TAXONOMY OF AUSTRALIAN MAMMALS

TAXONOMY OF AUSTRALIAN MAMMALS

Stephen Jackson and Colin Groves

PUBLISHING

National Library of Australia Cataloguing-in-Publication entry

Jackson, Stephen M., author.

Taxonomy of Australian mammals / Stephen Jackson and Colin Groves.

9781486300129 (hardback)
9781486300136 (epdf)
9781486300143 (epub)

Includes bibliographical references and index.

Mammals – Australia – Classification.

Groves, Colin P. (Colin Peter), author.

599.0994

Published by

CSIRO Publishing
36 Gardiner Road, Clayton VIC 3168
Private Bag 10, Clayton South VIC 3169
Australia

Telephone: +61 3 9545 8400
Email: publishing.sales@csiro.au
Website: www.publish.csiro.au

Front cover (from top left): Koala (*Phascolarctos cinereus*), Torresian Striped Possum (*Dactylopsila trivirgata*) and Eastern Quoll (*Dasyurus viverrinus*). Photos by Stephen Jackson.

Cover design by James Kelly
Typeset by Thomson Digital
Printed by Ingram Lightning Source

Contents

Introduction

The Australian continent, associated islands and territories oceans comprise 411 species (which includes one undescribed native species and 33 introduced species) of mammals that are recognised here. The scientific naming of the mammals of Australia began on 1 January 1758 when many of the cetaceans and animals that would subsequently be introduced into Australia were described by Carl Linnaeus in the 10th edition of the *Systema Naturae*, which is the starting point for all modern scientific names. The taxonomy of mammals from Australian soil began in 1785 when the Eastern Ring-tailed Possum was described, which was followed by the Eastern Grey Kangaroo in 1790 and the Yellow-bellied Glider in 1791.

After the colonisation of Australia by Europeans in 1788 there was a rapid increase in the number of native species described, with 165 new species recognised by 1860 and 230 species by 1900 (Fig. 1). Since this time each decade has resulted in the descriptions of many new species, with the most recent decades showing an increase in the rate of new species being discovered. These recent discoveries have included many native dasyurids, bats and rodents as might be expected because of their small size and more cryptic nature, but has also included five macropods. Incredibly, since 2000 a total of four new species of cetaceans have also been described from the waters off the Australian coast, with several others identified and awaiting formal description. With the development of new genetic technology and the reassessment of previously described populations, subspecies or synonyms that had not been adequately recognised (apparently unjustly) there has increasingly been a recognition of many of these as distinct species or subspecies. So it is likely that there are dozens of new species awaiting formal description or re-classification, which include not only the small cryptic species but also other larger species such as the cetaceans.

The rate at which species of the different orders of Australian mammals have been described by science has varied considerably over time (Fig. 2). Some groups such as the monotremes, bandicoots, seals, and exotic species were all described by 1900, while other groups such as the diprotodontians, chiropterans, rodents and dasyurids have continued to steadily increase in number in the last 30 years. Given the current knowledge of the taxonomy of Australian mammals it is clear that groups such as the dasyurids, chiropterans, rodents and the cetaceans will continue to reveal further species well into the future.

Taxonomy is a continually evolving process where the work of each author is built upon as more species are described and as taxonomic studies of species, genera and larger groups reveal a clearer picture of their phylogenetic relationships. With this in mind this present research acknowledges the important and ground-breaking work of its predecessors, especially those by Waterhouse (1841a, 1846), Gould (1845–1863), Thomas (1888a), Lydekker (1894a), Iredale and Troughton (1934), Ride (1970), Walton (1988a), Strahan (1983, 1995), Van Dyck and Strahan (2008) and the enormous number of taxonomic revisions of specific groups that have been undertaken.

Despite the production of various excellent books on Australian mammals over the last few decades there is no current comprehensive checklist that provides a full list of scientific names and a list of agreed common names – indeed, there is no formal system for setting up an agreed list of common names. Therefore an effort has been made here to set and follow several rules in the formation of common names. With respect to higher taxonomic ranks, the absence of an updated checklist in Walton (1988a) has resulted in some authors such as Van Dyck and Strahan (2008) perforce retaining a somewhat outdated higher taxonomy, while Menkhorst and Knight (2011) included only the families within each order without any further division. Therefore an exhaustive effort has been made to include all higher ranks and their associated synonyms.

The taxonomy, common names and phylogenetic sequence used here typically follow Van Dyck and Strahan (2008) for accepted species names and ranks

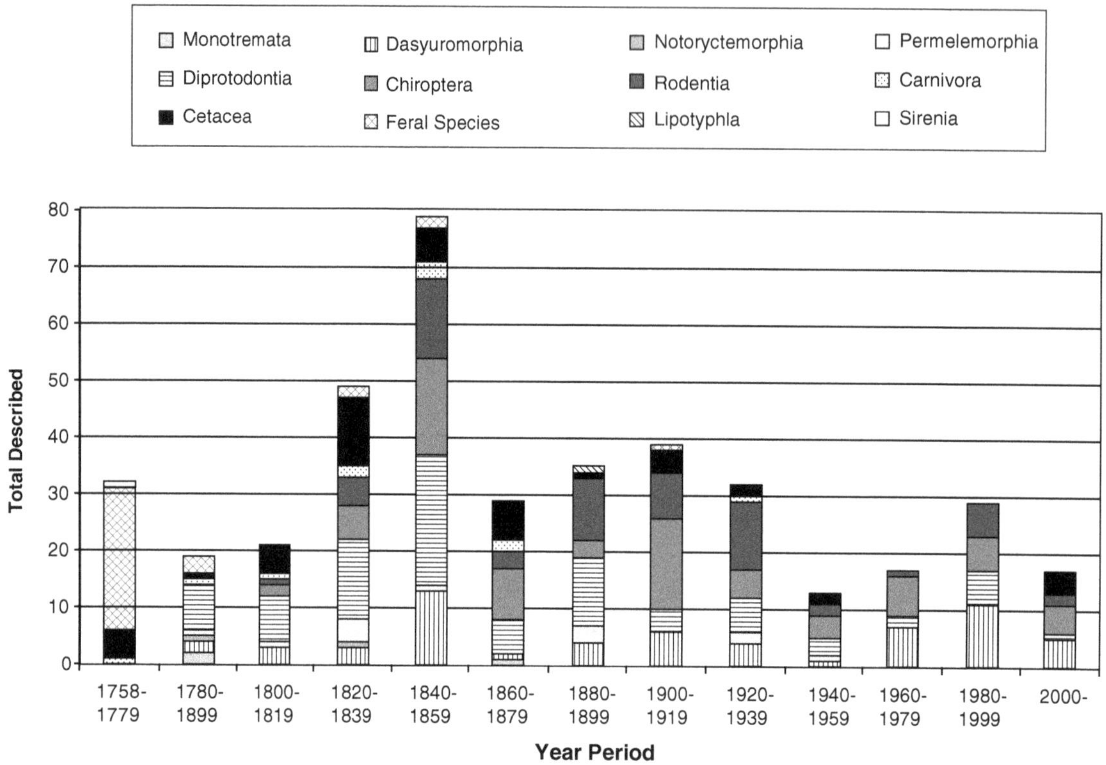

Fig. 1. The rate of descriptions of Australian mammals over time.

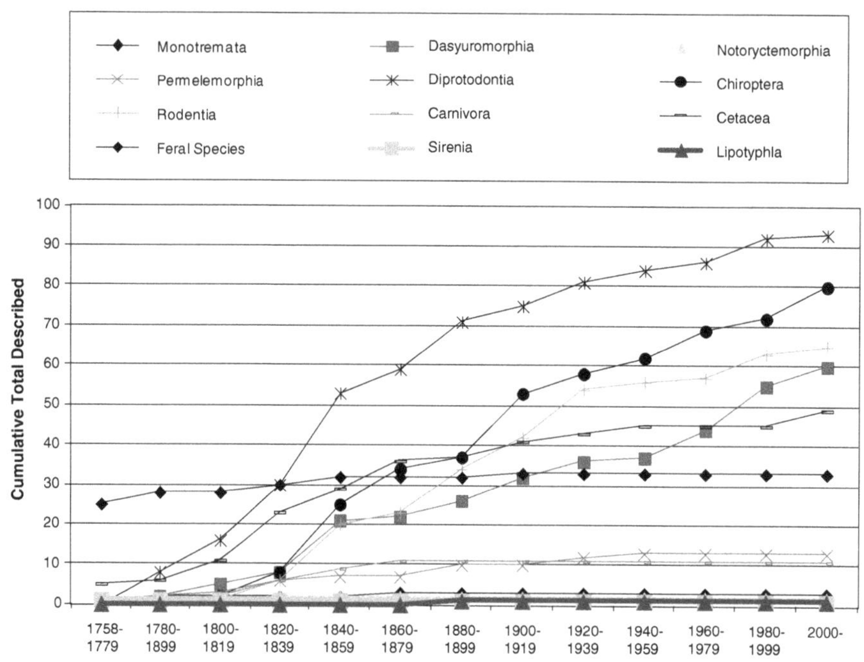

Fig. 2. Cumulative description of the different groups of Australian mammals.

up to family, in order to maximise consistency with the most current treatment of all Australian mammals. Ranks above family typically follow Van Dyck and Strahan (2008), although there are some notable exceptions as a result of significant recent research. These include the acceptance of the subordinal division of Chiroptera into Yinpterochiroptera and Yangochiroptera in preference to the old division into Megachiroptera and Microchiroptera (the latter of which is now known to be non-monophyletic), and the placement of the cetaceans within the Artiodactyla. The synonymies of species typically follow Wilson and Reeder (2005), those of genera following Wilson and Reeder (2005) and McKenna and Bell (1997), and those above the generic level following McKenna and Bell (1997), but we have ventured to register our disagreement in many places due to the recognition of appropriate recent research, and we give reasons for this deviation.

Unlike previous taxonomic treatments of Australian mammals, a full list of all known homonyms is provided including their full citation and taxonomic hierarchy. This review is also unique in that it includes subspecies and associated synonyms that are extralimital in order to give a full account of the distribution of species that occur within Australia and its territories. To rationalise the taxa covered, and the volume of the final text, we do not include synonyms at any rank for those species that are introduced to Australia unless the taxa concerned have synonyms based on specimens collected within Australia.

A deliberate attempt has been made to undertake this revision from first principles by making every effort to personally view and obtain a copy of every citation used. This process has revealed many discrepancies within the published literature so it is hoped that this treatment has helped to rectify many of these inconsistencies while at the same time it is hoped to not create many new ones. We also hope that this review helps to identify priority areas for future research so that the cycle of taxonomic research and review can be expedited.

In order to assist in the accuracy of the dates of the references used, and in turn the date priority of names used, an effort has been made to give each reference an exact or approximate date of publication depending on the available information. To undertake this process the primary source or numerous secondary sources were used including the Biodiversity Heritage Library website, journal websites, and numerous secondary papers and books that have been published. Appendix 1 includes a table of references included in the text and sources that have been used to include their dates of publication. Some publications were released over multiple years, and have been referred to in the text by the year the particular information was published along with the range of years in which the book was published. For example, the dates of John Gould's *The Mammals of Australia* spanned from 1845 to 1863 and it was released in 13 parts in three volumes. Taxa that were first described by Gould in that book have been given their specific year of publication using the publication dates proposed by Waterhouse (1885). References to this publication are cited with the year that the plate and associated text were published with [1845–1863] placed immediately afterwards so there is no confusion within the reference list.

The most recent complete review of Australian mammal taxonomy by Walton (1988a) is now well out of date by the description of new species and the major rearrangements of genera, families and even orders that has occurred as a result of subsequent research. Therefore the aims of this publication are to:

- Present an up to date taxonomic list of all Australia mammals and every rank from species (and subspecies and synonyms) up to the Class Mammalia.
- Provide a history of the major taxonomic changes that have occurred during the life of each taxon's name from the time of its initial description, and therefore provide a justification of the taxonomy used in this review.
- Provide an indication of the current stability of names by showing how names have changed over time, and show whether each name has stabilised over time or still shows fluctuation.
- Make taxonomic decisions where appropriate to assist in the stabilisation of names used.
- Include all taxa that occur outside Australia, but which have representative native species within Australia, in order to give a greater context to the Australian taxa.
- Include all species that occur within island territories of Australia including Christmas Island, Torres Strait islands, Lord Howe Island, Norfolk Island, Macquarie Island, Heard Island and McDonald Islands, and also the Australian Antarctic Territory, for all that this is not formally recognised internationally.

- Include all species occurring within the territorial waters of Australia and the island territories listed above, and the Australian Antarctic Territory.
- Provide details of all homonyms including non-mammalian names.
- Identify areas where future taxonomic research should be focussed.

Taxonomic decisions

The aim of this work is to provide a taxonomy that reflects the current state of knowledge and to document the enormous literature that has been published on studies focussed at all ranks within the Class Mammalia down to subspecies, and the synonyms for each rank. For each taxonomic rank an effort has been made to review the primary literature to either support its rank or synonymise it within other taxa. As part of this process the major taxonomic decisions have been identified from the primary literature. The construction of synonyms has endeavoured to follow that developed by Gardner and Hayssen (2004).

Common names

The first attempt to establish a recommended list of common names, or vernacular names, was published by Strahan (1980a) on behalf of a committee of the Australian Mammal Society. The general principles of the names proposed by the committee were to (where possible) be descriptive, indicate relationships, be memorable, be euphonious and be as short as possible. They also recognised the value of Aboriginal names such as Kultarr, Mulgara, Quoll, Kowari, Dunnart, Numbat, Bilby, Bettong, Pademelon, Potoroo and Quokka.

Common names used here are generally derived from those of Strahan (1980a) and more recently by Van Dyck and Strahan (2008) but with several differences. The names used are typically well known and commonly referred to, ideally descriptive, short and pronounceable. The noun-part of the name should correspond with the genus and the adjectival part should, if possible, refer to a diagnostic external character but may refer to the distribution. Whenever possible, names that are parochial and convey little or no information, for example 'Mitchell's Hopping-mouse', should be avoided. Names such as Monjon for *Petrogale burbidgei* and Nabarlek for *Petrogale concinna* are also not encouraged as they should include 'Rock-wallaby' in the name. With this in mind the names Kakarratul for *Notoryctes caurinus* and Itjaritjari for *Notoryctes typhlops* have been replaced by the Northern Marsupial Mole and Southern Marsupial Mole respectively following Groves (2005a). Familiarity may make terms such as Ringtail and Brushtail, as used by Groves (2005b), acceptable but for clarity are improved by the inclusion of 'Possum' in the group name.

Though common names can cause confusion on occasion, a perhaps unexpected advantage is that they typically remain constant when scientific names are changed. Importantly, common names and scientific names perform different functions, with common names reflecting the phenotype while the scientific names reflect the genotype (Andrew 2008). An example of this is the term Marsupial Mole that is used for the marsupial Order Notoryctemorphia as opposed to the true placental golden moles of the Order Afrosoricida. Common names have generally been restricted to the species rank, although they have been allocated to subspecies in some cases where these have been recognised in the modern literature – usually because the subspecies have extralimital distributions with recognised regional names or have been proposed to potentially warrant being elevated to species rank.

The protocols for the construction, hyphenation and capitalisation of common names used here draws on Duckworth and Pine (2003), Armstrong and Reardon (2006) and Andrew (2008) in an endeavour to give the names a consistent construction. For example, the free-tail bats have been spelt Freetail Bat by Menkhorst and Knight (2011), Free-tailed Bat by Simmons (2005a) and Van Dyck and Strahan (2008), and free-tailed bat by Armstrong and Reardon (2006).

The protocol of compounding group names is well established in some Australian taxa, with names including Rock-wallaby and Tree-kangaroo in common usage. Elsewhere, retaining the adjectival form, where possible, keeps groups large for unfamiliar taxa, for example Bent-winged Bat (rather than Bentwing-bat, but Horseshoe-bat because the adjectival form is too clumsy as in Horse-shoe-marked Bat). Thus the generalised rule adopted by Andrew (2008) and followed here is to compound group names comprising two nouns, but to retain two words when the first part of a group name is adjectival. Thus the group of possums with 'ringtails' (the 'ringtail possum' of Strahan, 1980a) should be compounded as

Ringtail-possum or become adjectival as in Ring-tailed Possum. The latter retains larger groups and is easier to find in an index for those less familiar with, for example, the Pseudocheiridae. This has the added advantage of being grammatically correct but we recognise there are still challenges in Cetacea.

Species concepts

In the light of the discussions of species concepts over the last 20 years or so, the failure of earlier checklists to specify what precisely they mean by species and subspecies seems a fault; but, given that the Biological Species Concept of Dobzhansky (1937) and Mayr (1942) held sway almost unchallenged for some 50 years since the period of the Evolutionary Synthesis in the 1930s, it was understandable. Today, however, it is mandatory for taxonomists to be clear about what species concept they adopt, and whether or not they recognise the rank of subspecies.

The Dobzhansky/Mayr concept defined species as not interbreeding with each other in nature. It has been pointed out many times that this gives no guidance for allopatric populations: under what circumstances should they be classified as distinct species, and when should they be combined into one species? For this reason alone, taxonomists should be aware that a simple claim to be working with the Biological Species Concept cannot be accurate: they are very evidently using some extra, unspoken criterion or multiple criteria.

We now know, however, that what we have always classed as distinct species, which remain discrete in sympatry, may actually be interbreeding on the sly – introgression, unbeknownst to the taxonomist and even usually to the field worker, may well be occurring or have occurred, its only evidence being the presence of the 'Wrong' mitochondrial or other genetic components in a given population. It would clearly be nonsense to unite two species on such grounds. The Biological Species Concept, defining species as not interbreeding, actually does not work.

Molecular data have been used to create what has become known as the Genetic Species Concept, based on the idea that the amount of genetic difference necessary for reproductive isolation can be measured (Mayden, 1997). Bradley and Baker (2001) tested this using Cytochrome-*b* sequences for several bat and rodent genera, and found that genetic distances do indeed tend to increase in the expected sequence from intrapopulational via intrasubspecific and intraspecific to intrageneric; moreover, distances between presumed sister species were less than between congeneric species in general. In a later paper the same authors amplified this, and offered a definition of the genetic species as 'a group of genetically compatible interbreeding natural populations that is genetically isolated from other such groups. Under our definition of the Genetic Species Concept, speciation is the accumulation of genetic changes in 2 lineages' (Baker & Bradley, 2006). In their accompanying table, they compared the Genetic Species Concept with other concepts, showing that most of the criteria for specific status under the concept were very close to those under the Phylogenetic Species Concept (see below), with this important difference: there must be a certain amount of difference in certain systems (nowadays, this would equate to DNA sequences) between two taxa in order to qualify, which, they argue, would constitute proof that they have entered separate evolutionary trajectories. This amount would have, of necessity, to vary across taxonomic groups, unless it is to be set arbitrarily.

Many authors working with DNA data have either explicitly or (more usually) tacitly used 'amount of difference' in making their taxonomic judgments. This raises the question of precisely what amount of difference would suffice, and, importantly, why. Baker and Bradley (2006) faced this problem and noted that it is a matter of probabilities: there is simply a greater probability that speciation has occurred with genetic distances of >5% than with distances of <2%. When faced with allopatric populations, the response of the Genetic Species Concept is similar to that of the Biological Species Concept: compare the amount of difference between two allopatric populations with that between two acknowledged congenerics (an argument which risks being circular!). There is also the question of how a difference in one DNA sequence (such as the Control Region) would equate to a difference in another (such as 12S or, even more problematic, a nuclear sequence such as an intron or pseudogene). Then there is of course the problem of species for which no genetic data are available, including fossils for which – except for rather recent specimens, preserved under appropriate conditions of fossilisation – no DNA sequencing will ever be possible. Given the arbitrariness of the 'amount of difference' criterion, we cannot follow the Genetic Species Concept, while recognising the enormous contributions that its adherents have made to the field. In effect, the Genetic Species Concept is less a species concept as such than a recipe for uncovering the existence of cryptic species; Baker and Bradley (2006), indeed, emphasise this role of the concept, and indeed we acknowledge that its considerable merit is its past history of, and future potential for, recognising cryptic evolutionary lineages within what has hitherto been considered a single species.

Our criterion for species status in this work is what has been called the Phylogenetic Species Concept: 'species are populations (or aggregations of populations) that are diagnosably distinct'. In this concept, we often have to assume that our species are populations (although this is often amenable to testing); apart from that, we need make no assumptions, noting only that on the evidence before us the ranges of variation of two species are separate. To put it another way, species have fixed heritable differences between them (Groves and Grubb, 2011). Such differences may be pelage characters, morphometric characters (either single measurements, or separation under Discriminant Analysis), or consistent differences in DNA sequences. It is in this latter area that genetic data are valuable for the delimitation of species.

Above all, the Phylogenetic Species Concept is testable, as a scientific proposition should be: 'the criterion of the scientific status of a theory is its falsifiability, or refutability, or testability' (Popper, 1963). Diagnosability (whether the differences are fixed or not) is testable; the question of whether the candidates for species status are populations (or metapopulations) is in principle testable; and the heritability of the differences is likewise in principle testable, although we must acknowledge that the heritability of morphological differences may well be unproven on the available evidence, but it is clearly best to act on the working hypothesis that they are heritable until breeding or rearing experiments prove otherwise.

In a significant clarification, De Queiroz (2007) proposed what he calls the Unified Species Concept – species are 'separately evolving metapopulation lineages (or, more properly, segments thereof)' (De Queiroz, 2007); there are particular properties, or lines of evidence, by which species may be delimited, but these are not themselves 'species concepts'. This has the effect of transferring the disputes from the question of which 'species concepts' are appropriate to the question of what evidence is appropriate for delimiting species, and it is relevant that De Queiroz goes on to argue that one consequence of this view is that 'any evidence of lineage separation is sufficient to infer the existence of separate species ... Although presence of a single property provides evidence for lineage separation, a highly corroborated hypothesis of lineage separation ... requires multiple lines of evidence' (De Queiroz, 2007: 884). That is to say, species delimitation relies on what has hitherto been called the Phylogenetic Species Concept.

The De Queiroz model significantly clarifies the discussion: there is only one way of defining species, and that is as evolutionary lineages. The 'phylogenetic species' is the minimal delimitation of species; while not necessarily 'highly corroborated', it offers clear evidence, falsifiable when fresh evidence comes to hand, that lineage separation has occurred.

In this way, the decision of whether two populations are specifically distinct or not is objective; it depends entirely on the evidence to hand, not on an arbitrary 'enough difference' nor on the basis of whether they would interbreed or not if given the chance.

Subspecies

Subspecies are geographic divisions within a species, which differ to a certain extent from each other in heritable features. The usual criterion is that 75% of individuals of a subspecies should be distinguishable from all individuals of other subspecies within that species, but this is a somewhat arbitrary criterion (Amadon, 1949). Essentially, a subspecies may be regarded as the point on a continuum at which it becomes convenient to dignify a population with a separate name (see, for example, Groves, 2001): the concept is, accordingly, a 'convenience category', and subspecies should on no account be reified. Nonetheless, as argued by Parkes (1982), subspecies serve as a conservative buffer for taxonomic uncertainty; this is especially true for cases where future taxonomic inquiry may demonstrate that currently recognised subspecies should be elevated to specific rank (e.g. Braby *et al.*, 2012). In contrast to the more recent support of subspecies, several authors have suggested that they be abandoned because they are often poorly diagnosed, their delimitation is difficult to determine (especially for parapatric populations), and taxonomic decisions made for a particular set of populations are often arbitrary, subjective, and based on too few characters (Wilson and Brown, 1953; Gillham, 1956).

Subspecies are not the only infraspecific category. Evolutionarily Significant Units ('ESU') (Moritz, 1994), are discrete populations of notably divergent evolutionary history and (consequently) genetic composition; very often, they are found to be equivalent to phylogenetic species.

Genera

A genus is a monophyletic group of species, but where to draw the line is at present arbitrary. Groves (2001) (see also Groves and Grubb, 2011) proposed that, to qualify as genera, two monophyletic groups should have been separated at least since the Miocene–Pliocene boundary. There seems little problem if molecular clock data (in the absence of the fossil record) indicate separation times, between candidates for generic status, of greatly above or below this level, but very often the inferred dates fall quite close to this boundary, and in such a case we would require much firmer data – more loci, better calculations of evolutionary rates, and so on – before proposing to change 'traditional' generic rankings. But it is mandatory that genera, and indeed other higher categories, be monophyletic, and if the evidence indicates otherwise, we have in this research ventured to alter the 'traditional' classification.

Families

A family is a monophyletic group of genera, but where to draw the line is at present arbitrary, as in the case of genera. Groves (2001, and elsewhere) proposed that, to qualify as families, two monophyletic groups should have been separated since at least the Oligocene–Miocene boundary. As in the case of genera, we have been conservative unless the evidence is strongly in favour of changing a 'traditional' arrangement.

Subgenera, subfamilies

When a genus or a family is too large and unwieldy in its contents, it may be convenient to divide it into two or more monophyletic groups called subgenera and subfamilies respectively. Their use is optional; for example, if a family contains just two known genera, however well differentiated, there is no need for subfamilies. Frequently the use of subfamilies is not sufficient, and tribes, and below that subtribes, may be used as subordinate (but always monophyletic) divisions. Subgenera, on the other hand, have been all too often used simply as a way of dividing a large and speciose genus without asking whether they are actually monophyletic. The informal term 'species group', which carries no implication other than sheer convenience, performs this function much better, and it is perhaps appropriate that the subgenus category seems to be falling out of use. Given the issues associated with subgenera they are not recognised here, so proposed subgenera including *Notamacropus* and *Osphranter* have been separated from *Macropus*, while *Micronomus*, *Ozimops* and *Setirostris* have been removed from *Mormopterus* as distinct genera.

Ranks above genus

In order to better understand the phylogeny of the different groups of Australian mammals, an effort has been made to include all ranks above genera and their associated synonyms. To better understand the original scope of each individual rank, the ranks above and below have been included. For family group names this includes all genera.

Given that the composition of most higher taxa has changed considerably since their original description, the current usage of the rank is also attributed to the author(s) from which this revision follows.

Higher non-Linnean ranks

The increasing research of the higher-order relationships of the different groups of mammals has resulted in a proliferation of names for ranks or clades above family group; as these names are not subject to the rules of priority that are required by the Code of Zoological Nomenclature there has been a degree of confusion as to which ones should be used. In an effort to provide stability to these names the principles of priority and stability as outlined by Simpson (1945), and followed by Brothers (1983) and enhanced by Asher and Helgen (2010), are followed. One of the main principles used is that the older name is retained if the proposed new name retains at least its original families. The senior and junior names of higher ranks of placental mammals advocated by Asher and Helgen (2010) are specifically followed.

Book outline

Each currently recognised taxon is identified with the relevant author and year of publication under which is the common name for species (and some subspecies) ranks. This is followed by a complete synonymy including the senior synonym and all known junior synonyms using the spelling and formation as described with the relevant author, year of publication and page number on which the name occurs. Sometimes two page numbers are included and these comprise the first listing of the name in the publication (often in an abstract) followed by the location where the name is formally described in the main text. The full citation for each author and year of publication is included within the reference section at the back of the text. Each synonym also includes the type genus for family group names, type species for genus group names and type locality for species group names. Within the comments section for ranks above the genus level an effort has been made to include the placement of the taxon within its taxonomic hierarchy by including the named rank above and ranks below, if they were provided, in order to give the reader a sense of the scope and context in which the proposed name was established. For convenience, the taxa that have individual entries do not typically have page numbers included when they are mentioned elsewhere in the text (only author and year). The comments section also outlines a broad taxonomic history of each taxon based on the major reviews. The synonymy for each taxon includes any unique name combination applied by any author for the first time to that taxon. This includes taxon names that have been made in error.

A future taxonomic research section is also included where appropriate. It very often happens that a particular genus or species has not been revised for many years, if ever, since its original description, and we point out if this has occurred and seems to be potentially a problem. We also take the liberty of pointing out, on occasion, when work of taxonomic relevance (especially a DNA phylogeny) has been published but the authors have not themselves ventured to tamper with the 'accepted classification' or where further confirmation of a taxonomic decision may be needed. Finally, a list of known homonyms is included within a taxonomic hierarchy for not only mammal groups but also other animal groups.

Acknowledgements

This work would not have been possible without significant assistance from numerous sources. We would like to thank those who helped read sections of this book and suggested numerous changes including Paul Andrew (Taronga Zoo), Ken Aplin (Smithsonian Institution), Joanne Burden (Muséum National d'Histoire Naturelle), Morgan Churchill (University of Wyoming), Mark Eldridge (Australian Museum), Erich Fitzgerald (Museum Victoria), Ewan Fordyce (University of Otago), Kris Helgen (Smithsonian Institution), Harry Parnaby (Australian Museum) and Terry Reardon (South Australian Museum). Stephen Jackson would like to thank several libraries and their associated staff that were very helpful in providing assistance during his visits. These include Carol Gokce, Paul Cooper, Eliza Walsh, Kirsten Marshall, John Rose and Emma Solway from the Natural History Museum in London who provided many of the references and assisted during the first visit to the library in 2003; Nicola Gamba, Paul Cooper, Lisa Di Tommaso, Samantha Gare, Nadja Noel, Kamila Reekie, Sarah Sworder, John Rose and Angela Thresher during the second visit in 2008; and Hellen Sharman, Sarah Sworder, Kamila Reekie, Harriet Campbell Longley, Lisa Atalla, Sarah Stewart, Goulven Keineg, Rachel Whittington, Elinor Skedgell, Lisa Di Tommaso and Lorraine Portch during the third visit in 2012. Thanks also to Therese Nouaille-Degorce and Evelyne Bremond-Hoslet from the Bibliothèque Centrale du Museum National d'Histoire Naturelle in Paris for providing several valuable references. Great thanks also to the staff at the National Museum of Natural History (Smithsonian Institution) libraries in Washington DC including Martha Rosen, Leslie Overstreet, Daria Wingreen-Mason and Kirstin van der Veen who helped him enormously in finding and copying references for this project. Thanks also to the staff of the National Natuurhistorisch Museum in Leiden, including Tom Gilissen, Marianne van der Wal and Agnes Bavelaar for all their help. Many thanks also to the Australian Museum and staff including Anja Divljan, Anina Hainsworth, Sandy Ingleby, Fiona Simpson and Fran Smith. Many thanks to Davide Molone who provided accommodation and interesting discussions during one of the visits to the Natural History Museum in London. Many thanks to Lindell Andrews and Dawn Roberts for providing several important references. Enormous thanks also go to the extraordinary Biodiversity Heritage Library that has allowed so many of the books and journals to be made available. Many thanks to Susan Bell (American Museum of Natural History) for answering various queries and providing several difficult-to-find citations. Ken Aplin provided important information on the taxonomy of various groups, which has been very much appreciated. The University of New South Wales and Professors Mike Archer and Sue Hand are gratefully acknowledged for their all their assistance. Finally, Stephen Jackson sincerely thanks Kerstin, Olivia and James for all their support during the writing of this book.

Definitions of nomenclatural and bibliographic terms

Absolute tautonomy: The identical spelling of a generic or subgeneric name and the specific or subspecific name of one of its originally included species or subspecies.

Abtheilung (Abth.): Part (German).

Available: Available to be used in nomenclature.

Available name: A scientific name that satisfies the nomenclatural requirements set forth in the International Code of Zoological Nomenclature, International Commission on Zoological Nomenclature (the current Code is the 4th edition, 1999; see ICZN in References). An available name may or may not be the valid name for a taxon.

Band (Bd.): Volume (German).

Berolini: of Berlin, Germany (Latin).

Binominal (binomial) name: The combination of two names, the first being the generic name and the second being the specific name, that together constitute the scientific name of a species. Any interpolated name (i.e. subgeneric name) is not counted as a component of a binominal name.

The Code: International Code of Zoological Nomenclature. First edition, 1961; 2nd edition, 1964; 3rd edition, 1985a; 4th edition, 1999.

Combination (or name combination): The association of a generic name and a specific name to form the name of a species, or of both with a subspecific name to form the name of the subspecies. In synonymy, new combinations do not constitute new names as such.

The Commission: International Commission on Zoological Nomenclature. Executive Secretary, The Natural History Museum, Cromwell Road, London SW7 5BD, United Kingdom; email: iczn@nhm.ac.uk.

Conditionally proposed names (or nomenclatural acts): Conditional proposal of a name means that the author used language to the effect that if the subject specimen or sample proved in the future to represent a different taxon, it should be known as (*Scientific name*). Names or nomenclatural acts proposed conditionally after 1960 are not thereby made available (ICZN, 1999: Art. 15.1).

Definition: A statement in words that purports to give those characters that, in combination, uniquely distinguishes a taxon.

Description: A statement in words of taxonomic characters of a specimen or a taxon. A description of a nominal taxon when it is established is called an original description.

Diagnosis: A statement in words that purports to give those characters that differentiates the taxon from other taxa with which it is likely to be confused.

Ed./Eds.: editor/editors

Editio: Edition (Latin).

Edn.: Edition

Emendation (emend.): Any intentional change in the original spelling of an available name that is not mandated by the Code.

emend. pro: emendation for [a particular name].

errore pro: inadvertent error for [a particular name].

extralimital: Of an occurrence or a taxon, beyond the geographic limits of this work.

***fide*:** According to.

First reviser: The 1st author to subsequently cite names (including different original spellings of the same name) or nomenclatural acts published on the same date and to select one of them to have precedence over the other(s).

First use of a name: The oldest use of a name.

Genus group: In the hierarchy of classification the group of taxa ranked between the family group and the species group. Contains taxa at the ranks of genus and subgenus. Holmiae: Stockholm, Sweden (Latin).

Haplotype: A combination of alleles that are closely linked on the same chromosome (or in mtDNA), and are [almost invariably] inherited together.

Holotype: A single specimen designated as the name-bearing type of a species or subspecies when it is established, or the single specimen on which such a taxon was based when no type was specified.

Homonym: Each of two or more available names having the same spelling, or differing only in suffix, and denoting different nominal taxa, whether in the family group, the genus group, or the species group.

Junior homonyms are unavailable in nomenclature. Homonyms in the species group apply only within the same genus, and may be either primary or secondary; homonyms in the genus and family groups apply throughout the animal kingdom. Variant spellings of what is essentially the same word are deemed to be identical (ICZN, 1999: Art. 58).

Homonymy: The relationship between homonyms; that is to say, the situation when one name is a homonym of another.

ICZN: International Commission on Zoological Nomenclature (see the Commission).

***Incertae sedis*:** Of uncertain taxonomic position.

Incorrect original spelling: An original spelling that is incorrect because of a misspelling or printer's error, one of two or more different spellings of the same name in the original description, or a spelling requiring a mandatory change.

Incorrect subsequent spelling: Any change in the spelling of an available name other than a mandatory change or emendation (ICZN, 1999: Art. 33.3). An inadvertent alteration of an available name. Incorrect subsequent spellings are not available names unless an author purposely uses the misspelling for a previously undescribed taxon.

Indication: A reference to a description, definition, or figure in another publication, including pre-Linnaean (pre-1758) works. Using a reference in place of a description was a common practice in the late 18th century and during the 19th century, but is not permissible in literature after 1930.

Invalid name: An available name that is not in current use (i.e. for any recognised taxon).

Junior: A more recent name, comparative to previous use or older name.

Junior objective synonym: Of two synonyms, the one that was established at a later date.

Justified emendation: The correction of an incorrect original spelling.

Lapsus calami (pl. lapsus calamarum): 'A slip of the pen'; when an author uses a name other than the name intended as opposed to a misspelling, typographical error, or printer's error.

Lectotype: A syntype designated as the single name-bearing type specimen subsequent to the establishment of a nominal species or subspecies. If the original description of a species group name was based on two or more specimens, and no specimen was identified as the holotype, these specimens are syntypes (previously referred to as cotypes). When one of these syntypes is selected as the name-bearing type specimen, it is referred to as the lectotype.

Lieferung (Lfg.): Fascicle (German).

Linnaean tautonomy: The identical spelling of a new generic or subgeneric name established before 1931 and a pre-1758, one-word name cited as a synonym of one of the species or subspecies originally included in that genus.

Livr.: Livraison. Part of a serial issued from time to time.

Logotype: The type by subsequent determination. The historical type of a genus, selected from two or more original species. A genus whose type is selected from two or more original species is logotypic (see O. Cook, 1914: 314).

Lugduni: Lyons, France (Latin).

Mandatory change: A change in the original spelling mandated by the Code: (1) a change in the ending of a specific or subspecific name (such as agreement in gender with the generic name or in number such as a change from *-ii* to *-orum* when the species-group name was intended to honour two or more individuals of the same name instead of a single individual); or (2) the dropping of hyphens or diacritical marks such as accents, the separation of diphthongs (e.g. æ or oe), or a change of spelling required (for names published before 1985) such as converting *-ü* - to *-ue-* (e.g. *mülleri* to *muelleri*).

Monotypic: Represented by a single taxon. A taxon that includes only one subordinate taxon of the next lower rank.

Monotypy: A situation where: (1) the author does not explicitly indicate a type species for a genus or subgenus but lists a single species by an available name (type species by monotypy), or (2) when a species description is based on a single specimen not explicitly designated the holotype (holotype by monotypy). This is not to be confused with the term 'monotypic', a term in taxonomy denoting that a genus has only one species, or a species is not divided into subspecies.

Multiple original spellings: Two or more different original spellings for the same name (ICZN 1999: Art. 32.2.1).

Nec: Preoccupied by.

Neotype: The single specimen designated as the name-bearing type of a nominal species or subspecies when a need arises to objectively define the nominal taxon and no name-bearing type is believed to exist.

New combination: The first combination of a generic name and a previously established species-group name.

Nomenclatural act: Any published act that affects the status of a scientific name or its type. Description of a taxon, revision of a species, designation of a type species, restriction of a type locality, designation of a neotype, selection of a lectotype, and so on, are examples of a nomenclatural act.

Nomenclature: A system of names and provisions for their formation and use. Binomial nomenclature is the system of nomenclature whereby species, but no other taxon, are denoted by a binomen, a combination of two names, the first being the generic name and the second the specific name. Nomenclature is a man-made system, designed to serve taxonomy, which is a reflection of the natural world. The two should not be confused.

***Nomen dubium* (pl. nomina dubia):** A Latin term meaning a name of unknown or doubtful application.

***Nomen novum* (pl. nomina nova):** A Latin term used for a new replacement name. A new name that is published to replace an earlier name (and valid only if the latter is preoccupied) and which is expressively proposed as a replacement name; a new name, not to be confused with a new species, or a new genus, etc., which represent new taxa. Commonly applied to names proposed to replace junior homonyms.

***Nomen novum pro*:** New name for [some other name, either a homonym or for some reason 'preferred', this being in the older literature].

***Nomen nudum* (nom. nud):** A naked name, a name that has not met the criteria for availability as outlined in the International Code of Zoological Nomenclature; a name that, if published before 1931, was not accompanied by a description, definition, or indication, or if published after 1930, is not accompanied by a statement that purports to give characters differentiating the taxon; or is not accompanied by a definitive bibliographic reference to such a statement; or is not proposed expressly as a replacement for a pre-existing name. A *nomen nudum* is not an available name, and therefore the same name may be made available later for the same or a different concept; in such a case it would take authorship and date from that act of establishment, not from any earlier publication as a *nomen nudem*.

***Nomen oblitum* (pl. nomina oblita):** A Latin term meaning a forgotten name. A name that has not been used since 1899; such a name is not allowed to displace a junior synonym or homonym in prevailing usage.

Nominal taxon: A taxon denoted by an available name.

Nominotypical: The nominal taxon at a subordinate rank within the family group, the genus group, or the species group that contains the name-bearing type of that group.

Non: Not.

nov. = *novus, novum*: New.

Objective synonym: A name whose synonymy with another is not merely a matter of opinion, because the two are based on the same type material.

Original designation: The designation of the type of a nominal taxon when it is established.

Original description: See Description.

Operational Taxonomic Unit: A group, or sample, or single specimen, for the time being assumed to be a valid taxon for purposes of an analysis.

***Pace*:** 'With peace', meaning with all due respect.

Polytypic: A taxon represented by two or more subordinate taxa of the next lower rank. This is not to be confused with its use in taxonomy, referring to a taxon which has two or more subordinate taxa.

Preoccupied name: A name that is unavailable because it is: (1) a generic name of the same spelling (junior homonym) as a generic name previously used for a different animal; or (2) a species or subspecies name of the same spelling originally combined with the same generic name as an earlier described name, even if that species-level name is now used with a different genus.

Preoccupied: A homonym.

Primary homonym: Each of two or more homonyms in the species group, originally combined with the same generic name.

Principle of Priority: The principle that the valid name of a taxon is the oldest available name applied to it provided that the name is not invalidated by any provision of the Code or by any ruling by the Commission.

Pro: Before.

***Recte*:** Correctly.

Renaming: The act of providing a replacement name for a preoccupied name. Note that a different name combination does not constitute a renaming.

Scientific name: A name treated as a Latinised name for a taxon (usually cited in combination with the author and date of publication).

Secondary homonym: Each of two or more homonyms in the species group, originally combined with different generic names but subsequently combined with the same generic name.

Senior: An older name, comparative with a more recent name

Senior synonym: Of two synonyms, the one that was established at an earlier date

***Sensu*:** In the sense of.

***Sensu lato*:** In the broad sense.

***Sensu stricto*:** in the strict (narrow) sense.

Separate: A copy (reprint or offprint) of an article printed separately from the journal or periodical in which the article appears. If separates are printed and distributed before the printing and distribution of the journal or periodical, they are referred to as preprints and usually bear an earlier date and different pagination. Today scientific publications normally print separates (usually called reprints) after the printing and distribution of the book, journal, or periodical in which the article appears. Separates qualifying as reprints have the same pagination found in the publication in which they appear. The advance distribution of separates after 1999 does not constitute publication for the purposes of zoological nomenclature.

Species group: In zoological classification, the lowest-ranking group of taxa, the names of which are regulated by the Code. The species group includes all taxa at the ranks of species and subspecies.

Subjective synonym: A name based on a type specimen thought to belong to the same taxon as another name.

Subsequent designation: Designation of type species for a genus group taxon in another publication after the genus-group taxon was described.

Subsequent spelling: A subsequent spelling that is not the same as the original spelling is an emendation, an incorrect subsequent spelling, or a mandated change.

Syntype: Each specimen of a type series from which a holotype (at time of publication) or a lectotype (subsequent to publication) has not been selected. The syntypes collectively constitute the name-bearing type. Syntypes are sometimes referred to as cotypes, a term that should not now be used in zoological nomenclature.

Tautonymy: The use of the same word for the name of a genus and of one of its originally included nominal species or subspecies. Linnean tautonomy is the identical spelling of a new generic or subgeneric name and pre-1758 name cited as a synonym of only one of the species or subspecies originally included in that genus.

Tautotype: The type of a genus designated because its [species-group] name is identical in spelling with the genus-group name.

Taxon (pl. taxa): A taxonomic unit, whether named or not, considered to comprise a population or group of populations of organisms usually inferred to be phylogenetically related and have characters in common that differentiate the unit from other such units. A taxon encompassing all included taxa of lower rank.

Taxonomy: The theoretical study of classification, including its bases, principles, procedures, and rules. It includes the classification and naming of animals.

Tome: Volume (French).

Tomus: Volume (Latin).

Type: A tern used alone, or forming part of a compound term, to denote a particular kind of specimen or taxon.

Type locality: The geographical place of capture or collection of the name-bearing type of a nominal species or subspecies.

Typotypical: [of a specimen] originating from the type locality of the species or subspecies to which it is thought to belong.

Unavailable: Not available for use in nomenclature.

Unavailable work: A publication in which names and nomenclatural acts are rejected for nomenclatural purposes. The work may be unavailable because it was published before 1758, or because the author was not consistently binomial, or (after 1950) was published anonymously, or contained a disclaimer, or because the Commission has ruled it to be unavailable.

Unjustified emendation: An intentional alteration of the original spelling of an available name that is not mandated.

Valid name: An available name that is used as the current name for a taxon. A valid name is a term applied in nomenclature to mean only the name by which a taxon is currently identified.

Vernacular name: A name of an animal or animals in a language used for general purposes as opposed to a Latinised name proposed only for zoological nomenclature.

Vide: See.

Virtual tautonomy: The nearly identical spelling, or of the same origin or meaning, of a generic or subgeneric name and the specific or subspecific name in a binomen or trinomen. The term virtual tautonomy is not regulated by the Code.

Abbreviations

†: Extinct.
Φ: Extralimital taxon.
Ω: Introduced taxon.

Taxonomy of the mammals of Australia

CLASS MAMMALIA Linnaeus, 1758

SUBCLASS PROTOTHERIA Gill, 1872

ORDER MONOTREMATA Bonaparte, 1832 *sensu* Bonaparte, 1838

Family Ornithorhynchidae J. Gray, 1825 *sensu* Burnett, 1830 [Platypus]

Ornithorhynchus anatinus (Shaw, 1799) Platypus

Family Tachyglossidae Gill, 1872 [Echidnas]

Tachyglossus aculeatus (Shaw, 1792) Short-beaked Echidna

† *Zaglossus bruijni* (Peters & Doria, 1876) Western Long-beaked Echidna

SUBCLASS THERIA Parker & Haswell, 1897

SUPERLEGION TRECHNOTHERIA McKenna, 1975

LEGION YANGOTHERIA Chow & Rich, 1982

SUBLEGION CLADOTHERIA McKenna, 1975

INFRALEGION ZATHERIA McKenna, 1975

INFRACLASS TRIBOSPHENIDA McKenna, 1975

SUPERCOHORT MARSUPIALIA Illiger, 1811 *sensu* Cuvier, 1816

COHORT AUSTRALIDELPHIA Szalay, 1982

ORDER DASYUROMORPHIA Gill, 1872 *sensu* Aplin & Archer, 1987

Superfamily Dasyuroidea Goldfuss, 1820 *sensu* Marshall *et al.*, 1990

Family Dasyuridae Goldfuss, 1820 *sensu* Owen, 1839 [Dasyurids]

Subfamily Dasyurinae Goldfuss, 1820 *sensu* Marshall *et al.*, 1990

Dasycercus blythi (Waite, 1904) Brush-tailed Mulgara

Dasycercus cristicauda (Krefft, 1867) Crest-tailed Mulgara

Dasykaluta rosamondae (Ride, 1964) Kaluta

Dasyuroides byrnei Spencer, 1896 Kowari

Dasyurus geoffroii Gould, 1841 Western Quoll

Dasyurus hallucatus Gould, 1842 Northern Quoll

Dasyurus maculatus (Kerr, 1792) Spotted-tailed Quoll [Tiger Quoll]

Dasyurus viverrinus (Shaw, 1800) Eastern Quoll

Parantechinus apicalis (J. Gray, 1842) Dibbler

Pseudantechinus bilarni (Johnson, 1954) Sandstone Pseudantechinus

Pseudantechinus macdonnellensis (Spencer, 1895) Fat-tailed Pseudantechinus

Pseudantechinus mimulus (Thomas, 1906) Carpentarian Pseudantechinus

Pseudantechinus ningbing Kitchener, 1988 Ningbing Pseudantechinus

Pseudantechinus roryi N. Cooper *et al.*, 2000 Tan Pseudantechinus

Pseudantechinus woolleyae Kitchener & Caputi, 1988 Woolley's Pseudantechinus

Sarcophilus harrisii (Boitard, 1841) Tasmanian Devil

Subfamily Phascogalinae Gill, 1872 *sensu* Marshall, 1990

Antechinus adustus (Thomas, 1923) Rusty Antechinus

Antechinus agilis Dickman *et al.*, 1998 Agile Antechinus

Antechinus argentus Baker *et al.*, 2013 Silver-headed Antechinus

Antechinus arktos Baker *et al.*, 2014 Black-tailed Antechinus

Antechinus bellus (Thomas, 1904) Fawn Antechinus

Antechinus flavipes (Waterhouse, 1838) Yellow-footed Antechinus
Antechinus godmani (Thomas, 1923) Atherton Antechinus
Antechinus leo Van Dyck, 1980 Cinnamon Antechinus
Antechinus minimus (É. Geoffroy, 1803) Swamp Antechinus
Antechinus mysticus Baker *et al.*, 2012 Buff-footed Antechinus
Antechinus stuartii Macleay, 1841 Brown Antechinus
Antechinus subtropicus Van Dyck & Crowther, 2000 Subtropical Antechinus
Antechinus swainsonii (Waterhouse, 1840) Dusky Antechinus
Phascogale calura Gould, 1844 Red-tailed Phascogale
Phascogale pirata Thomas, 1904 Northern Phascogale
Phascogale tapoatafa (Meyer, 1793) Brush-tailed Phascogale

Subfamily Planigalinae Archer, 1982 *sensu* Marshall *et al.*, 1990
Planigale gilesi Aitken, 1972 Gile's Planigale
Planigale ingrami (Thomas, 1906) Long-tailed Planigale
Planigale maculata (Gould, 1851) Common Planigale
Planigale tenuirostris Troughton, 1928 Narrow-nosed Planigale

Subfamily Sminthopsinae Archer, 1982 *sensu* Marshall *et al.*, 1990
Antechinomys laniger (Gould, 1856) Kultarr
Ningaui ridei Archer, 1975 Wongai Ningaui
Ningaui timealeyi Archer, 1975 Pilbara Ningaui
Ningaui yvonnae Kitchener *et al.*, 1983 Southern Ningaui
Sminthopsis archeri Van Dyck, 1986 Chestnut Dunnart
Sminthopsis bindi Van Dyck *et al.*, 1994 Kakadu Dunnart
Sminthopsis butleri Archer, 1979 Butler's Dunnart
Sminthopsis crassicaudata (Gould, 1844) Fat-tailed Dunnart
Sminthopsis dolichura Kitchener, *et al.*, 1984 Little Long-tailed Dunnart
Sminthopsis douglasi Archer, 1979 Julia Creek Dunnart
Sminthopsis fuliginosus (Gould, 1852) Grey-bellied Dunnart
Sminthopsis gilberti Kitchener *et al.*, 1984 Gilbert's Dunnart
Sminthopsis granulipes Troughton, 1932 White-tailed Dunnart
Sminthopsis hirtipes Thomas, 1898 Greater Hairy-footed Dunnart
Sminthopsis leucopus (J. Gray, 1842) White-footed Dunnart
Sminthopsis longicaudata Spencer, 1909 Large Long-tailed Dunnart
Sminthopsis macroura (Gould, 1845) Stripe-faced Dunnart
Sminthopsis murina (Waterhouse, 1838) Common Dunnart
Sminthopsis ooldea Troughton, 1965 Ooldea Dunnart
Sminthopsis psammophila Spencer, 1895 Sandhill Dunnart
Sminthopsis virginiae (de Tarragon, 1847) Red-cheeked Dunnart
Sminthopsis youngsoni McKenzie & Archer, 1982 Lesser Hairy-footed Dunnart

Family Myrmecobiidae Waterhouse, 1841 [Numbat]
Myrmecobius fasciatus Waterhouse, 1836 Numbat

Family Thylacinidae Bonaparte, 1838 [Thylacine]
† *Thylacinus cynocephalus* (Harris, 1808) Thylacine

ORDER NOTORYCTEMORPHIA Kirsch, 1977 *sensu* Aplin & Archer, 1987

Family Notoryctidae J. Ogilby, 1892 [Marsupial Moles]
Notoryctes caurinus Thomas, 1920 Northern Marsupial Mole
Notoryctes typhlops (Stirling, 1889) Southern Marsupial Mole

ORDER PERAMELEMORPHIA Ameghino, 1889 *sensu* Aplin & Archer, 1987

Superfamily Perameloidea J. Gray, 1825 *sensu* Van Dyck & Strahan, 2008

Family Chaeropodidae Gill, 1872 *sensu* Groves, 2005 [Pig-footed Bandicoot]
† *Chaeropus ecaudatus* (W. Ogilby, 1838) Pig-footed Bandicoot

Family Peramelidae J. Gray, 1825 *sensu* Van Dyck & Strahan, 2008 [Bandicoots]

Subfamily Echymiperinae McKenna & Bell, 1997 *sensu* Van Dyck & Strahan, 2008

Echymipera rufescens (Peters & Doria, 1875) Long-nosed Echymipera

Subfamily Peramelinae J. Gray, 1825 *sensu* Kirsch *et al.*, 1997

Isoodon auratus (Ramsay, 1887) Golden Bandicoot

Isoodon macrourus (Gould, 1842) Northern Brown Bandicoot

Isoodon obesulus (Shaw, 1797) Southern Brown Bandicoot

Isoodon peninsulae Thomas, 1922 Cape York Brown Bandicoot

Perameles bougainville Quoy & Gaimard, 1824 Western Barred Bandicoot

† *Perameles eremiana* Spencer, 1897 Desert Bandicoot

Perameles gunnii J. Gray, 1838 Eastern Barred Bandicoot

Perameles nasuta É. Geoffroy, 1804 Southern Long-nosed Bandicoot

Perameles pallescens Thomas, 1923 Northern Long-nosed Bandicoot

Family Thylacomyidae Bensley, 1903 *sensu* Archer & Kirsch, 1977 [Bilbies]

Macrotis lagotis (Reid, 1837) Greater Bilby

† *Macrotis leucura* (Thomas, 1887) Lesser Bilby

ORDER DIPROTODONTIA Owen, 1877

SUBORDER VOMBATIFORMES Woodburne, 1984 *sensu* Aplin & Archer, 1987

INFRAORDER PHASCOLARCTOMORPHIA Aplin & Archer, 1987

Family Phascolarctidae Owen, 1839 [Koala]

Phascolarctos cinereus (Goldfuss, 1817) Koala

INFRAORDER VOMBATOMORPHIA Aplin & Archer, 1987

Family Vombatidae Burnett, 1830 *sensu* Dawson, 1983 [Wombats]

Lasiorhinus krefftii (Owen, 1872) Northern Hairy-nosed Wombat

Lasiorhinus latifrons (Owen, 1845) Southern Hairy-nosed Wombat

Vombatus ursinus (Shaw, 1800) Bare-nosed Wombat

SUBORDER PHALANGERIDA Aplin & Archer, 1987

Superfamily Burramyoidea Broom, 1898 *sensu* Aplin & Archer, 1987

Family Burramyidae Broom, 1898 *sensu* Aplin & Archer, 1987 [Pygmy-possums]

Burramys parvus Broom, 1895 Mountain Pygmy-possum

Cercartetus caudatus (Milne-Edwards, 1877) Long-tailed Pygmy-possum

Cercartetus concinnus (Gould, 1845) Western Pygmy-possum

Cercartetus lepidus (Thomas, 1888) Little Pygmy-possum

Cercartetus nanus (Desmarest, 1817) Eastern Pygmy-possum

Superfamily Petauroidea Bonaparte, 1832 *sensu* Aplin & Archer, 1987

Family Petauridae Bonaparte, 1832 *sensu* Baverstock, 1984 [Striped Possum, Leadbeater's Possum, and Lesser Gliders]

Subfamily Dactylopsilinae Kirsch, 1977 *sensu* Edwards & Westerman, 1992

Dactylopsila trivirgata J. Gray, 1858 Torresian Striped Possum

Gymnobelideus leadbeateri McCoy, 1867 Leadbeater's Possum

Subfamily Petaurinae Bonaparte, 1832 *sensu* Edwards & Westerman, 1992

Petaurus australis Shaw, 1791 Yellow-bellied Glider

Petaurus breviceps (Waterhouse, 1838) Sugar Glider

Petaurus gracilis (De Vis, 1883) Mahogany Glider

Petaurus norfolcensis (Kerr, 1792) Squirrel Glider

Family Pseudocheiridae Winge, 1893 *sensu* Kirsch *et al.*, 1997 [Ring-tailed Possums and Greater Gliders]

Subfamily Hemibelideinae Kirsch *et al.*, 1997

Hemibelideus lemuroides (Collett, 1884) Lemuroid Ring-tailed Possum

Petauroides armillatus Thomas, 1923 Central Greater Glider

Petauroides minor (Collett, 1887) Northern Greater Glider

Petauroides volans (Kerr, 1792) Southern Greater Glider

Subfamily Pseudocheirinae Winge, 1893 *sensu* Kirsch *et al.*, 1997

Pseudocheirus occidentalis (Thomas, 1888) Western Ring-tailed Possum

Pseudocheirus peregrinus (Boddaert, 1785) Eastern Ring-tailed Possum

Pseudochirulus cinereus (Tate, 1945) Daintree River Ring-tailed Possum

Pseudochirulus herbertensis (Collett, 1884) Herbert River Ring-tailed Possum

Subfamily Pseudochiropsinae Kirsch *et al.*, 1997 *sensu* Meredith *et al.*, 2010

Petropseudes dahlii (Collett, 1895) Rock Ring-tailed Possum

Pseudochirops archeri (Collett, 1884) Green Ring-tailed Possum

Family Tarsipedidae Gervais & Verreaux, 1842 [Honey Possum]

Tarsipes rostratus Gervais & Verreaux, 1842 Honey Possum

Family Acrobatidae Aplin (in Aplin and Archer), 1987 [Feather-tailed Gliders]

Acrobates frontalis (De Vis, 1887) Broad-toed Feather-tailed Glider

Acrobates pygmaeus (Shaw, 1794) Narrow-toed Feather-tailed Glider

Superfamily Phalangeroidea Thomas, 1888 *sensu* Aplin & Archer, 1987

Family Phalangeridae Thomas, 1888 *sensu* Flannery *et al.*, 1987 [Cuscuses and Brush-tailed Possums]

Subfamily Phalangerinae Thomas, 1888 *sensu* Flannery *et al.*, 1987

Tribe Phalangerini Thomas, 1888 *sensu* Flannery *et al.*, 1987

Phalanger mimicus Thomas, 1922 Southern Common Cuscus

Spilocuscus nudicaudatus (Gould, 1850) Australian Spotted Cuscus

Tribe Trichosurini Flynn, 1911 *sensu* Flannery *et al.*, 1987

Trichosurus caninus (W. Ogilby, 1836) Short-eared Brush-tailed Possum

Trichosurus cunninghami Lindenmayer *et al.*, 2002 Mountain Brush-tailed Possum

Trichosurus vulpecula (Kerr, 1792) Common Brush-tailed Possum

Wyulda squamicaudata Alexander, 1919 Scaly-tailed Possum

SUBORDER MACROPODIFORMES Kirsch *et al.*, 1997

Superfamily Macropodoidea J. Gray, 1821 *sensu* Kear & Cooke, 2001

Family Hypsiprymnodontidae Collett, 1887 [Musky Rat-kangaroo]

Subfamily Hypsiprymnodontinae Collett, 1887 [Musky Rat-kangaroo]

Hypsiprymnodon moschatus Ramsay, 1876 Musky Rat-kangaroo

Family Potoroidae J. Gray, 1821 *sensu* Kear & Cooke, 2001 [Bettongs and Potoroos]

Subfamily Potoroinae J. Gray, 1821 *sensu* Flannery, 1989 [Potoroos]

Tribe Bettongini Flannery & Archer, 1987 [Bettongs]

Aepyprymnus rufescens (J. Gray, 1837) Rufous Bettong

† *Bettongia anhydra* Finlayson, 1957 Desert Bettong

Bettongia gaimardi (Desmarest, 1822) Eastern Bettong

Bettongia lesueur (Quoy & Gaimard, 1824) Burrowing Bettong

Bettongia penicillata J. Gray, 1837 Brush-tailed Bettong

† *Bettongia pusilla* McNamara, 1997 Nullarbor Dwarf Bettong

Bettongia tropica Wakefield, 1967 Northern Bettong

† *Caloprymnus campestris* (Gould, 1843) Desert Rat-kangaroo

Tribe Potoroini J. Gray, 1821 *sensu* Flannery, 1989 [Potoroos]

Potorous gilbertii (Gould, 1841) Gilbert's Potoroo

Potorous longipes Seebeck & Johnston, 1980 Long-footed Potoroo

† *Potorous platyops* (Gould, 1844) Broad-faced Potoroo

Potorous tridactylus (Kerr, 1792) Long-nosed Potoroo

Family Macropodidae J. Gray, 1821 *sensu* Kear & Cooke, 2001 [Kangaroos and Wallabies]

Subfamily Macropodinae J. Gray, 1821 *sensu* Kear & Cooke, 2001

Tribe Dendrolagini Flannery, 1989 *sensu* Baverstock *et al.*, 1989

Dendrolagus bennettianus De Vis, 1887 Bennett's Tree-kangaroo

Dendrolagus lumholtzi Collett, 1884 Lumholtz's Tree-kangaroo
Petrogale assimilis Ramsay, 1877 Allied Rock-wallaby
Petrogale brachyotis (Gould, 1841) Western Short-eared Rock-wallaby
Petrogale burbidgei Kitchener & Sanson, 1978 Monjon Rock-wallaby
Petrogale coenensis Eldridge & Close, 1992 Cape York Rock-wallaby
Petrogale concinna Gould, 1842 Nabarlek Rock-wallaby
Petrogale godmani Thomas, 1923 Godman's Rock-wallaby
Petrogale herberti Thomas, 1926 Herbert's Rock-wallaby
Petrogale inornata Gould, 1842 Unadorned Rock-wallaby
Petrogale lateralis Gould, 1840 Black-footed Rock-wallaby
Petrogale mareeba Eldridge & Close, 1992 Mareeba Rock-wallaby
Petrogale penicillata (J. Gray, 1827) Brush-tailed Rock-wallaby
Petrogale persephone Maynes, 1982 Proserpine Rock-wallaby
Petrogale purpureicollis Le Souef, 1924 Purple-necked Rock-wallaby
Petrogale rothschildi Thomas, 1904 Rothschild's Rock-wallaby
Petrogale sharmani Eldridge & Close, 1992 Mount Claro Rock-wallaby
Petrogale wilkinsi Thomas, 1926 Eastern Short-eared Rock-wallaby
Petrogale xanthopus J. Gray, 1855 Yellow-footed Rock-wallaby
Thylogale billardierii (Desmarest, 1822) Rufous-bellied Pademelon
Thylogale stigmatica (Gould, 1860) Red-legged Pademelon
Thylogale thetis (Lesson, 1827) Red-necked Pademelon

Tribe Macropodini J. Gray, 1821 *sensu* Flannery, 1989
†*Lagorchestes asomatus* Finlayson, 1943 Central Hare-wallaby
Lagorchestes conspicillatus Gould, 1842 Spectacled Hare-wallaby
Lagorchestes hirsutus Gould, 1844 Rufous Hare-wallaby
†*Lagorchestes leporides* (Gould, 1841) Eastern Hare-wallaby
Macropus fuliginosus (Desmarest, 1817) Western Grey Kangaroo
Macropus giganteus Shaw, 1790 Eastern Grey Kangaroo
Notamacropus agilis (Gould, 1842) Agile Wallaby
Notamacropus dorsalis (J. Gray, 1837) Black-striped Wallaby
Notamacropus eugenii (Desmarest, 1817) Tammar Wallaby
† *Notamacropus greyi* (Waterhouse, 1846) Toolache Wallaby
Notamacropus irma (Jourdan, 1837) Western Brush Wallaby
Notamacropus parma (Waterhouse, 1846) Parma Wallaby
Notamacropus parryi (Bennett, 1835) Whip-tailed Wallaby
Notamacropus rufogriseus (Desmarest, 1817) Red-necked Wallaby
Onychogalea frenata (Gould, 1840) Bridled Nail-tailed Wallaby
† *Onychogalea lunata* (Gould, 1840) Crescent Nail-tailed Wallaby
Onychogalea unguifera (Gould, 1840) Northern Nail-tailed Wallaby
Osphranter antilopinus (Gould, 1842) Antilopine Wallaroo
Osphranter bernardus (Rothschild, 1904) Black Wallaroo
Osphranter robustus (Gould, 1840) Common Wallaroo
Osphranter rufus (Desmarest, 1822) Red Kangaroo
Wallabia bicolor (Desmarest, 1804) Swamp Wallaby

Tribe Setonichini Jackson & Groves, 2015 New
Setonix brachyurus (Quoy & Gaimard, 1830) Quokka

Subfamily Lagostrophinae Flannery, 1989
Lagostrophus fasciatus (Péron & Lesueur, 1807) Banded Hare-wallaby

SUPERCOHORT PLACENTALIA Bonaparte, 1838

COHORT AFROTHERIA Stanhope *et al.*, 1998

ORDER PAENUNGULATA Simpson, 1945

SUBORDER SIRENIA Iliger, 1811
Family Dugongidae J. Gray, 1821 [Dugong]
Subfamily Dugonginae J. Gray, 1821
Dugong dugon (P. Müller, 1776) Dugong

COHORT EUARCHONTOGLIRES Murphy *et al.*, 2001

SUPERORDER ARCHONTA Gregory, 1910 *sensu* Waddell *et al.*, 1999

ORDER PRIMATES Linnaeus, 1758
Family Hominidae J. Gray, 1825
Ω *Homo sapiens* Linnaeus, 1758 Human

SUPERORDER GLIRES Linnaeus, 1758

ORDER RODENTIA Bowdich, 1821

SUBORDER MYOMORPHA Brandt, 1855
Superfamily Muroidea Illiger, 1811
Family Muridae Illiger, 1811 [Rats and Mice]
Subfamily Murinae Illiger, 1811
Tribe Hydromyini Alston, 1876 *sensu* Lecompte *et al.*, 2008
† *Conilurus albipes* (M. Lichtenstein, 1829) White-footed Rabbit-rat
† *Conilurus capricornensis* Cramb & Hocknull, 2010 Capricorn Rabbit-rat
Conilurus penicillatus (Gould, 1842) Brush-tailed Rabbit-rat
Hydromys chrysogaster É. Geoffroy, 1804 Water Rat
Leggadina forresti (Thomas, 1906) Central Short-tailed Mouse
Leggadina lakedownensis Watts, 1976 Northern Short-tailed Mouse
† *Leporillus apicalis* (Gould, 1853) Lesser Stick-nest Rat
Leporillus conditor (Gould, 1848) Greater Stick-nest Rat
Mastacomys fuscus Thomas, 1882 Broad-toothed Rat
Melomys burtoni (Ramsay, 1887) Grassland Melomys
Melomys capensis Tate, 1951 Cape York Melomys
Melomys cervinipes (Gould, 1852) Fawn-footed Melomys
† *Melomys rubicola* Thomas, 1924 Bramble Cay Melomys
Mesembriomys gouldii (J. Gray, 1843) Black-footed Tree-rat
Mesembriomys macrurus (Peters, 1876) Golden-backed Tree-rat
Notomys alexis Thomas, 1922 Spinifex Hopping-mouse
† *Notomys amplus* Brazenor, 1936 Short-tailed Hopping-mouse
Notomys aquilo Thomas, 1921 Northern Hopping-mouse
Notomys cervinus (Gould, 1853) Fawn Hopping-mouse
Notomys fuscus (Wood Jones, 1925) Dusky Hopping-mouse
† *Notomys longicaudatus* (Gould, 1844) Long-tailed Hopping-mouse
† *Notomys macrotis* Thomas, 1921 Big-eared Hopping-mouse
Notomys mitchellii (W. Ogilby, 1838) Mitchell's Hopping-mouse
† *Notomys mordax* Thomas, 1922 Darling Downs Hopping-mouse
† *Notomys robustus* Mahoney *et al.*, 2008 Broad-cheeked Hopping-mouse
Pogonomys sp. undescribed Tree Mouse
Pseudomys albocinereus (Gould, 1845) Ash-grey Mouse
Pseudomys apodemoides Finlayson, 1932 Silky Mouse
† *Pseudomys auritus* Thomas, 1910 Long-eared Mouse
Pseudomys australis J. Gray, 1832 Plains Mouse
Pseudomys bolami Troughton, 1932 Bolam's Mouse
Pseudomys calabyi Kitchener & Humphreys, 1987 Kakadu Pebble-mouse
Pseudomys chapmani Kitchener, 1980 Western Pebble-mouse
Pseudomys delicatulus (Gould, 1842) Delicate Mouse
Pseudomys desertor Troughton, 1932 Desert Mouse
Pseudomys fieldi (Waite, 1896) Shark Bay Mouse
Pseudomys fumeus Brazenor, 1934 Smoky Mouse

† *Pseudomys glaucus* Thomas, 1910 Blue-grey Mouse
† *Pseudomys gouldii* (Waterhouse, 1839) Gould's Mouse
Pseudomys gracilicaudatus (Gould, 1845) Eastern Chestnut Mouse
Pseudomys hermannsburgensis (Waite, 1896) Sandy Inland Mouse
Pseudomys higginsi (Trouessart, 1897) Long-tailed Mouse
Pseudomys johnsoni Kitchener, 1985 Central Pebble-mouse
Pseudomys nanus (Gould, 1858) Western Chestnut Mouse
Pseudomys novaehollandiae (Waterhouse, 1843) New Holland Mouse
Pseudomys occidentalis Tate, 1951 Western Mouse
Pseudomys oralis Thomas, 1921 Hastings River Mouse
Pseudomys patrius (Thomas & Dollman, 1909) Eastern Pebble-mouse
Pseudomys shortridgei (Thomas, 1907) Heath Mouse
Uromys caudimaculatus (Krefft, 1867) Giant White-tailed Rat
Uromys hadrourus (Winter, 1984) Pygmy White-tailed Rat
Xeromys myoides Thomas, 1889 Water Mouse
Zyzomys argurus (Thomas, 1889) Common Rock-rat
Zyzomys maini Kitchener, 1989 Arnhem Land Rock-rat
Zyzomys palatalis Kitchener, 1989 Carpentarian Rock-rat
Zyzomys pedunculatus (Waite, 1896) Central Rock-rat
Zyzomys woodwardi (Thomas, 1909) Kimberley Rock-rat

Tribe Murini Illiger, 1811 *sensu* Lecompte *et al.*, 2008
Ω *Mus musculus* Linnaeus, 1758 House Mouse

Tribe Rattini Burnett, 1830 *sensu* Lecompte *et al.*, 2008
Rattus colletti (Thomas, 1904) Dusky Rat
Ω *Rattus exulans* (Peale, 1848) Pacific Rat
Rattus fuscipes (Waterhouse, 1839) Bush Rat
Rattus leucopus (J. Gray, 1867) Cape York Rat
Rattus lutreolus (J. Gray, 1841) Swamp Rat
† *Rattus macleari* (Thomas, 1887) Maclear's Rat
† *Rattus nativitatis* (Thomas, 1889) Bulldog Rat
Ω *Rattus norvegicus* (Berkenhout, 1769) Brown Rat
Ω *Rattus rattus* (Linnaeus, 1758) Black Rat
Rattus sordidus (Gould, 1858) Canefield Rat
Ω *Rattus tanezumi* Temminck, 1844 Oriental House Rat
Rattus tunneyi (Thomas, 1904) Pale Field Rat
Rattus villosissimus (Waite, 1898) Long-haired Rat

SUBORDER SCIUROMORPHA Brandt, 1855
Family Sciuridae G. Fischer, 1814 [Squirrels]
Subfamily Sciurinae G. Fischer, 1814
Tribe Sciurini G. Fischer, 1814
Ω *Funambulus pennantii* Wroughton, 1905 Northern Palm Squirrel
Ω *Sciurus carolinensis* Gmelin, 1788 Eastern Grey Squirrel

ORDER LAGOMORPHA Brandt, 1855
Family Leporidae G. Fischer, 1814 [Hares and Rabbits]
Ω *Lepus europaeus* Pallas, 1778 European Brown Hare
Ω *Oryctolagus cuniculus* (Linnaeus, 1758) European Rabbit

COHORT LAURASIATHERIA Waddell *et al.*, 1999

ORDER LIPOTYPHLA Haeckel, 1866

SUBORDER SORICOMORPHA Gregory, 1910
Superfamily Soricoidea G. Fischer, 1814
Family Soricidae G. Fischer, 1814 [Shrews]
Subfamily Crocidurinae Milne-Edwards, 1872
† *Crocidura trichura* Dobson, 1889 Christmas Island Shrew

SUBCOHORT SCROTIFERA Waddell *et al.*, 1999

ORDER CHIROPTERA Blumenbach, 1779

SUBORDER YINPTEROCHIROPTERA Springer *et al.*, 2001
Superfamily Pteropodoidea J. Gray, 1821

Family Pteropodidae J. Gray, 1821 [Old World Fruit Bats]

Subfamily Macroglossinae J. Gray, 1866

Macroglossus minimus (É. Geoffroy, 1810) Northern Blossom-bat

Syconycteris australis (Peters, 1867) Eastern Blossom-bat

Subfamily Nyctimeninae Miller, 1907

Nyctimene robinsoni Thomas, 1904 Eastern Tube-nosed Bat

Subfamily Pteropodinae J. Gray, 1821

Dobsonia magna Thomas, 1905 Bare-backed Fruit-bat

Pteropus alecto Temminck, 1837 Black Fruit-bat

† *Pteropus brunneus* Dobson, 1878 Percy Island Fruit-bat

Pteropus conspicillatus Gould, 1850 Spectacled Fruit-bat

Pteropus macrotis Peters, 1867 Large-eared Fruit-bat

Pteropus natalis Thomas, 1887 Christmas Island Fruit-bat

Pteropus poliocephalus Temminck, 1825 Grey-headed Fruit-bat

Pteropus scapulatus Peters, 1862 Little Red Fruit-bat

Superfamily Rhinolophoidea J. Gray, 1825

Family Megadermatidae H. Allen, 1864 [Ghost Bat]

Macroderma gigas (Dobson, 1880) Ghost Bat

Family Rhinolophidae J. Gray, 1825 [Horseshoe Bats]

Rhinolophus megaphyllus J. Gray, 1834 Eastern Horseshoe-bat

Rhinolophus robertsi Tate, 1952 Large-eared Horseshoe-bat

Family Hipposideridae Flower & Lydekker, 1891 [Leaf-nosed Bats]

Hipposideros ater Templeton, 1848 Dusky Leaf-nosed Bat

Hipposideros cervinus (Gould, 1854) Fawn Leaf-nosed Bat

Hipposideros diadema (É. Geoffroy, 1813) Diadem Leaf-nosed Bat

Hipposideros inornatus McKean, 1970 Arnhem Leaf-nosed Bat

Hipposideros semoni Matschie, 1903 Semon's Leaf-nosed Bat

Hipposideros stenotis Thomas, 1913 Northern Leaf-nosed Bat

Family Rhinonycteridae J. Gray, 1866 [Orange Leaf-nosed Bat]

Rhinonicteris aurantia (J. Gray, 1845) Orange Leaf-nosed Bat

SUBORDER YANGOCHIROPTERA Koopman, 1985

Family Emballonuridae Gervais, 1855 [Sheath-tailed Bats]

Subfamily Taphozoinae Jerdon, 1867

Saccolaimus flaviventris (Peters, 1867) Yellow-bellied Sheath-tailed Bat

Saccolaimus mixtus Troughton, 1925 Cape York Sheath-tailed Bat

Saccolaimus saccolaimus (Temminck, 1838) Bare-rumped Sheath-tailed Bat

Taphozous australis Gould, 1854 Coastal Sheath-tailed Bat

Taphozous georgianus Thomas, 1915 Common Sheath-tailed Bat

Taphozous hilli Kitchener, 1980 Hill's Sheath-tailed Bat

Taphozous kapalgensis McKean & Friend, 1979 Arnhem Sheath-tailed Bat

Taphozous troughtoni Tate, 1952 Troughton's Sheath-tailed Bat

Family Molossidae Gervais, 1855 [Free-tailed Bats]

Subfamily Molossinae Gervais, 1855

Austronomus australis (J. Gray, 1838) White-striped Free-tailed Bat

Chaerephon jobensis (Miller, 1902) Greater Northern Free-tailed Bat

Micronomus norfolkensis (J. Gray, 1839) Eastern Coastal Free-tailed Bat

Ozimops cobourgianus (Johnson, 1959) Northern Coastal Free-tailed Bat

Ozimops halli (Reardon *et al.*, 2014) Cape York Free-tailed Bat

Ozimops kitcheneri (McKenzie *et al.* 2014) Western Free-tailed Bat

Ozimops lumsdenae (Reardon *et al.* 2014) Northern Free-tailed Bat

Ozimops petersi (Leche 1884) Inland Free-tailed Bat

Ozimops planiceps (Peters, 1866) Southern Free-tailed Bat

Ozimops ridei (Felten, 1964) Ride's Free-tailed Bat
Setirostris eleryi (Reardon & McKenzie, 2008) Bristle-faced Free-tailed Bat

Family Miniopteridae Dobson, 1875 [Bent-winged Bats]
Miniopterus australis (Tomes, 1858) Little Bent-winged Bat
Miniopterus orianae Thomas, 1922 Large Bent-winged Bat

Family Vespertilionidae J. Gray, 1821 [Vespertilionid Bats]

Subfamily Kerivoulinae Miller, 1907
Phoniscus papuensis (Dobson, 1878) Golden-tipped Bat

Subfamily Murininae Miller, 1907
Murina florium Thomas, 1908 Flute-nosed Bat

Subfamily Nyctophilinae Peters, 1865
Nyctophilus arnhemensis Johnson, 1959 Arnhem Long-eared Bat
Nyctophilus bifax Thomas, 1915 Eastern Long-eared Bat
Nyctophilus corbeni Parnaby, 2009 Corben's Long-eared Bat
Nyctophilus daedalus Thomas, 1915 Pallid Long-eared Bat
Nyctophilus geoffroyi Leach, 1821 Lesser Long-eared Bat
Nyctophilus gouldi Tomes, 1858 Gould's Long-eared Bat
† *Nyctophilus howensis* McKean, 1975 Lord Howe Long-eared Bat
Nyctophilus major J. Gray, 1844 Greater Long-eared Bat
Nyctophilus sherrini Thomas, 1915 Tasmanian Long-eared Bat
Nyctophilus walkeri Thomas, 1892 Pygmy Long-eared Bat

Subfamily Vespertilioninae J. Gray, 1821

Tribe Vespertilionini J. Gray, 1821
Chalinolobus dwyeri Ryan, 1966 Large-eared Wattled Bat
Chalinolobus gouldii (J. Gray, 1841) Gould's Wattled Bat
Chalinolobus morio (J. Gray, 1841) Chocolate Wattled Bat
Chalinolobus nigrogriseus (Gould, 1856) Hoary Wattled Bat
Chalinolobus picatus (Gould, 1852) Little Pied Wattled Bat
Falsistrellus mackenziei Kitchener *et al.*, 1986 Western Falsistrelle
Falsistrellus tasmaniensis (Gould, 1858) Eastern Falsistrelle

Tribe Pipistrellini Tate, 1942
Pipistrellus adamsi Kitchener *et al.*, 1986 Forest Pipistrelle
† *Pipistrellus murrayi* Andrews, 1900 Christmas Island Pipistrelle
Pipistrellus westralis Koopman, 1984 Northern Pipistrelle

Tribe Nycticeiini Gervais, 1855
Scoteanax rueppellii (Peters, 1866) Greater Broad-nosed Bat
Scotorepens balstoni (Thomas, 1906) Inland Broad-nosed Bat
Scotorepens greyii (J. Gray, 1843) Little Broad-nosed Bat
Scotorepens orion (Troughton, 1937) Eastern Broad-nosed Bat
Scotorepens sanborni (Troughton, 1937) Northern Broad-nosed Bat
Vespadelus baverstocki (Kitchener *et al.*, 1987) Inland Forest-bat
Vespadelus caurinus (Thomas, 1914) Northern Cave-bat
Vespadelus darlingtoni (G. Allen, 1933) Large Forest-bat
Vespadelus douglasorum (Kitchener, 1976) Yellow-lipped Cave-bat
Vespadelus finlaysoni (Kitchener *et al.*, 1987) Finlayson's Cave-bat
Vespadelus pumilus (J. Gray, 1841) Eastern Forest-bat
Vespadelus regulus (Thomas, 1906) Southern Forest-bat
Vespadelus troughtoni (Kitchener *et al.*, 1987) Eastern Cave-bat
Vespadelus vulturnus (Thomas, 1914) Little Forest-bat

Subfamily Myotinae Tate, 1942
Myotis macropus (Gould, 1855) Large-footed Myotis

SUPERORDER FEREUUNGULATA Waddell *et al.*, 1999

ORDER CARNIVORA Bowdich, 1821

SUBORDER CANIFORMIA Kretzoi, 1943

INFRAORDER CYNOIDEA Flower, 1869
Family Canidae G. Fischer, 1817 [Dogs]
Ω *Canis familiaris* Linnaeus, 1758 Domestic Dog and Dingo
Ω *Vulpes vulpes* (Linnaeus, 1758) Red Fox

INFRAORDER ARCTOIDEA Flower, 1869 *sensu* Tedford, 1976
Superfamily Phocoidea J. Gray, 1821
Family Otariidae J. Gray, 1825 [Eared Seals]
Arctocephalus pusillus (Schreber, 1775) Cape Fur Seal
Arctophoca forsteri (Lesson, 1828) Long-nosed Fur Seal
Arctophoca gazella Peters, 1875 Antarctic Fur Seal
Arctophoca tropicalis (J. Gray, 1872) Subantarctic Fur Seal
Neophoca cinerea (Péron, 1816) Australian Sea-lion
Phocarctos hookeri (J. Gray, 1844) New Zealand Sea-lion
Family Phocidae J. Gray, 1821 [Earless Seals]
Subfamily Monachinae J. Gray, 1869
Hydrurga leptonyx (de Blainville, 1820) Leopard Seal
Leptonychotes weddellii (Lesson, 1826) Weddell Seal
Lobodon carcinophaga (Hombron & Jacquinot, 1842) Crabeater Seal
Mirounga leonina (Linnaeus, 1758) Southern Elephant Seal
Ommatophoca rossii J. Gray, 1844 Ross Seal

INFRAORDER MUSTELIDA Tedford, 1976
Family Mustelidae G. Fischer, 1814 [Weasels, Badgers, Skunks & Otters]
Subfamily Mustelinae G. Fischer, 1814 [Grison, Polecats, Weasels and Ferrets]
Ω *Mustela putorius* Linnaeus, 1758 European Polecat

SUBORDER FELIFORMIA Kretzoi, 1945
Family Felidae G. Fischer, 1817 [Cats]
Subfamily Felinae G. Fischer, 1817
Ω *Felis catus* Linnaeus, 1758 Domestic Cat

SUPERORDER EUUNGULATA Waddell *et al.*, 2001

ORDER PERISSODACTYLA Owen, 1848
Family Equidae J. Gray, 1821 [Horses and Asses]
Ω *Equus asinus* Linnaeus, 1758 Donkey
Ω *Equus caballus* Linnaeus, 1758 Horse

ORDER ARTIODACTYLA Owen, 1848 *sensu* Montgelard *et al.*, 1997

SUBORDER SUINA J. Gray, 1868
Family Suidae J. Gray, 1821 [Pigs]
Ω *Sus scrofa* Linnaeus, 1758 Pig

SUBORDER TYLOPODA Illiger, 1811
Family Camelidae J. Gray, 1821 [Camels and relatives]
Ω *Camelus dromedarius* Linnaeus, 1758 One-humped Camel
Ω *Lama glama* (Linnaeus, 1758) Llama
Ω *Lama pacos* (Linnaeus, 1758) Alpaca

SUBORDER RUMINANTIA Scopoli, 1777
Family Bovidae J. Gray, 1821 [Cattle, Sheep and Goats]
Subfamily Bovinae J. Gray, 1821
Ω *Bos bison* Linnaeus, 1758 American Bison
Ω *Bos javanicus* d'Alton, 1823 Banteng
Ω *Bos taurus* Linnaeus, 1758 Cattle
Ω *Bubalus bubalis* (Linnaeus, 1758) Swamp Buffalo
Subfamily Antilopinae J. Gray, 1821
Tribe Antilopini J. Gray, 1821
Ω *Antilope cervicapra* (Linnaeus, 1758) Blackbuck
Tribe Caprini J. Gray, 1821
Ω *Capra hircus* Linnaeus, 1758 Goat
Ω *Ovis aries* Linnaeus, 1758 Sheep
Family Cervidae Goldfuss, 1820 [Deer]
Subfamily Cervinae Goldfuss, 1820
Ω *Axis axis* (Erxleben, 1777) Chital Deer
Ω *Axis porcinus* (Zimmermann, 1780) Hog Deer
Ω *Cervus elaphus* Linnaeus, 1758 Red Deer
Ω *Cervus timorensis* de Blainville, 1822 Rusa Deer
Ω *Cervus unicolor* Kerr, 1792 Sambar Deer
Ω *Dama dama* (Linnaeus, 1758) Fallow Deer

SUBORDER WHIPPOMORPHA Waddell *et al.*, 1999

INFRAORDER CETACEA Brisson, 1762

PARVORDER MYSTICETI J. Gray, 1864

Family Neobalaenidae J. Gray, 1873 [Pygmy Right Whale]

Caperea marginata (J. Gray, 1846) Pygmy Right Whale

Family Balaenidae J. Gray, 1821 [Right Whales]

Eubalaena australis (Desmoulins, 1822) Southern Right Whale

Family Balaenopteridae J. Gray, 1864 [Rorquals]

Balaenoptera acutorostrata Lacépède, 1804 Common Minke Whale

Balaenoptera bonaerensis Burmeister, 1867 Antarctic Minke Whale

Balaenoptera borealis Lesson, 1828 Sei Whale

Balaenoptera brydei Olsen, 1913 Bryde's Whale

Balaenoptera edeni Anderson, 1879 Eden's Whale

Balaenoptera musculus (Linnaeus, 1758) Blue Whale

Balaenoptera omurai Wada *et al.*, 2003 Omura's Whale

Balaenoptera physalus (Linnaeus, 1758) Fin Whale

Megaptera novaeangliae (Borowski, 1781) Humpback Whale

PARVORDER ODONTOCETI Flower, 1867

Superfamily Physeteroidea J. Gray, 1821 *sensu* Bianucci & Landini, 2006

Family Physeteridae J. Gray, 1821 [Sperm Whales]

Physeter macrocephalus Linnaeus, 1758 Sperm Whale

Family Kogiidae Gill, 1871 [Pygmy and Dwarf Sperm Whales]

Kogia breviceps (de Blainville, 1838) Pygmy Sperm Whale

Kogia sima (Owen, 1866) Dwarf Sperm Whale

Superfamily Ziphioidea J. Gray, 1865 *sensu* Rice, 2009

Family Ziphiidae J. Gray, 1865 [Beaked Whales]

Berardius arnuxii Duvernoy, 1851 Arnoux's Beaked Whale

Hyperoodon planifrons Flower, 1882 Southern Bottle-nosed Whale

Indopacetus pacificus (Longman, 1926) Longman's Beaked Whale

Mesoplodon bowdoini R. Andrews, 1908 Andrews' Beaked Whale

Mesoplodon densirostris (de Blainville, 1817) Blainville's Beaked Whale

Mesoplodon ginkgodens Nishiwaki & Kamiya, 1958 Ginkgo-toothed Beaked Whale

Mesoplodon grayi Haast, 1876 J. Gray's Beaked Whale

Mesoplodon hectori (J. Gray, 1871) Hector's Beaked Whale

Mesoplodon layardii (J. Gray, 1865) Strap-toothed Beaked Whale

Mesoplodon mirus True, 1913 True's Beaked Whale

Tasmacetus shepherdi Oliver, 1937 Tasman Beaked Whale

Ziphius cavirostris G. Cuvier, 1823 Cuvier's Beaked Whale

Superfamily Delphinoidea J. Gray, 1821 *sensu* Rice, 2009

Family Delphinidae J. Gray, 1821 [Dolphins and Killer Whales]

Delphinus capensis Gray, 1828 Long-beaked Common Dolphin

Delphinus delphis Linnaeus, 1758 Short-beaked Common Dolphin

Feresa attenuata J. Gray, 1874 Pygmy Killer Whale

Globicephala macrorhynchus (J. Gray, 1846) Short-finned Pilot Whale

Globicephala melas (Traill, 1809) Long-finned Pilot Whale

Grampus griseus (G. Cuvier, 1812) Risso's Dolphin

Lagenodelphis hosei Fraser, 1956 Fraser's Dolphin

Lagenorhynchus cruciger (Quoy & Gaimard, 1824) Hourglass Dolphin

Lagenorhynchus obscurus (J. Gray, 1828) Dusky Dolphin

Lissodelphis peronii (Lacépède, 1804) Southern Rightwhale Dolphin

Orcaella heinsohni Beasley *et al.*, 2005 Australian Snub-finned Dolphin

Orcinus orca (Linnaeus, 1758) Killer Whale

Peponocephala electra (J. Gray, 1846) Melon-headed Whale

Pseudorca crassidens (Owen, 1846) False Killer Whale

Sousa sahulensis Jefferson & Rosenbaum, 2014 Australian Hump-backed Dolphin

Stenella attenuata (J. Gray, 1846) Pantropical Spotted Dolphin

Stenella coeruleoalba (Meyen, 1833) Striped Dolphin
Stenella longirostris (J. Gray, 1828) Spinner Dolphin
Steno bredanensis (G. Cuvier, 1828) Rough-toothed Dolphin
Tursiops aduncus (Ehrenberg, 1833) Indo-Pacific Bottle-nosed Dolphin
Tursiops australis Charlton-Robb *et al.*, 2011 Burrunan Bottle-nosed Dolphin
Tursiops truncatus (Montagu, 1821) Common Bottle-nosed Dolphin

Family Phocoenidae J. Gray, 1825 [Porpoises]
Phocoena dioptrica Lahille, 1912 Spectacled Porpoise

Class Mammalia Linnaeus, 1758

Class Mammalia Linnaeus, 1758: 12.

COMMENTS: When originally proposed, this rank included the orders Primates (Linnaeus 1758: 16, 20) (including *Homo*, *Simia*, *Lemur*, *Vespertilio*), Bruta (Linnaeus 1758: 16, 33) (including *Elephas*, *Trichechus*, *Bradypus*, *Myrmecophaga*, *Manis*), Ferae (Linnaeus 1758: 16, 37) (including *Phoca*, *Canis*, *Felis*, *Viverra*, *Mustela*, *Ursinus*), Bestiae (Linnaeus 1758: 16, 49) (including *Sus*, *Dasypus*, *Erinaceus*, *Talpa*, *Sorex* and *Didelphis*), Glires (Linnaeus 1758: 16, 56) (including *Rhinoceros*, *Hystrix*, *Lepus*, *Castor*, *Mus*, *Sciurus*), Pecora (Linnaeus 1758: 16, 65) (including *Camellus*, *Moschatus*, *Cervinus*, *Capra*, *Ovis* and *Bos*), Belluae (Linnaeus 1758: 16, 73) (including *Equus* and *Hippopotamus*) and Cete (Linnaeus 1758: 16, 75) (including *Monodon*, *Balaena*, *Physeter* and *Delphinus*). McKenna and Bell (1997: 35) identified several pre-Linnean synonyms including Zootoka (Aristotle 330BC; see Gill 1873a: 458) and Vivipera (Ray 1693: 53), along with several other names including Mastodologie (Rafinesque 1814a: 9), Mastodia (Rafinesque 1814b: 47), Thricozoa (Oken 1847: 638), Aistheseozoa (Oken 1847: 563), Pilifera (Bonnet, 1892: 236) and Mammalea (Kinman 1994). Mammalia recognised at cohort rank by Gardiner (1982: 229) but at class rank by most authors including Iredale and Troughton (1934: vii, 1), Simpson (1945: 39) and McKenna and Bell (1997: 35). The definition of the Class Mammalia was reviewed by Rowe and Gauthier (1992: 372) and Lucas (1992: 370).

Klasse Säugthiere Duméril, 1806a: 4.

COMMENTS: When originally proposed, this rank included the families Bimanen (Duméril, 1806a: 6 [=Hominidae (J. Gray, 1825a)]); Quadrumanen [=Quadrumana (Blumenbach, 1791: 49) [=Primates (Linnaeus, 1758: 16, 20 part)]); Chiropteren (Duméril, 1806a [=Chiroptera (Blumenbach, 1779) and Dermoptera (Illiger, 1811: 63, 116]); Digigraden (Duméril, 1806a: 4, 12 [=Carnivora (Bowdich, 1821 part)]); Plantigraden (Duméril, 1806a: 4, 14 [=Carnivora (Bowdich, 1821 part) and Order Lipotyphla (Haeckel, 1866 part)]); Pedimanen (Duméril, 1806a [=Marsupialia (Illiger, 1811 part)]); Nagethiere (Duméril, 1806a: 4, 18 [=Marsupialia (Illiger, 1811 part), Rodentia (Bowdich, 1821 part) and Primates (Linnaeus, 1758 part)]); Zahnlose (Duméril, 1806a: 4, 20 [=Pilosa (Flower, 1883: 184 part), Tubulidentata (Huxley, 1872: 288), Pholidota (Weber, 1904: 411), Cingulata (Illiger, 1811: 110 part), Monotremata (Bonaparte, 1832)]); Tardigraden (Duméril, 1806a: 4, 22 [=Pilosa (Flower, 1883: 184 part)]); Pachydermen (Duméril, 1806a: 4, 22 [=Hyracoidea (Huxley, 1869: 101), Artiodactyla (Owen, 1848 part), Perissodactyla (Owen, 1848 part), and Proboscidea (Illiger, 1811: 96)]); Wienerkäuer (Duméril, 1806a: 4, 24 [=Artiodactyla (Owen, 1848 part)]); Solipeden (Duméril, 1806a: 4, 26 [=Perissodactyla (Owen, 1848 part)]); Amphibien (Duméril, 1806a [=Carnivora (Bowdich, 1821 part) and Sirenia (Illiger, 1811)]); and Cetaceen (Duméril, 1806a [=Cetacea (Brisson, 1762)]).

Family Zahnlose Duméril, 1806a: 4, 20.

TYPE GENUS: Not based on a genus group name.

COMMENTS: When originally proposed, this rank was placed in the Klasse Säugthiere (Duméril, 1806a [=Mammalia (Linnaeus, 1758)]) and included the genera *Myrmecophaga* Linnaeus, 1758: 35; *Orycteropus* É. Geoffroy Saint-Hilaire, 1796a: 102; *Manis* Linnaeus, 1758: 36; *Dasypus* Linnaeus, 1758: 50; *Echidna* G. Cuvier, 1797 [=*Tachyglossus* (Illiger, 1811)], and *Ornithorhynchus* Blumenbach, 1800a.

Order Pollicata Illiger, 1811: 58, 66.

COMMENTS: When originally proposed, this rank was placed in the Class Mammalia (Linnaeus, 1758) and included the families Quadrumana (Blumenbach, 1779: 49 [=Primates (Linnaeus, 1758 part)]), Prosimii (Illiger, 1811: 72) [=Suborder Strepsirrhini (É. Geoffroy, 1812: 156)]), Macrotarsi (Illiger, 1811: 73 [=Family Tarsiidae (J. Gray, 1825a: 338)]), Leptodactyla (Illiger, 1811: 75 [=Family Daubonentoniidae (J. Gray, 1863a: 151)]) and Marsupialia (Illiger, 1811). Subsequently recognised by Illiger (1815: 64). Synonymised within the Subclass Marsupialia by McKenna and Bell (1997: 51).

Order Pedimana G. Fischer, 1813a: 14.

COMMENTS: When originally proposed, this rank was placed in the Unguiculata (Linnaeus, 1766 [=Placentalia (Bonaparte, 1838 part)]) and included the genera *Didelphis* Linnaeus, 1758: 54; *Sipalus* G. Fischer, 1813b [=*Phalanger* Storr, 1780]; *Dasyurus* É. Geoffroy, 1796b) and *Cheiromys* G. Cuvier, 1800: Table 1 [=*Daubentonia* É. Geoffroy, 1795: 195; Order Primates]. Name also recognised by G. Fischer (1813b: xxiii, 569) and G. Fischer (1813b: 569; 1817: 372).

Order Metatarsigrada G. Fischer, 1813a: 14.

COMMENTS: When originally proposed, this rank was placed in the Unguiculata (Linnaeus, 1766 [=Placentalia (Bonaparte, 1838 part)]) and included the genera *Perameles* É. Geoffroy, 1803d; *Wombatus* Desmarest, 1804a [=*Vombatus* É. Geoffroy, 1803b]; *Phascolomys* Duméril, 1806a [=*Vombatus* É. Geoffroy, 1803b]; *Kangurus* É. Geoffroy and G. Cuvier, 1795 [=*Macropus* Shaw, 1790]; *Dipus* Zimmermann, 1780: 354; *Sciurus* Linnaeus, 1758; *Myoxus* Zimmermann, 1780: 351; *Lepus* Linnaeus, 1758; *Lagomys* Storr, 1780: Table B [=*Marmota* Blumenbach, 1779: 79]; *Cavia* Pallas, 1766: 30; *Paca* G. Fischer, 1813a: 14 [=*Cuniculus* Brisson, 1762: 13, 98]; *Mus* Linnaeus, 1758; *Brachyurus* G. Fischer, 1813a: 14, 24 [=*Lemmus* Link, 1795: 75]; *Talpoides* Lacépède, 1799a: 10 [=*Spalax* Güldenstädt, 1770: 409]; *Spalax* Güldenstädt, 1770: 409; *Cricetus* Leske, 1779: 168; *Arctomys* Schreber, 1780: Plate 207 [=*Marmota* Blumenbach, 1779: 79]; *Hystrix* Linnaeus, 1758: 56; and *Coandu* [=*Coendou*] Lacépède, 1799a. Name is equivalent to Metatarsii (G. Fischer, 1814).

Order Plantigrada G. Fischer, 1813a: 14.

COMMENTS: When originally proposed, this rank was placed in the Unguiculata (Linnaeus, 1766 [=Placentalia (Bonaparte, 1838 part)]) and included the genera *Manis* Linnaeus, 1758: 36; *Dasypus* Linnaeus, 1758: 50; *Myrmecophaga* Linnaeus, 1758: 35; *Oryctopus* G. Cuvier, 1797: 144, † *Onychotherium* G. Fischer, 1809: 253; *Bradypus* Linnaeus, 1758: 18, 34; † *Megatherium* G. Cuvier, 1796: 303, 308; *Echinopus* G. Fischer, 1813a [=*Tachyglossus* Illiger, 1811]; *Erinaceus* Linnaeus, 1758: 52; *Tenrecus* [=*Tenrec*] Lacépède, 1799a: 7; *Sorex* Linnaeus, 1758: 53; *Talpa* Linnaeus, 1758: 52; *Scalops* Illiger, 1811: 126 [sic=*Scalopus* É. Geoffroy, 1803c: 77]; *Chrysochloris* Lacépède, 1799a: 7; and *Ursus* Linnaeus, 1758: 47. Rank also included within G. Fischer (1814: ix, 106; 1817: 372).

HOMONYMS:

Family Plantigrada Illiger, 1811, are carnivores of the Class Mammalia (Order Carnivora). See individual entry.

Order Metatarsii G. Fischer, 1814: v, 3.

COMMENTS: When originally proposed, this rank was placed in the Unguiculata (Linnaeus, 1766 [=Placentalia (Bonaparte, 1838 part)]) and included the families Marsupio [sic] [=Marsupialia (Illiger, 1811)], Sciuriorum [=Sciuridae (G. Fischer, 1814)], Cricetinorum [=Cricetidae (G. Fischer, 1814: vi, 42)], Murinorum (G. Fischer, 1814 [=Muridae (Illiger, 1811)]), Spalacoidum (G. Fischer, 1814: viii, 71 [=Spalacidae (Gray, 1821: 303]), Caviarum [=Caviidae (G. Fischer, 1814: viii, 81)], Leporinorum [=Leporidae (G. Fischer, 1814)], and Hystricinorum [=Hystricidae (G. Fischer, 1814: viii, 99)]. Name also referred to by G. Fischer (1817: 372).

Class Mastodologie Rafinesque, 1814a: 9.

COMMENTS: Name used in preference to Mammalia (Linnaeus, 1758). Synonymised within the Class Mammalia by McKenna and Bell (1997: 35).

Class Mastodia Rafinesque, 1814b: 47.

COMMENTS: Name used in preference to Mammalia (Linnaeus, 1758). Also referred to by Rafinesque (1814a: 12) who gives the spelling as Mastodologie on page 9. Reviewed by Paclt (1960: 47). Synonymised within the Class Mammalia by McKenna and Bell (1997: 35).

Order Carnassiers G. Cuvier, 1816a: xxx, 119.

COMMENTS: When originally proposed, this rank was placed in the Class Mammifères (G. Cuvier, 1816a: xxix, 70 [=Mammalia (Linnaeus, 1758)]) and included the families Cheiroptères (G. Cuvier, 1816a: xxx, 121 [=Chiroptera (Blumenbach, 1779)]), Insectivores (G. Cuvier, 1816a: xxx, 131 [=Order Lipotyphla (Haeckel, 1866)]), Carnivores (G. Cuvier, 1816a: xxx, 138 [=Carnivora (Bowdich, 1821)]), and Marsupiaux (G. Cuvier, 1816a: xxxi, 169 [=Marsupialia (Illiger, 1811)]).

HOMONYMS:

Order Carnassiers de Blainville, 1816a, mammals of the Class Mammalia in part (Subclass Placentalia). See individual entry.

Order Sarcophaga Bowdich, 1821: 7, 23.

COMMENTS: When originally proposed, this rank was placed in the Class Mammalia (Linnaeus, 1758) and included the families Cheiroptera (J. Gray, 1821 [=Chiroptera (Blumenbach, 1779)]), Insectivora (Bowdich, 1821 [=Lipotyphla (Haeckel, 1866)]), Carnivora (Bowdich, 1821), and genus *Didelphis* Linnaeus, 1758: 54.

HOMONYMS:

Sarcophaga Owen, 1839a, carnivorous marsupials of the Class Mammalia (Order Dasyuromorphia). Synonymised within the Superfamily Dasyuroidea (Goldfuss, 1820a) in this work. See individual entry.

Sub-Kingdom Vertebrosa Gray, 1821: 297.

COMMENTS: When originally proposed this rank included the classes Bimanes (J. Gray, 1821: 297 [=Primates (Linnaeus, 1758 part)]), Quadrumanes (J. Gray, 1821: 297 [=Primates (Linnaeus, 1758 part)]), Cheiroptera (J. Gray, 1821 [=Chiroptera Blumenbach, 1779])), Quadripedes (J. Gray, 1821 [=Placentalia (Bonaparte, 1838 part)]), Pedimanes (J. Gray, 1821 [=Mammalia (Linnaeus, 1758 part)]), and Cetacea (Brisson, 1762).

Class Pedimanes J. Gray, 1821: 308.

COMMENTS: When originally proposed, this rank was placed in the Sub-kingdom Vertebrosa (J. Gray, 1821: 297 [=Mammalia (Linnaeus, 1758)]) and included the orders Ferae (J. Gray, 1821 [=Marsupialia (Illiger, 1811 part)]), Brutae (J. Gray, 1821 [=Marsupialia (Illiger, 1811 part)]), Glires (Linnaeus, 1758) and Rosores (J. Gray, 1821 [=Rodentia (Bowdich, 1821 part)]).

Series Ovovivipara Bonaparte, 1838: 113.

COMMENTS: When originally proposed, this rank was placed in the Class Mammalia (Linnaeus, 1758) and included the orders Marsupialia (Illiger, 1811) and Monotremata (Bonaparte, 1832). Name also recognised and further described by Bonaparte (1840: 249).

Class Thricozoa Oken, 1847: xi, 563, 566.

COMMENTS: Constituents of this class not clear. Synonymised within the Class Mammalia by McKenna and Bell (1997: 35).

Class Aistheseozoa Oken, 1847: 563.

COMMENTS: Constituents of this class not clear. Synonymised within the Class Mammalia by McKenna and Bell (1997: 35).

Subclass Lyencephala Owen, 1858a: 14.

COMMENTS: When originally proposed, this rank was placed within the Class Mammalia (Linnaeus, 1758) and included the orders Marsupialia (Illiger, 1811) and Monotremata (Bonaparte, 1832). Followed by Owen (1859: 24).

Subclass Ditremata Ameghino, 1889: xvii, 43.

COMMENTS: When originally proposed, this rank was placed in the Class Mammalia (Linnaeus, 1758) and included the 'grandes ramas' Heterodonta (Ameghino, 1889 [=Mammalia (Linnaeus, 1758 part)]) and Homalodonta (Ameghino, 1889 [=Mammalia (Linnaeus, 1758 part)]).

Grandes Ramas Heterodonta Ameghino, 1889: xvii, 43.

COMMENTS: When originally proposed, this rank was placed in the Subclass Ditremata (Ameghino, 1889 [=Mammalia (Linnaeus, 1758 part)]) and included the seccions Planungulata (Ameghino, 1889: xvii, 44 [=Mammalia (Linnaeus, 1758 part)]), Unguiculata (Linnaeus, 1766 [=Placentalia (Bonaparte, 1838 part)]), Ungulata (Linnaeus, 1766 part), Ptética (Ameghino, 1889 [=Chiroptera (Blumenbach, 1779)]) and † Hydrothereuta (Ameghino, 1889: xxi, 44, 353 [=† Archaeoceti (Flower, 1883: 182)]).

Grandes Ramas Homalodonta Ameghino, 1889: xxiv, 43, 653.

COMMENTS: When originally proposed, this rank was placed in the Subclass Ditremata (Ameghino, 1889 [=Mammalia (Linnaeus, 1758 part)]) and included the grand seccions Bruta (Ameghino, 1889: xxiv, 653) [non Linnaeus, 1758: 16, 33 and Bruta Ameghino, 1889: xxiv, 653] and Cetacea (Brisson, 1762).

Grand Group Sarcobora Ameghino, 1889: xx, 105, 276.

COMMENTS: When originally proposed, this rank was placed in the Seccion Unguiculata (Linnaeus, 1766 [=Placentalia (Bonaparte, 1838 part)]), and included the orders Phonoctonia (Ameghino, 1889: xx, 276), Pedimana (Van der Hoeven, 1855: xiii, 902 [=Didelphimorphia (Gill, 1872: vi, 26)]), Dasyura (Ameghino, 1889 [=Dasyuromorphia (Gill, 1872)]), † Creodonta (Cope, 1875: 446), Carnivora (Bowdich, 1821) and Pinnipedia (Illiger, 1811 [=Phocoidea (J. Gray, 1821)]).

Grand Group Alloidea Ameghino, 1889, xx, 105, 263.

COMMENTS: When originally proposed, this rank was placed in the Grand Section Unguiculata (Linnaeus, 1766 [=Placentalia (Bonaparte, 1838 part)]) and included the orders Insectivora (Bowdich, 1821 [=Lipotyphla (Haeckel, 1866)]), Microbiotheria (Ameghino, 1889: xx, 263), Peramelia (Ameghino, 1889 [=Peramelemorphia (Ameghino, 1889)]), Macropoda (Ameghino, 1889 [=Diprotodontia (Owen, 1877 part)]) and † Plagiaulacoidea (Ameghino, 1889: xx, 263, 268 [=† Multituberculata (Cope, 1884: 687 part)]).

Pilifera Bonnett, 1892: 236.

COMMENTS: Rank unknown. Constituents of this rank not clear. Synonymised within the Class Mammalia by McKenna and Bell (1997: 35).

Subclass Palaeotherida Broom, 1935: 36.

COMMENTS: When originally proposed, this rank was placed in the Class Mammalia (Linnaeus, 1758) and included the Monotremata (Bonaparte, 1832), Marsupialia (Illiger, 1811), Edentata (G. Cuvier, 1797: 142 [=Order Cingulata (Illiger, 1811: 110) and Pilosa (Flower, 1883: 184)]), Artiodactyla (Owen, 1848), Perissodactyla (Owen, 1848), Proboscidea (Illiger, 1811: 96), Sirenia (Illiger, 1811) and Chrysochloridea (Broom, 1915: 353) based on having no mesethmoid. Rank criticised by Parrington (1974: 425), who suggested it had no credence whatsoever, and not recognised by subsequent authors.

HOMONYMS:

† Family Palaeotherida Haeckel, 1866: clviii, of the Class Mammalia (Order Perissodactyla). Synonymised within the † Family Paleotheridae (Bonaparte, 1850a: unpaginated chart) by McKenna and Bell (1997: 472).

Subclass Marsupionta Gregory, 1947: 46.

COMMENTS: When originally proposed, this rank was placed in the Class Mammalia (Linnaeus, 1758) and included the Orders Marsupialia (Illiger, 1811) and Monotremata (Bonaparte, 1832). The Marsupionta hypothesis was supported by Kühne (1973: 63; 1975: 585; 1977: 225), and subsequently by several genetic studies, though at different degrees of confidence, including Janke *et al.* (1996: 153;

1997: 1276; 2002: 71), Penny and Hasegawa (1997: 550), and Zardoya and Mayer (1998: 14229, 14231). Weak support was provided by Toyosawa *et al.* (1998: 13060) using amino acid sequences and Kirsch and Mayer (1998: 1126) using DNA hybridisation. The Marsupionta hypothesis was reviewed by Musser (2003: 936). The Marsupionta was not supported by Parrington (1974: 425), Marshall (1979: 400), Kuhn and Zeller (1987: 68), Luckett and Zeller (1989: 193, 202), McKenna and Bell (1997: 51), M. Phillips and Penny (2003: 171) or Kullberg *et al.* (2008: 115). Rather than a link between monotremes and marsupials, Kullberg (2008: 115) found unequivocal support for the Monotremata as a sister group to the Marsupialia and Placentalia. Synonymised within the Subclass Marsupialia by McKenna and Bell (1997: 51).

Mammaliamorpha Rowe, 1988: 245, 249.

COMMENTS: When originally proposed as a new name it was at unknown rank and included the last common ancestor of † Family Tritylodontidae (Cope, 1884: 687) and Class Mammalia (Linnaeus, 1758), and all its descendants.

Mammaliaformes Rowe, 1988: 245, 250.

COMMENTS: When originally proposed as a new name it was at unknown rank as the sister taxon of the † Family Tritylodontidae (Cope, 1884: 687) within the Mammaliamorpha (Rowe, 1988: 249) and comprises the last common ancestor of † Morganucodontidae (Kühne, 1958: 197, 222) and Mammalia (Linnaeus, 1758: 12, 14), and all descendants. Name reviewed by McKenna and Bell (1997: 507, 511).

Class Mammalea Kinman, 1994: 37.

COMMENTS: When originally proposed, this rank included most current mammalian orders with the addition of a 'formes' suffix. Synonymised within the Class Mammalia by McKenna and Bell (1997: 35).

Subclass Prototheria Gill, 1872

Subclass Prototheria Gill, 1872: vi.

COMMENTS: When originally proposed, this rank was placed in the Class Mammalia (Linnaeus, 1758), with Ornithodelphia (de Blainville, 1834 [=Monotremata (Bonaparte, 1832)]), and included the Order Monotremata (Bonaparte, 1832). On page 27 of Gill (1872) only the Subclass Ornithodelphia is referred to. The terms Prototheria, Metatheria and Eutheria used by Huxley (1881: 654, 657) were theoretical terms to designate stages of evolution rather than as taxonomic names (see Gregory, 1910: 94; Aplin & Archer, 1987: xxvi; Simpson, 1945: 164). Rank synonymised within the Monotremata, which was recognised at the subclass rank by Iredale and Troughton (1934: vii, 1). Prototheria recognised at subclass rank by most authors including Simpson (1945: 39), Strahan (1983: xxi, 1; 1995: 6, 31), McKenna and Bell (1997: 35), and Van Dyck and Strahan (2008: 9, 29).

Order Reptantia Illiger, 1811: 63, 113.

COMMENTS: When originally proposed, this rank was placed in the Class Mammalia (Linnaeus, 1758) and included the Family Reptantia Illiger, 1811 [=Prototheria (Gill, 1872)]. Synonymised within the Subclass Prototheria by McKenna and Bell (1997: 35).

Family Reptantia Illiger, 1811: 114.

TYPE GENUS: Not based on a genus group name.

COMMENTS: When originally proposed, this rank was placed in the Order Reptantia (Gill, 1872 [=Prototheria (Gill, 1872)]) and included the genera *Tachyglossus* Illiger, 1811; *Ornithorhynchus* Blumenbach, 1800a; and ?†*Pamphractus* Illiger, 1811: 115. Reptantia is also a taxon name used in the higher classification of tetrapods (McKenna and Bell (1997: 35). Synonymised within the Subclass Prototheria by McKenna and Bell (1997: 35).

Suborder Didelphes de Blainville, 1822a: Table 3.

COMMENTS: When originally proposed, this rank was placed in the Class Mammifères (G. Cuvier, 1816a: xxix, 70 [=Mammalia (Linnaeus, 1758)]) that included the 'Normaux' containing the marsupials as 'les Sarigues' and 'les Phalangers', and the 'Anomaux' containing the monotremes as 'L'Echidné' [=Tachyglossidae Gill, 1872] and 'L'Ornithorhynchque' [=Ornithorhynchidae (J. Gray, 1825)].

Amasta Haeckel, 1866: cxlii.

COMMENTS: Rank not provided but placed in brackets after Ornithodelphia (de Blainville, 1834 [=Monotremata (Bonaparte, 1832)]), and included the genera *Ornithorhynchus* Blumenbach, 1800a; and *Echidna* G. Cuvier, 1797. Synonymised within the Subclass Prototheria by McKenna and Bell (1997: 35).

Sauropsidelphia Roger, 1887: 4.

COMMENTS: Rank not given but placed above Order Monotremata (Bonaparte, 1832). Synonymised within the Subclass Prototheria by McKenna and Bell (1997: 35).

Order Ornithostomi Cope, 1889b: 874.

COMMENTS: When originally proposed, this rank was placed in the Prototheria (Gill, 1872) and included the families Ornithorhynchidae (J. Gray, 1825a) and Echidnidae (Burnett, 1830a [=Tachyglossidae (Gill, 1872)]). Synonymised within the Subclass Prototheria by McKenna and Bell (1997: 35).

Subclass Atheria Kermack *et al.*, 1973: 108.

COMMENTS: When originally proposed, this rank was placed in the Class Mammalia (Linnaeus, 1758) contained the non-therian mammals including the orders † Triconodonta (Osborn 1888: 251), † Multituberculata (Cope, 1884: 687) and Monotremata (Bonaparte 1832. Placed within the Prototheria by Kemp (1983: 354) and subsequently not recognised.

Order Monotremiformes Kinman, 1994: 37.

COMMENTS: When originally proposed, this rank was placed in the Class Mammalea (Kinman, 1994 [=Mammalia (Linnaeus, 1758)]). Synonymised within the Subclass Prototheria by McKenna and Bell (1997: 35).

Infraclass Australosphenida Luo *et al.*, 2001a: 53, 56.

COMMENTS: When originally proposed as a new rank it was placed in the Subclass Holotheria (Wible *et al.*, 1995: 10, 11 [=Theria (Parker & Haswell, 1897)]) and included the Monotremata (Bonaparte, 1832), † Ausktribosphenida (Rich *et al.*, 1997: 1439) and † *Ambondro* J. Flynn *et al.*, 1999: 58. Recognised at the rank of Superdivision by Benton (2005: 300, 301). The recognition of Australosphenida remains controversial as most taxonomists maintain the name Prototheria as a fitting contrast to the other group of living mammals, the Theria. In theory, the Prototheria is taxonomically redundant, since Monotremata is currently the only order which can still be confidently included, but its retention might be justified if new fossil evidence, or a re-examination of known fossils, enables extinct relatives of the monotremes to be identified and placed within a wider grouping. Clade has since been recognised by several authors including Luo *et al.* (2002: 1, 22), Martin and Rauhut (2005: 414) and Rougier *et al.* (2007: 1, 5). In contrast to these suggestions other authors have criticised the inclusion of Monotremata within Australosphenida and placed this clade close to, or inside, Placentalia (e.g. Rich *et al.*, 2002: 467; Woodburne, 2003: 195, 235; Woodburne *et al.*, 2003: 360, 372).

Order Monotremata Bonaparte, 1832 *sensu* Bonaparte, 1838

Class Monotremata Bonaparte, 1832: 76.

COMMENTS: When originally proposed, this rank was placed in the Class Mammalia (Linnaeus, 1758) and included the genus *Echidna* G. Cuvier, 1797 [=*Tachyglossus* Illiger, 1811]. Order rank recognised, with the current spelling, by Bonaparte (1838: 113) who included the families Echidnidae and Ornithorhynchidae. Palmer (1904: 888) gave 'Monotreme' É. Geoffroy (1803a: 226) as the earliest form and use of this name. Recognised as the Suborder Monotremata by J. Gray (1869a: 393) who referred to 'Monotrema, Geoff' [=Geoffroy (1803a: 226)]. Monotremata recognised at family rank by Waterhouse (1841a: 60), within the Order Marsupialia, 'Section' rank by Waterhouse (1846: 18), subclass rank by Ameghino (1889: xvii, 43), suborder rank by Iredale and Troughton (1934: vii, 1) and ordinal rank by Bonaparte (1838: 113; 1840: 258), Gill (1872: vi, 27), Gregory (1947: 46), Strahan (1983: xxi, 3; 1995: 6, 32), and Van Dyck and Strahan, 2008: 9, 29). Synonymised within Subclass Prototheria by McKenna and Bell (1997: 35). Typically recognised at ordinal rank though recognised at the Subcohort rank by Gardiner (1982: 229). Order reviewed by M. Griffiths (1978).

Class Monotrema Bonaparte, 1831: 28.

COMMENTS: When originally proposed, this rank was placed in the Class Mammalia (Linnaeus, 1758) and included the orders Tachyglossa (Bonaparte, 1831 [=Tachyglossidae (Gill, 1872)]) and Platypoda (Bonaparte, 1831 [=Ornithorhynchidae (J. Gray, 1825a)]).

Order Herpornitherae Burnett, 1830a: 362, 365.

COMMENTS: When originally proposed, this rank was included the 'types' Fissipeda (including the Kind [=Family] Echidnidae (Burnett, 1830a)) and Remipeda (including the Kind [=Family] Ornithorhynchidae (J. Gray, 1825a)).

Subclass Ornithodelphes de Blainville, 1834: 82.

COMMENTS: When originally proposed, this rank was placed in the Class Mammalia (Linnaeus, 1758) and included the unranked terms 'Échidné' [=Tachyglossidae (Gill, 1872)] and 'Ornithorhynqué' [=Ornithorhynchidae (J. Gray, 1825a)]. According to Simpson (1954: 356) the classification used by de Blainville (1834) was copied in Gervais (1836: 619), which has been referred to here. Originally referred to as 'Didelphes – Anomaux' by de Blainville (1816a: 117). Recognised at subclass rank by Haeckel (1866: xi, cxlii), subordinal rank by Gill (1871a: 533; 1872: vi, 27, 46) and infraclass by Hopson (1970: 7). Synonymised within the Monotremata, which was recognised at the subclass rank, by Iredale and Troughton (1934: vii, 1), and within the Subclass Prototheria by McKenna and Bell (1997: 35).

Order Ornithodelphie Lesson, 1842: 195.

COMMENTS: When originally proposed, this rank was placed in the Class Mammalia (Linnaeus, 1758) and included the families Echidneae (Lesson, 1842 [=Tachyglossidae (Gill, 1872)]) and Paradoxideae (Lesson, 1842 [=Ornithorhynchidae J. Gray, 1825a]). Name does not appear to have been recognised by subsequent authors.

Family Biclavulata Wagner, 1844: vii, 226.

TYPE GENUS: Not based on a genus group name.

COMMENTS: When originally proposed, this rank included the genera *Tachyglossus* Illiger, 1811 and *Ornithorhynchus* Blumenbach, 1800a.

Order Monotrèmes Gervais, 1854: xx.

COMMENTS: When originally proposed, this rank was placed in the Class Mammalia (Linnaeus, 1758) and included the suborders Échidnés (Gervais, 1854 [=Tachyglossidae (Gill, 1872)]) and Ornithorhynqués (Gervais, 1854 [=Ornithorhynchidae (J. Gray, 1825a)].

Family Ornithorhynchidae J. Gray, 1825 *sensu* Burnett, 1830

Tribe Ornithorhyncina J. Gray, 1825a: 343.

TYPE GENUS: *Ornithorhynchus* Blumenbach, 1800a.

COMMENTS: When originally proposed, this rank was placed in the Family Dasypidae [=Family Dasypodidae (J. Gray, 1821: 305)] and included the genera *Ornithorhynchus* Blumenbach, 1800a and *Echidna* G. Cuvier, 1797 [=*Tachyglossus* Illiger, 1811]. Family rank first recognised by Burnett (1830a: 365), who restricted the family to the genus *Ornithorhynchus*, which was followed by Bonaparte (1838: 113; 1840: 258; 1845: 6) and subsequent authors. Family name often attributed to Burnett (1830a: 365), but J. Gray (1825a) was recognised as the valid family author by McKenna and Bell (1997: 35).

Type Remipeda Burnett, 1830a: 365.

TYPE GENUS: Not based on a genus group name.

COMMENTS: When originally proposed, this rank was placed in the Order Herpornitherae (Burnett, 1830a [=Monotremata (Bonaparte, 1832)]) and included the Kind [=Family] Ornithorhynchidae (Burnett, 1830a [=Ornithorhynchidae (J. Gray, 1825a)]).

Kind Ornithorhynchidae Burnett, 1830a: 365.

TYPE GENUS: *Ornithorhynchus* Blumenbach, 1800a.

COMMENTS: When originally proposed, this rank was placed in the Type Remipeda (Burnett, 1830a [=Ornithorhynchidae (J. Gray, 1825a)]) and included the genus *Ornithorhynchus* Blumenbach, 1800a. Name synonymised within the Family Ornithorhynchidae (J. Gray, 1825a) by McKenna and Bell (1997: 35).

Order Platypoda Bonaparte, 1831: 28.

COMMENTS: When originally proposed, this rank was placed in the Class Monotrema (Bonaparte, 1831 [=Monotremata (Bonaparte, 1832)]) and included the Family Ornithorhynchidae (J. Gray, 1825a). Author of this name has typically been forgotten in favour of its use by Gill (1872: vi, 27; see individual entry below).

Tribe Ornithorhynchina Bonaparte, 1838: 113.

TYPE GENUS: *Ornithorhynchus* Blumenbach, 1800a.

COMMENTS: When originally proposed, this rank was placed in the Family Ornithorhynchidae (J. Gray, 1825a). Usage subsequently followed by Bonaparte (1840: 258). Synonymised within the Family Ornithorhynchidae (J. Gray, 1825a) by McKenna and Bell (1997: 35).

Family Paradoxideae Lesson, 1842: 196.

TYPE GENUS: Not based on a genus group name.

COMMENTS: When originally proposed, this rank was placed in the Order Ornithodelphie (Lesson, 1842 [=Monotremata (Bonaparte, 1832)]) and included the genus *Ornithorhynchus* Blumenbach, 1800a. Family name not derived from a genus but *Ornithorhynchus paradoxus* Blumenbach, 1800a. Does not appear to have been recognised by other authors.

Suborder Ornithorhynqués Gervais, 1854: xx.

COMMENTS: When originally proposed, this rank was placed in the Order Monotrèmes (Gervais, 1854 [=Monotremata (Bonaparte, 1832)]). Synonymised within Platypoda by McKenna and Bell (1997: 35) but not by other authors.

Family Ornithorhynchidés Gervais, 1855a: 291, 292.

TYPE GENUS: *Ornithorhynchus* Blumenbach, 1800a.

COMMENTS: When originally proposed, this rank was placed in the Order Monotrèmes (Gervais, 1854 [=Monotremata (Bonaparte, 1832)]) and included the genus *Ornithorhynchus* Blumenbach, 1800a.

Suborder Platypoda Gill, 1872: vi, 27.

COMMENTS: When originally proposed, this rank was placed in the Order Monotremata (Bonaparte, 1832) and included the Family Ornithorhynchidae (J. Gray, 1825a). Elevated to the ordinal rank by McKenna, in Stucky and McKenna (1993: 740) and followed by McKenna and Bell (1997: 35) but not other authors.

HOMONYMS:

Section Pladypoda J. Gray, 1865a: 102, mustelids of the Class Mammalia (Order Carnivora, Family Mustelidae). Invalid family name.

Ornithorhynchus Blumenbach, 1800

Ornithorhynchus Blumenbach, 1800a: 205.

TYPE SPECIES: *Nomen novum* for *Platypus* Shaw, 1799.

COMMENTS: Also described by Blumenbach (1800b: 609). Genus recognised by Waterhouse (1841a: 309; 1846: 24), Thomas (1888a: xiii, 387), Iredale and Troughton (1934: vii, 2) and subsequent authors. The taxonomy of this genus has not been examined since Thomas (1923a: 176).

FUTURE TAXONOMIC RESEARCH: It is possible that this genus consists of more than one species, or that *O. anatinus* is divisible into subspecies. In the absence of evidence either way, we have placed all platypus in a single taxon, and listed the available names in synonymy.

Platypus Shaw, 1799: Text to Plates 385–386.

TYPE SPECIES: *Platypus anatinus* Shaw, 1799 [=*Ornithorhynchus anatinus* (Shaw, 1799)] by monotypy.

COMMENTS: Junior homonym of *Platypus* Herbst, 1793: vii, 128. Synonymised within *Ornithorhynchus* by Waterhouse (1846: 24), Thomas (1888a: 387), Iredale and Troughton

(1934: 1) as *Platypus* was preoccupied and *Ornithorhynchus* is the next available name.

HOMONYMS:

Platypus Herbst, 1793: vii, 128, weevil beetles of the Class Insecta (Order Coleoptera, Family Platypodidae). Currently accepted genus. See Beaver (1998: 182), and Bright and Skidmore (2002: 169).

Platypus C. Brehm, 1824a: 805, 828 and 1824b: 28, eider ducks of the Class Aves (Order Anseriformes, Family Anatidae). Genus is junior synonym of *Somateria* Leach, 1819: 61. See Hellmayr and Conover (1948: 385).

Platypus Marshall, 1868: 281, bugs of the Class Insecta (Order Hemiptera, Family Pentatomidae). Name proposed as an alternative spelling of *Platynopus* Amyot and Audinet-Serville, 1843: 79.

Platypus Faxon, 1878: 13, spiders of the Class Arachnida (Order Acarina, Family Demodecidae). Name arose from incorrectly arranging *Macrogaster platypus* (Miescher, 1943: 198). Genus is a synonym of *Macrogaster* Miescher, 1843: 198.

Dermipus Wiedemann, 1800: 180, Plate 4 (as Plate 3).

TYPE SPECIES: *Nomen novum* for *Platypus* Shaw, 1799.

COMMENTS: Synonymised within *Ornithorhynchus* by Waterhouse (1846: 24), Thomas (1888a: 387), and Iredale and Troughton (1934: 1).

Ornithorinchus Artaud, 1803: 162.

TYPE SPECIES: Unjustified emendation of *Ornithorhynchus* Blumenbach, 1800a.

COMMENTS: Mahoney (1988a: 8) proposed that this spelling was an unjustified emendation of *Ornithorhynchus* Blumenbach, 1800a. Spelling not recognised by other authors.

Ornithorincus Péron, 1807: Plate 34.

TYPE SPECIES: Unjustified emendation of *Ornithorhynchus* Blumenbach, 1800a.

COMMENTS: Does not appear to have been recognised subsequently.

Ornithorhynchi Meckel, 1826: 1.

TYPE SPECIES: Incorrect subsequent spelling of *Ornithorhynchus* Blumenbach, 1800a.

COMMENTS: Does not appear to have been recognised subsequently.

Ornithorynchus Macgillivray, 1827: 127.

TYPE SPECIES: Incorrect subsequent spelling of *Ornithorhynchus* Blumenbach, 1800a.

COMMENTS: Spelling not recognised by other authors.

Ornithorhynchus anatinus (Shaw, 1799)

Platypus

Platypus Anatinus Shaw, 1799: Text to Plates 385–386.

TYPE LOCALITY: Sydney, New South Wales, Australia.

COMMENTS: Discussion of the description made by Anderson (1802: 562). Species transferred to *Ornithorhynchus* by Waterhouse (1841a: 315; 1846: 25), J. Gray (1843a: xxviii, 191), Gould (1855 [1845–1863]: Text to Plate 1) and followed by most authors including Thomas (1888a: 388), Iredale and Troughton (1934: vii, 1) and subsequent authors. No subspecies are generally recognised.

FUTURE TAXONOMIC RESEARCH: The taxonomy of the platypus needs to be revisited as several studies have found large genetic differences between mainland and Tasmanian animals, suggesting long-term isolation (Furlan *et al.* 2010: 319; Gongora *et al.* 2012: 110).

Ornithorhynchus paradoxus Blumenbach, 1800a: 205.

TYPE LOCALITY: Replacement name for *Platypus anatinus* Shaw, 1799.

COMMENTS: Also described by Blumenbach (1800: 609). Name recognised by Waterhouse (1838a: 68; 1841a: 309) but appears to have been synonymised within *anatinus* on page 315. Synonymised within *anatinus* by Waterhouse (1846: 25), Thomas (1888a: 388), Iredale and Troughton (1934: 1) and subsequent authors.

Ornithorhynchus novae Hollandiae Lacépède, 1800: 78.

TYPE LOCALITY: Sydney, New South Wales, Australia.

COMMENTS: Generic name spelt two ways by Lacépède (1800: 78), including *Ornithorhynchus* by itself and *Ornithorynchus* in conjunction with the specific name *novaehollandiae*. Not included with Iredale and Troughton (1934: 1–2). Description reviewed by Husson and Holthius (1953: 211).

Ornithorincus Fuscus Péron, 1807: Plate 34, Fig. 1.

TYPE LOCALITY: Sydney, New South Wales, Australia.

COMMENTS: Recognised at species rank by Waterhouse (1841a: 315), but synonymised within *anatinus* by Waterhouse (1846: 25), Thomas (1888a: 389), Iredale and Troughton (1934: 1) and subsequent authors.

Ornithorincus Rufus Péron, 1807: Plate 34, Fig. 2.

TYPE LOCALITY: Sydney, New South Wales, Australia.

COMMENTS: Synonymised within *anatinus* by Waterhouse (1841a: 315; 1846: 25), Thomas (1888a: 389), Iredale and Troughton (1934: 1) and subsequent authors.

Ornithorhynchi paradoxi Meckel, 1826: 1.

TYPE LOCALITY: Incorrect subsequent spelling of *O. paradoxus* Blumenbach, 1800a.

COMMENTS: Synonymised within *anatinus* by Pasitschniak-Arts and Marineli (1998: 1).

Ornithorynchus crispus Macgillivray, 1827: 128.

TYPE LOCALITY: Tasmania, Australia.

COMMENTS: Genus name misspelt as *Ornithorynchus*. Synonymised within *fuscus* by Waterhouse (1841a: 315), and within *anatinus* by Waterhouse (1846: 25) and Thomas (1888a: 389). Iredale and Troughton (1934: 2) elevated this taxon to a subspecies of *anatinus* but this has not been subsequently accepted.

Ornithorynchus laevis Macgillivray, 1827: 132.

TYPE LOCALITY: *Nomen novum* for *Ornithorhynchus rufus* Péron, 1807.

COMMENTS: Genus name misspelt as *Ornithorynchus*. Synonymised within *anatinus* by Waterhouse (1841a: 315; 1846: 25), Thomas (1888a: 389) and Iredale and Troughton (1934: 1) who gave the year of publication as 1832. Mahoney (1988a: 9) correctly established dated the publication date after I. Geoffroy (1827: 408).

Ornithorhynchus brevirostris W. Ogilby, 1832: 150.

TYPE LOCALITY: Swan River, Tasmania, Australia.

COMMENTS: Synonymised within *anatinus* by Waterhouse (1841a: 31; 1846: 25), Thomas (1888a: 387), Iredale and Troughton (1934: 1) and subsequent authors.

Ornithorhynchus agilis De Vis, 1885a: 2, col. 6.

TYPE LOCALITY: King Creek, in the vicinity of Pilton, Darling Downs, Queensland, Australia.

COMMENTS: Initial description was an abstract but was described in greater detail by De Vis (1885b: 35). Subjective synonym of *Ornithorhyncus anatinus* according to M. Archer *et al.* (1978: 9, 18).

Ornithorhynchus anatinus phoxinus Thomas, 1923a: 176.

TYPE LOCALITY: Dinner Creek, Ravenshoe, Queensland, Australia. 2900 feet.

COMMENTS: Recognised as a subspecies of *anatinus* by Iredale and Troughton (1934: 2), but not subsequently recognised as a subspecies by subsequent authors.

Ornithorhynchus anatinus triton Thomas, 1923a: 178.

TYPE LOCALITY: Victorian side of the Murray River, opposite Deniliquin, SW New South Wales, Australia.

COMMENTS: Recognised as a subspecies of *anatinus* by Iredale and Troughton (1934: 1), but not subsequently recognised as a subspecies.

Family Tachyglossidae Gill, 1872

Family Tachyglossidae Gill, 1872: 27.

TYPE GENUS: *Tachyglossus* Illiger, 1811.

COMMENTS: When originally proposed, this rank was placed in the Suborder Tachyglossa (Gill, 1872 [=Tachyglossidae (Gill, 1872)]) and included the genus *Tachyglossus* Illiger, 1811. Family name used by Thomas (1888a: 374, 375), Iredale and Troughton (1934: vii, 2) and subsequent authors.

Type Fissipeda Burnett, 1830a: 365.

TYPE GENUS: Not based on a genus group name.

COMMENTS: When originally proposed, this rank was placed in the Order Herpornitherae (Burnett, 1830a [=Monotremata (Bonaparte, 1832)]) and included the Kind [=Family] Echidnidae (Burnett, 1830a [=Tachyglossidae (Gill, 1872)]).

HOMONYMS:

Fissipeda Blumenbach, 1791, carnivores of the Class Mammalia (Order Carnivora). See individual entry.

Kind Echidnidae Burnett, 1830a: 365.

TYPE GENUS: *Echidna* G. Cuvier, 1797.

COMMENTS: When originally proposed, this rank was placed in the Type Fissipeda (Burnett, 1830a [=Tachyglossidae (Gill, 1872)]) and included the genus *Echidna* G. Cuvier, 1797. Used by Bonaparte (1838: 113; 1840: 258; 1845: 6). Family rank recognised by Bonaparte (1840: 258) and Thomas (1888a: xiii, 375). Synonymised within the Family Tachyglossidae by Simpson (1945: 39), and McKenna and Bell (1997: 35).

HOMONYMS:

Echidnae Hübner, 1806: 1, moths of the Class Insecta (Order Lepidoptera, Family Saturniidae).

Order Tachyglossa Bonaparte, 1831: 28.

COMMENTS: When originally proposed, this rank was placed in the Class Monotrema (Bonaparte, 1831 [=Monotremata (Bonaparte, 1832)]) and included the Family Echidnidae (Burnett, 1830a [=Tachyglossidae (Gill, 1872)]). Author of this name has typically been forgotten in favour of its use by Gill (1872: vi, 27; see individual entry below).

Tribe Echidnina Bonaparte, 1838: 113.

TYPE GENUS: *Echidna* G. Cuvier, 1797.

COMMENTS: When originally proposed, this rank was placed in the Family Echidnidae (Burnett, 1830a [=Tachyglossidae (Gill, 1872)]). Usage subsequently followed by Bonaparte (1840: 258). Synonymised within the Family Tachyglossidae by McKenna and Bell (1997: 35).

Family Echidneae Lesson, 1842: 196.

TYPE GENUS: *Echidna* G. Cuvier, 1797.

COMMENTS: When originally proposed, this rank was placed in the Order Ornithodelphie (Lesson, 1842 [=Monotremata (Bonaparte, 1832)]) and included the genus *Echidna* (G. Cuvier, 1797 [=*Tachyglossus* Illiger, 1811]). Does not appear to have been recognised by other authors.

Suborder Ėchidnés Gervais, 1854: xx.

COMMENTS: When originally proposed, this rank was placed in the Order Monotrèmes [=Monotremata (Bonaparte, 1832)]. Synonymised within Tachyglossa by McKenna and Bell (1997: 35) but not by other authors.

Family Échidnidés Gervais, 1855a: 291.

TYPE GENUS: *Echidna* G. Cuvier, 1797.

COMMENTS: When originally proposed, this rank was placed in the Order Monotrèmes (Gervais, 1855a: 288 [=Monotremata (Bonaparte, 1832)]) and included the genus *Echidna* (G. Cuvier, 1797).

Family Echidnida Haeckel, 1866: clvii.

TYPE GENUS: *Echidna* G. Cuvier, 1797.

COMMENTS: When originally proposed, this rank was placed in the Order Monotremata (Bonaparte, 1832) and included the genus *Echidna* G. Cuvier, 1797. Synonymised within the Family Tachyglossidae by McKenna and Bell (1997: 35).

Suborder Tachyglossa Gill, 1872: vi, 27.

COMMENTS: When originally proposed, this rank was placed in the Order Monotremata (Bonaparte, 1832) and included the Family Tachyglossidae (Gill, 1872). Elevated to the rank of order by McKenna, in Stucky and McKenna (1993: 740). Recognised as the order for Family Tachyglossidae by McKenna and Bell (1997: 35) but not followed by other authors.

Tachyglossus Illiger, 1811

Tachyglossus Illiger, 1811: 114.

TYPE SPECIES: *Nomen novum* for *Echidna* G. Cuvier, 1797.

COMMENTS: Illiger (1811: 114), immediately under the heading *Tachyglossus*, lists *Echidna* G. Cuvier, implying perhaps that his new name substitutes for Cuvier's, and gives the species in the genus as '*Myrmecophaga aculeata* Shaw [1792] et *Echidna setosa* Cuvier [=É. Geoffroy, 1803a]'. As *Echidna novaehollandiae* is the only species listed by Lacépède (1799a: 11) under *Echidna*, it could be maintained that this is the type species of *Tachyglossus*, but Illiger specifically cites a species. The type was given as *Echidna aculeata* by Thomas (1888a: 377), and Palmer (1904: 658) says of the Thomas citation, '(type fixed)'; hence we regard *aculeatus* as the type species of the genus. Genus synonymised within *Echidna* by Waterhouse (1846: 40) and Thomas (1888a: 377), but recognised by Thomas (1897a: 621), Iredale and Troughton (1934: vii, 2) and subsequent authors.

FUTURE TAXONOMIC RESEARCH: It seems likely that several distinct taxa are involved in this genus, but whether at species or subspecies level has not been determined, nor is it known how many taxa, and whether there are names for all recognisable forms. Here, we list as subspecies those that have been recognised in the most recent revisions.

Myrmecophaga Shaw, 1792: Text to Plate 109.

TYPE SPECIES: *Myrmecophaga aculeata* Shaw, 1792 [=*Tachyglossus aculeatus* (Shaw, 1792)] by monotypy.

COMMENTS: Name in use until replaced by *Tachyglossus* Illiger, 1811.

HOMONYMS:

Myrmecophaga Linnaeus, 1758: 35, the Giant Anteater of the Class Mammalia (Order Pilosa, Family Myrmecophagidae). Currently used name. See Gardner (2005a: 102).

Myrmecophaga Lacépède, 1799b: 6, ant-thrush birds of the Class Aves (Order Passeriformes, Family Formicariidae). This taxon appears to be a *nomen nudum* and has been listed as a junior synonym of *Formicarius* Boddaert, 1783: 43. See Ridgeway (1893: 669; 1911: 115).

aculeata [sic] É. Geoffroy, 1796a: 103.

TYPE SPECIES: *Nomen novum* for *Myrmecophaga* Shaw, 1792.

COMMENTS: Discussed by Thomas (1897a: 621) who suggested that the name was referrable to a species rather than a genus, but was considered at generic rank by Palmer (1904: 77). Name synonymised within *Tachyglossus* by Iredale and Troughton (1934: 2), but was not considered by Mahoney (1988b: 4). Synonymised within the Family Tachyglossidae by McKenna and Bell (1997: 35).

Echidna G. Cuvier, 1797: 143.

TYPE SPECIES: *Echidna novaehollandiae* Lacépède, 1799a [=*Tachyglossus aculeatus* (Shaw, 1792)] by subsequent monotypy.

COMMENTS: Recognised as the genus for *hystrix* by Waterhouse (1838a: 68; 1841a: 303) and *aculeatus* by J. Ogilby (1892: 3). Recognised by Waterhouse (1846: 40) and Thomas (1888a: 377) who suggested that *Echidna* Forster, 1788: 81 was not the senior synonym as it is virtually a *nomen nudum*. Discarded with 'regret' in favour of *Tachyglossus* by Thomas (1897a: 621) who noted that the name was previously occupied by *Echidna* Forster (1788: 81). Synonymised within *Tachyglossus* by Palmer (1904: 248), Iredale and Troughton (1934: 2) and subsequent authors. An effort was made to suspend the name *Echidna*, but this was denied by Opinion 90 of the ICZN (1926a: 103; see also Cleave, 1943: 231).

HOMONYMS:

Echidna Forster, 1788: 81, moray eels of the Superclass Pisces (Order Anguilliformes, Family Muraenidae). Currently recognised name. See G. Allen *et al.* (2006: 245).

Echidna Link, 1806: 290, 296, 299, vipers of the Class Reptilia (Order Squamata, Family Viperidae). Genus is a junior synonym of *Bitis* J. Gray, 1842a: 69.

Echidna Hübner, 1806: 1 and 1807: Plate 172, moths of the Class Insecta (Order Lepidoptera, Family Saturniidae). Included in a work rejected for nomenclatural purposes by the ICZN (1926b: 19; 1954a: 140). Genus is junior synonym of *Aglia* Ochsenheimer, 1810: 11.

Acanthonotus Goldfuss, 1809: 308.

TYPE SPECIES: *Nomen novum* for *Echidna* G. Cuvier, 1797.

COMMENTS: Synonymised within *Tachyglossus* by Iredale and Troughton (1934: 2). Junior homonym of *Acanthonotus* Bloch, 1797: 113, pl. ccccxxxi, which is an unneeded substitute name for *Notacanthus* Bloch, 1788: 278.

HOMONYMS:

Acanthonotus Bloch, 1797: 113, pl. ccccxxxi, spiny eels of the Superclass Pisces (Class Actinopterygii, Order Notacanthiformes, Family Notacanthidae). Genus is an objective synonym of *Notacanthus* Bloch, 1788: 278. See Goode (1881: 535) and Paxton *et al.* (2006: 233).

Acanthonotus G. Cuvier, 1800: Table 4, fish of the Class Pisces (Order Perciformes). Genus is a *nomen nudum*.

Acanthonotus J. Gray, 1830a: Plate 85, Fig. 1, glass catfish of the Class Actinopterygii (Order Siluriformes, Family Schilbeidae). Genus is a synonym of *Ailia* J. Gray, 1830a: Plate 85, Fig. 2. See Ferraris (2007: 356).

Acanthonotus Swainson & Richardson, 1832: 168, cuckooshrikes of the Class Aves (Order Passeriformes, Family Campephagidae). Genus is a synonym of *Coracina* Vieillot, 1816: 37.

Acanthonotus Ross, 1835: xc, amphipods of the Subphylum Crustacea (Order Amphipoda, Family Acanthonotozomatidae). Name is a synonym within *Acanthonotozoma* Boeck, 1876: 237. See Costello and Bellan-Santini (2011).

Acanthonotus Koch, 1839: 36, harvestmen the Class Arachnida (Order Opiliones, Family Sclerosomatidae). Junior synonym of *Syleus* Thorell, 1876: 112. See Roewer (1929: 114).

Acanthonotus Taczanowski, 1872: 85, crab spiders of the Class Arachnida (Order Araneae, Family Thomisidae). Synonym of *Acentroscelus* Simon, 1886: 185. See Platnick (2013).

Acanthonotus Tickell in Day, 1889: 807, carp fish of the Superclass Pisces (Order Cypriniformes, Family Cyprinidae). Genus is a synonym of *Mystacoleucus* Günther, 1868: 206. See Roberts (1989: 45).

Acanthonotus Nalepa, 1889: 116, mites of the Class Arachnida (Order Acarina, Family Eriophyidae). Genus is a synonym of *Tegonotus* Nalepa, 1890: 213. See Xiao-Feng and Zhi-Qiang (2009: 60).

Acanthonotus Buckton, 1901: 81, tree hoppers of the Class Insecta (Order Hemiptera, Family Membracidae). Incorrect subsequent spelling of *Acanthonota* Buckton, 1901: 77, which is a junior synonym of *Cladonota* Stål, 1869: 273.

Echinopus G. Fischer, 1813a: 14.

TYPE SPECIES: *Nomen novum* for *Echidna* G. Cuvier, 1797.

COMMENTS: Also referred to by G. Fischer, 1814: 691. Synonymised within *Echidna* by Thomas (1888a: 377) and within *Tachyglossus* by Iredale and Troughton (1934: 2) and subsequent authors.

HOMONYMS:

Echinopus Schönherr, 1836: 457, true weevils of the Class Insecta (Order Coleoptera, Family Curculionidae). This genus is considered to be a junior synonym of *Peloropus* Schönherr, 1836: 456. See Gattolliat and Jacobus (2010: 159).

Echinopus Gattolliat, 2002: 143, 149, mayflies of the Class Insecta (Order Ephemeroptera, Family Baetidae). Genus is a junior synonym of *Madaechinopus* Gattolliat and Jacobus, 2010: 159.

Syphomia Rafinesque, 1815: 57, 219.

TYPE SPECIES: *Nomen novum* for *Echidna* G. Cuvier, 1797.

COMMENTS: Listed as an available name by Iredale and Troughton (1934: 2) but this name has no status in nomenclature as Rafinesque withdrew it in the '*Additions et Correction*' to his work (p. 219). Synonymised within *Tachyglossus* by McKenna and Bell (1997: 36).

Tachyglossus aculeatus (Shaw, 1792)

Short-beaked Echidna

Tachyglossus aculeatus aculeatus (Shaw, 1792)

Myrmecophaga Aculeata Shaw, 1792: Text to Plate 109.

TYPE LOCALITY: Sydney, New South Wales, Australia.

COMMENTS: Synonymised within *Echidna Hystrix* by Waterhouse (1841a: 303). Recognised within *Echidna* by Waterhouse (1846: 41) and *Tachyglossus* by Illiger (1811: 114). See comments under *Echidna* and *Tachyglossus* above.

Echidna novae Hollandiae Lacépède, 1799a: 11.

TYPE LOCALITY: Unknown, presumably New South Wales, Australia.

COMMENTS: Not considered by Iredale and Troughton (1934: 2–3) and considered a synonym of *aculeatus* by Mahoney (1988b: 4), Groves (1993a: 13; 2005c: 1) and subsequent authors.

Ornithorhynchus Hystrix Home, 1802: 348; Plates 10–13.

TYPE LOCALITY: *Nomen novum* for *Myrmecophaga aculeata* Shaw, 1792.

COMMENTS: Species recognised within *Echidna* by Waterhouse (1838a: 68; 1841a: 303) and Gould (1852 [1845–1863]: Text to Plate 2). Considered a synonym of *aculeatus* by Waterhouse (1846: 41), Thomas (1888a: 379), Iredale and Troughton (1934: 2), Mahoney (1988b: 5), Groves (1993a: 13; 2005c: 1) and subsequent authors.

E. [chidna] longiaculeata Tiedemann, 1808: 592.

TYPE LOCALITY: *Nomen novum* for *Myrmecophaga aculeata* Shaw, 1792.

COMMENTS: Synonymised within *aculeatus* by Waterhouse (1846: 41), Thomas (1888a: 379), Iredale and Troughton (1934: 2), Mahoney (1988b: 5), Groves (1993a: 13; 2005c: 1) and subsequent authors.

Acanthonotus myrmecophagus Goldfuss, 1809: xix, 309.

TYPE LOCALITY: *Nomen novum* for *Myrmecophaga aculeata* Shaw, 1792.

COMMENTS: Considered a synonym of *aculeatus* by Iredale and Troughton (1934: 2), Mahoney (1988b: 5), Groves (1993a: 13; 2005c: 1) and subsequent authors.

Echidna novae Hollandiae G. Fischer, 1813b: 444.

TYPE LOCALITY: *Nomen nudum.*

COMMENTS: Synonymised within *aculeatus* by Iredale and Troughton (1934: 2) and not considered by other authors including Mahoney (1988b: 4–6) or Groves (1993a: 13; 2005c: 1). Here placed (arbitrarily) as a synonym of *aculeatus*.

Echidna australiensis Lesson, 1827a: 318.

TYPE LOCALITY: *Nomen novum* for *Myrmecophaga aculeata* Shaw, 1792.

COMMENTS: Species recognised by Gervais (1835a: 623). Considered a synonym of *aculeatus* by Thomas (1888a: 379), Iredale and Troughton (1934: 2), Mahoney (1988b: 5), Groves (1993a: 13; 2005c: 1) and subsequent authors.

ornithorhynchus eracinius Mudie, 1829: 180.

TYPE LOCALITY: Unknown.

COMMENTS: Considered a synonym of *aculeatus* by Iredale and Troughton (1934: 3), Mahoney (1988b: 5), Groves (1993a: 13; 2005c: 1) and subsequent authors.

Echidna Australis Lesson, 1836: Plate 52.

TYPE LOCALITY: *Nomen novum* for *Myrmecophaga aculeata* Shaw, 1792.

COMMENTS: The full title of Lesson's work is recorded as *Complement des Oeuvres de Buffon, ou histoire naturelle des animalaux rares decouverts par les naturalistes et les voyageurs depuis la mort de Buffon*. Tome V. *Suite des mammifères*, on an added title page. Year recorded as 1838 by Iredale and Troughton (1934: 3). Considered a synonym of *aculeatus* by Thomas (1888a: 379), Iredale and Troughton (1934: 3), Groves (2005c: 1), though not considered by Mahoney (1988b: 5) or Groves (1993a: 13).

Echidna Histrix [sic] Waterhouse, 1838: 68.

TYPE LOCALITY: Incorrect subsequent spelling of *Ornithorhynchus [=Echidna] Hystrix* Home, 1802.

COMMENTS: Name not subsequently used.

Echidna brevicaudata J. Gray, 1865b: 386.

TYPE LOCALITY: Unknown.

COMMENTS: Considered a synonym of *setosus* by Iredale and Troughton (1934: 3) but not considered by other authors including Mahoney (1988b: 5) or Groves (1993a: 13; 2005c: 1). Here placed (arbitrarily) as a synonym of *aculeatus*.

Echidna orientalis Krefft, 1872a: 808.

TYPE LOCALITY: 'Eastern Australia'. Locality given as Cape York, Queensland, Australia by Mahoney (1988b: 5).

COMMENTS: Was not considered by Iredale and Troughton (1934: 3) and a synonym of *aculeatus* by Mahoney (1988b: 5) and Groves (1993a: 13; 2005c: 1).

Echidna corealis Krefft, 1872a: 808.

TYPE LOCALITY: Cape York, Queensland, Australia.

COMMENTS: Was not considered by Iredale and Troughton (1934: 3) and a synonym of *aculeatus* by Mahoney (1988b: 5), Groves (1993a: 13; 2005c: 1) and subsequent authors.

Echidna typica Thomas, 1885: 338; Plate 23, Fig. B.

TYPE LOCALITY: Australia.

COMMENTS: Synonymised within *aculeatus* by Groves (1993a: 13; 2005c: 1).

E. [chidna] sydneiensis Kowarzik, 1909: 214.

TYPE LOCALITY: *Nomen novum* for *Echidna hystrix multiaculeata* Rothschild, 1905a [=*Tachyglossus aculeatus multiaculeata* (Rothschild, 1905)].

COMMENTS: Considered a subspecies of *aculeatus* by Iredale and Troughton (1934: 3) and a synonym by Mahoney (1988b: 6), Groves (1993a: 13; 2005c: 1) and subsequent authors.

Tachyglossus aculeatus setosus (É. Geoffroy, 1803)

Echidna setosa É. Geoffroy, 1803a: 226 as 126.

TYPE LOCALITY: Adventure Bay, Bruny Island, Tasmania, Australia.

COMMENTS: Recognised at the species rank within *Echidna* by Gould (1849 [1845–1863]: Text to Plate 3), Waterhouse (1846: 47) Krefft (1868a: 94), and within *Tachyglossus* by Iredale and Troughton (1934: vii, 3) and Troughton (1967: 10). Synonymised within *aculeatus* by Ride (1970: 231), Mahoney (1988b: 5) and Groves (1993a: 13). Recognised as a subspecies by Thomas (1888a: 381), Strahan (1983: 8), Flannery (1990: 40; 1995a: 68), Groves (2005c: 1) and Clayton *et al.* (2006: 100). A future revision will almost certainly include this presumed subspecies as a valid taxon, given that Tasmanian echidnas are instantly recognisable by pelage features.

FUTURE TAXONOMIC RESEARCH: A morphometric and/or genetic examination of the Tasmanian echidnas appear to be warranted.

E. [chidna] breviaculeata Tiedemann, 1808: 592.

TYPE LOCALITY: Neighbourhood of Adventure Bay, Bruny Island, Tasmania, Australia.

COMMENTS: Considered a synonym of *setosa* by Waterhouse (1846: 47) and Iredale and Troughton (1934: 3) and a synonym of *aculeatus* by Mahoney (1988b: 5) and Groves (1993a: 13), and synonym of *setosus* by Groves (2005c: 1).

Platypus Longirostra Perry, 1810: Second page of text associated with Plate 10.

TYPE LOCALITY: Unknown.

COMMENTS: Considered a synonym of *setosus* by Iredale and Troughton (1934: 3) and a synonym of *aculeatus* by Mahoney (1988b: 5) and Groves (1993a: 13), and a synonym of *setosus* by Groves (2005c: 1).

Echidna (Tachyglossus) hobartensis Kowarzik, 1909: 215.

TYPE LOCALITY: Hobart, Tasmania, Australia.

COMMENTS: Considered a subspecies of *setosus* by Iredale and Troughton (1934: 3), synonym of *aculeatus* by Mahoney (1988b: 6) and Groves (1993a: 13), and synonym of *setosus* by Groves (2005c: 1).

Φ *Tachyglossus aculeatus lawesii* Ramsay, 1877

Φ *T. [achyglossus] Lawesii* Ramsay, 1877a: 32.

TYPE LOCALITY: Port Moresby, Papua New Guinea.

COMMENTS: Not considered by Iredale and Troughton (1934: 3) and a synonym of *aculeatus* by Mahoney (1988b: 5) and Groves (1993a: 13). Recognised as a subspecies by Thomas (1888a: 377), Strahan (1983: 8), Flannery (1990: 40; 1995a: 68) and Groves (2005c: 1).

Tachyglossus aculeatus acanthion (Collett, 1884)

Echidna acanthion Collett, 1884a: 1.

TYPE LOCALITY: Gracemore, near Rockhampton, Queensland, Australia.

COMMENTS: Synonymised within *aculeatus* by Thomas (1888a: 379) and Mahoney (1988b: 5). Considered a subspecies of *aculeatus* by Iredale and Troughton (1934: 3), D. Johnson (1964: 433), Strahan (1983: 8), Flannery (1990: 40; 1995a: 68), Groves (2005c: 1) and Clayton *et al.* (2006: 100).

Tachyglossus aculeatus multiaculeatus (Rothschild, 1905)

Echidna hystrix multiaculeata Rothschild, 1905a: 306.

TYPE LOCALITY: Extreme south of South Australia, Australia. The type locality given as Kangaroo Island, South Australia by Griffiths (1978: 61).

COMMENTS: Considered a subspecies of *aculeatus* by Iredale and Troughton (1934: 3) and a synonym by Mahoney (1988b: 5) and Groves (1993a: 13). Recognised as a subspecies by Flannery (1990: 40; 1995a: 68), Groves (2005c: 1) and Clayton *et al.* (2006: 100).

Tachyglossus aculeatus ineptus Thomas, 1906a: 2.

TYPE LOCALITY: Parker Range, SE of Southern Cross, Western Australia, Australia. (About 220 miles E. of Perth, 1163 feet).

COMMENTS: Considered a subspecies of *aculeatus* by Iredale and Troughton (1934: 3) and a synonym by Mahoney (1988b: 6) and synonym of *acanthion* by Groves (1993a: 13; 2005c: 1). Subspecies rank recognised by Flannery (1990: 40; 1995a: 68).

Zaglossus Gill, 1877

Zaglossus Gill, 1877: clxxi.

TYPE SPECIES: *Tachyglossus bruijni* Peters & Doria, 1876 [=*Zaglossus bruijni* (Peters & Doria, 1876)] by monotypy.

COMMENTS: Genus recognised in preference to other names by authors including Gill (1885: 642), Palmer (1895a: 518), Coues (1895: 610), Rothschild (1905a: 305), Toldt (1906: 1), who established its priority, G. Allen (1912: 253) and Kerbert (1913: 162). Genus reviewed by Toldt (1906: 1), G. Allen (1912: 253), Rothschild (1913: 188), Van Deusen and George (1969: 1), and Flannery and Groves (1998: 367).

Acanthoglossus Gervais, 1877a: 838.

TYPE SPECIES: *Tachyglossus bruijni* Peters & Doria, 1876 (as *Acanthoglossus bruijni*) [=*Zaglossus bruijni* (Peters & Doria, 1876)] by original designation.

COMMENTS: Taxon further described by Gervais (1877b: 991) and discussed by Gervais (1877c: 377). Date priority given to *Zaglossus* Gill, 1877. Name recognised by various authors including Thomas (1907a: 293) but not recognised or reduced to a synonym of *Zaglossus* by authors including Gill (1885: 642), Palmer (1895a: 518) and Coues (1895: 610), and Rothschild (1905a: 305).

HOMONYMS:

Acanthoglossa Kraatz, 1859: 144, rove beetles of the Class Insecta (Order Coleoptera, Family Staphylinidae).

Proechidna Gervais, 1877d: 43.

TYPE SPECIES: New name for *Acanthoglossus* Gervais, 1877a.

COMMENTS: Date priority given to *Zaglossus* Gill, 1877. Genus recognised by Toldt (1905: 5), but synonymised within *Zaglossus* by Kerbert (1913: 162) and McKenna and Bell (1997: 36).

Bruynia Dubois 1882: 267.

TYPE SPECIES: *Bruynia tridactyla* Dubois 1882: 267 [=*Zaglossus bruijni* (Peters & Doria, 1876)] by original designation.

COMMENTS: New name for *Acanthoglossus* Gervais, 1877a. Synonymised within *Zaglossus* by McKenna and Bell (1997: 36).

Bruijnia Thomas, 1883: 40.

TYPE SPECIES: New name for *Acanthoglossus* Gervais, 1877a.

COMMENTS: Synonymised within *Zaglossus* by McKenna and Bell (1997: 36).

Prozaglossus Kerbert, 1913: 166.

TYPE SPECIES: *Acanthoglossus bruijnii bartoni* Thomas, 1907a: 294 (as *Prozaglossus bartoni*) [=*Zaglossus bartoni* (Thomas, 1907a: 294)] by monotypy.

COMMENTS: Synonymised within *Zaglossus* by McKenna and Bell (1997: 36).

† ***Zaglossus bruijni*** (Peters & Doria, 1876)

Western Long-beaked Echidna

Tachyglossus bruijni Peters & Doria, 1876: 183.

TYPE LOCALITY: Vogelkop, Manokwari Division, Arfak Mountains, Province of Papua (=Irian Jaya), Indonesia.

COMMENTS: Recent occurrence within Australia suggested by Helgen *et al.* (2012: 103) who describe a specimen apparently collected from Mount Anderson in the Western Kimberley region of northern Western Australia in 1901 by John Tunney that is now located within the Natural History Museum, London. Though this species appears to be extinct in Australia there may be some chance that it still occurs in the Western Kimberley, so efforts should be made to search for it as a matter of urgency. Species is still extant in western New Guinea, though is considered an endangered species under the IUCN (see Groves, 2005c: 2).

Subclass Theria Parker & Haswell, 1897

Subclass Theria Parker & Haswell, 1897: 448.

COMMENTS: When originally proposed, this rank was placed in the Class Mammalia (Linnaeus, 1758) and included the sections Metatheria (Marsupialia) (Huxley, 1881 [=Marsupialia (Illiger, 1811)]) and Eutheria (Huxley, 1881 [=Placentalia (Bonaparte, 1838)]). Subclass rank recognised by authors including Gregory (1910: 230), Simpson (1945: xi, 40), Turnbull (1971: 176), McKenna (1975: 27), Prothero (1981: 281, 286), Aplin and Archer (1987: xxi) and Szalay (1994: 40). Recognised at subcohort rank by Gardiner (1982: 229) and supercohort by McKenna and Bell (1997: 49). Strong support for the Marsupialia and Placentalia as sister groups has been provided by Killian *et al.* (2001: 513), Kullberg (2008: 115) and Warren *et al.* (2008: 176). The ranks Trechnotheria (McKenna, 1975), Yangotheria (Chow & Rich, 1982), Cladotheria (McKenna, 1975), Zatheria (McKenna, 1975) and Tribosphenida (McKenna, 1975) are listed below with ascribed ranks for completeness sake following Aplin and Archer (1987: xxi), although they are of limited relevance in a work on modern mammals since in each case their sister groups are known only as fossils.

† Order Pappotherida Butler, 1978: 1, 25.

COMMENTS: When originally proposed as a new rank it was placed in the † Infraclass Tribotheria (Butler, 1978) and included the † Family Pappotheriidae (Slaughter, 1965: 4). Recognised as an Order by Szalay (1994: 40) but synonymised within the Supercohort Theria by McKenna and Bell (1997: 49).

† Infraclass Tribotheria Butler, 1978: 1, 25.

COMMENTS: When originally proposed as a new rank it was placed in the Subclass Theria (Parker & Haswell, 1897) and included the orders † Aegialodintidae (Kermack *et al.*, 1968: 421) and † Pappotherida (Butler, 1978). Recognised as an infraclass by Szalay (1994: 40) but synonymised within the Supercohort Theria by McKenna and Bell (1997: 49).

Infraclass Holotheria Hopson, 1994: 205, 208.

COMMENTS: Name formalised by Wible *et al.* (1995: 10: 11) who restricted it only to the common ancestor of † *Kuehneotherium* (Kermack *et al.* (1968: 407, 408) and living therians plus all its descendants, which would exclude the monotremes and multituberculates (see McKenna and Bell, 1997: 43). The term Holotheria represents a slightly modified replacement for 'Theria' *sensu lato*, conceived as one of two clades representing an early, fundamental dichotomy in mammalian history (Luo *et al.*, 2002: 4). Recognised at infraclass by McKenna and Bell (1997: 43). Given the very incomplete data for *Kuehneotherium*, Luo *et al.* (2002: 17) suggested that any clade defined using this taxon was unstable.

HOMONYMS:

Holotheria Jaeckel, 1911, mammals of the Class Mammalia. Name is a junior synonym of Placentalia (part) (Bonaparte, 1838). See individual entry.

Superlegion Trechnotheria McKenna, 1975

Superlegion Trechnotheria McKenna, 1975: 27, 40.

COMMENTS: When originally proposed as a new rank it was placed in the Subclass Theria (Parker & Haswell, 1897) and included the legions † Symmetrodonta (Simpson, 1925: 560) and Cladotheria (McKenna, 1975: 27, 40). Recognised at subclass rank by McKenna in Stucky and McKenna (1993:

742) and at superlegion rank by Aplin and Archer (1987: xxi), and McKenna and Bell (1997: 43).

Legion Yangotheria Chow & Rich, 1982

Legion Yangotheria Chow & Rich, 1982: 129.

COMMENTS: When originally proposed, this rank was placed in the Superlegion Trechnotheria (McKenna, 1975) and included the sublegions † Symmetrodonta (Simpson, 1925: 560) and Cladotheria (McKenna, 1975). Recognised at legion rank by Aplin and Archer (1987: xxi), but synonymised within the Superlegion Trechnotheria (McKenna, 1975) by McKenna and Bell (1997: 43).

Sublegion Cladotheria McKenna, 1975

Legion Cladotheria McKenna, 1975: 27, 40.

COMMENTS: When originally proposed, this rank was introduced as new and placed in the Superlegion Trechnothera (McKenna, 1975) and included the sublegions † Dryolestoidea (Butler, 1939: 353) and Zatheria (McKenna, 1975). Recognised at infraclass rank by McKenna in Stucky and McKenna (1993: 742), sublegion rank by Chow and Rich (1982: 129), and Aplin and Archer (1987: xxi), and legion rank by McKenna and Bell (1997: 45).

Infralegion Zatheria McKenna, 1975

Sublegion Zatheria McKenna, 1975: 27, 40.

COMMENTS: When originally proposed as a new rank it was placed in the Legion Cladotheria and included the infraclasses † Peramura (McKenna, 1975: 27, 40) and Tribosphenida (McKenna, 1975). Recognised at infralegion rank by Chow and Rich (1982: 129), and Aplin and Archer (1987: xxi), and sublegion rank by McKenna and Bell (1997: 48).

Infraclass Prototribosphenida Rougier, 1993: 455.

COMMENTS: When originally proposed, this rank was placed in the Sublegion Zatheria (McKenna, 1975) and included the † Family Vincelestidae (J. Bonaparte, 1986: 58). Name subsequently recognised by authors including Wible *et al.* (1995: 1, 2) and Rougier *et al.* (1996: 1, 23), but synonymised within the Sublegion Zatheria by McKenna and Bell (1997: 48).

Infraclass Tribosphenida McKenna, 1975

Infraclass Tribosphenida McKenna, 1975: 27, 40.

COMMENTS: When originally proposed as a new rank it was placed in the Sublegion Zatheria and included the supercohorts Marsupialia (Iliger, 1811) and Eutheria (Gill, 1872) as modified by Huxley (1881: 657). Recognised at infraclass rank by Aplin and Archer (1987: xxi) and infralegion rank by McKenna and Bell (1997: 48).

Supercohort Marsupialia Illiger, 1811 *sensu* Cuvier, 1816

Family Marsupialia Illiger, 1811: 75.

TYPE GENUS: Not based on a genus group name.

COMMENTS: When originally proposed, this rank was placed in the Order Pollicata (Illiger, 1811 [=Mammalia (Linnaeus, 1758 part)]) and included the genera *Didelphys* [sic=*Didelphis*] Linnaeus, 1758: 54; *Chironectes* Illiger, 1811: 76; *Thylacis* Illiger, 1811 [=*Perameles* É. Geoffroy, 1803d]; *Dasyurus* É. Geoffroy, 1796b; *Amblotis* Illiger, 1811 [=*Vombatus* É. Geoffroy, 1803b]; *Balantia* Illiger, 1811 [=*Phalanger* Storr, 1780]; *Phalangista* É. Geoffroy and G. Cuvier, 1795 [=*Phalanger* Storr, 1780]; and *Phascolomys* Duméril, 1806a [=*Vombatus* É. Geoffroy, 1803b]. Name appears to be derived from *Animalia crumenata* (or purse-bearing animals) of Scaliger (1557: 277b). Recognised as the Order 'Marsupiaux' by G. Cuvier (1816a: 169) who better defined the rank and included the genera *Didelphis*, *Chironectes*, *Dasyurus*, *Perameles*, *Phalangista*, *Petaurus*, *Hypsiprymnus*, *Macropus*, 'Les Koala' and *Phascolomys*. Elevated to ordinal rank by G. Cuvier (1829a: 172), Bonaparte (1838: 113; 1850a: Unpaginated table), Owen (1859: 52), Gill (1871a: 533; 1872: vi, 25) and Gregory (1947: 46), and class by Newman (1843: 120). Superorder rank recognised by Ride (1964a: 99) and Kirsch (1968a: 420; 1977a: 111) and supercohort by McKenna (1975: 27, 40), Aplin and Archer (1987: xxi) and Wroe (1999: 512), while Gardiner (1982: 229) recognised it at infracohort rank and Kirsch *et al.* (1997: 244) recognised it at infraclass rank. McKenna and Bell (1997: 51) created the new rank of cohort. Subclass rank recognised by Iredale and Troughton (1934: vii, 4), Strahan (1983: xxi, 11; 1995: 6, 45), and Van Dyck and Strahan (2008: 9, 41).

Family Pedimanen Duméril, 1806a: 4, 16.

TYPE GENUS: Not based on a genus group name.

COMMENTS: When originally proposed, this rank was placed in the Klasse Säugthiere (Duméril, 1806a [=Mammalia (Linnaeus, 1758)]) and included the genera *Phalangista* É. Geoffroy and G. Cuvier, 1795 [=*Phalanger* Storr, 1780]; *Coescoes* Lacépède, 1799a [=*Phalanger* Storr, 1780]; *Wombatus* Desmarest, 1804a [=*Vombatus* É. Geoffroy, 1803b]; *Dasyurus* É. Geoffroy, 1796b; *Didelphis* Linnaeus, 1758: 54; and *Perameles* É. Geoffroy, 1803d.

Family Nagethiere Duméril, 1806a: 4, 18.

TYPE GENUS: Not based on a genus group name.

COMMENTS: When originally proposed, this rank was placed in the Klasse Säugthiere (Duméril, 1806a [=Mammalia (Linnaeus, 1758)]) and included the genera *Kangurus* É.

Geoffroy and G. Cuvier, 1795 [=*Macropus* Shaw, 1790]; *Phascolomys* Duméril, 1806a [=*Vombatus* É. Geoffroy, 1803b]; *Cheiromys* G. Cuvier, 1800: Table 1 [=*Daubentonia* É. Geoffroy, 1795: 195; Order Primates]; rodents of the genera *Coendus* É. Geoffroy Saint-Hilaire, 1803c: 157 [=*Coendou* Lacépède, 1799a:]; *Hystrix* Linnaeus, 1758: 56; *Cavia* Pallas, 1766: 30; *Sphalax* [=*Spalax* Güldenstaedt, 1770: 409]; *Cricetus* Leske, 1779: 168; *Arctomys* Schreber, 1780: Plate 207 [=*Marmota* Blumenbach, 1779: 79]; *Sciurus* Linnaeus, 1758: 63; *Myoxus* Zimmermann, 1780: 351; *Arvicola* Lacépède, 1799a, 10; *Mus* Linnaeus, 1758; *Ondatra* Link, 1795: 76, *Hydromys* É. Geoffroy, 1804a; and *Fibre* Duméril, 1806a: 18 [=*Castor* Linnaeus, 1758: 58], and the lagomorph genus *Lepus* Linnaeus, 1758.

Subclass Didelphes de Blainville, 1816a: 117.

COMMENTS: When originally proposed, this rank was placed in the Mammifères (G. Cuvier, 1816a: xxix, 70 [=Mammalia (Linnaeus, 1758)]) that included the 'Normaux' containing the Carnassiers (G. Cuvier, 1816a [=Mammalia (Linnaeus, 1758 part)]) and Rongeurs (de Blainville, 1816a: 117 [=Glires (Linnaeus, 1758)]) and the 'Anomaux' containing the 'L'Echidné' [=Tachyglossidae Gill, 1872] and 'L'Ornithorhynchque' [=Ornithorhynchidae (J. Gray, 1825)]. The name was also introduced by de Blainville (1816b: 251) and subsequently recognised by de Blainville (1834), whose taxonomy was copied in Gervais (1836: 619), according to Simpson (1954: 356). Name recognised by Haeckel (1866: xi, cxlii) and Gill (1871a: 532) who both included only the marsupials within the Didelphia, which appears to have been adopted by subsequent authors. Name synonymised within Marsupialia by Simpson (1945: 41) and McKenna and Bell (1997: 51), who give the author as de Blainville (1916a: 109) and recognised at division rank by Aplin and Archer (1987: xxi) who give the author as de Blainville (1834).

Family Marsupiaux G. Cuvier, 1816a: xxxi, 169.

TYPE GENUS: Not based on a genus group name.
COMMENTS: When originally proposed, this rank was placed in the Order Carnassiers (G. Cuvier, 1816a [=Mammalia (Linnaeus, 1758 part)]) and included the genera *Didelphis* Linnaeus, 1758: 54; *Dasyurus* É. Geoffroy, 1796b; *Perameles* É. Geoffroy, 1803d; *Phalangista* É. Geoffroy and G. Cuvier, 1795 [=*Phalanger* Storr, 1780]; *Hypsyprymnus* [sic=*Hypsiprymnus*] Illiger, 1811 [=*Potorous* Desmarest, 1804]; *Macropus* Shaw, 1790; Les Koala [=*Phascolarctos* de Blainville, 1816a] and *Phascolomys* Duméril, 1806a [=*Vombatus* É. Geoffroy, 1803b].

Order Ferae J. Gray, 1821: 308.

COMMENTS: When originally proposed, this rank was placed in the Class Pedimanes (J. Gray, 1821 [=Mammalia (Linnaeus, 1758 part)]) and included the families Didelphidae (J. Gray, 1821: 308) and Phalangistadae (J. Gray, 1821 [=Phalangeridae (Thomas, 1888a)]). Synonymised within the Subclass Marsupialia by McKenna and Bell (1997: 51).

HOMONYMS:
Ferae Linnaeus, 1758, mammals of the Class Mammalia (Order Carnivora). Recognised as grandorder by McKenna and Bell (1997: 211) and as a superorder here. See individual entry.

Family Marsupiata J. Gray, 1827a: 53, 185.

TYPE GENUS: Not based on a genus group name.
COMMENTS: When originally proposed, this rank was placed in the Order Carnassiers (G. Cuvier, 1816a [=Mammalia (Linnaeus, 1758 part)]) and included the genera *Didelphis* Linnaeus, 1758: 54; *Dasyurus* É. Geoffroy, 1796b; *Parameles* J. Gray, 1827a [=*Perameles* É. Geoffroy, 1803d]; *Phalangista* É. Geoffroy and G. Cuvier, 1795 [=*Phalanger* Storr, 1780]; *Petaurista* Desmarest, 1821 [=*Petauroides* Thomas, 1888a]; *Potorous* Desmarest, 1804a; *Kangurus* É. Geoffroy and G. Cuvier, 1795 [=*Macropus* Shaw, 1790]; *Phascolarctos* de Blainville, 1816a; and *Phascolomys* Duméril, 1806a [=*Vombatus* É. Geoffroy, 1803b]. Recognised at unknown rank by Owen (1834: 333) and Anon (1839: 450), and at ordinal rank by Bell (1829: 121), Richardson (1837: 138, 149) and Waterhouse (1838a: 64; 1841a: 60; 1846: 1). The term does not appear to appear to have been used much by subsequent authors except Turnbull (1971: 176) who elevated it to cohort rank. Synonymised within the Subclass Marsupialia by McKenna and Bell (1997: 51). The author of the name was given as Richardson (1837: 149) by McKenna and Bell (1997: 51).

Race Feridentiae Burnett, 1830b: 351.

COMMENTS: When originally proposed, this rank included the kinds [=families] Didelphidae (J. Gray, 1821: 308), Dasyuridae (Goldfuss, 1820a) and Phalangistadae (J. Gray, 1821 [=Phalangeridae (Thomas, 1888a)]).

Family Genuina Eichwald, 1831: 373.

TYPE GENUS: Not based on a genus group name.
COMMENTS: When originally proposed, this rank was placed in the Order Marsupialia (Illiger, 1811) and included the genera *Didelphis* Linnaeus, 1758: 54; and *Phalangista* É. Geoffroy and G. Cuvier, 1795 [=*Phalanger* Storr, 1780]. Synonymised within the Family Phalangeridae by Marshal (1981: 27).

Tribe Didelphina Bonaparte, 1838: 113.

TYPE GENUS: *Didelphis* Linnaeus, 1758: 54.
COMMENTS: When originally proposed, this rank was placed in the Family Didelphidae (J. Gray, 1821: 308). Synonymised within the Subclass Marsupialia by McKenna and Bell (1997: 51).

Tribe Entomophaga Owen, 1839a: 8, 19.

TYPE GENUS: Not based on a genus group name.
COMMENTS: When originally proposed, this rank was placed in the Marsupialia (Illiger, 1811) and included the families Ambulatoria (Owen, 1839a [=Myrmecobiidae

(Waterhouse, 1841a)]), Saltatoria (Owen, 1839a [=Peramelemorphia (Ameghino, 1889)]) and Scansoria (Owen, 1839a: 11 [=Didelphidae (J. Gray, 1821: 308)]). Tribe rank recognised by Owen (1840: 318) and family rank recognised by Giebel (1855: xi, 703) and Owen (1859: 52). Recognised as a rank between the Suborder Syndactyli and Family Peramelidae by Gill (1872: 26), but synonymised within the Order Didelphimorphia (Gill, 1872: vi, 26) by McKenna and Bell (1997: 68).

HOMONYMS:

Entomophaga Owen, 1859: 52, 'opossums' of the Class Mammalia (Order Didelphimorphia). Name is a synonym of the Order Didelphimorphia. See McKenna and Bell (1997: 68).

Entomophaga A. Murray, 1866, bats of the Class Mammalia (Order Chiroptera, Suborder Microchiroptera). Name is a synonym of the Suborder Yangochiroptera (Koopman, 1985). See individual entry.

Order Mastodidelphie Lesson, 1842: 185.

COMMENTS: When originally proposed, this rank was placed in the Class Mammalia (Linnaeus, 1758) and included the families Didelphisideae (Lesson, 1842 [=Marsupialia (Illiger, 1811 part)]), Phalangistae (Lesson, 1842 [=Phalangeridae (Thomas, 1888a)]), Petaurusideae (Lesson, 1842 [=Petauridae (Bonaparte, 1832 part) and Pseudocheiridae (Winge, 1893 part)]), Dasyurideae (Lesson, 1842 [=Dasyuridae (Goldfuss, 1820 part) and Thylacinidae (Bonaparte, 1838)]), Myrmecobineae (Lesson, 1842 [=Myrmecobiidae (Waterhouse, 1841a)]), Peramelisideae (Lesson, 1842 [=Peramelemorphia Ameghino, 1889]), Phascolarctideae (Lesson, 1842 [=Phascolarctidae (Owen, 1839a)]), Phascolomysideae (Lesson, 1842 [=Vombatidae (Burnett, 1830)]) and Macropodineae (Lesson, 1842 [=Macropodoidea (J. Gray, 1821)]). Does not appear to have been recognised by other authors.

Family Didelphisideae Lesson, 1842: 186 (part).

TYPE GENUS: *Didelphis* Linnaeus, 1758: 54.

COMMENTS: When originally proposed, this rank was placed in the Order Mastodidelphie (Lesson, 1842 [=Marsupialia (Illiger, 1811)]) and included the genera *Didelphis* Linnaeus, 1758: 54; *Micoureus* Lesson, 1842: 186; *Peramys* Lesson, 1842: 187 [=*Monodelphis* Burnett, 1830b: 351]; *Tarsipes* Gervais and Verreaux, 1842b; and *Chironectes* Illiger, 1811: 76.

Order? Rapacia Wagner, 1843: v, 16.

COMMENTS: When originally proposed, this rank included the families Dasyurina (J. Gray, 1825a [=Dasyuroidea (Goldfuss, 1820)]) (containing the genera *Thylacinus* Temminck, 1824; *Dasyurus* É. Geoffroy, 1796b and *Phascogale* Temminck, 1824); and Opossina (Wagner, 1843 [=Marsupialia (Illiger, 1811 part)]). See entry below.

Family Opossina Wagner, 1843: v, 31.

TYPE GENUS: Not based on a genus group name.

COMMENTS: When originally proposed, this rank was placed in the Rapacia (Wagner, 1843 [=Marsupialia (Illiger, 1811 part)]) and included the genera *Myrmecobius* Waterhouse, 1836a; *Didelphys* [sic=*Didelphis*] Linnaeus, 1758: 54; *Chironectes* Illiger, 1811: 76; *Perameles* É. Geoffroy, 1803d; and *Choeropus* J. Gray, 1838a [=*Chaeropus* W. Ogilby, 1838a]. Synonymised within Myrmecobiidae and Peramelidae by Marshall (1981: 27) and Marshal *et al.* (1990: 489).

Order Marsupiaux Gervais, 1855a: 263.

COMMENTS: When originally proposed, this rank was placed in the Class Mammifères (G. Cuvier, 1816a: xxix, 70 [=Mammalia (Linnaeus, 1758)]) and included the families Phascolomydés (Gervais, 1855a [=Vombatidae (Burnett, 1830b)]), Macropodés (Gervais, 1855a [=Macropodoidea (J. Gray, 1821)]), Phalangeridés (Gervais, 1855a [=Phalangerida (Aplin & Archer, 1987)]), Tarsipédidés (Gervais, 1855a [=Tarsipedidae (Gervais & Verreaux, 1842a)]), Péramélidés (Gervais, 1855a [=Perameloidea (J. Gray, 1825)]), Dasyuridés (Gervais, 1855a [=Dasyuromorphia (Gill, 1872)]), Myrmécobidés (Gervais, 1855a [=Myrmecobiidae (Waterhouse, 1841a)]) and Didelphidés (Gervais, 1855a: 285 [=Didelphimorphia (Gill, 1872: vi, 26)]).

Family Kangeroidae J. Gray, 1858a: 108.

TYPE SPECIES: Not based on a genus group name, although both *Kanguroo* and *Kangurus* are synonyms of *Macropus*.

COMMENTS: When originally proposed, this rank included the genera *Cuscus* Lesson, 1827b [= *Phalanger* Storr, 1780]; *Belideus* Waterhouse, 1839a [= *Petaurus* Shaw, 1791]; *Dactylopsila* J. Gray, 1858a; *Myoictis* J. Gray, 1858a: 111; *Perameles* É. Geoffroy, 1803b; and *Paradoxurus* F. Cuvier, 1821: 5 of Plate 186. Synonymised within the Family Macropodidae by Marshall (1981: 29), Marshall *et al.* (1990: 492) and McKenna and Bell (1997: 62), but given the constituents it should not be placed within that family.

Order Cantharophaga Haeckel, 1866: clvii.

COMMENTS: When originally proposed, this rank was placed in the Subclass Didelphia (Haeckel, 1866 [=Marsupialia (Illiger, 1811)]) and included the families † Phascolotherida (Haeckel, 1866: clvii [=† Amphilestidae (Osborn, 1888: 228)]), Myrmecobida (Haeckel, 1866 [=Myrmecobiidae (Waterhouse, 1841a)]) and Peramelida (Haeckel, 1866 [=Peramelidae (J. Gray, 1825a)]).

Subclass Didelphia Haeckel, 1866: cxlii.

COMMENTS: When originally proposed, this rank was placed in the Mammalia (Linnaeus, 1758) and included the Order Marsupialia (Illiger, 1811). Name recognised by Gill (1871a: 532). Synonymised within Marsupialia by McKenna and Bell (1997: 51) who attributed the name to de Blainville (1816a: 117).

Suborder Syndactyli Gill, 1871a: 533.

COMMENTS: When originally proposed, this rank was placed in the Order Marsupialia (Illiger, 1811) and included the families Macropodidae (J. Gray, 1821), Tarsipedidae (Gervais & Verreaux, 1842a), Phalangistidae (Burnett, 1830b) [=Phalangerida (Aplin & Archer, 1987)], Phascolarctidae (Owen, 1839a), † Diprotodontidae (Gill, 1872: 26), † Thylacoleonidae (Gill, 1872: 26) and Peramelidae (J. Gray, 1825a). The seven families referred to with the original description were provided by Gill (1872: 25). Preference for the terms Syndactyla and Diadactyla ahead of Diprotodontia and Polyprotodontia was made by Bensley (1903: 208). The division of the marsupials into the Di[a]dactyla and Syndactyla was pre-empted by de Blainville (1834; see Gervais, 1836: 619). Recognised as superorder by McKenna, in Stucky and McKenna (1993: 743) and at grandorder rank by McKenna and Bell (1997: 56). Not recognised by other authors.

Subclass (Eutheria) Didelphia Gill, 1872: v.

COMMENTS: Recorded as 'Sub-Class Placentalia' on page 1. When originally proposed, this rank was placed in the Class Mammalia (Linnaeus, 1758) and contained the Order Marsupialia (Illiger, 1811). The use of Eutheria by Gill (1872: v, vi) is the first usage of the term and clearly included both the Placentalia and Marsupialia under the one lineage. The history and confusion relating to this name was discussed by Aplin and Archer (1987: xxvi) who advised that it should be considered available and an appropriate taxon for the superdivisional level that unites the marsupials and placental mammals and expressing an understanding of their monophyly. Eutheria recognised at supercohort rank by McKenna (1975: 27, 40).

HOMONYMS:

Eutheria Huxley, 1881, placental mammals of the Class Mammalia. Synonymised within the Infraclass Placentalia (Bonaparte, 1838). See individual entry.

Suborder Polyprotodontia Owen, 1877a: xii, 105.

COMMENTS: When originally proposed, this rank was placed within the Order Marsupialia and included the Entomophaga (Owen, 1839a [=Marsupialia (Illiger, 1811 part)]) and Sarcophaga (Owen, 1839a [=Mammalia (Linnaeus, 1758 part)]) (see Owen, 1877b: 360; Nicholson, 1880: 661). The rank is derived from Owen (1868: 293) who introduced the term 'Polyprotodont' without a rank, but suggests that marsupial dentition shows them to be divisible into 'two classes: one 'polyprotodont,' or characterised by several pairs of mandibular incisors; the other 'diprotodont,' or by a single pair: these are large, more or less procumbent, and ever growing; the incisors of the first group are small, and of the usual limited growth.' Authorship of the term Polyprotodontia was first given by Gregory (1910: 199) who suggested that 'Owen [1866a], as a result of important studies of *Diprotodon*, *Thylacoleo* and *Nototherim*, proposes the terms 'diprotodont' and 'polyprotodont'', which was further highlighted on page 492 which gives the reference as Owen (1866a) and underneath refers to this as the source of the terms 'polyprotodont' and 'diprotodont'. It appears that this error was subsequently repeated. Recognised at suborder rank by Nicholson (1880: 661) who gives credit for the names to Owen but does not give the year of his work directly but in the reference list he refers to 'Fossil mammalia of Australlia. Owen. 1877a', which appears to be an abbreviated title for this work. The name was subsequently used at subordinal rank by Lydekker (1887: xi, xxiii, 254), Thomas (1888a: xii, 219), Cope (1889b: 876; 1891: 69; 1898: 108, 133), Flower and Lydekker (1891: x, 133), J. Ogilby (1892: 4), Gregory (1910: 197, 200), Osborn (1910: 515), Weber (1904: 348; 1928: xiii) and Iredale and Troughton (1934: vii, 4). Recognised at ordinal rank by Haeckel (1895: 466) and Kirsch (1968a: 420; 1977a: 111; 1977b: 45) (as Polyprotodonta) and Strahan (1983: xxi) (as Polyprotodonta). This order has historically included the bandicoots, dasyurids and marsupial moles. The spelling Polyprotodonta has been used by various authors including Kirsch (1977a: 111; 1977b: 45), Strahan (1983: xxi) and M. Archer (1984: 784). Rank not recognised by Krefft (1871a: 3, Introduction), Gill (1872), Flower (1883: 184), Bensley (1903: 208) (who preferred the terms Syndactyla and Diadactyla), Simpson (1945: 43) or subsequent authors. The author has always been given as Owen (1866a), never with page number, by all authors that include the citation for the rank including Gregory (1910: 98, 464, 492), Marshall (1981: 17) and McKenna and Bell (1997: 54). Synonymised within Dasyuromorphia by McKenna and Bell (1997: 54).

Metatheria Huxley, 1881: 654.

COMMENTS: Name does not appear to have been used as a formal taxon but appears to be placed in the Class Mammalia (Linnaeus, 1758) and included all marsupials though specific lower ranks not listed. The terms Prototheria, Metatheria and Eutheria used by Huxley (1881: 654, 657) were theoretical terms to designate stages of evolution rather than as taxonomic names (see Gregory, 1910: 94; Aplin & Archer, 1987: xxvi; Simpson, 1945: 164). Recognised as a subclass by J. Ogilby (1892: 4), infraclass by Simpson (1945: 41) and Szalay (1994: 40), and supercohort or infraclass by Shoshani (1992: 108). Synonymised within the Subclass Marsupialia by McKenna and Bell (1997: 51). This name has been recognised as a clade to include the Marsupialia and all extinct mammals that are more closely related to extant marsupials than to extant placentals according by Rougier *et al.* (1998: 459), Luo *et al.* (2003: 1934) and Black *et al.* (2012: 986).

Suborder Polyprotodonta Thomas, 1896a: 876.

COMMENTS: When originally proposed, this rank was placed in the Order Marsupialia (Illiger, 1811) and included the families Peramelidae (J. Gray, 1825a), Didelphyidae [=Didelphidae] (J. Gray, 1821: 308), Dasyuridae (Goldfuss, 1820a) and Notoryctidae (J. Ogilby, 1892). Name is an incorrect emendation of Polyprotodontia (Owen, 1877a). The spelling Polyprotodonta has been used by various authors at the ordinal rank including Kirsch (1968a: 420; 1977a: 111; 1977b: 45), Strahan (1983: xxi) and M. Archer (1984: 784), and at unknown rank by Deberer (1909: 614).

Family Properamelidae Bensley, 1903: 192.

TYPE GENUS: Not based on a genus group name.
COMMENTS: When originally proposed this rank was a hypothetical group that included common ancestors of syndactylous Australasian taxa. Discussed by Marshall (1981: 18).

Syndactyla Bensley, 1903: 208, 210.

COMMENTS: When originally proposed the term was not given a rank but included the families Peramelidae (J. Gray, 1825a), Notoryctidae (J. Ogilby, 1892), Phalangeridae (Thomas, 1888a), † Diprotodontidae (Gill, 1872: 26), Phascolomidae (J. Gray, 1821 [=Vombatidae Burnett, 1830]), Macropodidae (J. Gray, 1821) and † Thylacoleontidae [=† Thylacoleonidae] (Gill, 1872: 26).

Suborder Syndactyla Wood Jones, 1923[1923–1925]: 82, 133.

COMMENTS: When originally proposed, this rank was placed in the Order Marsupialia (Illiger, 1811) and included the sections Syndactyla Polyprotodontia (Wood Jones, 1923[1923–1925] [=Peramelemorphia (Ameghino, 1889)]) and Syndactyla Diprotodontia (Wood Jones, 1923[1923–1925] [=Diprotodontia (Owen, 1877 part)]). Recognised at ordinal rank by Szalay (1982: 631; 1994: 42), who also included the Superfamily Notoryctoidea. Synonymised within Diprotodontia by Aplin and Archer (1987: xliii) and within Syndactyli (Gill, 1871a) by McKenna and Bell (1997: 56).

Suborder Didactyla Wood Jones, 1923 [1923–1925]: 82, 84.

COMMENTS: When originally proposed, this rank was placed in the Order Marsupialia (Illiger, 1811) and included the families Didelphidae (J. Gray, 1821: 308), Dasyuridae (Goldfuss, 1820a), † Family Thylacinidae (Bonaparte, 1838), Myrmecobiidae (Waterhouse, 1841a) and Notoryctidae (J. Ogilby, 1892).

Simplicicommissurala Abbie, 1937: 432.

COMMENTS: Rank not specified but placed below the Marsupialia (Illiger, 1811) at a rank similar to Polyprotodontia. Rank not subsequently recognised.

Suborder Eometatheria Simpson, 1970: 38.

COMMENTS: When originally proposed, this rank was placed within the Metatheria (Huxley, 1881 [=Marsupialia (Illiger, 1811)]) and included the superfamilies Dasyuroidea (Goldfuss, 1820), Perameloidea (J. Gray, 1825) and Phalangeroidea (Thomas, 1888a). Recognised at ordinal rank by M. Archer (1984: 786), magnorder by McKenna in Stucky and McKenna (1993: 743), cohort by Kirsch *et al.* (1997: 245, 261) and superorder by McKenna and Bell (1997: 53). Kirsch *et al.* (1997: 245) included the orders Dasyuromorphia (Gill, 1872), Notoryctemorphia (Kirsch, 1977b), Microbiotheria (Ameghino, 1889: xx, 263), Diprotodontia (Owen, 1877a) and † Yalkaparidontia (M. Archer *et al.*, 1988: 1528). Composition of this rank discussed by Asher *et al.* (2004: 240) and Beck *et al.* (2014: 5).

Cohort Metadelphia M. Archer, 1984: 786.

COMMENTS: When originally proposed, this rank was placed in the Suborder Marsupialia (Illiger, 1811) and included three options for the lower taxa. Synonymised within the Subclass Marsupialia by McKenna and Bell (1997: 51).

Cohort Eomarsupialia M. Archer, 1984: 787.

COMMENTS: When originally proposed, this rank was placed in the Subclass Marsupialia (Illiger, 1811) and included all extant Australian marsupial orders. Support for the monophyly of this clade was provided by various authors including Beck (2012: 715). Synonymised within Eometatheria by McKenna and Bell (1997: 53).

Supercohort Notometatheria Kirsch *et al.*, 1997: 244.

COMMENTS: When originally proposed as a new rank it was placed in the Infraclass Marsupialia (Illiger, 1811) and included the cohorts Didelphidia (J. Gray, 1821: 308), Pseudiprotodontia (Kirsch *et al.*, 1997: 244), Perametatheria (Kirsch *et al.*, 1997 [=Peramelemorphia (Ameghino, 1889)]) and Eometatheria (Simpson, 1970 [=Marsupialia (Illiger, 1811 part)]).

Clade Euaustralidelphia Nilsson, 2010: 4.

COMMENTS: When originally proposed, this clade was placed in the Subclass Marsupialia (Illiger, 1811) and included the four Australasian orders Notoryctemorphia (Kirsch, 1977b), Dasyuromorphia (Gill, 1872), Peramelemorphia (Ameghino, 1889), and Diprotodontia (Owen, 1877a). Name synonymised within Eomarsupialia by Beck (2012: 716).

Clade Agreodontia Beck *et al.* 2014: 127, 132.

COMMENTS: When originally proposed, this clade was placed in the Marsupialia (Illiger, 1811) and included the Dasyuromorphia (Gill, 1872), Peramelemorphia

(Ameghino, 1889), and Notoryctemorphia (Kirsch, 1977b) to the exclusion of Diprotodontia (Owen, 1877a). This clade is supported by the molecular studies of Amrine-Madsen *et al.* (2003a: 190); Phillips *et al.* (2006: 129); Beck (2008: 179, 180, 183; 2012: 723); Meredith *et al.* (2008a: 8–9; 2009a: 559, 564; 2009b: 388, 392) and was recovered in the morphological and total evidence analyses of Beck *et al.* (2008: 3).

Cohort Australidelphia Szalay, 1982

Cohort Australidelphia Szalay, 1982: 629.

COMMENTS: When originally proposed as a new rank it was placed in the Infraclass Metatheria (Huxley, 1881 [=Marsupialia (Illiger, 1811)]) and included the orders Dromiciopsia (Szalay, 1982: 631 [=Microbiotheria Ameghino, 1889: 263]), Dasyurida (Szalay, 1982 [=Dasyuromorphia (Gill, 1872)]) and Syndactyla (Wood Jones, 1923[1923–1925] [=Marsupialia (Illiger, 1811 part)]). Term used by Aplin and Archer (1987: xxi), Szalay (1994: 42), Marshall *et al.* (1990: 459), McKenna and Bell (1997: 53) and Wroe (1999: 512). Australidelphia recognised at the rank of magnorder by McKenna and Bell (1997: 53). Szalay (1982: 623) divided the marsupials into two cohorts: the Australidelphia and the Ameridelphia (Szalay, 1982: 623).

Order Gondwanadelphia Szalay, 1993: 237, 240.

COMMENTS: When originally proposed, this rank was placed in the Cohort Australidelphia (Szalay, 1982) and included the suborders Microbiotheria (Ameghino, 1889: xx, 263) and Dasyuromorphia (Gill, 1872). Reiterated as new by Szalay (1994: 42, 348), but synonymised within Australidelphia by McKenna and Bell (1997: 53).

Order Dasyuromorphia Gill, 1872 *sensu* Aplin & Archer, 1987

Suborder Dasyuromorphia Gill, 1872: vi, 26.

COMMENTS: When originally proposed, this rank was placed in the Order Marsupialia (Illiger, 1811) and included the families Dasyuridae (Goldfuss, 1820a), Myrmecobiidae (Waterhouse, 1841a), Didelphidae (J. Gray, 1821: 308), † Plagiaulacidae (Gill, 1872: 27) and † Dromatheriidae (Gill, 1872: 27). Suborder Polyprotodontia used in preference by Iredale and Troughton (1934: vii, 4). Recognised at subordinal rank by Kirsch (1968a: 420), Strahan (1983: xxi) and Szalay (1994: 42), and ordinal rank by Aplin and Archer (1987: xxi, xxxviii) who revised the composition of this rank. Subsequently given superordinal rank by McKenna in Stucky and McKenna (1993: 743). Placed in Grandorder Dasyuromorphia by McKenna and Bell (1997: 54). Ordinal rank recognised by Marshall *et al.* (1990: 459), Strahan (1995: 6, 50), Kirsch *et al.* (1997: 245, 261) and Groves (2005d: 23), and Van Dyck and Strahan (2008: 9, 44). McKenna and Bell (1997: 54) noted that although Creatophaga and Creophaga antedate Dasyuromorphia, Dasyura and Dasyurida, priority need not be applied above the family group level.

Family Dasyuridés Gervais, 1855a: 280.

TYPE GENUS: *Dasyurus* É. Geoffroy, 1796b.

COMMENTS: When originally proposed, this rank was placed in the Order Marsupiaux (Gervais, 1855a [=Marsupialia (Illiger, 1811)]) and included the genera *Thylacynus* [=*Thylacinus*] Temminck, 1824; *Sarcophilus* F. Cuvier, 1837a; *Dasyurus* É. Geoffroy, 1796b; *Phascogale* E. Geoffr. [=Temminck, 1824]; and *Antechinus* Macleay, 1841.

Family Creatophaga Giebel, 1855: xi, 723.

TYPE GENUS: Not based on a genus group name.

COMMENTS: When originally proposed, this rank was placed within the Marsupialia (Illiger, 1811) and included the genera *Myrmecobius* Waterhouse, 1836a; *Thylacotherium* Lund, 1839: 233 [=*Didelphis* Linnaeus, 1758: 54]; † *Phascolotherium* Owen, 1838a: 9; *Phascogale* Temminck, 1824; *Dasyurus* É. Geoffroy, 1796b; and *Thylacinus* Temminck, 1824. Antedates Dasyuromorphia of Gill (1872: vi, 26), but priority need not apply above the family group level (Art. 1.2.2). Synonymised with the Order Dasyuromorphia by McKenna and Bell (1997: 54).

Order Creophaga Haeckel, 1866: clvii.

COMMENTS: When originally proposed, this rank was placed in the Class Mammalia (Linnaeus, 1758) and included the Family Dasyurida (Szalay, 1982 [=Dasyuromorphia (Gill, 1872)]). Antedates Dasyuromorphia of Gill (1872: vi, 26); but priority need not apply above the family group level (see Article 1.2.2 of the Code; ICZN, 1999: 3). Synonymised within the Order Dasyuromorphia by McKenna and Bell (1997: 54).

Family Dasyurida Haeckel, 1866: clvii.

TYPE GENUS: *Dasyurus* É. Geoffroy, 1796b.

COMMENTS: When originally proposed, this rank was placed in the Order Creophaga (Haeckel, 1866 [=Dasyuromorphia (Gill, 1872)] and included the genera *Dasyurus* É. Geoffroy, 1796b; and *Thylacinus* Temminck, 1824.

HOMONYMS:

Order Dasyurida Szalay, 1982, carnivorous marsupials of the Class Mammalia (Order Dasyuromorphia). Synonymised within the Order Dasyuromorphia here.

Order Dasyura Ameghino, 1889: xx, 276, 284.

COMMENTS: When originally proposed, this rank was placed in the Grand Group Sarcobora (Ameghino, 1889

[=Mammalia (Linnaeus, 1758)]) and included all of the carnivorous marsupials. Rank also discussed by Ameghino (1916: 481). Synonymised within the Order Dasyuromorphia by McKenna and Bell (1997: 54).

Order Marsupicarnivora Ride, 1964a: 97, 99, 101.

COMMENTS: When originally proposed as a new rank it was placed in the Superorder Marsupialia (Illiger, 1811) and included the superfamilies Didelphoidea (J. Gray, 1821: 308), † Borhyaenoidea (Ameghino, 1893: 371) and Dasyuroidea (Goldfuss, 1820a) (including the families Dasyuridae (Goldfuss, 1820a) and Thylacinidae (Bonaparte, 1838)). Taxon subsequently recognised by Turnbull (1971: 176), but not Honacki *et al.* (1982: 18) and most other authors.

Order Dasyurida Szalay, 1982: 631.

COMMENTS: When originally proposed as a new rank it was placed within the Cohort Australidelphia (Szalay, 1982). Rank recognised by Woodburne (1984a: 71) but synonymised within the Order Dasyuromorphia by McKenna and Bell (1997: 54).

HOMONYMS:

Family Dasyurida Haeckel, 1866, carnivorous marsupials of the Class Mammalia (Order Dasyuromorphia). Synonymised within the Superfamily Dasyuroidea (Goldfuss, 1820a) in this work. See individual entry.

Order Dasyuriformes Kinman, 1994: 37.

COMMENTS: When originally proposed, this rank was placed in the Class Mammalea (Kinman, 1994 [=Mammalia (Linnaeus, 1758)]). Synonymised within the Order Dasyuromorphia by McKenna and Bell (1997: 54).

Superfamily Dasyuroidea Goldfuss, 1820 *sensu* Marshall *et al.*, 1990

Family Dasyurini Goldfuss, 1820a: xxiii, 447.

TYPE GENUS: *Dasyurus* É. Geoffroy, 1796b.

COMMENTS: When originally proposed, this rank was placed in the Order Marsupialia (Illiger, 1811) and included the genera *Thylacis* Illiger, 1811 [=*Perameles* É. Geoffroy, 1803d]; and *Dasyurus* É. Geoffroy, 1796b. Family name spelt Dasyurini on page xxiii and Dosyurini on page 447. Superfamily rank recognised by Simpson (1930: 9) and subsequently by various authors, though these early authors invariably included the Family Notoryctidae. Kirsch (1977a: 112) separated the Notoryctidae into its own order, but he did not include the Family Thylacinidae, which was included within the Superfamily Borhyaenoidea (Ameghino, 1893: 371). Superfamily rank recognised within the Order Dasyurida by M. Archer (1984: 787), and within the Order Dasyuromorphia by Marshall *et al.* (1990: 489) (who placed it in the Order Dasyuromorphia and included the families Dasyuridae, Thylacinidae and Myrmecobiidae), Strahan (1995: 6, 51), Wroe (1996: 1033), and Van Dyck and Strahan (2008: 9, 44). Synonymised within the Order Dasyuromorphia by Groves (1993b: 29; 2005d: 23).

Tribe Dasyurina J. Gray, 1825a: 340.

TYPE GENUS: *Dasyurus* É. Geoffroy, 1796b.

COMMENTS: When originally proposed, this rank was placed in the Family Didelphidae (J. Gray, 1821: 308) and included the genera *Peracyon* J. Gray, 1825a [=*Thylacinus* Temminck, 1824]; *Dasyurus* É. Geoffroy, 1796b; and *Phascogale* Temminck, 1824. Recognised at tribe rank by J. Gray (1842e: 16; 1843a: xxii, 97) and at family rank by Wagner (1843: v, 17) and Van der Hoeven (1855: xiii, 900; 1858: 619). Synonymised within the Family Dasyuridae by Waterhouse (1846: 393). Family rank recognised by Bonaparte (1845: 6) and Van der Hoeven (1855: xiii, 900; 1858: 619) but not typically by subsequent authors including McKenna and Bell (1997: 55) who synonymised it within Dasyurinae and Dasyuridae.

Tribe Sarcophaga Owen, 1839a: 6, 19.

TYPE GENUS: Not based on a genus group name.

COMMENTS: When originally proposed, this rank was placed in the Marsupialia (Illiger, 1811) and included the Family Dasyuridae (Goldfuss, 1820a) (including the genera *Thylacinus* Temminck, 1824; *Dasyurus* É. Geoffroy, 1796b; and *Phascogale* Temminck, 1824. Tribe rank recognised Owen (1840: 316) and as a suborder by Gill (1871a: 533).

HOMONYMS:

Sarcophaga Bowdich, 1821, mammals of the Class Mammalia. Synonymised within the Class Mammalia (Linnaeus, 1758) in this work. See individual entry.

Family Dasyuridae Owen, 1839a: 19.

TYPE GENUS: *Dasyurus* É. Geoffroy, 1796b.

COMMENTS: Placed in Tribe Sarcophaga (Owen, 1839a [=Dasyuroidea (Goldfuss, 1820 part)]) and included the genera *Thylacinus* Temminck, 1824; *Dasyurus* É. Geoffroy, 1796b; and *Phascogale* Temminck, 1824. Name also described by Owen (1840: 332) and subsequently used by Waterhouse (1841a: 60, 117; 1846: 11, 393) and other authors, though typically credited with Goldfuss (1820a: xxxiii, 447) as the author. Synonymised within the Family Dasyuridae (Goldfuss, 1820a) by McKenna and Bell (1997: 54).

Family Dasyurideae Lesson, 1842: 190.

TYPE GENUS: *Dasyurus* É. Geoffroy, 1796b.

COMMENTS: When originally proposed, this rank was placed in the Order Mastodidelphie (Lesson, 1842 [=Marsupialia (Illiger, 1811)]) and included the genera *Thylacinus* Temminck, 1824; *Dasyurus* É. Geoffroy, 1796b; *Tapoa* Lesson, 1842 [=*Phascogale* Temminck, 1824]; *Sarcophilus* F. Cuvier, 1837a; and *Phascogale* Temminck, 1824.

Family Dasyuridés Gervais, 1855a: 280.

TYPE GENUS: *Dasyurus* É. Geoffroy, 1796b.

COMMENTS: When originally proposed, this rank was placed in the Order Marsupiaux (Gervais, 1855a [=Marsupialia (Illiger, 1811)]) and included the genera *Thylacinus* Temminck, 1824; *Sarcophilus* F. Cuvier, 1837a; *Dasyurus* É. Geoffroy, 1796b; *Phascogale* Temminck, 1824; and *Antechinus* Macleay, 1841.

Family Dasyurida Haeckel, 1866: clvii.

TYPE GENUS: *Dasyurus* É. Geoffroy, 1796b.

COMMENTS: When originally proposed, this rank was placed in the Order Creophaga (Haeckel, 1866 [=Dasyuromorphia (Gill, 1872)] and included the genera *Dasyurus* É. Geoffroy, 1796b; and *Thylacinus* Temminck, 1824. Synonymised within the Family Dasyuridae by McKenna and Bell (1997: 54).

HOMONYMS:

Order Dasyurida Szalay, 1982, carnivorous marsupials of the Class Mammalia (Order Dasyuromorphia). Synonymised within the Order Dasyuromorphia (Gill, 1872) here. See individual entry.

Subfamily Dasyurinae Thomas, 1888a: xii, 253.

TYPE GENUS: *Dasyurus* É. Geoffroy, 1796b.

COMMENTS: When originally proposed, this rank was placed in the Family Dasyuridae (Goldfuss, 1820a) and included the genera *Thylacinus* Temminck, 1824; *Sarcophilus* F. Cuvier, 1837a; *Dasyurus* É. Geoffroy, 1796b; *Phascogale* Temminck, 1824; *Sminthopsis* Thomas, 1887a; and *Antechinomys* Krefft, 1867a. Synonymised within the Subfamily Dasyurinae by McKenna and Bell (1997: 55).

Tribe Dasyurini Winge, 1893a: 88, 93.

TYPE GENUS: *Dasyurus* É. Geoffroy, 1796b.

COMMENTS: When originally proposed, this rank was placed in the Family Dasyuridae (Goldfuss, 1820a) and included the genera *Phascologale* Thomas, 1888a [=*Antechinus* Macleay, 1841]; *Podabrus* Gould, 1845 [1845–1863] [=*Sminthopsis* Thomas, 1887a]; *Antechinomys* Krefft, 1867a; *Dasyurus* É. Geoffroy, 1796b; *Sarcophilus* F. Cuvier, 1837a; and *Myrmecobius* Waterhouse, 1836a. Tribe subsequently recognised by Winge (1941: 69) and more recently, with a modified composition, by authors including Krajewski *et al.* (2000a: 98; 2000b: 375; 2000c: 423), Krajewski and Westerman (2003: 14) and Groves (2005d: 24).

Superfamily Dasyuroidea Simpson, 1930: 9.

TYPE GENUS: *Dasyurus* É. Geoffroy, 1796b.

COMMENTS: When originally proposed, this rank was placed in the Order Marsupialia (Illiger, 1811) and included the families Dasyuridae (Goldfuss, 1820a) and Notoryctidae (J. Ogilby, 1892). Superfamily rank recognised by Simpson (1931: 262; 1945: 43; 1970: 38), Ride (1964a: 99), Kirsch (1977a: 112; 1977b: 4, 45), Strahan (1983: xxi; 15; 1995: 7, 51), Marshall *et al.* (1990: 459), and Van Dyck and Strahan (2008: 9, 44). Included as a synonym of Dasyuromorphia by Groves (1993b: 29; 2005d: 23).

Family Dasyuridae Goldfuss, 1820 *sensu* Owen, 1839

Family Dasyurini Goldfuss, 1820a: xxiii, 447.

TYPE GENUS: *Dasyurus* É. Geoffroy, 1796b.

COMMENTS: When originally proposed, this rank was placed in the Order Marsupialia (Illiger, 1811) and included the genera *Thylacis* Illiger, 1811 [=*Perameles* É. Geoffroy, 1803d]; and *Dasyurus* É. Geoffroy, 1796b. Family name spelt Dasyurini on page xxiii and Dosyurini on page 447. Winge (1941: 69) recognised the family and included the tribes Dasyurini (Goldfuss, 1820a), Thylacinini [sic] (Bonaparte, 1838), and Notoryctini (J. Ogilby, 1892). Simpson (1945: 43) gives Waterhouse (1838a) as the citation but this name does not appear in this publication, similarly Cramb and Hocknull (2010a: 128) gives the citation as *sensu* Waterhouse (1838b) but the name does not occur in that publication either. The confusion appears to be because Waterhouse (1841a: 60) suggests that the arrangement adopted in that work follows Waterhouse (1838b), even though it is not actually listed in this earlier work. The current spelling of the Family Dasyuridae was used by Waterhouse (1838c: 152; 1841a: 60, 117; 1846: 11, 393), Owen (1839a: 19; 1840: 332), Bonaparte (1845: 6), Krefft (1866a: 6), Thomas (1888a: xii, 253), Iredale and Troughton (1934: vii, 4), Marshall (1981: 26), Mahoney and Ride (1988a: 14) and subsequent authors. The current subfamilies follow Marshall *et al.* (1990: 459).

Family Macropidae J. Gray, 1842b: 261.

TYPE GENUS: Usage of the name appears to be in error.

COMMENTS: Included the description of *Phascogale leucopus* J. Gray, 1842b: 261 [=*Sminthopsis leucopus*] with no other taxa discussed under the family name.

HOMONYMS:

Family Macropidae J. Gray, 1821, macropods of the Class Mammalia (Order Diprotodontia). Rank is equivalent to the Superfamily Macropodoidea J. Gray, 1821. See individual entry.

Subfamily Macropodinae Liem, 1963: 47, fish of the Class Osteichthyes, Perciformes, Family Osphronemidae, Belontiidae or Anabantidae). The spelling of this name was emended to Macropodusinae by Opinion 2058 (Case 2661) to remove homonymy with Macropodinae J. Gray, 1821 (ICZN, 2001: 297; 2003a: 253).

Diadactyla [sic] Bensley, 1903: 208, 210.

TYPE GENUS: Not based on a genus group name.

COMMENTS: When originally proposed the term was not given a rank but included the Family Dasyuridae (Goldfuss, 1820a). The division of the marsupials into the Di[a]dactyla and Syndactyla was pre-empted by de Blainville (1834; see Gervais, 1836: 619). The terms Syndactyla and Diadactyla [sic] were given favour ahead of the terms Diprotodontia and Polyprotodontia by Bensley (1903: 208).

Subfamily Dasyurinae Goldfuss, 1820 *sensu* Marshall *et al.*, 1990

Family Dasyurini Goldfuss, 1820a: xxiii, 447.

TYPE GENUS: *Dasyurus* É. Geoffroy, 1796b.

COMMENTS: When originally proposed, this rank was placed in the Order Marsupialia (Illiger, 1811) and included the genera *Thylacis* Illiger, 1811 [=*Perameles* É. Geoffroy, 1803d]; and *Dasyurus* É. Geoffroy, 1796b. Simpson (1945: 43) designated Thomas (1888a: xii, 253) as the author of this subfamily. Subfamily rank attributed to Goldfuss (1820a) by M. Archer (1982a: 439) who also introduced the tribes Parantechini (nova) and Dasyurini (Goldfuss, 1820a). Subfamily rank recognised by Gill (1872: 26), Bensley (1903: 91), Osborn (1910: 516), Troughton (1967: 39), Marshall (1981: 26), Marshall *et al.* (1990: 459, 488), M. Archer (1982a: 439; 1984: 635), Strahan (1995: 6, 54), Kirsch *et al.* (1997: 245) and followed by most subsequent authors. Tribe Dasyurini recognised within the Family Dasyuridae by Marshall (1984: 83) and within the Subfamily Dasyurinae by Marshall (1981: 26), M. Archer (1982a: 439; 1984: 635), Marshall *et al.* (1990: 488), Kirsch *et al.* (1997: 245), Krajewski and Westerman (2003: 15) and Groves (2005d: 24). Tribe synonymised within the Subfamily Dasyurinae by McKenna and Bell (1997: 55). Subfamily classification reviewed by Krajewski *et al.* (1993: 158) and Krajewski and Westerman (2003: 3).

Tribe Antechini A. Murray, 1866: vx, 286, 362.

TYPE GENUS: *Antechinus* Macleay, 1841.

COMMENTS: When originally proposed, this rank was placed in the Entomophaga (Owen, 1839a [=Marsupialia (Illiger, 1811 part)]), at unknown rank, and included the genus *Antechinus* Macleay, 1841. Synonymised within the Subfamily Dasyurinae by Marshall (1981: 26), and McKenna and Bell (1997: 55).

Subfamily Sarcophilinae Gill, 1872: 26.

TYPE GENUS: *Sarcophilus* F. Cuvier, 1837a.

COMMENTS: When originally proposed, this rank was placed in the Family Dasyuridae (Goldfuss, 1820a). Synonymised within the Subfamily Dasyurinae by Marshall (1981: 26), and McKenna and Bell (1997: 55). Tribe rank recognised by Marshall (1981: 26; 1984: 83).

Tribe Dasyurini Moeller, 1973: 300.

TYPE GENUS: *Dasyurus* É. Geoffroy, 1796b.

COMMENTS: When originally proposed as a new rank it was placed in the Subfamily Dasyurinae (Thomas, 1888a) [=Goldfuss, 1820a) and included the genera *Dasyurus* É. Geoffroy, 1796b and *Myoictis* J. Gray, 1858a: 111.

Tribe Sarcophilini Moeller, 1973: 300.

TYPE GENUS: *Dasyurus* É. Geoffroy, 1796b.

COMMENTS: When originally proposed as a new rank it was placed in the Subfamily Dasyurinae (Thomas, 1888a [=Goldfuss, 1820a] and included the genus *Sarcophilus* F. Cuvier, 1837a. Tribe rank recognised by Marshall (1981: 26; 1984: 83) but typically not other authors.

Subfamily Murexinae M. Archer, 1982a: 438.

TYPE GENUS: *Murexia* Tate & Archbold, 1937: 331, 335, 339.

COMMENTS: When originally proposed as a new rank it was placed in the Family Dasyuridae (Goldfuss, 1820a) and included the genus *Murexia* Tate and Archbold, 1937: 331, 335, 339. Synonymised within the Subfamily Dasyurinae by McKenna and Bell (1997: 55).

Subfamily Phascolosorexinae M. Archer, 1982a: 438.

TYPE GENUS: *Phascolosorex* Matschie, 1916: 263.

COMMENTS: When originally proposed as a new rank it was placed in the Family Dasyuridae (Goldfuss, 1820a) and included the genera *Phascolosorex* Matschie, 1916: 263; and *Neophascogale* Stein, 1933: 87. Synonymised within the Subfamily Dasyurinae by McKenna and Bell (1997: 55).

Tribe Parantechini M. Archer, 1982a: 439.

TYPE GENUS: *Parantechinus* Tate, 1947.

COMMENTS: When originally proposed as a new rank it was placed in the Subfamily Dasyurinae (Goldfuss, 1820a) and included the genera *Parantechinus* Tate, 1947; *Dasykaluta* M. Archer, 1982a; and *Pseudantechinus* Tate, 1947. Tribe rank recognised by Marshall *et al.* (1990: 488). Synonymised within Dasyurinae by McKenna and Bell (1997: 55) and confirmed by Westerman *et al.* (2008: 207).

Subfamily Phascolosoricinae M. Archer, 1989: 67.

TYPE GENUS: *Phascolosorex* Matschie, 1916: 263.

COMMENTS: When originally proposed this rank was introduced as a replacement name for Phascolosorexinae M. Archer (1982a). Synonymised within the Subfamily Dasyurinae by McKenna and Bell (1997: 55).

Dasycercus Peters, 1875

Dasycercus Peters, 1875a: 73.

TYPE SPECIES: *Nomen novum* for *Chaetocercus* Krefft, 1867a.

COMMENTS: Synonymised within *Phascogale* by Thomas (1888a: 273), but recognised by Iredale and Troughton (1934: vii, 8), Simpson (1945: 43), Troughton (1967: 28), Honacki *et al.* (1982: 28), Marshall (1984: 83), and McKenna and Bell (1997: 56). Genus recognised in preference to *Dasyuroides* Mahoney and Ride (1988a: 18) and reviewed by Adams *et al.* (2000: 2).

Chaetocercus Krefft, 1867a: 434.

TYPE SPECIES: *Chaetocercus cristicauda* Krefft, 1867a [=*Dasycercus cristicauda* (Krefft, 1867a)] by monotypy.

COMMENTS: Synonymised within *Phascogale* by Thomas (1888a: 273) and *Dasycercus* by Iredale and Troughton (1934: 8), Simpson (1945: 43), Wood Jones (1949: 409), Ride (1970: 242), Marshall (1981: 26), Marshall *et al.* (1990: 488), Groves (1993b: 31) and other authors.

HOMONYMS:

Chaetocercus G. Gray, 1855: 22, hummingbirds of the Class Aves, Order Apodiformes, Family Trochilidae). Currently accepted name. See Freyman and Schuchmann (2005: 61).

Amperta Cabrera, 1919: 65.

TYPE SPECIES: *Nomen novum* for *Chaetocercus* Krefft, 1867a.

COMMENTS: Synonymised within *Dasycercus* by Iredale and Troughton (1934: 8), Wood Jones (1949: 409), Marshall (1981: 26), Marshall *et al.* (1990: 488), Groves (1993b: 31) and subsequent authors.

Dasycercus blythi (Waite, 1904)

Brush-tailed Mulgara

Phascogale blythi Waite, 1904: 123.

TYPE LOCALITY: Pilbara district, Western Australia, Australia.

COMMENTS: Recognised as a full species within *Dasycercus* by Iredale and Troughton (1934: vii, 8), Tate (1947: 142) and Troughton (1967: 29). Synonymised within *cristicauda* by Ride (1970: 242), Mahoney and Ride (1988a: 19) and Groves (1993b: 31; 2005d: 24). Recognised as a species by Woolley (2005: 213; 2006: 117), Van Dyck and Strahan (2008: 47) and Menkhorst and Knight (2011: 15).

Phascogale blighi Woodward, 1902: 213.

TYPE LOCALITY: Pilbara district, Western Australia, Australia.

COMMENTS: *Nomen nudum*. Incorrectly designated to Shortridge (1910: 804, 840) by Iredale and Troughton (1934: 8). Synonymised within *Dasycercus blythi* by Iredale and Troughton (1934: 8). Synonymised within *cristicauda* by Mahoney and Ride (1988a: 19) and Groves (1993b: 31; 2005d: 24), but synonymised within *blythi* by Woolley (2005: 216).

Dasycercus cristicauda (Krefft, 1867)

Crest-tailed Mulgara

Chaetocercus cristicauda Krefft, 1867a: 435; Plate 36.

TYPE LOCALITY: Lake Alexandrina, South Australia, Australia.

COMMENTS: The accuracy of Krefft's description of *cristicauda* was questioned by Spencer (1896a: 20). Placed within *Phascologale* by Thomas (1887a: 509; 1888a: xiii, 276) and *Dasycercus* by Iredale and Troughton (1934: vii, 8) and most subsequent authors.

Phascogale Hillieri Thomas, 1905a: 427.

TYPE LOCALITY: Killalpaninna, east of Lake Eyre, South Australia, Australia.

COMMENTS: Synonymised within *cristicauda* by Iredale and Troughton (1934: 8), Wood Jones (1949: 418), Ride (1970: 242), Mahoney and Ride (1988a: 19) and Groves (1993b: 31; 2005d: 24). Subspecies rank, within *cristicauda*, recognised by Finlayson (1933a: 200) and Strahan (1983: 26; 1995: 56). Elevated to species status by Maxwell *et al.* (1996: 2) based on unpublished information (received information from M. Adams of the South Australian Museum) for the species status of *hillieri*. Subsequently Adams *et al.* (2000: 2) confirmed the reasons for the species rank of *hillieri*. Synonymised within *cristicauda* by Woolley (2005: 216) who proposed that Adams *et al.* (2000) misplaced *cristicauda* in the clade containing the syntype *blythi*. Taxon subsequently synonymised within *cristicauda* by Van Dyck and Strahan (2008: 50).

Dasykaluta M. Archer, 1982

Dasykaluta M. Archer, 1982a: 434, 435.

TYPE SPECIES: *Antechinus rosamondae* Ride, 1964b [=*Dasykaluta rosamondae* (Ride, 1964b)] by original designation.

COMMENTS: Synonymised within *Antechinus* by Strahan (1983: 44) and McKenna and Bell (1997: 55). Synonymised within *Parantechinus* by Mahoney and Ride (1988a: 23), but M. Archer (1982a: 428, 439) and Krajewski *et al.* (2000a: 102) showed that they are distinct groups. Recognised as a valid genus by Marshall *et al.* (1990: 488), Groves (1993b: 31; 2005d: 24), Strahan (1995: 57), Van Dyck and Strahan (2008: 51) and Menkhorst and Knight (2011: 15).

Dasykaluta rosamondae (Ride, 1964)

Kaluta

Antechinus rosamondae Ride, 1964b: 58.

TYPE LOCALITY: Pullcunah Hill, Woodstock Station, via Marble Bar, Western Australia, Australia.

COMMENTS: Included in the genus *Antechinus* by Troughton (1967: 20), Ride (1970: 119), Honacki *et al.* (1982: 28) and Strahan (1983: 44). Transferred to *Parantechinus* by Mahoney and Ride (1988a: 24) and *Dasykaluta* by Groves (1993b: 31; 2005d: 24), Strahan (1995: 57), Van Dyck and Strahan (2008: 51) and Menkhorst and Knight (2011: 15).

Dasyuroides Spencer, 1896

Dasyuroides Spencer, 1896b: 5.

TYPE SPECIES: *Dasyuroides byrnei* Spencer, 1896b by monotypy.

COMMENTS: Recognised as a valid genus by Cabrera (1919: 65), Iredale and Troughton (1934: vii, 8), Simpson (1945: 43), Tate (1947: 142), Mack (1961: 214), Troughton (1967: 29), Marshall (1981: 26), Baverstock *et al.* (1982: 648), Honacki *et al.* (1982: 28) and Strahan (1983: 24; 1995: 59). Synonymised within *Dasycercus* by Mahoney and Ride (1988a: 18) and Groves (1993b: 31), but Kirsch *et al.* (1997: 262) considered this premature. Further support to synonymise *Dasyuroides* came from N. Cooper *et al.* (2000: 134) who favoured making them congeneric. Despite these studies the genus was recognised for *byrnei* by Groves (2005d: 24), McKenna and Bell (1997: 55), Van Dyck and Strahan (2008: 52) and Menkhorst and Knight (2011: 15).

FUTURE TAXONOMIC RESEARCH: The relationships of the species comprising *Dasyuroides* and *Dasycercus* need to be assessed to confirm whether *Dasyuroides* is a valid genus.

Dasyuroides byrnei Spencer, 1896

Kowari

Dasyuroides byrnei Spencer, 1896b: 6.

TYPE LOCALITY: Charlotte Waters, Northern Territory, Australia.

COMMENTS: Taxon further described by Spencer (1896a: 36). Type designation by Dixon (1970: 106). Included as species within *Dasyuroides* by Iredale and Troughton (1934: vii, 9), Tate (1947: 142), Mack (1961: 214) and Ride (1970: 112) and other authors until Mahoney and Ride (1988a: 18) transferred it to *Dasycercus*. Moved back to *Dasyuroides* by Strahan (1995: 59), McKenna and Bell (1997: 55), Groves (2005d: 24), and Van Dyck and Strahan (2008: 52).

Dasyuroides Byrnei pallidior Thomas, 1906b: 330.

TYPE LOCALITY: Killalpaninna, South Australia, Australia.

COMMENTS: Subspecies recognised by Finlayson (1933a: 202). Synonymised within *byrnei* by Iredale and Troughton (1934: 9), Mack (1961: 214) and subsequent authors. Strahan (1983: 24; 2005: 60), and Van Dyck and Strahan (2008: 54) considered this a doubtful subspecies.

Dasyurus É. Geoffroy, 1796

dasyurus [sic] É. Geoffroy, 1796b: 469.

TYPE SPECIES: *Didelphis maculata* Anon, 1791 (as 'Spotted Opossum' of Phillip, 1789: 147) [=*Dasyurus viverrinus* (Shaw, 1800)] by subsequent monotypy. See Thomas (1888a: 261).

COMMENTS: Also described by É. Geoffroy (1796c: 106), which is given as the original description by some authors including Thomas (1888a: 261), who gives the types species as *D. viverrinus*. Genus recognised by Waterhouse (1846: 432) and subsequent authors. See Haltenorth (1958: 20). Krajewski *et al.* (1997a: 217) confirmed that *Dasyurus* is monophyletic using molecular data. The work of Firestone (2000: 1) on the mtDNA Control Region has shown that there are more diagnosable units in this genus than have been hitherto recognised. These include the six species traditionally listed, plus what have been classified as the Queensland population of *D. hallucatus* and the Queensland and Tasmanian populations of *D. maculatus*.

Viverra Kerr, 1792: 170.

TYPE SPECIES: *Didelphis maculata* Anon, 1791 (as *Viverra maculata* Kerr, 1792) [=*Dasyurus maculatus* (Kerr, 1792)] by original designation.

COMMENTS: Not typically associated with *maculatus*, or other members of the genus *Dasyurus* since Shaw (1800: 433).

HOMONYMS:

Viverra Linnaeus, 1758: 43, civet's of the Subclass Eutheria (Order Carnivora, Family Viverridae). Currently used name. See Wozencraft (2005: 558).

Mustela F. Meyer, 1793: 27.

TYPE SPECIES: *Didelphis maculata* Anon, 1791 (as *Mustela NovaeHollandiae* F. Meyer, 1793) [=*Dasyurus maculatus* (Kerr, 1792)].

COMMENTS: Not typically associated with *Mustela*, or other members of the genus *Dasyurus* since its description.

HOMONYMS:

Mustela Linnaeus, 1758, mustelids of the Class Mammalia (Order Carnivora, Family Mustelidae). Currently used name. See Wozencraft (2005: 558).

Nasira Harvey, 1841: 210.

TYPE SPECIES: *Didelphis viverrina* Shaw, 1800 [=*Dasyurus viverrinus* (Shaw, 1800)] by subsequent designation. See Zietz (1904: 180).

COMMENTS: Appears to be an incorrect subsequent spelling of *Naisira* by Harvey (1904: 177). Synonymised within *Dasyurus* by Iredale and Troughton (1934: 12), Marshall (1981: 26), Marshall *et al.* (1990: 488) and subsequent authors.

Dasurus Anon, 1845: 743.

TYPE SPECIES: Incorrect subsequent spelling of *Dasyurus* É. Geoffroy, 1796b.

COMMENTS: Synonymised within *Dasyurus* by Palmer (1904: 218).

Dasyurinus Matschie, 1916: 262.

TYPE SPECIES: *Dasyurus geoffroii* Gould, 1841a (as *Dasyurus geoffroyi*) by original designation.

COMMENTS: Described as a subgenus of *Dasyurus*. Originally made available as a subgenus *Dasyurus* É. Geoffroy, 1796b. Used as a valid genus for *geoffroii* by Iredale and Troughton (1934: 13) and Tate (1947: 143). Synonymised within *Dasyurus* by Simpson (1945: 43),

Marshall (1981: 26), Marshall *et al.* (1990: 488) and subsequent other authors.

Dasyurops Matschie, 1916: 262.

TYPE SPECIES: *Viverra maculata* Kerr, 1792 [=*Dasyurus maculatus* (Kerr, 1792)] by original designation.

COMMENTS: Described as a subgenus of *Dasyurus*. Used as a valid genus for *maculatus* by Iredale and Troughton (1934: 14), Tate (1947: 144) and Troughton (1967: 42). Synonymised within *Dasyurus* by Simpson (1945: 43), Ride (1970: 242), Marshall (1981: 26), Marshall *et al.* (1990: 488) and subsequent authors.

Notoctonus Pocock, 1926a: 1082.

TYPE SPECIES: *Dasyurus geoffroii* Gould, 1841a (as *Dasyurus geoffroyi*) by original designation.

COMMENTS: Objective synonym of *Dasyurinus*. Was included as a synonym within *Dasyurinus* by Iredale and Troughton (1934: 13). Synonymised within *Dasyurus* by Mahoney and Ride (1988a: 19), Marshall (1981: 26), Marshall *et al.* (1990: 488) and subsequent other authors.

Satanellus Pocock, 1926a: 1083.

TYPE SPECIES: *Dasyurus hallucatus* Gould, 1842a by original designation.

COMMENTS: Recognised, with *hallucatus*, by Iredale and Troughton (1934: 13), Tate (1947: 142), D. Johnson (1964: 443), Troughton (1967: 41), M. Archer (1982a: 412, 439), Kirsch and Archer (1982: 597, 604), Baverstock *et al.* (1990a: 278) and Marshall *et al.* (1990: 488). Synonymised within *Dasyurus* by Simpson (1945: 43), Ride (1970: 247), Marshall (1981: 26), Baverstock *et al.* (1982: 643), Mahoney and Ride (1988a: 19), Van Dyck (1988: 145), Kirsch *et al.* (1990a: 673, 692), Krajewski *et al.* (1994: 26), and Wroe and Mackness (1998: 605).

Stictophonus Pocock, 1926a: 1083.

TYPE SPECIES: *Viverra maculata* Kerr, 1792 [=*Dasyurus maculatus* (Kerr, 1792)] by original designation.

COMMENTS: Objective synonym of *Dasyurops*. Synonymised within *Dasyurops* by Iredale and Troughton (1934: 14) and within *Dasyurus* by Marshall (1981: 26), Mahoney and Ride (1988a: 19), Marshall *et al.* (1990: 488) and subsequent other authors.

Dasyurus geoffroii Gould, 1841

Western Quoll

Dasyurus geoffroii geoffroii Gould, 1841

Dasyurus Geoffroii Gould, 1841a: 151.

TYPE LOCALITY: Liverpool Plains, New South Wales, Australia.

COMMENTS: Placed within *Dasyurinus* by Iredale and Troughton (1934: vii, 13) and Tate (1947: 144), but within *Dasyurus* by most other authors including Thomas (1888a: 268). The New Guinea records of this species refer to *Dasyurus spartacus* (Groves, 2005d: 25).

FUTURE TAXONOMIC RESEARCH: The taxa *fortis* and *spartacus* need to be assessed to determine their relationships to the nominate subspecies of this species, and to each other.

Dasyurus Geoffroyi Waterhouse, 1841a: 132.

TYPE LOCALITY: Alternative spelling of *Dasyurus geoffroii* Gould, 1841a.

COMMENTS: Though the two spellings, *geoffroii* and *geoffroyi*, were published in the same year, that of Waterhouse (1841a: 132) gives the author as Gould (1841a: 151). The spelling *geoffroyi* has been used by most authors including Waterhouse (1846: 437), Gould (1851 [1845–1863]: Text to Plate 51), Thomas (1888a, xiii, 268; 1906c: 476), Serventy (1954: 141), Whittell (1954: 105), Finlayson (1961a: 152), Marlow (1965: 40; 1981: 40), Lyne (1967: 19), L. Collins (1973: 124) and Helgen (2007a: 713), while the spelling *geoffroyii* was used by authors including J. Gray (1841: 400) and Parton (1952: 93). McAllan and Bruce (1989: 449) note that this taxon is a junior objective synonym of *geoffroii*.

Dasyurus geoffroii fortis Thomas, 1906

Dasyurus geoffroyi fortis Thomas, 1906c: 476.

TYPE LOCALITY: Arthur River, near Wagin, Western Australia, Australia.

COMMENTS: Included as a subspecies of *geoffroii* by Iredale and Troughton (1934: 13) and Tate (1947: 144). Synonymised within *geoffroii* by Mahoney and Ride (1988a: 20), Serena *et al.* (1991: 1), Groves (2005d: 25) and Burbidge *et al.* (2014: 17). Recognised as a subspecies of *geoffroii* by Strahan (1983: 22; 2005: 62), Clayton *et al.* (2006: 100), and Van Dyck and Strahan (2008: 56). The validity of this putative subspecies needs to be carefully re-examined.

Φ ***Dasyurus geoffroii spartacus*** Van Dyck, 1988

Φ *Dasyurus spartacus* Van Dyck, 1988: 145.

TYPE LOCALITY: Morehead, Trans-Fly Plains, Papua New Guinea. (8°41′S, 141°39′E)

COMMENTS: When the first specimens of this taxon were first discovered in New Guinea in 1972–1973 (Waithman, 1979: 315) it was recognised as *geoffroii* by M. Archer (1979a: 32; 1982a: 416), Honacki *et al.* (1982: 29) and Strahan (1983: 22) until it was formally described as a distinct species by Van Dyck (1988: 145). It has typically been recognised as a distinct species since its description by authors including Flannery (1990: 64; 1995a: 86) and Groves (1993b: 32; 2005d: 25). However Firestone (2000: 11, 18) questioned its specific status, which was supported by the close relationship between *spartacus* and *geoffroii* observed by Krajewski *et al.* (2000b: 388) and the suggestion that it is probably better recognised as a subspecies of *geoffroii* by Helgen (2007a: 713); we do not agree, as it is still diagnosably distinct, but we list it here because of the controversy.

Dasyurus hallucatus Gould, 1842

Northern Quoll

Dasyurus hallucatus Gould, 1842a: 41.

TYPE LOCALITY: Port Essington, Northern Territory, Australia.

COMMENTS: Included within the genus *Satanellus* by Iredale and Troughton (1934: vii, 13), Tate (1947: 143) and Troughton (1967: 41). Included within *Dasyurus* by most authors including Waterhouse (1846: 434), Thomas (1888a: xiii, 269; 1906d, 540), Baverstock *et al.* (1982: 643), Van Dyck (1988: 145), Kirsch *et al.* (1990a: 673), Krajewski *et al.* (1994: 26), and Wroe and Mackness (1998: 605). The once recognised subspecies *nesaeus*, *exilis* and *predator* are no longer recognised by recent authors (e.g. Strahan, 1995: 66; Maxwell *et al.*, 1996: 2)

FUTURE TAXONOMIC RESEARCH: Preliminary results suggest strongly divergent lineages between Northern Territory and Queensland populations (at least as great as between *D. geoffroii* and *D. spartacus*) (Firestone, 2000: 18) so the question of whether the species actually consists of two diagnosably distinct species should be examined.

Mustela Quoll Zimmermann, 1783: 181.

TYPE LOCALITY: North Queensland, Australia.

COMMENTS: It was the quoll referred to in Hawkesworth (1773: 626). Although accepted as an available name by Mahoney and Ride (1984: 57) they suggested it should be suppressed because its acceptance would upset a stable nomenclature. The specific name *quoll* was suppressed under Article 80 of the Code (ICZN, 1985a: 175) for *Dasyurus hallucatus* (see Groves, 1993b: 32; 2005d: 25). An application to suppress the name *Mustella quoll* Zimmermann, 1783 was also made by Mahoney and Ride (1986: 50). Synonymised within *hallucatus* by Mahoney and Ride (1988a: 20) and Groves (2005d: 25).

HOMONYMS:

Mustela quoll Zimmermann, 1777, Eastern Quolls of the Class Mammalia (Order Dasyuromorphia: Family Dasyuridae). Name is a synonym of *Dasyurus viverrinus* (Shaw, 1800). See individual entry.

Dasyurus hallucatus exilis Thomas, 1909a: 152.

TYPE LOCALITY: Parry's Creek, near Wyndham, Western Australia, Australia.

COMMENTS: Subspecies rank recognised by Iredale and Troughton (1934: 13), Tate (1947: 143) and tentatively by Strahan (1983: 23). Synonymised within *hallucatus* by Mahoney and Ride (1988a: 20) and Groves (1993b: 32; 2005d: 25).

Dasyurus hallucatus predator Thomas, 1926a: 543.

TYPE LOCALITY: Utingu, Cape York Peninsula, Queensland, Australia.

COMMENTS: Subspecies recognised by Iredale and Troughton (1934: 13), Tate (1947: 143) and tentatively by Strahan (1983: 23). Synonymised within *hallucatus* by Mahoney and Ride (1988a: 20) and Groves (1993b: 32; 2005d: 25).

Dasyurus hallucatus nesaeus Thomas, 1926a: 544.

TYPE LOCALITY: Groote Eylandt, Gulf of Carpentaria, Northern Territory, Australia.

COMMENTS: Subspecies recognised by Iredale and Troughton (1934: 14), Tate (1947: 143), D. Johnson (1964: 445) and tentatively by Strahan (1983: 23). Synonymised within *hallucatus* by Mahoney and Ride (1988a: 20) and Groves (1993b: 32; 2005d: 25).

Dasyurus maculatus (Kerr, 1792)

Spotted-tailed Quoll

Dasyurus maculatus maculatus (Kerr, 1792)

Viverra maculata Kerr, 1792: 170.

TYPE LOCALITY: Port Jackson, New South Wales, Australia.

COMMENTS: The holotype is described and figured under the name Spotted Martin and Martin Cat of Phillip (1789: 276, pl.). Recognised within *Viverra* by Shaw (1800: 433). Placed in *Dasyurus* by G. Fischer (1813b: 584), J. Gray (1841: 400; 1843a: 98), Waterhouse (1846: 439), Gould (1851 [1845–1863]: Text to Plate 49), Krefft (1868a: 94) and Thomas (1888a: xii, 262, 263). Included within genus *Dasyurops* by Iredale and Troughton (1934: vii, 14), Tate (1947: 145) and Troughton (1967: 42). See Haltenorth (1958: 20). Placed in the genus *Dasyurus* by Mahoney and Ride (1988a: 21). Recent genetic analysis by Firestone *et al.* (1999: 1621) suggests the taxonomic status of tiger quolls in Tasmania should be reassessed and elevated to at least subspecies rank to more accurately reflect the significant divergence between the two evolutionary species units. This species was recorded on King Island by Campbell (1888: 136) but is now considered extinct (Hope, 1972: 177).

FUTURE TAXONOMIC RESEARCH: The research of Firestone *et al.* (1999: 1621) suggests that the existing taxonomic status of *maculata* should be reassessed and the Tasmanian population perhaps elevated to the specific level. Therefore studies are required to formally assess the taxonomy of the Tasmanian and King Island quolls to determine if these groups are distinct. If these prove distinct from those from the mainland, but similar to each other, then the Tasmanian taxon may be given the name *bowlingi*. If the Tasmanian and King Island forms prove to be distinct from each other than the Tasmanian form will need to be given its own name.

[Mustela] Novae Hollandiae F. Meyer, 1793: 27.

TYPE LOCALITY: *Nomen novum* for *Viverra maculata* Kerr, 1792.

COMMENTS: Synonymised within *maculatus* by Iredale and Troughton (1934: 14) and Tate (1947: 145).

Dasyurus macrourus É. Geoffroy, 1803e: 259 as 159.

TYPE LOCALITY: *Nomen novum* for *Viverra maculata* Kerr, 1792.

COMMENTS: Recognised by Temminck (1824: 69) and Waterhouse (1838a: 65; 1841a: 130). Synonymised within *maculatus* by Waterhouse (1846: 439), Thomas (1888a: 263), Iredale and Troughton (1934: 14), Tate (1947: 145) and Mahoney and Ride (1988a: 21).

Dasyurus affinis McCoy, 1865: Note. Quarter Sheet 7.

TYPE LOCALITY: Bone Cave, Gisborne, central Victoria, Australia.

COMMENTS: Synonymised within *maculatus* by Mahoney (1964: 525). It is possible, given the findings of Firestone *et al.* (1999), that this could represent a distinct species; but larger samples need to be studied and more research is needed.

Dasyurus ursinus Giebel, 1874: Key to Plate 18, Fig. 4.

TYPE LOCALITY: Unknown.

COMMENTS: This name is not invalidated by *Didelphis ursinus* Harris, 1808. Synonymised within *maculatus* by Thomas (1888a: 264), Groves (1993b: 32; 2005d: 25) and subsequent authors.

† *Dasyurus bowlingi* Spencer & Kershaw, 1910a: 29, 33; Plate 8, Figs. 1–2, 4–5.

TYPE LOCALITY: King Island, Tasmania, Australia.

COMMENTS: Synonymised within *maculatus* by Marshall and Hope (1973: 225) and Hope (1972: 178). Firestone *et al.* (1999) sequenced the mitochondrial control region, and also some microsatellites, from 91 Spotted-tailed Quolls. They found reciprocal monophyly between the control region haplotypes of mainland and Tasmanian samples, but did not propose any taxonomic changes; there are 13 base-pair substitutions between Tasmanian Spotted-tailed Quolls and the nearest mainland populations.

Dasyurus maculatus gracilis Ramsay, 1888

Dasyurus gracilis Ramsay, 1888: 1296.

TYPE LOCALITY: Bellenden-Kerr Range, Queensland, Australia.

COMMENTS: Elevated to species rank within *Dasyurops* by Iredale and Troughton (1934: vii, 14) and Tate (1947: 145). Synonymised within *maculatus* by Ride (1970: 242). More recent genetic analysis has proposed that the designation of this subspecies does not reflect actual genetic subdivisions present within this species; nonetheless it was proposed that this form be treated as a separate management unit (Firestone *et al.*, 1999: 1621). The smaller size was suggested by Firestone *et al.* (1999: 1621) to reflect clinal adaptations to climatic differences (Bergmann's Rule). Recognised as a subspecies of *maculatus* by Tate (1952a: 580), which was accepted by subsequent authors including Strahan (1983: 18; 1995: 68), Clayton *et al.* (2006: 100), and Van Dyck and Strahan (2008: 61).

FUTURE TAXONOMIC RESEARCH: Given the uniqueness of the mitochondrial control region haplotypes as sequenced by Firestone *et al.* (1999: 1613) further research needs to be undertaken to determine the status of this taxon and whether it should be either reduced to a synonym of *maculatus* or else split up into distinct taxa.

Dasyurus viverrinus (Shaw, 1800)

Eastern Quoll

Didelphis Viverrina Shaw, 1800: 491.

TYPE LOCALITY: Sydney region, New South Wales, Australia.

COMMENTS: The holotype is described and figured under the name 'Spotted Opossum' of Phillip (1789: 147) and Tapoa Tafa by Anon (in White, 1790: 285). Note the Tapoa Tafa or Tapha referred to on page 281 is *Phascogale tapoatafa*. Placed in *Dasyurus* by É. Geoffroy (1804b: 360), Temminck (1824: 72), Waterhouse (1838a: 65; 1846: 442), Gould (1851 [1845–1863]: Text to Plate 50), Krefft (1868a: 94) and followed by subsequent authors including Thomas (1888a: xii, 265), who reviewed its historic taxonomy. Synonymised within *Dasyurus quoll* by Iredale and Troughton (1934: 12) and Tate (1947: 145) but not subsequent authors.

Mustela quoll Zimmermann, 1777: 489.

TYPE LOCALITY: Bare Island, Botany Bay, Sydney, New South Wales, Australia.

COMMENTS: *Dasyurus quoll* Zimmermann, 1777 (not *Mustella quoll* Zimmermann, 1783). Recognised at the species rank within *Dasyurus* by Iredale and Troughton (1934: vii, 12), and Tate (1947: 144). This work was rejected by Opinion 257 of the ICZN (1954b: 231) so the name *Dasyurus quoll* Zimmermann, 1777 is therefore invalid.

HOMONYMS:

Mustella quoll Zimmermann, 1783, Northern Quolls of the Class Mammalia (Order Dasyuromorphia: Family Dasyuridae). Name is a synonym of *Dasyurus hallucatus* Gould, 1842). See individual entry.

Didelphis maculata Anon, 1791: 186.

TYPE LOCALITY: Port Jackson, Sydney, New South Wales, Australia.

COMMENTS: Referred to as the 'Spotted Martin' by Phillip (1789: 276). Suppressed under Article 80 of the Code

(ICZN, 1985a: 175) in favour of *Didelphis viverrina* Shaw, 1800 (see Groves, 1993b: 32; 2005d: 25).

Dasyurus Maugei É. Geoffroy, 1803e: 259 as 159.

TYPE LOCALITY: Tasmania, Australia.

COMMENTS: Name further discussed by É. Geoffroy (1804b: 359). Recognised by Temminck (1824: 71), Waterhouse (1838a: 65; 1841a: 133). Synonymised within *Dasyurus quoll* by Iredale and Troughton (1934: 12), and within *viverrinus* by Waterhouse (1846: 444), Thomas (1888a: 266), and Mahoney and Ride (1988a: 21).

Dasyurus guttatus Desmarest, 1804b: 10.

TYPE LOCALITY: Australia.

COMMENTS: Synonymised within *Dasyurus quoll* by Iredale and Troughton (1934: 12), and within *viverrinus* by Thomas (1888a: 266) and Mahoney and Ride (1988a: 22).

Did. [elphys] [sic] alboguttata Burmeister, 1854: 340.

TYPE LOCALITY: Brazil [error=Sydney, Australia]

COMMENTS: Name recognised as described by Thomas (1888: 366), who noted that he has never seen a specimen. Synonymised within *Dasyurus quoll* by Iredale and Troughton (1934: 13) and within *viverrinus* by Mahoney and Ride (1988a: 22). Information about the type locality can be found in Goeldi (1894: 466) and also discussed by Pine (1980: 396–397) and Gardner (2007: 83), who questioned whether this was a *Dasyurus*.

Parantechinus Tate, 1947

Parantechinus Tate, 1947: 137.

TYPE SPECIES: *Phascogale apicalis* J. Gray, 1842c [=*Parantechinus apicalis* (J. Gray, 1842c)] by original designation.

COMMENTS: Synonymised within *Antechinus* by Ride (1964b: 62), which was followed by M. Archer (1976a: 250), Honacki *et al.* (1982: 27), Marshall (1981: 26) and McKenna and Bell (1997: 55). After the publication of new information on genetics by Kirsch (1977a: 54), Kirsch and Archer (1982: 618), and Baverstock *et al.* (1982: 648), and penile anatomy by Woolley and Webb (1977: 316) and Woolley (1982: 780), M. Archer (1982a: 434, 439) reinstated the genus *Parantechinus*. Subsequently recognised as a valid genus by Strahan (1983: 29), Mahoney and Ride (1988a: 23), Marshall *et al.* (1990: 488), Groves (1993b: 33; 2005d: 26) and Strahan (1995: 72).

Parantechinus apicalis (J. Gray, 1842)

Dibbler

Phascogale apicalis J. Gray, 1842c: 518.

TYPE LOCALITY: Australia.

COMMENTS: Recognised within *Phascogale* by Waterhouse (1846: 413) and *Phascologale* by Thomas (1888a: 274, 277), who reviewed its early taxonomic history. Included in the subgenus *Antechinus* by Waterhouse (1846: 413) and genus *Antechinus* by J. Gray (1843a: 99), Gould (1845 [1845–1863]: Text to Plate 39), Krefft (1867a: 432), Iredale and Troughton (1934: vii, 6) and Troughton (1967: 23). Placed in *Parantechinus* by Tate (1947: 137). It was however subsequently often included in the genus *Antechinus* by authors including Ride (1970: 118), Woolley (1971: 99) and Honacki *et al.* (1982: 27). Placed in the genus *Parantechinus* by M. Archer (1982a: 434) and Strahan (1995: 72), where its position was confirmed by N. Cooper *et al.* (2000: 135) and subsequently followed by Clayton *et al.* (2006: 100), Van Dyck and Strahan (2008: 65), and Menkhorst and Knight (2011: 15).

Pseudantechinus Tate, 1947

Pseudantechinus Tate, 1947: 139.

TYPE SPECIES: *Phascogale macdonnellensis* Spencer, 1895 [=*Pseudantechinus macdonnellensis* (Spencer, 1895)] by original designation.

COMMENTS: Synonymised within *Antechinus* by Ride (1964b: 62), which was followed by M. Archer (1976a: 250), Honacki *et al.* (1982: 27), Marshall (1981: 26) and McKenna and Bell (1997: 55). After the publication of new information on genetics by Kirsch (1977a: 54) and Baverstock *et al.* (1982: 648), and penile anatomy by Woolley and Webb (1977: 320) and Woolley (1982: 767), the genus was reinstated by M. Archer (1982a: 434, 439) and followed by Strahan (1983: 33). Despite this it was again synonymised within *Parantechinus* by Mahoney and Ride (1988a: 23), but then recognised as a valid genus by Marshall *et al.* (1990: 488), Groves (1993b: 34), Strahan (1995: 76) and subsequent authors.

Pseudantechinus bilarni (Johnson, 1954)

Sandstone Pseudantechinus

Antechinus bilarni D. Johnson, 1954: 77.

TYPE LOCALITY: Oenpelli, East Alligator River, Northern Territory, Australia.

COMMENTS: Taxonomic decision of Kirsch and Calaby (1977: 15) to separate this species from *A. macdonnellensis* (within *Antechinus*), where it was previously placed by Ride (1970: 241). Recognised within *Antechinus* by Troughton (1967: 22), Calaby and Taylor (1981: 329), and Honacki *et al.* (1982: 27). Species placed in *Parantechinus* by M. Archer (1982a: 434), Mahoney and Ride (1988a: 23), Woolley (1995: 83) and Strahan (1995: 74) and Krajewski *et al.* (1997a: 226). Tentatively placed in the genus *Pseudantechinus* by Kitchener and Caputi (1988: 35, 56), which was followed by Maxwell *et al.* (1996: 2), N. Cooper *et al.* (2000: 115), Groves (2005d: 27), Clayton *et al.* (2006: 100), Van Dyck and Strahan (2008: 67), and Menkhorst and Knight (2011: 15).

Pseudantechinus macdonnellensis (Spencer, 1895)

Fat-tailed Pseudantechinus

Phascogale macdonnellensis Spencer, 1895: 222.

TYPE LOCALITY: Alice Springs, Northern Territory, Australia.

COMMENTS: Discussed further by Spencer (1896c: 84, 85). Included in the genus *Antechinus* as a species by Iredale and Troughton (1934: vii, 6), Troughton (1967: 22) and Honacki *et al.* (1982: 28). Placed in the genus *Pseudantechinus* by Tate (1947: 139). Placed in the genus *Parantechinus* by Mahoney and Ride (1988a: 24) then in the genus *Pseudantechinus* by Kitchener and Caputi (1988: 35), N. Cooper *et al.* (2000: 115), Groves (2005d: 27), Van Dyck and Strahan (2008: 69), and Menkhorst and Knight (2011: 15).

Pseudantechinus mimulus (Thomas, 1906)

Carpentarian Pseudantechinus

Phascogale mimulus Thomas, 1906d: 540.

TYPE LOCALITY: Alexandria Station, Northern Territory, Australia.

COMMENTS: Included in the genus *Antechinus* as a species by Iredale and Troughton (1934: vii, 6) and Troughton (1967: 23). Placed in *Pseudantechinus* by Tate (1947: 139), while subsequent authors including Ride (1964b: 62; 1970: 241), Strahan (1983: 33), Mahoney and Ride (1988a: 24), and Groves (1993b: 35) synonymised it with *macdonnellensis*. Recognised as a distinct species within *Pseudantechinus* by Tate (1947: 139) and within *Antechinus* by M. Archer (1982a: 431). Doubt was placed on that taxonomic placement of *mimulus* within *macdonnellensis* by Kitchener and Caputi (1988: 46) and was subsequently separated by Kitchener (1991: 192) who placed it in the genus *Pseudantechinus*, which was followed by N. Cooper *et al.* (2000: 115), Groves (2005d: 27), Van Dyck and Strahan (2008: 71), and Menkhorst and Knight (2011: 15).

Pseudantechinus ningbing Kitchener, 1988

Ningbing Pseudantechinus

Pseudantechinus ningbing Kitchener, 1988: 61, 62.

TYPE LOCALITY: Mitchell Plateau, Kimberley region, Western Australia, Australia. 220 m elevation. (14°53′40″S, 125°45′20″E)

COMMENTS: The uniqueness of the 'ningbing' antechinus was first identified by M. Archer (1979a: 37) and recognised as an undescribed species by Baverstock *et al.* (1982: 648) who included it in the false antechinus group. It was confirmed as a new species by D. Cooper and Woolley (1983: 743); although the specific name was not formalised until Kitchener (1988: 61, 62). It was subsequently included in the genus *Pseudantechinus* by Kitchener and Caputi (1988: 35), N. Cooper *et al.* (2000: 135) and later authors.

Pseudantechinus roryi N. Cooper *et al.*, 2000

Tan Pseudantechinus

Pseudantechinus roryi N. Cooper *et al.*, 2000: 117, 125; Figs. 6–11.

TYPE LOCALITY: Woodstock Station, 500 m north of homestead, Pilbara Region, Western Australia, Australia. (21°36′42″S, 118°57′20″E)

COMMENTS: Studies by Westerman *et al.* (2008: 201) failed to demonstrate reciprocal monophyly between *P. macdonnellensis* and *P. roryi* haplotypes and they suggested that recognition of the later taxon may be premature. As a result of this research Burbidge *et al.* (2014: 17) synonymised *roryi* within *macdonnellensis*. Nonetheless, the morphological distinctions of this species as described by N. Cooper *et al.* (2000: 25) seem cogent, and may still warrant species recognition.

FUTURE TAXONOMIC RESEARCH: Further research is required to confirm whether this taxon should be recognised as a distinct or synonymised within *Pseudantechinus macdonnellensis*.

Pseudantechinus woolleyae Kitchener & Caputi, 1988

Woolley's Pseudantechinus

Pseudantechinus woolleyae Kitchener & Caputi, 1988: 35, 39.

TYPE LOCALITY: Near Newlingunn bore, 10 km 117° from Errabiddy Homestead, Western Australia, Australia. (25°33′00″S, 117°08′00″E)

COMMENTS: Subsequently placed in *Pseudantechinus* by N. Cooper *et al.* (2000: 115) who admitted that the inclusion of this species (and of *P. bilarni*) in *Pseudantechinus* might make the genus paraphyletic.

Sarcophilus F. Cuvier, 1837

Sarcophilus F. Cuvier, 1837a: Text to Plate 70.

TYPE SPECIES: *Didelphis ursina* Harris, 1808 (as *Sarcophilus ursinus*) [=*Sarcophilus harrisii* (Boitard, 1841)] by monotypy.

COMMENTS: Not recognised by Waterhouse (1841a: 128), but placed as a subgenus of *Dasyurus* by Waterhouse (1846: 448). Description attributed to É. Geoffroy and Cuvier by Iredale and Troughton (1934: 14). Genus recognised by Gould (1855: [1845–1863]: Text to Plate 40), Krefft (1868a: 94), Thomas (1888a: xii, 258), Iredale and Troughton (1934: vii, 14), Simpson (1945: 44) and all subsequent authors. Taxonomic decision of Thomas (1912a: 116) for the specific name of the Tasmanian devil to be *S. harrisii* instead of *satanicus*. This genus includes the fossil taxon † *Sarcophilus laniarius* (Owen, 1838b: 369) from Wellington Caves in New South Wales. The species *laniarius* was placed in *Sarcophilus* by Lydekker (1887: 265; 1894a: 267) and recognised as a separate larger species than modern *harrisii* by authors including Bartholomai (1977: 44), M. Archer (1978: 75), Dawson (1982a: 517, 523) and P. Murray (1991: 1090). It was proposed to be conspecific with *harrisii*, with the specific name *laniarius* proposed ahead of *harrisii*, by Stephenson (1963: 618) and Werdelin (1987: 9). This proposal has been followed by various authors, especially those directly associated with research on the species, such as Groves (1993b: 35), Gerdtz and Archbold (2003: 45), Jones *et al.* (2003: 277), McGlashan *et al.* (2006: 95), Piper (2007: 494) and Cramb *et al.* (2009: 665). The taxon *laniarius* was placed as a subspecies of *harrisii* by Roberts *et al.* (2001: 1889) and Blumstein and Daniel (2003: 591). Groves (2005d: 28), however, separated *laniarius* and *harrisii* as distinct species, which has been followed here.

Diabolus J. Gray, 1841: 400.

TYPE SPECIES: *Nomen novum* for *Sarcophilus* F. Cuvier, 1837a.

COMMENTS: Recognised by J. Gray (1843a: xxii, 97). Synonymised within *Sarcophilus* by Thomas (1888a: 258), Iredale and Troughton (1934: 14) and subsequent authors.

HOMONYMS:

Diabolus Karawajew, 1926: 424, ants of the Class Insecta (Order Hymenoptera, Family Formicidae). Genus is a synonym of *Dolichoderus* Lund, 1831: 130. See Brown (1973: 180).

Ursinus Boitard, 1841: 290.

TYPE SPECIES: *Ursinus harrisii* Boitard, 1841 [= *Sarcophilus harrisii* (Boitard, 1841)] by monotypy.

COMMENTS: Synonymised within *Sarcophilus* by Iredale and Troughton (1934: 14) and subsequent authors.

Sacrophilus Boitard, 1842: 204.

TYPE SPECIES: Incorrect subsequent spelling of *Sarcophilus* F. Cuvier, 1837a.

COMMENTS: He also used the spelling *Sacrophius* in the paragraph above. Synonymised within *Sarcophilus* by Palmer (1904a: 615, 618).

Sarcophilus harrisii (Boitard, 1841)

Tasmanian Devil

Sarcophilus harrisii harrisii (Boitard, 1841)

Ursinus Harrisii Boitard, 1841: 290.

TYPE LOCALITY: Tasmania, Australia.

COMMENTS: A replacement name for *Didelphis ursina* Harris, 1808, which is preoccupied by the common wombat that was originally described as *Didelphis ursina* Shaw, 1800 (e.g. see Thomas, 1912a: 116). Taxonomic decision of Thomas (1912a: 116) for the specific name of the Tasmanian devil to be *S. harrisii* instead of *satanicus*. Werdelin (1987: 9) argued that the Pleistocene taxon *Sarcophilus laniarius* (Owen, 1838b: 369) and the recent *Sarcophilus harrisii* (Boitard, 1841) were only subspecifically distinct, and as Owen's name antedates Boitard's by three years, *laniarius* would take precedence. The specific name *laniarius* was subsequently used by Groves (1993b: 35) for the living form, but Groves (2005d: 28) argued that the information used by Werdelin (1987: Tables 1 and 2) shows that the recent and fossil ranges do not overlap in many variables, so they may be retained as different species. Groves wrote that the Victorian subfossil *dixonae* remains as a subfossil of *S. harrisii*, noting that though it is distinctive its measurements overlap with those of the living form. Recognised as a subspecies of *laniarius* by Werdelin (1987: 1, 10), Gerdtz and Archbold (2003: 45) and Piper (2007: 494), but typically recognised as a distinct taxon.

Didelphis ursina Harris, 1808: 176; Plate 19, Fig. 2.

TYPE LOCALITY: Tasmania, Australia.

COMMENTS: Recognised within *Dasyurus* by Waterhouse (1841a: 128; 1846: 448), within *Diabolus* by J. Gray (1841: 400; 1843a: xxii, 97) and within *Sarcophilus* by F. Cuvier (1837a: Text to Plate 70), Gould (1851 [1845–1863]: Text to Plate 48), Krefft (1868a: 94) and Thomas (1888a: xii, 259). Synonymised within *harrisii* by Iredale and Troughton (1934: 15).

HOMONYMS:

Didelphis ursina Shaw, 1800, the Common Wombat of the Class Mammalia (Order Diprotodontia, Family Vombatidae). Taxon now recognised as *Vombatus ursinus* (Shaw, 1800). See individual entry.

Sarcophilus satanicus Thomas, 1903a: 289.

COMMENTS: *Nomen novum* for *Didelphis ursina* Harris, 1808. Synonymised within *harrisii* by Iredale and Troughton (1934: 15). Taxonomic decision of Thomas (1912a: 116) for the specific name of the Tasmanian devil to be *S. harrisii* instead of *satanicus*.

† ***Sarcophilus harrisii dixonae*** Werdelin, 1987

† *Sarcophilus harrisii dixonae* Werdelin, 1987: 1, 10.

TYPE LOCALITY: Mount Hamilton, Victoria, Australia.

COMMENTS: Described from fossil deposits. Regarded as a synonym of *harrisii* by Groves (1993b: 35) but recognised as a subspecies of *harrisii* by Groves (2005d: 28), who suggested that though it is distinctive its measurements overlap with those of the living Tasmanian form. Taxon recognised as a subspecies of *laniarius* by Piper (2007: 494).

Subfamily Phascogalinae Gill, 1872 *sensu* Marshall, 1990

Subfamily Phascogalinae Gill, 1872: 26.

TYPE GENUS: *Phascogale* Temminck, 1824.

COMMENTS: When originally proposed, this rank was placed in the Family Dasyuridae (Goldfuss, 1820a). Synonymised within the Subfamily Dasyurinae (Goldfuss, 1820a) by Marshall (1981: 26), Marshall *et al.* (1990: 488) and McKenna and Bell (1997: 55). Treated as a distinct subfamily by Osborn (1910: 516), Simpson (1945: 43), Troughton (1967: 17), M. Archer (1982a: 439; 1984: 635), Marshall *et al.* (1990: 459, 488), Krajewski *et al.* (1996: 81), Strahan (1995: 6, 85), Wroe (1997: 23; 1999: 512), and Van Dyck and Strahan (2008: 9, 81).

Phascogalina Bonaparte, 1850a: Unpaginated table.

TYPE GENUS: *Phascogale* Temminck, 1824.

COMMENTS: Rank unknown. Synonymised within the Subfamily Phascogalinae by Simpson (1945: 43), and within the Subfamily Dasyurinae by Marshall (1981: 26) and McKenna and Bell (1997: 55).

Tribe Phascogalini Kirsch *et al.* 1997: 245.

TYPE GENUS: *Phascogale* Temminck, 1824.

COMMENTS: When originally proposed this rank was attributed to Gill (1872: 26) and placed in the Subfamily Dasyurinae (Goldfuss, 1820). Rank recognised by Krajewski *et al.* (Krajewski *et al.*, 2000a: 98; 2000b: 375; 2000c: 423), Krajewski and Westerman (2003: 14) and Groves (2005d: 28).

Antechinus Macleay, 1841

Antechinus Macleay, 1841: 242.

TYPE SPECIES: *Antechinus stuartii* Macleay, 1841 by monotypy.

COMMENTS: Recognised as a subgenus of *Phascogale* by Waterhouse (1846: 411) and synonym of *Phascologale* by Thomas (1888a: 273) and within *Phascogale* by Simpson (1945: 43). Genus recognised by Iredale and Troughton (1934: vii, 4) and most subsequent authors. The genera *Parantechinus* and *Pseudantechinus* were separated by Haltenorth (1958: 18), and despite Ride's (1964b: 62) misgivings of these genera they have become subsequently recognised. The genus *Dasykaluta* was removed from *Antechinus* by M. Archer (1982a: 428).

Antechinus adustus (Thomas, 1923)

Rusty Antechinus

Phascogale flavipes adusta Thomas, 1923a: 175.

TYPE LOCALITY: Dinner Creek (now Charmillan Creek), near Ravenshoe, Queensland, Australia. 885 m elevation. (17°42′S, 145°31′E)

COMMENTS: The year after its description Thomas (1924a: 528) assigned *adusta* to a subspecies of *Phascogale unicolor* [=*A. stuartii*]. Despite this it was subsequently recognised as a subspecies of *flavipes* by Iredale and Troughton (1934: 4), Troughton (1941: 24), Tate (1947: 127; 1952a: 572, 579), Brass (1953: 199), Horner and Taylor (1959: 3), Marlow (1965: 19) and Troughton (1967: 18). Wakefield and Warneke (1967: 69) however could not resolve its status because of a lack of material, but demonstrated that it belonged with *A. stuartii*, not *A. flavipes*, and established the name *A. stuartii adustus*. In his detailed taxonomic review Van Dyck (1982: 730) regarded *adustus* as distinctive, but did not formally elevate it to species status. Strahan (1983: 40) recognised it as a subspecies of *stuartii* while Mahoney and Ride (1988a: 17) and Strahan (1995: 96) considered it a synonym of *stuartii*, which was followed by Groves (1993b: 30). Elevated to full species rank by Van Dyck and Crowther (2000: 611, 620) and followed by Groves (2005d: 28), Van Dyck and Strahan (2008: 81), and Menkhorst and Knight (2011: 15).

Antechinus agilis Dickman *et al.*, 1998

Agile Antechinus

Antechinus agilis Dickman *et al.*, 1998: 1, 5.

TYPE LOCALITY: Corner of Warkes Road and Blundells Creek Road, Near Lees Creek, Brindabella Range, Australian Capital Territory, Australia. 740 m elevation. (35°21′45″S, 148°50; 17″E)

COMMENTS: After it was shown by Dickman *et al.* (1988: 455) that *A. stuartii* in eastern New South Wales is composed of two species, this species was formally described and separated from the *Antechinus stuartii* complex by Dickman *et al.* (1998: 5) and followed by subsequent authors.

Antechinus argentus Baker *et al.*, 2013

Silver-headed Antechinus

Antechinus argentus Baker *et al.*, 2013: 201, 208.

TYPE LOCALITY: The plateau, west and north-west of The Lookout on the eastern escarpment of Kroombit Tops NP, 400 km NNW of Brisbane, 60 km SSW of Gladstone, south-east Queensland, Australia. (24°23′36″S, 151°02′34″E)

COMMENTS: Species sympatric with *Antechinus flavipes*.

Antechinus arktos Baker *et al.*, 2014

Black-tailed Antechinus

Antechinus arktos Baker *et al.*, 2014: 101, 106.

TYPE LOCALITY: A small gully near Best of All Lookout, Mount Mumdjin, Springbrook National Park, south-east Queensland, Australia. 950 m elevation. (28°14′29.6″S, 153°15′50.6″E)

COMMENTS: Taxon previously known as a northern outlier of the Dusky Antechinus.

Antechinus bellus (Thomas, 1904)

Fawn Antechinus

Phascogale bella Thomas, 1904a: 229.

TYPE LOCALITY: South Alligator River, Northern Territory, Australia.

COMMENTS: Transferred to *Antechinus* by Iredale and Troughton (1934: vii, 5) and followed by Tate (1947: 133), Troughton (1967: 19) Baverstock *et al.* (1982: 648), Mahoney and Ride (1988a: 15) and subsequent authors. Species reviewed and redescribed by Baker and Van Dyck (2013a: 201).

Antechinus flavipes (Waterhouse, 1838)

Yellow-footed Antechinus

Antechinus flavipes flavipes (Waterhouse, 1838)

Phascogale flavipes Waterhouse, 1838b: 75.

TYPE LOCALITY: North of Hunters River, New South Wales, Australia.

COMMENTS: Recognised within *Phascogale* by Waterhouse (1838a: 65; 1841a: 138) and within *Phascologale* by Thomas (1888a: xiii, 289) and subgenus *Antechinus* by Waterhouse (1846: 415). Placed in *Antechinus* by J. Gray (1843a: 99), Gould (1854 [1845–1863]: Text to Plate 40), Krefft (1866a: 9; 1967a: 432), and Iredale and Troughton (1934: vii, 4). Taxonomic decision of Wakefield and Warneke (1967: 69) and Ride in Mahoney and Ride (1988a: 15). Species and subspecies reviewed and redescribed by Baker and Van Dyck (2013b: 1).

Phascogale rufogaster J. Gray, 1841: 407.

TYPE LOCALITY: Vicinity of Adelaide, South Australia, Australia.

COMMENTS: Reduced to a subspecies of *flavipes* by Iredale and Troughton (1934: 4), Tate (1947: 128) and Marlow (1965: 19). Synonymised within *flavipes* by Mahoney and Ride (1988a: 15) and subsequent authors.

Antechinus flavipes leucogaster (J. Gray, 1841)

Phascogale leucogaster J. Gray, 1841: 407.

TYPE LOCALITY: Banks of the Canning River, Western Australia, Australia.

COMMENTS: Recognised at species rank within *Antechinus* by J. Gray (1843a: 99), Gould (1854 [1845–1863]: Text to Plate 38) and Krefft (1867a: 432), and within *Phascogale* (*Antechinus*) by Waterhouse (1846: 417). Synonymised within *flavipes* by Mahoney and Ride (1988a: 15) and Groves (2005d: 29). Recognised as a subspecies of *flavipes* by Thomas (1888a: 291), Iredale and Troughton (1934: 4), Tate (1947: 128), Marlow (1965: 19), Wakefield and Warneke (1967: 69), Van Dyck (1982: 730) and Strahan (1983: 38; 1995: 87), Maxwell *et al.* (1996: 2), Crowther *et al.* (2002: 627), Clayton *et al.* (2006: 101), and Van Dyck and Strahan (2008: 88). Subspecies reviewed and redescribed by Baker and Van Dyck (2013b: 1).

Antechinus flavipes rubeculus Van Dyck, 1982

Antechinus flavipes rubeculus Van Dyck, 1982: 727.

TYPE LOCALITY: Flaggy Creek Forestry barracks, Black Mountain Road, Kuranda, north Queensland, Australia.

COMMENTS: The available measurements overlap widely with specimens from New South Wales and Victoria; and there are no electrophoretic differences between it and southern populations (Baverstock *et al.*, 1982). Recognised as a subspecies of *flavipes* by Strahan (1983: 38), Maxwell *et al.* (1996: 2), Groves (2005d: 29), Clayton *et al.* (2006: 101), and Van Dyck and Strahan (2008: 88). Subspecies reviewed and redescribed by Baker and Van Dyck (2013b: 1).

Antechinus godmani (Thomas, 1923)

Atherton Antechinus

Phascogale godmani Thomas, 1923a: 174.

TYPE LOCALITY: Dinner Creek, Ravenshoe, Queensland, Australia. 2900 ft (884m) elevation. (17°40′S, 145°30′E)

COMMENTS: Transferred to *Antechinus* by Iredale and Troughton (1934: vii, 5) and followed by subsequent authors including Tate (1947: 128), although Tate (1952a: 572, 579) subsequently reduced *godmani* to a subspecies of *A. flavipes*, which was followed by Horner and Taylor (1959: 3). Also included within *flavipes* by Haltenorth (1958: 18). Separated from *flavipes* by Wakefield and Warneke (1967: 73, 95) as a distinct species, which was followed by Troughton (1967: 19), Kirsch and Calaby (1977: 15), Honacki *et al.* (1982: 27), Van Dyck (1982: 726) and subsequent authors. Taxon redescribed by Baker and Van Dyck (2013c: 401).

Antechinus leo Van Dyck, 1980

Cinnamon Antechinus

Antechinus leo Van Dyck, 1980: 5.

TYPE LOCALITY: Nesbit River, Buthen Buthen, Cape York Peninsula, north Queensland, Australia. (13°21′S, 143°28′E)

COMMENTS: Species has been recognised since its description.

Antechinus minimus (É. Geoffroy, 1803)

Swamp Antechinus

Antechinus minimus minimus (É. Geoffroy, 1803)

Dasyurus minimus É. Geoffroy, 1803e: 259 as 159.

TYPE LOCALITY: Tasmania, Australia. Probably Waterhouse Island, Bass Strait. See Wakefield and Warneke (1963: 209–210).

COMMENTS: Description extended by É. Geoffroy (1804b: 362). Included within *Phascogale* by Waterhouse (1841a: 140; 1846: 419), within *Phascologale* by Thomas (1888a: xiii, 287) and within *Antechinus* by J. Gray (1843a: 99), Iredale and Troughton (1934: vii, 5) and subsequent authors.

Phascogale affinis J. Gray, 1841: 406.

TYPE LOCALITY: Tasmanian Peninsula, Tasmania, Australia.

COMMENTS: Species recognised within *Antechinus* by Krefft (1867a: 432). Synonymised within *minimus* by Thomas (1888a: 287), Iredale and Troughton (1934: 5), Wakefield and Warneke (1963: 210) and subsequent authors.

Antechinus Rolandensis Higgins & Petterd, 1883: 171.

TYPE LOCALITY: Near Mt Roland, Tasmania, Australia.

COMMENTS: Synonymised within *minimus* by Thomas (1888a: 287), Iredale and Troughton (1934: 5), Mahoney and Ride (1988a: 16) and subsequent authors.

Antechinus Concinnus Higgins & Petterd, 1884a: 184.

TYPE LOCALITY: Tasmania, Australia. 'Exact locality unknown'.

COMMENTS: Synonymised within *minimus* by Thomas (1888a: 287), Iredale and Troughton (1934: 5), Mahoney and Ride (1988a: 16) and subsequent authors.

Antechinus minimus maritimus (Finlayson, 1958)

Phascogale (Antechinus) swainsoni maritima Finlayson, 1958a: 148; Plates 1–2.

TYPE LOCALITY: Port MacDonnell, South Australia, Australia. See Aitken (1976: 197) for details about the holotype.

COMMENTS: Recognised as a subspecies of *minimus* by Wakefield and Warneke (1963: 194), A. Smith (1983: 753) and Strahan (1983: 48). Synonym of *minimus* by Mahoney and Ride (1988a: 16) and Groves (2005d: 30). Recognised as a subspecies by Strahan (1983: 48; 1995: 93), Clayton *et al.* (2006: 101), and Van Dyck and Strahan (2008: 94).

Antechinus mysticus Baker *et al.*, 2012

Buff-footed Antechinus

Antechinus mysticus Baker *et al.*, 2012: 1, 8.

TYPE LOCALITY: Collected from the kitchen cupboard of Steve Van Dyck, Cedar Creek Road, Samford, south-east Queensland, Australia. (27°19′56″S, 152°48′26″E)

COMMENTS: Taxon found in sympatry with *Antechinus flavipes* and *A. subtropicus*.

Antechinus stuartii Macleay, 1841

Brown Antechinus

Antechinus Stuartii Macleay, 1841: 242.

TYPE LOCALITY: Sydney. Neotype from Waterfall, Royal National Park, New South Wales, Australia. Neotype selected by Wakefield and Warneke (1967: 69).

COMMENTS: This species was initially synonymised with *Antechinus flavipes* by most authors including Waterhouse (1846: 415), Thomas (1888a: 289), and Iredale and Troughton (1934: 4). Wakefield and Warneke (1967: 69) elevated this taxon to species rank from *Antechinus flavipes*. What was thought to be one species was shown to consist of a northern species (*A. stuartii*) and southern species now known as *A. agilis* by Dickman *et al.* (1988) and Dickman *et al.* (1998: 5). The two species have been found together at Kiola in southern New South Wales.

Antechinus unicolor Gould, 1854 [1845–1863]: Text to Plate 37.

TYPE LOCALITY: Coast of district north of Sydney, New South Wales, Australia. Type designated by Thomas (1922a: 128).

COMMENTS: Species recognised within *Antechinus* by Krefft (1867a: 432). Synonymised within *flavipes* by Iredale and Troughton (1934: 4). Synonymised within *stuartii* by Mahoney and Ride (1988a: 17) and Groves (1993b: 30; 2005d: 30).

Phascogale flavipes burrelli Le Souef & Burrell, 1926: 344.

TYPE LOCALITY: Highlands of northern New South Wales, Australia.

COMMENTS: Iredale and Troughton (1934: 4) recognised this taxon as a subspecies of *flavipes*. Synonymised within *stuartii* by Mahoney and Ride (1988a: 17) and Groves (1993b: 30; 2005d: 30).

Antechinus subtropicus Van Dyck & Crowther, 2000

Subtropical Antechinus

Antechinus subtropicus Van Dyck & Crowther, 2000: 611, 613.

TYPE LOCALITY: Emu Creek, SE Queensland, Australia. 38 km east of Warwick (23°13′03″S, 152°24′43″E). This species was separated from *Antechinus stuartii* complex by Van Dyck and Crowther (2000), with the species status subsequently confirmed by Crowther *et al.* (2003: 443, 455).

COMMENTS: Related to *stuartii*, with which it is sympatric in SE Queensland at Wallangarra (28°55′S, 151°55′E) and Pyramid Creek, Wyberba (28°50S, 151°57S).

Antechinus swainsonii (Waterhouse, 1840)

Dusky Antechinus

Antechinus swainsonii swainsonii (Waterhouse, 1840)

Phascogale Swainsonii Waterhouse, 1840a: 300.

TYPE LOCALITY: Tasmania, Australia.

COMMENTS: Synonymised within *Phascogale minimus* by Waterhouse (1841a: 140) but recognised at species rank within *Phascogale* (*Antechinus*) by Waterhouse (1846: 411) and within *Phascologale* by Thomas (1888a: xiii, 285). Species recognised within *Antechinus* by Gould (1854 [1845–1863]: Text to Plate 34), Krefft (1867a: 432) and Iredale and Troughton (1934: vii, 5), and followed by subsequent authors.

Antechinus niger Higgins & Petterd, 1883: 172.

TYPE LOCALITY: Unknown.

COMMENTS: Synonymised within *swainsonii* by Thomas (1888a: 285), Iredale and Troughton (1934: 5) and Groves (1993b: 31; 2005d: 30).

Antechinus Moorei Higgins & Petterd, 1884b: 182.

TYPE LOCALITY: Long Plains, Tasmania, Australia.

COMMENTS: Synonymised within *swainsonii* by Thomas (1888a: 285), Iredale and Troughton (1934: 5) and Groves (1993b: 31; 2005d: 30).

Antechinus Moorei Var. *Assimilis* Higgins & Petterd, 1884a: 185.

TYPE LOCALITY: Northern Tasmania, Australia.

COMMENTS: Synonymised within *swainsonii* by Iredale and Troughton (1934: 5) and Groves (1993b: 31; 2005d: 30).

Antechinus swainsonii mimetes (Thomas, 1924)

Phascogale swainsonii mimetes Thomas, 1924a: 528.

TYPE LOCALITY: Guy Fawkes district, New South Wales, Australia.

COMMENTS: Synonymised within *swainsonii* by Mahoney and Ride (1988a: 17) and Groves (1993b: 31; 2005d: 30). Considered a subspecies of *swainsonii* by Iredale and Troughton (1934: 5), Tate (1947: 132), Wakefield and Warneke (1963: 194), A. Smith (1983: 753) and Strahan (1983: 46; 1995: 99), Clayton *et al.* (2006: 101), and Van Dyck and Strahan (2008: 100).

Antechinus swainsonii insulanus Davison, 1991

Antechinus swainsonii insulanus Davison, 1991: 103, 104.

TYPE LOCALITY: Silverband Falls, Grampians National Park, western Victoria, Australia.

COMMENTS: Recognised as a subspecies by Clayton *et al.* (2006: 101), and Van Dyck and Strahan (2008: 100).

Phascogale Temminck, 1824

Phascogale Temminck, 1824: 23 footnote, 56.

TYPE SPECIES: *Didelphis penicillata* Shaw, 1800 (as *Dasyurus penicillatus*) [=*Phascogale tapoatafa* (F. Meyer, 1793)] by monotypy.

COMMENTS: Taxonomic decision of Iredale and Troughton (1934: vii, 7), Tate (1947: 126, 135) and Ride in Mahoney and Ride (1988a: 25) to recognise only the species *tapoatafa* and *calura* within *Phascogale*. Synonymised in part within *Antechinus* by Van Dyck (2002: 243).

Phascologale Lenz, 1831: 156.

TYPE SPECIES: Invalid emendation of *Phascogale* Temminck, 1824.

COMMENTS: Not considered by Iredale and Troughton (1934: 7) and synonymised within *Phascogale* by Palmer (1904: 529), Marshall (1981: 26), Mahoney and Ride (1988a: 24), Marshall *et al.* (1990: 489) and McKenna and Bell (1997: 55).

Phascalogale Reichenbach, 1837: xiv.

TYPE SPECIES: Misprint of *Phascogale* Temminck, 1824.

COMMENTS: Synonymised within *Phascogale* by Palmer (1904: 529).

Ascogăle Gloger, 1841: xxx, 83.

TYPE SPECIES: *Nomen novum* for *Phascogale* Temminck, 1824.

COMMENTS: Synonymised within *Phascogale* by Thomas (1888a: 273; 1895a: 190), Iredale and Troughton (1934: 7), Marshall (1981: 26), Mahoney and Ride (1988a: 25), Marshall *et al.* (1990: 489) and McKenna and Bell (1997: 55).

Tapoa Lesson, 1842: 190.

TYPE SPECIES: *Dasyurus tafa* White, 1803 (as *Tapoa tafa*) [=*Phascogale tapoatafa* (F. Meyer, 1793)] by monotypy.

COMMENTS: Synonymised within *Phascogale* by Iredale and Troughton (1934: 7), Marshall (1981: 26), Mahoney and Ride (1988a: 25), Marshall *et al.* (1990: 489) and McKenna and Bell (1997: 55).

HOMONYMS:

Tapoa Owen, 1839a, possums of the Class Mammalia (Order Diprotodontia, Family Phalangeridae). Described as a subgenus of *Phalangista*. Relegated to a *nomen nudum* by Palmer (1904: 663). See individual entry.

Phascogalea S. Müller & Schlegel, 1845a: 149; Plate 25.

TYPE SPECIES: Invalid emendation of *Phascogale* Temminck, 1824.

COMMENTS: Was not considered by Iredale and Troughton (1934: 7) and synonymised within *Phascogale* by Palmer (1904: 529), and McKenna and Bell (1997: 55).

Phascologale Thomas, 1888a: 273.

TYPE SPECIES: *Didelphis penicillata* Shaw, 1800 (as *Phascologale penicillata*) [=*Phascogale tapoatafa* (F. Meyer, 1793)] by original designation.

COMMENTS: Synonymised in part within *Antechinus* by Van Dyck (2002: 243).

Phascoloictis Matschie, 1916: 263.

TYPE SPECIES: *Phascogale calura* Gould, 1844a by original designation.

COMMENTS: Described as a subgenus of *Phascogale* Temminck, 1824. Synonymised within *Phascogale* by Iredale and Troughton (1934: 7), Mahoney and Ride (1988a: 25) and McKenna and Bell (1997: 55).

Phascogale calura Gould, 1844

Red-tailed Phascogale

Phascogale calurus Gould, 1844a: 104.

TYPE LOCALITY: Military Station, Williams River, Western Australia, Australia.

COMMENTS: Described further by Gould (1845 [1845–1863]: Text to Plate 32). Recognised within *Phascologale* by Thomas (1888a: xiii, 296) and within *Phascogale* by Waterhouse (1846: 409), Krefft (1866a: 8; 1867a: 431), Iredale and Troughton (1934: vii, 8), Tate (1947: 136) and subsequent authors.

Phascogale pirata Thomas, 1904

Northern Phascogale

Phascogale penicillata pirata Thomas, 1904a: 228.

TYPE LOCALITY: South Alligator River, Northern Territory, Australia.

COMMENTS: Considered a subspecies of *tapoatafa* by Iredale and Troughton (1934: 8), Finlayson (1934: 236), Tate (1947: 135), Strahan (1983: 34; 1995: 104, 106), Maxwell *et al.* (1996: 2), Clayton *et al.* (2006: 101) and most subsequent authors, though was synonymised within *tapoatafa* by Mahoney and Ride (1988a: 25) and Groves (1993b: 34; 2005d: 32). Elevated to species rank by Van Dyck and Strahan (2008: 103), though not Menkhorst and Knight (2011: 15), and confirmed at species rank by Aplin *et al.* (2015).

Phascogale tapoatafa (Meyer, 1793)

Brush-tailed Phascogale

Phascogale tapoatafa tapoatafa (Meyer, 1793)

[Viverra] Tapoatafa F. Meyer, 1793: 28, 177.

TYPE LOCALITY: Sydney, New South Wales, Australia.

COMMENTS: Based on the 'Tapoa Tafa, or Tapha' of White (1790: 281). Included within the genus *Phascogale* by Iredale and Troughton (1934: vii, 8) and subsequent authors. The phascogales from south-west Western Australia were proposed to be highly distinct by Spencer *et al.* (2001: 374) and as a potential new taxon by Rhind *et al.* (2001: 345, 366), although they did not propose a name. One of us (CPG) has seen the Western Australian specimens concerned, and concurs that a new species is warranted. Another population from the Kimberley region of northern Western Australia was identified as an undescribed subspecies by Van Dyck and Strahan (2008: 105). These two populations are recognised provisionally as distinct subspecies below, with a potential fourth subspecies/species being found on Cape York but it is not sufficiently known for adequate characterisation.

Didelphis Penicillata Shaw, 1800: 502.

TYPE LOCALITY: Australia.

COMMENTS: Recognised within *Phascogale* by Temminck (1824: 58), Waterhouse (1838a: 65; 1841a: 136; 1846: 407), Gould (1845 [1845–1863]: Text to Plate 31), Krefft (1867a: 431; 1868b: 4) and within *Phascologale* by Thomas (1888a: xiii, 294). Synonymised within *tapoatafa* by Iredale and Troughton (1934: 7), Tate (1947: 135), Ride (1970: 246), Mahoney and Ride (1988a: 25) and subsequent authors.

Dasyurus tafa White, 1803: 259 as 159.

TYPE LOCALITY: Sydney region, New South Wales, Australia.

COMMENTS: Synonymised within *tapoatafa* by Iredale and Troughton (1934: 7), Tate (1947, 135), Mahoney and Ride (1988a: 25) and subsequent authors.

Phascogale tapoatafa kimberleyensis Aplin & Rhind, in Aplin *et al.*, 2015

Phascogale tapoatafa kimberleyensis Aplin & Rhind, in Aplin *et al.*, 2015.

TYPE LOCALITY: Pago Mission, Napier Bay, near Broome, Western Australia, Australia. 14°8′S, 126°43′E.

COMMENTS: Taxon represents the geographically isolated north-western Western Australian population.

Phascogale tapoatafa wambenger Rhind & Aplin, in Aplin *et al.*, 2015

Phascogale tapoatafa wambenger Rhind & Aplin, in Aplin *et al.*, 2015

TYPE LOCALITY: Quindalup, Western Australia, Australia. 33°40S′, 115°00′E.

COMMENTS: Taxon represents the geographically isolated south-western Western Australian population.

Subfamily Planigalinae M. Archer, 1982 *sensu* Marshall *et al.*, 1990

Subfamily Planigalinae M. Archer, 1982a: 439.

TYPE GENUS: *Planigale* Troughton, 1928.

COMMENTS: When originally proposed, this rank was placed in the Family Dasyuridae (Goldfuss, 1820a) and included the genus *Planigale* Troughton, 1928. Revised by M. Archer (1976b: 341) on the basis of morphology and by Painter *et al.* (1995: 406) on the basis of mitochondrial Cytochrome *b* sequencing. Synonymised within Dasyuridae by McKenna and Bell (1997: 55). Subfamily rank not recognised by Strahan (1983: xxi), but recognised by Marshall *et al.* (1990: 459, 489), Strahan (1995: 5, 107), and Van Dyck and Strahan (2008: 9, 107). This subfamily included the genus *Ningaui* by Strahan (1995: 5), and Van Dyck and Strahan (2008: 9, 107), though it has been placed within the Subfamily Sminthopsinae by Marshall *et al.* (1990: 489) and Krajewski *et al.* (1997b: 246; 2000b: 375).

Tribe Planigalini M. Archer, 1984: 635.

TYPE GENUS: *Planigale* Troughton, 1928.

COMMENTS: When originally proposed, this rank was placed in the Subfamily Sminthopsinae (M. Archer, 1982a) and included the genus *Planigale* Troughton, 1928. Synonymised within Dasyuridae by McKenna and Bell (1997: 55). Rank recognised by several authors including Krajewski and Westerman (2003: 15) and Krajewski *et al.* (1997b: 246; 2000a: 98; 2000b: 375; 2000c: 423; 2012: 266).

Planigale Troughton, 1928

Planigale Troughton, 1928: 282.

TYPE SPECIES: *Planigale ingrami brunnea* Troughton, 1928 [=*Planigale ingrami* (Thomas, 1906e)] by original description.

COMMENTS: Recognised as a genus by Iredale and Troughton (1934: vii, 6) but synonymised within *Phascogale* by Simpson (1945: 43). Genus recognised by Tate (1947: 133), Troughton (1967: 24), Ride (1970: 119), M. Archer (1976b: 341) who reviewed the genus, and subsequent authors. A review by Blacket *et al.* (2000: 443, 453) found that the currently accepted geographical ranges of many planigale species requires careful re-evaluation and many specimens in collections are apparently misidentified. Blacket *et al.* (2000: 446, 453) also showed that an animal collected from Mount Tom Price in the Pilbara of Western Australia is not closely related to any other sampled population and appears to be an undescribed species. This seems especially true for *P. ingrami*, which appears to have a much greater range than is currently recognised, being present in South Australia. A further undescribed distinct taxon from the Pilbara in Western Australia was referred to as *Planigale* species 1 by Painter *et al.* (1995: 407, 411), with its species distinctiveness confirmed by Blacket *et al.* (2000: 444).

FUTURE TAXONOMIC RESEARCH: The genus *Planigale* needs revision, with the forms Mount Tom Price [Pilbara] (*sensu* Blacket *et al.* (2000: 446, 453) and *Planigale* species 1 [Pilbara] (Painter *et al.*, 1995: 407, 411) needing to be formally described and named.

Planigale gilesi Aitken, 1972

Gile's Planigale

Planigale gilesi Aitken, 1972a: 1.

TYPE LOCALITY: No. 3 Bore, Pastoral Property of Anna Creek, South Australia, Australia. (28°18′S, 136°19′40″E)

COMMENTS: Recognised since its description.

Planigale ingrami (Thomas, 1906)

Long-tailed Planigale

Phascogale ingrami Thomas, 1906e: 6.

TYPE LOCALITY: Buchanan, Alexandria Station, Northern Territory, Australia.

COMMENTS: This taxon was included in *Planigale* by Troughton (1928: 282), Iredale and Troughton (1934: vii, 6) and followed by subsequent authors.

Phascogale subtilissima Lönnberg, 1913: 9.

TYPE LOCALITY: Noonkambah, Western Australia, Australia.

COMMENTS: Included in *Planigale* by Troughton (1928: 282), which was followed by Iredale and Troughton (1934: vii, 7). Tate (1947: 134) considered this taxon be a subspecies of *Planigale ingrami* which was followed by Marlow (1965: 23). Troughton (1967: 25), Ride (1970: 120) and Woolley (1974: 11) regarded it as a full species within *Planigale*. M. Archer (1976b: 351) included this taxon within *ingrami*. Recognised as a form of *ingrami* by Strahan (1983: 76; 1995: 110) and subspecies by Blacket *et al.* (2000: 454), which was followed by Groves (2005d: 36) and Clayton *et al.* (2006: 101). Taxon synonymised within *ingrami* by Van Dyck and Strahan (2008: 111).

Planigale ingrami brunneus Troughton, 1928: 282; Plate 39, Figs 1a-h.

TYPE LOCALITY: Wyangarie, Flinders River, Richmond district, Western Australia, Australia.

COMMENTS: Subspecies recognised, as described, by Iredale and Troughton (1934: 7). M. Archer (1976b: 351) included this taxon within *ingrami*. Recognised as a form by Strahan (1983: 76; 1995: 110) and subspecies of *ingrami* by Blacket *et al.* (2000: 454) that was followed by Groves (2005d: 36) and Clayton *et al.* (2006: 101). Taxon synonymised within *ingrami* by Van Dyck and Strahan (2008: 111).

Planigale maculata (Gould, 1851)

Common Planigale

Planigale maculata maculata (Gould, 1851)

Antechinus maculata Gould, 1851 [1845–1863]: Text to Plate 44.

TYPE LOCALITY: Clarence River, near Clarence, New South Wales, Australia.

COMMENTS: Taxon also described by Gould (1854: 284). Recognised within *Antechinus* by Krefft (1867a: 432) and most modern authors including Iredale and Troughton (1934: vii, 6), Tate (1947: 131), Troughton (1967: 21) and Ride (1970: 120), until it was placed in the genus *Planigale* by M. Archer (1976b: 346), which was followed by subsequent authors. Considered the most distinctive species of the genus by Painter *et al.* (1995: 410), who thought it may be closer to *Sminthopsis*.

FUTURE TAXONOMIC RESEARCH: The subspecies *sinualis* needs to be assessed to determine if it is a valid taxon.

Antechinus minutissimus Gould, 1851: 923.

TYPE LOCALITY: Cressbrook, near Moreton Bay, Queensland, Australia.

COMMENTS: Name subsequently described by Gould (1852 [1845–1863]: Text to Plate 45) that was published on 1 December 1852 and Gould (1854: 284), which was read before the Zoological Society of London on 9 December 1851 (see discussion of McAllan & Bruce, 1989: 452). Recognised as a species within *Antechinus* by Krefft (1867a: 432), within *Phascologale* by Thomas (1888a: xiii, 292) and within *Phascogale* by Finlayson (1934: 227). Synonymised within *maculata* by Iredale and Troughton (1934: 6), Tate (1947: 131) and M. Archer (1976b: 346), who placed *maculata* within *Planigale*. Subsequently considered as synonym of *Planigale maculata*.

Podabrus minutus Schmeltz, 1879: 3.

TYPE LOCALITY: *Nomen nudum*.

COMMENTS: Error for *minutisimus*. Iredale and Troughton (1852: 6) included it as a synonym of *maculatus*. Not typically considered by subsequent authors.

Planigale maculata sinualis (Thomas, 1926)

Phascogale minutissima sinualis Thomas, 1926b: 634.

TYPE LOCALITY: Groote Eylandt, Northern Territory, Australia.

COMMENTS: Included as a subspecies of *Antechinus maculatus* by Iredale and Troughton (1934: 6) and Tate (1947: 131). Taxon synonymised within *maculata* by M. Archer (1976b: 346), which was followed by Burbidge *et al.* (2014: 18). Recognised as a subspecies of *maculata* by D. Johnson (1964: 440), in *Antechinus*, Strahan (1983: 75; 1995: 112), Painter *et al.* (1995: 410), who thought that it may be specifically distinct, Blacket *et al.* (2000: 454), Groves (2005d: 36), Clayton *et al.* (2006: 101), and Van Dyck and Strahan (2008: 113) who suggested the subspecies need revision.

Planigale tenuirostris Troughton, 1928

Narrow-nosed Planigale

Planigale tenuirostris Troughton, 1928: 285; Plate 285, Figs. 2a-g.

TYPE LOCALITY: Bourke or Wilcannia, near Darling River, New South Wales, Australia.

COMMENTS: Species recognised since its description.

Subfamily Sminthopsinae M. Archer, 1982 *sensu* Marshall *et al.*, 1990

Subfamily Sminthopsinae M. Archer, 1982a: 439.

TYPE GENUS: *Sminthopsis* Thomas, 1887a.

COMMENTS: When originally proposed as a new rank it was placed in the Family Dasyuridae (Goldfuss, 1820a) and included the genera *Sminthopsis* Thomas, 1887a; *Antechinomys* Krefft, 1867a; and *Ningaui* M. Archer, 1975. Rank synonymised within Dasyuridae by McKenna and Bell (1997: 55). Subfamily rank recognised by M. Archer (1984: 635), Marshall *et al.* (1990: 459), Strahan (1995: 6, 121), Kirsch *et al.* (1997: 245), Wroe (1997: 23; 1999: 512), Groves (2005d: 32), Krajewski and Westerman (2003: 14), and Van Dyck and Strahan (2008: 9, 122). Though *Antechinomys* has consistently been placed with *Sminthopsis* (e.g. M. Archer, 1981: 65; 1982a: 439; Baverstock *et al.*, 1982: 646), the placement of the genus *Ningaui* has varied. Strahan (1995: 5, 116), and Van Dyck and Strahan (2008: 9, 107) included *Ningaui* in the Subfamily Planigalinae, although it has historically normally been recognised as sister to a *Sminthopsis-Antechinomys* clade by authors including M. Archer (1975: 242–243; 1981: 65, 77; 1982a: 439) and Baverstock *et al.* (1982: 646) or even more closely to *Sminthopsis* than *Antechinomys* (e.g. Krajewski *et al.*, 1997b: 246; Blacket *et al.*, 1999: 140, 150; Krajewski & Westerman, 2003: 12, 14). As a result of this relationship, *Ningaui* has been placed in the Subfamily Sminthopsinae by Marshall *et al.* (1990: 489) or in the Tribe Sminthopsini (within the Subfamily Sminthopsinae) by several authors (see individual entry below).

Tribe Sminthopsini M. Archer, 1984: 635.

TYPE GENUS: *Sminthopsis* Thomas, 1887a.

COMMENTS: When originally proposed, this rank was placed in the Subfamily Sminthopsinae (M. Archer, 1982a) and included the genus *Sminthopsis* Thomas, 1887a.

Synonymised within Dasyuridae by McKenna and Bell (1997: 55) but recognised by several authors including Krajewski and Westerman (2003: 15) and Krajewski *et al.* (1997b: 246; 2000a: 98; 2000b: 375; 2000c: 423; 2012: 265).

Antechinomys Krefft, 1867

Antechinomys Krefft, 1867a: 434.

TYPE SPECIES: *Phascogale lanigera* Gould, 1856 [=*Antechinomys laniger* (Gould, 1856)] by monotypy.

COMMENTS: Considered a subgenus of *Sminthopsis* by M. Archer (1979b: 329; 1981: 63, 187) and Honacki *et al.* (1982: 31). Genus rank recognised by Iredale and Troughton (1934: vii, 11), Simpson (1945: 43), Tate (1947: 125), Troughton (1967: 37), Ride (1970: 124, 126), M. Archer (1977a: 17), Kirsch and Calaby (1977: 15), Baverstock *et al.* (1982: 643), Lidicker (1983: 1317), Marshall (1984: 83), Woolley (1984: 481), Mahoney and Ride (1988a: 14) and most subsequent authors. Genus confirmed by genetic evidence of Krajewski *et al.* (1997b: 236) and Blacket *et al.* (1999: 150) and followed by Groves (2005d: 32). Synonymised within *Sminthopsis* by Groves (1993b: 35) and McKenna and Bell (1997: 56), but recognised as a valid genus by Strahan (1995: 121) and other authors. Genus reviewed by Lidicker and Marlow (1970: 212).

Antechinomys laniger (Gould, 1856)

Kultarr

Antechinomys laniger laniger (Gould, 1856)

Phascogale lanigera Gould, 1856 [1845–1863]: Text to Plate 33.

TYPE LOCALITY: Eastern Australia.

COMMENTS: Recognised within *Phascogale* by Krefft (1866a: 9) but placed in *Antechinomys* with species name spelt *lanigera* by Krefft (1867a: 434). Spelling of the specific name changed to *laniger* by Thomas (1888a: xiii, 309) and followed by subsequent authors when using *Antechinomys*. Placed within *Antechinomys* by Thomas (1888a: 309), Iredale and Troughton (1934: vii, 12), Tate (1947: 125) and most subsequent authors including the review of *Antechinomys* by M. Archer (1977a: 17) and Mahoney and Ride (1988a: 14), though was recognised as a subgenus of *Sminthopsis* by M. Archer (1981: 63, 187) and Honacki *et al.* (1982: 32).

FUTURE TAXONOMIC RESEARCH: The subspecies *spenceri* needs to be assessed to determine if it is a valid taxon.

Antechinomys laniger spenceri Thomas, 1906

Antechinomys Spenceri Thomas, 1906b: 331.

TYPE LOCALITY: Charlotte Waters, Northern Territory, Australia.

COMMENTS: Considered a valid species by Iredale and Troughton (1934: vii, 12), Tate (1947: 125), Troughton (1967: 38), Lidicker and Marlow (1970: 222), Ride (1970: 126) and Watts and Aslin (1974: 64), but was synonymised within *laniger* by M. Archer (1977a: 19), Honacki *et al.* (1982: 32), Mahoney and Ride (1988a: 14) and Burbidge *et al.* (2014: 18). Subspecies rank recognised by Strahan (1983: 73; 1995: 122), Clayton *et al.* (2006: 101), and Van Dyck and Strahan (2008: 123).

Ningaui M. Archer, 1975

Ningaui M. Archer, 1975: 237, 239.

TYPE SPECIES: *Ningaui timealeyi* M. Archer, 1975 by original designation.

COMMENTS: An undescribed species of *Ningaui* occurs in the Northern Territory, Australia (K. Johnson and Roff, 1980: 127). According to recent authors, including Blacket *et al.* (1999: 140) and Krajewski *et al.* (1997b: 246) the genus *Ningaui* is most closely related to *Sminthopsis*.

Ningaui ridei M. Archer, 1975

Wongai Ningaui

Ningaui ridei M. Archer, 1975: 246.

TYPE LOCALITY: 38.6 km along White Cliffs Road, east-north-east of Laverton, White Cliffs Road, Western Australia, Australia. 28°30′S, 122°47′E.

COMMENTS: Species recognised since its description.

Ningaui timealeyi M. Archer, 1975

Pilbara Ningaui

Ningaui timealeyi M. Archer, 1975: 243.

TYPE LOCALITY: 32.2 km SE of Mt Robinson, north-western Western Australia, Australia.

COMMENTS: Species recognised since its description. The taxon referred to as *Planigale* 2, from the Pilbara in Western Australia, by Painter *et al.* (1995: 407, 411) was indicated to be *Ningaui timealeyi* by Blacket *et al.* (2000: 444).

Ningaui yvonneae Kitchener *et al.*, 1983

Southern Ningaui

Ningaui yvonneae Kitchener *et al.*, 1983: 366.

TYPE LOCALITY: Mt. Manning, Western Australia, Australia. (29°58′S, 119°32′E)

COMMENTS: Species recognised since its description.

Sminthopsis Thomas, 1887

Sminthopsis Thomas, 1887a: 503.

TYPE SPECIES: *Noven novum* for *Podabrus* Gould, 1845 [1845–1863].

COMMENTS: Taxonomic decision of Thomas (1887a: 503) to synonymise *Podabrus* within *Sminthopsis*. Genus reviewed by Troughton (1965: 307), M. Archer (1981, 61), Blacket *et al.* (1999: 140; 2001: 149; 2006: 125) and Krajewski *et al.* (2012: 265). Species groups proposed by Blacket *et al.* (1999: 140), and adopted by Groves (2005d: 33), include the *S. crassicaudata* group (monotypic); *S. macroura* group (containing *S. bindi*, *S. butleri*, *S. douglasi*, *S. macroura* and *S. virginiae*); *S. granulipes* group (monotypic); *S. fuliginosus* group (containing *S. aitkeni*, *S. boullangerensis*, *S. fuliginosus*); *S. longicaudata* group (monotypic); *S. murina* group (*S. archeri*, *S. dolichura*, *S. fuliginosus*, *S. gilberti*, *S. leucopus*, *S. murina*); and *S. psammophila* group (*S. hirtipes*, *S. ooldea*, *P. psammophila*, *S. youngsoni*). These species groups are strongly distinct, and some of them may ultimately be given generic rank (Groves, 2005d: 33). Krajewski *et al.* (2012: 265, 272) suggest *Sminthopsis* is polyphyletic with *Ningaui* nested within *Sminthopsis*.

Podabrus Gould, 1845 [1845–1863]: Text to Plate 47.

TYPE SPECIES: *Phascogale crassicaudata* Gould, 1844a [=*Sminthopsis crassicaudata* (Gould, 1844a)] by monotypy.

COMMENTS: Genus also described by Gould (1845a: 79) and subsequently recognised by Krefft (1867a: 432). Thomas (1888a: xiii, 298) replaced this genus with *Sminthopsis* and subsequently included as a synonym of *Sminthopsis* by Palmer (1904: 553), Iredale and Troughton (1934: 9), Marshall (1981: 26), Marshall *et al.* (1990: 489) and subsequent authors.

HOMONYMS:

Podabrus G. Fischer, 1821: 36, beetles of the Class Insecta (Order Coleoptera, Family Cantharidae). *Nomen nudum*. See Arnett *et al.* (2002: 211).

Podabrus Westwood, 1840: 27, beetles of the Class Insecta (Order Coleoptera, Family Cantharidae). Currently recognised subgenus. See Arnett *et al.* (2002: 211).

Podabrus Richardson, 1848: 11, sculpin fish of the Superclass Pisces (Order Scorpaeniformes, Family Cottidae). Genus is a synonym of *Vellitor* Jordan and Starks, 1904: 318. See Iwata (1983: 1).

Sminthopsis archeri Van Dyck, 1986

Chestnut Dunnart

Sminthopsis archeri Van Dyck, 1986: 111, 112.

TYPE LOCALITY: Morehead, Trans-Fly Plains, Papua New Guinea. (8°04′S, 141°39′E)

COMMENTS: Separated from the *Sminthopsis murina* species group. Also known to occur within New Guinea (Flannery, 1990: 65; 1995a: 101).

Sminthopsis bindi Van Dyck *et al.*, 1994

Kakadu Dunnart

Sminthopsis bindi Van Dyck *et al.*, 1994: 312.

TYPE LOCALITY: Eva Valley Station, Stage 3, Kakadu National Park, Northern Territory, Australia.

COMMENTS: Separated from the *Sminthopsis macroura* species group.

Sminthopsis butleri M. Archer, 1979

Butler's Dunnart

Sminthopsis butleri M. Archer, 1979b: 329.

TYPE LOCALITY: Kalumburu Mission, northern Western Australia, Australia. (14°15′S, 126°40E). The coordinates for the type locality were recorded as 14°18′S, 126°38′E by Kitchener and Vicker (1981: 26).

COMMENTS: Though the name *Sminthopsis butleri* is mentioned numerous times in the text of Kirsch (1977a: 47, 49, 50, 51, 55) without an author, it has been generally attributed to M. Archer (1977b: 137). Despite the mention (without description, hence a *nomen nudum*) of the name in M. Archer (1977b: 137), which McAllan and Bruce (1989: 455) argued to be the first description of the name, there seems no doubt that the formal description dates to M. Archer (1979b: 329). M. Archer (1981: 210) noted that its affinities are probably with the *S. macroura* species group.

Sminthopsis crassicaudata (Gould, 1844)

Fat-tailed Dunnart

Phascogale crassicaudata Gould, 1844a: 105.

TYPE LOCALITY: Military Station, Williams River, Western Australia, Australia.

COMMENTS: Transferred to *Podabrus* by Gould (1845 [1845–1863]: Text to Plate 47) and Krefft (1866a: 11; 1867a: 433). Transferred to *Sminthopsis* by Thomas (1888a: xiii, 306), Iredale and Troughton (1934: vii, 9), Troughton (1967: 31), M. Archer (1979b: 329; 1981: 65, 176) and subsequent authors. Included in its own species group. Proposed subspecies were not recognised by Morton and Alexander (1982: 698), Hope and Godfrey (1988: 451) or S. Cooper *et al.* (2000: 461).

Sminthopsis crassicaudata centralis Thomas, 1902a: 492.

TYPE LOCALITY: Kilalpannina, Lake Eyre, South Australia, Australia.

COMMENTS: Recognised as a subspecies within *S. crassicaudata* by Finlayson (1933a: 197), Iredale and Troughton (1934: 9), M. Archer (1981: 185) and Strahan (1983: 61; 1995: 130). Considered a subspecies of *macrura* [=*Sminthopsis macroura*] by Tate (1947: 122).

Synonymised within *S. crassicaudata* by M. Archer (1979b: 329), Honacki *et al.* (1982: 31), Morton and Alexander (1982: 698), Mahogany and Ride (1988a: 29) and S. Cooper *et al.* (2000: 461).

Sminthopsis crassicaudata ferruginea Finlayson, 1933a: 199.

TYPE LOCALITY: Macdonnell Range, Northern Territory, Australia.

COMMENTS: Lectotype designation by Morton and Alexander (1982: 698). Synonymised within *S. crassicaudata* by M. Archer (1979b: 329; 1981: 176), Honacki *et al.* (1982: 31), Morton and Alexander (1982: 698), Mahogany and Ride (1988a: 29) and subsequent authors.

Sminthopsis dolichura Kitchener *et al.*, 1984

Little-Long-tailed Dunnart

Sminthopsis dolichura Kitchener *et al.*, 1984a: 201, 214.

TYPE LOCALITY: 6 km SSE of Buningonia, Western Australia, Australia. 250 m elevation. (32°28′10″S, 123°36′00″E)

COMMENTS: Separated from the *Sminthopsis murina* species group and recognised by subsequent authors.

Sminthopsis douglasi M. Archer, 1979

Julia Creek Dunnart

Sminthopsis douglasi M. Archer, 1979b: 337.

TYPE LOCALITY: Julia Creek near Cloncurry River, north central, Queensland, Australia. (20°40′S, 141°40′E)

COMMENTS: From the *Sminthopsis macroura* species group.

Sminthopsis fuliginosus (Gould, 1852)

Grey-bellied Dunnart

Sminthopsis fuliginosus fuliginosus (Gould, 1852)

Antechinus fuliginosus Gould, 1852 [1845–1863]: Text to Plate 41.

TYPE LOCALITY: 'R. Avon' [=Hill sides on the Banks of the Avon, W.A.], Western Australia, Australia. See M. Archer (1981: 85), and Woinarski *et al.* (2014: 899–900). Type locality refined to be King George Sound, near Albany, Western Australia by Woinarski *et al.* (2014: 899–900).

COMMENTS: No type specimen was assigned to this species, but a young male specimen was designated as lectotype by Thomas (1888a: 305). Type locality often given as King Georges Sound, Western Australia, but this appears to be in error (see M. Archer, 1981: 85). The taxon was transferred to *Podabrus* by Krefft (1867a: 433). Considered a subspecies of *Sminthopsis murina* by Iredale and Troughton (1934: 10), Tate (1947: 121), M. Archer (1979b: 329; 1981: 95) and Strahan (1983: 52) and synonym of *murina* by Honacki *et al.* (1982: 32). Separated from *murina* by Kitchener *et al.* (1984a: 245), Mahoney and Ride (1988a: 29), and Groves (1993b: 36; 2005d: 34). Taxon reviewed by Crowther *et al.* (1999: 226–227) who also did not use the name *S. fuliginosus* or *S. griseoventer fuliginosus* for a variety of reasons. Taxon not recognised by Clayton *et al.* (2006: 102) or Woolley *et al.* (2007: 1381) who suggested this taxon was generally not recognised and assumed it to be synonymous with S. *griseoventer*. Synonymised within *dolichura* by Van Dyck and Strahan (2008: 135) but suggested to have an unclear taxonomic status by Burbidge *et al.* (2014: 18). A comprehensive review of this taxon by Woinarski *et al.* (2014: 899–900) suggested that, as the type locality of *S. fuliginosa* is King George Sound, near Albany, on the basis of known distributions, *S. griseoventer* is likely to be a synonym as neither *S. dolichura* nor *S. gilberti* occurs there. Despite this conclusion, this taxon was not recognised by Woinarski *et al.* (2014: 899–900), but based on the information provided in that review it is recognised here as the senior synonym of *Sminthopsis griseoventer* Kitchener *et al.*, 1984a.

FUTURE TAXONOMIC RESEARCH: A genetic and/or morphological assessment is required of the type series of this taxon in the Natural History Museum to identify whether it is the same as *S. griseoventer*. Unfortunately Woinarski *et al.* (2014: 900) report there is little likelihood of DNA being obtained from the London specimens; DNA might be extracted from the Leiden specimens, but this would be useful only if they are part of the same series as the London skulls.

Sminthopsis griseoventer Kitchener *et al.*, 1984a: 201, 225.

TYPE LOCALITY: 13.2 km and 68° from Bindoon, Western Australia, Australia. 150 m elevation. (31°18′15″S, 116°01′00″E)

COMMENTS: Separated from the *Sminthopsis murina* species group and is now its own species group. Though historically confined to Western Australia, specimens allocated to this taxon have been collected on the Eyre Peninsula, South Australia (Brandle, 2010: 50; Kemper *et al.*, 2011: 136).

Sminthopsis caniventer Baverstock *et al.*, 1984: 823.

COMMENTS: *Nomen novum* for *Sminthopsis griseoventer* Kitchener *et al.*, 1984a. Synonymised within *griseoventer* by Mahogany and Ride (1988a: 30) and not recognised by subsequent authors.

Sminthopsis griseoventer boullangerensis Crowther *et al.*, 1999: 215, 220.

TYPE LOCALITY: Boullanger Island, Western Australia, Australia. (30°19′S, 115°E)

COMMENTS: This subspecies was split from the *S. murina* species group, with the subspecies designation considered 'conservative'. The population on Boullanger Island, Western Australia, first indicated as distinct from genetic analysis by Lynam (1987). Recognised as a full species by Groves (2005d: 33) but as a subspecies of *griseoventer* by Clayton *et al.* (2006: 102), and Van Dyck and Strahan (2008: 143). Start *et al.* (2006: 51) explored new taxonomic evidence relating to this taxon and concluded that neither molecular nor the morphological data support differentiation at taxonomic or evolutionarily significant levels. Not recognised as distinct from mainland *griseoventer* by Labrinidis *et al.* (1998: 293) or Start *et al.* (2006: 51), which was followed by Woolley *et al.* (2007: 1381) who suggested that this taxon should no longer be recognised. As a result this taxon is not recognised here.

Sminthopsis fuliginosus aitkeni Kitchener *et al.*, 1984

Sminthopsis aitkeni Kitchener *et al.*, 1984a: 201, 230.

TYPE LOCALITY: Section 146 Hundreds of Cassini, Kangaroo Island, South Australia, Australia. (35°35′S, 137°19′E)

COMMENTS: Separated from the *Sminthopsis murina* species group. Originally described from Kangaroo Island as *Sminthopsis murina* (Aitken, 1972b: 36). More recent genetic and morphological research by Kemper *et al.* (2011: 138) suggested that *aitkeni* and *griseoventer* were conspecific. As a result they suggested that *aitkeni* perhaps warrants separate subspecies status based on its size and pelage colour differences.

FUTURE TAXONOMIC RESEARCH: Research is required to resolve the affinities and nomenclature of the *S. fuliginosus*/*S. aitkeni* complex, including re-evaluating the morphology of specimens from Western Australia and South Australia (see Kemper *et al.*, 2011: 138). In the Kemper *et al.* (2011) study, control region sequences of *S. aitkeni* were nested within those of Eyre Peninsula *S. griseoventer* (here *S. fuliginosus*), whereas there were fixed allelic differences in three allozyme loci (but there were only two available specimens of *S. aitkeni*). Discrepancies of this nature indicate that further research is needed, although it is admitted that the two species (if they are indeed distinct) are closely related.

Sminthopsis gilberti Kitchener *et al.*, 1984

Gilbert's Dunnart

Sminthopsis gilberti Kitchener *et al.*, 1984a: 201, 221.

TYPE LOCALITY: 10 km and 260° from Mt Saddleback, Western Australia, Australia. 240 m elevation. (32°58′15″S, 116°20′20″E)

COMMENTS: Separated from the *Sminthopsis murina* species group. Blacket *et al.* (2006: 136) found that specimens from the south-west and the extreme south-east of Western Australia are genetically divergent, and this needs further investigation.

Sminthopsis granulipes Troughton, 1932

White-tailed Dunnart

Sminthopsis granulipes Troughton, 1932a: 350; Fig. 1.

TYPE LOCALITY: King Georges sound, Western Australia, Australia.

COMMENTS: Recognised as species by Iredale and Troughton (1934: vii, 9) and subsequent authors. Forms a species group by itself.

Podabrus albocaudatus Krefft, 1872b: 598.

TYPE LOCALITY: King Georges Sound, Western Australia.

COMMENTS: This is a valid name that appears to have been forgotten after its description, with the name *Sminthopsis granulipes* Troughton, 1932 being recognised instead. It was proposed by Parnaby *et al.* (2015: 283) that the holotype of *granulipes* is also likely to be the holotype of *albocaudatus* and therefore an objective synonym. In order to maintain nomenclatural stability Parnaby *et al.* (2015: 283, 288) proposed that *Sminthopsis granulipes* should be declared a *nomen protectum* and *Podabrus albocaudatus* be considered a *nomen oblitum*, as required by Article 23.9.2 of the Code (ICZN, 1999: 28) because Article 23.9.1 has been complied with, giving evidence that the condition of 23.9.1.1 applies, and because they were not aware of any published citations of *albocaudatus*.

Sminthopsis hirtipes Thomas, 1898

Greater Hairy-footed Dunnart

Sminthopsis hirtipes Thomas, 1898a: 3.

TYPE LOCALITY: Station Point, Charlotte Waters, Northern Territory, Australia. Considered a subspecies of *murina* by Iredale and Troughton (1934: 10).

COMMENTS: Included in the *P. psammophila* species group.

Sminthopsis leucopus (J. Gray, 1842)

White-footed Dunnart

Sminthopsis leucopus leucopus (J. Gray, 1842)

Phascogale leucopus J. Gray, 1842b: 261.

TYPE LOCALITY: Tasmania, Australia.

COMMENTS: Recognised within *Phascogale* (*Antechinus*) by Waterhouse (1846: 423), within *Antechinus* by Gould (1860 [1845–1863]: Text to Plate 35) and *Podabrus* by

Krefft (1867a: 433). Reduced to a subspecies of *murina* by Tate (1947: 121) and as an apparent synonym of *murina* by Troughton (1965: 315), but recognised as a distinct species within *Sminthopsis* by Thomas (1888a: xiii, 302), Iredale and Troughton (1934: vii, 11), and M. Archer (1979b: 329; 1981: 63, 102), which has been followed by subsequent authors. Two subspecies are currently recognised from both morphological (M. Archer, 1981: 105, 106) and biochemical (A. Smith, 1983: 753; Blacket *et al.*, 1999: 143) studies. The *leucopus* population from the Atherton Tablelands is distinct from the south-eastern Australian populations both morphologically (Van Dyck, 1985: 53) and genetically (Baverstock *et al.*, 1984: 823), Reinforced by Blacket *et al.* (2006: 134), who indicated from their mtDNA data that it is strongly divergent from other populations, although they did not indicate whether there are fixed differences.

FUTURE TAXONOMIC RESEARCH: Morphometric and taxonomic studies are required to confirm the taxonomic status of the Atherton Tablelands population and formally name it if warranted.

Antechinus (Podabrus) leucogenys Higgins & Petterd, 1883: 172.

TYPE LOCALITY: Ringarooma, Tasmania, Australia.

COMMENTS: Synonymised within *leucopus* by M. Archer (1979b: 329; 1981: 102), Honacki *et al.* (1982: 32), and Mahogany and Ride (1988a: 31).

Sminthopsis leucopus ferruginifrons (Gould, 1854)

Antechinus ferruginifrons Gould, 1854 [1845–1863]: Text to Plate 36.

TYPE LOCALITY: Sydney, New South Wales, Australia. Type designated by Thomas (1922a: 128).

COMMENTS: Species recognised within *Antechinus* by Krefft (1867a: 432). Synonymised within *leucopus* by M. Archer (1979b: 329), Honacki *et al.* (1982: 32), Mahogany and Ride (1988a: 31), Maxwell (1996: 3) and Clayton *et al.* (2006: 102). Considered a subspecies of *murina* by Tate (1947: 121), and subsequently a subspecies of *leucopus* by Iredale and Troughton (1934: 11), M. Archer (1981: 106), A. Smith (1983: 753), Strahan (1983: 55; 1995: 144), Blacket *et al.* (1999: 143), and Van Dyck and Strahan (2008: 147). Usually ranked as a subspecies of *S. leucopus*, but morphological characters, admittedly on small samples ($n = 7$ for both species), do not overlap in several craniodental and bodily ratios, e.g. the tail is shorter than the head and body in Tasmania, longer in Victoria (M. Archer, 1981: Table 2, 105, 106) and in biochemical characters they are also distinct (A. Smith, 1983: 753; Blacket *et al.*, 1999: 143).

FUTURE TAXONOMIC RESEARCH: Morphometric and taxonomic studies are required to confirm the taxonomic status of this taxon and determine if differentiation at species rank is warranted.

P. [odabrus] mitchelli Krefft, 1867a: 433.

TYPE LOCALITY: Interior of New South Wales, Australia.

COMMENTS: Considered a subspecies of *ferruginifrons* by Iredale and Troughton (1934: 11). Synonymised within *leucopus* by M. Archer (1979b: 329; 1981: 102), Honacki *et al.* (1982: 32), and Mahogany and Ride (1988a: 31).

Antechinus ferrugineifrons Thomas, 1888a: 302.

TYPE LOCALITY: Invalid emendation for *Antechinus ferruginifrons* Gould, 1954.

COMMENTS: Synonymised within *leucopus* by Thomas (1888a: 302), M. Archer (1981: 102), Mahogany and Ride (1988a: 31) and subsequent authors.

Sminthopsis longicaudata Spencer, 1909

Large Long-tailed Dunnart

Sminthopsis longicaudatus Spencer, 1909: 449.

TYPE LOCALITY: Pilbara and the desert to east of Western Australia, Australia.

COMMENTS: Considered a species by Iredale and Troughton (1934: vii, 11) and subsequent authors. Forms a species group by itself.

Sminthopsis macroura (Gould, 1845)

Stripe-faced Dunnart

Sminthopsis macroura macroura (Gould, 1845)

Podabrus macrourus Gould, 1845a: 79.

TYPE LOCALITY: Open plains, Darling Downs, Queensland, Australia.

COMMENTS: Placed in *Phascogale* (*Antechinus*) by Waterhouse (1846: 426), but transferred to *Podabrus* by Gould (1849 [1845–1863]: Text to Plate 46) and Krefft (1867a: 433). Included as a subspecies of *S. crassicaudata* by Iredale and Troughton (1934: 9) and Finlayson (1934: 227). Elevated to species rank by Troughton (1967: 32) (as *macrura*), M. Archer (1979b: 329; 1981: 63, 148) and Mahoney and Ride (1988a: 31). This appears to be a species complex according to Blacket *et al.* (2001: 162–163) who recognised *froggatti* and *stalkeri* as distinct species. These subspecies were not recognised by Clayton *et al.* (2006: 102), but were accepted by Van Dyck and Strahan (2008: 152).

Podabrus macrurus Thomas, 1888a: 306.

TYPE LOCALITY: Invalid emendation of *Podarus macrourus* Gould, 1845a.

COMMENTS: Not considered by Iredale and Thomas (1934). Synonymised within *macroura* by M. Archer (1981: 148), and Mahoney and Ride (1988a: 31).

Sminthopsis larapinta Spencer, 1896a: 33.

TYPE LOCALITY: Charlotte Waters, Northern Territory, Australia.

COMMENTS: Species also described by Spencer (1896b: 8), though it is unclear which description was published first. Recognised at species rank by Thomas (1906d: 542), Finlayson (1933a: 199), Iredale and Troughton (1934: 10), Tate (1947: 123), Troughton (1965: 317; 1967: 33). Synonymised with *froggatti* by Ride (1970: 226) and within *macroura* by M. Archer (1979b: 329; 1981: 148), Honacki *et al.* (1982: 32), Mahoney and Ride (1988a: 31), Strahan (1983: 63; 1995: 148), and Van Dyck and Strahan (2008: 152).

Sminthopsis monticola Troughton, 1965: 311.

TYPE LOCALITY: Lawson, 56 road miles west of Sydney, New South Wales, Australia. 2403 feet.

COMMENTS: Species rank recognised by Troughton (1967: 32) and as a subspecies of *macroura* by Strahan (1983: 63). Synonymised within *macroura* by Ride (1970: 248), M. Archer (1979b: 329; 1981: 148), Honacki *et al.* (1982: 32) and Mahoney and Ride (1988a: 32).

Sminthopsis macroura froggatti (Ramsay, 1887)

Antechinus (Podabrus) froggatti Ramsay, 1887a: 552.

TYPE LOCALITY: From near the beach in the 'dindan'[sic] [=pindan] scrub bordering King Sound near Derby, Western Australia, Australia.

COMMENTS: Abstract of description by Ramsay (1887b: vi). Synonymised with *crassicaudata* the year after its description by Thomas (1888a: 306). Recognised as a subspecies of *macroura* by Tate (1947: 122) and Strahan (1983: 63; 1995: 148). Troughton (1932a: 352) restored this taxon as a valid species, as did Iredale and Troughton (1934: vii, 10), Troughton (1967: 33), Ride (1970: 124) and Watts and Aslin (1974: 63). Synonymised within *macroura* by M. Archer (1979b: 329), Honacki *et al.* (1982: 32), and Mahoney and Ride (1988a: 31). Recognised at the species rank using mitochondrial 12S rRNA by Blacket *et al.* (2001: 149) and followed by Clayton *et al.* (2006: 102), but reduced to subspecies rank by Van Dyck and Strahan (2008: 152).

FUTURE TAXONOMIC RESEARCH: More research is required, but given the geographic separation a specific distinction seems likely.

Sminthopsis macroura stalkeri Thomas, 1906

Sminthopsis stalkeri Thomas, 1906d: 543.

TYPE LOCALITY: SW of Alroy, Western Australia, Australia. 800 ft.

COMMENTS: Included at species rank by Iredale and Troughton (1934: vii, 10) and Troughton (1965: 318). Lowered to a subspecies of *larapinta* by Tate (1947: 123). Synonymised within *macroura* by M. Archer (1979b: 329; 1981: 148), Honacki *et al.* (1982: 32), and Mahoney and Ride (1988a: 32). Recognised as a species using mitochondrial 12S rRNA by Blacket *et al.* (2001: 149) and Clayton *et al.* (2006: 102) but reduced to subspecies rank by Van Dyck and Strahan (2008: 152).

FUTURE TAXONOMIC RESEARCH: As in the case of *froggatti*, more research is needed; at any rate some kind of taxonomic differentiation seems plausible.

Sminthopsis murina (Waterhouse, 1838)

Common Dunnart

Sminthopsis murina murina (Waterhouse, 1838)

Phascogale murina Waterhouse, 1838b: 76.

TYPE LOCALITY: Hunter River, New South Wales, Australia.

COMMENTS: Species rank recognised within *Phascogale* by Waterhouse (1838a: 65; 1841a: 143; 1846: 425), *Antechinus* by Gould (1852 [1945–1863]: Text to Plate 43) and *Podabrus* by Krefft (1867a: 433). Transferred to *Sminthopsis* by Thomas (1888a: xiii, 303), Iredale and Troughton (1934: vii, 10), Troughton (1967: 35), M. Archer (1979b: 329; 1981: 63, 94–99), Honacki *et al.* (1982: 32), Kitchener *et al.* (1984a: 201) and subsequent authors. A genetic assessment by Blacket *et al.* (2006: 135) suggested there are at least two genetically distinct lineages within what is currently recognised as *S. murina murina*. Interestingly, *S. m. murina* from west of the Great Dividing range is distinguishable morphology from eastern and southern *S. m. murina* (Kitchener *et al.*, 1984a: 201).

Phascogale albipes Waterhouse, 1842: 48.

TYPE LOCALITY: Port Adelaide, South Australia, Australia.

COMMENTS: Species recognised within *Phascogale* (*Antechinus*) by Waterhouse (1846: 421), *Antechinus* by Gould (1852 [1845–1863]: Text to Plate 42) and Krefft (1866a: 10), and *Podabrus* by Krefft (1867a: 433). Considered a subspecies of *murina* by Iredale and Troughton (1934: 10) and Tate (1947: 121). Synonymised within *murina* by Thomas (1888a: 304), M. Archer (1979b: 329; 1981: 94), Honacki *et al.* (1982: 32), and Mahoney and Ride (1988a: 32). Kitchener *et al.* (1984a: 241) found that what they referred to as *S. murina* on the western side of the Great Dividing Range (north-eastern Queensland and the Murray/Darling basin district) differed from that on the eastern side, for example having longer tails and more closely spaced palatine vacuities; Blacket *et al.* (2006: 135) found that they were also distinct using two regions of mitochondrial DNA (the Control Region and 12 S) and the nuclear ω-globin gene; those from east of the Divide (that is to say, the true *S. murina*) form a clade with what is here called *S. tatei*, sister to the present species.

FUTURE TAXONOMIC RESEARCH: Further assessment is required to determine the distinctiveness of this taxon and determine whether those from west and east of the Dividing Range are consistently different, hence distinct species.

Sminthopsis murina tatei Troughton, 1965

Sminthopsis murina tatei Troughton, 1965: 316.

TYPE LOCALITY: Tolga, Atherton Tableland, Queensland, Australia. Approx. 2460 ft.

COMMENTS: Baverstock *et al.* (1984: 831) were unable to find genetic differences between this taxon and *Sminthopsis murina*. Subsequently synonymised within *murina* by M. Archer (1979b: 329), Honacki *et al.* (1982: 32) and Mahoney and Ride (1988a: 32), but made a subspecies of *murina* by M. Archer (1981: 99), Strahan (1983: 52; 1995: 150), Clayton *et al.* (2006: 102), and Van Dyck and Strahan (2008: 153). It is nonetheless morphologically fully distinct (Kitchener *et al.*, 1984a: 208, 241).

FUTURE TAXONOMIC RESEARCH: Further assessment is required to determine the distinctiveness of this taxon and determine where it needs to be elevated to species rank as Kitchener *et al.*'s (1984a: 208, 241) findings would seem to require.

Sminthopsis ooldea Troughton, 1965

Ooldea Dunnart

Sminthopsis murina ooldea Troughton, 1965: 316.

TYPE LOCALITY: Ooldea, South Australia, Australia.

COMMENTS: Separated as a distinct species by M. Archer (1975: 243; 1979b: 329; 1981: 63, 108) and Kirsch and Calaby (1977: 15) and followed by subsequent authors. Included in the *S. psammphila* species group.

Sminthopsis psammophila Spencer, 1895

Sandhill Dunnart

Sminthopsis psammophilus Spencer, 1895: 223.

TYPE LOCALITY: Between Kurtitina Well and Ayers Rock, Northern Territory, Australia.

COMMENTS: Taxon further described by Spencer (1896a: 35; 1896c, 84). Recognised as a species, with the specific name *psammophila*, by Iredale and Troughton (1934: vii, 10) and subsequent authors with the exception of Tate (1947: 123), who reduced it to a subspecies of *macrura* [=*Sminthopsis macroura*]. Included in the *S. psammophila* species group. Rediscovered in February 1969 after not having been seen since the collection of the type specimen (Aitken, 1971: 103).

Sminthopsis virginiae (de Tarragon, 1847)

Red-cheeked Dunnart

Sminthopsis virginiae virginiae (de Tarragon, 1847)

Phascogale Virginiae de Tarragon, 1847: 177.

TYPE LOCALITY: Herbert Vale, Herbert River, Queensland, Australia. 43 m elevation. (18°30′S, 145°50′E). Type locality designated by M. Archer (1981: 132).

COMMENTS: Species recognised within *Phascogale* by Collett (1887a: 548) and then transferred to *Sminthopsis* by Thomas (1888a: xiii, 300) and followed by subsequent authors. Also recorded in New Guinea and the Aru Islands by Flannery (1990: 66; 1995a: 102; 1995b: 63; Helgen, 2007a: 711).

Sminthopsis virginiae nitela **Collett, 1897**

Sminthopsis nitela Collett, 1897: 334.

TYPE LOCALITY: Daly River, about miles from the coast, Northern Territory, Australia.

COMMENTS: Synonymised within *larapinta* by Iredale and Troughton (1934: 10), recognised as a subspecies of *larapinta* by Tate (1947: 124), and a valid species by Troughton (1965: 318; 1967: 34) and Ride (1970: 122). Recognised as a subspecies of *S. virginiae* by M. Archer (1981: 141), Strahan (1983: 56; 1995: 157) and Flannery (1990: 67; 1995a: 103; 1995b: 63). Synonymised within *virginiae* by Honacki *et al.* (1982: 33) and Mahoney and Ride (1988a: 33). Subspecies rank recognised by Blacket *et al.* (2001: 149, 159) who suggested that it is quite distinct from the two other *S. virginiae* subspecies and may warrant recognition as a separate species. Subsequently recognised as a subspecies of *virginiae* by Groves (2005d: 36), Clayton *et al.* (2006: 102), and Van Dyck and Strahan (2008: 159).

Sminthopsis lumholtzi Iredale & Troughton, 1934: vii, 11.

TYPE LOCALITY: Herbert Vale, Queensland, Australia.

COMMENTS: Name introduced as a *nomen novum* for *Phascogale virginiae* specimen referred to by Collett (1887a: 548). Species recognised by Troughton (1941: 40), Tate (1947: 120), who noted that it may be a synonym of *rufigenis*, and Troughton (1965: 319; 1967: 36). Synonymised within *rufigenis* by Ride (1970: 247) and within *virginiae* by M. Archer (1981: 132), Honacki *et al.* (1982: 33), Mahoney and Ride (1988a: 33), Flannery (1990: 67; 1995a: 103; 1995b: 63) and Groves (1993b: 37) and within *nitela* by Groves (2005d: 36).

Φ ***Sminthopsis virginiae rufigenis*** Thomas, 1922

Φ *Sminthopsis rufigenis* Thomas, 1922b: 265.

TYPE LOCALITY: Aru Islands, Indonesia.

COMMENTS: Considered a valid species by Tate (1947: 120) and Ride (1970: 122). Recognised as a subspecies of *S. virginiae* by M. Archer (1981: 139) and Strahan (1983:

56; 1995: 157). Synonymised within *virginiae* by Honacki *et al.* (1982: 33), Mahoney and Ride (1988a: 33) and Clayton *et al.* (2006: 102). Subspecies rank recognised by Flannery (1990: 67; 1995a: 103; 1995b: 63), Blacket *et al.* (2001: 161), Groves (2005d: 36), and Van Dyck and Strahan (2008: 159).

Φ *Phascogale rona* Tate and Archbold, 1936: 2.

TYPE LOCALITY: Rona, Laloki River, Central Division of Papua New Guinea.

COMMENTS: Synonymised within *virginiae* by M. Archer (1981: 132), Honacki *et al.* (1982: 33), Groves (1993b: 37), Flannery (1990: 67; 1995a: 103; 1995b: 63) and within *rufigenis* by Groves (2005d: 36).

Sminthopsis youngsoni McKenzie & Archer, 1982

Lesser Hairy-footed Dunnart

Sminthopsis youngsoni McKenzie & Archer, 1982: 267.

TYPE LOCALITY: Edgar Ranges, Northern edge of the Great Sandy Desert, Western Australia, Australia. (18°49′50″S, 123°04′30″E)

COMMENTS: Included in the *S. psammophila* species group.

Incertae Sedis

Sminthopsis murina var. *constricta* Spencer, 1896a: 33.

TYPE LOCALITY: Oodnadatta, South Australia, Australia.

COMMENTS: Considered a subspecies of *murina* by Iredale and Troughton (1934: 11) and a subspecies of *macrorura* by Tate (1947: 122). Type designation by Dixon (1970: 107). Available for homonymy only, see M. Archer (1976c: 127) and ICZN (1981: 274).

Family Myrmecobiidae Waterhouse, 1841

Family Myrmecobiidae Waterhouse, 1841a: 60, 144.

TYPE GENUS: *Myrmecobius* Waterhouse, 1836a.

COMMENTS: When originally proposed, this rank was placed in the Order Marsupialia (Illiger, 1811) and included the genus *Myrmecobius* Waterhouse, 1836a. Recognised as a subfamily within the Family Dasyuridae by Thomas (1888a: xiii, 311), Bensley (1903: 91, 99), Simpson (1945: 44), Ride (1964a: 97, 111) and McKenna and Bell (1997: 56). Family rank recognised by Gill (1872: 26), Osborn (1910: 516), Iredale and Troughton (1934: vii, 15), Troughton (1967: 46), M. Archer and Kirsch (1977: 20), Kirsch (1977a: 112; 1977b: 4, 45), Kirsch and Calaby (1977: 15), Marshall (1981: 27), Szalay (1994: 42), Strahan (1983: xxi, 84; 1995: 6, 159), Kirsch *et al.* (1997: 245) and subsequent authors. Groves (1993b: 29) noted that Waterhouse (1838a) is often referred to as the citation for this family, but Myrmecobiidae is not used in this catalogue. This error appears to have arisen as Waterhouse (1841a: 60) stated that the arrangement adopted in that work was originally published in 1838a (Waterhouse), but this is not the case.

HOMONYMS:

Subfamily Myrmecobiinae de Mello-Leitão, 1923: 523=3, corinnid sac spiders of the Class Arachnida (Order Araneae, Family Corinnidae) is a junior homonym of the Subfamily Attacobiinae. See Platnick and Baptista (1995: 2).

Family Ambulatoria Owen, 1839a: 9, 19.

TYPE GENUS: Not based on a genus group name.

COMMENTS: When originally proposed, this rank was placed in the Tribe Entomophaga (Owen, 1839a [=Marsupialia (Illiger, 1811 part)]) and included the genus *Myrmecobius* Waterhouse, 1836a. Recognised by Owen (1840: 318, 332). Synonymised within Myrmecobiidae by Gill (1872: 26) and within Myrmecobiidae by Marshall *et al.* (1990: 489) and within the Subfamily Myrmecobiinae by McKenna and Bell (1997: 56).

Family Myrmecobineae Lesson, 1842: 191.

TYPE GENUS: *Myrmecobius* Waterhouse, 1836a.

COMMENTS: When originally proposed, this rank was placed in the Order Mastodidelphie (Lesson, 1842 [=Marsupialia (Illiger, 1811)]) and included the genus *Myrmecobius* Waterhouse, 1836a. Does not appear to have been recognised by other authors.

Tribe Myrmecobina Bonaparte, 1845: 6.

TYPE GENUS: *Myrmecobius* Waterhouse, 1836a.

COMMENTS: When originally proposed, this rank was placed within the Family Dasyuridae (Goldfuss, 1820a).

Family Myrmécobidés Gervais, 1855a: 284.

TYPE GENUS: *Myrmecobius* Waterhouse, 1836a.

COMMENTS: When originally proposed, this rank was placed in the Order Marsupiaux (Gervais, 1855a [=Marsupialia (Illiger, 1811)]) and included the genus *Myrmecobius* Waterhouse, 1836a.

Family Myrmecobiidae Gill, 1872: 26.

TYPE GENUS: *Myrmecobius* Waterhouse, 1836a.

COMMENTS: When originally proposed, this rank was placed in the Suborder Dasyuromorphia (Gill, 1872). Recognised as the author, at subfamily rank, by Kirsch (1968a: 420). Synonymised within the Subfamily Myrmecobiinae by McKenna and Bell (1997: 56).

Order Mirmecobia Ameghino, 1889: xxi, 347, 348.

COMMENTS: When originally proposed, this rank was placed in the Grand Group Archaeoidea (Ameghino, 1889: xxi, 105, 346) and included the genus *Mirmecobius* [sic] [=*Myrmecobius*] Waterhouse, 1836a. Rank also discussed by Ameghino (1916: 586) as Myrmecobia. Synonymised within the Subfamily Myrmecobiinae by McKenna and Bell (1997: 56).

Myrmecobius Waterhouse, 1836

Myrmecobius Waterhouse, 1836a: 69.

TYPE SPECIES: *Myrmecobius fasciatus* Waterhouse, 1836a by monotypy.

COMMENTS: Also described by Waterhouse (1836b: 520; 1838c: 149).

HOMONYMS:

Myrmecobius H. Lucas, 1846: 234, round fungus beetles of the Class Insecta (Order Coleoptera, Family Leiodidae) is a junior homonym. Genus is a synonym of *Ptomaphagus* Hellwig, 1795: 358.

Myrmecobius de Mello-Leitão, 1923: 524=4, corinnid sac spiders of the Class Arachnida (Order Araneae, Family Corinnidae) is a junior homonym. Genus is a synonym of *Attacobius* de Mello-Leitão, 1925: 455. See Platnick and Baptista (1995: 2).

Mirmecobius Ameghino, 1889: 348.

TYPE SPECIES: Incorrect subsequent spelling of *Myrmecobius fasciatus* Waterhouse, 1836a.

COMMENTS: Not subsequently recognised.

Myrmecobius fasciatus Waterhouse, 1836

Numbat

Myrmecobius fasciatus fasciatus Waterhouse, 1836

Myrmecobius fasciatus Waterhouse, 1836a: 69.

TYPE LOCALITY: My Kokeby, S of Beverley, Western Australia, Australia.

COMMENTS: Described in further detail by Waterhouse (1838c: 151). Species recognised by Gould (1845 [1845–1863]: Text to Plate 4). There appears to be significant doubt over the authorship of the subspecies *rufus* as different authors recognise either Wood Jones (1923 [1923–1925]: 123) or Finlayson (1933b: 203). Species reviewed by C. Cooper (2011: 129).

M. [yrmecobius] Diemensis J. Gray, 1843a: 100.

TYPE LOCALITY: Swan River, Western Australia, Australia.

COMMENTS: Synonymised within *fasciatus* by Waterhouse (1846: 396), Thomas (1888a: 313) and Iredale and Troughton (1934: 16). Subspecies rank recognised by Tate (1951a: 5). Not considered by Mahoney and Ride (1988b).

Myrmecobius fasciatus rufus Wood Jones, 1923

Myrmecobius rufus Wood Jones, 1923 [1923–1925]: 123.

TYPE LOCALITY: South Australia, Australia.

COMMENTS: Synonymised within *fasciatus* by Mahoney and Ride (1988b: 34) who note its difference from the name *rufus* described by Finlayson (1933b: 203), and there appears to be significant doubt over the authorship of the subspecies *rufus* as different authors recognise either Wood Jones (1923[1923–1925]: 123) or Finlayson (1933b: 203). Wood Jones (1923:123) described 'a typical South Australian example, which I proposed to describe in a scientific journal under the name *Myrmecobius rufus*', but later Finlayson (1933b: 203) described 'the central animal which he' 'proposed to separate under the name *Myrmecobius fasciatus* var. *rufus*' from 'mulga and sand dunes south and south-west of the Everard Ranges', inferentially admitting that it was the same as that described by Wood Jones, and thanking the latter for loan of specimens. There seems no doubt that Wood Jones's description makes his use of the name available, and it is not even clear that Finlayson was actually intending to describe a new taxon, rather he was probably validating Wood Jones's name and providing it with a type series (he nominated two specimens in the South Australian Museum as 'co-types', *recte* lectotypes). Aitken (1976: 197) found that the localities of the two lectotypes are different: one is from south of the Musgrave and north of the Everard Ranges, the other is Oolarrinna, north of the Everard Ranges – neither is from south or south-west of the Everard, *contra* Finlayson. Species rank recognised by Iredale and Troughton (1934: vii, 16) and Troughton (1967: 48), with subspecies rank recognised by Tate (1951a: 5), Groves (2005d: 23), Clayton *et al.* (2006: 102), who attributed the name to Finlayson (1933b: 203), and C. Cooper (2011: 129) who correctly attributed the subspecies to Wood Jones (1923[1923–1925]: 123). The name *rufus* was synonymised within *fasciatus* by Ride (1970: 128, 245) and recognised as a subspecies by Van Dyck and Strahan (2008: 165).

FUTURE TAXONOMIC RESEARCH: The status of this taxon needs to be confirmed.

Myrmecobius? rufus Mitchell, 1838a: xvii.

TYPE LOCALITY: Australia.

COMMENTS: Species was referred to as the 'red shrew-mouse' by Mitchell (see Waterhouse (1841a: 150) who described it with considerable doubt because he neglected to note any generic characters. Waterhouse (1841a: 149) referred to this description and reiterated hesitation in naming the animal. Doubt also cast by Macleay (1841: 241) who stated that 'the marsupial called by the colonists "the Red Shrew Mouse," and which has been supposed by Sir T.L. Mitchell to be a *Myrmecobius*, proves now to be a new and minute species of *Perameles* ; that is, if I may be allowed to judge from the feet of the two stuffed specimens in the Colonial Museum, the only ones I have seen (the teeth are not visible in either of the specimens).' Because a description of the species was not provided the name was not reconsidered by subsequent authors including Waterhouse (1846: 396) and Thomas (1888a: 312). Name considered a *nomen nudum* by authors including Tate (1951a: 5) and Friend (1989: 583) as well as, inferentially, by Finlayson (1933b: 205).

† **Family Thylacinidae** Bonaparte, 1838

† Family Thylacinidae Bonaparte, 1838: 113.

TYPE GENUS: *Thylacinus* Temminck, 1824.

COMMENTS: When originally proposed, this rank was placed in the Order Marsupialia (Illiger, 1811) and included the Tribe Thylacinina (Bonaparte, 1838). Recognised as the Subfamily Thylacininae (Bensley, 1903: 91, 107), within the Family Dasyuridae, by Simpson (1945: 44). Family rank recognised by Ride (1964a: 105), Kirsch (1968a: 420; 1977a: 112; 1977b: 45), M. Archer and Kirsch (1977: 21), Kirsch and Calaby (1977: 15), Marshall (1981: 27), Strahan (1983: xxi, 81), Mahoney and Ride (1988c: 11) and subsequent authors. The relationship of the Australasian thylacinids and the South American borhyaenids has long been the subject of debate from Bensley (1903: 107, 209) onward. Family Thylacinidae placed in the † Superfamily Borhyaenoidea (Ameghino, 1893: 371) by Kirsch (1977a: 112), and within the Superfamily Didelphoidea (J. Gray, 1821: 308) by Kirsch (1977b: 45). This relationship was examined by Marshall (1977: 410) who suggested the thylacinids have evolved from a common ancestor shared with the Australian dasyuroids, while the borhyaenoids evolved from a common ancestor shared with didelphoids. Though the conclusions of Marshall were criticised by M. Archer (1982b: 445), the borhyaenids were placed in their own superfamily and the thylacinids included within the Family Dasyuridae by Marshall (1981: 9), which has in this broad sense, been subsequently followed. See review by M. Archer (1982b: 449). The phylogeny of *Thylacinus* was assessed by Krajewski *et al.* (1997c: 911) who showed that they are unambiguously members of the Dasyuromorphia.

† Tribe Thylacinina Bonaparte, 1838: 113.

TYPE GENUS: *Thylacinus* Temminck, 1824.

COMMENTS: When originally proposed, this rank was placed in the Family Thylacinidae (Bonaparte, 1838). Not used by subsequent authors.

† Tribe Thylacinini Winge, 1893a: 88, 93.

TYPE GENUS: *Thylacinus* Temminck, 1824.

COMMENTS: When originally proposed, this rank was placed in the Family Dasyuridae (Goldfuss, 1820a) and included the genus *Thylacinus* Temminck, 1824. Tribe rank subsequently recognised by Winge (1941: 69).

† Subfamily Thylacininae Bensley, 1903: 91, 107.

TYPE GENUS: *Thylacinus* Temminck, 1824.

COMMENTS: When originally proposed, this rank was placed in the Family Dasyuridae (Goldfuss, 1820a) and included the genus *Thylacinus* Temminck, 1824. Subfamily rank, within Dasyuridae, recognised by Troughton (1967: 44). Synonymised within Thylacinidae by Marshall (1981: 27), Marshall *et al.* (1990: 489) and McKenna and Bell (1997: 54).

† ***Thylacinus*** Temminck, 1824

† *Thylacinus* Temminck, 1824: 23 (footnote), 60.

TYPE SPECIES: *Didelphis cynocephala* Harris, 1808 [=*Thylacinus cynocephalus* (Harris, 1808)] by monotypy.

COMMENTS: Genus recognised by Waterhouse (1841a: 60, 123), Gould (1851 [1845–1863]: Text to Plates 53–54), Thomas (1888a: xii, 255), Iredale and Troughton (1934: vii, 15) and subsequent authors.

† *Peracyon* J. Gray, 1825a: 340.

TYPE SPECIES: *Nomen nudum*. See Palmer (1904: 512).

COMMENTS: Spelling used by J. Gray (1843a: xxii) but he then uses *Paracyon* on page 97. Synonymised within *Thylacinus* by Waterhouse (1846: 453), Thomas (1888a: 255), and Iredale and Troughton (1934: 15). Not considered by Mahoney and Ride (1988c: 11).

† *Paracyon* Brook[e]s, 1827: 192.

TYPE SPECIES: *Nomen novum* for *Thylacinus* Temminck, 1824.

COMMENTS: Genus recognised by J. Gray (1843a: 97). Synonymised within *Thylacinus* by Iredale and Troughton (1934: 15), Marshall (1981: 27), Mahoney and Ride (1988c: 11), Marshall *et al.* (1990: 489), and McKenna and Bell (1997: 54). Confusion exists over the author of this name as the description states 'Mr. Brooks, it is understood; proposed to make this species a type of a new genus, to be named *Paracyon*. Temminck has since done so, and applied to it the name *Thylacynus*'. Author of this name is given as J. Gray, in Griffiths *et al.* (1827: 192) by Marshall (1981: 27), Marshall *et al.* (1990: 489) and McKenna and Bell (1997: 54); as Griffith *et al.* (*ex* Brookes MS.)(1827: 192) by Iredale and Troughton (1934: 15); and to J. Gray (1843a: 97) by Mahoney and Ride (1988c: 11) and Groves (2005d: 23).

HOMONYMS:

† *Paracyon* Arredondo, 1981: 5, Holocene dog fossils of the Class Mammalia (Order Carnivora, Family Canidae). Synonymised within *Canis* Linnaeus, 1758 by McKenna and Bell (1997: 247).

† *Thylacynus* Temminck, 1824: xxiii; 1827: 267.

TYPE SPECIES: Incorrect subsequent spelling of *Thylacinus* Temminck, 1824.

COMMENTS: The correct spelling *Thylacinus* was used on page 60 that contains the main description of the genus. Not considered by Iredale and Troughton (1934: 15). Synonymised within *Thylacinus* by Marshall (1981: 27),

Mahoney and Ride (1988c: 11), Marshall *et al.* (1990: 489) and McKenna and Bell (1997: 54).

† *Lycaon* Wagler, 1830: 24.

TYPE SPECIES: *Didelphis cynocephalus* Harris, 1808 (as *Didelphys cynocephala*) [=*Thylacinus cynocephalus* (Harris, 1808)] by original designation.

COMMENTS: Synonymised within *Thylacinus* by Thomas (1888a: 255), Iredale and Troughton (1934: 15), and Mahoney and Ride (1988c: 11).

HOMONYMS:

Lycaon J. Gray, 1825a: 339, the African Wild Dog of the Class Mammalia (Order Carnivora, Family Canidae). Name attributed to Brook[e]s, 1827: 151. Currently used name. See Wozencraft (2005: 581).

Lycaon de Bonvouloir, 1870: Suppl. 718, beetles of the Class Insecta (Order Coleoptera, Family Eucnemidae). Genus is a junior synonym of *Hemiopsida* Macleay, 1872: 261.

† *Peralōpex* Gloger, 1841: xxx, 82.

TYPE SPECIES: *Nomen novum* for *Thylacinus* Temminck, 1824.

COMMENTS: Synonymised within *Thylacinus* by Thomas (1888a: 255; 1895a: 190), Iredale and Troughton (1934: 15), Marshall (1981: 27), Mahoney and Ride (1988c: 11), Marshall *et al.* (1990: 489) and McKenna and Bell (1997: 54).

† ***Thylacinus cynocephalus*** (Harris, 1808)

Thylacine

† *Didelphis cynocephala* Harris, 1808: 174; Plate 19, Fig. 1.

TYPE LOCALITY: Neighbourhood of the highest mountainous pars of Van Diemen's Land (Tasmania, Australia).

COMMENTS: Transferred to *Thylacinus* by J. Fischer (1829: 270), Gould (1851 [1845–1863]: Text to Plates 53–54) and followed by subsequent authors. See comments under the Family Thylacinidae.

† *Thylacinus Harrisii* Temminck, 1824: 23–24 (footnote), 63.

TYPE LOCALITY: Tasmania, Australia.

COMMENTS: Synonymised within *cynocephalus* by Waterhouse (1841a: 123; 1846: 456), Thomas (1888a: 256), Iredale and Troughton (1934: 15) and Mahoney and Ride (1988c: 12).

† *Dasyurus Lucocephalus* Grant, 1831: 177.

TYPE LOCALITY: Tasmania, Australia.

COMMENTS: Synonymised within *cynocephalus* by Thomas (1888a: 256), Iredale and Troughton (1934: 15) and Mahoney and Ride (1988c: 12).

† *Thylacinus striatus* Warlow, 1833: 97.

TYPE LOCALITY: Tasmania, Australia.

COMMENTS: *Nomen novum* for *Didelphis cynocephala* Harris, 1808, as *Didelphis cynocephalus*. Synonymised within *cynocephalus* by Thomas (1888a: 256), Iredale and Troughton (1934: 15) and Mahoney and Ride (1988c: 12).

† *Thylacinus communis* Anon, 1859: 147.

TYPE LOCALITY: Name attributed to Temminck, but the reference is unknown. Name may be a *nomen novum* for *Thylacinus harrisii* Temminck, 1824, or *Didelphis cynocephala* Harris, 1808 according to Mahoney and Ride (1988c: 12).

COMMENTS: Not considered by Iredale and Troughton (1934: 15) and synonymised *cynocephalus* by Mahoney and Ride (1988c: 12).

† *T. [hylacinus] breviceps* Krefft, 1868c: 296; Plate 17.

TYPE LOCALITY: Ouse River, Tasmania, Australia.

COMMENTS: Synonymised within *cynocephalus* by Thomas (1888a: 256), Iredale and Troughton (1934: 15) and Mahoney and Ride (1988c: 12).

† *Thylacinus major* Owen, 1877a: 106, 107.

TYPE LOCALITY: Wellington Caves, Wellington, New South Wales, Australia.

COMMENTS: Synonymised within *cynocephalus* by Dawson (1982b: 527, 534).

† *Thylacinus rostralis* De Vis, 1893: v.

TYPE LOCALITY: Ellangowan, near Cambooya, Darling Downs, Queensland, Australia (see De Vis, 1894: 444).

COMMENTS: Taxon described in greater detail by De Vis (1894: 446). Synonymised within *cynocephalus* by Dawson (1982b: 527, 534).

† *Thylacinus spelaeus* Owen, 1845a: 335.

TYPE LOCALITY: 'Wellington Valley', New South Wales, Australia.

COMMENTS: Synonymised within *cynocephalus* by Lydekker (1887: xxiii, 264), Ride (1964a: 105) and Dawson (1982b: 527, 533).

Order Notoryctemorphia Kirsch, 1977 *sensu* Aplin & Archer, 1987

Suborder Notoryctemorphia Kirsch, 1977b: 45.

COMMENTS: When originally proposed, this rank was placed in the Order Polyprotodonta (Thomas, 1896a [=Marsupialia (Illiger, 1811 part)]) and included the Superfamily Notoryctoidea (Osborn, 1910 [=Notoryctidae (J. Ogilby, 1892)]) with the Family Notoryctidae (J. Ogilby,

1892). Suborder rank also recognised by Kirsch (1977a: 112), Marshall (1981: 17) and followed by Strahan (1983: xxi). Reduced to a family within the Semiorder Peramelina by Szalay (1994: 42), but elevated to ordinal rank by Aplin and Archer (1987: xxi, xli), which was followed by Marshall *et al.* (1990: 459), Westerman (1991: 529), Strahan (1995: 8, 409), Kirsch *et al.* (1997: 245), McKenna and Bell (1997: 54) and Groves (2005a: 22). The placement of *Notoryctes* has been controversial since its description; it has typically been placed next to the Dasyuridae and Peramelidae, although it was placed adjacent to the Diprotodontia by Strahan (1995: 8, 409), and Van Dyck and Strahan (2008: 10, 409). The Family Notoryctidae, and hence the Order Notoryctemorphia, was placed next to Dasyuromorphia by Kirsch *et al.* (1997: 245), Cardillo *et al.* (2004: 16) and Beck (2008: 179). Nilsson *et al.* (2004: 191) placed *Notoryctes* adjacent to members of the Dasyuridae, while Thylacinidae was not included in the analysis. The exact position of the marsupial moles has remained difficult.

Order Zalambdadonta [sic] Turnbull, 1971: 175, 176.

COMMENTS: When originally proposed, this rank was placed in the Cohort Tribosphenata *nova* and included *Notoryctes* Stirling, 1891a; and the eutherian zalambdodonts from Gill (1883: 119) in the sense of Vandebroek (1961: 308). The highly unusual zalambdodont dentition of the marsupial moles led Turnbull (1971: 175) to propose that *Notoryctes*, † *Necrolestes* Ameghino, 1891: 303 and several species of eutherians belong to the Zalambdodonta, though this has not been accepted as it does not acknowledge the obvious marsupial relationship of *Notoryctes* (K. Johnson & Walton, 1989: 600; Long *et al.* (2002: 66).

HOMONYMS:

Zalambdodonta Gill, 1883: 119, of the Class Mammalia. Order is a synonym of the Superfamily Tenrecoidea J. Gray, 1821: 301, according to McKenna and Bell (1989: 293).

Suborder Syndactyliformes Szalay, 1982: 631.

COMMENTS: When originally proposed as a new rank it was placed in the Order Syndactyla (Wood Jones, 1923[1923–1925] [=Marsupialia (Illiger, 1811 part)]) and included an unknown superfamily and the Superfamily Notoryctoidea (Osborn, 1910 [=Notoryctidae (J. Ogilby, 1892)]). Does not appear to have been used by subsequent authors. Synonymised within Notoryctemorphia by McKenna and Bell (1997: 54).

Order Notoryctiformes Kinman, 1994: 37.

COMMENTS: When originally proposed, this rank was placed in the Class Mammalea (Kinman, 1994 [=Mammalia (Linnaeus, 1758)]). Synonymised within Notoryctemorphia by McKenna and Bell (1997: 54).

Family Notoryctidae J. Ogilby, 1892

Family Notoryctidae J. Ogilby, 1892: 5.

TYPE GENUS: *Notoryctes* Stirling, 1891a.

COMMENTS: When originally proposed, this rank was placed in the Suborder Polyprotodontia (Owen, 1877a [=Marsupialia (Illiger, 1811 part)]) and included the genus Notoryctes Stirling, 1891a. Family rank recognised by Gadow (1892: 370), Bensley (1903: 117), Iredale and Troughton (1934: vii, 16) and most subsequent authors.

Tribe Notoryctini Winge, 1893a: 88, 93.

TYPE GENUS: *Notoryctes* Stirling, 1891a.

COMMENTS: When originally proposed, this rank was placed in the Family Dasyuridae (Goldfuss, 1820a) and included the genus *Notoryctes* Stirling, 1891a. Tribe rank subsequently recognised by Winge (1941: 69).

[Family] Pronotoryctidae Gregory, 1910: 204.

TYPE GENUS: Not based on a genus group name.

COMMENTS: Name was written as 'Pronotoryctids'. When originally proposed this rank was placed in the Superfamily Notoryctoidea (Osborn, 1910 [=Notoryctidae (J. Ogilby, 1892)]) and was proposed to have evolved from the Properamelidae (Bensley, 1903 [=Marsupialia (Illiger, 1811)]) and gave rise to the Family Notoryctidae (J. Ogilby, 1892). Discussed by Marshall (1981: 18).

Superfamily Notoryctoidea Osborn, 1910: 516.

TYPE GENUS: *Notoryctes* Stirling, 1891a.

COMMENTS: When originally proposed, this rank was placed in the Suborder Polyprotodontia (Owen, 1877a [=Marsupialia (Illiger, 1811 part)]) and included the Family Notoryctidae (J. Ogilby, 1892). Superfamily rank recognised, and typically attributed to Ogilby (1892: 5), by Gregory (1910: 204), Calaby *et al.* (1974: 532), Kirsch (1977a: 112; 1977b: 45), Szalay (1982: 631), Strahan (1983: xxi) and Marshall *et al.* (1990: 459), but not by subsequent authors including Strahan (1995: 8, 409), McKenna and Bell (1997: 54) and Groves (2005a: 22).

Notoryctes Stirling, 1891

Notoryctes Stirling, 1891a: 154, Plates 2–9.

TYPE SPECIES: *Nomen novum* for *Psammoryctes* Stirling, 1889a.

COMMENTS: Taxonomic decision of Stirling (1891a: 154, 186) to synonymise *Psammoryctes* within *Notoryctes*. Further described by Stirling (1891b: 283). Subsequently recognised by Trouessart (1891: 290), Lydekker (1895: 276) and subsequent authors.

Psammoryctes Stirling, 1889a: 158.

TYPE SPECIES: *Psammoryctes typhlops* Stirling, 1889a [=*Notoryctes typhlops* (Stirling, 1889a)] by original designation.

COMMENTS: Synonymised within *Notoryctes* by Stirling (1891a: 154, 186), Palmer (1904: 464), Iredale and Troughton (1934: 16), Walton (1988b: 46), Groves (1993c: 43), Marshall *et al.* (1990: 490) and subsequent authors.

HOMONYMS:

Psammoryctes Poeppig, 1835: 252, coruro of the Class Mammalia (Order Rodentia, Family Octodontidae). Genus is a synonym of *Spalacopus* Wagler, 1832a: 1219. See Woods and Kilpatrick (2005: 1573).

Psammoryctes Vejdovsky, 1876a: 194 and 1876b: 137, oligochaete worms of the Phylum Annelida (Class Clitellata, Order Haplotaxida, Family Tubificidae). Genus is a synonym of *Psammoryctides* Hrabe, 1964: 107.

Neoryctes Stirling, 1891a: 186.

TYPE SPECIES: *Nomen novum* for *Psammoryctes* Stirling, 1889a.

COMMENTS: Stirling suggested the name was 'previously proposed by Dr. Sclater.' Synonymised within *Notoryctes* by Iredale and Troughton (1934: 16), Walton (1988b: 46), Groves (1993c: 43), Marshall *et al.* (1990: 490) and subsequent authors.

Notoryctes caurinus Thomas, 1920

Northern Marsupial Mole

Notoryctes caurinus Thomas, 1920a: 111.

TYPE LOCALITY: Wollal [=Wallal, Western Australia], Ninety Mile Beach, north-west Australia.

COMMENTS: Synonymised within *typhlops* by Ride (1970: 245), Strahan (1983: 88) and Strahan (1995: 410) who suggested that it warranted separate treatment. Recognised at species rank by Iredale and Troughton (1934: vii, 16), Troughton (1967: 50), Walton (1988b: 47) and Maxwell *et al.* (1996: 3). Since 1996 this species has received widespread support based on, in part, unpublished morphological studies by Ken Aplin, formerly of the Western Australian Museum, and Penny van Oosterzee at Alice Springs (Benshemesh, 2004: 7). Recognised as a distinct species by Van Dyck and Strahan (2008: 410). The relationships and distributions of the two species remain to be worked out.

Notoryctes typhlops (Stirling, 1889)

Southern Marsupial Mole

Psammorytes typhlops Stirling, 1889a: 158; Plates 2–5.

TYPE LOCALITY: Indracowie Station, 100 miles from Charlotte Waters Telegraph Station, central Australia.

COMMENTS: Preliminary notes of this species, without giving it a name, were described by Stirling (1888: 588; 1889b: 21). A more detailed description was outlined by Stirling (1891a: 154; Stirling, 1894: 1).

Order Peramelemorphia Ameghino, 1889 *sensu* Aplin & Archer, 1987

Order Peramelia Ameghino, 1889: xx, 263, 266.

COMMENTS: When originally proposed, this rank was placed in the Grand Group Alloidea (Ameghino, 1889, xx, 105, 263) and included the genus *Perameles* É. Geoffroy, 1803d. Recognised at the suborder rank Peramelemorphia by Kirsch (1968a: 420), Strahan (1983: xxi) and at order rank by Aplin and Archer (1987: xxi, xxxix). Order Peramelemorphia synonymised within Peramelia by McKenna and Bell (1997: 56), but recognised by Strahan (1995: 6, 166), Wroe (1999: 512), Groves (2005e: 38), Van Dyck and Strahan (2008: 9, 169), Meredith *et al.* (2008b: 1, 4), and Menkhorst and Knight (2011: 16). The intergeneric relationships were explored by Kirsch *et al.* (1997: 229, 233, 237). Two superfamilies are recognised: Superfamily Perameloidea (J. Gray, 1825) and † Superfamily Yaraloidea (Muirhead, 2000: 512).

Tribe Peramelina J. Gray, 1825a: 340.

COMMENTS: When originally proposed, this rank was placed in the Family Didelphidae (J. Gray, 1821: 308) and included the genera *Perameles* É. Geoffroy, 1803d; and *Isodon* [sic=*Isoodon*] Desmarest, 1817a. Recognised within the Family Dasyuridae by Bonaparte (1845: 6). Synonymised within the Family Peramelidae by Simpson (1945: 44) and within the Marsupialia by Honacki *et al.* (1982: 18). Recognised at infraorder rank by M. Archer (1984: 786), semiorder rank by Szalay (1994: 42) and ordinal rank name by Ride (1964a: 97, 112), Turnbull (1971: 176), M. Archer (1984: 787), Marshall *et al.* (1990: 459) and Kirsch *et al.* (1997: 245, 261). Unlike other authors that included only bandicoots within this rank, the Semisuborder Peramelina used by Szalay (1994: 42) included the Family Notoryctidae as well as the Family Peramelidae.

Family Peramelisideae Lesson, 1842: 191.

TYPE GENUS: *Perameles* É. Geoffroy, 1803d.

COMMENTS: When originally proposed, this rank was placed in the Order Mastodidelphie (Lesson, 1842 [=Marsupialia (Illiger, 1811)]) and included the genera *Chaeropus* W. Ogilby, 1838a, *Perameles* É. Geoffroy, 1803d; and *Echymipera* Lesson, 1842. Does not appear to have been recognised by other authors.

Section Syndactyla Polyprotodontia Wood Jones, 1924 [1923–1925]: 136.

COMMENTS: When originally proposed, this rank was placed in the Suborder Syndactyla (Wood Jones, 1923 [1923–1925] [=Marsupialia (Illiger, 1811 part)]) and included the Family Peramelidae (J. Gray, 1825a), which included all bandicoots. Recognised at ordinal rank by Szalay (1982: 631; 1994: 42). Synonymised within Diprotodontia by Aplin and Archer (1987: xliii) and other authors. Synonymised within Syndactyli (Gill, 1872: 25)[=Gill, 1871a: 533] by McKenna and Bell (1997: 56).

Order Peramelina Ride, 1964a: 99.

COMMENTS: When originally proposed, this rank was placed in the Superorder Marsupialia (Illiger, 1811) and included the Family Peramelidae (J. Gray, 1825a) that contains all the bandicoots. Authorship of rank attributed to J. Gray, 1825a by Ride (1964a: 99) and Kirsch *et al.* (1997: 245). Used as the ordinal name rank by Kirsch *et al.* (1997: 245) and McKenna and Bell (1997: 56).

Suborder Peramelemorphia Kirsch, 1968a: 420.

COMMENTS: When originally proposed as a new rank it was placed in the Suborder Polyprotodonta (given as Owen, 1866) [=Polyprotodontia (Owen, 1877a [=Marsupialia (Illiger, 1811 part)]) and included the Superfamily Perameloidea (Waterhouse, 1838a *fide* Waterhouse, 1841a [= Perameloidea (J. Gray, 1825)]) that contains all the bandicoots. Synonymised within Peramelina by McKenna and Bell (1997: 56). Recognised at ordinal rank by Long *et al.* (2002: 68) and subordinal rank by Kirsch (1977a: 112; 1977b: 45) and Aplin and Archer (1987: xxi).

Suborder Perameliformes Szalay, 1982: 631.

COMMENTS: When originally proposed as a new rank it was placed in the Order Syndactyla (Wood Jones, 1923 [1923–1925] [=Marsupialia (Illiger, 1811 part)]) with no lower ranks specified. Synonymised within Peramelina by McKenna and Bell (1997: 56) and within Peramelemorphia by Groves (2005e: 38).

Suborder Peramelomorphia Strahan, 1983: xix, xxi.

COMMENTS: Incorrect subsequent spelling of Peramelemorphia. When originally proposed, this rank was placed in the Order Polyprotodonta (Thomas, 1896a [=Marsupialia (Illiger, 1811 part)]) and included the Superfamily Perameloidea (J. Gray, 1825) containing the families Peramelidae (J. Gray, 1825a) and Thylacomyidae (Bensley, 1903).

Order Perameliformes Kinman, 1994: 37.

COMMENTS: When originally proposed, this rank was placed in the Class Mammalea (Kinman, 1994 [=Mammalia (Linnaeus, 1758)]). Synonymised within Peramelina by McKenna and Bell (1997: 56).

Cohort Perametatheria Kirsch *et al.*, 1997: 245, 261.

COMMENTS: When originally proposed as a new rank it was placed in the Supercohort Notometatheria (Kirsch *et al.*, 1997 [=Marsupialia (Illiger, 1811 part)]) and included the Order Peramelida (Haeckel, 1866 [=Peramelidae (J. Gray, 1825)]) that contains all bandicoots in the families Peramelidae (J. Gray, 1825a) and Thylacomyidae (Bensley, 1903).

Superfamily Perameloidea J. Gray, 1825 *sensu* Van Dyck & Strahan, 2008

Tribe Peramelina J. Gray, 1825a: 340.

TYPE GENUS: *Perameles* É. Geoffroy, 1803d.

COMMENTS: When originally proposed, this rank was placed in the Family Didelphidae (J. Gray, 1821: 308) and included the genera *Perameles* É. Geoffroy, 1803d; and *Isodon* [sic=*Isoodon*] Desmarest, 1817a. The author of the superfamily (and family rank) has been unstable and been attributed to 'Waterhouse, 1838' by authors including Kirsch (1968a: 420; 1977a: 112) and Long *et al.* (2002: 69), but this appears to be incorrect. The confusion appears to be because Waterhouse (1841a: 60) wrote that the arrangement adopted in that work follows Waterhouse (1838b), even though it is not actually listed in this earlier work. Rank was attributed to Osborn (1910: 516) by Simpson (1945: 44) and McKenna and Bell (1997: 56), but to J. Gray (1825a: 340) by Marshall (1981: 18). Recognised as a superfamily by Simpson (1945: 44; 1970: 38), Kirsch (1977a: 112; 1977b: 45), but synonymised within the Order Peramelemorphia by Groves (1993d: 39; 2005e: 38). Groves and Flannery (1990: 1) proposed there were only two families, the Peroryctidae and Peramelidae. Rank not recognised by Groves (2005e: 39). Superfamily rank recognised by Marshall *et al.* (1990: 459), Strahan (1995: 6, 167), and Van Dyck and Strahan (2008: 9, 169), who recognised the families Chaeropodidae, Peramelidae and Thylacomyidae.

Family Saltatoria Owen, 1839a: 10, 19.

TYPE GENUS: Not based on a genus group name.

COMMENTS: When originally proposed, this rank was placed in the Tribe Entomophaga (Owen, 1839a [=Marsupialia (Illiger, 1811 part)]) and included the genera *Chaeropus* W. Ogilby, 1838a; and *Perameles* É. Geoffroy, 1803d. Also used by Owen (1840: 319, 332). Name does not appear to have been previously recognised.

Family Pérамélidés Gervais, 1855a: 278.

TYPE GENUS: *Perameles* É. Geoffroy, 1803d.

COMMENTS: When originally proposed, this rank was placed in the Order Marsupiaux (Gervais, 1855a [=Marsupialia (Illiger, 1811)]) and included the genera *Chaeropus* W. Ogilby, 1838a; *Peragalea* [=*Paragalea*] J. Gray, 1843a; and *Perameles* É. Geoffroy, 1803d.

Family Syndactylina Wagner, 1855: xiii, 209.

TYPE GENUS: Not based on a genus group name.

COMMENTS: When originally proposed, this rank included the genera *Perameles* É. Geoffroy, 1803d and *Choeropus* J.

Gray, 1838a. Synonymised within Peramelidae by Marshall (1981: 27), Marshall *et al.* (1990: 489) and McKenna and Bell (1997: 56).

Syndactylina Haeckel, 1866: clvii.

TYPE GENUS: Not based on a genus group name.

COMMENTS: Rank not listed but placed in the Family Peramelida (Haeckel, 1866) above the genera *Perameles* É. Geoffroy, 1803d; and *Choeropus* J. Gray, 1838a [=*Chaeropus* W. Ogilby, 1838a].

Superfamily Perameloidea Osborn, 1910: 516.

TYPE GENUS: *Perameles* É. Geoffroy, 1803d.

COMMENTS: When originally proposed, this rank was placed in the Suborder Polyprotodontia (Owen, 1877a [=Marsupialia (Illiger, 1811 part)]) and included the Family Peramelidae (J. Gray, 1825a) that contains the genera *Perameles* É. Geoffroy, 1803d; *Chaeropus* W. Ogilby, 1838a; and *Thylacomys* Blyth, 1840. Name recognised by Simpson (1930: 9; 1931: 262; 1945: 44), Strahan (1983: xxi), Baverstock *et al.* (1990b: 14), Kirsch *et al.* (1990b: 434) and Westerman *et al.* (1999: 271). Synonymised within the Family Peramelidae by McKenna and Bell (1997: 56) and within Peramelemorphia by Groves (1993d: 39; 2005e: 38).

Family Chaeropodidae Gill, 1872 *sensu* Groves, 2005

Subfamily Choeropodinae Gill, 1872: 26.

TYPE GENUS: *Chaeropus* W. Ogilby, 1838a.

COMMENTS: When originally proposed, this rank was placed in the Family Peramelidae (J. Gray, 1825a). Incorrect original spelling, which was subsequently corrected in line with the spelling of the type genus. The single genus and species *Chaeropus ecaudatus* has typically been included within either the Family Peramelidae or with *Macrotis* within the Family Thylacomyidae. Name not recognised at any rank by Strahan (1983: 104; 2005: 170) or considered by most authors until it was synonymised within Peramelidae by Marshall (1981: 27) and Marshall *et al.* (1990: 489). Recognised as a subfamily within Peramelidae by McKenna and Bell (1997: 57), but considered a sister-group of all other Peramelemorphia by Westerman *et al.* (1999: 277; 2001: 98), though they did not give it a rank. Elevated to the family rank by Groves (2005e: 38), which was followed by Meredith *et al.* (2008b: 1, 4), and Van Dyck and Strahan (2008: 9, 172), but not Menkhorst and Knight (2011: 16).

Tribe Chaeropini Szalay, 1994: 42.

TYPE GENUS: *Chaeropus* W. Ogilby, 1838a.

COMMENTS: When originally proposed, this rank was attributed to Gill (1872: 26) and placed in the Subfamily Peramelinae (J. Gray, 1825a). Synonymised within the Family Chaeropodidae by Groves (2005e: 38).

† ***Chaeropus*** W. Ogilby, 1838

† *Chaeropus* W. Ogilby, 1838a: 26.

TYPE SPECIES: *Perameles ecaudatus* W. Ogilby, 1838a [=*Chaeropus ecaudatus* (W. Ogilby, 1838a)] by monotypy.

COMMENTS: Genus recognised by Mitchell (1838b: 132, Plate 27), Thomas (1888a: xii, 250), who spelt it *Chœropus*, Palmer (1904: 884) and subsequent authors.

† *Choeropus* J. Gray, 1838a: xvii.

TYPE SPECIES: Incorrect subsequent spelling of *Chaeropus* W. Ogilby, 1838a. Name correctly spelt in the second volume (see Mitchell, 1838b: 132, Plate 27) when it discusses the description by W. Ogilby (1838a: 26).

COMMENTS: Name also recognised by Waterhouse (1841a: 163). Name attributed to J. Gray by authors including Palmer (1904: 177), Simpson (1945: 44), Marshall (1981: 27), Marshall *et al.* (1990: 489), and McKenna and Bell (1997: 57); and to Waterhouse (1841a: 163) by Ride (1988b: 36) and Groves (2005e: 38). Synonymised within *Chaeropus* by Palmer (1904: 177), Simpson (1945: 44), Marshall (1981: 27), Ride (1988b: 36), Marshall *et al.* (1990: 489), and McKenna and Bell (1997: 57) and subsequent authors.

† ***Chaeropus ecaudatus*** (W. Ogilby, 1838)

Pig-footed Bandicoot

† *Per. [ameles] ecaudatus* W. Ogilby, 1838a: 25.

TYPE LOCALITY: Murray River, South of junction with Murrimbigee River between Mitchell's camp of Passage and the 2nd camp (as 'banks of the River Murray … in the interior of New South Wales'), Victoria, Australia. See also Mitchell (1838b: 132, Plate 27).

COMMENTS: Placed within *Choeropus* by Waterhouse (1841a: 163). Reduced to a synonym of *Choeropus castanotis* by Waterhouse (1846: 391) and Thomas (1888a: 250). Recognised as the senior synonym by Iredale and Troughton (1934: 21) and followed by subsequent authors. The large difference of *Chaeropus*, and *Macrotis*, to all other bandicoots was noted by Tate (1948a: 320), though he did not provide them with any formal higher taxonomic status. Considered a sister group to all other living bandicoots by Westerman *et al.* (1999: 277; 2001: 98). Taxon included within the Tribe Chaeropini by Szalay (1994: 42) and within the Subfamily Peramelinae by Strahan (1995: 170). Placed in its own family by Groves (2005e: 39), Meredith *et al.* (2008b: 1, 4), and Van Dyck and Strahan (2008: 9, 172).

† *Chaeropus castanotis* J. Gray, 1842d: 42.

TYPE LOCALITY: Murray River, South Australia, Australia.

COMMENTS: Species recognised by Gould (1845 [1845–1863]: Text to Plate 6), Waterhouse (1846: 391), Thomas (1888a: 251) and Lydekker (1894a: 146). Synonymised within *ecaudatus* by Iredale and Troughton (1934: 21), Tate

(1948a: 341), Ride (1970: 242), Mahoney and Ride (1988d: 36), Groves (1993d: 39; 2005e: 39) and subsequent authors.

† *[Chaeropus] occidentalis* Gould, 1845 [1845–1863]: Text to Plate 6.

TYPE LOCALITY: Walyormourning near Goomallig, Western Australia, Australia (as Walyemara district ~45 miles NE of Townsite of Northam, Western Australia. White Gum forest).

COMMENTS: Name recognised by Krefft (1866a: 12), but synonymised within *castanotis* by Thomas (1888a: 251) and Lydekker (1894a: 146). Subspecies rank recognised by Iredale and Troughton (1934: 21) and species status recognised by Tate (1948a: 341). Synonymised within *ecaudatus* by Mahoney and Ride (1988d: 36), Groves (1993d: 39; 2005e: 39).

Family Peramelidae J. Gray, 1825 *sensu* Van Dyck & Strahan, 2008

Tribe Peramelina J. Gray, 1825a: 340.

TYPE GENUS: *Perameles* É. Geoffroy, 1803d.

COMMENTS: When originally proposed, this rank was placed in the Family Didelphidae (J. Gray, 1821: 308) and included the genera *Perameles* É. Geoffroy, 1803d; and *Isodon* [sic=*Isoodon*] Desmarest, 1817a. Family rank recognised by Waterhouse (1841a: 150; 1846: 11, 354), Krefft (1866a: 6), Thomas (1888a: xii, 219) and subsequent authors. Simpson (1945: 44) and Kirsch (1968a: 420) proposed Waterhouse (1838a) as the author of the family but this does not appear to be correct. Kirsch *et al.* (1997: 245), McKenna and Bell (1997: 56) and Long *et al.* (2002: 69) recognise J. Gray (1825a) as the author. Current arrangement recognised by Van Dyck and Strahan (2008: 9, 174) and Meredith *et al.* (2008b: 4).

Family Peramelida Haeckel, 1866: clvii.

TYPE GENUS: *Perameles* É. Geoffroy, 1803d.

COMMENTS: When originally proposed, this rank was placed in the Order Cantharophaga (Haeckel, 1866 [=Marsupialia (Illiger, 1811 part)]) and included the genera *Perameles* É. Geoffroy, 1803d; and *Choerupus* [sic=*Choeropus*] J. Gray, 1838a [=*Chaeropus* W. Ogilby, 1838a]. Synonymised within Peramelidae by McKenna and Bell (1997: 56).

Superfamily Perameloidea Osborn, 1910: 516.

TYPE GENUS: *Perameles* É. Geoffroy, 1803d.

COMMENTS: When originally proposed, this rank was placed in the Suborder Polyprotodontia (Owen, 1877a [=Marsupialia (Illiger, 1811 part)]) and included the Family Peramelidae (J. Gray, 1825a) containing the genera *Perameles* É. Geoffroy, 1803d; *Chaeropus* W. Ogilby, 1838a; and *Thylacomys* Blyth, 1840. Superfamily name recognised by Simpson (1945: 44). Synonymised within the Family Peramelidae by McKenna and Bell (1997: 56) and Groves (2005e: 38).

Family Peroryctidae Groves & Flannery, 1990: 1.

TYPE GENUS: *Peroryctes* Thomas, 1906c: 476.

COMMENTS: Rank above the family name was not stated but included the genera *Peroryctes* Thomas, 1906c, 476; and *Microperoryctes* Stein, 1932: 256. Recognised at the Family rank by Strahan (1995: 7, 191) and subfamily rank by Kirsch *et al.* (1997: 245) and Groves (2005e: 40) who only included extralimital *Peroryctes* Thomas, 1906c: 476. Genetic studies by Westerman *et al.* (1999: 275; 2001: 93) found no support for the Family Peroryctidae. They suggested that a major morphological reassessment of New Guinean bandicoot relationships was needed. As a result the rank was synonymised within Peramelidae by Groves (2005e: 39) and Helgen (2007a: 713). Author attributed to M. Archer *et al.* (1989: 32) by McKenna and Bell (1997: 57), but Groves (2005e: 40) noted that is a *nomem nudum*, referring to the manuscript of Groves and Flannery (1990: 1).

Subfamily Echymiperinae McKenna & Bell, 1997 *sensu* Van Dyck and Strahan (2008)

Subfamily Echymiperinae McKenna & Bell, 1997: 57.

TYPE GENUS: *Echymipera* Lesson, 1842.

COMMENTS: When originally proposed, this rank was placed in the Family Peroryctidae (Groves & Flannery, 1990) and included the genera *Echymipera* Lesson, 1842; *Rhynchomeles* Thomas, 1920b: 429; and *Microperoryctes* Stein, 1932: 256. Subfamily recognised by Groves (2005e: 41), Van Dyck and Strahan (2008: 9, 174) and Meredith *et al.* (2008b: 1, 4).

Echymipera Lesson, 1842

Echymipera Lesson, 1842: 192.

TYPE SPECIES: Φ *Perameles kalubu* J. Fischer, 1829: 274 (as *Echymipera Kalubu*) [= Φ *Echymipera kalubu* (J. Fischer, 1829: 274)] by monotypy.

COMMENTS: Genus recognised by Thomas (1920b: 430) and most subsequent authors. Reviewed by Tate (1948a: 329), and Groves and Flannery (1990: 1).

Brachymelis de Miklouho-Maclay, 1884: 713.

TYPE SPECIES: Φ [*Perameles*] *Brachymelis garagassi* de Miklouho-Maclay, 1884: 715 [= Φ *Echymipera kalubu* (J. Fischer, 1829: 274)] by monotypy.

COMMENTS: Described as a subgenus of *Perameles*. Synonymised within *Echymipera* by Tate (1948a: 329), Marshall (1981: 27), Marshall *et al.* (1990: 489), and McKenna and Bell (1997: 57).

HOMONYMS:
Brachymeles Duméril & Bibron, 1839: 776, lizards of the Class Reptilia (Order Squamata: Family Scincidae).

Peramelopsis Heude, 1896: 143, footnote.

TYPE SPECIES: *Peramelopsis welsianus* Heude, 1896 [=*Echymipera rufescens* (Peters & Doria, 1875)] by monotypy.
COMMENTS: Synonymised within *Echymipera* by Groves (2005e: 41).

Anuromeles K. Heller, 1897: 5.

TYPE SPECIES: Φ *Anuromeles rufiventris* K. Heller, 1897: 5 [=Φ *Echymipera kalubu* J. Fischer, 1829: 274] by original designation.
COMMENTS: Synonymised within *Echymipera* by Tate (1948a: 329), Marshall (1981: 27), Marshall *et al.* (1990: 489) and McKenna and Bell (1997: 57).

Suillomeles G. Allen & Barbour, 1909: 43.

TYPE SPECIES: Φ *Suillomeles hispida* G. Allen & Barbour, 1909: 44 [= Φ *Echymipera kalubu* (J. Fischer, 1829: 274)] by original designation.
COMMENTS: Synonymised within *Echymipera* by Simpson (1945: 44), Tate (1948a: 329), Marshall (1981: 27), Marshall *et al.* (1990: 489), McKenna and Bell (1997: 57) and Groves (2005e: 41).

Echymipera rufescens (Peters & Doria, 1875)

Long-nosed Echymipera

Φ *Echymipera rufescens rufescens* (Peters & Doria, 1875)

Φ *Perameles rufescens* Peters & Doria, 1875: 541.

TYPE LOCALITY: 'Insulae Kei' [=Kai Islands], New Guinea.
COMMENTS: Species first found to occur in Australia by Tate (1948a: 334) from a specimen collected from the McIlwaith Range on Cape York Peninsula in 1932, which was given the name *Echymipera rufescens australis* Tate, 1948a.

Φ *Perameles aruensis* Peters & Doria, 1875: 542, footnote.

TYPE LOCALITY: Aru Islands, New Guinea.
COMMENTS: Synonymised within *rufescens* by Tate (1948a: 333) and subsequent authors including Laurie and Hill (1954: 12), Groves (1993d: 41) and Flannery (1990: 80; 1995a: 111; 1995b: 70).

Φ *Peramelopsis welsianus* Heude, 1896: 143 footnote.

TYPE LOCALITY: Kei Islands, New Guinea.
COMMENTS: Synonymised within *rufescens* by Laurie and Hill (1954: 12), Groves (1993b: 41) and Flannery (1990: 80; 1995a: 111; 1995b: 70).

Φ *Perameles keiensis* Cohn, 1910: 727.

TYPE LOCALITY: Kei Island, New Guinea.
COMMENTS: Synonymised within *rufescens* by Laurie and Hill (1954: 12) and subsequent authors including Groves (1993d: 41) and Flannery (1990: 80; 1995a: 111; 1995b: 70).

Φ *Echymipera gargantua* Thomas, 1914a: 443.

TYPE LOCALITY: Wahatuna, Mimika River, south-west Netherlands, New Guinea.
COMMENTS: Synonymised within *rufescens* by Tate (1948a: 333) and subsequent authors including Laurie and Hill (1954: 12), Groves (1993d, 41) and Flannery (1990: 80; 1995a: 111; 1995b: 70).

Echymipera rufescens australis Tate, 1948

Echymipera rufescens australis Tate, 1948a: 334.

TYPE LOCALITY: 'Rocky Scrub', Rocky River, east slope of McIlwraith Range, near Coen, Cape York, north Queensland, Australia.
COMMENTS: The presence of this species within Australia was confirmed by Hulbert *et al.* (1971: 331). Synonymised within *rufescens* by Mahoney and Ride (1988d: 37). Recognised as a subspecies of *rufescens* by Troughton (1967: 67), Strahan (1983: 103; 1995: 192), Flannery (1990: 80; 1995a: 111; 1995b: 70), Groves (2005e: 42), Clayton *et al.* (2006: 103), and Van Dyck and Strahan (2008: 175).

Subfamily Peramelinae J. Gray, 1825 *sensu* Kirsch, 1997

Tribe Peramelina J. Gray, 1825a: 340.

TYPE GENUS: *Perameles* É. Geoffroy, 1803d.
COMMENTS: When originally proposed, this rank was placed in the Family Didelphidae (J. Gray, 1821: 308) and included the genera *Perameles* É. Geoffroy, 1803d; and *Isodon* [sic=*Isoodon*] Desmarest, 1817a. Subfamily status recognised by Bensley (1903: 110), Kirsch (1968a: 420), Kirsch *et al.* (1997: 345) and followed by McKenna and Bell (1997: 57), Groves (2005e: 39) and Meredith *et al.* (2008b: 1, 4).

Subfamily Peramelinae Bensley, 1903: 110, 111.

TYPE GENUS: *Perameles* É. Geoffroy, 1803d.
COMMENTS: When originally proposed, this rank was placed in the Family Peramelidae (J. Gray, 1825a), and included the genera *Perameles* É. Geoffroy, 1803d; and *Chaeropus* W. Ogilby, 1838a. Synonymised within the Subfamily Peramelinae J. Gray (1825a: 340) by McKenna and Bell (1997: 57).

Tribe Peramelini Szalay, 1994: 42.

TYPE GENUS: *Perameles* É. Geoffroy, 1803d.

COMMENTS: When originally proposed, this rank was attributed to J. Gray (1825a: 340) and placed in the Subfamily Peramelinae (J. Gray, 1825a). Synonymised within the Subfamily Peramelinae by McKenna and Bell (1997: 57).

Isoodon Desmarest, 1817

Isoodon Desmarest, 1817a: 409.

TYPE SPECIES: *Didelphis obesula* Shaw, 1797 [=*Isoodon obesulus* (Shaw, 1797)] by monotypy.

COMMENTS: Synonymised within *Perameles* by Waterhouse (1846: 356) and Thomas (1888a: 227) but recognised by Iredale and Troughton (1934: vii, 18). Taxonomic decision to use *Isoodon* rather than *Thylacis* (Illiger, 1811) was made by Lidicker and Follett (1968: 251). Current taxonomy at the species level follows that of Ride (1970: 96), with subspecies designated by Strahan (1995). Genus revised by Lyne and Mort (1981: 107).

FUTURE TAXONOMIC RESEARCH: Further taxonomic examination of this continental genus is critical to assess all recognised, and unrecognised, taxa given the catastrophic declines observed in regional populations of many of its constituent taxa (Westerman *et al.*, 2012: 104).

Isodon Agassiz, 1842: 16.

TYPE SPECIES: Incorrect subsequent spelling of *Isoodon* Desmarest, 1817a.

COMMENTS: Not subsequently recognised.

HOMONYMS:

Isodon Say, 1822: 332, hutia rodents of the Class Mammalia (Order Rodentia, Family Capromyidae). Name is a synonym of *Capromys* Desmarest, 1822a: 185. See Woods and Kilpatick (2005: 1594).

Thylacis Haltenorth, 1958: 24.

TYPE SPECIES: *Nomen novum* for *Perameles* É. Geoffroy, 1803d.

COMMENTS: Recognised as a valid genus by Simpson (1945: 44), Marlow (1958: 79, 88) and Ziegler and Lidicker (1968: 43). Synonymised within *Isoodon* by most other authors including Iredale and Troughton (1934: 18), Lidicker and Follett (1968: 251), Honacki *et al.* (1982: 34), Mahoney and Ride (1988d: 39), Marshall *et al.* (1990: 489) and Groves (1993d: 39; 2005e: 39). Van Deusen and Jones (1967: 74) note that the name *Thylacis* is an objective synonym of *Perameles* so the use for *Isoodon* is incorrect.

HOMONYMS:

Thylacis Illiger, 1811, long-nosed bandicoots of the Class Mammalia (Order Peramelemorphia, Family Peramelidae). Genus is synonym of *Perameles* É. Geoffroy, 1803d. See individual entry.

Isoodon auratus (Ramsay, 1887)

Golden Bandicoot

Isoodon auratus auratus (Ramsay, 1887)

Perameles auratus Ramsay, 1887a: 551.

TYPE LOCALITY: Derby, Western Kimberley, Western Australia, Australia.

COMMENTS: Abstract of description by Ramsay (1887b: vi). Synonymised within *Perameles macroura* by Thomas (1888a: 234) but elevated to species rank within *Perameles* by Thomas (1904a: 227). Species rank recognised within *Isoodon* by Iredale and Troughton (1934: vii, 17) and lowered to a subspecies of *obesulus* by Tate (1948a: 338). Ride (1970: 96) separated *auratus*, which included *barrowensis*, from *obesulus* as distinct species within *Isoodon*. Elevated to species rank by Troughton (1967: 54), Lyne and Mort (1981: 107, 129), Honacki *et al.* (1982: 34), Mahoney and Ride (1988d: 37) and Strahan (1995: 172). One study has shown that this taxon is morphologically distinct from *obesulus*, although it was not genetically distinct on the DNA segments sequenced by Pope *et al.* (2001: 411, 424–425) who proposed that the most parsimonious compromise from both genetics and morphology from the current data would be to consider *I. auratus* as a subspecies of *I. obesulus*. These results were supported by Westerman and Krajewski *et al.* (2000: 6), and Zenger *et al.* (2005: 200) who suggested that the maintenance of *I. obesulus* and *I. auratus* as different species could not be supported and that only three subspecies should be recognised (ie. *I. o. obesulus*, *I. o. peninsulae* and *I. o. fusciventer*, with the exact position of *affinis* from Tasmania currently unknown but likely to be placed within *I. o. obesulus*. More recently *auratus* and *obesulus* were found to be genetically distinct from each other by Westerman *et al.* (2012: 106), which appears to have been followed by Van Dyck and Strahan (2008: 177) who recognised *auratus* at the species rank with the subspecies *arnhemensis* and *barrowensis*.

Isoodon auratus arnhemensis Lyne & Mort, 1981

Isoodon arnhemensis Lyne & Mort, 1981: 107, 128.

TYPE LOCALITY: Melville Bay, Cape Arnhem area, Northern Territory, Australia.

COMMENTS: Species status was recognised by Mahoney and Ride (1988d: 37), but lowered to a subspecies of *auratus* by Strahan (1995: 172), Groves (2005e: 39), and Van Dyck and Strahan (2008: 177). The population designated by this name needs to be re-examined.

Isoodon auratus barrowensis (Thomas, 1901)

Perameles barrowensis Thomas, 1901a: 396.

TYPE LOCALITY: Barrow Island, north-west Western Australia, Australia.

COMMENTS: Recognised at the species rank by Wood Jones (1922a: 39), Iredale and Troughton (1934: vii, 17), Tate (1948a: 337) and Troughton (1967: 54). It was included within *auratus* by Ride (1970: 96), though it is not clear at what rank. Synonymised within *auratus* by Honacki *et al.* (1982: 35). Elevated to species rank by Lyne and Mort (1981: 107, 123) but recognised as a subspecies of *auratus* by Strahan (1983: 98; 1995; 173). In reference to Lyne and Mort, Mahoney and Ride (1988d: 38) suggested it was possible that *barrowensis* is sufficiently distinct to be treated taxonomically as a separate subspecies, but synonymised it within *auratus*. Similarly Groves (1993d: 39) synonymised this taxon with *I. auratus*. Close *et al.* (1990: 26) suggested that *barrowensis* was strikingly similar *I. o. obesulus*, despite their geographical separation, but recognised *barrowensis* as a subspecies of *auratus*. These observations appear to be supported by other genetic studies by Pope *et al.* (2001: 425) and morphological observations of Thomas (1901a: 396) and Lyne and Mort (1981: 123) that showed closer resemblance of Barrow Island skulls to *obesulus* than *auratus*. The research by Pope *et al.* (2001: 424) also showed that this taxon is not genetically distinct from *obesulus*. Despite these findings this taxon was recognised as a subspecies of *auratus* by Groves (2005e: 39), Clayton *et al.* (2006: 103), and Van Dyck and Strahan (2008: 177).

FUTURE TAXONOMIC RESEARCH: Further taxonomic studies are required to assess the validity of this taxon, and its true affinities.

Isoodon macrourus (Gould, 1842)

Northern Brown Bandicoot

Isoodon macrourus macrourus (Gould, 1842)

Perameles macrourus Gould, 1842a: 41.

TYPE LOCALITY: Port Essington, Northern Territory, Australia.

COMMENTS: Recognised as a species within *Perameles* by Waterhouse (1846: 366) and Thomas (1888a: xii, 234) who spelt it *macrura*. Transferred to *Isoodon* by Iredale and Troughton (1934: vii, 18) and followed by subsequent authors. Three subspecies were recognised by Tate (1948a: 338), Strahan (1995: 175), and Van Dyck and Strahan (2008: 179) who suggest the subspecific taxonomy is in need of revision. Lyne and Mort (1981: 129–130) included *torosus* and *moresbyensis* (extralimital) within *macrourus* as synonyms, which was also done by Mahoney and Ride (1988d: 38).

P. [erameles] macrura Wagner, 1855: xiii, 211.

TYPE LOCALITY: Invalid emendation of *Perameles macrourus* Gould, 1842a.

COMMENTS: Synonymised within *macrourus* by Flannery (1990: 86; 1995a: 123) and Groves (1993d: 39: 2005e: 39).

Perameles wombeyensis Broom, 1896a: 56.

TYPE LOCALITY: Broom Cave, Wombeyan Caves, near Taralga, New South Wales, Australia.

COMMENTS: Synonymised within *Isoodon macrourus* by Wakefield (1972a: 20).

Φ *Isoodon macrourus moresbyensis* (Ramsay, 1877)

Φ *Perameles moresbyensis* Ramsay, 1877b: 14.

TYPE LOCALITY: Port Moresby, Papua New Guinea.

COMMENTS: Recognised as a species, within *Isoodon*, by Tate and Archbold (1937: 361) and synonymised within *macrourus* by Groves (1993d: 39) and Flannery (1990: 86; 1995a: 123), but recognised as a subspecies of *macrourus* by Tate (1948a: 340), Strahan (1983: 96; 1995: 175), Groves (2005e: 39), and Van Dyck and Strahan (2008: 179).

Isoodon macrourus torosus (Ramsay, 1877)

Perameles macroura var. *torosus* Ramsay, 1877b: 12.

TYPE LOCALITY: Near Cooktown, north Queensland, Australia.

COMMENTS: Synonymised within *macrourus* by Thomas (1888a: 234). Transferred to *Isoodon* and elevated to species status by Iredale and Troughton (1934: vii, 18) and Troughton (1967: 55). Synonymised within *macrourus* by Ride (1970: 243), and Lyne and Mort (1981: 131), Mahoney and Ride (1988d: 38) and Flannery (1990: 86; 1995a: 123). Recognised as a subspecies of *macrourus* by Finlayson (1934: 229), Tate (1948a: 339), Strahan (1983: 96; 1995: 175), Close *et al.* (1990: 19) and supported as a subspecies by Westerman and Krajewski (2000: 6). Despite this designation, significant doubt was cast on the validity of *torosus* using genetic and morphological analysis by Pope *et al.* (2001: 420). Groves (2005e: 39) synonymised *torosus* within *moresbyensis*. Subspecies rank recognised by Clayton *et al.* (2006: 102), and Van Dyck and Strahan (2008: 179). Spelling of this name is sometimes given as '*torosa*' but this is in error.

FUTURE TAXONOMIC RESEARCH: Further taxonomic revision is required to confirm the status of this taxon, in the context of a fresh revision, using both morphology and DNA, of the entire genus.

Isoodon obesulus (Shaw, 1797)

Southern Brown Bandicoot

Isoodon obesulus obesulus (Shaw, 1797)

Didelphis Obesula Shaw, 1797: Text to Plate 298.

TYPE LOCALITY: Ku-Ring-Gai Chase National Park, Sydney, New South Wales, Australia (33°36′S, 151°16′E). Designation of type by Dixon (1981: 132).

COMMENTS: Included within *Perameles* by Waterhouse (1838a: 65; 1841a: 159; 1846: 368), Gould (1856 [1845–1863]: Text to Plate 12), Krefft (1866a: 16; 1868a: 94), and Thomas (1888a: xii, 231), who reviewed its taxonomic history. Transferred to *Isoodon* by Desmarest (1817a: 409), F. Cuvier (1825: 416), Desmarest (1847: 579), and Iredale and Troughton (1934: vii, 17). Six subspecies (*obesulus*, *affinis*, *fusciventer*, *nauticus*, *peninsulae* and *auratus*) were recognised by Tate (1948a: 338). Mahoney and Ride (1988d: 39) suggested that it is possible that *fusciventer*, *nauticus* and *peninsulae* are sufficiently distinct to be treated as subspecies. Lyne and Mort (1981: 122–123) treated *nauticus* and *peninsulae* as separate species. Subsequently Strahan (1995: 176) synonymised *affinis*, *fusciventer* and *nauticus* and recognised *peninsulae* as a subspecies and elevated *auratus* to species rank.

FUTURE TAXONOMIC RESEARCH: The taxonomic status of all subspecies of *obesulus* needs to be confirmed.

Isoodon obesulus fusciventer (J. Gray, 1841)

Perameles fusciventer J. Gray, 1841: 407, locality 401.

TYPE LOCALITY: King George Sound, Western Australia, Australia.

COMMENTS: Synonymised within *obesulus* by Waterhouse (1846: 368) and Thomas (1888a: 231). Designated a subspecies of *obesulus* by Iredale and Troughton (1934: 17) and Tate (1948a: 338) and synonymised within *obesulus* by Lyne and Mort (1981: 129). Taxon treated as a subspecies of *obesulus* by Strahan (1983: 94), Mahoney and Ride (1988d: 39) and Maxwell *et al.* (1996: 3). Genetic research has proposed that *obesulus* from Western Australia and *auratus* could be placed within *fusciventer* (Pope *et al.*, 2001: 425). Synonymised within *obesulus* by Groves (2005e: 39). Recognised as a subspecies of *obesulus* by Clayton *et al.* (2006: 103), Van Dyck and Strahan (2008: 181) and Burbidge *et al.* (2014: 19, 28). Westerman *et al.* (2012: 103) suggested this taxon was genetically highly distinct and not related to *obesulus*.

Isoodon obesulus affinis (Waterhouse, 1846)

Perameles affinis Waterhouse, 1846: 373.

TYPE LOCALITY: Kangaroos Point near Hobart, Tasmania, Australia.

COMMENTS: Waterhouse (1846: 268) published the name as a junior synonym of *obesula* and apparently as a distinct taxon on page 373 (though this later discussion clearly suggests it is the same as *obesula*). *P. affinis* became available through its adoption for a taxon, *Perameles obesulus*, by Iredale and Troughton (1934: 17). Synonymised within *obesulus* by Thomas (1888a: 231), Lyne and Mort (1981: 129), Mahoney and Ride (1988d: 39), Groves (1993d: 39; 2005e: 39), Burbidge *et al.* (2014: 19), and not recognised by Strahan (1995: 176). Recognised as a subspecies of *obesulus* by Iredale and Troughton (1934: 17), Tate (1948a: 338), Strahan (1983: 94), Clayton *et al.* (2006: 103), and Van Dyck and Strahan (2008: 181).

Perameles affinis J. Gray, 1843a: 96.

TYPE LOCALITY: Tasmania, Australia.

COMMENTS: *Nomen nudum*. J. Gray recognised as the author of the name by Thomas (1888a: 231), but shown to be a *nomen nudum* by Iredale and Troughton (1934: 17), which was followed by subsequent authors (in giving priority to Waterhouse, 1846: 373 as the original description) including Tate (1948a: 338), Mahoney and Ride (1988d: 39) and Groves (2005e: 39).

Isoodon obesulus nauticus Thomas, 1922

Isoodon nauticus Thomas, 1922c: 678.

TYPE LOCALITY: Franklin Island, Nuyts Archipelago, South Australia, Australia.

COMMENTS: Recognised at the species rank by Iredale and Troughton (1934: vii, 17) and Troughton (1967: 54). Lowered to a subspecies of *obesulus* by Tate (1948a: 338), elevated to species by Lyne and Mort (1981: 122), synonymised within *obesulus* by Ride (1970: 243), Burbidge *et al.* (2014: 19), and not recognised by Strahan (1995: 172). Recognised as a subspecies of *obesulus* by Mahoney and Ride (1988d: 39), Groves (2005e: 39) and Clayton *et al.* (2006: 103), and Van Dyck and Strahan (2008: 181).

Isoodon peninsulae Thomas, 1922

Cape York Brown Bandicoot

Isoodon peninsulae Thomas, 1922c: 679.

TYPE LOCALITY: Utingu, northern Cape York, Queensland, Australia. 10 m elevation.

COMMENTS: The taxonomic position of *peninsulae* has been unstable. It was placed within *obesulus* by Ride (1970: 243), lowered to subspecies rank (within *obesulus*) by Tate (1948a: 338), Strahan (1983: 94; 1995: 176), and Mahoney and Ride (1988d: 39). Taxon also recognised at the species rank by Iredale and Troughton (1934: vii, 17), Troughton (1967: 55), Lyne and Mort (1981: 107, 123) and Close *et al.* (1990: 26). Recent research has been inconsistent, with Westerman and Krajewski (2000: 6) proposing there was little support for the continued recognition of *I. o. peninsulae* as a distinct subspecies; in contrast it was found to be morphologically and genetically distinct from *obesulus* and was strongly supported as a subspecies by Pope *et al.* (2001: 424–425). Synonymised within *obesulus* by Groves (2005e: 39), but recognised as a subspecies of *obesulus* by Clayton *et al.* (2006: 102), Van Dyck and Strahan (2008:

181), and Menkhorst and Knight (2011: 80). Most recently Westerman *et al.* (2012: 104) suggested this taxon was more closely related to *auratus* than *obesulus*, though it is highly distinct and should be treated as a distinct species, which was followed by Burbidge *et al.* (2014: 28).

Perameles É. Geoffroy, 1803

Perameles É. Geoffroy, 1803d: 249 as 149.

TYPE SPECIES: *Perameles nasuta* É. Geoffroy, 1804c by subsequent designation (see Lesson, 1828a: 197).

COMMENTS: *Perameles* also described by É. Geoffroy (1804c: 56), but Groves (2005e: 40) noted that the description which has priority is not known as the dates are uncertain. For variants of *Perameles* see Palmer (1904: 884).

HOMONYMS:

Perameles J. Gray, 1839a: 333, reptiles of the Class Reptilia (Order Squamata, Family Scincidae). Appears to be an incorrect subsequent spelling of *Peromeles* Wiegmann, 1834: 11, which is a junior synonym of *Saiphos* J. Gray, 1831a: 72.

Thylacis Illiger, 1811: 76.

TYPE SPECIES: Unnecessary *nomen novum* for *Perameles* É. Geoffroy, 1803d.

COMMENTS: Genus recognised by Simpson (1945: 44) and followed by Laurie and Hill (1954: 12) and Marlow (1958: 79, 88). Synonymised within *Perameles* by Thomas (1888a: 227), Iredale and Troughton (1934: 18), Tate (1948a: 323), Mackerras and Mackerras (1960: 52), Mahoney and Ride (1988d: 39) and Marshall *et al.* (1990: 490). Taxonomic decision to use *Isoodon* rather than *Thylacis* was made by Lidicker and Follett (1968: 251).

HOMONYMS:

Thylacis Haltenorth, 1958, short-nosed bandicoots of the Class Mammalia (Order Peramelemorphia, Family Peramelidae). Genus is a synonym of *Isoodon* Desmarest, 1817a. See individual entry.

Thylax Oken, 1816: xvi, 1128.

TYPE SPECIES: *Nomen novum* for *Perameles* É. Geoffroy, 1803d.

COMMENTS: Synonymised within *Perameles* by Iredale and Troughton (1934: 18), Marshall (1981: 27) and Marshall *et al.* (1990: 490). Not considered by Mahoney and Ride (1988d: 39–40) or Groves (2005e: 40).

HOMONYMS:

† *Thylax* Hag en, 1866: 172, bark louse of the Class Insecta (Order Psocoptera, Family Lepidopsocidae).

Parameles J. Gray, 1827a: 194.

TYPE SPECIES: Not designated from *Perameles nasuta* É. Geoffroy, 1804c; *Perameles bougainville* Quoy & Gaimard, 1824 and *P. obesula* Shaw, 1797.

COMMENTS: Not considered by Iredale and Troughton (1934: 18), and synonymised within *Perameles* by Mahoney and Ride (1988d: 39) who listed the author as 'Anon [J. Gray, J.E]'.

Perimeles Lenz, 1831: 158.

TYPE SPECIES: Unjustified emendation of *Perameles* É. Geoffroy, 1803d.

COMMENTS: Not considered by Iredale and Troughton (1934: 18) or Groves (2005e: 40). Synonymised within *Perameles* by Mahoney and Ride (1988d: 40).

HOMONYMS:

Perimeles Godman in Godman and Salvin, 1900: 542, butterflies of the Class Insecta (Order Lepidoptera, Family Hesperiidae). Genus is a synonym of *Remella* Hemming, 1939: 39.

Peromeles Winge, 1893a: 124.

TYPE SPECIES: Unjustified emendation of *Perameles* É. Geoffroy, 1803d.

COMMENTS: Not considered by Iredale and Troughton (1934: 18) or Groves (2005e: 40). Synonymised within *Perameles* by Mahoney and Ride (1988d: 40).

Perameles bougainville Quoy & Gaimard, 1824

Western Barred Bandicoot

Perameles bougainville bougainville Quoy & Gaimard, 1824

Perameles Bougainville Quoy & Gaimard, 1824: 56; Plate 5.

TYPE LOCALITY: Péron Peninsula, Shark Bay, Western Australia, Australia.

COMMENTS: Consistently recognised since its description including Waterhouse (1841a: 162; 1846: 385) and Thomas (1888a: xii, 246). Occurs on both Bernier and Dorre Islands in Shark Bay, Western Australia, but is extinct on mainland Australia.

† *Perameles myosuros* Wagner, 1841a: 293.

TYPE LOCALITY: Swan River District and surrounds of King Georges Sound, Western Australia, Australia.

COMMENTS: Multiple original spellings of *myosuros*; *myosuros* in text and table of contents, *myosurus* in title only; erroneous masculine gender of *mysurus*, as compared with the indeclinable *myosuros* when combined with *Perameles* indicates lapsus; confirmed by Wagner's subsequent usage. Recognised at species rank by Gould (1845 [1845–1863]: Text to Plate 10), Waterhouse (1846: 381) and Iredale and Troughton (1934: vii, 19) and Troughton (1967: 59), as

P. myosura. Synonymised within *bougainville* by Thomas (1888a: 246), Tate (1948a: 324), Ride (1970: 246), Mahoney and Ride (1988d: 40) and Groves (2005e: 40). Recognised as a subspecies of *bougainville* by Strahan (1983: 101; 1995: 179), Clayton *et al.* (2006: 103), and Van Dyck and Strahan (2008: 184). Synonymised within *bougainville* by Helgen and Flannery (2003: 209) and followed here.

† *Perameles arenaria* Gould, 1844a: 104.

TYPE LOCALITY: Near York, Western Australia, Australia.

COMMENTS: Synonymised within *myosuros* by Waterhouse (1846: 381) and as *myosura* by Iredale and Troughton (1934: 19). Synonymised within *bougainville* by Thomas (1888a: 247), Tate (1948a: 324), Mahoney and Ride (1988d: 40) and subsequent authors.

† ***Perameles bougainville fasciata*** J. Gray, 1841

† *Perameles fasciata* J. Gray, 1841: 407.

TYPE LOCALITY: Liverpool Plains, New South Wales, Australia. Type designated by Thomas (1922a: 128).

COMMENTS: Recognised at species rank by Waterhouse (1846: 379), Gould (1849 [1845–1963], Text to Plate 8), Krefft (1866a: 15), Iredale and Troughton (1934: vii, 19) and Troughton (1967: 58). Included within *Perameles bougainville* by Thomas (1888a: 246) and Ride (1970: 245) but at unknown status. Recognised as a subspecies of *bougainville* by Strahan (1983: 101), Helgen and Flannery (2003: 209) and Clayton *et al.* (2006: 103). Synonymised within *bougainville* by Mahoney and Ride (1988d: 40), Groves (2005e: 40), and Van Dyck and Strahan (2008: 184). Subspecies rank recognised by Helgen and Flannery (2003: 209) and followed here.

† *Perameles myosura notina* Thomas, 1922d: 144.

TYPE LOCALITY: St Vincent Gulf, South Australia, Australia.

COMMENTS: The species name *myosura* is a modified subsequent spelling of *mysuros* Wagner, 1841a: 293, but Thomas did not explain why he adopted this spelling. Considered a subspecies of *myosura* by Iredale and Troughton (1934: 19), subspecies of *bougainvillea* by Tate (1948a: 324) and Freedman and Rightmire (1971: 31), and synonymised within *bougainville* by Mahoney and Ride (1988d: 40) and Groves (2005e: 40). Recognised as a subspecies of *bougainville* by Strahan (1983: 101; 1995: 179), Clayton *et al.* (2006: 103), and Van Dyck and Strahan (2008: 184). Synonymised within *fasciata* by Helgen and Flannery (2003: 209) and followed here.

† ***Perameles eremiana*** Spencer, 1897

Desert Bandicoot

† *Perameles eremiana* Spencer, 1897: 9; Plate 2, Figs. 5–7.

TYPE LOCALITY: Burt Plain, north of Alice Springs, Northern Territory, Australia.

COMMENTS: Extinct. See Ride (1970: 200).

Perameles gunnii J. Gray, 1838

Eastern Barred Bandicoot

Perameles Gunnii J. Gray, 1838b: 107.

TYPE LOCALITY: Tasmania, Australia.

COMMENTS: Species recognised by Gould (1859 [1845–1863]: Text to Plate 9), Krefft (1868a: 94) and subsequent authors. A genetic analysis by Westerman and Krajewski (2000: 5) found little genetic difference between Victorian and Tasmanian samples. The Victorian population was considered an unnamed subspecies by N. Robinson *et al.* (1993: 203) and Maxwell *et al.* (1996: 7), although the analysis of Westerman and Krajewski (2000: 5) failed to support this differentiation. Clayton *et al.* (2006: 103) recognised an unnamed Victorian subspecies but this is not recognised here.

FUTURE TAXONOMIC RESEARCH: The Victorian population [*sensu* N. Robinson *et al.* (1993: 203)] needs to be revised and formally described if appropriate.

Perameles nasuta É. Geoffroy, 1804

Southern Long-nosed Bandicoot

Perameles nasuta É. Geoffroy, 1804c: 62; Plate 44.

TYPE LOCALITY: Unknown.

COMMENTS: Species recognised by Waterhouse (1838a: 65), Gould (1855 [1845–1863]: Text to Plate 11) and Thomas (1888a: xii, 242), who reviewed its early taxonomic history.

perameles Lawson Quoy & Gaimard, 1824 [1824–1826]: 711.

TYPE LOCALITY: Bathurst, New South Wales, Australia.

COMMENTS: Synonymised within *nasuta* by Waterhouse (1846: 374) and Iredale and Troughton (1934: 18) and subsequent authors including Mahoney and Ride (1988d: 42) and Groves (1993d: 40; 2005e: 40).

Perameles major Schinz, 1825: 375.

TYPE LOCALITY: Blue Mountains, New South Wales, Australia.

COMMENTS: Synonymised within *nasuta* by Iredale and Troughton (1934: 18) and subsequent authors including Mahoney and Ride (1988d: 42) and Groves (1993d: 40; 2005e: 40).

Isoodon Musei Boitard, 1841: 291.

COMMENTS: Replacement name for *Perameles nasuta* É. Geoffroy, 1804c. Tentatively included within *obesulus* by Iredale and Troughton (1934: 17) and synonymised within *nasuta* by Mahoney and Ride (1988d: 42) and Groves (1993d: 40; 2005e: 40).

Perameles auritus Waterhouse, 1841a: 156.

TYPE LOCALITY: Port Jackson, Sydney, New South Wales, Australia.

COMMENTS: Specimen referred to by Waterhouse (1841: 156) was seen at the Paris Museum, and he turn synonymised it within *nasuta*, which was followed by Iredale and Troughton (1934: 18). Not considered by Mahoney and Ride (1988d: 42).

Perameles tenuirostris Owen, 1877a: 107.

TYPE LOCALITY: Wellington area, New South Wales, Australia (see Mackness 2000: 133).

COMMENTS: Synonymised within *Perameles nasuta* by Lydekker (1887: 255).

Perameles pallescens Thomas, 1923

Northern Long-nosed Bandicoot

Perameles nasuta pallescens Thomas, 1923a: 173.

TYPE LOCALITY: Vine Creek, Ravenshoe, Queensland, Australia. 3000ft.

COMMENTS: Recognised at subspecies rank by Iredale and Troughton (1934: 18) and Tate (1948a: 326), and synonymised within *nasuta* by Mahoney and Ride (1988d: 42) and Groves (1993d: 40; 2005e: 40). Often considered a subspecies of *nasuta* by authors including Strahan (1983: 99; 1995: 185), which was supported by Westerman and Krajewski (2000: 5), Clayton *et al.* (2006: 103), and Van Dyck and Strahan (2008: 189). Westerman and Krajewski (2000: 4) and Pope *et al.* (2001: 415) found that Control Region haplotypes of this presumed taxon from Iron Range and from the Carbine Tableland are extremely different, suggesting that more detailed taxonomic investigation is in order. Most recently Westerman *et al.* (2012: 104) suggested this taxon is highly distinct and should be recognised as a separate species, which was followed by Burbidge *et al.* (2014: 19, 28) and is recognised here.

Family Thylacomyidae Bensley, 1903 *sensu* M. Archer & Kirsch, 1977

Subfamily Thylacomyinae Bensley, 1903: 110, 115.

TYPE GENUS: *Thylacomys* Blyth, 1840 [=*Macrotis* Reid, 1837].

COMMENTS: When originally proposed, this rank was placed in the Family Peramelidae (J. Gray, 1825a) and included the genus *Thylacomys* Blyth, 1840 [=*Macrotis* Reid, 1837]. Synonymised within Peramelidae by Groves and Flannery (1990: 1) and Groves (1993d: 39). *Macrotis* placed by itself in Subfamily Thylacomyinae by Szalay (1994: 42) and then in Subfamily Chaeropodinae, with *Chaeropus*, by McKenna and Bell (1997: 57). The Family Thylacomyidae was synonymised within the Subfamily Chaeropodinae by McKenna and Bell (1997: 57). Recognised as the Subfamily Thylacomyinae by Kirsch (1968a: 420) and Strahan (1995: 6, 186). Family rank recognised by M. Archer and Kirsch (1977: 21, 23), Kirsch (1977a: 112; 1977b: 4, 45), Kirsch and Calaby (1977: 16), Marshall (1981: 27), Honacki *et al.* (1982: 36), Strahan (1983: xxi, 106), Marshall (1984: 71, 91), Mahoney and Ride (1988e: 43), Kirsch *et al.* (1990b: 434; 1997: 245), Wroe (1999: 512), Groves (2005e: 38), Meredith *et al.* (2008b: 1, 4), and Van Dyck and Strahan (2008: 9, 191). Family name not recognised by Menkhorst and Knight (2011: 16). McAllan and Bruce (1989: 454) argued that the assertion of M. Archer and Kirsch (1977: 24) to adopt the family- name Thylacomyidae because of the general acceptance of the Subfamily Thylacomyinae was incorrect as the Article of the ICZN makes no mention of the issue (ICZN, 1985a: 81). McAllan and Bruce (1989: 454) argued that the generic name *Thylacomys* (a *nomen nudum*) was corrected to *Macrotis* before 1961, so Article 40 (ICZN, 1985a: 81) does not apply, and that the family-group name was not corrected before 1961, so section (b) of Article 40 does not apply either. As a result of these issues McAllan and Bruce (1989: 454) proposed a new family name, Macrotidae (see below). In fact, all these assertions are based on a misreading of the Code, and the name Thylacomyidae stands.

Family Macrotidae McAllan & Bruce, 1989: 454.

TYPE GENUS: *Macrotis* Reid, 1837.

COMMENTS: When originally proposed, this rank included the genus *Macrotis* Reid, 1837, and was developed as a replacement name for the Family Thylacomyidae (Bensley, 1903). An argument for the use of Macrotidae compared with Thylacomyidae was made by McAllan and Bruce (1989: 454) (see discussion above) but the name Macrotidae has not been recognised since its description.

Macrotis Reid, 1837

Macrotis Reid, 1837: 131.

TYPE SPECIES: *Perameles lagotis* Reid, 1837 [=*Macrotis lagotis* (Reid, 1837)] by monotypy.

COMMENTS: Proposed as a subgenus of *Perameles* É. Geoffroy, 1803d. Not preoccupied by *Macrotis* Dejean, 1833: 186, a *nomen nudum* (Sherborn, 1928: 3777; Troughton, 1932b: 223; Groves, 1993d: 40). Recognised as a subgenus of *Perameles* by Waterhouse (1846: 358). Synonymised within *Thylacomys* by Simpson (1945: 44). Taxonomic decision of Iredale and Troughton (1934: vii, 19) and Ride in Mahoney and Ride (1988e: 43). Groves and Flannery (1990: 11) placed *Macrotis* back in Peramelidae.

HOMONYMS:

Macrotis, Dejean, 1833: 186, flower beetles of the Class Insecta (Order Coleoptera, Family Cetoniidae). A *nomen nudum* (Sherborn, 1928: 3777; Troughton, 1932b: 223). Genus is a junior synonym of *Macrotina* Strand, 1934: 277.

Macrotis Wagner, 1855: xvi, 350, 368, white-tailed deer and mule deer of the Class Mammalia (Order Artiodactyla,

Family Cervidae). Genus is a junior synonym of *Odocoileus* Rafinesque, 1832: 109. See W. Smith (1991: 1) and Grubb (2005a: 657).

Macrotis Schürhoff, 1933: 92, scarab beetles of the Class Insecta (Order Coleoptera, Family Scarabaeidae). Genus is a junior synonym of *Macrotina* Strand, 1934: 277. See Krikken (1984: 51).

Thylacomys Owen, 1838c: 747.

TYPE SPECIES: *Nomen nudum*. See Thomas (1888a: 221–222, footnote) and Palmer (1904: 677).

COMMENTS: Synonymised within *Macrotis* by McKenna and Bell (1997: 57). Palmer (1899a: 302) suggested that as *Macrotis* is preoccupied, *Paragalia* or *Peragale* is antedated and *Thylacomys* is the first tenable name for the genus that it should be adopted on the ground of priority. Palmer (1899a: 301–302) discussed the history of this name and suggested it is 'better not to hold Owen responsible for *Thylacomys*, but to adopt it on the authority of Blyth by whom it was published as a perfectly valid name one year prior to the appearance of *Paragalia*, J. Gray'.

HOMONYMS:

Thylacomys Waite, 1898, hopping mice of the Class Mammalia (Order Rodentia, Family Muridae). Genus is a synonym of *Notomys* Lesson, 1842: 129. See individual account.

Thalacomys Blyth, 1840: 104.

TYPE SPECIES: *Perameles lagotis* Reid, 1837 [=*Macrotis lagotis* (Reid, 1837)] by monotypy.

COMMENTS: Blyth used two spellings *Thalacomys* (p. 104) and *Thylacomys* (p. 107). Palmer (1904: 884) chose *Thylacomys* as the original spelling. The spelling *Thalacomys* was used subsequently by Wood Jones (1923: 339). Synonymised within *Macrotis* by Iredale and Troughton (1934: 19), Tate (1948a: 341), Ride (1970: 248), Mahoney and Ride (1988e: 43), Marshall (1981: 27) and Marshall *et al*. (1990: 490).

Thylacomys Blyth, 1840: 107.

TYPE SPECIES: *Perameles lagotis* Reid, 1837 [=*Macrotis lagotis* (Reid, 1837)] by monotypy.

COMMENTS: Blyth used two spellings *Thalacomys* (p. 104) and *Thylacomys* (p. 107). Palmer (1904: 884) chose *Thylacomys* as the original spelling. Synonymised within *Macrotis* by Iredale and Troughton (1934: 19), and Mahoney and Ride (1988e: 43).

Paragalia J. Gray, 1841: 401.

TYPE SPECIES: *Perameles lagotis* Reid, 1837 [=*Macrotis lagotis* (Reid, 1837)] by monotypy.

COMMENTS: Described as a subgenus of *Perameles*. Synonymised within *Peragale* by Thomas (1888a: 221), within *Thylacomys* by Simpson (1945: 44) and within *Macrotis* by Waterhouse (1846: 358), Iredale and Troughton (1934: 19), Tate (1948a: 341), Marshall (1981: 27), Mahoney and Ride (1988e: 43) and Marshall *et al*. (1990: 490).

Perigalea J. Gray, 1842e: 15, 16.

TYPE SPECIES: Incorrect subsequent spelling of *Paragalia* J. Gray, 1841.

COMMENTS: Recognised at generic rank. Also recognised by J. Gray (1843a: xxii, 96). Synonymised within *Paragalia* by Palmer (1899a: 301).

Paragalea J. Gray, 1843a: 96.

TYPE SPECIES: Incorrect subsequent spelling of *Paragalia* J. Gray, 1841.

COMMENTS: Synonymised within *Paragalia* by Palmer (1899a: 301).

Peragalea Gould, 1845 [1845–1863]: Text to Plate 7.

TYPE SPECIES: Emendation of *Paragalia* J. Gray, 1841.

COMMENTS: Recognised at generic rank. Emendation of *Paragalia* originally used by J. Gray as an incorrect subsequent spelling of *Paragalia*, but adopted by Gould as an emendation. Synonymised within *Paragalia* by Palmer (1899a: 301). Not considered by Iredale and Troughton (1934: 19–20) and Mahoney and Ride (1988e: 43).

Phalacomys Anon, 1854: 382.

TYPE SPECIES: *Perameles lagotis* Reid, 1837 [=*Macrotis lagotis* (Reid, 1837)] by monotypy.

COMMENTS: Incorrect subsequent spelling of *Thylacomys* Owen, 1838c. Synonymised within *Macrotis* by Iredale and Troughton (1934: 20) and Mahoney and Ride (1988e: 43).

Thalaconus J. Richardson *et al*., 1862: 214.

TYPE SPECIES: *Perameles lagotis* Reid, 1837 [=*Macrotis lagotis* (Reid, 1837)] by monotypy.

COMMENTS: Not considered by Iredale and Troughton (1934: 19–20) and synonymised within *Macrotis* by Mahoney and Ride (1988e: 43).

Peragale Lydekker, 1887: 256, footnote 5.

TYPE SPECIES: Emendation of *Paragalia* J. Gray, 1841.

COMMENTS: Recognised by Thomas (1887b: 397; 1888a: xii, 221) ahead of *Macrotis*. Not considered by Iredale and Troughton (1934: 20) and synonymised within *Macrotis* by Marshall (1981: 27), Mahoney and Ride (1988e: 43) and Marshall *et al*. (1990: 490).

Macrotis lagotis (Reid, 1837)

Greater Bilby

Perameles lagotis Reid, 1837: 129.

TYPE LOCALITY: Swan River, Western Australia, Australia.

COMMENTS: Recognised within *Perameles* by Waterhouse (1838a: 65; 1841a: 153) and *Perameles* (*Macrotis*) *lagotis* by Waterhouse (1846: 360). Transferred to *Peragalea* by Gould (1845 [1845–1863]: Text to Plate 7) and Krefft (1866a: 14), *Macrotis* by Jentink (1887: 305) and *Peragale* by Thomas (1888a: xii, 223). Transferred to *Macrotis* by Iredale and Troughton (1934: vii, 20) included all described taxa as subspecies (see below as synonyms).

FUTURE TAXONOMIC RESEARCH: The validity of each of the reported subspecies (including *sagitta*, *cambrica*, *nigripes*, *grandis* and *interjecta*) needs to be resolved.

Thylacomys sagitta Thomas, 1905a: 426.

TYPE LOCALITY: Killalpaninna, east of Lake Eyre, South Australia, Australia.

COMMENTS: Subspecies rank recognised by Iredale and Troughton (1934: 20), Finlayson (1935a: 233), Tate (1948a: 343) and Troughton (1967: 63). Synonymised within *lagotis* by Ride (1970: 244), Mahoney and Ride (1988e: 44) and most subsequent authors including Groves (2005e: 38). Recognised as a subspecies by Strahan (1983: 107; 1995: 187), Clayton *et al.* (2006: 103), and Van Dyck and Strahan (2008: 192). Subspecies not recognised by Burbidge *et al.* (2014: 19, 28), as the species is thought to have had a largely continuous distribution, which is followed here.

Thalacomys nigripes Wood Jones, 1923: 347.

TYPE LOCALITY: Ooldea, South Australia, Australia.

COMMENTS: Subspecies rank recognised by Iredale and Troughton (1934: 20), Tate (1948a: 343) and Troughton (1967: 63). Synonymised within *lagotis* by Ride (1970: 244), Mahoney and Ride (1988e: 44) and most subsequent authors including Groves (2005e: 38). Recognised as a subspecies by Strahan (1983: 107; 1995: 187), Clayton *et al.* (2006: 103), and Van Dyck and Strahan (2008: 192). Subspecies not recognised by Burbidge *et al.* (2014: 19, 28), as the species is thought to have had a largely continuous distribution, which is followed here.

Macrotis lagotis interjecta Troughton, 1932b: 227.

TYPE LOCALITY: Rawlinna, Western Australia, Australia.

COMMENTS: Subspecies rank recognised by Iredale and Troughton (1934: 20), Tate (1948a: 343) and Troughton (1967: 63). Synonymised within *lagotis* by Mahoney and Ride (1988e: 44) and Groves (2005e: 38). Recognised as a subspecies by Strahan (1983: 107; 1995: 187), Clayton *et al.* (2006: 103), and Van Dyck and Strahan (2008: 192). Subspecies not recognised by Burbidge *et al.* (2014: 19, 28), as the species is thought to have had a largely continuous distribution, which is followed here.

Macrotis lagotis grandis Troughton, 1932b: 229.

TYPE LOCALITY: Nalpa, Lake Alexandrina district, South Australia, Australia.

COMMENTS: Subspecies rank recognised by Iredale and Troughton (1934: 20), Tate (1948a: 344) and Troughton (1967: 63). Synonymised within *lagotis* by Mahoney and Ride (1988e: 44) and most subsequent authors including Groves (2005e: 38). Recognised as a subspecies by Strahan (1983: 107; 1995: 187), Clayton *et al.* (2006: 103), and Van Dyck and Strahan (2008: 192). Subspecies not recognised by Burbidge *et al.* (2014: 19, 28), as the species is thought to have had a largely continuous distribution, which is followed here.

Macrotis lagotis cambrica Troughton, 1932b: 230.

TYPE LOCALITY: Bathurst, New South Wales, Australia.

COMMENTS: Subspecies rank recognised by Iredale and Troughton (1934: 20), Tate (1948a: 344) and Troughton (1967: 63). Synonymised within *lagotis* by Mahoney and Ride (1988e: 44) and most subsequent authors including Groves (2005e: 38). Recognised as a subspecies by Strahan (1983: 107; 1995: 187), Clayton *et al.* (2006: 103), and Van Dyck and Strahan (2008: 192). Subspecies not recognised by Burbidge *et al.* (2014: 19, 28), as the species is thought to have had a largely continuous distribution, which is followed here.

† ***Macrotis leucura*** (Thomas, 1887)

Lesser Bilby

† *Peragale leucura* Thomas, 1887b: 397.

TYPE LOCALITY: Undesignated. The specimen was sent by the South Australian Museums taxidermist to London (Groves, 2005e: 38). Thought to have originated near Adelaide or in the northern part of South Australia, where other specimens in the same collection had come from, by Thomas (1887b: 397).

COMMENTS: Included within *Peragale* by Thomas (1888a: xii, 225). Transferred to *Macrotis* by Iredale and Troughton (1934: vii, 21) and followed by subsequent authors.

† *Peragale minor* Spencer, 1897: 6; Plate 2, Figs. 1–4.

TYPE LOCALITY: Sand hills ~40 miles NE of Charlotte Waters, Northern Territory, Australia.

COMMENTS: Holotype identified by Troughton (1932b: 233). Considered a valid species by Iredale and Troughton (1934: vii, 20), which also included the taxon *miselius* as a subspecies, and was followed by Tate (1948a: 343). Synonymised within *leucura* by Ride (1970: 244), and Mahoney and Ride (1988e: 44).

† *Thalacomys Minor Miselius* Finlayson, 1932: 168.

TYPE LOCALITY: Cooncherie, on the lower Diamantina, South Australia, Australia.

COMMENTS: Recognised as a subspecies of *minor* by Iredale and Troughton (1934: 20) and Finlayson (1935a:

227). Synonymised within *leucura* by Troughton (1941: 73) and Mahoney and Ride (1988e: 44).

† *Macrotis minor miseliae* Tate, 1948a: 343.

TYPE LOCALITY: Emendation of *Thalacomys minor miselius* Finlayson, 1932.

COMMENTS: Synonymised within *leucura* by Mahoney and Ride (1988e: 45).

Order Diprotodontia Owen, 1877

Suborder Diprotodontia Owen, 1877a: xii, 107.

COMMENTS: When originally proposed, this rank was placed within the Order Marsupialia and included the sections Rhizophaga (Owen, 1839a [= Vombatidae (Burnett, 1830)]), Poëphaga (Owen, 1839a [= Macropodoidea (J. Gray, 1821)]) and Carpophaga (Owen, 1839a [= Diprotodontia (Owen, 1877a part)]) (see Owen, 1877b: 360; Nicholson, 1880: 661). The rank appears to be derived from Owen (1868: 293) when he suggested that marsupial dentition shows them to be divisible into 'two classes: one "polyprotodont," or characterised by several pairs of mandibular incisors; the other "diprotodont," or by a single pair: these are large, more or less procumbent, and ever growing; the incisors of the first group are small, and of the usual limited growth.' Authorship of the term Polyprotodontia was given to Owen (1866a) by Gregory (1910: 199) who suggested that 'Owen [1866a], as a result of important studies of *Diprotodon*, *Thylacoleo* and *Nototherim*, proposes the terms "diprotodont" and "polyprotodont"' and it appears that this error was subsequently repeated. Name not recognised by Krefft (1871a: 3, Introduction), Gill (1872), Flower (1883: 184), Bensley (1903: 207) or Simpson (1945: 45). Recognised at suborder rank by Nicholson (1880: 661) who gave credit for the name to Owen but did not give the year of his work directly but in the reference list he referred to 'Fossil Mammalia of Australia. Owen. 1877', which appears to be an abbreviated title for this work. The name was subsequently used at subordinal rank by Lydekker (1887: xi, xxi, 146), Thomas (1888a: xi, 3), Cope (1889b: 876; 1891: 69; 1898: 108, 133), Flower and Lydekker (1891: xi, 144), J. Ogilby (1892: 24), Weber (1904: 348; 1928: xiii, 75), Gregory (1910: 197, 215), Osborn (1910: 517) and Iredale and Troughton (1934: vii, 21). Recognised at ordinal rank by Haeckel (1895: 466), Kirsch (1968a: 420; 1977a: 112; 1977b: 45) (as Diprotodonta), Strahan (1983: xxi) (as Diprotodonta), M. Archer (1984: 787) (as Diprotodonta), Woodburne (1984a: 71) (as Diprotodonta), and Aplin and Archer (1987: xxi) who proposed a new classification that included two suborders, the Vombatiformes and Phalangerida. Infraorder rank recognised by M. Archer (1984: 786) and semiorder by Szalay (1993: 239, 240; 1994: 42). Kirsch *et al.* (1997: 245–246) recognised three suborders including the Vombatiformes, Macropodiformes and Phalangeriformes. The author is always given as Owen, 1866a, never with page number, by all authors that include the author and/or citation for the rank including Gregory (1910: 98, 199, 464, 492), Marshall (1981: 17), McKenna and Bell (1997: 58) and Long *et al.* (2002: 77). The author and year of the desciption of Diprotodontia follows Jackson and Thorington (2012: 9). The recognition of the suborders Vombatiformes and Phalangerida was supported by Munemasa *et al.* (2006: 181).

HOMONYMS:

† Diprotodontia Haeckel & Lankester, 1876: 239, includes the extinct hoofed marsupial animals. This usage of the term does appear to have been widely adopted.

Order Brutae J. Gray, 1821: 308.

COMMENTS: When originally proposed, this rank was placed in the Class Pedimanes (J. Gray, 1821 [= Mammalia (Linnaeus, 1758 part)]) and included the families Potoridae (J. Gray, 1821 [= Potoroidae (J. Gray, 1821)]), Macropidae (J. Gray, 1821 [= Macropodidae (J. Gray, 1821)]) and Koladae (J. Gray, 1821 [= Phascolarctidae (Owen, 1839a)]).

HOMONYMS:

Order Bruta Linnaeus, 1758: 16, 33, mammals of the Class Mammalia. Synonymised in part within Xenarthra (Cope, 1889a: 657 [= Order Cingulata (Illiger, 1811: 110) and Pilosa (Flower, 1883: 184)]) and Tethytheria (McKenna, 1975: 42). See McKenna and Bell (1997: 82, 492).

Grand Seccion Bruta Ameghino, 1889: xxiv, 653, mammals of the Class Mammalia.

Race Semidentiae Burnett, 1830b: 351.

COMMENTS: When originally proposed, this rank included the Family Koalidae (Burnett, 1830b [= Phascolarctidae (Owen, 1839) and Potoroidae (J. Gray, 1821)]).

Family Koalidae Burnett, 1830b: 351.

TYPE GENUS: *Koala* Schinz, 1821 [=*Phascolarctos* de Blainville, 1816].

COMMENTS: When originally proposed, this rank was placed in the Race Semidentiae (Burnett, 1830b [= Diprotodontia (Owen, 1877a part)]) and included the genera *Potorus* J. Gray, 1821 [= *Potorous* Desmarest, 1804] and *Koala* Schinz, 1821 [= *Phascolarctos* de Blainville, 1816]. Synonymised within the Family Phascolarctidae by Marshall (1981: 31), Marshall *et al.* (1981: 490), and McKenna and Bell (1997: 66).

Race Gliridentiae Burnett, 1830b: 351.

COMMENTS: Included the kinds Macropidae (J. Gray, 1821 [= Macropodidae (J. Gray, 1821)]) and Vombatidae (Burnett, 1830b).

Family Marsupidae Swainson, 1835: 391.

TYPE GENUS: Not based on a genus group name.

COMMENTS: When originally proposed, this rank was placed in the Order Glires (Linnaeus, 1758) and included the genera *Halmaturus* Illiger, 1811 [= *Macropus* Shaw, 1790]; *Hypsiprymnus* Illiger, 1811 [= *Potorous* Desmarest, 1804]; *Phalangista* É. Geoffroy and G. Cuvier, 1795 [= *Phalanger* Storr, 1780]; *Petaurista* Desmarest, 1821 [= *Petauroides* Thomas, 1888a]; and *Petaurus* Shaw, 1791. Synonymised within Macropodidae by Marshall (1981: 27) and Marshall *et al.* (1990: 492).

Tribe Carpophaga Owen, 1839a: 12, 19.

TYPE GENUS: Not based on a genus group name.

COMMENTS: When originally proposed, this rank was placed in the Marsupialia (Illiger, 1811) and included the genera *Phalangista* É. Geoffroy and G. Cuvier, 1795 [= *Phalanger* Storr, 1780]; *Petaurus* Shaw, 1791; and *Phascolarctus* Owen, 1839a [= *Phascolarctos* de Blainville, 1816a]. Recognised at tribe by Owen (1840: 322), at family rank by Giebel (1855: xi, 691) and Owen (1859: 52), and as an order by Haeckel (1866: clvii). Recognised at a rank between the Suborder Syndactyli and the families Tarsipedidae, Phalangeridae and Phascolarctidae by Gill (1872: 25). Synonymised within Phalangeroidea by Marshall *et al.* (1990: 493) and McKenna and Bell (1997: 61).

Tribe? Macropodina J. Gray, 1842e: 16.

TYPE GENUS: *Macropus* Shaw, 1790.

COMMENTS: When originally proposed, this rank was placed in the Family Macropodidae (J. Gray, 1821) and included the genera *Dendrolegus* [= *Dendrolagus*] S. Müller, 1840; *Macropus* Shaw, 1790; *Halmaturus* Illiger, 1811 [= *Macropus* Shaw, 1790]; *Osphranter* Gould, 1842e; *Petrogale* J. Gray, 1837; *Bettongia* J. Gray, 1837; *Hypsiprymnus* Illiger, 1811 [= *Potorous* Desmarest, 1804]; *Lagorchestes* Gould, 1841 [1841–1842]; and *Phascolomys* Duméril, 1806a [= *Vombatus* É. Geoffroy, 1803b].

Order? Phytophaga Wagner, 1843: vi, 64.

COMMENTS: When originally proposed, this rank included the Scandentia (Wagner, 1843 [= Diprotodontia (Owen, 1877a part)]), Macropoda (Wagner, 1843 [= Macropodoidea (J. Gray, 1821)]) and Glirina (Wiegmann, 1832 [= Vombatidae (Burnett, 1830)]).

HOMONYMS:

Phytophaga Huxley, 1872: 283, sloths of the Class Mammalia (Order Pilosa, Family Bradypodidae). Synonymised within the Suborder Phyllophaga (Owen, 1842: 167) by McKenna and Bell (1997: 93).

Scandentia Wagner, 1843: vi, 65.

COMMENTS: Rank unknown but is either at tribe or family rank. When originally proposed, this rank was placed in the Order? Phytophaga (Wagner, 1843 [= Diprotodontia (Owen, 1877a part)]) and included the genera *Phalangista* É. Geoffroy and G. Cuvier, 1795 [= *Phalanger* Storr, 1780]; *Petaurus* Shaw, 1791; and *Phascolarctos* de Blainville, 1816a.

HOMONYMS:

Scandentia G. Fischer, 1817: 372, anteaters of the Class Mammalia (Order Pilosa, Family Myrmecophagidae). See McKenna and Bell (1997: 92).

Scandentia Newman, 1843: 34, primates of the Class Mammalia (Order Primates). Synonymised with the Order Primates by McKenna and Bell (1997: 323).

Scandentia Wagner, 1855: xix, 524, tree shrews of the Class Mammalia. Recognised as a valid order by McKenna and Bell (1997: 356).

Order Macropoda Ameghino, 1889: xx, 263, 266.

COMMENTS: When originally proposed, this rank was placed in the Grand Group Alloidea (Ameghino, 1889 [= Mammalia (Linnaeus, 1758 part)]) and included the families Macropodidae (J. Gray, 1821), Phalangistidae (Owen, 1839a [= Phalangerida (Aplin & Archer, 1987 part)]) and Phascolomyidae (Owen, 1839a [= Vombatidae (Burnett, 1830b)]). Rank also discussed by Ameghino (1916: 448, 452). Synonymised within Diprotodontia by McKenna and Bell (1997: 58).

HOMONYMS:

Macropoda Illiger, 1811, rodents of the genera *Dipus* Zimmermann, 1780: 354; *Pedetes* Illiger, 1811: 81; and *Meriones* Illiger, 1811: 82 of the Class Mammalia (Order Rodentia, families Dipodidae, Muridae & Pedetidae). See D. Wilson and Reeder (2005: 882, 1234, 1535). See individual entry.

Macropoda Wagner, 1843, macropods of the Class Mammalia (Order Diprotodontia, Superfamily Macropodoidea). Synonymised within the Superfamily Macropodoidea J. Gray, 1821 here. See individual entry.

Suborder Diprotodonta Thomas, 1896a: 876.

COMMENTS: When originally proposed, this rank was placed in the Order Marsupialia (Illiger, 1811) and include the families † Epanorthidae (Ameghino, 1889: 268, 270), Phalangeridae (Thomas, 1888a), Phascolomyidae (Owen, 1839a [= Vombatidae (Burnett, 1830b)]) and Macropodidae (J. Gray, 1821). Name is an incorrect emendation of Diprotodontia. Spelling used by a number of authors including at the suborder rank by Deberer (1909: 614), infraordinal rank by M. Archer (1984: 786), and ordinal rank by Ride (1964a: 97, 99), Kirsch (1968a: 420; 1977a: 112; 1977b: 45), Turnbull (1971: 176), Strahan (1983: xxi), M. Archer (1984: 787), Baverstock (1984a: 2) and Woodburne (1984a: 71). The correct spelling was reviewed by Aplin and Archer (1987: xliii) who concluded that Diprotodonta is an invalid emendation of Diprotodontia.

Section Syndactyla Diprotodontia Wood Jones, 1924 [1923–1925]: 171.

COMMENTS: When originally proposed, this rank was placed in the Suborder Syndactyla (Wood Jones, 1923 [1923–1925]) and included the families Phalangeridae (Thomas, 1888a) and Macropodidae (J. Gray, 1821). Recognised at ordinal rank by Szalay (1982: 631; 1994: 42). Synonymised within Diprotodontia by Aplin and Archer (1987: xliii), and within Syndactyli Gill (1871a: 533) by McKenna and Bell (1997: 56).

Duplicicommissurala Abbie, 1937: 432.

COMMENTS: Rank not specified but placed below the Marsupialia (Illiger, 1811) at a rank equivalent to Diprotodontia. Synonymised within Diprotodontia by Aplin and Archer (1987: xliii) and McKenna and Bell (1997: 58).

Semisuborder Diprotodontiformes Szalay, 1993: 240.

COMMENTS: When originally proposed, this rank was placed in the Suborder Vombatiformes (Woodburne, 1984a) with the Semisuborder Vombatomorphia (Aplin & Archer, 1987) was separated. Synonymised within Diprotodontia by McKenna and Bell (1997: 58).

Order Diprotodontiformes Kinman, 1994: 37.

COMMENTS: When originally proposed, this rank was placed in the Class Mammalea (Kinman, 1994 [= Mammalia (Linnaeus, 1758)]). Synonymised within Diprotodontia by McKenna and Bell (1997: 58).

Suborder Vombatiformes Woodburne, 1984 *sensu* Aplin & Archer, 1987

Suborder Vombatiformes Woodburne, 1984a: 71.

COMMENTS: When originally proposed as a new rank it was placed in the Order Diprotodonta [sic] (Owen, 1966a [= Diprotodontia (Owen, 1877a)]) and included the superfamilies Vombatoidea (Kirsch, 1977b) and Phascolarctoidea (Woodburne, 1984a). Suborder recognised by Aplin and Archer (1987: xxi), Szalay (1994: 43), Strahan (1995: 7, 194), Kirsch *et al.* (1997: 245), Long *et al.* (2002: 77), Groves (2005b: 43), and Van Dyck and Strahan (2008: 9, 197). Author for rank attributed to Woodburne (1984a: 71) by Aplin and Archer (1987: xxi), Marshall *et al.* (1990: 490), Szalay (1994: 43), Kirsch *et al.* (1997: 245) and to Burnett (1830a: 351) by Groves (2005b: 43). Synonymised within Diprotodontia and Vombatoidea by McKenna and Bell (1997: 58, 59). This suborder included the new infraorders Phascolarctomorphia and Vombatomorphia by Aplin and Archer (1987: xxi), which was followed by Long *et al.* (2002: 77, 85) and is recognised here.

Superfamily Vombatoidea Kirsch, 1968: 420.

TYPE GENUS: *Vombatus* É. Geoffroy, 1803b.

COMMENTS: When originally proposed this rank was attributed to Iredale and Troughton (1934: viii, 33) and placed Order Diprotodonta [sic] (Owen, 1966a=Thomas, 1896a) and included the Families Vombatidae (Iredale & Troughton, 1939 [= 1934] [= Burnett, 1830b] and Phascolarctidae (Owen, 1839a). The recognition of the superfamily name with Kirsch (1968: 420) as the author appears to have only been recognised by Kirsch (1977a: 112) and Marshall (1990: 490). Kirsch *et al.* (1997: 245) and McKenna and Bell (1997: 59) recognised the author of the superfamily as Burnett (1830b: 351). Name also recognised at superfamily rank with unnamed author by Kirsch (1977b: 45), Szalay (1982: 631) and Strahan (1983: xxi).

Infraorder Phascolarctomorphia Aplin & Archer, 1987

Infraorder Phascolarctomorphia Aplin & Archer, 1987: xxi, xvli.

COMMENTS: When originally proposed as a new rank it was placed in the Suborder Vombatiformes (Woodburne, 1984a) and included the Family Phascolarctidae (Owen, 1839a). Synonymised within Diprotodontia by McKenna and Bell (1997: 58), but was subsequently recognised by Black (1999: 16) and Long *et al.* (2002: 77).

Family Phascolarctidae Owen, 1839

Family Phascolarctidae Owen, 1839a: 19.

TYPE GENUS: *Phascolarctos* de Blainville, 1816a.

COMMENTS: When originally proposed, this rank was placed in the Tribe Carpophaga (Owen, 1839a [= Diprotodontia (Owen, 1877a part)]) and included the genus *Phascolarctus* Owen, 1839a [= *Phascolarctos* de Blainville, 1816a]. Winge (1893: 88) recognised the family and included the tribes Phascolarctini (Winge, 1893a), Thylacoleontini (Winge, 1893a: 88), Diprotodontini (Winge, 1893a: 88), and Phascolomyini (Winge, 1893a). Rank also recognised by Owen (1940: 332) but it was included within the Subfamily Phascolarctinae, in the Family Phalangeridae, by Thomas (1888a: xii, 209), Bensley (1903: 125), Simpson (1945: 46) and Tate (1945a: 11) who also included *Schoinobates* [= *Petauroides*]. Recognised at the family rank by Sonntag (1922: 894–895), who included the subfamilies Phascolarctinae and Phascolomyinae, Gill (1872: 25), Iredale and Troughton (1934: viii, 33), Troughton (1967: 111), Kirsch (1968a: 420), Ride (1970: 225), Honacki *et al.* (1982: 50), McKay (1988a: 51), Groves (1993e: 43) and subsequent authors.

Family Koladae J. Gray, 1821: 309.

TYPE GENUS: *Kola* J. Gray, 1821 [= *Phascolarctos* de Blainville, 1816].

COMMENTS: When originally proposed, this rank was placed in the Order Brutae (J. Gray, 1821 [= Marsupialia (Illiger, 1811 part)]) and included the genus *Kola* [sic] Schinz, 1821 [= *Phascolarctos* de Blainville, 1816a]. Synonymised within the Family Phascolarctidae by McKenna and Bell (1997: 66).

Family Phascolarctideae Lesson, 1842: 192.

TYPE GENUS: *Phascolarctos* de Blainville, 1816a.

COMMENTS: When originally proposed, this rank was placed in the Order Mastodidelphie (Lesson, 1842 [= Marsupialia (Illiger, 1811)]) and included the genus *Phascolarctos* de Blainville, 1816a. Does not appear to have been recognised by other authors.

Tribe Phascolarctins Gervais, 1855a: 273.

TYPE GENUS: *Phascolarctos* de Blainville, 1816a.

COMMENTS: When originally proposed, this rank was placed in the Family Phalangeridés (Gervais, 1855a [= Phalangerida (Aplin & Archer, 1987)]) and included the genus *Phascolarctos* de Blainville, 1816a.

Subfamily Phascolarctinae Thomas, 1888a: xii, 209.

TYPE GENUS: *Phascolarctos* de Blainville, 1816a.

COMMENTS: When originally proposed, this rank was placed in the Family Phalangeridae (Thomas, 1888a) and included the genus *Phascolarctus* Owen, 1839a [= *Phascolarctos* de Blainville, 1816a]; which was followed by Osborn (1910: 517). Synonymised within the Family Phascolarctidae by McKenna and Bell (1997: 66).

Tribe Phascolarctini Winge, 1893a: 88.

TYPE GENUS: *Phascolarctos* de Blainville, 1816a.

COMMENTS: When originally proposed, this rank was placed in the Family Phascolarctidae (Owen, 1839a) and included the genus *Phascolarctos* de Blainville, 1816a. Tribe rank subsequently recognised by Winge (1941: 69).

Superfamily Phascolarctoidea Woodburne, 1984a: 71.

TYPE GENUS: *Phascolarctos* de Blainville, 1816a.

COMMENTS: When originally proposed, this rank was placed in the Suborder Vombatiformes (Woodburne, 1984a) and included the Family Phascolarctidae (Owen, 1839a). Superfamily rank recognised by Marshall *et al.* (1990: 459, 490) who proposed the author as Woodburne (1984a: 71), and Szalay (1994: 43) and Kirsch *et al.* (1997: 245) who both recognised the author as Owen (1839a: 19). Synonymised within the Family Phascolarctidae by McKenna and Bell (1997: 66) but recognised by Black (1999: 16).

Phascolarctos de Blainville, 1816

Phascolarctos de Blainville, 1816a: 116.

TYPE SPECIES: *Lipurus cinereus* Goldfuss, 1817 [= *Phascolarctos cinereus* (Goldfuss, 1817)] by monotypy.

COMMENTS: Description of *Lipurus cinereus* was seen by de Blainville as a manuscript before it was published in 1817.

Lipurus Goldfuss, 1817: Plate 155, Aa, Ab.

TYPE SPECIES: *Lipurus cinereus* Goldfuss, 1817 [= *Phascolarctos cinereus* (Goldfuss, 1817)] by monotypy.

COMMENTS: The description of the name was by two plates only, with further information provided by Goldfuss (1818: col. 1081; 1819a: col. 271). Year and author of description confirmed by Sherborn (1927: 3608). Synonymised within *Phascolarctos* by J. Gray (1843a: xxii), Waterhouse (1846: 258) and subsequent authors including Thomas (1888a: 209), Iredale and Troughton (1934: 33), Marshall (1981: 32), McKay (1988a: 51) and Marshall *et al.* (1990: 490).

HOMONYMS:

Lipura Illiger, 1811: 95, marmots of the Class Mammalia (Order Rodentia, Family Sciuridae). Genus is a synonym of *Marmota* Blumenbach (1779: 79). See Palmer (1904: 378) and McKenna and Bell (1997: 124). *Lipura* has been attributed to Storr (1780) by authors including Thorington and Hoffmann (2005: 799) but this appears to be incorrect.

Lipurus Agassiz, 1846: 213, louse of the Class Insecta (Order Phthiraptera, Family Philopteridae). Name is an emendation of *Lipeurus* Nitzsch, 1818: 292.

Morodactylus Goldfuss, 1820a: xxxiii, 445.

TYPE SPECIES: *Nomen novum* for *Lipurus* Goldfuss, 1817.

COMMENTS: Synonymised within *Phascolarctos* by J. Gray (1843a: xxii), Thomas (1888a: 209), Iredale and Troughton (1934: 33), Marshall (1981: 32), McKay (1988a: 51) and Marshall *et al.* (1990: 490).

Koala Schinz, 1821: 265.

TYPE SPECIES: *Lipurus cinereus* Goldfuss, 1817 [= *Phascolarctos cinereus* (Goldfuss, 1817)] by monotypy.

COMMENTS: Based on 'Les Koala' of G. Cuvier (1816a: 184). Genus also recognised by Burnett (1830a: 351) but with a different type species. Synonymised within *Phascolarctos* by J. Gray (1843a: xxii), Iredale and Troughton (1934: 33) and McKay (1988a: 51).

Kola J. Gray, 1821: 309.

TYPE SPECIES: Incorrect subsequent spelling of *Koala* Schinz, 1821.

COMMENTS: May also be derived from 'Les Koala' G. Cuvier (1816a: 184). Synonymised within *Phascolarctos* by McKenna and Bell (1997: 66).

Draximenus Lay, 1825: 744.

TYPE SPECIES: *Nomen novum* for *Lipurus* Goldfuss, 1817. Cited as 'new name only'.

COMMENTS: Date of publication uncertain as Palmer (1904: 244) give the date as 1845, while Iredale and Troughton

(1934: 33) and Groves (2005b: 43) give the date as 1825. Cited as 'new name only' by Iredale and Troughton (1934: 33). Synonymised within *Phascolarctos* by Iredale and Troughton (1934: 33) and McKay (1988a: 51).

Phascolaretus Gray, 1825: 340.

TYPE SPECIES: Incorrect subsequent spelling of *Phascolarctos* de Blainville, 1816a.

COMMENTS: Spelling not subsequently used.

Koala Burnett, 1830b: 351.

TYPE SPECIES: *Koala subiens* Burnett, 1830b [= *Phascolarctos cinereus* (Goldfuss, 1817)] by monotypy.

COMMENTS: Synonymised within *Phascolarctos* by Marshall (1981: 32), Marshall *et al.* (1990: 490) and McKenna and Bell (1997: 66).

Liscurus McMurtrie, 1834: 78.

TYPE SPECIES: Incorrect subsequent spelling of *Lipurus* Goldfuss, 1817.

COMMENTS: Synonymised within *Phascolarctos* by Iredale and Troughton (1934: 33), McKay (1988a: 51) and Groves (2005b: 43).

Phascolarctus Owen, 1839a: 15, 19.

TYPE SPECIES: Incorrect subsequent spelling of *Phascolarctos* de Blainville, 1816a.

COMMENTS: Spelling recognised by Thomas (1888a: 209) but synonymised within *Phascolarctos* by Palmer (1904: 529).

Phascolarctos cinereus (Goldfuss, 1817)

Koala

Lipurus cinereus Goldfuss, 1817: Plate 155, Aa, Ab.

TYPE LOCALITY: New South Wales, Australia.

COMMENTS: Described in inaccurate detail by Perry (1810 [1810–1811]: Text to Plate 17), which appears to be the first published illustration of the species, but it was not given a scientific name and only referred to by the common names Koalo and Kaolo. The first formal description of the species name included two plates with no associated text, but further information was provided by Goldfuss (1818: col. 1081; 1819a: col. 271, 274), the latter of which was acknowledged as the original description by Thomas (1888a: 209) who reviewed the early taxonomic history. Author and year of description confirmed by Sherborn (1927: 3608). Species name placed within *Phascolarctos* by J. Fischer (1829: 285) and most authors since. Iredale and Troughton (1934: 33) included *cinereus* and *adustus* as subspecies. One year later *victor* was described, and since then there have traditionally been three subspecies recognised within *P. cinereus*: *cinereus*, *adustus* and *victor*. No subspecies were recognised by Lee and Martin (1988: 22) or McKay (1988a: 52), though they were recognised as races or sub-species by Martin and Handasyde (1999: 7). An assessment of the mitochondrial DNA variation confirmed that the geographically distinct populations of koalas represented a single evolutionary significant unit and that the morphological differences may be interpreted as clinal variation and not subspecies (Houlden *et al.*, 1999: 1009), which is supported by the observations of Takami *et al.* (1998: 1163). As a result no subspecies are recognised here.

phascolarctos fuscus Desmarest, 1821: 276.

TYPE LOCALITY: Australia.

COMMENTS: Recognised by Waterhouse (1838a: 68; 1841a: 295). Synonymised within *cinereus* by Waterhouse (1846: 259), Thomas (1888a: 210), Iredale and Troughton (1934: 33), McKay (1988a: 51) and subsequent authors.

Phascolarctos Flindersii Lesson, 1827a: 221.

TYPE LOCALITY: Australia.

COMMENTS: Possibly based on same animal as *Phascolarctos fuscus* Desmarest, 1821. Synonymised within *cinereus* by Waterhouse (1846: 259), McKay (1988a: 52) and subsequent authors.

[Phascolarctos] koala J. Gray, 1827a: 205.

TYPE LOCALITY: Replacement name only.

COMMENTS: Although the combination J. Gray used was *K. Koala* it is listed under the generic heading: '*Phascolarctos*' the 'K' is undoubtedly a misprint. Synonymised within *cinereus* by Thomas (1888a: 210), Iredale and Troughton (1934: 33) and McKay (1988a: 52) and subsequent authors.

Koala Subiens Burnett, 1830b: 351.

TYPE LOCALITY: *Nomen nudum*.

COMMENTS: Synonymised within *cinereus* by Iredale and Troughton (1934: 33), McKay (1988a: 52) and subsequent authors.

K. [oala] cinerea McMurtrie, 1834: 78.

TYPE LOCALITY: *Nomen novum* only.

COMMENTS: *Nomen nudum*. Synonymised within *cinereus* by McKay (1988a: 52).

Phascolarctos cinereus adustus Thomas, 1923b: 246.

TYPE LOCALITY: O'Bil O'Bil near Mundubbera, Queensland, Australia.

COMMENTS: Subspecies status accepted by Iredale and Troughton (1934: 33), Finlayson (1934: 220), and Troughton (1967: 114). Synonymised within *cinereus* by McKay (1988a: 52). Recognised as a subspecies by Strahan (1983: 113; 1995: 197), Clayton *et al.* (2006: 103), and Van Dyck and Strahan (2008: 201) who suggested that the subspecies may represent arbitrary selections from a cline.

Phascolarctos cinereus victor Troughton, 1935a: 139.

TYPE LOCALITY: 'Booral', Victoria, Australia.

COMMENTS: Synonymised within *cinereus* by McKay (1988a: 52). Recognised as a subspecies by Troughton (1967: 114), Strahan (1983: 113; 1995: 197), Clayton *et al.* (2006: 103), and Van Dyck and Strahan (2008: 201) who suggested that the subspecies may represent arbitrary selections from a cline.

Infraorder Vombatomorphia Aplin & Archer, 1987

Infraorder Vombatomorphia Aplin & Archer, 1987: xxi.

COMMENTS: When originally proposed, this rank was placed in the Suborder Vombatiformes (Woodburne, 1984a) and included the families † Diprotodontidae (Gill, 1872: 26), † Palorchestidae (Tate, 1948b: 338), † Wynyardiidae (Osgood, 1921: 138), † Ilariidae (Tedford & Woodburne, 1987: 401), Vombatidae (Burnett, 1830b) and † Thylacoleonidae (Gill, 1872: 26). Name recognised as a semisuborder by Szalay (1994: 43). Synonymised within Diprotodontia and Vombatoidea by McKenna and Bell (1997: 58, 59) but recognised as an infraorder by Long *et al.* (2002: 85).

Infraorder Vombatimorphia Aplin & Archer, 1987: xlvi.

COMMENTS: McKenna and Bell (1997: 58, 59) suggested the name is a *lapsus calami* as the correct spelling should be Vombatomorphia (Aplin and Archer, 1987: xxi), but this spelling was used at semisuborder rank by Szalay (1993: 240).

Family Vombatidae Burnett, 1830 *sensu* Dawson, 1983

Family Vombatidae Burnett, 1830b: 351.

TYPE GENUS: *Vombatus* É. Geoffroy, 1803b.

COMMENTS: When originally proposed, this rank was placed in the Race Gliridentiae (Burnett, 1830b [= Diprotodontia (Owen, 1877a part)]) and included the genus *Phascolomys* Duméril, 1806a [= *Vombatus* É. Geoffroy, 1803b]. Phascolomyidae (Goldfuss, 1820a), is based on *Phascolomis*, a junior synonym (Haltenorth, 1958: 32; Groves, 1993e: 45). It was noted by Groves (1993e: 45; 2005b: 43) that as Phascolomyidae was replaced with Vombatidae before 1961, and because Vombatidae has won general acceptance, it is to be maintained following Article 40.2 of the Code (ICZN, 1999: 46). Family rank name recognised by most authors including Iredale and Troughton (1934: viii, 33), Troughton (1967: 116), Dawson (1983: 101; 1988: 48) and subsequent authors.

Family Phascolomyda Goldfuss, 1820a: xxii, 444.

TYPE GENUS: *Phascolomys* Duméril, 1806a [= *Vombatus* É. Geoffroy, 1803b].

COMMENTS: When originally proposed, this rank was placed in the Order Marsupialia (Illiger, 1811) and included the genus *Phascolomys* Duméril, 1806a [= *Vombatus* É. Geoffroy, 1803b]. Synonymised within Phascolomidae by Simpson (1945: 46) and within Vombatidae by Marshall (1981: 30), Marshall *et al.* (1990: 491) and also within the Superfamily Vombatoidea by McKenna and Bell (1997: 59).

Order Glires J. Gray, 1821: 309.

COMMENTS: When originally proposed, this rank was placed in the Class Pedimanes (J. Gray, 1821 [= Mammalia (Linnaeus, 1758 part)]) and included the Family Phascolomidae (J. Gray, 1821 [= Vombatidae Burnett, 1830]).

HOMONYMS:

Order Glires Linnaeus, 1758, Rodentia (Bowdich, 1821) and Lagomorpha (Brandt, 1855) of the Class Mammalia. Recognised as a superorder here. See individual entry.

Family Phascolomidae J. Gray, 1821: 309.

TYPE GENUS: *Phascolomys* Duméril, 1806a [= *Vombatus* É. Geoffroy, 1803b].

COMMENTS: When originally proposed, this rank was placed in the Order Glires (J. Gray, 1821 [= Vombatidae (Burnett, 1830b)]) and included the genus *Phascolomys* Duméril, 1806a [= *Vombatus* É. Geoffroy, 1803b]. Recognised at family rank by Bonaparte (1845: 6) and Simpson (1945: 46). Synonymised within Vombatidae by McKenna and Bell (1997: 60).

Tribe Phascolomina J. Gray, 1825a: 340.

TYPE GENUS: *Phascolomys* Duméril, 1806a [= *Vombatus* É. Geoffroy, 1803b].

COMMENTS: When originally proposed, this rank was placed in the Family Didelphidae (J. Gray, 1821: 308) and included the genus *Phascolomys* Duméril, 1806a [= *Vombatus* É. Geoffroy, 1803b]. Recognised as a within the Family Phascolomidae by Bonaparte (1845: 6). Synonymised within Phascolomidae by Waterhouse (1846: 241) and within Vombatidae by McKenna and Bell (1997: 60).

Family Glirina Wiegmann, 1832: 51.

TYPE GENUS: Not based on a genus group name.

COMMENTS: When originally proposed, this rank was placed in the Order Marsupialia (Illiger, 1811) and included the genus *Phascolomys* Duméril, 1806a [= *Vombatus* É. Geoffroy, 1803b]. Synonymised within Vombatidae by Marshall (1981: 30) and Marshall *et al.* (1990: 491).

Family Phascolomyidae Owen, 1839a: 19.

TYPE GENUS: *Phascolomys* Duméril, 1806a [= *Vombatus* É. Geoffroy, 1803b].

COMMENTS: When originally proposed, this rank was placed in the Tribe Rhizophaga (Owen, 1839a [= Vombatidae (Burnett, 1830)]) and included the genera *Phascolomys* Duméril, 1806a [= *Vombatus* É. Geoffroy, 1803b]; and † *Diprotodon* Owen, 1838b: 362. Family name recognised by

Owen (1840: 332), Waterhouse (1841a: 60; 1846: 241), Gill (1872: 25), Thomas (1888a: xii, 212) and Bensley (1903: 158). Synonymised within Phascolomidae by Simpson (1945: 46) and within Vombatidae by Marshall (1981: 30), Marshall *et al.* (1990: 491) and McKenna and Bell (1997: 60).

Tribe Rhizophaga Owen, 1839a: 18, 19.

TYPE GENUS: Not based on a genus group name.

COMMENTS: When originally proposed, this rank was placed in the Marsupialia (Illiger, 1811) and included the Family Phascolomyidae (Owen, 1839a [= Vombatidae (Burnett, 1830b)]). Recognised at tribe rank by Owen (1840: 329) and at family rank by Giebel (1855: xi, 668) and Owen (1859: 52), and as an order by Haeckel (1866: clviii). Recognised as a suborder by Gill (1871a: 533; 1872: vi, 25) who included the Family Phascolomyidae. Synonymised within Superfamily Vombatoidea by McKenna and Bell (1997: 59).

Subfamily Phascolomydina Bonaparte, 1840: 257.

TYPE GENUS: *Phascolomys* Duméril, 1806a [= *Vombatus* É. Geoffroy, 1803b].

COMMENTS: When originally proposed, this rank was placed in the Family Halmaturidae (Bonaparte, 1831 [= Macropodidae (J. Gray, 1821)]). Synonymised within Vombatidae by Marshall (1981: 30), Marshall *et al.* (1990: 491) and McKenna and Bell (1997: 60).

Family Phascolomysideae Lesson, 1842: 192.

TYPE GENUS: *Phascolomys* Duméril, 1806a [= *Vombatus* É. Geoffroy, 1803b].

COMMENTS: When originally proposed this rank was placed in the Order Mastodidelphie (Lesson, 1842 [= Marsupialia (Illiger, 1811)]) and included the genus *Phascolomys* Duméril, 1806a [= *Vombatus* É. Geoffroy, 1803b]. Does not appear to have been recognised by subsequent authors.

Family Phascolomidae Bonaparte, 1845: 6.

TYPE GENUS: *Phascolomys* Duméril, 1806a [= *Vombatus* É. Geoffroy, 1803b].

COMMENTS: When originally proposed, this rank was placed in the Order Marsupialia (Illiger, 1811) and included the Tribe Phascolomina (J. Gray, 1825a [= Vombatidae (Burnett, 1830b)]). Synonymised within Vombatidae by Marshall (1981: 30) and Marshall *et al.* (1990: 491).

Family Phascolomydés Gervais, 1855a: 267.

TYPE GENUS: *Phascolomys* Duméril, 1806a [= *Vombatus* É. Geoffroy, 1803b].

COMMENTS: When originally proposed, this rank was placed in the Order Marsupiaux (Gervais, 1855a [= Marsupialia (Illiger, 1811)]) and included the genus *Phascolomys* Duméril, 1806a [= *Vombatus* É. Geoffroy, 1803b].

Family Phascolomyida Haeckel, 1866: clvii.

TYPE GENUS: *Phascolomys* Duméril, 1806a [= *Vombatus* É. Geoffroy, 1803b].

COMMENTS: When originally proposed, this rank was placed in the Order Rhizophaga (Owen, 1839a [= Vombatidae Burnett, 1830]) and included the genus *Phascolomys* Duméril, 1806a [= *Vombatus* É. Geoffroy, 1803b]. Synonymised within Vombatidae by McKenna and Bell (1997: 60).

Glirina Haeckel, 1866: clvii.

TYPE GENUS: Not based on a genus group name.

COMMENTS: Rank not listed but placed between the Family Phascolomyida (Haeckel, 1866 [= Vombatidae (Burnett, 1830b)]) and the genus *Phascolomys* Duméril, 1806a [= *Vombatus* É. Geoffroy, 1803b]. Synonymised within *Vombatus* by McKenna and Bell (1997: 60).

Tribe Fossoria Owen, 1877b: 360.

TYPE GENUS: Not based on a genus group name.

COMMENTS: When originally proposed, this rank was placed in the Poëphaga (Owen, 1839a [= Macropodoidea J. Gray, 1821]) and included the genus *Phascolomys* Duméril, 1806a [= *Vombatus* É. Geoffroy, 1803b] and extinct taxa including the † *Phascolonus gigas* Owen, 1858b: 447, 450. This appears to be the first usage of this rank but this needs to be confirmed.

HOMONYMS:

Fossoria Voigt, 1839: xxii, 475, digger wasps of the Class Insecta (Order Hymenoptera). Rank is a synonym of the Family Sphecidae.

Family Fossoria Alfken, 1891: 120, digger wasps of the Class Insecta (Order Hymenoptera). Rank is a synonym of the Family Sphecidae.

Tribe Phascolomyini Winge, 1893a: 88.

TYPE GENUS: *Phascolomys* Duméril, 1806a [= *Vombatus* É. Geoffroy, 1803b].

COMMENTS: When originally proposed, this rank was placed in the Family Phascolarctidae (Owen, 1839a) and included the genus *Phascolomys* Duméril, 1806a [= *Vombatus* É. Geoffroy, 1803b]. Tribe rank subsequently recognised by Winge (1941: 68).

Lasiorhinus J. Gray, 1863

Lasiorhinus J. Gray, 1863b: 458.

TYPE SPECIES: *Lasiorhinus mcoyi* J. Gray, 1863b [= *Lasiorhinus latifrons* (Owen, 1845b)] by monotypy.

COMMENTS: Synonymised within *Phascolomys* by Thomas (1888a: 213), but recognised by Iredale and Troughton (1934: viii, 35), Dawson (1988: 48) and subsequent authors. Groves (1993e: 45) noted that this genus needs revision.

HOMONYMS:

Lasiorhinus Broun, 1880: 532, true weevils of the Class Insecta (Order Coleoptera, Family Curculionidae). Genus a synonym of *Pogonorhinus* Hutton, 1904: 368, as 218.

Wombatula Troughton, 1941: 145, 148.

TYPE SPECIES: *Phascolomys gillespei* De Vis, 1900 [= *Lasiorhinus krefftii* (Owen, 1872)] by monotypy.

COMMENTS: Name was first mentioned by Iredale & Troughton (1934: viii, 35), but this was noted to be a *nomen nudum* by McKenna and Bell (1997: 61), who proposed the correct citation to be Troughton (1941: 145, 148). Synonymised within *Lasiorhinus* by Ride (1970: 249), Marshall (1981: 30), Dawson (1988: 48) and Marshall *et al.* (1990: 491).

Lasiorhinus krefftii (Owen, 1872)

Northern Hairy-nosed Wombat

† *Lasiorhinus krefftii krefftii* (Owen, 1872)

† *Phascolomys Krefftii* Owen, 1872: 178; Plates 17, 20.

TYPE LOCALITY: Breccia Cavern, Wellington Caves, New South Wales, Australia.

COMMENTS: Described as a Pleistocene fossil. Not recognised by Ride (1970: 94) who recognised *L. gillespiei* and *L. barnardi* in preference. Wilkinson (in Merrilees, 1973: 181) argued that *L. krefftii* is conspecific with living and recently extinct wombats of Epping Forest and Moonie River (Qld.) and recently extinct wombats of Deniliquin, hence the name should take preference over *L. gillespiei*, the next available name. Kirsch and Calaby (1977: 23), Honacki *et al.* (1982: 50) and Dawson (1983: 115; 1988: 48) synonymised *gillespiei* and *barnardi* within *L. krefftii*.

FUTURE TAXONOMIC RESEARCH: It was proposed by Groves (2005b: 43) that *krefftii* may be better restricted to the Pleistocene species and that neither *barnardi* nor *gillespiei* may belong to it. This conclusion is derived from an unpublished MSc thesis by Scott (1988: iii, 76), who proposed that the living population from Epping Forest (*L. k. barnardi*) and the recently extinct ones from Moonie River (*L. k. gillespiei*) and Deniliquin are different from the Pleistocene *L. krefftii* and from each other, and merit status as distinct species. This proposition should be tested again, in light of the additional material from Epping Forest and from Pleistocene deposits which has become available since then, and from DNA sequencing. A fourth population, related to that from Moonie River, is known from fossil (subfossil?) material from the Willandra Lakes, and may have survived until the early 19th century.

† *Lasiorhinus krefftii gillespiei* (De Vis, 1900)

† *Phascolomys gillespiei* De Vis, 1900: 16; Plates 9–10.

TYPE LOCALITY: Bullanon Station, Moonie River, South East Queensland, Australia.

COMMENTS: Recognised as a species within a new genus *Wombatula* by Iredale and Troughton (1934: viii, 35) and Troughton (1967: 123) and within *Lasiorhinus* by Crowcroft (1967: 383, 397) and Ride (1970: 94). Synonymised within *L. krefftii* by Kirsch and Calaby (1977: 23), Honacki *et al.* (1982: 50), Dawson (1983: 115; 1988: 48) and Strahan (1983: 122; 1995: 200). Subspecies recognised by Groves (2005b: 43), but not Van Dyck and Strahan (2008: 203).

Lasiorhinus krefftii barnardi Longman, 1939

Lasiorhinus latifrons barnardi Longman, 1939: 283; Plate 26.

TYPE LOCALITY: Epping Forest Station, 75 km west of Clermont, Queensland, Australia.

COMMENTS: Subspecies rank, within *latifrons*, recognised by Troughton (1967: 123). Species rank recognised by Crowcroft (1967: 383, 397) and Ride (1970: 94). Synonymised within *L. krefftii* by Kirsch and Calaby (1977: 23), Honacki *et al.* (1982: 50), Strahan (1983: 122; 1995: 200) and Dawson (1988: 48). Subspecies rank recognised by Groves (2005b: 43) but not Van Dyck and Strahan (2008: 203). As noted above, on the limited available samples, there appear to be consistent differences between the fossil *Lasiorhinus* from Wellington Caves (to which the name *L. krefftii* belongs) and the three populations that are known to have survived into historical times at Moonie River (Qld), Epping Forest (Qld) and Deniliquin (NSW); and these latter appear to differ consistently from each other.

Lasiorhinus latifrons (Owen, 1845)

Southern Hairy-nosed Wombat

Phascolomys latifrons Owen, 1845b: 82.

TYPE LOCALITY: South Australia, Australia.

COMMENTS: Recognised within *Phascolomys* by Gould (1859 [1845–1863]: Text to Plates 57–58), Angus (1861: 268) and Thomas (1888a: xii, 217). Taxonomic decision of Iredale and Troughton (1934: 35) to include no subspecies and use the genus *Lasiorhinus*. Taxon recognised by Dawson (1988: 49) and subsequent authors.

Phascolomys lasiorhinus Gould, 1863 [1845–1863]: Text to Plates 59–60.

TYPE LOCALITY: South Australia, Australia.

COMMENTS: Synonymised within *latifrons* by Iredale and Troughton (1934: 35), Dawson (1988: 49) and subsequent authors.

Lasiorhinus M'Coyi J. Gray, 1863b: 458.

TYPE LOCALITY: *Nomen novum* for *Phascolomys lasiorhinus* Gould, 1863 [1845–1863] [= *Lasiorhinus latifrons* (Owen, 1845b)].

COMMENTS: Described as a replacement name for *Phascolomys lasiorhinus* Gould, 1863 [1845–1863]. Synonymised within *latifrons* by Thomas (1888a: 217), Iredale and Troughton (1934: 35), Dawson (1988: 49) and subsequent authors.

Phascolomys niger Krefft, 1871a: Text to Plate 5.

TYPE LOCALITY: Port Lincoln, South Australia, Australia.
COMMENTS: Not *Phascolomys niger* Gould (1863 [1845–1863]: Text to Plate 60). Synonymised within *ursinus* by Thomas (1888a: 217) but typically not considered by subsequent authors so placed within *latifrons* here due to the type locality.
HOMONYMS:
Phascolomys niger Gould, 1863 [1845–1863], the Common Wombat of the Class Mammalia (Order Diprotodontia, Family Vombatidae). Name is a synonym of *Vombatus ursinus hirsutus* (Perry, 1810). See individual entry.

Vombatus É. Geoffroy, 1803

Vombatus É. Geoffroy, 1803b: 185.

TYPE SPECIES: *Didelphis ursina* Shaw, 1800 [= *Vombatus ursinus* (Shaw, 1800)] by monotypy.
COMMENTS: Synonymised within *Phascolomys* by Waterhouse (1846: 242) and Thomas (1888a: 213), but resurrected by Iredale and Troughton (1934: viii, 33), Troughton (1967: 118), Ride (1970: 94) and subsequent authors.

phascolomis [sic] É. Geoffroy, 1803f: 365.

TYPE SPECIES: *Didelphis ursina* Shaw, 1800 [= *Vombatus ursinus* (Shaw, 1800)] by monotypy.
COMMENTS: Name also introduced by É. Geoffroy (1803d: 249, as 149). Synonymised within *Vombatus* by Iredale and Troughton (1934: 33), Marshall (1981: 30), Dawson (1988: 49) and Marshall *et al.* (1981: 491).

Wombatus Desmarest, 1804a: 19.

COMMENTS: Unjustified emendation of *Vombatus* É. Geoffroy, 1803b *fide* Neave (1940: 659).
COMMENTS: Name also referred to by Sonnini (1804: 480). Not considered by Iredale and Troughton (1934: 33–34) and synonymised within *Vombatus* by Dawson (1988: 49) and subsequent authors.

phascolomys [sic] Duméril, 1806a: 19.

TYPE SPECIES: *Nomen novum* for *Phascolomis* É. Geoffroy, 1803f.
COMMENTS: Name recognised as a valid genus by Illiger (1811: 78), Waterhouse (1846: 242) and Thomas (1888a: xii, 213) who referred to the author as É. Geoffroy (1803f: 365). Not considered by Iredale and Troughton (1934: 33–34) and synonymised within *Vombatus* by Dawson (1988: 49) and subsequent authors. Name attributed to Haeckel (1866: clvii) by McKenna and Bell (1997: 60).

Uomlatus Rees, 1807: Sign. Zz5 [= unpaginated - p. 5, Classification].

TYPE SPECIES: *Error pro Wombatus* Desmarest, 1804a.
COMMENTS: Not subsequently recognised.

Opossum Perry, 1810 [1810–1811]: Text to Plate 21.

TYPE SPECIES: *Opossum hirsutum* Perry, 1810 [1810–1811] [= *Vombatus ursinus hirsutus* (Perry, 1810)] by monotypy.
COMMENTS: Synonymised within *Vombatus* by Iredale and Troughton (1934: 34), Dawson (1988: 49) and subsequent authors. Name spelt *Opposum* by Groves (1993e: 45).
HOMONYMS:
Opossum Perry, 1810 [1810–1811], the Feathertail Glider of Class Mammalia (Order Diprotodontia, Family Acrobatidae). Genus is a synonym of *Acrobates* Desmarest, 1817b. See individual entry.

Amblotis Illiger, 1811: 77.

TYPE SPECIES: *Nomen novum* for *Wombatus* Desmarest, 1804a.
COMMENTS: Synonymised within *Phascolomys* by Thomas (1888a: 213) and within *Vombatus* by Iredale and Troughton (1934: 34), Dawson (1988: 49) and subsequent authors.

Phascolomus Rafinesque, 1815: 55.

TYPE SPECIES: *Errore pro Phascolomis* É. Geoffroy, 1803f.
COMMENTS: Refer to Neave (1940: 699). Not considered by Iredale and Troughton (1934: 33–34) and synonymised within *Vombatus* by Dawson (1988: 49).

Amblyotis Agassiz, 1846: 16.

TYPE SPECIES: Incorrect subsequent spelling of *Amblotis* Illiger, 1811.
COMMENTS: Emendation pro for *Amblotis* Illiger, 1811.

Vombatus ursinus (Shaw, 1800)

Bare-nosed Wombat

Vombatus ursinus ursinus (Shaw, 1800)

Didelphis Ursina Shaw, 1800: 504.

TYPE LOCALITY: From an island in the Furneaux Groups, Bass Strait, Australia (probably Clarke or Cape Barren Island). See Spencer and Kershaw (1910b: 37).
COMMENTS: Referred as 'Wom-bat' and 'Womback' by D. Collins (1802: 153, plate). Distribution restricted to Flinders Island by Van Dyck and Strahan (2008: 208). Synonymised within *Phascolomys wombat* by Waterhouse

(1846: 246), though recognising J. Gray (1843a: 95) as the author. Species status recognised within *Phascolomys* by Thomas (1888a: xii, 215), who reviewed its early taxonomic history. Transferred to *Vombatus* by Iredale and Troughton (1934: viii, 34) and subsequent authors. The geographic variation of the mainland, Tasmanian and Flinders Island wombats was assessed by Young (1980: 201) who found the Flinders Island animals clearly smaller than from Tasmania and the mainland. The vernacular name of this species has typically been known as the 'Common Wombat', but an important treatise on wombats by Triggs (2009: 6) refers to this species as the 'Bare-nosed Wombat', which has been followed by several subsequent authors including Borchard and Wright (2010: 16), and Crook *et al.* (2012: 23).

HOMONYMS:

Didelphis ursina Harris, 1808, the Tasmanian Devil of the Class Mammalia (Order Dasyuromorphia, Family Dasyuridae). Synonymised within *Sarcophilus laniarius* (Owen, 1838b). See individual entry.

wombatus fossor Desmarest, 1804a: 20.

TYPE LOCALITY: *Nomen novum* for *Didelphis ursina* Shaw, 1800.

COMMENTS: Name recognised within *Wombatus* by Sevastianof (1809: 445), but synonymised within *Phascolomys wombat* by Waterhouse (1846: 246), who gave the author as Sevastianof (1809: 445). Synonymised within *ursinus* by Thomas (1888a: 215), Iredale and Troughton (1934: 34), Dawson (1988: 49), Groves (1993e: 46; 2005b: 44) and subsequent authors.

Phascolomis wombat Lesueur & Petit, 1807: Plate 28.

TYPE LOCALITY: Probably King Island, Bass Strait, Australia.

COMMENTS: Species recognised by Waterhouse (1838a: 68; 1846: 246) and Gould (1855 [1845–1863]: Text to Plates 55–56). Recognised as *Phascolomys wombat* by Owen (1841: 408). Not considered by Iredale and Troughton (1934: 34) and synonymised within *ursinus* by Thomas (1888a: 215), Dawson (1988: 50) and subsequent authors.

Phascolomis vombatus Leach, 1815: 102; Plate 96.

TYPE LOCALITY: Australia.

COMMENTS: Emendation of *Phascolomis wombat* Lesueur and Petit, 1807. Synonymised within *ursinus* by Waterhouse (1846: 246), Thomas (1888a: 215), Dawson (1988: 50) and Groves (1993e: 46; 2005b: 44).

Phascolomys Bassii Lesson, 1827a: 229.

TYPE LOCALITY: None designated, from King Island, Furneaux Group, Bass Strait, Australia.

COMMENTS: Synonymised within *Phascolomys wombat* by Waterhouse (1846: 246), and within *ursinus* by Thomas (1888a: 216), Iredale and Troughton (1934: 34), Dawson (1988: 50) and Groves (1993e: 46; 2005b: 44).

Vombatus ursinus hirsutus (Perry, 1810)

Opossum hirsutum Perry, 1810 [1810–1811]: Text to Plate 21.

TYPE LOCALITY: Botany Bay, Sydney, Australia.

COMMENTS: Considered a valid species within *Vombatus* by Iredale and Troughton (1934: viii, 34), who also included the taxon *niger* Gould, 1863 as a subspecies, and Troughton (1967: 120). Synonymised within *ursinus* by Thomas (1888a: 215), Ride (1970: 248), as *hirsutus*, and provisionally by Dawson (1988: 50). Not recognised by Groves (1993e: 46; 2005b: 44), Strahan (1983: 117; 1995: 205), Van Dyck and Strahan (2008: 208), but recognised as a subspecies by Clayton *et al.* (2006: 103) and Burbidge *et al.* (2014: 19, 28).

Phascolomys platyrhinus Owen, 1853a: 334.

TYPE LOCALITY: Australia.

COMMENTS: Recognised by Lydekker (1887: 155), but synonymised within *mitchelli* by Thomas (1888a: 213). Synonymised within *hirsutus* by Iredale and Troughton (1934: 34) and synonymised within *ursinus* by Dawson (1988: 50) and Groves (1993e: 46; 2005b: 44). Subspecies not recognised by Clayton *et al.* (2006: 104) but accepted by Van Dyck and Strahan (2008: 208).

Didelphis wombat Voigt, 1802: 681.

TYPE LOCALITY: Unknown.

COMMENTS: Recognised within *Phascolomys* by Desmarest (1821: 276) and Waterhouse (1841a: 300; 1846: 246). Synonymised within *ursinus* by Thomas (1888a: 215), Iredale and Troughton (1934: 34), Dawson (1988: 49) and Groves (1993e: 46; 2005b: 44).

P. [hascolomys] fuscus Tiedemann, 1808: 437.

TYPE LOCALITY: Australia.

COMMENTS: Apparently a synonym of *Didelphis ursina* Shaw, 1800 (via Dawson, 1988: 49). Spelling '*fuscus*' is correct as masculine ending for *mys*. Synonymised within *ursinus* by Thomas (1888a: 215), Iredale and Troughton (1934: 34), Dawson (1988: 49) and Groves (1993e: 46; 2005b: 44).

Phascolomys mitchellii Owen, 1838b: 368.

TYPE LOCALITY: Wellington Valley (Wellington Caves), New South Wales, Australia.

COMMENTS: Designation by Dawson (1983: 99). Species recognised by Thomas (1888a: xii, 213) and Finlayson (1961b: 207), but synonymised within *hirsutus* by Iredale and Troughton (1934: 34). Recognised as a subspecies of *ursinus* by Tate (1951b: 5) and Dawson (1983: 111), but synonymised within *ursinus* by Dawson (1988: 50), Groves (1993e: 46; 2005b: 44) and subsequent authors.

Phascolomys niger Gould, 1863 [1845–1863]: Text to Plate 60.

TYPE LOCALITY: Australia.

COMMENTS: Synonymised within *mitchelli* by Thomas (1888a: 213). Recognised as a subspecies of *hirsutus* by Iredale and Troughton (1934: 34) but synonymised within *ursinus* by Dawson (1988: 50) and Groves (1993e: 46; 2005b: 44).

HOMONYMS:

Phascolomys niger Krefft, 1871a, the Southern Hairy-nosed Wombat of the Class Mammalia (Order Diprotodontia, Family Vombatidae). Name is a synonym of *Lasiorhinus latifrons* (Owen, 1845b). See individual entry.

Phascolomys Angasii J. Gray, 1863b: 458.

TYPE LOCALITY: Australia.

COMMENTS: Synonymised within *mitchelli* by Thomas (1888a: 214). Synonymised within *niger* by Iredale and Troughton (1934: 34) and within *ursinus* by Dawson (1988: 50) and Groves (1993e: 46; 2005b: 44).

Phascolomys setosus J. Gray, 1863b: 459.

TYPE LOCALITY: Australia.

COMMENTS: Synonymised within *mitchelli* by Thomas (1888a: 214). Synonymised within *niger* by Iredale and Troughton (1934: 34) and within *ursinus* by Dawson (1988: 50) Groves (1993e: 46; 2005b: 44).

Phascolomys assimilis Krefft, 1873a: 796.

TYPE LOCALITY: Probably New South Wales, Australia.

COMMENTS: Synonymised within *mitchelli* by Thomas (1888a: 214). Synonymised within *hirsutus* by Iredale and Troughton (1934: 34) and synonymised into *ursinus* by Dawson (1988: 50) and Groves (1993e: 46; 2005b: 44).

† *S. [arcophilus] prior* De Vis, 1883a: 189.

TYPE LOCALITY: Wellington Caves, New South Wales, Australia. Pleistocene fossil.

COMMENTS: Name has been a *nomen dubium* since its first description, but was associated with *Vombatus* by Bartholomai and Marshall (1973: 369).

Vombatus ursinus tasmaniensis (Spencer & Kershaw, 1910)

Phascolomys tasmaniensis Spencer & Kershaw, 1910b: 58.

TYPE LOCALITY: Tasmania, Australia.

COMMENTS: Recognised as a subspecies of *ursinus* by Iredale and Troughton (1934: 34) and Troughton (1967: 119), but synonymised within *ursinus* provisionally by Dawson (1988: 50) and by Groves (1993e: 46; 2005b: 44). Elevated to subspecies rank by Strahan (1983: 117; 19`95: 205), Clayton *et al.* (2006: 104), Van Dyck and Strahan (2008: 208) and Burbidge *et al.* (2014: 19, 28).

Suborder Phalangerida Aplin & Archer, 1987

Suborder Phalangerida Aplin & Archer, 1987: xxii, xlix.

COMMENTS: When originally proposed as a new rank it was placed in the Order Diprotodontia (Owen, 1877a) and included the superfamilies Phalangeroidea (Thomas, 1888a), Macropodoidea (J. Gray, 1821), Burramyoidea (Broom, 1898), Petauroidea (Bonaparte, 1832) and Tarsipedoidea (Gervais & Verreaux, 1842a). Synonymised within Diprotodontia by McKenna and Bell (1997: 58). Recognised by Strahan (1995: 7, 206), Kear and Cooke (2001: 84), Roberts *et al.* (2007: 2), and Van Dyck and Strahan (2008: 9, 209). Recent evidence from both nuclear genes (Meredith *et al.*, 2008c: 395) and mitochondrial RNA (M. Phillips & Pratt, 2008: 594) has provided strong support for a close relationship between the macropods and possums.

Kind Phalangistidae Burnett, 1830b: 351.

TYPE GENUS: *Phalangista* É. Geoffroy & G. Cuvier, 1795 [= *Phalanger* Storr, 1780].

COMMENTS: When originally proposed, this rank was placed in the Race Feridentiae (Burnett, 1830b [= Marsupialia (Illiger, 1811 part)]) and included the genera *Phalangista* É. Geoffroy and G. Cuvier, 1795 [= *Phalanger* Storr, 1780]; *Balantia* Illiger, 1811 [= *Phalanger* Storr, 1780]; *Petaurista* Desmarest, 1821 [= *Petauroides* Thomas, 1888a]; and *Acrobata* Desmarest, 1821 [= *Acrobates* Desmarest, 1817b].

Family Phalangistidae Owen, 1839a: 19.

TYPE GENUS: *Phalangista* É. Geoffroy & G. Cuvier, 1795 [= *Phalanger* Storr, 1780].

COMMENTS: When originally proposed, this rank was placed in the Tribe Carpophaga (Owen, 1839a [= Diprotodontia (Owen, 1877a part)]) and included the genera *Phalangista* É. Geoffroy and G. Cuvier, 1795 [= *Phalanger* Storr, 1780]; and *Petaurus* Shaw, 1791. Followed by Owen (1840: 332) and Krefft (1866a: 6) who included the genera *Phalangista* and *Belideus*. This spelling was used in preference to spelling Phalangistadae of J. Gray (1821: 308) by Bonaparte (1845: 6) and most authors subsequently who recognised this name (and in accordance with the rules of nomenclature). Family rank recognised and included the tribes Pseudochirini and Phalangistini. Synonymised within Phalangeridae by Marshall (1981: 27), Marshall *et al.* (1990: 493), and McKenna and Bell (1997: 61).

Family Phalangeridés Gervais, 1855a: 272.

TYPE GENUS: *Phalangista* É. Geoffroy & G. Cuvier, 1795 [= *Phalanger* Storr, 1780].

COMMENTS: When originally proposed, this rank was placed in the Order Marsupiaux (Gervais, 1855a [=

Marsupialia (Illiger, 1811)]) and included the tribes Phascolarctins (Gervais, 1855a [= Phascolarctidae (Owen, 1839a)]), Phalangistins (Gervais, 1855a [= Phalangerida (Aplin & Archer, 1987 part)]) and Pétauristins (Gervais, 1855a [= Phalangerida (Aplin & Archer, 1987 part)]).

Tribe Phalangistins Gervais, 1855a: 274.

TYPE GENUS: *Phalangista* É. Geoffroy & G. Cuvier, 1795 [= *Phalanger* Storr, 1780].

COMMENTS: When originally proposed, this rank was placed in the Family Phalangeridés (Gervais, 1855a [= Phalangerida (Aplin & Archer, 1987)]) and included the genera *Phalangista* É. Geoffroy and G. Cuvier, 1795 [= *Phalanger* Storr, 1780]; *Trichosurus* Lesson, 1828b; *Pseudochirus* W. Ogilby, 1836a [= *Pseudocheirus* W. Ogilby, 1837a]; and *Dromicia* J. Gray, 1841 [= *Cercartetus* Gloger, 1841].

Tribe Pétauristins Gervais, 1855a: 276.

TYPE GENUS: *Petaurus* F. Cuvier [= Shaw, 1791].

COMMENTS: When originally proposed, this rank was placed in the Family Phalangeridés (Gervais, 1855a [= Phalangerida (Aplin & Archer, 1987 part)]) and included the genera *Petaurus* F. Cuvier [= Shaw, 1791]; *Belideus* Waterhouse, 1839a [= *Petaurus* Shaw, 1791]; and *Acrobates* Desmarest, 1817b.

Volitantia A. Murray, 1866: vx, 363.

TYPE GENUS: Not based on a genus group name.

COMMENTS: When originally proposed, this rank was placed in the Marsupialia (Illiger, 1811) at unknown rank and included the genus *Petaurus* Shaw, 1791; which included species from the currently recognised genera *Petaurus*, *Petauroides* and *Acrobates*.

Suborder Phalangeriformes Szalay, 1982: 631.

COMMENTS: When originally proposed as a new rank it was placed in the Order Syndactyla (Wood Jones, 1923 [1923–1925] [= Marsupialia (Illiger, 1811)]) and included the superfamilies Phalangeroidea (Thomas, 1888a), Vombatoidea (Kirsch, 1977b), † Diprotodontoidea (Gill, 1872: 26) and Macropodoidea (J. Gray, 1821). Synonymised within Diprotodontia by Aplin and Archer (1987: xliii), and McKenna and Bell (1997: 58). Subordinal rank recognised by Woodburne (1984a: 71), who introduced the term as new, and Marshall *et al.* (1990: 459) and Kirsch *et al.* (1997: 245), who both attributed the name to Woodburne (1984a: 71). Subsequently both Szalay (1994: 42) and Groves (2005b: 44) recognised the suborder with Szalay (1982) as the author.

Suborder Phalangeriformes Woodburne, 1984a: 71.

COMMENTS: When originally proposed as a new rank it was placed in the Order Diprotodonta [sic] (Owen, 1877a) and included the superfamilies Phalangeroidea (Thomas, 1888a) (that contains the families Phalangeridae (Thomas, 1888a), † Ektopodontidae (Stirton *et al.*, 1967: 437), Petauridae (Bonaparte, 1832), Burramyidae (Broom, 1898) and Macropodidae (J. Gray, 1821)) and Tarsipedoidea (Gervais & Verreaux, 1842a).

Superfamily Burramyoidea Broom, 1898 *sensu* Aplin & Archer, 1987

Subfamily Burramyinae Broom, 1898: 73.

TYPE GENUS: *Burramys* Broom, 1895a.

COMMENTS: When originally proposed, this rank was placed in the Family Phalangeridae (Thomas, 1888a) and included the genus *Burramys* Broom, 1895a. Rank not recognised by Kirsch *et al.* (1997: 245) who included the Family Burramyidae in the Superfamily Phalangeroidea. Groves (2005b: 44) included the Family Burramyidae within the Superfamily Phalangeroidea. Superfamily rank recognised by Aplin and Archer (1987: xxii), Strahan (1995: 7, 206), Kear and Cooke (2001: 84), Long *et al.* (2002: 123), and Van Dyck and Strahan (2008: 9, 209).

Family Burramyidae Broom, 1898 *sensu* Aplin & Archer, 1987

Subfamily Burramyinae Broom, 1898: 73.

TYPE GENUS: *Burramys* Broom, 1895a.

COMMENTS: When originally proposed, this rank was placed in the Family Phalangeridae (Thomas, 1888a) and included the genus *Burramys* Broom, 1895a. Subfamily rank recognised, within the Family Phalangeridae, by Simpson (1945: 46) who included only the genus *Burramys*, while *Cercartetus* (and most of the other possums) were placed within the Subfamily Phalangerinae. Gunson *et al.* (1968: 41) suggested that *Acrobates, Burramys* and *Cercartetus* should be separated from the Subfamily Phalangerinae. Family rank was recognised by Kirsch (1968a: 420; 1968b: 45; 1977a: 2, 113), who initially included the genera *Burramys*, *Cercartetus* and *Acrobates*; which was followed until *Acrobates* was placed in its own family by Aplin (1987: xxii, lvii). This composition has typically been followed with the exception of Szalay (1994: 43) who placed the Subfamily Burramyinae (Broom, 1898) in the Family Petauridae and included the Tribes Burramyini and Acrobatini.

Tribe Burramyini Szalay, 1994: 43.

TYPE GENUS: *Burramys* Broom, 1895a.

COMMENTS: When originally proposed, this rank was attributed to Broom (1898: 63) and placed in the Subfamily Burramyinae (Broom, 1898). Rank not subsequently recognised.

Burramys Broom, 1895

Burramys Broom, 1895a: ii.

TYPE SPECIES: *Burramys parvus* Broom, 1895a by monotypy.

COMMENTS: The name was also published by Broom (1895b: 371; 1895c: 373; 1896b: 47), with a full description being made by Broom (1896a: 51; 189`6c: 563).

Burramys parvus **Broom, 1895**

Mountain Pygmy-possum

Burramys parvus Broom, 1895a: ii.

TYPE LOCALITY: Fossil in Wombeyan Caves, near Taralga, New South Wales, Australia.

COMMENTS: Abstract of the description was also published by Broom (1895b: 371; 1895c: 373; 1896a: 47), with a full description being made by Broom (1896a: 51; 1896c: 563). Thought by the describer to be a missing link between the phalangerid possums and the rat kangaroos, which appears to have influenced Tate (1948b: 260) who included it within the macropod Subfamily Hypsiprymnodontinae. Broom (1898: 63) concluded that *Burramys* should be classified in a monotypic subfamily of Phalangeridae. The historically recognised link with *Hypsiprymnodon* was proposed to be due to convergence by Ride (1956a: 428), who suggested that *Burramys* should be included with *Cercartetus* and *Petaurus* in the same subfamily instead of being placed in a separate monotypic subfamily. The link to other possums of the Burramyidae was confirmed when live specimens were found in 1966 (Anon, 1966: 225). The relationship between *Burramys* and *Cercartetus*, which initially also included *Acrobates*, was confirmed by Kirsch (1968b: 45) and has been subsequently followed.

Cercartetus Gloger, 1841

Cercartētus Gloger, 1841: xxx, 85.

TYPE SPECIES: *Phalangista nana* Desmarest, 1817c [= *Cercartetus nanus* (Desmarest, 1817c)] by monotypy.

COMMENTS: Synonymised within *Pseudochirus* by Thomas (1888a: 166; 1895a: 190). Genus recognised by Iredale and Troughton (1934: viii, 22) and subsequent authors. Genus reviewed by Wakefield (1963a: 99) and McKay (1988b: 99).

Dromicia J. Gray, 1841: 401, 407.

TYPE SPECIES: *Phalangista nana* Desmarest, 1817c (author cited as É. Geoffroy, 1803f) [= *Cercartetus nanus* (Desmarest, 1817c)] by monotypy. See Thomas (1888a: 140).

COMMENTS: Recognised as a subgenus of *Phalangista* É. Geoffroy & G. Cuvier, 1795 [= *Phalanger* Storr, 1780] by Waterhouse (1846: 307) and at the genus rank by Thomas (1888a: xii, 140). Synonymised within *Cercartetus* by Iredale and Troughton (1934: 22), Marshall (1981: 28), McKay (1988b: 100) and subsequent authors.

Eudromicia Mjöberg, 1916: 13.

TYPE SPECIES: *Eudromicia macrura* Mjöberg, 1916 [= *Cercartetus caudatus macrurus* (Mjöberg, 1916)], by subsequent designation. See Matschie (1916: 260).

COMMENTS: Recognised as a genus by Iredale and Troughton (1934: viii, 23) for *macrura* and *lepida*, which was followed by Tate (1945a: 3) who also included *caudata*. Synonymised within *Cercartetus* by Kirsch and Calaby (1977: 16), Marshall (1981: 28), McKay (1988b: 100) and subsequent authors.

Dromiciola Matschie, 1916: 260.

TYPE SPECIES: *Dromicia lepida* Thomas, 1888a [= *Cercartetus lepidus* (Thomas, 1888a)] by original designation.

COMMENTS: Described as a subgenus of *Dromicia* J. Gray, 1841. Synonymised within *Eudromicia* by Iredale and Troughton (1934: 23) and within *Cercartetus* by Marshall (1981: 29), McKay (1988b: 100) and subsequent authors.

Dromiciella Matschie, 1916: 260.

TYPE SPECIES: *Dromicia concinna* Gould, 1845b [= *Cercartetus concinnus* (Gould, 1845b)] by original designation.

COMMENTS: Described as a subgenus of *Dromicia* J. Gray, 1841. Synonymised within *Cercartetus* by Iredale and Troughton (1934: 22), Marshall (1981: 28), McKay (1988b: 100) and subsequent authors.

Cercartetus caudatus (Milne-Edwards, 1877)

Long-tailed Pygmy-possum

Φ ***Cercartetus caudatus caudatus*** (Milne-Edwards, 1877)

Φ *Dromicia caudata* Milne-Edwards, 1877: 1079.

TYPE LOCALITY: Arfak Mountains, Irian Jaya, Indonesia. Designation by de Beaufort (1966: 530)

COMMENTS: Recognised in the genus *Dromicia* by Thomas (1888a: xii, 143), not considered by Iredale and Troughton (1934: 22). Placed in *Eudromicia* by Tate (1945a: 3) and *Cercartetus* by Wakefield (1963a: 101), Honacki *et al.* (1982: 39), McKay (1988b: 100) and subsequent authors.

FUTURE TAXONOMIC RESEARCH: There appears to have been no recent revision testing whether there is a single species in different parts of New Guinea and in Australia.

Cercartetus caudatus macrurus (Mjöberg, 1916)

Eudromicia macrura Mjöberg, 1916: 14.

TYPE LOCALITY: Cedar Creek, Atherton, near Cairns, north Queensland, Australia.

COMMENTS: Recognised within *Eudromicia* by Iredale and Troughton (1934: viii, 23), Tate (1945a: 3) and Troughton (1967: 77). Synonymised within *caudatus* by Ride (1970: 224, 242) and Honacki *et al.* (1982: 39). Considered a subspecies within *Cercartetus caudatus* by Wakefield (1963a: 114), Strahan (1983: 166; 1995: 212), McKay (1988b: 100), Flannery (1990: 142; 1994: 42, 46; 1995a: 193), Osborne and Christidis (2002: 33), Groves (2005b: 44), Clayton *et al.* (2006: 104), and Van Dyck and Strahan (2008: 214).

Cercartetus concinnus (Gould, 1845)

Western Pygmy-possum

Dromicia concinna Gould, 1845b: 2.

TYPE LOCALITY: Swan River, Western Australia, Australia. Type designated by Thomas (1922a: 128).

COMMENTS: Recognised within *Phalangista* (*Dromicia*) by Waterhouse (1846: 314) and *Dromicia* by Gould (1845 [1845–1863]: Text to Plate 30) and Thomas (1888a: xii, 146). Transferred to *Cercartetus* by Iredale and Troughton (1934: viii, 22), Wakefield (1963a: 100), Troughton (1967: 76), Honacki *et al.* (1982: 39) and subsequent authors. Taxonomy reviewed by Harris (2009a: 1).

Phalangista (Dromicia) neillii G. Waterhouse, 1846: 315.

TYPE LOCALITY: King Gorge Sound, Western Australia, Australia.

COMMENTS: Synonymised within *concinna* by Thomas (1888a: 146), Iredale and Troughton (1934: 23), Wakefield (1963a: 100), McKay (1988b: 101), Flannery (1994: 50) and subsequent authors.

Cercartetus concinnus minor Wakefield, 1963a: 100.

TYPE LOCALITY: Nurcoung 16 km north west of Natimuk, Victoria, Australia.

COMMENTS: Recognised as subspecies of *concinnus* by McKay (1988b: 101), Flannery (1994: 42), Groves (2005b: 45) and Clayton *et al.* (2006: 104), but not Osborne and Christidis (2002: 34), Van Dyck and Strahan (2008: 216) or Harris (2009a: 2).

Cercartetus lepidus (Thomas, 1888)

Little Pygmy-possum

Dromicia lepida Thomas, 1888a: xii, 142.

TYPE LOCALITY: Tasmania, Australia.

COMMENTS: Taxon placed in the genus *Eudromicia* by Iredale and Troughton (1934: viii, 23), Tate (1945a: 3) and Troughton (1967: 77). Transferred to *Cercartetus* by Wakefield (1963a: 101) and followed by Honacki *et al.* (1982: 39), Strahan (1983: 164), McKay (1988b: 101) and subsequent authors. Taxonomic history reviewed by (Harris (2009b: 1). The potential taxonomic distinctiveness of the Victorian and South Australian populations compared to that found in Tasmania was raised by Bennett and Lumsden (1995: 100), which was subsequently confirmed genetically by Osborne and Christidis (2002: 33).

FUTURE TAXONOMIC RESEARCH: The mainland population (*sensu* Bennett & Lumsden, 1995: 100) needs to be taxonomically reviewed to determine if it should be recognised as a distinct taxon.

Cercartetus nanus (Desmarest, 1817)

Eastern Pygmy-possum

Cercartetus nanus nanus (Desmarest, 1817)

Phalangista nana Desmarest, 1817c: 477.

TYPE LOCALITY: Maria Island, Tasmania, Australia. Designation by de Beaufort (1966: 530).

COMMENTS: There has been a lot of debate regarding the author and year of publication of this species. Some authors recognise Desmarest, others É. Geoffroy, while others include both Desmarest and É. Geoffroy (see Harris, 2006: 108 for review and Van Dyck and Strahan, 2008: 219). The name was initially cited by Desmarest as '*Phalangista nana* Geoff.' supposedly based on É. Geoffroy's (1803f) unpublished manuscript, but the name does not occur there (see Julien-Laferrière, 1994: 18; Harris, 2006: 108). The year of publication is often given as 1818; this was reviewed by Harris (2006: 108) who concluded the year of publication is 1817. Recognised in *Phalangista* by Waterhouse (1838a: 68; 1841a: 279) and *Dromicia* by J. Gray (1841: 407) and Thomas (1888a: xii, 144), who reviewed the taxonomic history. Transferred to *Cercartetus* by Iredale and Troughton (1934: viii, 22) and followed by Tate (1945a: 3) and recognised by most subsequent authors. Taxonomic decision for subspecific arrangement by Wakefield (1963a: 103).

Phalangista gliriformis T. Bell, 1829: 121; Plates 13–14.

TYPE LOCALITY: Tasmania (as Australia).

COMMENTS: Species recognised within *Dromicia* by Gould (1845 [1845–1863]: Text to Plate 29) and Krefft (1868a: 94). Synonymised within *nanus* by Waterhouse (1846: 309), Thomas (1888a: 144), Iredale and Troughton (1934: 22), McKay (1988b: 101), Flannery (1994: 54) and subsequent authors.

Cercartetus nanus unicolor (Krefft, 1863)

Dromicia unicolor Krefft, 1863: 49.

TYPE LOCALITY: St. Leonards, Sydney, Australia.

COMMENTS: Synonymised within *nanus* by Thomas (1888a: 144). Considered a subspecies of *nanus* by Iredale and Troughton (1934: 22), Wakefield (1963a: 103), Strahan (1983: 160; 1995: 218), McKay (1988b: 101), Flannery (1994: 42, 54), Groves (2005b: 45), Clayton *et al.* (2006: 104), Harris (2006: 107; 2008: 1), and Van Dyck and Strahan (2008: 221).

Dromicia britta Wood Jones, 1925a: 97.

TYPE LOCALITY: Millicent, South Australia, Australia.

COMMENTS: Considered a subspecies of *nanus* by Iredale and Troughton (1934: 22) and a synonym of *unicolor* by McKay (1988b: 101), Flannery (1994: 54) and subsequent authors.

Superfamily Petauroidea Bonaparte, 1832 *sensu* Aplin & Archer, 1987

Tribe Petaurina Bonaparte, 1832: 69.

TYPE GENUS: *Petaurus* Shaw, 1791.

COMMENTS: When originally proposed, this rank was placed in the Family Halmaturidae (Bonaparte, 1831 [= Macropodidae (J. Gray, 1821)]). Tribe rank recognised by Bonaparte (1838: 113; 1840: 257; 1845: 6) and subfamily rank by Gill (1872: 25). Author of name attributed to Gill (1872: 25) by authors including Aplin and Archer (1987: xxii), Szalay (1994: 43), Kirsch *et al.* (1997: 245) and Kear and Cooke (2001: 84), while McKenna and Bell (1997: 65) attributed it to Szalay (1994: 43). Recognised at the superfamily rank by Aplin and Archer (1987: xxii), who included the families Petauridae and Pseudocheiridae, Marshall *et al.* (1990: 460), Szalay (1994: 43), Kirsch *et al.* (1997: 245) and Long *et al.* (2002: 126). Synonymised within Family Petauridae by McKenna and Bell (1997: 65).

Family Petaurusideae Lesson, 1842: 189.

TYPE GENUS: *Petaurus* Shaw, 1791.

COMMENTS: When originally proposed, this rank was placed in the Order Mastodidelphie (Lesson, 1842 [= Marsupialia (Illiger, 1811)]) and included the genera *Petaurus* Shaw, 1791 (and subgenera *Belideus* Waterhouse, 1839a [= *Petaurus* Shaw, 1791] and *Acrobata* Desmarest, 1821 [= *Acrobates* Desmarest, 1817b]); and *Schoinobates* Lesson, 1842 [= *Petauroides* Thomas, 1888a]. Synonymised within Petauridae by Marshall *et al.* (1990: 495).

Family Petauridae Bonaparte, 1832 *sensu* Baverstock, 1984

Tribe Petaurina Bonaparte, 1832: 69.

TYPE GENUS: *Petaurus* Shaw, 1791.

COMMENTS: When originally proposed, this rank was placed in the Family Halmaturidae (Bonaparte, 1831 [= Macropodidae (J. Gray, 1821)]). Tribe rank recognised by Bonaparte (1838: 113; 1840: 257; 1845: 6). Family rank recognised by Kirsch (1968a: 420), Smith (1984: xiii), Aplin and Archer (1987: xxii), Marshall *et al.* (1990: 460), Szalay (1994: 43) and Kirsch *et al.* (1997: 245). The family name has often been attributed to Gill (1872: 25) by authors including Kirsch (1968a: 420; 1977a: 113), Marshall (1981: 28), Aplin and Archer (1987: xxii), Szalay (1994: 43) and Kirsch *et al.* (1997: 245). More recently the family name has been attributed to Bonaparte (1838: 113) by McKenna and Bell (1997: 65) and Groves (2005b: 53). The family historically included species now recognised within the Pseudocheiridae with the subfamilies Petaurinae, Pseudocheirinae (and sometimes Dactylopsilinae) usually being recognised (e.g. Kirsch, 1968a: 420; 1977a: 113; Marshall, 1981: 28; Strahan, 1983: 124; McKay, 1988c: 87; McKenna and Bell, 1997: 65). In contrast Szalay (1994: 43) recognised the subfamilies Petaurinae and Burramyinae, with the former including the tribes Petaurini, Dactylopsilini and Pseudocheirini and the latter including the tribes Burramyini and Acrobatini. Separate families, i.e. Petauridae and Pseudocheiridae, were recognised by M. Archer (1984: 710, 719, 786) and Baverstock (1984a: 4–5) and followed by Aplin and Archer (1987: xxii), Marshall *et al.* (1990: 460, 494), Groves (1993e: 58, 60), Flannery (1994: 60, 102), Kirsch *et al.* (1997: 245) and most subsequent authors.

Subfamily Petaurinae Gill, 1872: 25.

TYPE GENUS: *Petaurus* Shaw, 1791.

COMMENTS: When originally proposed, this rank was placed in the Phalangistidae (Owen, 1839a [= Phalangerida (Aplin & Archer, 1987 part)]). Recognised at family, subfamily and tribe (without *Dactylopsila*) ranks by Szalay (1994: 43). Author recognised at family rank by Long *et al.* (2002: 132). Synonymised within the Subfamily Petaurinae (Bonaparte, 1832) by McKenna and Bell (1997: 66).

Subfamily Dactylopsilinae Kirsch, 1977 *sensu* Edwards & Westerman, 1992

Subfamily Dactylopsilinae Kirsch, 1977a: 113.

TYPE GENUS: *Dactylopsila* J. Gray, 1858a.

COMMENTS: When originally proposed as a new rank it was placed in the Family Petauridae (Bonaparte, 1832) and included the genus *Dactylopsila* J. Gray, 1858a. Subfamily status recognised by Marshall (1981: 28; 1984: 102), Marshall *et al.* (1990: 460), Edwards and Westerman (1992: 563), Flannery (1994: 60), Strahan (1995: 7, 222), Kirsch *et al.* (1997: 245), and Van Dyck and Strahan (2008: 9, 224). Rank synonymised within the Subfamily Petaurinae (Bonaparte, 1832) by McKenna and Bell (1997: 66). When originally proposed this rank did not contain the genus *Gymnobelideus*, but genetic studies by Baverstock *et al.* (1990a: 283), Edwards and Westerman (1992: 563), Kirsch *et al.* (1997: 228, 245), Osborne and Christidis (2001: 221) and Meredith *et al.* (2009a: 559, 560), placed this genus closer to *Dactylopsila* than *Petaurus*. Despite these observations authors including Flannery (1994: 60), and Van Dyck and Strahan (2008: 224, 226) included *Gymnobelideus* within the Subfamily Petaurinae.

Tribe Dactylopsilini Szalay, 1994: 43.

TYPE GENUS: *Dactylopsila* J. Gray, 1858a.

COMMENTS: When originally proposed, this rank was attributed to Kirsch (1977a: 113) and placed in the Subfamily Petaurinae (Bonaparte, 1832). Synonymised within the Subfamily Petaurinae (Bonaparte, 1832) by McKenna and Bell (1997: 66).

Dactylopsila J. Gray, 1858

Dactylopsila J. Gray, 1858a: 109.

TYPE SPECIES: *Dactylopsila trivirgata* J. Gray, 1858a by monotypy.

COMMENTS: Recognised since its description.

Dactylonax Thomas, 1910a: 610.

TYPE SPECIES: Φ *Dactylopsila palpator* Milne-Edwards, 1888: 174 (as *Dactylonax palpator*) by original designation.

COMMENTS: Genus recognised by Tate and Archbold (1937: 394), Tate (1945a: 5), and Laurie and Hill (1954: 18). Synonymised within *Dactylopsila* by Simpson (1945: 46), Haltenorth (1958: 28), Wakefield (1963a: 100), Marshall (1981: 28), Marshall *et al.* (1990: 495), McKenna and Bell (1997: 66) and Groves (2005b: 53).

Dactylopsila trivirgata J. Gray, 1858

Torresian Striped Possum

Φ ***Dactylopsila trivirgata trivirgata*** J. Gray, 1858

Φ *Dactylopsila trivirgata* J. Gray, 1858a: 111; Plate 63.

TYPE LOCALITY: Aru Islands, Indonesia.

COMMENTS: Early taxonomic history reviewed by Thomas (1888a: 160).

Φ *Dactylopsila Albertisii* Peters & Doria, 1875: 542.

TYPE LOCALITY: Sorong, west Vogelkop, north-west Netherlands, Papua, Indonesia.

COMMENTS: Synonymised within *trivirgata* by Thomas (1888a: 160), Laurie and Hill (1954: 17), McKay (1988c: 88), Flannery (1990: 150; 1994: 98; 1995a: 203; 1995b: 115), Groves (1993e: 61; 2005b: 54) and subsequent authors.

Φ *Phalangista (Dactylopsila) angustivittis* Peters & Doria, 1880: 674.

TYPE LOCALITY: Sorong, west Vogelkop, north-west Netherlands, Papua, Indonesia.

COMMENTS: Synonymised within *trivirgata* by Thomas (1888a: 160), Laurie and Hill (1954: 17), McKay (1988c: 88), Flannery (1990: 150; 1994: 98; 1995a: 203; 1995b: 115), Groves (1993e: 61; 2005b: 54) and subsequent authors.

Φ *Dactylopsila occidentalis* Matschie, 1916: 302.

TYPE LOCALITY: Waigeu Islands, north-west of Papua, Indonesia.

COMMENTS: Synonymised within *trivirgata* by McKay (1988c: 88), Laurie and Hill (1954: 17), Flannery (1990: 150; 1994: 98; 1995a: 203; 1995b: 115), Groves (1993e: 61; 2005b: 54) and subsequent authors.

Φ *Dactylopsila arfakensis* Matschie, 1916: 302.

TYPE LOCALITY: Hatam, Arfak Mountains, Vogelkop, north-west Netherlands, Papua, Indonesia.

COMMENTS: Synonymised within *trivirgata* by McKay (1988c: 88), Laurie and Hill (1954: 17), Flannery (1990: 150; 1994: 98; 1995a: 203; 1995b: 115), Groves (1993e: 61; 2005b: 54) and subsequent authors.

Φ ***Dactylopsila trivirgata melampus*** Thomas, 1908

Φ *Dactylopsila melampus* Thomas, 1908a: 122.

TYPE LOCALITY: Kokoda, Mambare River, New Guinea. 1000 feet.

COMMENTS: Synonymised within *trivirgata* Groves (1993e: 61). Recognised as a subspecies of *trivirgata* by Tate and Archbold (1937: 393), Laurie and Hill (1954: 17), Flannery (1990: 150; 1994: 60, 98; 1995a: 203; 1995b: 115) and Groves (2005b: 54).

Φ *Dactylopsila hindenburgi* Ramme, 1914: 413.

TYPE LOCALITY: Sattelberg, north-east New Guinea, New Guinea.

COMMENTS: Synonymised within *melampus* by Laurie and Hill (1954: 17) and within *trivirgata* by Flannery (1990: 150; 1994: 98; 1995a: 203; 1995b: 115), Groves (1993e: 61; 2005b: 54) and subsequent authors.

Φ *Dactylopsila biedermanni* Matschie, 1916: 303.

TYPE LOCALITY: Upper Aroa River, south east Papua New Guinea.

COMMENTS: Recognised as a synonym of *melampus* by Laurie and Hill (1954: 18). Synonymised within *trivirgata* by Groves (1993e: 61) and Flannery (1990: 150; 1994: 98; 1995a: 203; 1995b: 115), and within *melampus* by Groves (2005b: 54).

Φ ***Dactylopsila trivirgata kataui*** Matschie, 1916

Φ *Dactylopsila kataui* Matschie, 1916: 271, 304.

TYPE LOCALITY: Katau, near mouth of Fly River, Papua New Guinea.

COMMENTS: Synonymised within *trivirgata* by Groves (1993e: 61). Recognised as a subspecies of *trivirgata* by Tate and Archbold (1937: 393), Laurie and Hill (1954: 18), Flannery (1990: 150; 1994: 60, 98; 1995a: 203; 1995b: 115) and Groves (2005b: 54).

Dactylopsila trivirgata picata Thomas, 1908

Dactylopsila trivirgata picata Thomas, 1908a: 123.

TYPE LOCALITY: Somerset (as Port Albany), Cape York, Queensland, Australia.

COMMENTS: Elevated to species status by Iredale and Troughton (1934: viii, 25) and Troughton (1967: 79). Synonymised within *trivirgata* by Ride (1970: 242) and Strahan (1983: 144). Recognised as a subspecies of *trivirgata* by McKay (1988c: 88), Flannery (1990: 150; 1994: 60, 98; 1995a: 203; 1995b: 115) and subsequent authors. This is the only subspecies that occurs within Australia.

Dactylopsila trivirgata infumata Tate, 1945a: 4.

TYPE LOCALITY: Lake Barrine, near Cairns, Queensland, Australia.

COMMENTS: Subspecies recognised with scepticism by Strahan (1983: 144). Synonymised within *trivirgata* by Groves (1993e: 61), Flannery (1990: 150; 1994: 98; 1995a: 203; 1995b: 115) and subsequent authors. Synonymised within *picata* by McKay (1988c: 88) and Groves (2005b: 54).

Gymnobelideus McCoy, 1867

Gymnobelideus McCoy, 1867a: 287.

TYPE SPECIES: *Gymnobelideus leadbeateri* McCoy, 1867a by monotypy.

COMMENTS: The placement of this genus within the subfamily is still contentious. It was considered to be more closely related to *Petaurus* by Kirsch (1977a: 122), Nelson and Stephan (1982: 703), M. Archer (1984: 720), Marshall (1990: 495) and Springer *et al.* (1994: 108). More recently however it has typically been placed closer to *Dactylopsila* than *Petaurus* by authors including Baverstock *et al.* (1990a: 283), Edwards and Westerman (1992: 563), Kirsch *et al.* (1997: 228), Osborne and Christidis (2001: 221) and Meredith *et al.* (2009a: 559, 560), which is followed here. Despite this, it was placed within the Subfamily Petaurinae rather than the Subfamily Dactylopsilinae by Van Dyck and Strahan (2008: 10, 226) so it appears that further resolution of the relationship of this species may be warranted.

Gymnobelides de Marschall, 1873: 6.

TYPE SPECIES: Incorrect subsequent spelling of *Gymnobelideus* McCoy, 1867a.

COMMENTS: Synonymised within *Gymnobelideus* by Palmer (1904: 303) and McKenna and Bell (1997: 66).

† *Palaeopetaurus* Broom, 1896d: 568.

TYPE SPECIES: *Palaeopetaurus elegans* Broom, 1895d [= *Gymnobelideus leadbeateri* McCoy, 1867a] by original designation.

COMMENTS: Pleistocene. Unknown if members of this genus could glide. Genus typically placed within the Family Petauridae by authors including Mahoney and Ride (1975: 45), but synonymised within the non-gliding extant *Gymnobelideus* McCoy, 1867a by Wakefield (1972a: 19, 20), Marshall (1981: 28), M. Archer *et al.* (1984: 1037) and McKenna and Bell (1997: 66).

Gymnobelideus leadbeateri McCoy, 1867

Leadbeater's Possum

Gymnobelideus leadbeateri McCoy, 1867a: 287; Plate 6.

TYPE LOCALITY: Bass River, Victoria, Australia. Type designation by Dixon (1970: 108).

COMMENTS: Genus has been recognised since its description with a re-examination undertaken by Brazenor (1932: 106). A recent genetic analysis has revealed that there are two evolutionarily significant units, Yellingbo and the central highlands (Hansen & Taylor 2008: 4039) and the relationship between these two, and the level of their distinctiveness, needs to be re-examined.

† *Palaeopetaurus elegans* Broom, 1896d: 568; Plate 46.

TYPE LOCALITY: Wombeyan Caves, near Taralga, New South Wales, Australia.

COMMENTS: Pleistocene. Synonymised within *leadbeateri* by Wakefield (1972a: 19, 20), M. Archer *et al.* (1984: 1037) and McKenna and Bell (1997: 66).

Subfamily Petaurinae Bonaparte, 1832 *sensu* Edwards & Westerman, 1992

Tribe Petaurina Bonaparte, 1832: 69.

TYPE GENUS: *Petaurus* Shaw, 1791.

COMMENTS: When originally proposed, this rank was placed in the Family Halmaturidae (Bonaparte, 1831 [= Macropodidae (J. Gray, 1821)]). Tribe rank recognised by Bonaparte (1838: 113; 1840: 257; 1845: 6). Subfamily rank recognised by Gill (1872: 25), Marshall (1981: 28; 1984: 102), Marshall *et al.* (1990: 460), Strahan (1995: 7, 224), Kirsch *et al.* (1997: 245), McKenna and Bell (1997: 66), and Van Dyck and Strahan (2008: 10, 226). The composition of the subfamily recognised here follows Edwards and Westerman (1992: 563), and Kirsch *et al.* (1997: 228, 245).

Subfamily Petaurinae Gill, 1872: 25.

TYPE GENUS: *Petaurus* Shaw, 1791.

COMMENTS: When originally proposed, this rank was placed in the Family Phalangistidae (Owen, 1839a [=

Phalangerida (Aplin & Archer, 1987 part)]). Author of the subfamily recognised by Marshall (1981: 28), Szalay (1994: 43), and Kirsch *et al.* (1997: 245).

Tribe Petaurini Szalay, 1994: 43.

TYPE GENUS: *Petaurus* Shaw, 1791.

COMMENTS: When originally proposed, this rank was attributed to Gill (1872: 25) and placed in the Subfamily Petaurinae (Gill, 1872 [= Petaurinae (Bonaparte, 1832)]). Synonymised within the Subfamily Petaurinae by McKenna and Bell (1997: 66).

Petaurus Shaw, 1791

Petaurus Shaw, 1791: Text to Plate 60.

TYPE SPECIES: *Petaurus australis* Shaw, 1791 by monotypy.

COMMENTS: Genus recognised by most subsequent authors including Waterhouse (1841a: xvi, 282).

Ptilotus G. Fischer, 1814: 512.

TYPE SPECIES: *Petaurus australis* Shaw, 1791 by subsequent designation. See Thomas (1888a: 150).

COMMENTS: Synonymised within *Petaurus* by J. Gray (1843a: xxii), Thomas (1888a: 150), Iredale and Troughton (1934: 23), Marshall (1981: 28), McKay (1988c: 91) and Marshall *et al.* (1990: 495).

HOMONYMS:

Ptilotis Swainson, 1837: 326, honeyeaters of the Class Aves (Order Passeriformes, Family Meliphagidae). Genus is a junior synonym of *Meliphaga* Lewin, 1808: Index for Plate 5. See Pizzey and Knight (1998: 378).

Belidea Waterhouse, 1838d: 880.

TYPE SPECIES: *Didelphis sciurea* Shaw, 1794 (as *Petaurista [Belidea] sciurea* [= *Petaurus norfolcensis* (Kerr, 1792)] by original designation.

COMMENTS: Proposed as a subgenus of *Petaurista* [= *Petauroides* Thomas, 1888a]. Recognised at generic rank by Gould (1842b: 11; 1843a: 404). Synonymised within *Belideus* by Palmer (1904: 135).

Belideus Waterhouse, 1839a: 149.

TYPE SPECIES: *Didelphis sciurea* Shaw, 1794 [= *Petaurus norfolcensis* (Kerr, 1792)] by original designation.

COMMENTS: Described as a subgenus of *Petaurus*. Generic rank recognised by various authors, including De Vis (1883b: 619) and as a subgenus of *Petaurus* by Waterhouse (1841a: 286; 1846: 325) and Gervais (1869: 574). Synonymised within *Petaurus* by J. Gray (1843a: xxii), Thomas (1888a: 150), Iredale and Troughton (1934: 23) Marshall (1981: 28), McKay (1988c: 91) and Marshall *et al.* (1990: 495).

Belidens Wiegmann, 1839: 418.

TYPE SPECIES: Incorrect subsequent spelling of *Belideus* Waterhouse, 1839a.

COMMENTS: Synonymised within *Belideus* by Palmer (1904: 135).

Xenochīrus Gloger, 1841: xxx, 85.

TYPE SPECIES: *Didelphis sciurea* Shaw, 1794 [= *Petaurus norfolcensis* (Kerr, 1792)] by monotypy.

COMMENTS: Synonymised within *Petaurus* by Thomas (1888a: 150; 1895a: 190), Iredale and Troughton (1934: 23), Marshall (1981: 28), McKay (1988c: 91) and Marshall *et al.* (1990: 495). Palmer (1904: 710) considered to be antedated by *Belideus*.

HOMONYMS:

Xenochirus Gilbert, 1890: 90, poacher fish or poachers of the Superclass Pisces (Order Scorpaeniformes, Family Agonidae). Junior synonym of *Xeneretmus* Gilbert, 1903: 360. See Sheiko and Mecklenburg (2004: 5).

Petaurula Matschie, 1916: 261.

TYPE SPECIES: *Petaurus breviceps* Waterhouse, 1839a by original designation.

COMMENTS: Described as a subgenus of *Petaurus* Shaw, 1791. Synonymised within *Petaurus* by Iredale and Troughton (1934: 23), Marshall (1981: 28), and McKay (1988c: 91) and Marshall *et al.* (1990: 495).

Petaurella Matschie, 1916: 261.

TYPE SPECIES: Φ *Petaurus breviceps papuanus* Thomas, 1888a (as *Petaurus papuanus*) by original designation.

COMMENTS: Described as a subgenus of *Petaurus* Shaw, 1791. Not considered by Iredale and Troughton (1934: 23) and synonymised within *Petaurus* by Marshall (1981: 28), McKay (1988c: 91) and Marshall *et al.* (1990: 495).

Petaurus australis Shaw, 1791

Yellow-bellied Glider

Petaurus Australis Shaw, 1791: Text to Plate 60.

TYPE LOCALITY: Sydney, New South Wales, Australia.

COMMENTS: Type based on 'Hepoona Roo' of Hunter, J. in White (1790: 288). Taxonomic history reviewed by Thomas (1888a: 151). Iredale and Troughton (1934: 24) refer to the author as Shaw and Nodder; Nodder is actually the publisher.

FUTURE TAXONOMIC RESEARCH: In M. Brown *et al.*'s (2006) minimum-spanning haplotype network diagram of the ND4 gene (M. Brown *et al.*, 2006: Figure 4), North Queensland animals differ from those from elsewhere by eight substitutions, and the authors also noted a few morphological and apparent socio-behavioural differences in addition. On any one of these grounds, the North Queensland animals could potentially be ranked as a distinct species so a full analysis of the differentiation of the North Queensland population needs to be undertaken to confirm or refute this. It should be noted also that M. Brown *et al.* (2006) found that the sample from western Victoria and southeastern South Australia different by two substitutions, and in their Principal Components diagram (Figure 5) they were largely, but not entirely, separate from other samples; it is possible that discriminant analysis might show complete separation (i.e. species status for this population as well), and this in turn needs to be tested in future.

[Sciurus] Novae Hollandiae F. Meyer, 1793: 11, 177 (part A).

TYPE LOCALITY: New South Wales, Australia.
COMMENTS: Combined description incorporating 'Hepoona Roo' of Hunter, J. in White (1790: 288) and 'Norfolk Island Flying Squirrel' on Anon in Philip (1789: 151). Included within *P. norfolcensis* by Iredale and Troughton (1934: 24). Split between *australis* and *norfolcensis* [as *sciureus*] by Thomas (1888a: 151, 153), McKay (1988c: 92, 94) and Flannery (1994: 74, 86).

Didelphis Petaurus Shaw, 1800: 496; Plate 112.

TYPE LOCALITY: Sydney, New South Wales, Australia.
COMMENTS: Replacement name for *Petaurus australis* Shaw, 1791, but part of description refers to the 'Black Flying Opossum' of Anon in Phillip (1789: 297) [= *Didelphis volans* Kerr, 1792]. Synonymised within *australis* by Thomas (1888a: 151), Iredale and Troughton (1934: 24), McKay (1988c: 92) and Flannery (1994: 74).

Voluccella nigra Bechstein, 1800: 351. (part A)

TYPE LOCALITY: Botany Bay, New South Wales, Australia.
COMMENTS: Combined description; see part B under *Petauroides volans volans* (Kerr, 1792). Synonymised within *australis* by McKay (1988c: 92) and Flannery (1994: 74).

P. [etaurus] Hepuna Ru Oken, 1816: 1118.

TYPE LOCALITY: *Nomen novum* for *Didelphis petaurus* Shaw, 1800.
COMMENTS: Synonymised within *australis* by Iredale and Troughton (1934: 24) and not included within McKay (1988c: 92).

Petaurus flaviventer Desmarest, 1817b: 403.

TYPE LOCALITY: New South Wales, Australia.
COMMENTS: Recognised as a valid species within *Petaurista* by Waterhouse (1838a: 68) [= *Petauroides* Thomas, 1888a] and *Petaurus* by Waterhouse (1841a: 286) and within *Belideus* by Gould (1845 [1845–1863]: Text to Plate 23). Synonymised within *australis* by Waterhouse (1846: 327), Thomas (1888a: 151), Iredale and Troughton (1934: 24), McKay (1988c: 92) and Flannery (1994: 74).

P. [etaurus] Cunninghami J. Gray, 1843a: 83.

TYPE LOCALITY: New South Wales, Australia.
COMMENTS: Synonymised within *australis* by Thomas (1888a: 152), Iredale and Troughton (1934: 24), McKay (1988c: 92) and Flannery (1994: 74).

petaurus opossum Falcimagne, 1854: 366, footnote.

TYPE LOCALITY: New South Wales, Australia.
COMMENTS: Synonymised within *australis* by Iredale and Troughton (1934: 24) and not included within McKay (1988c: 92).

Petaurus australis reginae Thomas, 1923b: 249.

TYPE LOCALITY: Gin Gin inland of Bundaberg, southeast Queensland, Australia.
COMMENTS: Subspecies rank recognised by Iredale and Troughton (1934: 24), Finlayson (1934: 222), Tate (1945a: 7), Troughton (1967: 86), Strahan (1983: 136; 1995: 228), McKay (1988c: 92) and Flannery (1994: 60, 74). Though the separate north Queensland population is usually allocated to this subspecies, the type locality at Gin Gin in southern Queensland lies within the continuous range of *Petaurus australis australis*. The north Queensland population was proposed to be treated as an undescribed subspecies by Maxwell *et al.* (1996: 5). More recently it was recognised as a subspecies of *australis* by Groves (2005b: 55) and Clayton *et al.* (2006: 104), but a detailed genetic analysis by M. Brown *et al.* (2006: 305) of the original type specimen from southern Queensland found no support for the classification of *reginae* as a separate taxon, but concurred that the north Queensland population represents a distinct 'evolutionary significant unit', using this term in preference to subspecies.
FUTURE TAXONOMIC RESEARCH: As noted above, the evidence strongly suggests that the north Queensland population represents a distinct species, yet to be described.

Petaurus breviceps (Waterhouse, 1838)

Sugar Glider

Petaurus breviceps breviceps (Waterhouse, 1838)

[Petaurista] [Belidea]Breviceps Waterhouse, 1838d: 880.

TYPE LOCALITY: Sydney, New South Wales, Australia.
COMMENTS: The original publication of this taxon appears to have been overlooked, with authors recognising Waterhouse (1839a: 152) as the publication which uses the name *Petaurus [Belideus] breviceps*. This publication arose from the paper being read before the Zoological Society of London on 13 November 1838 and subsequently published in May 1839. McAllan and Bruce (1989: 447) argued that the original publication of the name was Waterhouse (1838d: 880) which was published on 8 December 1838; and this has been followed here. Included within *Petaurus* by Waterhouse (1839a: 152; 1841a: 290; 1846: 334), J. Gray (1843a: 83) and most subsequent authors, except Gould (1849 [1845–1863]: Text to Plate 5) who placed it within *Belideus*. Taxonomic history reviewed by Thomas (1888a:

156). Flannery (1990: 146) recognised '*alba* Gervais, 1869' as a synonym of *breviceps*, but this appears to be an error (Flannery pers. comm.) and was not referred to by Flannery (1994: 80; 1995a: 207; 1995b: 118) or by any other author. Taxonomy reviewed by M. Smith (1973: 1).

FUTURE TAXONOMIC RESEARCH: The molecular study by Malekian *et al.* (2007: 24; 2010a: 122; 2010b: 165) recovered two divergent mtDNA clades in this species, whose distribution did not in the least correspond to morphologically described taxa. Evidently, this species needs close re-examination. The distinctiveness of the Tasmanian sugar glider also needs to be assessed to confirm whether or not it was introduced into Tasmania (as proposed by Gunn, 1846: 458; 1851: 253) or occurred there naturally.

Petaurus sciureus Gunn, 1851: 253.

TYPE LOCALITY: Tasmania, Australia.

COMMENTS: Recognised as a species within *Belideus* by Gould (1845 [1845–1863]: Text to Plate 24) and within *Petaurus* by Finlayson (1934: 222), who attributed the name to Shaw without citation. Synonymised within *Petaurus breviceps* by Thomas (1888a: 156).

HOMONYMS:

Didelphis Sciurea Shaw, 1794, Squirrel Gliders of the Class Mammalia (Order Diprotodontia, Family Petauridae). Name is a synonym of *Petaurus norfolcensis* (Kerr, 1792). See individual entry.

Petaurus (Belideus) notatus Peters, 1859a: 14.

TYPE LOCALITY: Port Philip, Victoria, Australia.

COMMENTS: Recognised within *Belideus* by Gould (1860 [1845–1863]: Text to Plate 26). Synonymised within *Petaurus breviceps* by Thomas (1888a: 156), Iredale and Troughton (1934: 25), McKay (1988c: 93), Groves (1993e: 61), Flannery (1990: 146; 1994: 80; 1995a: 207; 1995b: 118) and subsequent authors.

Petaurus breviceps ariel (Gould, 1842)

Belidea ariel Gould, 1842b: 11.

TYPE LOCALITY: Port Essington, Northern Territory, Australia. Type designated by Thomas (1922a: 128).

COMMENTS: Recognised within *Petaurus (Belideus)* by Waterhouse (1846: 336), and within *Belideus* by Gould, 1849 [1845–1863]: Text to Plate 27). Synonymised within *papuanus* by Thomas (1888a: 156). Subspecies rank recognised by Iredale and Troughton (1934: 25), Finlayson (1934: 223), Tate (1945a: 9), D. Johnson (1964: 452), Troughton (1967: 83), M. Smith (1973: 1), Strahan (1983: 138; 1995: 230), McKay (1988c: 93) and subsequent authors.

Petaurus [Belideus] arul Gervais, 1869: 574.

TYPE LOCALITY: Incorrect subsequent spelling of *ariel* Gould, 1842b.

COMMENTS: Not considered by Iredale and Troughton (1934: 25), but synonymised within *ariel* by M. Smith (1973: 1), McKay (1988c: 93), Flannery (1990: 146; 1994: 80; 1995a: 207; 1995b: 118) and subsequent authors.

Petaurus breviceps longicaudatus Longman, 1924

Petaurus breviceps longicaudatus Longman, 1924a: ix.

TYPE LOCALITY: Mapoon Mission, Gulf of Carpentaria, north Queensland, Australia.

COMMENTS: Not considered by Iredale and Troughton (1934: 25), but the subspecies rank was recognised by Tate (1945a: 8), Troughton (1967: 83), M. Smith (1973: 1), Strahan (1983: 138; 1995: 230), McKay (1988c: 93), Flannery (1990: 146; 1994: 60, 80) and confirmed by Colgan and Flannery (1992: 247) and followed by subsequent authors.

Φ *Petaurus breviceps papuanus* Thomas, 1888

Φ *Petaurus breviceps* var. *papuanus* Thomas, 1888a: xii, 158.

TYPE LOCALITY: Huon Gulf, north eastern New Guinea.

COMMENTS: Elevated to species rank by Matschie (1916: 261) who also erected the subgenus *Petaurella*. Tate and Archbold (1935a: 1) also recognised the specific status. Subspecies recognised by Tate (1945a: 9), Laurie and Hill (1954: 19) and M. Smith (1973: 1). Again synonymised within *breviceps* by Groves (1993e: 61), but elevated to a subspecies of *breviceps* by Strahan (1983: 138), Flannery (1990: 146; 1994: 60, 80; 1995a: 207; 1995b: 118) and Groves (2005b: 55).

Φ *Petaurus* (*Petaurella*) *papuensis flavidus* Tate & Archbold, 1935a: 2.

TYPE LOCALITY: Dogwa, Oriomo River, Western Division, Papua New Guinea. 30 metres.

COMMENTS: Recognised at the subspecies rank by Tate (1945a: 9), Laurie and Hill (1954: 19), M. Smith (1973: 1), Strahan (1983: 138) and Flannery (1990: 146; 1995b: 118). Synonymised within *breviceps* by Groves (1993e: 61) and Flannery (1994: 80; 1995a: 207), and within *papuanus* by Groves (2005b: 55).

Φ *Petaurus* (*Petaurella*) *papuensis tafa* Tate & Archbold, 1935a: 1.

TYPE LOCALITY: Eastern ridge, Mount Tafa, Central Division, Papua New Guinea. 2000m.

COMMENTS: Subspecies rank recognised by Tate (1945a: 10), Laurie and Hill (1954: 19), M. Smith (1973: 1) and Flannery (1990: 146). Synonymised within *breviceps* by Groves (1993e: 61) and Flannery (1994: 80; 1995a: 207; 1995b: 118), and within *papuanus* by Groves (2005b: 55).

Petaurus gracilis (De Vis, 1883)

Mahogany Glider

Belideus gracilis De Vis, 1883c: 27.

TYPE LOCALITY: 'North of Cardwell', north Queensland, Australia.

COMMENTS: Though the description by De Vis (1883b: 619) was published in April that year, the abstract of the description was published by De Vis (1883c: 27) in January. History of description given by Van Dyck (1990: 329; 1993: 77). Taxon synonymised within *Petaurus norfolcensis* (as *P. sciureus*) by Thomas (1888a: 154) and D. Elliot (1907: 14). Elevated to subspecies of *norfolcensis* by Iredale and Troughton (1934: 24), which was accepted by subsequent authors including Tate (1945a: 8), Fleay (1947: 111; 1954: 210), Marlow (1965: 75), Troughton (1967: 84), Alexander (1981: 64), Strahan (1983: 140), and Colgan and Flannery (1992: 245, 255) who used limited allozyme data. Taxon synonymised within *norfolcensis* by McKay (1988c: 93). With the exception of the studies by Van Dyck (1990; 1993) and Colgan and Flannery (1992) these classifications were not based on samples or inspections of specimens as the species had not been recorded between 1886 and 6 December 1989. Upon its rediscovery and the assessment of new samples it was resurrected from synonymy with *P. norfolcensis* to species rank by Van Dyck (1991: 350), with a formal reappraisal in Van Dyck (1993: 84) who found *gracilis* and *norfolcensis* to be distinctly different morphologically. Subsequently recognised as a species by all authors including Flannery (1994: 60, 84), Strahan (1995: 232), and Van Dyck and Strahan (2008: 233). Species rank retained by Malekian *et al.* (2010a: 122, 130), though it was recognised that mitochondrial divergence from *norfolcensis* was less than within *P. australis*. Taxonomic history discussed by Jackson (2011: 141). This species is clearly distinct from *norfolcensis* in all its body measurements, skull measurements and mass (Van Dyck and Strahan, 2008: 234, 236).

Petaurus norfolcensis (Kerr, 1792)

Squirrel Glider

S. [ciurus] Petaurus norfolcensis Kerr, 1792: 270.

TYPE LOCALITY: Sydney, New South Wales, Australia. Norfolk Island in error.

COMMENTS: Type based on the 'Norfolk Island Flying-Squirrel' of Anon in Phillip (1789: 151, Plate 17).

[Sciurus] NovaeHollandiae F. Meyer, 1793: 11, 177 (part).

TYPE LOCALITY: New South Wales, Australia.

COMMENTS: Synonymised within *norfolcensis* by Thomas (1888a: 153) and Iredale and Troughton (1934: 24) and noted by McKay (1988c: 93) as a combined description; see Part A under *Petaurus australis australis* Shaw, 1791. Also synonymised within *norfolcensis* by Flannery (1994: 86).

Didelphis Sciurea Shaw, 1794: 29; Plate 11.

TYPE LOCALITY: Sydney, New South Wales, Australia.

COMMENTS: Recognised as a valid species within *Didelphis* by Shaw (1800: 498), within *Petaurista* (as *Petaurista sciureus* Geoff.) by Waterhouse (1838a: 68) and within *Petaurus* by Desmarest (1817b: 403), Waterhouse (1841a: 289; 1846: 331), J. Gray (1843a: 83) and Thomas (1888a: xii, 153). Synonymised within *norfolcensis* by Iredale and Troughton (1934: 24), McKay (1988c: 93), Flannery (1994: 86) and subsequent authors.

HOMONYMS:

Petaurus sciureus Gunn, 1851, Sugar Gliders of the Class Mammalia (Order Diprotodontia, Family Petauridae). Name is a synonym of *Petaurus breviceps* (Waterhouse, 1838). See individual entry.

Petaurus Leucogaster Mitchell, 1838a: xvii.

TYPE LOCALITY: Banks of the Murray River, New South Wales?, Australia.

COMMENTS: *Nomen nudum*. It was placed as a synonym of *Petauroides volans* by Iredale and Troughton (1934: 29), but this species does not occur near the Murray River. Considered *incertae sedis* by McKay (1988c: 97), but the type locality suggests it is most likely that it should be placed within *Petaurus norfolcensis*.

Family Pseudocheiridae Winge, 1893 *sensu* Kirsch *et al.*, 1997

Tribe Pseudochirini Winge, 1893a: 89, 100.

TYPE GENUS: *Pseudocheirus* W. Ogilby, 1837a.

COMMENTS: When originally proposed, this rank was placed in the Family Phalangistidae (Owen, 1839a [= Phalangerida (Aplin & Archer, 1987 part)]) and included the genera *Pseudocheirus* W. Ogilby, 1837a; and *Petaurista* Desmarest, 1821 [= *Petauroides* Thomas, 1888a]. Tribe rank subsequently recognised by Winge (1941: 69). Name recognised as the Subfamily Pseudocheirinae, within the Family Petauridae, by Marshall (1984: 102), but both groups were combined within the Family Petauridae by McKay (1988c: 87) and Szalay (1994: 43). Separate families recognised by M. Archer (1984: 710, 719, 786) and Baverstock (1984a: 4–5) and followed by Aplin and Archer (1987: xxii), Marshall *et al.* (1990: 460, 494), Baverstock *et al.* (1990c: 519), Groves (1993e: 58, 60), Flannery (1994: 60, 102), Edwards and Westerman (1995: 231), Kirsch *et al.* (1997: 245), Osborne and Christidis (2001: 211), though they did question the separation of the families, and subsequent authors accepted the family except McKenna and Bell (1997: 65) who reduced it to a subfamily of the Family Petauridae. Tribe rank recognised by Szalay (1994: 43) but not subsequent authors. The recognition of the subfamilies followed here was pre-empted by Baverstock *et al.* (1990c: 519) and formally recognised by Kirsch *et al.* (1997: 245).

Tribe Pseudocheirini Szalay, 1994: 43.

TYPE GENUS: *Pseudocheirus* W. Ogilby, 1837a.

COMMENTS: When originally proposed, this rank was attributed to Winge (1893: 89, 100) and placed in the Subfamily Petaurinae (Gill, 1872 [= Petaurinae (Bonaparte, 1832)]).

Subfamily Hemibelideinae Kirsch *et al.*, 1997

Subfamily Hemibelideinae Kirsch *et al.*, 1997: 245.

TYPE GENUS: *Hemibelideus* Collett, 1884b.

COMMENTS: When originally proposed as a new rank it was placed in the Family Pseudocheiridae (Winge, 1893a) and included the genera *Hemibelideus* Collett, 1884b; and *Petauroides* Thomas, 1888a. Relationship between these genera was recognised by Baverstock *et al.* (1990c: 519). Subfamily rank recognised by Groves (2005b: 50) and Meredith *et al.* (2010: 75) but not by Van Dyck and Strahan (2008: 10, 238).

Hemibelideus Collett, 1884

Hemibelideus Collett, 1884b: 385.

TYPE SPECIES: *Phalangista (Hemibelideus) lemuroides* Collett, 1884b [= *Hemibelideus lemuroides* (Collett, 1884b)] by monotypy.

COMMENTS: Proposed as a new subgenus of *Phalangista* É. Geoffroy and G. Cuvier, 1795 [= *Phalanger* Storr, 1780]. Recognised as a subgenus of *Pseudocheirus* by Tate (1945b: 1, 22) and Kirsch and Calaby (1977: 16) and a synonym of it by Simpson (1945: 46), Marshall (1981: 28), Honacki *et al.* (1982: 41) and McKenna and Bell (1997: 65). Distinct genus status recognised by Iredale and Troughton (1934: viii, 28), Troughton (1967: 98), Ride (1970: 76), McKay (1988c: 89) and subsequent authors.

Hemibelideus lemuroides (Collett, 1884)

Lemuroid Ring-tailed Possum

Phalangista (Hemibelideus) lemuroides Collett, 1884b: 385; Plate 31, Figs. 3–4.

TYPE LOCALITY: Spur between Gowry Creek and Herbert River, north Queensland, Australia.

COMMENTS: Included within *Pseudochirus* by Thomas (1888a: 168, 170). Taxonomic decision of Iredale and Troughton (1934: viii, 28) to designate this taxon to *Hemibelideus* and followed by most subsequent authors including Troughton (1967: 98), Ride (1970: 76) and McKay (1988c: 89).

Pseudochirus (Hemibelideus) cervinus Longman, 1915: 22.

TYPE LOCALITY: Atherton Tableland, north Queensland, Australia.

COMMENTS: Synonymised within *lemuroides* by Iredale and Troughton (1934: 28), Troughton (1967: 99), McKay (1988c: 89), Flannery (1994: 146) and subsequent authors.

Petauroides Thomas, 1888

Petauroides Thomas, 1888a: xii, 163.

TYPE SPECIES: *Nomen novum* for *Voluccella* Bechstein, 1800.

COMMENTS: The names *Volucella* and *Petaurista* were both preoccupied. *Schoinobates* was used in preference to *Petauroides* by Iredale and Troughton (1934: viii, 29), Simpson (1945: 46), Tate (1945a: 11) and Ride (1970: 80). Taxonomic decision of McKay (1988c: 89) to use *Petauroides*, and was followed by subsequent authors with the exception of McKenna and Bell (1997: 66) who used *Schoinobates*. Taxonomy reviewed by Harris and Maloney (2010: 207).

Voluccella Bechstein, 1800: 352.

TYPE SPECIES: *Didelphis volans* Kerr, 1792 [= *Petauroides volans* (Kerr, 1792)] by original designation.

COMMENTS: Synonymised within *Petauroides* by Thomas (1888a: 163), Palmer (1904: 526) and McKay (1988c: 89). Synonymised within *Schoinobates* by Iredale and Troughton (1934: 28) and McKenna and Bell (1997: 66).

HOMONYMS:

Volucella E.L. Geoffroy, 1762: 540, hoverflies of the Class Insecta (Order Diptera, Family Syrphidae).

Voluccella Fabricius, 1794: 55, hoverflies of the Class Insecta (Order Diptera, Family Syrphidae). Emendation of *Volucella* E.L. Geoffroy, 1762: 540.

Phalanger Lacépède, 1801: 491.

TYPE SPECIES: *Phalanger* [= *Didelphis*] *volans* Kerr, 1792 [= *Petauroides volans* (Kerr, 1792)] by original designation.

COMMENTS: Genus is a junior homonym of *Phalanger* Storr, 1780.

HOMONYMS:

Phalanger Storr, 1780, cuscuses of the Class Mammalia (Order Diprotodontia, Family Phalangeridae). Currently recognised genus. See individual entry.

Petaurista Rafinesque, 1815: 55.

TYPE SPECIES: *Nomen nudum*.

COMMENTS: Synonymised within *Petauroides* by Palmer (1904: 526).

HOMONYMS:

Petaurista Link, 1795: 52, 78, giant flying squirrels of the Class Mammalia (Order Rodentia, Family Sciuridae). Currently recognised genus. See Thorington and Hoffman (2005: 770).

Petaurista Meigen, 1800: 15, flies of the Class Insecta (Order Diptera, Family Trichoceridae). Name suppressed by

Opinion 678 of the ICZN (1963: 339). Genus is a synonym of *Trichocera* Meigen, 1803: 262.

Petaurista Desmarest, 1821, the Greater Glider of the Class Mammalia (Order Diprotodontia, Family Pseudocheiridae). Synonymised within *Petauroides* (Thomas, 1888a). See individual entry.

Petaurista Latreille, 1827: 400, leaf beetles of the Class Insecta (Order Coleoptera, Family Chrysomelidae). Genus is a synonym of *Lema* Fabricius, 1798: 4.

Petaurista H. Reichenbach, 1862: 105, guenon monkeys on the Class Mammalia (Order Primates, Family Cercopithecidae). Genus is a junior synonym of *Cercopithecus* (Linnaeus, 1758: 26). See Groves (2005f: 154).

petaurista Desmarest, 1821: 268.

TYPE SPECIES: *Petaurus taguanoides* Desmarest, 1817b [= *Petauroides volans* (Kerr, 1792)] by subsequent designation.

COMMENTS: Recognised as a subgenus of *Petaurus* by Waterhouse (1846: 322) for *taguanoïdes*. Designation of *P. taguanoides* Desmarest, 1817b as type species antedates that of *Petaurus australis* Shaw, 1791 by Iredale and Troughton (1934: 29). Included as a synonym of *Petauroides* by Thomas (1888a: 163), both *Petaurus* and *Schoinobates* by Iredale and Troughton (1934: 23, 29), and *Petauroides* by McKay (1988c: 89) and Groves (2005b: 50).

HOMONYMS:

See homonyms discussed above.

Petaurides Ramsay, 1890a: 77.

TYPE SPECIES: Incorrect subsequent spelling of *Petauroides* Thomas, 1888a.

COMMENTS: Not recognised by subsequent authors.

Schoinobates Iredale & Troughton, 1934: viii, 28.

TYPE SPECIES: *Didelphis volans* Kerr, 1782 [= *Petauroides volans* (Kerr, 1792)] by monotypy.

COMMENTS: Iredale and Troughton (1934: 28) gave the author as Lesson (1842: 190), but they, like Palmer (1904: 886) were wrong in assuming that the name *Schoinobates* initially referred to a marsupial (though the name was mistakenly placed with other marsupials). *Schoinobates* was applied by Lesson (1842), as *Petaurus* (*Schoinobates*) *leucogenys*, which is now known as the Japanese giant flying squirrel *Petaurista leucogenys* (Temminck, 1824: xxvii). *Schoinobates* (Lesson, 1842 is therefore a junior subjective synonym of *Petaurista* (Link, 1795: 52, 78) (McKay, 1982: 38). *Schoinobates* has commonly been used in preference to *Petauroides* by authors including Simpson (1945: 46), Tate (1945a: 11), Troughton (1967: 87), Ride (1970: 80), Kirsch and Calaby (1977: 16), Marshall (1981: 28), Honacki *et al.* (1982: 42), and McKenna and Bell (1997: 66).

HOMONYMS:

Schoinobates Lesson, 1842: 190, giant flying squirrels of the Class Mammalia (Order Rodentia, Family Sciuridae). Name is a synonym of *Petaurista* Link, 1795: 52, 78. Type specimen is given as *Schoinobates* [= *Petaurista*] *leucogenys* (Temminck, 1824: xxvii).

Petauroides armillatus Thomas, 1923

Central Greater Glider

Petauroides volans armillatus Thomas, 1923b: 248.

TYPE LOCALITY: Coomooboolaroo Station, 25km south west of Duaringa, central Queensland, Australia.

COMMENTS: Subspecies discussed by Troughton (1967: 90). Synonymised within *volans* by Iredale and Troughton (1934: 30), McKay (1988c: 91), Flannery (1994: 148) and subsequent authors. Recently Aplin (pers. comm.) and Arbogast *et al.* (2011) recognised its distinctiveness, both morphologically and genetically, from *P. volans*.

Petauroides minor (Collett, 1887)

Northern Greater Glider

Petaurista volans var. *minor* Collett, 1887b: 926.

TYPE LOCALITY: Herbert Vale, Queensland, Australia.

COMMENTS: Subspecies status recognised by Thomas (1888a: 166), Iredale and Troughton (1934: 30), Troughton (1967: 91), Strahan (1983: 134; 1995: 240), McKay (1988c: 91), Groves (2005b: 51), Flannery (1994: 103, 148), Clayton *et al.* (2006: 104), and Van Dyck and Strahan (2008: 241). Recently Aplin (pers. comm.) and Arbogast *et al.* (2011) recognised its distinctiveness, both morphologically and genetically, from *P. volans*.

Petaurides cinereus Ramsay, 1890a: 77.

TYPE LOCALITY: Bellenden-Ker Range, northeast Queensland, Australia.

COMMENTS: McKay (1988c: 91) suggested the type locality was probably the Atherton Tablelands (as Bellenden-Ker Range), Queensland. In the same year this species was also exhibited by Ramsay (1890b: 1030) who suggested that 'The *Belideus* will be described under the name of *B. cinereus*'. Synonymised within *volans* by Iredale and Troughton (1934: 30), McKay (1988c: 91), Flannery (1994: 148) and subsequent authors.

Petauroides volans (Kerr, 1792)

Southern Greater Glider

Petauroides volans volans (Kerr, 1792)

Didelphis volans Kerr, 1782: 199.

TYPE LOCALITY: Sydney, New South Wales, Australia.

COMMENTS: Type species based on 'Black Flying Opossum' of Anon in Philip (1789: 297). Confused in the early years with *Petaurus australis* Shaw, 1791 (McKay, 1982: 38). Between Desmarest (1817b: 400) and Thomas (1879: 397), the name *Didelphis volans* was either synonymised or ignored (McKay, 1982: 38). Early taxonomic history reviewed by Thomas (1888a: 164) and recently by Maloney and Harris (2008: 39), who disentangled the confusion in the early records between this and other gliding mammals, both marsupial and placental. Thomas (1879: 297) revived this taxon along with other names first used by Kerr (McKay, 1982: 38). Included within *Schoinobates* by Iredale and Troughton (1934: viii, 29) and Ride (1970: 80). Transferred to *Petauroides* by Thomas (1888a: xii, 164), Strahan (1983: 134), McKay (1988c: 90) and subsequent authors. Taxonomy reviewed by Maloney and Harris (2008: 39).

FUTURE TAXONOMIC RESEARCH: The alpha taxonomy of greater gliders has not been reviewed in detail since Thomas (1923b: 246), so could be revisited.

[Didelphis] Voluccella F. Meyer, 1793: 26, 174.

TYPE LOCALITY: New South Wales, Australia.

COMMENTS: Synonymised within *volans* by Thomas (1888a: 164), Iredale and Troughton (1934: 29), McKay (1988c: 90), Flannery (1994: 148) and subsequent authors.

Didelphis Macroura Shaw, 1794: 33; Plate 12.

TYPE LOCALITY: New South Wales, Australia.

COMMENTS: Recognised by Shaw (1800: 500). Considered to be the young of *Petaurus australis* by Waterhouse (1841a: 288; 1846: 330). Synonymised within *volans* by Thomas (1888a: 164), Iredale and Troughton (1934: 29), McKay (1988c: 90), Flannery (1994: 148) and subsequent authors.

Voluccella nigra Bechstein, 1800: 351. (part B)

TYPE LOCALITY: Botany Bay, New South Wales, Australia. Type species based on 'Black Flying Opossum' of Anon in Philip (1789: 297) and 'Hepoona Roo' of Hunter in White (1790: 288).

COMMENTS: Synonymised within *volans* by Thomas (1888a: 164), Iredale and Troughton (1934: 29), McKay (1988c: 90), Flannery (1994: 148) and subsequent authors.

Phal. [anger] petaurista É. Geoffroy, 1803c: 150.

TYPE LOCALITY: Australia.

COMMENTS: Synonymised within *volans* by Thomas (1888a: 164).

P. [etaurus] niger Oken, 1816: 1119.

TYPE LOCALITY: Australia.

COMMENTS: Synonymised within *volans* by Iredale and Troughton (1934: 29).

Petaurus taguanoïdes Desmarest, 1817b: 400.

TYPE LOCALITY: Sydney, New South Wales, Australia. Type designation by de Beaufort (1966: 534).

COMMENTS: Recognised as a valid species by Waterhouse (1838a: 68; 1841a: 283; 1846: 322), J. Gray (1843a: 84) and Gould (1853 [1845–1863]: Text to Plate 22). Synonymised within *volans* by Thomas (1879: 397), Thomas (1888a: 164), Iredale and Troughton (1934: 29), McKay (1988c: 90), Flannery (1994: 148) and subsequent authors.

Petaurus Peronii Desmarest, 1817b: 404.

TYPE LOCALITY: Sydney, New South Wales, Australia. Type designation by de Beaufort (1966: 534).

COMMENTS: Recognised as a valid species by Waterhouse (1841a: 284). Synonymised within *taguanoides* by Waterhouse (1846: 322) and within *volans* by Thomas (1888a: 164), Iredale and Troughton (1934: 29), McKay (1988c: 90), Flannery (1994: 148) and subsequent authors.

[Petaurus] didelphoides G. Cuvier, 1825: 129.

TYPE LOCALITY: Australia.

COMMENTS: Synonymised within *volans* by Thomas (1888a: 164), Iredale and Troughton (1934: 29), McKay (1988c: 90), Flannery (1994: 148) and subsequent authors. Note that the author of this taxon has been confused; authors such as Thomas (1888a: 164) and Iredale and Troughton (1934: 29) gave the author as F. Cuvier, whereas McKay (1988c: 90) and Groves (2005b: 51) attributed the author to G. Cuvier. The confusion appears to have arisen because the title page highlights that the publication of F. Cuvier is based on the cabinet of anatomy formed by G. Cuvier.

P. [etaurus] maximus Partington, 1837: 424.

TYPE LOCALITY: Sydney, New South Wales, Australia.

COMMENTS: McKay (1988c: 90) incorrectly referred to the abbreviation used by Iredale and Troughton (1934: 29) as Partington (1837 [=1838]) and noted that no reference to *maximus* or any other animal can be found. Synonymised within *volans* by Iredale and Troughton (1934: 29), Flannery (1994: 148) and subsequent authors.

Petauroides volans incanus (Thomas, 1923)

Petaurus volans incanus Thomas, 1923b: 247.

TYPE LOCALITY: Eidsvold, South-eastern Queensland, Australia.

COMMENTS: Subspecies discussed by Troughton (1967: 90) and recognised by Iredale and Troughton (1934: 29), Finlayson (1934: 219) and resurrected by Aplin pers. comm. Synonymised within *volans* by McKay (1988c: 90), Flannery (1994: 148) and subsequent authors. Recently Aplin (pers. comm.) and Arbogast *et al.* (2011) have recognised its distinctiveness, both morphologically and genetically, from *P. volans volans*.

Subfamily Pseudocheirinae Winge, 1893 *sensu* Kirsch *et al.*, 1997

Tribe Pseudochirini Winge, 1893a: 89, 100.

TYPE GENUS: *Pseudocheirus* W. Ogilby, 1837a.

COMMENTS: When originally proposed, this rank was placed in the Family Phalangistidae (Owen, 1839a [= Phalangerida (Aplin & Archer, 1987 part)]) and included the genera *Pseudocheirus* W. Ogilby, 1837a; and *Petaurista* Desmarest, 1821 [= *Petauroides* Thomas, 1888a]. Tribe rank subsequently recognised by Winge (1941: 69). Subfamily rank recognised, within the Family Phascolarctidae, by Turnbull and Lundelius (1970: 26), within the Family Petauridae by Marshall (1981: 28; 1984: 102) and within the Family Pseudocheiridae by Kirsch *et al.* (1997: 245). The subfamily recognised by Groves (2005b: 51) included the genera *Petropseudes*, *Pseudocheirus* and *Pseudochirulus*, while Meredith *et al.* (2010: 75) included only *Pseudochirulus* and *Pseudocheirus*. Van Dyck and Strahan (2008: 10) did not recognise any subfamilies.

Pseudocheirus W. Ogilby, 1837

Pseudocheirus W. Ogilby, 1837a: 457.

TYPE SPECIES: *Phalangista cookii* Desmarest, 1817c [= *Pseudocheirus peregrinus* (Boddaert, 1785)] by subsequent designation (see Thomas, 1888a: 166).

COMMENTS: Genus further described by W. Ogilby (1838b: 131). Recognised as 'Section 3 *Pseudocheirus*' within *Phalangista* by Waterhouse (1841a: 273) and subsequently ignored in preference to *Pseudochirus* W. Ogilby, 1836a until resurrected by Iredale and Troughton (1934: viii, 25) and followed by Simpson (1945: 46), Tate (1945b: 1, 6), Troughton (1967: 90) and subsequent authors.

Pseudochirus W. Ogilby, 1836a: 26.

TYPE SPECIES: *Phalangista cookii* Desmarest, 1817c [= *Pseudocheirus peregrinus* (Boddaert, 1785)], by subsequent designation. See Thomas (1888a: 166).

COMMENTS: *Nomen nudum*. Incorrect original spelling that was later corrected by W. Ogilby (1837a: 457) when he spelt it *Pseudocheirus*. Recognised as a subgenus of *Phalangista* by Waterhouse (1846: 297) and genus by Thomas (1888a: xii, 166). Not considered by Iredale and Troughton (1934: 25). Taxon synonymised within *Pseudocheirus* by McKay (1988c: 94), Groves (1993e: 59) and subsequent authors.

Hepoona J. Gray, 1841: 402, 407.

TYPE SPECIES: *Phalangista cookii* Desmarest, 1817c [= *Pseudocheirus peregrinus* (Boddaert, 1785)], by subsequent designation. See Thomas (1888a: 166).

COMMENTS: *Hepoona* of W. Ogilby (1838c: 218–219). Synonymised within *Pseudocheirus* by Thomas (1888a: 166), Iredale and Troughton (1934: 25), Marshall (1981: 28), McKay (1988c: 94) and Marshall *et al.* (1990: 495).

Ptenos J. Gray, 1843a: xxii.

TYPE SPECIES: First used as a synonym of *Hepoona* J. Gray, 1841.

COMMENTS: A *nomen nudum* as noted by Palmer (1904: 594, 886). Synonymised within *Pseudocheirus* by Iredale and Troughton (1934: 25) and McKay (1988c: 94).

HOMONYMS:

Ptenos Norton, 1872: 77, sawflies of the Class Insecta (Order Hymenoptera, Family Argidae). Genus is a synonym of *Ptenus* Kirby, 1882: 51.

Pseudocheirus occidentalis (Thomas, 1888)

Western Ring-tailed Possum

Pseudochirus occidentalis Thomas, 1888a: xii, 174.

TYPE LOCALITY: King George Sound, Western Australia, Australia.

COMMENTS: Included at species rank within *Pseudocheirus* by Iredale and Troughton (1934: viii, 27), Tate (1945b: 15) and Troughton (1967: 95). Synonymised within *peregrinus* by Ride (1970: 246) and McKay (1988c: 96) and listed as a subspecies of *peregrinus* by Strahan (1983: 126), Flannery (1994: 103, 122), Strahan (1995: 252) and Groves (2005b: 51). The sex chromosomes were revealed to be markedly distinct by McKay (1984: 11). Elevated to species status by Maxwell *et al.* (1996: 5), and Van Dyck and Strahan (2008: 253).

Pseudocheirus peregrinus (Boddaert, 1785)

Eastern Ring-tailed Possum

Pseudocheirus peregrinus peregrinus (Boddaert, 1785)

[Didelphis] Peregrinus Boddaert, 1785: 78.

TYPE LOCALITY: Endeavour River, Queensland, Australia.

COMMENTS: Based on the 'Opossum' referred to by Hawkesworth (1773: 586), 'New Holland Opossum' of Pennant (1781: 310), J. Cook and King (1784: 55, Plate 8) and the 'White-tailed Opossum' of Shaw (1800: 504). Placed within *Pseudochirus* by Thomas (1888a: xii, 172) and transferred to *Pseudocheirus* by Iredale and Troughton (1934: viii, 25) and subsequent authors. Smeenk (2009: 723) has shown that a specimen in the Leiden Museum (Naturalis) is very likely the type of this species.

FUTURE TAXONOMIC RESEARCH: This species is badly in need of taxonomic revision. The subspecies as presently recognised have distributions that seem to make little geographic sense, and further research is warranted.

Didelphis caudivolvula Kerr, 1792: 196.

TYPE LOCALITY: Endeavour River, Queensland, Australia.

COMMENTS: On same basis as *Didelphis peregrinus* Boddaert, 1785. Synonymised within *peregrinus* by Thomas (1888a: 172), Iredale and Troughton (1934: 25), McKay (1988c: 95), Flannery (1994: 122) and subsequent authors.

Didelphis Novae Hollandiae Bechstein, 1800: 348.

TYPE LOCALITY: Endeavour River, Queensland, Australia.

COMMENTS: On same basis as *Didelphis peregrinus* Boddaert, 1785. Synonymised within *peregrinus* by Thomas (1888a: 172), Iredale and Troughton (1934: 25), McKay (1988c: 95), Flannery (1994: 122) and subsequent authors.

B. [alantia] Banksii Oken, 1816: 1125.

TYPE LOCALITY: Novae Hollandiae [= Australia].

COMMENTS: Based on 'New Holland Opossum' of Pennant (1781: 310). Name is unavailable. See Opinion 417 of the ICZN (1956: 1).

Phalangista Cookii Schinz, 1821: 258.

TYPE LOCALITY: 'North coast of New Holland' [= Australia].

COMMENTS: Synonymised within *peregrinus* by Thomas (1888a: 172), Iredale and Troughton (1934: 25), McKay (1988c: 95), Flannery (1994: 122) and subsequent authors.

HOMONYMS:

Phalangista cookii Desmarest, 1817c, the Eastern Ring-tailed Possum of the Class Mammalia (Order Diprotodontia, Family Pseudocheiridae). Recognised as a subspecies of *Pseudocheirus peregrinus* (Boddaert, 1785). See individual entry.

Phalangista cookii G. Cuvier, 1824, the Common Brush-tailed Possum of the Class Mammalia (Order Diprotodontia, Family Phalangeridae). Synonym of *Trichosurus vulpecula* (Kerr, 1792). See individual entry.

Phalangista cookii Gould, 1856 [1845–1863], the Eastern Ring-tailed Possum of the Class Mammalia (Order Diprotodontia, Family Pseudocheiridae). Synonym of *Pseudocheirus peregrinus cookii* (Desmarest, 1817c). See individual entry.

Phalangista Banksii J. Gray, 1838b: 107.

TYPE LOCALITY: Endeavour River, Queensland, Australia.

COMMENTS: Based on 'New Holland Opossum' of Pennant (1781: 310). *Balantia banksii* Oken, 1816: 1125 is not available (see ICZN, 1956: 1). Synonymised within *cookii* Desmarest, 1817c by Waterhouse (1846: 299) and *peregrinus* by Thomas (1888a: 172), Iredale and Troughton (1934: 26), McKay (1988c: 96), Flannery (1994: 122) and subsequent authors.

Pseudochirus laniginosus incanens Thomas, 1923b: 249.

TYPE LOCALITY: Vine Creek, near Ravenshoe, north Queensland, Australia.

COMMENTS: Subspecies status recognised, as *P. peregrinus incanens*, by Iredale and Troughton (1934: 26). Synonymised within *peregrinus* by McKay (1988c: 96), Flannery (1994: 122) and subsequent authors.

Pseudochirus laniginosus notialis Thomas, 1923c: 158.

TYPE LOCALITY: Aldgate, Mt. Lofty, near Adelaide, South Australia, Australia.

COMMENTS: Recognised as a subspecies of *laniginosus* by Iredale and Troughton (1934: 26) and Troughton (1967: 94). Synonymised within *peregrinus* by McKay (1988c: 96), Flannery (1994: 122) and subsequent authors.

Pseudochirus laniginosus oralis Thomas, 1926b: 631.

TYPE LOCALITY: Bloomsbury, Queensland, Australia.

COMMENTS: Subspecies status recognised, as *P. peregrinus oralis*, by Iredale and Troughton (1934: 26) and Finlayson (1934: 218). Synonymised within *peregrinus* by McKay (1988c: 96), Flannery (1994: 122) and subsequent authors.

Pseudochirus rubidus Troughton & Le Souef, 1929a: 294.

TYPE LOCALITY: North of Mt Mowbullan, Bunya Range, Queensland, Australia.

COMMENTS: Species rank recognised by Troughton (1967: 93). Recognised as a subspecies of *laniginosus* by Iredale and Troughton (1934: 26) and of *peregrinus* by J. Murray *et al.* (1980: 73). Synonymised within *peregrinus* by Ride (1970: 246), McKay (1988c: 96), Flannery (1994: 122) and within *pulcher* by Groves (2005b: 51).

Pseudocheirus peregrinus cookii
(Desmarest, 1817)

Phalangista Cookii Desmarest, 1817c: 476.

TYPE LOCALITY: Sydney, New South Wales, Australia.

COMMENTS: Described from two specimens collected by Péron; although reference is made to 'opossum' plate in Cooks 3rd Voyage (J. Cook & King, 1784: 109), description is not from that plate. Type designation by de Beaufort (1966: 532). Recognised as a species, within *Phalangista* (*Pseudochirus*), by Waterhouse (1846: 299). Considered a synonym within *convolutor* by Iredale and Troughton (1934: 27) and within *peregrinus* by Ride (1970: 246) and McKay (1988c: 95). Recognised as a subspecies by of *peregrinus* by J. Murray *et al.* (1980: 73), Strahan (1983: 126; 1995: 255), Flannery (1994: 103), Clayton *et al.* (2006: 104), and Van Dyck and Strahan (2008: 255).

HOMONYMS:

Phalangista cookii Schinz, 1821, the Eastern Ring-tailed Possum of the Class Mammalia (Order Diprotodontia,

Family Pseudocheiridae). Synonym of *Pseudocheirus peregrinus* (Boddaert, 1785). See individual entry.

Phalangista cookii G. Cuvier, 1824, the Common Brush-tailed Possum of the Class Mammalia (Order Diprotodontia, Family Phalangeridae). Synonym of *Trichosurus vulpecula* (Kerr, 1792). See individual entry.

Phalangista cookii Gould, 1856 [1845–1863], the Eastern Ring-tailed Possum of the Class Mammalia (Order Diprotodontia, Family Pseudocheiridae). Synonym of *Pseudocheirus peregrinus cookii* (Desmarest, 1817c). See individual entry.

Pseudochirus antiquus Broom, 1896a: 55.

TYPE LOCALITY: Breccia deposit near Wombeyen Caves New South Wales, Australia.

COMMENTS: Synonymised within *Pseudocheirus peregrinus* by Wakefield (1972a: 21).

Pseudochirus victoriae Matschie, 1915a: 85.

TYPE LOCALITY: Cape Otway, Victoria, Australia.

COMMENTS: Recognised as a subspecies of *laniginosus* by Iredale and Troughton (1934: 26) and Troughton (1967: 94). Synonymised within *peregrinus* by McKay (1988c: 96), Flannery (1994: 122) and subsequent authors.

Pseudochirus laniginosus modestus Thomas, 1926b: 632.

TYPE LOCALITY: St. Georges River, New South Wales, Australia, but no such place exists. Possibly Georges River, Sydney, Australia.

COMMENTS: Recognised as a subspecies of *laniginosus* by Iredale and Troughton (1934: 26) and Troughton (1967: 94). Synonymised within *peregrinus* by McKay (1988c: 96), Flannery (1994: 122) and subsequent authors.

Pseudocheirus peregrinus convolutor (Schinz, 1821)

Phal. [angista] convolutor Schinz, 1821: 258.

TYPE LOCALITY: Adventure Bay, Tasmania, Australia.

COMMENTS: *Balantia convolutor* Oken, 1816 was used in preference by Iredale and Troughton (1934: viii, 27), who synonymised *P. convolutor* Schinz, 1821 within it. *Balantia convolutor* Oken, 1816 which Schinz's name is undoubtedly based, is not available, as it was suppressed by the ICZN (1956: 1). Recognised as a synonym of *peregrinus* by Ride (1970: 76), McKay (1988c: 95) and Flannery (1994: 122), but as a subspecies of it by Maxwell *et al.* (1996: 5), Groves (2005b: 51), Clayton *et al.* (2006: 104), and Van Dyck and Strahan (2008: 255).

B. [alantia] Convolutor Oken, 1816: 1126.

TYPE LOCALITY: Adventure Bay, Tasmania, Australia.

COMMENTS: Synonymised within *peregrinus* by Thomas (1888a: 172). Taxon name with Oken, 1816 as the author recognised as a full species within *Pseudocheirus* by Iredale and Troughton (1934: viii, 27) and Troughton (1967: 95). Name suppressed by the ICZN (1956: 1) and not subsequently considered by McKay (1988c) or Flannery (1994: 122).

P. [halangista] viverrina W. Ogleby [sic=Obilby], 1837b: 833.

TYPE LOCALITY: Tasmania, Australia.

COMMENTS: Though the author and date of this reference is typically recorded as W. Ogilby (1838b: 131), the first date of publication adopted here was highlighted by McAllan and Bruce (1989: 445). Recognised by Waterhouse (1838a: 67; 1841a: 277), Gould (1856 [1845–1863]: Text to Plate 19) and Krefft (1866a: 17; 1868a: 94). Synonymised within *cooki* Desmarest, 1817c, by Waterhouse (1846: 299) and Thomas (1888a: 176). Considered a synonym of *convolutor* by Iredale and Troughton (1934: 27), a subspecies of *peregrinus* by Strahan (1983: 126) and Flannery (1994: 103, 122). Synonymised within *peregrinus* by McKay (1988c: 95) and *convolutor* by Groves (2005b: 51).

Ph. [alangista] incana Schinz, 1844: 530.

TYPE LOCALITY: Tasmania, Australia. See Iredale and Troughton (1934: 27).

COMMENTS: Synonymised within *cookii* Desmarest, 1817c, by Thomas (1888a: 176). Considered a synonym of *peregrinus* by McKay (1988c: 96) and Flannery (1994: 122) and *convolutor* by Iredale and Troughton (1934: 27) and Groves (2005b: 51).

Pseudochirus cooki bassianus Le Souef, 1929: 330.

TYPE LOCALITY: Flinders Island, Bass Strait, Australia.

COMMENTS: This taxon was included as a subspecies of *Pseudocheirus convolutor* by Iredale and Troughton (1934: 27) and Troughton (1967: 95). Synonymised within *peregrinus* by McKay (1988c: 96) and Flannery (1994: 122) and within *convolutor* by Groves (2005b: 51).

Pseudocheirus peregrinus pulcher (Matschie, 1915)

Pseudochirus pulcher Matschie, 1915a: 85.

TYPE LOCALITY: Probably from northern rivers region of New South Wales, Australia.

COMMENTS: Intended as a replacement name for *Phalangista cookii* Gould, 1856 [1845–1863], despite the slightly different citation of the type locality. Recognised as a subspecies of *laniginosus* by Iredale and Troughton (1934: 26) and Troughton (1967: 94). Synonymised within *peregrinus* by McKay (1988c: 96). Recognised as a subspecies by J. Murray *et al.* (1980: 73), Strahan (1983: 126; 1995: 255), Flannery (1994: 103, 122), Groves (2005b: 51), Clayton *et al.* (2006: 104), and Van Dyck and Strahan (2008: 255).

Phalangista cookii Gould, 1856 [1845–1863]: Text to Plate 18.

TYPE LOCALITY: 'Brushes of Clarence' River, New South Wales, Australia.

COMMENTS: Label of lectotypes in Gould's hand: '*Phalangista cookii* Brushes of Clarence'. Considered a synonym of *Pseudochirus peregrinus* (under *P. convolutor*) by Iredale and Troughton (1934: 27). Recognised as a tentative subspecies of *peregrinus* by J. Murray *et al.* (1980: 73) but synonymised within *peregrinus* by McKay (1988c: 96) and Flannery (1994: 122) and subsequent authors.

HOMONYMS:

Phalangista cookii Desmarest, 1817c, the Eastern Ring-tailed Possum of the Class Mammalia (Order Diprotodontia, Family Pseudocheiridae). Recognised as a subspecies of *Pseudocheirus peregrinus* (Boddaert, 1785). See individual entry.

Phalangista cookii Schinz, 1821, the Eastern Ring-tailed Possum of the Class Mammalia (Order Diprotodontia, Family Pseudocheiridae). Synonym of *Pseudocheirus peregrinus* (Boddaert, 1785). See individual entry.

Phalangista cookii G. Cuvier, 1824, the Common Brush-tailed Possum of the Class Mammalia (Order Diprotodontia, Family Phalangeridae). Synonym of *Trichosurus vulpecula* (Kerr, 1792). See individual entry.

Phalangista laniginosa Gould, 1858 [1845–1863]: Text to Plate 20.

TYPE LOCALITY: Dartbrook, near Muswellbrook, upper Hunter River, New South Wales, Australia.

COMMENTS: Type designation by Thomas (1888a: 174). Recognised as a species within *Pseudocheirus* by Iredale and Troughton (1934: viii, 26) and Troughton (1967: 93) and within *Pseudochirus* by Finlayson (1935b: 222). Synonymised within *peregrinus* by Thomas (1888a: 172), Ride (1970: 246) and McKay (1988c: 96).

Pseudochirulus Matschie, 1915

Pseudochirulus Matschie, 1915a: 91.

TYPE SPECIES: Φ *Phalangista (Pseudochirus) canescens* Waterhouse, 1846: 305 [= Φ *Pseudochirulus canescens* (Waterhouse, 1846: 305)] by original designation.

COMMENTS: Described as a subgenus of *Pseudochirus* W. Ogilby, 1837a. Not considered by Iredale and Troughton (1934). Synonymised within *Pseudocheirus* by Marshall (1981: 28), McKay (1988c: 94), Marshall *et al.* (1990: 495) and McKenna and Bell (1997: 65). Genus rank resurrected by Flannery (1994: 103) and followed by Flannery (1995a: 211), Strahan (1995: 247), Groves (2005b: 51), Van Dyck and Strahan (2008: 248).

Pseudochirulus cinereus (Tate, 1945)

Daintree River Ring-tailed Possum

Pseudocheirus herbertensis cinereus Tate, 1945b: 17.

TYPE LOCALITY: Mt. Spurgeon, northwest of Cairns, north Queensland, Australia.

COMMENTS: Recognised at the subspecies rank by Troughton (1967: 97) and Strahan (1983: 128) but synonymised within *herbertensis* by McKay (1988c: 95) and Groves (1993e: 59) who proposed that it may be a distinct species. Elevated to species status within *Pseudocheirus* by J. Murray *et al.* (1989: 1119) and Baverstock *et al.* (1990a: 284), but placed within *Pseudochirulus* at species rank by Flannery (1994: 103, 134), Strahan (1995: 247), Maxwell *et al.* (1996: 5) and subsequent authors.

Pseudochirulus herbertensis (Collett, 1884)

Herbert River Ring-tailed Possum

Phalangista herbertensis Collett, 1884b: 383; Plate 30, Figs. 3–4.

TYPE LOCALITY: Top of ranges, approximately 25 km west of Cardwell (as Herbert Vale), north Queensland, Australia. See Lumholtz (1884: 407).

COMMENTS: Placed within *Pseudochirus* by Thomas (1888a: xii, 170). Species status, within *Pseudocheirus*, recognised by Iredale and Troughton (1934: viii, 27), Troughton (1967: 96), Ride (1970: 76), Kirsch and Calaby (1977: 16), Honacki *et al.* (1982: 42), McKay (1988c: 94) and Groves (1993e: 59). Transferred to *Pseudochirulus* by Flannery (1994: 103, 140), Maxwell *et al.* (1996: 5) and subsequent authors.

Pseudochirus mongan De Vis, 1887a: 1130.

TYPE LOCALITY: Herbert Gorge, near Herberton, north Queensland, Australia.

COMMENTS: Type designation by McKay (1988c: 94). Synonymised within *herbertensis* by Thomas (1888a: 170), Iredale and Troughton (1934: 27), McKay (1988c: 94), Flannery (1994: 140) and subsequent authors.

Pseudochirus herbertensis var. *colletti* Waite, 1899: 92.

TYPE LOCALITY: Herberton district Queensland, Australia.

COMMENTS: Designation by McKay (1988c: 94). Subspecies rank recognised by Iredale and Troughton (1934: 27), Tate (1945b: 16) and Troughton (1967: 97). Synonymised within *herbertensis* by McKay (1988c: 94), Flannery (1994: 140) and subsequent authors.

Subfamily Pseudochiropsinae Kirsch *et al.*, 1997 *sensu* Meredith *et al.*, 2010

Subfamily Pseudochiropsinae Kirsch *et al.*, 1997: 245.

TYPE GENUS: *Pseudochirops* Matschie, 1915a.

COMMENTS: When originally proposed as a new rank it was placed in the Family Pseudocheiridae (Winge, 1893a) and included the genus *Pseudochirops* Matschie, 1915a.

Distinctiveness of this group was identified by Baverstock *et al.* (1990c: 519). Subfamily rank recognised by Groves (2005b: 53) who included only the genus *Pseudochirops*, following Kirsch *et al.* (1997) who did not specifically assess the affinities of *Petropseudes*. A further assessment of the Family Pseudocheiridae by Meredith *et al.* (2010: 75), which included all genera, modified the composition of the subfamily and placed *Petropseudes* within it. Subfamily rank not recognised by Van Dyck and Strahan (2008: 10).

Petropseudes Thomas, 1923

Petropseudes Thomas, 1923b: 250.

TYPE SPECIES: *Pseudochirus dahlii* Collett, 1895 [= *Petropseudes dahlii* (Collett, 1895)] by monotypy.

COMMENTS: Proposed as subgenus of *Pseudocheirus* (W. Ogilby, 1837a). Genus rank recognised by Iredale and Troughton (1934: viii, 28), D. Johnson (1964: 453), Troughton (1967: 99) and Ride (1970: 76). Subgenus rank, within *Pseudocheirus*, was recognised by Tate (1945b: 2, 22). Synonymised within *Pseudocheirus* by Simpson (1945: 46), Kirsch and Calaby (1977: 16), Marshall (1981: 28), Honacki *et al.* (1982: 41), Strahan (1983: 132), and McKenna and Bell (1997: 65). Separated from *Pseudocheirus* at generic rank by McKay (1988c: 94), Strahan (1995: 242), Flannery (1994: 103), Van Dyck and Strahan (2008: 243) and most modern authors. An assessment by Meredith *et al.* (2009a: 567) found *Pseudochirops* to be paraphyletic with respect to *Petropseudes*, grouping the latter more closely to *Pseudochirops cupreus* than to *Pseudochirops archeri*. As a result they suggested subsuming *Petropseudes dahlii* into *Pseudochirops* as *Pseudochirops (Petropseudes) dahlii* (Meredith *et al.* (2009: 567). We here retain the distinctive genus *Petropseudes*, but recommend revision of the entire subfamily to determine whether the monophyletic clades require recognition at generic level.

Petropseudes dahlii (Collett, 1895)

Rock Ring-tailed Possum

Pseudochirus dahlii Collett, 1895: 464.

TYPE LOCALITY: Mary River, Northern Territory, Australia.

COMMENTS: Recognised within *Pseudocheirus* by Honacki *et al.* (1982: 41), Kirsch and Calaby (1977: 16) and Strahan (1983: 132). Placed in the genus *Petropseudes* by Iredale and Troughton (1934: viii, 28), D. Johnson (1964: 453), Troughton (1967: 99), Ride (1970: 76), McKay (1988c: 94) and most subsequent authors.

Pseudochirops Matschie, 1915

Pseudochirops Matschie, 1915a: 86.

TYPE SPECIES: Φ *Phalangista (Pseudochirus) albertisii* Peters, 1874a: 303 [= Φ *Pseudochirops albertisii* (Peters, 1874a: 303)] by original designation.

COMMENTS: Described as a subgenus of *Pseudochirus* W. Ogilby, 1837a. Synonymised within *Pseudocheirus* by Simpson (1945: 46), Ride (1970: 246), Marshall (1981: 28), Honacki *et al.* (1982: 41) and McKenna and Bell (1997: 65). Recognised as subgenus of *Pseudocheirus* by Tate (1945b: 2, 17) and Kirsch and Calaby (1977: 16). Separated from *Pseudocheirus* by Iredale and Troughton (1934: viii, 28), Troughton (1967: 97), McKay (1988c: 97), Marshall *et al.* (1990: 494), Groves (1993e: 60), Flannery (1994: 103, 112) and subsequent authors.

Pseudochirops archeri (Collett, 1884)

Green Ring-tailed Possum

Phalangista (*Pseudochirus*) *archeri* Collett, 1884b: 381; Plate 29, Figs. 1–2.

TYPE LOCALITY: Herbert River, as 'Herbert Vale', north Queensland, Australia.

COMMENTS: Recognised within *Pseudochirus* by Thomas (1888a: xii, 177), Ride (1970: 76), Honacki *et al.* (1982: 41) and Strahan (1983: 130). Transferred to the genus *Pseudochirops* by Iredale and Troughton (1934: viii, 28), Troughton (1967: 97), McKay (1988c: 97), Flannery (1994: 103, 112) and subsequent authors.

Family Tarsipedidae Gervais & Verreaux, 1842

Family Tarsipedidae Gervais & Verreaux, 1842a: 1.

TYPE GENUS: *Tarsipes* Gervais & Verreaux, 1842b.

COMMENTS: When originally proposed, this rank was placed in the Marsupialia (Illiger, 1811) and included the genus *Tarsipes* Gervais and Verreaux, 1842b. Correct authorship of the single species was reviewed by Mahoney (1981: 135). The monotypic genus has previously been included within the Phalangeridae as the Subfamily Tarsipedinae by Thomas (1888a: xii, 132), Bensley (1903: 125) and Simpson (1945: 46). Family status recognised by Marshall (1981: 30; 1984: 109), Honacki *et al.* (1982: 51), Aplin and Archer (1987: xxii), Marshall *et al.* (1990: 460), Szalay (1994: 43), Kirsch *et al.* (1997: 245), Kavanagh *et al.* (2004: 207, 210) and subsequent authors.

Family Tarsipédidés Gervais, 1855a: 277.

TYPE GENUS: *Tarsipes* Gervais & Verreaux, 1842b.

COMMENTS: When originally proposed, this rank was placed in the Order Marsupiaux (Gervais, 1855a [= Marsupialia (Illiger, 1811)]) and included the genus *Tarsipes* Gervais and Verreaux, 1842b. Synonymised within Tarsipedidae by McKenna and Bell (1997: 59).

Order Edentula Haeckel, 1866: clvii.

COMMENTS: When originally proposed, this rank was placed in the Class Mammalia (Linnaeus, 1758) and included the Family Tarsipedina (Haeckel, 1866 [= Tarsipedidae (Gervais & Verreaux, 1842a)]).

Family Tarsipedina Haeckel, 1866: clvii.

TYPE GENUS: *Tarsipes* Gervais & Verreaux, 1842b.

COMMENTS: When originally proposed, this rank was placed in the Order Edentula (Haeckel, 1866 [= Tarsipedidae (Gervais & Verreaux, 1842a)]) and included the genus *Tarsipes* Gervais and Verreaux, 1842b. Synonymised within Tarsipedidae by McKenna and Bell (1997: 59).

Family Tarsipedidae Gill, 1872: 25.

TYPE GENUS: *Tarsipes* Gervais & Verreaux, 1842b.

COMMENTS: When originally proposed, this rank was placed in the Suborder Syndactyli (Gill, 1871a [= Diprotodontia (Owen, 1877a part)]). Synonymised within Tarsipedidae Gervais and Verreaux, 1842a by Marshall (1981: 30), Marshall *et al.* (1981: 495) and McKenna and Bell (1997: 59).

Subfamily Tarsipedinae Thomas, 1888a: 130.

TYPE GENUS: *Tarsipes* Gervais & Verreaux, 1842b.

COMMENTS: When originally proposed, this rank was placed in the Family Phalangeridae (Thomas, 1888a) and included the genus *Tarsipes* Gervais and Verreaux, 1842b. Subfamily rank, within the Family Phalangeridae, recognised by Troughton (1967: 70). Synonymised within Tarsipedidae by Marshall (1981: 30), Marshall *et al.* (1981: 495) and McKenna and Bell (1997: 59).

Superfamily Tarsipedoidea Kirsch, 1968: 420.

TYPE GENUS: *Tarsipes* Gervais & Verreaux, 1842b.

COMMENTS: When originally proposed, this rank was attributed to Gill (1872: 25), and placed in the Order Diprotodont[i]a (Owen, 1877a) and included the Family Tarsipedidae (Gervais & Verreaux, 1842a). Superfamily rank recognised by Kirsch (1977a: 113; 1977b: 45), Strahan (1983: xxi), Aplin and Archer (1987: xxii), Strahan (1995: 7, 257) and Long *et al.* (2002: 143). Aplin and Archer (1987: xxii) revised this rank and included the families Acrobatidae and Tarsipedidae, which was followed by Kear and Cooke (2001: 84), Long *et al.* (2002: 143) and Crosby *et al.* (2004: 171). Rank not recognised by McKenna and Bell (1997: 59), Kirsch *et al.* (1997: 245), Kavanagh *et al.* (2004: 207) or Groves (2005b: vii, 55). Van Dyck and Strahan (2008: 10, 258) placed the families Tarsipedidae and Acrobatidae within the Superfamily Petauroidea. The phylogenetic position of *Tarsipes* was considered by Kavanagh *et al.* (2004: 207) who placed it within the Superfamily Petauroidea.

Tarsipes Gervais & Verreaux, 1842

Tarsipes Gervais & Verreaux, 1842b: 75.

TYPE SPECIES: *Tarsipes rostratus* Gervais & Verreaux, 1842b by monotypy.

COMMENTS: Type designation by de Beaufort (1966: 532). This species was also described by Gervais and Verreaux (1842a: 1; 1842c: 1) and Gervais (1842a: 19; 1842b: 94; 1842c: 631). There has been great confusion over the author of the genus and species names. See comments below.

Tarsipes rostratus Gervais & Verreaux, 1842

Honey Possum

Tarsipes rostratus Gervais & Verreaux, 1842b: 75.

TYPE LOCALITY: Swan River, Western Australia, Australia.

COMMENTS: This species was also described by Gervais and Verreaux (1842a: 1; 1842c: 1) and Gervais (1842a: 19; 1842b: 94; 1842c: 631). The paper by Gervais and Verreaux (1842b: 75) was published via an abstract from a presentation given to the Société Philomatique de Paris held on 19 February 1842, which was published on 3 March 1842. The publication by Gervais (1842b: 94) was proposed to be a *nomen nudum* by Mahoney (1981: 136). Species name recognised by Gould (1845 [1845–1863]: Text to Plate 5), Waterhouse (1846: 345), Thomas (1888a: xii, 133), A. Lucas and Le Souëf (1909: 110) and de Beaufort (1966: 532). Publication date and taxonomic decision taken from Mahoney (1981: 136) to accept *rostratus* as the senior synonym.

Tarsipes spenserae J. Gray, 1842d: 40.

TYPE LOCALITY: King Gorge Sound, Western Australia, Australia.

COMMENTS: Originally described by J. Gray in reference to a manuscript of Gervais. The name *spenserae* is a misspelling as it was named after Lady Spencer, the daughter of Captain Spencer. Recognised as the valid species name by various authors because the publication of J. Gray (1842d: 40) predates that of Gervais and Verreaux (1842a: 1) describing *rostratus* [but it does not predate Gervais & Verreaux, 1842b: 75]. Examples of *spenserae* being given preference to *rostratus* include Palmer (1904: 664), Thomas (1906c: 475), Troughton (1923a: 148), Le Souef and Burrell (1926: 241), Iredale and Troughton (1934: vii, 21), Glauert (1950: 122), Troughton (1967: 71), Marlow (1965: 68), Ride (1970: 88), with the spelling *spencerae*, Vose (1973: 245), Renfree (1980: 81), and Honacki *et al.* (1982: 51). Synonymised within *rostratus* by Waterhouse (1846: 345), Thomas (1888a: 133), most authors after the taxonomic review of Mahoney (1981: 135), including Strahan (1983: 173; 1995: 258), McKay (1988d: 103) and subsequent authors.

Tarsipes spencerae Ride, 1970: 88.

TYPE LOCALITY: Invalid emendation for *Tarsipes spenserae* J. Gray, 1842d.

COMMENTS: Synonymised within *rostratus* by Strahan (1983: 173), Groves (1993e: 62; 2005b: 55) and Flannery (1994: 30).

Family Acrobatidae Aplin (in Aplin and Archer), 1987

Family Acrobatidae Aplin (in Aplin and Archer), 1987: xxii, lvii.

TYPE GENUS: *Acrobates* Desmarest, 1817b.

COMMENTS: When originally proposed, this rank was placed in the Superfamily Tarsipedoidea (Gervais & Verreaux, 1842a) and included the genera *Acrobates* Desmarest, 1817b; and *Distoechurus* Peters, 1874a: 303. Recognition of this family being distinct from the Burramyidae followed by Strahan (1987: v, 113) and supported by Baverstock *et al.* (1990a: 273), Marshall *et al.* (1990: 460) and subsequent authors.

Tribe Acrobatini Szalay, 1994: 43.

TYPE GENUS: *Acrobates* Desmarest, 1817b.

COMMENTS: When originally proposed, this rank was attributed to Aplin (*in* Aplin and Archer, 1987: xxii, lvii) and placed in the Subfamily Burramyinae (Broom, 1898). Synonymised within Acrobatidae by McKenna and Bell (1997: 67).

Acrobates Desmarest, 1817

Acrobates Desmarest, 1817b: 405.

TYPE SPECIES: *Didelphis pygmaea* Shaw, 1794 [= *Acrobates pygmaeus* (Shaw, 1794)] by monotypy.

COMMENTS: Described as a subgenus of *Petaurus*, which was followed by Waterhouse (1841a: 293; 1846: 337). Genus recognised by Waterhouse (1838a: 68), J. Gray (1841: 402; 1843a: xxii, 83), Gould (1849 [1845–1863]: Text to Plate 28), Krefft (1871a: 3, Text to Plate 7) and Thomas (1888a: xii, 136).

HOMONYMS:

Acrobates Bonaparte, 1850b: 284, scrub robins or bush chats of the Class Aves (Order Passériformes, Family Muscicapidae). Appears to be an incorrect subsequent spelling of *Agrobates* Swainson, 1837: 63, 241 or *Agrabates* Swainson, 1837: 63, 64. These appear to be synonyms of *Cercotrichas* F. Boie, 1831: col. 542.

Opossum Perry, 1810 [1810–1811]: Text to Plate 32.

TYPE SPECIES: *Opossum opossum* Perry, 1810 [1810–1811] [= *Acrobates pygmaeus* (Shaw, 1794)] by monotypy.

COMMENTS: Genus not recognised by subsequent authors.

HOMONYMS:

Opossum Perry, 1810 [1810–1811], the Common Wombat of the Class Mammalia (Order Diprotodontia, Family Vombatidae). Name is a junior synonym of *Vombatus* É. Geoffroy, 1803b. See individual entry.

acrobata Desmarest, 1821: 270.

TYPE SPECIES: In error for *Acrobates* Desmarest, 1817b.

COMMENTS: Does not appear to have been previously recognised. Included here as a synonym.

Ascobates Anon, 1839: 454.

TYPE SPECIES: In error for *Acrobates* Desmarest, 1817b.

COMMENTS: Synonymised within *Acrobates* by Iredale and Troughton (1934: 21) and McKay (1988b: 98).

Cercoptēnus Gloger, 1841: xxx, 85.

TYPE SPECIES: *Didelphis pygmaea* Shaw, 1794 [= *Acrobates pygmaeus* (Shaw, 1794)] by monotypy.

COMMENTS: Synonymised within *Acrobates* by Thomas (1888a: 136; 1895a: 190), Iredale and Troughton (1934: 22), Marshall (1981: 28) and McKay (1988b: 98).

Acrobates frontalis (De Vis, 1887)

Broad-toed Feather-tailed Glider

Dromicia frontalis De Vis, 1887a: 1134.

TYPE LOCALITY: Herbert district, north Queensland, Australia.

COMMENTS: Abstract of description provided in De Vis (1886a: vi). Considered a subspecies by Iredale and Troughton (1934: 22) and Van Deusen (1960: 263), but synonymised within *pygmaeus* by Thomas (1888a: 137), McKay (1988b: 99), Flannery (1994: 40) and subsequent authors until it was recognised as being both morphologically and genetically distinct by Aplin (pers. comm.) and formally recognised by Van Dyck *et al.* (2013: 25, 86).

Acrobates pygmaeus (Shaw, 1794)

Narrow-toed Feather-tailed Glider

Didelphis Pygmaea Shaw, 1794: 5; Plate 1.

TYPE LOCALITY: Sydney, New South Wales, Australia. Type designated by Thomas (1922a: 128).

COMMENTS: Placed within *Petaurus* (*Acrobata*) by Waterhouse (1841a: 293; 1846: 339), but transferred to *Acrobates* by Desmarest (1817b: 405) and followed by subsequent authors including Waterhouse (1838a: 68), J. Gray (1841: 402; 1843a: 83), Gould (1849 [1845–1863]: Text to Plate 28), Krefft (1871a: 3, Text to Plate 7) and Thomas (1888a: xii, 136), who described the taxonomic history. Tate (1938: 60) believed the single specimen (see under *A. pulchellus* below) obtained in NW New Guinea was probably an introduction as a pet. Species included within Burramyidae by Kirsch and Calaby (1977: 16) and McKay (1988b: 99).

Opossum opossum Perry, 1810 [1810–1811]: Text to Plate 32.

TYPE LOCALITY: New Holland [= Australia].

COMMENTS: Species not recognised by subsequent authors.

Φ *Acrobates pulchellus* Rothschild, 1893: 546.

TYPE LOCALITY: Unknown island north of Dutch New Guinea [= West Papua].

COMMENTS: Species recognised by Laurie and Hill (1954: 19), but not by other authors including Tate (1938: 60), who believed the single specimen obtained in NW New Guinea was probably an introduction as a pet from Australia, and Van Deusen (1960: 264), which was followed by subsequent authors. The possibility of this species occurring in New Guinea was further explored by Helgen (2003a: 107) who examined several specimens from Stockholm and London of which one was collected in 1899–1900 and received from a dealer with the only information being 'Nya Guinee'. Helgen (2003a: 108) suggested that both specimens (from London and Stockholm) are left open to doubt due to imprecise locality data but lent some credibility by their appearance. It seems likely that the specimens associated with this taxon are the result of a major locality error rather than being introduced as a pet (K. Aplin pers. comm.).

Superfamily Phalangeroidea Thomas, 1888 *sensu* Aplin & Archer, 1987

Family Phalangeridae Thomas, 1888a: xii, 126.

TYPE GENUS: *Phalanger* Storr, 1780.

COMMENTS: When originally proposed, this rank was placed in the Suborder Diprotodontia (Owen, 1877a) and included the subfamilies Tarsipedinae (Thomas, 1888a [= Tarsipedidae (Gervais & Verreaux, 1842a)]), Phalangerinae (Thomas, 1888a) and Phascolarctinae (Thomas, 1888a [= Phascolarctidae (Owen, 1839a)]). Simpson (1945: 45) designated the author of this superfamily as Weber (1928: xiii). Superfamily status recognised by Osborn (1910: 517), Simpson (1930: 10; 1931: 262; 1945: xi, 45; 1970: 38), Kirsch (1968a: 420; 1977a: 112; 1977b: 45), Szalay (1982: 631; 1994: 42), Strahan (1983: xxi, 123), Aplin and Archer (1987: xxii), Marshall *et al.* (1990: 493), Kirsch *et al.* (1997: 245), McKenna and Bell (1997: 61), Kear and Cooke (2001: 84), Groves (2005b: 44) and subsequent authors. The composition of the superfamily has reduced due to the removal of the other superfamilies.

Family Phalangeridae Weber, 1928: xiii, 76.

TYPE GENUS: *Phalanger* Storr, 1780.

COMMENTS: When originally proposed, this rank was placed in the Diprotodontia (Owen, 1877a). Synonymised within Phalangeroidea by McKenna and Bell (1997: 61).

Family Phalangeridae Thomas, 1888 *sensu* Flannery *et al.*, 1987

Family Phalangeridae Thomas, 1888a: xii, 126.

TYPE GENUS: *Phalanger* Storr, 1780.

COMMENTS: When originally proposed, this rank was placed in the Suborder Diprotodontia (Owen, 1877a) and included the subfamilies Tarsipedinae (Thomas, 1888a [= Tarsipedidae (Gervais & Verreaux, 1842a)]), Phalangerinae (Thomas, 1888a) and Phascolarctinae (Thomas, 1888a [= Phascolarctidae (Owen, 1839a)]). Family name recognised by most authors since its description, although the composition has varied considerably as different species have been removed into distinct families including the Pseudocheiridae, Petauridae, Burramyidae and Acrobatidae. Two subfamilies are currently recognised: the Subfamily Ailuropinae (Flannery *et al.*, 1987: 477, 503), whose single representative does not occur within Australia, and the Subfamily Phalangerinae, which includes the tribes Phalangerini and Trichosurini. Ruedas and Morales (2005: 353) proposed that the Family Phalangeridae should be composed of the subfamilies Phalangerinae (including *Phalanger* and *Spilocuscus*), Ailuropinae (including *Ailurops* Wagler, 1830: 26 and *Strigocuscus* Gray, 1862a: 319) and Trichosurinae (including *Trichosurus* and *Wyulda*).

Family Phalangistadae J. Gray, 1821: 308.

TYPE GENUS: *Phalangista* É. Geoffroy & G. Cuvier, 1795 [= *Phalanger* Storr, 1780].

COMMENTS: When originally proposed, this rank was placed in the Order Ferae (J. Gray, 1821 [= Diprotodontia (Owen, 1877a)]) and included the genera *Balantia* Illiger, 1811 [= *Phalanger* Storr, 1780]; *Petaurus* Shaw, 1791; *Phalanger* Storr, 1780; and *Coesiodes* J. Gray, 1821 [= *Phalanger* Storr, 1780]. Synonymised within the Family Phalangeridae by Marshall (1981: 27), Marshall *et al.* (1990: 493) and McKenna and Bell (1997: 61).

Tribe Phalangistina J. Gray, 1825a: 340.

TYPE GENUS: *Phalangista* É. Geoffroy & G. Cuvier, 1795 [=*Phalanger* Storr, 1780].

COMMENTS: When originally proposed, this rank was placed in the Family Didelphidae (J. Gray, 1821: 308) and included the genera *Acrobata* Desmarest, 1821 [=*Acrobates* Desmarest, 1817b]; *Petaurus* Shaw, 1791; *Phalangista* É. Geoffroy and G. Cuvier, 1795 [= *Phalanger* Storr, 1780]; *Balantia* Illiger, 1811 [= *Phalanger* Storr, 1780]; and *Phascolaretus* [sic=*Phascolarctos*] de Blainville, 1816a. Recognised as a tribe within the Family Macropodidae by J. Gray (1842e: 16) and as a tribe within the Family Phalangistidae by Bonaparte (1845: 6). Synonymised within the Family Phalangeridae by McKenna and Bell (1997: 61).

Family Phalangistae Lesson, 1842: 188.

TYPE GENUS: *Phalangista* É. Geoffroy & G. Cuvier, 1795 [= *Phalanger* Storr, 1780].

COMMENTS: When originally proposed, this rank was placed in the Order Mastodidelphie (Lesson, 1842 [= Marsupialia (Illiger, 1811)]) and included the genera *Cuscus* Lesson, 1827b [= *Phalanger* Storr, 1780]; *Phalangista* É. Geoffroy and G. Cuvier, 1795 [= *Phalanger* Storr, 1780]; and *Trichosurus* Lesson, 1828b.

Family Phalangistida Haeckel, 1866: clvii.

TYPE GENUS: *Phalangista* É. Geoffroy & G. Cuvier, 1795 [= *Phalanger* Storr, 1780].

COMMENTS: When originally proposed, this rank was placed in the Carpophaga (Owen, 1839a [= Diprotodontia (Owen, 1877a part)]) and included the genera *Phalangista* É. Geoffroy and G. Cuvier, 1795 [= *Phalanger* Storr, 1780]; and *Petaurus* Shaw, 1791. Synonymised within Phalangeridae by McKenna and Bell (1997: 61).

Subfamily Phalangerinae Thomas, 1888 *sensu* Flannery *et al.*, 1987

Subfamily Phalangerinae Thomas, 1888a: xii, 135.

TYPE GENUS: *Phalanger* Storr, 1780.

COMMENTS: When originally proposed, this rank was placed in the Suborder Diprotodontia (Owen, 1877a) and included the subfamilies Tarsipedinae (Thomas, 1888a [= Tarsipedidae (Gervais & Verreaux, 1842a)]), Phalangerinae (Thomas, 1888a) and Phascolarctinae (Thomas, 1888a [= Phascolarctidae (Owen, 1839a)]). The inclusion of the possums, gliders and koala within the Family Phalangeridae was followed by Bensley (1903: 125) and Simpson (1945: 46), although the taxa representing the Families Phascolarctidae, Acrobatidae, Burramyidae, Petauridae, Pseudocheiridae and Tarsipedidae were subsequently removed (see individual entries). The only other currently accepted subfamily is the Ailuropinae (Flannery *et al.*, 1987: 477, 503) which includes only the bear cuscus of Sulawesi, genus *Ailurops* (Wagler, 1830: 26). Subfamilies not recognised by Strahan (1983: xxi; 1995: 7, 265) or Van Dyck and Strahan (2008: 10). Subfamily recognised by Kirsch (1968a: 420), Marshall (1981: 27; 1984: 98), Szalay (1994: 42) and Groves (2005b: 45).

Tribe Phalangistini Winge, 1893a: 89, 103.

TYPE GENUS: *Phalangista* É. Geoffroy & G. Cuvier, 1795 [= *Phalanger* Storr, 1780].

COMMENTS: When originally proposed, this rank was placed in the Family Phalangistidae (Owen, 1839a [= Phalangerida (Aplin & Archer, 1987 part)]) and included the genera *Pseudocheirus* W. Ogilby, 1837a; and *Petaurista* Desmarest, 1821 [= *Petauroides* Thomas, 1888a]. Tribe rank subsequently recognised by Winge (1941: 69) but not typically by subsequent authors.

Tribe Phalangerini Thomas, 1888 *sensu* Flannery *et al.*, 1987

Subfamily Phalangerinae Thomas, 1888a: xii, 135.

TYPE GENUS: *Phalanger* Storr, 1780.

COMMENTS: When originally proposed, this rank was placed in the Family Phalangeridae and included the genera *Acrobates* Desmarest, 1817b; *Distoechurus* Peters, 1874a: 303; *Dromicia* J. Gray, 1841 [= *Cercartetus* Gloger, 1841]; *Gymnobelideus* McCoy, 1867a; *Petaurus* Shaw, 1791; *Dactylopsila* J. Gray, 1858a; *Petauroides* Thomas, 1888a; *Pseudochirus* W. Ogilby, 1836a [= *Pseudocheirus* W. Ogilby, 1837a]; *Trichosurus* Lesson, 1828b; and *Phalanger* Storr, 1780. Tribe Phalangerini recognised by Flannery *et al.* (1987: 477, 503), Norris (1994: 93), Szalay (1994: 42), Marshall *et al.* (1990: 494), Kirsch *et al.* (1997: 245) and Groves (2005b: 46). Tribe not recognised by Strahan (1983: xxi; 1995: 7), and Van Dyck and Strahan (2008: 10) or McKenna and Bell (1997: 61) who synonymised it within the Subfamily Phalangerinae.

Phalanger Storr, 1780

Phalanger Storr, 1780: 33.

TYPE SPECIES: Φ *Didelphis orientalis* Pallas, 1766: 59 [= Φ *Phalanger orientalis* (Pallas, 1766: 59)] by monotypy.

COMMENTS: Genus reviewed by Tate (1945c: 1), Feiler (1978a: 1; 1978b: 385) and George (1979: 97). Placed in the Tribe Phalangerini (Thomas, 1888a) by Flannery *et al.* (1987: 504) and many current authors (see above).

HOMONYMS:

Phalanger Lacépède, 1801, the Greater Glider of the Class Mammalia (Order Diprotodontia, Family Phalangeridae). Genus is a synonym of *Petauroides* Thomas, 1888a. See individual entry.

Coes-Coes Pallas, 1766: 60.

TYPE SPECIES: Φ *Didelphis orientalis* Pallas, 1766: 59 [= Φ *Phalanger orientalis* (Pallas, 1766: 59)] by monotypy.

COMMENTS: *Nomen oblitum*. *Coescoes* has often been attributed to Lacépède (1799a: 5; 1801: 491) by authors including McKenna and Bell (1997: 61) and Groves (2005b: 46). Synonymised within *Phalanger* by Thomas (1888a: 193), McKenna and Bell (1997: 61), Norris (1999: 1) and Groves (2005b: 46).

Phalangista É. Geoffroy & G. Cuvier, 1795: 187.

TYPE SPECIES: Φ *Didelphis orientalis* Pallas, 1766: 59 [= Φ *Phalanger orientalis* (Pallas, 1766: 59)] by monotypy.

COMMENTS: Genus recognised by Waterhouse (1846: 265), but synonymised within *Phalanger* by Thomas (1888a: 193), Marshall (1981: 28), McKay (1988e: 81), Marshall *et al.* (1990: 494), McKenna and Bell (1997: 61) and Norris (1999: 1).

HOMONYMS:

Phalangista G. Cuvier, 1829b: 163, poacher fish of the Class Actinopterygii (Order Scorpaeniformes, Family Agonidae). Genus appears to be an incorrect subsequent spelling of *Phalangistes* Pallas, 1811: 263 and 1831: 110. Genus is a synonym of *Agonus* Bloch and Schneider, 1801: xxx.

Coescoes Lacépède, 1799a: 5.

TYPE SPECIES: Φ *Didelphis orientalis* Pallas, 1766: 59 (as *Coescoes amboinensis*) [= Φ *Phalanger orientalis* (Pallas, 1766: 59)] by monotypy.

COMMENTS: Synonymised within *Phalanger* by Palmer (1904: 194), Marshall (1981: 28), Marshall *et al.* (1990: 494) and McKenna and Bell (1997: 61).

Balantia Illiger, 1811: 77.

TYPE SPECIES: Φ *Didelphis orientalis* Pallas, 1766: 59 [= Φ *Phalanger orientalis* (Pallas, 1766: 59)] by subsequent designation by Thomas (1888a: 193).

COMMENTS: Synonymised within *Phalanger* by Waterhouse (1846: 265), Thomas (1888a: 193), Palmer (1904: 132), Marshall (1981: 28), Marshall *et al.* (1990: 494), McKenna and Bell (1997: 61) and Norris (1999: 1).

Cuscus Illiger, 1811: 77.

TYPE SPECIES: Emendation of *Coescoes* Lacépède, 1799a.

COMMENTS: Ascription of this name to Illiger, 1811 not typically recognised. Taxon recognised as a subgenus of *Phalangista* by Waterhouse (1846: 266), but typically not by subsequent authors.

HOMONYMS:

Cuscus Lesson, 1827b, cuscuses of the Class Mammalia (Order Diprotodontia, Family Phalangeridae). See individual entry.

Sipalus G. Fischer, 1813b: xxiii, 581.

TYPE SPECIES: Φ *Didelphis orientalis* Pallas, 1766: 59 [= Φ *Phalanger orientalis* (Pallas, 1766: 59)] by subsequent designation by Thomas (1888a: 193).

COMMENTS: *Nomen novum* for *Phalanger* Storr (1780) and *Coescoes* Lacépède, 1799a, which are not Latin or Greek names. Synonymised within *Phalanger* by Thomas (1888a: 193), Marshall (1981: 28), Marshall *et al.* (1990: 494), McKenna and Bell (1997: 61) and Norris (1999: 1).

HOMONYMS:

Sipalus Schönherr, 1825: 587, true weevils of the Class Insecta (Order Coleoptera, Family Curculionidae).

Coesiodes J. Gray, 1821: 308.

TYPE SPECIES: Φ *Didelphis orientalis* Pallas, 1766: 59 [= Φ *Phalanger orientalis* (Pallas, 1766: 59)] by monotypy.

COMMENTS: Emendation of *Coescoes* (Pallas, 1766: 60). Not generally recognised.

Cuscus Lesson, 1827b: 150.

TYPE SPECIES: Φ *Didelphis orientalis* Pallas, 1766: 59 [= Φ *Phalanger orientalis* (Pallas, 1766: 59)] by subsequent designation. See Thomas (1888a: 193).

COMMENTS: Name attributed to Lacépède with reference. Taxon recognised as a subgenus of *Phalangista* by Waterhouse (1846: 266). Synonymised within *Phalanger* by Thomas (1888a: 193), Marshall (1981: 28), Marshall *et al.* (1990: 494), McKenna and Bell (1997: 61) and Norris (1999: 1).

HOMONYMS:

Cuscus Illiger, 1811, cuscuses of the Class Mammalia (Order Diprotodontia, Family Phalangeridae). See individual entry.

Ceonix Temminck, 1824: 10, footnote.

TYPE SPECIES: Φ *Phalangista ursina* Temminck, 1824: 10 [= Φ *Ailurops ursinus* (Temminck, 1824: 10)] by monotypy.

COMMENTS: *Nomen provis* (see Thomas, 1888a: 193). Synonymised within *Phalanger* by Thomas (1888a: 193), Marshall (1981: 28), Marshall *et al.* (1990: 494) and McKenna and Bell (1987: 61). Not recognised by Norris (1999: 1).

Tapoa Owen, 1839a: 19.

TYPE SPECIES: Described as a subgenus of *Phalangista* É. Geoffroy & G. Cuvier, 1795 [= *Phalanger* Storr, 1780].

COMMENTS: Recognised as a *nomen nudum* by Palmer (1904: 663) and synonymised within *Trichosurus* by Iredale and Troughton (1934: 30) and McKay (1988e: 83).

HOMONYMS:

Tapoa Lesson, 1842, phascogales of the Class Mammalia (Order Dasyuromorphia, Family Dasyuridae). Genus is a synonym of *Phascogale* Temminck, 1824. See individual entry.

Ceonyx Agassiz, 1842: 6.

TYPE SPECIES: Emendation of *Ceonix* Temminck, 1824.

COMMENTS: Date established from Palmer (1904: 168). Also incorrectly spelt by Thomas (1888a: 193).

Cursus J. Gray, 1849: 20.

TYPE SPECIES: Φ *Didelphis orientalis* Pallas, 1766: 59 [= Φ *Phalanger orientalis* (Pallas, 1766: 59)] by monotypy.

COMMENTS: Pro *Coescoes* Lacépède (1799a: 5). Misprint of *Cuscus* Lesson, 1827b.

Eucuscus J. Gray, 1862a: 316.

TYPE SPECIES: Φ *Phalangista ursina* Temminck, 1824: 10 [= Φ *Ailurops ursinus* (Temminck, 1824: 10)] by subsequent designation. See Thomas (1888a: 193).

COMMENTS: Described as a subgenus of *Cuscus*. Synonymised within *Phalanger* by Thomas (1888a: 193), Marshall *et al.* (1990: 494) and McKenna and Bell (1997: 61). Not recognised by Norris (1999: 1).

Phalanger mimicus Thomas, 1922

Southern Common Cuscus

Phalanger orientalis mimicus Thomas, 1922e: 680.

TYPE LOCALITY: Parimau, Mimika River, Nassau Range, New Guinea. [4°31′S 136°36′E]

COMMENTS: Recognised as described by Tate (1945c: 12). Recorded in Australia on Cape York Peninsula, as *Phalanger orientalis peninsulae*, by Brass (1953: 199). Included at species rank within *Strigocuscus* Gray, 1862a: 319 by Flannery *et al.* (1987: 481). Synonymised within *Phalanger orientalis* by Groves (1993e: 46) and *Phalanger intercastellanus* Thomas, 1895b: 165 by Colgan *et al.* (1993: 375) and Flannery (1994: 190; 1995a: 170; 1995b: 93). Placed as a subspecies of *Phalanger orientalis* (Pallas, 1766: 59) by Feiler (1978b: 388), and Menzies and Pernetta (1986: 586). Recognised as a species within *Phalanger* by Norris and Musser (2001: 5) and followed by Groves (2005b: 47), Van Dyck and Strahan (2008: 268) and Menkhorst and Knight (2011: 16).

Φ *Phalanger microdon* Tate & Archbold, 1935a: 8.

TYPE LOCALITY: Dogwa, Oriomo River, Western Division of Papua New Guinea. 30 metres.

COMMENTS: Synonymised within *mimicus* by Tate (1945c: 12), Menzies and Pernetta (1986: 586), Norris and Musser (2001: 5) and Groves (2005b: 47).

Phalanger orientalis peninsulae Tate, 1945c: 2.

TYPE LOCALITY: Rocky scrub, 30 miles north of Coen, north Queensland, Australia.

COMMENTS: Recognised as described by Brass (1953: 199) and Troughton (1967: 110). Synonymised within *orientalis* by Groves (1993e: 46), within *intercastellanus* by Colgan *et al.* (1993: 375) and Flannery (1994: 190), but synonymised within *mimicus* by Menzies and Pernetta (1986: 586) and Norris and Musser (2001: 5). Recognised as a subspecies within *mimicus* by Groves (2005b: 47) but not by Van Dyck and Strahan (2008: 270) who recognised *Phalanger mimicus* within Australia with no distinct subspecies.

Spilocuscus J. Gray, 1862

Spilocuscus J. Gray, 1862a: 316.

TYPE SPECIES: Φ *Phalangista maculata* É. Geoffroy, 1803c: 149 [= Φ *Spilocuscus maculatus* (É. Geoffroy, 1803c: 149)] by subsequent designation. See Thomas (1888a: 193).

COMMENTS: Described as a subgenus of *Cuscus*. Synonymised within *Phalanger* by Thomas (1888a: 193), Ride (1970: 248), Honacki *et al.* (1982: 37), McKenna and Bell (1997: 61) and Norris (1999: 1). Recognised at genus rank by Simpson (1945: 46), Troughton (1967: 108), Flannery *et al.* (1987: 481), Flannery (1994: 153) and Groves (1993e: 47). Placed in the Tribe Phalangerini (Thomas, 1888a) by Flannery *et al.* (1987: 504). The Australian *Spilocuscus nudicaudatus* has typically been included as a subspecies of the *Spilocuscus maculatus* (É. Geoffroy, 1803c: 149), which has also included as subspecies *Spilocuscus maculatus chrysorrhous* (Temminck, 1824: 12) from the Moluccas and *Spilocuscus maculatus goldiei* (Ramsay, 1877c: 395) from New Guinea. However, the morphological attributes of *nudicaudatus* had not previously been the subject of a close review until this was done by Helgen (2007b: 232, 273) who concluded that it was distinct.

Spilocuscus nudicaudatus (Gould, 1850)

Australian Spotted Cuscus

Phalangista (Pseudocheirus) nudicaudata Gould, 1850: 110.

TYPE LOCALITY: Cape York, north Queensland, Australia.

COMMENTS: Recognised within the genus *Spilocuscus* at the species rank by Iredale and Troughton (1934: viii, 32). Synonymised in *maculatus* within the genus *Phalanger* by Thomas (1888a: 198) and Ride (1970: 246), but elevated to a subspecies of *Phalanger maculatus* by Tate (1945c: 26), Feiler (1978a: 14) and McKay (1988e: 82). Recognised at species rank, within *Spilocuscus*, by Troughton (1967: 109) and as a subspecies of *maculatus* by Flannery (1990: 130; 1994: 153, 220; 1995a: 182; 1995b: 106), Groves (2005b: 48), and Van Dyck and Strahan (2008: 267). In contrast to this arrangement, Helgen and Flannery (2004: 829) and Helgen (2007b: 232, 273) recognised this taxon as a distinct species, which is followed here.

Cuscus brevicaudatus J. Gray, 1858b: 102.

TYPE LOCALITY: Cape York, north Queensland, Australia.

COMMENTS: Replacement name for *Phalangista (Pseudocheirus) nudicaudata* Gould, 1850. Synonymised within *nudicaudatus* by Iredale and Troughton (1934: 32) and McKay (1988e: 82). Synonymised within *maculatus* by Thomas (1888a: 198), Groves (1993e: 47) and Flannery (1990: 130; 1994: 220; 1995a: 182; 1995b: 106). Taxon synonymised within *nudicaudatus* by Feiler (1978a: 14) and Groves (2005b: 48).

Cuscus maculatus var. *ochropus* J. Gray, 1866a: 220.

TYPE LOCALITY: Port Albany, Cape York, north Queensland, Australia. Type designated by Thomas (1922a: 128).

COMMENTS: Synonymised within *nudicaudatus* by Troughton (1934: 32) and McKay (1988e: 82). Synonymised within *maculatus* by Thomas (1888a: 198), Groves (1993e: 47) and Flannery (1990: 130; 1994: 220; 1995a: 182; 1995b: 106). Taxon synonymised within *nudicaudatus* by Groves (2005b: 48).

Tribe Trichosurini T. Flynn, 1911 *sensu* Flannery *et al.*, 1987

Family Trichosuridae T. Flynn, 1911: 120.

TYPE GENUS: *Trichosurus* Lesson, 1828b.

COMMENTS: When originally proposed, this rank was placed in the Marsupialia (Illiger, 1811) and included the genus

Trichosurus Lesson, 1828b. Not recognised subsequently at family rank. Subfamily Trichosurinae recognised by Kirsch (1968a: 420) and Marshall (1981: 28; 1984: 98) but not other authors. Tribe Trichosurini recognised by Flannery *et al.* (1987: 477, 503), Marshall *et al.* (1990: 494), Norris (1994: 93), Kirsch *et al.* (1997: 245) and Groves (2005b: 49), but not by Strahan (1983: xxi; 1995: 7, 265) or Van Dyck and Strahan (2008: 10, 265). Synonymised within Phalangeridae by Marshall *et al.* (1990: 493) and McKenna and Bell (1997: 61).

Subfamily Trichosurinae Kirsch & Wolman, 2001: 23, 29.

TYPE GENUS: *Trichosurus* Lesson, 1828b.

COMMENTS: When originally proposed, this rank was correctly attributed to Flynn (1911: 120) and placed in the Family Phalangeridae and included the genera *Trichosurus* Lesson, 1828b, *Wyulda* W. Alexander, 1919 and fossil *Strigocuscus* Gray, 1862a: 319. Subfamily Trichosurinae recognised by Kirsch (1968a: 420) and Marshall (1981: 28; 1984: 98) but not other authors. When recognising this rank Kirsch and Wolman (2001: 23, 28) also recognised the subfamily Phalangeridae, but included within it the tribes Phalangerini (Thomas, 1888a) and Ailuropini (Flannery *et al.*, 1987: 477, 503).

Trichosurus Lesson, 1828

Trichosurus Lesson, 1828b: 333.

TYPE SPECIES: *Didelphis vulpecula* Kerr, 1792 [= *Trichosurus vulpecula* (Kerr, 1792)] by subsequent designation. See Waterhouse (1846: 298).

COMMENTS: Described as a subgenus of *Phalangista*, which was followed by Lesson (1828b: 333) and Waterhouse (1846: 283). Elevated to generic rank by Lesson (1842: 189) and Thomas (1888a: xii, 184). Placed in the Tribe Trichsurini (T. Flynn, 1911) by Flannery *et al.* (1987: 503).

HOMONYMS:

Trichosurus Rafinesque, 1815: 99, crustaceans of the Subphylum Crustacea (Order Notostraca). Name is a *nomen nudum*. See Holthuis (1954: 7).

Cercaërtus Burmeister, 1837: 814.

TYPE SPECIES: *Didelphis vulpina* F. Meyer, 1793 [= *Trichosurus vulpecula* (Kerr, 1792)] by original designation.

COMMENTS: Wakefield (1963a: 114) suggested the reference to Gloger in the description was due a reference drawn from an unpublished manuscript by Constantine Gloger, but when the work was published in Gloger (1841: xxx, 85) the name was not mentioned, but rather *Psilogrammũrus* was proposed. Taxon synonymised within *Pseudochirus* by Thomas (1888a: 166) and *Trichosurus* by Iredale and Troughton (1934: 30), Wakefield (1963a: 114), McKay (1988e: 83), Marshall (1981: 28), Marshall *et al.* (1990: 494) and subsequent authors.

Psilogrammũrus Gloger, 1841: xxx, 85.

TYPE SPECIES: *Didelphis vulpecula* Kerr, 1792 [= *Trichosurus vulpecula* (Kerr, 1792)] by subsequent designation. See Thomas (1888a: 184).

COMMENTS: Synonymised within *Trichosurus* by Thomas (1888a: 184; 1895a: 190), Iredale and Troughton (1934: 30), McKay (1988e: 83), Marshall (1981: 28) and Marshall *et al.* (1990: 494).

Trichurus Wagner, 1843: vi, 74.

TYPE SPECIES: *Didelphis vulpecula* Kerr, 1792 [= *Trichosurus vulpecula* (Kerr, 1792)] by subsequent designation.

COMMENTS: Proposed as subgenus of *Phalangista* Ė. Geoffroy and G. Cuvier, 1795 [= *Phalanger* Storr, 1780]. Synonymised within *Trichosurus* by Thomas (1888a: 184), McKay (1988e: 83), Marshall (1981: 28) and Marshall *et al.* (1990: 494).

HOMONYMS:

Trichuris Röderer, 1761: 243, whipworms of the Phylum Nematoda (Class Adenophorea, Order Trichurida, Family Trichuridae). Valid genus name.

Trichurus Donndorff, 1798: 244, cutlass fishes of the Class Actinopterygii (Order Perciformes, Family Trichiuridae). Incorrect subsequent spelling of *Trichiurus* Linnaeus, 1758: 242, 246.

Trichura Hübner, 1819 [1816–1826]: 126, moths of the Class Insecta (Order Lepidoptera, Family Arctiidae).

Trichurus Fagge and Pye-Smith, 1902: 465, whipworms of the Phylum Nematoda (Class Adenophorea, Order Trichurida, Family Trichuridae). Incorrect subsequent spelling of *Trichuris* Roederer, 1761: 243.

Trichosurus caninus (W. Ogilby, 1836)

Short-eared Brush-tailed Possum

Phalangista Canina W. Ogilby, 1836b: 191.

TYPE LOCALITY: Beyond the Hunter River, north of Sydney, New South Wales, Australia. Designation by Thomas (1888a: 191).

COMMENTS: Recognised within *Phalangista* by Waterhouse (1838a: 67; 1841a: 271; 1846: 296) and Gould (1856 [1845–1863]: Text to Plate 17). Placed in the genus *Trichosurus* by Thomas (1888a: xii, 191) and followed by subsequent authors.

Trichosurus caninus nigrans Le Souef, 1916: 64.

TYPE LOCALITY: Tweed River, New South Wales, Australia.

COMMENTS: Synonymised within *caninus* by Iredale and Troughton (1934: 32), McKay (1988e: 84), Flannery (1994: 172) and subsequent authors.

Trichosurus cunninghami Lindenmayer *et al.*, 2002

Mountain Brush-tailed Possum

Trichosurus cunninghami Lindenmayer *et al.*, 2002: 369, 385.

TYPE LOCALITY: Tommy's Bend Road, Cambarville regions, central Victoria, Australia. (37°30′S, 145°49′E)

COMMENTS: Formerly regarded as the Victorian population of *T. caninus*.

Trichosurus vulpecula (Kerr, 1792)

Common Brush-tailed Possum

Trichosurus vulpecula vulpecula (Kerr, 1792)

Didelphis vulpecula Kerr, 1792: 198.

TYPE LOCALITY: Sydney, New South Wales, Australia.

COMMENTS: Derived from the Vulpine Opossum of Phillip (1789: 150) and Wha Tapoau Roo of White (1790: 278). Included in *Phalangista* by Waterhouse (1841a: 265) and Krefft (1866a: 17). Placed in the genus *Trichosurus* by Gervais (1855a: 275), as *vulpina*, Jentink (1885: 25), Thomas (1888a: xii, 187), who reviewed its taxonomic history, and subsequent authors. A recent genetic analysis of *vulpecula* by A. Taylor and Foulkes (2004: 466) suggested the presence of a previously unrecognised phylogeographic break splitting *T. v. vulpecula* into distinct northeastern and southeastern clades. Morphological analysis by Kerle *et al.* (1991: 313) and A. Taylor and Foulkes (2004: 467) failed to separate populations on the basis of morphology, with the exception of A. Taylor and Folkes (2004: 467) finding significant separation between *fuliginosus* from Tasmania and other *Trichosurus*. No subspecies were recognised by Groves (2005b: 50), but Van Dyck and Strahan (2008: 274) recognised six including the nominate subspecies which is followed here.

FUTURE TAXONOMIC RESEARCH: There has been limited published taxonomic information on this species since Kerle *et al.* (1991: 313) and A. Taylor and Foulkes (2004: 455). The contrasting unpublished genetic studies of Collins (2003: v, 49) and morphological studies of Kerr (2011: 18, 95) suggest a fresh study, using both morphology and DNA, of samples throughout the species range is surely overdue. Observations reveal that some of the geographic populations ascribed to the species appear distinctive in pelage and proportions, as was highlighted by Kerle *et al.* (1991: 313). Taxonomic status of each subspecies needs clarification.

[Didelphis] Vulpina F. Meyer, 1793: 23, 174.

TYPE LOCALITY: Sydney, New South Wales, Australia.

COMMENTS: Type based on 'Vulpine Opossum' of Anon in Philip, 1789). Name recognised by Shaw (1800: 503) and transferred to *Phalangista* by Lesson (1828b: 335), Waterhouse (1838a: 67), Gould (1856 [1845–1863]: Text to Plate 12) and Krefft (1868a: 94). Synonymised within *vulpecula* by Thomas (1888a: 187), Iredale and Troughton (1934: 30), McKay (1988e: 84), Flannery (1994: 176) and Groves (1993e: 48; 2005b: 50).

[Didelphis] Tapouaru F. Meyer, 1793: 24, 174.

TYPE LOCALITY: Sydney, New South Wales, Australia. Type based on 'Wha Tapoau Roo' of Hunter, J. in White (1790: 278).

COMMENTS: Synonymised within *vulpecula* by Thomas (1888a: 187), Iredale and Troughton (1934: 30), McKay (1988e: 85), Groves (1993e: 48; 2005b: 50) and Flannery (1994: 176).

Didelphis Lemurina Shaw, 1800: 487.

TYPE LOCALITY: Sydney, New South Wales, Australia. Type based on 'Wha Tapoau Roo' of Hunter, J. in White (1790: 278).

COMMENTS: Synonymised within *vulpecula* by Waterhouse (1846: 284), Thomas (1888a: 187), Iredale and Troughton (1934: 30), McKay (1988e: 85), Groves (1993e: 48; 2005b: 50) and Flannery (1994: 176) who misspelt it '*lemorina*'.

Ursus novae Hollandiae Bechstein, 1800: 337.

TYPE LOCALITY: Sydney, New South Wales, Australia. Type based on 'New Holland Bear' in Pennant (1783: 13).

COMMENTS: Synonymised within *vulpecula* by Thomas (1888a: 187), Iredale and Troughton (1934: 30), McKay (1988e: 85), Groves (1993e: 48; 2005b: 50) and Flannery (1994: 176).

Phalangista cookii G. Cuvier, 1824: unpaginated, Plate.

TYPE LOCALITY: Australia.

COMMENTS: Synonymised within *vulpecula* by Thomas (1888a: 187), McKay (1988e: 85), Groves (1993e: 48; 2005b: 50) and Flannery (1994: 176).

HOMONYMS:

Phalangista cookii Desmarest, 1817c, the Eastern Ring-tailed Possum of the Class Mammalia (Order Diprotodontia, Family Pseudocheiridae). Recognised as a subspecies of *Pseudocheirus peregrinus* (Boddaert, 1785). See individual entry.

Phalangista cookii Schinz, 1821, the Eastern Ring-tailed Possum of the Class Mammalia (Order Diprotodontia, Family Pseudocheiridae). Synonym of *Pseudocheirus peregrinus* (Boddaert, 1785). See individual entry.

Phalangista cookii Gould, 1856 [1845–1863], the Eastern Ring-tailed Possum of the Class Mammalia (Order Diprotodontia, Family Pseudocheiridae). Synonym of *Pseudocheirus peregrinus cookii* (Desmarest, 1817c). See individual entry.

Ph. [alangistae] Bougainvillei J. Fischer, 1829: 583, misprint as 383.

TYPE LOCALITY: Australia. Type based on 'Le Phalanger de Bougainville' of G. Cuvier (1829a: 183).

COMMENTS: Synonymised within *vulpecula* by Thomas (1888a: 187), Iredale and Troughton (1934: 31), McKay (1988e: 85), Groves (1993e: 48; 2005b: 50) and Flannery (1994: 176).

Phalangista xanthopus W. Ogilby, 1831: 135.

TYPE LOCALITY: Glenelg River, Victoria, Australia. See Waterhouse (1846: 294).

COMMENTS: Recognised by Waterhouse (1838a: 67; 1841a: 269; 1846: 294). Synonymised within *vulpecula* by Thomas (1888a: 188), Iredale and Troughton (1934: 31), McKay (1988e: 85), Groves (1993e: 48; 2005b: 50) and Flannery (1994: 176).

Ph. [alangista] felina Wagner, 1843: 76.

TYPE LOCALITY: Australia.

COMMENTS: Considered a synonym of *fuliginosus* by Iredale and Troughton (1934: 31) and synonym of *vulpecula* by Waterhouse (1846: 284), McKay (1988e: 85; 2005b: 50), Groves (1993e: 48) and Flannery (1994: 176).

Ph. [alangista] melanura Wagner, 1843: 81.

TYPE LOCALITY: Australia.

COMMENTS: Replacement name for *Phalangista cookii* G. Cuvier, 1824. Synonymised within *vulpecula* by Waterhouse (1846: 284), Thomas (1888a: 188), Iredale and Troughton (1934: 31), McKay (1988e: 85), Groves (1993e: 48; 2005b: 50) and Flannery (1994: 176).

Phal. [angista] selma Gervais, 1847: 704.

TYPE LOCALITY: Australia.

COMMENTS: In error for *P. felina* Wagner, 1843. Considered a synonym of *fuliginosus* by Iredale and Troughton (1934: 31) and synonym of *vulpecula* by McKay (1988e: 85), Groves (1993e: 48; 2005b: 50) and Flannery (1994: 176).

Trichosurus vulpecula mesurus Thomas, 1926b: 633.

TYPE LOCALITY: Inkerman, near Ayr, Queensland, Australia.

COMMENTS: Subspecies status recognised by Iredale and Troughton (1934: 31) and Troughton (1967: 102), and synonym of *vulpecula* by McKay (1988e: 86), Groves (1993e: 48; 2005b: 50) and Flannery (1994: 176).

Trichosurus vulpecula raui Finlayson, 1963: 18.

TYPE LOCALITY: Rocky River, Flinders Chase National park, Kangaroo Island, South Australia, Australia.

COMMENTS: Synonym of *vulpecula* by McKay (1988e: 86), Groves (1993e: 48; 2005b: 50) and Flannery (1994: 176).

Trichosurus vulpecula fuliginosus (W. Ogilby, 1831)

Phalangista fuliginosa W. Ogilby, 1831: 135.

TYPE LOCALITY: Tasmania, Australia.

COMMENTS: 'Brought from Sydney'. Recognised within *Phalangista* by Waterhouse (1841a: 267), Gould (1856 [1845–1849]: Text to Plate 15) and Krefft (1868a: 94). Transferred to *Phalangista* (*Trichosurus*) by Waterhouse (1838a: 67; 1846: 284) and to *Trichosurus vulpecula* by Thomas (1888a: xii, 190). See also Waterhouse (1846: 288). Considered a subspecies of *vulpecula* by Thomas (1888a: 190), Iredale and Troughton (1934: 31) and Strahan (1983: 147; 1995: 273). Species rank, within *Trichosurus*, by Troughton (1967: 106). Reduced to a synonym of *vulpecula* by Ride (1970: 248), McKay (1988e: 85), Groves (1993e: 48; 2005b: 50) and Flannery (1994: 176). Morphological and/or genetic analysis by Triggs (1990: 545), Kerle *et al.* (1991: 328), Collins (2003: 42), and A. Taylor and Foulkes (2004: 466) did not distinguish *fuliginosus* from *vulpecula* from southeastern Australia. Nonetheless Kerle *et al.* (1991: 329) suggested the subspecific status of *fuliginosus* may be retained on the grounds of their geographic isolation and morphological divergence, which was supported by the morphological component of the study of Taylor and Foulkes (2004: 467) which showed a significant separation of Tasmanian *Trichosurus* from other *Trichosurus*. Recognised as a subspecies by Clayton *et al.* (2006: 105), and Van Dyck and Strahan (2008: 274) but not by Burbidge *et al.* (2014: 20, 28).

Phalangista cuvieri Waterhouse, 1841a: 268.

TYPE LOCALITY: Probably Tasmania, Australia.

COMMENTS: Synonymised within *vulpecula* by Waterhouse (1846: 284). Considered a synonym of *fuliginosus* by Iredale and Troughton (1934: 31) and synonym of *vulpecula* by Waterhouse (1846: 284), McKay (1988e: 85), Groves (1993e: 48; 2005b: 50) and Flannery (1994: 176).

Phalangista fuliginosa grisea J. Gray, 1841: 401.

TYPE LOCALITY: Circular Head, near Stanley, Tasmania, Australia.

COMMENTS: *Nomen nudum*. Considered a synonym of *fuliginosus* by Iredale and Troughton (1934: 31) and synonym of *vulpecula* by McKay (1988e: 85), Groves (1993e: 48; 2005b: 50) and Flannery (1994: 176).

Trichosurus vulpecula ruficollis Schwarz, 1909: 626.

TYPE LOCALITY: 'Murchison District', probably Tasmania, Australia. But there is also a Murchison District in New South Wales and Victoria.

COMMENTS: Recognised as a subspecies of *vulpecula* by Iredale and Troughton (1934: 32) and Troughton (1967: 103), and synonym of *vulpecula* by McKay (1988e: 86), Groves (1993e: 48; 2005b: 50) and Flannery (1994: 176).

Trichosurus vulpecula hypoleucus (Wagner, 1855)

Ph. [alangista] hypoleucus Wagner, 1855: xiv, 273.

TYPE LOCALITY: Stirling Range, Western Australia, Australia. Type designated by Schwarz (1909: 625).

COMMENTS: Recognised as a subspecies of *vulpecula* by Iredale and Troughton (1934: 32) and Troughton (1967: 103). Synonymised within *vulpecula* by Thomas (1888a: 188), McKay (1988e: 85), Groves (1993e: 48; 2005b: 50) and Flannery (1994: 176). Recognised as a subspecies by Clayton *et al.* (2006: 105), Van Dyck and Strahan (2008: 274) and Burbidge *et al.* (2014: 20, 28). Phylogenetic support for this taxon being recognised at subspecies rank was provided by Collins (2003: v, 60).

Trichosurus vulpecula johnstonii (Ramsay, 1888)

Phalangista johnstonii Ramsay, 1888: 1297.

TYPE LOCALITY: Probably Atherton Tablelands, North Queensland, Australia. Designated by McKay (1988e: 85).

COMMENTS: Recognised as a subspecies of *vulpecula* by Iredale and Troughton (1934: 31), Troughton (1967: 102) and Strahan (1983: 147; 1995: 273). Synonymised within *vulpecula* by McKay (1988e: 85), Groves (1993e: 48) and Collins (2003: v, 49). Kerle *et al.* (1991: 313, 329) had insufficient data to provide adequate resolution of the status of the Atherton Tablelands population. Groves (1993e: 48) noted that this taxon might be a distinct species, which was followed by Flannery (1994: 153, 174) and Groves (2005b: 50). Recognised as a subspecies by Clayton *et al.* (2006: 105), and Van Dyck and Strahan (2008: 274), but not by Burbidge *et al.* (2014: 20, 28). In contrast to the genetic studies of Collins (2003: v, 49), Kerr (2011: 18, 95) found this taxon to be morphologically distinct from *T. v. vulpecula*, both in fur colour and body shape but not in body mass, though fur colour was proposed to be a phenotypically plastic trait that may develop due to maternal diet rather than an inherited trait. She also found that the broadly sympatric *T. v. vulpecula* and *T. v. johnstonii* had strongly differing habitat affinities, the former taxon typically occupying dry sclerophyll forest while the latter was typically associated with neighbouring rainforest. Given the uncertainty of this taxon there is clearly a need to re-examine it more closely.

Trichosurus vulpecula arnhemensis Collett, 1897

[Trichosurus vulpecula] var. arnhemensis Collett, 1897: 328.

TYPE LOCALITY: Daly River and Katherine River, North Territory, Australia.

COMMENTS: Recognised as a subspecies of *vulpecula* by Thomas (1906d: 540), Iredale and Troughton (1934: 31), D. Johnson (1964: 450) and Troughton (1967: 103). Phylogenetic support for this taxon being recognised at subpecies rank was provided by Collins (2003: v, 60). Elevated to species rank by Ride (1970: 72), Kirsch and Calaby (1977: 16), Honacki *et al.* (1982: 38), Strahan (1983: 149), McKay (1988e: 84), Groves (1993e: 48; 2005b: 49) and Flannery (1994: 153, 170). Recognised as a subspecies of *vulpecula* by Kerle *et al.* (1991: 328), Strahan (1995: 273), Collins (2003: v, 42), Clayton *et al.* (2006: 105), Van Dyck and Strahan (2008: 274) and Burbidge *et al.* (2014: 20, 28).

Trichosurus vulpecula eburacensis Lönnberg, 1916

Trichosurus vulpecula eburacensis Lönnberg, 1916: 9.

TYPE LOCALITY: Olen Creek, between Coleman River and Mitchell River, Cape York, Queensland, Australia.

COMMENTS: Recognised as a subspecies of *vulpecula* by Iredale and Troughton (1934: 31) and Troughton (1967: 103), and synonym of *vulpecula* by McKay (1988e: 86), Groves (1993e: 48; 2005b: 50) and Flannery (1994: 176). Recognised as a subspecies of *vulpecula* by Clayton *et al.* (2006: 105), and Van Dyck and Strahan (2008: 274), but not by Burbidge *et al.* (2014: 20, 28).

Wyulda Alexander, 1919

Wyulda W. Alexander, 1919: 31

TYPE SPECIES: *Wyulda squamicaudata* Alexander, 1919.

Wyulda squamicaudata Alexander, 1919

Scaly-tailed Possum

Wyulda squamicaudata Alexander, 1919: 31.

TYPE LOCALITY: Violet Valley (Station), Aboriginal Reserve near Wyndham, Western Australia, Australia.

COMMENTS: Placed in *Trichosurus* by Flannery *et al.* (1987: 503), which has not been followed by other authors.

Suborder Macropodiformes Kirsch *et al.*, 1997

Suborder Macropodiformes Kirsch *et al.*, 1997: 246.

COMMENTS: When originally proposed, this rank was placed in the Order Diprotodontia (Owen, 1877a) and included the Superfamily Macropodoidea (J. Gray, 1821) that included the families Macropodidae (J. Gray, 1821) and Hypsiprymnodontidae (Collett, 1887b). Groves (2005b: 56) recognised this suborder with the author as Ameghino (1889: xx, 263, 266). Suborder not recognised by Strahan (1995: 7),

McKenna and Bell (1997: 62), and Van Dyck and Strahan (2008: 10) who placed the macropods within the Suborder Phalangerida. Strong support for monophyly of this suborder arises from various studies including Kavanagh *et al.* (2004: 210, 217) and Meredith *et al.* (2008c: 395).

Superfamily Macropodoidea J. Gray, 1821 *sensu* Kear and Cooke, 2001

Family Macropidae J. Gray, 1821: 308.

TYPE GENUS: *Macropus* Shaw, 1790.

COMMENTS: When originally proposed, this rank was placed in the Order Brutae (J. Gray, 1821 [= Marsupialia (Illiger, 1811 part)]) and included the genus *Macropus* Shaw, 1790. Name corrected and changed to Macropodidae by Owen (1839a: 16, 19). The family name included all marsupials when used by J. Gray (1843a: xxii). Spelling Macropidae (J. Gray, 1821) synonymised within the Family Macropodidae by McKenna and Bell (1997: 62). Given superfamily status by M. Archer and Bartholomai (1978: 12) and followed by Szalay (1982: 631; 1994: 43), Marshall *et al.* (1990: 459), Kirsch *et al.* (1997: 246), Kear and Cooke (2001: 84) and Long *et al.* (2002: 146). Authors often include the family rank with two extant subfamilies, the Macropodinae and Potoroinae (e.g. Kirsch, 1968a: 420; McKenna and Bell, 1997: 62, 63). Three subfamilies were recognised by Bensley (1903: 143): Macropodinae, Potoroinae and Bettongiinae. More recently three families or subfamilies have been recognised, namely Hypsiprymnodontidae, Potoroidae and Macropodidae by authors including Szalay (1994: 43), Burk *et al.* (1998: 469), Burk and Springer (2000: 228), and Van Dyck and Strahan (2008: 10). Kavanagh *et al.* (2004: 210) recognised the Family Hypsiprymnodontidae and included the potoroids within the Family Macropodidae, while Kirsch *et al.* (1997: 246) also recognised these families, with the latter including the subfamilies Macropodinae and Potoroinae. A revised classification of the Macropodoidea was undertaken by Kear and Cooke (2001: 84), which is followed here.

HOMONYMS:

Family Macropidae J. Gray, 1842b, dunnarts of the Class Mammalia (Order Dasyuromorphia, Family Dasyuridae). Usage of the name appears to be in error. Rank is a synonym of the Family Dasyuridae Goldfuss, 1820a. See individual entry.

Subfamily Macropodinae Liem, 1963: 47, fish of the Class Osteichthyes, Perciformes, Family Osphronemidae, Belontiidae or Anabantidae). The spelling of this name was emended to Macropodusinae by Opinion 2058 (Case 2661) to remove homonymy with the Subfamily Macropodinae J. Gray, 1821 (ICZN, 2001: 297; 2003a: 253).

Family Halmaturini Goldfuss, 1820a: xxiii, 445.

TYPE GENUS: *Halmaturus* Illiger, 1811 [=*Macropus* Shaw, 1790].

COMMENTS: When originally proposed, this rank was placed in the Order Marsupialia (Illiger, 1811) and included the genera *Halmaturus* Illiger, 1811 [= *Macropus* Shaw, 1790]; and *Hypsiprymnus* Illiger, 1811 [= *Potorous* Desmarest, 1804]. Synonymised within Macropodidae by McKenna and Bell (1997: 62) who referred to Article 37(b) of the Code (ICZN, 1985a: 77). Synonymised within the Family Macropodidae by McKenna and Bell (1997: 62).

Tribe Macropina J. Gray, 1825a: 340.

TYPE GENUS: *Macropus* Shaw, 1790.

COMMENTS: When originally proposed, this rank was placed in the Family Didelphidae (J. Gray, 1821: 308) and included the genera *Macropus* Shaw, 1790; *Halmaturus* Illiger, 1811 [= *Macropus* Shaw, 1790]; and *Potorous* Desmarest, 1804a. Synonymised within Macropodidae by McKenna and Bell (1997: 62).

Tribe Poëphaga Owen, 1839a: 16, 19.

TYPE GENUS: Not based on a genus group name.

COMMENTS: When originally proposed, this rank was placed in the Marsupialia (Illiger, 1811) and included the Family Macropodidae (J. Gray, 1821), which included the genera *Hypsiprymnus* Illiger, 1811 [= *Potorous* Desmarest, 1804] and *Macropus* Shaw, 1790. Recognised at tribe rank by Owen (1840: 326) and at family rank by Giebel (1855: xi, 670) and Owen (1859: 52). Synonymised within Phalangeriformes by McKenna and Bell (1997: 58).

Family Macropodineae Lesson, 1842: 193.

TYPE GENUS: *Macropus* Shaw, 1790.

COMMENTS: When originally proposed, this rank was placed in the Order Mastodidelphie (Lesson, 1842 [= Marsupialia (Illiger, 1811)]) and included the genera *Hypsiprymnus* Illiger, 1811 [= *Potorous* Desmarest, 1804]; *Macropus* Shaw, 1790; *Setonix* Lesson, 1842; *Petrogale* J. Gray, 1837; *Conoyces* Lesson, 1842 [= *Thylogale* J. Gray, 1837]; *Heteropus* Jourdan, 1837a [= *Petrogale* J. Gray, 1837]; and *Halmaturus* Illiger, 1811 [= *Macropus* Shaw, 1790]. Does not appear to have been recognised by subsequent authors.

Family Macropoda Wagner, 1843: vi, 96.

TYPE GENUS: Not based on a genus group name.

COMMENTS: Rank unknown. When originally proposed, this rank was placed in the Phytophaga (Wagner, 1843 [= Diprotodontia (Owen, 1877a part)]), and included the genera *Hypsiprymnus* Illiger, 1811 [= *Potorous* Desmarest, 1804] and *Halmaturus* Illiger, 1811 [= *Macropus* Shaw, 1790]. Recognised as an order by Haeckel (1866: clviii).

HOMONYMS:

Macropoda Illiger, 1811, rodents of the Class Mammalia (Order Rodentia, Families Dipodidae, Muridae & Pedetidae). See D. Wilson and Reeder (2005: 882, 1234: 1535). See individual entry.

Order Macropoda Ameghino, 1889, diprotodont marsupials of the Class Mammalia (Order Diprontodontia). Synonymised within the Order Diprotodontia (Owen, 1877a) here. See individual entry.

Family Macropodés Gervais, 1855a: 268.

TYPE GENUS: Not based on a genus group name.

COMMENTS: When originally proposed, this rank was placed in the Order Marsupiaux (Gervais, 1855a [= Marsupialia (Illiger, 1811)]) and included the genera *Macropus* Shaw, 1790; *Onychogales* [sic=*Onychogalea*] J. Gray, 1841; *Lagorchestes* Gould, 1841 [1841–1842]; Halmatures [sic=*Halmaturus*] Illiger, 1811 [= *Macropus* Shaw, 1790]; *Heteropes* [= *Heteropus* Jourdan, 1837a] [= *Petrogale* J. Gray, 1837]; *Dendrolagus* S. Müller, 1840; *Hypsiprymnus* Illiger, 1811 [= *Potorous* Desmarest, 1804]; and *Bettongia* J. Gray, 1837.

Family Halmaturida Haeckel, 1866: clvii.

TYPE GENUS: *Halmaturus* Illiger, 1811 [=*Macropus* Shaw, 1790].

COMMENTS: When originally proposed, this rank was placed in the Order Macropoda (Wagner, 1843) and included the genera *Halmaturus* Illiger, 1811 [= *Macropus* Shaw, 1790]; *Macropus* Shaw, 1790 and *Hypsiprymnus* Illiger, 1811 [= *Potorous* Desmarest, 1804]). Synonymised within Macropodidae by McKenna and Bell (1997: 62).

Tribe Saltigrada Owen, 1877b: 360.

TYPE SPECIES: Not based on a genus group name.

COMMENTS: When originally proposed, this rank was placed in the Poëphaga (Owen, 1839a [= Macropodoidea J. Gray, 1821]) and included the extinct and living macropods.

Family Hypsiprymnodontidae Collett, 1887

Family Hypsiprymnodontidae Collett, 1887b: 833, 906.

TYPE GENUS: *Hypsiprymnodon* Ramsay, 1876a.

COMMENTS: When originally proposed, this rank was placed in the Order Marsupialia (Illiger, 1811) and included the genus *Hypsiprymnodon* Ramsay, 1876a. Rank synonymised within Macropodidae by Simpson (1945: 46) and not recognised by Strahan (1983: xxi), Marshall (1984: 107) or Woods (1960: 211). Tribe within the Subfamily Potoroinae recognised by Marshall (1981: 30) and Marshall *et al.* (1990: 493). Recognised as the Subfamily Hypsiprymnodontinae within the Family Macropodidae by Thomas (1888a: xii, 122), Cabrera (1919: 132), Raven (1929: 255), Iredale and Troughton (1934: viii, 35), Troughton (1967: 125), Baverstock *et al.* (1989: 46). Considered a subfamily of the Family Potoroidae by Pearson (1950: 211), M. Archer and Bartholomai (1978: 13), Strahan (1995: 7, 282) and Flannery (1989: 25). Synonymised within the Subfamily Potoroinae by McKenna and Bell (1997: 62), but recognised as a distinct family by Ride (1993: 442), Szalay (1994: 43), Kirsch *et al.* (1997: 246), Burk *et al.* (1998: 469), Burk and Springer (2000: 228), Kear and Cooke (2001: 84), Long *et al.* (2002: 149), Kavanagh *et al.* (2004: 210), and Van Dyck and Strahan (2008: 10, 281). Comprises two subfamilies, the Hypsiprymnodontinae and † Propleopinae (M. Archer & Flannery, 1985), according to Ride (1993: 442) and followed by Kirsch *et al.* (1997: 246), Kear and Cooke (2001: 84) and Long *et al.* (2002: 150–151).

Family Hypsiprymnidae Owen, 1849: 933.

TYPE GENUS: *Hypsiprymnodon* Ramsay, 1876a.

COMMENTS: Placed within Marsupialia (Illiger, 1811) with no details of lower taxa included. Family recognised by Ameghino (1889: 268). Synonymised within Macropodidae by Marshall *et al.* (1990: 493) and McKenna and Bell (1997: 62).

Family Pleopodidae Owen, 1878a: 574.

TYPE GENUS: *Pleopus* Owen, 1877c [= *Hypsiprymnodon* Ramsay, 1876].

COMMENTS: When originally proposed, this rank was placed in the Marsupialia (Illiger, 1811) and included the genus *Pleopus* Owen, 1877c [= *Hypsiprymnodon* Ramsay, 1876a]. Objective synonym of Hypsiprymnodontidae. Recognised as a subfamily of Hypsiprymnodontidae by Kirsch *et al.* (1997: 246). Synonymised within Potorini J. Gray, 1821 by McKenna and Bell (1997: 62).

Superfamily? Hypsiprymnoidea Ameghino, 1893: 331.

TYPE GENUS: *Hypsiprymnodon* Ramsay, 1876a.

COMMENTS: When originally proposed as a new rank it was placed in the Diprotodonta [sic] (= Diprotodontia Owen, 1877a). Synonymised within Macropodidae (J. Gray, 1821) by Marshall (1981: 29) and Marshall *et al.* (1990: 492).

† Subfamily Propleopinae M. Archer & Flannery, 1985: 1331, 1332.

TYPE GENUS: † *Propleopus* Longman, 1924b: 20.

COMMENTS: When originally proposed as a new rank it was placed in the Family Potoroidae (J. Gray, 1821) and included the genera † *Propleopus* Longman, 1924b: 20; and † *Ekaltadeta* M. Archer & Flannery, 1985: 1331, 1332. Subfamily recognised for the extinct genus † *Propleopus* by Ride (1993: 442) and followed by Long *et al.* (2002: 151).

Subfamily Hypsiprymnodontinae Collett, 1887

Family Hypsiprymnodontidae Collett, 1887b: 833, 906.

TYPE GENUS: *Hypsiprymnodon* Ramsay, 1876a.

COMMENTS: When originally proposed, this rank was placed in the Order Marsupialia (Illiger, 1811) and included the genus *Hypsiprymnodon* Ramsay, 1876a. Subfamily rank, within the Family Hypsiprymnodontidae (Collett, 1887b) recognised by Ride (1993: 442) and followed by Kirsch *et al.* (1997: 246), Kear and Cooke (2001: 84) and Long *et al.* (2002: 150).

Hypsiprymnodon Ramsay, 1876

Hypsiprymnodon Ramsay, 1876a: 33.

TYPE SPECIES: *Hypsiprymnus moschatus* Ramsay, 1876a by monotypy.

COMMENTS: Taxonomic decision of Owen (1878a: 573; 1878b: 103) to use *Hypsiprymnodon* in preference to *Pleopus*.

Pleopus Owen, 1877c: 542.

TYPE SPECIES: *Pleopus nudicaudatus* Owen, 1877c [= *Hypsiprymnodon moschatus* Ramsay, 1876a] by monotypy.

COMMENTS: Synonymised within *Hypsiprymnodon* by Thomas (1888a: 123), Iredale and Troughton (1934: 35), Marshall (1981: 30), Calaby and Richardson (1988a: 56) and Marshall *et al.* (1990: 493).

Hypsiprymnodon moschatus Ramsay, 1876

Musky Rat-kangaroo

Hypsiprymnodon moschatus Ramsay, 1876a: 33, 34.

TYPE LOCALITY: Rockingham Bay, Queensland, Australia.

COMMENTS: Acknowledgement of the original publication was made by Owen (1878b: 103). Baverstock *et al.* (1989: 46) found *Hypsiprymnodon* was distinct from the other potoroids and suggested it be placed in a separate subfamily, the Hypsiprymnodontinae.

[Pleopus] nudicaudatus Owen, 1877c: 542.

TYPE LOCALITY: North Queensland, Australia. Type designated by Thomas (1922a: 128).

COMMENTS: Synonymised within *moschatus* by Thomas (1888a: 123), Iredale and Troughton (1934: 35) and Calaby and Richardson (1988a: 57).

Family Potoroidae J. Gray, 1821 *sensu* Kear & Cooke, 2001

Family Potoridae J. Gray, 1821: 308.

TYPE GENUS: *Potorous* Desmarest, 1804a.

COMMENTS: When originally proposed, this rank was placed in the Order Brutae (J. Gray, 1821 [= Marsupialia (Illiger, 1811 part)]) and included the genus *Potorus* [sic] J. Gray, 1821 [= *Potorous* Desmarest, 1804a]. Simpson (1945: 47) allocated the author of family and subfamily rank to Trouessart (1899: 1195). Family rank recognised by Pearson (1950: 211), M. Archer and Bartholomai (1978: 12), Aplin and Archer (1987: xxii), Marshall *et al.* (1990: 459) and most recent authors. Synonymised within Subfamily Potoroinae by McKenna and Bell (1997: 62). *Hypsiprymnodon* was separated and placed in the Family Hypsiprymnodontidae by Szalay (1994: 43), Kirsch *et al.* (1997: 246), Burk *et al.* (1998: 469), Burk and Springer (2000: 228), and Van Dyck and Strahan (2008: 284). See comments under Family Hypsiprymnodontidae. The Family Potoroidae, with Flannery *et al.* (1984: 1087) as the author, was synonymised within Macropodidae by McKenna and Bell (1997: 58) but this is not followed here. The separation of *Hypsiprymnodon* into its own family, and the recognition of the tribes Potoroinini and Bettongini, was recognised by Kear and Cooke (2001: 84), and is followed here.

Subfamily Bettongiinae Bensley, 1903: 143.

TYPE GENUS: *Bettongia* J. Gray, 1837.

COMMENTS: When originally proposed, this rank was placed in the Family Macropodidae (J. Gray, 1821) and included the genera *Hypsiprymnodon* Ramsay, 1876a; *Bettongia* J. Gray, 1837; and *Aepyprymnus* Garrod, 1875. Subfamily recognised by Osborn (1910: 517). Synonymised within the Subfamily Potoroinae by Simpson (1945: 47), Marshall (1981: 30), Marshall *et al.* (1990: 492) and McKenna and Bell (1997: 62).

Family Potoroidae Pearson, 1950: 211.

TYPE GENUS: *Potorous* Desmarest, 1804a.

COMMENTS: When originally proposed, this rank was placed in the Marsupialia (Illiger, 1811) and included the subfamilies Hypsiprymnodontinae (Collett, 1887) and Potoroinae (J. Gray, 1821). Synonymised within Macropodidae by Marshall (1981: 29) and Marshall *et al.* (1990: 492).

Subfamily Potoroinea Szalay, 1994: 43.

TYPE GENUS: *Potorous* Desmarest, 1804a.

COMMENTS: When originally proposed, this rank was attributed to J. Gray (1821: 308) and placed in the Family Macropodidae (J. Gray, 1821) and included the tribes Potoroini (J. Gray, 1821: 308) and † Bulungamayini (Flannery *et al.*, 1983: 287, 288). Name attributed to J. Gray (1821: 308). Appears to be an incorrect subsequent spelling of the Subfamily Potoroinae.

Subfamily Potoroinae J. Gray, 1821 *sensu* Flannery, 1989

Family Potoridae J. Gray, 1821: 308.

TYPE GENUS: *Potorous* Desmarest, 1804a.

COMMENTS: When originally proposed, this rank was placed in the Order Brutae (J. Gray, 1821 [= Marsupialia (Illiger, 1811 part)]) and included the genus *Potorus* [sic] J. Gray, 1821 [= *Potorous* Desmarest, 1804a]. Recognised as the Subfamily Potoroinae (spelling corrected) within Family Macropodidae by Thomas (1888a: xi, 102), Trouessart

(1899: 1195), Bensley (1903: 143), Raven (1929: 255), Iredale and Troughton (1934: viii, 35), Simpson (1945: 47), Troughton (1967: 126), M. Archer and Bartholomai (1978: 13), Marshall (1984: 107), Flannery (1989: 25), Kirsch *et al.* (1997: 246) and McKenna and Bell (1997: 62). Flannery and Archer (1987: 759), and Flannery (1989: 25) introduced the tribes Bettongini and Potoroini, which was accepted by Kear and Cooke (2001: 84), and is followed here.

Tribe Potoroini Flannery, 1989: 25.

TYPE GENUS: *Potorous* Desmarest, 1804a.

COMMENTS: When originally proposed as a new rank it was placed in the Subfamily Potoroinae (J. Gray, 1821) and included the genus *Potorous* Desmarest, 1804a. Tribe attributed to J. Gray (1821: 308) by Marshall (1981: 30), Marshall *et al.* (1990: 493), Szalay (1994: 43), Kirsch *et al.* (1997: 246) and McKenna and Bell (1997: 62).

Subtribe Potoroina Szalay, 1994: 43.

TYPE GENUS: *Potorous* Desmarest, 1804a.

COMMENTS: When originally proposed, this rank was attributed to J. Gray (1821: 308) and placed in the Tribe Potoroini (J. Gray, 1821). Name attributed to J. Gray (1821: 308). Synonymised within Potoroini (J. Gray, 1821) by McKenna and Bell (1997: 62).

Tribe Bettongiini Flannery & Archer, 1987

Tribe Bettongini Flannery & Archer, 1987: 759.

TYPE GENUS: *Bettongia* J. Gray, 1837.

COMMENTS: When originally proposed as a new rank it was placed in the Subfamily Potoroinae (J. Gray, 1821) and included the genera *Bettongia* J. Gray, 1837; *Caloprymnus* Thomas, 1888a; and *Aepyprymnus* Garrod, 1875. Tribe rank recognised by Flannery (1989: 25), who redescribed it and included fossil genera, and Kear and Cooke (2001: 84). The tribe name is here corrected as above (the stem of Bettongia would be 'Bettongi-').

Tribe Bettongini Flannery, 1989: 25.

TYPE GENUS: *Bettongia* J. Gray, 1837.

COMMENTS: When originally proposed as a new rank it was placed in the Subfamily Potoroinae (J. Gray, 1821) and included the genera † *Gumardee* Flannery *et al.*, 1983: 292; † *Wakiewakie* Woodburne, 1984b: 1062, 1063; *Bettongia* J. Gray, 1837; † *Caloprymnus* Thomas, 1888a; and *Aepyprymnus* Garrod, 1875. Also recognised at subtribe rank by Szalay (1994: 43).

Subtribe Bettongina Szalay, 1994: 43.

TYPE GENUS: *Bettongia* J. Gray, 1837.

COMMENTS: When originally proposed, this rank was attributed to Flannery (1990: 25) and placed in the Tribe Potoroini (J. Gray, 1821). Synonymised within the Tribe Potoroini (J. Gray, 1821) by McKenna and Bell (1997: 62).

Aepyprymnus Garrod, 1875

Aepyprymnus Garrod, 1875: 59.

TYPE SPECIES: *Bettongia rufescens* J. Gray, 1837 [= *Aepyprymnus rufescens* (J. Gray, 1837)] by monotypy.

COMMENTS: Genus recognised by Garrod (1875: 59), Lydekker (1887: 205), and Thomas (1888a: xi, 102), and followed by subsequent authors.

Aepyprymnus rufescens (J. Gray, 1837)

Rufous Bettong

Bettóngia rufèscens J. Gray, 1837: 584.

TYPE LOCALITY: New South Wales, Australia.

COMMENTS: Recognised within *Hypsiprymnus* by Waterhouse (1841a: 188; 1846: 196), but transferred to *Bettongia* by Gould (1841 [1841–1842]: Unnumbered Plate; Gould, 1855 [1845–1863]: Text to Plate 65) and Krefft (1866a: 20). Subsequently transferred to *Aepyprymnus* by Garrod (1875: 59), Lydekker (1887: 205), Thomas (1888a: xi, 103) and followed by subsequent authors.

Hyp. [siprymnus] melanotis W. Ogilby, 1838d: 62.

TYPE LOCALITY: New South Wales, Australia. Based on same specimen as *Bettongia rufescens* J. Gray, 1837.

COMMENTS: Also listed within Waterhouse (1838a: 65). Synonymised within *rufescens* by Waterhouse (1841a: 188; 1846: 196), Thomas (1888a: 103), Iredale and Troughton (1934: 37) and Calaby and Richardson (1988a: 53).

Bettongia J. Gray, 1837

Bettóngia J. Gray, 1837: 584.

TYPE SPECIES: *Bettongia setosa* J. Gray, 1837 [= *Bettongia gaimardi* (Desmarest, 1822b)] (not *Hypsiprymnus setosus* W. Ogilby, 1832) by subsequent designation. See Thomas (1888a: xi, 104).

COMMENTS: Genus reviewed by Wakefield (1967a: 8).

Bettongiops Matschie, 1916: 264, footnote.

TYPE SPECIES: *Hypsiprymnus lesueur* Quoy and Gaimard, 1824 (as *B. leseurii*) [= *Bettongia lesueur* (Quoy & Gaimard, 1824)] by original designation.

COMMENTS: Synonymised within *Bettongia* by Iredale and Troughton (1934: 36), Marshall (1981: 30), Calaby and Richardson (1988a: 54), Rose and Rose (1998: 1) and Marshall *et al.* (1990: 493).

† ***Bettongia anhydra*** Finlayson, 1957

Desert Bettong

† *Bettongia penicillata anhydra* Finlayson, 1957: 552.

TYPE LOCALITY: McEwin Hills, Lake MacKay area, Northern Territory, Australia.

COMMENTS: Synonymised within *lesueur* by Calaby and Richardson (1988a: 55) and subsequent authors until McDowell (2013: 57) proposed that it should be treated as a distinct, and extinct, species. As a result of this Burbidge *et al.* (2014: 21, 29) recognised this taxon as a distinct species.

Bettongia gaimardi (Desmarest, 1822)

Eastern Bettong

† ***Bettongia gaimardi gainardi*** (Desmarest, 1822)

† *kangurus Gaimardi* Desmarest, 1822b: 542.

TYPE LOCALITY: Port Jackson, New South Wales, Australia.

COMMENTS: Placed within *Hypsiprymnus* (*Bettongia*) by Waterhouse (1846: 207). Included within *Bettongia* at genus rank by Flower and Garson (1884: 726) and Thomas (1888a: xi, 108), who reviewed its early taxonomic history, and subsequent authors.

† *Macropus minor* G. Cuvier, 1816a: 181.

TYPE LOCALITY: New South Wales, Australia.

COMMENTS: Synonymised within *gaimardi* by Waterhouse (1846: 207).

HOMONYMS:

Macropus minor Shaw, 1800, the Long-nosed Potoroo of the Class Mammalia (Order Diprotodontia, Family Potoridae). Name is a synonym of *Potorous tridactylus* (Kerr, 1792). See individual entry.

† *Hypsiprymnus White* Quoy & Gaimard, 1824: 62; Plate 10.

TYPE LOCALITY: Blue Mountains, New South Wales, Australia.

COMMENTS: *Kangarus gaimardi* Desmarest, 1822b. Recognised by Waterhouse (1841a: 181) and J. Gray (1841: 403), who placed it in *Bettongia*, and both authors spelt the specific name *whitei*. Synonymised within *gaimardi* by Waterhouse (1846: 207), Thomas (1888a: 108), Iredale and Troughton (1934: 36), Calaby and Richardson (1988a: 54) and Rose and Rose (1998: 1).

† *kangurus lepturus* Quoy & Gaimard, 1824: 64, footnote; Plate 4.

TYPE LOCALITY: Blue Mountains, New South Wales, Australia.

COMMENTS: *Nomen novum* for *Hypsiprymnus whitei* Quoy and Gaimard, 1824. Synonymised within *gaimardi* by Waterhouse (1846: 207), Thomas (1888a: 108), Iredale and Troughton (1934: 36), Calaby and Richardson (1988a: 54) and Rose and Rose (1998: 1).

† *Bettóngia setòsus* J. Gray, 1837: 584.

TYPE LOCALITY: Tasmania, Australia.

COMMENTS: J. Gray (1837: 584) wrote '*Bettóngia setòsus Hypsiprýmnus setòsus* W. Ogilby. Grey grizzled. End of tail blackish, slightly tufted. Ears moderate. Inhabits Swan River. British Museum.' Although placed by Calaby and Richardson (1988a: 54) in the synonymy of *B. gaimardi*, the brief description by J. Gray, and the locality, indicate *B. penicillata*, despite the fact that J. Gray himself described this in the next two lines. Junior secondary homonym to *Hypsiprymnus setosus* W. Ogilby, 1832 [= *Potorous tridactylus* (Kerr, 1792)]. Synonymised within *Hypsiprymnus cuniculus* by Waterhouse (1846: 200). Not considered by Iredale and Troughton (1934: 36). Synonymised within *gaimardi* by Calaby and Richardson (1988a: 54), and followed by Groves (1993e: 49) and Rose and Rose (1998: 1).

HOMONYMS:

Hypsiprymnus setosus W. Ogilby, 1832, the Long-nosed Potoroo of the Class Mammalia (Order Diprotodontia, Family Potoroidae). Name is synonym of *Potorous tridactylus* (Kerr, 1792). See individual entry.

† *Hyp. [siprymnus] formosus* W. Ogilby, 1838d: 62.

TYPE LOCALITY: Unknown. Locality given as New South Wales, Australia by Iredale and Troughton (1934: 36).

COMMENTS: Synonymised within *gaimardi* by Waterhouse (1846: 207), Thomas (1888a: 108), Iredale and Troughton (1934: 36), Calaby and Richardson (1988a: 54) and Rose and Rose (1998: 1).

† *Hyp. [siprymnus] Phillippi* W. Ogilby, 1838d: 62.

TYPE LOCALITY: Unknown. Locality given as New South Wales, Australia by Iredale and Troughton (1934: 36).

COMMENTS: Synonymised within *whitei* by Waterhouse (1841a: 181) and *gaimardi* by Waterhouse (1846: 207), Thomas (1888a: 108), Iredale and Troughton (1934: 36), Calaby and Richardson (1988a: 54) and Rose and Rose (1998: 1).

† *Hypsiprymnus hunteri* Owen, 1841: 408.

TYPE LOCALITY: Unknown. Locality given as New South Wales, Australia by Iredale and Troughton (1934: 36).

COMMENTS: Synonymised within *gaimardi* by Thomas (1888a: 108), by Iredale and Troughton (1934: 36), Calaby and Richardson (1988a: 54) and Rose and Rose (1998: 1).

† *Bett. [ongia] Whitei* J. Gray, 1841: 403.

COMMENTS: Emendation for *B. white* Quoy and Gaimard, 1824. Recognised within *Hypsiprymnus* by Waterhouse (1841a: 181). Synonymised within *gaimardi* by Thomas (1888a: 108), Iredale and Troughton (1934: 36), Calaby and Richardson (1988a: 54) and Rose and Rose (1998: 1).

† *Potorous minimus* Boitard, 1842: 207.

COMMENTS: As of 'Desm' (= Desmarest), in synonymy, so *nomen nudum*. Synonymised within *gaimardi* by Iredale and Troughton (1934: 36), Calaby and Richardson (1988a: 54) and Rose and Rose (1998: 1), who give the publication date as 1842, and other authors.

Bettongia gaimardi cuniculus (W. Ogilby, 1838)

Hyp.[siprymnus] Cuniculus W. Ogilby, 1838d: 63.

TYPE LOCALITY: Hunter's River [= Tasmania, Australia]. See Iredale and Troughton (1934: 37).

COMMENTS: Recognised at species rank in *Hypsiprymnus* by Waterhouse (1838a: 65; 1841a: 186; 1846: 200), and within *Bettongia* by Gould (1842 [1841–1842]: Unnumbered Plate; Gould, 1854 [1845–1863]: Text to Plate 63), Krefft (1868a: 94), Thomas (1888a: xi, 106), Iredale and Troughton (1934: viii, 37), Tate (1948b: 268) and Troughton (1967: 130). Synonymised within *gaimardi* by Ride (1970: 242), Calaby and Richardson (1988a: 54), and Groves (2005b: 57). Recognised as a subspecies by Strahan (1983: 186; 1995: 288), Rose and Rose (1998: 1), Clayton *et al.* (2006: 105), and Van Dyck and Strahan (2008: 287).

Bettongia lesueur (Quoy & Gaimard, 1824)

Burrowing Bettong

Bettongia lesueur lesueur (Quoy & Gaimard, 1824)

hypsiprymnus [sic] Lesueur Quoy & Gaimard, 1824: 64.

TYPE LOCALITY: Dirk Hartog's Island, Shark Bay, Western Australia, Australia.

COMMENTS: Synonymised within *graii* by Waterhouse (1846: 203). Species recognised within *Hypsiprymnus* by J. Gray (1841: 403), and within *Bettongia* by Thomas (1888a: xi, 112), who reviewed its historic taxonomy, and subsequent authors. Groves (1993e: 49; 2005b: 57) noted that it is commonly spelt '*lesueuri*', but that the original spelling is *lesueur*, with no indication that it is an error. An unnamed subspecies from Barrow Island and Boodie Island, Western Australia, was recorded by Clayton *et al.* (2006: 105).

FUTURE TAXONOMIC RESEARCH: The identity of the different subspecies needs to be resolved. These include *graii* from the Swan River, *harveyi* from the Eyre Peninsula, South Australia, and a potentially undescribed subspecies from Barrow and Boodie Islands (see Maxwell *et al.* 1996; Richardson, 2007: 1).

† *Bettongia lesueur graii* (Gould, 1841)

† *Hypsiprymnus Graii* Gould, 1841b: 178.

TYPE LOCALITY: Swan River, Western Australia, Australia.

COMMENTS: The original spelling of the species name appears to have been in error, and was subsequently changed by various authors (including Gould) to either '*grayi*' (e.g. Thomas, 1888a: 112), '*grayii*' (e.g. J. Gray, 1841: 403; Gould, 1863 [1845–1863]: Plate 64) or '*greyi*' (e.g. Krefft, 1871a: 5, Text to Plate 11). Taxon recognised by Waterhouse (1841a: 190; 1846: 203) within *Hypsiprymnus (Bettongia)* and Krefft (1866a: 22) who placed it within *Bettongia*. Synonymised within *lesueur* by Thomas (1888a: 112) and Calaby and Richardson (1988a: 55), recognised as a subspecies of *lesueur* by Iredale and Troughton (1934: 37), Troughton (1967: 129), Strahan (1983: 187; 1995: 290), Clayton *et al.* (2006: 105) and Burbidge *et al.* (2014: 21, 29) but not by Van Dyck and Strahan (2008: 290).

† *Perameles Harveyi* Waterhouse, 1842: 47.

TYPE LOCALITY: South Australia, Australia.

COMMENTS: Synonymised within *graii* by Waterhouse (1846: 388) and *lesueur* by Thomas (1888a: 112) and Calaby and Richardson (1988a: 55). Recognised as a subspecies of *lesueur* by Iredale and Troughton (1934: 37), Troughton (1967: 130), Strahan (1983: 187; 1995: 290), but not by Van Dyck and Strahan (2008: 290) or Burbidge *et al.* (2014: 21, 29).

Bettongia penicillata J. Gray, 1837

Brush-tailed Bettong

† *Bettongia penicillata penicillata* J. Gray, 1837

† *Bettóngia penicillàta* J. Gray, 1837: 584.

TYPE LOCALITY: No locality given; type in Natural History Museum, London, from New South Wales, Australia (see Jenkins & Knutson, 1983: 20).

COMMENTS: Recognised within *Hypsiprymnus* by Waterhouse (1841a: 183) and within subgenus *Bettongia* by Waterhouse (1846: 212). Placed within *Bettongia* by Gould (1841 [1841–1842]: Unnumbered Plate; Gould, 1852 [1845–1863]: Text to Plate 61) and Thomas (1888a: xi, 110), who also describes its historic taxonomy, Iredale and Troughton (1934: viii, 36) and subsequent authors. Taxa *penicillata*, *gouldii* and *ogilbyi* recognised as subspecies by Iredale and Troughton (1934: 36).

FUTURE TAXONOMIC RESEARCH: It is unclear whether representatives of this taxon in southeastern Australia (where it is now extinct) and southwest Western Australia (where it is threatened) are different and should be taxonomically separated. The distinctiveness of the two taxa described from South Australia (*gouldii* and *francisca*) also need to be assessed to determine if they have any taxonomic validity.

† *Hyp. [siprymnus] murinus* W. Ogilby, 1838d: 63.

TYPE LOCALITY: New South Wales, Australia.

COMMENTS: Species recognised within *Hypsiprymnus* by Gould (1854 [1845–1863]: Text to Plate 67). Synonymised within *penicillata* by Waterhouse (1841a: 183; 1846: 212).

† *Bettongia Gouldii* J. Gray, 1843a: 94.

TYPE LOCALITY: *Nomen nudum*.

COMMENTS: Not considered by Iredale and Troughton (1934: 36). Synonymised within *penicillata* by Waterhouse (1846: 212), Thomas (1888a: 110), Calaby and Richardson (1988a: 56) and Groves (1993e: 49).

† *Bettongia Gouldii* Waterhouse, 1846: 219.

TYPE LOCALITY: 40 miles north of Adelaide, Gulf St Vincent, South Australia, Australia.

COMMENTS: Waterhouse (1946: 219) noted that this specimen referred to by J. Gray (1843a: 94) appeared to represent an immature *Hypsiprymnus penicillatus* [= *Bettongia penicillata*]. Recognised as a subspecies of *penicillata* by Iredale and Troughton (1934: 36). Synonymised within *penicillata* by Waterhouse (1846: 219), Calaby and Richardson (1988a: 56) and Groves (1993e: 49).

† *Bettongia penicillata francisca* Finlayson, 1957: 552.

TYPE LOCALITY: St Francis Island, Nuyt's Archipelago, South Australia, Australia.

COMMENTS: Synonymised within *penicillata* by Calaby and Richardson (1988a: 56) and Groves (1993e: 49).

Bettongia penicillata ogilbyi (Waterhouse, 1841)

Hypsiprymnus Ogilbyi Waterhouse, 1841a: 185.

TYPE LOCALITY: York, Swan River, Western Australia, Australia.

COMMENTS: Recognised as a species of *Bettongia* by Gould (1852 [1845–1863]: Text to Plate 62). Synonymised within *penicillata* by Waterhouse (1846: 212), Thomas (1888a: 110), Calaby and Richardson (1988a: 55), and Groves (1993e: 49). More recently recognised as a subspecies of *penicillata* by Iredale and Troughton (1934: 36), Tate (1948b: 267), Strahan (1983: 184; 1995: 292), Clayton *et al.* (2006: 106), and Van Dyck and Strahan (2008: 292).

† ***Bettongia pusilla*** McNamara, 1997

Nullarbor Dwarf Bettong

† *Bettongia pusilla* McNamara, 1997: 98.

TYPE LOCALITY: Late Holocene deposits, Nullarbor Plain, South Australia, Australia.

COMMENTS: Specimens were originally allocated to † *Caloprymnus campestris* by Lundelius and Turnbull (1984: vii, 14). Species recognised from Holocene deposits by Long *et al.* (2002: 174) and as being present at the onset of European settlement and subsequently extinct by Burbidge *et al.* (2008: 412).

Bettongia tropica Wakefield, 1967

Northern Bettong

Bettongia tropica Wakefield, 1967a: 15.

TYPE LOCALITY: Mt Spurgeon, eastern Queensland, Australia.

COMMENTS: Ride (1970: 68) stated that *B. tropica* was not distinguishable externally from *B. penicillata* or *B. gaimardi* but recognised it as a valid species. Subsequently Sharman *et al.* (1980: 62) found no chromosomal basis for the distinction currently afforded to what was then referred to as the North Queensland Brush-tailed Bettong. Recognised as a subspecies of *penicillata* by Strahan (1983: 184) and a species by Calaby and Richardson (1988a: 56) and Seebeck and Rose (1989: 716). Synonymised within *penicillata* by Honacki *et al.* (1982: 43) and Groves (1993e: 49). Species rank recognised by Strahan (1995: 294), Maxwell *et al.* (1996: 4) and subsequent authors.

† ***Caloprymnus*** Thomas, 1888

† *Caloprymnus* Thomas, 1888a: xi, 114.

TYPE SPECIES: † *Bettongia campestris* Gould, 1843b (as *Caloprymnus campestris*) [= † *Caloprymnus campestris* (Gould, 1843b)] by original designation.

COMMENTS: Genus name accepted since description.

† ***Caloprymnus campestris*** (Gould, 1843)

Desert Rat-kangaroo

† *Bettongia campestris* Gould, 1843b: 81.

TYPE LOCALITY: South Australia, Australia.

COMMENTS: Placed in *Hypsiprymnus* (*Bettongia*) by Waterhouse (1846: 221) and *Caloprymnus* by Thomas (1888a: xi, 115). No live specimens recorded since 1935 (Carr & Robinson, 1997: 5; Groves, 2005b: 58).

Tribe Potoroini J. Gray, 1821 *sensu* Flannery, 1989

Family Potoridae J. Gray, 1821: 308.

TYPE GENUS: *Potorous* Desmarest, 1804a.

COMMENTS: When originally proposed, this rank was placed in the Order Brutae (J. Gray, 1821 [= Marsupialia (Illiger, 1811 part)]) and included the genus *Potorus* [sic] J. Gray, 1821 [= *Potorous* Desmarest, 1804a]. Tribe rank recognised within the Subfamily Potoroinae and to include all potoroids, with *Hypsiprymnodon* being placed in a separate tribe by various authors including Marshall (1981: vi, 30), Woodburne (1984a: 71), Case (1984: 1074, 1078), Flannery and Rich (1986: 444) and Marshall *et al.* (1990: 493). Flannery (1989: 25) was the first to recognise the tribes Potoroini and Bettongini within the Subfamily Potoroinae and *Hypsiprymnodon* it its own family, while Kirsch *et al.* (1997: 246) placed the Tribe Potoroini within the Subfamily Potoroinae, with two extinct tribes, and placed *Hypsiprymnodon* in its own family. Tribe rank recognised as distinct from the Tribe Bettongini by Flannery (1989: 25) and Kear and Cooke (2001: 84) who recognised the author as Trouessart (1899: 1195). Frankham *et al.* (2012: 592) calculated the separation time of potoroos from bettongs at 13.41 Ma (range of estimates about 10-19).

Potorous Desmarest, 1804

Potoroüs Desmarest, 1804a: 20.

TYPE SPECIES: *Didelphis murinus* G. Cuvier, 1797 [= *Potorous tridactylus* (Kerr, 1792)] by monotypy.

COMMENTS: Recognised as a subgenus of *Hypsiprymnus* by Waterhouse (1946: 223) and genus rank by Desmarest (1821: 38, 271), Thomas (1888a: xi, 116), Iredale and Troughton (1934: viii, 37), Simpson (1945: 47), Troughton (1967: 131), Calaby and Richardson (1988a: 57) and other authors.

Hypsiprymnus Illiger, 1811: 79.

TYPE SPECIES: *Nomen novum* for *Potorous* Desmarest, 1804a.

COMMENTS: Genus recognised by Waterhouse (1846: 190), but synonymised within *Potorous* by Waterhouse (1846: 223), Thomas (1888a: 116), Iredale and Troughton (1934: 37), Marshall (1981: 30), Calaby and Richardson (1988a: 57), Marshall *et al.* (1990: 493) and subsequent authors.

Potoroiis Rafinesque, 1815: 55.

TYPE SPECIES: *Errore pro Potorous* Desmarest, 1804a.

COMMENTS: Not considered by Iredale and Troughton (1934: 37–38) and synonymised within *Potorous* by Calaby and Richardson (1988a: 57) and subsequent authors.

Potorus J. Gray, 1821: 308.

TYPE SPECIES: *Errore pro Potorous* Desmarest, 1804a.

COMMENTS: Spelling used by Burnett (1830a: 351). Not considered by Iredale and Troughton (1934: 37–38) and synonymised within *Potorous* by Calaby and Richardson (1988a: 57) and subsequent authors.

Myorthius Lay, 1825: 743.

TYPE SPECIES: *Didelphis tridactyla* Kerr, 1792 [= *Potorous tridactylus* (Kerr, 1792)] by subsequent designation. See Iredale and Troughton (1934: 37).

COMMENTS: Date of publication uncertain as Palmer (1904: 440) gives the date as 1845, while Iredale and Troughton (1934: 33) and Groves (2005b: 43) give the date as 1825. Synonymised within *Potorous* by Iredale and Troughton (1934: 37) and not considered by Calaby and Richardson (1988a: 57) and subsequent authors.

Potoroo Berthold, 1827: 50.

TYPE SPECIES: *Errore pro Potorous* Desmarest, 1804a.

COMMENTS: Synonymised within *Potorous* by Iredale and Troughton (1934: 38) and Calaby and Richardson (1988a: 57).

Patoroo Partington, 1837: 12.

TYPE SPECIES: *Macropus minor* Shaw, 1800 [= *Potorous tridactylus* (Kerr, 1792)] by monotypy.

COMMENTS: Synonymised within *Potorous* by Iredale and Troughton (1934: 38) and Calaby and Richardson (1988a: 58).

Hypsiprimnus de Serres, 1838: 201.

TYPE SPECIES: *Errore pro Hypsiprymnus* Illiger, 1811.

COMMENTS: Not considered by Iredale and Troughton (1934: 37–38) and Calaby and Richardson (1988a: 57).

Hypsirymnus Jäger, 1850: 904.

TYPE SPECIES: *Error Pro Hypsiprymnus* Illiger, 1811.

COMMENTS: Does not appear to have been previously recognised. Synonymised here.

Potoroops Matschie, 1916: 264, footnote.

TYPE SPECIES: † *Hypsiprymnus platyops* Gould, 1844a [= † *Potorous platyops* (Gould, 1844a)] by original designation.

COMMENTS: Described as a subgenus of *Potorous*. Synonymised within *Potorous* by Iredale and Troughton (1934: 38), Calaby and Richardson (1988a: 57) and Marshall *et al.* (1990: 493).

Potorous gilbertii (Gould, 1841)

Gilbert's Potoroo

Hypsiprymnus Gilbertii Gould, 1841 [1841–1842]: Unnumbered Plate.

TYPE LOCALITY: King George Sound, Western Australia, Australia.

COMMENTS: Also described by Gould (1841c: 14) and also recognised within Gould (1854 [1845–1863]: Text to Plate 69). Recognised at species rank within *Hypsiprymnus (Potorous)* by Waterhouse (1846: 229) and within *Potorous* by Thomas (1888a: xi, 120), Iredale and Troughton (1934: viii, 38), Tate (1948b: 264) and Troughton (1967: 133). Subsequently synonymised within *tridactylus* by Ride (1970: 224), Johnston and Sharman (1976: 573; 1977: 733), Strahan (1983: 181), Calaby and Richardson (1988a: 59) and Groves (1993e: 50). Recognised as a subspecies of *tridactylus* by Calaby (1971: 18), Seebeck and Rose (1989: 729) and Seebeck *et al.* (1989: 68). This form was considered extinct by Shortridge (1910: 824) and subsequent authors as live animals had not been seen since the 1870s. After live specimens were rediscovered in December 1994 at Two Peoples Bay Nature Reserve, south west Western Australia (Start *et al.*, 1995: 29; Sinclair *et al.*, 1996: 69), it was resurrected from synonymy by Maxwell *et al.* (1996: 8). Species rank confirmed by Sinclair and Westerman (1997: 147), Sinclair *et al.* (2000: 285) and Frankham *et al.* (2012: 597). Recent mtDNA and nDNA research (Frankham *et al.*, 2012: 592) has shown that *P. gilbertii* forms a clade with what the authors called *P. tridactylus trisulcatus* (Victoria and Carrington Falls, NSW) and *P. t. apicalis* (Tasmania), and this clade in turn is sister to *P. t. tridactylus* (northern NSW and Queensland). The separation between these two major clades was put at 7.2 Ma; that between *P. gilbertii* and the two southern 'subspecies' of *P. tridactylus* was placed at 5.77 Ma.

Hypsiprymnus micropus Waterhouse, 1841a: 180.

TYPE LOCALITY: King George's Sound, Western Australia, Australia.

COMMENTS: Same as for *Hypsiprymnus gilberti* Gould, 1841 [1841–1842]. Synonymised within *gilberti* by Waterhouse (1846: 229), Thomas (1888a: 120), Iredale and Troughton (1934: 38) and within *tridactylus* by Calaby and Richardson (1988a: 59) and Groves (2005b: 58), but the type locality suggests *gilberti* is the correct species allocation.

Potorous longipes Seebeck & Johnson, 1980

Long-footed Potoroo

Potorous longipes Seebeck & Johnson, 1980: 119, 121.

TYPE LOCALITY: Princes Highway at Bellbird Creek, 32km east of Orbost, Victoria, Australia.

COMMENTS: Known from very few specimens since it was first collected in 1968 (Groves, 2005b: 58). Subfossil specimens have been collected from New South Wales (Seebeck, 1992: 173). According to Frankham *et al.* (2012: 592), this species separated from the *tridactylus/gilbertii* complex about 9.4 Ma, which suggests, if this deep separation is confirmed in further research, that this species should be placed in a distinct genus.

† ***Potorous platyops*** (Gould, 1844)

Broad-faced Potoroo

† *Hypsiprymnus platyops* Gould, 1844a: 103.

TYPE LOCALITY: Walyema Swamps near Northam, Western Australia, Australia [= Lake Walyormouning, Western Australia]. Locality discussed by Calaby (1954: 148).

COMMENTS: Placed within *Hypsiprymnus (Potoroüs)* by Waterhouse (1846: 231) and recognised within *Hypsiprymnus* by Gould (1854 [1845–1863]: Text to Plate 70). Recognised as a species within *Potorous* by Thomas (1888a: xi, 121), Iredale and Troughton (1934: viii, 39), Ride (1970: 68), Honacki *et al.* (1982: 49) and Calaby and Richardson (1988a: 58). No records of living animals after 1875 (Groves, 2005b: 58).

† *Potoroüs morgani* Finlayson, 1938: 135.

TYPE LOCALITY: Kelly's Hill Caves, Kangaroo Island, South Australia, Australia.

COMMENTS: Recognised as a subspecies of *platyops* by Troughton (1967: 134). Not considered by Iredale and Troughton (1934: 39) and synonymised within *platyops* by Ride (1970: 224), Calaby and Richardson (1988a: 58) and Groves (1993e: 50; 2005b: 58).

Potorous tridactylus (Kerr, 1792)

Long-nosed Potoroo

Potorous tridactylus tridactylus (Kerr, 1792)

Didelphis tridactyla Kerr, 1792: 198.

TYPE LOCALITY: Sydney, New South Wales, Australia. Locality of neotype given as Gosford, New South Wales, Australia (Frankham *et al.*, 2012: 600).

COMMENTS: Based on the Kanguroo Rat (and Kangooroo Rat) of Phillip (1789: 277) and the 'Poto Roo, or Kangaroo Rat' of White (1790: 286). Transferred to *Potorous* by Thomas (1888a: xi, 117), who also reviewed its taxonomic history. Johnston and Sharman (1976: 573) proposed that on morphological grounds they could only recognise one highly variable species from south-eastern or south-western Australia. Frankham *et al.* (2012: 592) found that, in a study including both mtDNA and nDNA, potoroos from Queensland and central/northern New South Wales (Mt Royal National Park, Mangrove Mountain) formed a sister clade to those from further south (Carrington Falls in NSW, Victoria and Tasmania) plus *P. gilbertii.* They considered that there is currently insufficient definitive evidence to raise the northern mainland *P. tridactylus* lineage to species status because of the limited samples available. As we explained in our introduction, however, this criterion does not determine species status; species are evolutionary lineages, and the evidence offered by Frankham *et al.* (2012: 592) shows that the 'species' *P. tridactylus* consists of three well-separated evolutionary units, two of which are actually closer to *P. gilbertii* than to the third. To resolve the nomenclatural problem, Frankham *et al.* (2012: 600) designated a neotype of *tridactylus* because the type appears to have been destroyed (see Calaby & Richardson, 1988a: 58) 'from a collection locality as near to Sydney as possible', namely Gosford, New South Wales central coast.

FUTURE TAXONOMIC RESEARCH: Research is required to confirm the limits of the distribution of each of the two mainland subspecies to confirm whether: 1) *tridactylus* and *trisulcatus* are the appropriate names to represent the two distinct mainland clades identified by Frankham *et al.* (2012: 592); or 2) *tridactylus* and *trisulcatus* are synonyms (with the former name having priority) because both occur within the same distribution and therefore the northern clade represents an unnamed taxon that needs to be formally described.

[Didelphis] Potoru F. Meyer, 1793: 13, 173.

TYPE LOCALITY: Sydney, New South Wales, Australia.

COMMENTS: Based on the 'Poto Roo, or Kangaroo Rat' of White (1790: 286). Synonymised within *tridactylus* by Thomas (1888a: 117), Iredale and Troughton (1934: 38) and Calaby and Richardson (1988a: 58).

Didelphis murina G. Cuvier, 1797: 126.

TYPE LOCALITY: Australia.

COMMENTS: Recognised as a species within *Hypsiprymnus* by Illiger (1811: 79) and Waterhouse (1841a: 175; 1846: 224). Synonymised within *tridactylus* by Thomas (1888a: 117), Iredale and Troughton (1934: 38) and Calaby and Richardson (1988a: 58).

HOMONYMS:

Didelphis murina Linnaeus, 1758: 55, Linnaeus's Mouse Opossum of the Class Mammalia (Order Didelphimorphia, Family Didelphidae). Taxon currently recognised as *Marmosa murina* (Linnaeus, 1758: 55). See Gardner (2005b: 9).

Macropus Minor Shaw, 1800: 513; Plate 116.

TYPE LOCALITY: Sydney, New South Wales, Australia.

COMMENTS: Presumably based on the Kangaroo Rat of Philip (1789: 277) and 'Poto Roo, or Kangaroo Rat' of White (1790: 286). Synonymised within *murina* (as *murinus*) by Waterhouse (1846: 224) and within *tridactylus* by Thomas (1888a: 117), Iredale and Troughton (1934: 38) and Calaby and Richardson (1988a: 58).

HOMONYMS:

Macropus minor G. Cuvier, 1816a, the Eastern Bettong of the Class Mammalia (Order Diprotodontia, Family Potoridae). Name is a synonym of *Bettongia gaimardi* (Desmarest, 1822b). See individual entry.

Dipus muscola Perry, 1810 [1810–1811]: Text to Plate 27.

TYPE LOCALITY: New South Wales, Australia.

COMMENTS: Based on the 'Poto Roo, or Kangaroo Rat' of White (1790: 286) and Philip (1789). Considered a synonym of *tridactyla* by Iredale and Troughton (1934: 38) and Calaby and Richardson (1988a: 58).

hypsiprymnus [sic] Peron Quoy & Gaimard, 1824: 64.

TYPE LOCALITY: New South Wales., Australia

COMMENTS: Synonymised within *Hypsiprymnus murinus* by Waterhouse (1841a: 175; 1846: 224) and within *tridactylus* by Thomas (1888a: 118), Iredale and Troughton (1934: 38) and Calaby and Richardson (1988a: 58).

Hypsiprymnus setosus W. Ogilby, 1832: 149.

TYPE LOCALITY: 'Swan River', New South Wales?, Australia

COMMENTS: Recognised by Waterhouse (1838a: 65), but synonymised within *Hypsiprymnus murinus* by Waterhouse (1841a: 175), within *penicillatus* by Waterhouse (1846: 212), and within *P. tridactylus* by Thomas (1888a: 118), Iredale and Troughton (1934: 38), and Calaby and Richardson (1988a: 59).

HOMONYMS:

Bettongia setosus J. Gray, 1837, the Eastern Bettong of the Class Mammalia. (Order Diprotodontia, Family Potoroidae). Name is a synonym of *Bettongia gaimardi* (Desmarest, 1822b). See individual entry.

Hypsiprymnus myosurus W. Ogilby, 1838d: 62.

TYPE LOCALITY: New South Wales, Australia.

COMMENTS: Also listed in Waterhouse (1838a: 65). Synonymised within *Hypsiprymnus murinus* by Waterhouse (1841a: 175; 1846: 224) and within *tridactylus* by Thomas (1888a: 118), Iredale and Troughton (1934: 38) and Calaby and Richardson (1988a: 59).

Perameles Tuckeri J. Gray, 1840a: 150.

TYPE LOCALITY: Australia.

COMMENTS: Considered a synonym of *Hypsiprymnus murinus* by Waterhouse (1846: 388) and *tridactylus* by Thomas (1888a: 118), Iredale and Troughton (1934: 38) and Calaby and Richardson (1988a: 59).

† *Potorous tridactylus var. antiquus* Broom, 1896a: 50.

TYPE LOCALITY: Broom Cave, Wombeyan Caves, near Taralga, New South Wales, Australia.

COMMENTS: Placed within *tridactylus* by Wakefield (1972a: 21) and Seebeck (1992: 175).

Potorous tridactylus apicalis (Gould, 1851)

Hypsiprymnus apicalis Gould, 1851 [1845–1863]: Text to Plate 68.

TYPE LOCALITY: New Norfolk, Tasmania, Australia.

COMMENTS: Recognised within *Hypsiprymnus* by Krefft (1868a: 94). Synonymised within *tridactylus* by Thomas

(1888a: 118), and Calaby and Richardson (1988a: 59). Populations of *Potorous* found on Bass Strait Islands (Flinders, Hunter and King) as well as some islands adjacent to Tasmania (Maria, Bruny and De Witt) have been assumed to belong to *P. t. apicalis* (Maxwell *et al.*, 1996: 8), but Johnston and Sharman (1996: 8) suggested that the Bass Strait populations are closer to *P. t. tridactylus*. Species rank recognised by Ride (1970: 68), but recorded as a subspecies of *tridactylus* by Iredale and Troughton (1934: 38), Troughton (1967: 133), Strahan (1983: 181; 1995: 301), Maxwell *et al.* (1996: 8), Groves (2005b: 58) and subsequent authors. Frankham *et al.* (2012: 592, 598) found that Tasmanian potoroos form a clade distinct from those from Victoria and southern New South Wales, and the two together cluster with *P. gilbertii* and separately from those from northern New South Wales and Queensland; consequently it may be appropriate to recognise these lineages as separate species.

FUTURE TAXONOMIC RESEARCH: The taxonomic status of *Potorous* from the Bass Strait islands needs to be confirmed.

Potorous Rufus Higgins & Petterd, 1884b: 181.

TYPE LOCALITY: Long's Plains, Tasmania, Australia.

COMMENTS: Synonymised within *apicalis* by Iredale and Troughton (1934: 38) and synonymised within *tridactylus* by Thomas (1888a: 118) and Calaby and Richardson (1988a: 59), but placed within *apicalis* by Groves (2005b: 58).

Potorous tridactylus benormi Courtney, 1963: 19.

TYPE LOCALITY: King Island, Bass Strait, Australia.

COMMENTS: Printing errors and the nominations of the type, which was omitted from the original article, are given on page 92. Synonymised within *tridactylus* by Calaby and Richardson (1988a: 59) and within *apicalis* by Groves (2005b: 58).

† *Potorous tridactylus trisulcatus* (McCoy, 1865)

† *Hypsiprymuus [sic] trisulcatus* McCoy, 1865: Quarter Sheet 7.

TYPE LOCALITY: Bone Cave, Gisborne, central Victoria, Australia. Locality of neotype given as Toolern Creek cave deposit near Gisborne, Victoria (Frankham *et al.*, 2012: 600).

COMMENTS: Synonymised within *tridactylus* by Mahoney (1964: 531), Calaby and Richardson (1988a: 59) and subsequent authors until Frankham *et al.* (2012: 592) revisited the phylogenetic relationships of *Potorous* and identified three major clades that were given the subspecific names *tridactylus*, *apicalis* and *trisulcatus*. As they showed that this taxon, as represented by samples from Victoria and as far north as Carrington Falls, NSW, forms a discrete evolutionary clade, we have no hesitation in recommending that *P. trisulcatus* be recognised as a distinct species. As the type specimen is lost, Frankham *et al.* (2012: 600) nominated a neotype (from a locality close to that of the original type specimen).

Family Macropodidae J. Gray, 1821 *sensu* Kear & Cooke, 2001

Family Macropidae J. Gray, 1821: 308.

TYPE GENUS: *Macropus* Shaw, 1790.

COMMENTS: When originally proposed, this rank was placed in the Order Brutae (J. Gray, 1821 [= Marsupialia (Illiger, 1811 part)]) and included the genus *Macropus* Shaw, 1790. Family recognised by Krefft (1866a: 6) and included the genera *Macropus*, *Onychogalea*, *Lagorchestes* and *Bettongia*. Family designated to Owen (1839a: 19) by Simpson (1945: 46). Recognised as the Family Macropodidae by Owen (1839a: 19; 1840: 332), Waterhouse (1841a: 165; 1846: 11, 50), Krefft (1866a: 6) with subfamilies Macropodinae and Potoroinae, by Thomas (1888a: xi, 3, 10) (who also recognised the Subfamily Hypsiprymnodontinae). Family rank recognised by Ameghino (1889: xx, 263, 266), Bensley (1903: 142), Simpson (1945: 46) and subsequent authors. The composition of the family, including extant and extinct, higher taxa was undertaken by Kear and Cooke (2001: 84), which is followed here.

Order Salienta Illiger, 1811: 58, 79.

COMMENTS: When originally proposed, this rank was placed in the Class Mammalia (Linnaeus, 1758) and included the Family Salienta (Illiger, 1811 [= Family Macropodidae (J. Gray, 1821: 308)]). Synonymised within the Subclass Marsupialia by McKenna and Bell (1997: 51).

HOMONYMS:

Order Salientia Laurenti, 1768: 24, amphibians of the Class Amphibia (Order Anura). Invalid synonym and/or a part of Order Anura G. Fischer, 1813a: 58. See Frost (2011) for discussion.

Family Salienta Illiger, 1811: 79.

TYPE GENUS: Not based on a genus group name.

COMMENTS: When originally proposed, this rank was placed in the Order Salienta (Illiger, 1811 [= Macropodidae (J. Gray, 1821)]) and included the genera *Hypsiprymnus* Illiger, 1811 [= *Potorous* Desmarest, 1804]; and *Halmaturus* Illiger, 1811 [= *Macropus* Shaw, 1790]. Synonymised within Macropodidae by McKenna and Bell (1997: 62).

Family Halmaturidae Bonaparte, 1831: 8, 19.

TYPE GENUS: *Halmaturus* Illiger, 1811 [=*Macropus* Shaw, 1790].

COMMENTS: When originally proposed, this rank was placed in the Order Marsupialia (Illiger, 1811). Subsequently used by Bonaparte (1832: 69; 1838: 113; 1845: 6). Synonymised within Macropodidae by Simpson (1945: 46), Marshall

(1981: 29), Marshall *et al.* (1990: 492), and McKenna and Bell (1997: 62).

Family Macropidae Burnett, 1830b: 351.

TYPE GENUS: *Macropus* Shaw, 1790.

COMMENTS: When originally proposed, this rank was placed in the Race Gliridentiae (Burnett, 1830b [= Diprotodontia (Owen, 1877a part)]) and included the genera *Kangurus* É. Geoffroy and G. Cuvier, 1795 [= *Macropus* Shaw, 1790]; *Macropus* Shaw, 1790; and *Halmaturus* Illiger, 1811 [= *Macropus* Shaw, 1790]. Synonymised within Macropodidae by McKenna and Bell (1997: 62).

Family Macropodidae Owen, 1839a: 19.

TYPE GENUS: *Macropus* Shaw, 1790.

COMMENTS: When originally proposed, this rank was placed in the Tribe Poëphaga (Owen, 1839a [= Macropodoidea J. Gray, 1821]) and included the genera *Hypsiprymnus* Illiger, 1811 [= *Potorous* Desmarest, 1804]; and *Macropus* Shaw, 1790. Also described by Owen (1840: 332). Synonymised within Macropodidae J. Gray, 1821 by Marshall (1981: 29) and Marshall *et al.* (1990: 493).

Subfamily Macropodinae J. Gray, 1821 *sensu* Kear & Cooke, 2001

Family Macropidae J. Gray, 1821: 308.

TYPE GENUS: *Macropus* Shaw, 1790.

COMMENTS: When originally proposed, this rank was placed in the Order Brutae (J. Gray, 1821 [= Marsupialia (Illiger, 1811 part)]) and included the genus *Macropus* Shaw, 1790. Subfamily designated to Thomas (1888a: 10) by Simpson (1945: 46). Recognised as the Family Macropodidae, with subfamilies Macropodinae and Potoroinae, by Thomas (1888a: xi, 3, 10) (who also recognised the Subfamily Hypsiprymnodontinae), Bensley (1903: 142), Cabrera (1919: 139), Simpson (1945: 46) and Szalay (1994: 43). Subfamily rank, within the Family Macropodidae, recognised by Troughton (1967: 136), Strahan (1995: 7, 306), Kirsch *et al.* (1997: 246), McKenna and Bell (1997: 63), Groves (2005b: 59), and Van Dyck and Strahan (2008: 10, 307). Names of the subfamily were reviewed by Ride (1962: 367). A revised classification of the Macropodoidea was undertaken by Kear and Cooke (2001: 84), which is followed here.

Family Halmaturini Goldfuss, 1820a: xxiii, 445.

TYPE GENUS: *Halmaturus* Illiger, 1811 [=*Macropus* Shaw, 1790].

COMMENTS: When originally proposed, this rank was placed in the Order Marsupialia (Illiger, 1811) and included the genera *Halmaturus* Illiger, 1811 [= *Macropus* Shaw, 1790]; and *Hypsiprymnus* Illiger, 1811 [= *Potorous* Desmarest, 1804]. Synonymised within Macropodidae by Marshall (1981: 29), Marshall *et al.* (1990: 492) and McKenna and Bell (1997: 62). McKenna and Bell (1997: 62) refer to Article 37(b) of the Code (ICZN, 1985a: 79), which states that if a name in use for a family group taxon, and hence for its nominotypical taxa, is invalid or unavailable and has been replaced by the name valid under Article 23e; the subordinate taxa bearing the valid name are the nominotypical taxa.

Tribe Macropina J. Gray, 1825a: 340.

TYPE GENUS: *Macropus* Shaw, 1790.

COMMENTS: When originally proposed, this rank was placed in the Family Didelphidae (J. Gray, 1821: 308) and included the genera *Macropus* Shaw, 1790; *Halmaturus* Illiger, 1811 [= *Macropus* Shaw, 1790]; and *Potorous* Desmarest, 1804a. Synonymised within the Subfamily Macropodinae by McKenna and Bell (1997: 63).

Tribe Halmaturina Bonaparte, 1832: 69.

TYPE GENUS: *Halmaturus* Illiger, 1811 [=*Macropus* Shaw, 1790].

COMMENTS: When originally proposed, this rank was placed in the Family Halmaturidae (Bonaparte, 1831 [= Macropodidae (J. Gray, 1821)]). Recognised by Bonaparte (1838: 113; 1840: 257; 1845: 6). Synonymised within the Subfamily Macropodinae by McKenna and Bell (1997: 63).

Tribe Poëphaga Owen, 1839a: 16, 19.

TYPE GENUS: Not based on a genus group name.

COMMENTS: When originally proposed, this rank was placed in the Marsupialia (Illiger, 1811) and included the Family Macropodidae (J. Gray, 1821). Tribe rank recognised by Owen (1840: 326) and family rank by Owen (1859: 52). McKenna and Bell (1997: 58) synonymised this taxon within Diprotodontia and attribute the author to Giebel (1855: xi, 670).

† Family Protemnodontidae De Vis, 1883d: 191.

TYPE GENUS: † *Protemnodon* Owen, 1873a: 128.

COMMENTS: Higher and lower ranks not recorded. Name also described by De Vis (1883e: 221). Synonymised within the Subfamily Macropodinae by Marshall (1981: 29), Marshall *et al.* (1990: 492) and McKenna and Bell (1997: 63). The type genus has commonly been used to include the larger extant wallabies (e.g. Tate, 1948: 256, 295), although it is currently considered to include only extinct species (e.g. Stirton, 1963: 97; Bartholomai, 1973a: 609; 360; Dawson & Flannery, 1985: 473; Long *et al.* 2002: 170).

Subfamily Macropodinae Thomas, 1888a: xi, 10.

TYPE GENUS: *Macropus* Shaw, 1790.

COMMENTS: When originally proposed, this rank was placed in the Family Macropodidae (J. Gray, 1821) and included the genera *Macropus* Shaw, 1790; *Petrogale* J. Gray, 1837; *Onychogale* Thomas, 1888a [= *Onychogalea* J. Gray, 1841]; *Lagorchestes* Gould, 1841 [1841–1842];

Dorcopsis Schlegel and Müller, 1845: 130; *Dendrolagus* S. Müller, 1840; and *Lagostrophus* Thomas, 1887c. Synonymised within the Subfamily Macropodinae (J. Gray, 1821) by McKenna and Bell (1997: 63).

Tribe Dendrolagini Flannery, 1989 *sensu* Baverstock *et al.*, 1989

Tribe Dendrolagini Flannery, 1989: 41.

TYPE GENUS: *Dendrolagus* S. Müller, 1840.

COMMENTS: When originally proposed as a new rank it was placed in the Subfamily Macropodinae (J. Gray, 1821) and included the genera † *Bohra* Flannery and Szalay, 1982: 83; and *Dendrolagus* S. Müller, 1840. Though Flannery (1989) placed *Dendrolagus* near the root of the macropodine tree, several subsequent studies suggest a relationship between *Dendrolagus* and *Petrogale*, with a close or sister relationship with *Thylogale* by authors including Baverstock *et al.* (1989: 38, 47; 1990d: 140), Kirsch *et al.* (1995: 309; 1997: 237), and Prideaux and Warburton (2010: 960, 969). Tribe recognised by Szalay (1994: 43), Long *et al.* (2002: 163), Kear and Cooke (2001: 84), and Prideaux and Warburton (2010: 954, 969) who expanded it to include *Petrogale*.

Tribe? Dendrolagina Bonaparte, 1850a: 1.

TYPE GENUS: *Dendrolagus* S. Müller, 1840.

COMMENTS: When originally proposed, this rank was placed in the Family Halmaturidae (Bonaparte, 1831 [= Macropodidae (J. Gray, 1821)]). Synonymised within the Subfamily Macropodinae by Marshall (1981: 29), Marshall *et al.* (1990: 492), and McKenna and Bell (1997: 63).

Dendrolagus S. Müller, 1840

Dendrolagus S. Müller, 1840: 20, footnote.

TYPE SPECIES: Φ *Dendrolagus ursinus* (S. Müller, 1840: 20, footnote) (*recte* Φ *Hypsiprymnus ursinus* Temminck, 1836: vi, footnote 2). Designated by Thomas (1888a: 92). See also Groves (1982: 157).

COMMENTS: Further described by Schlegel and Müller (1845: 138).

Dendrolegus J. Gray, 1842e: 16.

TYPE SPECIES: Incorrect subsequent spelling of *Dendrolagus* S. Müller, 1840.

COMMENTS: Spelling used by J. Gray (1843a: xxii, 87), but typically not recognised by subsequent authors.

Drendrolagus Fletcher, 1882: 801.

TYPE SPECIES: Incorrect subsequent spelling of *Dendrolagus* S. Müller, 1840.

COMMENTS: Fletcher (1991: 801) used the same species name with the spelling *Dendrolagus* earlier on the same page, but again used *Drendrolagus* on page 805.

Dendrolagus bennettianus De Vis, 1887

Bennett's Tree-kangaroo

D. [endrolagus] Bennettianus De Vis, 1887b: 13.

TYPE LOCALITY: Daintree River, Queensland, Australia.

COMMENTS: Species rank recognised by Iredale and Troughton (1934: viii, 39), Rothschild and Dollman (1936: 499) and subsequent authors. Considered a subspecies of *Dendrolagus dorianus* (from New Guinea) by Haltenorth (1958: 38). Subsequently recognised as a species by Ride (1970: 62), Groves (1982: 157; 1993e: 50), Calaby and Richardson (1988b: 60), Flannery *et al.* (1996: 94) and subsequent authors.

Dendrolagus Bennettianus De Vis, 1886b: v.

TYPE LOCALITY: Daintree River, Queensland, Australia.

COMMENTS: *Nomen nudum*. Synonymised within *bennettianus* by Calaby and Richardson (1988b: 60).

Dendrolagus lumholtzi Collett, 1884

Lumholtz's Tree-kangaroo

Dendrolagus lumholtzi Collett, 1884b: 387; Plate 32, Figs. 7–8.

TYPE LOCALITY: Herbert Vale, north Queensland, Australia.

COMMENTS: Species recognised by Thomas (1888a: xi, 96), Iredale and Troughton (1934: viii, 39) and subsequent authors.

Dendrolagus fulvus De Vis, 1888: 132.

TYPE LOCALITY: Herberton, north Queensland, Australia.

COMMENTS: *Nomen nudum*. Synonymised within *lumholtzi* by Tate (1948b: 294), Iredale and Troughton (1934: 39), Calaby and Richardson (1988b: 60) and Groves (1993e: 51; 2005b: 60).

Petrogale J. Gray, 1837

Petrógale J. Gray, 1837: 583.

TYPE SPECIES: *Kangurus penicillatus* J. Gray, 1827a [= *Petrogale penicillata* (J. Gray, 1827a)] by monotypy.

COMMENTS: Recognised as a subgenus within *Macropus* by Waterhouse (1841a: 240) but synonymised within *Heteropus* by Waterhouse (1846: 165). Genus recognised by Gould (1841 [1841–1842]: Plates 42, 46), Thomas (1888a: xi, 62), Iredale and Troughton (1934: viii, 42) and subsequent authors. Genus reviewed by Eldridge (1997: 113).

FUTURE TAXONOMIC RESEARCH: Meredith *et al.* (2008c: 399, 405) placed rock wallabies in a clade with *Dendrolagus*, *Thylogale* and *Dorcopsis*. The only two species that they examined in this genus were *P. concinna* (which they referred to as *Peradorcas*) and *P. xanthopus*, which they found separated at about 5 Ma. More recently Potter *et al.*

(2012a: 2254) examined mtDNA and nDNA sequences, finding that *P. concinna*, *burbidgei* and *brachyotis* form a clade separate from other species, and, that among the others, *P. persephone*, *P. xanthopus* and the *P. penicillata/lateralis* group form three distinct clades; and that the four major clades within the genus separated some 7.5–9 Ma. It may be therefore, that the genus *Peradorcas* may need to be reinstated (but with different composition from the way it was formerly envisaged) and that new genera should be contemplated for *P. persephone* and *P. xanthopus*.

Heteropus Jourdan, 1837a: 522.

TYPE SPECIES: *Kangurus penicillatus* J. Gray, 1827a [= *Petrogale penicillata* (J. Gray, 1827a)] by monotypy.

COMMENTS: Recognised as a subgenus within *Macropus* by Waterhouse (1846: 165), but synonymised within *Petrogale* by Thomas (1888a: 62), Iredale and Troughton (1934: 42) and subsequent authors.

HOMONYMS:

Heteropus Palisot de Beauvois, 1820: 230, pygmy mole crickets of the Class Insecta (Order Orthoptera, Family Tridactylidae). Genus is a synonym of *Tridactylus* Oliver, 1789: 26.

Heteropus Fitzinger, 1826: 23, skinks of the Class Reptilia (Order Squamata, Family Scincidae). Genus is a synonym of *Carlia* J. Gray, 1845a: 271.

Heteropus Spinola, 1837: 337, stink bugs of the Class Insecta (Order Hemiptera, Family Pentatomidae). Genus is a synonym of *Montrouzieriellus* Kirkaldy (1908: 124).

Heteropus Germar, 1839: 206, click beetles of the Class Insecta (Order Coleoptera, Family Elateridae).

Heteropus de Laporte, 1840: 221, darkling beetles of the Class Insecta (Order Coleoptera, Family Tenebrionidae). Genus is a synonym of *Blaptinus* Sturm, 1826: 101.

Heteropus Hodgson, 1843: 127, birds of prey of the Class Aves (Order Falconiformes, Family Accipitridae). Genus is a synonym of *Ictinäetus* Blyth, 1843: 128.

Heteropus Schönherr, 1845: 1, weevils of the Class Insecta (Order Coleoptera, Family Curculiodidae). Genus is a synonym of *Sclerocardius* Schönherr, 1848: 132.

Heteropus Newport, 1850a: 71 and 1850b: 396, mites of the Class Arachnida (Order Acarina, Family Pyemotidae). Genus is a synonym of *Pyemotes* Amerling, 1861: 54.

Peradorcas Thomas, 1904a: 226.

TYPE SPECIES: *Petrogale concinna* Gould, 1842c by monotypy.

COMMENTS: Genus recognised by Cabrera (1919: 142), Iredale and Troughton (1934: viii, 41), Tate (1948b: 280), Marlow (1965: 114), Troughton (1967: 150), Ride (1970: 62), Kitchener and Sanson (1978: 280), Marshall (1981: 29), Honacki *et al.* (1982: 48), Strahan (1983: 223; 1995: 371), Marshall (1984: 107), Marshall *et al.* (1990: 492) and McKenna and Bell (1997: 64). Synonymised within *Petrogale* by Simpson (1945: 47), Calaby and Richardson (1988b: 74), Groves (1993e: 56; 2005b: 66), and Van Dyck and Strahan (2008: 370).

Petrogale assimilis Ramsay, 1877

Allied Rock-wallaby

Petrogale assimilis Ramsay, 1877d: 360.

TYPE LOCALITY: Palm Island, north of Townsville, Queensland, Australia.

COMMENTS: Synonymised within *P. penicillata* by Thomas (1888a: 67), then within *Petrogale inornata* by Iredale and Troughton (1934: 43), Tate (1948b: 274) and Marlow (1965: 113). Subspecies rank within *inornata* recognised by Troughton (1967: 147). Subspecies rank within *penicillata* recognised by Poole (1979: 21) but transferred to *inornata* as a race by Briscoe *et al.* (1982: 74) and subspecies by Strahan (1983: 213). Elevated to species rank by Calaby and Richardson (1988b: 74), Sharman *et al.* (1990: 353), Eldridge and Close (1992: 605, 618) and subsequent authors.

Petrogale puella Thomas, 1926b: 627.

TYPE LOCALITY: Flinders River and Torrens Creek, north Queensland, Australia.

COMMENTS: Considered a subspecies of *Petrogale inornata* by Iredale and Troughton (1934: 43), Tate (1948b: 274), Marlow (1965: 113), Troughton (1967: 147) and Strahan (1983: 214). Subspecies rank within *penicillata* recognised by Poole (1979: 21) and as a race within *inornata* by Briscoe *et al.* (1982: 79). Synonymised within *P. assimilis* by Calaby and Richardson (1988b: 74), Eldridge and Close (1992: 612) and subsequent authors.

Petrogale brachyotis (Gould, 1841)

Western Short-eared Rock-wallaby

Petrogale brachyotis brachyotis (Gould, 1841)

Macropus (Petrogale) brachyotis Gould, 1841d: 128.

TYPE LOCALITY: Swan River, Western Australia, Australia (= Hanover Bay, North West Weste rn Australia (see Calaby & Richardson, 1988b: 74). Type designated by Thomas (1922a: 128).

COMMENTS: Included within the genus *Petrogale* and further described by Gould (1841 [1841–1842]: no page numbers). Placed in *Macropus* in the subgenus *Petrogale* by Waterhouse (1841a: 247) and within subgenus *Heteropus* by Waterhouse (1846: 176). Placed in *Petrogale* by Gould (1841 [1841–1842]: Plate 6; Gould, 1859 [1845–1863]: Text to Plate 47), J. Gray (1841: 403) and Thomas (1888a: xi, 69), who reviewed its taxonomic history. Taxon consistently recognised as a species since its description. Strahan (1995: 368) suggested that it is possible that *P. brachyotis* is divisible into three races from the Kimberley Region (Western

Australia), Victoria River District (Northern Territory) and Arnhem Land Region (Northern Territory). This was supported by Potter *et al.* (2012a: 2262; 2012b: 645) who found multiple biographic barriers and the identification of eight geographically discrete and genetically distinct lineages within the *brachyotis* group (including the three mentioned above), five of which are separated by major river valleys.

Petrogale brachyotis victoriae Potter *et al.*, 2014

Petrogale brachyotis victoriae Potter *et al.*, 2014: 401, 409.

TYPE LOCALITY: Lobby Creek, 'Bradshaw', Victoria River region of the western Northern Territory. (15°20′S, 130°06″E)

COMMENTS: Differences in the Kimberley, Victoria River and Northern Territory populations were identified by Strahan (1995: 367). Subsequently Potter *et al.* (2012b: 645; 2014: 401, 409) found that together they form a non-monophyletic cluster, with the Northern Territory population (now recognised as *P. wilkinsi*) forming a clade with *P. burbidgei*, with the Kimberley and the Victoria River populations forming another clade.

Petrogale burbidgei Kitchener & Sanson, 1978

Monjon Rock-wallaby

Petrogale burbidgei Kitchener & Sanson, 1978: 269.

TYPE LOCALITY: Crystal Creek, Mitchell Plateau, Western Australia, Australia. (14°30′00″S, 125°47′20″E)

COMMENTS: Has consistently been recognised as species since its description.

Petrogale coenensis Eldridge & Close, 1992

Cape York Rock-wallaby

Petrogale coenensis Eldridge & Close, 1992: 605, 621.

TYPE LOCALITY: 'Twin Humps', north of Coen, north Queensland, Australia. (13°47′27″S, 143°04′24″E)

COMMENTS: Initially recognised as the 'Cape York race' of *P. godmani* by Briscoe *et al.* (1982: 74) and Strahan (1983: 215), proposed to be a separate species by Eldridge *et al.* (1989: 935), but kept as the 'Cape York race' until Eldridge and Close (1992: 605, 621) formally elevated it to species rank, which has been followed by subsequent authors including Groves (2005b: 67), Clayton *et al.*, 2006: 107), and Van Dyck and Strahan (2008: 368).

Petrogale concinna Gould, 1842

Nabarlek Rock-wallaby

Petrogale concinna concinna Gould, 1842

Petrogale concinna Gould, 1842c: 57.

TYPE LOCALITY: North west coast of Australia [= Victoria River, Northern Territory; See Eldridge, 1997: 113 for discussion]

COMMENTS: Placed in *Macropus* (*Heteropus*) by Waterhouse (1846: 177) and *Petrogale* by Gould (1856 [1845–1863]: Text to Plate 48). Placed within *Peradorcas* by Thomas (1909b: 198; 1926b: 630), Cabrera (1919: 142), Iredale and Troughton (1934: viii, 42), Tate (1948b: 280), Marlow (1965: 114), Troughton (1967: 150), Ride (1970: 62), Honacki *et al.* (1982: 48), Strahan (1983: 223; 1995: 371), Marshall (1984: 107) and McKenna and Bell (1997: 64). Recognised as a species within *Petrogale* by Thomas (1888a: xi, 71), Poole (1979: 21), Calaby and Richardson (1988b: 75), Groves (2005b: 67), Clayton *et al.* (2006: 107), and Van Dyck and Strahan (2008: 370).

Petrogale concinna canescens (Thomas, 1909)

Peradorcas concinna canescens Thomas, 1909b: 198.

TYPE LOCALITY: Nellie Creek, Arnhem Land, Northern Territory, Australia.

COMMENTS: Synonymised within *concinna* by Calaby and Richardson (1988b: 75) and Groves (2005b: 67). Recognised as a subspecies of *concinna* by Iredale and Troughton (1934: 42), Tate (1948b: 284), D. Johnson (1964: 456), Marlow (1965: 114), Troughton (1967: 151), Poole (1979: 21), Briscoe *et al.* (1982: 74), Strahan (1983: 223; 1995: 372), Sharman *et al.* (1990: 354), Clayton *et al.* (2006: 107), and Van Dyck and Strahan (2008: 371).

Petrogale concinna monastria (Thomas, 1926)

Peradorcas concinna monastria Thomas, 1926b: 630.

TYPE LOCALITY: Napier Broome Bay, Western Australia, Australia.

COMMENTS: Synonymised within *concinna* by Calaby and Richardson (1988b: 75) and Groves (2005b: 67). Recognised as a subspecies of *concinna* by most authors including by Iredale and Troughton (1934: 42), Tate (1948b: 284), Marlow (1965: 114), Troughton (1967: 151), Poole (1979: 21), Strahan (1983: 223; 1995: 372), Clayton *et al.* (2006: 107), and Van Dyck and Strahan (2008: 371).

Petrogale godmani Thomas, 1923

Godman's Rock-wallaby

Petrogale godmani Thomas, 1923d: 13.

TYPE LOCALITY: Black Mountain, 16 Miles south west of Cooktown, north Queensland, Australia.

COMMENTS: A fuller description occurs in Thomas (1923e: 177). Reduced to a subspecies of *Petrogale inornata* by Iredale and Troughton (1934: 43), Tate (1948b: 274), Marlow (1965: 113), Troughton (1967: 147); and as a subspecies of *P. penicillata* by Poole (1979: 21). Recognised at species rank by Ride (1970: 61), Briscoe *et al.* (1982: 74), Honacki *et al.* (1982: 48), Strahan (1983: 215), Calaby and Richardson (1988b: 75), Sharman *et al.* (1990: 356) and Eldridge and Close (1992: 605, 612) and subsequent authors.

Petrogale herberti Thomas, 1926

Herbert's Rock-wallaby

Petrogale herberti Thomas, 1926b: 626.

TYPE LOCALITY: Eidsvold, Burnett River, southern Queensland, Australia.

COMMENTS: Recognised as a subspecies of *Petrogale inornata* by Iredale and Troughton (1934: 43) and Troughton (1967: 147). Transferred to *P. penicillata* at the subspecies rank by Finlayson (1931: 82), Tate (1948b: 273), Marlow (1965: 113), Ride (1970: 223), Poole (1979: 21), Briscoe *et al.* (1982: 78) and Strahan (1983: 211) and Eldridge *et al.* (1990: 798). Synonymised within *penicillata* by Calaby and Richardson (1988b: 76) and Groves (1993e: 56). Elevated again to subspecies of *penicillata* by Sharman *et al.* (1990: 356) and to species rank by Eldridge and Close (1992: 605, 615) and followed by subsequent authors.

Petrogale inornata Gould, 1842

Unadorned Rock-wallaby

Petrogale inornata Gould, 1842 [1841–1842]: Plate 46.

TYPE LOCALITY: North coast of Australia [= Cape Upstart, Queensland, Australia]. See Stokes (1846: 336).

COMMENTS: Name also described by Gould (1842d: 5). Placed in *Macropus* (*Heteropus*) by Waterhouse (1846: 175) and *Petrogale* by Gould (1842 [1841–1842]: Unnumbered Plate; Gould, 1860 [1845–1863]: Text to Plates 45–46). Name misspelt as '*Petregale Inanata*' by Stokes (1846: 336). Synonymised within *penicillata* by Ride (1970: 246) and Honacki *et al.* (1982: 48). Recognised as a species within *Petrogale* by Thomas (1888a: xi, 70), Iredale and Troughton (1934: viii, 42), Tate (1948b: 273), Marlow (1965: 112), Troughton (1967: 146), Poole (1979: 21), Strahan (1983: 213), Calaby and Richardson (1988b: 75), Eldridge and Close (1992: 605, 616) and subsequent authors.

Petrogale lateralis Gould, 1840

Black-footed Rock-wallaby

Petrogale lateralis lateralis Gould, 1840

Petrogale lateralis Gould, 1840a: 685.

TYPE LOCALITY: Swan River, Western Australia, Australia. Type designated by Thomas (1922a: 128).

COMMENTS: Name also described by Gould (1840b: 561), which was also published on 29 August 1840. McAllan and Bruce (1989: 450) proposed that they act as first revisers in giving priority to the publication of Gould (1840a: 685). Taxon further described by Gould (1841 [1841–1842]: no pages number). Placed in *Macropus* (*Heteropus*) by Waterhouse (1846: 172) and again within *Petrogale* by Gould (1857 [1845–1863]: Text to Plates 41–42). Reduced to a subspecies of *P. penicillata* by Tate (1948b: 273) and Poole (1979: 21). Placed in *Petrogale* at species rank by Thomas (1888a: xi, 68), Iredale and Troughton (1934: viii, 44), Troughton (1967: 144) and Marlow (1965: 112), Briscoe *et al.* (1982: 74), Strahan (1983: 209), Calaby and Richardson (1988b: 76) and Sharman *et al.* (1990: 355). Two races are known of this species, which were given the names MacDonnell Range race and West Kimberley race by Briscoe *et al.* (1982: 74), and were subsequently recognised by Sharman *et al.* (1990: 355), Strahan (1983: 209; 1995: 378), Clayton *et al.* (2006: 107), and Van Dyck and Strahan (2008: 376). Potter *et al.* (2012b: 645) found that the West Kimberley race is sister to a clade uniting *P. l. lateralis* and *P. l. pearsoni*, while the MacDonnell Ranges race is sister to a clade containing all those three.

FUTURE TAXONOMIC RESEARCH: The MacDonnell Range and West Kimberley races (*sensu* Briscoe *et al.*, 1982: 74) need to be reviewed and formally described if appropriate.

HOMONYMS:

Petrogale lateralis J. Gray, 1841: 403, rock-wallaby of the Class Mammalia (Order Diprotodontia, Family Macropodidae). A *nomen nudum*. See Thomas (1888a: 68) and Iredale and Troughton (1934: 44).

Petrogale lateralis hacketti Thomas, 1905

Petrogale lateralis Hacketti Thomas, 1905a: 425.

TYPE LOCALITY: Mondrain Island, Recherche Group, south west Australia.

COMMENTS: Elevated to species rank by Iredale and Troughton (1934: viii, 44). Reduced to a subspecies of *P. penicillata* by Tate (1948b: 273), then transferred to a subspecies of *P. lateralis* by Marlow (1965: 113). Troughton (1967: 145) elevated it to species status, but it was reduced to a synonym of *penicillata* by Ride (1970: 246), before being elevated again to a subspecies of *P. penicillata* by Poole (1979: 21) and subsequent authors with the exception of Calaby and Richardson (1988b: 76) who synonymised it within *P. lateralis*. Recognised as a race of *lateralis* by Briscoe *et al.* (1982: 76) and as a subspecies of *lateralis* by Strahan (1983: 209; 1995: 379), Sharman *et al.* (1990: 355),

Groves (2005b: 68), Clayton *et al.* (2006: 107), and Van Dyck and Strahan (2008: 376). Potter *et al.* (2012b: 645) found that it is sister to *P. l. lateralis*.

Petrogale lateralis pearsoni Thomas, 1922

Petrogale pearsoni Thomas, 1922f: 682.

TYPE LOCALITY: Pearson Island, Investigator Group, South Australia, Australia.

COMMENTS: Species rank recognised by Iredale and Troughton (1934: viii, 45) and Troughton (1967: 146). Subspecies within *P. lateralis* recognised by Marlow (1965: 113) and within *penicillata* by Ride (1970: 246) and Poole (1979: 21). Synonymised within *lateralis* by Briscoe *et al.* (1982: 74), Strahan (1983: 209) and Calaby and Richardson (1988b: 76). It was subsequently elevated to subspecies rank by Sharman *et al.* (1990: 355), Strahan (1995: 379), Groves (2005b: 68), Clayton *et al.* (2006: 107), and Van Dyck and Strahan (2008: 376).

Petrogale mareeba Eldridge & Close, 1992

Mareeba Rock-wallaby

Petrogale mareeba Eldridge & Close, 1992: 605, 619.

TYPE LOCALITY: Mungana trucking yards, 16km west of Chillagoe, north Queensland, Australia. (17°06′S, 144°23′E)

COMMENTS: Initially recognised as *P. penicillata* subspecies nova 3 by Poole (1979: 21), then moved to *Petrogale inornata* as 'Mareeba race' by Briscoe *et al.* (1982: 74), which was followed by Strahan (1983: 214) and Eldridge *et al.* (1988: 228). Synonymised within *P. assimilis* by Sharman *et al.* (1990: 353). Finally given a name and elevated to species rank by Eldridge and Close (1992: 605, 619), which has been followed by subsequent authors.

Petrogale penicillata (J. Gray, 1827)

Brush-tailed Rock-wallaby

Kangurus Penicillatus J. Gray, 1827a: 204; Plate.

TYPE LOCALITY: Australia.

COMMENTS: Recognised within *Macropus* by Waterhouse (1838a: 66), Gould (1840a: 685) (incorrectly as *penecillatus*), subgenus *Petrogale* by Waterhouse (1841a: 243) and subgenus *Heteropus* by Waterhouse (1846: 167). Placed within genus *Petrogale* by J. Gray (1837: 583), Gould (1842 [1841–1842]: Unnumbered Plate; Gould, 1853 [1845–1863]: Text to Plates 39–40), Thomas (1888a: xi, 66), who reviewed its early taxonomic history, and subsequent authors.

heteropus [sic] albogularis Jourdan, 1837a: 522.

TYPE LOCALITY: Mountains south east of Sydney, New South Wales, Australia.

COMMENTS: Recognised within *Heteropus* by Waterhouse (1841a: 246). Synonymised within *penicillata* by Waterhouse (1846: 167), Thomas (1888a: 66), Iredale and Troughton (1934: 42) and Calaby and Richardson (1988b: 76).

Petrogale longicauda Krefft, 1865: 324.

TYPE LOCALITY: Rylstone, 250 miles northwest of Sydney, New South Wales, Australia.

COMMENTS: Synonymised within *penicillata* by Thomas (1888a: 66), Iredale and Troughton (1934: 42) and Calaby and Richardson (1988b: 76).

Petrogale persephone Maynes, 1982

Proserpine Rock-wallaby

Petrogale persephone Maynes, 1982: 47.

TYPE LOCALITY: Base of Mt. Dryander, 9.6km of Proserpine, Queensland, Australia. (20°19′41″S, 148°33′21″E)

COMMENTS: Recognised as *Petrogale* sp. nov. 1 by Poole (1979: 21). Species has consistently been recognised since its description.

Petrogale purpureicollis Le Souef, 1924

Purple-necked Rock-wallaby

Petrogale purpureicollis Le Souef, 1924: 274.

TYPE LOCALITY: Dajarra, north west Queensland, Australia.

COMMENTS: Reduced to a subspecies of *Petrogale inornata* by Iredale and Troughton (1934: 43), Tate (1948b: 275), Marlow (1965: 113), Troughton (1967: 147). Tate (1948b: 275), upon examining a single skull of *purpureicollis*, noted that it differed rather sharply from *inornata* and suggested that if the differences were typical of the taxon then it should be recognised as distinct species. Species rank recognised by Ride (1970: 61) and Honacki *et al.* (1982: 48). Reduced to a subspecies of *P. penicillata* by Poole (1979: 21) and as a race of *lateralis* by Briscoe *et al.* (1982: 74). Subsequently placed as a subspecies of *P. lateralis* by Strahan (1983: 209; 1995: 379), Sharman *et al.* (1990: 355) and Eldridge *et al.* (1991a: 625; 1991b: 631). Synonymised within *P. lateralis* by Calaby and Richardson, 1988b: 76) and Groves (1993e: 56). Finally elevated to full species status by Eldridge *et al.* (2001: 323), which has been followed by subsequent authors; see especially Potter *et al.* (2012b: 640).

Petrogale rothschildi Thomas, 1904

Rothschild's Rock-wallaby

Petrogale rothschildi Thomas, 1904b: 366.

TYPE LOCALITY: Cossack, north west Australia. (not Cossack River).

COMMENTS: Consistently recognised as a species since its description. It was recently shown by Potter *et al.* (2012b: 645) to be sister to the clade containing the *P. penicillata* and *P. lateralis* groups.

Petrogale sharmani Eldridge & Close, 1992

Mount Claro Rock-wallaby

Petrogale sharmani Eldridge & Close, 1992: 605, 618.

TYPE LOCALITY: Mt Claro, north Queensland, Australia. (18°52′05″S, 145°44′05″E)

COMMENTS: Initially recognised as *P. penicillata* subspecies nova 4 by Poole (1979: 21), moved to *Petrogale inornata* as 'Mt Claro race' by Briscoe *et al.* (1982: 74), which was followed by Strahan (1983: 214) and Eldridge *et al.* (1988: 228). Synonymised within *P. assimilis* by Sharman *et al.* (1990: 353) who no longer regarded the race as a subspecies of *assimilis*. Finally given a name and elevated to species rank by Eldridge and Close (1992: 605, 618), which has been followed by subsequent authors.

Petrogale wilkinsi Thomas, 1926

Eastern Short-eared Rock-wallaby

Petrogale wilkinsi Thomas, 1926c: 185.

TYPE LOCALITY: Roper River, Northern Territory, Australia. 200 feet. (15°S, 135°E)

COMMENTS: Taxon recognised as a species by Iredale and Troughton (1934: viii, 43), Marlow (1965: 112) and Troughton (1967: 149). Recognised as a subspecies of *penicillata* by Ride (1970: 246) and synonymised within *P. brachyotis* by Poole (1979: 22), Briscoe *et al.* (1982: 74), Strahan (1983: 221; 1995: 368), Calaby and Richardson (1988b: 74), and Van Dyck and Strahan (2008: 366). The distinctiveness of this taxon was identified by Potter *et al.* (2010: 56) who suggested *P. brachyotis* (*sensu lato*) from the Kimberley and Northern Territory represents two highly divergent species with *P. wilkinsi*, therefore, potentially being a distinct taxon. The distinctiveness of this taxon at species rank was confirmed by Potter *et al.* (2014: 401, 411), which has been adopted here.

Petrogale longmani Thomas, 1926c: 186.

TYPE LOCALITY: Groote Eylandt, Northern Territory, Australia.

COMMENTS: Considered a distinct species by Iredale and Troughton (1934: viii, 43), Marlow (1965: 112) and Troughton (1967: 148). Synonymised within *penicillata* by Ride (1970: 246) and within *brachyotis* by Poole (1979: 22), Briscoe *et al.* (1982: 74), Strahan (1983: 221), Calaby and Richardson (1988b: 74) and subsequent authors until it was synonymised within *wilkinsi* by Potter *et al.* (2014: 411).

Petrogale venustula Thomas, 1926b: 628.

TYPE LOCALITY: King River, upper Daly River, Northern Territory, Australia.

COMMENTS: Recognised as a subspecies of *longmani* by Iredale and Troughton (1934: 43), Marlow (1965: 113) and Troughton (1967: 149). Species rank recognised by D. Johnson (1964: 454). Reduced to a synonym of *P. brachyotis* by Poole (1979: 22), Briscoe *et al.* (1982: 74), Strahan (1983: 221), Calaby and Richardson (1988b: 74) and subsequent authors until it was synonymised within *wilkinsi* by Potter *et al.* (2014: 411).

Petrogale brachyotis signata Thomas, 1926b: 629.

TYPE LOCALITY: Mary River, Northern Territory, Australia.

COMMENTS: Recognised as subspecies of *Petrogale brachyotis* by Iredale and Troughton (1934: 44), Marlow (1965: 113), Troughton (1967: 148) and Strahan (1983: 221), but doubt over its validity was raised by Strahan (1995: 368). Synonymised within *brachyotis* by Briscoe *et al.* (1982: 74), and Calaby and Richardson (1988b: 75). Taxon not recognised by Groves (2005b: 67), and Van Dyck and Strahan (2008: 366), but synonymised within *wilkinsi* by Potter *et al.* (2014: 411).

Petrogale xanthopus J. Gray, 1855

Yellow-footed Rock-wallaby

Petrogale xanthopus xanthopus J. Gray, 1855

Petrogale xanthopus J. Gray, 1855: 249; Plate 39.

TYPE LOCALITY: Flinders Ranges, South Australia, Australia. Type designated by Thomas (1922a: 128).

COMMENTS: Species recognised within *Petrogale* by subsequent authors.

Phalangista xanthopygus Giebel, 1874: Key to Plate 20, Fig. 4.

TYPE LOCALITY: Error for *Petrogale xanthopus*.

COMMENTS: Synonymised within *xanthopus* by Thomas (1888a: 65), Iredale and Troughton (1934: 44) and Calaby and Richardson (1988b: 77).

Petrogale xanthopus celeris Le Souef, 1924

Petrogale celeris Le Souef, 1924: 273.

TYPE LOCALITY: Adavale, Bulloo River, south west Queensland, Australia.

COMMENTS: Recognised as a subspecies of *Petrogale xanthopus* by Iredale and Troughton (1934: 44) and most subsequent authors including Tate (1948b: 275), Marlow (1965: 113), Troughton (1967: 150), Poole (1979: 19), Briscoe *et al.* (1982: 74), Strahan (1983: 217; 1995: 392), Sharman *et al.* (1990: 354), Clayton *et al.* (2006: 108), and

Van Dyck and Strahan (2008: 394), with the exception of Calaby and Richardson (1988b: 77) who synonymised it within *P. xanthopus*.

Thylogale J. Gray, 1837

Thylógale J. Gray, 1837: 583.

TYPE SPECIES: *Halmaturus (Thylogale) eugenii* J. Gray, 1837 [= *Thylogale thetis* (Lesson, 1827a)] by monotypy.

COMMENTS: Described as a subgenus of *Halmaturus*. Synonymised within *Macropus* Shaw (1790: Text to Plate 33) by Thomas (1888a: 10). Recognised at subgeneric rank by Cabrera (1919: 146). Included in *Macropus* by Van Gelder (1977: 6) but recognised as a genus by Simpson (1945: 47), Kirsch and Calaby (1977: 17) and subsequent authors. Genetic differentiation amongst populations of *T. thetis* and *T. stigmatica* in eastern Australia was explored by Eldridge *et al.* (2011: 103), with the phylogenetics and historical biogeography being explored by Macqueen *et al.* (2010: 1134).

Conoyces Lesson, 1842: 194.

TYPE SPECIES: Φ *Didelphis brunii* Schreber, 1778: 551 (as *Macropus brunii*) [= Φ *Thylogale brunii* (Schreber, 1778: 551)] by monotypy.

COMMENTS: Described as a subgenus of *Macropus*. Synonymised within *Thylogale* by Tate (1948b: 312) and Groves (1993e: 57) but not recognised by Groves (2005b: 69).

Thylacogale Agassiz, 1846: 370.

TYPE SPECIES: Replacement name for *Thylogale* J. Gray, 1837.

COMMENTS: Not subsequently recognised.

Thylogale billardierii (Desmarest, 1822)

Rufous-bellied Pademelon

kangurus Billardierii Desmarest, 1822b: 542.

TYPE LOCALITY: Tasmania, Australia.

COMMENTS: Recognised within *Halmaturus* by Gould (1841 [1841–1842]: Unnumbered Plate), Waterhouse (1846: 159) and Krefft (1868a: 94), and within *Macropus* by Waterhouse (1841a: 227) and Thomas (1888a: xi, 58), who reviewed the early taxonomic history. Taxon transferred to *Thylogale* by Iredale and Troughton (1934: viii, 47) and subsequent authors.

Halmaturus (Thylogale) Tasmanei J. Gray, 1838b: 108.

TYPE LOCALITY: Tasmania, Australia.

COMMENTS: Synonymised within *billardierii* by Waterhouse (1841a: 227; 1846: 159), Thomas (1888a: 58), Iredale and Troughton (1934: 47) and Calaby and Richardson (1988b: 78).

Macropus (Halmaturus) rufiventer W. Ogilby, 1838c: 220.

TYPE LOCALITY: Tasmania, Australia.

COMMENTS: Also described by W. Ogilby, 1838e: 23. Taxon recognised by Waterhouse (1838a: 67). Synonymised within *billardierii* by Waterhouse (1841a: 227; 1846: 159), Thomas (1888a: 58), Iredale and Troughton (1934: 47) and Calaby and Richardson (1988b: 78).

H. [almaturus] brachytarsus Wagner, 1843: 121.

TYPE LOCALITY: Tasmania, Australia.

COMMENTS: Synonymised within *billardierii* by Waterhouse (1846: 159), Thomas (1888a: 59), Iredale and Troughton (1934: 47) and Calaby and Richardson (1988b: 78).

Thylogale stigmatica (Gould, 1860)

Red-legged Pademelon

Thylogale stigmatica stigmatica (Gould, 1860)

Halmaturus stigmatica Gould, 1860 [1845–1863]: Text to Plates 33–34.

TYPE LOCALITY: Point Cooper, north of Rockingham Bay, Queensland, Australia.

COMMENTS: Placed within *Macropus* by Thomas (1888a: xi, 47) and *Thylogale* and Iredale and Troughton (1934: viii, 45), Tate (1948b: 315) and subsequent authors. Groves (2005b: 70) noted that some authors give the citation of the original description as Gould, 1860 [= 1861a: 375], but this was dated 13 November 1860 [and published in March 1861], while *Mammals of Australia*, Part 12 was published 1 November 1860. Tate (1948b: 315–316) recognised *wilcoxi*, *coxeni* and *oriomo* as subspecies. Genetic studies by Eldridge *et al.* (2011: 103) found a broad zone of introgression, for both nuclear and mtDNA markers, between *T. s. stigmatica* (Wet Tropics) and *T. s. wilcoxi* (south-east Queensland). In addition they found that individuals sampled from around Proserpine were genetically *T. s. stigmatica* rather than *T. s. wilcoxi*, as had previously been assumed.

FUTURE TAXONOMIC RESEARCH: This species requires revision using both morphological and molecular evidence. In particular, it is of considerable biogeographic significance to find out whether those animals from the tropical rainforest and the southeast are really conspecific.

Thylogale stigmatica coxeni (J. Gray, 1866)

Halmaturus coxeni J. Gray, 1866a: 220; Plate 25.

TYPE LOCALITY: Port Albany, Cape York, Queensland, Australia. Type designated by Thomas (1922a: 128).

COMMENTS: Recognised at the species rank, within *Macropus*, by Thomas (1888a: xi, 44) and within *Thylogale* by Iredale and Troughton (1934: viii, 45) and Troughton (1967: 157). Taxon recognised as a subspecies of *stigmatica* by Tate (1948b: 316). Synonymised within *stigmatica* by Ride (1970: 248), Calaby and Richardson (1988b: 78) and Groves (1993e: 57). Recognised as a subspecies by Strahan (1983: 225; 1995: 398), Flannery (1990: 113; 1995a: 159), Groves (2005b: 70), Clayton *et al.* (2006: 108), and Van Dyck and Strahan (2008: 399).

Halmaturus Gazella De Vis, 1884a: 110.

TYPE LOCALITY: Somerset, Cape York, Queensland, Australia.

COMMENTS: Synonymised within *stigmatica* by Calaby and Richardson (1988b: 79), Flannery (1990: 113; 1995a: 159) and Groves (1993e: 57). Synonymised within *coxeni* by Thomas (1888a: 44) and Iredale and Troughton (1934: 45) and Groves (2005b: 70).

Thylogale stigmatica wilcoxi (McCoy, 1866)

Halmaturus wilcoxi McCoy, 1866a: 583.

TYPE LOCALITY: Richmond River, New South Wales, Australia. Designation by Dixon (1970: 107).

COMMENTS: Subsequent full description by McCoy (1866b: 322). Species rank within *Macropus* recognised by Thomas (1888a: xi, 48), Finlayson (1931: 81) (in the subgenus *Thylogale*), and within the genus *Thylogale* by Troughton (1967: 156). Lowered to a subspecies of *Thylogale stigmatica* by Iredale and Troughton (1934: 45) and Tate (1948b: 315). Synonymised within *stigmatica* by Ride (1970: 248), Calaby and Richardson (1988b: 79) but more recently recognised as a subspecies of *stigmatica* by Strahan (1983: 225; 1995: 398), Flannery (1990: 113; 1995a: 159), Groves (2005b: 70), Clayton *et al.* (2006: 108), and Van Dyck and Strahan (2008: 399).

Φ ***Thylogale stigmatica oriomo*** (Tate & Archbold, 1935)

Φ *Macropus coxenii oriomo* Tate & Archbold, 1935b: 1.

TYPE LOCALITY: Wuroi, Oriomo River, Western Division, Papua New Guinea.

COMMENTS: Recognised as a subspecies of *stigmatica* by Laurie and Hill (1954: 28) and Tate (1948b: 316). Synonymised within *stigmatica* by Groves (1993e: 57). Subspecies rank recognised by Strahan (1983: 225; 1995: 398), Flannery (1990: 113; 1995a: 159) and Groves (2005b: 70).

Φ *Halmaturus temporalis* De Vis, 1884a: 111.

TYPE LOCALITY: Moreton Bay, Queensland, Australia.

COMMENTS: Synonymised within *wilcoxi* by Thomas (1888a: 48) and Iredale and Troughton (1934: 45) and within *stigmatica* by Calaby and Richardson (1988b: 79) and Flannery (1990: 113; 1995a: 159). Taxon synonymised within *oriomo* by Groves (2005b: 70).

Thylogale thetis (Lesson, 1827)

Red-necked Pademelon

Halmaturus Thetis Lesson, 1827a: 229.

TYPE LOCALITY: Port Jackson, New South Wales, Australia.

COMMENTS: Species name spelt *thetidis* and included within *Halmaturus* by Gould (1857 [1845–1863]: Text to Plates 31–32) and with similar spelling in *Macropus* by Thomas (1888a: xi, 52). Placed in *Thylogale* by Iredale and Troughton (1934: viii, 46), Tate (1948b: 313) and followed by subsequent authors. Genetic studies by Eldridge *et al.* (2011: 103) found limited evidence of introgression between sympatric populations of *T. thetis* and *T. s. wilcoxi* in south-east Queensland.

[Halmaturus] thetidis F. Cuvier, 1829a: 2; Text to Plate 225.

TYPE LOCALITY: *Nomen novum* for *Halmaturus thetis* Lesson, 1827a.

COMMENTS: Taxon recognised as a species within *Halmaturus* by Gould (1842 [1841–1842]: Unnumbered Plate; Gould, 1857 [1845–1863]: Text to Plates 31–32). Synonymised within *Macropus eugenii* by Waterhouse (1841a: 232) and recognised within *Macropus (Halmaturus)* by Waterhouse (1846: 144) and within *Macropus* by Thomas (1888a: xi, 52) and Lydekker (1894a: 38), but typically not by subsequent authors. Taxon placed within *thetis* by Julien-Laferrière (1994: 30).

Halmaturus [Thylogale] Eugènii J. Gray, 1837: 583.

TYPE LOCALITY: Unknown.

COMMENTS: Synonymised within *thetis* by Calaby and Richardson (1988b: 79) and Groves (1993e: 57; 2005b: 70).

HOMONYMS:

Kangurus Eugenii Desmarest, 1817d, the Tammar Wallaby of the Class Mammalia (Order Diprotodontia, Family Macropodidae). Name is now recognised as *Notamacropus eugenii* (Desmarest, 1817d). See individual entry.

H. [almaturus] nuchalis Wagner, 1843: 128.

TYPE LOCALITY: New South Wales, Australia.

COMMENTS: Synonymised within *thetidis* by Thomas (1888a: 52) and within *thetis* by Iredale and Troughton (1934: 46), Tate (1948b: 313), Calaby and Richardson (1988b: 79) and subsequent authors.

Tribe Macropodini J. Gray, 1821 *sensu* Flannery, 1989

Family Macropidae J. Gray, 1821: 308.

TYPE GENUS: *Macropus* Shaw, 1790.

COMMENTS: When originally proposed, this rank was placed in the Order Brutae (J. Gray, 1821 [= Marsupialia (Illiger, 1811 part)]) and included the genus *Macropus* Shaw,

1790. Tribe rank has been recognised as Macropodini by L. Marshall (1981: 29), Flannery (1989: 41), Szalay (1994: 43), Kear and Cooke (2001: 84), Long *et al.* (2002: 164), Prideaux and Warburton (2010: 954, 969) and is followed here.

Tribe Macropodini L. Marshall, 1981: 29.

TYPE GENUS: *Macropus* Shaw, 1790.

COMMENTS: When originally proposed, this name was attributed to J. Gray (1821: 308) and introduced as a new rank in the Subfamily Macropodinae (J. Gray, 1821) and included the genera *Dendrolagus* S. Müller, 1840; *Dorcopsis* Schlegel and Müller, 1845: 130; † *Dorcopsoides* Woodburne, 1967: 43; † *Fissuridon* Bartholomai, 1973b: 365; † *Hadronomas* Woodburne, 1967: 83; *Lagorchestes* Gould, 1841 [1841–1842]; *Lagostrophus* Thomas, 1887c; *Macropus* Shaw, 1790; *Onychogalea* J. Gray, 1841; *Peradorcas* Thomas, 1904a [= *Petrogale* J. Gray, 1837]; *Petrogale* J. Gray, 1837; † *Prionotemnus* Stirton, 1955: 252; † *Protemnodon* Owen, 1873a: 128; *Setonix* Lesson, 1842; † *Synaptodon* De Vis, 1889: 158; *Thylogale* J. Gray, 1837; † *Troposodon* Bartholomai, 1967: 21; † *Wabularoo* M. Archer, 1979c: 299; and *Wallabia* Trouessart, 1905. Rank followed by Marshall *et al.* (1990: 492).

Tribe Macropodini Flannery, 1989: 41.

TYPE GENUS: *Macropus* Shaw, 1790.

COMMENTS: When originally proposed as a new rank it was placed in the Subfamily Macropodinae (J. Gray, 1821) and included the genera *Setonix* Lesson, 1842; *Thylogale* J. Gray, 1837; *Petrogale* J. Gray, 1837; † *Baringa* Flannery and Hann, 1984: 193, 195; *Lagorchestes* Gould, 1841 [1841–1842]; *Onychogalea* J. Gray, 1841; † *Kurrabi* Flannery and Archer, 1984: 357, 364; *Wallabia* Trouessart, 1905; and *Macropus* Shaw, 1790. Tribe recognised by authors including Szalay (1994: 43), Kirsch *et al.* (1997: 246), Kear and Cooke (2001: 84) and Long *et al.* (2002: 164).

Lagorchestes Gould, 1841

Lagorchestes Gould, 1841 [1841–1842]: Text to Plate 12.

TYPE SPECIES: † *Macropus leporides* Gould, 1841e [= † *Lagorchestes leporides* (Gould, 1841e)] by monotypy.

COMMENTS: Recognised as a subgenus of *Macropus* by Waterhouse (1846: 81). Genus recognised by Thomas (1888a: xi, 79), Iredale and Troughton (1934: viii, 39) and subsequent authors.

Lagocheles Owen, 1847a: 330.

TYPE SPECIES: *Nomen nudum*.

COMMENTS: Synonymised within *Lagorchestes* by Iredale and Troughton (1934: 39), Calaby and Richardson (1988b: 61) and Groves (1993e: 52).

† *Lagorchestes asomatus* Finlayson, 1943

Central Hare-wallaby

† *Lagorchestes asomatus* Finlayson, 1943: 319; Plates 33–34.

TYPE LOCALITY: Between Mt Farewell and Lake McKay, Northern Territory, Australia.

COMMENTS: Known only from a single unsexed skull (Kirsch & Calaby, 1977: 22).

Lagorchestes conspicillatus Gould, 1842

Spectacled Hare-wallaby

Lagorchestes conspicillatus conspicillatus Gould, 1842

Lagorchestes conspicillatus Gould, 1842e: 82.

TYPE LOCALITY: Barrow Island, Western Australia, Australia. Type designated by Thomas (1922a: 128).

COMMENTS: Groves (1993e: 52; 2005b: 63) proposed that it might consist of two or three distinct species. Included within *Macropus* (*Lagorchestes*) by Waterhouse (1846: 85) and *Lagorchestes* by Gould (1842 [1841–1842]: Unnumbered Plate), Gould (1860 [1845–1863]: Text to Plate 59), Thomas (1888a: xi, 80) and subsequent authors. This species was collected in July 1997 from the upper Torassi or Bensbach River in the Trans Fly River region in south-west Papua New Guinea (Hitchcock, 1997: 47).

FUTURE TAXONOMIC RESEARCH: This species needs revision. Courtenay (1993: 133, 135) found evidence of the existence of two different taxa on the mainland of tropical Australia, apparently in partial sympatry; neither is identical to the form on Barrow Island.

Lagorchestes conspicillatus leichardti Gould, 1853

Lagorchestes leichardti [sic] Gould, 1853 [1845–1863]: Text to Plate 60.

TYPE LOCALITY: Between Port Essington and Gulf of Carpentaria [= probably Valley of Lagoons, Queensland, Australia (Calaby & Richardson, 1988b: 61)].

COMMENTS: Recognised as a subspecies of *conspicillatus* by Thomas (1888a: xi, 82), Iredale and Troughton (1934: 40), Troughton (1967: 139) and Strahan (1983: 197; 1995: 314), but synonymised within *conspicillatus* by Calaby and Richardson (1988b: 61) and Groves (1993e: 52; 2005b: 63). Recognised as a subspecies by Clayton *et al.* (2006: 106), and Van Dyck and Strahan (2008: 316).

Lagorchestes conspicillatus pallidior Thomas & Dollman, 1909: 793.

TYPE LOCALITY: Inkerman, north Queensland, Australia.

COMMENTS: Recognised as a subspecies of *conspicillatus* by Iredale and Troughton (1934: 40), Tate (1948b: 280) Troughton (1967: 139), and Strahan (1983: 197; 1995: 314), but synonymised within *conspicillatus* by Calaby and Richardson (1988b: 61) and Groves (1993e: 52) and subsequent authors.

Lagorchestes hirsutus Gould, 1844

Rufous Hare-wallaby

† ***Lagorchestes hirsutus hirsutus*** Gould, 1844

† *Lagorchestes hirsutus* Gould, 1844b: 32.

TYPE LOCALITY: York District, Western Australia, Australia.

COMMENTS: Included within *Macropus* (*Lagorchestes*) by Waterhouse (1846: 92) and *Lagorchestes* by Gould (1849 [1845–1863]: Text to Plate 58), Thomas (1888a: xi, 84), which was followed by Thomas (1907b: 775) and subsequent authors. An unnamed subspecies from the central mainland was identified by Courtenay (1993: 95) and Clayton *et al.* (2006: 106).

FUTURE TAXONOMIC RESEARCH: The unnamed subspecies from central Australia identified by Courtenay (1993: 95), and followed by Clayton *et al.* (2006: 106), is very different from the type series from southwestern Western Australia, and from the surviving animals from the Shark Bay islands. A study is needed, using specimens that have become available more recently, to determine whether the surviving populations from Central Australia really are conspecific with *L. hirsutus* from Western Australian localities.

Lagorchestes hirsutus bernieri Thomas, 1907

Lagorchestes hirsutus bernieri Thomas, 1907b: 775.

TYPE LOCALITY: Bernier Island, Shark Bay, Western Australia, Australia.

COMMENTS: Recognised as a subspecies of *hirsutus* by Iredale and Troughton (1934: 40), Troughton (1967: 139) and Strahan (1983: 199; 1995: 316). Taxon synonymised within *hirsutus* by Calaby and Richardson (1988b: 61) and Groves (2005b: 63). The Bernier and Dorre Islands populations were found to have substantially lower genetic diversity than the remnant mainland population by Eldridge *et al.* (2004: 329). Recognised as a subspecies of *hirsutus* by Clayton *et al.* (2006: 106), and Van Dyck and Strahan (2008: 319).

Lagorchestes hirsutus dorreae Thomas, 1907b: 775.

TYPE LOCALITY: Dorre Island, Shark Bay, Western Australia, Australia.

COMMENTS: Recognised as a race of *hirsutus* by Troughton (1967: 139), and as a subspecies by Iredale and Troughton (1934: 40), Strahan (1983: 199; 1995: 316), How *et al.* (2001: 93), who noted unpublished research suggesting that the typically recognised subspecies are only weakly differentiated, Clayton *et al.* (2006: 106), and Van Dyck and Strahan (2008: 319). Synonymised within *hirsutus* by Calaby and Richardson (1988b: 62), Groves (2005b: 63) and Burbidge *et al.* (2014: 21, 29) who noted unpublished research that showed very little difference between animals from Bernier and Dorre Islands, Shark Bay, in Western Australia, suggesting that this taxon should be treated as a junior synonym of the former (*bernieri*).

† ***Lagorchestes leporides*** (Gould, 1841)

Eastern Hare-wallaby

† *Macropus leporides* Gould, 1841e: 93.

TYPE LOCALITY: Interior of Australia [= New South Wales?].

COMMENTS: Included within *Macropus* by Waterhouse (1841a: 204), *Macropus* (*Lagorchestes*) by Waterhouse (1846: 82) and *Lagorchestes* by Gould (1841 [1841–1842]: Unnumbered Plate); Gould, 1859 [1845–1863]: Text to Plate 57), Thomas (1888a: xi, 82), Iredale and Troughton (1934: viii, 40), Ride (1970: 58) and subsequent authors.

† *Lagorchestes leporoïdes* Gould (1841 [1841–1842]: Unnumbered Plate.

TYPE LOCALITY: Unjustified emendation of *leporides*.

COMMENTS: Spelling used by Krefft (1866a: 20), Wood Jones (1924 [1923–1925]: 222), Troughton (1941: 168; 1967: 138) and Finlayson (1958b: 270). Synonymised within *leporides* by McAllan and Bruce (1989: 450).

† *L. [agorchestes] gymnotis* Blyth, 1859a: 276.

TYPE LOCALITY: Unknown.

COMMENTS: Synonymised within *leporides* by Thomas (1888a: 83), Iredale and Troughton (1934: 40) and Calaby and Richardson (1988b: 62).

Macropus Shaw, 1790

Macropus Shaw, 1790: Text to Plate 33.

TYPE SPECIES: *Macropus giganteus* Shaw, 1790 by monotypy. Type species validated by ICZN (1966: 292).

COMMENTS: Van Gelder (1977: 5–6) included *Thylogale* and *Wallabia* within *Macropus*, but Kirsch and Calaby (1977: 17) and all subsequent authors have separated them from the genus. For an alternative taxonomic arrangement see Dawson and Flannery (1985: 473). The genera † *Prionotemnus* Stirton, 1955: 252 and † *Protemnodon* Owen, 1873a: 128 have been recognised as distinct genera, subgenera or synonyms of *Macropus* (e.g. Kirsch & Calaby, 1977: 17, 22; Honacki *et al.*, 1982: 46; Dawson & Flannery, 1985: 476; Calaby & Richardson, 1988b: 64; Groves, 1993e: 53, 2005b: 63), but these are both currently recognised as distinct genera for fossil taxa (e.g. Long *et al.*, 2002: 170). Using sequences from five nuclear genes, Meredith *et al.* (2008c: 399, 405) found *Macropus* to be paraphyletic with respect to *Wallabia*, and they proposed to subsume the latter into *Macropus* as a fourth subgenus (in addition to the subgenera *Macropus*, *Notamacropus* and *Osphranter*). The subgenera of *Macropus* depicted by Meredith *et al.* (2008c: 402) show them separating some 8–9 million years ago, very shortly after the genus itself separated from its closest relatives, *Lagorchestes* and *Setonix*. In contrast to these results using nuclear DNA, subsequent research using mitochondrial DNA by Phillips *et al.* (2013: 1, 2) strongly favoured the topological placement of *bicolor* as sister to *Macropus*. We have indicated above that we urge an objective standard for the recognition of genera, and that the only one that seems readily applicable is time depth; and our preferred time depth for a genus is 4–5 million years. Under these circumstances, each of the three subgenera of *Macropus* (*Macropus*, *Notamacropus* and *Osphranter*) would be split into distinct genera, with *Wallabia* also being recognised as a distinct genus, and this is what has been done here.

FUTURE TAXONOMIC RESEARCH: Further research is required to assess relationships of the *Macropus*, *Notamacropus*, *Osphranter* and *Wallabia* complex.

HOMONYMS:

Macropus Latreille, 1802: 27, crabs of the Subphylum Crustacea (Order Decapoda, Family Inachoidae). Genus is a synonym of *Inachus* Weber, 1795: 93. See Ng *et al.* (2008: 111).

Macropus Thunberg, 1805: 282, long-horned beetles of the Order Insecta (Class Coleoptera, Family Cerambycidae). Genus a synonym of *Macropophora* Thomson, 1864: 15.

Macropus G. Fischer, 1811: 12, galagos of the Class Mammalia (Order Primates, Family Galagidae). Emend. Pro of *Galago* É. Geoffroy, 1796d: 49. See Palmer (1904: 393–394).

Macropus de Spix, 1824: 53, pheasant cuckoos of the Class Aves (Order Cuculiformes, Family Cuculidae). Genus is a synonym of *Morococcyx* Wied-Neuwied, 1832: 351. See J. Peters (1964: 60) and Payne (2005: 187).

Macropus Nuttal, 1834: 450, Eider Ducks of the Class Aves (Order Anseriformes, Family Anatidae). Genus is a synonym of *Polysticta* Eyton, 1838: 55, 150.

Macropus Günther, 1861: 381, paradise fishes of the Superclass Pisces (Order Perciformes, Family Osphronemidae). Genus is an unjustifiable emendation of *Macropodus* Lacépède, 1802a: 416.

Macropus Birula, 1893: 387, arachnids of the Class Arachnida (Order Acarina, Family Erythraeidae). Genus is a synonym of *Eatoniana* Cambridge, 1898: 348.

Gigantomys Link, 1794: 70.

TYPE SPECIES: *Gigantomys canguru* Link, 1794 by monotypy.

COMMENTS: Synonymised within *Macropus* by Iredale and Troughton (1934: 53), Tate (1948b: 321), Marshall (1981: 29), Calaby and Richardson (1988b: 63), Marshall *et al.* (1990: 492) and Groves (1993e: 53; 2005b: 63).

Kangurus É. Geoffroy & G. Cuvier, 1795: 188.

TYPE SPECIES: *Macropus giganteus* Shaw, 1790 by monotypy.

COMMENTS: Name also described by É. Geoffroy (1896c: 106). Synonymised within *Macropus* by Thomas (1888a: 10), Iredale and Troughton (1934: 53), Tate (1948b: 321), Marshall (1981: 29), Calaby and Richardson (1988b: 63), Marshall *et al.* (1990: 492) and Groves (1993e: 53; 2005b: 63).

Kanguroo Lacépède, 1799a: 6.

TYPE SPECIES: *Kanguroo gigas* Lacépède, 1799a [= *Macropus giganteus* Shaw, 1790] by monotypy.

COMMENTS: Synonymised within *Macropus* by J. Gray (1843a: xxii), Iredale and Troughton (1934: 53), Calaby and Richardson (1988b: 63) and Groves (1993e: 53; 2005b: 63).

Halmaturus Illiger, 1811: 80.

TYPE SPECIES: *Nomen novum* for '*Kanguru*' [sic] Lacépède, 1799a and *Kangurus* É. Geoffroy and G. Cuvier, 1795. Palmer (1904: 308) says type is '*Didelphis giganteus* Gmelin [= *Yerboa gigantea* Zimmermann, type]'.

COMMENTS: Recognised as a subgenus of *Macropus* by Waterhouse (1841a: 205; 1846: 94) but it has the same type species. Synonymised within *Macropus* by J. Gray (1843a, xxii), Thomas (1888a: 10), Iredale and Troughton (1934: 54), Simpson (1945: 47), Tate (1948b: 321), Marshall (1981: 29), Calaby and Richardson (1988b: 63), Marshall *et al.* (1990: 492) and Groves (1993e: 53; 2005b: 63).

Kalmaturus Gervais, 1835b: 533.

TYPE SPECIES: *Errore pro Halmaturus* Illiger, 1811.

COMMENTS: Synonymised within *Macropus* by Iredale and Troughton (1934: 54), Calaby and Richardson (1988b: 63) and Groves (1993e: 53; 2005b: 63).

Zèbua J. Gray, 1837: 582.

TYPE SPECIES: *Jaculus giganteus* Erxleben, 1777 (as *Zèbua gigantèa* Erxl.) by monotypy.

COMMENTS: Error only.

Halmatopus Wagner, 1841b: 8.

TYPE SPECIES: *Errore pro Halmaturus* Illiger, 1811.

COMMENTS: Synonymised within *Macropus* by Iredale and Troughton (1934: 54), Calaby and Richardson (1988b: 63) and Groves (1993e: 53; 2005b: 63).

† *Leptosiagon* Owen, 1873c: 386.

TYPE SPECIES: † *Leptosiagon gracilis* Owen, 1874b: 785 [= † *Macropus ferragus* Owen, 1874b: 874] by monotypy.

COMMENTS: Abstract of paper. An expanded description included in Owen (1874b: 783, 785). Suggested to be a *nomen nudum* by Palmer (1904: 373). Synonymised within *Macropus* by Marshall (1981: 29), Marshall *et al.* (1990: 492), and McKenna and Bell (1997: 64).

HOMONYMS:

Leptosiagon Trask, 1856: 99, fossil worms of the Phylum Annelida (Class Polychaeta). *Incertae sedis*.

Macropus fuliginosus (Desmarest, 1817)

Western Grey Kangaroo

Macropus fuliginosus fuliginosus (Desmarest, 1817)

Kangurus fuliginosus Desmarest, 1817d: 35; Plate 22.

TYPE LOCALITY: Kangaroo Island, South Australia, Australia.

COMMENTS: Placed within *Macropus* by Lesson (1827a: 225), Waterhouse (1841a: 200; 1846: 73), Gould (1842 [1841–1842]: Unnumbered Plate; Gould, 1858 [1845–1863]: Text to Plate 5) and as a subspecies of *Macropus giganteus* by Thomas (1888a: xi, 19), who reviewed the early taxonomic history. Recognised as a subspecies within *canguru* [= *giganteus*] by Tate (1948b: 333). The taxonomy of the grey kangaroos was explored using serology by Kirsch and Poole (1967: 1098) who concluded that the greys down the east coast, and in Tasmania, were *Macropus giganteus* while those further west were *Macropus fuliginosus*, within which the animals from Kangaroo Island constituted the first named form. Species rank subsequently recognised by Troughton (1967: 177), and was confirmed by Kirsch and Poole (1972: 315). Subspecific variation reviewed by Poole *et al.* (1990: 159). Species placed into the subgenus *Macropus* by Dawson and Flannery (1985: 482) and followed by Groves (1993e: 54).

FUTURE TAXONOMIC RESEARCH: The craniometric differentiation between *fuliginosus*, *melanops* and *ocydromus* is striking and, on the face of it, absolute (Poole *et al.*, 1990: 159), and gives rise to speculation that they may not be 'merely subspecies'. In contrast a molecular study by Neaves *et al.* (2012: 1558) suggested the Kangaroo Island population is not highly divergent genetically from mainland animals and does not represent a major component of the genetic diversity present within *M. fuliginosus*.

Macropus fuliginosus melanops Gould, 1842

Macropus melanops Gould, 1842b: 10.

TYPE LOCALITY: 'Port Essington' according to the holotype label. It is very like *M. fuliginosus* specimens from southern South Australia, and the real locality appears to be somewhere in South Australia.

COMMENTS: Synonymised within *giganteus* by Waterhouse (1846: 62), and within *M. fuliginosus* by Calaby and Richardson (1988b: 67). Recognised as a subspecies of *giganteus* by Thomas (1888a: xi, 20), with doubt by Iredale and Troughton (1934: ix, 54), Kirsch and Poole (1972: 335), Poole *et al.* (1990: 159), Strahan (1983: 248; 1995: 334), Groves (2005b: 64), Clayton *et al.* (2006: 106), and Van Dyck and Strahan (2008: 334), but not by Burbidge *et al.* (2014: 21, 29).

Macropus ocydromus Gould, 1842f: 1.

TYPE LOCALITY: Swan River, Western Australia, Australia. Type designated by Thomas (1922a: 128).

COMMENTS: Species recognised by Gould (1860 [1845–1863]: Text to Plates 3–4). Synonymised within *giganteus* by Waterhouse (1846: 62) and Thomas (1888a: 17). Recognised as a subspecies within *canguru* [= *giganteus*] by Tate (1948b: 332) and as a species by Iredale and Troughton (1934: ix, 54) and Troughton (1967: 175). Subspecies status within *fuliginosis* recognised by Kirsch and Poole (1972: 335), Strahan (1983: 248; 1995: 334) and Groves (2005b: 64). Poole *et al.* (1990: 159) considered it as much closer to *M. f. melanops* than to *M. f. fuliginosus*, possibly synonymous, and it was not recognised subsequently within *M. fuliginosus* by Clayton *et al.* (2006: 106), Van Dyck and Strahan (2008: 334) or Burbidge *et al.* (2014: 21, 29).

FUTURE TAXONOMIC RESEARCH: As noted above, the craniometric differentiation between *fuliginosus*, *melanops* and *ocydromus* is striking and, on the face of it, absolute (Poole *et al.*, 1990: 159), and gives rise to speculation that they may not be 'merely subspecies'. Nonetheless, the three share Y-chromosome haplotypes (Neaves *et al.*, 2013: 1552, 1558).

Macropus giganteus Shaw, 1790

Eastern Grey Kangaroo

Macropus Giganteus Shaw, 1790: Text to Plate 33.

TYPE LOCALITY: Kings Plains, Cooktown, Queensland, Australia.

COMMENTS: Early taxonomic history reviewed by Thomas (1888a: 15). Opinion 760 of the ICZN (1966: 292) validated the nomenclature presented here (see Calaby *et al.*, 1963: 376 for discussion). Designation of type specimens was made by Calaby *et al.* (1962: 29). Recognised within *Macropus* by Waterhouse (1838a: 66; 1841a: 192; 1846: 62). Proposed to make *Macropus giganteus* the appropriate name by Kirkpatrick and Woods (1964: 250), and Ride and Calaby (1964: 254). Taxonomic decision of Kirsch and Poole (1972: 315) confirmed the split of the grey kangaroos into two species, namely *Macropus giganteus* from *M. fuliginosus*. Placed into the subgenus *Macropus* by Dawson and Flannery (1985: 481) and followed by Groves (1993e: 54).

Kanguroo gigas Lacépède, 1799a: 6.

TYPE LOCALITY: None given; presumably error for *giganteus*.

COMMENTS: Synonymised within *giganteus* by Poole (1982: 1), but not typically discussed by subsequent authors.

Macropus Major Shaw, 1800: 505; Plate 115.

TYPE LOCALITY: Sydney, New South Wales, Australia.

COMMENTS: Recognised at the species rank by Gould (1841 [1841–1842]: no pages number); Gould, 1857 [1845–1863]: Text to Plates 1–2), Krefft (1866a: 19; 1868a: 94), and Iredale and Troughton (1934: ix, 54). Recognised as a subspecies of *canguru* [= *giganteus*] by Tate (1948b: 331). The name *Macropus major* Shaw, 1800 was ruled under the plenary powers of the International Commission on Zoological Nomenclature not to be an objective synonym of *Macropus giganteus* Shaw, 1790, and the two names may have separate type-specimens and type-localities by Opinion 760 of the ICZN (1966: 292) and subsequently added to the Official List of Specific Names (Melville & Smith, 1987: 255). Recognised as a subspecies of *giganteus* by Kirsch and Poole (1972) on p. 335; however on page 336 suggested it should not be recognised. Recognised with possible subspecies by Strahan (1983: 244; 1995: 338) who suggested that it has not been demonstrated as distinct. Synonymised within *Macropus giganteus* by Waterhouse (1846: 62), Thomas (1888a: 16), Ride (1970: 244), Calaby and Richardson (1988b: 67) and Groves (1993e: 54; 2005b: 64).

Dipus tridactylus Perry, 1811 [1810–1811]: Text to Plate 73.

TYPE LOCALITY: New South Wales, Australia.

COMMENTS: Synonymised within *Macropus major* by Iredale and Troughton (1934: 54) and synonymised within *M. giganteus* by Calaby and Richardson (1988b: 67).

Kangurus labiatus Desmarest, 1817d: 33.

TYPE LOCALITY: Vicinity of Botany Bay and Port Jackson, New South Wales, Australia.

COMMENTS: Synonymised within *giganteus* by Thomas (1888a: 16). Synonymised within *Macropus major* by Waterhouse (1846: 62), Iredale and Troughton (1934: 54) and synonymised within *M. giganteus* by Calaby and Richardson (1988b: 67).

Halmaturus griseo-fuscus Goldfuss, 1819b: 266.

TYPE LOCALITY: New South Wales, Australia.

COMMENTS: Synonymised *within giganteus* by Thomas (1888a: 16). Synonymised within *Macropus major* by Iredale and Troughton (1934: 54) and synonymised within *M. giganteus* by Waterhouse (1846: 62) and Calaby and Richardson (1988b: 67).

Macropus giganteus tasmaniensis Le Souef, 1923: 145.

TYPE LOCALITY: Tasmania, Australia.

COMMENTS: Recognised as a distinct species by Iredale and Troughton (1934: ix, 54) and Troughton (1967: 178), and as a subspecies within *canguru* [= *giganteus*] by Tate (1948b: 334). Synonymised within *M. giganteus* by Ride (1970: 244), and Calaby and Richardson (1988b: 67). Recognised as a subspecies within *giganteus* by Kirsch and Poole (1972: 335), Strahan (1983: 244; 1995: 338), Groves (2005b: 64) and Clayton *et al.* (2006: 107), but not by Zenger *et al.* (2003: 160), Van Dyck and Strahan (2008: 337) or Burbidge *et al.* (2014: 21, 29).

Notamacropus Dawson & Flannery, 1985

Notamacropus Dawson & Flannery, 1985: 473, 489.

TYPE SPECIES: *Halmaturus agilis* Gould, 1842e [= *Notamacropus agilis* (Gould, 1842e)] by original designation.

COMMENTS: Proposed as a subgenus of *Macropus*. Synonymised within *Macropus* by Calaby and Richardson (1988b: 64), Groves (1993e: 53; 2005b: 63) and McKenna and Bell (1997: 64). See discussion under *Macropus* for the elevation of this taxon to generic rank.

Notamacropus agilis (Gould, 1842)

Agile Wallaby

Notamacropus agilis agilis (Gould, 1842)

Halmaturus agilis Gould, 1842e: 81.

TYPE LOCALITY: Port Essington, Northern Territory, Australia. Type designated by Thomas (1922a: 128).

COMMENTS: Species further described by Gould (1842 [1841–1842]: Unnumbered Plate; Gould, 1857 [1845–1863]: Text to Plate 24–25). Placed in *Macropus (Halmaturus)* by Waterhouse (1846: 108) and Thomas (1888a: xi, 42), who reviewed its early taxonomic history. Included in the genus *Wallabia* as a distinct species by Iredale and Troughton (1934: viii, 47) and Troughton (1967: 169), and within *Protemnodon* by Tate (1948b: 301). Included in the genus *Macropus* by Finlayson (1931: 71), Ride (1970: 46), Honacki *et al.* (1982: 46), Calaby and Richardson (1988b: 64). Placed in the subgenus *Notamacropus* by Dawson and Flannery (1985: 473, 489) and followed by Groves (1993e: 53; 2005b: 63).

Halmaturus Binoë Gould, 1842c: 58.

TYPE LOCALITY: Port Essington, Northern Territory, Australia.

COMMENTS: Recognised as a subspecies of *eugenii* (within *Thylogale*) by Iredale and Troughton (1934: 47) and Tate (1948b: 303). Synonymised within *agilis* by Waterhouse (1846: 108), Thomas (1888a: 42), Calaby and Richardson (1988b: 64), Groves (1993e: 53; 2005b: 63) and Flannery (1990: 114; 1995a: 154; 1995b: 81).

Halmaturus siva De Vis, 1895: 113.

TYPE LOCALITY: Darling Downs, Queensland, Australia. Described from a fossil specimen.

COMMENTS: Subspecies rank recognised by Flannery (1990: 114; 1995a: 154; 1995b: 81). Synonymised within *agilis* by Groves (1993e: 53) but not considered by Groves (2005b: 63).

Macropus agilis aurescens Schwarz, 1910a: 166.

TYPE LOCALITY: Fitzroy River, Western Australia, Australia.

COMMENTS: Recognised as a subspecies of *agilis*, within *Wallabia*, by Iredale and Troughton (1934: 48) and Troughton (1967: 170). Synonymised within *agilis* by Calaby and Richardson (1988b: 64), Groves (1993e: 53; 2005b: 63) and Flannery (1990: 114; 1995a: 154; 1995b: 81).

FUTURE TAXONOMIC RESEARCH: There has been no recent revision of geographic variation within this species; all populations appear to be closely related, but this cannot be assumed.

Notamacropus agilis jardinii (De Vis, 1884)

Halmaturus Jardinii De Vis, 1884a: 109.

TYPE LOCALITY: Cape York, Queensland, Australia.

COMMENTS: Erroneously spelt *Halmaturus jardinei* in Schwarz (1910a: 165), who recognised it as a subspecies of *agilis*. Included as a subspecies of *agilis* by Iredale and Troughton (1934: 48), Tate (1948b: 302), Troughton (1967: 170) and Groves (2005b: 63). Synonymised within *agilis* by Thomas (1888a: 42), Calaby and Richardson (1988b: 64), Groves (1993e: 53), Clayton *et al.* (2006: 106) and Burbidge *et al.* (2014: 21, 29). Subspecies rank within *Protemnodon agilis*, by Tate (1948b: 301) and within *Macropus agilis* recognised by Strahan (1983: 242; 1995: 323), Flannery (1990: 114; 1995a: 154; 1995b: 81), Groves (2005b: 63), and Van Dyck and Strahan (2008: 323).

Φ *Notamacropus agilis papuanus* (Peters & Doria, 1875)

Φ *Macropus papuanus* Peters & Doria, 1875: 544.

TYPE LOCALITY: Eastern mainland (Papua) New Guinea, opposite Yule Island.

COMMENTS: Synonymised within *agilis* by Thomas (1888a: 42) and Groves (1993e: 53). Recognised as a subspecies of *Protemnodon agilis* by Tate (1948b: 302) and Laurie and Hill (1954: 27), and within *Macropus agilis* by Strahan (1983: 242; 1995: 323), Flannery (1990: 114; 1995a: 154; 1995b: 81) and Groves (2005b: 63).

Φ *Macropus papuensis* P. Sclater, 1876: 532, footnote.

COMMENTS: Errore pro *Macropus papuanus* Peters and Doria, 1875. Synonymised within *agilis* by Flannery (1990: 114; 1995a: 154; 1995b: 81) and within *papuanus* by Groves (2005b: 63).

Φ *Halmaturus crassipes* Ramsay, 1876b: 162.

TYPE LOCALITY: Southern New Guinea.

COMMENTS: Synonymised within *agilis* by Thomas (1888a: 42), Groves (1993e: 53) and Flannery (1990: 114; 1995a: 154; 1995b: 81), and within *papuanus* by Laurie and Hill (1954: 27) and Groves (2005b: 63).

Φ *Dorcopsis* (?) *aurantiacus* Rothschild & Rothschild, 1898: 513.

TYPE LOCALITY: New Guinea.

COMMENTS: Synonymised within *agilis*, in the genus *Wallabia*, by Iredale and Troughton (1934: 47). Synonymised within *agilis* by Calaby and Richardson (1988b: 64), Groves (1993e: 53) and Flannery (1990: 114; 1995a: 154; 1995b: 81). Synonymised within *papuanus* by Laurie and Hill (1954: 27) and Groves (2005b: 63),

Notamacropus agilis nigrescens (Lönnberg, 1913)

Macropus agilis nigrescens Lönnberg, 1913: 8.

TYPE LOCALITY: Broome, Western Australia, Australia.

COMMENTS: Synonymised within *agilis* by Calaby and Richardson (1988b: 64), Groves (1993e: 53), Clayton *et al.* (2006: 106) and Burbidge *et al.* (2014: 21, 29). Subspecies rank recognised, within *Protemnodon agilis*, by Tate (1948b: 301). Recognised as a subspecies of *agilis*, within *Wallabia*, by Iredale and Troughton (1934: 48), Troughton (1967: 169), and within *Macropus* by Strahan (1983: 242; 1995: 323), Flannery (1990: 114; 1995a: 154; 1995b: 81), Groves (2005b: 63), and Van Dyck and Strahan (2008: 323).

Notamacropus dorsalis (J. Gray, 1837)

Black-striped Wallaby

Halmatùrus dorsàlis J. Gray, 1837: 583.

TYPE LOCALITY: Unknown. [= Namoi Hills, *fide* Iredale & Troughton, 1934: 50].

COMMENTS: Species recognised within *Halmaturus* by Gould (1841 [1841–1842]: no pages number; Gould, 1857 [1845–1863]: Text to Plates 26–27), within *Wallabia* by Iredale and Troughton (1934: viii, 50) and Troughton (1967: 166), and within *Protemnodon* by Tate (1948b: 311).

Included in *Macropus* by Waterhouse (1838a: 67; 1841a: 230; 1846: 152), Thomas (1888a: xi, 37), Finlayson (1931: 72), Ride (1970: 47), Honacki *et al.* (1982: 46), and Calaby and Richardson (1988b: 65). Early taxonomic history reviewed by Thomas (1888a: 37). Placed into the subgenus *Notamacropus* by Dawson and Flannery (1985: 473, 491) and followed by Groves (1993e: 53).

Notamacropus eugenii (Desmarest, 1817)

Tammar Wallaby

Notamacropus eugenii eugenii (Desmarest, 1817)

† *Kangurus Eugenii* Desmarest, 1817d: 38.

TYPE LOCALITY: St. Peters Island, Nuyt's Archipelago, South Australia, Australia (as L'île Eugène, Josephine Archipelago). Population is now extinct.

COMMENTS: Included in the genus *Thylogale* by Iredale and Troughton (1934: viii, 46) and Troughton (1967: 158). Species placed within *Protemnodon* by Tate (1948b: 303). Included within *Macropus* by Lesson (1827a: 227), Waterhouse (1838a: 66; 1841a: 232; 1846: 140), Thomas (1888a: xi, 54), Ride (1970: 48), Honacki *et al.* (1982: 46), and Calaby and Richardson (1988b: 65). Placed into the subgenus *Notamacropus* by Dawson and Flannery (1985: 473, 490) and followed by Groves (1993e: 53). Taxonomic history reviewed by Thomas (1888a: 54). Morphometric research supports three groupings: 1) Western Australia including islands, 2) Kangaroo Island, and 3) New Zealand (introduced population) that appears to be most closely related to extinct South Australian mainland animals (Poole *et al.*, 1991: 625). The samples studied by them were very strongly differentiated, and it may be that further study, with more complete geographic representation as well as molecular sequences, would separate them at full species level (M. Eldridge, pers. comm.). A comparison of microsatellite DNA was made between animals from Kangaroo Island and Kawau Island, New Zealand, by A. Taylor and Cooper (1999: 41, 47) who concluded that these two populations were distinct and that the Kawau Island animals represented an undescribed taxon extinct in Australia.

HOMONYMS:

Halmaturus eugenii J. Gray, 1837, the Red-necked Pademelon of the Class Mammalia (Order Diprotodontia, Family Macropodidae). Name is a junior synonym of *Thylogale thetis* (Lesson, 1827a). See individual entry.

FUTURE TAXONOMIC RESEARCH: Research is required to assess the distinctiveness of each population of this species in order to determine the validity of potential species or subspecies including *derbianus*, *decres* and a possibly unnamed taxon formerly from mainland South Australia.

Thylogale eugenii decres Troughton, 1941: 194.

TYPE LOCALITY: Kangaroo Island, South Australia, Australia.

COMMENTS: Recognised as a subspecies of *eugenii* by Tate (1948b: 304), Troughton (1967: 159) and Maxwell *et al.* (1996: 4) and Clayton *et al.* (2006: 106). Taxon synonymised within *eugenii* by Calaby and Richardson (1988b: 66) and Burbidge *et al.* (2014: 21, 29).

Notamacropus eugenii derbianus (J. Gray, 1837)

Halmatùrus Derbiànus J. Gray, 1837: 583.

TYPE LOCALITY: Western Australia, Australia.

COMMENTS: Recognised as a species within *Macropus* by Waterhouse (1838a: 67; 1841a: 234; 1846: 154) and within *Halmaturus* by Gould (1841 [1845–1863]: Unnumbered Plate; Gould, 1859 [1845–1863]: Text to Plate 29–30). Designated a subspecies of *eugenii*, within *Thylogale*, by Iredale and Troughton (1934: 46). Recognised as subspecies within *eugenii* by Tate (1948b: 303) and Maxwell *et al.* (1996: 4). Taxon synonymised within *eugenii* by Thomas (1888a: 54), and Calaby and Richardson (1988b: 65). Recognised as a subspecies of *eugenii* by Clayton *et al.* (2006: 106) but not other authors.

Hal. [maturus] Derbianus Var. obscurior J. Gray, 1841: 403.

TYPE LOCALITY: Rottnest Island [sic] and Gardens Islands, Western Australia, Australia.

COMMENTS: *Nomen nudum*, proposed as a variety. Synonymised within *derbianus* by Iredale and Troughton (1934: 46). Synonymised within *eugenii* by Calaby and Richardson (1988b: 65).

Halmaturus Emiliae J. Gray, 1843a: 90.

TYPE LOCALITY: Houtman's Abrolhos, Western Australia, Australia.

COMMENTS: *Nomen nudum* (see Thomas, 1888a: 54). Synonymised with *binoe* (within *Thylogale*) by Iredale and Troughton (1934: 47) and Tate (1948b: 303). Taxon synonymised within *Macropus eugenii* by Thomas (1888a: 54) and Calaby and Richardson (1988b: 66).

Halmaturus Houtmannii Gould, 1844b: 31.

TYPE LOCALITY: Houtman Abrolhos, Western Australia, Australia. Type designated by Thomas (1922a: 128).

COMMENTS: Synonymised with *binoe* (within *Thylogale*) by Iredale and Troughton (1934: 47) and Tate (1948b: 303). Synonymised within *eugenii* (within *Macropus*) by Waterhouse (1846: 154), Thomas (1888a: 54) and Calaby and Richardson (1988b: 66).

Halmaturus Dama Gould, 1844b: 32.

TYPE LOCALITY: Unknown.

COMMENTS: Synonymised within *Thylogale eugenii binoe* (Gould, 1842c) by Iredale and Troughton (1934: 47), and within *Macropus eugenii* by Waterhouse (1846: 140), Thomas (1888a: 54) and Calaby and Richardson (1988b: 66).

Macropus gracilis Gould, 1844a: 103.

TYPE LOCALITY: Walyema Swamps, 40 miles NE of Northam, Western Australia, Australia [= Lake Walyormouring, Western Australia].

COMMENTS: Synonymised within *dama* by Waterhouse (1846: 141, 143), *eugenii derbianus* by Iredale and Troughton (1934: 46) and Tate (1948b: 303), and within *eugenii* by Thomas (1888a: 54) and Calaby and Richardson (1988b: 66).

Thylogale bedfordi Thomas, 1900a: 112.

TYPE LOCALITY: 'Queensland or north Australia'.

COMMENTS: Included in the genus *Thylogale* as full species by Iredale and Troughton (1934: viii, 47) and in *Protemnodon* as a full species by Tate (1948b: 305). Synonymised within *eugenii* by Calaby and Richardson (1988b: 66).

Thylogale flindersi Wood Jones, 1924: 12.

TYPE LOCALITY: Flinders Island, Investigator Group, South Australia, Australia.

COMMENTS: Included in the genus *Thylogale* as full species by Iredale and Troughton (1934: viii, 46) and Troughton (1967: 159). Synonymised within *eugenii* by Ride (1970: 244), Strahan (1983: 232; 1995: 329) and Calaby and Richardson (1988b: 66).

† ***Notamacropus greyi*** (Waterhouse, 1846)

Toolache Wallaby

† *Macropus (Halmaturus) greyi* Waterhouse, 1846: 122.

TYPE LOCALITY: Coorong [= South Australia, Australia]. Type designated by Thomas (1922a: 128).

COMMENTS: Included in *Macropus* by Thomas (1888a: xi, 36), Ride (1970: 47) and Strahan (1983: 234). Species recognised within *Halmaturus* by Gould (1852 [1845–1863]: Text to Plate 18-19) and within *Wallabia* by Iredale and Troughton (1934: ix, 50) and Troughton (1967: 166). Reduced to a subspecies of *irma* by Tate (1948b: 307), but as a species within *Macropus* by all other authors. Placed into the subgenus *Wallabia*, within *Macropus*, by Finlayson (1927: 366) and *Notamacropus* by Dawson and Flannery (1985: 473, 491) and followed by Groves (1993e: 54).

† *Halmaturus Greyii* J. Gray, 1843a: 90.

TYPE LOCALITY: *Nomen nudum*.

COMMENTS: Not considered by Iredale and Troughton (1934: 50). Taxon synonymised within *greyi* by Calaby and Richardson (1988b: 68).

Notamacropus irma (Jourdan, 1837)

Western Brush Wallaby

Halmaturus irma Jourdan, 1837a: 523.

TYPE LOCALITY: Swan River, Western Australia, Australia.

COMMENTS: Placed in *Macropus (Halmaturus)* by Waterhouse (1841a: 222; 1846: 117) and followed by Thomas (1888a: xi, 40) who reviewed its early taxonomic history. Recognised within *Wallabia* by Iredale and Troughton (1934: ix, 50) and Troughton (1967: 168). Synonymised within *manicatus* by J. Gray (1841: 402). Transferred to *Macropus* and Ride (1970: 47) and followed by subsequent authors. Placed into the subgenus *Notamacropus* by Dawson and Flannery (1985: 473, 491) and followed by Groves (1993e: 54), but unexpectedly placed in the subgenus *Osphranter* by Phillips (2013: 1).

Macropus melanopus Gould, 1840c: 876.

TYPE LOCALITY: Unknown.

COMMENTS: The history of this species was discussed by McAllan and Bruce (1989: 450). Name not typically discussed by subsequent authors until McAllan and Bruce (1989: 450) placed it within *irma*.

Macropus (Halmaturus) manicatus Gould, 1841d: 127.

TYPE LOCALITY: Swan River, Western Australia, Australia.

COMMENTS: Species recognised within *Halmaturus* by Gould (1841 [1841–1842]: Unnumbered Plate) and J. Gray (1841: 402) and within *Macropus* by Waterhouse (1841a: 223), but transferred back to *Halmaturus* by Gould (1852 [1845–1863]: Text to Plates 20–21). Synonymised within *irma* by Iredale and Troughton (1934: 50), Tate (1948b: 306) and Calaby and Richardson (1988b: 68).

Notamacropus parma (Waterhouse, 1846)

Parma Wallaby

Macropus (Halmaturus) parma Waterhouse, 1846: 149; Plate 5, Fig. 7.

TYPE LOCALITY: New South Wales, Australia.

COMMENTS: Taxon included in the genus *Halmaturus* by Gould (1856 [1845–1863]: Text to Pate 28), and within *Macropus* by Thomas (1888a: xi, 57). Placed in *Thylogale* by Iredale and Troughton (1934: viii, 46) and *Protemnodon* by Tate (1948b: 305), which was followed by Ride (1957: 327). Transferred to *Wallabia* by Troughton (1967: 165) and then into *Macropus* by Ride (1970: 48) and followed by subsequent authors. Placed into the subgenus *Notamacropus* by Dawson and Flannery (1985: 473, 491) and followed by Groves (1993e: 54).

Hal. [maturus] Parma J. Gray, 1841: 403.

TYPE LOCALITY: Unknown. [= New South Wales?, Australia].

COMMENTS: *Nomen nudum* (see Thomas, 1888a: 57). Synonymised within *parma* by Thomas (1888a: 57).

Halmaturus Parma J. Gray, 1843a: 91.

TYPE LOCALITY: Unknown. New South Wales, Australia.

COMMENTS: *Nomen nudum* (see Thomas, 1888a: 57). Says '*Halmaturus Parma*, Gould, P.Z.S.' Gould (in J. Gray, 1843a: 91) was recognised as the correct citation by Waterhouse (1846: 149) who states on page 150 that Mr Gould informed him that he had doubts whether he had described it in the *Proceedings of the Zoological Society of London*. Synonymised within *Macropus (Halmaturus) parma* Waterhouse 1846: 149 by Iredale and Troughton (1934: 46) and Calaby and Richardson (1988b: 69).

Notamacropus parryi (Bennett, 1835)

Whip-tailed Wallaby

Macropus parryi E. Bennett, 1835: 151; Plate 27.

TYPE LOCALITY: Stroud, New South Wales, Australia.

COMMENTS: Species recognised within *Halmaturus* by Gould (1842 [1841–1842]: Unnumbered Plate) and *Macropus* by Waterhouse (1838a: 66; 1841a: 206; 1846: 113), Thomas (1888a: xi, 39), who reviewed its early taxonomic history, and Finlayson (1931: 75). Placed within *Osphranter* by Gould (1852 [1845–1863]: Text to Plates 12–13). Taxon synonymised within *Wallabia elegans* (Lambert, 1807: 318) by Iredale and Troughton (1934: 50). Included within *Protemnodon* by Haltenorth (1958: 39), as a subspecies of *canguru* and Tate (1948b: 308), and within *Macropus* by Ride (1970: 47), Kirsch and Calaby (1977: 17), Strahan (1983: 236; 1995: 344), and Calaby and Richardson (1988b: 69). Included in the subgenus *Notamacropus* by Dawson and Flannery (1985: 473, 490) and followed by Groves (1993e: 54).

[Halmatùrus Párryi] Var. pállida J. Gray, 1837: 583.

TYPE LOCALITY: 'Swan River Wallaroo' (error=New South Wales, Australia), See Iredale and Troughton (1934: 50).

COMMENTS: Taxon synonymised within *Wallabia elegans* (Lambert, 1807: 318) by Iredale and Troughton (1934: 50). Synonymised within *Macropus parryi* by Calaby and Richardson (1988b: 69) and Groves (1993e: 54; 2005b: 65).

Notamacropus rufogriseus (Desmarest, 1817)

Red-necked Wallaby

Notamacropus rufogriseus rufogriseus (Desmarest, 1817)

Kangurus rufogriseus Desmarest, 1817d: 36.

TYPE LOCALITY: King Island, Bass Strait, Australia, Australia.

COMMENTS: Placed within *Macropus* by Lesson (1827a: 226), Waterhouse (1841a: 217), within *Wallabia* by Iredale and Troughton (1934: viii, 49) and Troughton (1967: 164), and within *Protemnodon* by Tate (1948b: 309). Taxon synonymised within *ruficollis* by Thomas (1888a: 32). Recognised as a species within *Macropus* by Ride (1970: 46), Honacki *et al.* (1982: 47), Strahan (1983: 239), Calaby and Richardson (1988b: 70) and subsequent authors. Placed into the subgenus *Wallabia*, within *Macropus*, by Finlayson (1930: 47) and *Notamacropus* by Dawson and Flannery (1985: 473, 490) and followed by Groves (1993e: 55; 2005b: 65).

Halmaturus rutilans Illiger, 1815: 102.

COMMENTS: *nomen nudum* (see Thomas 1888a: 16). Taxon synonymised within *rufogriseus* by Calaby and Richardson (1988b: 70) and Groves (2005b: 65).

Halmaturus Kingii Illiger, 1815: 102.

COMMENTS: *nomen nudum*. Synonymised within *ruficollis* by Thomas (1888a: 32) and *rufogriseus* (in *Wallabia*) by Iredale and Troughton (1934: 49). Taxon synonymised within *rufogriseus* (in *Macropus*) by Calaby and Richardson (1988b: 70) and Groves (2005b: 65).

Kangurus ruficollis Desmarest, 1817d: 37.

TYPE LOCALITY: King Island, Bass Strait, Tasmania, Australia.

COMMENTS: Recognised as a species within *Macropus* by Lesson (1827a: 226); Waterhouse (1838a: 66; 1841a: 216; 1846: 125) and Thomas (1888a: xi, 32). Synonymised within *rufogriseus* (in *Wallabia*) by Iredale and Troughton (1934: 49). Recognised within *Halmaturus* by Gould (1842 [1841–1842]: Unnumbered Plate; Gould, 1854 [1845–1863]: Text to Plates 14–15). Synonymised within *rufogriseus* (in *Macropus*) by Calaby and Richardson (1988b: 70) and Groves (1993e: 55).

Halmaturus rutilus M. Lichtenstein, 1818: 17.

TYPE LOCALITY: Unknown.

COMMENTS: *nomen nudum*. Synonymised within *rufogriseus* by Calaby and Richardson (1988b: 70) and Groves (1993e: 55; 2005b: 65).

H. [almaturus] griseo-rufus Goldfuss, 1819b: 267.

TYPE LOCALITY: Australia.

COMMENTS: Synonymised within *rufogriseus* (in *Wallabia*) by Iredale and Troughton (1934: 49). Synonymised within *rufogriseus* (in *Macropus*) by Calaby and Richardson (1988b: 71) and Groves (2005b: 65).

K. [angurus] Griseus J. Gray, 1827a: 202.

TYPE LOCALITY: New Holland [= Australia].

COMMENTS: Taxon synonymised within *ruficollis* by Thomas (1888a: 33). Synonymised within *rufogriseus* (in *Wallabia*) by Iredale and Troughton (1934: 49) and within *rufogriseus* (in *Macropus*) by Calaby and Richardson (1988b: 71) and Groves (1993e: 55; 2005b: 65).

Macropus (Halmaturus) fruticus W. Ogilby, 1838c: 219.

TYPE LOCALITY: Tasmania, Australia.

COMMENTS: Recognised as a subspecies of *rufogriseus* by Iredale and Troughton (1934: 49), Tate (1948b: 310) and Troughton (1967: 165). Subspecies status recognised within *rufogriseus* by Groves (2005b: 65). Synonymised within *rufogriseus* by Strahan (1983: 239; 1995: 351), Calaby and Richardson (1988b: 71), Groves (1993e: 55).

Kangurus vinosus Boitard, 1842: 209.

TYPE LOCALITY: King Island, Tasmania, Australia. See Iredale and Troughton (1934: 49).

COMMENTS: Synonymised within *rufogriseus* (in *Wallabia*) by Iredale and Troughton (1934: 49) and within *rufogriseus* (in *Macropus*) by Calaby and Richardson (1988b: 71) and Groves (1993e: 55; 2005b: 65).

H. [almaturus] leptonyx Wagner, 1843: 116.

TYPE LOCALITY: Unknown.

COMMENTS: Recognised as a species within *Macropus (Halmaturus)* by Waterhouse (1846: 134). Taxon synonymised within Macropus *ruficollis* var. *bennettii* by Thomas (1888a: 35) and within *Wallabia rufogriseus fruticus* by Iredale and Troughton (1934: 49). Taxon synonymised within *Macropus rufogriseus* by Calaby and Richardson (1988b: 71) and Groves (1993e: 55). Groves (2005b: 65) gave this taxon an 'unassigned' status.

Notamacropus rufogriseus banksianus (Quoy & Gaimard, 1825)

Kangurus banksianus Quoy & Gaimard, 1825: 481, footnote.

TYPE LOCALITY: Blue Mountains, New South Wales, Australia.

COMMENTS: Synonymised within *ruficollis* by Thomas (1888a: 33). Recognised as a subspecies of *rufogriseus* by Iredale and Troughton (1934: 49) and Tate (1948b: 309). Taxon synonymised within *rufogriseus* by Calaby and Richardson (1988b: 71) and Groves (1993e: 55), but recognised as a subspecies of *rufogriseus* by Strahan (1983: 239; 1995: 351), Groves (2005b: 65) and Clayton *et al.* (2006: 107).

Macropus Bennetti Waterhouse, 1838e: 103.

TYPE LOCALITY: New South Wales, Australia (error=Tasmania, Australia)

COMMENTS: Species rank recognised within *Halmaturus* by Gould (1841 [1841–1842]: no pages number; Gould, 1856 [1845–1863]: Text to Plate 16–17) and Krefft (1868a: 94), and within *Macropus* by Waterhouse (1838a: 66; 1841a: 211). Subspecies rank within *M. ruficollis* recognised by Waterhouse (1846: 130) and Thomas (1888a: xi, 34) who reviewed its taxonomic history. Reduced to a synonym of *fruticus* by Iredale and Troughton (1934: 49) and Tate (1948b: 310). Synonymised within *rufogriseus* (in *Macropus*) by Calaby and Richardson (1988b: 71) and Groves (1993e: 55) and within *fruticus* by Groves (2005b: 65).

Onychogalea J. Gray, 1841

Onychogalea J. Gray, 1841: 402.

TYPE SPECIES: *Macropus unguifer* Gould, 1841e [= *Onychogalea unguifera* (Gould, 1840a)] by subsequent designation. See Thomas (1888a: 73).

COMMENTS: Described as a subgenus of *Macropus* and recognised at this rank by Waterhouse (1846: 75). Elevated to generic rank by J. Gray (1843a: xxii, 88), who misspelt it *Onichogalea*, and most subsequent authors. Groves (1993e: 55) attributed the citation to J. Gray (1841: 402). Simpson (1945: 47) also stated that it is generally but incorrectly written as *Onychogale*.

Onichogalea J. Gray, 1843a: xxii, 88.

TYPE SPECIES: Incorrect subsequent spelling of *Onychogalea* J. Gray, 1841.

COMMENTS: Synonymised with *Onychogalea* by Calaby and Richardson (1988b: 73).

Onychogale Thomas, 1888a: xi, 73.

TYPE SPECIES: Incorrect subsequent spelling of *Onychogalea* J. Gray, 1841.

COMMENTS: Spelling used by several authors including J. Ogilby (1892: 47), and Le Souef and Burrell (1926: 13, 210). Synonymised with *Onychogalea* by Calaby and Richardson (1988b: 73).

HOMONYMS:

Onychogale J. Gray, 1865c: 570, mongoose of the Class Mammalia (Order Carnivora, Family Herpestidae). Genus is a synonym of *Herpestes* Illiger, 1811: 135. See Wozencraft (2005: 567).

Onychogalea frenata (Gould, 1840)

Bridled Nail-tailed Wallaby

[Macropus] Frenatus Gould, 1840a: 685.

TYPE LOCALITY: Interior of New South Wales, Australia.

COMMENTS: McAllan and Bruce (1989: 449) argued that the spelling *frenatus* (published on 29 August 1840) should be used in preference to *fraenatus* (published in April 1841) and proposed that they act as first revisers in giving priority to the publication of Gould (1840a: 685). McAllan and Bruce (1989: 449) suggested that Gould's (1841e: 92) emendation of *frenatus* to *fraenatus* is 'unjustified' (citing Art. 33(b), ICZN, 1985a), and noted that the spelling *frenatus* (or *frenata* when used in combination with *Onychogalea* or *Onychogale*) has been used within the last 50 years by authors. The spelling *frenata* has been used by J. Gray (1841: 402; 1843a: 88), Thomas (1888a: xi, 75), J. Ogilby (1892: 47), Le Souef and Burrell (1926: 13, 212), Troughton (1945a: 347), Frith and Calaby (1969: 22) and Hyett and Shaw (1980: 120). As a result of these points the name *fraenata* was proposed to be a junior objective synonym of *frenata* by McAllan and Bruce (1989: 449), but this was not adopted by Groves (1993e: 55; 2005b: 66) or subsequent authors who have recognised the spelling as *fraenata*.

Macropus fraenatus Gould, 1841e: 92.

TYPE LOCALITY: Interior of New South Wales, Australia.

COMMENTS: Taxon further described by Gould (1841 [1841–1842]: no pages number). Included within *Macropus* by Waterhouse (1841a: 202 and in the subgenus *Onychogalea* by Waterhouse (1946: 77). Placed in the genus *Onychogalea* by J. Gray (1843a: 88) (as *Onichogalea*), Gould (1849 [1845–1863]: Text to Plate 54), Thomas (1888a: xi, 75) and subsequent authors. The species name for this taxon has been spelt *fraenatus* (or *fraenata* when used in combination with *Onychogalea*) by many authors including Iredale and Troughton (1934: 41), Troughton (1941: 186; 1967: 152), Ride (1970: 54), Honacki *et al.* (1982: 47), Strahan (1983: 205; 1995: 356), Caughley *et al.* (1987: 32), Calaby and Richardson (1988b: 73), Groves (1993e: 55; 2005b: 66) and Van Dyck and Strahan (2008: 355). This species was thought to be extinct until it was rediscovered in 1973 (G. Gordon & Lawrie, 1980: 339).

† *Onychogalea lunata* (Gould, 1840)

Crescent Nail-tailed Wallaby

† *Macropus lunatus* Gould, 1840a: 685.

TYPE LOCALITY: West coast of Australia.

COMMENTS: Name also described by Gould (1840b: 561), which was also published on 29 August 1840. McAllan and Bruce (1989: 449) proposed that they act as first revisers in giving priority to the publication of Gould (1840a: 685). Taxon further described by Gould (1841 [1841–1842]: no pages number). Included within *Macropus* by Waterhouse (1841a: 203) and in the subgenus *Onychogalea* by Waterhouse (1946: 79). Placed in the genus *Onichogalea* by J. Gray (1843a: 88), *Onychogalea* by Gould (1849 [1845–1863]: Text to Plate 55), and *Onychogale* by Thomas (1888a: xi, 77), who reviewed the species' taxonomic history. Placed in *Onychogalea* by Iredale and Troughton (1934: viii, 41), Tate (1948b: 278), and followed by most subsequent authors.

Onychogalea unguifera (Gould, 1840)

Northern Nail-tailed Wallaby

Onychogalea unguifera unguifera (Gould, 1840)

[Macropus] Unguifer Gould, 1840a: 685.

TYPE LOCALITY: North west coast of Australia.

COMMENTS: Name also described by Gould (1840b: 561), which was also published on 29 August 1840. McAllan and Bruce (1989: 449) proposed that they act as first revisers in giving priority to the publication of Gould (1840a: 685). Taxon further described by Gould (1841e: 93; 1841 [1841–1842]: no pages number). Included within *Macropus* by Waterhouse (1841a: 201) and in subgenus *Onychogalea* by Waterhouse (1946: 75). Placed in the genus *Onichogalea* by J. Gray (1843a: 88), *Onychogalea* by Gould (1849 [1845–1863]: Text to Plates 52–53), and *Onychogale* by Thomas (1888a: xi, 74). Specific name has been spelt *unguifer* by several authors including Tate (1948b: 276) and Troughton (1967: 153). Placed in *Onychogalea* by Iredale and Troughton (1934: viii, 41), Tate (1948b: 276), and followed by most subsequent authors.

Onychogalea unguifera annulicauda De Vis, 1884

Onychogalea annulicauda De Vis, 1884b: 157.

TYPE LOCALITY: Kimberley, Norman River, Gulf of Carpentaria, Queensland, Australia.

COMMENTS: Taxon synonymised within *unguifera* by Thomas (1888a: 74). Recognised as a subspecies of *unguifera* by Iredale and Troughton (1934: 41), Troughton (1967: 153), Strahan (1983: 204; 1995: 362), Clayton *et al.* (2006: 107) and subsequent authors.

FUTURE TAXONOMIC RESEARCH: The validity of the subspecies ascribed to this species needs to be examined. Interrelationships between the three species of the genus are poorly understood, although the monophyly of the two living species was corroborated by Meredith *et al.* (2008c: 399).

Osphranter Gould, 1842

Osphranter Gould, 1842e: 80.

TYPE SPECIES: *Osphranter antilopinus* Gould, 1842e by monotypy.

COMMENTS: Synonymised within *Macropus* Shaw, 1790 by Thomas (1888a: 10). Recognised at generic rank by Iredale and Troughton (1934: ix, 50), D. Johnson (1964: 460) and Troughton (1967: 180). Synonymised within *Macropus* by Simpson (1945: 47), Ride (1970: 245), Marshall (1981: 29), Calaby and Richardson (1988b: 63), Marshall *et al.* (1990: 492) and Groves (1993e: 53; 2005b: 63). Recognised as a subgenus of *Macropus* by Haltenorth (1958: 39), Tate (1948b: 325) and Dawson and Flannery (1985: 473, 485). See discussion under *Macropus* for the elevation of this taxon to generic rank.

Megaleia Gistel, 1848: ix.

TYPE SPECIES: *Kangurus laniger* Gaimard, 1823 [= *Osphranter rufus* (Desmarest, 1822b)] by monotypy.

COMMENTS: Described as a subgenus of *Halmaturus*. Recognised as valid genus (for *M. rufus*) by Iredale and Troughton (1934: ix, 52), Sharman (1961: 38), Troughton (1967: 178), Ride (1970: 44) and McKenna and Bell (1997: 64). Tate (1948b: 334) recognised *Megaleia* as a subgenus of *Macropus* for *rufus*, as did Haltenorth (1958: 39). Synonymised within *Macropus* by Simpson (1945: 47), Kirsch (1977a: 60), Kirsch and Calaby (1977: 17, 22), Marshall (1981: 29), Honacki *et al.* (1982: 46), Calaby and Richardson (1988b: 63), Marshall *et al.* (1990: 492) and Groves (1993e: 53; 2005b: 63).

Gerboides Gervais, 1855a: 271.

TYPE SPECIES: *Kangurus rufus* Desmarest, 1822b [= *Osphranter rufus* (Desmarest, 1822b)] by monotypy.

COMMENTS: Synonymised within *Macropus* Shaw, 1790 by Thomas (1888a: 10) and within *Megaleia* by Iredale and Troughton (1934: 52) and McKenna and Bell (1997: 64). Synonymised within *Macropus* by Marshall (1981: 29), Calaby and Richardson (1988b: 63), Marshall *et al.* (1990: 492) and Groves (1993e: 53; 2005b: 63).

† *Boriogale* Owen, 1874a: 247.

TYPE SPECIES: † *Macropus (Boriogale) magnus* Owen, 1874a [= *Osphranter rufus* (Desmarest, 1822b)] by monotypy.

COMMENTS: Described as a subgenus of *Macropus*. Synonymised within *Macropus* Shaw, 1790 by Thomas (1888a: 10). Synonymised within *Megaleia* by Iredale and Troughton (1934: 52) and McKenna and Bell (1997: 64). Synonymised within *Macropus* by Marshall (1981: 29), Calaby and Richardson (1988b: 64), Marshall *et al.* (1990: 492) and Groves (1993e: 53; 2005b: 63).

Phascolagus Owen, 1873a: 128.

TYPE SPECIES: *Macropus erubescens* P. Sclater, 1870 [= *Osphranter robustus erebescens* (P. Sclater, 1870)] by monotypy.

COMMENTS: Abstract of paper. Also introduced by Owen (1873b: 255) with an expanded description by Owen (1874a: 245, 261–262). Synonymised within *Macropus* Shaw, 1790 by Thomas (1888a: 10) and within *Osphranter* by Iredale and Troughton (1934: 50) and Tate (1948b: 325). Synonymised within *Macropus* by Marshall (1981: 29), Calaby and Richardson (1988b: 63), Marshall *et al.* (1990: 492) and Groves (1993e: 53; 2005b: 63).

Dendrodorcopsis Rothschild, 1903: 414.

TYPE SPECIES: *Dendrodorcopsis woodwardi* Rothschild, 1903 [= *Osphranter bernardus* (Rothschild, 1904)] by monotypy.

COMMENTS: Synonymised in *Osphranter* by Iredale and Troughton (1934: 50) and Tate (1948b: 325). Synonymised within *Macropus* by Marshall (1981: 29), Calaby and Richardson (1988b: 64), Marshall *et al.* (1990: 492) and Groves (1993e: 53; 2005b: 63).

Osphranter antilopinus (Gould, 1842)

Antilopine Wallaroo

Osphranter Antilopinus Gould, 1842e: 80.

TYPE LOCALITY: Port Essington, Northern Territory, Australia. Type designated by Thomas (1922a: 128).

COMMENTS: Placed in *Macropus (Halmaturus)* by Waterhouse (1846: 95) and *Macropus* by Thomas (1888a: xi, 21), who reviewed the early taxonomic history. Transferred to *Osphranter* by Gould (1842 [1841–1842]; Unnumbered Plate; Gould, 1858 [1845–1863]: Text to Plates 8–9), Iredale and Troughton (1934: ix, 51), D. Johnson (1964: 460) and Troughton (1967: 182), and as a subspecies of *robustus* by Tate (1948b: 329). Included in *Macropus* by Ride (1970: 44), Richardson and Sharman (1976: 508), Honacki *et al.* (1982: 46), Calaby and Richardson (1988b: 65). Placed into the subgenus *Osphranter* by Dawson and Flannery (1985: 486) and followed by Groves (1993e: 53; 2005b: 63).

Osphranter bernardus (Rothschild, 1904)

Black Wallaroo

Macropus bernardus Rothschild, 1904: 543.

TYPE LOCALITY: Alligator River, Northern Territory, Australia.

COMMENTS: *Nomen novum* for *Macropus woodwardi* Thomas, 1901a. Included within *Osphranter* by Iredale and Troughton (1934: ix, 52) and Troughton (1967: 184) and as a subspecies of *robustus* by Tate (1948b: 328). Placed in *Macropus* by Ride (1970: 47), Richardson and

Sharman (1976: 508), Honacki *et al.* (1982: 46), Calaby and Richardson (1988b: 65). Placed in the subgenus *Osphranter* by Groves (1993e: 53).

Dendrodorcopsis woodwardi Rothschild, 1903: 414.

TYPE LOCALITY: Granite Ranges, Head of South Alligator River, Northern Territory, Australia.

COMMENTS: Name unavailable as preoccupied (secondary homonym) within both *Macropus* and *Osphranter* by *Macropus robustus woodwardi* Thomas, 1901a. Synonymised within *Osphranter bernardus* by Iredale and Troughton (1934: 52). Synonymised within *bernardus* by Calaby and Richardson (1988b: 65).

HOMONYMS:

Macropus robustus woodwardi Thomas, 1901a, the Common Wallaroo of the Class Mammalia (Order Diprotodontia, Family Macropodidae). Name is now recognised as *Osphranter robustus woodwardi* (Thomas, 1901). See individual entry.

Osphranter robustus (Gould, 1840)

Common Wallaroo

Osphranter robustus robustus (Gould, 1840)

[Macropus] Robustus Gould, 1840a: 685.

TYPE LOCALITY: Interior of New South Wales, Australia. Type designated by Thomas (1922a: 128).

COMMENTS: Name also described by Gould (1840b: 561), which was also published on 29 August 1840. McAllan and Bruce (1989: 449) proposed that they act as first revisers in giving priority to the publication of Gould (1840a: 685). Taxon further described by Gould (1841e: 92), who included it in the subgenus *Petrogale*, and Gould (1841 [1841–1842]: no pages number). Early taxonomic history reviewed by Thomas (1888a: xi, 22). Included within *Osphranter* by Gould (1842 [1841–1842]: Unnumbered Plate; Gould, 1858 [1845–1863]: Text to Plates 10-11), Iredale and Troughton (1934: ix, 52) and Troughton (1967: 181). Included in *Macropus (Halmaturus)* by Waterhouse (1841a: 241; 1846: 100), Thomas (1888a: xi, 22), Finlayson (1931: 69), Ride (1970: 44), Honacki *et al.* (1982: 47), and Calaby and Richardson (1988b: 69). Placed into the subgenus *Osphranter* by Tate (1948b: 327), Dawson and Flannery (1985: 487) and followed by Groves (1993e: 54). Species reviewed by Richardson and Sharman (1976: 499, 509) who designated the subspecies recognised here.

FUTURE TAXONOMIC RESEARCH: The validity of each of the typically recognised subspecies is in urgent need of revision to confirm their validity.

Macropus robustus reginae Schwarz, 1910b: 103.

TYPE LOCALITY: Mt Abbott, SW of Townsville, Queensland, Australia.

COMMENTS: Recognised as a species, within *Osphranter*, by Iredale and Troughton (1934: ix, 52) and Troughton (1967: 184) and as a subspecies of *robustus* by Tate (1948b: 328). Synonymised in *robustus* (in *Macropus*) by Ride (1970: 244), Richardson and Sharman (1976: 509), and Calaby and Richardson (1988b: 70).

Osphranter robustus erubescens (Sclater, 1870)

Macropus erubescens P. Sclater, 1870: 126; Plate 10.

TYPE LOCALITY: Port Augusta, South Australia, Australia.

COMMENTS: Synonymised within *robustus* by Thomas (1888a: 23). Recognised as a species, within *Osphranter*, by Iredale and Troughton (1934: ix, 51) and Troughton (1967: 182) and as a subspecies of *robustus* by Tate (1948b: 327) and D. Johnson (1964: 464). Taxon synonymised within *robustus* (in *Macropus*) by Calaby and Richardson (1988b: 69). Recognised as a subspecies of *robustus* by Richardson and Sharman (1976: 509), Strahan (1983: 250; 1995: 348), Groves (2005b: 65), Clayton *et al.* (2006: 107), and Van Dyck and Strahan (2008: 346), but not Burbidge *et al.* (2014: 21, 29).

O. [sphranter] crebescens Waterhouse, 1876: 284.

TYPE LOCALITY: *Errore pro Osphranter erubescens* P. Sclater, 1870.

COMMENTS: Synonymised within *robustus* by Thomas (1888a: 23). Synonymised within *erebescens*, within *Osphranter*, by Iredale and Troughton (1934: 51) and within *robustus* (in *Macropus*) by Calaby and Richardson (1988b: 69).

Macropus robustus cervinus Thomas, 1900b: 113.

TYPE LOCALITY: Yalgoo, Murchison District, Western Australia, Australia.

COMMENTS: Subspecies of *antilopinus* recognised by Iredale and Troughton (1934: 51) and Tate (1948b: 330). Species rank recognised, within *Osphranter*, by Troughton (1967: 183). Synonymised within *Macropus robustus* by Calaby and Richardson (1988b: 69) and Groves (1993e: 54), and within *erubescens* by Richardson and Sharman (1976: 509) and Groves (2005b: 65).

Macropus argentatus Rothschild, 1905b: 509.

TYPE LOCALITY: Northern Territory, Australia.

COMMENTS: Synonymised within *erebescens* by Iredale and Troughton (1934: 51) and Tate (1948b: 327). Synonymised within *Macropus robustus* by Calaby and Richardson (1988b: 70) and Groves (1993e: 54), and within *erubescens* by Richardson and Sharman (1976: 509) and Groves (2005b: 65).

Macropus magnus Rothschild, 1905b: 509.

TYPE LOCALITY: Murchison River, Western Australia, Australia.

COMMENTS: Primary junior homonym *Macropus magnus* Owen, 1874a; believed by some to be a hybrid between *Macropus robustus* and *Macropus rufus* (Calaby & Richardson, 1988b: 69).

HOMONYMS:

Macropus (Boriogale) magnus Owen, 1874a, the Red Kangaroo of the Class Mammalia (Order Diprotodontia, Family Macropodidae). Name is a synonym of *Osphranter rufus* (Desmarest, 1822b). See individual entry.

Macropus hagenbecki Rothschild, 1907: 333.

TYPE LOCALITY: Northern or north-western Australia, Australia.

COMMENTS: *Nomen novum* for *Macropus magnus* Rothschild, 1905b. Recognised as a species, within *Megaleia*, by Iredale and Troughton (1934: ix, 52). Synonymised in *robustus* (in *Macropus*) by Ride (1970: 244), and Calaby and Richardson (1988b: 70).

Macropus robustus rubens Schwarz, 1910b: 99.

TYPE LOCALITY: Box Soak, Shaw River, Western Australia, Australia.

COMMENTS: Synonymised within *cervinus* (*Osphranter*) by Iredale and Troughton (1934: 51), within *Macropus robustus* by Calaby and Richardson (1988b: 70) and (Groves (1993e: 54), and within *erubescens* by Richardson and Sharman (1976: 509) and Groves (2005b: 65).

Macropus robustus alexandriae Schwarz, 1910b: 102.

TYPE LOCALITY: Alexandria, Northern Territory, Australia.

COMMENTS: Recognised as a subspecies of *erubescens* by Iredale and Troughton (1934: 52), Tate (1948b: 327) and Troughton (1967: 182). Synonymised within *Macropus robustus* by Calaby and Richardson (1988b: 70) and Groves (1993e: 54), and within *erubescens* by Richardson and Sharman (1976: 509) and Groves (2005b: 65).

Osphranter robustus isabellinus (Gould, 1842)

Osphranter isabellinus Gould, 1842e: 81.

TYPE LOCALITY: Barrow Island, Western Australia, Australia.

COMMENTS: Further description provided by Waite (1901: 131). Recognised as a species within *Macropus (Halmaturus)* by Waterhouse (1846: 99) and within *Macropus* by Thomas (1888a: xi, 25). Recognised as a species, within *Osphranter*, by Iredale and Troughton (1934: ix, 51) and Troughton (1967: 183), and as a subspecies of *robustus* by Tate (1948b: 329). Synonymised in *robustus* (in *Macropus*) by Ride (1970: 244), and Calaby and Richardson (1988b: 69). Recognised as a subspecies of *robustus* by Richardson and Sharman (1976: 510), Strahan (1983: 250; 1995: 248), Groves (2005b: 65), Clayton *et al.* (2006: 107), Van Dyck and Strahan (2008: 346), and Burbidge *et al.* (2014: 21, 29).

Osphranter robustus woodwardi (Thomas, 1901)

Macropus robustus woodwardi Thomas, 1901a: 395.

TYPE LOCALITY: Grant range, near Fitzroy River, Western Australia, Australia.

COMMENTS: Recognised as a subspecies within *Osphranter antilopinus* by Iredale and Troughton (1934: 51), within *Osphranter cervinus* by Troughton (1967: 183), and within *Macropus robustus* by Richardson and Sharman (1976: 510), Strahan (1983: 250; 1995: 248), Groves (2005b: 65), Clayton *et al.* (2006: 107), and Van Dyck and Strahan (2008: 346), but not Burbidge *et al.* (2014: 21, 29).

HOMONYMS:

Dendrodorcopsis woodwardi Rothschild, 1903, the Black Wallaroo of the Class Mammalia (Order Diprotodontia, Family Macropodidae). Currently recognised as *Osphranter bernardus* (Rothschild, 1904). See individual entry.

Macropus robustus alligatoris Thomas, 1904a: 224.

TYPE LOCALITY: South Alligator River, Northern Territory, Australia.

COMMENTS: Synonymised in *woodwardi* (within *Osphranter*) by Iredale and Troughton (1934: 51) and Groves (2005b: 65) within *Macropus*. Recognised as a subspecies of *Osphranter robustus* by D. Johnson (1964: 464), within *Macropus robustus* by Calaby and Richardson (1988b: 69) and Groves (1993e: 54), and within *woodwardi* by Richardson and Sharman (1976: 510) and Groves (2005b: 65).

Macropus robustus bracteator Thomas, 1911a: 609.

TYPE LOCALITY: McClintock Range, east Kimberley, Western Australia, Australia.

COMMENTS: Synonymised within *Macropus robustus* by Calaby and Richardson (1988b: 70) and Groves (1993e: 54). Synonymised within *Osphranter antilopinus woodwardi* by Iredale and Troughton (1934: 51) and within *woodwardi* by Richardson and Sharman (1976: 510) and Groves (2005b: 65).

Osphranter rufus (Desmarest, 1822)

Red Kangaroo

kangurus rufus Desmarest, 1822b: 541.

TYPE LOCALITY: West of the Blue Mountains, New South Wales, Australia.

COMMENTS: Synonymised within *laniger* by Waterhouse (1841a: 198). Recognised at the species rank within *Macropus* by G. Bennett (1837: 6), Waterhouse (1846: 104) and Thomas (1888a: xi, 25), who reviewed the early taxonomic history. Placed in *Osphranter* by Gould (1853 [1845–1863]: Text to Plates 6-7) and Krefft (1864: 48; 1866a: 18). Included within *Megaleia*, as *rufa*, by Iredale and Troughton (1934: ix, 52), Troughton (1967: 178), Ride (1970: 44) and McKenna and Bell (1997: 64). Included in the genus *Macropus* by Honacki *et al*. (1982: 47), Strahan (1983: 255; 1995: 353), Dawson and Flannery (1985: 488), Calaby and Richardson (1988b: 71) and subsequent authors with the exception of McKenna and Bell (1997: 64). Placed in the subgenus *Osphranter* by Dawson and Flannery (1985: 488) and followed by Groves (1993e: 55).

Kangurus laniger Gaimard, 1823: 138.

TYPE LOCALITY: Port Macquarie, New South Wales, Australia.

COMMENTS: Recognised within *Macropus* by Waterhouse (1841a: 198) and Gould (1841 [1841–1842]: no pages number). Synonymised within *rufus* by Waterhouse (1846: 104), Thomas (1888a: 25), Iredale and Troughton (1934: 52), Calaby and Richardson (1988b: 71) and subsequent authors.

Kangurus griseo-lanosus Quoy & Gaimard, 1825: 482.

TYPE LOCALITY: Blue Mountains, New South Wales, Australia.

COMMENTS: Synonymised within *rufus* by Thomas (1888a: 25), Iredale and Troughton (1934: 52), Calaby and Richardson (1988b: 71) and subsequent authors

Macropus lanigerus J. Gray, 1826a: 49; Plate.

TYPE LOCALITY: Bathurst Plains, New South Wales, Australia.

COMMENTS: Synonymised within *rufus* by Thomas (1888a: 25), Iredale and Troughton (1934: 53), Calaby and Richardson (1988b: 71) and subsequent authors

K. [angurus] Lanosus J. Gray, 1827a: 202.

COMMENTS: As of Gaimard, in synonymy. Synonymised within *rufus* by Iredale and Troughton (1934: 53), Calaby and Richardson (1988b: 71) and subsequent authors.

Kangurus lanosus J. Gray, 1843a: 88.

COMMENTS: *Nomen nudum* as of Gaimard. Synonymised within *rufus* by Iredale and Troughton (1934: 53), Calaby and Richardson (1988b: 72) and subsequent authors.

Macropus (Osphranter) pictus Gould, 1861b: 373.

COMMENTS: *Nomen nudum*; error vide Gould (1845–1863). Synonymised within *rufus* by Thomas (1888a: 26), Iredale and Troughton (1934: 53), Calaby and Richardson (1988b: 72) and subsequent authors.

M. [acropus] ruber Crisp, 1862: 135.

COMMENTS: Errore pro *M. rufa*. Synonymised within *rufus* by Thomas (1888a: 26), Iredale and Troughton (1934: 53), Calaby and Richardson (1988b: 72) and subsequent authors.

Macropus (Boriogale) magnus Owen, 1874a: 247.

TYPE LOCALITY: Far north of South Australia, Australia.

COMMENTS: Recognised as a species by Thomas (1888a: xi, 27), but synonymised within *rufus* by Iredale and Troughton (1934: 53), Calaby and Richardson (1988b: 72) and subsequent authors.

HOMONYMS:

Macropus magnus Rothschild, 1905b, the Common Wallaroo of the Class Mammalia (Order Diprotodontia, Family Macropodidae). Name is a synonym of *Osphranter robustus* (Gould, 1840a). See individual entry.

Macropus rufus dissimulatus Rothschild, 1905c: 508.

TYPE LOCALITY: North West Australia [= Gascoyne River, Western Australia, Australia]. Established from Iredale and Troughton (1934: 53).

COMMENTS: Recognised as a subspecies by Cahn (1906: 381). Included as a subspecies of *rufus* by Iredale and Troughton (1934: 53), Tate (1948b: 336) and Troughton (1967: 180). Synonymised within *Macropus rufus* by Calaby and Richardson (1988b: 72) and subsequent authors.

Macropus rufus dissimulator Lydekker, 1906: 47.

COMMENTS: *Errore pro M. rufus dissimulatus* Rothschild, 1905c. Synonymised within *Macropus rufus* by Calaby and Richardson (1988b: 72) and subsequent authors.

M. [acropus] rufus occidentalis Cahn, 1906: 381.

TYPE LOCALITY: Murchison River, Western Australia, Australia. Established from Cahn (1907: 3–4).

COMMENTS: Included as a synonym of *dissimulator* by Iredale and Troughton (1934: 53). Synonymised within *Macropus rufus* by Calaby and Richardson (1988b: 72).

Macropus rufus pallidus Schwarz, 1910b: 89.

TYPE LOCALITY: Shaw River, Western Australia, Australia.

COMMENTS: Included as a synonym of *dissimulator* by Iredale and Troughton (1934: 53) and as described by Tate (1948b: 336). Synonymised within *Macropus rufus* by Calaby and Richardson (1988b: 72).

Wallabia Trouessart, 1905

Wallabia Trouessart, 1905 [1904–1905]: 834, footnote.

TYPE SPECIES: *Kangurus ualabatus* Lesson, 1827b [= *Wallabia bicolor* (Desmarest, 1804c)] by subsequent designation. See Iredale and Troughton (1934: 47).

COMMENTS: Recognised at subgeneric rank by Cabrera (1919: 148). Included in *Macropus* by Van Gelder (1977: 5), but recognised by Iredale and Troughton (1934: viii, 47), Ride (1970: 47), Kirsch and Calaby (1977: 17) and subsequent authors.

Wallabia bicolor (Desmarest, 1804)

Swamp Wallaby

Wallabia bicolor bicolor (Desmarest, 1804)

Kangurus bicolor Desmarest, 1804c: 357.

TYPE LOCALITY: Unknown.

COMMENTS: Not recognised under this name by Waterhouse (1841a; 1846), Thomas (1888a), J. Ogilby (1892) or Lydekker (1994a), each of whom called it *ualabatus* and placed it within *Macropus*. Placed within *Wallabia* by Iredale and Troughton (1934: viii, 48), Troughton (1941: 202), who also included all of the 'Wallabies' within the genus, and was followed by Troughton (1967: 163). Transferred to *Protemnodon* by Tate (1948b: 297). Recognised as the only extant species within *Wallabia* by Ride (1970: 47) and all subsequent authors until Meredith *et al.* (2008c: 405) undertook a genetic analysis using nearly every extant *Macropus* species as well as *bicolor*. Meredith *et al.* (2008c: 405) suggested that the majority of evidence favours *Macropus* paraphyly, with *Wallabia* as a distinct lineage within this clade and they recommended subsuming the genus *Wallabia* within *Macropus* as a new subgenus, i.e. *Macropus* (*Wallabia*). This arrangement (whereby *bicolor* is not, or hardly, more distinct from the three clades traditionally recognised as subgenera of *Macropus*) appears to be supported by most molecular studies including Kirsch (1977a: 64), Baverstock *et al.* (1989: 39, 46) and Kirsch *et al.* (1995: 309; 1997: 228, 237). The placement of *bicolor* within *Macropus* is also intriguing given the ecological, karyotypic, and reproductive distinctiveness of this species compared to the other members of the genus *Macropus* (Van Dyck & Strahan, 2008: 404). As noted under '*Macropus*... Future taxonomic directions', we would prefer instead to keep *Wallabia* as a full genus and raise the present three subgenera of genus *Macropus* to generic rank, given the Late Miocene dates of their divergence. No subspecies were recognised by Groves (2005b: 70) or Clayton *et al.* (2006: 108). Van Dyck and Strahan (2008: 406) tentatively listed *bicolor*, *apicalis*, *mastersii*, *ingrami* and *welsbyi* as subspecies, although they suggested that the species needs revision at infraspecific level and that no great confidence can be placed on the subspecies listed. Placed in the subgenus *Wallabia* by Meredith *et al.* (2008c: 405).

FUTURE TAXONOMIC RESEARCH: This species needs revision to determine the validity of the supposed subspecies that typically include *W. b. apicalis*, *W. b. mastersii*, *W. b. ingrami* and *W. b. welsbyi*.

Kangurus ualabatus Lesson, 1827b: 161; Plate 7.

TYPE LOCALITY: Port Jackson, New South Wales, Australia.

COMMENTS: Recognised as a species within *Macropus* by Waterhouse (1838a: 66; 1841a: 219; 1846: 136), *Macropus* by Lydekker (1887: 219), Thomas (1888a: xi, 30) and J. Ogilby (1892: 57), but was placed within *Halmaturus* by Gould (1842 [1841–1842]: Unnumbered Plate) and Gould (1857[1845–1863]: Plates 22–23). Synonymised within *bicolor* by Iredale and Troughton (1934: 48), Ride (1970: 244), Calaby and Richardson (1988b: 79) and Groves (1993e: 57; 2005b: 70).

H. [almaturus] Lessònii J. Gray, 1837: 583.

TYPE LOCALITY: New South Wales?, Australia.

COMMENTS: Synonymised within *ualabatus* by Lydekker (1887: 219) and Thomas (1888a: 30). Synonymised within *bicolor* by Iredale and Troughton (1934: 48), Calaby and Richardson (1988b: 80) and Groves (1993e: 57; 2005b: 70).

H. [almaturus] nemoralis Wagner, 1843: 114.

TYPE LOCALITY: New South Wales, Australia.

COMMENTS: Synonymised within *ualabatus* by Lydekker (1887: 219) and Thomas (1888a: 31). Synonymised within *bicolor* by Iredale and Troughton (1934: 48), Calaby and Richardson (1988b: 80) and Groves (1993e: 57; 2005b: 70).

Wallabia bicolor mastersii (Krefft, 1871)

Halmaturus Mastersii Krefft, 1871a: 3; Text to Plate 11.

TYPE LOCALITY: Burnett River, Queensland, Australia.

COMMENTS: Considered a *nomen nudum* by Thomas (1888a: 32), but sufficient description is in fact given to validate the name, especially in the footnote, according to Calaby and Richardson (1988b: 80). Taxon synonymised within *bicolor* by Calaby and Richardson (1988b: 80) and Groves (1993e: 57; 2005b: 70). Recognised as a subspecies of *bicolor* by Iredale and Troughton (1934: 48), Tate (1948b: 299), Strahan (1983: 261; 1995: 405), and Van Dyck and Strahan (2008: 406), but not Burbidge *et al.* (2014: 29).

Halmaturus mastersii Krefft, 1867b: 96.

TYPE LOCALITY: Burnett River, Queensland, Australia.

COMMENTS: *nomen nudum*, made available by Krefft (1871a: 3, Text to Plate 11). Recognised as a subspecies of *W. bicolor* by Iredale and Troughton (1934: 48) and Troughton (1967: 163). Synonymised within *bicolor* by Calaby and Richardson (1988b: 80) and Groves (1993e: 57; 2005b: 70).

Wallabia bicolor apicalis (Günther, 1875)

Halmaturus apicalis Günther, 1875: 653; Plate 77.

TYPE LOCALITY: Cape Grafton, north Queensland, Australia.

COMMENTS: Recognised as a subspecies within *bicolor* by Tate (1948b: 300). Taxon synonymised within *bicolor* by Calaby and Richardson (1988b: 80) and Groves (1993e:

57; 2005b: 70). Recognised as subspecies within *ualabatus* by Thomas (1888a: xi, 32) and subspecies of *bicolor* by Troughton (1967: 163), Strahan (1983: 261; 1995: 405), and Van Dyck and Strahan (2008: 406), but not Burbidge *et al.* (2014: 29).

Wallabia bicolor ingrami Thomas & Dollman, 1909

Macropus ualabatus ingrami Thomas & Dollman, 1909: 792; Plate 42.

TYPE LOCALITY: Inkerman, Queensland, Australia.

COMMENTS: Synonymised within *mastersii* by Iredale and Troughton (1934: 48). Taxon synonymised within *bicolor* by Calaby and Richardson (1988b: 80) and Groves (1993e: 57; 2005b: 70). Recognised as a subspecies of *bicolor* by Tate (1948b: 299), Strahan (1983: 261; 1995: 405), and Van Dyck and Strahan (2008: 406), but not Burbidge *et al.* (2014: 29).

Wallabia bicolor welsbyi Longman, 1922

Macropus welsbyi Longman, 1922: 298.

TYPE LOCALITY: Stradbroke Island, Queensland, Australia.

COMMENTS: Taxon synonymised within *bicolor* by Ride (1970: 248), Calaby and Richardson (1988b: 80) and Groves (1993e: 57; 2005b: 70). Recognised as a subspecies of *bicolor* by Iredale and Troughton (1934: 48), Tate (1948b: 299), Strahan (1983: 261; 1995: 405), and Van Dyck and Strahan (2008: 406), but not Burbidge *et al.* (2014: 29).

Captain Cook's Kangaroo

During Captain James Cook's first voyage, from 1768 to 1771, he visited Australia in 1770. While repairing his ship the *Endeavour* from late June to early August at what is now the Endeavour River near Cooktown, north Queensland, he and his crew made a number of observations of one or more species of kangaroo and collected several specimens that have since been lost. The name *Mus canguru*, P. Müller, 1776, was based on it. The species seen has typically been assumed to be the Eastern Grey Kangaroo *Macropus giganteus*; but doubt has been cast on this assumption by some authors including Iredale and Troughton (1925: 311) who suggested that the description does not correspond with *Macropus giganteus*, it does not occur there (which is incorrect) and that it was more likely to be the Common Wallaroo (*Macropus robustus*). A further review by the same authors (Iredale & Troughton, 1937: 67, 70) argued that it was probably a whiptail wallaby *Macropus parryi*, which they thought should be called Captain Cook's Whiptail *Wallabia canguru* (P. Müller, 1776). They also proposed that the name *giganteus* should be a junior synonym of *major* (Iredale & Troughton, 1937: 70).

Several drawings and paintings of the animals seen by Cook and members of the *Endeavour* exist which appear to have been based on sketches by Sydney Parkinson who accompanied Cook. Subsequently it appears that these drawings were used by George Stubbs in his 1771–1772 painting and for an engraving in Hawkesworth's (1773: Plate 20) edition of Cook's Voyage. These drawings and paintings were reviewed by Lysaght (1957: 17) but no clear species identification was made.

Raven (1939: 56) held that the evidence is decidedly against Cook's kangaroo having been a whiptail or pretty-face wallaby, but rather that it was a grey kangaroo *Macropus giganteus*. Tate (1948b: 323) proposed that the specimens collected by Captain James Cook could not have been *M. parryi*, *M. rufus* or *M. robustus* and concluded that it must be *Macropus giganteus*. Morrison-Scott and Sawyer (1950: 49) noted that Captain Cook's first expedition to Australia obtained three specimens, all from the Endeavour River, Queensland, July 1770 and recorded that the skull of one of these was destroyed by bombs during World War II and that no trace of the others has been found. They suggested that a photograph of the specimen before it was destroyed in the Museum of the Royal College of Surgeons is clearly the skull of a young eastern grey kangaroo and they (Morrison-Scott & Sawyer, 1950: 49) then designated it as the photo-lectotype of *Macropus canguru* (P. Müller, 1776) or Captain Cook's kangaroo. This was subsequently supported by Calaby *et al.* (1962: 29), though they recognised *Macropus major* as the senior synonym. Shortly afterwards Kirkpatrick (1963: 541) reviewed these photos and the molar eruption and concluded that *Mus canguru* must be *M. robustus*, which was followed by Kirkpatrick and Woods (1964: 249). In contrast to these reviews Iredale and Troughton (1963: 182) suggested that *Macropus giganteus* can be excluded from consideration based on the rhinarium that is relatively naked, while the third upper premolar is simply bilobate instead of having a 'double-ridge' anteriorly. Instead, as noted above, they considered that Cook's Kangaroo may have been *Macropus parryi* and so that the valid name for the latter is *Wallabia canguru*.

In 1966, the ICZN issued Opinion 760 (ICZN, 1966: 292), which read, in part, as follows:

(1) Under the plenary powers:
   (a) the specific name *canguru* P. Müller, 1776, as published in the binomen *Mus canguru*, together with all usages of *canguru* (and its various spellings *kangaru*, *kanguro*, *kanguru*, *caenguru*, *cangaru*, *cangura*) in combination with *Mus*, *Yerboa*, *Jaculus*, *Zerbua*, *Didelphis*, *Didelphys* and *Macropus*, is hereby suppressed for the purposes of the Law of Priority but not for those of the Law of Homonymy;
   (b) the specific name *giganteus* Erxleben, 1777, as published in the binomen *Jaculus giganteus*,

and all usages of *giganteus* in combination with *Yerboa*, *Jaculus*, *Didelphis* and *Didelphys* prior to that by Shaw in 1790, is hereby suppressed for the purposes of both the Law of Priority and the Law of Homonymy;....

(2) The following specific names are hereby placed on the Official List of Specific Names in Zoology with the Name Numbers specified:
   (a) *giganteus* Shaw, 1790, as published in the binomen *Macropus giganteus*, as interpreted by the neotype designated by Ride and Calaby (1964: 254) (type-species of *Macropus* Shaw, 1790) (Name No. 2109);
   (b) *major* Shaw, 1800, as published in the binomen *Macropus major* (Name No. 2110).

The Gordian knot was thereby cut, but interest in the identity of the macropods seen/collected by Cook's expedition remains, and is of course of considerable historic significance.

Nomenclature Relating to the Cook Expedition

Mus Canguru P. Müller, 1776: 62.

TYPE LOCALITY: Endeavour River, north Queensland, Australia.

COMMENTS: Taxon based on the description and plate of 'Kanguroo' by Hawkesworth (1773: Plate 20, 578). This description antedates *Macropus giganteus* (Zimmermann, 1777: 526). Also note that Zimmermann (1777: 526) was rejected by Opinion 257 of the ICZN (1954b: 231). Recognised as the valid species name within *Macropus* by Tate (1948b: 330) and within *Wallabia* by Troughton (1967: 161). The specific name as published in the binomen *Mus canguru*, together with all usages of *canguru* (and its various spellings *kangaru*, *kanguro*, *kanguru*, *caenguru*, *cangaru* and *cangura*) in combination with *Mus*, *Yerboa*, *Jaculus*, *Zerbua*, *Didelphis*, *Didelphys* [sic =*Didelphis*] Linnaeus 1758: 54 and *Macropus* was suppressed by Opinion 760 of the ICZN (1966: 292). Synonymised within *parryi*, as *cangaru*, by Ride (1970: 248).

[Jaculus] giganteus Erxleben, 1777: 409.

TYPE LOCALITY: Endeavour River, north Queensland, Australia.

COMMENTS: Synonymised within *Macropus giganteus* by Thomas (1888a: 15). The specific name *giganteus* Erxleben, 1777 as published in the binomen *Jaculus giganteus*, and in all uses of *giganteus* in combination *Yerboa*, *Jaculus*, *Didelphis* and *Didelphys* [sic =*Didelphis*] Linnaeus 1758: 54 prior to that by Shaw in 1790, was suppressed by Opinion 760 of the ICZN (1966: 292).

Jerboa gigantea Zimmermann, 1777: 526.

TYPE LOCALITY: Endeavour River, north Queensland, Australia.

COMMENTS: Zimmermann (1777: 526) was rejected by Opinion 257 of the ICZN (1954b: 231). Synonymised within *Macropus giganteus* by Thomas (1888a: 15). The specific name *giganteus* Erxleben, 1777 as published in the binomen *Jaculus giganteus*, and all uses of *giganteus* in combination *Yerboa*, *Jaculus*, *Didelphis* and *Didelphys* [sic =*Didelphis*] Linnaeus 1758: 54 prior to that by Shaw in 1790 was suppressed by Opinion 760 of the ICZN (1966: 292). The effect of this suppression was discussed by Ride (1963: 126).

Didelphys giganteus Schreber, 1777: Plate 154; 1878: 552.

TYPE LOCALITY: Endeavour River, north Queensland, Australia.

COMMENTS: The specific name *giganteus* Erxleben, 1777 as published in the binomen *Jaculus giganteus*, and all uses of *giganteus* in combination *Yerboa*, *Jaculus*, *Didelphis* and *Didelphys* [sic =*Didelphis*] Linnaeus 1758: 54 prior to that by Shaw in 1790 was suppressed by Opinion 760 of the ICZN (1966: 292).

[Yerbua] Kanguru Forster, 1778: 113.

TYPE LOCALITY: Endeavour River, north Queensland, Australia.

COMMENTS: The specific name as published in the binomen *Mus canguru*, together with all usages of *canguru* (and its various spellings *kangaru*, *kanguro*, *kanguru*, *caenguru*, *cangaru* and *cangura*) in combination with *Mus*, *Yerboa*, *Jaculus*, *Zerbua*, *Didelphis*, *Didelphys* [sic =*Didelphis*] Linnaeus 1758: 54 and *Macropus* was suppressed by Opinion 760 of the ICZN (1966: 292). The genus *Yerbua* Forster, 1778: 108, 111 was also supressed by Opinion 730 of the ICZN (1965: 84).

Didelphis kenguru Zimmermann, 1780: 231.

TYPE LOCALITY: Australia.

COMMENTS: Synonymised within *Macropus giganteus* by Thomas (1888a: 15).

Mus Caenguru Höslin, 1781: 125.

TYPE LOCALITY: Australia.

COMMENTS: The specific name as published in the binomen *Mus canguru*, together with all usages of *canguru* (and its various spellings *kangaru*, *kanguro*, *kanguru*, *caenguru*, *cangaru* and *cangura*) in combination with *Mus*, *Yerboa*, *Jaculus*, *Zerbua*, *Didelphis*, *Didelphys* [sic =*Didelphis*] Linnaeus 1758: 54 and *Macropus* was suppressed by Opinion 760 of the ICZN (1966: 292).

[Didelphis] Kanguro Boddaert, 1785: 78.

TYPE LOCALITY: Australia.

COMMENTS: Synonymised within *Macropus giganteus* by Thomas (1888a: 15).

Names from later accounts.

[Didelphis] Giganteus F. Meyer, 1793: 16.

TYPE LOCALITY: Sydney, New South Wales, Australia.

Gigantomys canguru Link, 1794: 70.

TYPE LOCALITY: Sydney, New South Wales, Australia.

Gigantomys gigantia Link, 1794: 70.

TYPE LOCALITY: Sydney, New South Wales, Australia.
COMMENTS: Name listed adjacent to *Gigantomys cangura*, so the taxon may have been intended to be recorded as *Gigantomys gigantia*.

Tribe Setonichini New Tribe

Tribe Setonichini Jackson & Groves, 2015: New.

TYPE GENUS: *Setonix* Lesson, 1842.
COMMENTS: Palmer (1904: 630) derived the generic name from a combination of the Latin *seta* (bristle) and the Greek *onyx* (claw), although there is also a Greek word *setao*, meaning to gnaw; in either case, Thomas's emendation of Lesson's name was justified etymologically, if not in terms of zoological nomenclature. The stem of *onyx* is *onych-*, hence the correct formation of the tribe name is Setonichini. Previous studies have placed *Setonix* within *Macropus* (Thomas (1888a, xi, 60), near *Dendrolagus* (Wood Jones, 1924 [1923–1925], near *Thylogale* (Raven & Gregory, 1946: 10; Flannery, 1989: 41), and near *Macropus* (Cabrera, 1919: 153; Tate, 1948b: 312; Baverstock *et al*., 1989: 38). The comprehensive DNA hybridisation studies of Kirsch *et al*. (1997: 237) found genus *Setonix* to be distinct from the other macropodines, and most closely related to the New Guinean *Dorcopsulus* (Matschie, 1917: 57), and presumably *Dorcopsis* Schlegel & Müller, 1845: 130. In order to be consistent with the groups, and tribes recognised above, by Kirsch *et al*. (1997: 237) a new tribe is recognised here that includes the genera *Setonix* Lesson, 1842; Φ *Dorcopsis* Schlegel and Müller, 1845: 130; and Φ *Dorcopsulus* Matschie, 1917: 57.
FUTURE TAXONOMIC RESEARCH: Further confirmation of the relationship of the members of the genera within this tribe is required.

Setonix Lesson, 1842

Setonix Lesson, 1842: 194.

TYPE SPECIES: *Kangurus brachyurus* Quoy & Gaimardi, 1830 [=*Setonix brachyurus* (Quoy & Gaimard, 1830)] by monotypy.

COMMENTS: Described as a subgenus of *Macropus*. Synonymised within *Macropus* Shaw, 1790 by Thomas (1888a: 10). Genus recognised by Iredale and Troughton (1934: viii, 45) and subsequent authors.

Setonyx Thomas, 1888a: 10.

COMMENTS: A nomenclaturally invalid (although etymologically correct) emendation of *Setonix*. This spelling also used by Simpson (1945: 47). Synonymised within *Setonix* by Palmer (1904: 630) and McKenna and Bell (1997: 64).

Setonix brachyurus (Quoy & Gaimard, 1830)

Quokka

Kangurus brachyurus Quoy & Gaimard, 1830: 114; Plate 19.

TYPE LOCALITY: King Gorge Sound, Western Australia, south west Australia.
COMMENTS: Included within *Halmaturus* by Gould (1855 [1845–1863]: Text to Plates 37–38) and Krefft (1968b: 2). Transferred to *Macropus* by Waterhouse (1841a: 236; 1846: 162) and Thomas (1888a: xi, 60), who reviewed its early taxonomic history. Placed within the subgenus *Setonix* by Lesson (1842: 194) and elevated to genus rank *Setonix* by Iredale and Troughton (1934: viii, 45) and followed by subsequent authors.

H. [almaturus] Thylogale brevicaudatus J. Gray, 1838b: 108.

COMMENTS: Unnecessary replacement name for *Setonix brachyurus* (Quoy and Gaimard, 1830). Synonymised within *brachyurus* by Waterhouse (1846: 162), Thomas (1888a: 60), Iredale and Troughton (1934: 44), Tate (1948b: 312), Calaby and Richardson (1988b: 77) and followed by subsequent authors.

Subfamily Lagostrophinae Flannery, 1989

Tribe Lagostrophini Flannery, 1989: 30.

TYPE GENUS: *Lagostrophus* Thomas, 1887c.
COMMENTS: When originally proposed as a new rank it was placed in the Subfamily Sthenurinae (Glauert, 1926: 71) and included the genus *Lagostrophus* Thomas, 1887c.

There has been a lot of debate regarding the placement of the living species *Lagostrophus fasciatus*. It was proposed to be the sole living member of the Subfamily Sthenurinae (Glauert, 1926: 71) by Flannery (1983: 15) before being placed in its own tribe by Flannery (1989: 30). Subsequently the genus *Lagostrophus* was recognised within the Tribe Sthenurini by Marshall *et al*. (1990: 492) and Kirsch *et al*. (1997: 246), and within the Subfamily Sthenurinae by Strahan (1995: 7, 406) and Groves (2005b:

59). The placement of *Lagostrophus* within the Subfamily Sthenurinae was not followed by McKenna and Bell (1997: 64) or Menkhorst and Knight (2011: 17) and was considered *incertae sedis* by Van Dyck and Strahan (2008: 10, 406). Westerman *et al.*'s (2002: 209) molecular results could neither corroborate nor refute the sthenurine hypothesis, but agreed that *Lagostrophus* does not have a sister relationship to the Subfamily Macropodinae, and argued on this basis that the overriding implication is that *Lagostrophus* is a living relic of the ancient sthenurine kangaroo lineage. Prideaux (2004: 230), however, argued strongly against such a relationship, agreeing that *Lagostrophus* is highly divergent, very likely sister to the Subfamily Macropodinae, but not a sthenurine. Meredith *et al.* (2008c: 404) again confirmed a sister-group relationship between *Lagostrophus* and the Subfamily Macropodinae, but wanted further evidence to determine its relationship to the Subfamily Sthenurinae, suggesting that ancient DNA information from one of the acknowledged sthenurines was necessary to solve the problem. The studies by Prideaux and Warburton (2010: 954) revealed strong support for the clade containing *Lagostrophus* and † *Troposodon* whereas the genera † *Sthenurus* Owen, 1873a: 128; † *Simosthenurus* Tedford, 1966: 10; and † *Procoptodon* Owen, 1873c: 386 formed a distinct clade, hence *Lagostrophus* and † *Troposodon* could not be included within and Subfamily Sthenurinae.

Subfamily Lagostrophinae Prideaux & Warburton, 2010: 954, 968.

TYPE GENUS: *Lagostrophus* Thomas, 1887c.

COMMENTS: When originally proposed as a new rank it was placed in the Family Macropodidae (J. Gray, 1821) and included the genera *Lagostrophus* Thomas, 1887c; and † *Troposodon* Bartholomai, 1967: 21.

Lagostrophus Thomas, 1887

Lagostrophus Thomas, 1887c: 544.

TYPE SPECIES: *Kangurus fasciatus* Péron & Lesueur, 1807 [=*Lagostrophus fasciatus* (Péron & Lesueur, 1807)] by monotypy.

COMMENTS: Traditionally placed as a sister-taxon to *Lagorchestes* since Bensley (1903: 201). Synonymised within *Lagorchestes* by Simpson (1945: 46) but recognised by most subsequent authors including Ride (1970: 58), Marshall (1981: 29) and Strahan (1983: 201; 1995: 406). See comments above under Subfamily.

Lagostrophus fasciatus (Péron & Lesueur, 1807)

Banded Hare-wallaby

Lagostrophus fasciatus fasciatus (Péron & Lesueur, 1807)

Kangurus fasciatus Péron & Lesueur, 1807: 114; Plate 27.

TYPE LOCALITY: Bernier Island, Shark Bay, Western Australia, Australia.

COMMENTS: Included within *Macropus* by Waterhouse (1841a: 237), within *Bettongia* by Gould (1842 [1845–1863]: Unnumbered Plate) and within *Macropus* (*Lagorchestes*) by Waterhouse (1846: 87). Placed within *Lagorchestes* by Gould (1849 [1945–1863]: Text to Plate 56) and transferred to *Lagostrophus* by Thomas (1887c: 544; 1888a: xi, 100), who reviewed its early taxonomic history, and most subsequent authors including Iredale and Troughton (1934: viii, 41), and Troughton (1967: 137). The animals on Dorre Island and Bernier Island show little morphological, chromosomal or blood allozyme difference, which supports the existing taxonomy where both of these populations are placed within the same taxon (Courtenay, 1993: 191; J. Richards *et al.*, 2001: 318).

Mac. [ropus] elegans G. Cuvier, 1816a: 183.

TYPE LOCALITY: Bernier Island, Western Australia (see Helgen & Flannery (2003: 200).

COMMENTS: Synonymised within *fasciatus* by Thomas (1888a: 100), Iredale and Troughton (1934: 41), Calaby and Richardson (1988b: 63), and Helgen and Flannery (2003: 200).

HOMONYMS:

Macropus elegans Lambert, 1807, wallaby of the Class Mammalia (Order Diprotodontia, Family Macropodidae). Taxon is considered *incertae sedis*.

† *Lagorchestes albipilis* Gould, 1842f: 2.

TYPE LOCALITY: Wongan Hills, Western Australia, Australia. Type designated by Thomas (1922a: 128).

COMMENTS: Recognised as a subspecies of *fasciatus* by Thomas (1907b: 769), Troughton (1967: 137), Strahan (1983: 201; 1995: 406), Courtenay (1993: 191), Clayton *et al.* (2006: 106), and Van Dyck and Strahan (2008: 408). Synonymised within *fasciatus* by Waterhouse (1846: 87), Thomas (1888a: 100), Iredale and Troughton (1934: 41), Calaby and Richardson (1988b: 63), Helgen and Flannery (2003: 200, 203), which was followed by Burbidge *et al.* (2014: 29).

† *Halmaturus striatus* Lesson, 1842: 195.

TYPE LOCALITY: *Nomen nudum* for *Lagorchestes albipilis* Gould, 1842f (see Thomas, 1888a: 101).

COMMENTS: Synonymised within of *fasciatus* by Thomas (1888a: 101), Iredale and Troughton (1934: 41), Calaby and Richardson (1988b: 63) and Helgen and Flannery (2003: 200).

† *Lagostrophus fasciatus baudinettei* Helgen & Flannery, 2003

† *Lagostrophus fasciatus baudinettei* Helgen & Flannery, 2003: 202, 206; Figs. 2–4.

TYPE LOCALITY: Buchsfelde [=Loos] 40km north of Adelaide, South Australia, Australia. (34°37′S, 138°42′E)

COMMENTS: Described from a specimen collected in 1863 by Richard Schomburgk. Appears to have become extinct soon after the holotype was collected. Recognised as a subspecies of *fasciatus* by Clayton *et al.* (2006: 106), Van Dyck and Strahan (2008: 408) and Burbidge *et al.* (2014: 29).

Incertae Sedis

Macropus albus J. Gray, 1830b: 10.

TYPE LOCALITY: Unknown.

COMMENTS: Recognised at the species rank by Waterhouse (1838a: 66; 1841a: 214). Tentatively placed as a synonym within *ruficollis* [=*M. rufogriseus*] by Thomas (1888a: 34). Included as a synonym of *Macropus fruticus rufogriseus*, as *Wallabia rufogrisea frutica*, by Iredale and Troughton (1934: 49). *Incertae sedis* by Calaby and Richardson (1988b: 80) and not considered by Groves (1993e; 2005b).

Macropus elegans Lambert, 1807: 318; Plate 16.

TYPE LOCALITY: New South Wales, Australia.

COMMENTS: Recognised as described, within *Macropus*, by Waterhouse (1841a: 209). Iredale and Troughton (1934: viii, 50), and Troughton (1967: 162) recognised it within *Wallabia* in preference to *parryi* E. Bennett, 1835. Placed as a subspecies of *parryi* by Tate (1948b: 308) and synonym of *parryi* by Ride (1970: 244). *Incertae sedis* by Thomas (1888a: 33, footnote) and Calaby and Richardson (1988b: 80), and not considered by Groves (1993e; 2005b). Raven (1939: 56) suggested that Thomas was right in discarding the name *elegans* Lambert 1807, because it is impossible to decide with certainty to which species Lambert's description applies.

HOMONYMS:

Macropus elegans G. Cuvier, 1816a, the Banded Hare-wallaby of the Class Mammalia (Order Diprotodontia, Family Macropodidae). Genus is a synonym of *Lagostrophus fasciatus* (Péron & Lesueur, 1807). See individual entry.

[Macropus] Psilopus Gould, 1840a: 685.

TYPE LOCALITY: Interior of Australia.

COMMENTS: Also described by Gould (1840b: 561) but not discussed by other authors until McAllan and Bruce (1989: 450) who suggest that it remains an unused senior synonym.

[Macropus] Nepeanensis Gould, 1840b: 561.

TYPE LOCALITY: Nepean Bay [may be Kangaroo Island, South Australia, Australia].

COMMENTS: Appears to have not been discussed again after its description until McAllan and Bruce (1989: 448).

Supercohort Placentalia Bonaparte, 1838

Subclass Placentalia Bonaparte, 1838: 107.

COMMENTS: When originally proposed, this rank was placed in the Class Mammalia (Linnaeus, 1758) and included the sectios Educabilia (Bonaparte, 1838) (containing the Primates, Ferae, Pinnipedia, Cete, Belluae and Pecora) and Ineducabilia (Bonaparte, 1838) (containing the Bruta, Cheiroptera, Bestiae and Glires). Name also described by Bonaparte (1840: 248), which was read before the *Linnean Society of London* on 7 November 1837. The recognition of this rank, author and year has been unstable. Authors including Simpson (1945: 47), Aplin and Archer (1987: xxvii), and McKenna and Bell (1997: 80) give the author as Owen (1837: 903 =1849: 903), while Asher and Helgen (2010: 4) give the author as Bonaparte (1837 [=1840]: 248). Taxon recognised as a 'Series' by Bonaparte (1840: 248), subclass rank by Gill (1872: v, 1) and Iredale and Troughton (1934: ix, 55), supercohort rank by Aplin and Archer (1987: xxvii) (within the Superdivision Eutheria) and cohort rank by McKenna and Bell (1997: 80) who suggested that there was confusion over the use of Eutheria because of its use by Gill (1872: v, 1) and Huxley (1881: 657). In a similar sense to Aplin and Archer (1987: xxvii), authors including Wible *et al.* (2007: 1005) and Archibald (2003: 350) placed Placentalia as a lower (unnamed) rank within Eutheria. In turn Eutheria has been considered a clade that contains the Placentalia (crown group) and all taxa that share a more recent common ancestor with it (stem eutherians) than they do with the marsupials, which is the sister taxon of Eutheria (see Archibald *et al.*, 2001: 63–64).

HOMONYMS:

Subclass Placentalia Owen, 1849, mammals of the Infraclass Placentalia. See individual entry.

Order Belluae Linnaeus, 1758: 16, 73.

COMMENTS: When originally proposed, this rank was placed in the Class Mammalia (Linnaeus, 1758) and included the genera *Equus* and *Hippopotamus*. Name spelt Belluae on page 17 and Bellua on page 73. In 1766, Linnaeus added the genera *Sus* and *Rhinoceros*, which were in different orders in his 1758 edition.

Unguiculata Linnaeus, 1766: 21, 24.

COMMENTS: When originally proposed, this rank was placed in the Class Mammalia (Linnaeus, 1758) at unknown rank and included the orders Bruta (Linnaeus, 1758: 16, 33 [=Mammalia (Linnaeus, 1758 part)]), Glires (Linnaeus, 1758), Primates (Linnaeus, 1758) and Ferae (Linnaeus, 1758). Strictly speaking, it seems doubtful whether this term was intended by Linnaeus as a formal category, rather than simply a descriptive one for the four orders mentioned, since he recognised only classes, orders, genera and species. Synonymised within Placentalia by McKenna and Bell (1997: 80).

Ungulata Linnaeus, 1766: 21, 24.

COMMENTS: When originally proposed, this rank was placed in the Class Mammalia (Linnaeus, 1758) and included the orders Belluae (Linnaeus, 1758 [=Placentalia (Bonaparte, 1838 part)]) and Pecora (Linnaeus, 1758). Name spelt Ungulati on page 21. Strictly speaking, it seems doubtful whether this term was intended by Linnaeus as a formal category, rather than simply a descriptive one for the two orders mentioned, since he recognised only classes, orders, genera and species. Recognised as an order by J. Gray (1825a: 342), Gill (1872: v, 8, 47, 70) and Flower and Lydekker (1891: xii, 273), and grandorder by McKenna and Bell (1997: 357).

Order Pinnipeda Storr, 1780: Table C.

COMMENTS: When originally proposed, this rank was placed in the Mammalia (Linnaeus, 1758) and included the genera *Phoca* Linnaeus, 1758: 37; *Rosmarus* Brünnich, 1772: 34, 38. [=*Odobenus* Brisson, 1762: 30]; *Trichechus* Linnaeus, 1758: 34; and *Manatus* Brünnich, 1772: 34, 38 [=*Trichechus* Linnaeus, 1758: 34]. Storr (1780: Table C) was recognised as the author of the Suborder Pinnipedia used by Simpson (1931: 277) but not by Simpson (1945: 121) who recognised Illiger (1811: 60, 138) as the author.

HOMONYMS:

Type Pinnipeda Burnett, 1830c, whales of the Class Mammalia (Order Artiodactyla). Name is a synonym of the Infraorder Cetacea (Brisson, 1762). See individual entry.

Family Plantigraden Duméril, 1806a: 4, 14.

TYPE GENUS: Not based on a genus group name.

COMMENTS: When originally proposed, this rank was placed in the Klasse Säugthiere (Duméril, 1806a [=Mammalia (Linnaeus, 1758)]) and included the genera *Ursus* Linnaeus, 1758: 47; *Kinkaju* [=*Kinkajou* Lacépède, 1799a: 7], *Taxus* E. Geoffroy Saint-Hilaire & G. Cuvier, 1795: 187 [=*Meles* Boddaert, 1785: 45]; *Nasua* Storr, 1780: 35, Table A; *Erinaceus* Linnaeus, 1758: 52; *Sorex* Linnaeus, 1758: 53; and *Talpa* Linnaeus, 1758: 52.

Family Amphibien Duméril, 1806a: 4, 26.

TYPE GENUS: Not based on a genus group name.

COMMENTS: When originally proposed, this rank was placed Klasse Säugthiere (Duméril, 1806a [=Mammalia (Linnaeus, 1758)]) and included the genera *Phoca* Linnaeus, 1758: 37; *Trichechus* Linnaeus, 1758: 34; *Dugong* Lacépède, 1799a; and *Manatus* Brünnich, 1772: 34, 38 [=*Trichechus* Linnaeus, 1758: 34].

Order Volitantia Illiger, 1811: 60, 116.

COMMENTS: When originally proposed, this rank was placed in the Class Mammalia (Linnaeus, 1758) and included the families Dermoptera (Illiger, 1811: 63, 116) and Chiroptera (Blumenbach, 1779). The unity of the bats and colugos has been raised many times since it was first proposed including Leche (1886: 78), Miller (1907: 6–7), and Pocock (1926b: 444), and revived more recently by morphological studies by Novacek (1986: 96), Novacek and Wyss (1986: 265), Thewissen and Babcock (1991: 934; 1993: 106), Szalay and Lucas (1993: 222; 1996: 38), Simmons (1993: 44; 1995: 27), Wible and Martin (1993: 144), Simmons and Geisler (1998: 43), Stafford and Thorington (1998: 135), Stafford and Szalay (2000: 380), Sargis (2002: 1565) and Silcox *et al.* (2005: 127). Despite the apparent strong morphological evidence numerous molecular studies completed since the early 1990s never group bats with dermopterans or primates regardless of the methods used (Simmons, 2005a: 171). Synonymised within the Grandorder Archonta by McKenna and Bell (1997: 295).

Order Falculata Illiger, 1811: 60, 123.

COMMENTS: When originally proposed, this rank was placed in the Mammalia (Linnaeus, 1758) and included the families Subterranea (including the genera *Erinaceus* Linnaeus, 1758: 52; *Centetes* Illiger, 1811: 124 [=*Tenrec* Lacépède, 1799a: 7]; *Sorex* Linnaeus, 1758: 53; *Mygale* G. Cuvier, 1800: Table 1 [=*Desmana* Güldenstädt, 1777: 108]; *Condylura* Illiger, 1811: 125; *Chrysochloris* Lacépède, 1799a: 7; *Scalops* Illiger, 1811: 126 [sic=*Scalopus* É. Geoffroy, 1803c: 77]; *Talpa* Linnaeus, 1758: 52); Plantigrada (Illiger, 1811 [=Carnivora (Bowdich, 1821 part)]); Sanguinaria (including the genera *Megalotis* Illiger, 1811: 131; *Canis* Linnaeus, 1758; *Hyaena* Brisson, 1762: 13, 168; *Felis* Linnaeus, 1758; *Viverra* Linnaeus, 1758: 43; *Ryzaena* G. Cuvier, 1816a: 158 [=*Suricata* Desmarest, 1804a: 15]; and Gracilia (including the genera *Herpestes* Illiger, 1811: 135; *Mephitis* É. Geoffroy Saint-Hilare & G. Cuvier, 1795: 187; *Mustela* Linnaeus, 1758; and *Lutra* Brisson, 1762: 13, 201).

Order Pinnipedia Illiger, 1811: 60, 138.

COMMENTS: When originally proposed, this rank was placed within the Class Mammalia (Linnaeus, 1758) and included the Family Pinnipedia (Illiger, 1811 [=Placentalia (Bonaparte, 1838 part)]). Rank recognised with the inclusion of the seals and Sirenia by Haeckel (1866: clix), but has typically been recognised without the Sirenia as either a suborder of the Carnivora (e.g. J. Gray, 1871a: 1; Simpson, 1931: 277; 1945: 121; Troughton, 1967: 194; Sarich, 1969: 416; Deméré *et al.*, 2003: 48), Adam (2005: 1), J. Flynn *et al.* (2005: 317), Arnason *et al.* (2002: 8153; 2006: 348; Jefferson *et al.*, 2008: 10), or at ordinal rank of the Carnivora (e.g. Iredale and Troughton, 1934: 87; Strahan, 1983: xx, 457).

Synonymised within Superfamily Phocoidea by McKenna and Bell (1997: 252), but not by Wozencraft (2005: 590, 595), who synonymised the Suborder Pinnipedia within the Family Phocidae. Fulton and Strobeck (2006: 165) used the name Pinnipedia in preference to Phocoidea, but this usage appears to be at superfamily rank, which is contrary to the rules of nomenclature. Subsequently Deméré (*et al.*, 2003: 48) recognised this name as a crown clade for modern seals, and various associated extinct taxa, which was followed by Berta (2009a: 863). Rank also recognised by Fulton and Strobeck (2010a: 816) as a clade between the Infraorder Arctoidea and the superfamilies Phocoidea and Otarioidea.

Family Pinnipedia Illiger, 1811: 138.

TYPE GENUS: Not based on a genus group name.

COMMENTS: When originally proposed, this rank was placed within the Order Pinnipedia (Illiger, 1811 [=Placentalia (Bonaparte, 1838 part)]) and included the genera *Phoca* Linnaeus, 1758: 37; and *Trichechus* Linnaeus, 1758: 34.

Order Natantia Illiger, 1811: 60, 139.

COMMENTS: When originally proposed, this rank was placed in the Class Mammalia (Linnaeus, 1758) and included the families Sirenia (Illiger, 1811) and Cete (Linnaeus, 1758 [=Cetacea (Brisson, 1762)]).

HOMONYMS:

Natantia Newman, 1843, whales of the Class Mammalia (Infraorder Cetacea Brisson, 1762). See individual entry.

Subclass Monodelphes de Blainville, 1816a: 117.

COMMENTS: When originally proposed, this rank was placed within the Class Mammifères (G. Cuvier, 1816a: xxix, 70 [=Mammalia (Linnaeus, 1758)]) and included the orders Quadrumanes (de Blainville, 1816a: 117 [=Order Primates (Linnaeus, 1758)]), Carnassiers (de Blainville, 1816a: 117 [=Soricomorpha (Gregory, 1910), Chiroptera (Blumenbach, 1779), Carnivora (Bowdich, 1821)]), Edentés (de Blainville, 1816a: 117 [=Order Cingulata (Illiger, 1811: 110) and Pilosa (Flower, 1883: 184)]), Rongeurs (Célérigrades) (de Blainville, 1816a: 117 [=Glires (Linnaeus, 1758)]); Gravigrades (de Blainville, 1816a: 117 [=Proboscidea (Illiger, 1811: 96)]) and Ongulograd (de Blainville, 1816a: 117 [=Perissodactyla (Owen, 1848) and Artiodactyla (Owen, 1848)]). The name was also introduced by de Blainville (1816b: 250). Name recognised as the Subclass Monodelphia by Haeckel (1866: xi, cxliv), Gill (1871a: 527, 528) and Gregory (1947: 46). Synonymised within Eutheria by Simpson (1945: 47) and within Placentalia by McKenna and Bell (1997: 80).

Order Carnassiers de Blainville, 1816a: 117.

COMMENTS: When originally proposed, this rank was placed within the Subclass Monodelphes (de Blainville, 1816a [=Placentalia Bonaparte, 1838]) and included the Plantigrade Omnivores, Digitigrade Carnivores [=?Carnivora Bowdich, 1821], Insectivores [=?Insectivora] (Bowdich, 1821 [=Order Lipotyphla (Haeckel, 1866 part)]), Cheiroptères [=Chiroptera (Blumenbach, 1779)], Taupes [=Order Lipotyphla (Haeckel, 1866 part)]) and Phoques [=Phocoidea (J. Gray, 1821)].

HOMONYMS:

Order Carnassiers G. Cuvier, 1816a, mammals of the Class Mammalia in part. See individual entry.

Class Quadripedes J. Gray, 1821: 300.

COMMENTS: When originally proposed, this rank was placed in the Sub-Kingdom Vertebrosa (J. Gray, 1821: 297 [=Mammalia (Linnaeus, 1758 part)]) and included the orders Pterophorae (J. Gray, 1821: 300 [=Dermoptera (Illiger, 1811: 63, 116)]), Plantigradae (J. Gray, 1821 [=Placentalia (Bonaparte, 1838 part)]), Digitigradae (J. Gray, 1821: 301 [=Carnivora (Bowdich, 1821 part)]), Amphibiae (J. Gray, 1821 [=Phocoidea (J. Gray, 1821 part)]), Rosores (J. Gray, 1821: 302 [=Rodentia (Bowdich, 1821)]), Tardigradae (J. Gray, 1821: 304 [=Phyllophaga (Owen, 1842: 167)]), Oligodontae (J. Gray, 1821: 305 [=Cingulata (Illiger, 1811: 110) and Pilosa (Flower, 1883: 184)]), Edentulae (J. Gray, 1821: 305 [=Vermilingua (Illiger, 1811: 112)]), Proboscidiae (J. Gray, 1821: 305 [=Proboscidea (Illiger, 1811: 96)]), Tesserachenae (J. Gray, 1821: 306 [=Suiformes (Jaekel, 1911: 233)]), Trichenae (J. Gray, 1821: 306 [=Altungulata (Prothero & Schoch, 1989: 510)]), Monochenae (J. Gray, 1821: 306 [=Perissodactyla (Owen, 1848)]), Hydrophorae (J. Gray, 1821: 307 [=Tylopoda (Illiger, 1811)]) and Ruminantes (J. Gray, 1821: 307 [=Ruminantia (Scopoli, 1777)]).

Order Plantigradae J. Gray, 1821: 300.

COMMENTS: When originally proposed, this rank was placed in the Class Quadripedes (J. Gray, 1821 [=Placentalia (Bonaparte, 1838 part)]) and included the families Erinacidae (J. Gray, 1821: 300 [=Erinaceidae (G. Fischer, 1814: ix, 143)]), Soricidae (J. Gray, 1821 [=Soricidae (G. Fischer, 1814: x, 143)]), Myaladae (J. Gray, 1821: 300 [=Desmaninae (Thomas, 1912b: 397)]), Tenrecidae (J. Gray, 1821: 301) and Ursinidae (J. Gray, 1821: 301 [=Ursidae (G. Fischer, 1814: x, 143)]).

Class Cetaceae J. Gray, 1821: 309.

COMMENTS: When originally proposed, this rank was placed in the Sub-Kingdom Vertebrosa (J. Gray, 1821: 297 [=Mammalia (Linnaeus, 1758 part)]) and orders the Herbivorae (J. Gray, 1821 [=Sirenia (Illiger, 1811)]) and Carnivorae (J. Gray, 1821 [=Cetacea (Brisson, 1762 part)]). Synonymised within the Suborder Cetacea by McKenna and Bell (1997: 368).

Placentaria Fleming, 1822a: xxxii; 1822b: 169.

COMMENTS: When originally proposed, this unranked name was placed in the Mammalia (Linnaeus, 1758) and included the Pedata (Fleming, 1822a: xxxii; 1822b: 170 [=Placentalia (Bonaparte, 1838)]) (skin with appendices in the form of hair, spines or scales and containing the Unguiculata (Linnaeus, 1766), Bimana (Fleming, 1822a: xxxii; 1822b: 171 [=Primates (Linnaeus, 1758 part)]), Quadrumana (Blumenbach, 1791: 49) [=Primates (Linnaeus, 1758 part)]), Cheiroptera (J. Gray, 1821 [=Chiroptera (Blumenbach, 1779)]), Ferae (Linnaeus, 1758), Plantigrada[e] (J. Gray, 1821 [=Placentalia (Bonaparte, 1838 part)]), Digitigrada[e] (J. Gray, 1821 [=Carnivora (Bowdich, 1821)]), Palmata (Blumenbach, 1779: 58, 136 [=Carnivora (Bowdich, 1821) and Sirenia (Illiger, 1811)]), Glires (Linnaeus, 1758), Ungulata (Linnaeus, 1766 [=Placentalia (Bonaparte, 1838 part)]), Pecora (Linnaeus, 1758), Belluae (Linnaeus, 1758 [=Placentalia (Bonaparte, 1838 part)]) and Apoda (Fleming, 1822a: xxxiii, 202) (including Herbivora[e] J. Gray, 1821 [=Sirenia (Illiger, 1811)]) and Cetacea (Brisson, 1762). Name further described by Fleming (1822b: 169). Synonymised within Eutheria by Simpson (1945: 47) and within Placentalia by McKenna and Bell (1997: 80).

Suborder Monodelphes de Blainville, 1822a: Table 3.

COMMENTS: When originally proposed, this rank was placed in the Class Mammifères and included all mammals except the marsupials the monotremes. See Subclass Monodelphes de Blainville, 1816a for contents.

Cetae J. Gray, 1825a: 340.

COMMENTS: When originally proposed, this rank was placed in the Class Mammalia (Linnaeus, 1758) and included the families Balaenidae (J. Gray, 1821), Delphinidae (J. Gray, 1821), Trichechidae (J. Gray, 1821: 302 [=Odobenidae (J. Allen, 1880: ix, 5)], Manatidae (J. Gray, 1821: 309 [=Trichechidae (Gill, 1872: 14, 91)]); and Halicoridae (J. Gray, 1825a: 341 [=Dugongidae (J. Gray, 1821: 309)]).

Order Cetetherae Burnett, 1830c: 360.

COMMENTS: When originally proposed, this rank was placed in the Class Mammalia (Linnaeus, 1758) and included the types Loripeda (Burnett, 1830c [=Placentalia (Bonaparte, 1838)]), Semipeda (Burnett, 1830c [=Sirenia (Illiger, 1811)]) and Pinnipeda (Burnett, 1830c [=Cetacea (Brisson, 1762)]).

Type Loripeda Burnett, 1830c: 360.

COMMENTS: When originally proposed, this rank was placed within the Order Cetetherae (Burnett, 1830c [=Placentalia (Bonaparte, 1838)]) and included the kinds (=families) Phocidae (J. Gray, 1821) and Trichecidae (J. Gray, 1821: 302 [=Odobenidae (Allen, 1880: ix, 5)]).

Sectio Educabilia Bonaparte, 1838: 107.

COMMENTS: When originally proposed, this rank was placed within the Subclass Placentalia (Bonaparte, 1838) and included the Primates *(Quadrumana)* (Linnaeus, 1758), Ferae *(Carnivora)* (Linnaeus, 1758), Pinnipedia *(Amphibia)* (Illiger, 1811), Cete *(Natantia)* (Linnaeus, 1758 [=Cetacea (Brisson, 1762)]), Belluae *(Pachydermata)* (Linnaeus, 1758 [=Placentalia (Bonaparte, 1838 part)]) and Pecora (*Ruminantia*) (Linnaeus, 1758). Name also recognised as a subclass by Bonaparte (1840: 248).

Subclass Ineducabilia Bonaparte, 1838: 108, 110.

COMMENTS: When originally proposed, this rank was placed within the Subclass Placentalia (Bonaparte, 1838) and included the Bruta (Linnaeus, 1758: 16, 33 [=Mammalia (Linnaeus, 1758 part)]), Chiroptera (Blumenbach, 1779), Bestiae (Linnaeus, 1758: 16, 49 [=Mammalia (Linnaeus, 1758 part)]) and Glires (Linnaeus, 1758). Name also recognised as a subclass by Bonaparte (1840: 249).

Cetina Newman, 1843: 148.

COMMENTS: Rank not specified but proposed to be the equivalent of Cete of Linnaeus (1758) and included the genera *Balaena* Linnaeus, 1758: 75; *Monodon* Linnaeus, 1758: 75; *Manatus* Brunnich, 1772: 34, 38 [=*Trichechus* Linnaeus, 1758: 34]; *Halicore* Illiger, 1811 [=*Dugong* Lacépède, 1799a]; and *Trichecus* Linnaeus, 1758: 34. Synonymised within Cete by McKenna and Bell (1997: 366).

Placentalia Owen, 1849: 903.

COMMENTS: Rank not specified by Owen. Year of publication given as 1837 by Simpson (1945: 47), which appears to have been followed by subsequent authors including Aplin and Archer (1987: xxvi), and McKenna and Bell (1997: 80), however the volume gives the year of publication as 1849–1852. Name synonymised within Eutheria by Simpson (1945: 47) and within Monodelphia by Gregory (1947: 46).

HOMONYMS:

Subclass Placentalia Bonaparte, 1838, mammals of the Infraclass Placentalia. See individual entry.

Subclass Lissencephala Owen, 1858: 14, 22.

COMMENTS: When originally proposed, this rank was placed in the Class Mammalia and included the orders Bruta (Linnaeus, 1758: 16, 33 [=Mammalia (Linnaeus, 1758 part)]), Cheiroptera (J. Gray, 1821 [=Chiroptera (Blumenbach, 1779)]), Insectivora (Bowdich, 1821 [=Order Lipotyphla (Haeckel, 1866)]) and Rodentia (Bowdich, 1821). Followed by Owen (1859: 25).

Subclass Gyrencephala Owen, 1858: 18.

COMMENTS: When originally proposed, this rank was placed in the Class Mammalia and included the orders Quadrumana (Blumenbach, 1791: 49 [=Primates (Linnaeus, 1758 part)]), Carnivora (Bowdich, 1821), Artiodactyla

(Owen, 1848), Perissodactyla (Owen, 1848), Proboscidia (J. Gray, 1821: 305 [=Proboscidea (Illiger, 1811: 96)]), † Toxodontia (Owen, 1853b: 291, 309), Sirenia (Illiger, 1811) and Cetacea (Brisson, 1762). Followed by Owen (1859: 25).

Sub-class (Eutheria) Placentalia Gill, 1872: v, 1.

COMMENTS: When originally proposed, this rank was placed in the Class Mammalia (Linnaeus, 1758) and contained the superorders Educabilia (Bonaparte, 1838 [=Placentalia (Bonaparte, 1838 part)]) (including the orders Primates (Linnaeus, 1758), Ferae (Linnaeus, 1758), Ungulata (Linnaeus, 1766 [=Placentalia (Bonaparte, 1838 part)]), † Toxodontia (Owen, 1853b: 291, 309), Hyracoidea (Huxley, 1869: 101), Proboscidea (Illiger, 1811: 96), Sirenia (Illiger, 1811) and Cete (Linnaeus, 1758 [=Cetacea (Brisson, 1762)]) and Ineducabilia (Bonaparte, 1838 [=Placentalia (Bonaparte, 1838 part)]) (including the orders Chiroptera (Blumenbach, 1779), Insectivora (Bowdich, 1821 [=Order Lipotyphla (Haeckel, 1866)]), Glires (Linnaeus, 1758) and Bruta (Linnaeus, 1758: 16, 33 [=Mammalia (Linnaeus, 1758 part)]). The use of Eutheria by Gill (1872: v, vi) is the first usage of the term and clearly includes both the Placentalia and Marsupialia under the one lineage. Taxon recognised as the Infraclass Eutheria by Simpson (1931: 262; 1945: 47) and Luo *et al.* (2001b: 442), infracohort by Gardiner (1982: 229), supercohort by McKenna (1975: 27, 40), superdivision by Aplin and Archer (1987: xxi) and supercohort or infraclass by Shoshani (1992: 108). The term Eutheria was synonymised within the Subclass Theria (Parker & Haswell, 1897) by Gregory (1910: 230) and Simpson (1945: 40), and in the Supercohort Theria by McKenna and Bell (1997: 49). Subclass rank recognised by authors including Osborn (1910: 515), who included the infraclasses Didelphia (Metatheria including the Order Marsupialia) and Monodelphia (Placentals), Strahan (1983: xxi, 269; 2005: 8, 412), and Van Dyck and Strahan (2008: 10, 415). The introduction of the term Eutheria and its inclusion of both marsupials and placental mammals was discussed by Aplin and Archer (1987: xxvi) and recognised as a distinct rank. Confusion however arose when Huxley (1881: 657) also used the term Eutheria, though not as a formal taxon.

HOMONYMS:

Eutheria Huxley, 1881, placental mammals of the Class Mammalia. Synonymised within the Infraclass Placentalia (Bonaparte, 1838). See individual entry.

Eutheria Huxley, 1881: 657.

COMMENTS: Name does not appear to have been used as a formal taxon but, when originally proposed, it was placed in the Class Mammalia (Linnaeus, 1758) and included all families of placental mammals. The terms Prototheria, Metatheria and Eutheria used by Huxley (1881: 657) were theoretical terms to designate stages of evolution rather than as taxonomic names (see Gregory, 1910: 94; Aplin & Archer, 1987: xxvi; Simpson, 1945: 47, 164). Though the authorship of Eutheria is typically attributed to Gill (1872) the rank is usually used in the sense established by Huxley (1881), which has created confusion. Synonymised within the Cohort Placentalia by McKenna and Bell (1997: 80).

HOMONYMS:

Eutheria Gill, 1872, placental and marsupial mammals of the Class Mammalia. Synonymised within the Infraclass Placentalia (Bonaparte, 1838). See two individual entries as Subclass (Eutheria) Placentalia Gill, 1872 and Subclass (Eutheria) Placentalia Gill, 1872.

Order Clinodactyla Marsh, 1886: 177.

COMMENTS: When originally proposed, this rank was placed in the Class Mammalia (Linnaeus, 1758) and included the Mesaxonia (Marsh, 1886: 9, 177 [=Perissodactyla (Owen, 1848)]) and Paraxonia (Marsh, 1886: 9, 177 [=Artiodactyla (Owen, 1848)]). Synonymised within the Order Artiodactyla by McKenna and Bell (1997: 391).

Legion Cetomorpha Haeckel, 1895: 166, 490, 562.

COMMENTS: When originally proposed, this rank was placed in the Placentalia (Bonaparte, 1838) and included the Orders Sirenia (Illiger, 1811), Denticeta (Haeckel, 1895 [=Odontoceti (Flower, 1867)]) and Mysticeta (Haeckel, 1895 [=Mysticeti (J. Gray, 1864c)]). Synonymised within Cetacea by McKenna and Bell (1997: 366).

Superfamily? Otarioidea Smirnov, 1908: 14.

TYPE GENUS: *Otaria* Péron, 1816: 37.

COMMENTS: When originally proposed, this rank included the families Trichechidae (Gill, 1872: 14, 91) and Otariidae (J. Gray, 1825). Synonymised within the Family Otariidae by Wozencraft (2005: 590).

Subclass Holotheria Jaeckel, 1911: viii, 227.

COMMENTS: When originally proposed, this rank was placed in the Class Mammalia (Linnaeus, 1758) and included the orders Carnivori [sic] [=Carnivora (Bowdich, 1821)], Cetacei [sic] [=Cetacea (Brisson, 1762)], Diungulati (Jaeckel, 1911: viii, 232 [=Suiformes (Jaeckel, 1911)]), Tubulidentati [=Tubulidentata (Huxley, 1872: 288)], Ulgulati [sic] [=Ungulata (Linnaeus, 1766 [=Placentalia (Bonaparte, 1838 part)])] and Sirenii [sic] [=Sirenia (Illiger, 1811)].

HOMONYMS:

Holotheria Hopson, 1994, mammals of the Class Mammalia. Name is a junior synonym of Theria (Parker & Haswell, 1897). See individual entry.

Subclass Neotherida Broom, 1935: 36.

COMMENTS: When originally proposed, this rank was placed in the Class Mammalia (Linnaeus, 1758) and included the

Pholidota (Weber, 1904: vi, 420), Tubulidentata (Huxley, 1872: 288), Hyracoidea (Huxley, 1869: 101), Rodentia (Bowdich, 1821), Insectivora (Bowdich, 1821 [=Lipotyphla (Haeckel, 1866)]), Carnivora (Bowdich, 1821), Menotyphla (Haeckel, 1866: clx), Chiroptera (Blumenbach, 1779), Dermoptera (Illiger, 1811: 63, 116), Primates (Linnaeus, 1758) and potentially the Cetacea (Brisson, 1762) based on having a mesethmoid. Rank criticised by Parrington (1974: 425), who suggested it had no credence whatsoever, and not recognised by subsequent authors.

Cohort Ferungulata Simpson, 1945: xiii, 105.

COMMENTS: When originally proposed as a new rank it was placed in the Class Mammalia (Linnaeus, 1758) and included the superorders Ferae (Linnaeus, 1758), Protungulata (Weber, 1904: 587 [=Ungulata (Linnaeus, 1766] [=Placentalia (Bonaparte, 1838 part)]), Paenungulata (Simpson, 1945), Mesaxonia (Marsh, 1886: 9, 177 [=Perissodactyla (Owen, 1848)]) and Paraxonia (Marsh, 1886: 9, 177 [=Artiodactyla (Owen, 1848)]). Synonymised within Placentalia by McKenna and Bell (1997: 80). Recognised as a cohort by Skinner and Chimimba (2005: vi, 353) but was discussed in detail by Asher and Helgen (2010: 4, 5) who noted the different concepts that have been associated with this term and argued that a new name is justified. It can also be argued that cladistic structure within the Laurasiatheria is so insecure that further divisions above the level of order are not called for, and that Laurasiatheria should itself be given the rank of cohort (just below infraclass Placentalia).

Cohort Placentata Turnbull, 1971: 176.

COMMENTS: When originally proposed, this rank was placed in the Infraclass Eutheria (Gill, 1872 [=Placentalia (Bonaparte, 1838)]) at a rank equivalent to the 'Old Eutheria or Placentalia' and included the Cohort Division Unguiculata (with the orders Insectivora (Bowdich, 1821 [=Order Lipotyphla (Haeckel, 1866)]), Dermoptera (Illiger, 1811: 63, 116) and Chiroptera (Blumenbach, 1779) along with the rest of the placental orders). Synonymised within Placentalia by McKenna and Bell (1997: 80).

Clade Epitheria McKenna, 1975: 28, 41.

COMMENTS: When originally proposed, this clade included all placental mammals except Xenarthra (Cope, 1889a: 657). Clade recognised as having potential support by Asher and Helgen (2010: 5).

Clade Cetferungulata Arnason *et al.*, 1999: 339, 344.

COMMENTS: When originally proposed, this clade included the orders Carnivora (Bowdich, 1821), Perissodactyla (Owen, 1848), Artiodactyla (Owen, 1848) and Cetacea (Brisson, 1762). Grouping previously recognised by Xu *et al.* (1996: 1167). Clade recognised as having potential support by Asher and Helgen (2010: 5).

Clade Pseudoungulata Waddell *et al.*, 1999a: 4.

COMMENTS: When originally proposed, this clade included the Tubulidentata (Huxley, 1872: 288) and Paenungulata (Simpson, 1945).

Clade Zooamata Waddell *et al.*, 1999a: 4.

COMMENTS: When originally proposed, this clade included the 'animal friends, as the group contains cats, dogs and horses etc', which was refined by Asher and Helgen (2010: 5) to include Perissodactyla (Owen, 1848) and Ferae (Linnaeus, 1758) and who suggested it has potential support.

Clade Atlantogenata Waddell *et al.*, 1999c: 120.

COMMENTS: When originally proposed, this clade contained the Afrotheria (Stanhope *et al.*, 1998) and Xenarthra (Cope, 1889a: 657). Rank subsequently recognised by Asher and Helgen (2010: 4).

Clade Boreoeutheria Springer & de Jong, 2001: 1709.

COMMENTS: When originally proposed, this clade included the Laurasiatheria (Waddell *et al.*, 1999a) and Euarchontoglires (Murphy *et al.*, 2001a). Name also proposed by Murphy *et al.* (2001a: 2349) but was published later. Recognised at cohort rank by Springer *et al.* (2007: 21). Recognised as the senior name ahead of clades Boreotheria (Waddell *et al.*, 2001) and Boreoplacentalia (Arnason *et al.*, 2008) by Asher and Helgen (2010: 4).

Clade Boreotheria Waddell *et al.*, 2001: 141, 148.

COMMENTS: When originally proposed, this clade was placed in the Clade Exafroplacentalia (Waddell *et al.*, 2001 [=Placentalia (Bonaparte, 1838 part)]) and included the Laurasiatheria (Waddell *et al.*, 1999a) and Surpraprimates (Waddell *et al.*, 2001: 141). Synonymised within Boreoeutheria by Asher and Helgen (2010: 4).

Clade Exafroplacentalia Waddell *et al.*, 2001: 148.

COMMENTS: When originally proposed, this clade was placed in the Placentalia (Bonaparte, 1838) and included the Boreotheria (Waddell *et al.*, 2001 [=Placentalia (Bonaparte, 1838 part)]). Clade recognised as having potential support by Asher and Helgen (2010: 5).

Clade Ostentoria Amrine-Madsen *et al.*, 2003b: 225.

COMMENTS: When originally proposed, this clade included the orders Carnivora (Bowdich, 1821) and Pholidota (Weber, 1904: 412). The description of this rank refers to the publication of Springer *et al.* (2007: 21) as being *in press*, which was subsequently introduced as a new name. Synonymised within Ferae, of McKenna and Bell (1997: 211), by Asher and Helgen (2010: 4).

Clade Notolegia Springer *et al.*, 2005: 39.

COMMENTS: When originally proposed, this clade included all placental mammals except the Afrotheria (Stanhope *et al.*, 1998). Name synonymised within the Clade Exafroplacentalia (Waddell *et al.*, 2001).

Clade Xenafrotheria Asher, 2005: 65.

COMMENTS: When originally proposed, this clade was placed in the Placentalia (Bonaparte, 1838) and included the Xenarthra (Cope, 1889a: 657) and Afrotheria (Stanhope *et al.*, 1998). Clade synonymised within the Atlantogenata by Asher and Helgen (2010: 5).

Clade Pegasoferae Nishihara *et al.*, 2006: 9929.

COMMENTS: When originally proposed, this clade included the Chiroptera (Blumenbach, 1779), Perissodactyla (Owen, 1848), Carnivora (Bowdich, 1821) and Pholidota (Weber, 1904: vi, 420). Clade recognised as having potential support by Asher and Helgen (2010: 5).

Superorder Notolegia Springer *et al.*, 2007: 21.

COMMENTS: When originally proposed as a new rank it was placed in the Infralegion Placentalia (Bonaparte, 1838) and included the most recent common ancestor of Xenarthra (Cope, 1889a: 657), Eulipotyphla (Waddell *et al.*, 1999a [=Lipotyphla (Haeckel, 1866)]), Chiroptera (Blumenbach, 1779), Cetartiodactyla (Montgelard *et al.*, 1997 [=Artiodactyla (Owen, 1848) and Cetacea (Brisson, 1762)]), Perissodactyla (Owen, 1848), Carnivora (Bowdich, 1821), Pholidota (Weber, 1904: vi, 420), Rodentia (Bowdich, 1821), Lagomorpha (Brandt, 1855), Primates (Linnaeus, 1758), Scandentia (Wagner, 1855: xix, 524), Dermoptera (Illiger, 1811: 63, 116) and all of its descendants.

Mirorder Ostentoria Springer *et al.*, 2007: 21.

COMMENTS: When originally proposed as a new rank it was placed in the Grandorder Fereuungulata (Waddell *et al.*, 1999a [=Placentalia (Bonaparte, 1838)]) and included the orders Carnivora (Bowdich, 1821) and Pholidota (Weber, 1904: 412).

Clade Boreoplacentalia Arnason *et al.*, 2008: 37, 39.

COMMENTS: When originally proposed, this clade was placed in the Placentalia (Bonaparte, 1838) and included the Archontoglires (Arnason *et al.*, 2008 [=Euarchontoglires (Murphy *et al.*, 2001a)]) and Laurasiaplacentalia (Arnason *et al.*, 2008 [=Laurasiatheria (Waddell *et al.*, 1999a)]). Name synonymised within Boreoeutheria (Springer & de Jong, 2001) by Asher and Helgen (2010: 4).

Clade Notoplacentalia Arnason *et al.*, 2008: 37, 40.

COMMENTS: When originally proposed, this clade was placed in the Placentalia (Bonaparte, 1838) and included the Afroplacentalia (Arnason *et al.*, 2008 [=Afrotheria (Stanhope *et al.*, 1998)]) and Xenarthra (Cope, 1889a: 657). Taxon synonymised within the Atlantogenata by Asher and Helgen (2010: 4).

Cohort Afrotheria, Stanhope *et al.,* 1998

Supercohort Afrotheria Stanhope *et al.*, 1998: 9970, 9971.

COMMENTS: When originally proposed as a new rank it included the Paenungulata (Simpson, 1945) and Afroinsectiphilia (Waddell *et al.*, 2001: 148). Recognised as a valid clade by Asher and Helgen (2010: 3, 4). Recognised at supercohort rank by Skinner and Chimimba (2005: v, 1) and Springer *et al.* (2007: 20).

Clade Afroplacentalia Arnason *et al.*, 2008: 37, 39.

COMMENTS: When originally proposed, this clade was placed in the Notoplacentalia (Arnason *et al.*, 2008 [=Placentalia (Bonaparte, 1838)]) and included the Paenungulata (Simpson, 1945) and Afroinsectiphilia (Waddell *et al.*, 2001: 148). Synonymised within Afrotheria by Asher and Helgen (2010: 3, 4).

Order Paenungulata Simpson, 1945

Superorder Paenungulata Simpson, 1945: xiii, 131.

COMMENTS: When originally proposed as a new rank it was placed in the Class Mammalia (Linnaeus, 1758) and included the Cohort Ferungulata (Simpson, 1945 [=Placentalia (Bonaparte, 1838 part)]) and included the orders † Pantodonta (Cope, 1873: 40, 67), † Dinocerata (Marsh, 1873: 117), † Pyrotheria (Ameghino, 1895: 608), Proboscidea (Illiger, 1811: 96), † Embrithopoda (C. Andrews, 1906: 224), Hyracoidea (Huxley, 1869: 101) and Sirenia (Illiger, 1811). Synonymised within the new Order Uranotheria by McKenna and Bell (1997: 490) but recognised as a superorder by Minkoff (1976: 153), cohort by Skinner and Chimimba (2005: v, 41), grandorder by Springer *et al.* (2007: 20) and as a clade by Gheerbrant *et al.* (2005: 84) and Asher and Helgen (2010: 4, 5). Goodman *et al.* (1998: 595) proposed that the rank of Order in mammals be applied to a crown-group clade with a time depth going back to the K/T boundary. The separation between Proboscidea, Sirenia and Hyracoidea would appear to go back to about the Palaeocene/Eocene boundary (Springer, 1997: 295), so their ordinal rank is questionable, and we have here taken the view that the Paenungulata should be downranked, as an order.

Order Uranotheria McKenna & Bell, 1997: 490.

COMMENTS: When originally proposed as a new rank it was placed in the Mirorder Altungulata (Prothera & Schoch, 1989: 510) and included the suborders Hyracoida (Huxley, 1869: 101), † Embrithopoda (C. Andrews, 1906: 224) and

Tethytheria (McKenna, 1975 [=Paenungulata (Simpson, 1945)]). Synonymised within Paenungulata Simpson (1945) by Asher and Helgen (2010: 4).

Mirorder Tethytheria McKenna, 1975: 42.

COMMENTS: When originally proposed as a new rank it was placed in the Grandorder Ungulata (Linnaeus, 1766 [=Placentalia (Bonaparte, 1838 part)]), and included the orders Proboscidea (Illiger, 1811: 96) and Sirenia (Illiger, 1811). Further support for Tethytheria was provided by Arnason *et al.* (2008: 40), Poux *et al.* (2008: 3) and Asher and Helgen (2010: 4). A clade containing Sirenia and Hyracoidea has also been recognised by several authors including Malia *et al.* (2002: 99) and Douady *et al.* (2003: 8326), while Kuntner *et al.* (2010: 1) found Sirenia to be sister to a clade containing Hyracoidea plus Proboscidea. In addition Springer (1997: 290) and Pardini *et al.* (2007: 1333) found a clade containing Proboscidea and Sirenia to the exclusion of Hyracoidea, hence the validity of Tethytheria could not be supported.

Suborder Sirenia Illiger, 1811

Family Sirenia Illiger, 1811: 140.

TYPE GENUS: Not based on a genus group name.

COMMENTS: When originally proposed, this rank was placed in the Order Natantia (Illiger, 1811 [=Placentalia (Bonaparte, 1838 part)]) that included the genera *Manatus* Brünnich, 1772: 34, 38 [=*Trichechus* Linnaeus, 1758: 34]; *Halicore* Illiger, 1811 [=*Dugong* Lacépède, 1799a]; and † *Rytina* Illiger, 1811: 141 [=† *Hydrodamalis* Retzius, 1794: 292]. Sirenia was recognised at family rank within Cetacea by J. Gray (1827a: 377), suborder rank within Cetacea by J. Gray (1866b: 356) and ordinal rank by Gill (1872: v, 13, 49, 91), Simpson (1945: 135) and most subsequent authors, except McKenna and Bell (1997: 493) who placed it at infraorder rank. Order reviewed by Domning (1996: 1).

Order Herbivorae J. Gray, 1821: 309.

COMMENTS: When originally proposed, this rank was placed in the Class Cetaceae (J. Gray, 1821 [=Cetacea Brisson, 1762]) and included the families Manatidae (J. Gray, 1821: 309 [=Trichechidae (Gill, 1872: 14, 91)]) and Dugongidae (J. Gray, 1821). Synonymised within the Order Sirenia by McKenna and Bell (1997: 493) and Shoshani (2005: 92).

Type Semipeda Burnett, 1830c: 360.

TYPE GENUS: Not based on a genus group name.

COMMENTS: When originally proposed, this rank was placed in the Order Cetetherae (Burnett, 1830c [=Mammalia (Linnaeus, 1758 part)]) and included the Kind [=Family] Manatidae (J. Gray, 1821: 309 [=Trichechidae (Gill, 1872: 14, 91)]) (including the genera *Manatus* Brünnich, 1772: 34, 38 [=*Trichechus* Linnaeus, 1758: 34]; *Halicore* Illiger, 1811 [=*Dugong* Lacépède, 1799a]; and *Rytina* Illiger, 1811: 141 [=† *Hydrodamalis* Retzius, 1794: 292]).

Tribe Phytophaga F. Cuvier, 1836a: 563.

TYPE GENUS: Not based on a genus group name.

COMMENTS: When originally proposed, this rank was placed in the Order Cetacea (Brisson, 1762) and included the genera *Manatus* Brünnich, 1772: 34, 38 [=*Trichechus* Linnaeus, 1758: 34]; *Halicore* Illiger, 1811 [=*Dugong* Lacépède, 1799a]; and *Rytina* Illiger, 1811: 141 [=† *Hydrodamalis* Retzius, 1794: 292].

Order? Sirénoïdes Van Bénéden, 1852: 339.

COMMENTS: When originally proposed, this rank was placed in the Class Mammalia (Linnaeus, 1758). Not given but appears to be at ordinal rank. See review of Agassiz (1859: 360), who Latinised it as Sirenoidea. Synonymised (in its Latinised form) within the Order Sirenia by McKenna and Bell (1997: 493) and Shoshani (2005: 92).

HOMONYMS:

Sirenoidea Oken, 1845: 870, African lungfishes of the Class Sarcopterygii (Order Lepidosireniformes; Family Protopteridae).

Sirenoidea Goodrich, 1930, amphibians of the Order Caudata. Name is a synonym of Caudata (G. Fischer, 1813a: 58). See Frost (2011).

Order Sirénides Gervais, 1855a: 308.

COMMENTS: When originally proposed, this rank was placed in Class Mammalia (Linnaeus, 1758) and included the genera *Rytina* Illiger, 1811: 141 [=† *Hydrodamalis* Retzius, 1794: 292]; *Halicore* Illiger, 1811 [=*Dugong* Lacépède, 1799a]; † *Halitherium* Kaup, 1838a: 319; 1838b: 536; and *Manatus* Brünnich, 1772: 34, 38 [=*Trichechus* Linnaeus, 1758: 34].

Subclass Mutilata Owen, 1859: 9.

COMMENTS: When originally proposed, this rank was placed in the Class Mammalia (Linnaeus, 1758) and included the Order Cetacea (Brisson, 1762), and contained the dugong and whales.

Suborder Phycoceta Haeckel, 1866: clix.

COMMENTS: When originally proposed, this rank was placed in the Order Cetacea (Brisson, 1762) and included the families Manatida (Haeckel, 1866: clix [=Trichechidae (Gill, 1872: 14, 91)]) and Rhyntinida (Haeckel, 1866: clix [=Dugongidae (Gray, 1821: 309 part)]). Synonymised within the Order Sirenia by McKenna and Bell (1997: 493) and Shoshani (2005: 92).

Grand Tribe Halobioidea Ameghino, 1889: xxiv, 354, 652.

COMMENTS: Placed on the Grand Seccion Ungulata (Linnaeus, 1766 [=Placentalia (Bonaparte, 1838 part)]), and included the orders Sirenia (Illiger, 1811) and † Prosirenia (Ameghino, 1889: 652). Synonymised within the Order Sirenia by McKenna and Bell (1997: 493) and Shoshani (2005: 92).

Suborder Trichechiformes Hay, 1923: 109.

COMMENTS: When originally proposed as a new rank it was placed in the Order Sirenia (Illiger, 1811) and included the genera *Trichechus* Linnaeus, 1758: 34; and *Dugong* Lacépède, 1799a. Synonymised within the Order Sirenia by McKenna and Bell (1997: 493) and Shoshani (2005: 92).

Order Sireniformes Kinman, 1994: 38.

COMMENTS: When originally proposed, this rank was placed in the Class Mammalea (Kinman, 1994 [=Mammalia (Linnaeus, 1758)]). Synonymised within the Order Sirenia by McKenna and Bell (1997: 493) and Shoshani (2005: 92).

Family Dugongidae J. Gray, 1821

Family Dugongidae J. Gray, 1821: 309.

TYPE GENUS: *Dugong* Lacépède, 1799a.

COMMENTS: When originally proposed, this rank was placed in the Order Herbivorae (J. Gray, 1821 [=Sirenia (Illiger, 1811)]) and included the genus *Dugong* Lacépède, 1799a. Family name recognised by Palmer (1895b: 450), Simpson (1945: 135) and subsequent authors. The recent genera were assigned to the subfamilies Dugonginae (J. Gray, 1821: 309) and the extinct Hydrodamalinae (Palmer, 1895b: 450) by Domning (1994: 189), and followed by McKenna and Bell (1997: 496), Rice (1998: 131), and Shoshani (2005: 92).

Family Halicoridae J. Gray, 1825a: 341.

TYPE GENUS: *Halicore* Illiger, 1811 [=*Dugong* Lacépède, 1799a].

COMMENTS: When originally proposed, this rank was placed in the Order Cetae (J. Gray, 1825a [=Placentalia (Bonaparte, 1838 part)]) and included the genus *Halicora* [=*Halicore*] Illiger, 1811 [=*Dugong* Lacépède, 1799a]. Synonymised within the Family Dugongidae by Simpson (1945: 135), D. Wilson (1993: 365), McKenna and Bell (1997: 493), and Rice (1998: 131).

† Tribe Rhytineae Brandt, 1833a: 115.

TYPE GENUS: † *Rhytina* [=*Rytina*] Illiger, 1811: 141 [=† *Hydrodamalis* Retzius, 1794: 292].

COMMENTS: When originally proposed, this rank was placed in the 'Cetacea Herbivora Cuv.' and included the genera † *Rytina* Illiger, 1811: 141 [=† *Hydrodamalis* Retzius, 1794: 292] and *Stellerus* 'Cuvier' [=Desmarest, 1822b: 510, which is derived from 'Les Stellères' of G. Cuvier, 1816a: 275] [=† *Hydrodamalis* Retzius, 1794: 292]. Rice (1998: 131) noted that this is an incorrect original spelling and that the type genus is a junior synonym of † *Hydrodamalis* Retzius, 1794: 292, so the family name is invalid because it was replaced prior to 1961 (see Article 40(b) of the Code (ICZN, 1985a: 81). Synonymised within the Family Dugongidae by Rice (1998: 131).

† Family Rytinadae J. Gray, 1843a: xxiii.

TYPE GENUS: † *Rytina* Illiger, 1811: 141 [=† *Hydrodamalis* Retzius, 1794: 292].

COMMENTS: When originally proposed, this rank was placed in the Order Cete (Linnaeus, 1758 [=Cetacea (Brisson, 1762)]) and included the genera *Rytina* Illiger, 1811: 141 [=† *Hydrodamalis* Retzius, 1794: 292]; † *Stellerus* 'F. Cuvier' [=Desmarest, 1822b: 510, which is derived from 'Les Stellères' of G. Cuvier, 1816a: 275] [=† *Hydrodamalis* Retzius, 1794: 292]; and † *Hydrodamalis* Retzius, 1794: 292. Synonymised within the Family Dugongidae by McKenna and Bell (1997: 493), and Rice (1998: 131).

Family Halitherida Carus, 1868: 168.

TYPE GENUS: † *Halitherium* Kaup, 1838a: 319; 1838b: 536.

COMMENTS: When originally proposed, this rank was placed in the Suborder Sirenia (Illiger, 1811) (in the Order Natantia (Illiger, 1811 [=Placentalia (Bonaparte, 1838 part)])) and included the genera *Manatus* Brünnich, 1772: 34, 38 [=*Trichechus* Linnaeus, 1758: 34]; *Halicore* Illiger, 1811 [=*Dugong* Lacépède, 1799a]; and † *Rhytina* [=*Rytina*] Illiger, 1811: 141 [=† *Hydrodamalis* Retzius, 1794: 292]. Synonymised within the Family Dugongidae by McKenna and Bell (1997: 493).

Family Halitheriidae Gill, 1872: 13.

TYPE GENUS: † *Halitherium* Kaup, 1838a: 319; 1838b: 536.

COMMENTS: When originally proposed, this rank was placed in the Order Sirenia (Illiger, 1811) and included the † *Halitherium* Kaup, 1838a: 319; 1838b: 536. Synonymised within the Family Dugongidae by McKenna and Bell (1997: 493).

Family Rhytinidae Gill, 1872: 14, 91.

TYPE GENUS: † *Rhytina* [=*Rytina*] Illiger, 1811: 141 [=† *Hydrodamalis* Retzius, 1794: 292].

COMMENTS: When originally proposed, this rank was placed in the Order Sirenia (Illiger, 1811) and included the † *Rhytina* [=*Rytina*] Illiger, 1811: 141 [=† *Hydrodamalis* Retzius, 1794: 292]. Synonymised within the Family Dugongidae by McKenna and Bell (1997: 493).

Family Hydrodamalidae Palmer, 1895b: 450.

TYPE GENUS: † *Hydrodamalis* Retzius, 1794: 292.

COMMENTS: When originally proposed, this rank included the genus † *Hydrodamalis* Retzius, 1794: 292. Synonymised within the Family Dugongidae by Rice (1998: 131).

† Subfamily Rhytiodinae Abel, 1914: 217.

TYPE GENUS: † *Rytiodus* Lartet, 1866: 682.

COMMENTS: When originally proposed, this rank was placed in the Family Haricoriden [=Halicoridae] (Gray, 1825a [=Dugongidae (J. Gray, 1821)]) and included the genus † *Rytiodus* Lartet, 1866: 682. Subfamily combined within the Subfamily Dugonginae by Domning (1994: 186, 189).

† Subfamily Eotherioidinae Kretzoi, 1941: 154.

TYPE GENUS: † *Eotheroides* Palmer, 1899b: 494.

COMMENTS: When originally proposed, this rank was placed in the Family Dugongidae (J. Gray, 1821) and included the genera † *Eotherioides* Palmer, 1899b: 494; † *Archaeosiren* Abel, 1913: 300 [=† *Eosiren* Andrews, 1902: 293]; † *Eosiren* Andrews, 1902: 293; † *Manatherium* Hartlaub, 1886: 369, 378 and † *Thalattosiren* Sickenberg, 1928: 293.

† Family Prototheriidae Kretzoi, 1941: 155.

TYPE GENUS: † *Prototherium* de Zigno, 1887: 731.

COMMENTS: When originally proposed, this rank was family included the genus † *Prototherium* de Zigno, 1887: 731.

Subfamily Dugonginae J. Gray, 1821

Subfamily Dugonginae J. Gray, 1821: 309.

TYPE GENUS: *Dugong* Lacépède, 1799a.

COMMENTS: When originally proposed, this rank was placed in the Order Herbivorae (J. Gray, 1821 [=Sirenia (Illiger, 1811)]) and included the genus *Dugong* Lacépède, 1799a. Subfamily rank recognised by Simpson (1932: 424), Kretzoi (1941: 154) (introduced as new), Rathbun (1984: 543), McKenna and Bell (1997: 496) and Shoshani (2005: 92). The second subfamily of the Family Dugongidae is the extinct Subfamily Hydrodamalinae (Palmer, 1895b: 450). Subfamily rank not recognised by Van Dyck and Strahan (2008: 10), and as long as both subfamilies are monotypic their recognition seems redundant.

Dugong Lacépède, 1799

Dugong Lacépède, 1799a: 17.

TYPE SPECIES: *Dugong indicus* Lacépède, 1799a [=*Dugong dugon* (P. Müller, 1776)] by monotypy.

COMMENTS: Taxonomic decision of Corbet (1978: 193) to include the genera *Dugong* Lacépède, 1799a and † *Hydrodamalis* Retzius, 1794: 292 in the family Dugongidae. Reviewed by Husar (1978: 1).

Platystomus G. Fischer, 1803: 353.

TYPE SPECIES: *Trichecus dugon* P. Müller, 1776 (as *Platystomus dugong*) [=*Dugong dugon* (P. Müller, 1776)] by monotypy.

COMMENTS: Synonymised within *Dugong* by Iredale and Troughton (1934: 68), Husar (1978: 1), D. Wilson (1993: 365), Domning (1996: 391), McKenna and Bell (1997: 496) and Shoshani (2005: 92).

HOMONYMS:

Platystomus Schönherr, 1833: 129, beetles of the Class Insecta (Order Coleoptera, Family Anthribidae). Genus is a synonym of *Platystomos* Schneider, 1791: 21. See György (2006: 66).

Platystomus Swainson, 1837: 261, flycatchers of the Class Aves (Order Passeriformes, Family Cracticidae). Genus is a synonym of *Peltops* Wagler, 1829a: 656.

Dugungus Tiedemann, 1808: 554.

TYPE SPECIES: *Trichechus dugon* P. Müller, 1776 [=*Dugong dugon* (P. Müller, 1776)] by haplotypy.

COMMENTS: Emendation by Latinisation of *Dugong*. Synonymised within *Dugong* by Palmer (1904: 246), Iredale and Troughton (1934: 68), Husar (1978: 1), D. Wilson (1993: 365), Domning (1996: 391), McKenna and Bell (1997: 496) and Shoshani (2005: 92).

Halicore Illiger, 1811: 140.

TYPE SPECIES: *Trichechus dugong* Gmelin, 1788 [=*Dugong dugon* (P. Müller, 1776)] by original designation.

COMMENTS: Genus recognised by J. Gray (1866b: 357, 360). Synonymised within *Dugong* by Palmer (1904: 306), Iredale and Troughton (1934: 69), Husar (1978: 1), D. Wilson (1993: 365), Domning (1996: 391), McKenna and Bell (1997: 496) and Shoshani (2005: 92).

Amblychilus G. Fischer, 1814: 638.

TYPE SPECIES: *Trichechus dugon* P. Müller, 1776 [=*Dugong dugon* (P. Müller, 1776)] by original designation.

COMMENTS: *Nomen novum* for *Platystomus* G. Fischer, 1803. Synonymised within *Dugong* by Domning (1996: 391) and Shoshani (2005: 92).

Dugongidus J. Gray, 1821: 309.

TYPE SPECIES: *Trichechus dugon* P. Müller, 1776 [=*Dugong dugon* (P. Müller, 1776)] by original designation.

COMMENTS: Synonymised within *Dugong* by Palmer (1904: 246), Husar (1978: 1), Domning (1996: 391), McKenna and Bell (1997: 496) and Shoshani (2005: 92).

Halicora Fleming, 1822b: 204.

TYPE SPECIES: Unjustified emendation of *Halicore* Illiger, 1811: 140.

COMMENTS: Synonymised within *Dugong* by Domning (1996: 391).

Dugong dugon (P. Müller, 1776)

Dugong

Trichecus Dugon P. Müller, 1776: 21.

TYPE LOCALITY: Cape of Good Hope to the Philippines (Husar, 1978: 1).

COMMENTS: Placed in the genus *Dugong* by Palmer (1895b: 450) and *Halicore* by Iredale and Troughton (1934: 69). Taxonomic decision of Corbet (1978: 193). Reviewed by Husar (1978: 1) and Domning (1996: 391).

Phoca manatus Brisson, 1762: 164.

TYPE LOCALITY: Amboine [=Ambon], Maluku Islands, Indonesia.
COMMENTS: Synonymised within dugong by Domning (1996: 391).

[Trichechus] dugung Erxleben, 1777: 599.

TYPE LOCALITY: Indian Ocean.
COMMENTS: Synonymised within *dugon* by Husar (1978: 1), D. Wilson (1993: 365), Domning (1996: 391) and Shoshani (2005: 92).

[Rosmarus] Indicus Boddaert, 1785: 169.

TYPE LOCALITY: Based on Buffon's 'Dugon' and Pennant's 'Indian Walrus'.
COMMENTS: Synonymised within *dugon* by Domning (1996: 391) and Shoshani (2005: 92).

[Trichechus] Dugong Gmelin, 1788: 60.

TYPE LOCALITY: Philippines.
COMMENTS: Synonymised within *dugon* by Domning (1996: 391) and Shoshani (2005: 92).

M.[anatus] australis Retzius, 1794: 291.

TYPE LOCALITY: Unknown.
COMMENTS: Synonymised within *dugon* by Domning (1996: 391) and Shoshani (2005: 92).

Dugong inducus Lacépède, 1799a: 17.

TYPE LOCALITY: Indian Ocean.
COMMENTS: Synonymised within *dugon* by Husar (1978: 1), D. Wilson (1993: 365) and Domning (1996: 391)

Halicore cetacea Illiger, 1815: 103.

TYPE LOCALITY: Unknown.
COMMENTS: Synonymised within *dugon* by Domning (1996: 391) and Shoshani (2005: 92).

Halicore Syren Brookes, 1828: 40.

TYPE LOCALITY: Unknown. 'A very fine cranium presented by Captain Jeffreys.'
COMMENTS: Synonymised within *dugon* by Domning (1996: 391) and Shoshani (2005: 92).

Halicore hemprichii Ehrenberg, 1830: unnumbered page, footnote.

TYPE LOCALITY: Barkan Island, Red Sea.
COMMENTS: Synonymised within *dugon* by Husar (1978: 1), D. Wilson (1993: 365), Domning (1996: 391) and Shoshani (2005: 92).

Halicore Lottum Ehrenberg, 1830: unnumbered page, footnote.

TYPE LOCALITY: Hauakal Island, southern part of Red Sea.
COMMENTS: Synonymised within *dugon* by Husar (1978: 1), D. Wilson (1993: 365), Domning (1996: 391) and Shoshani (2005: 92).

Halicore tabernaculi Rüppell, 1834: 113.

TYPE LOCALITY: Red Sea.
COMMENTS: Species recognised by J. Gray (1866b: 364). Synonymised within *dugon* by Husar (1978: 1), D. Wilson (1993: 365) and Domning (1996: 391).

Halicore Australis Owen, 1847b: 323.

TYPE LOCALITY: Endeavour Strait, Cape York, north Queensland, Australia.
COMMENTS: Recognised as described by Iredale and Troughton (1934: 69) and within *Dugong* by Troughton (1967: 191). Synonymised within *dugon* by Husar (1978: 1), D. Wilson (1993: 365) and Domning (1996: 391).

Halicore malayana Owen, 1875a: 560.

TYPE LOCALITY: *Nomen nudum*; *lapsus*?
COMMENTS: Synonymised within *dugon* by Domning (1996: 391) and Shoshani (2005: 92).

Halicore cetacea Heuglin, 1877: 135.

TYPE LOCALITY: Red Sea.
COMMENTS: Synonymised within *dugon* by Husar (1978: 1) and D. Wilson (1993: 365).

Cohort Euarchontoglires Murphy *et al.*, 2001

Clade Euarchontoglires Murphy *et al.*, 2001a: 2348.

COMMENTS: When originally proposed, this clade was placed in the Clade Boreoeutheria (Springer & de Jong, 2001 [=Placentalia (Bonaparte, 1838 part)]) and included Glires (Linnaeus, 1758) and Archonta (Gregory, 1910). Clade recognised as the senior name ahead of clades Archontoglires (Arnason *et al.*, 2008) and Supraprimates (Waddell *et al.*, 2001) by Asher and Helgen (2010: 4). Given the rank of supercohort by Skinner and Chimimba (2005: v, 63) and magnorder by Springer *et al.* (2007: 21).

Clade Supraprimates Waddell *et al.*, 2001: 141, 148.

COMMENTS: When originally proposed, this clade was placed in the Placentalia (Bonaparte, 1838) and included the Euarchonta (Waddell *et al.*, 1999a [=Archonta (Gregory, 1910)]) and Glires (Linnaeus, 1758). Synonymised within Euarchontoglires (Murphy *et al.*, 2001a) by Asher and Helgen (2010: 4).

Clade Archontoglires Arnason *et al.*, 2008: 37, 39.

COMMENTS: When originally proposed, this clade was placed in the Boreoplacentalia (Arnason *et al.*, 2008) and included the Archonta (Gregory, 1910) (including the Orders Primates (Linnaeus, 1758), Dermoptera (Illiger, 1811: 63, 116), Scandentia (Wagner, 1855: xix, 524)) and Glires (Linnaeus, 1758) (including Lagomorpha (Brandt, 1855) and Rodentia (Bowdich, 1821)). Name is equivalent to the Clade Supraprimates (Waddell *et al.*, 2001) and was synonymised within Euarchontoglires (Murphy *et al.*, 2001a) by Asher and Helgen (2010: 4).

Superorder Archonta Gregory, 1910 *sensu* Waddell *et al.*, 1999

Superorder Archonta Gregory, 1910: 322, 465.

COMMENTS: When originally proposed, this rank was placed in the Infraclass Eutheria (Huxley, 1881 [=Placentalia (Bonaparte, 1838)]) and included the orders Menotyphla (Haeckel, 1866: clx), Dermoptera (Illiger, 1811: 63, 116), Chiroptera (Blumenbach, 1779) and Primates (Linnaeus, 1758). Recognised at grandorder rank by McKenna and Bell (1997: 295) and as a valid clade (without the Chiroptera) in favour of Euarchonta (Waddell *et al.*, 1999a), by Asher and Helgen (2010: 4).

Clade Euarchonta Waddell *et al.*, 1999a: 4.

COMMENTS: When originally proposed, this clade included the orders Primates (Linnaeus, 1758), Dermoptera (Illiger, 1811: 63, 116) and Scandentia (Wagner, 1855: xix, 524). Name recognised by many authors including Scally *et al.* (2002: 239), Silcox *et al.* (2005: 127), who reviewed it, and Springer *et al.* (2007: 21), who gave it the rank of grandorder, but synonymised within Archonta (Gregory, 1910) by Asher and Helgen (2010: 4).

Order Primates Linnaeus, 1758

Primates Linnaeus, 1758: 18, 20.

COMMENTS: When originally proposed, this rank was placed in the Class Mammalia (Linnaeus, 1758) and included the genera *Homo* Linnaeus, 1758; *Simia* Linnaeus, 1758: 18, 25; *Lemur* Linnaeus, 1758: 29, 30; and *Vespertilio* Linnaeus, 1758: 31. Order reviewed by Groves (2001; 2005f: 111).

Family Hominidae J. Gray, 1825

Family Hominidae J. Gray, 1825a: 338.

TYPE GENUS: *Homo* Linnaeus, 1758.

COMMENTS: When originally proposed, this rank was placed in the order Primates (Linnaeus, 1758) and included the Hominina (J. Gray, 1825a: 338) [=Family Hominidae (J. Gray, 1825a)]), Simiina (J. Gray, 1825a: 338 [=Family Hominidae (J. Gray, 1825a)]), Presbytina (J. Gray, 1825a: 338 [=Family Cercopithecidae (J. Gray, 1821: 297)]), Cercopithecina (J. Gray, 1825a: 338 [=Family Cercopithecidae (J. (Gray, 1821: 297)]) and Cynocephalina (J. Gray, 1825a: 338 [=Family Cercopithecidae (J. Gray, 1821: 297)]). Reviewed by Groves (2001: 298; 2005f: 182).

Homo Linnaeus, 1758

Homo Linnaeus, 1758: 7, 20.

TYPE SPECIES: Ω *Homo sapiens* Linnaeus, 1758.

COMMENTS: Reviewed by Groves (2001: 308; 2005f: 182).

Ω ***Homo sapiens*** Linnaeus, 1758

Human

Ω *[Homo] Sapiens* Linnaeus, 1758: 7, 20.

TYPE LOCALITY: Sweden.

COMMENTS: Reviewed by Groves (2001: 308; 2005f: 182).

Superorder Glires Linnaeus, 1758

Order Glires Linnaeus, 1758: 56.

COMMENTS: When originally proposed, this rank was placed in the Class Mammalia (Linnaeus, 1758) and included the genera *Rhinoceros*, *Hystrix*, *Lepus*, *Castor*, *Mus* and *Sciurus*. *Rhinoceros* was removed by Linnaeus (1766: 76), but included the bat genus *Noctilio* (Linnaeus, 1766: 88), with the notion of Glires linking rodents and lagomorphs persisting today. Significant support has been obtained for the recognition of Glires by recent research giving increasing support for monophyly between the Rodentia and Lagomorpha. Some authors suggest limited or no support for Glires including Wood (1957: 424), Honeycutt and Adkins (1993: 279), Graur *et al.* (1996: 333), Arnason *et al.* (2002: 8154, 8155), Adkins *et al.* (2003: 413) and Misawa and Janke (2003: 320). In contrast, support for the acceptance of Glires includes Minkoff (1976: 153), though it did not include the Order Lagomorpha, Landry (1999: 283), Madsen *et al.* (2001: 610), Meng and Wyss (2001: 1), Scally *et al.* (2002: 239), Meng *et al.* (2003: 1), Liu *et al.* (2001: 1786), Murphy *et al.* (2001b: 614), Huchon *et al.* (2002: 1053), Lin *et al.* (2002: 119), Douzery and Huchon (2004: 922), Meng (2004: 93), Asher *et al.* (2005: 1091), Meng and Wyss (2005: 145), Kriegs *et al.* (2007: 161) and Nikolaev *et al.* (2007: 5). Reviewed by Meng (2004: 93) and Meng and Wyss (2005: 145). Not recognised by McKenna and Bell (1997), but it was given the rank of cohort by Skinner and Chimimba (2005: v, 63).

HOMONYMS:

Order Glires J. Gray, 1821, wombats of the Class Mammalia (Order Diprotodontia, Family Vombatidae). Name is a synonym of the Family Vombatidae. See individual entry.

Order Rosores Storr, 1780: Table B.

COMMENTS: When originally proposed, this rank was placed in the Class Mammalia (Linnaeus, 1758) and included the genera *Hystrix* Linnaeus, 1758: 56; *Cavia* Pallas, 1766: 30; *Mus* Linnaeus, 1758; *Glis* Brisson, 1762: 113; *Sciurus* Linnaeus, 1758; *Lagomys* Storr, 1780: Table B [=*Marmota* Blumenbach, 1779: 79]; *Cavia* Pallas, 1766: 30; *Procavia* Storr, 1780: Table B; and *Lepus* Linnaeus, 1758.

HOMONYMS:

Order Rosores J. Gray, 1821, mammals of the Class Mammalia (Infraclass Placentalia Bonaparte, 1838). See individual entry.

Order Prensiculantia Illiger, 1811: 80.

COMMENTS: When originally proposed, this rank included the families Macropoda (Illiger, 1811: 81 [=Muroidea (Illiger, 1811)]), Aglia (Illiger, 1811: 82 [=Myoxidae (J. Gray, 1821: 303)] and Sciuridae (G. Fischer, 1814)]), Murina [=Muridae (Illiger, 1811)], Cunicularia (Illiger, 1811: 86 [=Bathyergidae (Waterhouse, 1841b: 81)]), Cricetidae (G. Fischer, 1817: 372, 410), Castoridae (Hemprich, 1820: 33), Palmipeda (Illiger, 1811: 88 [=Muridae (Illiger, 1811)]), Castoridae (Hemprich, 1820: 33), Aculeata (Illiger, 1811: 89 [=Hystricidae (G. Fischer, 1814: viii, 99)]), Echimyidae (J. Gray, 1825a: 341), Duplicidentata (Illiger, 1811 [=Lagomorpha (Brandt, 1855)] and Sub-Ungulata (Illiger, 1811: 92 [=Cavioidea (G. Fischer, 1814: viii, 81)]).

Rongeurs F. Cuvier, 1813: 268.

COMMENTS: Not given a rank. When originally proposed, this rank included Les Écureuils, La Marmotte des Aples, Les Rats, Les Loirs, Les Hamsters, Les Hydromis, Les Plus Petite Taupe du Cap, Le Zemni, Les Gerboises, Les Échimis, Les Castors, Les Pacas, La Grande Taupe du Cap, Les Porc-Épics, Les Agoutis, Le Cabiai, Le Cochon d'Inde, Le Campagnoi, Les Lièvres and Les Lagotis. Though often referred to by various authors this term does not appear to have been given a formal rank. Also referred to in a general sense by G. Cuvier (1809: 394).

Order Rosores J. Gray, 1821: 302.

COMMENTS: Placed in Class Quadrumanes (de Blainville, 1816a: 117) and included the families Castoridae (Hemprich, 1820: 33), Arvicolidae (J. Gray, 1821: 303), Myosidae (J. Gray, 1821: 303 [=Myoxidae (J. Gray, 1821: 303)]), Dipsidae (J. Gray, 1821: 303 [=Dipodidae (G. Fischer, 1817: 372, 407)]), Muridae (Illiger, 1811), Spalacidae (J. Gray, 1821: 303), Halamydae (J. Gray, 1821: 303 [=Pedetidae J. Gray, 1825a: 342)]), Arctomydae (J. Gray, 1821: 303 [=Sciuridae (G. Fischer, 1814)]), Sciuridae (G. Fischer, 1814), Hystricidae (G. Fischer, 1814: viii, 99), Leporidae (G. Fischer, 1814), Caviadae (J. Gray, 1821: 304 [=Caviidae (G. Fischer, 1814: viii, 81)]) and Agoutidae (J. Gray, 1821: 304). Later in the same paper J. Gray employed Order Rosores a second time to include the Family Cheiromydae (p. 309), which are primates that are now included in the Family Daubentoniadae (J. Gray, 1863a: 151). Synonymised within Rodentia by McKenna and Bell (1997: 114).

HOMONYMS:

Order Rosores Storr, 1780, mammals of the Class Mammalia (Infraclass Placentalia Bonaparte, 1838). See individual entry.

Order Rodentia Bowdich, 1821

Order Rodentia Bowdich, 1821: 7, 51.

COMMENTS: When originally proposed, this rank was placed in the Class Mammalia (Linnaeus, 1758) and included Group 1 containing the genera *Castor* Linnaeus, 1758: 58; *Mus* Linnaeus, 1758; *Spalax* Güldenstädt, 1770: 409; *Bathyergus* Illiger, 1811: 86; *Helamys* G. Cuvier, 1816a: 202 [=*Pedetes* Illiger, 1811: 81]; *Arctomys* Schreber, 1780: Plate 207 [=*Marmota* Blumenbach, 1779: 79]; *Sciurus* Linnaeus, 1758; and *Cheiromys* G. Cuvier, 1800: Table 1 [=*Daubentonia* É. Geoffroy, 1795: 195]; and Group 2 containing the genera *Hystrix* Linnaeus, 1758: 56; *Lepus* Linnaeus, 1758; *Hydrochoerus* Brisson, 1762: 12, 80; *Cavia* Pallas, 1766: 30; *Dasyprocta* Illiger, 1811: 93; and *Coelogenus* F. Cuvier, 1807: 203 [=*Agouti* Lacépède, 1799a: 9].

Class Rodentes Vicq-d'Azyr, 1792: xcvii.

COMMENTS: When originally proposed, this rank included the genera *Sciuriens* Vicq-d'Azyr, 1792: xcvii; *Ecureuils* Vicq-d'Azyr, 1792: xcviii; *Glirins* Vicq-d'Azyr, 1792: xcviii; *Murins* Vicq-d'Azyr, 1792: xcix; *Surmurins* Vicq-d'Azyr, 1792: c; *Essorillés* Vicq-d'Azyr, 1792: c; *Planiqueues* Vicq-d'Azyr, 1792: c; *Sauteurs* Vicq-d'Azyr, 1792: c; *Double-Dents* Vicq-d'Azyr, 1792: c; and *Épineux* Vicq-d'Azyr, 1792: ci. Synonymised within Rodentia by McKenna and Bell (1997: 114).

Clavicules G. Cuvier, 1816a: xxxi, 189.

COMMENTS: Rank unknown but placed within the Rongeurs (F. Cuvier, 1813 [=Glires (Linnaeus, 1758)]) and included the genera *Castor* Linnaeus, 1758: 58; *Mus* Linnaeus, 1758; *Arvicola* Lacépède, 1799a: 10; *Echimys* G. Cuvier, 1809: 394; *Myoxus* Zimmermann, 1780: 351; *Cricetus* Leske, 1779: 168; *Dipus* Zimmermann, 1780: 354; *Spalax* Güldenstädt, 1770: 409; *Bathygerus* Illiger, 1811: 86; *Pedetes* Illiger, 1811: 81; *Arctomys* Schreber, 1780: Plate 207 [=*Marmota* Blumenbach, 1779: 79]; *Sciurus* Linnaeus, 1758; *Pteromys* G. Cuvier, 1800: Table 1; and *Cheiromys* G. Cuvier, 1800: Table 1 [=*Daubentonia* É. Geoffroy, 1795: 195]. Recognised as the Family Claviculata within the Order Rodentia by Owen (1859: 52).

Sans Clavicules G. Cuvier, 1816a: xxxi, 208.

COMMENTS: Rank unknown but placed within the Rongeurs (F. Cuvier, 1813 [=Glires (Linnaeus, 1758)]) and included the genera *Hystrix* Linnaeus, 1758: 56; *Lepus* Linnaeus, 1758; *Lagomys* G. Cuvier, 1800: Table 1 [=*Ochotona* Link, 1795: 74]; *Cavia* Pallas, 1766: 30; *Hydrochoerus* Brisson, 1762: 12, 80; *Dasyprocta* Illiger, 1811: 93; and *Coelogenus* F. Cuvier, 1807: 203 [=*Agouti* Lacépède, 1799a: 9]. Recognised as the Family Claviculata within the Order Rodentia by Owen (1859: 52).

Order Rodentiformes Kinman, 1994: 37.

COMMENTS: When originally proposed, this rank was placed in the Class Mammalea (Kinman, 1994 [=Mammalia (Linnaeus, 1758)]). Synonymised within the Mirorder Simplicidentata (Weber, 1904: x, 495) by McKenna and Bell (1997: 113).

Clade Rodentiaformes Wyss & Meng, 1996: 562.

COMMENTS: When originally proposed, this clade was placed in the clade Simplicidentata (Weber, 1904: 495), within Glires (Linnaeus, 1758), and includes the genus † *Tribosphenomys* (Meng *et al*., 1994: 134) and Rodentia (Bowdich, 1821). Not currently recognised at a specific rank.

Suborder Myomorpha Brandt, 1855

Suborder Myomorpha Brandt, 1855: 152, 292, 300.

COMMENTS: When originally proposed, this rank was placed in the 'Order Glires seu Rodentia' and included the families Myoxoïdes [=Myoxidae (J. Gray, 1821: 303)], Castoroïdes [=Castoridae (Hemprich, 1820: 33)], Sciurospalacoïdes (Brandt, 1855: 301 [=Geomorpha (Thaler, 1966: 11)], Myoïdes [=Superfamily Muroidea (Illiger, 1811)], Spalacoïdes (Brandt, 1855: 306), and Dipodoïdes ([=Dipodoidea] G. Fischer, 1817: 372, 407). The Family Muridae was placed in the Suborder Sciurognathi (Tullberg, 1899: 43) by Musser and Carleton (1993: 501) and Skinner and Chimimba (2005: v, 99). Suborder rank recognised by Strahan (1983: xxi) and Musser and Carleton (2005: 871), and intraordinal rank by Carleton (1984: 258). Called Phaneraulata by Landry (1999: 283, 312), with a slightly different composition. Superfamily rank recognised by McKenna and Bell (1997: 131) but not by Van Dyck and Strahan (2008: 10).

Clade Phaneraulata Landry, 1999: 283, 312.

COMMENTS: When originally proposed, this clade was placed in the Sciuromorpha (Brandt, 1855) and included the old Myomorpha (Brandt, 1855), minus the anomalurids and included the geomyids, glirids and theridomyids.

Superfamily Muroidea Illiger, 1811

Family Murina Illiger, 1811: 84.

TYPE GENUS: *Mus* Linnaeus, 1758.

COMMENTS: When originally proposed, this rank was placed in the Order Prensiculantia (Illiger, 1811 [=Glires (Linnaeus, 1758 part)]) and included the genera *Arctomys* Schreber, 1780: Plate 207 [=*Marmota* Blumenbach, 1779: 79]; *Cricetus* Leske, 1779: 168; *Mus* Linnaeus, 1758; and *Bathyergus* Illiger, 1811: 86. Superfamily rank recognised by Ellerman (1940: 35), Carleton (1984: 258), McKenna and Bell (1997: 135) and Musser and Carleton (2005: 894), but not by Van Dyck and Strahan (2008: 10).

Family Macropoda Illiger, 1811: 61, 81.

TYPE GENUS: Not based on a genus group name.

COMMENTS: When originally proposed, this rank was placed in the Order Prensiculantia (Illiger, 1811 [=Glires (Linnaeus, 1758 part)]) and included the genera *Dipus* Zimmermann, 1780: 354; *Pedetes* Illiger, 1811: 81; and *Meriones* Illiger, 1811: 82. Each of these genera are currently recognised within separate families, i.e. the families Dipodidae for *Dipus* (Holden & Musser, 2005: 882), Pedetidae for *Pedetes* (Dieterlen, 2005: 1535) and Muridae for *Meriones* (Musser & Carleton, 2005: 1234).

HOMONYMS:

Macropoda Wagner, 1843, macropods of the Class Mammalia (Order Diprotodontia, Superfamily Macropodoidea). Synonymised within the Superfamily Macropodoidea J. Gray, 1821 here. See individual entry.

Order Macropoda Ameghino, 1889, diprotodont marsupials of the Class Mammalia (Order Diprontodontia). Synonymised within the Order Diprotodontia Owen, 1877a here. See individual entry.

Family Cricetinorum G. Fischer, vi, 42.

TYPE GENUS: *Cricetus* Leske, 1779: 168.

COMMENTS: When originally proposed, this rank was placed in the Order Metatarsii (G. Fischer, 1817 [=Mammalia (Linnaeus, 1758 part)]) and included the genera *Arctomys* Schreber, 1780: Plate 207 [=*Marmota* Blumenbach, 1779: 79]; *Cricetus* Leske, 1779: 168; and *Brachyurus* G. Fischer, 1813a: 14, 24 [=*Lemmus* Link, 1795: 75]. Name is also referred to by G. Fischer (1817: 410) and is equivalent to Cricetini (G. Fischer, 1817).

Family Cricetini G. Fischer, 1817: 372.

TYPE GENUS: *Cricetus* Leske, 1779: 168.

COMMENTS: When originally proposed, this rank was placed in the Order Metatarsii (G. Fischer, 1817 [=Mammalia (Linnaeus, 1758 part)]) and included the genera *Cricetus* Leske, 1779: 168; *Arctomys* Schreber, 1780: Plate 207 [=*Marmota* Blumenbach, 1779: 79]; *Hystrix* Linnaeus, 1758: 56; *Eucritus* G. Fischer, 1817: 372; and *Loncheres* Illiger, 1811: 90 [=*Echimys* G. Cuvier, 1809: 394]. Rank not stated but appears to be the same as the Family Cricetinorum on page 410. Synonymised within Muridae by McKenna and Bell (1997: 135).

Family Murini G. Fischer, 1817: 372.

TYPE GENUS: *Mus* Linnaeus, 1758.

COMMENTS: When originally proposed, this rank was placed in the Order Metatarsii (G. Fischer, 1817 [=Mammalia (Linnaeus, 1758 part)]) and included the genera *Mus* Linnaeus, 1758; *Brachyurus* G. Fischer, 1813a: 14, 24 [=*Lemmus* Link, 1795: 75]; and *Spalax* Güldenstädt, 1770: 409. Rank not stated but appears to be the same as the Family Murinorum on page 410. Synonymised within Muridae by McKenna and Bell (1997: 136).

Suborder Simplicidentati Lilljeborg, 1866a: 5, 9.

COMMENTS: When originally proposed, this rank was placed in the Order Glires (Linnaeus, 1758) and included the families Muridae (Illiger, 1811), Spalacidae (J. Gray, 1821: 303), Dipodidae [=G. Fischer, 1817: 372, 407], Myoxidae (J. Gray, 1821: 303), Saccomyidae (Baird, 1857: 235 [=Geomyidae (Bonaparte, 1845: 5)]), Castoridae (Hemprich, 1820: 33), Sciuridae (G. Fischer, 1814), Haploodontidae (Lilljeborg, 1866a: 41 [=Aplodontidae (Brandt, 1855: 148, 150)]), Chinchillidae (Bennett, 1833: 57), Spalacopodidae (Lilljeborg, 1866a: 44 [=Octodontidae (Waterhouse, 1839b: 172)]) and Hystricidae (G. Fischer, 1814: viii, 99). Suborder rank recognised by Gill (1872: v, 20).

Myoidea Gill, 1872: 20.

COMMENTS: When originally proposed, this rank was placed in the Suborder Simplicidentati (Lilljeborg, 1866a: 5, 9 [=Muroidea (Illiger, 1811 part)]) and included the families Pedetidae (J. Gray, 1825a: 342), Dipodidae (G. Fischer, 1817: 372, 407), Jaculidae (Gill, 1872: 20 [=Dipodidae (G. Fischer, 1817: 372, 407)]), Muridae (Illiger, 1811), Myoxidae (J. Gray, 1821: 303), Saccomyidae (Baird, 1857: 235 [=Geomyidae (Bonaparte, 1845: 5)]), Geomyidae (Bonaparte, 1845: 5), Castoridae (Hemprich, 1820: 33), Sciuridae (G. Fischer, 1814), Anomaluridae (Gervais, 1849a: 203), Haploodontidae (Lilljeborg, 1866a: 41 [=Aplodontidae (Brandt, 1855: 148, 150)]), Spalacopodidae (Lilljeborg, 1866a: 44 [=Octodontidae (Waterhouse, 1839b: 172)]), Hystricidae (G. Fischer, 1814: viii, 99), Dasyproctidae (J. Gray, 1825a: 341), Caviidae (G. Fischer, 1814: viii, 81), Hydrochoeridae (J. Gray, 1825a: 341) and Chinchillidae (Bennett, 1833: 57). Rank not specified by Gill, 1872. Synonymised within Muroidea by McKenna and Bell (1997: 135).

† Lophiomyoidea Gill, 1872: 20.

COMMENTS: When originally proposed, this rank was placed in the Suborder Simplicidentati (Lilljeborg, 1866a: 5, 9 [=Muroidea (Illiger, 1811 part)]) and included the Family Lophiomyidae (Milne-Edwards, 1878: 114). Synonymised within Muroidea by McKenna and Bell (1997: 135).

Superfamily Muroidae Miller & Gidley, 1918: 435.

TYPE GENUS: *Mus* Linnaeus, 1758.

COMMENTS: When originally proposed, this rank was placed in the Order Rodentia (Bowdich, 1821) and included the families Muscardinidae (Palmer, 1899c: 413), † Ischyromyidae (Alston, 1876: 67, 78), Cricetidae (G. Fischer, 1817: 372, 410), Platacanthomyidae (Alston, 1876: 81), Rhizomyidae (Winge, 1887: 109), Spalacidae (J. Gray, 1821: 303) and Muridae (Illiger, 1811). Rank recognised by Ellerman (1940: 35), but synonymised within Muroidea (Illiger, 1811) by McKenna and Bell (1997: 135).

Suborder Cricetomorpha Thaler, 1966: 137.

COMMENTS: When originally proposed as a new rank it was placed in the Order Rodentia (Bowdich, 1821) and included the Superfamily Cricetoidea (Thaler, 1966: 137). Synonymised within Muroidea by McKenna and Bell (1997: 135).

Superfamily Cricetoidea Thaler, 1966: 137.

TYPE GENUS: *Cricetus* Leske, 1779: 168.

COMMENTS: When originally proposed as a new rank it was placed in the Suborder Cricetomorpha (Thaler, 1966: 137) and contained the families Cricetidae (G. Fischer, 1817: 372, 410) and Microtidae (Cope, 1891: 90 [=Arvicolidae (J. Gray, 1821: 303)]). Synonymised within Muroidea by McKenna and Bell (1997: 135).

Family Muridae Illiger, 1811

Family Murina Illiger, 1811: 84.

TYPE GENUS: *Mus* Linnaeus, 1758.

COMMENTS: When originally proposed, this rank was placed in the Order Prensiculantia (Illiger, 1811 [=Glires (Linnaeus, 1758 part)]) and included the genera *Arctomys* Schreber, 1780: Plate 207 [=*Marmota* Blumenbach, 1779: 79]; *Cricetus* Leske, 1779: 168; *Mus* Linnaeus, 1758; and *Bathyergus* Illiger, 1811: 86. Subsequently recorded by Illiger (1815: 46, 129). The priority of Illiger (1811: 84) was recognised by McKenna and Bell (1997: 136), and Musser and Carleton (2005: 1189), while J. Gray (1821: 303) was recognised as the author by Simpson (1945: 88).

Family Murinorum G. Fischer, 1814: vii, 62.

TYPE GENUS: *Mus* Linnaeus, 1758.

COMMENTS: When originally proposed, this rank was placed in the Order Metatarsii (G. Fischer, 1814) and included the genus *Mus* Linnaeus, 1758. Name also referred to by G. Fischer (1817: 410).

Family Muridae J. Gray, 1821: 303.

TYPE GENUS: *Mus* Linnaeus, 1758.

COMMENTS: When originally proposed, this rank was placed in the Order Rosores (J. Gray, 1821) and included the genera *Hydromys* É. Geoffroy, 1804a; *Rattus* G. Fischer, 1803; *Mus* Linnaeus, 1758; and *Cricetus* Leske, 1779: 168. Synonymised within Muridae Illiger, 1811 by McKenna and Bell (1997: 136).

Kind Rattidae Burnett, 1830b: 350.

TYPE GENUS: *Rattus* G. Fischer, 1803.

COMMENTS: When originally proposed, this rank was placed in the Race 'Glires, Rodentia, Liberae' with the Muridae (Illiger, 1811) (mouse-kind) to include rat-kind animals, with both families including the genera *Sphermophilus* [sic =*Spermophilus*] F. Cuvier, 1825 [1821–1825]: 255; *Arctomys* Schreber, 1780: Plate 207 [=*Marmota* Blumenbach, 1779: 79]; *Pedetes* Illiger, 1811: 81; *Bathyergus* [=*Bathygerus*] Illiger, 1811: 86; *Dipus* Zimmermann, 1780: 354; *Gerbillus* Desmarest, 1804a: 22; *Aspalax* Desmarest, 1804a: 24 [=*Spalax* Güldenstädt, 1770: 409]; *Cricetus* Leske, 1779: 168; *Mus* Linnaeus, 1758; *Hydromys* É. Geoffroy, 1804a; *Myoxus* Zimmermann, 1780: 351; *Echimys* G. Cuvier, 1809: 394; *Lemmus* Link, 1795: 75; and *Arvicola* Lacépède, 1799a: 10. Synonymised within Muridae by McKenna and Bell (1997: 136) and the Subfamily Murinae by Musser and Carleton (2005: 1247).

Family Dipodineae Lesson, 1842: 129.

TYPE GENUS: *Dipus* Zimmermann, 1780: 354.

COMMENTS: When originally proposed, this rank was placed in the Suborder Rodentia (Bowdich, 1821) and included the genera *Pedetes* Illiger, 1811: 81; *Lagostomus* Brookes, 1829: 96; *Notomys* Lesson, 1842; *Dipus* Zimmermann, 1780: 354; *Alactaga* F. Cuvier, 1837b: 141, *Gerbillus* Desmarest, 1804a: 22; *Meriones* Illiger, 1811: 82; and *Eligmodontia* F. Cuvier, 1837c: 168. Does not appear to have been recognised by other authors.

Musideaee Lesson, 1842: 134.

TYPE GENUS: *Mus* Linnaeus, 1758.

COMMENTS: When originally proposed, this rank was placed in the Suborder Rodentia (Bowdich, 1821) and included the genus *Mus* (Linnaeus, 1758). Synonymised within Muridae by McKenna and Bell (1997: 136).

Family Murini Giebel, 1855: xi, 531.

TYPE GENUS: *Mus* Linnaeus, 1758.

COMMENTS: When originally proposed, this rank was placed in the Order Glires (Linnaeus, 1758) and included the genera *Acomys* I. Geoffroy, 1838: 126; *Sminthus* Nathusius, 1839: 49; *Reithrodon* Waterhouse, 1837a: 29; *Neotoma* Say and Ord, 1825a: 345; *Sigmodon* Say and Ord, 1825b: 352; *Hesperomys* Waterhouse, 1839c: 75; *Mus* Linnaeus, 1758; *Steatomys* Peters, 1846: 258; *Pseudomys* J. Gray, 1832; *Dendromys* Smuts, 1832: iii, 39; *Akodon* Meyen, 1833: 599; *Drymomys* Tschudi, 1844: 9; *Saccomys* F. Cuvier, 1823[1821–1825]: 186; *Perognathus* Wied-Neuwied, 1839: 449; *Saccostomus* Peters, 1846: 258; *Cricetomys* Waterhouse, 1840b: 2; *Cricetus* Leske, 1779: 168; *Hydromys* É. Geoffroy, 1804a; *Phloeomys* Waterhouse, 1839d: 108; and *Hapalotis* M. Lichtenstein, 1829 [=*Conilurus* W. Ogilby, 1838f]. Synonymised within Muridae by McKenna and Bell (1997: 136).

Family Murina Haeckel, 1866: clx.

TYPE GENUS: *Mus* Linnaeus, 1758.

COMMENTS: When originally proposed, this rank was placed in the Suborder Myomorpha and included the genera *Mus* Linnaeus, 1758; *Cricetus* Leske, 1779: 168; and *Hypudaeus* Illiger, 1811: 87 [=*Lemmus* Link, 1795: 75]. Synonymised within Muridae by McKenna and Bell (1997: 136).

Subfamily Murinae Illiger, 1811

Family Murina Illiger, 1811: 84.

TYPE GENUS: *Mus* Linnaeus, 1758.

COMMENTS: When originally proposed, this rank was placed in the Order Prensiculantia (Illiger, 1811 [=Glires (Linnaeus, 1758 part)]) and included the genera *Arctomys* Schreber, 1780: Plate 207 [=*Marmota* Blumenbach, 1779: 79]; *Cricetus* Leske, 1779: 168; *Mus* Linnaeus, 1758; and *Bathyergus* Illiger, 1811: 86. Subsequently recorded by Illiger (1815: 46, 129). Recognised at family rank by J. Gray (1825a: 341) and as a subfamily within the Family Muridae by A. Murray (1866: xv, 359), Longman (1916: 9), Iredale and Troughton (1934: 71), Tate (1951c: 237), Troughton (1967: 212), Carleton and Musser (1984: 294, 352), Strahan (1995: 9, 646), and Musser and Carleton (2005: 1247). The recognition of the tribes used here follows Lecompte *et al.* (2008: 8) with the informal rank of Division being included within each tribe following Musser and Carleton (2005: 902–905) and Lecompte *et al.* (2008: 8–9).

Tribe Murina J. Gray, 1825a: 341.

TYPE GENUS: *Mus* Linnaeus, 1758.

COMMENTS: When originally proposed, this rank was placed in the Family Muridae (Illiger, 1811) and included the genera *Mus* Linnaeus, 1758; *Otomys* F. Cuvier, 1824 [1821–1825]: 255; and *Capromys* Desmarest, 1822a: 185). Synonymised within the Subfamily Murinae by McKenna and Bell (1997: 161).

Tribe Hydromina J. Gray, 1825a: 341.

TYPE GENUS: *Hydromys* É. Geoffroy, 1804a.

COMMENTS: When originally proposed, this rank was placed in the Family Muridae (Illiger, 1811) and included the genus *Hydromys* É. Geoffroy, 1804a. Synonymised within the Subfamily Murinae by McKenna and Bell (1997: 161) and Musser and Carleton (2005: 1247).

Kind Rattidae Burnett, 1830b: 350.

TYPE GENUS: *Rattus* G. Fischer, 1803.

COMMENTS: When originally proposed, this rank was placed in the Race 'Glires, Rodentia, Liberae' with the Muridae (Illiger, 1811) (mouse-kind) to include rat-kind animals, with both families including the genera *Sphermophilus* [sic =*Spermophilus*] F. Cuvier, 1825 [1821–1825]: 255; *Arctomys* Schreber, 1780: Plate 207 [=*Marmota* Blumenbach, 1779: 79]; *Pedetes* Illiger, 1811: 81; *Bathyergus* [=*Bathygerus*] Illiger, 1811: 86; *Dipus* Zimmermann, 1780: 354; *Gerbillus* Desmarest, 1804a: 22; *Aspalax* Desmarest, 1804a: 24 [=*Spalax* Güldenstädt, 1770: 409]; *Cricetus* Leske, 1779: 168; *Mus* Linnaeus, 1758; *Hydromys* É. Geoffroy, 1804a; *Myoxus* Zimmermann, 1780: 351; *Echimys* G. Cuvier, 1809: 394; *Lemmus* Link, 1795: 75; and *Arvicola* Lacépède, 1799a: 10. Synonymised within the Subfamily Murinae by McKenna and Bell (1997: 161) and Musser and Carleton (2005: 1247).

Subfamily Murinae A. Murray, 1866: xv, 359.

TYPE GENUS: *Mus* Linnaeus, 1758.

COMMENTS: When originally proposed, this rank was placed in the Family Muridae (Illiger, 1811). Synonymised within the Subfamily Murinae by McKenna and Bell (1997: 161).

Subfamily Phloeomyinae Alston, 1876: 81.

TYPE GENUS: *Phloeomys* Waterhouse, 1839d: 108.

COMMENTS: When originally proposed, this rank was placed in the Family Muridae (Illiger, 1811) and included the genera *Phloeomys* Waterhouse, 1839d: 108; and *Nesokia* J. Gray, 1842b: 264. Synonymised within the Subfamily Murinae by McKenna and Bell (1997: 161) and Musser and Carleton (2005: 1247).

Murini Winge, 1887: 126.

TYPE GENUS: *Mus* Linnaeus, 1758.

COMMENTS: When originally proposed, this rank was placed in the Family Muridae (Illiger, 1811) above the unknown ranks Mures (Winge, 1887) and Gerbilli (J. Gray, 1825a: 342). Synonymised within the Subfamily Murinae by Musser and Carleton (2005: 1247).

Mures Winge, 1887: 125.

TYPE GENUS: *Mus* Linnaeus, 1758.

COMMENTS: When originally proposed, this rank was placed within the unknown rank Murini above the genera *Mus* Linnaeus, 1758; *Acomys* I. Geoffroy, 1838: 126; *Cricetomys* Waterhouse, 1840b: 2; *Dasymys* Peters, 1875b: 12; *Dendromys* Smuts, 1832: iii, 39; *Isomys* Sundevall, 1843: 219 [=*Arvicanthus* Lesson, 1842: 147]; *Lophuromys* Peters, 1874b: 234; *Pelomys* Peters, 1852a: 275; *Saccostomus* Peters, 1846: 258; *Steatomys* Peters, 1846: 258; *Chiropodomys* Peters, 1868a: 448; *Phloeomys* Waterhouse, 1839d: 108; *Spalacomys* Peters, 1860a: 139 [=*Nesokia* J. Gray, 1842b: 264]; *Uromys* Peters, 1867a; *Echiothrix* J. Gray, 1867: 599; *Hapalotis* M. Lichtenstein, 1829 [=*Conilurus* W. Ogilby, 1838f]; and *Mastacomys* Thomas, 1882. Synonymised within the Subfamily Murinae by Musser and Carleton (2005: 1247).

Subfamily Rhynchomyinae Thomas, 1897b: 1017.

TYPE GENUS: *Rhynchomys* Thomas, 1895c: 160.

COMMENTS: When originally proposed, this rank was placed in the Family Muridae (Illiger, 1811) and included the genus *Rhyncomys* Thomas, 1895c: 160. Synonymised within the Subfamily Murinae by McKenna and Bell (1997: 161) and Musser and Carleton (2005: 1247).

Subfamily Otomyinae Thomas, 1897b: 1017.

TYPE GENUS: *Otomys* F. Cuvier, 1824 [1821–1825]: 255.

COMMENTS: When originally proposed, this rank was placed in the Family Muridae (Illiger, 1811) and included the genera *Otomys* F. Cuvier, 1824 [1821–1825]: 255; and *Oreinomys* Trouessart, 1881: 111. Synonymised within the Subfamily Murinae by McKenna and Bell (1997: 161).

Subfamily Phloeomyini Tullberg, 1899: 274.

TYPE GENUS: *Phloeomys* Waterhouse, 1839d: 108.

COMMENTS: When originally proposed, this rank was placed in the Family Muridae (Illiger, 1811) and included the genus *Phloeomys* Waterhouse, 1839d: 108. Synonymised within the Subfamily Murinae by Musser and Carleton (2005: 1247).

Subfamily Otomyini Tullberg, 1899: 274.

TYPE GENUS: *Otomys* F. Cuvier, 1824 [1821–1825]: 255.

COMMENTS: When originally proposed, this rank was placed in the Family Muridae (Illiger, 1811) and included the genus *Otomys* F. Cuvier, 1824 [1821–1825]: 255. Tribe rank recognised by Ducroz *et al.* (2001: 199). Synonymised within the Subfamily Murinae by McKenna and Bell (1997: 161) and McKenna and Bell (1997: 161).

Group Anisomyes *sensu* Ellerman, 1941: 9, 76.

TYPE GENUS: *Anisomys* Thomas, 1904c: 199.

COMMENTS: When originally proposed, this rank was placed in the Subfamily Murinae (Illiger, 1811) and included the genus *Anisomys* Thomas, 1904c: 199. Synonymised within the Subfamily Murinae by Musser and Carleton (2005: 1247).

Tribe Anisomyini Lidicker & Brylski, 1987: 635.

TYPE GENUS: *Anisomys* Thomas, 1904c: 199.

COMMENTS: When originally proposed as a new rank it was placed in the Subfamily Hydromyinae (Alston, 1876) and included the genera *Anisomys* Thomas, 1904c: 199;

Pogonomys Milne-Edwards, 1877; *Chiruromys* Thomas, 1888b: 237; *Hyomys* Thomas, 1904c: 198; *Lorentzimys* Jentink, 1911: 174; *Mallomys* Thomas, 1898a: 1; and *Macruromys* Stein, 1933: 94. Synonymised within the Subfamily Murinae by McKenna and Bell (1997: 161) and Musser and Carleton (2005: 1247).

Tribe Arvicanthini Ducroz *et al.*, 2001: 173, 200.

TYPE GENUS: *Arvicanthis* Lesson, 1842: 147.

COMMENTS: When originally proposed as a new rank it was placed in the Subfamily Murinae (Illiger, 1811) and included the genera *Arvicanthis* Lesson, 1842: 147; *Aethomys* Thomas, 1915a: 477; *Dasymys* Peters, 1875b: 12; *Grammomys* Thomas, 1915b: 150; *Hybomys* Thomas, 1910b: 85; and probably *Golunda* J. Gray, 1837: 586. *Nomen nudum*. Synonymised within the Subfamily Murinae by Musser and Carleton (2005: 1247).

Tribe Hydromyini Alston, 1876 *sensu* Lecompte *et al.*, 2008

Subfamily Hydromyinae Alston, 1876: 80.

TYPE GENUS: *Hydromys* É. Geoffroy, 1804a.

COMMENTS: When originally proposed, this rank was placed in the Family Muridae (Illiger, 1811) and included the genus *Hydromys* É. Geoffroy, 1804a. Author to the subfamily given as Thomas (1897b: 1017) by authors including Ellerman (1941: 297), and Watts and Baverstock (1994: 303). Subfamily rank recognised by Longman (1916: 9), Ellerman (1940: 39; 1941: 297), Tate (1951c: 222), Iredale and Troughton (1934: 69), Simpson (1945: 91; 1961: 434), Troughton (1967: 206), Misonne (1969: iii, 154), Lee *et al.* (1981: 1530), Watts and Aslin (1981: 6), Strahan (1983: xxi; 1995: 9, 550), Baverstock (1984b: 917), Lidicker and Brylski (1987: 621), Watts and Kemper (1989: 948), Watts and Baverstock (1994: 303) and Strahan (1995: 9, 550). Subfamily rank synonymised within the Subfamily Murinae by McKenna and Bell (1997: 161) and Musser and Carleton (2005: 1247). Tribe rank within the Subfamily Murinae recognised by Strahan (1983: 366; 1995: 9, 628), Lidicker and Brylski (1987: 621, 635), Watts and Kemper (1989: 948), Flannery (1995a: 10, 233) and LeCompte *et al.* (2008: 7). Recognised as the '*Hydromys* Division' by Musser and Carleton (2005: 903) and '*Hydromys* Group' by Breed and Ford (2007: 9), and Van Dyck and Strahan (2008: 662).

TAXONOMIC COMMENTS: Ken Aplin (personal communication) has commented that, according to our time-depth criterion for genera (see above), most of the genera of Hydromyini are probably not valid.

Tribe Hydromina J. Gray, 1825a: 341.

TYPE GENUS: *Hydromys* É. Geoffroy, 1804a.

COMMENTS: When originally proposed, this rank was placed in the Family Muridae (Illiger, 1811) and included the genus *Hydromys* É. Geoffroy, 1804a. Synonymised within the Subfamily Murinae by Musser and Carleton (2005: 1247).

Hydromyes Winge, 1887: 126.

TYPE GENUS: *Hydromys* É. Geoffroy, 1804a.

COMMENTS: When originally proposed, this rank was placed within the unknown rank Murini (Illiger, 1811) above the genus *Hydromys* É. Geoffroy, 1804a. Synonymised within the Subfamily Murinae by Musser and Carleton (2005: 1247).

Family Coniluridae K. Dahl, 1897: 194.

TYPE GENUS: *Conilurus* W. Ogilby, 1838f.

COMMENTS: When originally proposed, the higher rank was not stated but included the genus *Conilurus* W. Ogilby, 1838f; and possibly *Hydromys* É. Geoffroy, 1804a. Synonymised within the Subfamily Murinae by Musser and Carleton (2005: 1247).

Subfamily Pseudomyinae Simpson, 1961: 433.

TYPE GENUS: *Pseudomys* J. Gray, 1832.

COMMENTS: When originally proposed, this rank was placed in the Family Muridae (Illiger, 1811) and included the genera *Pseudomys* J. Gray, 1832 (including *Thetomys* Thomas, 1910c; and *Gyomys* Thomas, 1910c), *Leggadina* Thomas, 1910c; *Zyzomys* Thomas, 1909c (including *Laomys* Thomas, 1909c), *Leporillus* Thomas, 1906f; *Mastacomys* Thomas, 1882; *Notomys* Lesson, 1842; *Mesembriomys* Palmer, 1906; and *Conilurus* W. Ogilby, 1838f. Subfamily rank synonymised within the Subfamily Murinae by Musser and Carleton (2005: 1247), but recognised as the '*Pseudomys* Division' by Musser and Carleton (2005: 904) and '*Pseudomys* Group' by Troughton (1967: 225), Breed and Ford (2007: 8), and Van Dyck and Strahan (2008: 10, 577).

Uromys Group *sensu* Misonne, 1969: iii, 144.

TYPE GENUS: *Uromys* Peters, 1867a.

COMMENTS: When originally proposed, this rank was placed in the *Rattus* Division (Misonne, 1969) and included the genera *Uromys* Peters, 1867a; *Solomys* Thomas, 1922g: 261; *Melomys* Thomas, 1922g; *Pogonomelomys* Rümmler, 1936: 248; *Xenuromys* Tate and Archbold, 1941: 3; and *Apomys* Mearns, 1905: 455.

Tribe Conilurini Lee *et al.*, 1981: 1530.

TYPE GENUS: *Conilurus* W. Ogilby, 1838f.

COMMENTS: When originally proposed, this rank was placed in the Subfamily Hydromyinae (Alston, 1876) and included the genera *Pseudomys* J. Gray, 1832; *Leggadina* Thomas, 1910c; *Mastacomys* Thomas, 1882; *Notomys* Lesson, 1842; *Conilurus* W. Ogilby, 1838f; *Mesembriomys* Palmer, 1906; *Leporillus* Thomas, 1906f; and *Zyzomys* Thomas,

1909c. Tribe rank recognised by Watts and Aslin (1981: 5), Strahan (1983: 380; 1995: 9, 551), Baverstock (1984b: 917), Lidicker and Brylski (1987: 621, 634), Watts and Kemper (1989: 948), Watts and Baverstock (1994: 303), Flannery (1995a: 11, 257) and Ford (2003: 15). Synonymised within the Subfamily Murinae by McKenna and Bell (1997: 161) and Musser and Carleton (2005: 1247).

Tribe Hydromyini Lee *et al.*, 1981: 1530.

TYPE GENUS: *Hydromys* É. Geoffroy, 1804a.

COMMENTS: When originally proposed, this rank was placed in the Subfamily Hydromyinae (Alston, 1876) and included the genera *Hydromys* É. Geoffroy, 1804a; and *Xeromys* Thomas, 1889a. Tribe rank recognised by Watts and Aslin (1981: 6), Baverstock (1984b: 917), Lidicker and Brylski (1987: 621, 635), Watts and Kemper (1989: 948), Watts and Baverstock (1994: 303), and Ford (2003: 15). Synonymised within the Subfamily Murinae by Musser and Carleton (2005: 1247) and McKenna and Bell (1997: 161).

Tribe Uromyini Lee *et al.*, 1981: 1530.

TYPE GENUS: *Uromys* Peters, 1867a.

COMMENTS: When originally proposed, this rank was placed in the Subfamily Hydromyinae (Alston, 1876) and included the genera *Uromys* Peters, 1867a; and *Melomys* Thomas, 1922g. Tribe rank recognised by Watts and Aslin (1981: 6), Strahan (1983: 370; 1995: 632), Baverstock (1984b: 917), Lidicker and Brylski (1987: 621, 635), Watts and Kemper (1989: 948), Watts and Baverstock (1994: 303), Flannery (1995a: 11, 260) and Ford (2003: 15). Synonymised within the Subfamily Murinae by Musser and Carleton (2005: 1247) and McKenna and Bell (1997: 161).

Hydromys Division *sensu* Musser & Carleton, 2005: 903.

TYPE GENUS: *Hydromys* É. Geoffroy, 1804a.

COMMENTS: When originally proposed, this rank was placed in the Subfamily Murinae (Illiger, 1811) and included the genera *Crossomys* Thomas, 1907c: 70; *Hydromys* É. Geoffroy, 1804a; *Microhydromys* Tate and Archbold, 1941: 2; *Parahydromys* Poche, 1906: 326; and *Paraleptomys* Tate and Archbold, 1941: 1. Recognised by Breed and Ford (2007: 9), and Van Dyck and Strahan (2008: 10, 662) as the '*Hydromys* Group', but recognised in the Tribe Hydromyini (Alston, 1876) as a division by Lecompte *et al.* (2008: 8). As this is a non Linnean rank it has not been recognised here.

Pogonomys Division *sensu* Musser & Carleton, 2005: 904.

TYPE GENUS: *Pogonomys* Milne-Edwards, 1877.

COMMENTS: When originally proposed, this rank was placed in the Subfamily Murinae and included the genera *Abeomelomys* Menzies, 1990: 133; *Anisomys* Thomas, 1904c: 199; *Chiruromys* Thomas, 1888b: 237; *Coccymys* Menzies, 1990: 132; *Coryphomys* Schaub, 1937: 2, 5; *Hyomys* Thomas, 1904c: 198; *Macruromys* Stein, 1933: 94; *Mallomys* Thomas, 1898a: 1; *Mammelomys* Menzies, 1996: 383; *Pogonomelomys* Rümmler, 1936: 248; *Pogonomys* Milne-Edwards, 1877; *Spelaeomys* Hooijer, 1957: 306; and *Xenuromys* Tate and Archbold, 1941: 3. Recognised as the '*Pogonomys* Group' by Breed and Ford (2007: 9), and Van Dyck and Strahan (2008: 10, 679), but in the Tribe Hydromyini as a division by Lecompte *et al.* (2008: 8). As this is a non Linnean rank it has not been recognised here.

Pseudomys Division *sensu* Musser & Carleton, 2005: 904.

GENUS: *Pseudomys* J. Gray, 1832.

COMMENTS: When originally proposed, this rank was placed in the Subfamily Murinae (Illiger, 1811) and included the genera *Conilurus* W. Ogilby, 1838f; *Leggadina* Thomas, 1910c; *Leporillus* Thomas, 1906f; *Mastacomys* Thomas, 1882; *Mesembriomys* Palmer, 1906; *Notomys* Lesson, 1842; *Pseudomys* J. Gray, 1832; and *Zyzomys* Thomas, 1909c. Recognised as the '*Pseudomys* Group' by Troughton (1967: 225), Breed and Ford (2007: 8), and Van Dyck and Strahan (2008: 10, 577), but in the Tribe Hydromyini as a division by Lecompte *et al.* (2008: 8). As this is a non Linnean rank it has not been recognised here.

Xeromys Division *sensu* Musser & Carleton, 2005: 905.

TYPE GENUS: *Xeromys* Thomas, 1889a.

COMMENTS: When originally proposed, this rank was placed in the Subfamily Murinae (Illiger, 1811) and included the genera *Leptomys* Thomas, 1897a: 610; *Pseudohydromys* Rümmler, 1934: 47; and *Xeromys* Thomas, 1889a. Placed in the Tribe Hydromyini as a division by Lecompte *et al.* (2008: 9). As this is a non Linnean rank it has not been recognised here.

Uromys Division *sensu* Musser & Carleton, 2005: 905.

TYPE GENUS: *Uromys* Peters, 1867a.

COMMENTS: When originally proposed, this rank was placed in the Subfamily Murinae and included the genera *Melomys* Thomas, 1922g; *Paramelomys* Rümmler, 1936: 248; *Protochromys* Menzies, 1996: 416; *Solomys* Thomas, 1922g: 261; and *Uromys* Peters, 1867a. Recognised as the '*Uromys* Group' by Breed and Ford (2007: 9), and Van Dyck and Strahan (2008: 10, 666). Placed in the Tribe Hydromyini by as a division by Lecompte *et al.* (2008: 8). As this is a non Linnean rank it has not been recognised here.

Conilurus W. Ogilby, 1838

Conilurus W. Ogilby, 1838f: 96.

TYPE SPECIES: Emendation of *Conylurus* W. Ogilby, 1837c.

COMMENTS: W. Ogilby (1838f: 96) is an abstract of W. Ogilby (1838g: 124). Synonymised within *Hapalotis* by Waterhouse (1840c: 173). Proposed to be used ahead of *Conylurus* (W. Ogilby, 1837c) by Mahoney (1982: 23) and Mahoney and Richardson (1988a: 154) as it is in general use for the taxon, which has been followed by subsequent authors. The valid name *Conylurus* W. Ogilby, 1837c is an unused senior synonym of *Conilurus* W. Ogilby, 1838f. Included within Hydromyini (incorporating Conilurini where it is usually kept).

Hapalotis M. Lichtenstein, 1829: two unnumbered pages of text, Plate 29.

TYPE SPECIES: † *Hapalotis albipes* M. Lichtenstein, 1829 [= † *Conilurus albipes* (M. Lichtenstein, 1829)] by monotypy.

COMMENTS: Publication date established from Mahoney (1982: 22). Recognised in preference to *Conilurus* by Waterhouse (1840c: 173), but synonymised within *Conilurus* by Iredale and Troughton (1934: 82), Ellerman (1941: 113), Watts and Aslin (1981: 129), Mahoney and Richardson (1988a: 154) and subsequent authors.

HOMONYMS:

Hapalotis Hübner, 1821 [1816–1826]: 254, owlet moths of the Class Insecta (Order Lepidoptera, Family Noctuidae). Genus is a synonym of *Elaphria* Hübner, 1818: 16. See Poole (1989).

Conylurus W. Ogilby, 1837c: 208.

TYPE SPECIES: † *Hapalotis albipes* M. Lichtenstein, 1829 (as *Conylurus constructor* W. Ogilby, 1837c) [= † *Conilurus albipes* (M. Lichtenstein, 1829)] by monotypy.

COMMENTS: W. Ogilby (1837c) is an abstract of W. Ogilby (1838g: 124). Synonymised within *Conilurus* by Mahoney (1982: 23), Mahoney and Richardson (1988a: 154) and subsequent authors.

† *Conilurus albipes* (M. Lichtenstein, 1829)

White-footed Rabbit-rat

† *Hapalotis albipes* M. Lichtenstein, 1829: Text to Plate 29.

TYPE LOCALITY: New South Wales, eastern Australia.

COMMENTS: Publication date established from Mahoney (1982: 15). Recognised within *Hapalotis* by Gould (1853 [1845–1863]: Text to Plate 1). Placed in *Conilurus* by Longman (1916: 23), Iredale and Troughton (1934: 82), Tate (1951c: 270), Troughton (1967: 242) and subsequent authors.

† *Conylurus constructor* W. Ogilby, 1837c: 208.

TYPE LOCALITY: South eastern New South Wales, Australia.

COMMENTS: W. Ogilby (1837c) reference above is an abstract of W. Ogilby (1838g: 126). Synonymised within *albipes* by Iredale and Troughton (1934: 82), Tate (1951c: 270), Watts and Aslin (1981: 129), and Mahoney (1982: 22).

† *Conilurus destructor* Palmer, 1897a: 259.

TYPE LOCALITY: New South Wales, Australia.

COMMENTS: Synonymised within *Conilurus albipes* by Iredale and Troughton (1934: 82) and Watts and Aslin (1981: 129).

† *Conilurus capricornensis* Cramb & Hocknull, 2010

Capricorn Rabbit-rat

† *Conilurus capricornensis* Cramb & Hocknull, 2010b: 41, 46.

TYPE LOCALITY: Dodgey's Cave surface collection, Broken River, north east Queensland, Australia.

COMMENTS: Based on Pleistocene and Holocene dental remains.

Conilurus penicillatus (Gould, 1842)

Brush-tailed Rabbit-rat

Conilurus penicillatus penicillatus (Gould, 1842)

Mus penicillatus Gould, 1842b: 12.

TYPE LOCALITY: Port Essington, Northern Territory, Australia.

COMMENTS: Species recognised within *Hapalotis* by Gould (1851 [1845–1863]: Text to Plate 5). Recognised at the species rank in *Conilurus* by J. Ogilby (1892: 116) and subsequent authors including Longman (1916: 23), Iredale and Troughton (1934: 82), Troughton (1967: 243), Mahoney and Richardson (1988a: 156).

Hapalotis melanura J. Gray, 1844a: 12c; Plate 29, Fig. 2.

TYPE LOCALITY: Port Essington, Northern Territory, Australia.

COMMENTS: Date established from Tomes (1857: 138), Dobson (1875e: 372) and Tate (1941a: 595). Synonymised within *C. penicillatus* by Iredale and Troughton (1934: 82), Watts and Aslin (1981: 132), Mahoney in Mahoney and Richardson (1988a: 155) and Flannery (1990: 192; 1995a: 258).

Hapalotis hemileucura Gould, 1858: 243.

TYPE LOCALITY: Northern Australia. Gould (1858) believed the specimen came from about midway between the Gulf of Carpentaria and Moreton Bay, but from what is known of the distribution of the species it was considered by Mahoney in Mahoney and Richardson (1988a: 156) to be an unlikely locality for the species.

COMMENTS: Species further described by Gould (1858 [1845–1863]: Text to Plate 3). Recognised at the species rank within *Conilurus* by J. Ogilby (1892: 116), Iredale and Troughton (1934: 82), Ellerman (1941: 114), and Troughton

(1967: 244) and as a subspecies of *penicillatus* by Tate (1951c: 271). Synonymised within *Conilurus penicillatus* by Ride (1970: 242), Watts and Aslin (1981: 132), Mahoney and Richardson (1988a: 156) and Kemper and Schmitt (1992: 437) and Flannery (1990: 192; 1995a: 258).

Conilurus penicillatus melibius Thomas, 1921

Conilurus melibius Thomas, 1921a: 431.

TYPE LOCALITY: Biro, Apsley Strait, Melville Island, Northern Territory, Australia. Sea Level.

COMMENTS: Synonymised within *penicillatus* by Watts and Aslin (1981: 132), Mahoney and Richardson (1988a: 156) and Flannery (1990: 192). Recognised at the species rank in *Conilurus* by Iredale and Troughton (1934: 82), Ellerman (1941: 114) and Tate (1951c: 271). Placed at subspecies rank within *penicillatus* by Troughton (1967: 243), Kemper and Schmitt (1992: 437), Flannery (1995a: 258), Clayton *et al*. (2006: 111), and Van Dyck and Strahan (2008: 581).

Φ *Conilurus penicillatus randi* Tate & Archbold, 1938

Φ *Conilurus randi* Tate & Archbold, 1938: 1.

TYPE LOCALITY: Original description states Penzara, between Waii Kussa and Morehead Rivers, Western Division, Papua New Guinea. 20 metres elevation.

COMMENTS: Synonymised within *penicillatus* by Musser and Carleton (1993: 586; 2005: 1309). Recognised as a subspecies of *penicillatus* by Tate (1951c: 272), Kemper and Schmitt (1992: 437), Flannery (1990: 192; 1995a: 258), and Van Dyck and Strahan (2008: 581).

Hydromys É. Geoffroy, 1804

Hydromys É. Geoffroy, 1804a: 353 as 253.

TYPE SPECIES: *Hydromys chrysogaster* É. Geoffroy, 1804a by subsequent designation. See Iredale and Troughton (1934: 69).

COMMENTS: Taxon spelt *Hydromis* by É. Geoffroy (1805a: 90). Genus recognised by subsequent authors.

Hydromis É. Geoffroy, 1805a: 81.

TYPE SPECIES: Incorrect subsequent spelling of *Hydromys* É. Geoffroy, 1804a.

COMMENTS: Spelling not typically recognised by subsequent authors.

Baiyankamys Hinton, 1943: 552.

TYPE SPECIES: Φ *Baiyankamys shawmeyeri* [sic] Hinton, 1943: 552 [=Φ *Hydromys shawmeyeri* (Hinton, 1943: 552)] by original designation.

COMMENTS: Synonymised within *Hydromys* by Mahoney (1968: 64).

Hydromys chrysogaster É. Geoffroy, 1804

Water Rat

Hydromys chrysogaster É. Geoffroy, 1804a: 354 as 254.

TYPE LOCALITY: Bruny Island, Tasmania, Australia.

COMMENTS: Also described by Geoffroy (1805a: 90). See Desmarest (1822b: 297) and Sherborn and Woodward (1906: 582) who list the page and plate numbers of Livr. 91. Taxonomic decision of Mahoney in Mahoney and Richardson (1988a: 158) to reduce *leucogaster* É. Geoffroy as a junior synonym of *chrysogaster* É. Geoffroy even though they have the same publication date. The specific name *leucogaster* was fixed by Lesson (1827a: 285) as the prior name to *chrysogaster* É. Geoffroy, 1804a, which is in general use for the taxon; *leucogaster* É. Geoffroy, 1804a: 354, is an unused senior synonym of *chrysogaster* É. Geoffroy, 1804a, and should not be resurrected as a name for the species. Species recognised within *Hydromys* by Gould (1853 [1845–1863]: Text to Plate 24). Flannery (1990: 188; 1995a: 236; 1995b: 126), and Van Dyck and Strahan (2008: 663) who suggested the subspecies need to be resolved and are of questionable validity. As a result all names are listed as synonyms.

FUTURE TAXONOMIC RESEARCH: Australian water rats are badly in need of revision, as acknowledged by several authors, most lately by Musser and Carleton (2005: 1333). For example, brief observations by Colin Groves in the Australian Museum collection suggest that two species exist in the south-east of Australia, distinguished by the size of the cheekteeth and the colour of the underparts; they are largely geographically segregated, but may overlap in the Tumut region of New South Wales.

Hydromys leucogaster É. Geoffroy, 1804a: 354 as 254.

TYPE LOCALITY: Maria Island, Tasmania, Australia

COMMENTS: Type designation by Rode (1945: 204). Species recognised within *Hydromys* by Gould (1853 [1845–1863]: Text to Plate 26). Synonymised with *Hydromys chrysogaster* by Iredale and Troughton (1934: 69), Ellerman (1941: 299), Brazenor (1936a: 63), Watts and Aslin (1981: 66), Flannery (1990: 188; 1995a: 236; 1995b: 126), and Musser and Carleton (2005: 1333).

Meriones apicalis Kuhl, 1820: 70.

TYPE LOCALITY: 'India oriantali'.

COMMENTS: Location given as Moluccas by Illiger (1815: 89) but as Brazil by Olfers (1818: 207 *fide* Hershkovitz, 1959: 339). Synonymised within *chrysogaster* by Trouessart (1898: 458), Ellerman (1941: 299), Hershkovitz (1959: 339), and Musser and Carleton (2005: 1333).

Hydromys fulvogaster Jourdan, 1837a: 523.

TYPE LOCALITY: 'les bords de la riviere des Cygnes (Australasie)' [Swan River, in error =Tasmania, see Iredale & Troughton, 1934: 69].

COMMENTS: Thomas and Dollman (1909: 790) believed that Jourdan's description of *H. fulvogaster* (as *H. fulvoventer*) is wholly inapplicable to any western Australian specimen. A review of Jourdan's manuscript was undertaken by F. Cuvier (1838: 2) with further discussion of Jourdan's paper and that of Cuvier's 'Rapport' made by Andersen (1909a: 21–22). The spelling *fulvoventer* (as *Fulvo-Venter*) found in F. Cuvier (1837d: 372; 1838: 5) is an incorrect subsequent spelling of *fulvogaster* according to Mahoney and Richardson (1988a: 157). This taxon was synonymised with *Hydromys chrysogaster* by Iredale and Troughton (1934: 69), Brazenor (1936a: 63), Ellerman (1941: 299), Watts and Aslin (1981: 66), Flannery (1990: 188; 1995a: 236; 1995b: 126), and Musser and Carleton (2005: 1333).

[Hydromis] Fulvo-Venter F. Cuvier, 1837d: 372.

TYPE LOCALITY: 'les bords de la riviere des Cygnes (Australasie)' [Swan River, in error =Tasmania, see Iredale & Troughton, 1934: 69].

COMMENTS: Appears to be an incorrect subsequent spelling, which was also used by F. Cuvier (1838: 5). Ex Jourdan MS (Iredale & Troughton, 1934: 69). Thomas and Dollman (1909: 790) believed the locality of the specimen described by F. Cuvier (1837d: 372), and assigned to Jourdan, was incorrectly assigned to any specimens from Western Australia. This taxon was synonymised with *Hydromys chrysogaster* by Iredale and Troughton (1934: 69), Brazenor (1936a: 63), Ellerman (1941: 299), Watts and Aslin (1981: 66), Flannery (1990: 188; 1995a: 236; 1995b: 126), and Musser and Carleton (2005: 1333).

Hydromys flaviventer Owen, 1840–1845: 26; Plate 105, Fig. 18.

TYPE LOCALITY: Australia.

COMMENTS: Synonymised within *chrysogaster* by Ellerman (1941: 299), and by Musser and Carleton (2005: 1333).

Hydromys fuliginosus Gould, 1853 [1845–1863]: Text to Plate 27.

TYPE LOCALITY: King George Sound, Western Australia, Australia.

COMMENTS: Type designation by Thomas (1921a: 432). Recognised as a species within *Hydromys* by Longman (1916: 21), Iredale and Troughton (1934: 70) and Troughton (1967: 211). Reduced to a subspecies of *chrysogaster* by Tate (1951c: 234) and Ellerman (1941: 299). Synonymised within *chrysogaster* by Ride (1970: 243), Watts and Aslin (1981: 66), Flannery (1990: 188; 1995a: 236; 1995b: 126), and Musser and Carleton (2005: 1333).

Hydromys fulvolavatus Gould, 1853 [1845–1863]: Text to Plate 25.

TYPE LOCALITY: Murray River, South Australia, Australia.

COMMENTS: Type designation by Thomas (1921a: 432). Recognised at the species rank by J. Ogilby (1892: 102). Reduced to a subspecies of *chrysogaster* by Iredale and Troughton (1934: 70) and Ellerman (1941: 299). Recognised as a species Iredale and Troughton (1934: 70) and as a subspecies of *chrysogaster* by Brazenor (1936a: 64) and Tate (1951c: 233). Synonymised with *Hydromys chrysogaster* by Watts and Aslin (1981: 66), Flannery (1990: 188; 1995a: 236; 1995b: 126), and Musser and Carleton (2005: 1333).

Hydromys Lutrilla Gould, 1863 [1845–1863]: xxxvi of Introduction in Volume 1.

TYPE LOCALITY: Elizabeth Bay, Sydney, New South Wales, Australia.

COMMENTS: Recognised as a subspecies of *H. chrysogaster* by Iredale and Troughton (1934: 70) and Brazenor (1936a: 64). Synonymised within *chrysogaster* by Tate (1951c: 232), Watts and Aslin (1981: 66), Flannery (1990: 188; 1995a: 236; 1995b: 126) and Musser and Carleton (2005: 1333).

Φ *Hydromys Beccarii* Peters, 1874a: 303.

TYPE LOCALITY: 'Key (Weri)' =Kei Island, Indonesia.

COMMENTS: Recognised as a species by Le Souef and Burrell (1926: 113) and subspecies of *chrysogaster* by Ellerman (1941: 300), Tate (1951c: 234) and D. Johnson (1964: 500). Synonymised within *Hydromys chrysogaster* by Watts and Aslin (1981: 66), Flannery (1990: 188; 1995a: 236; 1995b: 126), and Musser and Carleton (2005: 1333).

Φ *Hydromys esox* Thomas, 1906b: 324.

TYPE LOCALITY: Southern New Guinea.

COMMENTS: Recognised as a species by Le Souef and Burrell (1926: 113) and subspecies rank by Ellerman (1941: 300). Synonymised within *beccarii* by Tate (1951c: 234). Synonymised within *chrysogaster* by Flannery (1990: 188; 1995a: 236; 1995b: 126), and Musser and Carleton (2005: 1333).

Hydromys chrysogaster reginae Thomas & Dollman, 1909: 789.

TYPE LOCALITY: Inkerman Station, Queensland, Australia.

COMMENTS: Recognised as a subspecies of *chrysogaster* by Iredale and Troughton (1934: 70), Brazenor (1936a: 64), Ellerman (1941: 300) and Tate (1951c: 234). Synonymised with *Hydromys chrysogaster* by Watts and Aslin (1981: 66), Flannery (1990: 188; 1995a: 236; 1995b: 126), and Musser and Carleton (2005: 1333).

Hydromys chrysogaster caurinus Thomas, 1909b: 197.

TYPE LOCALITY: Parry Creek, 5 miles west of Trig Station, near Wyndham, eastern Kimberley, Western Australia,

Australia. Trig Station HJ9 is located at 15°34′24.69″S, 128°20′47.01″E.

COMMENTS: Recognised as a species by Iredale and Troughton (1934: 70) and Troughton (1967: 210) but reduced to subspecies rank by Ellerman (1941: 299). Synonymised within *beccarii* by Tate (1951c: 234). Synonymised within *chrysogaster* by Ride (1970: 243), Watts and Aslin (1981: 66), Flannery (1990: 188; 1995a: 236; 1995b: 126), and Musser and Carleton (2005: 1333).

Φ *Hydromys nauticus* Thomas, 1921a: 429.

TYPE LOCALITY: Aru Islands, New Guinea.

COMMENTS: Subspecies rank recognised by Ellerman (1941: 300). Synonymised within *beccarii* by Tate (1951c: 235) and within *chrysogaster* by Flannery (1990: 188; 1995a: 236; 1995b: 126), and Musser and Carleton (2005: 1333).

Hydromys melicertes Thomas, 1921a: 430.

TYPE LOCALITY: Biro, Apsley Strait, Melville Island, Northern Territory, Australia. Sea level.

COMMENTS: Recognised as a species by Iredale and Troughton (1934: 70) and Troughton (1967: 210). Subspecies rank recognised by Ellerman (1941: 300). Synonymised within *beccarii* by Tate (1951c: 235) and within *chrysogaster* by Ride (1970: 243), Watts and Aslin (1981: 66), Flannery (1990: 188; 1995a: 236; 1995b: 126), and Musser and Carleton (2005: 1333).

Φ *Hydromys esox illuteus* Thomas, 1922b: 264.

TYPE LOCALITY: Prauwen Bivak, Idenburg River, Papua New Guinea.

COMMENTS: Recognised as a subspecies of *chrysogaster* by Tate (1951c: 236), Flannery (1990: 188; 1995a: 236; 1995b: 126), and Musser and Carleton (2005: 1333).

Hydromys longmani Thomas, 1923a: 171.

TYPE LOCALITY: Vine Creek, Ravenshoe, north Queensland, Australia. 3000 feet.

COMMENTS: Recognised as a species by Iredale and Troughton (1934: 70) and Troughton (1967: 209) but reduced to subspecies rank by Ellerman (1941: 300). Synonymised within *beccarii* by Tate (1951c: 235) and within *chrysogaster* by Ride (1970: 243), Watts and Aslin (1981: 66), Flannery (1990: 188; 1995a: 236; 1995b: 126), and Musser and Carleton (2005: 1333).

Hydromys grootensis Troughton, 1935b: 252.

TYPE LOCALITY: Groote Eylandt, Northern Territory, Australia.

COMMENTS: Species rank recognised by Troughton (1967: 209) and subspecies rank by Ellerman (1941: 300). Synonymised within *beccarii* by Tate (1951c: 235) and within *chrysogaster* by Ride (1970: 243), Watts and Aslin (1981: 66), Flannery (1990: 188; 1995a: 236; 1995b: 126), and Musser and Carleton (2005: 1333).

Hydromys lawnensis Troughton, 1935b: 253.

TYPE LOCALITY: Lawn Hill Creek, Adel's Grove, about 12 miles south of Lawn Hill Station and 100 miles south of Burketown, Queensland, Australia.

COMMENTS: Species rank recognised by Troughton (1967: 210) and subspecies by Ellerman (1941: 300). Synonymised within *reginae* by Tate (1951c: 234) and within *chrysogaster* by Ride (1970: 243), Watts and Aslin (1981: 66), Flannery (1990: 188; 1995a: 236; 1995b: 126), and Musser and Carleton (2005: 1333).

Hydromys moae Troughton, 1935b: 254.

TYPE LOCALITY: Moa Island (as Moa or Banka Island), Torres Strait, north Queensland, Australia.

COMMENTS: Species rank recognised by Troughton (1967: 209) and subspecies rank recognised by Ellerman (1941: 300). Synonymised within *beccarii* by Tate (1951c: 235) and within *chrysogaster* by Flannery (1990: 188; 1995a: 236; 1995b: 126), and Musser and Carleton (2005: 1333).

Φ *Hydromys oriens* Troughton, 1937a: 127.

TYPE LOCALITY: Mount Lamington district, southern border of the Northern Division, Papua New Guinea.

COMMENTS: Synonymised within *chrysogaster* by Tate (1951c: 236), Flannery (1990: 188; 1995a: 236; 1995b: 126), and Musser and Carleton (2005: 1333).

Leggadina Thomas, 1910

Leggadina Thomas, 1910c: 606.

TYPE SPECIES: *Mus forresti* Thomas, 1906e (as *Pseudomys* (*Leggadina*) *forresti*) [=*Leggadina forresti* (Thomas, 1906e)] by original designation.

COMMENTS: Originally described as a subgenus of *Pseudomys* J. Gray, 1832, which was followed by Troughton (1932c: 289). Raised to generic status by Iredale and Troughton (1934), and followed by subsequent authors including Ellerman (1941: 255), Tate (1951c: 249), D. Johnson (1964: 493) and Troughton (1967: 231). Synonymised within *Pseudomys* by Ride (1970: 244). Morphological and genetic variation reviewed by N. Cooper *et al.* (2003a: 333).

Leggadina forresti (Thomas, 1906)

Central Short-tailed Mouse

Mus forresti Thomas, 1906e: 6.

TYPE LOCALITY: Alexandria Station, Northern Territory, Australia. (About 19°S, 137°E)

COMMENTS: Full description in Thomas (1906d: 538). Recognised as a species within *Pseudomys* by Longman (1916: 22) and Finlayson (1939a: 101), and as a species within *Leggadina* by Iredale and Troughton (1934: 78), Ellerman (1941: 256), Tate (1951c: 250) and Troughton (1967: 231). Transferred to *Pseudomys* by Ride (1970: 154) and finally to *Leggadina* by Mahoney and Richardson (1988a: 158) and subsequent authors.

Pseudomys (Leggadina) messorius Thomas, 1925: 670.

TYPE LOCALITY: Melrose, Spencer Gulf, South Australia, Australia.

COMMENTS: Recognised at the species rank within *Leggadina* by Iredale and Troughton (1934: 79), Ellerman (1941: 256), Tate (1951c: 251) and Troughton (1967: 232). Synonymised within *forresti* by Ride (1970: 246), Watts and Aslin (1981: 208), and Mahoney and Richardson (1988a: 158).

Pseudomys (Leggadina) waitei Troughton, 1932c: 290.

TYPE LOCALITY: Alice Springs, Northern Territory, Australia.

COMMENTS: Recognised as described by Finlayson (1941: 220), and at the species rank within *Leggadina* by Iredale and Troughton (1934: 79), Ellerman (1941: 256), Tate (1951c: 251) and Troughton (1967: 232). Synonymised within *forresti* by Ride (1970: 247), Watts and Aslin (1981: 208), and Mahoney and Richardson (1988a: 158).

Gyomys berneyi Troughton, 1936: 15.

TYPE LOCALITY: Timbered sand ridge country, Barcarolle Station, 135 miles south of Longreach, Queensland, Australia.

COMMENTS: Species rank recognised within *Gyomys* by Ellerman (1941: 221), within *Pseudomys* by Tate (1951c: 245) and within *Leggadina* by Troughton (1967: 232). Synonymised within *forresti* by Ride (1970: 246) and Watts and Aslin (1981: 208) and subsequent authors.

Leggadina lakedownensis Watts, 1976

Northern Short-tailed Mouse

Leggadina lakedownensis Watts, 1976: 105.

TYPE LOCALITY: Lakeland Downs, 70 miles south of Cooktown, Queensland, Australia.

COMMENTS: Specific status confirmed by Baverstock *et al.* (1976: 109, 112).

Leporillus Thomas, 1906

Leporillus Thomas, 1906f: 83.

TYPE SPECIES: † *Hapalotis apicalis* Gould, 1853 [=† *Leporillus apicalis* (Gould, 1853)] by original designation.

COMMENTS: Genus typically recognised since its description.

† ***Leporillus apicalis*** (Gould, 1853)

Lesser Stick-nest Rat

† *Hapalotis apicalis* Gould, 1853: 126.

TYPE LOCALITY: South Australia, Australia (see Mahoney & Richardson, 1988a: 159).

COMMENTS: Species further described by Gould (1853 [1845–1863]: Text to Plate 2). Lectotype selected by Thomas (1921a: 433) but Mahoney (1975: 102) discussed this further. Recognised within *Hapalotis* by Krefft (1866a: 4). Placed in *Conilurus* by J. Ogilby (1892: 116) and Waite (1898: 115). It subsequently placed in the genus *Leporillus* by Thomas (1906f: 83), Troughton (1923b: 32), Wood Jones (1925 [1923–1925]: 334) and subsequent authors.

Leporillus conditor (Gould, 1848)

Greater Stick-nest Rat

Mus conditor Gould, in Sturt, 1848: Vol 1, Pl. opposite p. 120; Vol. 2. Appendix: 4, 7.

TYPE LOCALITY: Polia area, about 45 miles (72 kilometres) from Laidley Ponds, New South Wales, Australia. See Mahoney and Richardson (1988a: 160).

COMMENTS: Species transferred to *Hapalotis* by Gould (1849 [1845–1863]: Text to Plate 6) and followed by Krefft (1864: 65; 1866a: 4). Recognised as a species within *Conilurus* by J. Ogilby (1892: 118), and *Leporillus* by Troughton (1923b: 24), Wood Jones (1925 [1923–1925]: 327) and subsequent authors. The describer of this species has been variable, with Gould recognised as the author by authors including J. Ogilby (1892: 118), Troughton (1923b: 24) and Wood Jones (1925 [1923–1925]). In contrast, Sturt was recognised as the author by Iredale and Troughton (1934: 81), Tate (1951c: 256), Mahoney (1975: 102), Watts and Aslin (1981: 146), Mahoney and Richardson (1988a: 160) and Musser and Carleton (2005: 1349). McAllan and Bruce (1989: 454) argued that the allocation of Sturt as the author of this taxon is incorrect, and that Gould's name is associated with it: volume 1 of Sturt (1848: 120–121, plate opposite page 120) included the name '*Mus conditor* Gould', while in the Appendix of Volume 2 of the same book (1848: Appendix, 4, 7–8) the taxon name is again mentioned as '*Mus conditor*, Gould'. Therefore Gould has been credited with as author of this name here.

Leporillus jonesi Thomas, 1921b: 618.

TYPE LOCALITY: Hut on northern side of Franklin Islands, Nuyt's Archipelago, South Australia, Australia.

COMMENTS: Recognised as a species by Wood Jones (1922b: 595), Iredale and Troughton (1934: 81), Ellerman (1941: 222), Tate (1951c: 256) and Troughton (1967: 246). Synonymised within *conditor* by Ride (1970: 244), Watts and Aslin (1981: 146), Strahan (1983: 424; 1995: 560) and Mahoney and Richardson (1988a: 160) and subsequent authors.

Mastacomys Thomas, 1882

Mastacomys Thomas, 1882: 413.

TYPE SPECIES: *Mastacomys fuscus* Thomas, 1882 by monotypy.

COMMENTS: Synonymised within *Pseudomys* by Watts *et al.* (1992: 81), Musser and Carleton (1993: 644) and McKenna and Bell (1997: 170). Recognised as a distinct genus by Longman (1916: 23), Iredale and Troughton (1934: 80), Ellerman (1941: 266), Tate (1951c: 256), Troughton (1967: 238), Watts and Aslin (1981: 213), Mahoney and Richardson (1988a: 160), Strahan (1995: 562) and subsequent authors.

Mastacomys fuscus Thomas, 1882

Broad-toothed Rat

Mastacomys fuscus fuscus Thomas, 1882

Mastacomys fuscus Thomas, 1882: 413.

TYPE LOCALITY: Tasmania, Australia.

COMMENTS: Recognised within *Mastacomys* by Finlayson (1933c: 125), Ellerman (1941: 267), Tate (1951c: 257), Mahoney and Richardson (1988a: 160) and subsequent authors.

Mus Castaneus Higgins & Petterd, 1884b: 183.

TYPE LOCALITY: Syntypes not preserved. Lord (1919: 25–26) and Lord and Scott (1924: 300) suggested they were from Long Plains, Tasmania, Australia.

COMMENTS: Synonymised within *Rattus velutinus* by Iredale and Troughton (1934: 74) and within *Mastacomys fuscus* by Lord (1923: 76), J. Taylor and Horner (1973: 68), Mahoney and Richardson (1988a: 160) and subsequent authors.

HOMONYMS:

Mus castaneus Waterhouse, 1843a: 145, the House Mouse of the Class Mammalia (Order Rodentia, Family Muridae). Taxon is a synonym of the *Mus musculus* Linnaeus, 1758. See individual entry.

Mastacomys fuscus mordicus Thomas, 1922

Mastacomys mordicus Thomas, 1922h: 551.

TYPE LOCALITY: A guano cave at Mount Gambier, South Australia, Australia.

COMMENTS: Synonymised within *fuscus* by Iredale and Troughton (1934: 80) and Watts and Aslin (1981: 214). Recognised at the species rank by Ellerman (1941: 267), Tate (1951c: 257) and subspecies of *fuscus* by Troughton (1967: 239), Wakefield (1972b: 21), Strahan (1983: 423; 1995: 563), Clayton *et al.* (2006: 111), and Van Dyck and Strahan (2008: 591).

† *Mastacomys wombeyensis* Ride, 1956b: 431.

TYPE LOCALITY: Wombeyan Caves, near Taralga, New South Wales, Australia.

COMMENTS: Synonymised within *fuscus* by Wakefield (1972a: 20), Watts and Aslin (1981: 214) and Musser and Carleton (2005: 1360). Synonymised within *mordicus* by Strahan (1983: 423; 1995: 563), and Van Dyck and Strahan (2008: 591).

Mastacomys fuscus brazenori Ride, 1956b: 436

TYPE LOCALITY: Olangolah, near Beech Forest, at the head of Gellibrand River, Victoria, Australia.

COMMENTS: Synonymised within *fuscus* by Watts and Aslin (1981: 214) and Musser and Carleton (2005: 1360), and within *mordicus* by Strahan (1983: 423; 1995: 563), and Van Dyck and Strahan (2008: 591).

Melomys Thomas, 1922

Melomys Thomas, 1922g: 261.

TYPE SPECIES: Φ *Uromys rufescens* Alston, 1877: 124 [=Φ *Melomys rufescens* (Alston, 1877: 124)] by original designation.

COMMENTS: Two species, *littoralis* and *burtoni*, were identified by Knox (1978: 276). Has previously included as synonyms *Mammelomys* Menzies (1996: 383), *Paramelomys* Rümmler (1936: 248) and *Protochromys* Menzies (1996: 416) (e.g. McKenna and Bell, 1997: 171), but these are now recognised as distinct genera (Musser & Carleton, 2005: 1358, 1432, 1451).

Melomys burtoni (Ramsay, 1887)

Grassland Melomys

Mus burtoni Ramsay, 1887a: 551, 553; Plate 17.

TYPE LOCALITY: Neighbourhood of Derby, Western Australia, Australia.

COMMENTS: Abstract of description by Ramsay (1887b: vi). Mahoney and Richardson (1988a: 161) and Musser and Carleton (1993: 615) treated populations from the Australian and New Guinea region as a single species. No subspecies recognised by Van Dyck and Strahan (2008: 668).

Uromys melicus Thomas, 1913a: 215.

TYPE LOCALITY: Biro, Apsley Strait, Melville Island, Northern Territory, Australia.

COMMENTS: Recognised as a species within *Uromys* by Longman (1916: 23) and Iredale and Troughton (1934: 87). Recognised as a subspecies of *cervinipes* by Ellerman (1941: 231) and Tate (1951c: 294). Synonymised within *Melomys lutillus* (Thomas, 1913a: 216) by Mahoney and Richardson (1988a: 161) and within *M. burtoni* by Watts and Aslin (1981: 83) and Musser and Carlton (1993: 615; 2005: 1375).

Uromys murinus Thomas, 1913a: 216.

TYPE LOCALITY: Maer Island (as Mer), Murray Islands, Torres Strait, north Queensland, Australia.

COMMENTS: Recognised as a species within *Uromys* by Longman (1916: 23) and Iredale and Troughton (1934: 86). Synonymised within *Melomys* by Watts and Aslin (1981: 83), Mahoney and Richardson (1988a: 161) and Musser and Carlton (1993: 615; 2005: 1375).

Φ *Uromys muscalis* Thomas, 1913a: 217.

TYPE LOCALITY: Lower Fly River, Papua New Guinea.

COMMENTS: Recognised as a subspecies of *Melomys lutillus* (Thomas, 1913a: 216) by Laurie and Hill (1954: 124), but synonymised within *burtoni* by Musser and Carlton (1993: 615; 2005: 1375).

Uromys littoralis Lönnberg, 1916: 5.

TYPE LOCALITY: Beach near the mouth of Russell (as Russel) River, Queensland, Australia.

COMMENTS: Recognised as a valid species, *Melomys littoralis*, by Iredale and Troughton (1934: 86), Troughton (1967: 255), Ride (1970: 156) and Knox (1978: 276). Synonymised within *Melomys* by Watts and Aslin (1981: 83), Mahoney and Richardson (1988a: 161) and Musser and Carlton (1993: 615; 2005: 1375).

Melomys australius Thomas, 1924b: 298.

TYPE LOCALITY: Piara, Queensland, Australia.

COMMENTS: Species recognised by Iredale and Troughton (1934: 86). Synonymised within *Melomys lutillus* (Thomas, 1913a: 216) by Ride (1970: 244) and within *burtoni* by Watts and Aslin (1981: 83), Mahoney and Richardson (1988a: 161) and Musser and Carlton (1993: 615; 2005: 1375).

Melomys littoralis insulae Troughton & Le Souef, 1929b: 96.

TYPE LOCALITY: Hinchinbrook Island, Queensland, Australia.

COMMENTS: Subspecies recognised, within *littoralis*, by Iredale and Troughton (1934: 86) and Troughton (1967: 255), but synonymised within *littoralis* by Ellerman (1941: 231), Synonymised within *burtoni* by Watts and Aslin (1981: 83), Mahoney and Richardson (1988a: 161) and Musser and Carlton (1993: 615; 2005: 1375).

Melomys mixtus Troughton, 1935b: 257.

TYPE LOCALITY: Groote Eylandt, Northern Territory, Australia.

COMMENTS: Recognised as a subspecies of *cervinipes* by D. Johnson (1964: 491). Tentatively synonymised within *cervinipes* by Ellerman (1941: 231), and within *burtoni* by Watts and Aslin (1981: 83), Mahoney and Richardson (1988a: 161) and Musser and Carlton (1993: 615; 2005: 1375).

Φ *Melomys muscalis froggatti* Troughton, 1937a: 123.

TYPE LOCALITY: Strickland River, 100 miles above junction with Fly River, south New Guinea.

COMMENTS: Recognised as a subspecies of *Melomys lutillus* (Thomas, 1913a: 216) by Laurie and Hill (1954: 125), but synonymised within *burtoni* by Musser and Carlton (1993: 615; 2005: 1375).

Melomys callopes Finlayson, 1942: 243; Plates 8–9.

TYPE LOCALITY: Approximately south west of Duaringa, Queensland, Australia.

COMMENTS: Synonymised within *burtoni* by Watts and Aslin (1981: 83), Mahoney and Richardson (1988a: 161) and Musser and Carlton (1993: 615; 2005: 1375).

Melomys cervinipes albiventer Kellogg, 1945: 69.

TYPE LOCALITY: Near either Brocks Creek or Douglas River, about 100 miles south of Port Darwin, Northern Territory, Australia.

COMMENTS: Recognised as a subspecies of *cervinipes* by Tate (1951c: 295) and D. Johnson (1964: 490). Synonymised within *burtoni* by Watts and Aslin (1981: 83), Mahoney and Richardson (1988a: 161) and Musser and Carlton (1993: 615; 2005: 1375).

Melomys capensis Tate, 1951

Cape York Melomys

Melomys cervinipes capensis Tate, 1951c: 295.

TYPE LOCALITY: Upper Nesbitt River, Rocky Scrub. East of Coen north Queensland, Australia. 1500 feet.

COMMENTS: Elevated to species rank by Baverstock *et al.* (1980: 553) and followed by Watts and Aslin (1981: 81) and subsequent authors.

Melomys cervinipes (Gould, 1852)

Fawn-footed Melomys

Mus cervinipes Gould, 1852 [1845–1863]: Text to Plate 14.

TYPE LOCALITY: Stradbroke Island, Queensland, Australia.

COMMENTS: Type designation by Thomas (1921a: 432). Recognised as a species within *Uromys* by J. Ogilby (1892: 121) and Longman (1916: 23). Transferred to *Melomys* by Ellerman (1941: 231), Tate (1951c: 292), Troughton (1967: 254) and subsequent authors.

Uromys banfieldi De Vis, 1907: 8.

TYPE LOCALITY: Dunk Island, north Queensland, Australia.

COMMENTS: Recognised as a species within *Uromys* by Longman (1916: 23) then as a species within *Melomys* by Iredale and Troughton (1934: 86) and Troughton (1967: 255). Recognised as a subspecies of *cervinipes* by Ellerman

(1941: 231) and Tate (1951c: 294). Synonymised within *cervinipes* by Watts and Aslin (1981: 77) and Mahoney and Richardson (1988a: 162).

Melomys cervinipes eboreus Thomas, 1924b: 297.

TYPE LOCALITY: Dinner Creek, Ravenshoe, Queensland, Australia. 2900 feet.

COMMENTS: Recognised as described by Iredale and Troughton (1934: 86), Ellerman (1941: 231), Tate (1951c: 293) and Strahan (1983: 374). Synonymised within *cervinipes* by Watts and Aslin (1981: 77) and Mahoney and Richardson (1988a: 162).

Melomys cervinipes pallidus Troughton & Le Souef, 1929b: 97.

TYPE LOCALITY: Hinchinbrook Island, Queensland, Australia.

COMMENTS: Recognised as a subspecies of *cervinipes* by Iredale and Troughton (1934: 86), Tate (1951c: 294) and Troughton (1967: 254). Synonymised within *littoralis* by Ellerman (1941: 231), and within *cervinipes* by Watts and Aslin (1981: 77), Strahan (1983: 374) and Mahoney and Richardson (1988a: 162).

Melomys limicauda Troughton, 1935b: 255.

TYPE LOCALITY: Hayman Island, Whitsunday Group, Queensland, Australia.

COMMENTS: Recognised as a subspecies of *cervinipes* by Tate (1951c: 294) and synonymised within *littoralis* by Ellerman (1941: 231), and within *cervinipes* by Watts and Aslin (1981: 77).

Melomys cervinipes bunya Tate, 1951c: 293.

TYPE LOCALITY: Rainforest bordering rest house, summit of Bunya Mountains, Queensland, Australia. 3500 feet.

COMMENTS: Subspecies rank recognised by Strahan (1983: 374). Synonymised within *cervinipes* by Watts and Aslin (1981: 77) and most subsequent authors.

† ***Melomys rubicola*** Thomas, 1924

Bramble Cay Melomys

† *Melomys rubicola* Thomas, 1924b: 298.

TYPE LOCALITY: Bramble Cay (as Bramble Key), Torres Strait, Queensland, Australia. (About 9°S, 144°E)

COMMENTS: Reduced to a subspecies of *cervinipes* by Ellerman (1941: 231). Species rank recognised by Iredale and Troughton (1934: 87), Tate (1951c: 296) and subsequent authors. A more recent revision of the taxonomy of this taxon was undertaken by Dennis and Storch (1998: 21) who confirmed its species rank, though they highlighted that the genus *Melomys* is in need of taxonomic revision. Species is possibly extinct (Woinarski *et al.*, 2014: 665).

Mesembriomys Palmer, 1906

Mesembriomys Palmer, 1906: 97.

TYPE SPECIES: *Nomen novum* for *Ammomys* Thomas, 1906f.

COMMENTS: Recognised by most authors since its description.

Ammomys Thomas, 1906f: 83, 84.

TYPE SPECIES: *Mus hirsutus* Gould, 1842b (as *Ammomys hirsutus*) [=*Mesembriomys gouldii* (J. Gray, 1843a)] by original designation.

COMMENTS: Synonymised within *Mesembriomys* by Iredale and Troughton (1934: 80), Ellerman (1941: 116), Tate (1951c: 267), Watts and Aslin (1981: 123) and subsequent authors.

HOMONYMS:

Ammomys Bonaparte, 1831: 20, meadow voles of the Class Mammalia (Order Rodentia, Family Cricetidae). Genus is a synonym of *Microtus* Schrank, 1798: 66, 72. See Musser and Carleton (2005: 989).

Mesembriomys gouldii (J. Gray, 1843)

Black-footed Tree-rat

Mesembriomys gouldii gouldii (J. Gray, 1843)

Hapalotis Gouldii J. Gray, 1843a: 116.

TYPE LOCALITY: Port Essington, Northern Territory, Australia.

COMMENTS: *Nomen novum* for *Mus hirsutus* Gould, 1842b. Recognised within *Mesembriomys* by Iredale and Troughton (1934: 80), Tate (1951c: 268), Troughton (1967: 241), Strahan (1983: 383; 1995: 564), Mahoney and Richardson (1988a: 163) and subsequent authors.

HOMONYMS:

Hapalotis gouldii Gould, 1855 [1845–1863], Mitchell's Hopping-mouse of the Class mammalia (Order Rodentia, Family Muridae). Name is now recognised as *Notomys mitchellii* (W. Ogilby, 1838f). See individual entry.

Mus hirsutus Gould, 1842b: 12.

TYPE LOCALITY: Port Essington, Northern Territory, Australia.

COMMENTS: Placed within *Hapalotis* by Gould (1853: 127; 1857 [1845–1863]: Text to Plate 4). Recognised at the species rank in *Conilurus* by J. Ogilby (1892: 117) and within *Mesembriomys* by Longman (1916: 23). Synonymised within *gouldii* by Iredale and Troughton (1934: 81), Tate (1951c: 268), Ride (1970: 244), Watts and Aslin (1981: 123), Strahan (1983: 383) and Mahoney and Richardson (1988a: 163).

HOMONYMS:

Mus hirsutus Elliot, 1839: 213, Indian bush rats of the Class Mammalia (Order Rodentia, Family Muridae). Name is a synonym of *Golunda ellioti* J. Gray, 1837: 586. See Musser and Cartleton (2005: 1322).

Mesembriomys gouldii rattoides Thomas, 1924

Mesembriomys hirsutus rattoides Thomas, 1924b: 296.

TYPE LOCALITY: Cooktown, Queensland, Australia.

COMMENTS: Synonymised within *gouldii* by Watts and Aslin (1981: 123), Mahoney and Richardson (1988a: 163), and Musser and Carlton (2005: 1381). Recognised as subspecies of subspecies of *hirsutus* by Finlayson (1961a: 158), and *gouldii* by Iredale and Troughton (1934: 81), Ellerman (1941: 117), Tate (1951c: 268), Troughton (1967: 242), Strahan (1983: 383; 1995: 564), Clayton *et al.* (2006: 111), and Van Dyck and Strahan (2008: 591).

Mesembriomys gouldii melvillensis Hayman, 1936

Mesembriomys hirsutus melvillensis Hayman, 1936: 366.

TYPE LOCALITY: Melville Island, Northern Territory, Australia.

COMMENTS: Synonymised within *gouldii* by Watts and Aslin (1981: 123), Mahoney and Richardson (1988a: 163), and Musser and Carlton (2005: 1381). Recognised as a subspecies of *hirsutus* by Finlayson (1961a: 157), and *gouldii* by Ellerman (1941: 117), Tate (1951c: 268), Troughton (1967: 242), Strahan (1983: 383; 1995: 564), Clayton *et al.* (2006: 112), and Van Dyck and Strahan (2008: 591).

Mesembriomys macrurus (Peters, 1876)

Golden-backed Tree-rat

Hapalotis macrura Peters, 1876a: 355; Plate 1.

TYPE LOCALITY: Small mainland creek, Mermaid Strait, Western Australia, Australia.

COMMENTS: Recognised as a species within *Mesembriomys* by Iredale and Troughton (1934: 81), Ellerman (1941: 117), Tate (1951c: 269), Strahan (1983: 385; 1995: 566), Mahoney and Richardson (1988a: 164) and subsequent authors.

Hapalotis boweri Ramsay, 1887c: 1153; Plate 18.

TYPE LOCALITY: Derby district, Western Australia, Australia.

COMMENTS: Abstract of description provided in Ramsay (1886: vi). Recognised at the species rank in *Conilurus* by J. Ogilby (1892: 116). Synonymised within *macrurus* by Iredale and Troughton (1934: 81), Tate (1951c: 269), Watts and Aslin (1981: 127), Mahoney and Richardson (1988a: 164) and subsequent authors.

Notomys Lesson, 1842

Notomys Lesson, 1842: 129.

TYPE SPECIES: *Dipus mitchellii* W. Ogilby, 1838f [=*Notomys mitchellii* (W. Ogilby, 1838f)] by monotypy.

COMMENTS: Genus reviewed by Thomas (1921c: 536), Brazenor (1934a: 77) and Musser and Carleton (2005: 1426).

FUTURE TAXONOMIC RESEARCH: The taxonomic diversity within this genus is probably much underestimated, according to Aplin (cited by Musser & Carleton, 2005: 1426). There also appear to be two undescribed species from the south-west Kimberley identified by Start *et al.* (2012: 36, 40) that need to be formally described.

Podanomalus Waite, 1898: 117.

TYPE SPECIES: *Hapalotis longicaudata* Gould, 1844a (as *Podanomalurus longicaudatus*) [=*Notomys longicaudatus* (Gould, 1844a)] by monotypy.

COMMENTS: Recognised by Brazener (1934a: 83). Synonymised within *Notomys* by Iredale and Troughton (1934: 83), Ellerman (1941: 263), Watts and Aslin (1981: 95), Mahoney and Richardson (1988a: 165) and Musser and Carleton (1993: 635; 2005: 1426).

Thylacomys Waite, 1898: 121.

TYPE SPECIES: *Hapalotis cervinus* Gould, 1853 (as *Thylacomys cervinus*) [=*Notomys cervinus* (Gould, 1853)] by monotypy.

COMMENTS: Synonymised within *Ascopharynx* by Iredale and Troughton (1934: 84). Synonymised within *Notomys* by Brazenor (1934a: 77), Ellerman (1941: 263), Watts and Aslin (1981: 95), Mahoney and Richardson (1988a: 166) and Musser and Carleton (1993: 635; 2005: 1426).

HOMONYMS:

Thylacomys Owen, 1838c, bilbies of the Class Mammalia (Order Peramelemorphia, Family Thylacomyidae). Genus is a synonym of *Macrotis* Reid, 1837: 131. See individual account.

Ascopharynx Waite, 1900: 223.

TYPE SPECIES: *Nomen novum* for *Thylacomys* Waite, 1898.

COMMENTS: Recognised as genus by Iredale and Troughton (1934: 84). Synonymised within *Notomys* by Brazenor (1934a: 77), Ellerman (1941: 263), Ride (1970: 241; Watts and Aslin (1981: 95), Mahoney and Richardson (1988a: 166) and Musser and Carleton (1993: 635; 2005: 1426).

Notomys alexis Thomas, 1922

Spinifex Hopping-mouse

Notomys alexis alexis Thomas, 1922

Notomys alexis Thomas, 1922i: 316.

TYPE LOCALITY: 35 miles (56.3 kilometres) south west of Alroy, Northern Territory, Australia. 800 feet elevation. (About 19°30′S, 135°40′E)

COMMENTS: Recognised by Iredale and Troughton (1934: x, 84), Finlayson (1940: 125) and subsequent authors.

Notomys alexis reginae Troughton, 1936

Notomys alexis reginae Troughton, 1936: 20.

TYPE LOCALITY: Timbered sand ridge country, Barcarolle Station, 135 miles of Longreach, Queensland, Australia.

COMMENTS: Synonymised within *alexis* by Mahoney and Richardson (1988a: 166), Musser and Carleton (1993: 635; 2005: 1426) and Burbidge *et al.* (2014: 24, 31), but recognised as a subspecies of *alexis* by Tate (1951c: 261), Watts and Aslin (1981: 108), Strahan (1983: 428; 1995: 570) and Clayton *et al.* (2006: 112).

Notomys alexis everardensis Finlayson, 1940

Notomys alexis everardensis Finlayson, 1940: 133; Plates 14–15.

TYPE LOCALITY: Near the waters of Chundrinna and Walthajalkanna, north of Everard Range, South Australia, Australia. (Approx. 26°50′S, 132°15′E)

COMMENTS: Synonymised within *alexis* by Mahoney and Richardson (1988a: 166), Musser and Carleton (1993: 635; 2005: 1426) and Burbidge *et al.* (2014: 24, 31). Recognised as a subspecies of *alexis* by Tate (1951c: 262), Watts and Aslin (1981: 108), and Clayton *et al.* (2006: 112).

† ***Notomys amplus*** Brazenor, 1936

Short-tailed Hopping-mouse

† *Notomys amplus* Brazenor, 1936b: 7; Plate 1, Figs. 2a-2e.

TYPE LOCALITY: Charlotte Waters, Northern Territory, Australia.

COMMENTS: Recognised at species rank by Tate (1951c: 265) and subsequent authors. Appears to be extinct.

Notomys aquilo Thomas, 1921

Northern Hopping-mouse

Notomys aquilo Thomas, 1921c: 540.

TYPE LOCALITY: Cape York, north Queensland, Australia.

COMMENTS: Taxon recognised by Ellerman (1941: 265), Tate (1951c: 262), Troughton (1967: 253), Mahoney and Richardson (1988a: 167) and subsequent authors.

Notomys carpentarius D. Johnson, 1959: 186.

TYPE LOCALITY: Umbakumba, Port Langdon, north east corner of Groote Eylandt, Northern Territory, Australia. (13°51′S, 136°45′E)

COMMENTS: Species recognised by D. Johnson (1964: 497) and Troughton (1967: 252). Synonymised within *aquilo* by Ride (1970: 245), Watts and Aslin (1981: 112), Mahoney and Richardson (1988a: 167), Musser and Carleton (1993: 636) and subsequent authors.

Notomys cervinus (Gould, 1853)

Fawn Hopping-mouse

Hapalotis cervinus Gould, 1853: 127.

TYPE LOCALITY: Interior of South Australia, Australia. The lectotype was collected by the Sturt Expedition to Central Australia in 1844–1846, with the inscription 26/3/45 on the label attached to the skin of the lectotype. If the lectotype was collected on that date the type locality would be the Depot Glen (Preservation Creek) area of New South Wales (29°40′14″S, 141°9′41″E), the location given by Sturt (1848: 324). The location 29°6′S, 141°E is however recorded for the lectotype by Thomas (1921a: 433) who places the locality in the area of Sturt's Depot at Pinaroo Lake, New South Wales.

COMMENTS: Species further described by Gould (1853 [1845–1863]: Text to Plate 10). Recognised as a species within *Conilurus* by J. Ogilby (1892: 120), within *Thylacomys* by Waite (1898: 122), *Notomys* by Thomas (1921c: 541) and *Ascopharynx* by Wood Jones (1925b: 3), Longman (1916: 23) and Iredale and Troughton (1934: 85). Recognised as a species within *Notomys* by Brazenor (1934a: 74, 82), Finlayson (1939a: 108), Ellerman (1941: 265), Tate (1951c: 262), Troughton (1967: 253) and subsequent authors including Mahoney and Richardson (1988a: 167).

Podanomalus aistoni Brazenor, 1934a: 84; Plate 5, Fig. 8 & Plate 6, Fig, 5.

TYPE LOCALITY: Mulka, east of Lake Eyre, South Australia, Australia.

COMMENTS: Species rank recognised with *Notomys* by Finlayson (1939a: 103; 1939b: 358), Ellerman (1941: 266), Synonymised within *cervinus* by Watts and Aslin (1981: 98), Strahan (1983: 432; 1995: 575), Mahoney and Richardson (1988a: 167), Musser and Carleton (1993: 636) and subsequent authors.

Notomys fuscus (Wood Jones, 1925)

Dusky Hopping-mouse

Ascopharynx fuscus Wood Jones, 1925b: 3.

TYPE LOCALITY: Ooldea District, South Australia, Australia. Designation by Finlayson (1960a: 81).

COMMENTS: Aitken (1976: 200) questioned the validity of Finlayson's lectotype selection, but it was accepted by Mahoney in Mahoney and Richardson (1988a: 168) as being a specimen eligible for selection as lectotype. Synonymised within *cervinus* by Iredale and Troughton (1934: 85) and within *alexis* by Brazener (1934a: 80). Recognised within *Notomys* by Tate (1951c: 263), Aitken (1968: 37), Mahoney and Richardson (1988a: 167) and subsequent authors.

Notomys fuscus eyreius Finlayson, 1960a: 81.

TYPE LOCALITY: Mulka (New Well), on the east side of Lake Eyre about 50 miles east north east of the Cooper (as Barcoo) inflow, South Australia, Australia.

COMMENTS: Synonymised within *fuscus* by Aitken (1968: 37), Watts and Aslin (1981: 114), Strahan (1983: 430; 1995: 576), Mahoney and Richardson (1988a: 168), Musser and Carleton (1993: 636) and subsequent authors.

Notomys filmeri Mack, 1961: 222.

TYPE LOCALITY: Near Birdsville, Queensland, Australia.

COMMENTS: Synonymised within *fuscus* by Aitken (1968: 37), Ride (1970: 245), Watts and Aslin (1981: 114), Strahan (1983: 430; 1995: 576), Mahoney and Richardson (1988a: 168), Musser and Carleton (1993: 636) and subsequent authors.

† ***Notomys longicaudatus*** (Gould, 1844)

Long-tailed Hopping-mouse

† *Hapalotis longicaudata* Gould, 1844a: 104.

TYPE LOCALITY: Moore River, Western Australia, Australia.

COMMENTS: Species further described by Gould (1845 [1845–1863]: Text to Plate 8). Designation by Thomas (1921a: 433). Recognised as a species within *Conilurus* by J. Ogilby (1892: 119) and within *Podanomalus* by Waite (1898: 117) and Brazenor (1934a: 84). Placed in *Notomys* by Thomas (1921c: 537), Wood Jones (1925b: 2), Iredale and Troughton (1934: 84), Finlayson (1940: 135), Ellerman (1941: 266), Tate (1951c: 264), Mahoney and Richardson (1988a: 168) and subsequent authors.

† *Notomys sturti* Thomas, 1921c: 537.

TYPE LOCALITY: Mt. Gipps area, adjacent to Broken Hill, New South Wales, Australia (as 'Coonbaralba Range about 85 miles from Laidley's Ponds', New South Wales).

COMMENTS: Recognised as a species by Iredale and Troughton (1934: 84), Brazener (1934a: 83), Ellerman (1941: 266) and Tate (1951c: 263). Synonymised within *longicauda* by Watts and Aslin (1981: 106), Mahoney and Richardson (1988a: 168), Musser and Carleton (1993: 636) and subsequent authors.

† ***Notomys macrotis*** Thomas, 1921

Big-eared Hopping-mouse

† *Notomys macrotis* Thomas, 1921c: 538.

TYPE LOCALITY: Moore River, Western Australia, Australia.

COMMENTS: Known only from the type locality. Species recognised within *Hapalotis* by Gould (1863 [1845–1863]: Volume 1, xxv) and Gerrard (1862: 171). Synonymised within *richardsoni* by Iredale and Troughton (1934: 84) and within *megalotis* by Ellerman (1941: 265) who suggested the name is invalid. Species rank recognised by Mahoney and Richardson (1988a: 168) and subsequent authors.

† *Notomys megalotis* Iredale & Troughton, 1934: x, 84.

TYPE LOCALITY: Moore's River, Western Australia, Australia.

COMMENTS: *Nomen novum* for *Notomys macrotis* Thomas, 1921c. Species rank recognised by Ellerman (1941: 265), Tate (1951c: 264), Troughton (1967: 249) and Ride (1970: 146). Synonymised within *macrotis* by Watts and Aslin (1981: 101), Mahoney and Richardson (1988a: 168), and Musser and Carleton (1993: 636).

Notomys mitchellii (W. Ogilby, 1838)

Mitchell's Hopping-mouse

Dipus Mitchellii W. Ogilby, 1838f: 96.

TYPE LOCALITY: Approximately 12km south east of Lake Boga, Victoria, Australia. Ellerman (1941: 265) give the locality as the junction of the Murray and Murrumbiidgee Rivers, New South Wales.

COMMENTS: Species had been described from a paper read on 15 December 1837 but not published until the following year (Ogilby, 1838g: 130), and Ogilby (1838f: 96) is always supposed to have priority. Recognised as described by Mitchell (1838a: xvii) and within *Hapalotis* by Gould (1845 [1845–1863]: Text to Plate 9) and Krefft (1866a: 5). Recognised as a species within *Conilurus* by J. Ogilby (1892: 119). Placed in *Notomys* by Thomas (1921c: 539), Wood Jones (1925b: 3), Iredale and Troughton (1934: 83) and subsequent authors including Troughton (1967: 249) and Mahoney and Richardson (1988a: 169).

Hapalotis Gouldii J. Gray, 1841: 404, 413.

TYPE LOCALITY: Perth, Western Australia, Australia.

COMMENTS: *Nomen nudum*. Species recognised by Gould (1853: 127), but synonymised within *mitchellii* by Watts and Aslin (1981: 117).

Hapalotis Richardsonii J. Gray, 1844a: 12d; Plate 28, Fig. 2.

TYPE LOCALITY: Swan River, Western Australia, Australia.

COMMENTS: Recognised at species rank within *Hapalotis gouldii* by Gould (1853: 127), and within *Notomys* by Iredale and Troughton (1934: 84) and Troughton (1967: 249). Tate (1951c: 262) and Troughton (1967: 249) recognised the author as Gould (1853: 127). Synonymised within *gouldii* by Brazenor (1934a: 77). Synonymised within *mitchellii* by Ride (1970: 245), Watts and Aslin (1981: 117), Mahoney and Richardson (1988a: 169) and Musser and Carleton (1993: 636; 2005: 1428).

Hapalotis Gouldii Gould, 1863 [1845–1863]: xxxv.

TYPE LOCALITY: South west Western Australia, Australia.

COMMENTS: Recognised as a species by Brazenor (1934a: 77) and Tate (1951c: 260). Synonymised within *mitchellii* by Mahoney and Richardson (1988a: 169) and Musser and Carleton (1993: 636; 2005: 1428).

HOMONYMS:

Hapalotis gouldii J. Gray, 1843a, the Black-footed Tree-rat of the Class Mammalia (Order Rodentia, Family Muridae). Name is now recognised as *Mesembriomys gouldii* (J. Gray, 1843a). See individual entry.

Notomys mitchelli macropus Thomas, 1921c: 540.

TYPE LOCALITY: South Australia, Australia. Believed to be Kangaroo Island, South Australia. Mahoney in Mahoney and Richardson (1988a: 170) believed the type locality was from the mainland where it was known to occur.

COMMENTS: Recognised as described by Iredale and Troughton (1934: 83), Brazenor (1936a: 79), Finlayson (1939b: 358), Ellerman (1941: 265) and Tate (1951c: 260). Synonymised within *mitchellii* by Watts and Aslin (1981: 117), and Musser and Carleton (1993: 636; 2005: 1428).

Notomys mitchelli alutacea Brazenor, 1934a: 79; Plate 5–6.

TYPE LOCALITY: Ooldea, South Australia, Australia.

COMMENTS: Recognised as described by Ellerman (1941: 265) and Tate (1951c: 260), but synonymised within *mitchellii* by Watts and Aslin (1981: 117), Mahoney and Richardson (1988a: 170) and Musser and Carleton (1993: 636; 2005: 1428).

† ***Notomys mordax*** Thomas, 1922

Darling Downs Hopping-mouse

† *Notomys mordax* Thomas, 1922i: 317.

TYPE LOCALITY: Darling Downs, Queensland, Australia.

COMMENTS: Known only from the type locality. Species rank confirmed by Ellerman (1941: 265), Tate (1951c: 264) and Mahoney (1977: 749).

† ***Notomys robustus*** Mahoney *et al.*, 2008

Broad-cheeked Hopping-mouse

† *Notomys robustus* Mahoney *et al.*, 2008: 117, 119.

TYPE LOCALITY: Late Holocene deposits, Upper Level, Cave D, Chambers Gorge northeastern Flinders Ranges, South Australia, Australia. (30°57′S, 139°15′E)

COMMENTS: The manuscript formally describing this species, above, was published in April 2008: it was preceded by a brief description by Medlin (2008: 609) that was published in March 2008, but this earlier description does not meet the formal requirements of the Article 16.4 of the Code (ICZN, 1999: 20). The distinctiveness of this species has been recognised for a number of years by various authors including Watts and Aslin (1981: 104) as '(undescribed) – *Notomys* sp.'; Tunbridge (1991: 15, 81) as 'Broad-cheeked Hopping-mouse *Notomys* sp.'; Medlin (1993: 78) as 'Broad-cheeked Hopping-mouse (undescribed species); Strahan (1995: 582) as 'Great Hopping-mouse *Notomys* sp.'; A. Robinson *et al.* (2000: 389) as '*Notomys* sp. (Broad-cheeked Hopping-mouse); and Flannery and Schouten (2001: 178) as the 'Great Hopping-mouse (*Notomys* sp.)'. Recognised as being present after the onset of European settlement by Burbidge *et al.* (2008: 412).

Pogonomys Milne-Edwards, 1877

Pogonomys Milne-Edwards, 1877: 1081.

TYPE SPECIES: Φ *Mus (Pogonomys) macrourus* Milne-Edwards, 1877: 1081 [=Φ *Pogonomys macrourus* (Milne-Edwards, 1877: 1081)] by monotypy.

COMMENTS: Proposed with generic and subgeneric rank as alternatives.

Pogonomys sp. Undescribed

Tree Mouse

Pogonomys sp. Undescribed.

TYPE LOCALITY: Not yet described.

COMMENTS: The Australian taxon has historically been considered to be *Pogonomys mollipilosus* (Peters & Doria, 1880: 698) by various authors including Watts and Aslin (1981: 63) and Strahan (1983: 439; 1995: 643). The Australian species is now not considered part of any currently described species and requires formal description as a distinct taxon, as argued by authors including Strahan (1995: 645), Musser and Carleton (2005: 1440), and Van Dyck and Strahan (2008: 680).

FUTURE TAXONOMIC RESEARCH: The Australian form of *Pogonomys* (*sensu* Strahan, 1995: 645), needs to be formally reviewed and described as a distinct taxon if appropriate.

Pseudomys J. Gray, 1832

Pseudomys J. Gray, 1832: 39.

TYPE SPECIES: *Pseudomys australis* J. Gray, 1832 by monotypy.

COMMENTS: This name was revived by Thomas (1910c: 605) when he separated the Australian mice from other forms and divided the genus *Pseudomys* into four subgenera including *Pseudomys*, *Thetomys*, *Gyomys* and *Leggadina*. A major review has been undertaken by Ford (2006: 117) who suggested the genus *Pseudomys* should be split into eight genera.

FUTURE TAXONOMIC DIRECTIONS: Extensive notes towards a taxonomic revision, which will involve dividing up the genus, were published by Ford (2006: 117). There are at least five clades, forming a polytomy with *Notomys*, and probably three others, one of which includes *P. nanus*, *P. gracilicaudatus* and *Mastacomys*. The final revision is anxiously awaited. There also appears to be an undescribed species from the south-west Kimberley that was identified by Start *et al.* (2012: 36, 40) that needs to be formally described.

HOMONYMS:

† *Pseudomys* J. Allen & Coues, 1877: 944, extinct rodents of the Class Mammalia (Order Rodentia, Family Ischyromyidae). Genus is a synonym of † *Pseudotomus* Cope, 1872a: 467. See Palmer (1904: 592) and McKenna and Bell (1997: 116).

Thetomys Thomas, 1910c: 606.

TYPE SPECIES: *Mus nanus* Gould, 1858 (as *Pseudomys (Thetomys) nanus*) [=*Pseudomys nanus* (Gould, 1858)] by original designation.

COMMENTS: Originally made available as a subgenus of *Pseudomys* J. Gray, 1832. This subgenus was raised to generic status by Iredale and Troughton (1934: 77), Troughton (1967: 229) and Wakefield (1963b: 131), but reduced to subgenus by Ellerman (1941: 224), Tate (1951c: 247). Synonymised within *Pseudomys* by Ride (1970: 248), Baverstock *et al.* (1981: 299), Watts and Aslin (1981: 155), Mahoney and Richardson (1988a: 171) and Musser and Carleton (1993: 644; 2005: 1453).

Gyomys Thomas, 1910c: 607.

TYPE SPECIES: *Mus novaehollandiae* Waterhouse, 1843a (as *Pseudomys (Gyomys) novaehollandiae*) [=*Pseudomys novaehollandiae* (Waterhouse, 1843a)] by original designation.

COMMENTS: Originally made available as a subgenus of *Pseudomys* J. Gray, 1832, which was followed by Troughton (1932c: 293), but raised to generic status by Iredale and Troughton (1934: 79), Ellerman (1941: 220), Troughton (1967: 235) and Wakefield (1963b: 131), but lowered to subgenus by Tate (1951c: 243). Synonymised within *Pseudomys* by Ride (1970: 243), Baverstock *et al.* (1981: 300), Watts and Aslin (1981: 155), Mahoney and Richardson (1988a: 171) and Musser and Carleton (1993: 644; 2005: 1453).

† *Paraleporillus* Martinez & Lidicker, 1971: 775.

TYPE SPECIES: † *Paraleporillus stirtoni* Martinez & Lidicker, 1971 [=*Pseudomys australis* J. Gray, 1832] by original designation.

COMMENTS: Synonymised within *Pseudomys* by Watts and Aslin (1981: 155), McKenna and Bell (1997: 170) and Musser and Carleton (1993: 644; 2005: 1453).

Pseudomys albocinereus (Gould, 1845)

Ash-grey Mouse

Pseudomys albocinereus albocinereus (Gould, 1845)

Mus albocinereus Gould, 1845a: 78.

TYPE LOCALITY: Scrubby plains near Perth, Western Australia, Australia.

COMMENTS: Recognised as a species within *Mus* by Gould (1849 [1845–1863]: Text to Plate 21) and *Pseudomys* by Longman (1916: 22). Placed in *Gyomys* by Iredale and Troughton (1934: 79), Ellerman (1941: 221) and Troughton (1967: 237), but transferred to *Pseudomys* by Tate (1951c: 244), Strahan (1983: 410; 1995: 583), Mahoney and Richardson (1988a: 171) and subsequent authors.

Pseudomys albocinereus squalorum (Thomas, 1907)

Mus albocinereus squalorum Thomas, 1907b: 776.

TYPE LOCALITY: Bernier Island, Shark Bay, Western Australia, Australia.

COMMENTS: Subspecies rank within *Gyomys albocinereus* recognised by Iredale and Troughton (1934: 79), Ellerman (1941: 221) and Troughton (1967: 237). Recognised as a subspecies of *Pseudomys albocinereus* (Tate, 1951c: 245) and Strahan (1983: 410; 1995: 584). Synonymised within *albocinereus* by Watts and Aslin (1981: 195), Mahoney and Richardson (1988a: 171), Musser and Carleton (1993: 644; 2005: 1454) and Burbidge *et al.* (2014: 24, 31). Recognised as a subspecies by Clayton *et al.* (2006: 112), and Van Dyck and Strahan (2008: 613).

Pseudomys apodemoides Finlayson, 1932

Silky Mouse

Pseudomys (gyomys)[sic] apodemoides Finlayson, 1932: 170.

TYPE LOCALITY: Coombe, South Australia, Australia.

COMMENTS: This species was placed in the genus *Gyomys* by Iredale and Troughton (1934: 79), Ellerman (1941: 221) and Troughton (1967: 236), but returned to *Pseudomys* by Tate (1951c: 245), Watts and Aslin (1981: 198) and Mahoney and Richardson (1988a: 172). Synonymised within in *albocinereus* by Ride (1970: 246), but species status confirmed by Baverstock *et al.* (1977: 471, 483).

† ***Pseudomys auritus*** Thomas, 1910

Long-eared Mouse

† *Pseudomys auritus* Thomas, 1910d: 607.

TYPE LOCALITY: Lake Albert, South Australia, Australia.

COMMENTS: Recognised as a species within in *Pseudomys* by Longman (1916: 22), Iredale and Troughton (1934: 76),

Ellerman (1941: 224), Tate (1951c: 243) and Troughton (1967: 227). Synonymised within *australis* by Ride (1970: 246), Watts and Aslin (1981: 159), Strahan (1983: 391; 1995: 586) and Musser and Carleton (1993: 645; 2005: 1455). Recognised as a distinct species by Breed and Ford (2007: 8), and Van Dyck and Strahan (2008: 615).

Pseudomys australis J. Gray, 1832

Plains Mouse

Pseudomys Australis J. Gray, 1832: 39.

TYPE LOCALITY: South west side of Liverpool Plains, New South Wales, Australia.

COMMENTS: Species recognised since its description.

Pseudomys Greyii J. Gray, 1843a: 113.

TYPE LOCALITY: South west side of Liverpool Plains, New South Wales, Australia.

COMMENTS: *Nomen novum* for *P. australis* J. Gray, 1832. Not recognised by Musser and Carleton (1993: 645; 2005: 1455).

Mus lineolatus Gould, 1845a: 77.

TYPE LOCALITY: Open plains, Darling Downs, Queensland, Australia (as New South Wales).

COMMENTS: Type designation by Thomas (1921a: 432). Recognised at the species rank in *Mus* by Gould (1857 [1845–1863]: Text to Plate 18) and J. Ogilby (1892: 104). Recognised as a species in *Pseudomys* by Longman (1916: 22) and Tate (1951c: 242). Synonymised within *australis* by Ellerman (1941: 223), Watts and Aslin (1981: 159), and Musser and Carleton (1993: 645; 2005: 1455).

Hapalotis murinus Gould, 1845a: 78.

TYPE LOCALITY: Liverpool Plains, New South Wales, Australia. As the plains of the Namoi, New South Wales by Gould (1845a: 78), but see Gould (1853: 127).

COMMENTS: Recognised within *Hapalotis* by Gould (1853 [1845–1863]: Text to Plate 7). Recognised as a species within *Conilurus* by J. Ogilby (1892: 118). Synonymised within *australis* by Ellerman (1941: 223), Watts and Aslin (1981: 159), and Musser and Carleton (1993: 645; 2005: 1455).

Pseudomys (Pseudomys) minnie Troughton, 1932c: 287.

TYPE LOCALITY: Minnie Downs Station, north eastern corner, South Australia, Australia.

COMMENTS: Recognised as a species by Iredale and Troughton (1934: 76), Finlayson (1939a: 94; 1939b: 354), Ellerman (1941: 224), and Troughton (1967: 227) and a subspecies of *australis* by Tate (1951c: 241). Synonymised within *australis* by Ride (1970: 246), Watts and Aslin (1981: 159), Musser and Carleton (1993: 645; 2005: 1455) and subsequent authors.

Pseudomys minnie flavescens Troughton, 1936: 19.

TYPE LOCALITY: Timbered sand ridge country, Barcarolle Station, 135 km south of Longreach, Queensland, Australia.

COMMENTS: Recognised by the name under which it was described by Ellerman (1941: 224), and as a subspecies of *australis* by Tate (1951c: 241) and Troughton (1967: 227). Synonymised within *australis* by Musser and Carleton (1993: 645; 2005: 1455).

† *Paraleporillus stirtoni* Martinez & Lidicker, 1971: 775, 778.

TYPE LOCALITY: Wellington Caves, New South Wales, Australia.

COMMENTS: Synonymised within *Pseudomys australis* by Watts and Aslin (1981: 159), and Musser and Carleton (1993: 645; 2005: 1455).

Pseudomys bolami Troughton, 1932

Bolam's Mouse

Pseudomys hermannsburgensis bolami Troughton, 1932c: 292.

TYPE LOCALITY: Ooldea, South Australia, Australia. (30°27′S, 131°50′E)

COMMENTS: Recognised as described by Finlayson (1941: 220), and as a subspecies of *hermannsburgensis* by Iredale and Troughton (1934: 78), Ellerman (1941: 256) and Tate (1951c: 253), but synonymised within *hermannsburgensis* by Watts and Aslin (1981: 189). Recognised as a subspecies of *Pseudomys hermannsburgensis* by Strahan (1983: 407) and elevated to species status by Kitchener *et al.* (1984b: 153) and followed by subsequent authors.

Pseudomys calabyi Kitchener & Humphreys, 1987

Kakadu Pebble-mouse

Pseudomys laborifex calabyi Kitchener & Humphreys, 1987: 285, 286.

TYPE LOCALITY: 3 km south east Uranium Development Project Falls, Northern Territory, Australia. Approx. 100 m elevation. (13°27′S, 132°26′E)

COMMENTS: Mahoney in Mahoney and Richardson (1988a: 177) also included this as a subspecies of *Pseudomys laborifex*. Synonymised within *Pseudomys laborifex* by Musser and Carleton (1993: 647), but elevated to species status by Lee (1995: 3) and Musser and Carleton (2005: 1455).

Pseudomys chapmani Kitchener, 1980

Western Pebble-mouse

Pseudomys chapmani Kitchener, 1980a: 405.

TYPE LOCALITY: 31 km 136° Mt Meharry (West Angelas Mine Site), Western Australia, Australia. (23°11′05″S, 118°47′31″E)

COMMENTS: Recognised by subsequent authors.

Pseudomys delicatulus (Gould, 1842)

Delicate Mouse

Mus delicatulus Gould, 1842b: 13.

TYPE LOCALITY: Port Essington, Northern Territory, Australia.

COMMENTS: Type designation by Thomas (1921a: 432). Recognised as a species within *Mus* by Gould (1855 [1845–1863]: Text to Plate 23) and *Pseudomys* by Longman (1916: 22). Placed in *Leggadina* at the species rank by Ellerman (1941: 256), Tate (1951c: 252), D. Johnson (1964: 493) and Troughton (1967: 233), but it was moved to *Pseudomys* by Baverstock *et al.* (1981: 299), Mahoney and Richardson (1988a: 173) and followed by subsequent authors.

Mus albirostris J. Gray, 1843a: 112.

TYPE LOCALITY: Unknown.

COMMENTS: Synonymised within *delicatulus* by Iredale and Troughton (1934: 78) and Flannery (1990: 193).

Pseudomys (Leggadina) delicatulus mimulus Thomas, 1926b: 634.

TYPE LOCALITY: Groote Eylandt, Northern Territory, Australia. Elevation 10 feet.

COMMENTS: Subspecies status within *delicatulus* recognised by Iredale and Troughton (1934: 78), Ellerman (1941: 256), Tate (1951c: 252) and D. Johnson (1964: 495), Strahan (1983: 406; 1995: 593), Flannery (1990: 193; 1995a: 258) and Clayton *et al.* (2006: 112). Synonymised within *delicatulus* by Watts and Aslin (1981: 187), Mahoney and Richardson (1988a: 173) and Musser and Carleton (1993: 645; 2005: 1456) and Burbidge *et al.* (2014: 24, 31).

Gyomys pumilus Troughton, 1936: 16.

TYPE LOCALITY: Byfield, north of Yeppoon, Queensland, Australia.

COMMENTS: Recognised as described by Ellerman (1941: 221) and Troughton (1967: 235). Placed within *Pseudomys* in the subgenus *Gyomys* by Tate (1951c: 244). Synonymised within *delicatulus* by Ride (1970: 247), Watts and Aslin (1981: 187), Strahan (1983: 406; 1995: 593), Mahoney and Richardson (1988a: 173), Flannery (1990: 193; 1995a: 258), Musser and Carleton (1993: 645; 2005: 1456) and Burbidge *et al.* (2014: 24, 31). Subspecies rank recognised within *delicatulus* by Van Dyck and Strahan (2008: 624).

Pseudomys pilligaensis B. Fox & Briscoe, 1980: 109, 112.

TYPE LOCALITY: Cumberdeen Road, 3km west of the Pilliga-Baradine Road, Merriwindi State Forest, New South Wales, Australia. (31°52′S, 148°59′E)

COMMENTS: Genetic analysis confirming the species status of *pilligaensis* was undertaken by Briscoe *et al.* (1981: 89) and was accepted by various subsequent authors including Strahan (1983: 418; 1995: 616), and Musser and Carleton (1993: 648; 2005: 1460). Species synonymised within *delicatulus* by Breed and Ford (2007: 8) which was followed by Van Dyck and Strahan (2008: 623) and Burbidge *et al.* (2014: 24, 31); this taxon needs further investigation.

Pseudomys desertor Troughton, 1932

Desert Mouse

Pseudomys (Gyomys) desertor Troughton, 1932c: 293.

TYPE LOCALITY: Central Australia.

COMMENTS: Placed in *Gyomys* by Iredale and Troughton (1934: 79), Ellerman (1941: 221) and Troughton (1967: 237), but transferred to *Pseudomys* by Tate (1951c: 246) and Mahoney and Richardson (1988a: 174). The specific name is recorded as *desertor* Troughton, 1932; as this is the name in general use for the taxon; *subrufus* Krefft, 1862 and *murrayensis* Krefft, 1862 are unused senior synonyms of *desertor* Troughton, 1932c and neither should be resurrected as a name for the species (Mahoney & Richardson, 1988a: 174; Musser & Carleton, 2005: 1456).

Mus subrufus Murrayensis Krefft, 1862: 2.

TYPE LOCALITY: Between Gol Gol Creek, Victoria and Darling River, New South Wales, Australia.

COMMENTS: The two names *subrufus* and *murrayensis* were introduced simultaneously in combination. Wakefield (1966: 387–388) discussed specimens that were given manuscript names by Krefft; Mahoney, in Mahoney and Richardson (1988a: 174), agreed that they were indeed very likely the specimens constituting the basis for the names *subrufus* and *murrayensis* Krefft (1862: 2). Synonymised within *desertor* by Wakefield (1966: 387), and Musser and Carleton (1993: 646; 2005: 1456). The specific name *Mus subrufus* was later recognised separately by Krefft (1866a: 5).

Pseudomys fieldi (Waite, 1896)

Shark Bay Mouse

Mus fieldi Waite, 1896: 403; Plate 26, Fig 4.

TYPE LOCALITY: Alice Springs, Northern Territory, Australia.

COMMENTS: Recognised as a species in *Leggadina* by Ellerman (1941: 256) and within *Pseudomys* by Longman (1916: 22), Troughton (1937b: 188), Tate (1951c: 248) and subsequent authors.

Pseudomys (Thetomys) praeconis Thomas, 1910d: 608.

TYPE LOCALITY: Péron Peninsula, Western Australia, Australia.

COMMENTS: Recognised as a species within *Pseudomys* by Longman (1916: 22). Placed in the genus *Thetomys* by Iredale and Troughton (1934: 77) and Troughton (1967: 230). Recognised within *Pseudomys* by Ellerman (1941: 224), Tate (1951c: 248), Mahoney and Richardson (1988a: 179), and Musser and Carleton (1993: 648). Synonymised within *fieldi* by Strahan (1995: 596), Musser and Carleton (2005: 1456), Breed and Ford (2007: 8), and Van Dyck and Strahan (2008: 628).

Pseudomys fumeus Brazenor, 1934

Smoky Mouse

Pseudomys (Gyomys) fumeus Brazenor, 1934b: 158; Plate 18.

TYPE LOCALITY: Turton's Pass, Otway Forest, Victoria, Australia.

COMMENTS: Recognised at species rank within *Pseudomys* by Tate (1951c: 245), within *Gyomys* by Ellerman (1941: 221), Wakefield (1963b: 131) and Troughton (1967: 236). Transferred back to *Pseudomys* by Ride (1970: 154), Strahan (1983: 413) and subsequent authors.

† *Pseudomys glaucus* Thomas, 1910

Blue-grey Mouse

† *Pseudomys (Gyomys) glaucus* Thomas, 1910d: 609.

TYPE LOCALITY: South Queensland, Australia.

COMMENTS: Recognised as a species in *Pseudomys* by Longman (1916: 22). Placed in *Gyomys* by Iredale and Troughton (1934: 79), Ellerman (1941: 221) and Troughton (1967: 236), but transferred to *Pseudomys* and recognised as a distinct species by Tate (1951c: 244), B. Fox and Briscoe (1980: 121), Mahoney and Richardson (1988a: 175) and Musser and Carleton (1993: 646). Doubts have existed over the taxonomic status of this species as it was synonymised within *albocinereus* by Ride (1970: 246), and within *apodemoides* by Watts and Aslin (1981: 198) and was not recognised by Lee (1995: 3). Subsequently recognised as a distinct species by Musser and Carleton (2005: 1457), Breed and Ford (2007: 8), and Van Dyck and Strahan (2008: 631).

† *Pseudomys gouldii* (Waterhouse, 1839)

Gould's Mouse

† *Mus gouldii* Waterhouse, 1839c: 67.

TYPE LOCALITY: Hunter River, New South Wales, Australia.

COMMENTS: The type locality given as New South Wales by Waterhouse (1839c: 67–68), but restricted to the Hunter River, New South Wales by Thomas (1921: 432). However Gould (1855 [1845–1863]: Text to Plate 19) notes that the original specimen from which the initial description was made probably came from either near the 'Upper Hunter' or the 'interior side of the Liverpool Range'. Recognised as a species in *Mus* by Gould (1855 [1845–1863]: Text to Plate 19) and within *Pseudomys* by Longman (1916: 22). Placed in the genus *Thetomys* by Iredale and Troughton (1934: 77) and Troughton (1967: 230). Transferred to *Pseudomys* by Ellerman (1941: 224), Tate (1951c: 249) and followed by most subsequent authors. Appears to be extinct.

† *Pseudomys (Pseudomys) rawlinnae* Troughton, 1932c: 289.

TYPE LOCALITY: Rawlinna, Western Australia, Australia.

COMMENTS: Recognised as a valid species by Iredale and Troughton (1934: 76), Finlayson (1939a: 101; 1939b: 354), Ellerman (1941: 224), Tate (1951c: 242) and Troughton (1967: 229). Synonymised within *gouldii* by Ride (1970: 247) and *australis* by Watts and Aslin (1981: 159) and Musser and Carleton (1993: 646; 2005: 1457).

Pseudomys gracilicaudatus (Gould, 1845)

Eastern Chestnut Mouse

Mus gracilicaudatus Gould, 1845a: 77.

TYPE LOCALITY: Oakley Creek, Darling Downs, Queensland, Australia.

COMMENTS: Placed in the genus *Thetomys* by Iredale and Troughton (1934: 77) and followed by Troughton (1967: 231). Recognised as a species within *Pseudomys* by Longman (1916: 22), Ellerman (1941: 224), Tate (1951c: 249), Strahan (1983: 400; 1995: 601), Mahoney and Richardson (1988a: 175) and subsequent authors.

Thetomys gracilicaudatus ultra Troughton, 1939: 281.

TYPE LOCALITY: Mackay, Queensland, Australia

COMMENTS: Recognised as a subspecies by *gracilicaudatus* by Tate (1951c: 249), Troughton (1967: 231), Strahan (1983: 400; 1995: 602), and Van Dyck and Strahan (2008: 635). Synonymised within *gracilicaudatus* by Watts and Aslin (1981: 179), Mahoney and Richardson (1988a: 176), Musser and Carleton (1993: 647) and Burbidge *et al.* (2014: 24, 31).

Pseudomys hermannsburgensis (Waite, 1896)

Sandy Inland Mouse

Mus hermannsburgensis Waite, 1896: 405.

TYPE LOCALITY: George Gill Range, Northern Territory, Australia.

COMMENTS: Designation by Troughton (1932c: 292). Recognised as a species in *Pseudomys* by Longman (1916: 22) and Finlayson (1941: 215). Placed in *Leggadina* by

Iredale and Troughton (1934: 78), Ellerman (1941: 256), Troughton (1941: 297), Tate (1951c: 252) and Troughton (1967: 234). Included within *Pseudomys* by Ride (1970: 154), Baverstock *et al.* (1981: 299), Mahoney and Richardson (1988a: 176) and subsequent authors.

Leggadina hermannsburgensis brazenori Troughton, 1937b: 187.

TYPE LOCALITY: Region of the junction of the Murray and Darling River, New South Wales, Australia. It was collected during the Blandowski Expedition to northwestern Victoria and could have come from New South Wales, Victoria or South Australia. Wakefield (1966: 387) discussed the localities of specimen collection by the expedition.

COMMENTS: Recognised as described by Tate (1951c: 253) and Troughton (1967: 234); and as a subspecies of *Pseudomys hermannsburgensis* by Strahan (1983: 407). Synonymised within *hermannsburgensis* by Watts and Aslin (1981: 189), Mahoney and Richardson (1988a: 176), Musser and Carleton (1993: 647; 2005: 1458) and Strahan (1995: 604).

Pseudomys higginsi (Trouessart, 1897)

Long-tailed Mouse

Mus (Epimys) higginsi Trouessart, 1897: 473.

TYPE LOCALITY: *Nomen novum* for *Mus leucopus* Higgins and Petterd, 1883.

COMMENTS: Recognised as a species within *Pseudomys* by Longman (1916: 22) and *Pseudomys* by Iredale and Troughton (1934: 76), Ellerman (1941: 224) and Tate (1951c: 243).

Mus leucopus Higgins & Petterd, 1883: 174.

TYPE LOCALITY: Kentishbury, Tasmania, Australia.

COMMENTS: Recognised as the unavailable senior synonym for *higginisi* by Ellerman (1941: 224), Watts and Aslin (1981: 172), Mahoney and Richardson (1988a: 176) and Musser and Carleton (1993: 647; 2005: 1458).

HOMONYMS:

Musculus leucopus Rafinesque, 1818a: 446, the White-footed Deer Mouse of the Class Mammalia (Order Rodentia, Family Cricetidae). Species is a junior synonym of *Peromyscus leucopus* (Rafinesque, 1818a: 446). See Musser and Carleton (2005: 1070).

Pseudomys higginsi australiensis Wakefield, 1972b: 15, 21.

TYPE LOCALITY: Pleistocene deposits, Pyramid Cave Deposits, Buchan, eastern Victoria, Australia.

COMMENTS: Description based on late Pleistocene fossils. Not considered by Mahoney and Richardson (1988a: 176–177). Synonymised within *higginsi* by Watts and Aslin (1981: 172), and Musser and Carleton (1993: 647; 2005: 1458).

Pseudomys johnsoni Kitchener, 1985

Central Pebble-mouse

Pseudomys johnsoni Kitchener, 1985: 207, 208.

TYPE LOCALITY: Kurinelli Mine, Kurundi Station, central Northern Territory, Australia. 150 m elevation. (20°37′S, 134°51′E)

Pseudomys laborifex Kitchener & Humphreys, 1986: 419, 420.

TYPE LOCALITY: Adjacent to Camp Creek, Mitchell Plateau, Kimberley Region, Western Australia, Australia. Approximately 270 m elevation. (14°49′00″S, 125°50′25″E)

COMMENTS: Recognised as a distinct species by Strahan (1995: 608) and Musser and Carleton (2005: 1458), but synonymised within *johnsoni* by Breed and Ford (2007: 8), and Van Dyck and Strahan (2008: 640).

Pseudomys nanus (Gould, 1858)

Western Chestnut Mouse

Pseudomys nanus nanus (Gould, 1858)

Mus nanus Gould, 1858: 242.

TYPE LOCALITY: Victoria Plains, Western Australia, Australia.

COMMENTS: Type designation by Thomas (1921a: 432). Described further by Gould (1858 [1845–1863]: Text to Plate 20). Recognised as a species within *Pseudomys* by Longman (1916: 22) and Finlayson (1941: 224). Placed in the genus *Thetomys* by Iredale and Troughton (1934: 77) and Troughton (1967: 229), but transferred back to *Pseudomys* by Ellerman (1941: 224), Tate (1951c: 247), Strahan (1983: 398; 1995: 609), Mahoney and Richardson (1988a: 177) and subsequent authors.

Pseudomys nanus ferculinus (Thomas, 1902)

Mus ferculinus Thomas, 1902a: 491.

TYPE LOCALITY: Barrow Island, Western Australia, Australia.

COMMENTS: Recognised as a species within *Pseudomys* by Longman (1916: 22). Recognised at the species rank within *Thetomys* by Iredale and Troughton (1934: 77) and Troughton (1967: 230), and within *Pseudomys* by Ellerman (1941: 224) and Tate (1951c: 248). Synonymised within *nanus* by Ride (1970: 246), Watts and Aslin (1981: 176), Mahoney and Richardson (1988a: 177) and Musser and Carleton (1993: 647; 2005: 1458). Recognised as a subspecies of *nanus* by Strahan (1983: 398; 1995: 610), Clayton *et al.* (2006: 113), and Van Dyck and Strahan (2008: 642). Synonymised within *nanus* in an Honours thesis by White (2006) and followed by Burbidge *et al.* (2014: 24, 31).

Pseudomys novaehollandiae (Waterhouse, 1843)

New Holland Mouse

Mus Novae-Hollandiae Waterhouse, 1843a: 146.

TYPE LOCALITY: Yarrundi, Upper Hunter River (as Upper Hunter), New South Wales, Australia. Type designation by Thomas (1921a: 432).

COMMENTS: Name also described by Waterhouse (1843b: 135). An extended description and comparison with *P. hermannsburgensis* was undertaken by Brazenor (1936c: 9). Recognised as a species in *Mus* by Gould (1856 [1845–1863]: Text to Plate 22) and within *Pseudomys* by Longman (1916: 22). Placed in *Gyomys* by Iredale and Troughton (1934: 79), Ellerman (1941: 221) and Troughton (1967: 235), but transferred to *Pseudomys* by Tate (1951c: 244), Strahan (1983: 409; 1995: 611), Mahoney and Richardson (1988a: 178) and subsequent authors.

Pseudomys occidentalis Tate, 1951

Western Mouse

Pseudomys (Gyomys) occidentalis Tate, 1951c: 246.

TYPE LOCALITY: Tambellup, Western Australia, Australia.

COMMENTS: Recognised within *Gyomys* by Troughton (1967: 238) but transferred back to *Pseudomys* by Ride (1970: 154), Strahan (1983: 414), Mahoney and Richardson (1988a: 178) and subsequent authors.

Pseudomys oralis Thomas, 1921

Hastings River Mouse

Pseudomys australis oralis Thomas, 1921b: 621.

TYPE LOCALITY: Likely to be north east New South Wales or south east Queensland, Australia. Proposed by Mahoney in Mahoney and Richardson (1988a: 179).

COMMENTS: Recognised as a subspecies of *australis* by Iredale and Troughton (1934: 76) and Ellerman (1941: 224). Elevated to species rank by Tate (1951c: 242) and Mahoney and Richardson (1988a: 179). Recognised at species rank by Musser and Carleton (1993: 648; 2005: 1459).

Pseudomys patrius (Thomas & Dollman, 1909)

Eastern Pebble-mouse

Mus patrius Thomas & Dollman, 1909: 791.

TYPE LOCALITY: Mount Inkerman, Queensland, Australia.

COMMENTS: Placed in the genus *Leggadina* by Iredale and Troughton (1934: 78), Ellerman (1941: 256), Tate (1951c: 252) and Troughton (1967: 233). Recognised as a species in *Pseudomys* by Longman (1916: 22), B. Fox and Briscoe (1980: 121), Mahoney and Richardson (1988a: 179) and Musser and Carleton (1993: 648). Judged a synonym of *P. delicatulus* by Ride (1970: 247), Watts and Aslin (1981: 187), Kitchener (1985: 218) and Strahan (1983: 406; 1995: 593). Returned to species status by Van Dyck (1997: 42) and confirmed by Breed (2000: 197), which has been followed by subsequent authors including Musser and Carleton (2005: 1459), and Van Dyck and Strahan (2008: 648).

Pseudomys shortridgei (Thomas, 1907)

Heath Mouse

Mus shortridgei Thomas, 1907b: 765.

TYPE LOCALITY: Woyerling Reserve (as Woyaline, east of Pinjelly), Western Australia, Australia. 973 feet.

COMMENTS: Recognised at species rank within *Pseudomys* by Longman (1916: 22), Ellerman (1941: 224), Tate (1951c: 242) and subsequent authors. The relationship between eastern and western population assessed by N. Cooper *et al.* (2003b: 367).

Uromys Peters, 1867

Uromys Peters, 1867a: 343; Plate.

TYPE SPECIES: *Mus macropus* J. Gray, 1866a [=*Uromys caudimaculatus* (Krefft, 1867c)] by monotypy.

COMMENTS: Reviewed by Tate (1951c: 308) and Groves and Flannery (1994: 145).

Gymnomys J. Gray, 1867: 597.

TYPE SPECIES: *Mus macropus* J. Gray, 1866a [=*Uromys caudimaculatus* (Krefft, 1867c)] by monotypy.

COMMENTS: Described as a subgenus of *Mus*. Synonymised within *Uromys* by Thomas (1922g: 260), Ellerman (1941: 232), Watts and Aslin (1981: 89), Musser and Carleton (1993: 671; 2005: 1513), and McKenna and Bell (1997: 171).

Cyromys Thomas, 1910e: 507.

TYPE SPECIES: Φ *Mus imperator* Thomas, 1888c: 157 (as *Cyromys imperator*) [= Φ *Uromys imperator* (Thomas, 1888c: 157)] by original designation.

COMMENTS: Synonymised within *Uromys* by Ellerman (1941: 232), Tate (1951c: 308), Musser and Carleton (1993: 671; 2005: 1513), and McKenna and Bell (1997: 171).

Melanomys Winter, 1983: 379.

TYPE SPECIES: *Melanomys hadrourus* Winter, 1983 [=*Uromys hadrourus* (Winter, 1984)] by monotypy.

COMMENTS: *Melanomys* was used in the reference list of Winter (1983: 379) to an in press description of '*Melanomys hadrourus*', which was published by Winter (1984: 519), but this publication did not use the genus name *Melanomys* but *Melomys*. McAllan and Bruce (1989: 455) suggested that *Melanomys* is a new generic name proposed by Winter (1983: 379), but it may be inadvertent and it is in any case preoccupied by Thomas (1902b: 248). Synonymised within *Uromys* by Musser and Carleton (1993: 671; 2005: 1513), and Groves and Flannery (1994: 167).

HOMONYMS:

Melanomys Thomas, 1902b: 248, rice rats of the Class Mammalia (Order Rodentia, Family Muridae). Currently recognised genus. See Musser and Carleton (2005: 1125).

Uromys caudimaculatus (Krefft, 1867)

Giant White-tailed Rat

Uromys caudimaculatus caudimaculatus (Krefft, 1867)

Hapalotis caudimaculata Krefft, 1867c: 316.

TYPE LOCALITY: Cape York Peninsula, north Queensland, Australia.

COMMENTS: Reviewed by Tate (1951c: 309) and Groves and Flannery (1994: 151).

Mus macropus J. Gray, 1866a: 221.

TYPE LOCALITY: Port Albany, Queensland, Australia.

COMMENTS: Recognised as a species within *Uromys* by J. Ogilby (1892: 121). Recognised as a species within *Uromys* by Longman (1916: 17). Synonymised within *caudimaculatus* by Ellerman (1941: 234), Tate (1951c: 310), Watts and Aslin (1981: 90), Musser and Carleton (1993: 671; 2005: 1514), Groves and Flannery (1994: 152) and Flannery (1990: 217; 1995a: 324; 1995b: 171).

HOMONYMS:

Mus macropus Hodgson, 1845: 268, the Greater Bandicoot Rat of the Class Mammalia (Order Rodentia, Family Muridae). Taxon is a synonym of *Bandicota indica* (Bechstein, 1800: 497). See Musser and Carleton (2005: 1294).

Φ *Uromys aruensis* J. Gray, 1873a: 418.

TYPE LOCALITY: Aru Islands, Indonesia.

COMMENTS: Subspecies rank, within *Uromys caudimaculatus*, recognised by Ellerman (1941: 234), Tate (1951c: 310), Laurie and Hill (1954: 129) and Flannery (1990: 217). Synonymised within *caudimaculatus* by Musser and Carleton (1993: 671; 2005: 1514), Groves and Flannery (1994: 152) and Flannery (1995a: 324; 1995b: 171).

Φ *Uromys papuanus* A. Meyer, 1876: 145, 146.

TYPE LOCALITY: Near Rubi, south point of Geelvinck Bay, Netherlands, New Guinea.

COMMENTS: *Nomen nudum*. Synonymised within *validus* by Ellerman (1941: 234), within *aruensis* by Laurie and Hill (1954: 129) and *caudimaculatus* by Flannery (1990: 217) and Musser and Carleton (1993: 671).

HOMONYMS:

Hapalotis papuanus Ramsay, 1883, Giant White-tailed Rat of the Class Mammalia (Order Rodentia, Family Muridae). Recognised as *Uromys caudimaculatus papuanus* (Ramsay, 1883). See individual entry.

Φ *Uromys validus* Peters and Doria, 1880: 703.

TYPE LOCALITY: Katau, near the mouth of the Oriomo River, south New Guinea.

COMMENTS: Subspecies rank recognised within *caudimaculatus* by Ellerman (1941: 234). Synonymised within *aruensis* by Tate (1951c: 310), and Laurie and Hill (1954: 129), and within *caudimaculatus* by Musser and Carleton (1993: 671; 2005: 1514), Groves and Flannery (1994: 152) and Flannery (1990: 217; 1995a: 324; 1995b: 171).

Uromys sherrini Thomas, 1923a: 171.

TYPE LOCALITY: Dinner Creek, Ravenshoe, Queensland, Australia. 2900 feet. In dense scrub.

COMMENTS: Recognised as a species by Iredale and Troughton (1934: 85) and Troughton (1967: 259). Subspecies of *caudimaculatus* recognised by Ellerman (1941: 234). Synonymised within *caudimaculatus* by Tate (1951c: 310), Ride (1970: 248), Watts and Aslin (1981: 90), Musser and Carleton (1993: 671; 2005: 1514), Groves and Flannery (1994: 152) and Flannery (1990: 217; 1995a: 324; 1995b: 171).

Uromys macropus exilis Troughton & Le Souef, 1929b: 98.

TYPE LOCALITY: Hinchinbrook Island, Queensland, Australia.

COMMENTS: Subspecies rank recognised by Iredale and Troughton (1934: 85) and species rank recognised by Troughton (1967: 259). Synonymised within *caudimaculatus* by Ellerman (1941: 234), Tate (1951c: 310), Ride (1970: 248), Watts and Aslin (1981: 90), Musser and Carleton (1993: 671; 2005: 1514), Groves and Flannery (1994: 152) and Flannery (1990: 218; 1995a: 324; 1995b: 171).

Φ *Uromys caudimaculatus papuanus* (Ramsay, 1883)

Φ *Hapalotis Papuanus* Ramsay, 1883: 18; Plate 11.

TYPE LOCALITY: Not known. Probably near Port Moresby or lower slopes of Astrolobe range, Papua New Guinea (Laurie & Hill, 1954: 129).

COMMENTS: Not preoccupied by *Uromys papuanus* A. Meyer, 1876: 145, 146 a *nomen nudum*. Synonymised within *aruensis* by Tate (1951c: 310) and Laurie and Hill (1954: 129), and within *caudimaculatus* by Flannery (1990: 217) and Musser and Carleton (1993: 671; 2005: 1514). Recognised as a subspecies by Groves and Flannery (1994: 153), Flannery (1995a: 324; 1995b: 171), and Van Dyck and Strahan (2008: 676).

HOMONYMS:

Uromys papuanus A. Meyer, 1876: 145, 146, Giant White-tailed Rat of the Class Mammalia (Order Rodentia, Family Muridae). Name is a *nomen nudum*. See individual entry.

Φ *Uromys ductor* Thomas, 1913a: 213.

TYPE LOCALITY: Avera, Aroa River, Papua New Guinea.

COMMENTS: Synonymised within *aruensis* by Tate (1951c: 311) and Laurie and Hill (1954: 129), and within *caudimaculatus* by Musser and Carleton (1993: 671; 2005: 1514) and Flannery (1990: 217; 1995a: 324; 1995b: 171).

Φ *Uromys prolixus* Thomas, 1913a: 213.

TYPE LOCALITY: Haveri. Papua New Guinea. 700 metres elevation. (9°25′S, 147°25′E)

COMMENTS: Synonymised within *validus* by Ellerman (1941: 234), within *aruensis* by Tate (1951c: 310) and Laurie and Hill (1954: 129), and within *caudimaculatus* by Musser and Carleton (1993: 671; 2005: 1514) and Flannery (1990: 217; 1995a: 324; 1995b: 171).

Φ *Uromys lamington* Troughton, 1937a: 126.

TYPE LOCALITY: Mount Lamington district, on the southern border of the Northern Division, Papua New Guinea.

COMMENTS: Synonymised within *aruensis* by Laurie and Hill (1954: 129), and within *caudimaculatus* by Musser and Carleton (1993: 671; 2005: 1514) and Flannery (1990: 218; 1995a: 324; 1995b: 171).

Φ *Uromys caudimaculatus multiplicatus* (Jentink, 1907)

Φ *Pogonomys multiplicatus* Jentink, 1907: 367; Plate 16, Figs. 4–6.

TYPE LOCALITY: Sentani Lake, north Netherlands, New Guinea.

COMMENTS: Recognised as a valid subspecies of *Uromys caudimaculatus* by Laurie and Hill (1954: 129) and Tate (1951c: 311). Synonymised within *validus* by Ellerman (1941: 234), and *caudimaculatus* by Musser and Carleton (1993: 671; 2005: 1514). Recognised as a subspecies by Groves and Flannery (1994: 153), Flannery (1990: 217; 1995a: 324; 1995b: 171), and Van Dyck and Strahan (2008: 676).

Φ *Uromys nero* Thomas, 1913b: 208.

TYPE LOCALITY: Camp No. 3, Utakwa River, south-west Netherlands, New Guinea. 2500 feet.

COMMENTS: Synonymised within *validus* by Ellerman (1941: 234), within *aruensis* by Laurie and Hill (1954: 129), and *caudimaculatus* by Tate (1951c: 310), Musser and Carleton (1993: 671; 2005: 1514) and Flannery (1990: 217; 1995a: 324; 1995b: 171).

Φ *Uromys scaphax* Thomas, 1913b: 209.

TYPE LOCALITY: Canoe Camp, lower Setakwa River, south-west Netherlands, New Guinea. 150 feet.

COMMENTS: Synonymised within *validus* by Ellerman (1941: 234), within *aruensis* by Tate (1951c: 310) and Laurie and Hill (1954: 129), and within *caudimaculatus* by Musser and Carleton (1993: 671; 2005: 1514) and Flannery (1990: 217; 1995a: 324; 1995b: 171).

Φ *Uromys waigeuensis* Frechkop, 1932: 1, 11.

TYPE LOCALITY: Waigeu Island, off coast of New Guinea.

COMMENTS: Synonymised within *validus* by Ellerman (1941: 234), within *multiplicatus* by Laurie and Hill (1954: 129) and Tate (1951c: 311), and within *caudimaculatus* by Musser and Carleton (1993: 671; 2005: 1514) and Flannery (1995a: 324; 1995b: 171).

Uromys hadrourus (Winter, 1984)

Pygmy White-tailed Rat

Melomys hadrourus Winter, 1984: 519.

TYPE LOCALITY: Thornton Peak summit area at altitude 1220m, North Queensland, Australia (16°09′30″S, 145°21′45″E).

COMMENTS: Species recognised in *Melomys* by Mahoney and Richardson (1988a: 163), but transferred to *Uromys* by Musser and Carleton (1993: 671; 2005: 1515), Groves and Flannery (1994: 153) and subsequent authors.

Melanomys hadrourus Winter, 1983: 379.

TYPE LOCALITY: *Nomen nudum*.

COMMENTS: This reference was given preference to Winter (1984: 519) by McAllan and Bruce (1989: 455); however, Winter (1984: 519) was considered the correct citation by authors including Mahoney and Richardson (1988a: 163) and Musser and Carleton (2005: 1515). It is clear from Winter (1983: 379) that the intended publication of the scientific name is Winter (1984: 519) as this latter publication was referred to as being 'in press'.

Xeromys Thomas, 1889

Xeromys Thomas, 1889a: 248.

TYPE SPECIES: *Xeromys myoides* Thomas, 1889a by monotypy.

COMMENTS: Taxon recognised since description.

Xeromys myoides Thomas, 1889

Water Mouse

Xeromys myoides Thomas, 1889a: 248.

TYPE LOCALITY: Port Mackay, Queensland, Australia.

COMMENTS: Recorded in New Guinea by Hitchcock (1998: 141).

Zyzomys Thomas, 1909

Zyzomys Thomas, 1909c: 372.

TYPE SPECIES: *Mus argurus* Thomas, 1889b [=*Zyzomys argurus* (Thomas, 1889b)] by original designation.

COMMENTS: Genus recognised by Iredale and Troughton (1934: 82), Tate (1951c: 192, 204) and Troughton (1967: 240). Taxonomic decision of Mahoney in Mahoney and Richardson (1988a: 190) for the currently recognised species. The priority of *Zyzomys* Thomas, 1909c over *Laomys* Thomas, 1909c was fixed by Ellerman (1949: 25, 26).

Laomys Thomas, 1909c: 373.

TYPE SPECIES: *Laomys woodwardi* Thomas, 1909c [=*Zyzomys woodwardi* (Thomas, 1909c)] by original designation.

COMMENTS: Recognised by Iredale and Troughton (1934: 80), Ellerman (1941: 115) and Troughton (1967: 239). Recognised as a subgenus of *Zyzomys* by Tate (1951c: 266) and synonymised within *Zyzomys* by Ellerman (1949: 25, 35), Ride (1970: 244), Watts and Aslin (1981: 136), Mahoney and Richardson (1988a: 190), Kitchener (1989: 332) and followed by Musser and Carleton (1993: 674).

Zyzomys argurus (Thomas, 1889)

Common Rock-rat

Mus argurus Thomas, 1889b: 433.

TYPE LOCALITY: Probably Northern Territory, Australia (as South Australia). The Northern Territory was brought under the control of South Australia on 6 July 1863 and remained under its control until 1 January 1911, when it was surrendered to the Commonwealth. Contemporary maps showing territorial divisions of Australia in 1863–1910 commonly have the South Australian boundary to include the Northern Territory (Mahoney & Richardson, 1988a: 191).

COMMENTS: Placed in *Zyzomys* by Longman (1916: 23), Iredale and Troughton (1934: 83), Tate (1951c: 192, 265), Troughton (1967: 240), Mahoney and Richardson (1988a: 191) and subsequent authors.

Mesembriomys argurus indutus Thomas, 1909a: 151.

TYPE LOCALITY: Parry Creek, Western Australia, Australia. 20 ft elevation.

COMMENTS: Recognised as a subspecies within *Zyzomys argurus* by Iredale and Troughton (1934: 83), Ellerman (1941: 115) and D. Johnson (1964: 484). Synonymised within *argurus* by Tate (1951c: 265), Watts and Aslin (1981: 137), Mahoney and Richardson (1988a: 191), Kitchener (1989: 353), Musser and Carleton (1993: 674) and subsequent authors.

Zyzomys maini Kitchener, 1989

Arnhem Land Rock-rat

Zyzomys maini Kitchener, 1989: 331, 357.

TYPE LOCALITY: Djawamba Massif, 1.5 km east of Ja Ja Billabong, Northern Territory, Australia. Approx. 150 m elevation. (12°31′S, 132°54′E)

Zyzomys palatalis Kitchener, 1989

Carpentarian Rock-rat

Zyzomys palatalis Kitchener, 1989: 331, 361.

TYPE LOCALITY: Echo Gorge, Wollogorang Station, Northern Territory, Australia. 180m elevation. (17°12′S, 137°41′E)

COMMENTS: Taxon accepted since its description.

Zyzomys pedunculatus (Waite, 1896)

Central Rock-rat

Conilurus pedunculatus Waite, 1896: 395; Fig. 1a-f.

TYPE LOCALITY: Alice Springs, Northern Territory, Australia.

COMMENTS: Recognised as a species within *Laomys* by Longman (1916: 23), Iredale and Troughton (1934: 80), Finlayson (1941: 223), Ellerman (1941: 116) and Troughton (1967: 239). Transferred to *Zyzomys* by Mahoney and Richardson (1988a: 191) and subsequent authors.

[Conilurus pedunculatus] var. *brachyotis* Waite, 1896: 397.

TYPE LOCALITY: Illamurta, James Range, Northern Territory, Australia.

COMMENTS: Recognised as a subspecies of *pedunculatus* by Ellerman (1941: 116). This has received little further examination, probably because of confusion surrounding the validity of the type specimens (Aitken, 1976: 200). Synonymised within *pedunculatus* by Mahoney and Richardson (1988a: 191), Kitchener (1989: 347), Musser and Carleton (1993: 675) and subsequent authors.

Zyzomys woodwardi (Thomas, 1909)

Kimberley Rock-rat

Laomys woodwardi Thomas, 1909c: 373.

TYPE LOCALITY: Parry Creek, 5 miles west of Trig Station, near Wyndham, Eastern Kimberley, Western Australia,

Australia. Trig Station HJ9 is located at 15°34′24.69″S, 128°20′47.01″E.

COMMENTS: Recognised as a species within *Laomys* by Longman (1916: 23), Iredale and Troughton (1934: 80), Ellerman (1941: 116) and Troughton (1967: 240). Transferred to *Zyzomys* by Tate (1951c: 267), Mahoney and Richardson (1988a: 191) and followed by subsequent authors.

Tribe Murini Illiger, 1811 *sensu* Lecompte *et al.*, 2008

Family Murina Illiger, 1811: 84.

TYPE GENUS: *Mus* Linnaeus, 1758.

COMMENTS: When originally proposed, this rank was placed in the Order Prensiculantia (Illiger, 1811 [=Glires (Linnaeus, 1758)]) and included the genera *Arctomys* Schreber, 1780: Plate 207 [=*Marmota* Blumenbach, 1779: 79]; *Cricetus* Leske, 1779: 168; *Mus* Linnaeus, 1758; and *Bathyergus* Illiger, 1811: 86. Subsequently recorded by Illiger (1815: 46, 129). The recognition of Illiger (1811: 84) as the author was made by McKenna and Bell (1997: 136), and Musser and Carleton (2005: 1189), while the reference of J. Gray (1821: 303) was recognised by Simpson (1945: 88). The Family Murini G. Fischer (1817: 372, 410) was synonymised within Muridae by McKenna and Bell (1997: 136) with Murini Winge (1887: 126) being synonymised within the Subfamily Murinae by Musser and Carleton (2005: 1247). Recognised by Breed and Ford (2007: 9), and Van Dyck and Strahan (2008: 10, 682) as the '*Rattus* Group' and Lecompte *et al.* (2008: 8) as the Tribe Murini within the Subfamily Murinae.

Mus Group *sensu* Misonne, 1969: iii, 147.

TYPE GENUS: *Mus* Linnaeus, 1758.

COMMENTS: When originally proposed, this rank was placed in the *Rattus* Division (Misonne, 1969) and included the genus *Mus* Linnaeus, 1758.

Mus Division *sensu* Musser & Carleton, 2005: 903.

TYPE GENUS: *Mus* Linnaeus, 1758.

COMMENTS: When originally proposed, this rank was placed in the Subfamily Murinae (Illiger, 1811) and included the genera *Muriculus* Thomas, 1903b: 314; and *Mus* Linnaeus, 1758. Placed in the tribe Murini as a division by Lecompte *et al.* (2008: 8).

Mus Linnaeus, 1758

Mus Linnaeus, 1758: 59.

TYPE SPECIES: Ω *Mus musculus* Linnaeus, 1758 by Linnean tautonymy. See Thomas (1911b: 146) and ICZN (1958a: 41).

COMMENTS: Genus reviewed by Musser and Carleton (2005: 1387).

Ω ***Mus musculus*** Linnaeus, 1758

House Mouse

Ω ***Mus musculus musculus*** Linnaeus, 1758

Ω *[Mus] Musculus* Linnaeus, 1758: 62.

TYPE LOCALITY: Unknown. Thomas (1911b: 147) suggests that it is 'Sweden (Upsala)'.

COMMENTS: Taxonomic decision of Mahoney in Mahoney and Richardson (1988a: 164). There are numerous extralimital synonyms and subspecies (e.g. Musser & Carleton, 1993: 625–626), but only those synonyms that have been described from Australia are included below. Species reviewed by Schwarz and Schwarz (1943: 59), Musser and Carleton (2005: 1398), and Suzuki *et al.* (2013: 375). History of introduction discussed by Long (2003: 204).

Ω ***Mus musculus domesticus*** Schwarz and Schwarz, 1943

Ω *Mus domesticus* Schwarz and Schwarz, 1943: 65.

TYPE LOCALITY: Dublin, Ireland.

COMMENTS: *Mus domesticus* Rutty, 1772: 281 is a *nomen nudum*, but the specific name *domesticus* Schwarz and Schwarz, 1943: 65 was conserved by Opinion 1607 of the ICZN (1990: 171). Taxon recognised at species rank by Marshall (1998: 10) and subspecies of *musculus* by ICZN (1990: 171), Musser and Carleton (2005: 1399) and Gabriel *et al.* (2011: 1). The *Mus* taxon that occurs within Australian was proposed to be *domesticus* by Schwarz and Schwarz (1943: 65) and confirmed by Gabriel *et al.* (2011: 1).

HOMONYMS:

Mus domesticus Rutty, 1772: 281, the House Mouse of the Class Mammalia (Order Rodentia, Family Muridae). A *nomen nudum* but is a synonym of *Mus musculus* Linnaeus, 1758. See Schwarz and Schwarz (1943: 66) and ICZN (1990: 171).

Ω *Mus Adelaidensis* J. Gray, 1841: 404.

TYPE LOCALITY: Adelaide, South Australia, Australia.

COMMENTS: See also Mahoney (1974: 223–224). Synonymised within *Mus musculus* by Iredale and Troughton (1934: 75), Watts and Aslin (1981: 262) and subsequent authors.

Ω *Mus Simsoni* Higgins & Petterd, 1883: 175.

TYPE LOCALITY: Ringarooma, Tasmania, Australia. Syntypes not preserved.

COMMENTS: See Lord (1919: 25) and Lord and Scott (1924: 300). Synonymised within *Mus musculus* by Lord (1923: 76), Iredale and Troughton (1934: 75) and subsequent authors.

Tribe Rattini Burnett, 1830 *sensu* Lecompte *et al.*, 2008

Kind Rattidae Burnett, 1830b: 350.

TYPE GENUS: *Rattus* G. Fischer, 1803.

COMMENTS: When originally proposed, this rank was placed in the Race 'Glires, Rodentia, Liberae' with the Muridae (Illiger, 1811) (mouse-kind) to include rat-kind animals, with both families including the genera *Sphermophilus* [sic =*Spermophilus*] F. Cuvier, 1825 [1821–1825]: 255; *Arctomys* Schreber, 1780: Plate 207 [=*Marmota* Blumenbach, 1779: 79]; *Pedetes* Illiger, 1811: 81; *Bathyergus* [=*Bathygerus*] Illiger, 1811: 86; *Dipus* Zimmermann, 1780: 354; *Gerbillus* Desmarest, 1804a: 22; *Aspalax* Desmarest, 1804a: 24 [=*Spalax* Güldenstädt, 1770: 409]; *Cricetus* Leske, 1779: 168; *Mus* Linnaeus, 1758; *Hydromys* É. Geoffroy, 1804a; *Myoxus* Zimmermann, 1780: 351; *Echimys* G. Cuvier, 1809: 394; *Lemmus* Link, 1795: 75; and *Arvicola* Lacépède, 1799a: 10. Recognised as a tribe within the Subfamily Murinae by Lecompte *et al.* (2008: 9).

Rattus Division *sensu* Misonne, 1969: iii, 117.

TYPE GENUS: *Rattus* G. Fischer, 1803.

COMMENTS: When originally proposed, this rank was placed in the Family Muridae and included the *Praomys* group (Misonne, 1969: iii, 123), *Maxomys* group (Misonne, 1969: iii, 127), *Rattus* group (Misonne, 1969), *Uromys* group (Misonne, 1969) and *Mus* group (Misonne, 1969). More recently this division was modified by Musser and Carleton (2005: 904) who placed it in the Subfamily Murinae (Illiger, 1811) and included the genera *Abditomys* Musser, 1982: 3; *Bandicota* J. Gray, 1873a: 418; *Berylmys* Ellerman, 1947: 261: *Bullimus* Mearns, 1905: 450; *Bunomys* Thomas, 1910e: 508; *Diplothrix* Thomas, 1916a: 404; *Kadarsanomys* Musser, 1981: 5; *Komodomys* Musser and Boeadi, 1980: 397; *Limnomys* Mearns, 1905: 451; *Nesokia* J. Gray, 1842b: 264; *Nesoromys* Thomas, 1922b: 263; *Palawanomys* Musser and Newcomb, 1983: 335; *Papagomys* Sody, 1941: 322; *Parumomys* Ellerman, 1954: 117; *Paulamys* Musser, 1986: 1, 2; *Rattus* G. Fischer, 1803; *Sundamys* Musser and Newcombe, 1983: 401; *Taeromys* Sody, 1941: 260; *Tarsomys* Mearns, 1905: 453; and *Tryphomys* Miller, 1910: 399. Placed in the Tribe Rattini by Lecompte *et al.* (2008: 9).

Rattus Group *sensu* Misonne, 1969: iii, 129.

TYPE GENUS: *Rattus* G. Fischer, 1803.

COMMENTS: When originally proposed, this rank was placed in the *Rattus* Division (Misonne, 1969) and included the genus *Rattus* G. Fischer, 1803.

Rattus G. Fischer, 1803

Ruttus [sic] G. Fischer, 1803: 128.

TYPE SPECIES: Ω *Mus decumanus* Pallas, 1779[1778–1779]: 91 [= Ω *Rattus norvegicus* (Berkenhout, 1769)] by monotypy. See Hollister (1916: 206). See discussion of type by Corbet and Hill (1992: 334).

COMMENTS: The original spelling is '*Ruttus*' and Corbet and Hill (1992: 334) suggested that there is no evidence within the publication that the spelling is an error; as the name with the spelling *Rattus* has been universally accepted, however, it would serve no useful purpose to revert back to *Ruttus*. The use of the name *Rattus* was also used by Frisch (1775: 7) and Zimmermann (1777: 344), both unavailable. Thomas (1916b: 240) mistakenly identified *Mus rattus* as the type species of this genus. Hollister (1916: 206) described the types and type species of *Rattus*. A review of the extralimital synonyms can be found in Ellerman (1941: 148), Musser and Carleton (1993: 649), and Musser and Carleton (1993 649; 2005: 1460).

HOMONYMS:

Rattus Frisch, 1775, rodents of the Class Mammalia (Order Rodentia, Family Muridae). Genus is an unavailable senior synonym (see below) of *Rattus* G. Fischer, 1803. See individual entry.

Rattus Zimmermann, 1777, rodents of the Class Mammalia (Order Rodentia, Family Muridae). Genus is an unavailable senior synonym (see below) of *Rattus* G. Fischer, 1803. See individual entry.

Rattus Donovan, 1825: Plate 73, text unnumbered, rodents of the Class Mammalia (Order Rodentia, Family Muridae). Genus is a synonym of *Arvicanthis* Lesson, 1842: 147. See McKenna and Bell (1997: 165).

Rattus Frisch, 1775: 7.

TYPE SPECIES: Ω *Mus domesticus* Rutty, 1772: 281 *nomen nudum* [= Ω *Mus musculus* Linnaeus, 1758]. See Marshall (1998: 14) and ICZN (1990: 171).

COMMENTS: This name is unavailable because it was not published in a consistently binomial work (McKenna & Bell, 1997: 166). Name initially rejected by Thomas (1905b: 463; 1916c: 70) and subsequently rejected for nomenclatorial purposes by Opinion 258 of the ICZN (1954c: 247) and Article 11(c) of the Code (ICZN, 1985a: 23). Synonymised within *Rattus* G. Fischer, 1803 by Musser and Carleton (2005: 1460).

Rattus Zimmermann, 1777: 344.

TYPE SPECIES: Ω *Rattus somnolentus* Zimmermann, 1777: 344 [= Ω *Glis glis* (Linnaeus, 1766: 87)] by monotypy. See Reuvens (1890: 61) and Holden (2005: 840).

COMMENTS: Publication rejected as unavailable because non-binomial (Opinion 257 of the ICZN (1954b: 231)).

Acanthomys J. Gray, 1867: 598.

TYPE SPECIES: *Acanthomys leucopus* J. Gray, 1867 [=*Rattus leucopus* (J. Gray, 1867)] by monotypy.

COMMENTS: Genus typically not recognised by subsequent authors but was proposed to be validated, as *Acanthomys leucopus*, by Calaby *et al.* (1966: 330). Synonymised within *Rattus* by Musser and Carleton (2005: 1460).

HOMONYMS:

Acanthomys Lesson, 1842: 135, spiny mice of the Class Mammalia (Order Rodentia, Family Muridae). Genus is a synonym of *Acomys* I. Geoffroy, 1838: 126. See Musser and Carleton (2005: 1193).

Acanthomys Tokuda, 1941: 95, Ryukyu spiny rats of the Class Mammalia (Order Rodentia, Family Muridae). Genus is a synonym of *Tokudaia* Kuroda, 1943: 61. See Musser and Carleton (2005: 1512).

Epimys Trouessart, 1881: 117.

TYPE SPECIES: Described as a subgenus of *Mus*.

COMMENTS: Recognised by Longman (1916: 12) but not by most subsequent authors including Ellerman (1941: 148), Corbet and Hill (1992: 334), and Musser and Carleton (2005: 1460).

Stenomys Thomas, 1910e: 507.

TYPE SPECIES: Φ *Mus verecundus* Thomas, 1904d: 598 (as *Stenomys verecundus*) [= Φ *Rattus verecundus* (Thomas, 1904d: 598)] by original designation.

COMMENTS: A new species from New Guinea. Synonymised within *Rattus* by Corbet and Hill (1992: 334) and Musser and Carleton (2005: 1460).

Christomys Sody, 1941: 260.

TYPE SPECIES: † *Mus macleari* Thomas, 1887d [=† *Rattus macleari* (Thomas, 1887d)] by original designation.

COMMENTS: Synonymised within *Rattus* by Corbet and Hill (1992: 334), and Musser and Carleton (2005: 1460).

Mollicomys Sody, 1941: 260.

TYPE SPECIES: Φ *Mus hoffmanni* Matschie, 1901a: 284 [=Φ *Rattus hoffmanni* (Matschie, 1901a: 284)] by original designation.

COMMENTS: Synonymised within *Rattus* by Corbet and Hill (1992: 334), and Musser and Carleton (2005: 1460).

Cironomys Sody, 1941: 260.

TYPE SPECIES: Φ *Rattus hoogerwerfi* Chasen, 1939: 496 by original designation.

COMMENTS: Synonymised within *Rattus* by Corbet and Hill (1992: 335), and Musser and Carleton (2005: 1460).

Pullomys Sody, 1941: 260.

TYPE SPECIES: Ω *Mus pulliventer* Miller, 1902a: 765 [= Ω *Rattus tanezumi* Temminck, 1844a] by original designation.

COMMENTS: Type was placed within *Rattus rattus* Linnaeus, 1758 by Corbet and Hill (1992: 335). Synonymised within *Rattus* by Corbet and Hill (1992: 335), and Musser and Carleton (2005: 1460).

Geromys Sody, 1941: 260.

TYPE SPECIES: *Mus gestri* Thomas, 1897a [=*Rattus sordidus* (Gould, 1858)] by original designation.

COMMENTS: Synonymised within *Rattus* by Musser and Carleton (2005: 1460).

Octomys Sody, 1941: 261.

TYPE SPECIES: Ω *Mus concolor* Blyth, 1859b [= Ω *Rattus exulans* (Peale, 1848)] by original designation.

COMMENTS: Synonymised within *Rattus* by Corbet and Hill (1992: 335), and Musser and Carleton (2005: 1460).

HOMONYMS:

Octomys Thomas, 1920c: 117, the Viscacha Rat of the Class Mammalia (Order Rodentia, Family Octodontidae). Currently recognised genus. See Woods and Kilpatrick (2005: 1572).

Togomys Dieterlen, 1986: 12.

TYPE SPECIES: Ω *Togomys melanoderma* Dieterlen, 1986 [= Ω *Rattus exulans* (Peale, 1848)].

COMMENTS: Synonymised within *Rattus* by Dieterlen (1989: 65), Corbet and Hill (1992: 335), and Musser and Carleton (2005: 1460).

Rattus colletti (Thomas, 1904)

Dusky Rat

Mus colletti Thomas, 1904d: 599.

TYPE LOCALITY: South Alligator River, Northern Territory, Australia. Open clay flats.

COMMENTS: Recognised as a species in the genus *Epimys* (Trouessart, 1881) by Longman (1916: 21) and as a species within *Rattus* by Iredale and Troughton (1934: 72), Ellerman (1941: 207) and Troughton (1967: 221). Subsequently it was made a subspecies of *R. sordidus* by Ellerman (1949: 67) and J. Taylor and Horner (1973: 85). Synonymised within *sordidus* by Ride (1970: 247) but recognised as a subspecies of *R. gestri* by Tate (1951c: 350) and D. Johnson (1964: 489). More recently recognised as a distinct species by Watts and Aslin (1981: 242), Mahoney and Richardson (1988a: 180), and Musser and Carleton (2005: 1468).

Ω *Rattus exulans* (Peale, 1848)

Pacific Rat

Ω *Mus exulans* Peale, 1848: 47; Plate 4, Fig. 1.

TYPE LOCALITY: Society Islands, Tahiti (France). Designation by Lyon and Osgood (1909: 148).

COMMENTS: Note that in the original description the text refers to the plate number as 'Plate XII. Fig. 1.' The original edition of the United States Exploring Expedition, in which Peale's description appeared, was suppressed and 10 years later another similar account of *M. exulans* with long quotes from Peale's account appeared under the authorship of Cassin (1858: 38); but Peale was the original describer. Included in the genus *Epimys* (Trouessart, 1881) by Longman (1916: 18). Numerous extralimital synonyms reviewed by Musser and Carleton (1993: 652; 2005: 1469). First described in Australia by Thomas (1926d: 309), from Adele Island specimens, which he allied to the *concolor-ephippium* group. Record in Australia confirmed by Raven (1935: 250) but he did not state its exact location and only stated its locality as 'Australia'. Reviewed by Tate (1935: 146) and J. Taylor and Horner (1973: 11) who confirmed its Australian distribution being Maer [=Mer] Island, Murray Islands, near Cape York Peninsula, Queensland and Adele Island, near Derby, Western Australia. Van Dyck and Strahan (2008: 704, 705) also recognised this species from Sunday Island, near Adele Island and Norfolk Island off the east coast of Australia. This species has not been recorded on the Australian mainland. Considered to be introduced to the Australian islands by Van Dyck and Strahan (2008: 705).

Rattus fuscipes (Waterhouse, 1839)

Bush Rat

Rattus fuscipes fuscipes (Waterhouse, 1839)

Mus fuscipes Waterhouse, 1839c: 66; Plate 25.

TYPE LOCALITY: 'Little Grove' on Princess Royal Harbour, approx. 4 miles south of Mount Melvile, Albany, Western Australia, Australia. Locality restricted by J. Taylor and Horner (1973: 19).

COMMENTS: Type designation by J. Taylor and Horner (1967: 8). Recognised as a species in *Mus* by Gould (1851 [1845–1863]: Text to Plate 11), and *Epimys* (Trouessart, 1881) by Longman (1916: 21). Recognised as a subspecies of *lutreolus* by Tate (1951c: 345) but separated as a distinct species by Iredale and Troughton (1934: 71), Ellerman (1941: 207), Troughton (1967: 217) and subsequent authors.

Rattus mondraineus Thomas, 1921a: 428.

TYPE LOCALITY: Mondrain Island, Recherche Archipelago, Western Australia, Australia.

COMMENTS: Recognised as a species within *Rattus* by Iredale and Troughton (1934: 73) and Ellerman (1941: 207) and recognised to a subspecies of *lutreolus* by Tate (1951c: 345). Synonymised within *fuscipes* by J. Taylor and Horner (1973: 19), J. Taylor and Calaby (1988a: 1), Watts and Aslin (1981: 222), and Mahoney and Richardson (1988a: 182), and Musser and Carleton (1993: 653; 2005: 1471).

Rattus murrayi Thomas, 1923f: 601.

TYPE LOCALITY: North Island, Pearson Group, South Australia, Australia.

COMMENTS: Recognised as a subspecies of *greyii* by Iredale and Troughton (1934: 72), Ellerman (1941: 207), Tate (1951c: 330) and Troughton (1967: 217). Synonymised within *greyii* by J. Taylor and Horner (1973: 24) and within *fuscipes* by Watts and Aslin (1981: 222), Mahoney and Richardson (1988a: 182), J. Taylor and Calaby (1988a: 1), and Musser and Carleton (1993: 653; 2005: 1471).

Rattus glauerti Thomas, 1926d: 308.

TYPE LOCALITY: East Wallabu (as East Wallaby Is.), Houtman Abrolhos, Western Australia, Australia. 'Rushes on sand Hills'.

COMMENTS: Recognised as a subspecies of *fuscipes* by Iredale and Troughton (1934: 71) and Ellerman (1941: 207), and full species rank by Tate (1951c: 325) and Troughton (1967: 218). Synonymised within *fuscipes* by Ride (1970: 247), J. Taylor and Horner (1973: 19), Watts and Aslin (1981: 222), J. Taylor and Calaby (1988a: 1), Mahoney and Richardson (1988a: 182), and Musser and Carleton (2005: 1471).

Rattus fuscipes greyii (J. Gray, 1841)

Mus Greyii J. Gray, 1841: 404.

TYPE LOCALITY: Adelaide, South Australia, Australia. Finlayson (1960b: 129–130, 138) recorded information which suggested to him that the type locality could lie in the Hills district to the south of the coastal Adelaide-Wakefield plain.

COMMENTS: Type designation by Thomas (1921a: 432). Recognised as a species in the genus *Epimys* (Trouessart, 1881) by Longman (1916: 21) and as a species within *Rattus* by Iredale and Troughton (1934: 72), Ellerman (1941: 207), Tate (1951c: 329) and Troughton (1967: 216). Synonymised within *fuscipes* by Ride (1970: 247), Watts and Aslin (1981: 222), Mahoney and Richardson (1988a: 182), and Musser and Carleton (2005: 1471). Recognised as a subspecies of *fuscipes* by J. Taylor and Horner (1973: 24), Strahan (1983: 443; 1995: 652), J. Taylor and Calaby (1988a: 1), Clayton *et al.* (2006: 113), and Van Dyck and Strahan (2008: 686).

Rattus greyi ravus Brazenor, 1936a: 69; Plate 13, Fig. 3.

TYPE LOCALITY: Portland, Victoria, Australia.

COMMENTS: Non *Epimys ravus* H. Robinson and Kloss, 1916; as *Rattus greyii ravus*. Type designation by Dixon (1970: 112). Synonymised within *greyii* by J. Taylor and Horner (1973: 24), Watts and Aslin (1981: 222), J. Taylor and Calaby (1988a: 1) and Mahoney and Richardson (1988a: 182).

HOMONYMS:

Epimys ravus H. Robinson and Kloss, 1916: 272, the Indomalayan Maxomys of the Class Mammalia (Order Rodentia, Family Muridae). Name is a synonym of *Maxomys surifer* (Miller, 1900: 148). See Musser and Carleton (2005: 1372).

[Rattus greyi] peccatus Troughton, 1937b: 189.

TYPE LOCALITY: *Nomen novum* for *Rattus greyii ravus* Brazener, 1936a; as *Rattus greyii peccatus*.

COMMENTS: Recognised as described by Tate (1951c: 329) and Troughton (1967: 217). Synonymised with *Rattus f. greyii* by Horner (1973: 24). Synonymised within *fuscipes* by Watts and Aslin (1981: 222), J. Taylor and Calaby (1988a: 1), Mahoney and Richardson (1988a: 182) and Musser and Carleton (2005: 1471).

[Rattus greyi] brazenori Tate, 1940: 6.

TYPE LOCALITY: *Nomen novum* for *Rattus greyii ravus* Brazener, 1936a; as *Rattus greyii peccatus*.

COMMENTS: Not considered by J. Taylor and Calaby (1988a: 1). Synonymised with *fuscipes* by Mahoney and Richardson (1988a: 182) and Musser and Carleton (2005: 1471).

Rattus greyi pelori Finlayson, 1960b: 140.

TYPE LOCALITY: Greenly Island, South Australia, Australia. (Approx. 34°39′S, 134°49′E)

COMMENTS: Synonymised within *greyii* by J. Taylor and Horner (1973: 24) and within *fuscipes* by Watts and Aslin (1981: 222), J. Taylor and Calaby (1988a: 1), Mahoney and Richardson (1988a: 182) and Musser and Carleton (2005: 1471).

Rattus fuscipes assimilis (Gould, 1858)

Mus assimilis Gould, 1858: 241.

TYPE LOCALITY: Clarence River, New South Wales, Australia.

COMMENTS: Described further by Gould (1858 [1845–1863]: Text to Plate 15). Type designation Thomas (1921a: 432). Recognised at the species rank in *Mus* by J. Ogilby (1892: 105). Recognised as a species within *Epimys* (Trouessart, 1881) by Longman (1916: 21). Recognised as a species of *Rattus* by Iredale and Troughton (1934: 71), Ellerman (1941: 207), Tate (1951c: 327) and Troughton (1967: 215). Synonymised within *fuscipes* by Ride (1970: 247), Watts and Aslin (1981: 222), and Mahoney and Richardson (1988a: 182). Recognised as a subspecies of *fuscipes* by J. Taylor and Horner (1973: 32), Strahan (1983: 443; 1995: 652), J. Taylor and Calaby (1988a: 1), Clayton *et al*. (2006: 113), and Van Dyck and Strahan (2008: 686).

Rattus fuscipes coracius Thomas, 1923

Rattus assimilis coracius Thomas, 1923a: 173.

TYPE LOCALITY: Dinner Creek, Ravenshoe, Queensland, Australia. 2900 feet.

COMMENTS: Recognised as a species of *Rattus* by Iredale and Troughton (1934: 71), but placed as a subspecies of *assimilis* by Ellerman (1941: 207) and Tate (1951c: 328). Synonymised within *Rattus fuscipes* by Watts and Aslin (1981: 222), Mahoney and Richardson (1988a: 182) and Musser and Carleton (2005: 1471). Recognised as a subspecies of *fuscipes* by J. Taylor and Horner (1973: 39), Strahan (1983: 443; 1995: 652), J. Taylor and Calaby (1988a: 1), Clayton *et al*. (2006: 113), and Van Dyck and Strahan (2008: 686).

Mus manicatus Gould, 1858: 242.

TYPE LOCALITY: ?Australia (as Port Essington).

COMMENTS: Described further by Gould (1858 [1845–1863]: Text to Plate 16). Recognised at the species rank in *Mus* by J. Ogilby (1892: 104). Recognised as a species in the genus *Epimys* (Trouessart, 1881) by Longman (1916: 21), then as a species within *Rattus* by Iredale and Troughton (1934: 72), Ellerman (1941: 207) and Troughton (1967: 225), then as a subspecies of *assimilis* by Tate (1951c: 328). Synonymised within *fuscipes* by Ride (1970: 247) and within *coracius* by J. Taylor and Horner (1973: 39–43, 120) who discussed the holotype's identity and listed the specimen as coming from an unknown Australian locality. Considered *incertae sedis* by Mahoney and Richardson (1988a: 192), but was synonymised within *R. fuscipes* by Watts and Aslin (1981: 222), and Musser and Carleton (1993: 653; 2005 1471).

Rattus leucopus (J. Gray, 1867)

Cape York Rat

Rattus leucopus leucopus (J. Gray, 1867)

Acanthomys leucopus J. Gray, 1867: 598.

TYPE LOCALITY: Cape York, Queensland, Australia.

COMMENTS: Type designation by Thomas (1920b: 424). Species transferred to *Mus* by Alston (1877: 124), who acknowledged that it was preoccupied by *Musculus leucopus* [=*Peromyscus leucopus*] Rafinesque (1818a: 446). Synonymised within *Mus terrareginae* by Alston (1879a: 646; 1979b: 211). Name reinstated and placed in *Rattus* by Thomas (1920b: 424) which was followed by Iredale and Troughton (1934: 74), Tate (1951c: 335), Troughton (1967: 223) and Mahoney and Richardson (1988a: 183). The specific name *leucopus* J. Gray, 1867, as published in the binomen *Acanthomys leucopus*, was placed on the Official List of Specific names in Zoology, ICZN (1979a: 180). By the ruling given in Opinion 1116 of the ICZN (1979a: 180), the original reference for *Acanthomys leucopus* is J. Gray (1867: 598). Species reviewed by J. Taylor and Horner (1973: 46) and J. Taylor *et al*. (1982: 232).

Hapalotis personata Krefft, 1867c: 318.

TYPE LOCALITY: Cape York, north Queensland, Australia.

COMMENTS: Lectotype designation by Mahoney (1972: 17). Recognised as a species within *Conilurus* by J. Ogilby (1892: 118). Placed as a subspecies of *Rattus rattus* by Iredale and Troughton (1934: 74). Synonymised within *leucopus* by J. Taylor and Horner (1973: 46), Watts and Aslin (1981: 227), Mahoney and Richardson (1988a: 183), Musser and Carleton (1993: 654; 2005: 1473), and Flannery (1990: 247; 1995a: 330; 1995b: 155).

Mus terrae-reginae Alston, 1879b: 211.

TYPE LOCALITY: Cape York, Queensland, Australia.

COMMENTS: *Nomen novum* for *Acanthomys leucopus* J. Gray, 1867. Description reference above is an abstract of Alston (1879a: 646). The specific name *terraereginae* Alston, 1879b, as published in the binomen *Mus terraereginae*, has been suppressed for the purposes of the Law of Priority, but not for the Law of Homonymy, and placed on the Official Index of Rejected and Invalid Specific Names, ICZN (1979a: 180). By the ruling given in Opinion 1116 (ICZN, 1979a) the original reference for *Mus terraereginae* Alston, 1879 is Alston (1879a: 646) in the *Proceedings of the Zoological Society of London* (published October, 1879) but this is predated by Alston (1879b) in *Nature* (published 26 June, 1879). Recognised as a species in the genus *Epimys* Trouessart, 1881 by Longman (1916: 21) and within *Rattus* by Wood Jones (1922b: 594). Synonymised within *leucopus* by Iredale and Troughton (1934: 74), Tate (1951c: 335), J. Taylor and Horner (1973: 46), Watts and Aslin (1981: 227) and subsequent authors.

Φ *Mus ringens* Peters & Doria, 1880: 700.

TYPE LOCALITY: Fly River, south New Guinea.

COMMENTS: Not recognised by Jentink (1909: 7). Recognised as a species within *Epimys* Trouessart, 1881 by Thomas (1914b: 319) then a species within *Rattus* by Thomas (1920b: 424). Transferred to a subspecies of *leucopus*, within *Stemonys*, by Rümmler (1938: 183). Then recognised as a subspecies of *Rattus leucopus* by Ellerman (1941: 204), species in *Rattus* by Ellerman (1949: 69), subspecies of *Rattus ruber* by Laurie and Hill (1954: 110) and subspecies of *Rattus leucopus* by Tate (1951c: 337), J. Taylor *et al*. (1982: 178, 234) and Flannery (1990: 247; 1995a: 330; 1995b: 155). Synonymised within *leucopus* by Musser and Carleton (1993: 654; 2005: 1473).

Φ *Mus ratticolor* Jentink, 1909: 3: 7; Plate 1, Fig. 11.

TYPE LOCALITY: Van Weel's Camp, junction of Reiger and Lorentz River, south-west Netherlands, New Guinea.

COMMENTS: Transferred to *Rattus* by Thomas (1920b: 424), then as a subspecies of *leucopus* within *Stenomys* by Rümmler (1938: 184). Transferred back to *Rattus* as a subspecies of *leucopus* by Ellerman (1941: 204) and then a subspecies of *ringens* by Ellerman (1949: 51), which was followed by Tate (1951c: 337). Synonymised within *Rattus ruber ringens* by Laurie and Hill (1954: 110). Recognised as a subspecies within *leucopus* by J. Taylor *et al*. (1982: 178, 238) and Flannery (1990: 247; 1995a: 330; 1995b: 155). Synonymised within *leucopus* by Musser and Carleton (1993: 654; 2005: 1473).

Φ *Mus doboensis* de Beaufort, 1911: 112.

TYPE LOCALITY: Aru Islands, New Guinea.

COMMENTS: Recognised as a subspecies of *leucopus* by Tate (1951c: 337). Synonymised within *leucopus* by Musser and Carleton (1993: 654; 2005: 1473).

Φ *Rattus ringens dobodurae* Troughton, 1946: 407.

TYPE LOCALITY: Dobodura, Northern Division, Papua New Guinea.

COMMENTS: Recognised as a subspecies of *Rattus ruber* by Laurie and Hill (1954: 111) and subspecies of *leucopus* by Tate (1951c: 338), J. Taylor *et al*. (1982: 178, 241) and Flannery (1990: 247; 1995a: 330; 1995b: 155). Synonymised within *leucopus* by Musser and Carleton (1993: 654).

Φ *Rattus owiensis* Troughton, 1945b: 374.

TYPE LOCALITY: Owi Island, Schouten Group, New Guinea.

COMMENTS: Recognised as a subspecies of *leucopus* by Tate (1951c: 338). Synonymised within *leucopus* by Musser and Carleton (1993: 654).

Rattus leucopus mcilwraithi Tate, 1951c: 335.

TYPE LOCALITY: Upper Nesbit River, Rocky Scrub, 20 miles east of Coen, Queensland, Australia. 1500 feet.

COMMENTS: Synonymised within *leucopus* by J. Taylor and Horner (1973: 46), Watts and Aslin (1981: 227), Mahoney and Richardson (1988a: 183) and Flannery (1990: 247; 1995a: 330; 1995b: 155).

Rattus leucopus cooktownensis Tate, 1951

Rattus leucopus cooktownensis Tate, 1951c: 336.

TYPE LOCALITY: Shipton's Flat, 30 miles south of Cooktown, Queensland, Australia. 900 feet.

COMMENTS: Synonymised within *leucopus* by Watts and Aslin (1981: 227) and Mahoney and Richardson (1988a: 183). Recognised as a subspecies by J. Taylor and Horner (1973: 50), Strahan (1983: 446; 1995: 654), Flannery (1990: 247; 1995a: 330; 1995b: 155), Clayton *et al*. (2006: 113), and Van Dyck and Strahan (2008: 688).

Rattus lutreolus (J. Gray, 1841)

Swamp Rat

Rattus lutreolus lutreolus (J. Gray, 1841)

Mus lutreola J. Gray, 1841: 409.

TYPE LOCALITY: Moscheto (as Mosquito) Island, Hunter River, New South Wales, Australia.

COMMENTS: Type designation by Thomas (1921a: 432). Synonymised within *fuscipes* by J. Gray (1843a: 111). Species rank, within *Rattus*, recognised by Iredale and Troughton (1934: 71), Finlayson (1935b: 224), Ellerman (1941: 208), Troughton (1967: 218), J. Taylor and Horner (1973: 3, 53), Mahoney and Richardson (1988a: 184) and subsequent authors.

Mus vellerosus J. Gray, 1847a: 5.

TYPE LOCALITY: Between the Murray River, South Australia and Glenelg River, Victoria, Australia.

COMMENTS: Type designation by Thomas (1921a: 432). Recognised at the species rank in *Mus* by Gould (1859 [1845–1863]: Text to Plate 12) and J. Ogilby (1892: 104). Recognised as a species in the genus *Epimys* by Longman (1916: 22) then as a species within *Rattus* by Iredale and Troughton (1934: 72), Ellerman (1941: 208) and Troughton (1967: 223). Recognised as a subspecies of *lutreolus* by Tate (1951c: 344). Synonymised within *lutreolus* by Ride (1970: 247) and within *lutreolus* by Mahoney and Richardson (1988a: 184), J. Taylor and Horner (1973: 57), Watts and Aslin (1981: 230), J. Taylor and Calaby (1988b: 1) and subsequent authors.

Rattus lutreolus cambricus Troughton, 1937c: 283.

TYPE LOCALITY: Booloombayt, Myall lakes District, New South Wales, Australia.

COMMENTS: Synonymised within *Rattus lutreolus* by Tate (1951c: 343), Mahoney and Richardson (1988a: 184), J. Taylor and Horner (1973: 57), J. Taylor and Calaby (1988b: 1) and subsequent authors. Synonymised within *lutreolus* by Watts and Aslin (1981: 230), Strahan (1983: 447; 1995: 656) and subsequent authors.

Rattus lutreolus velutinus (Thomas, 1882)

Mus velutinus Thomas, 1882: 415.

TYPE LOCALITY: Tasmania, Australia.

COMMENTS: Type designation by Thomas (1921a: 432). Recognised at the species rank in *Mus* by J. Ogilby (1892: 106). Recognised as a species in the genus *Epimys* by Longman (1916: 22) and as a species within *Rattus* by Iredale and Troughton (1934: 73) and Ellerman (1941: 208). Synonymised within *Rattus lutreolus* by Ride (1970: 247), Watts and Aslin (1981: 230), Mahoney and Richardson (1988a: 184) and Musser and Carleton (1993: 655; 2005: 1474). Recognised as a subspecies of *lutreolus* by Troughton (1937c: 284) and Tate (1951c: 344), J. Taylor and Horner (1973: 64), Strahan (1983: 447; 1995: 656), J. Taylor and Calaby (1988b: 1) and Clayton *et al.* (2006: 113).

Mus Pachyurus Higgins & Petterd, 1884b: 182.

TYPE LOCALITY: Long Plains, Tasmania, Australia.

COMMENTS: Synonymised within *Mus musculus* by Watts and Aslin (1981: 262) and *Rattus lutreolus* by Mahoney and Richardson (1988a: 184) and Musser and Carleton (1993: 655; 2005: 1474).

Mus Tetragonurus Higgins & Petterd, 1884c: 195.

TYPE LOCALITY: Mount Cameron and Springfield, Tasmania, Australia.

COMMENTS: Synonymised within *Rattus lutreolus* by Mahoney and Richardson (1988a: 184) and Musser and Carleton (1993: 655; 2005: 1474).

HOMONYMS:

Mus tetragonurus Kelaart, 1850a: 217, the Black Rat of the Class Mammalia (Order Rodentia, Family Muridae). Name is a synonym of *Rattus rattus* Linnaeus, 1758. See Musser and Carleton (2005: 1485).

[Mus] petterdi Trouessart, 1904 [1904–1905]: 373.

TYPE LOCALITY: *Nomen novum* for *Mus tetragonurus* Higgins and Petterd, 1884c.

COMMENTS: Synonymised within *Rattus lutreolus* by Mahoney and Richardson (1988a: 184) and Musser and Carleton (1993: 655; 2005: 1474).

Rattus lutreolus lacus Tate, 1951

Rattus lacus Tate, 1951c: 347.

TYPE LOCALITY: Lake Barrine, Queensland, Australia. 2400 feet.

COMMENTS: Recognised as a full species by Tate (1951c: 347). Synonymised within *Rattus lutreolus* by Ride (1970: 247), Watts and Aslin (1981: 230), Mahoney and Richardson (1988a: 184) and Musser and Carleton (1993: 655; 2005: 1474). Recognised as a subspecies of *Rattus lureolus* by J. Taylor and Horner (1967: 11), J. Taylor and Horner (1973: 69), Strahan (1983: 447; 1995: 656), J. Taylor and Calaby (1988b: 1) and Clayton *et al.* (2006: 113).

Rattus lutreolus imbil Troughton, 1937c: 283.

TYPE LOCALITY: Imbil, Queensland, Australia.

COMMENTS: Recognised as a subspecies of *lutreolus* by Tate (1951c: 344). Synonymised within *lutreolus* by Mahoney and Richardson (1988a: 184), and J. Taylor and Horner (1973: 57), Watts and Aslin (1981: 230), Strahan (1983: 447; 1995: 656) and subsequent authors.

† *Rattus macleari* (Thomas, 1887)

Maclear's Rat

† *Mus macleari* Thomas, 1887d: 513; Plate 42.

TYPE LOCALITY: Christmas Island, Indian Ocean.

COMMENTS: Species rank recognised by Ellerman (1941: 172) and subsequent authors. Thought to be extinct.

† ***Rattus nativitatis*** (Thomas, 1889)

Bulldog Rat

† *Mus nativitatis* Thomas, 1889c: 533.

TYPE LOCALITY: Christmas Island, Indian Ocean.

COMMENTS: Species rank recognised by Ellerman (1941: 172) and subsequent authors. Thought to be extinct.

Ω ***Rattus norvegicus*** (Berkenhout, 1769)

Brown Rat

Ω *Mus Norvegicus* Berkenhout, 1769: 5.

TYPE LOCALITY: England.

COMMENTS: Included in *Epimys* by Longman (1916: 15). Synonyms reviewed by Ellerman (1941: 183) and Musser and Carleton (1993: 657; 2005: 1478). The one synonym recorded from Australia is listed below. History of introduction discussed by Long (2003: 185).

Ω *Mus Tamarensis* Higgins & Petterd, 1884a: 185.

TYPE LOCALITY: ?Tamer River, northern Tasmania, Australia.

COMMENTS: Synonymised within *Rattus rattus* by Lord (1923: 76) and Iredale and Troughton (1934: 74) and Watts and Aslin (1981: 250), and within *Rattus norvegicus* by Mahoney in Mahoney and Richardson (1988a: 185) and Musser and Carleton (1993: 657; 2005: 1478).

Ω ***Rattus rattus*** (Linnaeus, 1758)

Black Rat

Ω *[Mus] Rattus* Linnaeus, 1758: 61.

TYPE LOCALITY: Sweden (Upsala).

COMMENTS: Included in *Epimys* by Longman (1916: 12). Taxonomic decision of Mahoney in Mahoney and Richardson (1988a: 186). Synonyms reviewed by Ellerman (1941: 174) and Musser and Carleton (1993: 658; 2005: 1484). Synonyms recorded from Australia are listed below. History of introduction discussed by Long (2003: 191).

Ω *Mus alexandrinus* É. Geoffroy, 1803c: 192.

TYPE LOCALITY: Australia. Recorded in Australia by Longman (1916: 13).

COMMENTS: Synonymised within *rattus* by Mahoney in Mahoney and Richardson (1988a: 186).

Ω *Mus arboricola* Gould, 1863 [1845–1863]: Intro to Volume 1, page xxxv.

TYPE LOCALITY: Elizabeth Bay, Sydney, New South Wales, Australia.

COMMENTS: Synonymised within *Rattus assimilis* by Tate (1951c: 327) and in *Rattus rattus* by Watts and Aslin (1981: 250), Mahoney in Mahoney and Richardson (1988a: 186) and Musser and Carleton (1993: 658).

Ω *Mu.[s] (Hapalotis?) Tompsoni* Ramsay, 1882: 763.

TYPE LOCALITY: Wagga Wagga, New South Wales, Australia.

COMMENTS: Recognised at the species rank in *Mus* by J. Ogilby (1892: 109). Synonymised within *rattus* by Iredale and Troughton (1934: 74), Watts and Aslin (1981: 250), Mahoney in Mahoney and Richardson (1988a: 186) and Musser and Carleton (1993: 658).

Ω *Mus griseocaeruleus* Higgins & Petterd, 1883: 173.

TYPE LOCALITY: Deloraine, Kentishbury & Launceston, Tasmania, Australia.

COMMENTS: Synonymised within *rattus* by Lord (1923: 76), Iredale and Troughton (1934: 74), Watts and Aslin (1981: 250), Mahoney in Mahoney and Richardson (1988a: 186) and Musser and Carleton (1993: 658).

Ω *Mus variabilis* Higgins & Petterd, 1883: 174.

TYPE LOCALITY: St Leonards, Tasmania, Australia.

COMMENTS: Recognised at the species rank in *Mus* by J. Ogilby (1892: 111). Synonymised within *rattus* by Lord (1923: 76), Iredale and Troughton (1934: 74), Watts and Aslin (1981: 250), Mahoney in Mahoney and Richardson (1988a: 186) and Musser and Carleton (1993: 658).

Ω *Epimys chionogaster* Lönnberg, 1916: 6.

TYPE LOCALITY: Tolga, Queensland, Australia.

COMMENTS: Synonymised with *Rattus rattus* by Iredale and Troughton (1934: 75). Synonymised within *rattus* by Watts and Aslin (1981: 250), Mahoney in Mahoney and Richardson (1988a: 186) and Musser and Carleton (1993: 658).

Ω *Rattus rattus keelingensis* Tate, 1950: 276.

TYPE LOCALITY: Pulo Tikus (Direction Island), Cocos-Keeling Islands.

COMMENTS: Synonymised with *Rattus rattus* by Musser and Carleton (2005: 1484).

Rattus sordidus (Gould, 1858)

Canefield Rat

Mus sordidus Gould, 1858: 242.

TYPE LOCALITY: Darling Downs, Queensland, Australia.

COMMENTS: Type designation by Thomas (1921a: 432). Described further by Gould (1858 [1845–1863]: Text to Plate 17). Recognised as a species in the genus *Epimys* by Longman (1916: 21). Transferred to *Rattus* by Iredale and Troughton (1934: 72), Troughton (1967: 219), Horner and Taylor (1973: 3, 72), J. Taylor *et al.* (1982: 178, 265),

Mahoney and Richardson (1988a: 187) and subsequent authors.

Φ *Mus gestri* Thomas, 1897a: 611.

TYPE LOCALITY: Kapa Kapa, Papua New Guinea. (9°50S, 147°30′E)

COMMENTS: Recognised as a species within *Rattus* by Tate (1951c: 349) and as a subspecies of *sordidus* by Laurie and Hill (1954: 109)(as *R. s. gestroi*), J. Taylor *et al.* (1982: 178, 269) and Flannery (1990: 253; 1995a: 335; 1995b: 161). Species rank recognised within *Rattus* by D. Johnson (1964: 489). Synonymised within *sordidus* by Strahan (1983: 449; 1995: 661), Musser and Carleton (1993: 659; 2005: 1489) and subsequent authors.

Φ *Rattus sordidus gestroi* Laurie & Hill, 1954: 109.

TYPE LOCALITY: Incorrect subsequent spelling of *Mus gestri* Thomas, 1897a.

COMMENTS: Synonymised within *sordidus* by Flannery (1990: 253; 1995a: 335; 1995b: 161), Musser and Carleton (1993: 659; 2005: 1489) and subsequent authors.

Rattus conatus Thomas, 1923c: 159.

TYPE LOCALITY: Annan River, Cooktown, north Queensland, Australia.

COMMENTS: Recognised as a species within *Rattus* by Iredale and Troughton (1934: 72), Ellerman (1941: 207) and Troughton (1967: 220). Lowered to a subspecies of *gestri* by Tate (1951c: 350). Synonymised within *sordidus* by Ride (1970: 247), J. Taylor and Horner (1973: 75), Watts and Aslin (1981: 238), Strahan (1983: 449; 1995: 661), Mahoney and Richardson (1988a: 187), Flannery (1990: 253; 1995a: 335; 1995b: 161), Musser and Carleton (1993: 659; 2005: 1489) and subsequent authors.

Rattus youngi Thomas, 1926d: 309.

TYPE LOCALITY: Cowan Cowan, Moreton Island, Queensland, Australia.

COMMENTS: Recognised as a subspecies within *culmorum* by Iredale and Troughton (1934: 74) and Ellerman (1941: 206). Synonymised within *sordidus* by J. Taylor and Horner (1973: 75), Strahan (1983: 449; 1995: 661), Mahoney and Richardson (1988a: 187), Flannery (1990: 253; 1995a: 335; 1995b: 161), Musser and Carleton (1993: 659; 2005: 1489) and subsequent authors.

Φ *Rattus brachyrhinus* Tate & Archbold, 1935c: 4.

TYPE LOCALITY: Baroka, near mouth of Angabunga River [=St. Joseph's River], Central Division, Papua New Guinea. 30 metres.

COMMENTS: Synonymised within *Rattus gestri* by Tate (1951c: 349). Synonymised within *Rattus sordidus gestroi* by Laurie and Hill (1954: 109), and within *sordidus* by Flannery (1990: 253; 1995a: 335; 1995b: 161), Musser and Carleton (1993: 659; 2005: 1489) and subsequent authors.

Φ *Rattus gestri aramia* Troughton, 1937a: 119.

TYPE LOCALITY: Aramia Lakes district, near mouth of Aramia River, Western Division, Papua New Guinea.

COMMENTS: Recognised as a subspecies of *Rattus gestri* by Tate (1951c: 350) and *Rattus sordidus* by Laurie and Hill (1954: 109), J. Taylor *et al.* (1982: 178, 272) and Flannery (1990: 253; 1995a: 335; 1995b: 161). Synonymised within *sordidus* by Musser and Carleton (1993: 659; 2005: 1489).

Φ *Rattus gestri bunae* Troughton, 1946: 408.

TYPE LOCALITY: Dobodura district, Northern District, Papua New Guinea.

COMMENTS: Recognised as a subspecies of *sordidus* by Laurie and Hill (1954: 110). Synonymised within *sordidus* by Flannery (1990: 253; 1995a: 335; 1995b: 161), Musser and Carleton (1993: 659; 2005: 1489) and subsequent authors.

Ω ***Rattus tanezumi*** Temminck, 1844

Oriental House Rat

Ω *Rattus tanezumi* Temminck, 1844a: 51.

TYPE LOCALITY: Japan, possibly from near Nagasaki on Kyushu Island.

COMMENTS: For review of the taxonomic history of this species, and its synonyms, see Musser and Carleton (2005: 1489). For a contemporary view that incorporates a rapidly emerging genetic perspective, see Aplin *et al.* (2011: 1); as explained therein, the current notion of *tanezumi* as an East Asian species with 42 chromosomes (contrasting with 38 in the Indian *R. rattus*; e.g. Yosida, 1980: 5, 32, 43, 157) is very likely an understatement of taxic diversity within this group (see also Robins *et al.*, 2007: 717; Pages *et al.* 2010: 17, 20). Nomenclature of the group is complex and the appropriate name for each of the taxa currently included under *R. tanezumi sensu lato* is uncertain (Aplin *et al.*, 2011: 2, 13). Within Australian territories, *R. tanezumi* has been suggested to be present on the Cocos (Keeling) Islands by Long (2003: 198), but this claim awaits genetic confirmation. Martin (1976; cited by Watts and Aslin 1981: 254) recorded individuals of '*Rattus rattus*' with both 38 and 42 chromosomes in Brisbane so this taxon may occur on the Australian mainland. The prior difficulty in identifying this species may be because *tanezumi* is morphologically similar to *rattus*, so confirmation of its identity is reliant upon molecular typing approaches (Mostert, 2009: v, 67, 69). Genetic studies of co-introduced populations of *R. rattus* and *R. tanezumi* (*sensu lato*) in other parts of the world suggest a high propensity for interbreeding and genetic introgression (Yosida, 1980: 105; Mostert, 2009: 64, 140; Bastos *et al.*, 2011: 14; Conroy *et al.*, 2012: 754; Lack *et al.*, 2012: 3545); hence, analysis using a suite of genetic markers will be needed to accurately identify original taxic contributions to any particular population.

Rattus tunneyi (Thomas, 1904)

Pale Field Rat

Rattus tunneyi tunneyi (Thomas, 1904)

Mus tunneyi Thomas, 1904a: 223.

TYPE LOCALITY: Mary River, Northern Territory, Australia.

COMMENTS: Recognised as a species in the genus *Epimys* by Longman (1916: 22) and a species within *Rattus* by Le Souef and Burrell (1926: 124) and Iredale and Troughton (1934: 73), Troughton (1967: 221) and Mahoney and Richardson (1988a: 187).

Mus woodwardi Thomas, 1908b: 374.

TYPE LOCALITY: Lagrange Bay, Western Australia, Australia.

COMMENTS: Recognised as a species within *Epimys* by Longman (1916: 22) and within *Rattus* by Iredale and Troughton (1934: 73) and Ellerman (1941: 207). Reduced to a subspecies of *tunneyi* by Tate (1951c: 346) and Troughton (1967: 221). Synonymised within *tunneyi* by J. Taylor and Horner (1973: 92), Watts and Aslin (1981: 235) and subsequent authors.

Rattus tunneyi culmorum (Thomas, 1909)

Mus culmorum Thomas & Dollman, 1909: 790.

TYPE LOCALITY: Beach Mountain, Inkerman, Queensland, Australia.

COMMENTS: Recognised as a species in the genus *Epimys* by Longman (1916: 21) and species in *Rattus* by Tate (1951c: 346), Ellerman (1941: 206) and Troughton (1967: 222). Recognised as a subspecies of *tunneyi* by Iredale and Troughton (1934: 74), and J. Taylor and Horner (1973: 97). Synonymised within *tunneyi* by Ride (1970: 247), Watts and Aslin (1981: 235), and Musser and Carleton (2005: 1493). Elevated to a subspecies of *tunneyi* by Strahan (1983: 451; 1995: 662), Clayton *et al.* (2006: 114), and Van Dyck and Strahan (2008: 698).

Rattus culmorum vallesius Thomas, 1921a: 426.

TYPE LOCALITY: Duck Creek, Macquarie River, Upper Darling, New South Wales, Australia. (31°10′S, 147°40′E)

COMMENTS: Recognised as a subspecies within *culmorum* by Iredale and Troughton (1934: 74) and Ellerman (1941: 206). Synonymised within *culmorum* by J. Taylor and Horner (1973: 97). Synonymised within *tunneyi* by Watts and Aslin (1981: 235), Musser and Carleton (1993: 661; 2005: 1493).

Rattus culmorum austrinus Thomas, 1921a: 427.

TYPE LOCALITY: South Australia, Australia (Probably Kangaroo Island). J. Taylor and Horner (1973: 103) believed Kangaroo Island and Port Lincoln, Eyre Peninsula are the two most probable choices for the type locality.

COMMENTS: Recognised as a subspecies within *culmorum* by Iredale and Troughton (1934: 74) and Ellerman (1941: 206), but as a subspecies of *greyii* by Tate (1951c: 330) and Troughton (1967: 216). Synonymised within *culmorum* by J. Taylor and Horner (1973: 97). Placed as a synonym of *tunneyi* by Watts and Aslin (1981: 235), Mahoney and Richardson (1988a: 188) and subsequent authors.

Rattus melvilleus Thomas, 1921a: 427.

TYPE LOCALITY: Biro, Apsley Strait, Melville Island, Northern Territory, Australia.

COMMENTS: Recognised as a species within *Rattus* by Le Souef and Burrell (1926: 123), Iredale and Troughton (1934: 73), Ellerman (1941: 206) and Troughton (1967: 222). Recognised as a subspecies of *tunneyi* by Tate (1951c: 346). Synonymised within *tunneyi* by Ride (1970: 247), J. Taylor and Horner (1973: 92), Watts and Aslin (1981: 235) and subsequent authors.

Rattus tunneyi dispar Brazenor, 1936b: 5; Plate 1.

TYPE LOCALITY: Alice Springs, Northern Territory, Australia.

COMMENTS: Recognised as a subspecies of *tunneyi* by Tate (1951c: 346) and Troughton (1967: 221). Synonymised within *tunneyi* by J. Taylor and Horner (1973: 92), Watts and Aslin (1981: 235) and subsequent authors.

Rattus culmorum apex Troughton, 1939: 280.

TYPE LOCALITY: Skull Creek, extreme north west of Cape York Peninsula, north Queensland, Australia. 20m.

COMMENTS: Recognised as a subspecies of *gestri* by Tate (1951c: 350). Synonymised within *culmorum* by J. Taylor and Horner (1973: 97). Synonymised within *tunneyi* by Watts and Aslin (1981: 235), Strahan (1983: 451; 1995: 662), and Musser and Carleton (2005: 1493).

Rattus villosissimus (Waite, 1898)

Long-haired Rat

Mus villosissimus Waite, 1898: 125.

TYPE LOCALITY: Probably Goorogooheeny Billabong, Cooper Creek, southwestern Queensland, Australia. Type locality identified by Calaby and Taylor (1974: 267).

COMMENTS: *Nomen novum* for *Mus longipilis* Gould, 1854[1845–1863], preoccupied by *Mus longipilis* Waterhouse, 1837b: 16. Recognised as a species in the genus *Epimys* by Longman (1916: 22), and within *Rattus* by Finlayson (1939a: 88), Ellerman (1941: 207), Tate (1951c: 351) and Troughton (1967: 224), but lowered to a subspecies of *sordidus* by J. Taylor and Horner (1973: 80) and Watts and Aslin (1974: 65). Species rank again recognised by

Mahoney and Richardson (1988a: 188) and subsequent authors.

Mus longipilis Gould, 1854 [1845–1863]: Text to Plate 13.

TYPE LOCALITY: Vicinity of Goonaghooheeny Billabong, Cooper Creek, Queensland, Australia. See Calaby and Taylor (1974: 267).

COMMENTS: Proposed to be the same as *M. vellerosus* J. Gray, 1847a by Krefft (1871a: 2, Text to Plate 4). The name *longipilis* was recognised as preoccupied by another member of the genus *Mus* Waterhouse, 1837b: 16 by Waite (1898: 125). Synonymised within *villosissimus* by Waite (1898: 125) and Tate (1951c: 351). Recognised at the species rank in *Mus* by J. Ogilby (1892: 106). Synonymised within *villosissimus* by J. Taylor and Horner (1973: 80), Watts and Aslin (1981: 244) and Musser and Carleton (1993: 662; 2005: 1494).

HOMONYMS:

Mus longipilis Waterhouse, 1837b: 16, the Long-haired Akodont of the Class Mammalia (Order Rodentia, Family Cricetidae). Genus is currently recognised as *Abrothrix longipilis* (this Waterhouse, 1837b: 16). See Musser and Carleton (2005: 1089).

Rattus villosissimus profusus Thomas, 1921b: 620.

TYPE LOCALITY: Liverpool Plains, New South Wales, Australia.

COMMENTS: Recognised as a subspecies of *Rattus villosissimus* by Iredale and Troughton (1934: 73), Ellerman (1941: 207), Tate (1951c: 352) and Troughton (1967: 224). Synonymised within *villosissimus* by J. Taylor and Horner (1973: 80), Watts and Aslin (1981: 244), and Musser and Carleton (1993: 662; 2005: 1494).

Incertae Sedis

Mus Platurus Mitchell, 1838a: xvii.

TYPE LOCALITY: River Darling, New South Wales, Australia.

COMMENTS: Described as not 'accurately located' by Iredale and Troughton (1934: 75).

Mus Hovellii Mitchell, 1838a: xvii.

TYPE LOCALITY: Near Bayunga River, New South Wales, Australia.

COMMENTS: Described as not 'accurately located' by Iredale and Troughton (1934: 75).

Mus fuscus J. Gray, 1843a: 112.

TYPE LOCALITY: South Australia, Australia.

COMMENTS: Does not appear to have been previously considered.

Mus Australasicus J. Gray, 1843a: 112.

TYPE LOCALITY: South Australia, Australia.

COMMENTS: Described as not 'accurately located' by Iredale and Troughton (1934: 75).

Mus tasmaniensis Krefft, 1868a: 93.

TYPE LOCALITY: Banks of Ouse River, Tasmania, Australia.

COMMENTS: Described as not 'accurately located' by Iredale and Troughton (1934: 75).

Myoxoïdes Australasiae Brookes, 1828: 52.

TYPE LOCALITY: Australia.

COMMENTS: Described as not 'accurately located' by Iredale and Troughton (1934: 75).

Suborder Sciuromorpha Brandt, 1855

Suborder Sciuromorphi Brandt, 1855: 144, 292.

COMMENTS: Subordinal rank recognised by McKenna and Bell (1997: 115) and Thorington and Hoffmann (2005: 753).

Family Sciuridae G. Fischer, 1814

Family Sciuriorum G. Fischer, 1814: vi, 29.

TYPE GENUS: *Sciurus* Linnaeus, 1758.

COMMENTS: When originally proposed this rank was placed in the Order Metatarsii (G. Fischer, 1817 [=Mammalia (Linnaeus, 1758 part)]) and included the genera *Sciurus* Linnaeus, 1758; and *Myoxus* Zimmermann, 1780: 351. Name also referred to by G. Fischer (1817: 408). Author of the family given as J. Gray (1821: 304) by Simpson (1945: 78), Hemprich (1820: 32) by Hoffmann *et al.* (1993: 419), and G. Fischer (1817: 372) by McKenna and Bell (1997: 121) and Thorington and Hoffmann (2005: 754).

Subfamily Sciurinae G. Fischer, 1814

Family Sciuriorum G. Fischer, 1814: vi, 29.

TYPE GENUS: *Sciurus* Linnaeus, 1758.

COMMENTS: See comments above.

Tribe Sciurini G. Fischer, 1814

Family Sciuriorum G. Fischer, 1814: vi, 29.

TYPE GENUS: *Sciurus* Linnaeus, 1758.

COMMENTS: See comments above.

Funambulus Lesson, 1835

Funambulus Lesson, 1835: Plate 43.

TYPE SPECIES: Φ *Sciurus indicus* Lesson, 1835: Plate 43 [= Φ *Funambulus palmarum* (Linnaeus, 1766: 86)] by original designation.

COMMENTS: Reviewed by Thorington and Hoffmann (2005: 781).

Ω ***Funambulus pennantii*** Wroughton, 1905

Northern Palm Squirrel

Ω *Funambulus pennantii* Wroughton, 1905: 411.

TYPE LOCALITY: 'Mandvi Taluka of Surat District,' Guzerath [=Gudjerat], India.

COMMENTS: Reviewed by Thorington and Hoffmann (2005: 782). History of introduction into Australia described by Seebeck (1989: 932) and Long (2003: 155).

Sciurus Linnaeus, 1758

Sciurus Linnaeus, 1758: 63.

TYPE SPECIES: Φ *Sciurus vulgaris* Linnaeus, 1758: 63 by tautonomy. See Thomas (1911b: 148).

COMMENTS: Reviewed by Thorington and Hoffmann (2005: 758).

Ω ***Sciurus carolinensis*** Gmelin, 1788

Eastern Grey Squirrel

Ω *[Sciurus] carolinensis* Gmelin, 1788: 148.

TYPE LOCALITY: 'Carolina'.

COMMENTS: Reviewed by Thorington and Hoffmann (2005: 760). History of introduction into Australia described by Seebeck (1989: 932) and Long (2003: 146).

Order Lagomorpha Brandt, 1855

Suborder Lagomorphi seu Lagomorpha Brandt, 1855: 295, 319.

COMMENTS: When originally proposed, this rank was placed in the Order Rodentia (Bowdich, 1821) and included the Family Lagoïdes (Brandt, 1855: 319).

Family Leporidae G. Fischer, 1814

Family Leporinorum G. Fischer, 1814: viii, 88.

TYPE GENUS: *Lepus* Linnaeus, 1758.

COMMENTS: When originally proposed, this rank was placed in the Order Metatarsii (G. Fischer, 1817 [=Mammalia (Linnaeus, 1758 part)]) and included the genera *Lepus* Linnaeus, 1758; and *Lagomys* G. Cuvier, 1800: Table 1 [=*Ochotona* Link, 1795: 74]. Name also referred to by G. Fischer (1817: 409). The name Leporinorum was synonymised within the Family Leporidae by Hoffmann and Smith (2005: 194).

Lepus Linnaeus, 1758

Lepus Linnaeus, 1758: 57.

TYPE SPECIES: Φ *Lepus timidus* Linnaeus, 1758: 57 by Linnaean tautonomy. See ICZN (1958a: 41).

COMMENTS: Taxonomic arrangement and extralimital synonyms can be found in Hoffmann and Smith (2005: 195).

Ω ***Lepus europaeus*** Pallas, 1778

European Brown Hare

Ω ***Lepus europaeus europaeus*** Pallas, 1778

Ω *Lepus europaeus* Pallas, 1778 [1778–1779]: 30.

TYPE LOCALITY: Not stated but restricted by Trouessart (1910: 219) to Poland, and Ognev (1940: 140), who further restricted it to northwest Poland.

COMMENTS: Australian species of hare recognised as *europaeus* by Van Dyck and Strahan (2008: 748). Species reviewed by Hoffman and Smith (2005: 198).

Ω ***Lepus europaeus occidentalis*** de Winton, 1898

Ω *Lepus europaeus occidentalis* de Winton, 1898: 152.

TYPE LOCALITY: Restricted to Herefordshire, England by Trouessart (1910: 220).

COMMENTS: Recognised as the subspecies that occurs Australia by Van Dyck and Strahan (2008: 749). History of introduction discussed by Long (2003: 122).

Oryctolagus Lilljeborg, 1874

Oryctolagus Lilljeborg, 1874: 417.

TYPE SPECIES: Ω *Lepus cuniculus* Linnaeus, 1758 [= Ω *Oryctolagus cuniculus* (Linnaeus, 1758)] by original designation.

COMMENTS: Originally made available as a subgenus of *Lepus* Linnaeus, 1758. See Hoffman and Smith (2005: 206) for a full list of subspecies and synonyms.

Ω ***Oryctolagus cuniculus*** (Linnaeus, 1758)

European Rabbit

Ω *[Lepus] Cuniculus* Linnaeus, 1758: 58.

TYPE LOCALITY: Europe. Miller (1912: 490) identified the type locality as Germany.

COMMENTS: See Hoffman and Smith (2005: 206) for a full list of subspecies and synonyms. History of introduction discussed by Long (2003: 92).

Cohort Laurasiatheria Waddell *et al.*, 1999

Clade Laurasiatheria Waddell *et al.*, 1999a: 4.

COMMENTS: When originally proposed, this clade was placed in the Clade Boreoeutheria (Springer & de Jong, 2001 [=Placentalia Bonaparte, 1838 part]) and included the Lipotyphla (Haeckel, 1866) and Scrotifera (Waddell *et al.*,

1999b [=Placentalia (Bonaparte, 1838 part)]). Recognised at magnorder rank by Springer *et al.* (2007: 21) and as a clade by Murphy *et al.* (2001a: 2349), Douady *et al.*, 2002: 206), Van Den Bussche *et al.* (2002: 333) and Asher and Helgen (2010: 4).

Clade Laurasiaplacentalia Arnason *et al.*, 2008: 37, 47.

COMMENTS: When originally proposed, this clade was placed in the Boreoplacentalia (Arnason *et al.*, 2008 [=Placentalia Bonaparte, 1838 part]) and included the Lipotyphla (Haeckel, 1866), Chiroptera (Blumenbach, 1779), Pholidota (Weber, 1904: vi, 420), Carnivora (Bowdich, 1821), Perissodactyla (Owen, 1848), Artiodactyla (Owen, 1848) and Cetacea (Brisson, 1762). Name synonymised within Laurasiatheria (Waddell *et al.*, 1999a) by Asher and Helgen (2010: 4).

Order Lipotyphla Haeckel, 1866

Suborder Lipotyphla Haeckel, 1866: clx.

COMMENTS: When originally proposed, this rank was placed in the Order Insectivora (Bowdich, 1821 [=Order Lipotyphla (Haeckel, 1866)]) and included the families Soricida (Haeckel, 1866 [=Soricidae (G. Fischer, 1814)]), Talpida[e] (G. Fischer, 1814: x, 143), Erinaceidea [=Erinaceidae] (G. Fischer, 1814: ix, 143) and Centetida (Haeckel, 1866: clx) [=Tenrecidae (J. Gray, 1821: 301)]. Recognised at Grandorder rank by McKenna and Bell (1997: 272) but not recognised by Simpson (1945: 176, 177) or D. Wilson and Reeder (2005). Asher and Helgen (2010: 4) recognised this as a valid clade that included the families Erinaceidae (G. Fischer, 1814: ix, 143), Talpidae (G. Fischer, 1814: x, 143), Soricidae (G. Fischer, 1814), and Solenodontidae (Gill, 1872: 19).

Family Subterranea Illiger, 1811: 123.

TYPE GENUS: Not based on a genus group name.

COMMENTS: When originally proposed, this rank was placed in the Order Falculata (Illiger, 1811: 123 [=Placentalia (Bonaparte, 1838 part)]) and included the genera *Erinaceus* Linnaeus, 1758: 52; *Centetes* Illiger, 1811: 124 [=*Tenrec* Lacépède, 1799a: 7]; *Sorex* Linnaeus, 1758: 53; *Mygale* G. Cuvier, 1800: Table 1 [=*Desmana* Güldenstädt, 1777: 108]; *Condylura* Illiger, 1811: 125; *Chrysochloris* Lacépède, 1799a: 7; *Scalops* Illiger, 1811: 126 [=*Scalopus* É. Geoffroy, 1803c: 77]; and *Talpa* Linnaeus, 1758: 52. Synonymised within Lipotyphla by McKenna and Bell (1997: 272).

Family Insectivores G. Cuvier, 1816a: xxx, 131.

TYPE GENUS: Not based on a genus group name.

COMMENTS: When originally proposed, this rank was placed in the Order Carnassiers (G. Cuvier, 1816a [=Mammalia (Linnaeus, 1758 part)]) and included the genera *Erinaceus* Linnaeus, 1758: 52; *Sorex* Linnaeus, 1758: 53; *Mygale* G. Cuvier, 1800: Table 1 [=*Desmana* Güldenstädt, 1777: 108]; *Scalops* Illiger, 1811: 126 [=*Scalopus* É. Geoffroy, 1803c: 77]; *Chrysochloris* Lacépède, 1799a: 7; *Centenes* [sic =*Centetes*] Illiger, 1811: 124 [=*Tenrec* Lacépède, 1799a: 7; and *Talpa* Linnaeus, 1758: 52. Synonymised within Lipotyphla by McKenna and Bell (1997: 272).

Family Insectivora Bowdich, 1821: 24, 31.

TYPE GENUS: Not based on a genus group name.

COMMENTS: When originally proposed, this rank was placed in the Order Sarcophaga (Bowdich, 1821 [=Mammalia (Linnaeus, 1758 part)]) and included the genera *Erinaceus* Linnaeus, 1758: 52; *Sorex* Linnaeus, 1758: 53; *Mygale* G. Cuvier, 1800: Table 1 [=*Desmana* Güldenstädt, 1777: 108]; *Scalops* Illiger, 1811: 126 [sic =*Scalopus* É. Geoffroy, 1803c: 77]; *Chrysochloris* Lacépède, 1799a: 7; *Centenes* [sic =*Centetes* Illiger, 1811: 124 [=*Tenrec* Lacépède, 1799a: 7]; and *Talpa* Linnaeus, 1758: 52. Not to be confused with Insectivorae of J. Gray (1821: 299), which a synonym of Chiroptera (Blumenbach (1779: 58, 74). Historically recognised as an order by authors including Gill (1871a: 527; 1872: v, 18), Simpson (1931: 262), Honacki *et al.* (1982: 58) and Yates (1984: 117), and many others. Synonymised within the Grandorder Lypotyphla by McKenna and Bell (1997: 272) and within the Order Soricomorpha by Hutterer (2005: 220). Used in preference to Soricomorpha by Van Dyck and Strahan (2008: 10, 417).

HOMONYMS:

Insectivorae J. Gray, 1821, bats of the Class Mammalia (Order Chiroptera). Name is a synonym of the Order Chiroptera Blumenbach, 1779. See individual entry.

Suborder Gradientia Gill, 1871a: 532.

COMMENTS: When originally proposed, this rank was placed in the Order Insectivora (Bowdich, 1821 [=Lipotyphla (Haeckel, 1866)]) and included nine unnamed families. Synonymised within Lipotyphla by McKenna and Bell (1997: 272).

Suborder Euinsectivora Heim de Balsac & Bourlière, 1955: 1656.

COMMENTS: When originally proposed, this rank was placed in the Order Insectivora (Bowdich, 1821 [=Lipotyphla (Haeckel, 1866)]) and included the superfamilies Tenrecoidea (J. Gray, 1821: 301), Erinacoidea (Gill, 1872: 18), and Soricoidea (Gill, 1872: 18). Synonymised within Lipotyphla by McKenna and Bell (1997: 272).

Suborder Erinaceota Van Valen, 1967: 261.

COMMENTS: When originally proposed, this rank was placed in the Order Insectivora (Bowdich, 1821 [=Lipotyphla (Haeckel, 1866)]) and included the superfamilies

Erinaceoidea (G. Fischer, 1817: 372, 414) and Soricoidea (G. Fischer, 1814). Synonymised within Lipotyphla by McKenna and Bell (1997: 272).

Order Lipotyphliformes Kinman, 1994: 37.

COMMENTS: When originally proposed, this rank was placed in the Class Mammalea (Kinman, 1994 [=Mammalia (Linnaeus, 1758)]) and included 15 families. Synonymised within Lipotyphla by McKenna and Bell (1997: 272).

Clade Eulipotyphla Waddell *et al*., 1999a: 1.

COMMENTS: When originally proposed, this clade was placed in the Placentalia (Bonaparte, 1838) and included the hedgehogs, shrews, solenodon, and moles though specific ranks were not listed. Name recognised in favour of Lipotyphla by Skinner and Chimimba (2005: v, 231), but synonymised within Lipotyphla by Asher and Helgen (2010: 4).

Suborder Soricomorpha Gregory, 1910

Section Soricomorpha Gregory, 1910: 465.

COMMENTS: When originally proposed, this rank was placed in the Suborder Lipotyphla (Haeckel, 1866) and included the families Soricidae (G. Fischer, 1814), and Talpididae (G. Fischer, 1814: x). Section roughly equals infraordinal or Superfamily rank. Recognised as at the subordinal rank by Saban (1954: 428), proposed it as a new rank, and ordinal rank by McKenna (1975: 41), McKenna and Bell (1997: 284) and Hutterer (2005: 220). Commonly included within the Insectivora or Lipotyphla (see Hutterer, 2005: 220).

Infraorder Solenodonta Kalandadze & Rautian, 1992: 54.

COMMENTS: When originally proposed, this rank was placed in the Suborder Soricomorpha (Gregory, 1910: 465) and included the families Solenodontidae (Gill, 1872: 19), † Palaeoryctidae (Winge, 1917: 161) and † Didelphodontidae (Matthew, 1818: 571 [=† Cimolestidae (Marsh, 1889: 89)]). Synonymised within the Soricomorpha by McKenna and Bell (1997: 284).

Infraorder Soricota Kalandadze & Rautian, 1992: 55.

COMMENTS: When originally proposed, this rank was placed in the Suborder Soricomorpha (Gregory, 1910) and included the superfamilies Talpoidea (G. Fischer, 1814: x) and Soricoidea (G. Fischer, 1814). Synonymised within Soricomorpha by McKenna and Bell (1997: 284).

Superfamily Soricoidea G. Fischer, 1814

Family Soricinorum G. Fischer, 1814: x.

TYPE GENUS: *Sorex* Linnaeus, 1758: 53.

COMMENTS: When originally proposed, this rank was placed in the Order Plantigrada (G. Fischer, 1813a [=Placentalia (Bonaparte, 1838 part)]) and included the genus *Sorex* Linnaeus, 1758: 53. Also described by G. Fischer (1817: 414). Superfamily rank recognised by Gill (1883: 119) and McKenna and Bell (1997: 285) who recognise the author as G. Fischer (1817: 372, 414).

Superfamily? Soricoidea Gill, 1872: 18.

TYPE GENUS: *Sorex* Linnaeus, 1758: 53.

COMMENTS: When originally proposed, this rank was placed in the Suborder Insectivora (Bowdich, 1821 [=Order Lipotyphla (Haeckel, 1866)]) and included the families Talpidae (G. Fischer, 1814: x, 143) and Soricidae (G. Fischer, 1814). Synonymised within Soricoidea by McKenna and Bell (1997: 285).

Family Soricidae G. Fischer, 1814

Family Soricinorum G. Fischer, 1814: x.

TYPE GENUS: *Sorex* Linnaeus, 1758: 53.

COMMENTS: See comments above. Recognised at the family rank by Honacki *et al*. (1982: 67), Yates (1984: 118) and subsequent authors. Rank appears to be the same as the Family Soricini (G. Fischer, 1814: 143, 144).

Family Soricini G. Fischer, 1814: 143, 144.

TYPE GENUS: *Sorex* Linnaeus, 1758: 53.

COMMENTS: When originally proposed, this rank was placed in the Order Plantigrada (G. Fischer, 1813a [=Placentalia (Bonaparte, 1838 part)]) and included the genus *Sorex* Linnaeus, 1758: 53. Name appear to be equivalent to Soricinorum (G. Fischer, 1814: x) and also mentioned within G. Fischer (1817: 372).

Family Soricidae J. Gray, 1821: 300.

TYPE GENUS: *Sorex* Linnaeus: 1758: 53.

COMMENTS: When originally proposed, this rank was placed in the Order Plantigradae (J. Gray, 1821 [=Placentalia (Bonaparte, 1838 part)]) and included genus *Sorex* Linnaeus, 1758: 53.

Tribe Sorinina J. Gray, 1825a: 339.

TYPE GENUS: *Sorex* Linnaeus, 1758: 53.

COMMENTS: When originally proposed, this rank was placed in the Family Talpidae (G. Fischer, 1814: x, 143) and included the genera *Sorex* Linnaeus, 1758: 53; and *Mygale* G. Cuvier, 1800: Table 1 [=*Desmana* Güldenstädt, 1777: 108]. Synonymised within Soricoidea by McKenna and Bell (1997: 285).

Family Sorexineae Lesson, 1842: 87.

TYPE GENUS: *Sorex* Linnaeus, 1758: 53.

COMMENTS: When originally proposed, this rank was placed in the Group Insectivora (Bowdich, 1821 [=Order Lipotyphla (Haeckel, 1866)]) and included the genera *Mygale* G. Cuvier, 1800: Table 1 [=*Desmana* Güldenstädt, 1777: 108]; *Galemys* Kaup, 1829: 119; *Solenodon* Brandt, 1833b: 459, 477; *Sorex* Linnaeus, 1758: 53; *Macroscelides* A. Smith, 1829: 435; *Tupaia* Raffles, 1821: 256; and *Gymnura* Lesson, 1827a: 171. Synonymised within the Family Soricidae by Hutterer (2005: 223). As the stem of the Latin sorex (=shrew) is soric-, Lesson's name is incorrectly formed.

Family Sorices Peters, 1863: 21.

TYPE GENUS: *Sorex* Linnaeus, 1758: 53.

COMMENTS: When originally proposed, this rank was placed in the Order Insectivora (Bowdich, 1821 [=Order Lipotyphla (Haeckel, 1866)]) and included the genus *Sorex* Linnaeus, 1758: 53. Rank recognised by Mivart (1868: 141). Synonymised within Soricoidea by McKenna and Bell (1997: 285).

Family Soricida Haeckel, 1866: clx.

TYPE GENUS: *Sorex* Linnaeus, 1758: 53.

COMMENTS: When originally proposed, this rank was placed in the Suborder Lipotyphla (Haeckel, 1866) and included the genera *Sorex* Linnaeus, 1758: 53; *Crossopus* Wagler, 1832b: 275; and *Crocidura* Wagler, 1832b.

Subfamily Crocidurinae Milne-Edwards, 1872

Subfamily Crocidurinae Milne-Edwards, 1872: 256.

TYPE GENUS: *Crocidura* Wagler, 1832b.

COMMENTS: When originally proposed, this rank was placed in the Family Soricidae (G. Fischer, 1814), and included the genera *Crocidura* Wagler, 1832b; and *Sorex* Linnaeus, 1758: 53. Subfamily rank within the Family Soricidae recognised by Repenning (1967: 15), Yates (1984: 134), Reumer (1989: 81) and Hutterer (1993: 81; 2005: 224), but not by Van Dyck and Strahan (2008: 10, 418).

Subfamily Scutisoricinae J. Allen, 1917a: 781.

TYPE GENUS: *Scutisorex* Thomas, 1913c: 321.

COMMENTS: When originally proposed, this rank was placed in the Family Soricidae (G. Fischer, 1814) and included the genus *Scutisorex* Thomas, 1913c: 321. Synonymised within the Subfamily Crocidurinae by Hutterer (2005: 224) and McKenna and Bell (1997: 292).

Tribe Crocidurini Pavlinov & Rossolimo, 1987: 25.

TYPE GENUS: *Crocidura* Wagler, 1832b.

COMMENTS: When originally proposed, this rank was placed in the Subfamily Crocidurinae (Milne-Edwards, 1872) and included the genera *Suncus* Ehrenberg, 1833: 4, within *Herpestes leucurus* entry; and *Crocidura* Wagler, 1832b. Synonymised within the Subfamily Crocidurinae by McKenna and Bell (1997: 292).

HOMONYMS:

Tribe Crocidurini Gureev, 1971: 8, 120, shrews of the Class Mammalia (Order Soricomorpha, Family Crociduridae). Appears to preoccupy Crocidurini Pavinov & Rossolimo, 1987. See McKenna and Bell (1997: 292).

Crocidura Wagler, 1832

Crocidura Wagler, 1832b: 275.

TYPE SPECIES: Φ *Sorex leucodon* Hermann, 1780: 382 (as *Crocidura leucodon*) [= Φ *Crocidura leucodon* (Hermann, 1780: 382)] by subsequent designation. See Palmer (1904: 204).

COMMENTS: Conserved by Opinion 91 of the ICZN (1926c: 1) and Direction 24 (ICZN, 1955: 222) of the ICZN. Taxonomy of the genus reviewed by G. Allen (1939: 29) and Heim de Balsac and Meester (1977: 9). Eurasian shrews reviewed by Jenkins (1976: 271).

Rhinomus A. Murray, 1860: 159.

TYPE SPECIES: Φ *Rhinomus soricoides* A. Murray, 1860: 159 [= Φ *Crocidura poensis* (L. Fraser, 1843: 200)] by monotypy.

COMMENTS: Synonymised within *Sorex* Linnaeus, 1758: 53 by J. Gray (1864a: 57), but synonymised within *Crocidura* by G. Allen (1939: 29) and Hutterer (2005: 224).

HOMONYMS:

Rhinomys M. Lichtenstein, 1831: Text to Plate 38, elephant shrews of the Class Mammalia (Order Macroscelidea, Family Macroscelididae). Genus is a synonym of *Macroscelides* (A. Smith, 1829: 435).

Leucodon Fatio, 1869: 132.

TYPE SPECIES: Φ *Leucodon microurus* Fatio, 1869: 137 [=Φ *Crocidura leucodon* (Hermann, 1780: 382)] by monotypy.

COMMENTS: Synonymised within *Crocidura* by Palmer (1904: 374), G. Allen (1939: 29) and Hutterer (2005: 224).

Paurodus E. Schulze, 1897: 90.

TYPE SPECIES: Φ *Sorex leucodon* Hermann, 1780: 382 [=Φ *Crocidura leucodon* (Hermann, 1780: 382)] by subsequent designation. See G. Allen (1939: 29).

COMMENTS: Described as a subgenus of *Crocidura*. Synonymised within *Crocidura* by G. Allen (1939: 29) and Hutterer (2005: 224).

HOMONYMS:

Paurodon Marsh, 1887: 342, mammals of the Class Mammalia († Order Dryolestida, † Family Paurodontidae). See McKenna and Bell (1997: 47).

Heliosorex E. Heller, 1910: 6.

TYPE SPECIES: Φ *Heliosorex roosevelti* E. Heller, 1910: 6 [= Φ *Crocidura roosevelti* (E. Heller, 1910: 6)] by original designation.

COMMENTS: A synonym of *Crocidura*, fide Hollister (1918: 68). Synonymised within *Crocidura* by G. Allen (1939: 29) and Hutterer (2005: 224).

Praesorex Thomas, 1913c: 320.

TYPE SPECIES: Φ *Crocidura goliath* Thomas, 1906g: 177 by monotypy.

COMMENTS: Not recognised by G. Allen (1939: 29). Synonymised within *Crocidura* by Hutterer (2005: 224).

Afrosorex Hutterer, 1986: 26.

TYPE SPECIES: Φ *Crocidura fischeri* Pagenstecher, 1885a: 34 by original designation.

COMMENTS: Described as a subgenus of *Crocidura*. Not recognised by G. Allen (1939: 29). Synonymised within *Crocidura* by Hutterer (2005: 224).

† *Crocidura trichura* Dobson, 1889

Christmas Island Shrew

† *Crocidura fuliginosa trichura* Dobson, 1889: 532.

TYPE LOCALITY: Christmas Island, Indian Ocean, Australia.

COMMENTS: Recognised as a variety within *fuliginosa* by authors including C. Andrews (1900: 27). Thought to be a relative of the southeast Asian white-toothed shrew *C. fuliginous* (Thomas, 1889c: 532) until Jenkins (1976: 297) redescribed the taxon as *C. attenuata trichura*, which was followed by Jenkins (1982: 276), Meek (2000: 43), Clayton *et al.* (2006: 108), and Van Dyck and Strahan (2008: 418). Synonymised within *Crocidura attenuata* Milne-Edwards (1872: 263) by Corbet and Hill (1992: 43). Species rank recognised by Ruedi (1995: 232) and Hutterer (2005: 252), which was subsequently confirmed by Eldridge *et al.* (2009: 3; 2013: 572, 576). The species is probably extinct (Meek, 2000: 43).

Subcohort Scrotifera Waddell *et al.*, 1999

Clade Scrotifera Waddell *et al.*, 1999b: 31, 50.

COMMENTS: When originally proposed, this clade was placed in the Placentalia (Bonaparte, 1838) and included the common ancestor of the extant orders Pholidota (Weber, 1904: vi, 420), Carnivora (Bowdich, 1821), Cetartiodactyla (Montgelard *et al.*, 1997 [=Artiodactyla (Owen, 1848) and Cetacea (Brisson, 1762)]), Perissodactyla (Owen, 1848) and Chiroptera (Blumenbach, 1779). Recognised as a valid clade ahead of Variamana (Springer *et al.*, 2005) by Asher and Helgen (2010: 4).

Clade Variamana Springer *et al.*, 2005: 41.

COMMENTS: When originally proposed, this clade was placed in the Laurasiatheria (Waddell *et al.*, 1999a) and included the Chiroptera (Blumenbach, 1779), Perissodactyla (Owen, 1848), Cetartiodactyla (Montgelard *et al.*, 1997 [=Artiodactyla (Owen, 1848) and Cetacea (Brisson, 1762)]), Pholidota (Weber, 1904: vi, 420) and Carnivora (Bowdich, 1821). Recognised at superordinal rank by Springer *et al.* (2007: 21), but synonymised within the Clade Scrotifera (Waddell *et al.*, 1999b) by Asher and Helgen (2010: 4).

Order Chiroptera Blumenbach, 1779

Order Chiroptera Blumenbach, 1779: 58, 74.

COMMENTS: When originally proposed, this rank was placed in the Mammalia (Linnaeus, 1758) and included the genus *Vespertilio* Linnaeus, 1758: 31. The Chiroptera were placed within the Archonta by Gregory (1910: 465) who used morphological comparisons. This superordinal group included the orders Chiroptera, Dermoptera, Primates and Menotyphla (that included the tree shrews and elephant shrews). Though various reviews support this conclusion (see Sargis, 2004: 56) some studies do not (e.g. Pumo *et al.*, 1998: 709). More recent molecular studies excluded the bats from this group under the new name of Euarchonta (Waddell *et al.*, 1999a). This group was pre-empted by Adkins and Honeycutt (1991: 10317) using mtDNA, with subsequent support given by various studies including Murphy *et al.* (2001b: 615), Kriegs *et al.* (2007: 160) and Bloch *et al.* (2007: 1159). Yet other studies have recognised the Chiroptera and Dermoptera as sister taxa in a group called the Volitantia (Illiger, 1811; see Volitantia entry). More recent phylogenetic assessments exclude Chiroptera from Euarchonta and place them with the Laurasiatheria (Zhou *et al.*, 2012: 150). So it appears that some work still needs to be done to resolve the higher relationships of the bats.

The phylogeny of bats was reviewed by Gunnell and Simmons (2005: 209) who agreed they are monophyletic. The classical subdivision of bats into the Megachiroptera and Microchiroptera has come under increasing question based on both body size (megabats are not necessarily large nor are microbats uniformly small) and, especially, molecular evidence. Though the relationships amongst the different families are not yet fully resolved, it is clear from molecular research over the past two decades that some previous microchiropteran families (Hipposideridae, Rhinolophidae, Megadermatidae, Rhinopomatidae and Craseonycteridae), those frequently included in the microchiropteran infraorder Yinochiroptera, are more closely related to the megachiropterans (Pteropodidae) than they are to the microchiropteran bats (Reardon, 2009a: 39; Teeling *et al.*, 2012: 2). A new phylogeny for the Chiroptera was proposed by Springer *et al.* (2001: 6243) and Teeling *et al.*

(2002: 1432) that abandons the suborders Megachiroptera and Microchiroptera and replaces them with the suborders Yinpterochiroptera and Yangochiroptera (the former being a combination of Yinochiroptera and Pteropodidae). The Yinpterochiroptera consists of the superfamilies Pteropodoidea and Rhinolophoidea, which included the families Pteropodidae, Rhinopomatidae, Megadermatidae, Hipposideridae and Rhinolophidae. The Yangochiroptera includes the superfamilies Emballonuroidea, Noctilionoidea and Vespertilionoidea and the remaining families of bats. Further support for the Yinpterochiroptera/ Yangochiroptera division has been given by Teeling *et al.* (2000: 189), Van Den Bussche and Hoofer (2004: 327), Eick *et al.* (2005: 1874), Miller-Butterworth *et al.* (2007: 1556, 1558) and Teeling *et al.*, 2012: 8). In contrast to the molecular data, morphological studies do not ally the Pteropodidae with the Yinochiroptera (Simmons, 2005b: 166–170). Subsequently new subordinal names have been proposed, the Vespertilioniformes (for the group including Emballonuridae, Nycteridae and the 'yangochiropterans') and Pteropodiformes (for the group comprised of the families Pteropodidae, Craseonycteridae, Hipposideridae, Megadermatidae, Rhinolophidae and Rhinopomatidae) (Hutcheon & Kirsch, 2006: 1).

Vespertiliones Pallas, 1767: 3.

COMMENTS: When originally proposed, this unknown rank included the genus *Vespertilio* Linnaeus, 1758: 31.

HOMONYMS:

Family Vespertiliones Peters, 1865a, bats of the Family Vespertilionidae J. Gray, 1821. See individual entry.

Group Vespertiliones Dobson, 1878, bats of the Subfamily Vespertilioninae J. Gray, 1821. See individual entry.

Family Chiropteren Duméril, 1806a: 4, 10.

TYPE GENUS: Not based on a genus group name.

COMMENTS: When originally proposed, this rank was placed in the Klasse Säugthiere (Duméril, 1806a [=Class Mammalia (Linnaeus, 1758)]) and included the genera *Galeopithecus* Pallas, 1780a: 208 [Order Dermoptera (Illiger, 1811: 63, 116)]; *Pteropus* Brisson, 1762; *Noctilio* Linnaeus, 1766: 88; *Vespertilio* Linnaeus, 1758: 31; *Rhinolophen* [=*Rhinolophus* Lacépède, 1799a]; and *Phyllostomen* [=*Phyllostomus* Lacépède, 1799a: 16].

Order Chiropteria Rafinesque, 1815: 54.

COMMENTS: When originally proposed, this rank was placed in the and included the families Galeopia (Rafinesque, 1815: 54 [=Galeopithecidae (Gray, 1821: 300)]) and Vespertilia (Rafinesque, 1815 [=Vespertilionidae (J. Gray, 1821)]). Name is an unjustified emendation of Chiroptera (Blumenbach, 1779).

Family Cheiroptères G. Cuvier, 1816a: xxx, 121.

TYPE GENUS: Not based on a genus group name.

COMMENTS: When originally proposed, this rank was placed in the Order Carnassiers (G. Cuvier, 1816a [=Mammalia (Linnaeus, 1758 part)])) and included the genera *Vespertilio* Linnaeus, 1758: 31; *Pteropus* Brisson, 1762; *Cephalotes* É. Geoffroy, 1810b [=*Nyctimene* Borkhausen, 1797]; *Dysopes* Illiger, 1811: 122 [=*Molossus* É. Geoffroy, 1805b: 151]; *Noctilio* Linnaeus, 1766: 88; *Phyllostoma* G. Cuvier, 1800: Table 1 [=*Phyllostomus* Lacépède, 1799a: 16]; Les Megadermes [=*Megaderma* É. Geoffroy, 1810a: 197]; *Rhinolophus* Lacépède, 1799a; *Nycteris* É. Geoffroy & G. Cuvier, 1795: 186; *Thaphozous* [=*Taphozous* É. Geoffroy, 1818]; and *Plecotus* É. Geoffroy, 1818: 112.

Class Cheiroptera J. Gray, 1821: 299.

COMMENTS: When originally proposed, this rank was placed in the Sub-Kingdom Vertebrosa (J. Gray, 1821: 297 [=Class Mammalia (Linnaeus, 1758 part)]) and included the orders Fructivorae (J. Gray, 1821 [=Pteropodidae (J. Gray, 1821)]) and Insectivorae (J. Gray, 1821 [=Yangochiroptera (Koopman, 1985 part)]). Synonymised within Chiroptera by McKenna and Bell (1997: 295).

Order Cheiroptera Fleming, 1822a: xxxii.

COMMENTS: When originally proposed, this rank was placed in the Unguiculata (Linnaeus, 1766 [=Placentalia (Bonaparte, 1838 part)]), and included the genera *Galeopithecus* Pallas, 1780a: 208 [Order Dermoptera (Illiger, 1811: 63, 116)]; *Pteropus* Brisson, 1762; *Cephalotes* É. Geoffroy, 1810b [=*Nyctimene* Borkhausen, 1797]; *Noctilio* Linnaeus, 1766: 88; *Phyllostoma* G. Cuvier, 1800: Table 1 [=*Phyllostomus* Lacépède, 1799a: 16]; *Molossus* É. Geoffroy, 1805b: 151; *Stenoderma* É. Geoffroy, 1818: 114; *Rhinopoma* É. Geoffroy, 1818: 113; *Rhinolophus* Lacépède, 1799a; *Nycteris* É. Geoffroy and G. Cuvier, 1795: 186; *Megaderma* É. Geoffroy, 1810a: 197; *Thaphozus* [sic =*Thaphozous*] Bowdich, 1821 [=*Taphozous* É. Geoffroy, 1818; *Vespertilio* Linnaeus, 1758: 31; and *Plecotus* É. Geoffroy, 1818: 112. Name further described by Fleming, 1822b: 175. Synonymised within Chiroptera by McKenna and Bell (1997: 295).

Suborder Nycterides Haeckel, 1866: clx.

COMMENTS: When originally proposed, this rank was placed in the Order Chiroptera (Blumenbach, 1779) and included the families Gymnorrhina [sic =Gymnorhina] (Giebel, 1855 [=Yangochiroptera (Koopman, 1985 part)]) and Histiorrhina [sic =Histiorhina] (Van der Hoeven, 1855: 1033 [=Rhinolophidae (J. Gray, 1825)]). *Nomen oblitum*. Synonymised within Microchiroptera by McKenna and Bell (1997: 301).

Suborder Animalivora Gill, 1872: v, 16.

COMMENTS: Placed within Order Chiroptera (Blumenbach, 1779) and included the families Desmodidae (Bonaparte, 1845: 5), Phyllostomidae (J. Gray, 1825b: 242), Mormopidae (Gill, 1872: 16 [=Mormoopidae (de Saussure, 1860: 286)]), Rhinolophidae (J. Gray, 1825a), Megadermidae (Gill, 1872 [=Megadermatidae (H. Allen, 1864)]), Vespertilionidae (J. Gray, 1821), Molossidae (Gervais, 1855b) and Noctilionidae (J. Gray, 1821: 299). Synonymised within Microchiroptera by Miller (1907: 78), Simpson (1945: 55) and subsequent authors.

Suborder Microchiroptera Dobson, 1875a: 346.

COMMENTS: When originally proposed, this rank was placed in the Order Chiroptera (Blumenbach, 1779) and included the families Rhinolophidae (J. Gray, 1825a), Nycteridae (Van der Hoeven, 1855: 1028), Vespertilionidae (J. Gray, 1821), Emballonuridae (Gervais, 1855b), and Phyllostomidae (J. Gray, 1825b: 242). Rank has historically been accepted by all authors until the revisions discussed above in Yangochiroptera.

Order Chiropteri de Rochebrune, 1883: 87.

COMMENTS: When originally proposed, this rank was attributed to Blumenbach (1779: 58, 74) and included the families Pteropidae [=Pteropodidae (J. Gray, 1821)], Megadermidae (Gill, 1872 [=Megadermatidae (H. Allen, 1864)], Nycteridae (Van der Hoeven, 1855: 1028), Rhinolophidae (J. Gray, 1825), Phyllorhinidae (de Rochebrune, 1883 [=Hipposideridae (Flower & Lydekker, 1891)]), Taphozoidae (Jerdon, 1867 [=Emballonuridae (Gervais, 1855b)]), Molossidae (Gervais, 1855b), and Vespertilionidae (J. Gray, 1821). Unjustified emendation of Chiroptera (Blumenbach, 1779).

Grand Seccion Ptética Ameghino, 1889: xxi, 44, 348.

COMMENTS: When originally proposed, this rank was placed in the Subclass Ditremata (Ameghino, 1889 [=Mammalia (Linnaeus, 1758 part)]) and included the orders Prochiroptera (Ameghino, 1889: 348) and Chiroptera (Blumenbach, 1779). Synonymised within Chiroptera by McKenna and Bell (1997: 295).

Infraorder Yinochiroptera Koopman, 1985: 26.

COMMENTS: When originally proposed, this rank was placed in the Suborder Microchiroptera (Dobson, 1875a [=Chiroptera (Blumenbach, 1779 part)]) and included the superfamilies Emballonuroidea (Gervais, 1855b) and Rhinolophoidea (J. Gray, 1825a). Yinochiroptera initially included the Family Emballonuridae but some subsequent authors either placed it outside both Yinochiroptera and Yangochiroptera, such as Simmons and Geisler (1998: 136) and Simmons (1998: 12), or more recently within the modified Yangochiroptera (*sensu* Hutcheon & Kirsch, 2004: 44). Yinochiroptera was not recognised by Simmons (2005a: 313), but it was noted that the data supporting novel clades such as Yinpterochiroptera are increasingly compelling and that it seems likely that a new consensus view of higher-level classification of bats that contradicts most traditional arrangements will soon emerge. Subsequently Giannini and Simmons (2007: 1) undertook a detailed study of the premaxillae of this group and found a lack of anatomical basis to support it, whereas the Yangochiroptera was supported. The Yinochiroptera was not allied with the Megachiroptera by authors including McKenna and Bell (1997: 304), Simmons and Geisler (1998: 1432), Simmons (1998: 12) and Göbbel (2002: 343), but this affiliation was subsequently strongly supported and was given the name Yinpterochiroptera or Pteropodiformes (see individual entries).

Order Chiropteriformes Kinman, 1994: 37.

COMMENTS: When originally proposed, this rank was placed in the Class Mammalea (Kinman, 1994 [=Class Mammalia (Linnaeus, 1758)]). Synonymised within Chiroptera by McKenna and Bell (1997: 295).

Suborder Vespertilioniformes Hutcheon & Kirsch, 2004: 44.

COMMENTS: When originally proposed, this rank was placed in the Order Chiroptera (Blumenbach, 1779) and included the families Phyllostomidae (J. Gray, 1825b: 242), Mormoopidae (de Saussure, 1860: 286), Noctilionidae (J. Gray, 1821: 299), Mystacinidae (Dobson, 1875a: 349), Emballonuridae (Gervais, 1855b), Vespertilionidae (J. Gray, 1821), Miniopteridae (Dobson, 1875a), Molossidae (Gervais, 1855b), Natalidae (J. Gray, 1866c: 90) and Nycteridae (Van der Hoeven, 1855: 1028). Created to include the Emballonuridae, Nycteridae and the 'yangochiropterans'. Name subsequently recognised by Eick *et al.* (2005: 1874) and Hutcheon and Kirsch (2006: 1).

Suborder Yinpterochiroptera Springer *et al.*, 2001

Suborder Yinpterochiroptera Springer *et al.*, 2001: 6243.

COMMENTS: When originally proposed as a new rank it was placed in the Order Chiroptera (Blumenbach, 1779) and included the superfamilies Pteropodoidea (including Pteropodidae) and Superfamily Rhinolophoidea (J. Gray, 1825a) (including Megadermatidae (H. Allen, 1864) and Rhinolophidae (J. Gray, 1825a). Name developed as a result of the revision of the Yinochiroptera to include the Pteropodidae. Suborder rank recognised and expanded by Teeling *et al.* (2002: 1432; 2003: 309), and subsequent given further support by Van Den Bussche and Hoofer (2004: 327), Teeling *et al.* (2005: 581) and Miller-Butterworth (2007: 1556, 1558). Name was rejected by Hutcheon and Kirsch (2006:

7) in favour of their proposed Pteropodiformes (Hutcheon & Kirsch, 2004: 44) as it is based on an incorporation of the Pteropodidae into the Yinochiroptera, rather than an equal fusion of the two taxa. The use of Pteropodiformes was supported by Eick *et al.* (2005: 1874) and Hutcheon and Kirsch (2006: 1) and Reardon (2009a: 40). This suborder includes the families Pteropodidae, Megadermatidae, Hipposideridae, Rhinolophidae and Rhinonycteridae, that occur within Australia, as well as the Craseonycteridae and Rhinopomatidae that are extralimital. It could be argued that the name ought to continue to be Yinochiroptera under the argumentation developed by Helgen (2003b: 1) and Asher & Helgen (2010: 1), although the fundamental change made by Springer *et al.* (2001: 6243) and modified by Teeling *et al.* (2002: 1432) should be recognised, with subsequent modifications on this being considered junior names, which has been adopted here. The recognition of Yinterochiroptera as a suborder was recognised by Yapa and Ratnavira (2013: 362).

Suborder Megachiroptera Dobson, 1875a: 346.

COMMENTS: When originally proposed, this rank was placed in the Order Chiroptera (Blumenbach, 1779) and included the groups Pteropi (containing the genera *Pteropus* Brisson, 1762; *Cynopterus* F. Cuvier, 1824 [1821–1825]: 248; *Cynonycteris* Peters, 1852b: 25 [=*Rousettus* J. Gray, 1821: 299]; *Harpyia* Illiger, 1811 [=*Nyctimene* Borkhausen, 1797]; *Epomophorus* E. Bennett, 1836: 149; and *Cephalotes* É. Geoffroy, 1810b [=*Nyctimene* Borkhausen, 1797]) and Macroglossi (containing the genera *Macroglossus* F. Cuvier, 1824 [1821–1825]; *Eonycteris* Dobson, 1873a: 148; and *Notopteris* J. Gray, 1859a: 36). Historically recognised since its description by all authors to contain the Pteropodidae. See comments above under Chiroptera.

Family Frugivora Giebel, 1855: xii, 991.

TYPE GENUS: Not based on a genus group name.
COMMENTS: When originally proposed, this rank was placed in the Order Chiroptera (Blumenbach, 1779) and included the genera *Hypoderma* É. Geoffroy, 1828 [= *Dobsonia* Palmer, 1898]; *Harpyia* Illiger, 1811 [= *Nyctimene* Borkhausen, 1797]; *Macroglossus* F. Cuvier, 1824 [1821–1825]; and *Pteropus* Brisson, 1762. Recognised at the family rank by Owen (1859: 52) and suborder by Gill (1872: v, 18). Synonymised within Megachiroptera by McKenna and Bell (1997: 295).

Suborder Pterocynes Haeckel, 1866: clx.

COMMENTS: When originally proposed, this rank was placed in the Order Chiroptera (Blumenbach, 1779) and included the families Pteropodida (Haeckel, 1866 [=Pteropodidae (J. Gray, 1821)]) and Hypodermida (Haeckel, 1866 [=Pteropodidae (J. Gray, 1821)]). *Nomen oblitum*. Synonymised within the Suborder Megachiroptera by McKenna and Bell (1997: 295).

Family Histiorrhina Haeckel, 1866: clx.

TYPE GENUS: Not based on a genus group name.
COMMENTS: When originally proposed, this rank was placed in the Suborder Nycterides (Haeckel, 1866 [=Chiroptera (Blumenbach, 1779 part)]) and included the genera *Rhinolophus* Lacépède, 1799a; *Megaderma* É. Geoffroy, 1810a: 197; and *Phyllostoma* G. Cuvier, 1800: Table 1 [=*Phyllostomus* Lacépède, 1799a: 16].

Megachiropteramorpha Simmons & Geisler, 1998: 135, 136.

COMMENTS: When originally proposed as a new taxon it was placed in the Order Chiroptera (Blumenbach, 1779) and included the † *Archaeopteropus* Meschinelli, 1903: 1344 and the Suborder Megachiroptera (Dobson, 1875a [=Yinpterochiroptera (Springer *et al.*, 2001 part)]). More broadly it included all chiropterans sharing a more recent common ancestor with Megachiroptera than with Microchiroptera. Does not appear to have been widely followed by subsequent authors.

Suborder Pteropodiformes Hutcheon & Kirsch, 2004: 44.

COMMENTS: When originally proposed, this rank was placed in the Order Chiroptera (Blumenbach, 1779) and included the families Pteropodidae (J. Gray, 1821), Hipposideridae (Flower & Lydekker, 1891), Rhinolophidae (J. Gray, 1825a), Megadermatidae (H. Allen, 1864), and Craseonycteridae (Hill, 1974: 303). Name subsequently recognised by Eick *et al.* (2005: 1874), and Hutcheon and Kirsch (2006: 1). This rank was subsequently modified to include Rhinopomatidae (Bonaparte, 1838: 112) by Hutcheon and Kirsch (2006: 1, 9), which was accepted as the most appropriate name ahead of Yinpterochiroptera (Springer *et al.*, 2001) by Reardon (2009a: 40).

Superfamily Pteropodoidea J. Gray, 1821

Family Pteropidae J. Gray, 1821: 299.

TYPE GENUS: *Pteropus* Brisson, 1762.
COMMENTS: When originally proposed, this rank was placed in the Order Fructivorae (J. Gray, 1821 [=Pteropodidae (J. Gray, 1821)]) and included the genera *Pteropus* Brisson, 1762; and *Rousettus* J. Gray, 1821: 299. Superfamily rank recognised by Teeling *et al.* (2002: 1432: 2003: 309).

Family Pteropodidae J. Gray, 1821

Family Pteropidae J. Gray, 1821: 299.

TYPE GENUS: *Pteropus* Brisson, 1762.

COMMENTS: When originally proposed, this rank was placed in the Order Fructivorae (J. Gray, 1821 [=Pteropodidae (J. Gray, 1821)]) and included the genera *Pteropus* Brisson, 1762; and *Rousettus* J. Gray, 1821: 299. The spelling Pteropidae was used by several authors including Dobson (1875a: 346), Miller (1907: vii, 45), Simpson (1945: 54) and Ellerman and Morrison-Scott (1951: 91), but the stem of the Greek *pous* (=foot) is *pod-*, hence that of the latinised *Pteropus* is 'pteropod-'. The spelling Pteropodidae was introduced by Bonaparte (1838: 112) and used by many authors including Gill (1872: 18). Spelling of the name reviewed by Handley (1980: 9) who concluded the correct spelling is Pteropodidae. The spelling of the name 'Pteropidae' was synonymised within Pteropodidae by McKenna and Bell (1997: 295). Subfamilies not recognised by Simmons (2005a: 313).

Order Fructivorae J. Gray, 1821: 299.

COMMENTS: When originally proposed, this rank was placed in the Class Cheiroptera (J. Gray, 1821 [=Chiroptera (Blumenbach, 1779)]) and included the families Pteropidae [=Pteropodidae (J. Gray, 1821)] and Cephalotidae (J. Gray, 1821 [=Pteropodidae (J. Gray, 1821)]). The term has long actual priority over Megachiroptera, but the Code does not govern above the family-group ranks. Included the families Pteropidae and Cephalotidae. Synonymised within Megachiroptera by Miller (1907: 44), Simpson (1945: 54) and McKenna and Bell (1997: 295).

Family Cephalotidae J. Gray, 1821: 299.

TYPE GENUS: *Cephalotes* É. Geoffroy, 1810b [= *Nyctimene* Borkhausen, 1797].

COMMENTS: When originally proposed, this rank was placed in the Order Fructivorae (J. Gray, 1821 [=Pteropodidae (J. Gray, 1821)]) and included the genus *Cephalotes* É. Geoffroy, 1810b [= *Nyctimene* Borkhausen, 1797]. Included within Pteropodidae by Simpson (1945: 54) and Simmons (2005a: 313). Also placed in Subfamily Nyctimeninae by Simpson (1945: 55) and the Subtribe Nyctimena (Miller, 1907) by McKenna and Bell (1997: 299).

Tribe Pteropina J. Gray, 1825a: 338.

TYPE GENUS: *Pteropus* Brisson, 1762.

COMMENTS: When originally proposed, this rank was placed in the Family Vespertilionidae (J. Gray, 1821) and included the genera *Pteropus* Brisson, 1762; *Cynopterus* F. Cuvier, 1824 [1821–1825]: 248; *Macroglossum* [sic =*Macroglossus*] F. Cuvier, 1824 [1821–1825]; *Cephalotis* [sic =*Cephalotes*] É. Geoffroy, 1810b [= *Nyctimene* Borkhausen, 1797]; and *Harpyia* Illiger, 1811 [= *Nyctimene* Borkhausen, 1797]. Name recognised at subfamily rank by J. Gray (1825b: 243). Tribe rank recognised by Bonaparte (1831: 6) and as a subtribe by Koopman and Jones (1970: 23). Synonymised within Pteropodidae by McKenna and Bell (1997: 295).

Family Pteropodidae Bonaparte, 1838: 112.

TYPE GENUS: *Pteropus* Brisson, 1762.

COMMENTS: When originally proposed, this rank was placed in the Order Chiroptera (Blumenbach, 1779) and included the Tribe Pteropodina (Bonaparte, 1831 [=Pteropodinae (J. Gray, 1821)]). Synonymised within Pteropodidae by McKenna and Bell (1997: 295).

Family Harpyidae C. Smith, 1842: 115.

TYPE GENUS: *Harpyia* Illiger, 1811 [= *Nyctimene* Borkhausen, 1797].

COMMENTS: When originally proposed, this rank was placed in the Order Chiroptera (Blumenbach, 1779) and included the genera *Pteropus* Brisson, 1762; *Pachystoma* [=*Pachysoma*] I. Geoffroy, 1828: 703 [=*Cynopterus* F. Cuvier, 1824 [1821–1825]: 248]; *Macroglossus* F. Cuvier, 1824 [1821–1825]; *Harpyia* Illiger, 1811 [= *Nyctimene* Borkhausen, 1797]; and *Cephalotes* É. Geoffroy, 1810b [= *Nyctimene* Borkhausen, 1797]. Synonymised within Pteropodidae by Simmons (2005a: 313).

Family Pteropi Peters, 1865a: 256.

TYPE GENUS: *Pteropus* Brisson, 1762.

COMMENTS: When originally proposed, this rank included the genera *Pteropus* Brisson, 1762; *Cynonycteris* Peters, 1852b: 25 [=*Rousettus* J. Gray, 1821: 299]; *Pterocyon* Peters, 1861a: 423 [=*Eidolon* Rafinesque, 1815: 54]; *Cynopterus* F. Cuvier, 1824 [1821–1825]: 248; *Megaerops* Peters, 1865a: 256; *Epomophorus* E. Bennett, 1836: 149; *Macroglossus* F. Cuvier, 1824 [1821–1825]; *Harpyia* Illiger, 1811 [= *Nyctimene* Borkhausen, 1797]; *Cephalotes* É. Geoffroy, 1810b [= *Nyctimene* Borkhausen, 1797]; and *Notopteris* J. Gray, 1859a: 36.

Family Pteropodida Haeckel, 1866: clx.

TYPE GENUS: *Pteropus* Brisson, 1762.

COMMENTS: When originally proposed, this rank was placed in the Suborder Pterocynes (Haeckel, 1866 [=Yinpterochiroptera (Springer *et al*., 2001 part)]) and included the genera *Pteropus* Brisson, 1762; and *Macroglossus* F. Cuvier, 1824 [1821–1825]. Synonymised within Pteropodidae by McKenna and Bell (1997: 295).

Family Hypodermida Haeckel, 1866: clx.

TYPE GENUS: *Hypoderma* É. Geoffroy, 1828 [=*Dobsonia* Palmer, 1898].

COMMENTS: When originally proposed, this rank was placed in the Suborder Pterocynes (Haeckel, 1866 [=Yinpterochiroptera (Springer *et al*., 2001 part)]) and included the genus *Hypoderma* É. Geoffroy, 1828 [= *Dobsonia* Palmer, 1898]. Synonymised within Pteropodidae by McKenna and Bell (1997: 295).

Tribe? Pteropi Winge, 1893b: 53.

TYPE GENUS: *Pteropus* Brisson, 1762.

COMMENTS: When originally proposed, this rank was placed in the Family Pteropodidae (J. Gray, 1821) and included the genera *Cynonycteris* Peters, 1852b: 25 [=*Rousettus* J. Gray, 1821: 299]; *Pteropus* Brisson, 1762; *Pteralopex* Thomas, 1888c: 155; *Epomophorus* E. Bennett, 1836: 149; *Cephalotes* É. Geoffroy, 1810b [=*Nyctimene* Borkhausen, 1797]; *Cynopterus* F. Cuvier, 1824 [1821–1825]: 248; and *Harpyia* Illiger, 1811 [=*Nyctimene* Borkhausen, 1797].

Subfamily Macroglossinae J. Gray, 1866

Tribe? Macroglossina J. Gray, 1866d: 64.

TYPE GENUS: *Macroglossus* F. Cuvier, 1824 [1821–1825].

COMMENTS: Described at unspecified rank but placed within the Family Pteropidae [=Pteropodidae (J. Gray, 1821)] and included the genera *Notopteris* J. Gray, 1859a: 36; and *Macroglossus* F. Cuvier, 1824 [1821–1825]. Subfamily Macroglossinae recognised by Koopman and Jones (1970: 23), Koopman (1984a: 153), Strahan (1995: 8, 421), McKenna and Bell (1997: 300), and Van Dyck and Strahan (2008: 10, 426).

Group Macroglossi Dobson, 1875a: 346.

TYPE GENUS: *Macroglossus* F. Cuvier, 1824 [1821–1825].

COMMENTS: When originally proposed, this rank was placed in the Family Pteropidae [=Pteropodidae (J. Gray, 1821)] and included the genera *Macroglossus* F. Cuvier, 1824 [1821–1825]; *Eonycteris* Dobson, 1873a: 148; and *Notopteris* J. Gray, 1859a: 36. Synonymised within the Subfamily Macroglossinae by Simpson (1945: 54).

Subfamily Carponycteriinae Flower & Lydekker, 1891: 654.

TYPE GENUS: *Carponycteris* Lydekker, 1891 [=*Macroglossus* F. Cuvier, 1824 [1821–1825]].

COMMENTS: When originally proposed, this rank was placed in the Family Pteropodidae (J. Gray, 1821) and included the genera *Notopteris* J. Gray, 1859a: 36; *Eonycteris* Dobson, 1873a: 148; *Carponycteris* Lydekker, 1891 [=*Macroglossus* F. Cuvier, 1824 [1821–1825]; *Nesonycteris* Thomas, 1887e: 147 [=*Melonycteris* Dobson, 1877a: 119]; *Callinycteris* Jentink, 1889: 209 [=*Eonycteris* Dobson, 1873a: 148]; and *Trygenycteris* Lydekker, 1891: xv, 655 [=*Megaloglossus* Pagenstecher, 1885b: 245]. Synonymised within the Subfamily Macroglossinae by Simpson (1945: 54).

Tribe? Macroglossi Winge, 1893b: 53.

TYPE GENUS: *Macroglossus* F. Cuvier, 1824 [1821–1825].

COMMENTS: When originally proposed, this rank was placed in the Family Pteropodidae (J. Gray, 1821) and included the genera *Notopteris* J. Gray, 1859a: 36; *Megaloglossus* Pagenstecher, 1885b: 245; *Macroglossus* F. Cuvier, 1824 [1821–1825]; *Melonycteris* Dobson, 1877a: 119; *Nesonycteris* Thomas, 1887e: 147 [=*Melonycteris* Dobson, 1877a: 119]; and *Eonycteris* Dobson, 1873a: 148.

Subfamily Macroglossinae Trouessart, 1897: v, 89.

TYPE GENUS: *Macroglossus* F. Cuvier, 1824 [1821–1825].

COMMENTS: When originally proposed, this rank was placed in the Family Pteropidae [=Pteropodidae (J. Gray, 1821)] and included the genera *Eonycteris* Dobson, 1873a: 148; *Megaloglossus* Pagenstecher, 1885b: 245; *Macroglossus* F. Cuvier, 1824 [1821–1825]; *Melonycteris* Dobson, 1877a: 119; *Callinycteris* Jentink, 1889: 209 [=*Eonycteris* Dobson, 1873a: 148]; *Nesonycteris* Thomas, 1887e: 147 [=*Melonycteris* Dobson, 1877a: 119]; and *Notopteris* J. Gray, 1859a: 36. Subfamily recognised by Simpson (1945: 54).

Subfamily Kiodotinae Palmer, 1898: 111.

TYPE GENUS: *Kiodotus* Blyth, 1840: 69.

COMMENTS: When originally proposed, this rank was placed in the Family Pteropodidae (J. Gray, 1821) and included the genera *Callinycteris* Jentink, 1889: 209 [=*Eonycteris* Dobson, 1873a: 148]; *Eonycteris* Dobson, 1873a: 148; *Kiodotus* Blyth, 1840: 69 [=*Macroglossus* F. Cuvier, 1824 [1821–1825]; *Melonycteris* Dobson, 1877a: 119; *Nesonycteris* Thomas, 1887e: 147 [=*Melonycteris* Dobson, 1877a: 119]; *Notopteris* J. Gray, 1859a: 36; and *Trygenycteris* Lydekker, 1891: xv, 655 [=*Megaloglossus* Pagenstecher, 1885b: 245]. Synonymised within the Subfamily Macroglossinae by Simpson (1945: 54) and McKenna and Bell (1997: 300).

Family Kiodotidae Iredale & Troughton, 1934: x, 92.

TYPE GENUS: *Kiodotus* Blyth, 1840.

COMMENTS: When originally proposed, this rank was placed in the Suborder Megachiroptera (Dobson, 1875a [=Yinpterochiroptera (Springer *et al.*, 2001 part)]) and included the *Odontonycteris* Jentink, 1902 [=*Macroglossus* F. Cuvier, 1824 [1821–1825]; and *Syconycteris* Matschie, 1899a. Recognised by Troughton (1967: 267), but not typically by subsequent authors.

Tribe Macroglossini Koopman & Jones, 1970: 23.

TYPE GENUS: *Macroglossus* F. Cuvier, 1824 [1821–1825].

COMMENTS: When originally proposed, this rank was placed in the Subfamily Macroglossinae (J. Gray, 1866d) and included the genera *Eonycteris* Dobson, 1873a: 148; *Megaloglossus* Pagenstecher, 1885b: 245; *Macroglossus* F. Cuvier, 1824 [1821–1825]; and *Syconycteris* Matschie, 1899a. Tribe rank recognised by McKenna and Bell (1997: 300).

Macroglossus F. Cuvier, 1824

macroglossus [sic] F. Cuvier, 1824 [1821–1825]: 40, 248.

TYPE SPECIES: *Pteropus minimus* É. Geoffroy, 1810b [=*Macroglossus minimus* (É. Geoffroy, 1810b)] by monotypy.

COMMENTS: Palmer (1904: 392), and McKenna and Bell (1997: 300) give the author as Schinz (1824: 71). The spelling *Macroglossa* was used by Lesson (1827a: 115). Taxonomic decision of Andersen (1912: 746) to recognise genus and place *minimus* within the genus.

HOMONYMS:

Macroglossus Neave, 1940: 13, moths of the Class Insecta (Order Lepidoptera, Family Sphingidae). An incorrect subsequent spelling of *Macroglossa* de Boisduval, 1833: 226. Genus is a junior synonym of *Macroglossum* Scopoli, 1777: 414.

Kiodotus Blyth, 1840: 69, footnote.

TYPE SPECIES: *Nomen novum* for *Macroglossus* F. Cuvier, 1824 [1821–1825].

COMMENTS: Name revived by Palmer (1898: 111). Misspelt as *Koidotus* by Waterhouse (1902: 188). Synonymised within *Macroglossus* by Andersen (1912: 746), Corbet and Hill (1992: 79) and Simmons (2005a: 325).

Rhynchocyon Gistel, 1848: ix.

TYPE SPECIES: Not stated.

COMMENTS: Synonymised within *Macroglossus* by Andersen (1912: 746), Corbet and Hill (1992: 79) and Simmons (2005a: 325).

HOMONYMS:

Rhynchocyon Peters, 1847: 36, elephant shrews of the Class Mammalia (Order Macroscelidea, Family Macroscelididae). Currently recognised genus. See Schlitter (2005: 84).

Carponycteris Lydekker, 1891: xv, 654.

TYPE SPECIES: *Nomen novum* for *Macroglossus* F. Cuvier, 1824 [1821–1825].

COMMENTS: Synonymised within *Macroglossus* by Andersen (1912: 746), Corbet and Hill (1992: 79) and Simmons (2005a: 325).

Odontonycteris Jentink, 1902: 131, 140.

TYPE SPECIES: Φ *Odontonycteris meyeri* Jentink, 1902 [= Φ *Macroglossus minimus lagochilus* Matschie, 1899a] by monotypy.

COMMENTS: Recorded from Australia by Iredale and Troughton (1934: 92). Synonymised within *Macroglossus* by Andersen (1912: 746), Corbet and Hill (1992: 79) and Simmons (2005a: 325).

Macroglossus minimus (É. Geoffroy, 1810)

Northern Blossom-bat

Φ *Macroglossus minimus minimus* (É. Geoffroy, 1810)

Φ *Pteropus minimus* É. Geoffroy, 1810b: 97.

TYPE LOCALITY: Java, Indonesia.

COMMENTS: Transferred to *Macroglossus* by F. Cuvier (1824 [1821–1825]: 248), Lesson (1836: 66) and accepted by most subsequent authors (see Andersen, 1912: 757).

Φ *Pteropus rostratus* Horsfield, 1822: No. III.

TYPE LOCALITY: Java, Indonesia.

COMMENTS: Synonymised within *minimus* by Corbet and Hill (1992: 79) and Simmons (2005a: 325).

Φ *Macroglossa Horsfieldii* Lesson, 1827a: 115.

TYPE LOCALITY: Java, Indonesia.

COMMENTS: Synonymised within *minimus* by Corbet and Hill (1992: 79), Koopman (1993: 154) and Simmons (2005a: 325).

Φ *Macroglossa kiodotes* Lesson, 1827a: 115.

TYPE LOCALITY: Java, Indonesia.

COMMENTS: Synonymised within *minimus* by Corbet and Hill (1992: 79), Koopman (1993: 154) and Simmons (2005a: 325).

Φ *Macroglossus minimus nanus* Matschie, 1899

Φ *[Macroglossus (Macroglossus)] nanus* Matschie, 1899a: viii, 98.

TYPE LOCALITY: Lamellana, Banda Island, New Britain, Bismarck Archipelago, N. Moluccas, Indonesia.

COMMENTS: Early taxonomic history discussed by Andersen (1912: 765), who recognised it as a subspecies of *lagochilus*, which was followed by Thomas (1914b: 317). Recognised as a subspecies within *Odontonycteris lagochilus*, by Iredale and Troughton (1934: 92) and subspecies within *Macroglossus minimus* Flannery (1990: 275). Synonymised within *minimus* by Koopman (1993: 154). Recognition of subspecies of *minimus*, and occurrence in Australia, by Corbet and Hill (1992: 80), Flannery (1995a: 356; 1995b: 211), Reardon (1999a: 11), Bergmans (2001: 128) and Simmons (2005a: 325) and Clayton *et al.* (2006: 108).

Φ *Macroglossus Novaeguineae* Matschie, 1899b: 78.

TYPE LOCALITY: New Guinea.

COMMENTS: *Nomen nudum*. Synonymised within *nanus* by Laurie and Hill (1954: 44) who suggested it was possibly a *lapsus* for *nanus*, and Simmons (2005a: 325).

Φ *Macroglossus lagochilus microtus* Andersen, 1911: 642.

TYPE LOCALITY: Aola, Guadalcanar Island, Solomon Islands.

COMMENTS: Recognised as a subspecies of *lagochilus* by Laurie and Hill (1954: 44). Synonymised within *minimus* by Koopman (1993: 154) and recognised as a subspecies by Flannery (1990: 275; 1995a: 356; 1995b: 211). Synonymised within *nanus* by Simmons (2005a: 325).

Φ *Macroglossus minimus lagochilus* Matschie, 1899

Φ *[Macroglossus (Macroglossus)] lagochilus* Matschie, 1899a: viii, 97.

TYPE LOCALITY: Buru Island, Moluccas, Indonesia.

COMMENTS: Recognised as a distinct species by Andersen (1912: 762), L. Hall and Richards (1979: 19) and Strahan (1983: 290). Synonymised within *minimus* by Lekagul and McNeely (1977: 50), Koopman (1993: 154) and subspecies by Hill (1983: 134), Corbet and Hill (1992: 80), Flannery (1990: 275; 1995a: 356; 1995b: 211), Bergmans (2001: 127) and Simmons (2005a: 325).

Φ *Odontonycteris meyeri* Jentink, 1902: 131, 140.

TYPE LOCALITY: Tabukan, Great Sanghir Island, Indonesia.

COMMENTS: Synonymised within *minimus* by Corbet and Hill (1992: 79), Koopman (1993: 154). Synonym within *lagochilus* by Simmons (2005a: 325).

Φ *Macroglossus fructivorus* E. Taylor, 1934: 125.

TYPE LOCALITY: Tatayan, Cotabato, Mindanao, Philippines.

COMMENTS: Synonymised within *minimus* Heaney and Rabor (1982: 12), Corbet and Hill (1992: 79), Koopman (1993: 154) and subsequent authors. Synonym within *lagochilus* by Simmons (2005a: 325).

Macroglossus minimus pygmaeus Andersen, 1911

Macroglossus lagochilus pygmaeus Andersen, 1911: 642.

TYPE LOCALITY: Maer Island (as Mer), Murray Islands, Torres Strait, north Queensland, Australia.

COMMENTS: Elevated to species rank, within *Odontonycteris*, by Iredale and Troughton (1934: 92) and as a subspecies of *Odontonycteris lagochilis* by Troughton (1967: 268). Subspecies rank within *minimus* recognised by Strahan (1995: 422) and Flannery (1990: 275; 1995a: 356; 1995b: 211). Synonymised within *nanus* by Simmons (2005a: 325) but recognised as the only subspecies to occur in Australia by Van Dyck and Strahan (2008: 427).

FUTURE TAXONOMIC RESEARCH: The relationship of this taxon with *Macroglossus minimus nanus* Matschie, 1899 needs to be assessed to confirm whether this taxon is distinct or synonymous.

Φ *Macroglossus minimus booensis* Kompanje & Moeliker, 2001

Φ *Macroglossus minimus booensis* Kompanje & Moeliker, 2001: 147, 156.

TYPE LOCALITY: Boo Besar, Boo Islands, Western Papuan Islands (Raya Ampat group), Papua (formerly called Irian Jaya), Indonesia. (129°22′10″E, 1°11′25″N).

COMMENTS: Recognised as a subspecies of *minimus* by Simmons (2005a: 325).

Syconycteris Matschie, 1899

Syconycteris Matschie, 1899a: viii, 98.

TYPE SPECIES: *Macroglossus minimus* var. *australis* Peters, 1867b [=*Syconycteris australis* (Peters, 1867b)] by original designation.

COMMENTS: Described as a subgenus of *Macroglossus* F. Cuvier, 1824 [1821–1825]. Elevated to genus rank by Andersen (1911: 641, 642) and followed by subsequent authors. Reviewed by Ziegler (1982: 1) and Hill (1983: 137).

Syconycteris australis (Peters, 1867)

Eastern Blossom-bat

Syconycteris australis australis (Peters, 1867)

Macroglossus minimus var. *australis* Peters, 1867b: 13, footnote.

TYPE LOCALITY: Rockhampton, Queensland, Australia.

COMMENTS: Included within *Macroglossus* by J. Ogilby (1892: 81). Elevated to species rank, within *Syconycteris*, by Iredale and Troughton (1934: 92), Tate (1942a: 346), Laurie and Hill (1954: 45) and Troughton (1967: 268). Species and subspecies reviewed by Koopman (1982: 8). The subspecies were proposed to be only slightly differentiated by Hill (1983: 139) but have been recognised by various authors since, including Simmons (2005a: 349). Taxon reviewed by Kitchener *et al*. (1994a: 485).

Φ *Syconycteris australis crassa* (Thomas, 1895)

Φ *Carponycteris crassa* Thomas, 1895b: 163.

TYPE LOCALITY: Fergusson Island, D'Entrecasteaux Archipelago.

COMMENTS: Recognised at species rank and transferred by *Synonycteris* by Andersen (1912: 775), Tate (1942a: 346), Laurie and Hill (1954: 44), McKean (1972: 9) and Ziegler

(1982: 1). Synonymised within *australis* by Koopman (1979: 8; 1993: 155), Hill (1983: 139) and Flannery (1990: 274; 1995a: 383; 1995b: 308). Subspecies status within *australis* recognised by Koopman (1982: 10), Kitchener *et al.* (1994a: 485) and Simmons (2005a: 349).

Φ *Syconycteris australis papuana* (Matschie, 1899)

Φ *[Macroglossus (Syconycteris)] papuanus* Matschie, 1899a: viii, 99.

TYPE LOCALITY: Andai, north-west Netherlands, Papua New Guinea.

COMMENTS: Recognised as a subspecies within *Syconycteris crassa* by Andersen (1912: 777), McKean (1972: 9), and Laurie and Hill (1954: 44). Synonymised within *australis* by Flannery (1990: 274) and Koopman (1993: 155). Recognised as a subspecies of *australis* by Koopman (1979: 8; 1982: 10), Hill (1983: 137), Kitchener *et al.* (1994a: 485), Flannery (1995a: 383; 1995b: 308), Bergmans (2001: 128) and Simmons (2005a: 349).

Φ *Syconycteris australis finschi* (Matschie, 1899)

Φ *[Macroglossus (Syconycteris)] finschi* Matschie, 1899a: viii, 100.

TYPE LOCALITY: New Pomerania, Bismarck Archipelago, Papua New Guinea.

COMMENTS: Recognised as a subspecies of *crassa* by Laurie and Hill (1954: 45). Synonymised within *papuana* by McKean (1972: 9), and within *australis* by Koopman (1993: 155) and Flannery (1990: 274; 1995a: 383; 1995b: 308). Recognised as a subspecies of *australis* by Koopman (1979: 8), Kitchener *et al.* (1994a: 485) and Simmons (2005a: 349).

Φ *Syconycteris australis keyensis* Andersen, 1911

Φ *Syconycteris keyensis* Andersen, 1911: 643.

TYPE LOCALITY: Key Island [=Kai or Kei Island], Maluku Islands, Indonesia.

COMMENTS: Recognised as a subspecies of *crassa* by Laurie and Hill (1954: 45). Synonymised within *papuana* by McKean (1972: 9) and within *australis* by Koopman (1993: 155) and Flannery (1990: 274; 1995a: 383; 1995b: 308). Recognised as a subspecies of *australis* by Kitchener *et al.* (1994a: 485) and Simmons (2005a: 349).

Φ *Syconycteris australis major* Andersen, 1911

Φ *Syconycteris crassa major* Andersen, 1911: 643.

TYPE LOCALITY: Ambon Island, Moluccas, Indonesia.

COMMENTS: Synonymised within *australis* by Corbet and Hill (1982: 81), Koopman (1993: 155) and Flannery (1990: 274). Recognised as a subspecies of *crassa* by Laurie and Hill (1954: 45), and as a subspecies of *australis* by Hill (1983: 139), Kitchener *et al.* (1994a: 485), Flannery (1995a: 383; 1995b: 308), Bergmans (2001: 129) and Simmons (2005a: 349).

Φ *Syconycteris australis naias* Andersen, 1911

Φ *Syconycteris naias* Andersen, 1911: 643.

TYPE LOCALITY: Woodlark Islands, New Guinea.

COMMENTS: Species recognised as described by Laurie and Hill (1954: 45). Synonymised within *australis* by McKean (1972: 9), Flannery (1990: 274) and Koopman (1993: 155). Suggested to be a synonym of *australis* by (Hill, 1983: 139), who also stated that it might prove to be another weakly separable subspecies. Subspecies rank within *australis* recognised by Lidicker and Ziegler (1968: 34), Koopman (1982: 10), Flannery (1995a: 383; 1995b: 308) and Simmons (2005a: 349).

Subfamily Nyctimeninae Miller, 1907

Subfamily Nyctymeninae Miller, 1907: viii, 75.

TYPE GENUS: *Nyctymene* Bechstein, 1800 [=*Nyctimene* Borkhausen, 1797].

COMMENTS: When originally proposed, this rank was placed in the Family Pteropidae [=Pteropodidae (J. Gray, 1821)] and included the genus *Nyctymene* Bechstein, 1800 [=*Nyctimene* Borkhausen, 1797]. Subfamily rank synonymised within Subtribe Nyctimenina Miller (1907: 75) by McKenna and Bell (1997: 299) and synonymised within Pteropodidae by Simmons (2005a: 313). Subfamily recognised by Simpson (1945: 55), Strahan (1995: 8, 426) and Bergmans (2001: 130), and Van Dyck and Strahan (2008: 10, 430).

Nyctimene Borkhausen, 1797

Nyctimene Borkhausen, 1797: 86.

TYPE SPECIES: Φ *Vespertilio cephalotes* Pallas, 1767: 10 [=Φ *Nyctimene cephalotes* (Pallas, 1767: 10)] by subsequent monotypy.

COMMENTS: Name subsequently recognised by Anon (1798a: 380). Spelling recognised by Thomas (1902c: 198) though the author was not stated. See Mahoney and Walton (1988a: 107) for discussion on validity of name and author. *Nyctimene albiventer* (J. Gray, 1863c: 262) was reported to occur within Australia by L. Hall and Richards (1979: 17), but this appears to have been made in error. Reviewed by J. Smith and Hood (1983: 7) and Bergmans (2001: 130). The species *Nyctimene vizcaccia* Thomas, 1914c: 436 was recorded within the Australian territory on Moa Island, off Cape York Peninsula, by Strahan (1995: 429), although no other authors suggest this species occurs within Australia (e.g. see Simmons, 2005a: 332). The species *Nyctimene cephalotes* (Pallas, 1767: 10) was recognised as occurring in Australia by Trouessart (1878: 207), Matschie (1899a: 83), Churchill (1998: 78), Reardon (1999a: 11), Van Dyck and Strahan (2008: 432), and Van Dyck *et al.* (2013: 120). However, the occurrence of this species on Cape York and the Torres Strait Islands was considered dubious by Churchill (2008: 68, 223), which was followed by Burbidge *et al.* (2014: 22). Genus partly reviewed by Newbound *et al.* (2008: 589).

HOMONYMS:

Nyctimene Morris, 1837: 123, true owls of the Class Aves (Order Strigiformes, Family Strigidae). Genus is a synonym of *Asio* Brisson, 1760: 28, but has been placed within *Aluco* Fleming 1922b: 236. See J. Allen (1908: 288).

Nyctimene Gistel & Bromme, 1847 [1847–1850]: 576, sea urchins of the Phylum Echinodermata (Class Echinoidea, Order Clypeasteroida, Family Clypeasteridae). Name is a synonym of *Clypeaster* Lamarck, 1801: 349. See Kroh (2010).

Nyctimene Thomson, 1857: 314, longicorn beetles of the Class Insecta (Order Coleoptera, Family Cerambycidae). Genus is a synonym of *Nyctimenius* Gressitt 1951: 629. See Heffern (2005: 47).

Nyctimene Heine & Reichenow, 1890: 252, owls of the Class Aves (Order Strigiformes, Family Strigidae). Synonym of the genus *Strix* Linnaeus, 1758: 92.

Nyctymene Bechstein, 1800: 615, 627–628, 736.

TYPE SPECIES: Unjustified emendation of *Nyctimene* Borkhausen, 1797.

COMMENTS: Taxon recognised by Miller (1907: 75), but synonymised within *Nyctimene* by Iredale and Troughton (1934: 91) and Mahoney and Walton (1988a: 107).

Cephalotes É. Geoffroy, 1810b: 104.

TYPE SPECIES: Φ *Vespertilio cephalotes* Pallas, 1767: 10 [= Φ *Nyctimene cephalotes* (Pallas, 1767: 10)] by absolute tautonymy.

COMMENTS: New genus name proposed for Φ *Vespertilio cephalotes* Pallas, 1767: 10. Recorded from Australia by Trouessart (1899). Synonymised within *Nyctimene* by Iredale and Troughton (1934: 91), Mahoney and Walton (1988a: 107), Corbet and Hill (1992: 77) and Simmons (2005a: 329).

HOMONYMS:

Cephalotes Latreille, 1802: 357, ants of the Class Insecta (Order Hymenoptera, Family Formicidae). Genus is currently recognised.

Cephalotes Bonelli, 1810: Tabula Synoptica, ground beetles of the Class Insecta (Order Coleoptera, Family Carabidae). Genus is a synonym of *Broscus* Panzer, 1813: 62. See Roig-Juñent (2000: 22).

Harpyia Illiger, 1811: 118.

TYPE SPECIES: Φ *Vespertilio cephalotes* Pallas, 1767: 10 [= Φ *Nyctimene cephalotes* (Pallas, 1767: 10)] by monotypy.

COMMENTS: Name misspelt *Harpyja* by Gloger (1841: xxviii, 49). Synonymised within *Nyctimene* by Iredale and Troughton (1934: 91), Mahoney and Walton (1988a: 107), Corbet and Hill (1992: 77) and Simmons (2005a: 329).

HOMONYMS:

Harpyia Ochsenheimer, 1810: 19, moths of the Class Insecta (Order Lepidoptera, Family Notodontidae). Genus is currently recognised.

Harpyia G. Cuvier, 1816a: 317, the Harpy Eagle of the Class Aves (Order Falconiformes, Family Accipitridae). Genus is a synonym of *Harpia* Vieillot, 1816: 24.

Harpyia Agassiz, 1846: 173, the Harpy Eagle of the Class Aves (Order Falconiformes, Family Accipitridae). Emendation pro of *Harpia* Vieillot, 1816: 24.

Gelasinus Temminck, 1837: 100.

TYPE SPECIES: Φ *Cephalotes pallasii* É. Geoffroy, 1810b: 107 (as *Harpyia pallasii*) [= Φ *Nyctimene cephalotes* (Pallas, 1767: 10)] by monotypy.

COMMENTS: Pre-occupied by *Gelasinus* Van der Hoeven, 1827: 446 (Crustacea). Synonymised within *Nyctimene* by Iredale and Troughton (1934: 91), Mahoney and Walton (1988a: 108), Corbet and Hill (1992: 77) and Simmons (2005a: 329).

HOMONYMS:

Gelasinus Van der Hoeven, 1827: 446, crabs of the Subphylum Crustacea (Order Decapoda, Family Ocypodidae). Error pro *Gelasimus* Latreille, 1817: 517. *Gelasimus* is a synonym of *Uca* Leach, 1814: 430. See Davie and Türkay (2011).

Uronycteris J. Gray, 1863c: 262.

TYPE SPECIES: Φ *Cynopterus (Uronycteris) albiventer* J. Gray, 1863c: 262 [=Φ *Nyctimene albiventer* (J. Gray, 1863c: 262)] by monotypy.

COMMENTS: Originally proposed as a subgenus of *Cynopterus* F. Cuvier, 1824 [1821–1825]: 248. Recorded from Australia by J. Ogilby (1892: 80). Synonymised within *Nyctimene* by Mahoney and Walton (1988a: 108), Corbet and Hill (1992: 77) and Simmons (2005a: 329).

Bdelygma Matschie, 1899a: viii, 84.

TYPE SPECIES: Φ *Harpyia major* Dobson, 1877a: 117 [=Φ *Nyctimene major* (Dobson, 1877a: 117)] by original designation.

COMMENTS: Described as a subgenus of *Gelasinus*. Synonymised within *Nyctimene* by Corbet and Hill (1992: 77) and Simmons (2005a: 329).

Nyctimene robinsoni Thomas, 1904

Eastern Tube-nosed Bat

Nyctimene Robinsoni Thomas, 1904e: 196.

TYPE LOCALITY: Cooktown, Queensland, Australia.

COMMENTS: Species recognised by Iredale and Troughton (1934: 91) and subsequent authors. No subspecies were recognised by Van Dyck and Strahan (2008: 431), although they suggested that recent DNA studies indicate a second species of *Nyctimene* in north eastern Queensland.

Nyctimene tryoni Longman, 1921: 179.

TYPE LOCALITY: Canungra, Queensland, Australia.

COMMENTS: Synonymised within *robinsoni* by Iredale and Troughton (1934: 92), Strahan (1983: 286; 1995: 428) and Simmons (2005a: 331).

Subfamily Pteropodinae J. Gray, 1821

Family Pteropidae J. Gray, 1821: 299.

TYPE GENUS: *Pteropus* Brisson, 1762.

COMMENTS: When originally proposed, this rank was placed in the Order Fructivorae (J. Gray, 1821 [=Pteropodidae (J. Gray, 1821)]) and included the genera *Pteropus* Brisson, 1762; and *Rousettus* J. Gray, 1821: 299. Subfamily name spelt Pteropinae by Simpson (1945: 54). Synonymised within the Subfamily Pteropinae by Simpson (1945: 54). Subfamily rank recognised by Simpson (1945: 54), Koopman and Jones (1970: 23), Koopman (1984a: 151) and Bergmans (2001: 120).

Tribe Pteropina J. Gray, 1825a: 338.

TYPE GENUS: *Pteropus* Brisson, 1762.

COMMENTS: When originally proposed, this rank was placed in the Family Vespertilionidae (J. Gray, 1821) and included the genera *Pteropus* Brisson, 1762; *Cynopterus* F. Cuvier, 1824 [1821–1825]: 248; *Macroglossum* [sic=*Macroglossus*] F. Cuvier, 1824 [1821–1825]; *Cephalotis* [sic=*Cephalotes*] É. Geoffroy, 1810b [=*Nyctimene* Borkhausen, 1797]; and *Harpyia* Illiger, 1811 [=*Nyctimene* Borkhausen, 1797]. Name also recognised at subfamily rank by J. Gray (1825b: 243). Synonymised within Pteropodini J. Gray, 1821 by McKenna and Bell (1997: 296).

Tribe Pteropodina Bonaparte, 1831: 6.

TYPE GENUS: *Pteropus* Brisson, 1762.

COMMENTS: When originally proposed, this rank was placed in the Family Vespertilionidae (Gray, 1821). Recognised as a subtribe by Koopman and Jones (1970: 23).

Family Pteropodae Bonaparte, 1835 [1832–1841]: Fascicolo 14 under *Dysopes cestonii*, second of four unnumbered pages.

TYPE GENUS: *Pteropus* Brisson, 1762.

COMMENTS: When originally proposed, this rank was placed in the Order Chiroptera (Blumenbach, 1779). Current spelling of the family group names and has been recognised by subsequent authors.

Family Pteropodidae Bonaparte, 1838: 112.

TYPE GENUS: *Pteropus* Brisson, 1762.

COMMENTS: When originally proposed, this rank was placed in the Order Chiroptera (Blumenbach, 1779) and included the Tribe Pteropodina (Bonaparte, 1831 [=Pteropodinae (J. Gray, 1821)]). Current spelling of the family group names and has been recognised by subsequent authors.

Subfamily Pteropodinae Flower & Lydekker, 1891: 650.

TYPE GENUS: *Pteropus* Brisson, 1762.

COMMENTS: When originally proposed, this rank was placed in the Family Pteropodidae (J. Gray, 1821) and included the genera *Epomophorus* E. Bennett, 1836: 149; *Pteropus* Brisson, 1762; *Xantharpyia* J. Gray, 1843a: xix, 37 [=*Rousettus* J. Gray, 1821: 299]; *Boneia* Jentink, 1879: 117 [=*Rousettus* J. Gray, 1821: 299]; *Cynopterus* F. Cuvier, 1824 [1821–1825]: 248; *Harpyia* Illiger, 1811 [=*Nyctimene* Borkhausen, 1797]; *Cephalotes* É. Geoffroy, 1810b [=*Nyctimene* Borkhausen, 1797]; and *Pteralopex* Thomas, 1888c: 155. Subfamily rank recognised by Koopman and Jones (1970: 23). Synonymised within the Subfamily Pteropodinae by McKenna and Bell (1997: 296).

Subfamily Pteropinae Trouessart, 1897: v, 77.

TYPE GENUS: *Pteropus* Brisson, 1762.

COMMENTS: When originally proposed, this rank was placed in the Family Pteropidae [=Pteropodidae (J. Gray, 1821)] and included the genera *Pteropus* Brisson, 1762; *Pteralopex* Thomas, 1888c: 155; *Cynonycteris* Peters, 1852b: 25 [= *Rousettus* J. Gray, 1821: 299]; *Harpyionycteris* Thomas, 1896b: 243; *Cynopterus* F. Cuvier, 1824 [1821–1825]: 248; *Scotonycteris* Matschie, 1894: 200; *Harpyia* Illiger, 1811 [=*Nyctimene* Borkhausen, 1797]; *Cephalotes* É. Geoffroy, 1810b [=*Nyctimene* Borkhausen, 1797]; *Epomophorus* E. Bennett, 1836: 149; *Hypsignathus* H. Allen, 1861: 156; and *Leiponyx* Jentink, 1881: 60 [=*Eidolon* Rafinesque, 1815: 54]. Spelling an author recognised by Simpson (1945: 54).

Subfamily Harpyionycterinae Miller, 1907: viii, 77.

TYPE GENUS: *Harpyionycteris* Thomas, 1896b: 243.

COMMENTS: When originally proposed, this rank was placed in the Family Pteropidae [=Pteropodidae (J. Gray, 1821)] and included the genus *Harpyionycteris* Thomas, 1896b: 243. Synonymised within the Subfamily Pteropodinae by McKenna and Bell (1997: 296).

Tribe Pteropini Koopman & Jones, 1970: 23.

TYPE GENUS: *Pteropus* Brisson, 1762.

COMMENTS: When originally proposed, this rank was placed in the Subfamily Pteropodinae (J. Gray, 1821) and included the subtribes Rousettina (=Koopman & Jones, 1970: 23) (containing the genera *Eidolon* Rafinesque, 1815: 54; *Rousettus* J. Gray, 1821: 299; *Myonycteris* Matschie, 1899a: 61, 63; and *Boneia* Jentink, 1879: 117 [=*Rousettus* J. Gray, 1821: 299]) and Pteropodina (Bonaparte, 1831 [=Pteropodinae (J. Gray, 1821)]) (containing the genera *Pteropus* Brisson, 1762; *Acerodon* Jourdan, 1837b: 156; *Neopteryx* Hayman, 1946: 569; *Pteralopex* Thomas, 1888c: 155; and *Styloctenium* Matschie, 1899a: 33). Synonymised within Pteropodini J. Gray, 1821 by McKenna and Bell (1997: 296).

Subtribe Dobsoniina Koopman & Jones, 1970: 23.

TYPE GENUS: *Dobsonia* Palmer, 1898.

COMMENTS: When originally proposed, this rank was placed in the Subfamily Pteropodinae (J. Gray, 1821) and included the genus *Dobsonia* Palmer, 1898. Recognised at subtribe rank by McKenna and Bell (1997: 297).

Subtribe Nyctimenina Koopman & Jones, 1970: 23.

TYPE GENUS: *Nyctimene* Borkhausen, 1797.

COMMENTS: When originally proposed, this rank was placed in the Tribe Cynopterini (J. Gray, 1866d: 64) and included the genera *Nyctimene* Borkhausen, 1797; and *Paranyctimene* Tate, 1942b: 1. Synonymised within Nyctimenina Miller (1907: 75) by McKenna and Bell (1997: 299).

Dobsonia Palmer, 1898

Dobsonia Palmer, 1898: 114.

TYPE SPECIES: Φ *Cephalotes peronii* É. Geoffroy, 1810b: 104 [= Φ *Dobsonia peronii* (É. Geoffroy, 1810b: 104)] by monotypy.

COMMENTS: Name adopted by Thomas (1902c: 198). Reviewed by Miller (1907: 63), de Jong and Bergmans (1981: 209) and Koopman (1994: 28).

Hypoderma É. Geoffroy, 1828: 28.

TYPE SPECIES: Φ *Pteropus paliatus* É. Geoffroy, 1810b: 99 (as *Pteropus palliatus*) [= Φ *Dobsonia peronii* (É. Geoffroy, 1810b: 104)] by monotypy.

COMMENTS: Species also described by I. Geoffroy (1828: 706). Publication date and author established from Andersen (1912: 448). Synonymised within *Dobsonia* by Andersen (1912: 448), Corbet and Hill (1992: 68) and Simmons (2005a: 318).

HOMONYMS:

Hypoderma Latreille, 1818: 272, bot flies of the Class Insecta (Order Diptera, Family Oestridae). Currently recognised genus.

Pteronotus Rafinesque, 1815: 54.

TYPE SPECIES: *Nomen novum* for *Pteropus* Brisson, 1762. Apparently based upon Φ *Pteropus paliatus* É. Geoffroy, 1810b: 99 [= Φ *Dobsonia peronii* (É. Geoffroy, 1810b: 104)]. See Andersen (1912: 448) and Corbet and Hill (1992: 68).

COMMENTS: *Nomen nudum*. Synonymised within *Dobsonia* by Andersen (1912: 448), Corbet and Hill (1992: 68) and Simmons (2005a: 318).

HOMONYMS:

Pteronotus J. Gray, 1838c: 500, naked-backed bats of the Class Mammalia (Order Chiroptera, Family Mormoopidae). Currently recognised genus. Name discussed by Gill (1901: 177). See also Simmons (2005a: 427).

Pteronotus Swainson, 1833: Text to Plate 122, murex snails or rock snails of the Phylum Mollusca (Order Neogastropoda, Family Muricidae). Incorrect subsequent spelling of *Pterynotus* Swainson, 1833: Text to Plate 100. See Houart (2010).

Pteronotus Swainson, 1839: 190, 309, catfishes of the Superclass Pisces (Order Siluriformes, Family Heptapteridae). Genus is a synonym of *Rhamdia* Bleeker, 1858: 197, 244. See Ferraris (2007: 197).

Pteronotus Ranzani, 1839: 234, chameleons of the Class Reptilia (Order Squamata, Family Chamaeleonidae).

Tribonophorus Burnett, 1829: 269

TYPE SPECIES: Φ *Tribonophorus desmarestii* Burnett, 1829: 269 (*nomen nudum*) [= Φ *Dobsonia peronii* (Geoffroy, 1810b: 104)] by monotypy.

COMMENTS: *Nomen nudum*. Synonymised within *Dobsonia* by Andersen (1912: 448), Corbet and Hill (1992: 68), and Simmons (2005a: 318).

Dobsonia magna Thomas, 1905

Bare-backed Fruit-bat

Dobsonia magna Thomas, 1905a: 423.

TYPE LOCALITY: Tamata, Papua New Guinea.

COMMENTS: Recognised as a subspecies of *Dobsonia moluccensis* (Quoy & Gaimard, 1830: 86) by Tate (1942a: 338), Laurie and Hill (1954: 42), McKean (1972: 7), Hill (1983: 114), Koopman (1979: 6; 1984a: 4) and Corbet and Hill (1992: 69). Recognised as occurring in Australia by G. Allen (1935: 151) and as a subspecies of *moluccensis* within Australia by Strahan (1983: 284; 1995: 430), Flannery (1990: 269) and Reardon (1999a: 11). Synonymised within *moluccensis* by Ride (1970: 242) (as *moluccense*), Mahoney and Walton (1988a: 106) and Koopman (1993: 140). Species rank recognised by Andersen (1912: 466), Troughton (1967: 266), Bergmans and Sarbini (1985: 183, 187), who raised concern over the validity of this taxon, Flannery (1995a: 351; 1995b: 200), Churchill (1998: 72), Simmons (2005a:

319), Clayton *et al.* (2006: 108), and Van Dyck and Strahan (2008: 433).

Pteropus Brisson, 1762

Pteropus Brisson, 1762: 13, 153.

TYPE SPECIES: Φ *Vespertilio vampyrus niger* Kerr, 1792: 90 [= Φ *Pteropus niger* (Kerr, 1792: 90) by designation under the plenary powers of the International Commission of Zoological Nomenclature.

COMMENTS: The name *Pteropus* Brisson, 1762 was identified as the original publication for *Pteropus* by Andersen (1912: 61, 77–78). Name has been considered not available because the names were published in a work that is not consistently binomial, with preference being given to Erxleben (1777: 130), by authors including Hopwood (1947: 535), Delson and Napier (1976: 47), Honacki *et al.* (1982: 120) and McKenna and Bell (1997: 297). Brisson is however given preference by Corbet and Hill (1992: 57), and the generic name *Pteropus* as described by Brisson was conserved by the Opinion 1894 of the ICZN (1998: 65) and added to the Official List of Generic Names (J. Smith, 2001: 67). Brisson's work was discussed by Merriam (1895: 375). Brisson recognised as the author by most recent authors including Simmons (2005a: 334). Genus reviewed by Andersen (1912: 61) and L. Hall (1987: 75).

HOMONYMS:

Pteropus Thunberg, 1815: 219, leaf insects of the Class Insecta (Order Phasmatodea, Family Phylliidae). Genus is a synonym of *Phyllium* Illiger, 1798: 499.

Pteropus Jennings, 1828: 390, finfoot birds of Class Aves (Order Gruiformes, Family Heliornithidae). Genus is a synonym of *Podica* Lesson, 1831: 596. See Mathews and Iredale (1922: 175) and Richmond (1927: 31).

Pteropus Canestrini and Fanzago, 1878: 2, mites of the Class Arachnida (Order Mesostigmata, Family Spinturnicidae). Genus is a synonym of *Spinturnix* Heyden, 1826: 612. See Radford (1943: 59).

Spectrum Lacépède, 1799a: 15.

TYPE SPECIES: Φ *Vespertilio vampyrus* Linnaeus, 1758: 31 (as *Spectrum vampirus* [sic]) [= Φ *Pteropus vampyrus* (Linnaeus, 1758: 31)] by monotypy.

COMMENTS: Andersen (1912: 61) identified *Pteropus niger* Kerr, 1792: 90 as the type species. Genus recognised by J. Gray (1870a: 99, 100) but recognised as a subgenus of *Pteropus* by Matschie (1899a: 19). Synonymised within *Pteropus* by J. Gray (1843a: xix), Miller (1907: 56), Andersen (1912: 61), Simpson (1945: 54), Corbet and Hill (1992: 57) and subsequent authors.

HOMONYMS:

Spectrum Scopoli, 1777: 413, moths of the Class Insecta (Order Lepidoptera, Family Sphingidae). Genus is a synonym of *Sphinx* Linnaeus, 1758: 489.

Spectrum Stoll, 1787: 3, stick insects of the Class Insecta (Order Orthoptera, Family Phasmatidae). Genus is a synonym of *Phasma* A. Lichtenstein, 1796: 77. See Uvarov (1940: 378).

Eunycteris J. Gray, 1866d: 64.

TYPE SPECIES: Φ *Pteropus phaiops* Temminck, 1825: 178 [= Φ *Pteropus rufus* É. Geoffroy, 1803c: 47] by monotypy. See Andersen (1912: 205) and Simmons (2005a: 334, 343).

COMMENTS: Recognised as a subgenus of *Pteropus* by Matschie (1899a: 7, 11). Synonymised within *Pteropus* by Miller (1907: 56), Andersen (1912: 61), Simpson (1945: 54), Corbet and Hill (1992: 57) and subsequent authors.

Pselaphon J. Gray, 1870a: 110.

TYPE SPECIES: Φ *Pteropus pselaphon* Lay, 1829: 457 by monotypy.

COMMENTS: Synonymised within *Pteropus* by Miller (1907: 56), Andersen (1912: 61), Corbet and Hill (1992: 57) and Simmons (2005a: 334).

HOMONYMS:

Pselaphus [=*Pselaphon*] Herbst, 1792: v, 106, beetles of the Class Insecta (Order Coleoptera, Family Psephalidae). Currently recognised genus. See Newton and Chandler (1989: 63).

Sericonycteris Matschie, 1899a: vii, 30.

TYPE SPECIES: Φ *Pteropus rubricollis* É. Geoffroy, 1810b: 93 [= Φ *Pteropus subniger* (Kerr, 1792: 91) by subsequent designation. See Palmer (1904: 629).

COMMENTS: Proposed as a subgenus of *Pteropus*. Synonymised within *Pteropus* by Miller (1907: 56), Andersen (1912: 61), Simpson (1945: 54) and Corbet and Hill (1992: 57).

Desmaplex Miller, 1907: vii, 60.

TYPE SPECIES: Φ *Pteropus leucopterus* Temminck, 1853: 60 by original designation.

COMMENTS: Synonymised within *Pteropus* by Andersen (1912: 61), Simpson (1945: 54), Corbet and Hill (1992: 57) and subsequent authors.

Pteropus alecto Temminck, 1837

Black Fruit-bat

Φ ***Pteropus alecto alecto*** Temminck, 1837

Φ *Pteropus alecto* Temminck, 1837: 75.

TYPE LOCALITY: Menado District, northern Sulawesi, Indonesia.

COMMENTS: There are records by Webb and Tidemann (1995: 19) of *P. alecto* hybridising with *P. poliocephalus*. Has been reported from mainland New Guinea on only two occasions in Western Province in southern New Guinea (Tate, 1942a: 337; Waithman, 1979: 321).

Φ *Pteropus nicobaricus* Heude, 1897: 176, footnote; Plate 5, Fig. 5.

TYPE LOCALITY: Macassar, S Sulawesi, Indonesia.

COMMENTS: Synonymised within *alecto* by Corbet and Hill (1992: 64), Flannery (1990: 264; 1995a: 370; 1995b: 247) and Simmons (2005a: 335).

HOMONYMS:

Pteropus nicobaricus Fitzinger, 1861, the Black-eared Fruit-bat of the Class Mammalia (Order Chiroptera, Family Pteropodidae). Taxon is a *nomen nudum* and synonym of *Pteropus melanotus* Blyth, 1863. See individual entry.

Pteropus nicobaricus Zelebor, 1869, the Christmas Island Fruit-bat of the Class Mammalia (Order Chiroptera, Family Pteropodidae). Taxon is a synonym of *Pteropus melanotus* Blyth, 1863. See individual entry.

Pteropus alecto gouldii Peters, 1867

Pteropus Gouldii Peters, 1867c: 703.

TYPE LOCALITY: Rockhampton, Queensland, Australia.

COMMENTS: Recognised as a species by Iredale and Troughton (1934: 91) and Troughton (1967: 264). Synonymised within *P. alecto* by Ride (1970: 247), Honacki *et al.* (1982: 120), Mahoney and Walton (1988a: 109) and Koopman (1993: 146). Recognised as a subspecies of *alecto* by Tate (1942a: 337), Corbet and Hill (1992: 64), Flannery (1990: 264; 1995a: 370; 1995b: 247), Reardon (1999a: 11), Simmons (2005a: 335), Clayton *et al.* (2006: 108), and Van Dyck and Strahan (2008: 436).

Pteropus banakrisi G. Richards & Hall, 2002: 69.

TYPE LOCALITY: St. Paul's Mission, east coast of Moa Island, Torres Strait. (10°11′S, 142°20′E)

COMMENTS: The true identity of this taxon appears to have been in doubt since its description. Specimens of this 'species' were subsequently reviewed by Helgen (2004: 1) who clearly found that this species is not valid as the type series consisted only of subadult individuals of *Pteropus alecto*, and *banakrisi* should be considered a junior synonym of *P. a. gouldii*. Despite this, the species rank was allocated to *banakrisi* by Simmons (2005a: 335), but she suggested that it might be conspecific with *alecto*, citing personal communication with Helgen. Subsequently the taxon has not been recognised by either Clayton *et al.* (2006: 108), Churchill (2008: 222) or Van Dyck and Strahan (2008: 436).

Φ ***Pteropus alecto aterrimus*** Matschie, 1899

Φ *Pt. [eropus] aterrimus* Matschie, 1899a: 17.

TYPE LOCALITY: Bawean Island, Kangean Islands, Indonesia.

COMMENTS: Subspecies rank recognised by Corbet and Hill (1992: 64), Flannery (1995b: 247) and Simmons (2005a: 335). Not recognised by Flannery (1995a: 370).

Φ *Pteropus aterrimus* Temminck, 1846: 333.

TYPE LOCALITY: Bawean Island, Kangean Islands, Indonesia.

COMMENTS: *Nomen nudum*. Subspecies rank within *alecto* recorded by Corbet and Hill (1992: 64). Synonymised within *alecto* by Koopman (1993: 146) and within *aterrimus* by Simmons (2005a: 335).

Φ *Pteropus baveanus* Miller, 1906a: 63.

TYPE LOCALITY: Bawean Island, Java Sea, Indonesia.

COMMENTS: Synonymised within *alecto* by Corbet and Hill (1992: 64) and Koopman (1993: 146) and within *aterrimus* by Simmons (2005a: 335).

Φ ***Pteropus alecto morio*** Andersen, 1908

Φ *Pteropus morio* Andersen, 1908: 369.

TYPE LOCALITY: Waingapo, Sumba, Indonesia.

COMMENTS: Synonymised within *alecto* by Koopman (1993: 146). Recognised as a subspecies of *alecto* by Laurie and Hill (1954: 37), Corbet and Hill (1992: 64), Flannery (1990: 264; 1995a: 370; 1995b: 247) and Simmons (2005a: 335).

† ***Pteropus brunneus*** Dobson, 1878

Percy Island Fruit-bat

† *Pteropus brunneus* Dobson, 1878: 37; Plate 3, Fig 4.

TYPE LOCALITY: Percy Island, off east central Queensland, Australia.

COMMENTS: This species is known only from the type specimen. The taxonomic status of *brunneus* remains unclear, though appears to be distinct (e.g. Koopman, 1984a: 2). Andersen (1912: xiv, 149) retained this form as a species and commented that it is an Australian representative of *Pteropus hypomelanus* Temminck, 1853: 61. Tate (1952a: 611) considered *P. brunneus* to be not truly Australian and was unable to allocate it with certainty to any described form. Synonymised within *hypomelanus* by Ride (1970: 247). L. Hall and Richards (1979: 13) and Corbet and Hill (1980: 36; 1986: 42) believed that it might be a vagrant *Pteropus hypomelanus*, while Koopman (1984a: 3) said that it was clearly distinct from any Australian *Pteropus* and placed *P. brunneus* as a member of the *subniger* group

[=*hypomelanus*] of Andersen (1912: 101). Species rank recognised by Simmons (2005a: 335), who noted however that it is not clear that this taxon represents a valid species.

Pteropus conspicillatus Gould, 1850

Spectacled Fruit-bat

Pteropus conspicillatus conspicillatus Gould, 1850

Pteropus conspicillatus Gould, 1850: 109.

TYPE LOCALITY: Fitzroy Island, Queensland, Australia.

COMMENTS: Further described by Gould (1851 [1845–1863]: Text to Plate 29). This species also occurs within New Guinea and several offshore islands (Flannery, 1995a: 371; Helgen, 2007a: 728).

Φ *Pteropus conspicillatus chrysauchen* Peters, 1862

Φ *Pteropus chrysauchen* Peters, 1862: 576, footnote.

TYPE LOCALITY: Batchian Island [=Bacan Island], Moluccas, Indonesia.

COMMENTS: Synonymised within *conspicillatus* by Koopman (1993: 147). Recognised as a subspecies of *conspicillatus* by Laurie and Hill (1954: 38), Tate (1942a: 337), Corbet and Hill (1992: 65), Flannery (1990: 265; 1995a: 372; 1995b: 257), Bergmans (2001: 125) and Simmons (2005a: 336). Subspecies also occurs off the north-eastern coast of New Guinea (S. Fox, 2011: 137).

Φ *Pteropus mysolensis* J. Gray, 1870a: 105.

TYPE LOCALITY: Mysol Island, off western New Guinea, Indonesia.

COMMENTS: Synonymised within *chrysauchen* by Laurie and Hill (1954: 38), and within *conspicillatus* by Koopman (1993: 147), Flannery (1990: 265; 1995a: 372; 1995b: 257) and Simmons (2005a: 336).

Pteropus macrotis Peters, 1867

Large-eared Fruit-bat

Φ *Pteropus macrotis macrotis* Peters, 1867

Φ *Pteropus macrotis* Peters, 1867d: 327.

TYPE LOCALITY: Wokam Island, Aru Islands, Indonesia.

Φ *Pteropus insignis* Rosenberg, 1867: 31.

TYPE LOCALITY: Zoogdieren der Aroe-Eilanden [=Aru Islands].

COMMENTS: Not recognised by Flannery (1995a: 375; 1995b: 264). Synonymised within *macrotis* by Simmons (2005a: 340).

Pteropus macrotis epularius Ramsay, 1877

Pteropus (Epomops?) epularius Ramsay, 1877b: 8.

TYPE LOCALITY: Katau, south New Guinea.

COMMENTS: Proposed to occur in the Torres Strait in the Boigu Island group by L. Hall and Richards (1991: 22). Recognised as a subspecies of *macrotis* by McKean (1972: 3), Flannery (1990: 268; 1995a: 375; 1995b: 264) and Strahan (1995: 439). Synonymised within *macrotis* by Churchill (1998: 86). A review of vouchered specimens by Helgen (2004: 1) of *P. macrotis* from Australia found they actually represent misidentified *P. scapulatus*. As a result of his findings Helgen (2004: 1) proposed that this taxon should be removed from the list of mammals known from Australia, which was followed by Churchill (2008: 223). Most recently Van Dyck and Strahan (2008: 441) discussed the presence of this species within Australia and suggested that photographic and general opinion by bat researchers does support its presence in Australia. The inclusion of this species within Australia and its surrounding islands is done here with caution but further research is required.

FUTURE TAXONOMIC RESEARCH: The presence of this taxon on Saibai and Boigu Islands in Torres Strait, which are Australian territories almost adjacent to the Papua New Guinea mainland, has not been confirmed but there is a good chance they may occasionally visit these islands. The presence of *Pteropus macrotis* in Australia, and its subspecies status, needs to be confirmed with the collection of voucher specimens.

Pteropus natalis Thomas, 1887

Christmas Island Fruit-bat

Pteropus natalis Thomas, 1887d: 511; Plate 41.

TYPE LOCALITY: Christmas Island, Indian Ocean, Australia.

COMMENTS: Synonymised within *melanotus* by Tidemann (1987a: 89), Corbet and Hill (1992: 62), Koopman (1993: 149) and Churchill (1998: 88). Recognised as a valid species by Andersen (1912: 93, 233), James *et al.* (2007: ii, 2), Churchill (2008: 74), Burbidge *et al.* (2014: 22, 30) and a subspecies of *melanotis* by Chasen (1940: 29), Reardon (1999a: 11), Simmons (2005a: 341), Clayton *et al.* (2006: 108), and Van Dyck and Strahan (2008, 443).

FUTURE TAXONOMIC RESEARCH: The status of this taxon, whether it is a true subspecies of extralimital *P. melanotus* or a distinct species, needs to be resolved.

Pteropus poliocephalus Temminck, 1825

Grey-headed Fruit-bat

Pteropus poliocephalus Temminck, 1825: 179.

TYPE LOCALITY: Eastern Australia.

COMMENTS: Name has been stable since its description.

Pteropus scapulatus Peters, 1862

Little Red Fruit-bat

Pteropus scapulatus Peters, 1862: 574.

TYPE LOCALITY: Cape York, north Queensland, Australia.

COMMENTS: Recognised within *Pteropus* by Andersen (1912: 403).

Pteropus elseyii J. Gray, 1866d: 67.

TYPE LOCALITY: Claremont Island, Queensland, Australia.

COMMENTS: Synonymised within *scapulatus* by Iredale and Troughton (1934: 91), who gave the author as Gould, and subsequent authors.

Superfamily Rhinolophoidea J. Gray, 1825

Tribe Rhinolophina J. Gray, 1825a: 338.

COMMENTS: When originally proposed, this rank was placed in the Family Vespertilionidae (J. Gray, 1821) and included the genera *Megaderma* É. Geoffroy, 1810a: 197; *Rhinolophus* Lacépède, 1799a; *Nycteris* É. Geoffroy & G. Cuvier, 1795: 186; *Mormoops* Leach, 1821a: 76; and *Nyctophilus* Leach, 1821a. Recognised at the superfamily rank by Weber (1928: 152), Simpson (1945: 56) and more recently by Simmons (1998: 12), Simmons and Geisler (1998: 136), Teeling *et al.* (2002: 1432; 2003: 309) and Miller-Butterworth *et al.* (2007: 1556, 1558).

Family Megadermatidae H. Allen, 1864

Family Megadermatidae H. Allen, 1864: iii, 1.

TYPE GENUS: *Megaderma* É. Geoffroy, 1810a: 197.

COMMENTS: When originally proposed, the higher rank not listed but included the genera *Macrotus* J. Gray, 1843b: 21; and *Megaderma* É. Geoffroy, 1810a: 197 (though not listed). Name used by authors including Simpson (1945: 56), Ellerman and Morrison-Scott (1951: 107) and Koopman and Cockrum (1967: 125) except several have used the spelling Megadermidae (see below). See Handley (1980: 10) for the correct formation of the family name. No subfamilies are currently recognised (Simmons, 2005: 379).

Family Megadermata Peters, 1865a: 256.

TYPE GENUS: *Megaderma* É. Geoffroy, 1810a: 197.

COMMENTS: When originally proposed, this rank was placed in the 'Chiropteren' and included the genera *Rhinopoma* É. Geoffroy, 1818: 113; *Megaderma* É. Geoffroy, 1810a: 197; *Nycteris* É. Geoffroy & G. Cuvier, 1795: 186; and *Nyctophilus* Leach, 1821a. Synonymised within Megadermidae [=Megadermatidae] by Miller (1907: 102) and within Megadermatidae by McKenna and Bell (1997: 304).

Family Megadermidae Gill, 1872: 17.

TYPE GENUS: *Megaderma* É. Geoffroy, 1810a: 197.

COMMENTS: When originally proposed, this rank was placed in the Suborder Animalivora (Gill, 1872 [=Chiroptera (Blumenbach, 1779 part)]) and included the subfamilies Vampyrinae (Peters, 1865b: 503 [=Family Phyllostomidae (Gray, 1825b: 242)]), Glossophaginae (Bonaparte, 1845: 5) and Stenoderminae (Gill, 1872: 17). Spelling utilised by Dobson (1878: 154) (as the Subfamily Megaderminae), Miller (1907: viii, 101), Cabrera (1922: 189) and Chasen (1940: iii, 34) but not by other authors as described under Megadermatidae. Correct spelling reviewed by Handley (1980 10) who concluded that the correct spelling for the family name is Megadermatidae. Synonymised within the Family Megadermatidae by Simpson (1945: 56) and McKenna and Bell (1997: 304).

Tribe? Megadermatini Winge, 1893b: 24.

TYPE GENUS: *Megaderma* É. Geoffroy, 1810a: 197.

COMMENTS: When originally proposed, this rank was placed in the Family Rhinolophidae (J. Gray, 1825a) and included the genera *Nycteris* É. Geoffroy and G. Cuvier, 1795: 186 and *Megaderma* É. Geoffroy, 1810a: 197.

Macroderma Miller, 1906

Macroderma Miller, 1906b: 84.

TYPE SPECIES: *Megaderma gigas* Dobson, 1880 by original designation.

COMMENTS: Genus recognised since its description.

Macroderma gigas (Dobson, 1880)

Ghost Bat

Megaderma gigas Dobson, 1880: 461; Plate 46.

TYPE LOCALITY: 'Mount Margaret', Wilson River, Queensland, Australia.

COMMENTS: Type locality discussed by Nelson (1988: 70). Taxonomic decision of Mahoney in Mahoney and Walton (1988b: 119) based on Koopman (1984a: 8) who argued that as no other characters correlate with the colour differences between *gigas* and *saturata* he could not recognise any subspecies. Species reviewed by Hudson and Wilson (1986: 1). The possible occurrence of this species in New Guinea was discussed by Filewood (1983: 35), but there are no known vouchered records to date (Helgen, 2007a: 721). Substantial genetic divergence has been observed between subpopulations by Worthington Wilmer *et al.* (1999: 1582) so the taxonomy of this species appears to need revisiting.

FUTURE TAXONOMIC RESEARCH: Surveys are needed within the Trans-Fly region of southern New Guinea to determine if this species occurs there. The genetic divergence observed between subpopulations by Worthington Wilmer *et al.*

(1999: 1582) needs to be explored further to determine if there are taxonomic implications.

Macroderma gigas saturata Douglas, 1962: 59, 60.

TYPE LOCALITY: Kalumburu, northwest Kimberley Division, Western Australia, Australia.

COMMENTS: Recognised as a subspecies of *gigas* by Strahan (1983: 292). Synonymised within *gigas* by Honacki *et al.* (1982: 135), Mahoney and Walton (1988b: 119), Koopman (1993: 163) and Simmons (2005a: 379).

Family Rhinolophidae J. Gray, 1825

Tribe Rhinolophina J. Gray, 1825a: 338.

TYPE GENUS: *Rhinolophus* Lacépède, 1799a.

COMMENTS: When originally proposed, this rank was placed in the Family Vespertilionidae (J. Gray, 1821) and included the genera *Megaderma* É. Geoffroy, 1810a: 197; *Rhinolophus* Lacépède, 1799a; *Nycteris* É. Geoffroy and G. Cuvier, 1795: 186; *Mormoops* Leach, 1821a: 76; and *Nyctophilus* Leach, 1821a. Name recognised at subfamily rank by J. Gray (1825b: 242) and family rank recognised by most authors including Troughton (1967: 270), Honacki *et al.* (1982: 136), Strahan (1983: xxi). Reviewed by Miller (1907: 106). T. Bell (1836: 599) reference used by Simpson (1945: 56), Corbet and Hill (1992: 90). Name also recognised by J. Gray (1866e: 81).

Family Rhinolophidae T. Bell, 1836: 599.

TYPE GENUS: *Rhinolophus* Lacépède, 1799a.

COMMENTS: When originally proposed, this rank was placed in the Cheiroptera (J. Gray, 1821 [=Chiroptera (Blumenbach, 1779)]) and included the genus *Rhinolophus* Lacépède, 1799a. Synonymised within Rhinolophidae by McKenna and Bell (1997: 305).

Tribe? Histiorhina Van der Hoeven, 1855: 1033.

TYPE GENUS: Not based on a genus group name.

COMMENTS: When originally proposed, this rank was placed within Vespertiliones (Pallas, 1767 [=Chiroptera (Blumenbach, 1779 part)]). Synonymised within Rhinolophidae by Simmons (2005a: 350).

Family Rhinolophi Peters, 1865a: 256.

TYPE GENUS: *Rhinolophus* Lacépède, 1799a.

COMMENTS: When originally proposed, this rank included the genera *Rhinolophus* Lacépède, 1799a; *Phyllorrhina* Bonaparte, 1837 [1832–1841] [=*Hipposideros* J. Gray, 1831b]; and *Coelops* Blyth, 1848: 251.

Subfamily Rhinolophinae Dobson, 1875a: 347.

TYPE GENUS: *Rhinolophus* Lacépède, 1799a.

COMMENTS: When originally proposed, this rank was placed in the Family Rhinolophidae (J. Gray, 1825a) and included the genus *Rhinolophus* Lacépède, 1799a. Subfamily rank recognised by Koopman (1984b: 164). Synonymised within Rhinolophidae by McKenna and Bell (1997: 305).

Family Phyllorhinidae de Rochebrune, 1883: 94.

TYPE GENUS: *Phyllorhina* Leach, 1816 [=*Rhinolophus* Lacépède, 1799a].

COMMENTS: When originally proposed, this rank was placed in the Order Chiropteri (de Rochebrune, 1883 [=Chiroptera (Blumenbach, 1779)]) and included the genus *Phyllorhina* Leach, 1816 [=*Rhinolophus* Lacépède, 1799a]. Unavailable. Synonymised within Rhinolophidae by McKenna and Bell (1997: 305).

Tribe? Rhinolophini Winge, 1893b: 24.

TYPE GENUS: *Rhinolophus* Lacépède, 1799a.

COMMENTS: When originally proposed, this rank was placed in the Family Rhinolophidae (J. Gray, 1825a) and included the genera *Phyllorhina* [sic] Bonaparte, 1837 [1832–1841] [=*Hipposideros* J. Gray, 1831b]; *Anthops* Thomas, 1888c: 156; *Rhinonicteris* J. Gray, 1847b; *Triaenops* Dobson, 1871a: 455; *Coelops* Blyth, 1848: 251; and *Rhinolophus* Lacépède, 1799a.

Rhinolophus Lacépède, 1799

Rhinolophus Lacépède, 1799a: 15.

TYPE SPECIES: Φ *Vespertilio ferrumequinum* Schreber, 1774: 190; Plate 62 [= Φ *Rhinolophus ferrumequinum* (Schreber, 1774: 190; Plate 62)] by monotypy. Conserved in the Opinion 91 of the ICZN (1926c: 2) and Direction 24 (ICZN, 1955: 222).

COMMENTS: Genus recognised by Desmarest (1803a: 108; 1803b: 383) and most subsequent authors. Genus reviewed by Corbet and Hill (1992: 91) and Csorba *et al.* (2003: 1). See Also Qumsiyeh *et al.* (1988: 326) and Bogdanowicz and Owen (1992: 142). There is also considered to be an undescribed 'intermediate' species by S. Cooper *et al.* (1998: 203), which was also recognised by Churchill (2008: 88). The taxon *robertsi* has typically been included as a subspecies within *Rhinolophus philippinensis* Waterhouse, 1843c: 68, which has also included *Rhinolophus philippinensis achilles* Thomas, 1900c: 145; *Rhinolophus philippinensis alleni* Lawrence, 1939: 46; *Rhinolophus philippinensis maros* Tate & Archbold, 1939: 9; and *Rhinolophus philippinensis sanborni* Chasen, 1940: 39. *Rhinolophus philippinensis* was recognised as occurring within Australia by recent authors including Simmons (2005a: 360), and Van Dyck and Strahan (2008: 454), but not Churchill (2008: 87, 224) which was followed by Reardon *et al.* (2010: 9) and Burbidge *et al.* (2014: 23, 30) who recognised *robertsi* and an unnamed species in preference. The unnamed 'intermediate' horseshoe-bat has been recognised by several earlier authors including S. Cooper *et al.* (1998: 203), Reardon (1999a: 10) and Kutt (2004: 629).

Phyllorhina Leach, 1816: 5.

TYPE SPECIES: Φ *Vespertilio minutus* Montagu, 1908: 163 (as *Phyllorhina minuta*) [= Φ *Rhinolophus hipposideros minutus* (Montagu, 1908: 163)] by subsequent designation. See Palmer (1904: 535), Hill (1963: 4), and Corbet and Hill (1992: 91).

COMMENTS: Synonymised within *Rhinolophus* by Corbet and Hill (1992: 91), McKenna and Bell (1997: 305) and Simmons (2005a: 350).

HOMONYMS:

Phyllorrhina Bonaparte, 1831: 16, horseshoe bats of the Class Mammalia (Order Chiroptera, Family Rhinolophidae). *Nomen nudum* according to Palmer (1904: 535).

Phyllorrhina Bonaparte, 1837 [1832–1841], of the class Mammalia (Order Chiroptera, Family Hipposideridae). Described as a subgenus of *Rhinolophus*. Genus is a synonym of *Hipposideros* J. Gray, 1831b. See Simmons (2005a: 367).

Rhinocrepis Gervais, 1836: 617.

TYPE SPECIES: Φ *Vespertilio ferrumequinum* Schreber, 1774: 190; Plate 62 [= Φ *Rhinolophus ferrumequinum* (Schreber, 1774: 190; Plate 62)] by subsequent designation. See Palmer (1904: 606).

COMMENTS: Synonymised within *Rhinolophus* by Corbet and Hill (1992: 91), McKenna and Bell (1997: 305) and Simmons (2005a: 350).

Aquias J. Gray, 1847b: 15.

TYPE SPECIES: Φ *Rhinolophus luctus* Temminck, 1834: 23 and Φ *R. trifoliatus*, Temminck, 1834: 24.

COMMENTS: Synonymised within *Rhinolophus* by Corbet and Hill (1992: 91), McKenna and Bell (1997: 305) and Simmons (2005a: 350).

Phyllotis J. Gray, 1866e: 81.

TYPE SPECIES: Φ *Rhinolophus philippinensis* Waterhouse, 1843c: 68 (as *P. philippinensis*) by monotypy.

COMMENTS: Synonymised within *Rhinolophus* by Corbet and Hill (1992: 91), McKenna and Bell (1997: 305) and Simmons (2005a: 350).

HOMONYMS:

Phyllotis Waterhouse, 1837a: 28, large eared mice of the Class Mammalia (Order Rodentia, Family Cricetidae). Currently recognised genus. See Musser and Carleton (2005: 1161).

Coelophyllus Peters, 1867e: 427.

TYPE SPECIES: Φ *Rhinolophus coelophyllus* Peters, 1867a: 426 by monotypy.

COMMENTS: Subsequently recognised by Peters (1871: 303) as a subgenus. Synonymised within *Rhinolophus* by Corbet and Hill (1992: 91), McKenna and Bell (1997: 305) and Simmons (2005a: 350).

Euryalus Matschie, 1901b: 225.

TYPE SPECIES: Φ *Rhinolophus mehelyi* Matschie, 1901b: 225 by monotypy.

COMMENTS: Described as a subgenus of *Rhinolophus*. Synonymised within *Rhinolophus* by Corbet and Hill (1992: 91), McKenna and Bell (1997: 305) and Simmons (2005a: 350).

Rhinophyllotis Iredale & Troughton, 1934: x, 92.

TYPE SPECIES: *Rhinolophus megaphyllus* J. Gray, 1834 (as *Rhinophyllotis megaphyllus*] by monotypy.

COMMENTS: Name validated by Troughton (1941: 342). Synonymised within *Rhinolophus* by Corbet and Hill (1992: 91), McKenna and Bell (1997: 305) and Simmons (2005a: 350).

Rhinomegalophus Bourret, 1951: 607.

TYPE SPECIES: Φ *Rhinomegalophus paradoxolophus* Bourret, 1951: 607 [= Φ *Rhinolophus paradoxolophus* (Bourret, 1951: 607)] by monotypy.

COMMENTS: Synonymised within *Rhinolophus* by Thonglongya (1973: 587), Honacki *et al.* (1982: 141), Corbet and Hill (1992: 91), Koopman (1993: 163), McKenna and Bell (1997: 305) and Simmons (2005a: 350).

Rhinolophus megaphyllus J. Gray, 1834

Eastern Horseshoe-bat

Rhinolophus megaphyllus megaphyllus J. Gray, 1834

Rhinolophus megaphyllus J. Gray, 1834: 52.

TYPE LOCALITY: 'New Holland'. Ryan (1965a: 259) proposed the type locality was New South Wales, County of Harden, 'Cave Flat', limestone cave on the north bank of the Murrumbidgee River at the junction with the Goodradigbee River. (35°00′S, 148°39′E)

COMMENTS: Species recognised by Gould (1856 [1845–1863]: Text to Plate 33). Various subspecies and synonyms recognised by Csorba *et al.* (2003: 74–75) were revised by Simmons (2005a: 359), with most taxa being removed from the species. Subspecies are based on Simmons (2005a: 359). Australian subspecies reviewed by S. Cooper *et al.* (1998: 213), which was followed by Van Dyck and Strahan (2008: 453).

Φ *Rhinolophus megaphyllus monachus* Andersen, 1905

Φ *Rhinolophus megaphyllus monachus* Andersen, 1905a: 80.

TYPE LOCALITY: St. Aignan's [=Misima] Island, Louisiade Archipelago, Papua New Guinea.

COMMENTS: Synonymised within *megaphyllus* by Koopman (1993: 167). Recognised as a subspecies of *megaphyllus* by Laurie and Hill (1954: 52), Flannery (1990: 329; 1995a: 434; 1995b: 360), Bonaccorso (1998: 309), Csorba *et al.* (2003: 75) and Simmons (2005a: 359).

Φ *Rhinolophus megaphyllus fallax* Andersen, 1906

Φ *Rhinolophus fallax* Andersen, 1906: 173; Plate 3.

TYPE LOCALITY: Ighibierei, Kemp Welch River, Papua New Guinea.

COMMENTS: Synonymised within *megaphyllus* by Koopman (1993: 167). Recognised as a subspecies of *megaphyllus* by Laurie and Hill (1954: 52), McKean (1972: 23), Bonaccorso (1998: 309), Flannery (1990: 329; 1995a: 434; 1995b: 360), Csorba *et al.* (2003: 75) and Simmons (2005a: 359).

Rhinolophus megaphyllus ignifer G. Allen, 1933

Rhinolophus megaphyllus ignifer G. Allen, 1933: 149.

TYPE LOCALITY: Coen, Cape York Peninsula, north Queensland, Australia.

COMMENTS: Subspecies rank recognised by Iredale and Troughton (1934: 93) and McKean and Price (1967: 108), who showed that colour is not a good diagnostic tool, but could distinguish the two subspecies on the basis of size (using forearm length). Koopman (1984a: 8) found *megaphyllus* was smaller than *ignifer* (*contra* McKean & Price, 1967: 109), though he acknowledged uncertainty as to whether he had a genuine *R. m. megaphyllus*. Nonetheless he suggested that they should not be granted subspecies status and that *ignifer* should be synonymised with *R. m. megaphyllus* (Koopman, 1984a: 9). Also synonymised within *megaphyllus* by Csorba *et al.* (2003: 75). Subspecies rank recognised by Strahan (1983: 295), Flannery (1990: 329; 1995a: 434; 1995b: 360), Bonaccorso (1998: 309), S. Cooper *et al.* (1998: 213), Reardon (1999a: 11), Simmons (2005a: 359), Clayton *et al.* (2006: 109), and Van Dyck and Strahan (2008: 453).

Φ *Rhinolophus megaphyllus vandeuseni* Koopman, 1982

Φ *Rhinolophus megaphyllus vandeuseni* Koopman, 1982: 13.

TYPE LOCALITY: Bululogon plantation, east coast New Ireland, Bismarck Archipelago, Papua New Guinea. (3°22′S, 152°8′E)

COMMENTS: Synonymised within *megaphyllus* by Koopman (1993: 167). Subspecies rank recognised by Flannery (1990: 329; 1995a: 434; 1995b: 360), Bonaccorso (1998: 309), Csorba *et al.* (2003: 75) and Simmons (2005a: 359).

Rhinolophus robertsi Tate, 1952

Large-eared Horseshoe-bat

Rhinolophus maros robertsi Tate, 1952b: 1.

TYPE LOCALITY: Phoenician Tin mine, near summit of Mt. Amos, 18 miles south of Cooktown, northeast Queensland, Australia. 2000 feet.

COMMENTS: Recognised as a subspecies of *philippinensis* Waterhouse, 1843c: 68 by Goodwin (1979: 113) and followed by subsequent authors including Strahan (1983: 297; 1995: 452), Flannery (1990: 330), Bonaccorso (1998: 313), Csorba *et al.* (2003: 93) and Simmons (2005a: 361). Synonymised within *philippinensis* by Mahoney and Walton (1988c: 122) and Flannery (1995a: 436; 1995b: 361) and within *achilles* by Van Dyck and Strahan (2008: 456). Recognised at the species rank by Churchill (2008: 87) who noted that S. Cooper *et al.* (1998: 203) found this taxon to be unrelated to *Rhinolophus philippinensis*. However the elevation of *robertsi* by Churchill (2008: 87) was refuted by Reardon (2009b: 53) who suggested there are other potential relationships not covered in S. Cooper *et al.* (1998: 203), which included only limited representation of the subspecies and allies of *philippinensis*. Subsequently this taxon has been recognised at the species rank by Reardon *et al.* (2010: 1, 36).

Family Hipposideridae Flower & Lydekker, 1891

Subfamily Hipposiderinae Flower & Lydekker, 1891: 657.

TYPE GENUS: *Hipposideros* J. Gray, 1831b.

COMMENTS: When originally proposed, this rank was placed in the Family Rhinolophidae (J. Gray, 1825a) and included the genera *Hipposiderus* [sic] J. Gray, 1834 [=*Hipposideros* J. Gray, 1831b]; *Anthops* Thomas, 1888c: 156; *Rhinonicteris* J. Gray, 1847b; *Triaenops* Dobson, 1871a: 455; and *Coelops* Blyth, 1848: 251. McKenna and Bell (1997: 306) recognised the Subfamily Rhinonycterinae J. Gray (1866e: 81) for part of the group but this has not been accepted by other authors including Simmons (2005a: 365). Included within Rhinolophidae by Honacki *et al.* (1982: 136) and Strahan (1983: xxi). Recognised as a subfamily within the Family Rhinolophidae by Tate (1952a: 609), Koopman and Jones (1970: 25), Koopman (1984b: 164), Koopman (1993: 169; 1994: 60), Simmons (1998: 12) and Teeling *et al.* (2002: 1432; 2003: 309). Recognised at family rank by Miller (1907: 109), Simpson (1945: 56), Troughton (1967: 271), Mahoney and Walton (1988d: 124), Corbet and Hill (1992: 104), Strahan (1995: 454), Bates and Harrison (1997: 80), Hand and Kirsch (1998: 72), Simmons (2005a: 365) and various other authors.

More recently the genera *Rhinonicteris* J. Gray, 1847b; *Cloeotis* Thomas 1901b: 28; *Triaenops* Dobson 1871a: 455; † *Brachipposideros* Sigé, 1968: 83; † *Brevipalatus* Hand and Archer 2005: 372; and *Paratriaenops* Benda and Vallo 2009: 31 were removed by Foley *et al.* (2015: 313, 319) into the Family Rhinonycteridae (J. Gray, 1866).

Phyllorrhina Koch, 1860: 26.

TYPE GENUS: *Phyllorrhina* Bonaparte, 1837 [1832–1841] [=*Hipposideros* J. Gray, 1831b].

COMMENTS: Unavailable. When originally proposed, this rank was placed in the Chiroptera (Blumenbach, 1779) and included the genera Phyllorrhina Bonaparte, 1837 [1832–1841] [=*Hipposideros* J. Gray, 1831b] and *Rhinolophus* Lacépède, 1799a. Synonymised within Rhinolophidae by McKenna and Bell (1997: 305); however, they also synonymised the currently recognised Family Hipposideridae within the Family Rhinolophidae.

Subfamily Phyllorhininae Dobson, 1875a: 347.

TYPE GENUS: *Phyllorrhina* Bonaparte, 1837 [1832–1841] [= *Hipposideros* J. Gray, 1831b].

COMMENTS: When originally proposed, this rank was placed in the Family Rhinolophidae (J. Gray, 1825a) and included the genera *Coelops* Blyth, 1848: 251; *Phyllorrhina* Bonaparte, 1837 [1832–1841] [=*Hipposideros* J. Gray, 1831b]; *Rhinonicteris* J. Gray, 1847b; and *Triaenops* Dobson, 1871a: 455. Though Dobson (1875a: 347) spells the generic name *Phyllorhina* he includes the author as 'Bonap.' referring to Bonaparte, 1837 [1832–1841] who spells the name *Phyllorrhina*, which is a synonym of *Hipposideros*. Synonymised within Rhinolophidae by McKenna and Bell (1997: 305); however, they also synonymised the currently recognised Family Hipposideridae within the family Rhinolophidae.

Family Hipposideridae Miller, 1907: viii, 109.

TYPE GENUS: *Hipposideros* J. Gray, 1831b.

COMMENTS: When originally proposed, this rank was placed in the Suborder Microchiroptera (Dobson, 1875a [=Chiroptera (Blumenbach, 1779 part)]) and included the genera *Hipposideros* J. Gray, 1831b; *Asellia* J. Gray, 1838c: 493; *Anthops* Thomas, 1888c: 156; *Coelops* Blyth, 1848: 251; *Cloeotis* Thomas, 1901b: 28; *Rhinonicteris* J. Gray, 1847b; and *Triaenops* Dobson, 1871a: 455. Recognised by Simpson (1945: 56), but synonymised within Rhinolophidae. McKenna and Bell (1997: 305).

Subfamily Coelopinae Tate, 1941b: 11.

TYPE GENUS: *Coelops* Blyth, 1848: 251.

COMMENTS: When originally proposed, this rank was placed in the Family Hipposideridae (Flower & Lydekker, 1891) and included the genus *Coelops* Blyth, 1848: 251. Synonymised within the Family Hipposideridae by Simmons (2005a: 365).

Tribe Hipposiderini Koopman & Jones, 1970: 25.

TYPE GENUS: *Hipposideros* J. Gray, 1831b.

COMMENTS: When originally proposed, this rank was placed in the Subfamily Hipposiderinae (Flower & Lydekker, 1891) and included the genera *Hipposideros* J. Gray, 1831b; *Aselliscus* Tate, 1941b: 2; *Cloeotis* Thomas, 1901b: 28; *Rhinonicteris* J. Gray, 1847b; and *Triaenops* Dobson, 1871a: 455. Synonymised within Rhinonycterina (J. Gray, 1866e: 81) by McKenna and Bell (1997: 306).

Subtribe Hipposiderina Koopman, 1994: 60.

TYPE GENUS: *Hipposideros* J. Gray, 1831b.

COMMENTS: When originally proposed, this rank was attributed to Flower and Lydekker (1891: 657) and placed in the Tribe Hipposiderini (Flower & Lydekker, 1891) and included genera *Hipposideros* J. Gray, 1831b; *Anthops* Thomas, 1888c: 156; and *Asellia* J. Gray, 1838c: 493. Synonymised within Hipposiderina (Flower & Lydekker, 1891) by McKenna and Bell (1997: 306).

Hipposideros J. Gray, 1831

Hipposideros J. Gray, 1831b: 37.

TYPE SPECIES: Φ *Vespertilio speoris* Schneider, 1800 (pl. 59b) [= Φ *Hipposideros speoris* (Schneider, 1800: pl. 59b)] by subsequent designation. See W. Sclater (1901: 116).

COMMENTS: Genus reviewed by Blanford (1888: 637), Hill (1963: 1), and Decher and Fahr (2005: 1). *Hipposideros galeritus* Cantor, 1846: 183 was considered to occur in Australia by authors including Ride (1970: 166) and L. Hall and Richards (1979: 29), although it has been proposed that this was in error (Jenkins & Hill, 1981: 279). *Hipposideros galeritus* was considered a synonym of *cervinus* by Churchill (1998: 118) but recognised as a distinct species by Simmons (2005a: 372).

Hipposiderus J. Gray, 1834: 53.

TYPE SPECIES: Invalid emendation of *Hipposideros* J. Gray, 1831b.

COMMENTS: Spelling recognised by Blanford (1888: 637) and W. Sclater (1901: 116). Synonymised within *Hipposideros* by Hill (1963: 3).

Phyllorrhina Bonaparte, 1837 [1832–1841]: Fascicolo 21, third unnumbered page of Φ *Rhinolophus ferrum-equinum*).

TYPE SPECIES: *Rhinolophus diadema* É. Geoffroy, 1813 [=*Hipposideros diadema* (É. Geoffroy, 1813)] by subsequent designation. See W. Sclater (1901: 116).

COMMENTS: Described as a subgenus of *Rhinolophus* Lacépède, 1799a. Elevated to generic rank by Peters (1852b: 31) and Dobson (1878: 127). Peters (1871: 312) and Trouessart (1897: 96) allocated the Australian forms to *Phyllorhina* [sic] Bonaparte, 1831 [*nomen nudum*]. Synonymised within *Hipposideros* by Blanford (1888: 638) and W. Sclater (1901: 116). Synonymised with

Hipposideros by J. Gray (1843a: xix), Miller (1907: 110), Simpson (1945: 56), Hill (1963: 3), Corbet and Hill (1992: 105) and Simmons (2005a: 367).

HOMONYMS:

Phyllorrhina Leach, 1816, horseshoe bats of the Class Mammalia (Order Chiroptera, Family Rhinolophidae). Genus is a synonym of *Rhinolophus* Lacépède, 1799a.

Phyllorrhina Bonaparte, 1831: 16, horseshoe bats of the Class Mammalia (Order Chiroptera, Family Rhinolophidae). *Nomen nudum* according to Palmer (1904: 535).

Macronycteris J. Gray, 1866e: 82.

TYPE SPECIES: Φ *Rhinolophus gigas* Wagner, 1845: 148 (as *M. gigas*) [= Φ *Hipposideros gigas* (Wagner, 1845: 148)] by monotypy.

COMMENTS: Synonymised within *Hipposideros* by Miller (1907: 110) and McKenna and Bell (1997: 306).

Gloionycteris J. Gray, 1866e: 82.

TYPE SPECIES: Φ *Rhinolophus armiger* Hodgson, 1835: 699 (as *G. armigera*) [= Φ *Hipposideros armiger* (Hodgson, 1835: 699)] by monotypy.

COMMENTS: Synonymised with *Hipposideros* by Miller (1907: 110), Hill (1963: 3), Corbet and Hill (1992: 105) and Simmons (2005a: 367).

Rhinophylla J. Gray, 1866e: 82.

TYPE SPECIES: Φ *Phyllorrhina labuanensis* Tomes, 1859a (as *R. labuanensis*) [= Φ *Hipposideros cervinus labuanensis* (Tomes, 1859a)] by monotypy.

COMMENTS: Synonymised with *Hipposideros* by Miller (1907: 110), Hill (1963: 3), Corbet and Hill (1992: 105) and Simmons (2005a: 367).

HOMONYMS:

Rhinophylla Peters, 1865b: 520; 1865c: 355, fruit bats of the Class Mammalia (Order Chiroptera, Family Phyllostomidae). Currently recognised genus. See Simmons (2005a: 413).

Speorifera J. Gray, 1866e: 82.

TYPE SPECIES: Φ *Rhinolophus vulgaris* Horsfield, 1823: No VI, p. 5 of Φ *R. larvatus* entry (as *S. vulgaris*) [= Φ *Hipposideros larvatus* (Horsfield, 1823: Plate and 10 unnumbered pages)] by monotypy.

COMMENTS: Synonymised with *Hipposideros* by Miller (1907: 110), Hill (1963: 4), Corbet and Hill (1992: 105) and Simmons (2005a: 367).

Chrysonycteris J. Gray, 1866e: 82.

TYPE SPECIES: Φ *Hipposideros fulvus* J. Gray, 1838c: 492 (as *C. fulva*) by monotypy.

COMMENTS: Synonymised with *Hipposideros* by Miller (1907: 110), Hill (1963: 4), Corbet and Hill (1992: 105) and Simmons (2005a: 367).

Doryrhina Peters, 1871: 314.

TYPE SPECIES: Φ *Phyllorhina cyclops* Temminck, 1853: 75 [=Φ *Hipposideros cyclops* (Temminck, 1853: 75)] by monotypy.

COMMENTS: Described as a subgenus of '*Phyllorhina*' [= *Phylorrhina*]. Synonymised within *Hipposideros* by Miller (1907: 110), Hill (1963: 4), and McKenna and Bell (1997: 306).

Sideroderma Peters, 1871: 324.

TYPE SPECIES: Φ *Phyllorhina fuliginosa* Temminck, 1853: 77 [= Φ *Hipposideros fuliginosus* (Temminck, 1853: 77)] by monotypy.

COMMENTS: Described as a subgenus of '*Phyllorhina*' [=*Phylorrhina*]. Synonymised within *Hipposideros* by Miller (1907: 110), Hill (1963: 4), and McKenna and Bell (1997: 306).

Ptychorhina Peters, 1871: 325.

TYPE SPECIES: Φ *Rhinolophus caffer* Sundevall, 1846a: 118 (as *Phyllorhina caffra* [sic]) [= Φ *Hipposideros caffer* (Sundevall, 1846a: 118)] by subsequent designation. See Palmer (1904: 597).

COMMENTS: Synonymised within *Hipposideros* by Miller (1907: 110), Hill (1963: 4) and Simmons (2005a: 368).

Cyclorhina Peters, 1871: 326.

TYPE SPECIES: Φ *Phyllorhina obscura* Peters, 1861b: 709 [= Φ *Hipposideros obscurus* (Peters, 1861b: 709)] and Φ *P. doriae* Peters, 1871: 326 [= Φ *Hipposideros doriae* (Peters, 1871: 326)] by subsequent designation. See Palmer (1904: 208), Tate (1941c: 354), and Corbet and Hill (1992: 105).

COMMENTS: Described as a section of a subgenus of '*Phyllorhina*' [=*Phyllorrhina*]. Synonymised within *Hipposideros* by Miller (1907: 110), Hill (1963: 4), Corbet and Hill (1992: 105) and Simmons (2005a: 367).

HOMONYMS:

† *Cyclorhina* J. Hall and Clarke, 1894: 206, brachiopods of the Phylum Brachiopoda (Class Rhynchonellata, Order Rhynchonellida, Family Machaerariidae. Name is a synonym of *Callipleura* G. Cooper, 1942: 228. See Doescher (1981: 8, 11).

Thyreorhina Peters, 1871: 327.

TYPE SPECIES: Φ *Phyllorhina coronata* Peters, 1871: 327 [= Φ *Hipposideros coronatus* (Peters, 1871: 327)] by monotypy.

COMMENTS: Described as a subgenus of '*Phyllorhina*' [=*Phylorrhina*]. Synonymised within *Hipposideros* by Miller (1907: 111), Hill (1963: 4) and Simmons (2005a: 367).

Syndesmotis Peters, 1871: 329.

TYPE SPECIES: Φ *Phyllorhina megalotis* Heuglin, 1861: 4, 8 [= Φ *Hipposideros megalotis* (Heuglin, 1861: 4, 8)] by monotypy.

COMMENTS: Described as a subgenus of *Phyllorhina*. Synonymised within *Hipposideros* by Miller (1907: 111), Hill (1963: 4) and Simmons (2005a: 367).

Syndesmotus C. Waterhouse, 1902: 362.

TYPE SPECIES: Incorrect subsequent spelling of *Syndesmotis* Peters, 1871.

COMMENTS: Objective synonym of *Syndesmotis* Peters, 1871. Synonymised within *Hipposideros* by McKenna and Bell (1997: 306) and Simmons (2005a: 367).

Hipposideros ater Templeton, 1848

Dusky Leaf-nosed Bat

Φ ***Hipposideros ater ater*** Templeton, 1848

Φ *Hipposideros ater* Templeton, 1848: 252.

TYPE LOCALITY: Colombo area, Sri Lanka.

COMMENTS: Taxonomic decision of Koopman (1984a: 9), for synonyms below.

Φ *Hipposideros Atratus* Kelaart, 1850b: 208.

TYPE LOCALITY: Colombo area, Sri Lanka.

COMMENTS: Synonymised within *ater* by Corbet and Hill (1992: 108), Flannery (1990: 311; 1995a: 408; 1995b: 338), Koopman (1993: 171), Bates and Harrison (1997: 81), and Simmons (2005a: 367).

Hipposideros ater aruensis J. Gray, 1858

Eastern Dusky Leaf-nosed Bat

Φ *Hipposideros Aruensis* J. Gray, 1858a: 107.

TYPE LOCALITY: Aru Islands.

COMMENTS: Subspecies and occurrence in Australia recognised by Reardon (1999a: 12). Placed as a subspecies of *Hipposideros ater* by Hill (1963: 33), McKean and Price (1967: 110), McKean (1972: 23), Flannery (1990: 310; 1995a: 408; 1995b: 338), Strahan (1995: 455), Bonaccorso (1998: 264) and recognised within Australia by Clayton *et al.* (2006: 109), and Van Dyck and Strahan (2008: 459).

Hipposideros albanensis J. Gray, 1866a: 220.

TYPE LOCALITY: Port Albany, Cape York Peninsula, Queensland, Australia.

COMMENTS: Species recognised by Iredale and Troughton (1934: 93). Synonymised within *ater* by Ride (1970: 243), Flannery (1990: 311; 1995a: 408; 1995b: 338) and Koopman (1993: 171). Subspecies rank within *bicolor* recognised by Troughton (1967: 272) and within *ater* by Strahan (1983: 298). Taxon synonymised within *aruensis* by Hill (1963: 33), Koopman (1984a: 9), and Simmons (2005a: 367).

Φ ***Hipposideros ater antricola*** Peters, 1861

Φ *Phyllorhina antricola* Peters, 1861b: 709.

TYPE LOCALITY: Paracali, Luzon Islands, Philippines. Flannery (1995b: 338) says Palawan, Philippines.

COMMENTS: Synonymised within *ater* by Koopman (1993: 171). Subspecies within *ater* recognised by Hill (1963: 31), Corbet and Hill (1992: 108), Flannery (1990: 310; 1995a: 408; 1995b: 338) and Simmons (2005a: 367).

Φ ***Hipposideros ater amboinensis*** (Peters, 1871)

Φ *Ph. [yllorhina] amboinensis* Peters, 1871: 323.

TYPE LOCALITY: Ambon, Moluccas, Indonesia.

COMMENTS: Synonymised within *ater* by Bates and Harrison (1997: 81). Subspecies within *ater* recognised by Hill (1963: 33), Corbet and Hill (1992: 108), Flannery (1990: 310; 1995a: 408; 1995b: 338), Bonaccorso (1998: 264) and Simmons (2005a: 367).

Φ ***Hipposideros ater nicobarulae*** Miller, 1902

Φ *Hipposideros nicobarulae* Miller, 1902a: 781.

TYPE LOCALITY: Little Nicobar Island, India.

COMMENTS: Synonymised within *ater* by Koopman (1993: 171) and Bates and Harrison (1997: 81). Subspecies within *ater* recognised by Hill (1963: 30), Corbet and Hill (1992: 108), Flannery (1990: 310; 1995a: 408; 1995b: 338) and Simmons (2005a: 367).

Φ ***Hipposideros ater saevus*** Andersen, 1918

Φ *Hipposideros albanensis saevus* Andersen, 1918: 380.

TYPE LOCALITY: Kai or Kei Island, Maluku Islands, Indonesia.

COMMENTS: Subspecies status recognised by Hill (1963: 30), Corbet and Hill (1992: 108), Flannery (1990: 310; 1995a: 408; 1995b: 338), Bonaccorso (1998: 264) and Simmons (2005a: 367).

Φ *Hipposideros gentilis toala* Shamel, 1940: 352.

TYPE LOCALITY: Toeare, Sulawesi, Indonesia.

COMMENTS: Synonymised within *ater* by Flannery (1990: 311; 1995a: 408; 1995b: 338) and Corbet and Hill (1992: 108). Synonymised within *saevus* by Simmons (2005a: 367).

Hipposideros ater gilberti D. Johnson, 1959

Western Dusky Leaf-nosed Bat

Hipposideros bicolor gilberti D. Johnson, 1959: 183.

TYPE LOCALITY: 'Oenpelli', Northern Territory, Australia. 12°21′S 133°04′E.

COMMENTS: Recognised as a subspecies of *ater* by Hill (1963: 33), but synonymised within *aruensis* by McKean and Price (1967: 111). Koopman (1984a: 9–10) was not certain that *gilberti* should be synonymised with *aruensis*, and proposed that rather than describing the Western Australian populations as a separate subspecies, he tentatively associated Western Australian specimens with *gilberti*, recognising that the typotypical population is in the intergrade area between *gilberti* and *aruensis*. Recognised as a subspecies of *ater* by Flannery (1990: 310; 1995a: 408; 1995b: 338), Strahan (1983: 298; 1995: 455), Simmons (2005a: 367), Clayton *et al*. (2006: 109), and Van Dyck and Strahan (2008: 459).

Hipposideros cervinus (Gould, 1854)

Fawn Leaf-nosed Bat

Hipposideros cervinus cervinus (Gould, 1854)

Rhinolophus? *cervinus* Gould, 1854 [1845–1963]: Text to Plate 34.

TYPE LOCALITY: Cape York and sandstone caves on Albany Island, north Queensland, Australia.

COMMENTS: Synonymised within *Hipposideros galeritus* Cantor, 1846 by Ride (1970: 243) and recognised as a subspecies of *galeritus* by McKean (1972: 26). Transferred to *Hipposideros* at species rank by Iredale and Troughton (1934: 93), Troughton (1967: 272), Jenkins and Hill (1981: 288) and most subsequent authors except Hall and Richards (1979: 29) who recognised *galeritus* at species rank and did not recognise *cervinus*.

Φ *Hipposideros cervinus labuanensis* (Tomes, 1859)

Φ *Phyllorhina labuanensis* Tomes, 1859a: 537.

TYPE LOCALITY: Labuan Island, Borneo, Malaysia.

COMMENTS: Described as *Phyllorhina labuanensis*. Synonymised within *cervinus* by Koopman (1993: 171). Subspecies within *cervinus* recognised by Jenkins and Hill (1981: 290), Corbet and Hill (1992: 111), Flannery (1990: 309; 1995a: 413; 1995b: 342) and Simmons (2005a: 369).

Φ *Hipposideros schneidersi* Thomas, 1904f: 722.

TYPE LOCALITY: Upper Langkat, Sumatra, Indonesia.

COMMENTS: Misprint of *schneideri*. Synonymised within *cervinus* by Jenkins and Hill (1981: 290), Corbet and Hill (1992: 111), Koopman (1993: 171) and Flannery (1995a: 413; 1995b: 342), and within *labuanensis* by Simmons (2005a: 369).

Φ *Hipposideros cervinus batchianus* Matschie, 1901

Φ *Hipposideros batchianus* Matschie, 1901a: 273.

TYPE LOCALITY: Batchian (Bacan) Island, Moluccas, Indonesia.

COMMENTS: Recognised as a subspecies of *cervinus* by Laurie and Hill (1954: 57) and Jenkins and Hill (1981: 292). Synonymised within *cervinus* by Corbet and Hill (1992: 111) and Koopman (1993: 171). Subspecies within *cervinus* recognised by Flannery (1990: 299; 1995a: 413; 1995b: 342) and Simmons (2005a: 369).

Φ *Hipposideros celebensis* Sody, 1936: 47.

TYPE LOCALITY: Mampoe (Mampu) Cave, 20 kilometres north of Watoe Pone (Watampone), south Sulawesi.

COMMENTS: Recognised as a subspecies of *cervinus* by Laurie and Hill (1954: 57). Synonymised within *cervinus* by Jenkins and Hill (1981: 289), Corbet and Hill (1992: 111), Koopman (1993: 171) and Simmons (2005a: 369).

Φ *Hipposideros cervinus misoriensis* (Peters, 1906)

Φ *Phyllorhina cervina* var. *misoriensis* Peters, 1906: Plate 5L, Fig. 4.

TYPE LOCALITY: Probably Misor [=Schouten] Island, New Guinea.

COMMENTS: Provisionally associated with this species by Jenkins and Hill (1981: 292). Synonymised within *cervinus* by Koopman (1993: 171). Subspecies within *cervinus* recognised by Corbet and Hill (1992: 111), Flannery (1990: 299; 1995a: 413; 1995b: 342) and Simmons (2005a: 369).

Hipposideros diadema (É. Geoffroy, 1813)

Diadem Leaf-nosed Bat

Φ *Hipposideros diadema diadema* (É. Geoffroy, 1813)

Φ *Rhinolophus diadema* É. Geoffroy, 1813: 263; Plate 6.

TYPE LOCALITY: Timor Island [or Timor Leste], Indonesia.

COMMENTS: Koopman (1984a: 11) proposed that although the subspecies taxonomy is chaotic the two Australian subspecies which he recognised, *reginae* and *inornatus*, seem well differentiated from each other (note *inornatus* here recognised as a distinct species). Simmons (2005a: 371) suggested that many subspecies are of dubious validity. Previously included *H. demissus* Andersen, 1909b: 268 by authors including Laurie and Hill (1954: 58), Hill (1963: 111), Hill (1971b: 576), Flannery (1990: 318) and Bonaccorso (1998: 280), but this taxon was removed from *diadema* by Kitchener *et al.* (1992: 3) and recognised as a distinct taxon by Simmons (2005a: 370).

Φ *Hipposideros diadema nobilis* (Horsfield, 1823)

Φ *Rhinolophus nobilis* Horsfield, 1823: No. VII; Plate.

TYPE LOCALITY: Java, Indonesia.

COMMENTS: Synonymised within *diadema* by Hill (1963: 109), Corbet and Hill (1992: 113) and Koopman (1993: 172). Recognised as a subspecies of *diadema* by Kitchener *et al.* (1992: 19), Flannery (1995a: 417; 1995b: 346) and Simmons (2005a: 370).

Φ *Hipposideros diadema vicarius* Andersen, 1905b: 499.

TYPE LOCALITY: Niah Cave, Sarawak, Borneo, Malaysia.

COMMENTS: Recognised as a synonym of *masoni* (Dobson, 1872) by Hill (1963: 109). Recognised as a subspecies of *diadema* by Corbet and Hill (1992: 113). Synonymised within *diadema* by Koopman (1993: 172) and Flannery (1995a: 417; 1995b: 346) but synonymised within *nobilis* by Kitchener *et al.* (1992: 19) and Simmons (2005a: 370).

Φ *Hipposideros diadema griseus* (Meyen, 1833)

Φ *Rhinolophus griseus* Meyen, 1833: 608; Plate 46.

TYPE LOCALITY: San Matheo Cave, Montalban, near Manila, Luzon, Philippines.

COMMENTS: Synonymised within *diadema* by Corbet and Hill (1992: 113) and Koopman (1993: 172). Recognised as a subspecies of *diadema* recognised by Hill (1963: 109), Kitchener *et al.* (1992: 33), Bonaccorso (1998: 279), Flannery (1995a: 417; 1995b: 346) and Simmons (2005a: 370).

Φ *Hipposideros diadema pullatus* Andersen, 1905b: 498.

TYPE LOCALITY: Haveri, Papua New Guinea. 700 metres.

COMMENTS: Species rank recognised by Thomas (1914b: 317), but reduced to a subspecies of *diadema* by Laurie and Hill (1954: 57), Hill (1963: 111), McKean (1972: 25), Koopman (1982: 16) and Flannery (1990: 318). Synonymised within *diadema* by Koopman (1993: 172) and Flannery (1995a: 417; 1995b: 346), and within *griseus* by Kitchener *et al.* (1992: 33) and Simmons (2005a: 370).

Φ *Hipposideros diadema anderseni* E. Taylor, 1934: 246.

TYPE LOCALITY: Novaliches, Rizal Provence, Luzon, Philippines.

COMMENTS: Recognised as a subspecies of *diadema* by Lawrence (1939: 53) and Corbet and Hill (1992: 113). Synonymised within *diadema* by Koopman (1993: 172) and Flannery (1995a: 417; 1995b: 346), and within *griseus* by Hill (1963: 109), Kitchener *et al.* (1992: 33) and Simmons (2005a: 370).

Φ *Hipposideros diadema nicobarensis* (Dobson, 1871)

Φ *Phyllorhina Nicobarensis* Dobson, 1871b: 262; Plate 20, Figs. 1–2.

TYPE LOCALITY: Nicobar Islands, India.

COMMENTS: Synonymised within *diadema* by Corbet and Hill (1992: 113), Koopman (1993: 172), and Bates and Harrison (1997: 101). Recognised as a subspecies of *diadema* by Hill (1963: 109; 1967: 5), Flannery (1995a: 417; 1995b: 346) and Simmons (2005a: 370).

Φ *Hipposideros diadema masoni* (Dobson, 1872)

Φ *Phyllorhina Masoni* Dobson, 1872: 338.

TYPE LOCALITY: Moulmein, Burma.

COMMENTS: The taxon *masoni* was considered by Andersen (1905b: 500), who associated it with *Hipposideros diadema*, and formally considered a subspecies of *diadema* by Andersen (1907: 6). Synonymised within *diadema* by Corbet and Hill (1992: 113) and Koopman (1993: 172). Recognised as a subspecies of *diadema* by Hill (1963: 109), Kitchener *et al.* (1992: 38), Flannery (1990: 318; 1995a: 417; 1995b: 346) and Simmons (2005a: 370).

Φ *Hipposideros diadema oceanitis* Andersen, 1905

Φ *Hipposideros diadema oceanitis* Andersen, 1905b: 497.

TYPE LOCALITY: Aola, Guadalcanal Island, Solomon Islands.

COMMENTS: Synonymised within *diadema* by Koopman (1993: 172). Recognised as a subspecies of *diadema* by Laurie and Hill (1954: 57), Hill (1963: 111), Hill (1971b: 575), Kitchener *et al.* (1992: 41), Bonaccorso (1998: 279), Flannery (1990: 318; 1995a: 417; 1995b: 346) and Simmons (2005a: 371).

Φ *Hipposideros diadema malaitensis* C. Phillips, 1967: 35.

TYPE LOCALITY: Malaita Island, Solomon Islands.

COMMENTS: Synonymised within *diadema* by Koopman (1993: 172) and Flannery (1995a: 417; 1995b: 346), and within *oceanitis* by Hill (1971b: 575), Kitchener *et al.* (1992: 41) and Simmons (2005a: 371).

Φ *Hipposideros diadema euotis* Andersen, 1905

Φ *Hipposideros euotis* Andersen, 1905b: 502.

TYPE LOCALITY: Batchian (Bacan) Island, Halmahera Group, Moluccas, Indonesia.

COMMENTS: Synonymised within *diadema* by Corbet and Hill (1992: 113) and Koopman (1993: 172). Subspecies of *diadema* recognised by Laurie and Hill (1954: 58), Hill (1963: 111), Flannery (1990: 318; 1995a: 417; 1995b: 346) and Simmons (2005a: 370).

Φ ***Hipposideros diadema enganus*** Andersen, 1907

Φ *Hipposideros diadema enganus* Andersen, 1907: 8.

TYPE LOCALITY: Kifa-juc, Bua Bua, Enggano Island, Sumatra, Indonesia.

COMMENTS: Synonymised within *diadema* by Koopman (1993: 172). Recognised as a subspecies of *diadema* by Hill (1963: 109), Corbet and Hill (1992: 113), Flannery (1995a: 417; 1995b: 346) and Simmons (2005a: 370).

Φ ***Hipposideros diadema mirandus*** Thomas, 1914

Φ *Hipposideros demissus mirandus* Thomas, 1914c: 437.

TYPE LOCALITY: Manus Island, Admiralty Islands.

COMMENTS: Synonymised within *diadema* by Koopman (1993: 172). Recognised as a subspecies of *diadema* by Laurie and Hill (1954: 58), Hill (1963: 111), Bonaccorso (1998: 279), Flannery (1990: 318; 1995a: 417; 1995b: 346) and Simmons (2005a: 370).

Φ ***Hipposideros diadema custos*** Andersen, 1918

Φ *Hipposideros diadema custos* Andersen, 1918: 381.

TYPE LOCALITY: Ara, Key Island [=Kai or Kei Island], Maluku Islands, Indonesia.

COMMENTS: Synonymised within *diadema* by Koopman (1993: 172). Recognised as a subspecies of *diadema* by Laurie and Hill (1954: 58), Hill (1963: 111), Corbet and Hill (1992: 113), Flannery (1990: 318; 1995a: 417; 1995b: 346) and Simmons (2005a: 370).

Φ ***Hipposideros diadema speculator*** Andersen, 1918

Φ *Hipposideros diadema speculator* Andersen, 1918: 381.

TYPE LOCALITY: Kalao Island, Flores Sea, south of Sulawesi, Indonesia.

COMMENTS: Synonymised within *diadema* by Koopman (1993: 172). Recognised as a subspecies of *diadema* by Laurie and Hill (1954: 58), Hill (1963: 111), Corbet and Hill (1992: 113), Flannery (1990: 318; 1995a: 417; 1995b: 346) and Simmons (2005a: 371).

Hipposideros diadema reginae Troughton, 1937

Hipposideros diadema reginae Troughton, 1937d: 275.

TYPE LOCALITY: Bloomfield River, Cooktown area, north Queensland, Australia.

COMMENTS: Synonymised within *diadema* by Mahoney and Walton (1988d: 126). Subspecies rank recognised by Tate (1952a: 609), Hill (1963: 111), Troughton (1967: 272), McKean and Price (1967: 112), Strahan (1983: 304; 1995: 461), Churchill (1998: 122), Kitchener *et al.* (1992: 36) and Reardon (1999a: 12), who noted that it is probably to be recognised as a species. Subsequently recognised as a subspecies of *diadema* by Bonaccorso (1998: 280), Flannery (1990: 318; 1995a: 417; 1995b: 346), Simmons (2005a: 371), Clayton *et al.* (2006: 109), and Van Dyck and Strahan (2008: 463).

FUTURE TAXONOMIC RESEARCH: The argument of its status as a full species (see above) needs to be tested in the future.

Φ ***Hipposideros diadema trobrius*** Troughton, 1937

Φ *Hipposideros diadema trobrius* Troughton, 1937d: 276.

TYPE LOCALITY: Kiriwina Island, Trobriand group, east of Papua New Guinea.

COMMENTS: Synonymised within *diadema* by Koopman (1993: 172). Recognised as a subspecies of *diadema* by Laurie and Hill (1954: 58), Hill (1963: 111), Koopman (1982: 16), Flannery (1990: 318; 1995a: 417; 1995b: 346), Bonaccorso (1998: 279) and Simmons (2005a: 371).

Φ ***Hipposideros diadema natunensis*** Chasen, 1940

Φ *Hipposideros diadema natunensis* Chasen, 1940: 43.

TYPE LOCALITY: Bunguran Island, North Natuan Islands, Indonesia.

COMMENTS: Synonymised within *diadema* by Koopman (1993: 172). Recognised as a subspecies of *diadema* by Hill (1963: 109), Corbet and Hill (1992: 113), Flannery (1995a: 417; 1995b: 346) and Simmons (2005a: 370).

Φ ***Hipposideros diadema ceramensis*** Laurie & Hill, 1954

Φ *Hipposideros diadema ceramensis* Laurie & Hill, 1954: 58.

TYPE LOCALITY: Teleoti Bay, south Ceram Island, Moluccas, Indonesia.

COMMENTS: Synonymised within *diadema* by Koopman (1993: 172). Recognised as a subspecies of *diadema* by Hill

(1963: 111), Corbet and Hill (1992: 113), Flannery (1990: 318; 1995a: 417; 1995b: 346) and Simmons (2005a: 370).

Hipposideros inornatus McKean, 1970

Arnhem Leaf-nosed Bat

Hipposideros diadema inornatus McKean, 1970: 138.

TYPE LOCALITY: Deaf Adder Creek, where it emerges from the Arnhem Land Plateau, 55 miles due south of Oenpelli, Northern Territory, Australia. (13°06′S, 132°56′E)

COMMENTS: Synonymised within *diadema* by Mahoney and Walton (1988d: 126). Recognised as a subspecies of *diadema* by Strahan (1983: 304; 1995: 461), Churchill (1998: 120), Flannery (1990: 318), L. Hall and Richards (2003: 114), Clayton *et al.* (2006: 109), Van Dyck and Strahan (2008: 461), and Kitchener *et al.* (1992: 56) who thought it to be not closely related to *diadema*, which was followed by Flannery (1995b: 345) and Reardon (1999a: 12), who proposed that it be recognised as a species. Hall and Richards (2003: 114) suggested that this taxon was likely to be recognised as a separate species based on genetic studies in progress. Species rank recognised by Simmons (2005a: 373) and Churchill (2008: 96), reduced to a subspecies of *diadema* by Van Dyck and Strahan (2008: 461), but again recognised at species rank by Burbidge *et al.* (2014: 23, 30).

Hipposideros semoni Matschie, 1903

Semon's Leaf-nosed Bat

Hipposideros semoni Matschie, 1903: 132 [=774].

TYPE LOCALITY: Cooktown, north Queensland, Australia.

COMMENTS: Species rank recognised by subsequent authors. Extraliminat distribution includes Papua New Guinea (Bonaccorso, 1998: 295).

Hipposideros stenotis Thomas, 1913

Northern Leaf-nosed Bat

Hipposideros stenotis Thomas, 1913b: 206.

TYPE LOCALITY: Mary River, Northern Territory, Australia.

COMMENTS: Species rank recognised by subsequent authors.

Family Rhinonycteridae J. Gray, 1866

Tribe? Rhinonycterina J. Gray, 1866e: 81.

TYPE GENUS: *Rhinonicteris* J. Gray, 1847b.

COMMENTS: When originally proposed, this rank was placed in the Family Rhinolophidae (J. Gray, 1825a) and included the genus *Rhinonicteris* J. Gray, 1847b. The Subtribe Rhinonycterina (within the Tribe Hipposiderini) was recognised by Koopman (1994: 68), while McKenna and Bell (1997: 306, 307) recognised the Subfamily Rhinonycterinae, Tribe Rhinonycterini and Subtribe Rhinonycterina within the Family Rhinolophidae. Tribe and subtribe ranks, within the Family Hipposideridae, were recognised by Hand and Kirsch (2003: 1148), Hand and Archer (2005, 372) and Armstrong (2006a: 128). Elevated to the new rank of family by Foley *et al.* (2015: 313, 319), who included the genera *Rhinonicteris* J. Gray, 1847b; *Cloeotis* Thomas 1901b: 28; *Triaenops* Dobson 1871a: 455; † *Brachipposideros* Sigé, 1968: 83; † *Brevipalatus* Hand and Archer 2005: 372; and *Paratriaenops* Benda and Vallo 2009: 31. Foley *et al.* (2015: Suppl. 2) also discussed the correct spelling of the family name, which follows Article 29.5 of the Code (ICZN, 1999: 33).

Tribe Triaenopini Benda and Vallo 2009: 2, 33.

TYPE GENUS: *Triaenops* Dobson 1871a: 455.

COMMENTS: When originally proposed, this rank was placed in the Family Hipposideridae and included the genera *Triaenops* Dobson 1871a: 455 and *Paratriaenops* Benda and Vallo 2009: 31. Rank synonymised within the Family Rhinonycteridae J. Gray, 1866 by Foley *et al.* (2015: 319).

Rhinonicteris J. Gray, 1847

Rhinonicteris J. Gray, 1847b: 16.

TYPE SPECIES: *Rhinolophus aurantius* J. Gray, 1845b [=*Rhinonicteris aurantia* (J. Gray, 1845b)] by original designation.

COMMENTS: Also described by J. Gray (1847c: 408). Error in Latinisation. *Rhinonicteris* is the correct spelling (see Koopman, 1993: 175); *Rhinonycteris* is sometimes used (e.g. McKenna & Bell, 1997: 307).

Rhinonycteris J. Gray, 1866e: 81.

TYPE SPECIES: Unjustified emendation of *Rhinonicteris* J. Gray, 1847b.

COMMENTS: Corrected spelling, but unjustified emendation of *Rhinonicteris* J. Gray, 1847b. Synonymised within *Rhinonicteris* J. Gray, 1847b by Palmer (1904: 606) and Tate (1941b: 3). Sometimes used as the preferred spelling including Hill (1982: 165).

Rhinonicteris aurantia (J. Gray, 1845)

Orange Leaf-nosed Bat

Rhinolophus aurantius J. Gray, 1845b: 405; Tab 1.

TYPE LOCALITY: Port Essington, Northern Territory, Australia.

COMMENTS: Recognised by Gould (1851 [1845–1863]: Text to Plate 35) and transferred to *Rhinonicteris* by Iredale and Troughton (1934: 93) and followed by Troughton (1967: 271), Honacki *et al.* (1982: 148) and subsequent authors. Often spelled '*aurantius*', but '*aurantia*' is the correct spelling in combination with *Rhinonicteris* according to Simmons (2005a: 378) and Armstrong (2006a: 126), which appears to have been followed by subsequent authors. Reardon (1999a: 12) recognised a 'Pilbara form', also Burbidge *et al.* (2014: 23, 30). Nomenclature of the species reviewed by Armstrong (2006a: 125).

FUTURE TAXONOMIC RESEARCH: Research is required to determine the taxonomy of the unnamed Pilbara form (*sensu* Reardon, 1999a: 12) and describe it if appropriate.

Suborder Yangochiroptera Koopman, 1985

Infraorder Yangochiroptera Koopman, 1985: 26.

COMMENTS: When originally proposed, this rank was placed in the Suborder Microchiroptera (Dobson, 1875a [=Chiroptera (Blumenbach, 1779 part)]) and included the superfamilies Phyllostomoidea (J. Gray, 1825b: 242) and Vespertilionoidea (J. Gray, 1821: 299). Infraorder rank within the Suborder Microchiroptera recognised by McKenna and Bell (1997: 307), Simmons and Geisler (1998: 136), Simmons (1998: 12) and Göbbel (2002: 343). Support for Yangochiroptera at the subordinal rank provided by Springer *et al.* (2001: 6243), Teeling *et al.* (2002: 1432; 2003: 309), Hoofer *et al.* (2003: 809), Van Den Bussche and Hoofer (2004: 327) and Teeling *et al.* (2005: 581). Taxon not recognised by Simmons (2005a: 313), but see comments under Yinpterochiroptera. The recognition of Yangochiroptera as a suborder was recognised by Yapa and Ratnavira (2013: 412).

Order Insectivorae J. Gray, 1821: 299.

COMMENTS: When originally proposed, this rank was placed in the Class Cheiroptera (J. Gray, 1821 [=Chiroptera (Blumenbach, 1779)]) and included the families Noctilionidae (J. Gray, 1821: 299) and Vespertilionidae (J. Gray, 1821). Synonymised within Microchiroptera by Miller (1907: 78), Simpson (1945: 55) and McKenna and Bell (1997: 301).

HOMONYMS:

Insectivora Bowdich, 1821, insectivores of the Class Mammalia (Order Lipotyphla). Name is a synonym of the Order Lipotyphla (Haeckel, 1866). See individual entry.

Family Gymnorhina Giebel, 1855: xii, 926.

TYPE GENUS: Not based on a genus group name.

COMMENTS: When originally proposed, this rank was placed in the Order Chiroptera (Blumenbach, 1779) and included the genera *Furia* F. Cuvier, 1828: 150 [=*Furipterus* Bonaparte, 1837 [1832–1841]: Fascicolo 21, third unnumbered page of *Plecotus auritus*]; *Nycticejus* [sic =*Nycticeius*] Rafinesque, 1819a: 417; *Vespertilio* Linnaeus, 1758: 31; *Thyroptera* de Spix, 1823: 61; *Dysopes* Illiger, 1811: 122 [=*Molossus* É. Geoffroy, 1805b: 151]; *Emballonura* Kuhl [=Temminck, 1838a: 1, 22]; *Diclidurus* Wied-Neuwied, 1820: 1629; *Noctilio* (Geoff) [=Linnaeus, 1766: 88]; *Taphozous* É. Geoffroy, 1818; and *Mormops* [sic =*Mormoops*] Leach, 1821a: 76]. Rank recognised by Van der Hoeven (1855: 1028) who attributed the author to Wagner. Synonymised within Microchiroptera by McKenna and Bell (1997: 301).

Entomophaga A. Murray, 1866: xiv.

COMMENTS: When originally proposed, this rank was placed within the Cheiroptera (J. Gray, 1821 [=Chiroptera (Blumenbach, 1779)]) and included the Istiophora (Giebel, 1855: xii, 967 [=Phyllostomidae (Gray, 1825b: 242)]) and Gymnorhini (Giebel, 1855 [=Yangochiroptera (Koopman, 1985 part)]) at unknown rank. Synonymised within Microchiroptera by McKenna and Bell (1997: 301).

HOMONYMS:

Entomophaga Owen, 1839a, marsupials of the Class Mammalia (Order Dasyuromorphia). Name is a synonym of the Infraclass Marsupialia (Illiger, 1811). See individual entry.

Entomophaga Owen, 1859: 52, opossums of the Class Mammalia (Order Didelphimorphia). Name is a synonym of the Order Didelphimorphia (see McKenna & Bell, 1997: 68).

Infraorder Vespertilionia Van Valen, 1979: 109.

COMMENTS: When originally proposed, this rank was placed in the Suborder Microchiroptera (Dobson, 1875a [=Chiroptera (Blumenbach, 1779 part)]) and included the Superfamily Vespertilionoidea (J. Gray, 1821: 299) containing the families Natalidae (J. Gray, 1866c: 90), Vespertilionidae (J. Gray, 1821), Mystacinidae (Dobson, 1875a: 349) and Molossidae (Gervais, 1855b). Synonymised within Microchiroptera by McKenna and Bell (1997: 301).

Infraorder Phyllostomatia Van Valen, 1979: 109.

COMMENTS: When originally proposed, this rank was placed in the Suborder Vespertilionia and included the superfamilies Rhinopomatoidea (Bonaparte, 1838: 112) (consisting of the families Rhinopomatidae (Bonaparte, 1838: 112), Emballonuridae (Gervais, 1855b) and Craseonycteridae (Hill, 1974: 303)), Rhinolophoidea (J. Gray, 1825a) (including the families Rhinolophidae (J. Gray, 1825a), Nycteridae (Van der Hoeven, 1855: 1028) and Megadermatidae (H. Allen, 1864) and Noctilionoidea (J. Gray, 1821: 299) (including the families Noctilionidae (J. Gray, 1821: 299), Mormoopidae (de Saussure, 1860: 286), Phyllostomatidae (Coues & Yarrow, 1875: 80) [=Phyllostomidae (J. Gray, 1825b: 242)] and Desmodontidae (Bonaparte, 1845: 5)). Synonymised within Microchiroptera by McKenna and Bell (1997: 301).

Microchiropteramorpha Simmons & Geisler, 1998: 135, 136.

COMMENTS: When originally proposed as a new rank it was placed below the Order Chiroptera (Blumenbach, 1779) and included the genus † *Australonycteris* Hand *et al.*, 1994: 375, 376; the families † Family Icaronycteridae (Habersetzer & Storch, 1987: 125) and † Archaeonycteridae (Revilliod, 1917: 190); and the unranked Microchiropteraformes. More broadly it included all chiropterans sharing a more recent common ancestor with Microchiroptera than with Megachiroptera.

Microchiropteraformes Simmons & Geisler, 1998: 136.

COMMENTS: When originally proposed as a new taxon it was placed in the Microchiropteramorpha (Simmons & Geisler, 1998) and included the genus † *Eppsinycteris* Hooker, 1996: 284; the families † Palaeochiropterygidae (Revilliod, 1917: 190) and † Hassianycteridae (Habersetzer & Storch, 1987: 129); and the Suborder Microchiroptera (Dobson, 1875a [=Chiroptera (Blumenbach, 1779 part)]).

Family Emballonuridae Gervais, 1855

Tribe Emballonurina Gervais, 1855b: 62, footnote.

TYPE GENUS: *Emballonura* Temminck, 1838a: 1, 22.

COMMENTS: When originally proposed this rank was placed in Family Vespertilionidae (J. Gray, 1821) and included the genus *Urocryptus* Temminck, 1838a: 1, 31 [=*Saccopteryx* Illiger, 1811: 121]; *Diclidurus* Wied-Neuwied, 1820: 1629; *Saccopteryx* Illiger, 1811: 121; *Emballonura* Temminck, 1838a: 1, 22; *Proboscidea* de Spix, 1823: 61 [=*Rhynchonycteris* Peters, 1867f: 477]; *Centronycteris* J. Gray, 1838c: 499; and *Furia* F. Cuvier, 1828: 150 [=*Furipterus* Bonaparte, 1837 [1832–1841]: Fascicolo 21, third unnumbered page of *Plecotus auritus*]. Section Emballonurina and Subfamily Emballonurinae recognised by Flower and Lydekker (1891: 666), with subfamily rank also recognised by Koopman (1984b: 157). Family rank recognised by Dobson (1875a: 349), Simpson (1945: 55) and subsequent authors. Reviewed by Miller (1907: 82), Robbins and Sarich (1988: 1), and T. Griffiths and Smith (1991: 62). The placement of this family has been somewhat unstable as it was placed in the Yinochiroptera when it was first established by Koopman (1985: 26), but placed outside both the Yinochiroptera and Yangochiroptera by McKenna and Bell (1997: 302), Simmons and Geisler (1998: 136) and Simmons (1998: 12). More recently it was placed within the Yangochiroptera (or its synonyms) by authors including Springer *et al.* (2001: 6243), Teeling *et al.* (2002: 1432; 2003: 309), Van Den Bussche and Hoofer (2004: 325) and Miller-Butterworth (2007: 1556, 1558).

Family Brachyura Peters, 1865a: 257.

TYPE GENUS: Does not appear to be based on a genus group name.

COMMENTS: When originally proposed this rank included the genera *Mystacina* J. Gray, 1843d: 296; *Noctilio* Linnaeus, 1766: 88; *Taphozous* É. Geoffroy, 1818; *Emballonura* Temminck, 1838a: 1, 22; *Diclidurus* Wied-Neuwied, 1820: 1629; and *Furia* F. Cuvier, 1828: 150 [=*Furipterus* Bonaparte, 1837 [1832–1841]: Fascicolo 21, third unnumbered page of *Plecotus auritus*].

Family Emballonuridae Dobson, 1875a: 349.

TYPE GENUS: *Emballonura* Temminck, 1838a: 1, 22.

COMMENTS: When originally proposed, this rank was placed in the Suborder Microchiroptera (Dobson, 1875a [=Chiroptera (Blumenbach, 1779 part)]) and included the subfamilies Emballonurinae (Gervais, 1855b) and Molossinae (Gervais, 1855b). Synonymised within Emballonuridae (Gervais, 1855b) by McKenna and Bell (1997: 302).

Tribe? Emballonurini Winge, 1893b: 24.

TYPE GENUS: *Emballonura* Temminck, 1838a: 1, 22.

COMMENTS: When originally proposed, this rank was placed in the Family Emballonuridae (Gervais, 1855b) and included the genera *Mosia* J. Gray, 1843c: 117; *Emballonura* Temminck, 1838a: 1, 22; *Coleura* Peters, 1867f: 479: *Saccopteryx* Illiger, 1811: 121; *Rhynchonycteris* Peters, 1867f: 477; † *Vespertiliavus* Schlosser, 1887: 70; *Diclidurus* Wied-Neuwied, 1820: 1629; and *Taphozous* É. Geoffroy, 1818.

Tribus Emballonuroidea Weber, 1928: 154.

TYPE GENUS: *Emballonura* Temminck, 1838a: 1, 22.

COMMENTS: When originally proposed, this rank was placed in the Suborder Microchiroptera (Dobson, 1875a [=Chiroptera (Blumenbach, 1779 part)]) and included the families Rhinopomatidae (Bonaparte, 1838: 112), Emballonuridae (Gervais, 1855b) and Noctilionidae (J. Gray, 1821: 299). Synonymised within Emballonuridae Gervais, 1855b by McKenna and Bell (1997: 302). Recognised at superfamily rank by Teeling *et al.* (2002: 1432; 2003: 309).

Subfamily Taphozoinae Jerdon, 1867

Subfamily Taphozoinae Jerdon, 1867: 30.

TYPE GENUS: *Taphozous* É. Geoffroy, 1818.

COMMENTS: When originally proposed, this rank was placed in the Family Noctilionidae (J. Gray, 1821: 299) and included the genus *Taphozous* É. Geoffroy, 1818. Subfamily rank recognised by McKenna and Bell (1997: 302) and Simmons (2005a: 381) but not Van Dyck and Strahan (2008: 10, 472). Equivalent to Tribe Taphozoini of McKenna and Bell (1997: 302) according to Simmons (2005a: 381).

Tribe Taphozoini Robbins & Sarich, 1988: 10.

TYPE GENUS: *Taphozous* É. Geoffroy, 1818.

COMMENTS: When originally proposed, this rank was attributed to Jerdon (1867: 30), placed in the Subfamily Taphozoinae (Jerdon, 1867) and included the genera

Taphozous É. Geoffroy, 1818; and *Saccolaimus* Temminck, 1838a. Synonymised within Taphozoini (Jerdon, 1874) by McKenna and Bell (1997: 302).

Saccolaimus Temminck, 1838

Saccolaimus Temminck, 1838a: 14.

TYPE SPECIES: *Taphozous saccolaimus*, Temminck, 1838a [= *Saccolaimus saccolaimus* (Temminck, 1838a)] by absolute tautonymy.

COMMENTS: *Taphozous saccolaimus* Temminck, 1838a satisfies the provisions for availability specified in Article 11(d) of the Code (ICZN, 1985a: 23). Troughton (1925: 313) placed the Australian emballonurids in the genera *Taphozous* and *Saccolaimus*. Subsequent reviews by Ellerman and Morrison-Scott (1951: 106) and Tate (1952a: 606) relegated *Saccolaimus* to a subgenus of *Taphozous* while still recognising the same species as Troughton (1925). Subgenus rank also recognised by Corbet and Hill (1992: 87) and Bates and Harrison (1997: 48). Taxon synonymised within *Taphozous* by Simpson (1945: 56) and Ride (1970: 247). Elevated to genus rank by Barghoorn (1977: 5), Robbins and Sarich (1988: 1), Chimimba and Kitchener (1991: 203), Strahan (1995: 467), Churchill (1998: 54; 2008: 207), Van Dyck and Strahan (2008: 472) and subsequent authors.

Taphonycteris Dobson, 1876a: 548, 555.

TYPE SPECIES: *Taphozous saccolaimus* Temminck, 1838a [= *Saccolaimus saccolaimus* (Temminck, 1838a)] by subsequent designation. See Palmer (1904: 661) and Iredale and Troughton (1934: 100).

COMMENTS: Described as a subgenus of *Taphozous* É. Geoffroy, 1818. Synonymised within *Taphozous* by Simpson (1945: 56) and Corbet and Hill (1992: 85), and within *Saccolaimus* by Troughton (1925: 314), Iredale and Troughton (1934: 100) and most subsequent authors.

Saccolaimus flaviventris (Peters, 1867)

Yellow-bellied Sheath-tailed Bat

Taphozous flaviventris Peters, 1867g: 430.

TYPE LOCALITY: Northern Territory, Australia.

COMMENTS: History of the type discussed by Troughton (1925: 318). Recognised at species rank in *Taphozous* by Waterhouse (1876: 283) and L. Hall and Richards (1979: 21). Placed in *Saccolaimus* by Troughton (1925: 315), Tate (1941d: 3), Troughton (1967: 290), Koopman (1984a: 7), D. Johnson (1964: 471), Reardon and Flavel (1987: 56), Mahoney and Walton (1988e: 114), Flannery (1990: 299; 1995a: 398), Parnaby (1992: 30), Koopman (1993: 159), Bonaccorso (1998: 238), Churchill (1998: 94) and subsequent authors.

Taphozous hargravei Ramsay, 1876c: 81.

TYPE LOCALITY: 'Stanwell' (?=Stanwell Park), New South Wales, Australia.

COMMENTS: Synonymised within *flaviventris* by Troughton (1925: 316), Iredale and Troughton (1934: 100), Honacki *et al.* (1982: 131), Mahoney and Walton (1988e: 114) and Flannery (1990: 299; 1995a: 398).

Taphozous affinis var. *insignis* Leche, 1884: 51.

TYPE LOCALITY: South Australia, Australia.

COMMENTS: Synonymised within *flaviventris* by Troughton (1925: 316), Iredale and Troughton (1934: 100) and Mahoney and Walton (1988e: 115) and Flannery (1990: 299; 1995a: 398).

Saccolaimus mixtus Troughton, 1925

Cape York Sheath-tailed Bat

Saccolaimus mixtus Troughton, 1925: 322, Plates 47–48.

TYPE LOCALITY: Port Moresby, Papua New Guinea.

COMMENTS: Recognised at the species rank within *Saccolaimus* by Tate (1941d: 3) and within *Taphozous* by Troughton (1967: 291), Ride (1970: 170), L. Hall and Richards (1979: 21, 23) and Strahan (1983: 314). Subsequently recognised within *Saccolaimus* by Honacki *et al.* (1982: 131), Strahan (1995: 468), Koopman (1984a: 7), Churchill (1998: 96), Mahoney and Walton (1988e: 115) and most other authors.

Saccolaimus saccolaimus (Temminck, 1838)

Bare-rumped Sheath-tailed Bat

Φ ***Saccolaimus saccolaimus saccolaimus*** (Temminck, 1838)

Φ *Taphozous Saccolaimus* Temminck, 1838a: 14.

TYPE LOCALITY: Hasselt, Java, Indonesia (see Goodwin, 1979: 101).

COMMENTS: Type designated by Tate (1941d: 2). See Medway (1977: 45) for discussion. Included in *Taphozous* by Goodwin (1979: 101). Placed in *Saccolaimus* by Troughton (1925: 328), Honacki *et al.* (1982: 131), Mahoney and Walton (1988e: 115), Bonaccorso (1998: 243) and Churchill (1998: 98). Subspecies recognised within Australia by Churchill (1998: 98) and Van Dyck and Strahan (2008: 476), but not Burbidge *et al.* (2014: 23, 30) who instead recognised *nudicluniatus* as the only subspecies occurring within Australia.

FUTURE TAXONOMIC RESEARCH: Genetic studies by Milne *et al.* (2009: 505) showed that the Northern Territory and

Queensland populations both occur within one clade, which does not support the division of these two geographic populations into separate subspecies. Because specimens from the type locality were not examined it is not clear whether the Australian specimens are the same as the nominate subspecies (which would relegate *nudicluniatus* as a junior synonym) or whether all bats of this species in Australia should be attributed to *nudicluniatus*.

Φ *Saccolaimus saccolaimus crassus* Blyth, 1844

Φ *T. [aphozous] crassus* Blyth, 1844: 491.

TYPE LOCALITY: Mirzapore, Allahabad, Uttar Pradesh, India.

COMMENTS: Synonymised within *saccolaimus* by Corbet and Hill (1992: 87), Flannery (1990: 298), Koopman (1993: 159), and Bates and Harrison (1997: 48). Subspecies rank recognised by Flannery (1995a: 401; 1995b: 328), and Simmons (2005a: 382).

Φ *T. [aphozous] pulcher* Blyth, 1844: 492.

TYPE LOCALITY: Madras, India.

COMMENTS: Synonymised within *saccolaimus* by Flannery (1990: 298; 1995a: 401; 1995b: 328), Corbet and Hill (1992: 87), Koopman (1993: 159), and Bates and Harrison (1997: 48). Included within *crassus* by Simmons (2005a: 382).

Φ *Saccolaimus saccolaimus affinis* (Dobson, 1875)

Φ *Taphozous affinis* Dobson, 1875b: 232.

TYPE LOCALITY: Labuan Island, Borneo, Malaysia.

COMMENTS: Species rank recognised by Tate (1941d: 3). Synonymised within *saccolaimus* by Troughton (1925: 328), Honacki *et al*. (1982: 131), Corbet and Hill (1992: 87), Koopman (1993: 159) and Flannery (1990: 298; 1995a: 401; 1995b: 328). Subspecies recognised by Simmons (2005a: 382).

Φ *Saccolaimus flavimaculatus* Sody, 1931: 355.

TYPE LOCALITY: Kutei, Kalimantan, Borneo, Indonesia.

COMMENTS: Synonymised within *saccolaimus* by Honacki *et al*. (1982: 131), Corbet and Hill (1992: 87), Koopman (1993: 159) and within *affinis* by Simmons (2005a: 382).

Saccolaimus saccolaimus nudicluniatus (De Vis, 1905)

Taphozous nudicluniatus De Vis, 1905: 39.

TYPE LOCALITY: Gowrie Creek, Cardwell, north Queensland, Australia.

COMMENTS: Publication date established from Mahoney and Ride (1975: 115). Species rank recognised within *Saccolaimus* by Troughton (1925: 325), Iredale and Troughton (1934: 98), Tate (1941d: 3) and within *Taphozous* by Ride (1970: 170) and L. Hall and Richards (1979: 21, 23). Synonymised within *saccolaimus* by Goodwin (1979: 102), Honacki *et al*. (1982: 131), Strahan (1983: 312; 1995: 469), Mahoney and Walton (1988e: 115), Flannery (1990: 298), Chimimba and Kitchener (1991: 218) and Churchill (1998: 98). Recognised as a subspecies of *saccolaimus* by Koopman (1984a: 7), Flannery (1995a: 401; 1995b: 328), Reardon (1999a: 12), Simmons (2005a: 382) and Clayton *et al*. (2006: 109). Recognised within Australia by Clayton *et al*. (2006: 109), Van Dyck and Strahan (2008: 476) and Burbidge *et al*. (2014: 23, 30).

Φ *Taphozous granti* Thomas, 1911c: 378.

TYPE LOCALITY: Paramau, Mimika River, south Netherlands, New Guinea.

COMMENTS: Synonymised within *saccolaimus* by Koopman (1993: 159) and Flannery (1990: 298; 1995a: 401; 1995b: 328). Synonymised within *nudicluniatus* by Troughton (1925: 325) and Simmons (2005a: 382).

Φ *Saccolaimus saccolaimus pluto* (Miller, 1910)

Φ *Taphozous pluto* Miller, 1910: 396.

TYPE LOCALITY: Mercedes, near Zamboanga, Mindanao Island, Philippines.

COMMENTS: Recognised as a synonym within *saccolaimus* by Corbet and Hill (1992: 87) and subspecies by Simmons (2005a: 382).

Φ *Taphonycteris capito* Hollister, 1913: 308

TYPE LOCALITY: Pandan, Catanduanes Islands, Philippines.

COMMENTS: Synonymised within *saccolaimus* by Corbet and Hill (1992: 87) and within *pluto* by Simmons (2005a: 382).

Taphozous É. Geoffroy, 1818

Taphozous É. Geoffroy, 1818: 113, 126; Plate 3, Fig. 1.

TYPE SPECIES: Φ *Taphozous perforatus* É. Geoffroy, 1818: 113, 126, by subsequent designation. See Miller (1907: 93).

COMMENTS: The date of publication of this work has been controversial as it has been given as 1813 by Anon (1838: 35) and followed by various authors including I. Geoffroy (1847: 425), Flourens (1853: xlviii–xlix), Miller (1907: 93), Tate (1941d: 1), and Mahoney and Walton (1988e: 116). The publication date of 1818 is derived from Sherborn (1897a: 288), which has typically been recognised by more recent authors including Miller and Rehn (1901: 271), Lyon (1914:

217), Corbet and Hill (1992: 85), Reardon (1999b: 22) and Simmons (2005a: 382). The name *Taphozous* was also described by Oken (1816: 926) but this work is considered non-binomial (see ICZN, 1956: 39). Troughton (1925: 313) placed the Australian emballonurids in the genera *Taphozous* and *Saccolaimus*.

Thaphozous Bowdich, 1821: 30.

TYPE SPECIES: Incorrect subsequent spelling of *Taphozous* É. Geoffroy, 1818.

COMMENTS: Synonymised within *Taphozous* by Palmer (1904: 661).

Thaphosores J. Gray, 1821: 300.

TYPE SPECIES: Not given.

COMMENTS: Appears to be an incorrect subsequent spelling of *Taphozous* as 'Geoff' is given as the author.

Liponycteris Thomas, 1922j: 267.

TYPE SPECIES: Φ *Taphozous nudiventris* Cretzschmar, 1830: 70 (as *L. nudiventris*) by original designation.

COMMENTS: Synonymised within *Taphozous* by Simpson (1945: 56), Corbet and Hill (1992: 85) and Simmons (2005a: 382).

Taphozous australis Gould, 1854

Coastal Sheath-tailed Bat

Taphozous australis Gould, 1854 [1845–1863]: Text Plate 32.

TYPE LOCALITY: Albany Island, Queensland, Australia.

COMMENTS: Recognised in *Taphozous* by J. Ogilby (1892: 96), Troughton (1925: 332), and Iredale and Troughton (1934: 99), and within *Saccolaimus* by Troughton (1967: 289). Placed back in *Taphozous* by Mahoney and Walton (1988e: 116) at the species rank and followed by subsequent authors.

Taphozous fumosus De Vis, 1905: 37.

TYPE LOCALITY: Gowrie Creek, Cardwell, north Queensland, Australia.

COMMENTS: Publication date established from Mahoney and Ride (1975: 157). Synonymised within *australis* by Troughton (1925: 332), Iredale and Troughton (1934: 99), Honacki *et al.* (1982: 132), Mahoney and Walton (1988e: 116) and subsequent authors.

Taphozous georgianus Thomas, 1915

Common Sheath-tailed Bat

Taphozous australis georgianus Thomas, 1915c: 62.

TYPE LOCALITY: King Gorge Sound, Western Australia, Australia.

COMMENTS: Species rank recognised within *Taphozous* by Troughton (1925: 336), Iredale and Troughton (1934: 99), Tate (1941d: 5), L. Hall and Richards (1979: 21, 23), and within *Saccolaimus* by Troughton (1967: 289). Tate (1952a: 607) questioned whether this taxon differed in any substantial way from *T. australis*, as did D. Johnson (1964: 470). McKean and Price (1967: 104) further examined the relationship between these two taxa and suggested that they should be separated, which was followed by D. Johnson (1964: 470), Chimimba and Kitchener (1991: 225, 229) and subsequent authors.

Taphozous hilli Kitchener, 1980

Hill's Sheath-tailed Bat

Taphozous hilli Kitchener, 1980b: 161, 162.

TYPE LOCALITY: Hamersley Range National Park, Western Australia, Australia.

COMMENTS: Species separated from *georgianus*.

Taphozous kapalgensis McKean & Friend, 1979

Arnhem Sheath-tailed Bat

Taphozous kapalgensis McKean & Friend, 1979: 239.

TYPE LOCALITY: 'Kapalga', at the edge of the western flood plain of the South Alligator River, near Rookery Point, Northern Territory, Australia. (12°32′S, 132°23′E)

COMMENTS: Taxon has been recognised by subsequent authorities.

Taphozous troughtoni Tate, 1952

Troughton's Sheath-tailed Bat

Taphozous troughtoni Tate, 1952a: 572, 607.

TYPE LOCALITY: Rifle Creek, 10 miles east of Mount Isa, north west Queensland, Australia.

COMMENTS: McKean and Price (1967: 104) tentatively decided that *troughtoni* should be retained as a subspecies of *T. georgianus*. Similarly Ride (1970: 170) did not recognise *T. troughtoni*. It was also considered a synonym of *Taphozous georgianus* by Honacki *et al.* (1982: 132), Koopman (1984a: 5; 1993: 160) and Mahoney and Walton (1988e: 117). Recognised at species rank by Chimimba and Kitchener (1991: 232), Koopman (1994: 43), Churchill (1998: 107; 2008: 175), Reardon (1999a: 12) and Reardon and Thomson (2002: 1).

Family Molossidae Gervais, 1855

Tribe Molossina Gervais, 1855b: 53, footnote.

TYPE GENUS: *Molossus* É. Geoffroy, 1805b: 151.

COMMENTS: When originally proposed, this rank was placed in the Family Vespertilionidae (J. Gray, 1821) and included the genera *Molossus* É. Geoffroy, 1805b: 151; *Promops* Gervais, 1855b: 58; and *Nyctinomus* É. Geoffroy, 1818: 114, 128 [=*Tadarida* Rafinesque, 1814a: 55]. Recognised as a subfamily within the Family Emballonuridae by Flower and Lydekker (1891: 669) and as the Superfamily Molossoidea by Simmons (1998: 12), but does not appear to have been recognised by other authors. Family reviewed by Freeman (1981: 1), Legendre (1984: 399) and Allison (1989: 892). Simpson (1945: 60) and Corbet and Hill (1992: 156) gave the family authorship to Gill (1872: 17) as he was the first to use the current spelling of the family name. The Australian species of the Family Molossidae have typically been placed within *Mormopterus* Peters, 1865a: 258, with the exception of the species *jobensis* and *australis*. The Australian '*Mormopterus*' were reviewed by Adams *et al.* (1988: 315) who proposed there are at least five undescribed species in Australia (with a sixth species of undetermined generic affinities also found), although they did not allocate any of the existing names to the identified taxa. Five of these were recognised as operational taxonomic units by Van Dyck and Strahan (2008: 493–499), of which *eleryi* has since been described, with a sixth unnamed species being recognised by Menkhorst and Knight (2011: 18). Churchill (2008: 193) referred to research by Terry Reardon who suggested that *Mormopterus* does not occur in Australia, and reinstated *Micronomus* for *norfolkensis*. Subsequently Reardon *et al.* (2014: 109) taxonomically reviewed the work of Adams *et al.* (1988: 315) and recognised 11 species in the family including *australis* within *Austronomus*, *jobensis* within *Chaerephon*, *norfolkensis* within *Mormopterus* (*Micronomus*), seven species in the new subgenus *Mormopterus (Ozimops)* – including three new species, and *eleryi* within *Mormopterus (Setirostris)*. Historically Australian representatives of this genus has included *Mormopterus beccarii astrolabiensis* (Meyer, 1899: iii, 19) and *Mormopterus loriae* (Thomas, 1897a: 609), but these were removed from the Australian fauna by Reardon *et al.* (2014: 110). The names *Micronomus*, *Ozimops* and *Setirostris* were placed as subgenera within the genus *Mormopterus* Peters 1865a: 258 by Reardon *et al.* (2014: 62, 117–118) but these are recognised here as distinct genera.

Family Molossi Peters, 1865a: 258.

TYPE GENUS: *Molossus* É. Geoffroy, 1805b: 151.
COMMENTS: When originally proposed, this rank included the genera *Dysopes* Illiger, 1811: 122 [= *Molossus* É. Geoffroy, 1805b: 151]; *Molossus* É. Geoffroy, 1805b: 151; *Promops* Gervais, 1855b: 58; *Mormopterus* Peters, 1865a: 258; *Nyctinomus* É. Geoffroy, 1818: 114, 128 [=*Tadarida* Rafinesque, 1814a: 55]; and *Chiromeles* [=*Cheiromeles*] Horsfield, 1824: No. VIII].

Family Molossidae Gill, 1872: 17.

TYPE GENUS: *Molossus* É. Geoffroy, 1805b: 151.
COMMENTS: When originally proposed, this rank was placed in the Suborder Animalivora (Gill, 1872 [=Chiroptera (Blumenbach, 1779)]) and included the genus *Molossus* É. Geoffroy, 1805b: 151. Recognised as the author of the family name by Simpson (1945: 60), but synonymised within Molossidae (Gervais, 1855b) by McKenna and Bell (1997: 313) and Simmons (2005a: 432).

Tribe? Molossini Winge, 1893b: 24.

TYPE GENUS: *Molossus* É. Geoffroy, 1805b: 151.
COMMENTS: When originally proposed, this rank was placed in the Family Vespertilionidae (J. Gray, 1821) and included the genera *Mystacina* J. Gray, 1843d: 296; *Nyctinomus* É. Geoffroy, 1818: 114, 128 [=*Tadarida* Rafinesque, 1814a: 55]; *Chiromeles* [=*Cheiromeles*] Horsfield, 1824: No. VIII; and *Molossus* É. Geoffroy, 1805b: 151.

Superfamily Molossoidea Simmons, 1998: 12.

TYPE GENUS: *Molossus* É. Geoffroy, 1805b: 151.
COMMENTS: When originally proposed as a new rank it was placed in the Infraorder Yangochiroptera (Koopman, 1985) and included the families Molossidae (Gervais, 1855b) and Antrozoidae (Simmons, 1998: 12) [=Subfamily Antrozoinae (Miller, 1897: 41)].

Subfamily Molossinae Gervais, 1855

Tribe Molossina Gervais, 1855b: 53, footnote.

TYPE GENUS: *Molossus* É. Geoffroy, 1805b: 151.
COMMENTS: When originally proposed, this rank was placed in the Family Vespertilionidae (J. Gray, 1821) and included the genus *Molossus* É. Geoffroy, 1805b: 151; *Promops* Gervais, 1855b: 58; and *Nyctinomus* É. Geoffroy, 1818: 114, 128 [=*Tadarida* Rafinesque, 1814a: 55]. Recognised as a new subfamily within the Family Molossidae by Legendre (1984: 399, 425), McKenna and Bell (1997: 314) and Simmons (2005a: 432). Described by Simmons (2005a: 432) as equivalent to Molossidae *sensu* Freeman (1981: 150), Legendre (1984: 425) and Koopman (1993: 232; 1994: 135).

Subfamily Cheiromelinae Legendre, 1984: 399, 425.

TYPE GENUS: *Cheiromeles* Horsfield, 1824: No. VIII.
COMMENTS: When originally proposed as a new rank it was placed in the Family Molossidae (Gervais, 1855b) and included the genus *Cheiromeles* Horsfield, 1824: No. VIII. Synonymised within the Subfamily Molossinae (Gervais, 1855) by McKenna and Bell (1997: 314) and Simmons (2005a: 432).

Subfamily Molossinae Legendre, 1984: 399, 425.

TYPE GENUS: *Molossus* É. Geoffroy, 1805b: 151.
COMMENTS: When originally proposed as a new rank it was placed in the Family Molossidae (Gervais, 1855b) and

included the genera *Molossus* É. Geoffroy, 1805b: 151; *Eumops* Miller, 1906b: 85; *Molossops* Peters, 1865d: 575 including the subgenera *Cynomops* Thomas, 1920d: 189 and *Neoplatymops* Peterson 1965: 3; *Myopterus* É. Geoffroy, 1818: 113 and *Promops* Gervais, 1855b: 58. Rank introduced as new by Legendre, 1984: 399, 425. Synonymised within Subfamily Molossinae (Gervais, 1855) by McKenna and Bell (1997: 314).

Subfamily Tadaridinae Legendre, 1984: 399, 426.

TYPE GENUS: *Tadarida* Rafinesque, 1814a: 55.

COMMENTS: When originally proposed as a new rank it was placed in the Family Molossidae (Gervais, 1855b) and included the genera *Tadarida* Rafinesque, 1814a: 55; *Mormopterus* Peters, 1865a: 258; *Nyctinomops* Miller, 1902b: 393; *Otomops* Thomas, 1913d: 90; and *Rhizomops* Legendre, 1984: 415, 427 [=*Tadarida* Rafinesque, 1814a: 55]. Synonymised within the Subfamily Molossinae by McKenna and Bell (1997: 314) and Simmons (2005a: 432).

Austronomus Troughton, 1941

Austronomus Troughton, 1941: 360.

TYPE SPECIES: *Molossus australis* J. Gray, 1838c [=*Austronomus australis* (J. Gray, 1838c)] by monotypy.

COMMENTS: Initially named by Iredale and Troughton, 1934: xi, 100 as a *nomen nudum*, but the name was validated by Troughton (1941: 360). Synonymised within *Nyctinomus* É. Geoffroy, 1818: 114, 128 [=*Tadarida* Rafinesque, 1814a: 55] by Mahoney and Walton (1988f: 149), Strahan (1995: 487) and Churchill (1998: 212). Synonymised within *Tadarida* Rafinesque, 1814a: 55 by Hill (1961: 30), Ride (1970: 241), Freeman (1981: 165), Koopman (1993: 240), McKenna and Bell (1997: 315), Simmons (2005a: 449) and Van Dyck and Strahan (2008: 501). Genus *Austronomus* recognised by Troughton (1967: 292) and Churchill (2008: 190) who referred to research by Terry Reardon suggesting *Nyctinomus* É. Geoffroy, 1818: 114, 128 is an invalid name and *Tadarida* Rafinesque, 1814a: 55 does not occur in Australia. But see Reardon (2009b: 52). Subsequently the genus was recognised for *australis* by Ammerman *et al.* (2012: 13), Van Dyck *et al.* (2013: 29, 135), and Reardon *et al.* (2014: 109).

Austronomus australis (Gray, 1838)

White-striped Free-tailed Bat

Molossus australis J. Gray, 1838c: 501.

TYPE LOCALITY: Syntypes from New South Wales, Australia. Troughton (1941: 360) believes the original specimen came from near the Camden home of Major Macarthur, New South Wales.

COMMENTS: Species recognised within *Molossus* by Gould (1858 [1845–1863]: Text to Plate 31). Placed in the genus *Nyctinomus* É. Geoffroy, 1818: 114, 128 [=*Tadarida* Rafinesque, 1814a: 55] by Mahoney and Walton (1988f: 149), Strahan (1995: 487) and Churchill (1998: 212). Placed in *Tadarida* Rafinesque, 1814a: 55 by Hill (1961: 35), Ride (1970: 168), L. Hall and Richards (1979: 33, 35), Freeman (1981: 165), Koopman (1982: 23; 1984a: 32; 1993: 240), Strahan (1983: 319), Reardon and Flavel (1987: 57), Adams *et al.* (1988: 315), Parnaby (1992: 28), Reardon (1999a: 12), Simmons (2005a: 449), Van Dyck and Strahan (2008: 501), and Menkhorst and Knight (2011: 19). Species included within *Austronomus* by Troughton (1967: 292), Churchill (2008: 190), Van Dyck *et al.* (2013: 29, 135) and Reardon *et al.* (2014: 110, 115).

Nyctinomus albidus Leche, 1884: 50.

TYPE LOCALITY: Syntypes from South Australia, Australia.

COMMENTS: Synonymised within *australis* by Iredale and Troughton (1934: 100), Laurie and Hill (1954: 63) and Hill (1961: 35).

Nyctinomus australis atratus Thomas, 1924c: 456.

TYPE LOCALITY: Ooldea, South Australia, Australia.

COMMENTS: Synonymised within *australis* by Laurie and Hill (1954: 63), Hill (1961: 35) and Koopman (1984a: 32). Considered a subspecies of *australis* by Iredale and Troughton (1934: 100), Strahan (1983: 319; 1995: 488), and Clayton *et al.* (2006: 110), but not Burbidge *et al.* (2014: 23, 30).

Chaerephon Dobson, 1874

Chaerephon Dobson, 1874: 144.

TYPE SPECIES: Φ *Molossus (Nyctinomus) johorensis* Dobson, 1873b: 22 (as *Nyctinomus johorensis*) [= Φ *Chaerephon johorensis* (Dobson, 1873b: 22)] by monotypy.

COMMENTS: Described as a subgenus of *Nyctinomus* É. Geoffroy, 1818: 114, 128 [=*Tadarida* Rafinesque, 1814a: 55]. Synonymised within *Tadarida* Rafinesque, 1814a: 55 by Simpson (1945: 60), Hill (1961: 30) and Ride (1970: 242). Subgenus within *Tadarida* Rafinesque, 1814a: 55 recognised by Ellerman and Morrison-Scott (1951: 135), Koopman (1982: 23), Hill (1983: 195), Legendre (1984: 399, 415), and Corbet and Hill (1992: 159). Recognised at genus rank by Andersen (1907: 5), Koopman (1993: 232) and Simmons (2005a: 432). The species *C. plicatus* is recorded as occurring on the Cocos (Keeling) Islands, which is a territory of Australia by Simmons (2005a: 434).

HOMONYMS:

Chaerephon Godman in Godman and Salvin, 1900: 462, 474, butterflies of the Class Insecta (Order Lepidoptera,

Family Hesperiidae). Genus is a synonym of *Polites* Scudder, 1872: 78. See Pelham (2008: 76).

Choerephon Dobson, 1878: 431.

TYPE SPECIES: Incorrect subsequent spelling of *Chaerephon* Dobson, 1874.

COMMENTS: Recognised as a subgenus of *Nyctinomus* É. Geoffroy, 1818: 114, 128 [=*Tadarida* Rafinesque, 1814a: 55] when described. Spelling does not appear to have subsequently used.

Lophomops J. Allen, 1917b: 460.

TYPE SPECIES: Φ *Chaerephon (Lophomops) chapini* J. Allen, 1817b: 461 [= Φ *Chaerephon chapini* J. Allen, 1817b: 461] by original designation.

COMMENTS: Described as a subgenus of *Chaerephon*. Synonymised within *Chaerephon* by Freeman (1981: 150), McKenna and Bell (1997: 315) and Simmons (2005a: 432).

Chaerephon jobensis (Miller, 1902)

Greater Northern Free-tailed Bat

Φ *Chaerephon jobensis jobensis* (Miller, 1902)

Φ *Nyctinomus jobensis* Miller, 1902c: 246.

TYPE LOCALITY: Ansus, Pulau Japen (as Jobie Island), western Irian Jaya, Indonesia.

COMMENTS: Placed in the genus *Tadarida* Rafinesque, 1814a: 55 as a subspecies of *plicata* by Laurie and Hill (1954: 63) and Felten (1964: 8), but recognised as a distinct taxon within *Tadarida* Rafinesque, 1814a: 55 by subsequent authors including L. Hall and Richards (1979: 33, 37), Hill (1983: 195), Koopman (1982: 23) (in subgenus *Chaerephon*) and Corbet and Hill (1992: 159) who also placed this taxon in the subgenus *Chaerephon*. Transferred to *Chaerephon* at genus rank by Koopman (1984a: 32), Adams *et al.* (1988: 315), Mahoney and Walton (1988f: 146) and subsequent authors. Has included *bregullae* Felten (1964: 9) and *solomonis* Troughton, 1931: 207 but these were treated as separate species by Flannery (1995b: 405–406), which was followed by Bonaccorso (1998: 411) and Simmons (2005a: 434).

Chaerephon jobensis colonicus (Thomas, 1906)

Nyctinomus plicatus colonicus Thomas, 1906d: 537.

TYPE LOCALITY: Alexandria, Northern Territory, Australia.

COMMENTS: Species rank recognised within *Chaerephon* by Iredale and Troughton (1934: 101) and Troughton (1967: 293). Synonymised within *jobensis* by Ride (1970: 248). Reduced to subspecies of *Chaerephon plicatus* by Tate (1941e: 2) and within *T. jobensis* by Hill (1961: 55). Synonymised within *Nyctinomus jobensis* by Mahoney and Walton (1988f: 146). Subspecies recognised by Flannery (1990: 370; 1995a: 476; 1995b: 405), Reardon (1999a: 12), Bonaccorso (1998: 411), Simmons (2005a: 434) and Clayton *et al.* (2006: 110). The Australian representative of this species was given as '*C. j. plicatus* (Thomas, 1906)' [=*C. plicatus* (Buchannan, 1800: 261); see Simmons, 2005a: 434] by Clayton *et al.* (2006: 110), which appears to be an error because of the authorship. Ingleby and Colgan (2003: 13), Van Dyck and Strahan (2008: 486) and Burbidge *et al.* (2014: 23, 30) give the Australian representative as *C. j. colonicus*, which has been followed here.

Micronomus Troughton, 1944

Micronomus Troughton, 1944: 360.

TYPE SPECIES: *Molossus norfolkensis* J. Gray, 1839b [=*Micronomus norfolkensis* (J. Gray, 1839b)] by original designation.

COMMENTS: When described by Iredale and Troughton (1934: xi, 100) it was a *nomen nudum*. Type description assigned to Troughton (1944: 360) by Freeman (1981: 160), Mahoney and Walton (1988f: 147), McKenna and Bell (1997: 314) and Simmons (2005a: 444). Genus proposed to be invalid by Tate (1941e: 4), synonymised within *Tadarida* Rafinesque, 1814a: 55 by Hill (1961: 30) and Ride (1970: 244), and synonymised within *Mormopterus* Peters, 1865a: 258 by Freeman (1981: 160), Mahoney and Walton (1988f: 147), McKenna and Bell (1997: 314) and Simmons (2005a: 444). *Micronomus* recognised as a subgenus of *Mormopterus* Peters, 1865a: 258 by Tate (1952a: 604), Laurie and Hill (1954: 63), Legendre (1984: 399, 411, 426) and Hand *et al.* (1999: 291), but as a valid genus by Troughton (1967: 292) and Churchill (2008: 193), but see Reardon (2009b: 52). Recognised as a genus to include *norfolkensis* by Van Dyck *et al.* (2013: 29, 136), which is followed here.

Micronomus norfolkensis (J. Gray, 1839)

Eastern Coastal Free-tailed Bat

Mol. [ossus] Norfolkensis J. Gray, 1839b: 7.

TYPE LOCALITY: Norfolk Island. Placed as Sydney, New South Wales, Australia by Iredale and Troughton (1934: 100).

COMMENTS The type locality was recorded as Norfolk Island, but this appears to be in error though considerable confusion exists over what might be the true type locality; it was first thought to occur on mainland Australia by Dobson (1877b: 733; see Strahan (1980b: 6) and Reardon *et al.* (2008) for discussion). It was proposed by Strahan (1980b: 6) that it was possible that this species had once indeed existed on Norfolk Island, but none have been caught on the island in modern times (M. Gordon, 1984: 11; Tidemann, 1987b: 33; Hoye, 2011: 297). Recognised as a distinct species within *Nyctinomys* [sic =*Nyctinomus* É. Geoffroy, 1818: 114, 128] [=*Tadarida* Rafinesque, 1814a: 55] by Tate (1952a: 604) and then placed within *Tadarida* Rafinesque, 1814a: 55 by Hill (1961: 44), which was subsequently followed by Felten (1964: 6), L. Hall and Richards (1979: 35), Winter and Allison (1980: 34) and M. Gordon (1984: 4). Considerable doubt as to the status of this species was raised by Koopman (1993: 238) and Simmons (2005a: 445). Placed within *Micronomus* by Troughton (1967: 292), Churchill (2008: 193) and Van Dyck *et al.* (2013: 29, 136). Included in the genus *Mormopterus* Peters, 1865a: 258 by Freeman (1981: 161), Mahoney and Walton (1988f: 147), Parnaby (1992: 29), Koopman (1993: 238), Churchill (1998: 202), Simmons (2005a: 444), Reardon *et al.* (2008: 1), and Van Dyck and Strahan (2008: 491). Reviewed by Reardon *et al.* (2008: 22) who noted the type locality was not established beyond doubt, but the species was defined and confirmed on mainland Australia. Placed within the subgenus *Micronomus* by Reardon *et al.* (2014: 117).

Ozimops Reardon et al., 2014

Ozimops Reardon *et al.*, 2014: 109, 118.

TYPE SPECIES: *Mormopterus planiceps* (Peters, 1866a) [=*Ozimops planiceps* (Peters, 1866a)] by original designation.

COMMENTS: Proposed as a subgenus of *Mormopterus* Peters, 1865a: 258. Includes the species *cobourgianus*, *lumsdenae*, *kitcheneri*, *halli*, *petersi*, *planiceps* and *ridei* from Australia and the Indo-Papuan species *Mormopterus beccarii* Peters, 1881: 484; *Mormopterus beccarii astrolabiensis* (Meyer, 1899: iii, 19); and *Mormopterus loriae* (Thomas, 1897a: 609).

Ozimops cobourgianus (Johnson, 1959)

Northern Coastal Free-tailed Bat

Tadarida loriae cobourgiana D. Johnson, 1959: 185.

TYPE LOCALITY: Black Rock Point, on north shore of Van Diemen Gulf, 15 miles south east of Cape Don lighthouse, Cobourg Peninsula, Northern Territory, Australia. (11°26′S, 131°56′E)

COMMENTS: Subspecies rank within *Tadarida planiceps* [=*Ozimops planiceps*] recognised by Hill (1961: 46), and within *planiceps* by Koopman (1984: 31) and as '*coburgiana*' by Koopman (1994: 136). Synonymised within *Mormopterus loriae* (Thomas, 1897a: 609) by Freeman (1981: 161) and *Mormopterus planiceps* [=*Ozimops planiceps*] by Mahoney and Walton (1988f: 148). Subspecies recognised within *Mormopterus loriae* (Thomas, 1897a: 609) by Felten (1964: 8), Hall and Richards (1979: 37), as '*coburgiana*', Strahan (1983: 324; 1995: 483), Flannery (1990: 368; 1995a: 479), Bonaccorso (1998: 416), Reardon (1999a: 12), who expected it to be recognised as full species, Simmons (2005a: 445), and Van Dyck and Strahan (2008: 488), but not by Clayton *et al.* (2006: 110) or Burbidge *et al.* (2014: 23, 30). Taxon listed as *Mormopterus* 'Species 5 (populations U and V)' by Adams *et al.* (1988: 317, 321, 323), which was subsequently recognised as this taxon by Churchill (1998: 207), but see Reardon (2009b: 52). Species recognised with current name by Churchill (2008: 197) and with amended species name gender, by Van Dyck *et al.* (2013: 29, 137) and confirmed by Reardon *et al.* (2014: 131) who placed it within the subgenus *Ozimops*.

Ozimops halli (Reardon, McKenzie & Adams in Reardon *et al.*, 2014)

Cape York Free-tailed Bat

Mormopterus halli Reardon, McKenzie & Adams in Reardon *et al.*, 2014: 112, 132.

TYPE LOCALITY: Ironbark Dam, Oyala Thumotang National Park, Queensland, 13.625°S 142.801°E.

COMMENTS: Population treated as belonging under *Mormopterus loriae* (Thomas, 1897a: 609), as *Mormopterus loriae ridei* Felten, 1964 [=*Ozimops ridei*], by authors including Reardon (1999a: 12), and Strahan (1995: 482; 2008: 490), or under *M. planiceps* [=*Ozimops planiceps*] (Koopman, 1994: 136). This species equates to 'Species 5 (populations S and T)' from Adams *et al.* (1988: 317, 321, 323) from which time it was recognised as an unnamed taxon by Churchill (1998: 206) or as potentially a subspecies *M. ridei* [=*Ozimops ridei*] (Churchill, 2008: 198). Placed in the subgenus *Ozimops* by Reardon et al. (2014: 132).

Ozimops kitcheneri (McKenzie, Reardon & Adams in Reardon *et al.*, 2014)

Western Free-tailed Bat

Mormopterus kitcheneri Reardon, McKenzie & Adams in Reardon *et al.* 2014: 112, 126.

TYPE LOCALITY: 20 km north-west of Balladonia, Western Australia. 32.252°S 123.431°E.

COMMENTS: This species equates to Species 4 (population O) from Adams *et al.* (1988: 317, 321, 324), which was recognised as an unnamed taxon by Churchill (1998: 210; 2008: 201 part), Reardon (1999a: 12), Van Dyck and Strahan (2008: 496) and Van Dyck *et al.* (2013: 139). Placed in the subgenus *Ozimops* by Reardon et al. (2014: 126).

Ozimops lumsdenae (Reardon, McKenzie & Adams in Reardon *et al.*, 2014)

Northern Free-tailed Bat

Mormopterus lumsdenae Reardon, McKenzie & Adams in Reardon *et al.* 2014: 112, 127.

TYPE LOCALITY: Roadside dam, Peninsula Developmental Road, approximately 16 km north of Coen, Queensland. 14.809°S, 143.146°E.

COMMENTS: This species equates to Species 1 from Adams *et al.* (1988: 316, 322, 323). Taxon initially attributed to *Mormopterus beccarii* Peters, 1881: 484 by Winter and Allison (1980: 34) and followed by subsequent authors including Freeman (1981: 160), Mahoney and Walton (1989: 147), Koopman (1993: 237; 1994: 136), Churchill (1998: 200; 2008: 195), Simmons (2005a: 444), Van Dyck and Strahan (2008: 486), Menkhorst and Knight (2010: 19) and Van Dyck *et al.* (2013: 136). Australian forms allocated to *Tadarida beccarii astrolabiensis* [=*Mormopterus beccarii astrolabiensis* (Meyer, 1899: iii, 19)] by Hill (1983: 196), *Mormopterus beccarii astrolabiensis* (Meyer, 1899: iii, 19) by Simmons (1995: 444) and Koopman (1984: 32)(inferred), and within *Mormopterus astrolabiensis* (Meyer, 1899: iii, 19) by Peterson (1985: 208). Placed in the subgenus *Ozimops* by Reardon *et al.* (2014: 127).

Ozimops petersi (Leche 1884)

Inland Free-tailed Bat

Nyctinomous petersi Leche, 1884: 49.

TYPE LOCALITY: Syntypes from South Australia, Australia.

COMMENTS: Recognised at the species rank by J. Ogilby (1892: 99). Taxon synonymised within *Nyctinomus planiceps* [=*Ozimops planiceps*] by Thomas (1907b: 765) and Troughton (1926: 86); *Micronomus planiceps* [=*Ozimops planiceps*] by Iredale and Troughton (1934: 101); *Tadarida planiceps* [=*Ozimops planiceps*] by Hill (1961: 45) and Felten (1964: 3); and *Mormopterus planiceps* [=*Ozimops planiceps*] by Freeman (1981: 161), Strahan (1983: 322; 1995: 485), Mahoney and Walton (1988: 148) and Simmons (2005a: 446). Recognised at species rank within *Nyctinomus* É. Geoffroy, 1818: 114, 128 [=*Tadarida* Rafinesque, 1814a: 55] by Wood Jones (1925 [1923–1925]): 393) and *Mormopterus* Peters, 1865a: 258 by Peterson *et al.* (1985: 1). Taxon recognised as Species 3 in Adams *et al.* (1988: 317, 321, 324), which was subsequently recognised by Churchill (1998: 204; 2008: 200), Reardon (1999a: 12), Van Dyck and Strahan (2008: 494), and Van Dyck *et al.* (2013: 138), until the name *petersi* was formally resurrected and lectotype established by Reardon *et al.* (2014: 124) who placed it in the subgenus *Ozimops*.

Ozimops planiceps (Peters, 1866)

Southern Free-tailed Bat

Nyctinomus planiceps Peters, 1866a: 23.

TYPE LOCALITY: Sydney, Australia.

COMMENTS: Sydney designated as the type locality by Mertens (1925: 20), but believed to be Western Australia by Iredale and Troughton (1934: 101). Placed in *Micronomus* by Iredale and Troughton (1934: 101). The taxonomy of this species has been problematical due to the uncertainty of the type locality (Koopman, 1984a: 29). Synonymised within *norfolcensis* [sic] by Dobson (1877b: 732) and Peters (1881: 484) but given species rank in *Nyctinomus* É. Geoffroy, 1818: 114, 128 [=*Tadarida* Rafinesque, 1814a: 55] by Mertens (1925: 20), in *Micronomus* by Troughton (1967: 293), and within *Mormopterus* Peters, 1865a: 258 by Strahan (1983: 322; 1995: 485), Koopman (1984a: 31), Legendre (1984: 408, 426), Reardon and Flavel (1987: 59), Parnaby (1992: 29), Simmons (2005a: 446) and Reardon *et al.* (2014: 119). Placed in the genus *Tadarida* Rafinesque, 1814a: 55 by Hill (1961: 45), Felten (1964: 3), L. Hall and Richards (1979: 33, 37), Koopman (1982: 24) in subgenus *Mormopterus* Peters, 1865a: 258. Taxon equates to Species 4 (populations P, Q, R) from Adams *et al.* (1988: 317, 321, 324), which was recognised by Churchill (1998: 208; 2008: 201 part), Reardon (1999a: 12), Van Dyck and Strahan (2008: 497), Van Dyck *et al.* (2013: 139). Species 4 (populations P, Q, R) attributed to *planiceps* by Reardon *et al.* (2014: 112). Placed in the subgenus *Ozimops* by Reardon *et al.* (2014: 119).

Molossus Wilcoxii Krefft, 1873b: 694.

TYPE LOCALITY: Syntypes from Clarence River, New South Wales to Rockhampton, Queensland, Australia.

COMMENTS: Name introduced by Krefft (1864: 6; 1871b: 4) but this was reduced to a *nomen nudum* by Iredale and Troughton (1934: 101). Lectotype designated by Reardon *et al.* (2014: 124), but as a junior synonym of *planiceps*. Synonymised within *norfolcensis* [sic] by Dobson (1877b: 732), *planiceps* by Freeman (1981: 161) and within *norfolkensis* by Iredale and Troughton (1934: 101), Hill (1961: 44), Mahoney and Walton (1988f: 147), Koopman (1993: 238), Simmons (2005a: 445) and most subsequent authors, though synonymised within *planiceps* by Felten (1964: 3) and Freeman (1981: 161). The status of this taxon was discussed in detail by Reardon *et al.* (2014: 124) who placed it as a junior synonym on *Mormopterus planiceps* [=*Ozimops planiceps*].

Ozimops ridei (Felten, 1964)

Ride's Free-tailed Bat

Tadarida loriae ridei Felten, 1964: 6.

TYPE LOCALITY: Cairns, Queensland, Australia.

COMMENTS: Taxon synonymised within *Mormopterus loriae* (Thomas, 1897a: 609) by Freeman (1981: 161) and within *Mormopterus planiceps* [=*Ozimops planiceps*] by Mahoney and Walton (1988f: 148) and Koopman (1993: 238). Subspecies rank recognised within *Mormopterus loriae* (Thomas, 1897a: 609) by McKean and Price (1967: 112), Hall and Richards (1979: 37), Strahan (1983: 324; 1995: 483), Flannery (1990: 368; 1995a: 479), Bonaccorso (1998:

416), Reardon (1999a: 12) who expected it to be recognised as full species, Simmons (2005a: 445), and Van Dyck and Strahan (2008: 490), but not by Clayton *et al.* (2006: 110) or Burbidge *et al.* (2014: 23, 30). Species rank, within *Mormopterus* Peters, 1865a: 258, recognised by Churchill (2008: 198), but see Reardon (2009b: 52). Taxon recognised at species rank by Churchill (2008: 198) who suggested that it included 'Species 2' and 'Species 5 (populations S and T)' of Adams *et al.* (1988: 316, 321, 323). Taxon recognised as Species 2 (only) of Adams *et al.* (1988: 316, 321, 323) by Churchill (1998: 203), Reardon (1999a: 12), Van Dyck and Strahan (2008: 493)(who also recognised the subspecies *Mormopterus loriae ridei* [=*Ozimops ridei*]), and Van Dyck *et al.* (2013: 29, 138) who recognised it at species rank, which was subsequently formalised by Reardon *et al.* (2014: 112, 129) who placed it in the subgenus *Ozimops*.

Setirostris Reardon *et al.*, 2014

Setirostris Reardon *et al.*, 2014: 109, 117.

TYPE SPECIES: *Mormopterus eleryi* Reardon and McKenzie 2008 [=*Setirostris eleryi* (Reardon & McKenzie, 2008)] by original designation.

COMMENTS: Proposed as a subgenus of *Mormopterus* Peters, 1865a: 258. Includes the species *eleryi*.

Setirostris eleryi (Reardon & McKenzie, 2008)

Bristle-faced Free-tailed Bat

Mormopterus eleryi Reardon & McKenzie, 2008: 1, 13.

TYPE LOCALITY: 1.1 km ESE of Eringa, South Australia, Australia. Approx. 240 m elevation. (26.29184°S, 134.739030°E)

COMMENTS: Separated as a distinct species from *Micronomus norfolkensis*. This taxon equates to Species 6 from Adams *et al.* (1988: 316, 321, 322). Species 6 was recognised as an undescribed taxon by subsequent authors including Churchill (1998: 211; 2008: 203), and Van Dyck and Strahan (2008: 499), with the described species being recognised by Menkhorst and Knight (2011: 19, 154), Van Dyck *et al.* (2013: 137) and Burbidge *et al.* (2014: 23, 30). Recognised as the sole member of the newly proposed subgenus *Setirostris* by Reardon *et al.* (2014: 117).

Family Miniopteridae Dobson, 1875

Group Miniopteri Dobson, 1875a: 349.

TYPE GENUS: *Miniopterus* Bonaparte, 1837.

COMMENTS: When originally proposed, this rank was placed in the Family Vespertilionidae (J. Gray, 1821) and included the genera *Natalus* J. Gray, 1838c: 496; *Thyroptera* de Spix, 1823: 61; and *Miniopterus* Bonaparte, 1837. Subfamily rank recognised by Miller (1907: xi, 227), Tate (1941a: 567), Simpson (1945: 60), Corbet and Hill (1992: 145), Volleth and Heller (1994: 25), Strahan (1995: 8, 492), Simmons (1998: 12) and Simmons (2005a: 519). Family rank recognised by Hoofer and Van Den Bussche (2003: 1), Van Den Bussche and Hoofer (2004: 327), Hutcheon and Kirsch (2004: 43), Eick *et al.* (2005: 1874), Miller-Butterworth *et al.* (2007: 1553, 1557) and followed by Van Dyck and Strahan (2008: 10, 503), and Menkhorst and Knight (2011: 19).

Subfamily Miniopterinae Miller, 1907: xi, 227.

TYPE GENUS: *Miniopterus* Bonaparte, 1837.

COMMENTS: When originally proposed, this rank was placed in the Family Vespertilionidae (J. Gray, 1821) and included the genus *Miniopterus* Bonaparte, 1837. Subfamily rank recognised by Koopman and Jones (1970: 27), Koopman (1984b: 180), and Volleth and Heller (1994: 25). Synonymised within the Subfamily Miniopterinae (Dobson, 1875a: 349) by McKenna and Bell (1997: 322).

Family Miniopteridae Mein & Tupinier, 1977: 207, 209.

TYPE GENUS: *Miniopterus* Bonaparte, 1837.

COMMENTS: When originally proposed as a new rank it included the genus *Miniopterus* Bonaparte, 1837. Synonymised within the Subfamily Miniopterinae (Dobson, 1875a: 349) by McKenna and Bell (1997: 322).

Miniopterus Bonaparte, 1837

Miniopterus Bonaparte, 1837 [1832–1841]: Fascicolo 20, under Φ *Vespertilio emarginatus* (first unnumbered, of two, pages of text) and Fascicolo 21 under Φ *Vespertilio ursinii* (first unnumbered, of two, pages of text).

TYPE SPECIES: Φ *Vespertilio ursinii* Bonaparte, 1837 [1832–1841] [= Φ *Miniopterus schreibersii* (Kuhl, 1817)] by monotypy.

COMMENTS: Described as a subgenus of *Vespertilio*. Genus reviewed in part by Goodwin (1979: 118), Peterson (1981: 828), Maeda (1982: 1), Hill (1983: 171), Corbet and Hill (1992: 145) and Peterson *et al.* (1995: 115). The revision of Maeda (1982: 1) was contested by P. Wilson (1985: 29). More recently Appleton *et al.* (2004: 431, 437) suggested that the currently recognised species is a gross underestimate of the number of actual species. Following Tian *et al.* (2004: 303) and Appleton *et al.* (2004: 431) we recognise the split of *Miniopterus schreibersii* (Kuhl, 1817: 6, 41) into three species: M. *schreibersii* (Europe/N. Africa/Near East), *M. fuliginosus* (Hodgson, 1835: 700; Asia), and *M. orianae* (Australasia).

Miniopteris Tomes, 1858a: 116, footnote.

TYPE SPECIES: Unjustified emendation of *Miniopterus* Bonaparte, 1837.

COMMENTS: This spelling appears to have been used by J. Gray (1866c: 91) but may be a *nomen nudum*. Synonymised within *Miniopterus* by Palmer (1904: 426).

HOMONYMS:

Miniopteris J. Gray, 1849: 9, little brown bats of the Class Mammalia (Order Chiroptera, Family Vespertilionidae). Genus name used for what is now *Myotis macrotarsus* (Waterhouse, 1845: 3, 5). See Simmons (2005a: 511).

Minyopterus Agassiz, 1846: 235.

TYPE SPECIES: Unjustified emendation of *Miniopterus* Bonaparte, 1837.

COMMENTS: Spelling also used by Winge (1893b: 35).

Minneopterus Lampe, 1900: 12.

TYPE SPECIES: Appears to be a *nomen nudum*.

COMMENTS: Unjustified emendation of *Miniopterus* Bonaparte, 1837. Synonymised within *Miniopterus* by Palmer (1904: 426).

Miniopterus australis (Tomes, 1858)

Little Bent-winged Bat

Miniopterus australis australis (Tomes, 1858)

M. [iniopteris] Australis Tomes, 1858a: 125; Plate 65.

TYPE LOCALITY: Loyalty Islands, South Pacific. Designation by Hill (1983: 171).

COMMENTS: Synonyms used here are from Koopman (1993: 230) who notes they come from a variety of sources, some of which are probably erroneous. *Miniopterus australis minor* Laurie and Hill, 1954: 72 is additional to this list. Included in *Miniopterus* by L. Hall and Richards (1979: 57) and Mahoney and Walton (1988g: 135). Koopman (1989: 9) briefly described its extra limital distribution. Species reviewed by Maeda (1982: 1), Hill (1983: 171), Corbet and Hill (1992: 146), Flannery (1995a: 373; 1995b: 443), Bonaccorso (1998: 387) and Kitchener and Suyanto (2002: 9). Does not include *paululus* Hollister (1913: 311), *shortridgei* Laurie and Hill (1957: 128) or *witkampi* Sody (1930: 272) as these are now recognised as distinct species (Kitchener & Suyanto, 2002: 27, 29). Australian subspecies recognised as *australis* by Strahan (1995: 492), and Van Dyck and Strahan (2008: 504).

Φ ***Miniopterus australis solomonensis*** Maeda, 1982

Φ *Miniopterus solomonensis* Maeda, 1982: 32.

TYPE LOCALITY: Malaita and San Christobal, Solomon Islands.

COMMENTS: Synonymised within *australis* by Koopman (1993: 230) but elevated to subspecies rank by Simmons (2005a: 519).

Φ ***Miniopterus australis tibialis*** (Tomes, 1858)

Φ *Vesp. [ertilio] tibialis* Tomes, 1858a: 126.

TYPE LOCALITY: Ambon Island, Maluku, Tenggara, Indonesia.

COMMENTS: Synonymised within *australis* by Laurie and Hill (1954: 71) and Koopman (1993: 230) but elevated to subspecies rank by Strahan (1983: 338), Corbet and Hill (1992: 146), Flannery (1990: 359; 1995a: 444; 1995b: 374), and subspecies rank by Bonaccorso (1998: 387) and Simmons (2005a: 519). Proposed to occur in Australia and the Solomon Islands by Bonaccorso (1998: 387).

Miniopterus orianae Thomas, 1922

Large Bent-winged Bat

Miniopterus orianae orianae Thomas, 1922

Northern Bent-winged Bat

Miniopterus orianae Thomas, 1922k: 616.

TYPE LOCALITY: Casurina Bay, 17 miles from Darwin, Northern Territory, Australia.

COMMENTS: Recognised as a subspecies of *Miniopterus blepotis* (Temminck, 1840: 212) by Troughton (1967: 287) but has typically been associated with *Miniopterus schreibersii* (Kuhl, 1817: 6, 41). It was synonymised within *schreibersii* by Mahoney and Walton (1988g: 136) but recognised as a subspecies of *schreibersii* by D. Johnson (1964: 476), Strahan (1983: 336; 1995: 495), Flannery (1990: 360; 1995a: 452; 1995b: 378), Reardon (1999a: 12), who noted that it should probably be recognised as a full species, Cardinal and Christidis (2000: 14), Simmons (2005a: 522), Clayton *et al.* (2006: 110), and Van Dyck and Strahan (2008: 509). Mitochondrial studies by Appleton *et al.* (2004: 435) revealed that Australian populations of *Miniopterus schreibersii* (Kuhl, 1817: 6, 41) differed 3.5% from each other and 13% from *australis*. Further support for the distinctiveness of the Australian '*schreibersii*' was provided by Tian *et al.* (2004: 303) who suggested that *schreibersii* should be separated into three distinct species, named *M. schreibersii*, *M. fuliginosus* and *M. oceanensis*. As a result of these studies this taxon has more recently been typically recognised as a distinct species by authors including Churchill (2008: 186), Van Dyck *et al.* (2013: 142) and Burbidge *et al.* (2014: 23, 30); but Reardon (2009b: 53) suggested that although this name may be the oldest name for *Miniopterus* in Australia there are concerns that the relationship of Australian '*schreibersii*' to Indonesian *Miniopterus schreibersii blepotis* (Temminck, 1840: 212) (with an older name) has not been resolved. The common name was proposed to be 'Large Bent-wing Bat' by Parnaby (1996: 35), which is followed here.

FUTURE TAXONOMIC RESEARCH: The relationship of Australian *Miniopterus orianae* to Indonesian *blepotis* (Temminck, 1840) needs to be resolved to determine the available name.

Miniopterus orianae oceanensis Maeda, 1982

Eastern Bent-winged Bat

Miniopterus oceanensis Maeda, 1982: 23.

TYPE LOCALITY: Cable Station, Cape York, Queensland, Australia.

COMMENTS: Synonymised within *Miniopterus schreibersii* (Kuhl, 1817: 6, 41) by Mahoney and Walton (1988g: 136) and Flannery (1990: 361; 1995a: 452; 1995b: 378). Recognised as a subspecies of *schreibersii* by Corbet and Hill (1992: 146), Reardon (1999a: 12), Cardinal and Christidis (2000: 15), Simmons (2005a: 522), Clayton *et al.* (2006: 110) and Van Dyck and Strahan (2008: 507). Molecular studies by Tian *et al.* (2004: 303) proposed that this taxon should be recognised as a distinct species, while Churchill (2008: 182) recognised it as a subspecies of *orianae* and suggested it is likely to be upgraded to full species status, which was followed by Van Dyck *et al.* (2013: 141) and Burbidge *et al.* (2014: 23, 30).

FUTURE TAXONOMIC RESEARCH: The status of this taxon, whether as subspecies of *M. orianae* or distinct species, needs to be confirmed.

Miniopterus orianae bassanii Cardinal & Christidis, 2000

Southern Bent-winged Bat

Miniopterus schreibersii bassanii Cardinal & Christidis, 2000: 13.

TYPE LOCALITY: Naracoorte, South Australia, Australia.

COMMENTS: Recognised as a subspecies of *Miniopterus schreibersii* (Kuhl, 1817: 6, 41) by Simmons (2005a: 521), Clayton *et al.* (2006: 110), and Van Dyck and Strahan (2008: 505). Churchill (2008: 182) recognised it as a subspecies of *orianae* and suggested it is likely to be upgraded to full species status, which was followed by Van Dyck *et al.* (2013: 141) and Burbidge *et al.* (2014: 23, 30).

FUTURE TAXONOMIC DIRECTIONS: The status of this taxon, whether as subspecies of *M. orianae* or distinct species, needs to be confirmed.

Family Vespertilionidae J. Gray, 1821

Family Vespertilionidae J. Gray, 1821: 299.

TYPE GENUS: *Vespertilio* Linnaeus, 1758: 31.

COMMENTS: When originally proposed, this rank was placed in the Order Insectivorae (J. Gray, 1821 [=Yangochiroptera (Koopman, 1985 part)]) and included the genera *Megadermes* [sic =*Megaderma*] É. Geoffroy, 1810a: 197; *Rhynolophus* [sic =*Rhinolophus*] Lacépède, 1799; *Nycterus* [sic =*Nycteris*] É. Geoffroy and G. Cuvier, 1795: 186; *Rhynopoma* [sic =*Rhinopoma*] É. Geoffroy, 1818: 113; *Thaphosores* [sic =*Taphozous*] É. Geoffroy, 1818; *Vespertilio* Linnaeus, 1758: 31; *Pecotus* [sic =*Plecotus*] É. Geoffroy, 1818: 112; and *Barbastella* J. Gray, 1821: 300. Family name conserved by Direction 98 (ICZN, 1958b: 131). Family reviewed by Miller (1907: 195). Also recognised as the Superfamily Vespertilionoidea by authors including Weber (1928: xiv, 155, as tribe but equal in rank), Van Valen (1979: 109), Koopman (1985: 27), McKenna and Bell (1997: 313), Springer *et al.* (2001: 6243) and Teeling *et al.* (2002: 1432; 2003: 309).

Family Vespertilia Rafinesque, 1815: 54.

TYPE GENUS: *Vespertilio* Linnaeus, 1758: 31.

COMMENTS: When originally proposed, this rank was placed in the Order Chiropteria (Rafinesque, 1815 [=Chiroptera (Blumenbach, 1779)]), and included the subfamilies Lophinia (Rafinesque, 1815: 54) that included the genera *Rhinolophus* Lacépède, 1799a; *Phyllostoma* G. Cuvier, 1800: Table 1 [=*Phyllostomus* Lacépède, 1799a: 16]; *Vampyrum* Rafinesque, 1815: 54; and *Megaderma* É. Geoffroy, 1810a: 197; and Nycteria (Rafinesque, 1815: 54) that included the genera *Pteropus* Brisson, 1762; *Eidolon* Rafinesque, 1815: 54; *Pteronotus* Rafinesque, 1815; *Cephalotes* É. Geoffroy, 1810b [=*Nyctimene* Borkhausen, 1797]; *Tadaris* [sic =*Tadarida*] Rafinesque, 1814a: 55; *Vespertilio* Linnaeus, 1758: 31; *Nycterus* [sic =*Nycteris*] É. Geoffroy and G. Cuvier, 1795: 186; *Noctilio* Linnaeus, 1766: 88; *Molossus* É. Geoffroy, 1805b: 151; and *Atalapha* Rafinesque, 1814a: 12 [=*Nyctalus* Bowdich, 1825: 36]. Correction of name to Vespertilionidae provided by Direction 98 (ICZN, 1958b: 131). Name added to the Official List of Family-Group names (Melville & Smith, 1987: 38).

Tribe Vespertilionina J. Gray, 1825a: 339.

TYPE GENUS: *Vespertilio* Linnaeus, 1758: 31.

COMMENTS: When originally proposed, this rank was placed in the Family Vespertilionidae (J. Gray, 1821) and included the genera *Vespertilio* Linnaeus, 1758: 31; *Plecotus* É. Geoffroy, 1818: 112; *Barbastellus* J. Gray, 1825b: 243 [=*Barbastella* J. Gray, 1821: 300]; *Proboscidea* de Spix, 1823: 61 [=*Rhynchonycteris* Peters, 1867f: 477; *Thyroptera* de Spix, 1823: 61; and *Caelano* [sic =*Celaeno*] Leach, 1821b: 69 [*Noctilio* Linnaeus, 1766: 88]. Name also described by J. Gray (1825b: 243). Synonymised within Subfamily Vespertilioninae (J. Gray, 1821: 299) by McKenna and Bell (1997: 316).

Family Vespertiliones Peters, 1865a: 258.

TYPE GENUS: *Vespertilio* Linnaeus, 1758: 31.

COMMENTS: When originally proposed, this rank included the genera *Synotus* Keyserling & Blasius, 1839: 305 [=*Barbastella* J. Gray, 1821: 300]; *Plecotus* É. Geoffroy, 1818: 112; *Histiotus* Gervais, 1855b: 77; *Otonycteris* Peters, 1859b: 223; *Miniopterus* Bonaparte, 1837; *Vespertilio* Linnaeus, 1758: 31; *Vesperugo* Keyserling and Blasius, 1839: 312 [=*Vespertilio* Linnaeus, 1758: 31]; *Vesperus* Keyserling and Blasius, 1839: 313 [=*Vespertilio* Linnaeus, 1758: 31]; *Murina* J. Gray, 1842b; *Harpiocephalus* J. Gray, 1842b: 259; *Nycticejus* [sic =*Nycticeius*] Rafinesque, 1819a: 417]; *Atalapha* Rafinesque, 1814a: 12 [=*Nyctalus* Bowdich, 1825: 36]; *Thyroptera* de Spix, 1823: 61; and *Antrozous* H. Allen, 1862: 248.

HOMONYMS:

Vespertiliones Pallas, 1767, bats of the Order Chiroptera. See individual entry.

Group Vespertiliones Dobson, 1878, bats of the Subfamily Vespertilioninae J. Gray, 1821. See individual entry.

Family Vespertilionidae Gill, 1872: 17.

TYPE GENUS: *Vespertilio* Linnaeus, 1758: 31.

COMMENTS: When originally proposed, this rank was placed in the Suborder Animalivora (Gill, 1872 [=Chiroptera (Blumenbach, 1779 part)]) and included the subfamilies Vespertilioninae (J. Gray, 1821: 299) and Nycticejinae (Gill, 1872: 17). Synonymised within Superfamily Vespertilionoidea (J. Gray, 1821: 299) by McKenna and Bell (1997: 313).

Tribe? Vespertilionini Winge, 1893b: 24.

TYPE GENUS: *Emballonura* Temminck, 1838a: 1, 22.

COMMENTS: When originally proposed, this rank was placed in the Family Vespertilionidae (J. Gray, 1821) and included the genera *Vespertilio* Linnaeus, 1758: 31; *Plecotus* É. Geoffroy, 1818: 112; *Minyopterus* Agassiz, 1846 [=*Miniopterus* Bonaparte, 1837]; *Lasionycteris* Peters, 1865e: 648; *Vesperugo* Keyserling and Blasius, 1839: 312 [=*Vespertilio* Linnaeus, 1758: 31]; *Harpyiocephalus* J. Gray, 1866c: 90 [=*Harpiocephalus* J. Gray, 1842b: 259]; *Synotus* Keyserling & Blasius, 1839: 305 [=*Barbastella* J. Gray, 1821: 300]; *Chalinolobus* Peters, 1865d; *Otonycteris* Peters, 1859b: 223; *Nyctophilus* Leach, 1821a; *Atalapha* Rafinesque, 1814a: 12 [=*Nyctalus* Bowdich, 1825: 36]; and *Antrozous* H. Allen, 1862: 248. The genus *Vespertilio* included the subgenera *Kerivoula* J. Gray, 1842b: 258; *Natalus* J. Gray, 1838c: 496; and *Nyctiellus* Gervais, 1855b: 84.

Tribe Vespertilionoidea Weber, 1928: xiv, 155.

TYPE GENUS: *Vespertilio* Linnaeus, 1758: 31.

COMMENTS: When originally proposed, this rank was placed in the Suborder Microchiroptera (Dobson, 1875a [=Chiroptera (Blumenbach, 1779 part)]) and included the families Vespertilionidae (J. Gray, 1821), Natalidae (J. Gray, 1866c: 90), Myzopodidae (Thomas, 1904g: 5) and Molossidae (Gervais, 1855b). Rank recorded as tribe but is similar to superfamily. Recognised at superfamily rank by Simpson (1945: 58), McKenna and Bell (1997: 313) (via J. Gray, 1821: 299), Simmons and Geisler (1998: 136) and Teeling *et al.* (2002: 1432; 2003: 309).

Subfamily Kerivoulinae Miller, 1907

Subfamily Kerivoulinae Miller, 1907: xi, 232.

TYPE GENUS: *Kerivoula* J. Gray, 1842b: 258.

COMMENTS: When originally proposed, this rank was placed in the Family Vespertilionidae (J. Gray, 1821) and included the genera *Kerivoula* J. Gray, 1842b: 258; and *Phoniscus* Miller, 1905. Subfamily rank recognised by Troughton (1929: 85), Tate (1941a: 583), Simpson (1945: 60), Koopman and Jones (1970: 27), Koopman (1984b: 182), Volleth and Heller (1994: 25), Strahan (1995: 8, 490), McKenna and Bell (1997: 322), Hoofer and Van Den Bussche (2003: 14, 38), Simmons (2005a: 525), and Van Dyck and Strahan (2008: 10, 511).

Phoniscus Miller, 1905

Phoniscus Miller, 1905: 229

TYPE SPECIES: Φ *Phoniscus atrox* Miller, 1905: 230 by original designation.

COMMENTS: The taxonomic distinctiveness of this taxon has been unstable as it was placed as a subgenus within *Kerivoula* J. Gray, 1842b: 258 by Koopman (1982: 22; 1994: 99), and a synonym of it by Koopman (1993: 196) and Ryan (1965b: 517–518). In contrast Iredale and Troughton (1934: xi, 98) and Hill (1965: 547) recognised it as distinct, as did Corbet and Hill (1980: 77; 1992; 155), Churchill (1998: 164; 2008: 147), Simmons (2005a: 528) and most subsequent authors including Reardon *et al.* (2010: 44).

Phoniscus papuensis (Dobson, 1878)

Golden-tipped Bat

Kerivoula papuensis Dobson, 1878: 339.

TYPE LOCALITY: Port Moresby, Papua New Guinea.

COMMENTS: The generic position of this species has been unstable. It was placed in *Kerivoula* by Mahoney and Walton (1988g: 134), Parnaby (1992: 27), Koopman (1993: 197), Reardon (1999a: 12), and Van Dyck and Strahan (2008: 511). Included within the genus *Phoniscus* Miller, 1905 by Iredale and Troughton (1934: 99), Tate (1941a: 590), Troughton (1967: 288), L. Hall and Richards (1979: 55), Flannery (1990: 355; 1995a: 466; 1995b: 392), Churchill

(1998: 164; 2008: 147), Simmons (2005a: 529) and Reardon *et al.* (2010: 44).

FUTURE TAXONOMIC RESEARCH: The generic placement of this taxon needs to be confirmed.

Subfamily Murininae Miller, 1907

Subfamily Murininae Miller, 1907: xi, 229.

TYPE GENUS: *Murina* J. Gray, 1842b.

COMMENTS: When originally proposed, this rank was placed in the Family Vespertilionidae (J. Gray, 1821) and included the genera *Murina* J. Gray, 1842b; and *Harpiocephalus* J. Gray, 1842b: 259. Subfamily recognised by Simpson (1945: 60), Koopman and Jones (1970: 27), Koopman (1984b: 180), Volleth and Heller (1994: 25), Strahan (1995: 8, 496), McKenna and Bell (1997: 322), Hoofer and Van Den Bussche (2003: 38), Simmons (2005a: 522), and Van Dyck and Strahan (2008: 10, 514).

Murina J. Gray, 1842

Murina J. Gray, 1842b: 258.

TYPE SPECIES: Φ *Vespertilio suillus* Temminck, 1840: 224 [= Φ *Murina suilla* (Temminck, 1840: 224)] by monotypy.

COMMENTS: Typically recognised since its description.

Ocypetes Lesson, 1842: 30.

TYPE SPECIES: Φ *Vespertilio suillus* Temminck, 1840: 224 (as *Ocypetes suilla*) [= Φ *Murina suilla* (Temminck, 1840: 224)] by subsequent designation. e.g. See Corbet and Hill (1992: 148).

COMMENTS: Described as a subgenus of *Vespertilio*. Synonymised within *Murina* by Miller (1907: 229) and Simmons (2005a: 523).

HOMONYMS:

Ocypetes Risso, 1826a: 184, mites of the Class Arachnida (Order Trombidiformes, Family Trombidiidae). Genus name is an emendation of *Ocypete* Leach, 1814: 434.

Ocypetes Wagler, 1829b: 762, seedsnipe birds of the Class Aves (Order Charadriiformes, Family Thinocoridae). Genus is a synonym of *Thinocorus* Eschscholtz, 1829: 2.

Ocypetes Saunders, 1871: 77, beetles of the Class Insecta (Order Coleoptera, Family Buprestidae). Currently recognised name. See Bellamy (1997: 218).

Harpiola Thomas, 1915d: 309.

TYPE SPECIES: Φ *Murina grisea* Peters, 1872a: 256, 258 by original designation.

COMMENTS: Synonymised within *Murina* by Corbet and Hill (1992: 148) and Simmons (2005a: 523).

Murina florium Thomas, 1908

Flute-nosed Bat

Murina florium florium Thomas, 1908

Murina florium Thomas, 1908b: 371.

TYPE LOCALITY: Pulau Flores, Lesser Sunda Inlands (Nusatenggara), Indonesia.

COMMENTS: First identified in Australia by G. Richards *et al.* (1982: 149) with the distribution expanded by Clague *et al.* (1999: 175). This subspecies was allocated to Australia by Burbidge *et al.* (2014: 24, 30).

Φ *Murina florium lanosa* Thomas, 1910

Φ *Murina lanosa* Thomas, 1910f: 534.

TYPE LOCALITY: Ceram Island, Moluccas, Indonesia.

COMMENTS: Recognised as a subspecies of *florium* by Laurie and Hill (1954: 75). Synonymised within *florium* by Koopman (1993: 229), but was elevated to subspecies of *florium* by Corbet and Hill (1992: 150), Flannery (1990: 343; 1995a: 454; 1995b: 380), Bonaccorso (1998: 349) and Simmons (2005a: 523). Synonymised within *florium* by Van Dyck and Strahan (2008: 515).

Φ *Murina florium toxopei* Thomas, 1923

Φ *Murina toxopei* Thomas, 1923g: 254.

TYPE LOCALITY: En-Biloro, Buru Island, Moluccas, Indonesia.

COMMENTS: Recognised as a subspecies of *florium* by Laurie and Hill (1954: 75) and Flannery (1990: 343). Synonymised within *florium* by Flannery (1995a: 454; 1995b: 380), Koopman (1993: 229) and Churchill (1998: 148) (as *M. toxopeusi*); was elevated to a subspecies of *florium* by Simmons (2005a: 523). Synonymised within *florium* by Van Dyck and Strahan (2008: 515), who misspelt it *toxopeusi* following Laurie and Hill (1954: 75).

Subfamily Nyctophilinae Peters, 1865

Nyctophili Peters, 1865b: 524.

TYPE GENUS: *Nyctophilus* Leach, 1821a.

COMMENTS: When originally proposed, the rank was unknown but placed in the Family Megadermata (Peters, 1865a [=Megadermatidae (H. Allen, 1864)]) and included the genera *Nyctophilus* Leach, 1821a; and *Antrosous* [sic =*Antrozous*] H. Allen, 1862: 248. Recognised as the Tribe Nyctophilini within the Subfamily Vespertilioninae by McKenna and Bell (1997: 322) and Simmons (2005a: 468). Volleth and Tidemann (1991: 321) suggested that *Nyctophilus* may belong within Vespertilionini. Subfamily rank recognised by Koopman and Jones (1970: 27), Strahan (1995: 8, 498), and Van Dyck and Strahan (2008: 10, 516).

Subfamily Nyctophilinae Miller, 1907: xi, 234.

TYPE GENUS: *Nyctophilus* Leach, 1821a.

COMMENTS: When originally proposed, this rank was placed in the Family Vespertilionidae (J. Gray, 1821) and included the genera *Antrozous* H. Allen, 1862: 248; and *Nyctophilus* Leach, 1821a. Subfamily name recognised by Simpson (1945: 60).

Tribe Nyctophilini Koopman & Jones, 1970: 27.

TYPE GENUS: *Nyctophilus* Leach, 1821a.

COMMENTS: When originally proposed, this rank was placed within the Subfamily Nyctophilinae (Peters, 1865b) and included the genera *Lamingtona* McKean and Calaby, 1968; *Nyctophilus* Leach, 1821a; and *Pharotis* Thomas, 1914d: 381. Synonymised within Nyctophilini by McKenna and Bell (1997: 322).

Nyctophilus Leach, 1821

Nyctophilus Leach, 1821a: 74, 78; Plate 7.

TYPE SPECIES: *Nyctophilus geoffroyi* Leach, 1821a by monotypy.

COMMENTS: Australian species reviewed by L. Hall and Richards (1979: 58) and Koopman (1984a: 25), with the genus reviewed by Tomes (1858b: 25), Peters (1860b: 123), Thomas (1915e: 493), Tate (1941a: 567) and Parnaby (2009: 39). Van Dyck and Strahan (2008: 525–528) suggested there are two undescribed species.

Barbastellus J. Gray, 1831b: 38.

TYPE SPECIES: *Barbastellus pacificus* J. Gray, 1831b [=*Nyctophilus geoffroyi pacificus* (J. Gray, 1831b)] by subsequent designation. See Dobson (1878: 175).

COMMENTS: Origin of genus name discussed by Mahoney and Walton (1988g: 139). Synonymised within *Nyctophilus* by J. Gray (1843a: xix), Miller (1907: 236) and Mahoney and Walton (1988g: 139).

HOMONYMS:

Barbastella J. Gray, 1821: 300, barbastelle bats of the Class Mammalia (Order Chiroptera, Family Vespertilionidae). Currently recognised genus. See Simmons (2005a: 480).

Barbastellus J. Gray, 1825b: 243, barbastelle bats of the Class Mammalia (Order Chiroptera, Family Vespertilionidae). Incorrect subsequent spelling of *Barbastella* J. Gray, 1821: 300.

Lamingtona McKean & Calaby, 1968: 372

TYPE SPECIES: Φ *Lamingtona lophorhina* McKean & Calaby, 1968: 373 [= Φ *Nyctophilus microtis* (Thomas, 1888d: 226)] by original designation.

COMMENTS: Synonymised within *Nyctophilus* by Hill and Koopman (1981: 278) and Simmons (2005a: 468).

Nyctophilus arnhemensis Johnson, 1959

Arnhem Long-eared Bat

Nyctophilus arnhemensis D. Johnson, 1959: 184.

TYPE LOCALITY: Rocky Bay, south of Yirrkala, Cape Arnhem Peninsula, Northern Territory, Australia. (12°13′S, 136°47′E)

COMMENTS: Taxon has been accepted by subsequent authors.

Nyctophilus bifax Thomas, 1915

Eastern Long-eared Bat

Nyctophilus bifax Thomas, 1915e: 496.

TYPE LOCALITY: Herberton district, Queensland, Australia.

COMMENTS: Considered a synonym of *Nyctophilus gouldi* Tomes, 1858b by Mahoney and Walton (1988g: 140). Koopman (1984a: 27; 1994: 131) considered *bifax* to be a subspecies of *Nyctophilus gouldi* while Koopman (1993: 218) included it as a synonym. Species rank recognised by Troughton (1967: 276), L. Hall and Richards (1979: 58), Parnaby (1987: 153; 2002: 115), Churchill (1998: 154) and Reardon (1999a: 12) and subsequent authors.

Nyctophilus corbeni Parnaby, 2009

Corben's Long-eared Bat

Nyctophilus corbeni Parnaby, 2009: 39, 46.

TYPE LOCALITY: Old Coghill Track, 0.7 km east of junction with track to main Gilgai Waterhole; formerly Gilgai Flora Reserve, Pilliga East State Forest, New South Wales, Australia. Approximate 235 m elevation (30°29′58″S, 149°20′53″E).

COMMENTS: This species was separated from the previously recognised *Nyctophilus timoriensis*. É. Geoffroy, 1806. Recognised as *Nyctophilus* species 2 by Churchill (2008: 143).

Nyctophilus daedalus Thomas, 1915

Pallid Long-eared Bat

Nyctophilus daedalus Thomas, 1915e: 498.

TYPE LOCALITY: Daly River, Northern Territory, Australia.

COMMENTS: Recognised at the species rank by Iredale and Troughton (1934: 95) and Troughton (1967: 277), but reduced to a synonym of *bifax* by Ride (1970: 245) and synonym of *gouldi* by Koopman (1993: 218). Subsequently placed as a subspecies of *gouldi* by Koopman (1984a: 27; 1994: 131) and Mahoney and Walton (1988g: 140). Included as a subspecies of *bifax* by D. Johnson (1964: 479), Parnaby (1987: 153) and Reardon (1999a: 12) who noted that it was probably a full species. Recognised as a subspecies of *bifax* by Strahan (1983: 333; 1995: 501), Flannery (1990: 351; 1995a: 457), Bonaccorso (1998: 357), Simmons (2005a: 469), Clayton *et al.* (2006: 110), and Van Dyck and Strahan (2008: 519) who suggested that it is possibly a distinct species. Species rank recognised by Churchill (2008: 134) and confirmed by Parnaby (2009: 39, 61).

FUTURE TAXONOMIC RESEARCH: Parnaby (2009) drew attention to three distinctive morphs in this species, suggesting that it might in fact be 'a composite of two, and possibly three distinct forms... [which] are most likely to be broadly sympatric throughout the current range'. Future research needs to close this proposition, perhaps with the aid of DNA sequencing.

Nyctophilus geoffroyi Leach, 1821

Lesser Long-eared Bat

Nyctophilus geoffroyi geoffroyi Leach, 1821

Nyctophilus Geoffroyi Leach, 1821a: 78; Plate 7.

TYPE LOCALITY: Australia.

COMMENTS: Species recognised within *Nyctophilus* by Gould (1853 [1845–1863]: Text to Plates 36–37). Thomas (1915e: 495) recognised three subspecies including *geoffroyi*, *pallescens* and *pacificus*. Taxonomic decision of Iredale and Troughton (1934: 94) for *N. geoffroyi* to include as subspecies *pacificus* J. Gray, 1831b and *pallescens* as subspecies which was followed by Tate (1941a: 594). Reviewed by Koopman (1984a: 28) and in part by Kitchener *et al.* (1991: 99). The three subspecies were considered poorly defined by Simmons (2005a: 469).

FUTURE TAXONOMIC RESEARCH: The validity of the recognised subspecies needs to be confirmed.

B. [arbastellus] Novae Hollandiae J. Gray, 1831b: 38.

TYPE LOCALITY: Australia.

COMMENTS: Considered as *incertae sedis* by Mahoney and Walton (1988g: 145). Iredale and Troughton (1934: 94) include *Barbastellus novaehollandiae* in *Nyctophilus geoffroyi* Leach, 1821a: but it seems unlikely that the holotype of *B. novaehollandiae* was seen by them and their identification of it as *N. geoffroyi* might not be correct (Mahoney & Walton, 1988g: 145). This species was placed in the genus *Nyctophilus* Leach, 1821a by Mahoney in Mahoney and Walton (1988g: 145) because J. Gray included it in the same genus as *Barbastellus pacificus* (J. Gray, 1831b), a form correctly placed by Iredale and Troughton (1934) in *Nyctophilus*. Placed within *geoffroyi* by Simmons (2005a: 469).

Nyctophilus australis Peters, 1860b: 125; Plate.

TYPE LOCALITY: Western Australia?, Australia.

COMMENTS: Synonymised within *geoffroyi* by Simmons (2005a: 469).

Nyctophilus leachii Dobson, 1878: 174.

TYPE LOCALITY: Unknown.

COMMENTS: Synonymised within *geoffroyi* by Iredale and Troughton (1934: 94) and Simmons (2005a: 469).

Nyctophilus geoffroyi pacificus (J. Gray, 1831)

Barbastellus Pacificus J. Gray, 1831b: 38.

TYPE LOCALITY: 'the islands of the Southern Pacific', Australia.

COMMENTS: Recognised as a subspecies of *geoffroyi* by Iredale and Troughton (1934: 94). Considered a junior synonym of *pallescens* by Koopman (1984a: 28). Recognised as a subspecies of *geoffroyi* by Strahan (1983: 331; 1995: 503), R. Taylor *et al.* (1987: 110), Simmons (2005a: 469), Clayton *et al.* (2006: 110), and Van Dyck and Strahan (2008: 521), but not Burbidge *et al.* (2014: 24, 30).

Nyctophilus unicolor Tomes, 1858b: 33.

TYPE LOCALITY: Tasmania, Australia.

COMMENTS: Species recognised by Gould (1859 [1845–1863]: Text to Plate 38) and Krefft (1868a: 93). Synonymised within *pacificus* by Iredale and Troughton (1934: 94) and Simmons (2005a: 469).

Nyctophilus Geayi Trouessart, 1915: 146.

TYPE LOCALITY: Nicholson River, Victoria, Australia.

COMMENTS: Synonymised within *pacificus* by Iredale and Troughton (1934: 94) and Simmons (2005a: 469).

Nyctophilus geoffroyi pallescens Thomas, 1913

Nyctophilus geoffroyi pallescens Thomas, 1913e: 79.

TYPE LOCALITY: Alexandria, Northern Territory, Australia. 800 feet.

COMMENTS: Recognised as a subspecies of *geoffroyi* by Iredale and Troughton (1934: 94), Strahan (1983: 331; 1995: 503), Simmons (2005a: 469), Clayton *et al.* (2006: 110), and Van Dyck and Strahan (2008: 521), but not Burbidge *et al.* (2014: 24, 30).

Nyctophilus gouldi Tomes, 1858

Gould's Long-eared Bat

Nyctophilus Gouldi Tomes, 1858b: 31.

TYPE LOCALITY: Morton Bay, Queensland and Bathurst, New South Wales, Australia.

COMMENTS: Recognised as a subspecies of *timoriensis* by Iredale and Troughton (1934: 95). Koopman (1984a: 26) noted that that this taxon also included as subspecies the taxa *daedalus* and *bifax*, but these were definitively separated by Parnaby (1987: 153; 2002: 115; 2009: 39, 61).

† ***Nyctophilus howensis*** McKean, 1975

Lord Howe Long-eared Bat

† *Nyctophilus howensis* McKean, 1975: 330.

TYPE LOCALITY: Calcarenite Cave, North Bay, Lord Howe Island, Australia.

COMMENTS: Known only from the holotype, but there is no doubt of its specific distinctness (Parnaby, 2009: 68). Appears to have survived into historic times based on the observation of Etheridge (1889: 6) who stated that a bat large than *Chalinolobus morio* was occasionally seen on the island.

FUTURE TAXONOMIC RESEARCH: It was suggested by McKean (1975: 332) that the placement of this species within *Nyctophilus* may not prove satisfactory when more material becomes available, which was supported by Parnaby (2009: 39, 70) who argued that there appears to be no specific reason for assigning the holotype to *Nyctophilus*, other than its superficial resemblance in dental and cranial structure compared to any other genera in the Australian region.

Nyctophilus major J. Gray, 1844

Greater Long-eared Bat

Nyctophilus major major J. Gray, 1844

Western Long-eared Bat

Nyctophilus major J. Gray, 1844a: 12b; Plate 21, Fig. 2.

TYPE LOCALITY: Perth, Western Australia, Australia.

COMMENTS: Taxon reviewed by Tate (1941a: 592) who recognised the author as J. Gray (1841: 400). Synonymised within *timoriensis* by Iredale and Troughton (1934: 95). This taxon considered a subspecies of *timoriensis* by Koopman (1984a: 27). Synonymised within *timoriensis* by Strahan (1983: 328), Mahoney and Walton (1988g: 141), Koopman (1993: 218), Flannery (1995a: 461) and Churchill (1998: 160). Recognised as subspecies of *timoriensis* by Reardon (1999a: 12), who noted that it should be recognised as a full species, Simmons (2005a: 470) and Clayton *et al.* (2006: 110). Synonymised in part within several undescribed species by Van Dyck and Strahan (2008: 526, 528, 529), but species rank was recognised by Churchill (2008: 139), Parnaby (2009: 39, 52) and Burbidge *et al.* (2014: 24, 30).

Nyctophilus Geoffroyii Var. *major* J. Gray, 1841: 400.

TYPE LOCALITY: Unknown.

COMMENTS: *Nomen nudum.* Citation recognised by Tate (1941a: 595) but not typically by other authors who recognised the author of *major* as J. Gray (1844a: 12b; Plate 21, Fig. 2).

Nyctophilus major tor Parnaby, 2009

Central Long-eared Bat

Nyctophilus major tor Parnaby, 2009: 39, 58.

TYPE LOCALITY: Johnnies Dam, Jaurdi Station, 125 km west of Kalgoorlie, Western Australia, Australia. Approx 435 m elevation (30°46′22″S, 120°07′55″E).

COMMENTS: This species was separated from the previously recognised *Nyctophylus timoriensis*. É. Geoffroy, 1806, and was recognised as the Central Long-eared Bat *Nyctophilus* Species 1 by Churchill (2008: 142). Its status was extensively discussed by Parnaby (2009: 58), who suspected that it is in fact a distinct species, broadly sympatric with *N.(m.) major* in the Western Australian wheat belt and southern subcoastal areas, but described it as a subspecies because he was 'unable to refute the simpler hypothesis of a variable species with environmentally induced size variation' (Parnaby, 2009: 58).

FURTHER TAXONOMIC RESEARCH: Because of the equivocal morphometric separation between *tor* and nominotypical *major* (females seem to be separable, males not completely so), DNA sequencing seems called for to test the reality of the distinction between the two.

Nyctophilus sherrini Thomas, 1915

Tasmanian Long-eared Bat

Nyctophilus sherrini Thomas, 1915e: 495.

TYPE LOCALITY: Tasmania, Australia.

COMMENTS: This taxon considered a subspecies rank by Iredale and Troughton (1934: 95) and Koopman (1984a: 27). Synonymised within *timoriensis* by Mahoney and Walton (1988g: 141) and Koopman (1993: 218). Recognised as subspecies of *timoriensis* by Strahan (1983: 328; 1995: 508), R. Taylor *et al.* (1987: 110) and Reardon (1999a: 12) who noted that it was probably a full species. Recognised as a subspecies by Flannery (1990: 352; 1995a: 461), Simmons (2005a: 470) and Clayton *et al.* (2006: 110) but as a distinct species by Churchill (2008: 141) and Parnaby (2009: 39, 65). Not recognised by Van Dyck and Strahan (2008) but was recognised by Burbidge *et al.* (2014: 24, 30).

Nyctophilus walkeri Thomas, 1892

Pygmy Long-eared Bat

Nyctophilus Walkeri Thomas, 1892: 406.

TYPE LOCALITY: Adelaide River, Northern Territory, Australia.

COMMENTS: Has been recognised by most subsequent authors.

nomen dubium

Vesp. [ertilio] timoriensis É. Geoffroy, 1806: 200; Plate 47.

TYPE LOCALITY: Timor Island, Indonesia. This locality is not considered beyond doubt by Desmarest (1819: 481), Tomes (1858b: 31) and Koopman (1984a: 28).

COMMENTS: Species reviewed by Kitchener *et al.* (1991: 97), with the taxonomic history reviewed by Parnaby (2009: 44). Placed in the genus *Nyctophilus* by Gould (1859 [1845–1863]: Text to Plate 39), Iredale and Troughton (1934: 95) and subsequent authors including Troughton (1967: 276), Ride (1970: 164), L. Hall and Richards (1979: 58, 59), Reardon and Flavel (1987: 63), and Mahoney and Walton (1988g: 141). Species provisionally synonymised within *major* by Thomas (1914d: 383), who suggested that the name be dropped for the present, as it is impossible to identify it with certainty among the Australian species. Taxonomic decision of Koopman (1984a: 27) to recognise *major* and *sherrini* as subspecies. Van Dyck and Strahan (2008: 525–528) and Parnaby (2009: 43) suggested that it is not yet clear whether several undescribed forms are separate species or subspecies of *timoriensis*. The species *timoriensis* was considered not to occur in Australia by Churchill (2008: 143) and subsequently considered a *nomen dubium* by Parnaby (2009: 39) due to uncertainty surrounding the provenance of the original specimens(s), the lack of a definite type specimen, and lack of sufficient detail in the original description and illustration to relate the name to a singular currently recognised species.

Subfamily Vespertilioninae J. Gray, 1821

Family Vespertilionidae J. Gray, 1821: 299.

TYPE GENUS: *Vespertilio* Linnaeus, 1758: 31.

COMMENTS: When originally proposed, this rank was placed in the Order Insectivorae (J. Gray, 1821 [=Yangochiroptera (Koopman, 1985 part)]) and included the genera *Megadermes* [sic =*Megaderma*] É. Geoffroy, 1810a: 197; *Rhynolophus* [sic =*Rhinolophus*] Lacépède, 1799; *Nycterus* [sic =*Nycteris*] É. Geoffroy and G. Cuvier, 1795: 186; *Rhynopoma* [sic =*Rhinopoma*] É. Geoffroy, 1818: 113; *Thaphosores* [sic =*Taphozous*] É. Geoffroy, 1818; *Vespertilio* Linnaeus, 1758: 31; *Pecotus* [sic =*Plecotus* É. Geoffroy, 1818: 112; and *Barbastella* J. Gray, 1821: 300. Recognised as a tribe by Koopman and Jones (1970: 27), and Volleth and Heller (1994: 25). Subfamily recognised by Koopman (1984b: 180), Strahan (1995: 8, 510), McKenna and Bell (1997: 316), Simmons (2005a: 484), and Van Dyck and Strahan (2008: 10, 531).

Tribe Vespertilionina J. Gray, 1825a: 339.

TYPE GENUS: *Vespertilio* Linnaeus, 1758: 31.

COMMENTS: When originally proposed, this rank was placed in the Family Vespertilionidae (J. Gray, 1821) and included the genera *Vespertilio* Linnaeus, 1758: 31; *Plectotus* É. Geoffroy, 1818: 112; *Barbastellus* J. Gray, 1825b: 243 [=*Barbastella* J. Gray, 1821: 300]; *Proboscidea* de Spix, 1823: 61 [=*Rhynchonycteris* Peters, 1867f: 477]; *Thyroptera* de Spix, 1823: 61]; and *Caelano* [sic =*Celaeno*] Leach, 1821b: 69 [=*Noctilio* Linnaeus, 1766: 88]. Name also described at subfamily rank by J. Gray (1825b: 243). Synonymised within the Subfamily Vespertilioninae by McKenna and Bell (1997: 316).

Group Vespertiliones Dobson, 1878: 168.

TYPE GENUS: *Vespertilio* Linnaeus, 1758: 31.

COMMENTS: When originally proposed, this rank was placed in the Family Vespertilionidae (J. Gray, 1821) and included the genera *Vesperugo* Keyserling and Blasius, 1839: 312 [=*Vespertilio* Linnaeus, 1758: 31]; *Chalinolobus* Peters, 1865d; *Scotophilus* Tomes, 1857; *Nycticejus* [sic =*Nycticeius*] Rafinesque, 1819a: 417]; *Atalapha* Rafinesque, 1814a: 12 [=*Nyctalus* Bowdich, 1825: 36]; *Harpiocephalus* J. Gray, 1842b: 259; *Vespertilio* Linnaeus, 1758: 31]; and *Kerivoula* J. Gray, 1842b: 258. Synonymised within the Subfamily Vespertiloninae by Simpson (1945: 59).

HOMONYMS:

Vespertiliones Pallas, 1767, bats of the Order Chiroptera. See individual entry.

Family Vespertiliones Peters, 1865a, bats of the Family Vespertilionidae J. Gray, 1821. See individual entry.

Subfamily Plecotinae Miller, 1897a: 41, 46.

TYPE GENUS: *Plecotus* É. Geoffroy, 1818: 112.

COMMENTS: When originally proposed, this rank was placed in the Family Vespertilionidae (J. Gray, 1821) and included the genera *Euderma* H. Allen, 1891: 467; and *Plecotus* É. Geoffroy, 1818: 112.

Subfamily Vespertilioninae Miller, 1897a: 41, 54.

TYPE GENUS: *Vespertilio* Linnaeus, 1758: 31.

COMMENTS: When originally proposed, this rank was placed in the Family Vespertilionidae (J. Gray, 1821) and included the genera *Dasypterus* Peters, 1870: 912 [=*Lasiurus* J. Gray, 1831b: 38]; *Rhogeëssa* H. Allen, 1866: 285; *Nycticeius* Rafinesque, 1819a: 417; *Lasiurus* J. Gray, 1831b: 38; *Vespertilio* Linnaeus, 1758: 31; *Pipistrellus* Kaup, 1829; *Lasionycteris* Peters, 1865e: 648; and *Myotis* Kaup, 1829. Subfamily rank recognised by Simpson (1945: 59).

Subfamily Nyctophilinae Miller, 1907: xi, 234.

TYPE GENUS: *Nyctophilus* Leach, 1821a.

COMMENTS: When originally proposed, this rank was placed in the Family Vespertilionidae (J. Gray, 1821) and included the genera *Antrozous* H. Allen, 1862: 248;

and *Nyctophilus* Leach, 1821a. Synonymised within the Subfamily Vespertilioninae by McKenna and Bell (1997: 316).

Leuconoïdes Menu, 1987: 77, 82.

TYPE GENUS: *Leuconöe* Boie, 1830.

COMMENTS: When originally proposed, the rank was unclear but placed in the Subfamily Vespertilioninae (J. Gray, 1821), above the 'les leuconoformes' (Menu, 1987) and included the genera *Leuconöe* Boie, 1830 [=*Myotis*, 1829]; *Pizonyx* Miller, 1906b [=*Myotis*, 1829]; and *Perimyotis* Menu, 1984 [=*Pipistrellus* Kaup, 1829]. Emended by McKenna and Bell (1997: 316) from the form as originally published as 'Les leuconöides', as it apparently meets the requirements of Article 11 of the Code (ICZN, 1985a: 19). Synonymised within the Subfamily Vespertilioninae by McKenna and Bell (1997: 316).

Tribe Vespertilionini J. Gray, 1821

Family Vespertilionini J. Gray, 1821: 299.

TYPE GENUS: *Vespertilio* Linnaeus, 1758: 31.

COMMENTS: When originally proposed, this rank was placed in the Order Insectivorae (J. Gray, 1821 [=Yangochiroptera (Koopman, 1985 part)]) and included the genera *Megadermes* [sic =*Megaderma*] É. Geoffroy, 1810a: 197; *Rhynolophus* [sic =*Rhinolophus*] Lacépède, 1799; *Nycterus* [sic =*Nycteris*] É. Geoffroy and G. Cuvier, 1795: 186; *Rhynopoma* [sic =*Rhinopoma*] É. Geoffroy, 1818: 113; *Thaphosores* [sic =*Taphozous*] É. Geoffroy, 1818; *Vespertilio* Linnaeus, 1758: 31; *Pecotus* [sic =*Plecotus* É. Geoffroy, 1818: 112; and *Barbastella* J. Gray, 1821: 300. Tribe rank recognised by Koopman and Jones (1970: 27), McKenna and Bell (1997: 318), Hoofer and Van Den Bussche (2003: 38) and Simmons (2005a: 484), but not Van Dyck and Strahan (2008: 10).

Tribe Vespertilionina J. Gray, 1825a: 339.

TYPE GENUS: *Vespertilio* Linnaeus, 1758: 31.

COMMENTS: When originally proposed, this rank was placed in the Family Vespertilionidae (J. Gray, 1821) and included the genera *Vespertilio* Linnaeus, 1758: 31; *Plectotus* É. Geoffroy, 1818: 112; *Barbastellus* J. Gray, 1825b: 243 [=*Barbastella* J. Gray, 1821: 300]; *Proboscidea* de Spix, 1823: 61 [=*Rhynchonycteris* Peters, 1867f: 477]; *Thyroptera* de Spix, 1823: 61; and *Caelano* [sic =*Celaeno*] Leach, 1821b: 69 [=*Noctilio* Linnaeus, 1766: 88]. Synonymised within the Subfamily Vespertilioninae by McKenna and Bell (1997: 318).

Tribe Vespertilionini Koopman & Jones, 1970: 27.

TYPE GENUS: *Vespertilio* Linnaeus, 1758: 31.

COMMENTS: When originally proposed, this rank was placed in the Subfamily Vespertilioninae (J. Gray, 1821) and included the genera † *Samonycteris* Revilliod, 1922: 139; *Eudiscopus* Conisbee, 1953: 30; *Pipistrellus* Kaup, 1829; *Nyctalus* Bowdich, 1825: 36; *Glischropus* Dobson, 1875c: 472; *Eptesicus* Rafinesque, 1820a: 2; *Vespertilio* Linnaeus, 1758: 31; *Laephotis* Thomas, 1901c: 460; *Histiotus* Gervais, 1855b: 77; *Philetor* Thomas, 1902d: 220; *Tylonycteris* Peters, 1872b: 703; *Mimetillus* Thomas, 1904h: 12; *Hesperoptenus* Peters, 1868b: 626; and *Chalinolobus* Peters, 1865d. Synonymised within the Subfamily Vespertilioninae by McKenna and Bell (1997: 318).

Néoeptésiformes Menu, 1987: 77, 123, 134.

TYPE GENUS: *Nyctalus* Bowdich, 1825: 36.

COMMENTS: When originally proposed, the rank was unclear but placed in the Subfamily Vespertilioninae (J. Gray, 1821), in the Les Nyctaloïdes (Menu, 1987) and included the genera *Hypsugo* Kolenati, 1856: 131, 167; *Tylonycteris* Peters, 1872b: 703; and *Mimetillus* Thomas, 1904h: 12. Emended by McKenna and Bell (1997: 318) as published as 'les néoeptésiformes' as it seems to meet the requirements of Article 11 of the Code (ICZN, 1985a: 19). Synonymised within Vespertilioninae by McKenna and Bell (1997: 318).

Eptésiformes Menu, 1987: 78, 104, 134.

TYPE GENUS: *Nyctalus* Bowdich, 1825: 36.

COMMENTS: When originally proposed, the rank was unclear but placed in the Subfamily Vespertilioninae (J. Gray, 1821), in the Les Nyctaliformes (Menu, 1987) and included the genera *Eptesicus* Rafinesque, 1820a: 2; and *Nycterikaupius* Menu, 1987. Emended by McKenna and Bell (1997: 318) as published as 'les eptésiformes' as it seems to meet the requirements of Article 11 of the Code (ICZN, 1985a: 19). Synonymised within the Subfamily Vespertilioninae by McKenna and Bell (1997: 318).

Chalinolobus Peters, 1867

Chalinolobus Peters, 1867h: 680.

TYPE SPECIES: Φ *Vespertilio tuberculatus* Forster, 1844: 62 [= Φ *Chalinolobus tuberculatus* (Forster, 1844: 62)] by monotypy, by ruling of the ICZN Opinion 1994 (2002: 63).

COMMENTS: Reviewed by Dobson (1875d: 381), Tate (1942c: 260), Ryan (1966: 86) and Koopman (1971: 1; 1984a: 17). Does not include *Glauconycteris* Dobson (1875d: 383) as this was removed by Simmons (1997: 486).

Vespertilio J. Gray, 1843d: 181.

TYPE SPECIES: Φ *Vespertilio tuberculatus* Forster, 1844: 62 [= Φ *Chalinolobus tuberculatus* (Forster, 1844: 62)] by monotypy. See Opinion 1994 of the ICZN (2002: 63).

COMMENTS: Author of type species given as Forster (1844: 62) by Simmons (2005a: 485). Synonymised within *Chalinolobus* by Chruszcz and Barclay (2002: 1) who give the author of the type as J. Gray (1843d: 181).

HOMONYMS:

Vespertilio Linnaeus, 1758: 31, particolored bats of the Class Mammalia (Order Chiroptera, Family Vespertilionidae). Currently recognised genus. See Simmons (2005a: 498).

Vespertilio J. Gray, 1843a, little brown bats of the Class Mammalia (Order Chiroptera, Family Vespertilionidae). Synonymised within *Myotis* Kaup, 1829. See individual entry.

Vespertilio Mörch, 1852: 123, sea snails of the Class Mollusca (Order Neogastropoda, Family Volutidae). Genus is a synonym of *Cymbiola* Swainson, 1831: Text to Plate 83. See Darragh (1988: 259).

Scotophilus Tomes, 1857: 135; Plates 53–54.

TYPE SPECIES: Φ *Vespertilio tuberculatus* Forster, 1844: 62 (as *Scotophilus tuberculatus*) [= Φ *Chalinolobus tuberculatus* (Forster, 1844: 62)] by monotypy.

COMMENTS: Synonymised within *Chalinolobus* by Chruszcz and Barclay (2002: 1), but not discussed by Simmons (2005a: 484).

HOMONYMS:

Scotophilus Leach, 1821b: 69, 71, house bats of the Class Mammalia (Order Chiroptera, Family Vespertilionidae). Genus currently recognised. See Simmons (2005a: 465).

Scotophilus Swainson, 1836: 326, owls of the Class Aves (Order Strigiformes, Family Strigidae). Described as a subspecies and is a synonym of the genus *Strix* Linnaeus 1758: 92.

Scotophilus Meinert, 1870: 40, myriapods of the Phylum Arthropoda (Class Chilopoda, Order Geophilomorpha, Family Dignathodontidae). Genus a synonym of *Henia* Koch, 1847: 181. See Barber and Minelli (2012).

Scotophilus Hesse, 1881: 4, small crustaceans of the Subphylum Crustacea (Class Maxillopoda, Order Cyclopodea, Family *incertae sedis*). Synonym of the genus *Hesseius* Özdikmen, 2008: 267.

Chalinolobus dwyeri Ryan, 1966

Large-eared Wattled Bat

Chalinolobus dwyeri Ryan, 1966: 89.

TYPE LOCALITY: Copeton, 14 miles south of Inverell, New South Wales, Australia.

COMMENTS: Recognised since its description.

Chalinolobus gouldii (J. Gray, 1841)

Gould's Wattled Bat

S. [cotophilus] Gouldii J. Gray, 1841: 400, 405.

TYPE LOCALITY: Launceston, Tasmania, Australia.

COMMENTS: Designation by Thomas (1905a: 422). Recognised within *Scotophilus* by Gould (1855 [1845–1863]: Text to Plate 40) and transferred to *Chalinolobus* by Dobson (1878: 250), Iredale and Troughton (1934: 97) and followed by subsequent authors. Specimens of this species have been collected from Norfolk Island (Troughton, 1922: 40), but it now appears likely to be extinct (Gordon, 1984: 12; Tidemann, 1986: 511; 1987b: 33). The population from Norfolk Island (as yet unnamed) may also represent a distinct species (Flannery, 1995b: 363). Reviewed by Tidemann (1986: 503), Koopman (1993: 199) and Chruszcz and Barclay (2002: 1) who included *neocaledonicus* Revilliod, 1914: 355 within this species, but *neocaledonicus* was recognised as a distinct species by Flannery (1995b: 364) who suggested the morphometric evidence of Tidemann (1986) that synonymised it with *gouldii* was weak.

Chalinolobus gouldii venatoris Thomas, 1908b: 372.

TYPE LOCALITY: Alexandria, Northern Territory, Australia.

COMMENTS: The status of this taxon has been unstable as it was recognised as a subspecies by Iredale and Troughton (1934: 97), D. Johnson (1964: 475), L. Hall and Richards (1979: 45), Strahan (1983: 340; 1995: 513), Flannery (1995b: 360) and Simmons (2005a: 484), but was synonymised within *gouldii* by Koopman (1993:199), and not considered by Chruszcz and Barclay (2002: 1), Clayton *et al.* (2006: 110) or Van Dyck and Strahan (2008: 534).

Chalinolobus morio (J. Gray, 1841)

Chocolate Wattled Bat

Scotophilus morio J. Gray, 1841: 400, 405.

TYPE LOCALITY: Tasmania, Australia.

COMMENTS: Designation by Thomas (1905a: 422). Species recognised within *Scotophilus* by Gould (1855 [1845–1863]: Text to Plate 41) and transferred to *Chalinolobus* by Iredale and Troughton (1934: 97) and followed by subsequent authors. Also suggested to occur on Lord Howe Island by Troughton (1920: 118).

S. [cotophilus] australis J. Gray, 1841: 400, 406.

TYPE LOCALITY: Sydney, Australia. Syntypes collected in Liverpool range New South Wales; Adelaide and its vicinity, South Australia; Canning River Western Australia and Hobart Tasmania.

COMMENTS: Synonymised within *morio* by Iredale and Troughton (1934: 97) and subsequent authors.

Scotophilus microdon Tomes, 1859b: 68.

TYPE LOCALITY: Syntypes from Tasmania and South Australia?, Australia.

COMMENTS: Recognised by Gould (1860 [1845–1863]: Text to Plate 42) and Krefft (1868a: 93). Synonymised within *morio* by Iredale and Troughton (1934: 97) and subsequent authors.

Vespertilio Muelleri Becker, 1860: 43; Plate.

TYPE LOCALITY: Melbourne, Victoria, Australia.

COMMENTS: Synonymised within *morio* by Mahoney and Walton (1988g: 130).

Chalinolobus signifer Dobson, 1876b: 289.

TYPE LOCALITY: Peak Downs, Queensland, Australia.

COMMENTS: Recognised at the species rank by J. Ogilby (1892: 90). Synonymised within *morio* by Iredale and Troughton (1934: 97), Mahoney and Walton (1988g: 130) and subsequent authors.

Chalinolobus nigrogriseus (Gould, 1856)

Hoary Wattled Bat

Chalinolobus nigrogriseus nigrogriseus (Gould, 1856)

Scotophilus nigrogriseus Gould, 1856 [1845–1863]: Text to Plate 44.

TYPE LOCALITY: Moreton Bay, Queensland, Australia.

COMMENTS: Included within *Chalinolobus*, but synonymised within *picatus*, by Iredale and Troughton (1934: 97) and of uncertain status by Ride (1970: 176). Recognised as a subspecies of *picatus* by D. Johnson (1964: 476). Troughton (1967: 281) and Van Deusen and Koopman (1971: 4) recognised *nigrogriseus* at the species rank in *Chalinolobus*, as did L. Hall and Richards (1979: 42, 43), Koopman (1984a: 17) and subsequent authors.

Chalinolobus nigrogriseus rogersi Thomas, 1909

Chalinolobus nigrogriseus rogersi Thomas, 1909a: 150.

TYPE LOCALITY: Parry Creek, Western Australia, Australia. 10 feet.

COMMENTS: Recognised at the species rank by Iredale and Troughton (1934: 97), Troughton (1967: 282) and Ride (1970: 176). Subspecies status recognised by Van Deusen and Koopman (1971: 4), Strahan (1983: 344; 1995: 516) and Flannery (1990: 334). Synonymised within *nigrogriseus* by Honacki *et al.* (1982: 172), Mahoney and Walton (1988g: 130), Koopman (1993: 200) and Flannery (1995a: 440; 1995b: 365). Elevated to subspecies of *nigrogriseus* by Bonaccorso (1998: 322), Simmons (2005a: 484), Clayton *et al.* (2006: 111), and Van Dyck and Strahan (2008: 538), but not Burbidge *et al.* (2014: 24, 30).

Chalinolobus picatus (Gould, 1852)

Little Pied Wattled Bat

Scotophilus picatus Gould, 1852 [1845–1863]: Text to Plate 43.

TYPE LOCALITY: Depot Glen Preservation Creek, about 13 km north west of Milparinka, New South Wales, Australia. Holotype collected on the Sturt Expedition (1848: 324).

COMMENTS: Placed in *Chalinolobus* by Iredale and Troughton (1934: 97) and followed by subsequent authors. Reviewed by Van Deusen and Koopman (1971: 3).

Falsistrellus Troughton, 1944

Falsistrellus Troughton, 1944: 349.

TYPE SPECIES: *Vespertilio tasmaniensis* Gould, 1858 [=*Falsistrellus tasmaniensis* (Gould, 1858)] by original designation.

COMMENTS: Synonymised within *Pipistrellus* by Ride (1970: 243), Hill and Harrison (1987: 238), Corbet and Hill (1992: 133), Koopman (1993: 219) and McKenna and Bell (1997: 319). Recognised at subgeneric rank (within *Pipistrellus*) by Koopman (1994: 115), and generic rank by Kitchener *et al.* (1986: 435, 442), Adams *et al.* (1987a: 168), Mahoney and Walton (1988g: 133), Volleth and Heller (1994: 25), Reardon (1999a: 12) and Simmons (2005a: 485).

Falsistrellus mackenziei Kitchener *et al.*, 1986

Western Falsistrelle

Falsistrellus mackenziei Kitchener *et al.*, 1986: 435, 451.

TYPE LOCALITY: Donelly, Western Australia, Australia.

COMMENTS: Recognised as a distinct species by Adams *et al.* (1987a: 168) and Churchill (1998: 140). Synonymised within *tasmaniensis* in *Pipistrellus* by Koopman (1993: 224) and as a subspecies of *tasmaniensis* by Koopman (1994: 115).

Falsistrellus tasmaniensis (Gould, 1858)

Eastern Falsistrelle

Vespertilio tasmanensis Gould, 1858 [1845–1863]: Text to Plate 48.

TYPE LOCALITY: Syntypes from Tasmania, Australia; Philippine Islands and the continent of India.

COMMENTS: Described as *Vespertilio tasmanensis* on the cover and 14th plate of Part 10 and as *Vespertilio tasmaniensis* on unnumbered page of text. Gould (1863 [1845–1863]: Part 10, xxxix) appears to be the first reviser and adopts the name *tasmaniensis*. Name recognised by Krefft (1868a: 93). Recognised within *Glischropus* by Iredale and Troughton (1934: 96) and within *Pipistrellus* by Thomas (1906c: 470),

Tate (1942c: 251), Troughton (1967: 280), Ride (1970: 176), L. Hall and Richards (1979: 39), Koopman (1993: 224) and Reardon (1999a: 12). Kitchener *et al.* (1986: 443) recognised its species status and placed it in *Falsistrellus*, which was supported by Adams *et al.* (1987a: 168), R. Taylor *et al.* (1987: 110), Parnaby (1992: 26) and Simmons (2005a: 486).

Vesperugo Krefftii Peters, 1869: 404.

TYPE LOCALITY: New South Wales, Australia.

COMMENTS: Recognised as a species by J. Ogilby (1892: 89). Recognised as a subspecies of *tasmaniensis* by Tate (1942c: 251). Synonymised within *tasmaniensis* in *Pipistrellus* by Thomas (1906c: 470), Iredale and Troughton (1934: 96), Kitchener *et al.* (1986: 443) and subsequent authors.

Tribe Pipistrellini Tate, 1942

Tribe Pipistrellini Tate, 1942c: 221, 232.

TYPE GENUS: *Pipistrellus* Kaup, 1829.

COMMENTS: When originally proposed, this rank was placed in the Subfamily Vespertilioninae (J. Gray, 1821) and included the genera *Pipistrellus* Kaup, 1829; and *Glischropus* Dobson, 1875c: 472. Synonymised within Vespertilionini by McKenna and Bell (1997: 318). Tribe rank recognised by Volleth and Heller (1994: 25), Hoofer and Van Den Bussche (2003: 38) and Simmons (2005a: 470), but not Van Dyck and Strahan (2008: 10).

Nyctaloïdes Menu, 1987: 77, 94.

TYPE GENUS: *Nyctalus* Bowdich, 1825: 36.

COMMENTS: When originally proposed, the rank was unclear but placed in the Subfamily Vespertilioninae (J. Gray, 1821) and included the Nyctaliformes (Menu, 1987), Eptésiformes (Menu, 1987) and Néoeptésiformes (Menu, 1987). Synonymised within the Subfamily Vespertilioninae by McKenna and Bell (1997: 318).

Nyctaliformes Menu, 1987: 77, 95, 133.

TYPE GENUS: *Nyctalus* Bowdich, 1825: 36.

COMMENTS: When originally proposed, the rank was unclear but placed in the Subfamily Vespertilioninae (J. Gray, 1821), the Les Nyctaloïdes (Menu, 1987) and included the genera *Pipistrellus* Kaup, 1829; *Glischropus* Dobson, 1875c: 472; *Philetor* Thomas, 1902d: 220; *Scotoecus* Thomas, 1901d: 263; and *Nyctalus* Bowdich, 1825: 36. Synonymised within the Subfamily Vespertilioninae by McKenna and Bell (1997: 318).

Pipistrellus Kaup, 1829

Pipistrellus Kaup, 1829: 98.

TYPE SPECIES: Φ *Vespertilio pipistrellus* Schreber, 1774: 189; Plate 54 [= Φ *Pipistrellus pipistrellus* (Schreber, 1774: 189; Plate 54)] by monotypy.

COMMENTS: Taxonomic decision of Miller (1897a: 15, 20; 1897b: 384; 1907: 204) to include *Vesperugo* Keyserling and Blasius, 1839: 312; and *Nannugo* Kolenati 1856, 131, 169 in synonymy as *Pipistrellus* Kaup, 1829. There has been considerable confusion over the allocation of species to the genus *Pipistrellus* in Australia (e.g. Tate, 1942c; Ride, 1970: 176; Koopman, 1984a: 12). See taxonomic decision of Ride (1970: 176, 243) for *Pipistrellus* and *Falsistrellus* in synonymy as *Pipistrellus*. For discussion of the synonyms of *Pipistrellus* see Ellerman and Morrison-Scott (1951: 161), Menu (1984: 409) and Kitchener *et al.* (1986: 435). Genus reviewed by Koopman (1973: 113), Horáček and Hanák (1985: 9), Hill and Harrison (1987: 238) and Simmons (2005a: 472).

HOMONYMS:

Pipistrellus Gistel, 1848: xi, dance flies of the Class Insecta (Order Diptera, Family Hybotidae). Genus is a synonym of *Trichina* Meigen, 1830: 335. See Poole and Gentili (1996: 161).

Romicia J. Gray, 1838c: 495.

TYPE SPECIES: Φ *Romicia calcarata* J. Gray, 1838c: 495 [= unassigned – *nomen dubium*)] by monotypy.

COMMENTS: Synonymised within *Pipistrellus* by Miller (1907: 204), Corbet and Hill (1992: 133), McKenna and Bell (1997: 319) and Simmons (2005a: 472).

Romicius Blyth, 1840: 75.

TYPE SPECIES: Invalid emendation of *Romicia* J. Gray, 1838c.

COMMENTS: Synonymised within *Pipistrellus* by Hill and Harrison (1987: 238) and Simmons (2005a: 472).

Nannugo Kolenati, 1856: 131, 169.

TYPE SPECIES: Described as a subgenus of *Vesperugo* Keyserling and Blasius, 1839: 312). Included Φ *Vespertilio nathusii* Keyserling and Blasius, 1839: 320; Φ *Vespertilio kuhlii* Kuhl, 1819: 199); Φ *Vespertilio pipistrellus* Schreber, 1774: 189; Plate 54 [= Φ *Pipistrellus pipistrellus* (Schreber, 1774: 189; Plate 54)].

COMMENTS: Originally made available as a subgenus of *Vesperugo* Keyserling and Blasius, 1839: 312. Synonymised within *Pipistrellus* by Miller (1897a: 15; 1897b: 384; 1907: 204), Hill and Harrison (1987: 238), Corbet and Hill (1992: 133), McKenna and Bell (1997: 319) and Simmons (2005a: 472). Not considered by Koopman (1993).

Alobus Peters, 1867c: 707.

TYPE SPECIES: Φ *Vespertilio (Alobus) temminckii* Cretzschmar, 1826: 17 [= Φ *Pipistrellus rueppellii* (J. Fischer, 1829: 109)] by monotypy.

COMMENTS: Described as a subgenus of *Vespertilio*. Synonymised within *Pipistrellus* by Hill and Harrison

(1987: 238), McKenna and Bell (1997: 319) and Simmons (2005a: 472).

HOMONYMS:

Alobus Le Conte, 1856: 273, scarab beetles of the Class Insecta (Order Coleoptera, Family Scarabaeidae). Genus is a synonym of *Diplotaxis* Kirby, 1837: 129. See A. Smith (2002: 45) and Vaurie (1958: 282).

Euvesperugo Acloque, 1899: 35.

TYPE SPECIES: When proposed, this genus contained six species including Φ *Vespertilio noctula* Schreber, 1774: 166 [= Φ *Nyctalus noctula* (Schreber, 1774: 166); Φ *Vespertilio leisleri* Kuhl, 1817: 14, 46 [= Φ *Nyctalus leisleri* (Kuhl, 1817; 14, 46); Φ *Vesperugo maurus* Blasius, 1853: 35 [= Φ *Hypsugo savii* (Bonaparte, 1837[1832–1841]: Fascicolo 20)]; Φ *Vespertilio kuhlii* Kuhl, 1819: 199 [= Φ *Pipistrellus kuhlii* (Kuhl, 1819: 199)]; Φ *Vespertilio pipistrellus* Schreber, 1774: 167 [= Φ *Pipistrellus pipistrellus* (Schreber, 1774: 167)]; and Φ *Vespertilio abramus* Temminck, 1840: 232 [= Φ *Pipistrellus abramus* (Temminck, 1840: 232). Of these Φ *Vespertilio pipistrellus* Schreber, 1774: 167 [= Φ *Pipistrellus pipistrellus* (Schreber, 1774: 167)] is here nominated as the type species.

COMMENTS: Described as a subgenus of *Vesperugo* Keyserling and Blasius, 1839: 312 [= *Vespertilio* Linnaeus, 1758: 31]. Synonymised within *Pipistrellus* by Hill and Harrison (1987: 238), McKenna and Bell (1997: 319) and Simmons (2005a: 472).

Eptesicops A. Roberts, 1926: 245.

TYPE SPECIES: Φ *Scotophilus rusticus* Tomes, 1861: 35 (as *V. rusticus*) [= Φ *Pipistrellus rusticus* (Tomes, 1861: 35)] by original designation.

COMMENTS: Synonymised within *Pipistrellus* by Simpson (1945: 59), Hill and Harrison (1987: 238), McKenna and Bell (1997: 319) and Simmons (2005a: 472). Not considered by Koopman (1993).

Vansonia A. Roberts, 1946: 304.

TYPE SPECIES: Φ *Pipistrellus vernayi* (Roberts, 1932: 16) by original designation.

COMMENTS: Synonymised within *Pipistrellus* by Hill and Harrison (1987: 238), Koopman (1993: 219) and Simmons (2005a: 472).

HOMONYMS:

Vansonia Schein, 1956: 27, scarab beetles of the Class Insecta (Order Coleoptera, Family Scarabaeidae). Genus name was renamed *Vanstaronia* by Kammerer, 2006: 270.

Perimyotis Menu, 1984: 409, 415.

TYPE SPECIES: Φ *Vespertilio subflavus* F. Cuvier, 1832: 17 [= Φ *Pipistrellus subflavus* (F. Cuvier, 1832: 17)] by monotypy

COMMENTS: Synonymised within *Pipistrellus* by Hill and Harrison (1987: 238), Koopman (1993: 219), McKenna and Bell (1997: 319) and Simmons (2005a: 472).

Attelepharca Menu, 1987: 78, 126.

TYPE SPECIES: No type species designated and therefore is not available.

COMMENTS: Synonymised within *Pipistrellus* by Corbet and Hill (1992: 133) and Simmons (2005a: 472).

Pipistrellus adamsi Kitchener *et al.*, 1986

Forest Pipistrelle

Pipistrellus adamsi Kitchener *et al.*, 1986: 435, 463.

TYPE LOCALITY: 40 km East of the Archer River Crossing, Cape York, Queensland, Australia (13°27′S, 143°18′E).

COMMENTS: Confirmed as a species by Adams *et al.* (1987a: 167). Included within *tenuis* by Koopman (1993: 224; 1994: 112). Elevated to species rank by Strahan (1995: 524) and Simmons (2005a: 473).

† ***Pipistrellus murrayi*** C. Andrews, 1900

Christmas Island Pipistrelle

† *Pipistrellus murrayi* C. Andrews, 1900: 26.

TYPE LOCALITY: Christmas Island, Indian Ocean, Australia.

COMMENTS: Recognised as a subspecies of *Pipistrellus tenuis* Temminck, 1840 by Simmons (2005a: 479). Species rank recognised by Tate (1942c: 240), Churchill (1998: 167), Van Dyck and Strahan (2007: 547) and Burbidge *et al.* (2014: 24, 30). A major revision of south-east Asian species was undertaken to determine the evolutionary significance of this taxon and it revealed that this taxon was distinct at the species rank (Helgen *et al.* 2009: 19). This species appears to have gone extinct in 2009 (Beeton *et al.*, 2010: 11, 51).

Pipistrellus westralis Koopman, 1984

Northern Pipistrelle

Pipistrellus tenuis westralis Koopman, 1984a: 13.

TYPE LOCALITY: Cape Bossut, Western Australia, Australia. (18°40′S, 121°30′E)

COMMENTS: Included as a subspecies or synonym within *tenuis* by Koopman (1993: 224; 1994: 112). Elevated to species status by Kitchener *et al.* (1986: 456), which was subsequently confirmed by Adams *et al.* (1987a: 167).

Tribe Nycticeiini Gervais, 1855

Tribe Nycticeina Gervais, 1855b: 71, footnote.

TYPE GENUS: *Nycticeius* Rafinesque, 1819a: 417.

COMMENTS: When originally proposed, this rank was placed in the Family Vespertilionidae (J. Gray, 1821) and included the genera *Atalapha* Rafinesque, 1814a: 12 [=*Nyctalus* Bowdich, 1825: 36] and *Nycticejus* [sic =*Nycticeius*]

Rafinesque, 1819a: 417]. Tribe rank recognised by Volleth and Heller (1994: 25), McKenna and Bell (1997: 320) and Simmons (2005a: 461), but not Van Dyck and Strahan (2008: 10).

Subfamily Nycticejinae Gill, 1872: 17.

TYPE GENUS: *Nycticeius* Rafinesque, 1819a: 417.

COMMENTS: When originally proposed, this rank was placed in the Family Vespertilionidae (J. Gray, 1821). Synonymised within the Tribe Nycticeiini by McKenna and Bell (1997: 320).

Tribe Nycticeini Tate, 1942c: 280

TYPE GENUS: *Nycticeius* Rafinesque, 1819a: 417.

COMMENTS: When originally proposed, this rank was placed in the Subfamily Vespertilioninae (J. Gray, 1821) and included the genera *Scoteinus* Dobson, 1875e: 371 [=*Scotomanes* Dobson, 1875e: 371], *Nycticeius* Rafinesque, 1819a: 417; *Rhogeëssa* H. Allen, 1866: 285; *Baeodon* Miller, 1906b: 85 [=*Rhogeëssa* H. Allen, 1866: 285]; *Scotoecus* Thomas, 1901d: 263; *Scotophilus* Leach, 1821b: 69, 71; *Scotomanes* Dobson, 1875e: 371; and *Otonycteris* Peters, 1859b: 223. Synonymised within the Tribe Nycticeiini by McKenna and Bell (1997: 320).

Section Scotophili A. Murray, 1866: 238.

TYPE GENUS: *Scotophilus* Leach, 1821b: 69, 71.

COMMENTS: When originally proposed, this rank was placed in the Family Vespertilionidae (J. Gray, 1821). Synonymised within the Tribe Nycticeiini by McKenna and Bell (1997: 320).

Tribe Scotophilini Hill & Harrison, 1987: 278.

TYPE GENUS: *Scotophilus* Leach, 1821b: 69, 71.

COMMENTS: When originally proposed, this rank was placed in the Subfamily Vespertilioninae (J. Gray, 1821) and included the genera *Scotomanes* Dobson, 1875e: 371 (including *Scoteinus* Dobson, 1875e: 371); and *Scotophilus* Leach, 1821b: 69, 71). Rank was recognised by Hoofer and Van Den Bussche (2003: 38), but synonymised within the Tribe Nycticeiini by McKenna and Bell (1997: 320) and followed by Simmons (2005a: 461, 465).

Scoteanax Troughton, 1944

Scoteanax Troughton, 1944: 353.

TYPE SPECIES: *Nycticejus rüppellii* Peters, 1866a (as *Nycteicejus ruppellii* Peters, 1866a) [=*Scoteanax rueppellii* (Peters, 1866a)] by original designation.

COMMENTS: Type species given as *Oligotomus australis* Iredale (ex MacGillivray), 1937: 45. This genus was erected by Troughton (1944: 353, 354) for *ruppellii* and *Scotorepens* for the other forms. Often included within *Nycticeius* (e.g. Laurie and Hill, 1954: 70; Strahan, 1983: 348), but separated by Troughton (1967: 284), Kitchener and Caputi (1985: 85), Mahoney and Walton (1988g: 143), Parnaby (1992: 26), Strahan (1995: 527), and Van Dyck and Strahan (2008: 551).

Oligotomus Iredale (ex MacGillivray), 1937: 45.

TYPE SPECIES: *Oligotomus australis* Iredale (ex MacGillivray), 1937 [=*Scoteanax rueppellii* (Peters, 1866a)] by monotypy.

COMMENTS: Synonymised within *Scoteanax* by Simmons (2005a: 464).

HOMONYMS:

Oligotomus Cope, 1882a: 182, fossil of the Class Mammalia († Order Condylarthra, † Family Phenacodontidae). Genus is a synonym of *Ectocion* (Cope, 1882b: 522; Archibald, 1998: 321).

Scoteanax rueppellii (Peters, 1866)

Greater Broad-nosed Bat

Nycticejus Rüppellii Peters, 1866a: 21.

TYPE LOCALITY: 'Sydney in Westaustralien' [=Sydney, Australia].

COMMENTS: Dobson (1875e: 371) referred *rueppellii* to *Scotophilus* by placing it in the subgenus *Scoteinus*. Included in the genus *Scotophilus* by J. Ogilby (1892: 92). Included in *Scoteinus* by Iredale and Troughton (1934: 98). McKean (1966: 25) placed this species within *Nycticeius* as did Calaby (1966: 19, 44), Ride (1970: 177) and L. Hall and Richards (1979: 50, 55). Reviewed by Kitchener and Caputi (1985: 85). The placement of *rueppellii* in *Scoteanax* was supported by Baverstock *et al.* (1987: 417). Placed in the genus *Nycticeius* by Strahan (1983: 348), Koopman (1984a: 22; 1993: 217). Included in *Scoteanax* by Troughton (1967: 284), Kitchener and Caputi (1985: 85, 94), Mahoney and Walton (1988g: 143), Parnaby (1992: 26), Strahan (1995: 527), Churchill (1998: 170), Reardon (1999a: 13) and subsequent authors.

Oligotomus australis Iredale (ex MacGillivray), 1937: 45.

TYPE LOCALITY: Clarence River, New South Wales, Australia.

COMMENTS: Synonymised within *rueppellii* by Simmons (2005a: 464).

Scotorepens Troughton, 1944

Scotorepens Troughton, 1944: 354.

TYPE SPECIES: *Scoteinus orion* Troughton, 1937d [= *Scotorepens orion* (Troughton, 1937d)] by original designation.

COMMENTS: Publication date proposed to be 1944 rather than 1943 by Mahoney and Walton (1988g: 143). Tate

(1952a: 599) relegated *Scotorepens* to a subgenus of *Scoteinus*, while Laurie and Hill (1954: 70) and Koopman (1978: 166) considered it a subgenus of *Nycticeius*. Ride (1970: 247) synonymised it within *Nycticeius*. Taxon elevated to genus rank by Kitchener and Caputi (1985: 85) and Volleth and Tidemann (1991: 321). An undescribed species, known as the Central-eastern Broad-nosed Bat, appears to occur in eastern Australia (see Parnaby, 1992: 25, 1995: A4, A7; Menkhorst & Knight, 2011: 170; Churchill, 2008: 161; Van Dyck & Strahan, 2008: 558).

FUTURE TAXONOMIC RESEARCH: The distinctiveness of the Central-eastern Broad-nosed Bat (*sensu* Parnaby, 1992: 25) needs to be confirmed and formally described if appropriate. More generally the whole genus needs revision.

Scotorepens balstoni (Thomas, 1906)

Inland Broad-nosed Bat

Scoteinus balstoni Thomas, 1906a: 2.

TYPE LOCALITY: North Pool, Laverston, Victoria, Australia, Australia. 1650 feet.

COMMENTS: More detailed description given by Thomas (1906c: 472). Troughton (1926: 79) and Iredale and Troughton (1934: 98) included *balstoni* within the genus *Scoteinus*. This taxon was subsequently merged with *Nycticeius greyii* by Ride (1970: 245). Various authors including D. Johnson (1964: 475), McKean (1966: 28), L. Hall and Richards (1979: 51, 53) and Koopman (1982: 19; 1984a: 21; 1993: 217) recognised *balstoni* as a distinct species within *Nycticeius*. Koopman (1984a: 21) recognised *orion*, *aquilo* and *sanborni* as subspecies. Placed in the genus *Scotorepens* by Troughton (1944: 354; 1967: 285), Kitchener and Caputi (1985: 111), Reardon and Flavel (1987: 72) and Baverstock *et al.* (1987: 421), Parnaby (1992: 25), Churchill (1998: 172) and Reardon (1999a: 13) and subsequent authors.

Scoteinus influatus Thomas, 1924d: 540.

TYPE LOCALITY: Prairie, central part of southern north Queensland, Australia. 1400 feet elevation.

COMMENTS: Troughton (1926: 79) and Iredale and Troughton (1934: 98) included *influatus* within the genus *Scoteinus*. McKean (1966: 28), Ride (1970: 177), L. Hall and Richards (1979: 50, 51) and Koopman (1978: 166; 1984a: 22) included *influatus* within *Nycticeius* at the species rank. Recognised at the species rank in *Scotorepens* by Troughton (1944: 354; 1967: 285). Synonymised within *balstoni* by Mahoney and Walton (1988g: 143), Kitchener and Caputi (1985: 111) and Koopman (1993: 217), but recognised as a subspecies of *balstoni* by Simmons (2005a: 468). Taxon not recognised by Clayton *et al.* (2006: 111), Van Dyck and Strahan (2008: 553) or Burbidge *et al.* (2014: 24, 30).

Scotorepens greyii (J. Gray, 1843)

Little Broad-nosed Bat

Scotophilus Greyii J. Gray, 1843a: 30.

TYPE LOCALITY: Port Essington, Northern Territory, Australia.

COMMENTS: Recognised in *Scotophilus* by Gould (1856 [1845–1863]: Text to Plate 12) and J. Ogilby (1892: 92). Thomas (1906d: 537) placed *greyii* in the genus *Scoteinus* which was followed by Miller (1907: 217), and Iredale and Troughton (1934: 98). Dobson (1875e: 371) referred *rueppellii* to *Scotophilus* by placing it in the subgenus *Scoteinus*. McKean (1966: 28) placed this species within *Nycticeius* as did Ride (1970: 177), L. Hall and Richards (1979: 51) and Koopman (1984a: 21; 1993: 217). Recognised at the species rank within *Scotorepens* by Troughton (1944: 355; 1967: 286), Kitchener and Caputi (1985: 119), Reardon and Flavel (1987: 71), Baverstock *et al.* (1987: 417) and Mahoney and Walton (1988g: 144), Parnaby (1992: 25), Churchill (1998: 174) and Reardon (1999a: 13).

Scoteinus orion aquilo Troughton, 1937d: 278.

TYPE LOCALITY: Bowen, north Queensland, Australia.

COMMENTS: This taxon was merged with *orion* by McKean (1966: 28) and Ride (1970: 245) and made a subspecies by McKean and Price (1967: 113). Placed as a subspecies of *balstoni* within *Nycticeius* by Koopman (1978: 166; 1984a: 21). Synonymised within *greyii* by Kitchener and Caputi (1985: 119), Mahoney and Walton (1988g: 144), Koopman (1993: 217) and Simmons (2005a: 468).

Scoteinus balstoni caprenus Troughton, 1937d: 279.

TYPE LOCALITY: Roebuck Bay, Western Australia, Australia.

COMMENTS: This taxon was merged with *balstoni* by McKean (1966: 28). Placed as a subspecies of *balstoni* by D. Johnson (1964: 475), Koopman (1978: 166; 1984a: 21) and Strahan (1983: 352). Synonymised within *greyii* by Kitchener and Caputi (1985: 119), Mahoney and Walton (1988g: 144), Koopman (1993: 217), Strahan (1995: 530) and Simmons (2005a: 468).

Scotorepens orion (Troughton, 1937)

Eastern Broad-nosed Bat

Scoteinus orion Troughton, 1937d: 277.

TYPE LOCALITY: All Saints Church, Hunters Hill, Sydney, New South Wales, Australia.

COMMENTS: This taxon was merged within *greyii* by Ride (1970: 245). Subsequently L. Hall and Richards (1979: 51, 53) recognised *orion* as a distinct species within *Nycticeius* but it was reduced to a subspecies of *balstoni* by Koopman (1984a: 21). Recognised at the species rank within *Nycticeius* (*Scotorepens*) by Koopman (1978: 166) and within *Scotorepens* by Troughton (1944: 355; 1967: 286), Kitchener and Caputi (1985: 107) and Baverstock *et al.* (1987: 417). Synonymised within *balstoni* by Koopman (1993: 217). Included in *Scotorepens* at the species rank by Parnaby (1992: 24), Strahan (1995: 531), Churchill (1998: 176), Reardon (1999a: 13), Simmons (2005a: 468), and Van Dyck and Strahan (2008: 555).

Scotorepens sanborni (Troughton, 1937)

Northern Broad-nosed Bat

Scoteinus sanborni Troughton, 1937d: 280.

TYPE LOCALITY: East Cape, Papua New Guinea.

COMMENTS: Recorded from Australia by Tate (1952a: 601). This taxon was placed in *Scoteinus* by Tate (1952a: 601), *Nycticeius* by McKean (1966: 28; 1972: 31), Koopman (1978: 166) and L. Hall and Richards (1979: 51, 53). Koopman (1984a: 21) placed it as a subspecies of *balstoni*. Recognised at the species rank within *Scotorepens* by Kitchener and Caputi (1985: 115); Baverstock *et al.* (1987: 417) and Kitchener *et al.* (1994b: 31). Placed in the genus *Nycticeius* by Koopman (1993: 217). Included in *Scotorepens* again by Flannery (1990: 345; 1995a: 473; 1995b: 400), Bonaccorso (1998: 345), Churchill (1998: 178), Reardon (1999a: 13) and Simmons (2005a: 468).

Vespadelus Troughton, 1944

Vespadelus Troughton, 1944: 348.

TYPE SPECIES: *Scotophilus pumilus* J. Gray, 1841 (as *Vespadelus pumilus*) [= *Vespadelus pumilus* (J. Gray, 1841)] by monotypy.

COMMENTS: The name within Iredale and Troughton (1934: xi, 95) is a *nomen nudum* (Mahoney & Walton, 1988g: 131; Simmons, 2005a: 496). Though the name was also used within Troughton (1941: 348), it was fixed within Troughton (1944: 348) when he wrote 'The above description absolutely defines the genus *Vespadelus*.' Date of publication attributed to Iredale and Troughton (1934: 95) by McAllan and Bruce (1989: 455). Genus considered a junior synonym of *Eptesicus* Rafinesque, 1820a by various authors including Ride (1970: 248), McKean *et al.* (1978: 532), Mahoney and Walton (1988g: 131) and Koopman (1993: 200). Subgenus rank within *Pipistrellus* recognised by Hill and Harrison (1987: 242) and Koopman (1994: 116). Synonymised within *Pipistrellus* by McKenna and Bell (1997: 319). *Vespadelus* at the generic level was recognised by Troughton (1967: 278), Volleth and Tidemann (1991: 215), Volleth and Heller (1994: 11), Strahan (1995: 534), Reardon (1999a: 13), Simmons (2005a: 496), and Van Dyck and Strahan (2008: 560). Hill (1966: 306) suggested that *Vespadelus* (at either subgeneric or generic rank) could represent Australian forms currently placed in *Eptesicus*. Australian species reviewed by Kitchener *et al.* (1987: 427). Adams *et al.* (1987b: 143) proposed there was a minimum of nine species (then reviewed as *Eptesicus*).

Vespadelus Iredale & Troughton, 1934: xi, 95.

TYPE SPECIES: *Nomen nudum.*

COMMENTS: Recognised as a *nomen nudum* by McKenna and Bell (1997: 219) and Simmons (2005a: 496).

Registrellus Troughton, 1944: 349.

TYPE SPECIES: *Pipistrellus regulus* Thomas, 1906c [= *Vespadelus regulus* (Thomas, 1906c)] by original designation.

COMMENTS: This genus considered a junior synonym of *Eptesicus* Rafinesque, 1820a by Mahoney and Walton (1988g: 131). Synonymised within *Pipistrellus* by Hill and Harrison (1987: 238) and McKenna and Bell (1997: 319). Synonymised within *Vespadelus* by Simmons (2005a: 496).

Nycterikaupius Menu, 1987: 78, 108.

TYPE SPECIES: *Scotophilus pumilus* J. Gray, 1841: 400 (as *Eptesicus pumilus*) [= *Vespadelus pumilus* (J. Gray, 1841)] by original designation.

COMMENTS: Proposed as a genus. Not recognised by Koopman (1993), which was noted by Corbet and Hill (1994: 241), but was synonymised within *Pipistrellus* by McKenna and Bell (1997: 319). Taxon not recognised by Simmons (2005a).

Vespadelus baverstocki (Kitchener *et al.*, 1987)

Inland Forest-bat

Eptesicus baverstocki Kitchener *et al.*, 1987: 481.

TYPE LOCALITY: Yuinmery area, Western Australian Goldfields, Australia. 450 m elevation. (28°28′30″S, 119°17′15″E)

COMMENTS: Included in *vulturnus* by Koopman (1994: 116). Transferred to *Vespadelus* by Queale (1997: 29), Churchill (1998: 180; 2008: 163) and Reardon (1999a: 13).

Vespadelus caurinus (Thomas, 1914)

Northern Cave-bat

Eptesicus pumilus caurinus Thomas, 1914a: 439.

TYPE LOCALITY: Drysdale River, Kimberley, Western Australia, Australia.

COMMENTS: Considered a subspecies of *Eptesicus pumilus* by Iredale and Troughton (1934: 96), D. Johnson (1964: 474), S. Carpenter *et al.* (1978: 631), McKean *et al.* (1978: 534), L. Hall and Richards (1979: 47), Strahan (1983: 359), Adams *et al.* (1987b: 154), Mahoney and Walton (1988g: 132) and Koopman (1994: 116). Synonymised within *Eptesicus pumilus* by Koopman (1993: 203). Recognised as a full species within *Eptesicus* by Kitchener *et al.* (1987: 447) and

Vespadelus by Strahan (1995: 536), Churchill (1998: 182; 2008: 164), Reardon (1999a: 13) and subsequent authors.

Vespadelus darlingtoni (G. Allen, 1933)

Large Forest-bat

Eptesicus darlingtoni G. Allen, 1933: 150.

TYPE LOCALITY: Macpherson Range, Queensland National Park, Queensland, Australia. Approx. 915 m.

COMMENTS: Considered a subspecies of *Eptesicus pumilus* by Iredale and Troughton (1934: 96) and Mahoney and Walton (1988g: 132). Synonymised within *Eptesicus pumilus* by Koopman (1993: 203) and McKean *et al.* (1978: 533). Elevated to species rank within *Eptesicus* by Kitchener *et al.* (1987: 462), Reardon and Flavel (1987: 76), Parnaby (1992: 18) and Koopman (1994: 116) who placed it within the subgenus *Vespadelus*. Transferred to *Vespadelus* by Queale (1997: 29), Churchill (1998: 184; 2008: 166), Reardon (1999a: 13) and subsequent authors.

Eptesicus sagittula McKean *et al.*, 1978: 535.

TYPE LOCALITY: Braidwood, New South Wales, Australia. (35°21′S, 149°44′E)

COMMENTS: Species rank recognised by S. Carpenter *et al.* (1978: 631), L. Hall and Richards (1979: 47, 49) and R. Taylor *et al.* (1987: 109). This name was used in preference to *darlingtoni* by Adams *et al.* (1987b: 149) and Koopman (1993: 203). Synonymised within *pumilus* by Flannery (1995b: 367). Species rank recognised within *Pipistrellus* (*Vespadelus*) by Volleth and Tidemann (1989: 215). Kitchener *et al.* (1987: 462) synonymised *sagittula* within *darlingtoni*, which appears to be followed by subsequent authors including Simmons (2005a: 497).

Vespadelus douglasorum (Kitchener, 1976)

Yellow-lipped Cave-bat

Eptesicus douglasi Kitchener, 1976: 295.

TYPE LOCALITY: Tunnel Creek, Napier Range, Western Australia, Australia. Approx 140 m elevation. (17°37′S, 125°09′E)

COMMENTS: Specific name emended from *douglasi* to *douglasorum* by Kitchener *et al.* (1987: 452). Recognised within *Eptesicus* until transferred to *Vespadelus* by Churchill (1998: 186; 2008: 168), Reardon (1999a: 13) and subsequent authors.

Vespadelus finlaysoni (Kitchener *et al.*, 1987)

Finlayson's Cave-bat

Eptesicus finlaysoni Kitchener *et al.*, 1987: 456.

TYPE LOCALITY: Cossack, Western Australia, Australia. Approx. 5m elevation. (20°41′S, 117°11′E)

COMMENTS: Synonymised within *Eptesicus pumilus* by Koopman (1993: 203) and *Pipistrellus (Vespadelus) pumilus* by Koopman (1994: 116). Recognised as a full species within *Vespadelus* by Queale (1997: 29), Churchill (1998: 188; 2008: 169) and Reardon (1999a: 13).

Vespadelus pumilus (J. Gray, 1841)

Eastern Forest-bat

S. [cotophilus] pumilus J. Gray, 1841: 400, 406.

TYPE LOCALITY: Yarrundi, Dartbrook River, a tributary of the Hunter River, New South Wales, Australia. (Approx 32°00′S, 150°46′E) See Kitchener *et al.* (1987: 442).

COMMENTS: Recorded on Lord Howe Island by Troughton (1922: 40). Included in *Scotophilus* by Gould (1857 [1845–1863]: Text to Plate 46) and *Vesperugo* by J. Ogilby (1892: 88). Recognised as a full species and included within *Eptesicus* by L. Hall and Richards (1979: 47), Koopman (1984a: 15) and Parnaby (1992: 18). Transferred to *Vespadelus* by Troughton (1944: 348; 1967: 278), Churchill (1998: 190; 2008: 171) and Reardon (1999a: 13). Does not include *caurinus*, *darlingtoni*, *finlaysoni* and *troughtoni* (see Kitchener *et al.*, 1987; Adams *et al.* 1987a; Queale, 1997).

Vespadelus regulus (Thomas, 1906)

Southern Forest-bat

Pipistrellus regulus Thomas, 1906c: 470.

TYPE LOCALITY: King River, King Gorge Sound, southern Western Australia, Australia.

COMMENTS: Included in the genus *Pipistrellus* by Iredale and Troughton (1934: 96), and *Registrellus* by Troughton (1967: 279). Synonymised within *pumilus* by Ride (1970: 246). Moved to *Eptesicus* by Hill (1966: 302, 307), McKean *et al.* (1978: 534), Kitchener and Halse (1978: 257) and followed by S. Carpenter *et al.* (1978: 638), L. Hall and Richards (1979: 47, 49), Kitchener *et al.* (1987: 427, 471), Reardon and Flavel (1987: 74), R. Taylor *et al.* (1987: 109), Mahoney and Walton (1988g: 132) and Parnaby (1992: 19). Transferred to *Vespadelus* by Queale (1997: 29), Churchill (1998: 192; 2008: 173) and Reardon (1999a: 13).

Vespadelus troughtoni (Kitchener *et al.*, 1987)

Eastern Cave-bat

Eptesicus troughtoni Kitchener *et al.*, 1987: 467.

TYPE LOCALITY: Yarramulla Lava Tunnels, Mt Surprise, Queensland, Australia. 840m elevation. (18°13′30″S, 144°40′30″E)

COMMENTS: Synonymised within *Eptesicus pumilus* by Koopman (1993: 203) and *Pipistrellus (Vespadelus) pumilus* by Koopman (1994: 116). Recognised as a full species and included within *Vespadelus* by Churchill (1998: 194; 2008: 175), Reardon (1999a: 13) and subsequent authors.

Vespadelus vulturnus (Thomas, 1914)

Little Forest-bat

Eptesicus pumilus vulturnus Thomas, 1914a: 440.

TYPE LOCALITY: Tasmania, Australia.

COMMENTS: Recognised as a subspecies of *pumilus* by Iredale and Troughton (1934: 96). Taxonomic decision of McKean *et al.* (1978: 534) to recognise it at species rank, which was followed by S. Carpenter *et al.* (1978: 631), L. Hall and Richards (1979: 47, 48) and R. Taylor *et al.* (1987: 109). Recognised as a full species within *Pipistrellus* (*Vespadelus*) by Volleth and Tidemann (1989: 215) and within *Vespadelus* by Queale (1997: 29), Churchill (1998: 196; 2008: 177), Reardon (1999a: 13) and subsequent authors.

Vespertilio pygmaeus Becker, 1858: 5, col. 5.

TYPE LOCALITY: Oakleigh, near Dandenong, Victoria, Australia.

COMMENTS: See also Becker (1859: 40) for the full description. Synonymised within *vulturnus* by McKean *et al.* (1978: 534), Kitchener *et al.* (1987: 476) and Simmons (2005a: 498).

HOMONYMS:

Vespertilio pygmaeus Leach, 1825: 559, the Soprano Pipistrelle of the Class Mammalia (Order Chiroptera, Family Vespertilionidae). Currently recognised as *Pipistrellus pygmaeus* (Leach, 1825: 559). See Simmons (2005a: 477).

Subfamily Myotinae Tate, 1942

Tribe Myotini Tate, 1942c: 221, 229.

TYPE GENUS: *Myotis* Kaup, 1829.

COMMENTS: When originally proposed as a new rank it was placed in the Subfamily Vespertilioninae (J. Gray, 1821) and included the genera *Myotis* Kaup, 1829; *Lasionycteris* Peters, 1865e: 648; *Plecotus* É. Geoffroy, 1818: 112; *Corynorhinus* H. Allen, 1865: 173; *Idionycteris* Anthony, 1923: 1; and *Euderma* H. Allen, 1891: 467. Described as the Tribe Myotini within the Subfamily Vespertilioninae. Tribe rank recognised by Koopman and Jones (1970: 27), Volleth and Heller (1994: 25, 31), who suggesting elevating the rank to subfamily, and McKenna and Bell (1997: 317). Raised to subfamily rank by Simmons (1998: 12), following the suggestion of Volleth and Heller (1994: 31), and supported by Hoofer and Van Den Bussche (2003: 14, 38) and Simmons (2005a: 499). Rank not recognised by Van Dyck and Strahan (2008: 10).

Leuconoformes Menu, 1987: 77, 82, 133.

TYPE GENUS: *Leuconöe* Boie, 1830.

COMMENTS: When originally proposed, the rank was unclear but placed in the Subfamily Vespertilioninae (J. Gray, 1821) and included the genera *Leuconöe* Boie, 1830 [=*Myotis*, 1829]; *Pizonyx* Miller, 1906b [=*Myotis*, 1829]; and *Perimyotis* Menu, 1984 [=*Pipistrellus* Kaup, 1829]. Emended by McKenna and Bell (1997: 316) as published as 'Les leuconoformes', it appeared to meet the requirements of Article 11 of the Code (ICZN, 1985a: 19). Synonymised within Myotini by McKenna and Bell (1997: 317).

Subfamily Myotinae Simmons, 1998: 12.

TYPE GENUS: *Myotis* Kaup, 1829.

COMMENTS: When originally proposed as a new rank it was placed in the Family Vespertilionidae (J. Gray, 1821), although the contents of the subfamily were not included. See comments above.

Myotis Kaup, 1829

Myotis Kaup, 1829: 106, 188.

TYPE SPECIES: Φ *Vespertilio myotis* Borkhausen, 1797: 80 [= Φ *Myotis myotis* (Borkhausen, 1797: 80)] by subsequent designation. See Miller (1912: 167) and discussion below.

COMMENTS: There has been complicated nomenclature surrounding the Linnaean name *Vespertilio murinus*, formerly applied to the type species of this genus, hence leading to the frequent application of the name *Vespertilio* to this genus, see Rydell and Baagøe (1994: 4) and D. Wilson (2007: 469). The application of the Linnaean name to the European Particoloured Bat, presently known as *Vespertilio murinus*, was mandated by Direction 98 of the ICZN (1958a: 130, 145–146). Reviews of *Myotis* include Findley (1972: 31), Hayman and Hill (1971: 33), C. Phillips and Birney (1968: 495), Simmons (2005a: 500) and D. Wilson (2007: 468). Koopman (1984a: 12) considered that only one subspecies of *Myotis* occurred in Australia (as *M. adversus macropus*). A revision of *Myotis* by Kitchener *et al.* (1995: 191) split *adversus* into three species: *M. adversus* from New South Wales and Indonesia, *M. moluccarum* from northern Australia and New Guinea, and *M. macropus* from Victoria and South Australia. Analysis of mitochondrial DNA by S. Cooper *et al.* (2001: 328) to test the hypothesis that three species of *Myotis* that occur in Australia (*adversus*, *macropus* and *moluccarum*) suggested that only a single species is present in Australia. As *M. macropus* is distinct from *M. adversus* in Indonesia, and because it remains doubtful whether *macropus* and *moluccarum* should be regarded as separate species, a conservative approach should be adopted and only *M. macropus* should be recognised as it is senior to *M. moluccarum*.

HOMONYMS:

Myotis J. Gray, 1842b: 258, bats of the Class Mammalia (Order Chiroptera). This was described as a new genus to include the species *M. murinus*, *M. bechsteinii* and *M. natereri*.

Nystactes Kaup, 1829: 108.

TYPE SPECIES: Φ *Vespertilio bechsteinii* Kaup, 1829: 108 [= Φ *Myotis bechsteinii* (Kuhl, 1819: 14, 30)]. See D. Wilson (2007: 468).

COMMENTS: Name is objective synonym of *Paramyotis* Bianchi 1917. Synonymised within *Myotis* by Miller (1897a: 16; 1897b: 382; 1907: 200) and subsequent authors including D. Wilson (2007: 468).

HOMONYMS:

Nystactes Gloger, 1827: 277, puffbirds of the Class Aves (Order Piciformes, Family Bucconidae). Currently recognised genus.

Nytctactes G. Gray, 1840: 10, puffbirds of the Class Aves (Order Piciformes, Family Bucconidae). Incorrect subsequent spelling of *Nystactes* Gloger, 1827: 277.

Leuconöe Boie, 1830: 256.

TYPE SPECIES: Φ *Vespertilio daubentonii* Kuhl, 1817: 6 [= Φ *Myotis daubentonii* (Kuhl, 1817: 6)] by subsequent designation. See Thomas (1904i: 382).

COMMENTS: Synonymised within *Myotis* by Miller (1907: 200). Recognised at genus rank by Menu (1987: 77, 83) and subgenus by Thomas (1904i: 382), Tate (1941f: 548; 1952a: 597) and Corbet and Hill (1992: 125). Synonymised within *Myotis* by Koopman (1993: 207), Simmons (2005a: 500) and Wilson (2007: 468).

Vespertilio Keyserling & Blasius, 1839: 307.

TYPE SPECIES: Not designated.

COMMENTS: Synonymised within *Myotis* by D. Wilson (2007: 468).

HOMONYMS:

See above.

Capaccinius Bonaparte, 1841 [1832–1841]: Fascicolo 30, first of four unnumbered pages in Introduzione Alla Classe 1. Mammiferi.

TYPE SPECIES: Φ *Vespertilio megapodius* Temminck, 1840: 189 [= Φ *Myotis capaccinii* (Bonaparte, 1837 [1832–1841]: Fascicolo 20)] by tautonomy.

COMMENTS: Synonymised within *Myotis* by Miller (1907: 201), Simmons (2005a: 500) and Wilson (2007: 468).

Selysius Bonaparte, 1841 [1832–1841]: Fascicolo 30, third of four unnumbered pages in Introduzione Alla Classe 1. Mammiferi.

TYPE SPECIES: Φ *Vespertilio mystacinus* Kuhl, 1817: 7 (as *V. mystacinus* Leisler) [= Φ *Myotis mystacinus* (Kuhl, 1817: 7)] by monotypy.

COMMENTS: Subgenus recognised by Corbet and Hill (1992: 121) and Tate (1952a: 597). Synonymised within *Myotis* by Miller (1897a: 17; 1897b: 382; 1907: 201), Mahoney and Walton (1988g: 138), Koopman (1993: 207), Simmons (2005a: 500) and Wilson (2007: 468).

Trilatitus J. Gray, 1842b: 258.

TYPE SPECIES: Φ *Vespertilio hasseltii* Temminck, 1840: 225 (as *Trilatitus hasseltii*) [= Φ *Myotis hasseltii* (Temminck, 1840: 225)] by subsequent designation. See Kretzoi and Kretzoi (2000: 415).

COMMENTS: Synonymised within *Myotis* by Miller (1907: 201), Corbet and Hill (1992: 119), Simmons (2005a: 500) and Wilson (2007: 468).

Vespertilio J. Gray, 1843a: xix, 26.

TYPE SPECIES: Φ *Vespertilio mystacinus* Kuhl, 1817: 7 [= Φ *Myotis mystacinus* (Kuhl, 1817: 7)] by monotypy.

COMMENTS: Placed here within *Myotis* as the type species is from this genus. Synonymised within *Myotis* by Wilson (2007: 468).

HOMONYMS:

Vespertilio Linnaeus, 1758: 31, particolored bats of the Class Mammalia (Order Chiroptera, Family Vespertilionidae). Currently recognised genus. See Simmons (2005a: 498).

Vespertilio J. Gray, 1843d, wattled bats of the Class Mammalia (Order Chiroptera, Family Vespertilionidae). Genus is a synonym of *Chalinolobus* Peters, 1867h: 680. See individual entry.

Vespertilio Mörch, 1852: 123, sea snails of the Class Mollusca (Order Neogastropoda, Family Volutidae). Genus is a synonym of *Cymbiola* Swainson, 1831: Text to Plate 83. See Darragh (1988: 259).

Tralatitus Gervais, 1849b: 213.

TYPE SPECIES: Incorrect subsequent spelling of *Trilatitus* J. Gray, 1842b: 258. See Palmer (1904: 692).

COMMENTS: Synonymised within *Myotis* by Simmons (2005a: 500).

Brachyotus Kolenati, 1856: 131, 174.

TYPE SPECIES: Φ *Vespertilio mystacinus* Kuhl, 1817: 7 [= Φ *Myotis mystacinus* (Kuhl, 1817: 7)] by subsequent designation. See Ellerman and Morrison (1951: 137).

COMMENTS: Described as a subgenus of *Vespertilio* Linnaeus, 1758: 31. Synonymised with *Myotis* by Miller (1897b: 382; 1907: 201), Corbet and Hill (1992: 119) and Simmons (2005a: 500).

HOMONYMS:

Brachyotus Gould, 1837: 10, 11, true owls of Class Aves (Order Strigiformes, Family Strigidae). Genus is a synonym of *Asio* Brisson, 1760: 28.

Isotus Kolenati, 1856: 131, 177.

TYPE SPECIES: Φ *Vespertilio nattereri* Kuhl, 1817: 14, 33 [= Φ *Myotis nattereri* (Kuhl, 1817)]. See Tate (1941f: 546).

COMMENTS: Described as a subgenus of *Vespertilio* Linnaeus, 1758: 31. Synonymised within *Myotis* by Miller

(1897a: 14; 1897b: 383; 1907: 201), Simmons (2005a: 500) and Wilson (2007: 468).

Myotus Kolenati, 1856: 131, 179.

TYPE SPECIES: Incorrect subsequent spelling of *Myotis* Kaup, 1829. See Palmer (1904: 442).

COMMENTS: Described as a subgenus of *Vespertilio* Linnaeus, 1758: 31. Emendation of *Myotis* Kaup, 1829. Synonymised within *Myotis* by Palmer (1904: 442).

Pternopterus Peters, 1867c: 706.

TYPE SPECIES: Φ *Vespertilio (Pternopterus) lobipes* Peters, 1867c: 706 [= Φ *Myotis muricola* (J. Gray, 1846a: 4)] by monotypy.

COMMENTS: Synonymised within *Myotis* by Miller (1907: 201), Corbet and Hill (1992: 119), Simmons (2005a: 500) and Wilson (2007: 468).

Exochurus Fitzinger, 1870a: 75.

TYPE SPECIES: Φ *Vespertilio macrodactylus* Temminck, 1840: 231 [= Φ *Myotis macrodactylus* (Temminck, 1840: 231)] by subsequent designation. See Kretzoi and Kretzoi (2000: 136).

COMMENTS: Synonymised within *Myotis* by Miller (1907: 201), Simmons (2005a: 500) and Wilson (2007: 468).

HOMONYMS:

Exochura Kolenati, 1858: 251, bats of the Class Mammalia (Order Chiroptera, Family Vespertilionidae). As Φ *Amblyotis atratus* Kolenati, 1858: 252 [= Φ *Eptesicus nilssonii* (Keyserling & Blasius, 1839: 315)]. Name is a synonym of *Eptesicus* Rafinesque, 1820a: 2. See Simmons (2005: 452, 456).

Aeorestes Fitzinger, 1870b: 427.

TYPE SPECIES: Φ *Vespertilio villosissimus* Ė. Geoffroy, 1806: 204 (as *Aeorestes villosissimus*) [= Φ *Lasiurus cinereus villosissimus* (Ė. Geoffroy, 1806: 204)], Φ *V. albescens* Geoffroy, 1806: 204 [= Φ *Myotis albescens* (Geoffroy, 1806: 204)], Φ *V. nigricans* Schinz, 1821: 179 [= Φ *Myotis nigricans* (Schinz, 1821: 179)], Φ *V. levis* I. Geoffroy, 1824: 444 [=Φ *Myotis levis* (I. Geoffroy, 1824: 444)]. No type designated. See Palmer (1904: 82).

COMMENTS: Synonymised within *Myotis* by Miller (1897a: 12; 1897b: 383; 1907: 201), Simmons (2005a: 500) and Wilson (2007: 469).

Comastes Fitzinger, 1870c: 565.

TYPE SPECIES: Φ *Vespertilio capaccinii* Bonaparte, 1837 [1832–1841]: Fascicolo 20 [= Φ *Myotis capaccinii* (Bonaparte, 1837 [1832–1841]: Fascicolo 20)] by subsequent designation. See Kretzoi and Kretzoi (2000: 85).

COMMENTS: Synonymised within *Myotis* by Miller (1897a: 13; 1897b: 383; 1907: 201), Simmons (2005a: 500) and Wilson (2007: 469).

HOMONYMS:

Comastes Jan, 1863: 99, 102, keelback snakes of the Class Reptilia (Order Squamata, Family Colubridae). There appears to be some confusion over the placement of this name but it appears to be a synonym of *Xenochrophis* Günther, 1864: xx, 222, 273.

Euvespertilio Acloque, 1899: 38.

TYPE SPECIES: Not designated from Φ *Vespertilio bechsteinii* Kuhl, 1817: 6, 22 [= Φ *Myotis bechsteinii* (Kuhl, 1817: 6, 22)]; Φ *Vespertilio emarginatus* Ė. Geoffroy, 1806: 198 [=Φ *Myotis emarginatus* Ė. Geoffroy, 1806: 198)]; Φ *Vespertilio mystacinus* Kuhl, 1817: 7, 58 [= Φ *Myotis mystacinus* (Kuhl, 1817: 7, 58)]; Φ *Vespertilio natteri* Kuhl, 1817: 6, 25 [=Φ *Myotis natteri* (Kuhl, 1817: 6, 25)]; and Φ *Vespertilio nigricans* Schinz, 1821: 179 [= Φ *Myotis nigricans* (Schinz, 1821: 179)]. See D. Wilson (2007: 469).

COMMENTS: Described as a subgenus of *Vespertilio* Linnaeus, 1758: 31. Synonymised within *Myotis* by Simmons (2005a: 500) and Wilson (2007: 469).

Pizonyx Miller, 1906b: 85.

TYPE SPECIES: Φ *Myotis vivesi* Menegaux, 1901: 323 by original designation.

COMMENTS: Synonymised within *Myotis* by Simpson (1945: 59), Koopman (1993: 207), Simmons (2005a: 500) and Wilson (2007: 469).

Chrysopteron Jentink, 1910: 74.

TYPE SPECIES: Φ *Chryopteron bartelsii* Jentink, 1910: 74 [=Φ *Myotis formosus bartelsii* (Jentink, 1910: 74)]. Corbet and Hill (1992: 119) give the type as Φ *Kerivoula weberi* Jentink, 1890: 96 [=Φ *Myotis formosus weberi* (Jentink, 1890: 96)] by monotypy. See D. Wilson (2007: 469).

COMMENTS: Subgeneric status recognised by Tate (1941f: 547) and Corbet and Hill (1992: 121). Synonymised within *Myotis* by Thomas (1923g: 253), Simmons (2005a: 500) and Wilson (2007: 469).

Megapipistrellus Bianchi, 1917: lxxvii.

TYPE SPECIES: Φ *Pipistrellus annectans* Dobson, 1871c: 213 [= Φ *Myotis annectans* (Dobson, 1871c: 213] by monotypy.

COMMENTS: Proposed as a subgenus of *Pipistrellus*. Synonymised within *Myotis* by Corbet and Hill (1992: 119) and Simmons (2005a: 500).

Rickettia Bianchi, 1917: lxxviii.

TYPE SPECIES: Φ *Vespertilio (Leuconoë) ricketti* Thomas, 1894: 300 [= Φ *Myotis ricketti* (Thomas, 1894: 300)] by monotypy.

COMMENTS: Generic rank recognised by G. Allen (1936: 168; 1938: 224) and subgenus rank by Corbet and Hill (1992: 127). Synonymised within *Myotis* by Corbet and Hill (1992: 119), Simmons (2005a: 500) and Wilson (2007: 469).

Dichromyotis Bianchi, 1917: lxxviii.

TYPE SPECIES: Φ *Vespertilio formosa* Hodgson, 1835: 700 [= Φ *Myotis formosus* (Hodgson, 1835: 700)] by monotypy.

COMMENTS: Proposed as a subgenus of *Myotis*. Synonymised within *Myotis* by Corbet and Hill (1992: 119), Simmons (2005a: 500) and Wilson (2007: 469).

Paramyotis Bianchi, 1917: lxxix.

TYPE SPECIES: Replacement name for *Nystactes* Kaup, 1829: 108.

COMMENTS: Synonymised within *Myotis* by Simmons (2005a: 500) and Wilson (2007: 469).

Anamygdon Troughton, 1929: 87.

TYPE SPECIES: Φ *Anamygdon solomonis* Troughton, 1929: 89 [= Φ *Myotis moluccarum solomonis* (Troughton, 1929: 89)] by original designation.

COMMENTS: Genus rank tentatively recognised by Ryan (1965b: 518) who noted that there was not yet sufficient material to permit a careful determination of its status. Synonymised within *Myotis* by C. Phillips and Birney (1968: 495), Corbet and Hill (1992: 119), Koopman (1993: 207), Simmons (2005a: 500) and Wilson (2007: 469).

Hesperomyotis Cabrera, 1958: 103.

TYPE SPECIES: Φ *Myotis simus* Thomas, 1901e: 541 by original designation.

COMMENTS: Synonymised within *Myotis* by Simmons (2005a: 500) and Wilson (2007: 469).

Myottis Alberico & Orejuela, 1982: 58.

TYPE SPECIES: Incorrect subsequent spelling of *Myotis* Kaup, 1829.

COMMENTS: Synonymised within *Myotis* by D. Wilson (2007: 469).

Myotis macropus (Gould, 1855)

Large-footed Myotis

Vespertilio macropus Gould, 1855 [1845–1863]: Text to Plate 47.

TYPE LOCALITY: South Australia, Australia.

COMMENTS: Not separated from *adversus* by Dobson (1878: 292), which was agreed to by Thomas (1915f: 171). Thomas (1915f: 171) incorrectly stated the type locality was in Western Australia. Species recognised within *Myotis* by Iredale and Troughton (1934: 98) and Troughton (1967: 283). Considered a subspecies of *Myotis adversus* by McKean and Price (1967: 112), Strahan (1983: 346; 1995: 521), Koopman (1984a: 12) and Flannery (1990: 335; 1995a: 456; 1995b: 382). Synonymised within *Myotis adversus* by Ride (1970: 245), Mahoney and Walton (1988g: 138), Koopman (1993: 207) and Churchill (1998: 150). Elevated to species rank by Kitchener *et al.* (1995: 191) who revised the complex, and followed by Reardon (1999a: 12). Recognised as the only *Myotis* to occur in Australia by S. Cooper *et al.* (2001: 328) and followed by Simmons (2005a: 510), Churchill (2008: 128), and Van Dyck and Strahan (2008: 544).

Vespertilio australis Dobson, 1878: 317.

TYPE LOCALITY: Sydney, New South Wales, Australia.

COMMENTS: Recognised at the species rank in the genus *Myotis* by Iredale and Troughton (1934: 98), Tate (1941f: 555), Troughton (1967: 284), Koopman (1984a: 11), Mahoney and Walton (1988g: 138) and Simmons (2005a: 502). Synonymised within *macropus* by Churchill (2008: 128), and Van Dyck and Strahan (2008: 545).

Myotis moluccarum richardsi Kitchener *et al.*, 1995: 191, 208.

TYPE LOCALITY: Gayundah Creek, Hinchinbrook Island, Queensland, Australia.

COMMENTS: Recognised by Simmons (2005a: 511) but not considered by other authors including Churchill (1998: 150; 2008: 128), and Van Dyck and Strahan (2008: 545).

Superorder Fereuungulata Waddell *et al.*, 1999

Clade Fereuungulata Waddell *et al.*, 1999a: 3.

COMMENTS: When originally proposed, this clade was placed in the Placentalia (Bonaparte, 1838) and included the orders Carnivora (Bowdich, 1821), Pholidota (Weber, 1904: vi, 420), Perissodactyla (Owen, 1848), Artiodactyla (Owen, 1848) and Cetacea (Brisson, 1762). Name also discussed by Waddell *et al.* (1999b: 31, 50) who introduced it as new, but the reference above has page priority. This clade was proposed as an alternative hypothesis to Scrotifera for the superordinal relationships of placentals (Waddell *et al.*, 1999b: 50). Name recognised at grandorder rank by Springer *et al.* (2007: 21).

Order Carnivora Bowdich, 1821

Order Carnivora Bowdich, 1821: 33.

COMMENTS: When originally proposed, this rank was placed in the Class Mammalia (Linnaeus, 1758) and included the groupes Plantigrada (Illiger, 1811 [=Carnivora (Bowdich, 1821 part)]), Digitigrada (G. Fischer, 1813a [=Carnivora (Bowdich, 1821 part)]) and Amphibiae (J. Gray, 1821 [=Phocoidea (J. Gray, 1821 part)]).

HOMONYMS:

Carnivora Lesson, 1842, toothed whales of the Class Mammalia (Order Cetacea, Parvorder Odontoceti). See individual entry.

Ferae Linnaeus, 1758: 16, 37.

COMMENTS: When originally proposed, this rank was placed in the Class Mammalia (Linnaeus, 1758) and included the genera *Phoca* Linnaeus, 1758: 37; *Canis* Linnaeus, 1758; *Felis* Linnaeus, 1758; *Viverra* Linnaeus, 1758: 43; *Mustela* Linnaeus, 1758; and *Ursus* Linnaeus, 1758: 47. Included the orders Carnivora and Pholidota (Weber, 1904: 412) by McKenna and Bell (1997: 211, 220). Recognised as a superorder by Minkoff (1976: 153), and Skinner and Chimimba (2005: vi, 353) but as a clade by Asher and Helgen (2010: 4) who both included the orders Carnivora and Pholidota (Weber, 1904: 412).

HOMONYMS:

Ferae J. Gray, 1821, marsupials of the Class Mammalia. Synonymised within the Infraclass Marsupialia (Illiger, 1811) here. See individual entry.

Fissipeda Blumenbach, 1791.

COMMENTS: Recognised as the Suborder Fissipedia by Gill (1872: v, 3, 56), Simpson (1931: 263) and as Fissipeda by Simpson (1945: 108), who suggested that it is often incorrectly spelt 'Fissipedia' and credited to G. Fischer (1813a: 13) who also spell the name Fissipeda. Synonymised within Carnivora by McKenna and Bell (1997: 226). The land carnivores, to which the term traditionally has been applied, are now known to be paraphyletic relative to the marine carnivores.

HOMONYMS:

Type Fissipeda Burnett, 1830a, echidnas of the Class Mammalia (Order Monotremata, Family Tachyglossidae). See individual entry.

Classe Carnivori Vicq d'Azyr, 1792: civ.

COMMENTS: When originally proposed, this rank included the Oursinins (Vicq d'Azyr, 1792: civ), Mustelins (Vicq d'Azyr, 1792: civ), Ichneumons (Vicq d'Azyr, 1792: civ), Felins (Vicq d'Azyr, 1792: cv), Canins (Vicq d'Azyr, 1792: cvi), and Loutrins (Vicq d'Azyr, 1792: cvi). Synonymised within the Order Carnivora by Simpson (1945: 105) and McKenna and Bell (1997: 226).

Family Digitigraden Duméril, 1806a: 4, 12.

TYPE GENUS: Not based on a genus group name.

COMMENTS: When originally proposed, this rank was placed in the Klasse Säugthiere (Duméril, 1806a [=Mammalia (Linnaeus, 1758)]) and included the genera *Mustela* Linnaeus, 1758; *Mephitis* É. Geoffroy Saint-Hilare & G. Cuvier, 1795: 187; *Ichneumon* Frisch, 1775: 11 [=*Herpestes* Illiger, 1811: 135]; *Lutris* Duméril, 1806a: 13 [=*Lutra* Brisson, 1762: 13, 201]; *Felis* Linnaeus, 1758; *Civetta* É. Geoffroy Saint-Hilaire & G. Cuvier, 1795: 187 [=*Viverricula* Hodgson, 1838: 152]; *Hyaena* Brisson, 1762: 13, 168; and *Canis* Linnaeus, 1758.

Digitigrada Illiger, 1811: xv.

COMMENTS: When originally proposed the position and contents of this group were not clear and appears to be referring to Digitigraden of Duméril (1806a: 4, 12).

HOMONYMS:

Order Digitigrada G. Fischer, 1813a, carnivores of the Order Mammalia (Order Carnivora). See individual entry.

Family Plantigrada Illiger, 1811: 63, 127.

TYPE GENUS: Not based on a genus group name.

COMMENTS: When originally proposed, this rank was placed in the Order Falculata (Illiger, 1811: 123 [=Mammalia (Linnaeus, 1758 part)]) that included the genera *Cercoleptes* Illiger, 1811: 127 [=*Potos* É. Geoffroy Saint-Hilare & G. Cuvier, 1795: 187]; *Nasua* Storr, 1780: 35, Table A; *Procyon* Storr, 1780: 35; *Gulo* Pallas, 1780b: 25; *Meles* Boddaert, 1785: 45; and *Ursus* Linnaeus, 1758: 47. Name approximates Plantitigradae (J. Gray, 1821: 300).

HOMONYMS:

Order Plantigrada G. Fischer, 1813a, are mammals of the Class Mammalia. See individual entry.

Family Carnivores G. Cuvier, 1816a: xxx, 138.

TYPE GENUS: Not based on a genus group name.

COMMENTS: When originally proposed, this rank was placed in the Order Carnassiers (G. Cuvier, 1816a [=Mammalia (Linnaeus, 1758)]) and included the genera *Ursus* Linnaeus, 1758: 47; *Procyon* Storr, 1780: 35; *Nasua* Storr, 1780: 35, Table A; *Meles* Boddaert, 1785: 45; *Gulo* Pallas, 1780b: 25; *Mephitis* É. Geoffroy Saint-Hilare & G. Cuvier, 1795: 187; *Lutra* Brisson, 1762: 13, 201; *Mustela* Linnaeus, 1758; *Canis* Linnaeus, 1758; *Viverra* Linnaeus, 1758: 43; *Genetta* G. Cuvier, 1816a: 156; *Herpestes* Illiger, 1811: 135; *Ryzaena* G. Cuvier, 1816a: 158 [=*Suricata* Desmarest, 1804a: 15]; *Hyaena* Brisson, 1762: 13, 168; and *Felis* Linnaeus, 1758.

Order Digitigrada G. Fischer, 1813a: 14.

COMMENTS: Unguiculata (Linnaeus, 1766 [=Placentalia (Bonaparte, 1838 part)]) and included the genera *Taxus* É. Geoffroy Saint-Hilaire & G. Cuvier, 1795: 187 [=*Meles* Boddaert, 1785: 45]; *Nasua* Storr, 1780: 35, Table A; *Lotor* É. Geoffroy Saint-Hilaire & G. Cuvier, 1795: 187 [=*Procyon* Storr, 1780: 35]; Kinkaschu [=*Kinkajou* Lacépède, 1799a] [=*Potos* É. Geoffroy Saint-Hilare & G. Cuvier, 1795: 187]; *Mustela* Linnaeus, 1758; *Mephitis* É. Geoffroy Saint-Hilare & G. Cuvier, 1795: 187; *Viverra* Linnaeus, 1758: 43; *Otolicnus* Illiger, 1811: 74 [=*Galago* É. Geoffroy, 1796d: 49]; *Felis* Linnaeus, 1758; *Canis* Linnaeus, 1758; and *Hyaena* Brisson, 1762: 13, 168. Name also referred to by G. Fischer (1814: xi, 165; 1817: 372).

HOMONYMS:

Digitigrada Illiger, 1811, carnivores of the Order Mammalia (Order Carnivora). See individual entry.

Order Digitigradae J. Gray, 1821: 301.

COMMENTS: When originally proposed, this rank was placed in the Class Quadripedes (J. Gray, 1821 [=Placentalia Bonaparte, 1838 part)]) and included the families Mustelladae (J. Gray, 1821: 301) [=Mustelidae (G. Fischer, 1814)]), Canidae (G. Fischer, 1817: 372), Viveridae [=Viverridae (J. Gray, 1821: 301)], Hyaenadae [=Hyaenidae (J. Gray, 1821: 302)], and Felidae (G. Fischer, 1817: 372).

Order Carnaria Haeckel, 1866: clix.

COMMENTS: When originally proposed, this rank was placed in the Class Mammalia (Linnaeus, 1758) and included the suborders Carnivora (Bowdich, 1821) and Pinnipedia (Illiger, 1811 [=Placentalia (Bonaparte, 1838 part)]). Synonymised within the Order Carnivora by McKenna and Bell (1997: 226).

Parvorder Ursida Tedford, 1976: 372.

COMMENTS: When originally proposed as a new rank it was placed in the Intraorder Arctoidea (Flower, 1869) and included the superfamilies Ursoidea (G. Fischer, 1814: x, 143) and Otarioidea [=Otariidae] Gill, 1866 [=Otariina J. Gray, 1825a].

Suborder Cynofeliformia Ginsburg, 1982: 256.

COMMENTS: When originally proposed, this rank was placed in the Order Carnivora (Bowdich, 1821) and included the superfamilies Canoidea (Simpson, 1931) (containing the Family Canidae) and Feloidea (Simpson, 1931: 277) (containing the families Viverridae, Hyaenidae and Felidae). Synonymised within Carnivora by McKenna and Bell (1997: 226).

Legion Carnivoramorpha Kalandadze & Rautian, 1992: 53.

COMMENTS: When originally proposed, this rank was placed within the Infraclass Eutheria (Huxley, 1881 [=Placentalia (Bonaparte, 1838)]) and included the orders Lipotyphla (Haeckel, 1866), † Creodonta (Cope, 1875: 446), Caniformia (Kretzoi, 1943) and Feliformia (Kretzoi, 1945). See McKenna and Bell (1997: 226).

HOMONYMS:

Carnivoramomorpha Wyss & Flynn, 1993, carnivores of the Class Mammalia (Order Carnivora). Synonymised within the Order Carnivora by McKenna and Bell (1997: 226). See individual entry.

Clade Carnivoramomorpha Wyss & Flynn, 1993: 37.

COMMENTS: When originally proposed, this clade included the Carnivora (Bowdich, 1821) and all forms more closely related to it than to † Creodonta (Cope, 1875: 446). Synonymised within the Order Carnivora by McKenna and Bell (1997: 226).

HOMONYMS:

Carnivoramorphia Kalandadze & Rautian, 1992, carnivores of the Class Mammalia (Order Carnivora). See McKenna and Bell (1997: 226). See individual entry.

Order Carnivoriformes Kinman, 1994: 37.

COMMENTS: When originally proposed, this rank was placed in the Class Mammalea (Kinman, 1994 [=Mammalia (Linnaeus, 1758)]). Synonymised within Carnivora by McKenna and Bell (1997: 226).

Suborder Caniformia Kretzoi, 1943

Suborder Caniformia Kretzoi, 1943: 194.

COMMENTS: When originally proposed, this rank included the families † Miacidae (Cope, 1889b: 876), Canidae (G. Fischer, 1817), † Agriotheriidae (Kretzoi, 1929 [=† Hemicyonidae (Frick, 1926: 12)]), Ursidae (G. Fischer, 1814: x, 143), Ailuropodidae (Pocock, 1921: 420 [=Ursidae (G. Fischer, 1814: x, 143)]), Ailuridae [=Ailurinae (J. Gray, 1843a: xxi)], Procyonidae (J. Gray, 1825a: 339), Mustelidae, (G. Fischer, 1814), Herpestidae (Bonaparte, 1845: 3), Viverridae (J. Gray, 1821: 301), and Hyaenidae (J. Gray, 1821: 302). Recognised by McKenna and Bell (1997: 242), J. Flynn *et al.* (2005: 317, 324), Fulton and Strobeck (2006: 165; 2010a: 816), and Wozencraft (2005: 573).

Tribe Hypomycteri Cope, 1882c: 473.

TYPE GENUS: Not based on a genus group name.

COMMENTS: When originally proposed, this rank was placed in the fissiped carnivora and included the families Cercoleptidae (Bonaparte, 1838: 111 [=Procyonidae (J. Gray, 1825a: 339)]), Procyonidae (J. Gray, 1825a: 339), Mustelidae (G. Fischer, 1814), Aeluridae [=Ailurinae (J. Gray, 1843a: xxi)], Ursidae (G. Fischer, 1814: x, 143) and Canidae (G. Fischer, 1817). Synonymised within Caniformia by McKenna and Bell (1997: 242).

Infraorder Cynoidea Flower, 1869

Cynoidea Flower, 1869: 24.

COMMENTS: When originally proposed, this rank was placed in the Fissipedia [=Fissipeda] Blumenbach, 1791 [=Carnivora (Bowdich, 1821 part)]), within the Order Carnivora (Bowdich, 1821), and included the Family Canidae (G. Fischer, 1817) (including *Lycaon* Brook[e]s, 1827: 151). Recognised at infraordinal rank by McKenna and Bell (1997: 243), and Fulton and Strobeck (2006: 165).

Family Canidae G. Fischer, 1817

Family Canini G. Fischer, 1817: 372.

TYPE GENUS: *Canis* Linnaeus, 1758.

COMMENTS: When originally proposed, this rank was placed in the Digitigrada (G. Fischer, 1813a [=Carnivora (Bowdich, 1821 part)]) and included the genera *Canis* Linnaeus, 1758; and *Hyaena* Brisson, 1762: 13, 168. Rank appears to be the same as the Family Caninorum listed on page 417. Family subsequently recognised by J. Gray (1821: 301) and most subsequent authors. Various subfamilies have been recognised by authors including Simpson (1945: 108) and McKenna and Bell (1997: 243), however various revisions give little support to these subfamilies (e.g. Wozencraft, 2005: 573).

Family Caninorum G. Fischer, 1817: 417.

TYPE GENUS: *Canis* Linnaeus, 1758.

COMMENTS: When originally proposed, this rank was placed in the Digitigrada (G. Fischer, 1813a [=Carnivora (Bowdich, 1821 part)]) and included the genera *Canis* Linnaeus, 1758; and *Hyaena* Brisson, 1762: 13, 168. Rank appears to be the same as the Canini listed on page 372.

Family Canidae J. Gray, 1821: 301.

TYPE GENUS: *Canis* Linnaeus, 1758.

COMMENTS: When originally proposed, this rank was placed in the Order Digitigradae (J. Gray, 1821 [=Carnivora (Bowdich, 1821)]) and included the genera *Canis* Linnaeus, 1758; *Vulpes* Frisch, 1775; and *Fennecus* Desmarest, 1804a: 18 [=*Vulpes* Frisch, 1775]. Synonymised within Canidae (G. Fischer, 1817) by McKenna and Bell (1997: 243).

Tribe Canina J. Gray, 1825a: 339.

TYPE GENUS: *Canis* Linnaeus, 1758.

COMMENTS: When originally proposed, this rank was placed in the Family Felidae (G. Fischer, 1817) and included the genera *Canis* Linnaeus, 1758; *Fennecus* Desmarest, 1804a: 18 [=*Vulpes* Frisch, 1775]; and *Lycaon* Brook[e]s, 1827: 151. Synonymised within Canidae by McKenna and Bell (1997: 243).

Family Canina Haeckel, 1866: clix.

TYPE GENUS: *Canis* Linnaeus, 1758.

COMMENTS: When originally proposed, this rank was placed in the Suborder Carnivora (Bowdich, 1821) and included the genera † *Palaeocyon* Lund, 1843: 79 [=† *Protocyon* Giebel, 1855: 851]; † *Cynodon* Aymard, 1848: 244 [=† *Amphicynodon* Filhol, 1882: 32]; and *Canis* Linnaeus, 1758. Synonymised within Canidae by McKenna and Bell (1997: 243).

Family Megalotidae J. Gray, 1869a: 210.

TYPE GENUS: *Megalotis* Illiger, 1811: 131.

COMMENTS: When originally proposed, this rank was placed in the Suborder Carnivora (Bowdich, 1821) and included the genus *Megalotis* Illiger, 1811: 131. Synonymised within Canidae by McKenna and Bell (1997: 243).

Subfamily Caninae Gill, 1872: 63.

TYPE GENUS: *Canis* Linnaeus, 1758.

COMMENTS: When originally proposed, this rank was placed in the Family Canidae (G. Fischer, 1817) and included the genera *Lycaon* Brook[e]s, 1827: 151; *Icticyon* Lund, 1843: 80 [=*Speothos* Lund, 1839: 224]; *Cyon* Agassiz, 1846: 113 [=*Cuon* Hodgson, 1838: 152]; *Canis* Linnaeus, 1758; *Lycalopex* Burmeister, 1854: 95; *Pseudalopex* Burmeister, 1856: 44; *Vulpes* Frisch, 1775; *Urocyon* Baird, 1857: 121, 138; and *Nyctereutes* Temminck, 1838b: 285. Subfamily recognised by Stains (1984: 496) and McKenna and Bell (1997: 245).

Family Vulpinae de Rochebrune, 1883: 140.

TYPE GENUS: *Vulpes* Frisch, 1775.

COMMENTS: When originally proposed, this rank was placed in the Order Carnivora (Bowdich, 1821) and included the genus *Vulpes* Frisch, 1775. Synonymised within Canidae by McKenna and Bell (1997: 243).

Family Lycaonidae de Rochebrune, 1883: 133.

TYPE GENUS: *Lycaon* Brook[e]s, 1827: 151.

COMMENTS: When originally proposed, this rank was placed in the Order Carnivora (Bowdich, 1821) and included the genus *Lycaon* Brook[e]s, 1827: 151. Synonymised within Canidae by McKenna and Bell (1997: 243).

Family Otocyonidae Trouessart, 1885: 6, 51.

TYPE GENUS: *Otocyon* J. Müller, 1836: L [=50].

COMMENTS: When originally proposed, this rank was placed in the Suborder Carnivora Pissipedia 'Cope' and included the genus *Otocyon* J. Müller, 1836: L [=50]. Synonymised within Canidae by McKenna and Bell (1997: 243).

Superfamily Canoidea Simpson, 1931: 276.

TYPE GENUS: *Canis* Linnaeus, 1758.

COMMENTS: When originally proposed, this rank was placed in the Suborder Fissipedia [=Fissipeda (Blumenbach, 1791 [=Carnivora (Bowdich, 1821)])] and included the families Canidae (G. Fischer, 1817), Procyonidae (J. Gray, 1825a: 339), Ursidae (G. Fischer, 1814: x, 143) and Mustelidae (G. Fischer, 1814). Superfamily recognised by various authors including Mitchell and Tedford (1973: 278) and Stains (1984: 495). Synonymised within Canidae by McKenna and Bell (1997: 243).

Canis Linnaeus, 1758

Canis Linnaeus, 1758: 38.

TYPE SPECIES: Ω *Canis familiaris* Linnaeus, 1758 by subsequent designation.

COMMENTS: Genus recognised by Direction 24 (ICZN, 1955: 237). Placed on the Official List of Generic Names in Zoology (Melville & Smith, 1987: 62). Taxonomic arrangement and extralimital synonyms can be found in Wozencraft (2005: 574).

HOMONYMS:

Canis Goüan, 1770: 14, fish. *Insertae sedis*.

Ω *Canis familiaris* Linnaeus, 1758

Domestic Dog and Dingo

Ω *[Canis] familiaris* Linnaeus, 1758: 38.

TYPE LOCALITY: Unknown. Thomas (1911b: 134) identified the type locality as 'Sweden (Upsala)'.

COMMENTS: Molecular and fossil evidence typically estimates domestic dogs separated from wolves approximately 10,000–16,000 years ago (Olsen, 1985), Wayne, 1993: 221; Freedman *et al.*, 2014: 8), but estimates of more than 100,000 years ago have been proposed (Vilà *et al.*, 1997: 1687). Opinion 2027 of the ICZN (2003b: 81–82) ruled that *Canis lupus* Linnaeus, 1758: 39 is not invalid by virtue of being predated by the name based on a domestic form, *Canis familiaris*, as the name for the wild animal takes precedence over that of the domestic animal. An implication of Opinion 2027 (ICZN, 2003b: 81), though not explicitly stated, is that domesticated forms do not fall into the definition of subspecies and that specific names should be retained names for wild and domesticated animals. The implications of the ruling of the ICZN (2003b: 82) were also discussed by Gentry *et al.* (2004: 645). This ruling reflects in nomenclature the biological reality that a domestic 'taxon' cannot, by its very nature, have subspecies, nor can it be one. Therefore *familiaris* could be treated as a synonym of *lupus*, a 'quasi-subspecies' of it (it could not be a true subspecies because it is not based on a geographically distinct unit), or else ranked as a species in its own right. This last option is more a result of convenience than anything else, but is commonly adopted by authors including Corbet and Clutton-Brock (1984: 434) and Newsome *et al.* (2013: 196). Wozencraft (2005: 576) placed *familiaris* as a subspecies of *Canis lupus* and included 'domestic dog' after the name. History of introduction discussed by Long (2003: 252). The synonyms included here include only those that relate to Australasia.

Ω *Canis antarticus* [sic] Kerr, 1792: 136.

TYPE LOCALITY: Sydney, New South Wales, Australia.

COMMENTS: This was the first name designated for the Dingo, although was overlooked in preference to *dingo* until Iredale (1947: 35) resurrected the name and proposed that the Dingo should be known as *Canis antarticus*. An application to validate the specific name '*dingo*' and suppress the specific name '*antarticus*' was made by Tate (1955: 121) and supported by Morrison-Scott (1955: 168). Despite this proposal the name was subsequently reviewed and made available for Law of Homonymy only, but subsequently suppressed by Opinion 451 of the ICZN (1957a: 331). Recognised in preference to *dingo* by Troughton (1967: 187), though synonymised within *familiaris* by Ride (1970: 242) and within *dingo* by Mahoney and Richardson (1988b: 217) and Wozencraft (2005: 576).

HOMONYMS:

Canis antarcticus Bechstein, 1799: 271, the Falkland Island's Wolf of the Class Mammalia (Order Carnivora, Family Canidae). Name is a junior synonym of *Dusicyon australis* (Kerr, 1792: 144). See Cabrera (1931: 66), Hemming (1955: 122) and Wozencraft (2005: 579).

Ω *[Canis] Dingo* F. Meyer, 1793: 33.

TYPE LOCALITY: Sydney, New South Wales, Australia.

COMMENTS: The taxonomic position of the Dingo is complicated. The Dingo is a commensal animal that has developed in association with humans but has not been line bred like other domestic species so its status as a domestic species has been unclear. The name *dingo* was recognised in preference to *antarticus* by Opinion 451 of the ICZN (1957a: 331) and subsequently placed on the Official List of Specific Names (Melville & Smith, 1987: 222). Historically recognised as a distinct species by authors including Gould (1859 [1845–1863]: Text to Plates 51–52), Krefft (1866a: 2; 1868a: 93), Lesèble (1890: 681), and Iredale and Troughton (1934: 90). The Dingo has more often been accepted recently as a subspecies of *Canis lupus* i.e. *Canis lupus dingo* (e.g. Corbett, 1995: 5, 45; Strahan, 1995: 696; Bino, 1996: 43; Daniels & Corbett, 2003: 213; Wozencraft, 2005: 576; Elledge *et al.*, 2008: 812; Van Dyck & Strahan, 2008: 739; Sillero-Zubiri, 2009: 413; Purcell, 2010: 1; Stephens, 2011: viii; Radford *et al.*, 2012: 73; Newsome *et al.*, 2013: 196), but the relationship between the Dingo and the domestic dog is complex and some authors prefer to treat this putative taxon as a synonym of *Canis familiaris* (e.g. Ride, 1970: 186; Clutton-Brock *et al.*, 1976: 144; Mahoney & Richardson, 1988b: 217; Corbet & Hill, 1992: 188; Gentry *et al.* (2004: 647) or subspecies of *Canis familiaris*, i.e. *Canis familiaris dingo* (e.g. Wood Jones, 1921: 263; Finlayson, 1939a: 115; Wakefield, 1966: 390; Green, 1978: 207; Shepherd, 1981: 255; Strahan, 1983: 483; Strahan, 1992: 157; Sheldon, 1992: 5; Vernes *et al.*, 2001: 339; Butler *et al.*, 2004: 369; Elledge *et al.*, 2006: 142). Interestingly Wozencraft (2005: 576) recognised the name *dingo* as a subspecies of *lupus*, but in brackets after this included 'domestic dog', similar to what he did for *familiaris*. Though there is no ruling that does not allow subspecies of domestic forms there has been a general acknowledgement that a domestic species cannot have subspecies because these could not be geographic segments of the species as they have no natural distribution, as is required by definition; instead, domestic species have artificially maintained breeds (see Groves, 1995: 137).

The Dingo has been described as a 'part-modified wolf' introduced from mainland Asia, and arrived into Australia by Asian seafarers approximately 3,500 years according to fossil evidence (Corbett, 1995: 17). The earliest archaeological records of the Dingo in Australia are from Fromm's Landing, on the Murray River (3,170 ± 90 BP), Wombah Midden, northern New South Wales (3230 ± 100), and Madura Cave, on the Nullarbor in south-eastern Western Australia (3450 ± 95) (Flood, 1989: 204). This approximate arrival time is supported by mitochondrial DNA evidence, which suggests the Dingo originated 5,000 years ago from a small sample of domesticated dogs from East Asia and lived isolated from other dogs until the introduction of domestic dogs at the time of European settlement (Savolainen *et al.*, 2004: 12387). Some authors have suggested that there were differences between camp and wild dingoes to Aboriginal tribes as the two groups had different names (A. Hamilton, 1972: 287) and dingoes from archeological burial sites in eastern Australia, as compared with wild populations, have suggested to some that some Aboriginal tribes selected certain traits in tamed camp dingoes (Gollan, 1984: 924–925). On the other hand, as argued by Walters (1995: 29), the writings of the earliest colonial settlers in New South Wales mentioned under the name 'Dingo' only camp dogs, never wild dingoes, and it may be that they became feral, at least in this part of Australia, only after the collapse of traditional Aboriginal society. Genetic and morphological analysis has been developed to assist in the indentification of hybrids from pure dingoes and dogs (see Elledge *et al.*, 2006: 142; Radford *et al.*, 2012: 73). Wozencraft (2005: 576) included *Canis hallstomi* from New Guinea (the New Guinea Singing Dog) as a synonym, but this taxon was reviewed by Kohler-Matznick *et al.* (2003: 109) who recognised it as a distinct species while suggesting that further studies were needed to clarify the exact level of taxonomic differentiation. There is significant hybridisation between dingoes and wild dogs resulting in genetic dilution of the Dingo (e.g. Daniels & Corbett, 2003: 213; Elledge *et al.*, 2008: 812; Jones, 2009: 1; Stephens, 2011: 1). A genome-wide analysis of most dog breeds and different subspecies of the wolf suggest that the Dingo and the New Guinea singing dog are ancient breeds of dog that were probably established over 4,000 years ago and have existed in isolation from wolves (vonHoldt *et al.*, 2010: 901). Further support for the recognition of the Dingo as *Canis familiaris* was provided by Ardalan *et al.* (2012: 65, 71) who suggested they arrived via New Guinea and that the Dingo is an isolated introduced dog.

Crowther *et al.* (2014: 10) undertook a morphological assessment of historic museum specimens that were theoretically collected prior to hybridisation with domestic dogs. They concluded that because the ancestry of the dogs and dingoes is unknown, and because the Dingo was first described as a distinctive wild form and differs from wolves, New Guinea singing dogs and domestic dogs in many behavioral, morphological and molecular characteristics (Macintosh, 1975: 87; Corbett, 1995; Wilton *et al.*, 1999: 108; Koler-Matznick *et al.*, 2003: 109), they are effectively a diagnosable species, isolated in undisturbed natural environments, and thus can be considered a distinct taxon. As a result of this they proposed that *Canis dingo* is the correct binomial. It appears, however, that this argument does not duly recognise the relationship of the Dingo with *Canis familiaris*: not only do their morphometric results show considerable overlap with domestic dog samples, they readily hybridise with domestic dogs, as discussed above. Finally, we urge caution about applying formal scientific names to introduced populations, because they are in many respects still essentially domestic animals, with the reduced cranial capacities typical of their domestic forebears (Hemmer, 1983: 128–129), with which, under the right circumstances, they still form a genetic continuum. Therefore the name *Canis dingo* is here synonymised within *C. familiaris* as this later name has date priority.

The names *Canis familiaris papuensis* Ramsay, 1879 from Papua New Guinea; *Canis familiaris tenggerana* Kohlbrugge, 1896 from Java; *Canis tenggeranus harappensis* Prashad, 1936 from Pakistan; and *Canis hallstromi* Troughton, 1957 from New Guinea were also synonymised within *dingo*, in addition to the names listed here, by Wozencraft (2005). These names have not been included within *dingo* here because neither *tenggerana* or *harappensis* closely resemble the dingo (see Gollan, 1982; Gonzalez, 2013), *papuensis* may refer to feral dogs, and *hallstromi* is considered another primitive breed.

Ω *C. [anis] F. [amiliaris] Australasiae* Desmarest, 1821: 191.

TYPE LOCALITY: Sydney, New South Wales, Australia.

COMMENTS: Synonymised within *dingo* by Iredale and Troughton (1934: 90), Mahoney and Richardson (1988b: 218), Wozencraft (2005: 576) and subsequent authors.

Ω *Canis Australiae* J. Gray, 1826b: 412.

TYPE LOCALITY: Replacement name for *Canis familiaris australasiae* Desmarest, 1821.

COMMENTS: Synonymised within *dingo* by Iredale and Troughton (1934: 90), Mahoney and Richardson (1988b: 218), Wozencraft (2005: 576) and subsequent authors.

Ω *C. [anis] f. [amiliaris] novae Hollandiae* Voigt, 1831: 154.

TYPE LOCALITY: Australia.

COMMENTS: Synonymised within *dingo* by Wozencraft (2005: 576).

Ω *Canis diago* Temminck, 1838b: 285.

TYPE LOCALITY: Australia.

COMMENTS: Appears to be an incorrect subsequent spelling of *dingo*.

Ω *Canis dingoides* Matschie, 1915b: 103.

TYPE LOCALITY: Southern Queensland, Australia.

COMMENTS: Synonymised within *dingo* by Mahoney and Richardson (1988b: 218), Wozencraft (2005: 576) and subsequent authors.

Ω *Canis macdonnellensis* Matschie, 1915b: 105.

TYPE LOCALITY: MacDonnell Ranges, Northern Territory, Australia.

COMMENTS: Synonymised within *dingo* by Mahoney and Richardson (1988b: 218), Wozencraft (2005: 576) and most subsequent authors. The exception to this was Corbett (1995: 48) who suggested it should be recognised as a subspecies of *lupus* for the central Australian populations if they found to be distinct from those of the Australian alpine Dingo which should be referred to as *Canis lupus dingo*. The notion of subspeciation within the Australian Dingoes was subsequently abandoned by Corbett (2004: 223; 2006: 745) who recognised all Dingoes as *Canis lupus dingo*.

Ω *Canis lupus cobourgensis* Corbett, 1995: 48.

TYPE LOCALITY: None.

COMMENTS: *Nomen nudum*. Corbett (1995: 48) advised that as there appears to be no scientific name for northern populations of the dingo which were frequently recorded by the first European settlers in tropical Australia at Port Essington on the Cobourg Peninsula of Arnhem Land, an appropriate name for the Australian tropical dingo might be *Canis lupus cobourgensis*, but that this name would need to be formally submitted and accepted by the International Commission on Zoological Nomenclature.

Vulpes Frisch, 1775

Vulpes Frisch, 1775: 15.

TYPE SPECIES: Ω *Canis vulpes* Linnaeus, 1758 [= Ω *Vulpes vulpes* (Linnaeus, 1758)] by designation under the plenary powers of the ICZN (1979b: 76) and Melville and Smith (1987: 311).

COMMENTS: Though works by Frisch published in 1775 were rejected by Opinion 258 of the ICZN (1954c: 245), *Vulpes* Frisch, 1775 is a valid name by Opinion 1129 of the ICZN (1979b: 76). Work of Frisch reviewed by Thomas (1905b: 461). Taxonomic arrangement and extralimital synonyms can be found in Wozencraft (1993: 287; 2005: 583).

Ω ***Vulpes vulpes*** (Linnaeus, 1758)

Red Fox

Ω *[Canis] Vulpes* Linnaeus, 1758: 40.

TYPE LOCALITY: Syntypes from 'Europa, Asia, Africa'. Thomas (1911b: 134) identified the type locality as 'Sweden (Uppsala)'.

COMMENTS: Placed in *Canis* by J. Fischer (1829: 186). Many references in the early 19th Century used *Vulpes* as a subgenus of *Canis*, though some exceptions include DeKay (1842: 44), and Gerrard (1862: 85). Subsequently Miller (1912: xii, 325) used *Vulpes* as a genus. Synonyms and subspecies reviewed by Lariviere and Pasitsxhniak-Arts (1996: 1–2). The Australian subspecies of fox is *V. v. vulpes* (Van Dyck and Strahan (2008: 741). History of introduction discussed by Long (2003: 240).

Infraorder Arctoidea Flower, 1869 *sensu* Tedford, 1976

Arctoidea Flower, 1869: 15.

COMMENTS: When originally proposed, this rank was placed in the Fissipedia [=Fissipeda (Blumenbach, 1791 [=Carnivora (Bowdich, 1821)])], within the Order Carnivora (Bowdich, 1821), and included the families Ailuridae (Flower, 1869: 15 [=Ailurinae (J. Gray, 1843a: xxi)], Mustelidae (G. Fischer, 1814), Procyonidae (J. Gray, 1825a: 339) and Ursidae (G. Fischer, 1814: x, 143). Though this term has commonly been used as a crown group within the Carnivora, its acceptance as a formal taxonomic rank has been unsettled. For example, it was considered a division by Hunt (2002: 9), J. Flynn *et al.* (2005: 317), and unknown rank Tomiya (2011: 17), but recognised at infraordinal rank by McKenna and Bell (1997: 247), and Fulton and Strobeck (2006: 165, 172; 2010a: 816). The currently accepted composition of the Infraorder Arctoidea is to include three distinct lineages that comprise the bears of the Family Ursidae, the seals of the Superfamily Phocoidea, and the Infraorder Mustelida (Tedford, 1976) that includes the families Mustelidae (G. Fischer, 1814); Procyonidae (J. Gray, 1825a: 339); Mephitidae (Bonaparte, 1845); and Ailuridae (J. Gray, 1843a: xxi) (e.g. Tedford, 1976: 372; McKenna & Bell, 1997: 247; Flynn *et al.* 2005: 317; Fulton & Strobeck 2006: 165, 172; Finarelli, 2008: 232), though the monophyly of Musteoidea is dubious.

Suborder Arctoiformia Ginsburg, 1982: 247, 257.

COMMENTS: When originally proposed, this rank was placed in the Order Carnivora (Bowdich, 1821) and included the infraorders Ursida (Tedford, 1976: 372) and Mustelida (Tedford, 1976: 372). Named as synonym at rank of suborder. Synonymised within Arctoidea by McKenna and Bell (1997: 247).

Clade Pinnipedimorpha Berta *et al.*, 1989: 61.

COMMENTS: When originally proposed, this clade included the Pinnipedia (Illiger, 1811 [=Phocoidea (J. Gray, 1821)]) and *Enaliarctos* Mitchell and Tedford, 1973: 203, 218. Synonymised within Phocoidea (J. Gray, 1821) by McKenna and Bell (1997: 252).

Arctomorpha Wolsan, 1993: 363.

COMMENTS: When originally proposed, this name was allocated as an unnamed order group taxon between infraorder and category of taxon Mustelida (Tedford, 1976: 372). Synonymised within Arctoidea by McKenna and Bell (1997: 247).

Suborder Procyonia Kalandadze & Rautian, 1992: 62.

COMMENTS: When originally proposed, this rank was placed in the Order Caniformia (Kretzoi, 1943) and included the Superfamily Musteloidea (Kretzoi, 1929: 1350 or Gregory & Hellman, 1939: 313). Synonymised within Arctoidea by McKenna and Bell (1997: 247).

Superfamily Phocoidea J. Gray, 1821

Family Phocadae J. Gray, 1821: 302.

TYPE GENUS: *Phoca* Linnaeus, 1758: 37.

COMMENTS: When originally proposed, this rank was placed in the Order Amphibiae (J. Gray, 1821 [=Phocoidea (J. Gray, 1821)]) and included the genera *Phoca* Linnaeus, 1758: 37; and *Otearus* [sic =*Otaria*] Péron, 1816: 37. Superfamily Phocoidea rank recognised by Smirnov (1908: 36), Wyss and Flynn (1993: 38), who described it as new, and McKenna and Bell (1997: 252). Strahan (1995: 9, 671), and Van Dyck and Strahan (2008: 11, 714) placed all pinnipeds as families within the Order Carnivora and did not recognise the seals as a distinct group. There has been significant debate as to whether the seals originated from one (monophyletic) or two (diphyletic) ancestral stocks. Authors that have suggested the possibility that the seals may be diphyletic, with the otariids being most related to the bears (Family Ursidae) and the phocids most closely related to the otters (Family Mustelidae, Subfamily Lutrinae) include Mivart (1885: 497), McLaren (1960: 26–27); Tedford (1976: 372), Mitchell and Tedford (1973: 278), de Muizon (1982: 176) and Barnes (1989: 23). Authors that have tended to suggest that the seals are monophyletic include Wyss (1988: 427), Berta *et al.* (1989: 60), Berta and Wyss (1994: 33), Vrana *et al.* (1994: 54), Bininda-Emonds and Russell (1996: 187), Arnason *et al.* (1995: 78), Flynn and Nedbal (1998: 317), Davis *et al.* (2004: 363, 372) who suggested that more recent studies are increasingly agreeing with the notion that the seals are monophyletic, and Sato *et al.* (2006: 125). Given the strong recent support for monophyly of the different families of seals, they are recognised here as forming a single taxonomic group. This is close to the taxonomy used by McKenna and Bell (1997: 252), Rice (2009: 234), and Berta and Churchill (2012: 207). Wyss and Flynn (1993: 38) included only the common ancestor of phocids and desmatophocids plus all its descendants, which was supported by Berta and Wyss (1994: 41). Deméré *et al.* (2003: 48, 62) used this rank to include only the Family Phocidae and the extinct genera † *Desmatophoca* Condon, 1906: 1 and † *Allodesmus* Kellogg, 1922: 26. Similarly Fulton and Strobuck (2010a: 816) included only the Family Phocidae within this superfamily, with the families Odobenidae and Otariidae being placed in the Superfamily Otarioidea. Different again Arnason *et al.* (2006: 348) placed Pinnipedia above the Family Phocidae and Superfamily Otarioidea (containing the families Odobenidae and Otariidae).

Order Amphibiae J. Gray, 1821: 302.

COMMENTS: When originally proposed, this rank was placed in the Class Quadripedes (Gray, 1821: 300 [=Placentalia (Bonaparte, 1838)]) and included the families Phocadae [=Phocidae (J. Gray, 1821)] and Trichecidae (J. Gray, 1821: 302 [=Odobenidae (J. Allen, 1880: ix, 5)]. Synonymised within Phocoidea (J. Gray, 1821) by McKenna and Bell (1997: 252) and within Phocidae by Wozencraft (2005: 595).

Superfamily Otarioidea F. Lucas, 1899: 1.

TYPE GENUS: *Otaria* Péron, 1816: 37.

COMMENTS: When originally proposed, this rank was placed within the Pinnipedia (Illiger, 1811 [=Phocoidea (J. Gray, 1821)]) and included the 'eared seals and walruses'. Recognised as an unnamed rank within Pinnipedia by Flynn and Wesley-Hunt (2005: 179, 186). Synonymised within Phocoidea (J. Gray, 1821) by McKenna and Bell (1997: 252).

Clade Pinnipediformes Berta, 1994: 2.

COMMENTS: When originally proposed, this clade included the genera † *Allodesmus* Kellogg, 1922: 26; † *Desmatophoca* Condon, 1906: 1; † *Pinnarctidion* Barnes, 1979: 2, 16; † *Pteronarctos* Barnes, 1989: 3; and the families Otariidae (J. Gray, 1825a), Odobenidae (J. Allen, 1880: ix, 5) and Phocidae (J. Gray, 1821). Synonymised within Phocoidea (J. Gray, 1821) by McKenna and Bell (1997: 252).

Family Otariidae J. Gray, 1825

Tribe Otariina J. Gray, 1825a: 340.

TYPE GENUS: *Otaria* Péron, 1816: 37.

COMMENTS: When originally proposed, this rank was placed in the Family Phocidae (J. Gray, 1821) and included the genera *Otaria* Péron, 1816: 37; and *Platyrhynchus* F. Cuvier, 1826a: 555 [=*Otaria* Péron, 1816: 37]. Tribe recognised by J. Gray (1869b: 269; 1871: 12). Reviewed by J. Allen (1880: 187), Repenning *et al.* (1971: 1), Mitchell and Tedford (1973: 201), J. King (1983: 15) and Barnes (1989: 15, 23). Family rank typically recognised by historic and modern authors including Gill (1866: 4), Iredale and Troughton (1934: 89) and Troughton (1967:

196). The subfamilies Otariinae and Arctocephalinae, have been recognised by various authors including Stains (1984: 502), Riedman (1990: 58), McKenna and Bell (1997: 254), Reynolds *et al.* (1999: 5) and Deméré *et al.* (2003: 48) and Fulton and Strobeck (2010a: 816). The use of subfamilies has not been supported by other authors including Simpson (1945: 121), Repenning *et al.* (1971: 3), Repenning and Tedford (1977: 9–10), Wynen *et al.* (2001: 270, 278), Brunner (2004: 383), or Berta and Churchill (2012: 211) so have not been adopted here.

Family Otariadae Brookes, 1828: 37.

TYPE GENUS: *Otaria* Péron, 1816: 37.

COMMENTS: When originally proposed, this rank was placed in the Section Carnivora (Bowdich, 1821) and included the genus *Otaria* Péron, 1816: 37]. Recognised by J. Gray (1871a: 6). Synonymised within the Family Otariidae by Wozencraft (2005: 590).

Tribe? Arctocephalina J. Gray, 1837: 582.

TYPE GENUS: *Arctocephalus* F. Cuvier, 1826a.

COMMENTS: When originally proposed, this rank was placed in the Family Phocidae (J. Gray, 1821) included the genera *Arctocephalus* F. Cuvier, 1826a; and *Otaria* Péron, 1816: 37. The name Arctocephalina recognised at tribe rank by J. Gray (1869b: 269), and subfamily rank by J. Gray (1866b: 44) and Mivart (1885: 485). The Subfamily Arctocephalinae recognised by Stains (1984: 502), McKenna and Bell (1997: 253) and Deméré *et al.* (2003: 48). Synonymised within the Family Otariidae by Wozencraft (2005: 590).

Otariarina J. Gray, 1843a: xxiii.

TYPE GENUS: *Otaria* Péron, 1816: 37.

COMMENTS: When originally proposed, the rank was not stated but placed in the Family Phocidae (J. Gray, 1821) and included the genera *Otaria* Péron, 1816: 37; and *Arctocephalus* F. Cuvier, 1826a. Synonymised within the Family Otariidae by Wozencraft (2005: 590).

Family Otariidae Gill, 1866: 4, 7.

TYPE GENUS: *Otaria* Péron, 1816: 37.

COMMENTS: When originally proposed, the rank was placed in the 'Pinnipedes' and included genera *Otaria* Péron, 1816: 37; *Arctocephalus* F. Cuvier, 1826a; *Eumetopias* Gill, 1866: 7; *Zalophus* Gill, 1866: 7; and *Halarctus* Gill, 1866. Synonymised within the Family Otariidae by Wozencraft (2005: 590).

Tribe Callorhinina J. Gray, 1869b: 269.

TYPE GENUS: *Callorhinus* J. Gray, 1859b: 359.

COMMENTS: When originally proposed, this rank was placed in the Family Otariadae [Otariidae (J. Gray, 1825a)] and included the genus *Callorhinus* J. Gray, 1859b: 359. Synonymised within the Family Otariidae by Wozencraft (2005: 590).

Tribe Eumetopiina J. Gray, 1869b: 269.

TYPE GENUS: *Eumetopias* Gill, 1866: 7.

COMMENTS: When originally proposed, this rank was placed in the Family Otariadae (Brookes, 1828 [Otariidae (J. Gray, 1825a)]) and included the genera *Eumetopias* Gill, 1866: 7; and *Arctophoca* Peters, 1866b. Synonymised within the Family Otariidae and Phocidae by Wozencraft (2005: 590, 595).

Tribe Zalophina J. Gray, 1869b: 269.

TYPE GENUS: *Zalophus* Gill, 1866: 7.

COMMENTS: When originally proposed, this rank was placed in the Family Otariadae (Brookes, 1828 [Otariidae (J. Gray, 1825a)]) and included the genera *Zalophus* Gill, 1866: 7; and *Neophoca* J. Gray, 1866f. Synonymised within the Family Otariidae by Wozencraft (2005: 590).

Subfamily Trichiphocinae J. Allen, 1870: 23, 44.

TYPE GENUS: Not based on a genus group name.

COMMENTS: When originally proposed, this rank was placed in the Family Otariidae (J. Gray, 1825a) and included the genera *Otaria* Péron, 1816: 37; *Eumetopias* Gill, 1866: 7; and *Zalophus* Gill, 1866: 7. Discussed further by J. Allen (1880: 188). Synonymised within the Family Otariidae by Wozencraft (2005: 590).

Subfamily Ouliphocinae J. Allen, 1870: 23, 44.

TYPE GENUS: Not based on a genus group name.

COMMENTS: When originally proposed, this rank was placed in the Family Otariidae (J. Gray, 1825a) and included the genera *Callorhinus* J. Gray, 1859b: 359; and *Arctocephalus* F. Cuvier, 1826a. Discussed further by J. Allen (1880: 188). Synonymised within the Family Otariidae by Wozencraft (2005: 590).

Tribe Gypsophocina J. Gray, 1874a: 27.

TYPE GENUS: *Gypsophoca* J. Gray, 1866f [=*Arctophoca* Peters, 1866b].

COMMENTS: When originally proposed, this rank was placed in the Family Otariadae (Brookes, 1828 [Otariidae (J. Gray, 1825a)]) and included the genus *Gypsophoca* J. Gray, 1866f [=*Arctophoca* Peters, 1866b]. Synonymised within the Family Otariidae by Wozencraft (2005: 590).

Trichophocacae J. Allen, 1880: 208.

TYPE GENUS: Not based on a genus group name.

COMMENTS: Emended spelling of the Subfamily Trichiphocinae (J. Allen, 1870). When originally proposed this rank was placed in the Family Otariidae (J. Gray, 1825a) and included the genera *Otaria* Péron, 1816: 37; *Phocarctos*

Peters, 1866b; *Eumetopias* Gill, 1866: 7; and *Zalophus* Gill, 1866: 7. Synonymised within Otariidae by Rice (1998: 22).

Ouliphocacae J. Allen, 1880: 210.

TYPE GENUS: Not based on a genus group name.

COMMENTS: Emended spelling of the Subfamily Ouliphocinae (J. Allen, 1870). When originally proposed this rank was placed in the Family Otariidae (J. Gray, 1825a) and included the genera *Callorhinus* J. Gray, 1859b: 359; and *Arctocephalus* F. Cuvier, 1826a. Synonymised within Otariidae by Rice (1998: 22).

Order? Arctocephalida Haeckel, 1895: 590.

COMMENTS: When originally proposed, this rank was placed in the Pinnipedia (Illiger, 1811 [=Placentalia (Bonaparte, 1838 part)]) and included the eared seals. Synonymised within the Subfamily Arctocephalinae (J. Gray, 1837) by McKenna and Bell (1997: 253).

Subfamily Arctocephalinae Boetticher, 1934: 359.

TYPE GENUS: *Arctocephalus* F. Cuvier, 1826a.

COMMENTS: When originally proposed, this rank was placed in the Family Otariidae (J. Gray, 1825a) and included the genera *Callotaria* Palmer, 1892: 156 [=*Callorhinus* J. Gray, 1859b: 359]; and *Arctocephalus* F. Cuvier, 1826a. Synonymised within the Subfamily Arctocephalinae (J. Gray, 1837) by McKenna and Bell (1997: 253).

Subfamily Phocarctinae Boetticher, 1934: 359.

TYPE GENUS: *Phocarctos* Peters, 1866b.

COMMENTS: When originally proposed, this rank was placed in the Family Otariidae (J. Gray, 1825a) and included the genera *Phocarctos* Peters, 1866b. Synonymised within the Subfamily Otariinae (J. Gray, 1825a) by McKenna and Bell (1997: 254).

Subfamily Otariinae Mitchell, 1968: 1897.

TYPE GENUS: *Otaria* Péron, 1816: 37.

COMMENTS: When originally proposed, this rank was placed in the Family Otariidae (J. Gray, 1825a) and included the tribes Otariini (Mitchell, 1968 [=Otariidae (J. Gray, 1825)]) and Arctocephalini (Mitchell, 1968 [=Otariidae (J. Gray, 1825)]). Synonymised within Subfamily Otariinae (J. Gray, 1825a) by McKenna and Bell (1997: 254) and not supported by Berta and Churchill (2012: 211).

Tribe Otariini Mitchell, 1968: 1897.

TYPE GENUS: *Otaria* Péron, 1816: 37.

COMMENTS: When originally proposed, this rank was placed in the Subfamily Otariinae (Mitchell, 1968 [=Otariidae (J. Gray, 1825a)]) and included the genera *Phocarctos* Peters, 1866b; *Otaria* Péron, 1816: 37; *Zalophus* Gill, 1866: 7; *Neophoca* J. Gray, 1866f; and *Eumetopias* Gill, 1866: 7. Synonymised within Subfamily Otariinae (J. Gray, 1825a) by McKenna and Bell (1997: 254).

Tribe Arctocephalini Mitchell, 1968: 1843, 1897.

TYPE GENUS: *Arctocephalus* F. Cuvier, 1826a.

COMMENTS: When originally proposed, this rank was placed in the Subfamily Otariinae (Mitchell, 1968 [=Otariidae (J. Gray, 1825a)]) and included the genera † *Pithanotaria* Kellogg, 1925: 74; *Callorhinus* J. Gray, 1859b: 359; and *Arctocephalus* F. Cuvier, 1826a. Synonymised within the Subfamily Arctocephalinae (J. Gray, 1837) by McKenna and Bell (1997: 253).

Subfamily Callorhinae de Muizon, 1978: 180, 183.

TYPE GENUS: *Callorhinus* J. Gray, 1859b: 359.

COMMENTS: When originally proposed as a new rank it was placed in the Family Otariidae (J. Gray, 1825a) and included the genus *Callorhinus* J. Gray, 1859b: 359. Synonymised within the Family Otariidae by Wozencraft (2005: 590).

Arctocephalus F. Cuvier, 1826

Arctocephalus F. Cuvier, 1826a: 554.

TYPE SPECIES: *Phoca pusilla* Schreber, 1775 [= *Arctocephalus pusillus* (Schreber, 1775)] by subsequent designation. See ICZN, 2000: 193.

COMMENTS: Genus recognised as 'Arctocéphale' by F. Cuvier (1824: 205). The author and year of publication of this taxon has been inconsistent with Brunner (2004: 340) giving it as 'Geoffroy Saint-Hilaire and Cuvier, 1824' and Wozencraft (2005: 590) gave it as 'É. Geoffroy Saint-Hilaire and F. Cuvier, 1826', while Berta and Churchill (2012: 211) gave it as 'Cuvier, 1827'. Recognised by J. Gray (1866b: 47) and J. Allen (1905: 120), who synonymised within it *Halarctos*, *Arctophoca, Euotaria* and *Gysophoca*. F. Wood Jones (1925c: 12) stated that he could see no reason to dissent from Beddard's (1890: 380) conclusion that the rest of the eared seals, except *Otaria jubata*, should be embraced in F. Cuvier's (1824) genus *Arctocephalus*. Genus name *Arctocephalus* was conserved by Opinion 1962 of the ICZN (2000: 193) and the name was placed on the Official List of Generic Names (ICZN, 2000: 193–195; see J. Smith, 2001: 6). Recent studies suggest that the species that typically constitute *Arctocephalus* (e.g. Wozencraft, 2005: 590) do not constitute a monophyletic clade (e.g. Wynen *et al*., 2001: 276; Yonezawa *et al*., 2009: 94–95; Berta & Churchill, 2012: 209, 211; Nyakatura & Bininda-Emonds, 2013: 3). As a result, Berta and Churchill (2012: 207, 211) retained only *pusillus* in *Arctocephalus*, following Gardner and Robbins (1999: 136), and transferred all the other species to *Arctophoca*. Berta and Churchill (2012: 211) also noted that some molecular data (such as that mentioned above) suggest a sister relationship between *pusillus* and *tropicalis*, and that if this is confirmed, then *A. tropicalis* should be returned to *Arctocephalus*.

Halarctus Gill, 1866: 7.

TYPE SPECIES: *Arctocephalus delalandii* J. Gray, 1859c: 107 [=*Arctocephalus pusillus* (Schreber, 1775)] by original designation. See Gardner and Robbins (1998: 548).

COMMENTS: Synonymised within *Arctocephalus* by authors including Ling (1988a: 223), Wozencraft (2005: 590) and Berta and Churchill (2012: 211).

Arctocephalus pusillus (Schreber, 1775)

Cape Fur Seal

Φ ***Arctocephalus pusillus pusillus*** (Schreber, 1775)

Cape Fur Seal

Φ *Phoca pusilla* Schreber, 1775: Plate 85; Text 314 [1776].

TYPE LOCALITY: 'Presumably South Africa'.

COMMENTS: The year was given as 1776 by Brunner (2004: 340). Placed in *Arctocephalus* by Peters (1877: 506).

Φ *[Phoca] Parva* Boddaert, 1785: 172.

TYPE LOCALITY: 'Mare Mediterranco'.

COMMENTS: Synonymised within *pusillus* by Wozencraft (2005: 591).

Φ *Phoca Antarctica* Thunberg, 1811: 321.

TYPE LOCALITY: 'mare australe inhabitat'.

COMMENTS: Synonymised within *pusillus* by Wozencraft (2005: 591).

Φ *Otaria Delalandii* Lesson, 1827a: 206.

TYPE LOCALITY: Unknown.

COMMENTS: Recognised within *Arctocephalus* by J. Gray (1859c: 107), but synonymised within *pusillus* by Wozencraft (2005: 591).

Φ *Arctocephalus nivosus* J. Gray, 1868a: 219.

TYPE LOCALITY: Cape of Good Hope.

COMMENTS: Synonymised within *pusillus* by Wozencraft (2005: 591).

Φ *Arcto-cephalus schist-hyperoës* W. Turner, 1868: 114.

TYPE LOCALITY: Cape of Good Hope.

COMMENTS: Synonymised within *pusillus* by Wozencraft (2005: 591).

Φ *Euotaria compressa* J. Gray, 1874a: 38.

TYPE LOCALITY: South Africa.

COMMENTS: Synonymised within *pusillus* by Wozencraft (2005: 591).

Arctocephalus pusillus doriferus Wood Jones, 1925

Australian Fur Seal

Arctocephalus doriferus Wood Jones, 1925c: 12.

TYPE LOCALITY: Lady Julia Percy Island, Victoria, Australia.

COMMENTS: Type designation by Repenning *et al.* (1971: 10). Recognised at the species rank within *Gysophoca* by Iredale and Troughton (1934: 89) and Troughton (1967: 197), and within *Arctocephalus* by Scheffer (1958: 70). The placement of *doriferus* as a subspecies of *pusillus* was proposed by Repenning *et al.* (1971: 1, 7), and followed by most subsequent authors including Warneke and Shaughnessy (1985: 53), Brunner (1998: 67; 2004: 383), Wozencraft (2005: 591), and Berta and Churchill (2012: 212). This subspecies is the only representative of this species in Australian waters. Curiously, differentiation between *doriferus* and *pusillus* is slight; even with Discriminant Analysis, Brunner (2004) could not separate them completely, and though *doriferus* is recognised as a subspecies within *Arctocephalus pusillus* by Strahan (1983: 462; 1995: 682), Ling (1988a: 224) and subsequent authors, its recognition as even a subspecies has been questioned by Wynen *et al.* (2001: 282) who found the divergence between *A. p. pusillus* and *A. p. doriferus* to be very low.

Arctocephalus tasmanicus Scott & Lord, 1926: 187, 189.

TYPE LOCALITY: Bass Strait, Australia.

COMMENTS: Recognised at the species rank within the genus *Gysophoca* by Iredale and Troughton (1934: 89) and Troughton (1967: 198). Synonymised within *doriferus* by J. King (1968: 633), Ride (1970: 241; Repenning *et al.* (1971: 7), Ling (1988a: 224) and subsequent authors.

Otaria Peronii Desmarest, 1817e: 598.

TYPE LOCALITY: Rottnest Island, Western Australia, Australia.

COMMENTS: Synonymised within *Neophoca cinerea* by Iredale and Troughton (1934: 89) and within *pusillus* by Wozencraft (2005: 591).

Arctophoca Peters, 1866

Arctophoca Peters, 1866b: 276.

TYPE SPECIES: Φ *Otaria philippii* Peters, 1866b: 276 [= Φ *Arctophoca philippii* (Peters, 1866b: 276)] by original designation. See Gardner and Robbins (1998: 548).

COMMENTS: Described as a subgenus of *Otaria* Péron, 1816: 37. Synonymised within *Arctocephalus* by Ling (1988a: 223) and most authors until resurrected by Brunner (2004: 381) for *philippii* (Peters, 1866b: 276) and Berta and Churchill (2012: 207, 212) who restricted *Arctocephalus* to include *pusillus* only. The priority of *Arctophoca* over

Euotaria and *Gypsophoca* was confirmed by Gardner and Robbins (1999: 138).

Gypsophoca J. Gray, 1866f: 236.

TYPE SPECIES: *Arctocephalus cinereus* J. Gray, 1866b: 56 [=*Arctophoca forsteri* (Lesson, 1828c)] by monotypy. See J. Gray (1869b: 269) and Gardner and Robbins (1998: 548).

COMMENTS: Described as a subgenus of *Arctocephalus*. Genus rank recognised by J. Gray (1872: 659; 1874a: 27), Iredale and Troughton (1934: 89) and Troughton (1967: 197), but synonymised within *Arctocephalus* by J. Allen (1880: 210), Ride (1970: 243), Strahan (1983: 462; 1995: 682) and Ling (1988a: 223).

Euotaria J. Gray, 1866f: 236.

TYPE SPECIES: Φ *Phoca australis* Zimmermann, 1783: 276 (as *Arctocephalus nigrescens* (J. Gray, 1859c: 109)) [= Φ *Arctophoca australis* (Zimmermann, 1783: 276)] by monotypy. See Gardner and Robbins (1998: 548).

COMMENTS: Described as a subgenus of *Arctocephalus*. Synonymised within *Arctocephalus* by Ling (1988a: 223).

Arctophoca forsteri (Lesson, 1828)

Long-nosed Fur Seal

Otaria Forsteri Lesson, 1828c: 421.

TYPE LOCALITY: Dusky Sound, New Zealand.

COMMENTS: Included within *Arctocephalus* at species rank by J. Gray (1871a: 25), J. Allen (1892: 375), Wood Jones (1925c: 13) and most subsequent authors until various studies showed a close relationship of *forsteri* to Φ *Arctophoca australis* (Zimmermann, 1783: 276) (Wynen *et al.*, 2001: 275; Higdon *et al.*, 2007: 9; Yonezawa *et al.*, 2009: 94). As a result of these and other studies, including morphology, *forsteri* was reduced to a subspecies of *australis* by Brunner (2004: 384), and Berta and Churchill (2012: 213). The craniometric data of Brunner (1998: 67; 2004: 366) suggest there is a major division between populations from New Zealand and those from Australia and Macquarie Island; New Zealand examples are much broader and shorter snouted, especially in males, and these differences appear to be consistent, suggesting that further study may confirm them as distinct species, albeit sister species. More recent research including the super-tree analysis by Nyakatura and Bininda-Emonds (2012: 1) and earlier phylogenetic analysis of Higdon *et al.* (2007, 216) suggests this taxon should be recognised as a full species. The presence of this taxon within Australia was confirmed by King (1969: 841).

HOMONYMS:

Arctocephalus forsteri Wood Jones, 1922c, the Australian Sea-lion of the Class Mammalia (Order Carnivora, Family Otariidae). Species is a synonym of *Neophoca cinerea* (Péron, 1816). See individual entry.

Arctocephalus cinereus Gray, 1866b: 56.

TYPE LOCALITY: 'South coast of Australia'.

COMMENTS: Taxon appeared to have been confused with *Otaria cinerea* Péron, 1816 [=*Neophoca cinerea* (Péron, 1816)]. Name not typically recognised by subsequent authors.

HOMONYMS:

Otaria Cinerea Péron, 1816, the Australian Sea-lion of the Class Mammalia (Order Carnivora, Family Otariidae). Species is recognised as *Neophoca cinerea* (Péron, 1816). See individual entry.

Arctophoca gazella Peters, 1875

Antarctic Fur Seal

Arctophoca gazella Peters, 1875c: 393, 396.

TYPE LOCALITY: 'von Seehunden aus Kerguelenland' Restricted by Scheffer (1958: 72) to Kerguelen Islands, southern Pacific Ocean (49°09′S, 70°11′E).

COMMENTS: Found on Macquarie, Heard and McDonald Islands that are administratively part of Tasmania (Shaughnessy, 1992: 77–78), and on Macquarie Island it hybridises with *A. tropicalis* and *A. australis* (Lancaster *et al.*, 2006: 3681, 3686). Placed within *Arctocephalus* by Peters (1877: 507). Taxon recognised as a subspecies of *tropicalis* by J. King (1959a: 381) and as a distinct species (that included *tropicalis*) by Siverton (1954: 49, 74), Scheffer (1958: 72) and J. King (1959b: 39). Repenning *et al.* (1971: 10) and J. King (1983: 10) recognised *gazella* and *tropicalis* as distinct species, which has been followed by subsequent authors.

Arctophoca tropicalis (J. Gray, 1872)

Subantarctic Fur Seal

Gypsophoca tropicalis J. Gray, 1872: 653, 659.

TYPE LOCALITY: North coast of Australia.

COMMENTS: Taxon kept in the genus *Gysophoca* by Iredale and Troughton (1934: 89). Not considered by Ling (1988a: 224) as it was not considered to occur in Australian waters, but it is found on Macquarie Island and Heard Island that are administratively part of Tasmania (Shaughnessy, 1992: 78). Synonymised within *gazella* by Siverton (1954: 49, 74) but separated as a distinct species by Repenning *et al.* (1971: 17) and J. King (1983: 10), which has been followed by most subsequent authors. Placed in *Arctocephalus* by J. King (1959a: 381) and most subsequent authors until Berta and Churchill (2012: 214) placed this species within *Arctophoca*. If the apparent sister relationship between *tropicalis* and *pusillus* is confirmed (see discussion under *Arctocephalus* above), *tropicalis* should be placed within *Arctocephalus* (see Berta & Churchill, 2012: 211). Hybrids between this species, *A. australis* and *A. gazella* have been recorded from Macquarie Island (Lancaster *et al.*, 2006: 3681, 3686).

Otaria (Arctophoca) elegans Peters, 1876b: 316.

TYPE LOCALITY: St. Paul and Amsterdam Islands.

COMMENTS: Synonymised within *tropicalis* by Wozencraft (2005: 592).

Neophoca J. Gray, 1866

Neophoca J. Gray, 1866f: 231.

TYPE SPECIES: *Otaria cinerea* Péron, 1816 (as *Arctocephalus lobatus* J. Gray, 1828) [= *Neophoca cinerea* (Péron, 1816)] by monotypy. See Palmer (1904: 454).

COMMENTS: Synonymised within *Zalophus* by J. Allen (1880: 275), but separated as a distinct genus by Sivertsen (1954: 28), Scheffer (1958: 64), J. King (1960: 445; 1983: 30) and subsequent authors including Troughton (1967: 199), Marlow and King (1974: 125), Rice (1977: 2), Barnes (1989: 20, 23) and Wynen *et al.* (2001: 270).

Neophoca cinerea (Péron, 1816)

Australian Sea-lion

Otaria Cinerea Péron, 1816: 54.

TYPE LOCALITY: Kangaroo Island, South Australia, Australia.

COMMENTS: Placed in *Arctocephalus* by J. Gray (1866b: 56) and Wood Jones (1925c: 12), *Euotaria* by McCoy (1879: Plate 31) and *Zalophus* by J. Allen (1880: 275). Also included *Otaria peronii* Desmarest, 1817e by Iredale and Troughton (1934: 89). Transferred to *Neophoca* by Carter *et al.* (1945: 101) and Sivertsen (1954: 28), which appears to have been followed by subsequent authors. Species restricted to Australian waters by Sivertsen (1954: 31) and reviewed by Ling (1992: 1). Some authors including Wozencraft (2005: 593), and Berta and Churchill (2012: 216) place *Otaria stelleri* Temminck, 1844b: 10 as a synonym of *cinerea*, but this taxon was reviewed by J. King (1961: 211–213) who concluded there is no doubt the specimens referred to are those of the probably extinct *Zalophus japonicus* (Peters, 1867i: 668).

HOMONYMS:

Arctocephalus cinereus Gray, 1866b, the Long-nosed Fur Seal of the Class Mammalia (Order Carnivora, Family Otariidae). Species is a synonym of *Arctophoca forsteri* (Lesson, 1828). See individual entry.

Otaria Albicollis Péron, 1816: 118.

TYPE LOCALITY: St. Peter Island, Nuyts Archipelago, South Australia, Australia.

COMMENTS: Synonymised within *cinerea* by Wood Jones (1925c: 12), Iredale and Troughton (1934: 89), and Ling (1988a: 224; 1992: 1).

Arctocephalus lobatus J. Gray, 1828: 1.

TYPE LOCALITY: Houtman Abrolhos, Western Australia, Australia.

COMMENTS: Type designation by J. Gray (1874a: 43). Recognised as a species within *Arctocephalus* by Gould (1860 [1845–1863]: Text to Plate 49), Krefft (1868a: 93), *Neophoca* by J. Gray (1874a: 43) and *Zalophus* by J. Ogilby (1892: 126). Synonymised within *cinerea* by Wood Jones (1925b: 12), Iredale and Troughton (1934: 89) and Ling (1988a: 224; 1992: 1).

Otaria australis Quoy & Gaimard, 1830: 95; Plates 14–15.

TYPE LOCALITY: King George Sound, Western Australia, Australia.

COMMENTS: Recognised within *Arctocephalus* by J. Gray (1866b: 57). Synonymised within *cinerea* by Wood Jones (1925c: 12), Iredale and Troughton (1934: 89), and Ling (1988a: 225; 1992: 1).

Arctocephalus williamsi McCoy, 1877: 7.

TYPE LOCALITY: Queenscliff, Victoria, Australia.

COMMENTS: Synonymised within *cinerea* by Marlow and King (1974: 126) and Ling (1992: 1).

Arctocephalus forsteri Wood Jones, 1922c: 193.

TYPE LOCALITY: Nuyts Archipelago, off South Australia, Australia.

COMMENTS: Synonymised within *cinerea* by Wood Jones (1925c: 12), Ling (1993: 1) and subsequent authors including Wozencraft (2005: 593) who spelt it *Neophoca fosteri* [sic].

HOMONYMS:

Otaria forsteri Lesson, 1828c, the New Zealand Fur Seal of the Class Mammalia (Order Carnivora, Family Otariidae). Species is recognised as *Arctophoca forsteri* (Lesson, 1828). See individual entry.

Phocarctos Peters, 1866

Phocartos Peters, 1866b: 269.

TYPE SPECIES: *Arctocephalus hookeri* J. Gray, 1844b (as *Otaria Hookeri*) [= *Phocarctos hookeri* (J. Gray, 1844b)] by monotypy.

COMMENTS: Described as a subgenus of *Arctocephalus*. Considered a synonym of *Neophoca* by Scheffer (1958: 65) but not by Sivertsen (1954: 28), King (1960: 449; 1983: 30, 33), Barnes (1989: 23) and subsequent authors.

Phocarctos hookeri (J. Gray, 1844)

New Zealand Sea-lion

Arctocephalus Hookeri J. Gray, 1844b: 4.

TYPE LOCALITY: 'Falkland Islands and Cape Horn'. Locality on Error. Fixed by Clark (1874: 750) as Auckland

Islands [New Zealand] between 800–900 miles south of Tasmania. (50°48′S, 166°42′E)

COMMENTS: Only known to haul out in Australian waters on Macquarie Island.

Family Phocidae J. Gray, 1821

Family Phocadae [sic] J. Gray, 1821: 302.

TYPE GENUS: *Phoca* Linnaeus, 1758: 37.

COMMENTS: When originally proposed, this rank was placed in the Order Amphibiae (J. Gray, 1821 [=Phocoidea (J. Gray, 1821)]) and included the genera *Phoca* Linnaeus, 1758: 37; and *Otearus* [sic =*Otaria*] Péron, 1816: 37. Emended to Phocidae by J. Gray (1825a: 340). Family name spelt Phocidae by Gill (1866: 4). Family reviewed by Ling (1988b: 226) and McKenna and Bell (1997: 252) who recognised the subfamilies Odobeninae (J. Allen, 1880: ix, 5) and Phocinae (J. Gray, 1821). The placement of the odobenids continues to be controversial because morphological evidence relates them to the phocids whereas molecular studies provide consistent robust support for the relationship between odobenids and otarids (e.g. Arnason *et al*., 2006: 348; Fulton & Strobuck 2006: 166; Berta 2009a: 862; 2009b: 880). Subfamilies and tribes sometimes not recognised by authors (e.g. Wozencraft, 2005: 595), but there is increasingly strong molecular support for the subfamilies Monachinae and Phocinae (e.g. Deméré *et al*. 2003: 62; Davis *et al*. 2004: 363; Arnason *et al*., 2006: 348; Berta 2009a: 863; 2009b: 881; Dasmahapatra *et al*., 2009: 170; Fulton & Strobeck 2010a: 816; 2010b: 1066; Berta & Churchill, 2012: 218).

Tribe Phocina J. Gray, 1825a: 340.

TYPE GENUS: *Phoca* Linnaeus, 1758: 37.

COMMENTS: When originally proposed, this rank was placed in the Family Phocidae (J. Gray, 1821) and included the genus *Phoca* Linnaeus, 1758: 37. Emendation of Phocadae. Recognised in preference to J. Gray (1821: 302) by Simpson (1945: 122). Synonymised within Phocidae by McKenna and Bell (1997: 254).

Tribe Stemmotopina J. Gray, 1825a: 340.

TYPE GENUS: *Stemmotopus* F. Cuvier, 1826a: 551 [=*Cystophora* Nilsson, 1820: 382].

COMMENTS: When originally proposed, this rank was placed in the Family Phocidae (J. Gray, 1821) and included the genera *Stemmotopus* F. Cuvier, 1826a: 551 [=*Cystophora* Nilsson, 1820: 382] and *Macrorhinus* J. Gray, 1825a [=*Mirounga* J. Gray, 1827a]. Synonymised within Phocini (J. Gray, 1821) by McKenna and Bell (1997: 257) and the Family Phocidae by Rice (1998: 34).

Tribe? Stenorynchìna J. Gray, 1837: 582.

TYPE GENUS: *Stenorhynchus* J. Gray, 1825a.

COMMENTS: When originally proposed, this rank was placed in the Family Phocidae (J. Gray, 1821) included the genera *Leptonyx* J. Gray, 1837 [=*Leptonychotes* Gill, 1872], *Pelagiàs* [sic =*Pelagios*] F. Cuvier, 1824: 196 [=*Monachus* Fleming, 1822b: 187] and *Stenorhýnchus* J. Gray, 1825a [=*Hydrurga* Gistel, 1848]. Synonymised within the Family Phocidae by Rice (1998: 34) and Wozencraft (2005: 595).

Tribe? Phocìna J. Gray, 1837: 582.

TYPE GENUS: *Phoca* Linnaeus, 1758: 37.

COMMENTS: When originally proposed, this rank was placed in the Family Phocidae (J. Gray, 1821) included the genera *Phoca* Linnaeus, 1758: 37; and *Callocephalus* [sic =*Calocephalus*] F. Cuvier, 1826a: 544 [=*Phoca* Linnaeus, 1758: 37]. Subfamily rank recognised by J. Gray (1866b: 20; 1871: 2).

Subfamily Stenorynchina [sic] J. Gray, 1843a: xxiii, 102.

TYPE GENUS: *Stenorhynchus* J. Gray, 1825a.

COMMENTS: When originally proposed, this rank included the genera *Pelagius* [=*Pelagios*] F. Cuvier, 1824: 196 [=*Monachus* Fleming, 1822b: 187]; *Stenorhynchus* J. Gray, 1825a; and *Leptonyx* J. Gray, 1837 [=*Leptonychotes* Gill, 1872]. Synonymised within Phocidae by Rice (1998: 34).

Pinnigrada Owen, 1858: 37.

TYPE GENUS: Not based on a genus name.

COMMENTS: When originally proposed, this rank was placed in the Order Carnivora (Bowdich, 1821) at unknown rank. Recognised at family rank by (Owen, 1959: 52). Synonymised within Phocoidea (J. Gray, 1821) by McKenna and Bell (1997: 252) and within Phocidae by Wozencraft (2005: 595).

Pinnigrades Owen, 1859: 45.

TYPE GENUS: Not based on a genus group name.

COMMENTS: When originally proposed, this rank was placed in the Order Carnivora (Bowdich, 1821) at unknown rank but included the 'walrus and seal tribe'. Synonymised within Phocoidea (J. Gray, 1821) by McKenna and Bell (1997: 252) and within Phocidae by Wozencraft (2005: 595).

Subfamily Phocinae Gill, 1866: 5, 8.

TYPE GENUS: *Phoca* Linnaeus, 1758: 37.

COMMENTS: When originally proposed, this rank was placed in the Family Phocidae (J. Gray, 1821) and included the genera *Phoca* Linnaeus, 1758: 37; *Pagomys* J. Gray, 1864b: 31 [=*Pagophilus* J. Gray, 1844b: 3]; *Pagophilus* J. Gray, 1844b: 3; *Erignathus* Gill, 1866: 5; and *Halichoerus* Nilsson, 1820: 376. Recognised by Simpson (1945: 122) but synonymised within the Subfamily Phocinae by McKenna and Bell (1997: 257).

Family Phocida Haeckel, 1866: clix.

TYPE GENUS: *Phoca* Linnaeus, 1758: 37.

COMMENTS: When originally proposed, this rank was placed in the Suborder Pinnipedia (Illiger, 1811 [=Placentalia (Bonaparte, 1838 part)]) and included the genera *Phoca* Linnaeus, 1758: 37; *Cystophora* Nilsson, 1820: 382; and *Otaria* Péron, 1816: 37. Synonymised within Phocidae by McKenna and Bell (1997: 254).

Tribe Halichoerina J. Gray, 1869c: 345.

TYPE GENUS: *Halichoerus* Nilsson, 1820: 376.

COMMENTS: When originally proposed, this rank was placed in the Family Phocidae (J. Gray, 1821) and included the genus *Halichoerus* Nilsson, 1820: 376. Synonymised within Phocidae by Wozencraft (2005: 595).

Superfamily? Rosmaroidea Gill, 1872: 7, 70.

TYPE GENUS: *Rosmarus* Brünnich, 1772: 34, 38. [=*Odobenus* Brisson, 1762: 30].

COMMENTS: Placed in Suborder Pinnipedia (Illiger, 1811 [=Phocoidea (J. Gray, 1821)]) and included the Family Rosmaridae (Gill, 1866: 7, 11 [=Family Odobenidae (J. Allen, 1880: ix, 5)]). Synonymised within Phocoidea (J. Gray, 1821) by McKenna and Bell (1997: 252).

Suborder Amphibia Trouessart, 1879: 109.

COMMENTS: When originally proposed, this rank was placed in the Order Pinnipedia (Illiger, 1811 [=Phocoidea (J. Gray, 1821)]) and included the Phocae (Trouessart, 1879 [=Phocidae J. Gray, 1821]). Synonymised within Phocoidea (J. Gray, 1821) by McKenna and Bell (1997: 252) and within Phocidae by Wozencraft (2005: 595).

HOMONYMS:

Amphibia Linnaeus, 1758: 194, frogs, toads, newts and salamanders of the Class Amphibia. Currently recognised Class. See Frost (2011).

Amphibia J. Gray, 1821, mammals of the Superfamily Phocoidea (J. Gray, 1821) and Order Sirenia (Illiger, 1811).

Suborder Phocae Trouessart, 1879: 109.

COMMENTS: When originally proposed, this rank was placed in the Order Pinnipedia (Illiger, 1811 [=Phocoidea (J. Gray, 1821)]). Placed in brackets after the name Suborder Amphibia. Synonymised within Phocidae by Wozencraft (2005: 595).

Thalattailurina Albrecht, 1879: 22.

TYPE GENUS: Not based on a genus group name.

COMMENTS: Synonymised within Phocidae by Rice (1998: 34) and Wozencraft (2005: 595).

Subfamily Hydrurginae Trouessart, 1907: 7.

TYPE GENUS: *Hydrurga* Gistel, 1848.

COMMENTS: When originally proposed, this rank was placed in the Family Phocidae (J. Gray, 1821) and included the genus *Hydrurga* Gistel, 1848. Synonymised within Phocidae by Rice (1998: 34) and Wozencraft (2005: 595).

Superfamily? Phocoidea Smirnov, 1908: 36.

TYPE GENUS: *Phoca* Linnaeus, 1758: 37.

COMMENTS: When originally proposed, this rank included the Family Phocidae (J. Gray, 1821). Synonymised within Phocoidea (J. Gray, 1821) and Phocidae by McKenna and Bell (1997: 252, 254).

Family Sibiricopusidae Dybowski, 1929: 412.

TYPE GENUS: '*Baicalopusa*' Dybowski, 1929: 412.

COMMENTS: Also spelt Sibirico-Baicalo-Pusidae (p. 413) and Sibirico-bicuspidato-baicalopusidae (p. 414). Not based on a genus group name. Names are not available as they did not consistently apply the Principles of Binomial Nomenclature (Articles 4.1 and 11.4 of the Code (ICZN, 1999: 4, 10). Higher and lower ranks not listed. Synonymised within Phocidae by Rice (1998: 34) and Wozencraft (2005: 595).

Family Europäopusidae Dybowski, 1929: 412.

TYPE GENUS: '*Caspiopusa*' Dybowski, 1929 (see Rice, 1998: 34).

COMMENTS: Also spelt Europäo-Caspio-Pusidae (p. 413) and Europäo-tricuspidato-caspiopusidae (p. 414). Not based on a genus group name. Names are not available (see comment above). Synonymised within Phocidae by Rice (1998: 34).

Tribe Erignathini Chapskii, 1955: 164, 165.

TYPE GENUS: *Erignathus* Gill, 1866: 5.

COMMENTS: When originally proposed, this rank was placed in the Subfamily Phocinae (J. Gray, 1821) and included the genus *Erignathus* Gill, 1866: 5. Recognised by Dasmahapatra *et al.* (2009: 170). Synonymised within Phocini (J. Gray, 1821) by McKenna and Bell (1997: 257) and within Phocidae by Rice (1998: 34) and Wozencraft (2005: 595).

Subtribe Histriophocina Chapskii, 1955: 164, 188.

TYPE GENUS: *Histriophoca* Gill, 1873b: 179.

COMMENTS: When originally proposed, this rank was placed in the Tribe Phocini (Chapskii, 1955 [=Phocidae (J. Gray, 1821)]) and included the genera *Histriophoca* Gill, 1873b: 179; and *Pagophoca* Trouessart, 1904: 287 [=*Pagophilus* J. Gray, 1844b: 3]. Synonymised within Phocidae by Rice (1998: 34) and Wozencraft (2005: 595).

Tribe Phocini Chapskii, 1955: 164, 169.

TYPE GENUS: *Phoca* Linnaeus, 1758: 37.

COMMENTS: When originally proposed, this rank was placed in the Subfamily Phocinae (J. Gray, 1821) and included the

Phocina (J. Gray, 1821) and Histriophocina (Chapskii, 1955 [=Phocidae (J. Gray, 1821)]). Synonymised within Phocini (J. Gray, 1821) by McKenna and Bell (1997: 257). Rank recognised by Dasmahapatra *et al.* (2009: 170).

Clade Phocoidea Wyss & Flynn, 1993: 38.

COMMENTS: When originally proposed, this clade was placed in the Pinnipedia (Illiger, 1811 [=Phocoidea (J. Gray, 1821)]) and included the families † Desmatophocidae (Hay, 1930: 557) and Phocidae (J. Gray, 1821). Synonymised within Phocoidea (J. Gray, 1821) by McKenna and Bell (1997: 252).

Clade Phocomorpha Berta & Wyss, 1994: 41.

COMMENTS: When originally proposed, this clade was placed in the Pinnipedia (Illiger, 1811 [=Phocoidea (J. Gray, 1821)]) and included the most recent common ancestor of the Family Odobenidae (J. Allen, 1880: ix, 5) and Phocoidea (J. Gray, 1821), and all their descendants. Synonymised within Phocidae by Wozencraft (2005: 595).

Subfamily Monachinae J. Gray, 1869

Tribe Monachina J. Gray, 1869c: 345.

TYPE GENUS: *Monachus* Fleming, 1822b: 187.

COMMENTS: When originally proposed as a new rank it was placed in the Family Phocidae (J. Gray, 1821) and included the genus *Monachus* Fleming, 1822b: 187. Synonymised within the Subfamily Monachinae (Trouessart, 1897) by Simpson (1945: 123) and within Phocidae by Wozencraft (2005: 595), but recognised as the Tribe Monachini (within the subfamily Phocinae) by McKenna and Bell (1997: 257). Subfamily rank recognised by various authors including Scheffer (1958: 111), King (1983: 18), Deméré *et al.* (2003: 63), Davis *et al.* (2004: 363); Arnason *et al.* (2006: 348); Fulton and Strobeck (2006: 176; 2010a: 816); Berta (2009a: 863; 2009b: 881); and Dasmahapatra *et al.* (2009: 170). Within this subfamily various tribes have been recognised with various degrees of certainty including Monachini (Scheffer, 1958), Lobodontini (Scheffer, 1958), Cystophorini (Burns & Fay, 1970), and Miroungini (de Muizon, 1982). Of these the Monachini, Lobodontini and Miroungini have been more often accepted by authors including Berta (2009b: 881), Dasmahapatra *et al.* (2009: 170, 171) and Fulton and Strobeck (2010a: 816).

Tribe Stenorhyncina J. Gray, 1825a: 340.

TYPE GENUS: *Stenorhynchus* J. Gray, 1825a.

COMMENTS: When originally proposed, this rank was placed in the Family Phocidae (J. Gray, 1821) and included the genus *Stenorhynchus* J. Gray, 1825a [=*Hydrurga* Gistel, 1848]. Subfamily rank recognised by J. Gray (1837: 582; 1866b: 8; 1871: 3). Synonymised within the Subfamily Lobodontinae (Hay, 1930: 562) by Simpson (1945: 122), and within Phocidae by Rice (1998: 34) and Wozencraft (2005: 595).

Tribe? Cystophorina J. Gray, 1837: 582.

TYPE GENUS: *Cystophorina* J. Gray, 1837: 582.

COMMENTS: When originally proposed, this rank was placed in the Family Phocidae (J. Gray, 1821) included the genera *Cystophora* Nilsson, 1820: 382 and *Morunga* [sic =*Mirounga*] J. Gray, 1827a. Recognised at Subfamily rank by J. Gray (1866b: 38, 1871: 4). Synonymised within the Subfamily Cystophorinae (Gill, 1866: 6, 9) by Simpson (1945: 123) and within Phocidae by Rice (1998: 34) and Wozencraft (2005: 595).

Subfamily Stenorhynchina J. Gray, 1844b: 2.

TYPE GENUS: *Stenorhynchus* J. Gray, 1825a.

COMMENTS: When originally proposed, this rank included the genera *Lobodon* J. Gray, 1844b; *Stenorhynchus* J. Gray, 1825a; *Leptonyx* J. Gray, 1837 [=*Leptonychotes* Gill, 1872]; *Pelagios* F. Cuvier, 1824: 196 [=*Monachus* Fleming, 1822b: 187] and *Ommatophoca* J. Gray, 1844b. Synonymised within Phocidae by Rice (1998: 34) and Wozencraft (2005: 595).

Subfamily Stenorhynchinae Gill, 1866: 6, 10.

TYPE GENUS: *Stenorhynchus* J. Gray, 1825a.

COMMENTS: When originally proposed, this rank was placed in the Family Phocidae (J. Gray, 1821) and included the genera *Lobodon* J. Gray, 1844b; *Stenorhynchus* J. Gray, 1825a; *Leptonyx* J. Gray, 1837 [=*Leptonychotes* Gill, 1872]; and *Ommatophoca* J. Gray, 1844b. Synonymised within the Subfamily Lobodontinae (Hay, 1930: 562) by Simpson (1945: 122) and within the Subfamily Phocinae by McKenna and Bell (1997: 257).

Subfamily Cystophorinae Gill, 1866: 6, 9.

TYPE GENUS: *Cystophorina* J. Gray, 1837: 582.

COMMENTS: When originally proposed, this rank was placed in the Family Phocidae (J. Gray, 1821) and included the genera *Cystophora* Nilsson, 1820: 382 and *Macrorhinus* J. Gray, 1825a [=*Mirounga* J. Gray, 1827a]. Recognised by authors including Simpson (1945: 123), Koretsky and Grigorescu (2002: 151), and Koretsky and Holec (2002: 175), but synonymised within the Subfamily Phocinae by McKenna and Bell (1997: 257).

Tribe Lobodontina J. Gray, 1869c: 345.

TYPE GENUS: *Lobodon* J. Gray, 1844b.

COMMENTS: When originally proposed, this rank was placed in the Family Phocidae (J. Gray, 1821) and included the genus *Lobodon* J. Gray, 1844b. Recognised at the subtribe rank by McKenna and Bell (1997: 259). Synonymised within Phocidae by Rice (1998: 34) and Wozencraft (2005: 595).

Subfamily Ogmorhininae W. Turner, 1888: 2, 63.

TYPE GENUS: *Ogmorhinus* Peters, 1875c [=*Hydrurga* Gistel, 1848].

COMMENTS: When originally proposed, this rank was placed in the Family Phocidae (J. Gray, 1821) and included the genera *Ogmorhinus* Peters, 1875c [=*Hydrurga* Gistel, 1848]; *Leptonychotes* Gill, 1872; *Ommatophoca* J. Gray, 1844b; and *Monachus* Fleming, 1822b: 187. Synonymised within the Subfamily Phocinae by McKenna and Bell (1997: 257) and within the Family Phocidae by Rice (1998: 34) and Wozencraft (2005: 595).

Subfamily Monachinae Trouessart, 1897: 378.

TYPE GENUS: *Monachus* Fleming, 1822b: 187.

COMMENTS: When originally proposed, this rank was placed in the Family Phocidae (J. Gray, 1821) and included the genus *Monachus* Fleming, 1822b: 187. Recognised by Simpson (1945: 123) and various subsequent authors including Scheffer (1958: 111), and Bininda-Emonds and Russell (1996: 180) but reduced to the 'Tribe Monachini J. Gray, 1869' within the Subfamily Phocinae by McKenna and Bell (1997: 258).

Subfamily Lobodoninae Kellogg, 1922: 84, 89.

TYPE GENUS: *Lobodon* J. Gray, 1844b.

COMMENTS: When originally proposed, this rank was placed in the Family Phocidae (J. Gray, 1821) and included the genera *Leptonychotes* Gill, 1872; *Hydrurga* Gistel, 1848; *Lobodon* J. Gray, 1844b; and *Ommatophoca* J. Gray, 1844b. Synonymised within the Subfamily Lobodontinae (Hay, 1930: 562) by Simpson (1945: 122) and within Phocidae by Rice (1998: 34) and Wozencraft (2005: 595).

Subfamily Lobodontinae Hay, 1930: 562.

TYPE GENUS: *Lobodon* J. Gray, 1844b.

COMMENTS: When originally proposed, this rank was placed in the Family Phocidae (J. Gray, 1821) and included the genus *Lobodon* J. Gray, 1844b. Subfamily rank recognised by Simpson (1945: 122).

Tribe Monachini Scheffer, 1958: 112.

TYPE GENUS: *Monachus* Fleming, 1922b: 187.

COMMENTS: When originally proposed as a new rank it was placed in the Subfamily Monachinae (Trouessart, 1897) and included the genus *Monachus* Fleming, 1922b: 187. Rank recognised by McKenna and Bell (1997: 258), Berta (2009a: 881) and Dasmahapatra *et al.* (2009: 170).

Tribe Lobodontini Scheffer, 1958: 115.

TYPE GENUS: *Lobodon* J. Gray, 1844b.

COMMENTS: When originally proposed as a new rank it was placed in the Subfamily Monachinae (Trouessart, 1897) and included the genus *Lobodon* J. Gray, 1844b; *Ommatophoca* J. Gray, 1844b; *Hydrurga* Gistel, 1848; and *Leptonychotes* Gill, 1872. Synonymised within Lobodontina J. Gray, 1869c by McKenna and Bell (1997: 259). Rank recognised by Berta (2009a: 881) and Dasmahapatra *et al.* (2009: 170).

Tribe Cystophorini Burns & Fay, 1970: 363, 390.

TYPE GENUS: *Cystophora* Nilsson, 1820: 382.

COMMENTS: When originally proposed, this rank was placed in the Subfamily Phocinae (J. Gray, 1821) and included the genus *Cystophora* (Nilsson, 1820: 382). Tribe rank recognised by Dasmahapatra *et al.* (2009: 170). Synonymised within Phocidae by Wozencraft (2005: 595).

Tribe Miroungini de Muizon, 1982: 175, 199.

TYPE GENUS: *Mirounga* J. Gray, 1827a.

COMMENTS: When originally proposed as a new rank it was placed in the Subfamily Monachinae (Trouessart, 1897) and included the genera *Mirounga* J. Gray, 1827a; and † *Callophoca* Van Bénéden, 1876: 798. Rank recognised by Berta (2009a: 881) and Dasmahapatra *et al.* (2009: 170). Synonymised within Phocidae by Rice (1998: 34) and Wozencraft (2005: 595).

Hydrurga Gistel, 1848

Hydrurga Gistel, 1848: xi.

TYPE SPECIES: *Novum numen* for *Stenorhinchus* F. Cuvier, 1826a.

COMMENTS: Taxonomic decision of Iredale and Troughton (1934: 87), which was followed by subsequent authors. For history of the name *Hydrurga* see J. Allen (1905: 86).

Stenorhynchus J. Gray, 1825a: 340.

TYPE SPECIES: *Novum nudum* ascribed to Cuvier vernacular.

COMMENTS: Described as 'Stenorhinque' by F. Cuvier (1824: 190). Ex Cuvier vernacular. Genus recognised by J. Gray (1866b: 15). Publication date established from Iredale and Troughton (1934: 87) who synonymised it within *Hydrurga*.

HOMONYMS:

Stenorhynchus Lamarck, 1818: 236, crabs of the Subphylum Crustacea (Class Malacostraca, Order Decapoda, Family Inachidae). Currently accepted name. See Davie and Türkay (2010).

Stenorhynchus Hemprich, 1820: 105, of the Class Reptilia.

Stenorhynchus Dejean, 1821: 98, straight-snouted weevils of the Class Insecta (Order Coleoptera, Family Brentidae).

Stenorhynchus G. Dahl, 1823: 53, true weevils of the Class Insecta (Order Coleoptera, Family Curculionidae).

Stenorhynchus Berthold, 1827: 384, straight-snouted weevils of the Class Insecta (Order Coleoptera, Family Brentidae). Status unknown as poorly described.

Stenorhynchus Lesson, 1827a, the Leopard Seal of the Class Mammalia (Order Carnivora, Family Phocidae). Pro *Stenorhinchus* F. Cuvier, 1826a. Synonymised within *Hydrurga* by McKenna and Bell (1997: 260). See individual entry.

Stenorhynchus Villa & Villa, 1833: 43, true weevils of the Class Insecta (Order Coleoptera, Family Curculionidae). Genus is a synonym of *Lignyodes* Dejean, 1835: 278. See Arnett *et al.* (2002: 739).

Stenorhynchus Gould, 1836: 185, trembler birds of the Class Aves (Order Passeriformes, Family Mimidae). Genus is a synonym of *Cinclocerthia* G. J. Gray, 1840: 17.

Stenorhynchus A. Smith, 1849: Appendix. 23, dwarf puddle frogs of the Class Amphibia (Order Anura, Family Phrynobatrachidae or Ranidae (disputed)). Genus is a synonym of *Phrynobatrachus* Günther, 1862: 190. See Frost (1985: 443; 2011) and Ulber (1999: 11).

Stenorhynchus Lacordaire, 1866: 507, fungus weevils of the Class Insecta (Order Coleoptera, Family Anthribidae). Genus is a synonym of *Stenocerus* Schönherr, 1826: 39.

Stenorhinchus F. Cuvier, 1826a: 549.

TYPE SPECIES: *Phoca leptonyx* de Blainville, 1820 (as *Stenorhinchus leptonyx*) [=*Hydrurga leptonyx* (de Blainville, 1820)] by original designation.

COMMENTS: Junior homonym of *Stenorhynchus* Lamarck, 1818: 236. Spelling used by McMurtrie (1834: 71) but not typically by other authors. Publication date established from Scheffer (1958: 120). Synonymised within *Hydrurga* by Simpson (1845: 122), Iredale and Troughton (1934: 87) and McKenna and Bell (1997: 260).

Stenorhynchus Lesson, 1827a: 199.

TYPE SPECIES: Emendation of *Stenorhinchus* F. Cuvier, 1826a.

COMMENTS: Synonymised within *Hydrurga* by McKenna and Bell (1997: 260).

HOMONYMS:

Stenorhynchus J. Gray, 1825a, the Leopard Seal of the Class Mammalia (Order Carnivora, Family Phocidae). Synonymised within *Hydrurga* by Iredale and Troughton (1934: 87). See individual entry.

Stenorhyncus F. Cuvier, 1829b: 463.

TYPE SPECIES: Incorrect subsequent spelling of *Stenorhinchus* F. Cuvier, 1826a.

COMMENTS: Synonymised within *Stenorhinchus* by Palmer (1904: 647).

Ogmorhinus Peters, 1875c: 393.

TYPE SPECIES: *Novum numen* for *Stenorhinchus* F. Cuvier, 1826a.

COMMENTS: Publication date established from Scheffer (1958: 120). Synonymised within *Hydrurga* by Simpson (1845: 122) and Iredale and Troughton (1934: 87).

Stenorhynchotes W. Turner, 1888: 63, footnote.

TYPE SPECIES: *Nomen novum* for *Stenorhynchus* J. Gray, 1825a.

COMMENTS: Publication date established from Iredale and Troughton (1934: 87–89). Synonymised within *Hydrurga* by Iredale and Troughton (1934: 87).

Hydrurga leptonyx (de Blainville, 1820)

Leopard Seal

P. [hoca] leptonyx de Blainville, 1820: 298.

TYPE LOCALITY: 'des environs des îles Falckland ou Malouines'. Falkland Islands (United Kingdom).

COMMENTS: Included in the genus *Stenorhynchus* by Gould (1860 [1845–1863]: Text to Plate 50) and Krefft (1868a: 93), and *Ogmorhinus* by J. Ogilby (1892: 129). Placed in *Hydrurga* by Scheffer (1958: 120) and followed by subsequent authors. Synonyms allocated according to Cabrera (1958: 305).

Phoca Homei Lesson, 1828c: 417.

TYPE LOCALITY: Southern Australia, Australia.

COMMENTS: *Nomen novum* for *Phoca leptonyx* de Blainville, 1820. Synonymised within *leptonyx* by Iredale and Troughton (1934: 88).

Leptonychotes Gill, 1872

Leptonychotes Gill, 1872: 70.

TYPE SPECIES: *Otaria weddellii* Lesson, 1826 [=*Leptonychotes weddellii* (Lesson, 1826)] by monotypy.

COMMENTS: Replacement name for *Leptonyx* J. Gray, 1837, which is preoccupied by *Leptonyx* Swainson, 1832: Plate 117.

Léptonyx J. Gray, 1837: 582.

TYPE SPECIES: *Otaria weddelli* Lesson, 1826 (as *Léptonyx Weddéllii* J. Gray) [=*Leptonychotes weddellii* (Lesson, 1826)] by monotypy.

COMMENTS: Recognised by J. Gray (1866b: 11). Synonymised within *Leptonychotes* by Simpson (1845: 122) and Iredale and Troughton (1934: 88).

HOMONYMS:

Leptonyx Swainson, 1832: Plate 117, babblers of Class Aves (Order Passeriformes, Family Rhinocryptidae). Genus is a synonym of *Pteroptochos* Kittlitz, 1830: 178. See Cory and Hellmayr (1924: 3).

Leptonyx Swainson, 1837: 124, 290, finches of the Class Aves (Order Passeriformes, Family Emberizidae). Genus is a synonym of *Coryphaspiza* G. J. Gray, 1840: 47.

Leptonyx Lesson, 1842: 72, clawless otters of the Class Mammalia (Order Carnivora, Family Mustelidae). Genus is a synonym of *Aonyx* (Lesson, 1827a: 157). See Wozencraft (2005: 601).

Leptonyx P. Carpenter, 1864a: 652 and 1864b: 176, turban snails of the Class Gastropoda (Family Colloniidae). Genus is a synonym of *Homalopoma* Carpenter, 1864a: 537.

Leptonyx Hitchcock, 1865: v, 8, extinct reptiles of the superorder Dinosauria (Order Theropoda, † Family Grallatoridae). Genus is a synonym of † *Stenonyx* Lull, 1904: 479, 498 which was introduced as a replacement name.

Leptosonyx Weise, 1886: 576, leaf beetles of the Class Insecta (Order Coleoptera, Family Chrysomelidae). Genus is a synonym of *Theone* Gistel, 1857: 585.

Leptonyx Jacobson, 1895: 555, leaf beetles of the Class Insecta (Order Coleoptera, Family Chrysomelidae). Genus is an emendation of *Leptosonyx* Weise, 1886: 576 and is a synonym of *Theone* Gistel, 1857: 585.

Poecilophoca Lydekker, 1891: 605.

TYPE SPECIES: *Nomen novum* for *Leptonyx* J. Gray, 1837.

COMMENTS: Synonymised within *Leptonychotes* by Simpson (1945: 122), and Iredale and Troughton (1934: 88).

Leptonychotes weddellii (Lesson, 1826)

Weddell Seal

Otaria Weddellii Lesson, 1826: 437.

TYPE LOCALITY: 'sur le côtes des Orcades australes, situées sour 60 degrés 37 minutes de lat' (South Orkney Islands).

COMMENTS: Spelt 'weddelii' on page 438. First recorded as the 'sea leopard' by Weddell (1825: 22–23), which was used as the basis for the description of the species by Lesson (1826: 438). The first proper account of *weddelli* however was not made until J. Gray (1837: 582). Transferred to *Leptonychotes* by J. Allen (1880: 467) and accepted by Scheffer (1958: 122). Reviewed by Stirling (1971: 1), Kooyman (1981: 13) and Bonner (1988: 75).

Ph. [oca] Leopardina R. Hamilton, 1839: 183; Plate 12.

TYPE LOCALITY: South Orkney Islands.

COMMENTS: Synonymised within *weddelli* by Iredale and Troughton (1934: 88) and Wozencraft (2005: 597).

L. [eptonyx] leopardinus Wagner, 1846: 38.

TYPE LOCALITY: Unknown.

COMMENTS: Synonymised within *weddelli* by Wozencraft (2005: 597).

Stenorhynchus Leptonyx Moseley, 1879: 200.

TYPE LOCALITY: Kerguelens Land, South America.

COMMENTS: Synonymised within *weddelli* by Wozencraft (2005: 597).

Lobodon J. Gray, 1844

Lobodon J. Gray, 1844b: 2.

TYPE SPECIES: *Phoca carcinophaga* Hombron and Jacuinot, 1842 [=*Lobodon carcinophaga* (Hombron & Jacquinot, 1842)] by original designation.

COMMENTS: Appears to have been recognised by most authors since its description.

Lobodon carcinophaga (Hombron & Jacquinot, 1842)

Crabeater Seal

Phoca carcinophaga Hombron & Jacquinot, 1842: 27; Plate 10.

TYPE LOCALITY: 'capturé sur les glaces du Pole Sud, entre les îles Sandwich et les îles Powels, à 150 lieues de distances de chacune de ces îles.' [Scotia Sea (midway between South Orkney and South Sandwich Islands].

COMMENTS: Placed in *Lobodon* by J. Gray (1844b: 2), *Stenorhynchus* by Flower (1884: 213), *Ogmorhinus* by W. Turner (1888: 63). Returned to *Lobodon* by Scheffer (1958: 116) and followed by subsequent authors.

Stenorhynchus serridens Owen, 1843: 331.

TYPE LOCALITY: 'From a high latitude in the Australian seas'.

COMMENTS: Placed in *Leptorhynchus* by J. Gray (1844b: 5). Synonymised within *Carcinophaga* by Iredale and Troughton (1934: 87) and subsequent author.

Lobodon cancrivora J. Gray, 1844b: 7.

TYPE LOCALITY: *Errore pro Lobodon carcinophaga* Hombron and Jacquinot, 1842.

COMMENTS: Synonymised within *Lobodon carcinophaga* by Iredale and Troughton (1934: 87). Not considered by Wozencraft (2005: 597).

Lobodon carcinophagus Berg, 1898: 15.

COMMENTS: Unjustified emendation of *Lobodon carcinophaga*. Synonymised within *carcinophaga* by Adam (2005: 1).

Mirounga J. Gray, 1827

Mirounga J. Gray, 1827a: 179.

TYPE SPECIES: *Phoca proboscidea* Péron, 1816 [=*Mirounga leonina* (Linnaeus, 1758)] by original designation.

COMMENTS: Described as a subgenus of *Phoca*. Synonymised within *Macrorhinus* (F. Cuvier, 1826a) by Iredale and Troughton (1934: 88). Genus reviewed by Davidson (1929: 229), J. King (1966: 385), Briggs and Morejohn (1976: 199), Ling (1988b: 228) and Ling and Bryden (1992: 1).

Macrorhinus J. Gray, 1825a: 340.

TYPE SPECIES: *Nomen nudum*.

COMMENTS: Genus recognised as 'Macrorhine' by F. Cuvier (1824: 200) as '*Macrorhinus*, F. Cuv.' by J. Gray

(1825a: 340). The author for the genus was attributed to F. Cuvier (1826a) by Palmer (1904: 394). Genus recognised by Iredale and Troughton (1934: 88) and Troughton (1967: 203). Synonymised within *Mirounga* by Ling (1988b: 228) and Ling and Bryden (1992: 1). Not considered by Wozencraft (2005: 597).

HOMONYMS:

Macrorhine Latreille, 1825: 395, true weevils of the Class Insecta (Order Coleoptera, Family Curculionidae). Vernacular name. Genus is a junior synonym of *Eurhinus* Illiger, 1807: 326. See O'Brien and Wibmer (1982: 181).

Macrorhinus Berthold, 1827: 390, true weevils of the Class Insecta (Order Coleoptera, Family Curculionidae). Unnecessary replacement name. Genus is a junior synonym of *Eurhin* Illiger, 1807: 326. See O'Brien and Wibmer (1982: 181).

Macrorhinus F. Cuvier, 1826a: 552.

TYPE SPECIES: *Phoca proboscidea* Péron, 1816 [=*Mirounga leonina* (Linnaeus, 1758)] by original designation.

COMMENTS: Also junior homonym of *Macrorhinus* J. Gray, 1825a. Publication date established from Scheffer (1958: 128). Synonymised within '*Morunga*' by J. Gray (1943a: xxiii) and *Mirounga* by Ling (1988b: 228), Ling and Bryden (1992: 1) and Wozencraft (2005: 597).

HOMONYMS:

See homonyms above.

Macrohyna J. Gray, 1827a: 180.

TYPE SPECIES: Incorrect subsequent spelling of *Macrorhinus* F. Cuvier, 1826a.

COMMENTS: Synonymised within *Mirounga* by Ling and Bryden (1992: 1).

Rhinophoca Wagler, 1830: 27.

TYPE SPECIES: *Nomen novum* for *Macrorhinus* F. Cuvier, 1826a.

COMMENTS: Publication date established from Iredale and Troughton (1934: 88). Synonymised within '*Morunga*' by J. Gray (1943a: xxiii) and *Mirounga* by Ling (1988b: 228), Ling and Bryden (1992) and Wozencraft (2005: 597).

Mirunga T. Brown, 1832: 89.

TYPE SPECIES: Incorrect subsequent spelling of *Mirounga* J. Gray, 1827a.

COMMENTS: Not previously considered.

Morúnga J. Gray, 1837: 582.

TYPE SPECIES: Incorrect subsequent spelling of *Mirounga* J. Gray, 1827a.

COMMENTS: Subsequently recognised by J. Gray (1842e: 16; 1843a: xxiii, 103; 1866b: 38). Publication date established from Iredale and Troughton (1934). Incorrect subsequent spelling of *Mirounga* J. Gray, 1827a. Synonymised within *Mirounga* by Ling (1988b: 228) and Ling and Bryden (1992: 1). Not considered by Wozencraft (2005: 597).

Cystophora S. Nilsson, 1838: 240.

TYPE SPECIES: *Phoca proboscidea* Péron, 1816 (as *Cystophora proboscidea*) [=*Mirounga leonina* (Linnaeus, 1758)] by original designation.

COMMENTS: Synonymised within *Mirounga* by Ling and Bryden (1992: 1). Not considered by Wozencraft (2005: 597).

HOMONYMS:

Cystophora Nilsson, 1820: 382, the Hooded Seal of the Class Mammalia (Order Carnivora, Family Phocidae). Currently accepted name. See Wozencraft (2005: 595).

† *Cystophora* Yabe and Hayasaka, 1916: 70(78), horn corals of the Class Anthozoa (Order Stauriida, Family Durhaminidae). Genus is synonym of *Pseudocystophora* Kossovaya, 1997: 88. See Fedorowski *et al.* (2007: 73).

Physorhīnus Gloger, 1841: xxxiv, 163.

TYPE SPECIES: *Phoca proboscidea* Péron, 1816 (as *Physorhinus proboscidea*) [=*Mirounga leonina* (Linnaeus, 1758)] by monotypy.

COMMENTS: Synonymised within *Macrorhinus* (as F. Cuvier, 1824) by Thomas (1895a: 191) and within *Mirounga* by Ling (1988b: 228) and Ling and Bryden (1992: 1). Not considered by Wozencraft (2005: 597).

HOMONYMS:

Physorhinus Eschscholtz, 1836: 4, Table, click beetles of the Class Insecta (Order Coleoptera, Family Elateridae). Currently accepted name. See Casari (2008: 182).

Mirounga leonina (Linnaeus, 1758)

Southern Elephant Seal

[Phoca] Leonina Linnaeus, 1758: 37.

TYPE LOCALITY: 'ad polum Antarctum' restricted by Thomas (1911b: 133) to 'Juan Fernandez', which was confirmed by J. Hamilton (1940: 35).

COMMENTS: Type designation by J. Hamilton (1940: 35). Reviewed by Scheffer (1958: 130–131), Ling (1988b: 228), and Ling and Bryden (1992: 1). Various studies indicate some variation in morphology, growth rate and body size between different islands and ocean basins (Carrick *et al.*, 1962: 161; Bryden, 1968: 1106; Laws, 1994: 49). Other genetic studies have revealed strong differences between island populations and those on Valdez Peninsula (Gales *et al.*, 1989: 57; Slade *et al.*, 1998: 1945; Hoelzel *et al.*, 2001: 325). See discussion in Berta and Churchill (2012: 220).

FUTURE TAXONOMIC RESEARCH: Research is required to determine the merit of possible subspecies.

Phoca Elephantina Molina, 1782: 280.

TYPE LOCALITY: *Nomen novum* for *Phoca leonina* Linnaeus, 1758

COMMENTS: Species recognised within *Morunga* by J. Gray (1866b: 39). Synonymised within *Mirounga* by Ling and Bryden (1992: 1).

Phoca proboscidea Péron, 1816: 34; Plate 32.

TYPE LOCALITY: Des regions Australes. Given as King Island, Bass Strait, Australia by Scheffer (1958: 129) and Ling (1988b: 229).

COMMENTS: Recognised as a species within *Macrorhinus* by Iredale and Troughton (1934: 88) and Troughton (1967: 203). Synonymised within *leonina* by Ling (1988b: 228) and Ling and Bryden (1992: 1).

Phoca Resima Péron, 1816: 66.

TYPE LOCALITY: 'Des iles St. Pierre et St. Paul d'Amsterdam'.

COMMENTS: Synonymised within *leonina* by Ling and Bryden (1992: 1).

Phoca Coxii Desmarest, 1817e: 559.

TYPE LOCALITY: 'Des iles St. Pierre et St. Paul d'Amsterdam'.

COMMENTS: Synonymised within *leonina* by Ling and Bryden (1992: 1).

phoca Ansonii Desmarest, 1821: 239.

TYPE LOCALITY: 'Lile Georgia, la Terre de Feu, les iles Malouines, la cote est de 'Amerique, despuis la Terre des Etats jusqu'an 40° degree, sur la Terre des Patagons; quelquetois l'ile Sainte-Helene, selon Dampier, la Terre de Kerguelen'.

COMMENTS: Synonymised within *leonina* by Ling and Bryden (1992: 1).

Phoca ansonina de Blainville, 1820: 299.

TYPE LOCALITY: 'Des iles Kelckland' [sic].

COMMENTS: Synonymised within *leonina* by Ling and Bryden (1992: 1).

Phoca ansonii de Blainville, 1820: 300.

TYPE LOCALITY: 'Des iles Kelckland' [sic].

COMMENTS: Alternative spelling to *Phoca ansonina*. Synonymised within *leonina* by Ling and Bryden (1992: 1).

Macrorhinus proboscideus F. Cuvier, 1826a: 552.

TYPE LOCALITY: Name combination.

COMMENTS: Synonymised within *leonina* by Ling and Bryden (1992: 1).

Macrorhinus Ansonii Lesson, 1827a: 202.

TYPE LOCALITY: Name combination.

COMMENTS: Synonymised within *leonina* by Ling and Bryden (1992: 1).

[Phoca] M. [irounga]. Proboscidea J. Gray, 1827a: 180.

TYPE LOCALITY: Name combination.

COMMENTS: Described as '*M[irounga]. Proboscidea*' and '*Phoca Proboscidea*'. Synonymised within *leonina* by Ling and Bryden (1992: 1).

[Phoca] M. [irounga] Patagonica J. Gray, 1827a: 180.

TYPE LOCALITY: Patagonia.

COMMENTS: Recognised as a subspecies of *leonina* by Rothschild (1910: 446). Not recognised by Ling and Bryden (1992: 1).

[Phoca] M. [irounga] Ansonii J. Gray, 1827a: 180.

TYPE LOCALITY: Name combination.

COMMENTS: Synonymised within *leonina* by Ling and Bryden (1992: 1).

Phoca dubia J. Fischer, 1829: 235.

TYPE LOCALITY: 'In insulis Maluinis'.

COMMENTS: Synonymised within *leonina* by Ling and Bryden (1992: 1).

Cystophora proboscidea Nilsson, 1838: 240.

TYPE LOCALITY: Name combination.

COMMENTS: Synonymised within *leonina* by Ling and Bryden (1992: 1).

Macrorhinus Coxii Boitard, 1842: 196.

TYPE LOCALITY: Name combination.

COMMENTS: Synonymised within *leonina* by Ling and Bryden (1992: 1).

Morunga Elephantina J. Gray, 1844b: 4.

TYPE LOCALITY: Southern Ocean.

COMMENTS: Synonymised within *leonina* by Ling and Bryden (1992: 1).

C. [ystophora] falklandica Peters, 1875c: 394, footnote.

TYPE LOCALITY: Falkland Islands. See Flower (1881: 146).

COMMENTS: Based on Pernety (1769: Plate 9, Fig. 1). Recognised as a subspecies by Lydekker (1909: 603) and Strahan (1983: 467; 1995: 687), and recognised with uncertain validity by Ling and Bryden (1992: 1). Doubt cast over it by Lönnberg (1910: 580). Synonymised within *leonina* by Wozencraft (2005: 598) and the Society for Marine Mammalogy (Committee on Taxonomy, 2011), but recognised as a subspecies by Van Dyck and Strahan (2008: 729).

[Cystophora] kerguelensis Peters, 1875c: 394, footnote.

TYPE LOCALITY: 'Kerguelenland'.

COMMENTS: Recognised as a subspecies of *leonina* by Rothschild (1910: 446). Synonymised within *leonina* by Ling and Bryden (1992: 1).

Cystophora elephantina A. Brehm, 1877: 638.

TYPE LOCALITY: Name combination.

COMMENTS: Synonymised within *leonina* by Ling and Bryden (1992: 1).

Macrorhinus leoninus J. Allen, 1880: 456.

TYPE LOCALITY: None.

COMMENTS: First use of the current combination. Synonymised within *leonina* by Ling and Bryden (1992: 1).

Mirounga leonina J. Allen, 1905: 95.

TYPE LOCALITY: None.

COMMENTS: First use of the current combination. Synonymised within *leonina* by Ling and Bryden (1992: 1).

M. [irounga] leoninus typicus Lydekker, 1909: 601.

TYPE LOCALITY: Not designated.

COMMENTS: Synonymised within *leonina* by Wozencraft (2005: 598).

M. [irounga] l. [eoninus] macquariensis Lydekker, 1909: 603.

TYPE LOCALITY: Macquarie [sic] Island and Chatham Islands.

COMMENTS: Recognised as a subspecies by Rothschild (1910: 446) and Strahan (1983: 467; 1995: 687) and recognised with uncertain validity by Ling and Bryden (1992: 1). Doubt cast over it by Lönnberg (1910: 580). Synonymised within *leonina* by Wozencraft (2005: 598) and the Society for Marine Mammalogy (Committee on Taxonomy, 2011), but recognised as a subspecies by Van Dyck and Strahan (2008: 729).

M. [irounga] l. [eoninus] crosetensis Lydekker, 1909: 606.

TYPE LOCALITY: Crozet. (?)Kerguelen and Heard Islands.

COMMENTS: Synonymised within *kerguelensis* by Rothschild (1910: 446) with doubt over it by Lönnberg (1910: 580). Recognised at subspecific rank by Strahan (1983: 467; 1995: 687) and with uncertain validity by Ling and Bryden (1992: 1). Synonymised within *leonina* by Wozencraft (2005: 598) and the Society for Marine Mammalogy (Committee on Taxonomy, 2011), but recognised as a subspecies by Van Dyck and Strahan (2008: 729).

Ommatophoca J. Gray, 1844

Ommatophoca J. Gray, 1844b: 3, 7.

TYPE SPECIES: *Ommatophoca rossii* J. Gray, 1844b by subsequent designation. See Barrett-Hamilton (1902: 2, 46).

COMMENTS: Genus appears to have been recognised by most authors since its description.

Ommatophora H. Turner, 1849: 88.

TYPE SPECIES: Incorrect subsequent spelling of *Ommatophoca* J. Gray, 1844b.

COMMENTS: Synonymised within *Ommatophoca* by Palmer (1904: 474) and McKenna and Bell (1997: 260).

HOMONYMS:

Ommatophora Guenée, 1852: 190, owlet moths of the Class Insecta (Order Lepidoptera, Family Noctuidae). Genus is an available name. Not preoccupied by *Ommatophora* H. Turner, 1849: 88, an incorrect subsequent spelling of *Ommatophoca* J. Gray, 1844b.

Ommatophoca rossii J. Gray, 1844

Ross Seal

Ommatophoca Rossii J. Gray, 1844b: 3, 8.

TYPE LOCALITY: Ross Sea, Antarctica.

COMMENTS: Type designation by Barrett-Hamilton (1902: 46). Reviewed by Bonner (1988: 75).

Infraorder Mustelida Tedford, 1976

Parvorder Mustelida Tedford, 1976: 372.

COMMENTS: When originally proposed, this rank was placed in the Suborder Caniformia (Kretzoi, 1943) and included the Family Mustelidae (G. Fischer, 1814). Taxon recognised at parvorder rank by McKenna and Bell (1997: 260).

Family Mustelidae G. Fischer, 1814

Family Mustelinorum G. Fischer, 1814: 182.

TYPE GENUS: *Mustela* Linnaeus, 1758.

COMMENTS: When originally proposed, this rank was placed in the Digitigrada (G. Fischer, 1813a [=Carnivora (Bowdich, 1821 part)]), and included the genera *Suricata* Desmarest, 1804a: 15; *Ichneumon* Lacépède, 1799: 7 [=*Herpestes* Illiger, 1811: 135]; *Mustela* Linnaeus, 1758; *Mephitis* É. Geoffroy Saint-Hilare & G. Cuvier, 1795: 187; *Viverra* Linnaeus, 1758: 43; and *Otolicnus* Illiger, 1811: 74 [=*Galago* É. Geoffroy, 1796d: 49]. Name also introduced as Mustelini (G. Fischer, 1817: 372) and as Mustelinorum on page 415, which is sometimes referred to as the author of this family name. Family reviewed by Wozencraft (2005: 601).

Subfamily Mustelinae G. Fischer, 1814

Family Mustelinorum G. Fischer, 1814: 182.

TYPE GENUS: *Mustela* Linnaeus, 1758.

COMMENTS: When originally proposed, this rank was placed in the Digitigrada (G. Fischer, 1813a [=Carnivora (Bowdich, 1821 part)]), and included the genera *Suricata* Desmarest, 1804a: 15; *Ichneumon* Lacépède, 1799: 7 [=*Herpestes* Illiger, 1811: 135]; *Mustela* Linnaeus, 1758; *Mephitis* É. Geoffroy Saint-Hilare & G. Cuvier, 1795: 187; *Viverra* Linnaeus, 1758: 43; and *Otolicnus* Illiger, 1811: 74 [=*Galago* É. Geoffroy, 1796d: 49].

Mustela Linnaeus, 1758

Mustela Linnaeus, 1758: 45.

TYPE SPECIES: Ω *Mustela putorius* Linnaeus, 1758 by subsequent designation.

COMMENTS: Reviewed by Wozencraft (2005: 613).

Ω *Mustela putorius* Linnaeus, 1758

European Polecat

Ω *[Mustela] Putorius* Linnaeus, 1758: 46.

TYPE LOCALITY: Scania S. Sweden. See Thomas (1911b: 139).

COMMENTS: Species not typically recorded in Australia, but it was discussed by Menkhorst (2011: 214), as *Mustela furo*, and its history introduction was discussed by Long (2003: 277). Reviewed by Wozencraft (2005: 617).

Suborder Feliformia Kretzoi, 1945

Order Feliformia Kretzoi, 1945: 62.

COMMENTS: When originally proposed, this rank included the families † Nimravidae (Cope, 1880: 835), Felidae (G. Fischer, 1817), † Megantereontidae (Kretzoi, 1929: 1337 [=Felidae (G. Fischer, 1817)]) and † Machaerodontidae (Woodward, 1898: 399 [=Felidae (G. Fischer, 1817)]). Recognised by McKenna and Bell (1997: 227), J. Flynn *et al.* (2005: 317, 328) and Wozencraft (2005: 532) at the subordinal rank.

Family Felidae G. Fischer, 1817

Family Felini G. Fischer, 1817: 372.

TYPE GENUS: *Felis* Linnaeus, 1758.

COMMENTS: When originally proposed, this rank was placed in the Order Digitigrada (G. Fischer, 1813a [=Carnivora (Bowdich, 1821 part)]) and included the genus *Felis* (Linnaeus, 1758). Rank not listed but appears to be the same as the Felinorum on page 417. Reviewed by Wozencraft (2005: 532).

Subfamily Felinae G. Fischer, 1817

Family Felini G. Fischer, 1817: 372.

TYPE GENUS: *Felis* Linnaeus, 1758.

COMMENTS: When originally proposed, this rank was placed in the Order Digitigrada (G. Fischer, 1813a [=Carnivora (Bowdich, 1821 part)]) and included the genus *Felis* Linnaeus, 1758. Rank not listed but appears to be the same as the Felinorum on page 417. Reviewed by Wozencraft (2005: 532).

Felis Linnaeus, 1758

Felis Linnaeus, 1758: 41.

TYPE SPECIES: Ω *Felis catus* Linnaeus, 1758 by subsequent designation. See ICZN (1958a: 41).

COMMENTS: Taxonomic arrangements and synonymies are given by Wozencraft (2005: 534).

Ω *Felis catus* Linnaeus, 1758

Domestic Cat

Ω *[Felis] Catus* Linnaeus, 1758: 42.

TYPE LOCALITY: 'Europeae australis sylvis'. Thomas (1911b: 136) identified the type locality as 'Upsala', Sweden.

COMMENTS: Species reviewed by Pocock (1951: 6) and Wozencraft (2005: 534). History of introduction into Australia discussed by Long (2003: 316).

Superorder Euungulata Waddell *et al.*, 2001

Clade Euungulata Waddell *et al.*, 2001: 141.

COMMENTS: When originally proposed, this clade was placed in the Clade Fereuungulata (Waddell *et al.*, 1999a [=Placentalia (Bonaparte, 1838)]) and included the Artiodactyla (Owen, 1848) and Perissodactyla (Owen, 1848). Rank recognised by Asher and Helgen (2010: 4) and given further support by Zhou *et al.* (2012: 150, 159).

Clade Cetungulata Irwin & Wilson, 1993: 264.

COMMENTS: When originally proposed, this clade included the orders Cetacea (Brisson, 1762), Artiodactyla (Owen, 1848) and Perissodactyla (Owen, 1848). Placed in the Mirorder Eparctocyona (McKenna, 1975) by McKenna and Bell (1997: 358) and recognised (emended to include the Eparctocyona (McKenna, 1975) and Altungulata (Prothero & Schoch, 1989: 510)) by Shoshani and McKenna (1998: 578).

Order Perissodactyla Owen, 1848

Order Perissodactyla Owen, 1848: 131.

COMMENTS: When originally proposed, this rank was placed in the Ungulata (Linnaeus, 1766 [=Placentalia (Bonaparte, 1838 part)]) and included the Horse, Tapir, Rhinoceros and Hyrax. The phylogeny of the Order Perissodactyla was explored by Graur *et al.* (1997: 195), who suggested that this group is closer to the Carnivora and Cetartiodactyla than it is to the Paenungulata.

Family Equidae J. Gray, 1821

Family Equidae J. Gray, 1821: 307.

TYPE GENUS: *Equus* Linnaeus, 1758.

COMMENTS: When originally proposed, this rank was placed in the Order Monochenae (J. Gray, 1821: 306 [=Perissodactyla (Owen, 1848)]), and included only the genus *Equus* Linnaeus, 1758.

Equus Linnaeus, 1758

Equus Linnaeus, 1758: 73.

TYPE SPECIES: Ω *Equus caballus* Linnaeus, 1758 by subsequent designation. See ICZN (1958a: 88).

COMMENTS: Different taxonomic decisions for subgeneric arrangement of *Equus* include Bennett (1980: 280–283), Groves (1974: 29), Groves and Willoughby (1981: 349), and Grubb (2005b: 629).

Ω ***Equus asinus*** Linnaeus, 1758

Donkey

Ω *[Equus] Asinus* Linnaeus, 1758: 73.

TYPE LOCALITY: 'Habitat in oriente' [=Middle East?].

COMMENTS: History of introduction discussed by Long (2003: 342).

Ω ***Equus caballus*** Linnaeus, 1758

Horse

Ω *[Equus] Caballus* Linnaeus, 1758: 73.

TYPE LOCALITY: 'Europa'. Lydekker (1916: 6) identified the type locality as 'Scandinavia'.

COMMENTS: Corbet (1978: 194) proposed *ferus* to replace *caballus*, objecting to the use of specific names based on domestic animals. Gromov and Baranova (1981: 333–334) continued to recognise *gmelini* and *przewalskii*. Synonyms and subspecies reviewed by Bennett and Hoffmann (1999: 1) and Grubb (2005b: 629). History of introduction discussed by Long (2003: 347).

Order Artiodactyla Owen, 1848 *sensu* Montgelard *et al.,* 1997

Order Artiodactyla Owen, 1848: 131.

COMMENTS: When originally proposed, this rank was placed in the Ungulata (Linnaeus, 1766 [=Placentalia (Bonaparte, 1838 part)]) and included the Ox, Hog, Peccary and Hippopotamus. The term Artiodactyla dates from Owen (1848), but the concept was first expressed informally by de Blainville (1816a: 117). There is increasing molecular and morphological support to place the cetaceans firmly within the Artiodactyla (e.g. Geisler & Uhen, 2005: 145; Geisler *et al.*, 2007: 28; Rice 2009: 236) so this has been followed here. The use of Artiodactyla here unites it with the cetaceans in the same sense as that used by Montgelard *et al.* (1997: 550, 556).

Order Pecora Linnaeus, 1758: 17, 65.

COMMENTS: When originally proposed, this rank was placed in the Class Mammalia (Linnaeus, 1758) and included the genera *Camelus* Linnaeus, 1758; *Moschus* Linnaeus, 1758: 66; *Cervus* Linnaeus, 1758; *Capra* Linnaeus, 1758; *Ovis* Linnaeus, 1758; and *Bos* Linnaeus, 1758. Synonymised within the Suborder Ruminantia (Scopoli, 1777) by McKenna and Bell (1997: 418).

Order Hoplopoda Goldfuss, 1820: xx, 360.

COMMENTS: When originally proposed, this rank was placed in the Class Mammalia (Linnaeus, 1758) and included the families Cavicornia (Illiger, 1811: 106 [=Bovidae (J. Gray, 1821)]); Tylopoda (Illiger, 1811); Cervina (Goldfuss, 1820: xx, 374 [=Cervidae (Goldfuss, 1820: xx, 374)]); and Solidungula (Blumenbach, 1779: 58, 109 [=Hippomorpha (Wood, 1937: 106)]). Does not appear to have been considered by subsequent authors.

Mirorder Eparctocyona McKenna, 1975: 41.

COMMENTS: When originally proposed as a new rank it was placed in the Grandorder Ungulata (Linnaeus, 1766 [=Placentalia (Bonaparte, 1838 part)]) and included the orders † Arctocyonia (Van Valen, 1969: 123), † Tillodontia (Marsh, 1875: 221), Tubulidentata (Huxley, 1872: 288), † Dinocerata (Marsh, 1873: 117), † Embrithopoda (C. Andrews, 1906: 224), and Artiodactyla (Owen, 1848). Synonymised within Artiodactyla by Asher and Helgen (2010: 4).

Clade Cetartiodactyla Montgelard *et al.*, 1997: 550, 556.

COMMENTS: When originally proposed, this clade included the Artiodactyla (Owen, 1848) and Cetacea (Brisson, 1762). This name has been recognised by various authors including Nikaido *et al.* (1999: 10261), Geisler and Uhen (2005: 145), Price *et al.* (2005: 445), Agnarsson and May-Collado (2008: 964), O'Leary and Gatesy (2008: 397), Ayoub *et al.* (2009: 550), Rice (2009: 235), Zhou *et al.* (2011: 255), Committee on Taxonomy (2011). Rice (2009: 235) recognised Cetartiodactyla (in part) at ordinal rank with the intraorder Cetacea while Yapa and Ratnavira (2013: 9, 599) recognised the name at superordinal rank and included the orders Artiodactyla and Cetacea. Spaulding *et al.* (2009: 3) suggested that the term Cetartiodactyla had gained some traction in the literature, especially among molecular workers, although they formally retained it within the Artiodactyla. Synonymised within Artiodactyla by Asher and Helgen (2010: 4), which is followed here.

Clade Cetruminantia Waddell *et al.*, 1999a: 2.

COMMENTS: When originally proposed, this clade included the Whippomorpha (Waddell *et al.*, 1999a) and Ruminantia (Scopoli, 1777). This arrangement was supported by Hassanin *et al.* (2012: 37), but given the shortness of any branch uniting them, we do not recognise it here.

Clade Artiofabula Waddell *et al.*, 1999a: 2.

COMMENTS: When originally proposed, this clade included the Suidae (J. Gray, 1821) and Cetruminantia (Waddell *et al.*, 1999a [=Artiodactyla (Owen, 1848)]).

Suborder Suina J. Gray, 1868

Suina J. Gray, 1868b: 20.

COMMENTS: When originally proposed, this rank was placed in the Family Suidae (J. Gray, 1821) and included the genera *Sus* Linnaeus, 1758; *Porcula* Hodgson, 1847: 423; and *Potamochoerus* Gray, 1854: 129. Following Groves and Grubb (2011: 28, 33), we use this name to designate one of the four suborders of Artiodactyla, preferring it to Suiformes, which usually includes the Hippopotamidae, now shown to belong to the suborder Whippomorpha.

Suborder Suiformes Jaeckel, 1911: 233.

COMMENTS: When originally proposed, this rank was placed in the Order Diungulati (Jaeckel, 1911: viii, 232) and included the families † Trigonolestidae (Schlosser, 1899: 349 [=† Dichobunidae (H. Turner, 1850: 158)]), † Dichobunidae (H. Turner, 1850: 158), † Anthracotheridae (Leidy, 1869: 202), Hippopotamidae (J. Gray, 1821: 306), † Archaenodontidae (Jaeckel, 1911: 234), † Entelodontidae (Lydekker, 1883: 5–146), Dicotylidae (J. Gray, 1868b: 21 [=Tayassuidae (Palmer, 1897b: 174)]); and Suidae (J. Gray, 1821: 306). Suborder recognised McKenna and Bell (1997: 391) who describe the various synonyms.

Family Suidae J. Gray, 1821

Family Suidae J. Gray, 1821: 306.

TYPE GENUS: *Sus* Linnaeus, 1758.

COMMENTS: When originally proposed, this rank was placed in the Order Tesserachenae (J. Gray, 1821: 306 [=Suina (J. Gray, 1868)]) and included the genera *Sus* Linnaeus, 1758; *Phacochaeres* J. Gray, 1821: 306 [=*Phacochoerus* F. Cuvier, 1826b: 383]; *Dicotyles* G. Cuvier, 1816a: 237 [=*Tayassu* Fischer, 1814: 284]; and *Babiroussus* J. Gray, 1821: 306 [=*Babyrousa* Perry, 1811[1810–1811]: Text to Plate 67]. All extant suids were placed in the Subfamily Suinae by McKenna and Bell (1997: 394–395) but in different subfamilies within Suidae by Grubb (2005a: 637).

Sus Linnaeus, 1758

Sus Linnaeus, 1758: 49.

TYPE SPECIES: Ω *Sus scrofa* Linnaeus, 1758 by subsequent designation. See ICZN (1958a: 23).

COMMENTS: Taxonomic arrangements and synonymies can be found in Grubb (2005a: 639).

Ω *Sus scrofa* Linnaeus, 1758

Pig

Ω *[Sus] Scrofa* Linnaeus, 1758: 49.

TYPE LOCALITY: 'Europa australiore'. Thomas (1911b: 140) identified the type locality as Germany.

COMMENTS: Taxonomic arrangements and synonymies can be found in Groves (1981: 29), Genov (1999: 227) and Grubb (2005a: 641). History of introduction discussed by Long (2003: 368).

Suborder Tylopoda Illiger, 1811

Family Tylopoda Illiger, 1811: 102.

TYPE GENUS: Not based on a genus group name.

COMMENTS: When originally proposed, this rank was placed in the Order Bisulca (Illiger, 1811: 102 [=Tylopoda (Illiger, 1811)]) and included the genera *Camelus* Linnaeus, 1758; and *Auchenia* Illiger, 1811: 103 [=*Lama* G. Cuvier, 1800: Table 1]. Recognised by McKenna and Bell (1997: 412). Following Groves and Grubb (2011: 28), we recognise this as one of four suborders of Artiodactyla.

Family Camelidae J. Gray, 1821

Family Camelidae J. Gray, 1821: 307.

TYPE GENUS: *Camelus* Linnaeus, 1758.

COMMENTS: When originally proposed, this rank was placed in the order Hydrophorae (J. Gray, 1821: 307 [=Tylopoda (Illiger, 1811)]) and included the genera *Camelus* Linnaeus, 1758 and *Lama* G. Cuvier, 1800: Table 1.

Camelus Linnaeus, 1758

Camelus Linnaeus, 1758: 65.

TYPE SPECIES: Ω *Camelus dromedarius* Linnaeus, 1758 by subsequent designation. See G. Allen (1939: 465).

COMMENTS: Type species has been widely cited as *C. bactrianus* (see Gentry *et al*., 1996: 34) and a change in designation was proposed by Erridge (1988: 141) but this has not been supported (Grubb, 2005a: 645). See discussion of ICZN (1910: 37). Taxonomic arrangements and synonymies in Grubb (2005a: 645).

Ω *Camelus dromedarius* Linnaeus, 1758

One-humped Camel

Ω *[Camelus] Dromedarius* Linnaeus, 1758: 65.

TYPE LOCALITY: 'Africae desertis arenosis siticulosis'. See Thomas (1911b: 150).

COMMENTS: Recognised within *Camelus* by J. Fischer (1829: 435). Reviewed by Kohler-Rollefson (1991: 1). History of introduction discussed by Long (2003: 390).

Lama G. Cuvier, 1800

Lama G. Cuvier, 1800: Table 1,

TYPE SPECIES: Ω *Camelus glama* Linnaeus, 1758 [=Ω *Lama glama* (Linnaeus, 1758)] by subsequent designation.

COMMENTS: Taxonomic arrangements and synonymies in Grubb (2005a: 646) and, for the wild species (ancestral to the domestic species now established in Australia), see Groves and Grubb (2011:29–32). We do not regard the genus *Vicugna* Lesson, 1842: 167 as valid following Groves and Grubb (2011:30).

Ω *Lama glama* (Linnaeus, 1758)

Llama

Ω *[Camelus] Glama* Linnaeus, 1758: 65.

TYPE LOCALITY: 'Habitat in America meridionali'.

COMMENTS: Taxonomic arrangements and synonymies in Grubb (2005a: 646).

Ω *Lama pacos* (Linnaeus, 1758)

Alpaca

Ω *[Camelus] Pacos* Linnaeus, 1758: 66.

TYPE LOCALITY: 'Habitat in America meridionali'.

COMMENTS: This domestic species is based on a stabilised hybrid between the Llama and the wild Vicuña *Lama vicugna* (Molina, 1782: 342) (see Hemmer, 1990:60–63).

Suborder Ruminantia Scopoli, 1777

Ruminantia Scopoli, 1777: 493.

COMMENTS: Rank unknown but placed within the Order Ungulata (Linnaeus, 1766 [=Placentalia (Bonaparte, 1838 part)]) and included the genera *Camelus* Linnaeus, 1758; *Giraffa* Brisson, 1762: 12, 37; *Cervus* Linnaeus, 1758; *Antilope* Pallas, 1766; *Capra* Linnaeus, 1758; *Ovis* Linnaeus, 1758; *Bos* Linnaeus, 1758; and *Moschus* Linnaeus, 1758: 66. Recognised as a suborder within the Order Artiodactyla by Simpson (1945: 151) and McKenna and Bell (1997: 418), and as a suborder of the Order Cetartiodactyla (within the Clade Cetruminantia) by Hassanin *et al.* (2012: 36). Following Groves and Grubb (2011: 28), we recognise this as one of four suborders of Artiodactyla.

Family Bovidae J. Gray, 1821

Family Bovidae J. Gray, 1821: 308.

TYPE GENUS: *Bos* Linnaeus, 1758.

COMMENTS: When originally proposed, this rank was placed in the Order Ruminantes (J. Gray, 1821: 307 [=Ruminantia (Scopoli, 1777)]) and included the genus *Bos* (Linnaeus, 1758). Family group names reviewed by Grubb (2001: 374) and subfamily and tribes of the family Bovidae reviewed by Gentry (1992: 1).

Subfamily Bovinae J. Gray, 1821

Family Bovidae J. Gray, 1821: 308.

TYPE GENUS: *Bos* Linnaeus, 1758.

COMMENTS: When originally proposed, this rank was placed in the Order Ruminantes (J. Gray, 1821: 307 [=Ruminantia (Scopoli, 1777)]) and included the genus *Bos* (Linnaeus, 1758). Subfamily rank recognised by Gill (1872: 8) and McKenna and Bell (1997: 442).

Bos Linnaeus, 1758

Bos Linnaeus, 1758: 71.

TYPE SPECIES: Ω *Bos taurus* Linnaeus, 1758 by subsequent designation. See ICZN (1958a: 22).

COMMENTS: Taxonomic arrangements and extralimital synonyms can be found in Grubb (2005a: 690), and Groves and Grubb (2011: 109–117).

Ω *Bos bison* Linnaeus, 1758

American Bison

Ω *Bos bison* Linnaeus, 1758: 72.

TYPE LOCALITY: 'Habitat of Mexico, Florida'. Locality restricted to Hernandez, Mexico by Thomas (1911b: 154); to central 'Quivera Region, Mexico' now known as central Kansas, USA by Hershkovitz (1957: 32); and eastern New Mexico, Canadian River valley by McDonald (1981: 62).

COMMENTS: Species has historically been placed within the genus *Bison* Smith, 1827: 373 (e.g. Meagher, 1986:1; Grubb, 2005a: 689), but transferred to *Bos* by Groves and Grubb (2011: 115). Taxonomic arrangements and extralimital synonyms can be found in Grubb (2005a: 689), and Groves and Grubb (2011: 115). History of introduction discussed by Long (2003: 470). This species is included within this work as it has been introduced as domestic stock on rural properties in southeastern Australia.

Ω *Bos javanicus* d'Alton, 1823

Banteng

Ω *B[os] Javanicus* d'Alton, 1823: 11; legend to Plate 8 fig c, page unnumbered.

TYPE LOCALITY: Java.

COMMENTS: See Hooijer (1956: 223) who discusses the valid name of this species. Taxonomic arrangements and extralimital synonyms can be found in Grubb (2005a: 691), and Groves and Grubb (2011: 112–113). History of introduction discussed by Long (2003: 465).

Ω ***Bos taurus*** Linnaeus, 1758

Cattle

Ω *[Bos] Taurus* Linnaeus, 1758: 71.

TYPE LOCALITY: 'Poloniae depressis graminosts ferus Urus'. This refers to the wild ancestor, the aurochs (extinct even in Linnaeus's day), and Thomas (1911b: 154) identified the type locality for domestic cattle as 'Sweden (Upsala)'.

COMMENTS: Taxonomic arrangements and extralimital synonyms can be found in Grubb (2005a: 692). Recognition of the domestic form's scientific name follows Gentry *et al.* (2004: 649). History of introduction discussed by Long (2003: 466).

Bubalus C. Smith, 1827

Bubalus C. Smith, 1827: 371.

TYPE SPECIES: Ω *Bos bubalis* Linnaeus, 1758 by subsequent designation.

COMMENTS: Described as a subgenus of *Bos* Linnaeus, 1758. Taxonomic arrangements and extralimital synonyms can be found in Grubb (2005a: 693).

Ω ***Bubalus bubalis*** (Linnaeus, 1758)

Swamp Buffalo

Ω *[Bos] Bubalis* Linnaeus, 1758: 72.

TYPE LOCALITY: 'Asia, cultus in Italia'.

COMMENTS: Thomas (1911b: 154) identified the type locality as 'Italy (Rome)'. There are two breed-groups of domestic buffalo (water buffalo): River Buffaloes, originating in South Asia and subsequently spread through suitable areas of the Middle East, southern Europe, and Brazil; and Swamp Buffalo, bred in southern China and Southeast Asia. These may actually descend from different wild ancestors; if so, the name *Bubalus bubalis* will have to be restricted to the River Buffaloes, while the name *Bos kerabau* Sundevall (1846b: 202) is available for the Swamp Buffaloes. The feral buffaloes occurring in Australia were acknowledged by Van Dyck and Strahan (2008: 763) to be the taxon *Bubulus bubalis kerabau* Fitzinger, 1860a: viii, 329. History of introduction discussed by Long (2003: 461).

Subfamily Antilopinae J. Gray, 1821

Family Antilopidae J. Gray, 1821: 307.

TYPE GENUS: *Antilope* Pallas, 1766.

COMMENTS: When originally proposed, this rank was placed in the Order Ruminantes (J. Gray, 1821: 307 [=Ruminantia (Scopoli, 1777)]) and included the genera *Dorcas* J. Gray, 1821: 307 [=*Gazella* de Blainville, 1816c: 75]; *Antilope* Pallas, 1766: 1; *Cuama* J. Gray, 1821: 307 [=*Alcelaphus* de Blainville, 1816c: 75], *Onyx* J. Gray, 1821: 307 [=*Oryx* de Blainville, 1816c: 75], *Tseiran* J. Gray, 1821: 307 [=*Hippotragus* Sundevall, 1845: 31], *Canna* J. Gray, 1821: 307 [=*Taurotragus* Wagner, 1855: xvii, 438], *Nylgau* [=*Boselaphus* de Blainville, 1816c: 75] and *Catablepas* J. Gray, 1821: 307 [=*Connochaetes* M. Lichtenstein, 1814: 152].

Tribe Antilopini J. Gray, 1821

Family Antilopidae J. Gray, 1821: 307.

TYPE GENUS: *Antilope* Pallas, 1766.

COMMENTS: See comments above.

Antilope Pallas, 1766

Antilope Pallas, 1766: 1.

TYPE SPECIES: Ω *Capra cervicapra* Linnaeus, 1758 [= Ω *Antilope cervicapra* (Linnaeus, 1758)] by subsequent designation. See W. Ogilby (1837d: 137).

Ω ***Antilope cervicapra*** (Linnaeus, 1758)

Blackbuck

Ω *[Capra] Cervicapra* Linnaeus, 1758: 69.

TYPE LOCALITY: India.

COMMENTS: Taxonomic arrangements and extralimital synonyms can be found in Grubb (2005a: 678), and Groves and Grubb (2011: 157–158). History of introduction discussed by Long (2003: 487). This species is included within this work as it has been introduced as semi-domestic stock on rural properties in southeastern Australia.

Tribe Caprini J. Gray, 1821

Family Capridae J. Gray, 1821: 307.

TYPE GENUS: *Capra* Linnaeus, 1758.

COMMENTS: When originally proposed, this rank was placed in the Order Ruminantes (J. Gray, 1821: 307 [=Ruminantia (Scopoli, 1777)]) and included the genera *Capra* Linnaeus, 1758; and *Ovis* Linnaeus, 1758. Subfamily rank recognised by Gill (1872: 9) and McKenna and Bell (1997: 442). Placed as a tribe in Antilopinae by Groves and Grubb (2011: 109).

Capra Linnaeus, 1758

Capra Linnaeus, 1758: 68.

TYPE SPECIES: Ω *Capra hircus* Linnaeus, 1758 by subsequent designation. See ICZN (1958a: 40).

COMMENTS: Taxonomic arrangements and extralimital synonyms can be found in Grubb (2005a: 700).

Ω *Capra hircus* Linnaeus, 1758

Goat

Ω *[Capra] Hircus* Linnaeus, 1758: 68.

TYPE LOCALITY: Unknown. Linnaeus gives no indication of geographic locations, saying only that it lives in mountainous areas, feeding on various twigs and fronds of trees, and lichens.

COMMENTS: Taxonomic arrangements and extralimital synonyms can be found in Grubb (2005a: 700), and Groves and Grubb (2011: 223–234). History of introduction discussed by Long (2003: 505).

Ovis Linnaeus, 1758

Ovis Linnaeus, 1758: 70.

TYPE SPECIES: Ω *Ovis aires* Linnaeus, 1758 by subsequent designation.

COMMENTS: Taxonomic arrangements and extralimital synonyms can be found in Grubb (2005a: 707).

Ω *Ovis aries* Linnaeus, 1758

Sheep

Ω *[Ovis] Aries* Linnaeus, 1758: 70.

TYPE LOCALITY: None given by Linnaeus, who says only that it lives in hot, dry, sunny places.

COMMENTS: Taxonomic arrangements and extralimital synonyms can be found in Grubb (2005a: 708), and Groves and Grubb (2011: 234–245). History of introduction discussed by Long (2003: 528).

Family Cervidae Goldfuss, 1820

Family Cervina Goldfuss, 1820a: xx, 374.

TYPE GENUS: *Cervus* Linnaeus, 1758.

COMMENTS: When originally proposed, this rank was placed in the Order Hoplopoda (Goldfuss, 1820a: xx, 360 [=Artiodactyla (Owen, 1848)]) and included the genera *Moschus* Linnaeus, 1758: 66; *Camelopardalis* Schreber, 1784: Plate 255 [=*Giraffa* Brisson, 1762: 12, 37]; and *Cervus* Linnaeus, 1758. Recognised at family rank by J. Gray (1821: 307). Reviewed by Whitehead (1972), and Groves and Grubb (1987: 21; 2011: 71–107).

Subfamily Cervinae Goldfuss, 1820

Family Cervidae Goldfuss, 1820a: xx, 374.

TYPE GENUS: *Cervus* Linnaeus, 1758.

COMMENTS: When originally proposed, this rank was placed in the Order Hoplopoda (Goldfuss, 1820a: xx, 360 [=Artiodactyla (Owen, 1848)]) and included the genera *Moschus* Linnaeus, 1758: 66; *Camelopardalis* Schreber, 1784: Plate 255 [=*Giraffa* Brisson, 1762: 12, 37]; and *Cervus* Linnaeus, 1758.

Axis C. Smith, 1827

Axis C. Smith, 1827: 312.

TYPE SPECIES: Ω *Cervus axis* Erxleben, 1777 [=Ω *Axis axis* (Erxleben, 1777)] by absolute tautonymy.

COMMENTS: Described as a subgenus of *Cervus*. Raised to generic rank by J. Gray (1843a: xxvii, 178). Originally made available as a subgenus of *Cervus* Linnaeus, 1758. Taxonomic arrangements and synonymies in Grubb (2005a: 661), and Groves and Grubb (2011: 91–92).

Ω *Axis axis* (Erxleben, 1777)

Chital Deer

Ω *[Cervus] axis* Erxleben, 1777: 312.

TYPE LOCALITY: Banks of the Ganges, Bihar, India.

COMMENTS: Taxonomic decision of Haltenorth (1963: 54). Taxonomic arrangements and synonymies in Grubb (2005a: 661), and Groves and Grubb (2011: 91). History of introduction discussed by Long (2003: 396).

Ω *Axis porcinus* (Zimmermann, 1780)

Hog Deer

Ω *Cervus porcinus* Zimmermann, 1780: 131.

TYPE LOCALITY: 'Indo-Gangetic Plain of India'. See Lydekker (1915: 56).

COMMENTS: Originally described in Zimmermann (1777: 532), but this publication has been suppressed (ICZN, 1954b: 231). Included in the genus *Cervus* by Mahoney and Richardson (1988c: 241). History of introduction discussed by Long (2003: 429).

Cervus Linnaeus, 1758

Cervus Linnaeus, 1758: 66.

TYPE SPECIES: Ω *Cervus elaphus* Linnaeus, 1758: 67 by Linnean tautonomy. See ICZN (1958a: 40).

COMMENTS: Phylogeny explored by Randi *et al.* (2001: 1). Taxonomic arrangements and synonymies in Grubb (2005a: 662).

Ω *Cervus elaphus* Linnaeus, 1758

Red Deer

Ω *Cervus elaphus elaphus* Linnaeus, 1758

Ω *[Cervus] Elaphus* Linnaeus, 1758: 67.

TYPE LOCALITY: 'Europa: Asia'. Miller (1912: 967) identified the type locality as 'Southern Sweden'.

COMMENTS: Taxonomic arrangements and synonymies in Grubb (2005a: 662), and Groves and Grubb (2011: 95–99).

Ω ***Cervus elaphus scoticus*** Lönnberg, 1906

Ω *[Cervus elaphus] scoticus* Lönnberg, 1906a: 11.

TYPE LOCALITY: Scotland.

COMMENTS: Recognised as the subspecies that occurs in Australia by Van Dyck and Strahan (2008: 774), but subspecies not recognised as distinct by Groves and Grubb (2011: 96). History of introduction discussed by Long (2003: 412).

Ω ***Cervus timorensis*** de Blainville, 1822

Rusa Deer

Ω ***Cervus timorensis timorensis*** de Blainville, 1822

Ω *C. [ervus] Timorensis* de Blainville, 1822b: 267; Plate.

TYPE LOCALITY: Timor.

COMMENTS: Taxonomic decision of Haltenorth (1963: 59), and see Groves and Grubb (2011: 107).

Ω ***Cervus timorensis moluccensis*** Quoy & Gaimard, 1830

Ω *Cervus moluccensis* Quoy & Gaimard, 1830: 133; Plate 24.

TYPE LOCALITY: Buru Island, Moluccas, Indonesia.

COMMENTS: Recognised as a subspecies that occurs in Australia by Van Dyck and Strahan (2008: 776), but as the species was itself introduced in historic times (probably by pre-Islamic rajahs) into the Moluccas, the validity of the subspecies is dubious. History of introduction discussed by Long (2003: 431).

Ω ***Cervus timorensis russa*** Müller & Schlegel, 1845

Ω *Cervus russa* S. Müller & Schlegel, 1845b: 212, 217; Plates 43, 45.

TYPE LOCALITY: Java.

COMMENTS: Recognised as a subspecies that occurs in Australia by Van Dyck and Strahan (2008: 776). History of introduction discussed by Long (2003: 431).

Ω ***Cervus unicolor*** Kerr, 1792

Sambar Deer

Ω *C. [ervus] Axis unicolor* Kerr, 1792: 300.

TYPE LOCALITY: Syntypes from 'The dry hilly forest of Ceylon, Borneo, Celebes and Java'. Lydekker (1915: 73) identified the type locality as Ceylon.

COMMENTS: Taxonomic decision of Haltenorth (1963: 59). Taxonomic arrangements and synonymies in Grubb (2005a: 670), and Groves and Grubb (2011: 106). History of introduction discussed by Long (2003: 435).

Dama Frisch, 1775

Dama Frisch, 1775: 3.

TYPE SPECIES: Though works by Frisch published in 1775 were rejected by Opinion 258 of the ICZN (1954c: 245), the name *Dama* was validated by the Commission in Opinion 581 (1960: 267–275). Ω *Cervus dama* Linnaeus, 1758 [=Ω *Dama dama* (Linnaeus, 1758)] was placed on the Official List of Specific Names in Zoology by the same Opinion (ICZN, 1960: 267).

COMMENTS: Work of Frisch reviewed by Thomas (1905b: 461). Taxonomic arrangements and synonymies in Grubb (2005a: 664).

Ω ***Dama dama*** (Linnaeus, 1758)

Fallow Deer

Ω *[Cervus] Dama* Linnaeus, 1758: 67.

TYPE LOCALITY: 'Europa'. Thomas (1911b: 151) identified the type locality as 'vivariis Regis & Magnatum'.

COMMENTS: Described as *Cervus dama* by Mahoney and Richardson (1988c: 240). Taxonomic arrangements and synonymies in Grubb (2005a: 665). Reviewed by Feldhamer *et al.* (1988: 1).

Suborder Whippomorpha Waddell *et al.*, 1999

Clade Whippomorpha Waddell *et al.*, 1999a: 2.

COMMENTS: When originally proposed as a new clade it included the Class Cetacea (Brisson, 1762) and the Family Hippopotamidae (J. Gray, 1821: 306). The relationship between the cetaceans and hippopotamids was recognised, at no rank, by several authors including Gatesy *et al.* (1996: 954; 1999: 6), Luo (2000: 236), Waddell *et al.* (2001: 141), Beck *et al.* (2006: 7), Geisler and Theodor (2009: E1), and Hassanin *et al.* (2012: 37). The Cetacea and hippopotamus relationship was also supported by Nikaido *et al.* (1999: 10261). Whippomorpha was recognised at ordinal rank by Skinner and Chimimba (2005: vi, 556), while Liu *et al.* (2001: 1786) and Thewissen *et al.* (2007: 1191) did not recognise the relationship. Skinner and Chimimba (2005: vi, 556) included the suborders Ancodonta (Matthew, 1929: 406), containing the Family Hippopotamidae, and Cetacea containing the infraorders Odontoceti and Mysteceti, which was supported by Hassanin *et al.* (2012: 37). Given that the splits between the four major groups of Artiodactyla are early Eocene, and they are fairly even, we cannot rank them as full orders, and here recognise them as suborders following

Groves and Grubb (2011: 28). Reviewed by Gatesy *et al.* (2002: 652) who found most studies support the clade.

Clade Cetancodonta Arnason *et al.*, 2002: 8153.

COMMENTS: When originally proposed, this clade included the Class Cetacea (Brisson, 1762) and the Family Hippopotamidae (J. Gray, 1821: 306). Synonymised with Whippomorpha by Asher and Helgen (2010: 4).

Infraorder Cetacea Brisson, 1762

Class Cetacea Brisson, 1762: 5, 217, 225.

COMMENTS: When originally proposed, this rank included the genera *Balaena* Linnaeus, 1758: 75; *Cetus* Brisson, 1762: 217, 225 [=*Physeter* Linnaeus, 1758]; *Ceratodon* Brisson, 1762: 218, 231 [=*Monodon* Linnaeus, 1758: 75]; *Delphinus* Linnaeus, 1758; *Phocaena* Brisson, 1762: 234 [=*nomen nudum*]; *Gladius* Brisson, 1762: 235 [=*Orcinus* Fitzinger, 1860b]; *Orca* J. Gray, 1846b [=*Orcinus* Fitzinger, 1860b]; and *Physeter* Linnaeus, 1758. Brisson's work was discussed by Merriam (1895: 375). Recognised as a class on page 217 and order on page 218. Taxon recognised as the Class Cetacea by J. Gray (1821: 309), Order Cetae by J. Gray (1825a: 340), Order Cetacea by J. Gray (1866b: 61) and followed by subsequent authors except McKenna and Bell (1997: 366) who reinstated the Order Cete and placed Cetacea as a suborder. The nomenclature of cetaceans in the Linnaeus's *Systema Naturae* was reviewed by True (1898: 617), with an extensive taxonomic review of the cetaceans by Hershkovitz (1966), Fordyce and Barnes (1994: 419) and Mead and Brownell (2005: 723). The evolutionary history of cetaceans was reviewed by Fordyce and de Muizon (2001: 169). The taxonomy used here typically follows that of Rice (1998) with adjustments mainly reflecting more recent literature and that adopted by the Society for Marine Mammalogy (Committee on Taxonomy, 2014), though names Cetacea, Mysticeti and Odontoceti were not recognised by them as they were considered partially unresolved (e.g. Spaulding *et al.*, 2009: 1; Price *et al.*, 2005: 445; Agnarsson & May-Collado (2008: 964). The shortcomings of cetacean taxonomy were described by Reeves *et al.* (2004: 26).

Order Cete Linnaeus, 1758: 75.

COMMENTS: When originally proposed, this rank was placed within the Class Mammalia (Linnaeus, 1758) and included the genera *Monodon* Linnaeus, 1758: 75; *Balaena* Linnaeus, 1758: 75; *Physeter* Linnaeus, 1758; and *Delphinus* Linnaeus, 1758. Recognised at the ordinal rank, with Cetacea as a suborder, by McKenna and Bell (1997: 366, 368), but not recognised by other authors.

Mutica Linnaeus, 1766: 24.

COMMENTS: When originally proposed, this rank was placed within the Class Mammalia (Linnaeus, 1758) and included the Order Cete (Linnaeus, 1758 [=Cetacea (Brisson, 1762)]). Recognised at cohort rank by Simpson (1945: xiii, 100) and superordinal rank by Minkoff (1976: 153), but synonymised within Cetacea by McKenna and Bell (1997: 368).

Family Cetaceen Duméril, 1806a: 4, 28.

TYPE GENUS: Not based on a genus group name.

COMMENTS: When originally proposed, this rank was placed in the Klasse Säugthiere (Duméril, 1806a [=Mammalia (Linnaeus, 1758)]), and included the genera *Balaena* Linnaeus, 1758: 75; *Balenoptera* [=*Balaenoptera*] Lacépède, 1804; *Narwhalus* [=*Narwalus*] Lacépède, 1804: xxxvii, 142 [=*Monodon* Linnaeus, 1758: 75]; *Anarcus* [=*Anarnak*] Lacépède, 1804 [=*Incertae Sedis*]; *Catodon* Linnaeus, 1761 [=*Physeter* Linnaeus, 1758]; *Physalus* Lacépède, 1804 [=*Physeter* Linnaeus, 1758]; *Physeterus* Duméril, 1806b [=*Physeter* Linnaeus, 1758]; *Delphinus* Linnaeus, 1758; *Delphinapterus* Lacépède, 1804: xli, 243; *Hyperodon* [=*Hyperoodon*] Lacépède, 1804. See Bertuch (1801: B.1. No. 4).

Order Carnivorae J. Gray, 1821: 309.

COMMENTS: When originally proposed, this rank was placed in the Class Cetaceae (Brisson, 1762) and included the families Delphinidae (J. Gray, 1821), Monodontidae (J. Gray, 1821: 310), Physeteridae (J. Gray, 1821) and Balanadae (=Balaenidae (J. Gray, 1821). Not to be confused with the Order Carnivora. This order was included within the Class Cetaceae of J. Gray (1821: 309). Synonymised within Cetacea by McKenna and Bell (1997: 368).

Type Pinnipeda Burnett, 1830c: 360.

COMMENTS: When originally proposed, this rank was placed in the Order Cetetherae (Burnett, 1830c [=Placentalia (Bonaparte, 1838)]) and included the kinds [=families] Delphinidae (J. Gray, 1821), Narvallidae (Burnett, 1830c: 360 [=Monodontidae (Gray, 1821: 310]), Physeteridae (J. Gray, 1821) and Balaenidae (J. Gray, 1821).

HOMONYMS:

Order Pinnipeda Storr, 1780, marine mammals of the Class Mammalia. See individual entry.

Tribe Zoophaga F. Cuvier, 1836a: 563.

COMMENTS: When originally proposed, this rank was placed in the Order Cetacea (Brisson, 1762) and included the families Delphinidae (J. Gray, 1821), Catodontidae (F. Cuvier, 1836a [=Physeteridae (J. Gray, 1821)]) and Balaenidae (J. Gray, 1821).

Order Cétacés Lesson, 1842: 197.

COMMENTS: When originally proposed, this rank was placed in the Subclass Hydromastologie ou Cétologie (Lesson, 1842 [=Cetacea (Brisson, 1762 part)]) and included the tribes Carnivora (Lesson, 1842 [=Cetacea (Brisson, 1762 part)]) and Vermivora (Lesson, 1842 [=Cetacea (Brisson, 1762 part)]).

Sub Classe Hydromastologie Lesson, 1842: 197.

COMMENTS: When originally proposed, this rank was placed in the Class Mammalia (Linnaeus, 1758) and included the Order Cétacés (Lesson, 1842 [=Cetacea (Brisson, 1762)]).

Sub Classe Cétologie Lesson, 1842: 197.

COMMENTS: When originally proposed, this rank was placed in the Class Mammalia (Linnaeus, 1758) and included the Order Cétacés (Lesson, 1842 [=Cetacea (Brisson, 1762)]).

Natantia Newman, 1843: 35.

COMMENTS: Rank not specified. Also included other aquatic animals. Synonymised within Cete by McKenna and Bell (1997: 366).

HOMONYMS:

Order Natantia Illiger, 1811, sirenia and whales of the Class Mammalia. See individual entry.

Suborder Autoceta Haeckel, 1866: clix.

COMMENTS: When originally proposed, this rank was placed in the Order Cetacea (Brisson, 1762) and included the families Delphinida (Haeckel, 1866 [=Delphinidae J. Gray, 1821]), Hyperoodonta (Haeckel, 1866 [=Ziphiidae J. Gray, 1865d]), Monodonta (Haeckel, 1866: clix [=Monodontidae J. Gray, 1821: 310]), Physeterida (Haeckel, 1866 [=Odontoceti (Flower, 1867)]) and Balaenida (Haeckel, 1866 [=Balaenidae (J. Gray, 1821)]). This name was revisited by Geisler and Sanders (2003: 27, 65) who recognised it at ordinal rank.

Grand Seccion Cetacea Ameghino, 1889: xxvi, 653, 883.

COMMENTS: When originally proposed, this rank was placed in the Grandes Ramas Homalodonta (Ameghino, 1889 [=Mammalia (Linnaeus, 1758 part)]) and included the orders † Proteroceta (Ameghino, 1889: xxvi, 883), Odontoceta (Ameghino, 1889) and Mystacoceta (Ameghino, 1889).

Clade Neoceti Fordyce & de Muizon 2001: 188.

COMMENTS: When originally proposed, this clade was equal to crown Cetacea (Brisson, 1762), and at equivalent rank to Autoceta (Haeckel, 1866), and included the Mysticeti (J. Gray, 1864c) and Odontoceti (Flower, 1867) only to the exclusion of the Archaeoceti (Flower, 1883: 182). Clade recognised within Cetacea by Bianucci and Landini (2007: 45), Fitzgerald (2010: 370) and Lambert *et al.* (2010: 105), and within the unranked clade Pelagiceti (Uhen, 2008a: 591; 2008b: 434) by Steeman (2010: 65). Clade also discussed by Uhen (2010: 190) and Fordyce (2009: 759) who used it as an alternative to Autoceta.

Clade Pelagiceti Uhen, 2008a: 589, 591.

COMMENTS: When originally proposed, this clade was placed in the Cetacea (Brisson, 1762) and included the common ancestor of † Basilosauridae (Cope, 1868a: 144 *sensu* Uhen, 1998: 30; Uhen, 2004: 11), Neoceti (Fordyce & de Muizon 2001) and all its descendants.

Parvorder Mysticeti J. Gray, 1864

Section Mysticete J. Gray, 1864c: 198.

COMMENTS: When originally proposed, this rank was placed in the Suborder Cete (Linnaeus, 1758 [=Cetacea (Brisson, 1762)]) and included the families Balaenidae (J. Gray, 1821) and Balaenopteridae (J. Gray, 1864c). The current spelling of Mysticeti was formalised by Cope (1869: 14). The original description of this name has typically been overlooked in preference to that of Flower (1865a: 388) by authors including Simpson (1945: 104) and Mead and Brownell (2005: 723); or Cope (1891: 69) by authors including McKenna and Bell (1997: 374) and Bouetel and de Muizon (2006: 323). J. Gray (1864c: 198) recognised as the author of the name by Fordyce (2009: 759) and Garrison *et al.* (2012: 4), which is followed here. This name has been recognised at various ranks including suborder (e.g. Flower, 1865a: 388; Cope, 1891; Simpson 1945: 104), infraorder (e.g. Boutel & de Muizon, 2006: 323), and parvorder (e.g. McKenna & Bell, 1997: 374).

Tribe Vermivora Lesson, 1842: 201.

TYPE GENUS: Not based on a genus group name.

COMMENTS: When originally proposed, this rank was placed in the Order Cétacés (Lesson, 1842 [=Cetacea (Brisson, 1762)]) and included the Family Balaenae (Lesson, 1842 [=Family Balaenidae (J. Gray, 1821)]).

Suborder Mysticete Flower, 1865a: 388.

COMMENTS: When originally proposed, this rank was placed in the Order Cetacea (Brisson, 1762), as an alternative to Balaenoidea (Flower, 1865a [=Mysticeti (J. Gray, 1864c)]) and included the families Balaenidae (J. Gray, 1821) and Balaenopteridae (J. Gray, 1864c). Recognised as the Section Mysticete by J. Gray (1866b: 61, 68). Placed at the rank Parvorder by McKenna and Bell (1997: 374), who acknowledge the author as Cope (1891: 69). Suborder Mysticeti rank with the author as Flower (1865a: 388) recognised by Mead and Brownell (2005: 723).

Suborder Mystacoceti Flower, 1867: 110.

COMMENTS: When originally proposed, this rank was placed in the Order Cetacea (Brisson, 1762), as an alternative to Balaenoidea (Flower, 1865a [=Balaenidae (J. Gray, 1821)]), and included the families Balaenidae (J. Gray, 1821) and Balaenopteridae (J. Gray, 1864c). Rank recognised by

Iredale and Troughton (1934: ix, 55). Synonymised within Mysticeti by McKenna and Bell (1997: 374).

Suborder Mysticeti Cope, 1869: 14.

COMMENTS: When originally proposed placed in the Order Cetacea (Brisson, 1762) and included the Family Balaenidae (J. Gray, 1821). Both Mysticeti and Mysticete are mentioned. Name listed by Fordyce (2009: 759) but not recognised at a particular rank.

Suborder Anodontocete A. Scott, 1873: 63.

COMMENTS: When originally proposed, this rank was placed in the Order Cetacea (Brisson, 1762) and included the Microzoophaga (A. Scott, 1873 [=Mysticeti (J. Gray, 1864c)]). Synonymised within Mysticeti by McKenna and Bell (1997: 374).

Microzoophaga A. Scott, 1873: 63.

COMMENTS: When originally proposed, this rank was placed in the Suborder Anodontocete (A. Scott, 1873 [=Mysticeti (J. Gray, 1864c)]) and included the families Balaenopteridae (J. Gray, 1864c), Megapteridae (J. Gray, 1868c [=Balaenopteridae (J. Gray, 1864c)]), Agaphelidae (J. Gray, 1870b [=Balaenopteridae (J. Gray, 1864c)]) and Balaenidae (J. Gray, 1821). Synonymised within Mysticeti by McKenna and Bell (1997: 374).

Order Mystacoceta Ameghino, 1889: xxvi, 883, 888.

COMMENTS: When originally proposed, this rank was placed in the Grand Seccion Cetacea (Ameghino, 1889) and included the Family Balaenidae.

Suborder Mysticeti Cope, 1891: 69.

COMMENTS: When originally proposed, this rank was placed in the Order Cetacea (Brisson, 1762). Recognised as Parvorder Mysticeti by McKenna and Bell (1997: 374) and infraorder by Bouetel and de Muizon (2006: 323).

Order Mysticeta Haeckel, 1895: 566, 572.

COMMENTS: When originally proposed, this rank was placed with Mystacoceta (Haeckel, 1895 [=Mysticeti (J. Gray, 1864c)]) in the Legion Cetomorpha (Haeckel, 1895 [=Cetacea (Brisson, 1762)]) and included the families † Archibalaenae (Protobalaenida) (Haeckel, 1895: 566 [=Balaenidae (Gray, 1821)]), † Ogmobalaenae (Balaenopterida) (Haeckel, 1895: 566 [=Balaenopteridae (J. Gray, 1864c)]) and Lionbalaenae (Eubalaenida) (Haeckel, 1895: 566 [=Balaenidae (J. Gray, 1821)]). Synonymised within Mysticeti by McKenna and Bell (1997: 374).

Order Mystacoceta Haeckel, 1895: 566.

COMMENTS: When originally proposed, this rank was placed with Mysticeta (Haeckel, 1895 [=Mysticeti (J. Gray, 1864c)]) in the Legion Cetomorpha (Haeckel, 1895 [=Cetacea (Brisson, 1762)]) and included the families † Archibalaenae (Protobalaenida) (Haeckel, 1895: 566 [=Balaenidae (J. Gray, 1821)]), † Ogmobalaenae (Balaenopterida) (Haeckel, 1895: 566 [=Balaenopteridae (J. Gray, 1864c)]) and Lionbalaenae (Eubalaenida) (Haeckel, 1895: 566 [=Balaenidae (J. Gray, 1821)]). Synonymised within Mysticeti by McKenna and Bell (1997: 374).

Suborder Mystacoceti Imamura, 1961: 135.

COMMENTS: The placement and composition of this rank were not observed. Synonymised within Mysticeti by McKenna and Bell (1997: 374).

Infraorder Crenaticeti Mitchell, 1989: 2219, 2220, 2232.

COMMENTS: When originally proposed as a new rank it was placed in the Suborder Mysticeti (J. Gray, 1864c) and included the Family † Llanocetidae (Mitchell, 1989: 2219, 2220). Name also spelt Crenataceti on page 2232. Rank synonymised within Mysticeti (Cope, 1891) by McKenna and Bell (1997: 374).

Infraorder Chaeomysticeti Mitchell, 1989: 2219, 2232.

COMMENTS: When originally proposed as a new rank it was placed in the Suborder Mysticeti (J. Gray, 1864c) and included the superfamilies Balaenopteroidea (J. Gray, 1868c), Eschrichtioidea (from Eschrichtiidae) (Ellerman & Morrison Scott, 1951: 713) and Balaenoidea (J. Gray, 1868c). Synonymised within Mysticeti by McKenna and Bell (1997: 374), but recognised at unknown rank within Mysticeti by Bisconti (2012: 879).

Order Mysticetiformes Kinman, 1994: 38.

COMMENTS: When originally proposed, this rank was placed in the Class Mammalea (Kinman, 1994 [=Mammalia (Linnaeus, 1758]). Synonymised within Mysticeti by McKenna and Bell (1997: 374).

Clade Balaenomorpha Geisler & Sanders, 2003: 23, 27.

COMMENTS: When originally proposed, this clade was placed in the Infraorder Chaeomysticeti (Mitchell, 1989 [=Mysticeti (J. Gray, 1864c)]) and included the † Family Cetotheriidae (Brandt, 1872: 116), and the superfamilies Balaenoidea (Gray, 1868 [=Balaenidae (J. Gray, 1821)]) and Balaenopteroidea (Gray, 1868c [=Balaenopteridae (J. Gray, 1864c)]). Name subsequently recognised by Geisler *et al.* (2011: 6).

Clade Plicogulae Geisler *et al.*, 2011: 6, 28.

COMMENTS: When originally proposed, this clade was placed in the Balaenomorpha (Geisler & Sanders, 2003) and included the superfamily Balaenopteroidea (J. Gray, 1868c), that contained the families Balaenopteridae (J. Gray, 1864c), Eschrichtiidae (Ellerman & Morrison-Scott, 1951: 713), and

Neobalaenidae (J. Gray, 1873c). Further support for this clade was provided by Sasaki *et al.* (2005: 82) and Gatesy *et al.* (2012: 488).

Family Neobalaenidae J. Gray, 1873

Family Neobalaenidae J. Gray, 1873c: 108.

TYPE GENUS: *Neobalaena* J. Gray, 1870d [=*Caperea* J. Gray, 1864c].

COMMENTS: When originally proposed, this rank was placed in the Cetacea (Brisson, 1762) and included the genera *Neobalaena* J. Gray, 1870d [=*Caperea* J. Gray, 1864c]; and *Caperea* J. Gray, 1864c. Synonymised within the Family Balaenidae by McKenna and Bell (1997: 378), but recognised at subfamily rank by Rice (1984: 465) and family rank by Miller (1923: 21), Mead and Brownell (1993: 351; 2005: 726), Barnes and Noble (1984: 17) and most recent authors until Fordyce and Marx (2012: 1) placed *Caperea* within the † Family Cetotheriidae (Brandt, 1872: 116). The placement of *Caperea* within the monotypic Family Neobalaenidae is done with the qualification that future work may corroborate the referral of *Caperea* to the † Family Cetotheriidae.

FUTURE TAXONOMIC RESEARCH: The hypothesis that *Caperea* is a cetotheriid requires additional testing and confirmation.

† Family Cetotheriidae Miller, 1923: 21, 40.

TYPE GENUS: † *Cetotherium* Brandt, 1843a: 241, 1843b: 270; 1843c: 148.

COMMENTS: When originally proposed, this rank was placed in the Suborder Mysticeti (J. Gray, 1864c) and included † *Cetotherium* Brandt, 1843a: 241, 1843b: 270; 1843c: 148; and allied extinct genera. Name synonymised within the † Family Cetotheriidae Brandt, 1872 by McKenna and Bell (1997: 375).

Caperea J. Gray, 1864

Caperea J. Gray, 1864c: 202.

TYPE SPECIES: *Balaena (Caperea) antipodarum* J. Gray, 1864c [=*Caperea marginata* (J. Gray, 1846c)] by monotypy.

COMMENTS: Described as a subgenus of *Balaena* Linnaeus, 1758: 75, but elevated to genus rank by J. Gray (1864d: 349; 1866b: 101). Placed in Balaenidae by Iredale and Troughton (1934: ix, 56), Hershkovitz (1966: 183), Wakefield (1967b: 274), Bannister (1988a: 215) and McKenna and Bell (1997: 379), but placed within the Family Neobalaenidae by Miller (1923: 21), Mead and Brownell (1993: 351; 2005: 726) and most authors until it was included in the † Family Cetotheriidae by Fordyce and Marx (2012: 1). Evolutionary history discussed by Fitzgerald (2012: 976).

Neobalaena J. Gray, 1870d: 154, 155; Figs. 1–2.

TYPE SPECIES: *Balaena marginata* J. Gray, 1846c [=*Caperea marginata* (J. Gray, 1846c)] by original designation.

COMMENTS: Synonymised within *Caperea* by Iredale and Troughton (1934: 56), Hershkovitz (1966: 183), and Mead and Brownell (2005: 726).

Caperea marginata (J. Gray, 1846)

Pygmy Right Whale

Balaena marginata J. Gray, 1846c: 48.

TYPE LOCALITY: 'Inhab. W. Australia', Australia. The type material was stated as three baleen plates from Swan River, Western Australia by Bannister (1988a: 215).

COMMENTS: Included in *Neobalaena* by J. Gray (1870d: 154). Transferred to *Caperea* by Iredale and Troughton (1934: ix, 56), Wakefield (1967b: 274), Bannister (1988a: 215) and Mead and Brownell (1993: 351). Species reviewed by Baker (1985: 345).

Balaena (Caperea) antipodarum J. Gray, 1864c: 202; Fig. 2.

TYPE LOCALITY: South Pacific, New Zealand, Otago.

COMMENTS: Recognised as a species within *Balaena* by Iredale and Troughton (1934: ix, 56). Not considered by Bannister (1988a: 215). Synonymised within *marginata* by Hershkovitz (1966: 185), and Mead and Brownell (1993: 351).

HOMONYMS:

Balaena antipodarum J. Gray, 1843d, the Southern Right Whale of the Class Mammalia (Order Artiodactyla, Family Balaenidae). Name is a junior synonym of *Eubalaena australis* (Desmoulins, 1822a). See individual entry.

Balaena antipodarum Tomilin, 1957, the Fin Whale of the Class Mammalia (Order Artiodactyla, Family Neobalaenidae). Name is a junior synonym of *Balaenoptera physalus* (Linnaeus, 1758). See individual entry.

Family Balaenidae J. Gray, 1821

Family Balanadae [sic] J. Gray, 1821: 310.

TYPE GENUS: *Balaena* Linnaeus, 1758: 75.

COMMENTS: When originally proposed, this rank was placed in the Order Carnivorae (J. Gray, 1821 [=Cetacea (Brisson, 1762)]) and included the genera *Balaena* Linnaeus, 1758: 75; *Physalus* J. Gray, 1821 [=*Balaenoptera* Lacépède, 1804]; and *Boops* J. Gray, 1821 [=*Balaenoptera* Lacépède, 1804]. Commonly included *Caperea*, which is placed in a separate family, Neobalaenidae, following Barnes and McLeod (1984: 25) and Mead and Brownell (1993: 351).

Family Balaenidae J. Gray, 1825a: 340.

TYPE GENUS: *Balaena* Linnaeus, 1758: 75.

COMMENTS: Corrected spelling of Balanadae (J. Gray, 1821). When originally proposed, this rank was placed in the Order Cetae (J. Gray, 1825a [=Placentalia (Bonaparte, 1838 part)]) and included the tribes Balaenina (J. Gray, 1825a [=Balaenidae (J. Gray, 1821)]) and Physeterina (J. Gray, 1825a [=Physeteridae (J. Gray, 1821)]). Synonymised within Balaenidae by McKenna and Bell (1997: 378).

Tribe Balaenina J. Gray, 1825a: 340.

TYPE GENUS: *Balaena* Linnaeus, 1758: 75.

COMMENTS: When originally proposed, this rank was placed in the Family Balaenidae (J. Gray, 1821) and included the genera *Balaena* Linnaeus, 1758: 75 and *Balaenoptera* Lacépède, 1804. Synonymised within Balaenidae by McKenna and Bell (1997: 378).

Family Balaenae Lesson, 1842: 201.

TYPE GENUS: *Balaena* Linnaeus, 1758: 75.

COMMENTS: When originally proposed, this rank was placed in the Order Cetaces (Lesson, 1842 [=Cetacea (Brisson, 1762)]) and included the genera *Balaenoptera* Lacépède, 1804; and *Balaena* Linnaeus, 1758: 75. Does not appear to have been recognised by other authors.

Family Balaenodea Giebel, 1855: ix, 76.

TYPE GENUS: *Balaena* Linnaeus, 1758: 75.

COMMENTS: When originally proposed, this rank was placed in the Order Cetacea (Brisson, 1762) and included the genera *Balaena* Linnaeus, 1758: 75; and *Balaenoptera* Lacépède, 1804. Synonymised within Balaenidae by McKenna and Bell (1997: 378), who gave the year of publication as 1855.

Suborder Balaenoidea Flower, 1865a: 388.

COMMENTS: When originally proposed, this rank was placed in the Order Cetacea (Brisson, 1762), as an alternative to Mysticeti (J. Gray, 1864c), and included the families Balaenidae (J. Gray, 1821) and Balaenopteridae (J. Gray, 1864c). Recognised as a suborder by Gray (1868c: 1) and as a superfamily by Mitchell (1989: 2231) (who gave the author as Flower, 1864 =1865), Steeman (2007: 880) (who gave the author as Brandt 1873a: i, 17) and Rice (2009: 235). Synonymised within Mysticeti, with the author as Cope (1891: 69), by McKenna and Bell (1997: 374).

HOMONYMS:

Balaenoidea J. Gray, 1868c, whales of the Class Mammalia (Order Artiodactyla). Rank is a synonym of the Family Balaenidae. See individual entry.

Family Balaenida Haeckel, 1866: clix.

TYPE GENUS: *Balaena* Linnaeus, 1758: 75.

COMMENTS: When originally proposed, this rank was placed in the Suborder Autoceta (Haeckel, 1895 [=Cetacea (Brisson, 1762)]) and included the genera *Balaena* Linnaeus, 1758: 75; and *Balaenoptera* Lacépède, 1804. Synonymised within Balaenidae by McKenna and Bell (1997: 378).

Suborder Balaenoidea J. Gray, 1868c: 1.

COMMENTS: When originally proposed, this rank was placed in the Section Mysticete (Flower, 1865a [=Mysticeti (J. Gray, 1864c)]) and included the Family Balaenidae (J. Gray, 1821). Synonymised within Balaenidae by McKenna and Bell (1997: 378).

HOMONYMS:

Suborder Balaenoidea Flower, 1865a, whales of the Class Mammalia (Order Artiodactyla). Rank is a synonym of the Infraorder Mysticeti. See individual entry.

Eubalaenida Haeckel, 1895: 566.

TYPE GENUS: *Eubalaena* J. Gray, 1864c.

COMMENTS: Placed, with Family Lionbalaenae (Haeckel, 1895), in the Order Mysticeta (Haeckel, 1895 [=Mysticeti (J. Gray, 1864c)]) and included the genera *Eubalaena* J. Gray, 1864c; † *Balaenotus* Van Bénéden, 1872: 13 [=† *Balaenula* Van Bénéden, 1872: 11]; and *Balaena* Linnaeus, 1758: 75. Synonymised within Balaenidae by McKenna and Bell (1997: 378).

† Family Archibalaenae (Protobalaenida) Haeckel, 1895: 566.

TYPE GENUS: † *Protobalaena* Du Bus, 1867: 573.

COMMENTS: When originally proposed, this rank was placed in the Order Mysticeta (Haeckel, 1895 [=Mysticeti (J. Gray, 1864c)]) and included the genera † *Protobalaena* Du Bus, 1867: 573; and † *Plesiocetus* Van Bénéden, 1859: 139. Synonymised within the Family Balaenidae by McKenna and Bell (1997: 378) who suggested it is a *nomen nudum*.

Family Lionbalaenae (Eubalaenida) Haeckel, 1895: 566.

TYPE GENUS: Not based on a genus group name.

COMMENTS: When originally proposed, this rank was placed in the Order Mysticeta (Haeckel, 1895 [=Mysticeti (J. Gray, 1864c)]) and included the genera † *Balaenotus* Van Bénéden, 1872: 13 and *Balaena* Linnaeus, 1758: 75.

Eubalaena J. Gray, 1864

Eubalaena J. Gray, 1864c: 201.

TYPE SPECIES: *Balaena australis* Desmoulins, 1822a [=*Eubalaena australis* (Desmoulins, 1822a)] by monotypy.

COMMENTS: Genus synonymised within *Balaena* Linnaeus, 1758: 75 by Iredale and Troughton (1934: 55), but recognised by Hershkovitz (1966: 185), who included *australis* (Desmoulins, 1822a) and *japonica* (Lacépède, 1818: 469, 473) as subspecies within *glacialis* (O. Müller, 1776: 7). More recently *Eubalaena* was synonymised within

Balaena by McKenna and Bell (1997: 378) and Rice (1998: 61), but has typically been recognised by other recent authors. The taxa *glacialis* and *australis* were recognised within *Eubalaena* by Mead and Brownell (1993: 349), while Rosenbaum *et al.* (2000: 1800) and Gaines *et al.* (2005: 540) also gave support for the recognition of *japonica*, which was followed by Mead and Brownell (2005: 723).

Hunterus J. Gray, 1864d: 349.

TYPE SPECIES: *Hunterus temminckii* J. Gray, 1864d [=*Eubalaena australis* (Desmoulins, 1822a)] by monotypy.
COMMENTS: Synonymised within *Balaena* by Iredale and Troughton (1934: 55), and McKenna and Bell (1997: 378), and within *Eubalaena* by Hershkovitz (1966: 186), and Mead and Brownell (1993: 349; 2005: 723).

Macleayius J. Gray, 1865e: 588, 589; Figs. 1–2.

TYPE SPECIES: *Macleayius australiensis* J. Gray, 1865e [=*Eubalaena australis* (Desmoulins, 1822a)] by monotypy.
COMMENTS: Synonymised within *Balaena* by Iredale and Troughton (1934: 55), and within *Eubalaena* by Hershkovitz (1966: 186) and subsequent authors.

Macleayanus de Marschall, 1873: 8.

TYPE SPECIES: Emendation of *Macleayius* J. Gray, 1865e.
COMMENTS: Synonymised within *Macleayius* by Palmer (1904: 391), within *Balaena* by Iredale and Troughton (1934: 55), and McKenna and Bell (1997: 378), and within *Eubalaena* by Hershkovitz (1966: 186).

Halibalaena J. Gray, 1873b: 140.

TYPE SPECIES: Φ *Balaena britannica* J. Gray, 1870c: 200. [=Φ *Eubalaena glacialis* O. Müller, 1776: 7] by monotypy.
COMMENTS: Synonymised within *Balaena* by Iredale and Troughton (1934: 55), and McKenna and Bell (1997: 378), and within *Eubalaena* by Hershkovitz (1966: 186) and Mead and Brownell (1993: 349; 2005: 723).

Eubalaena australis (Desmoulins, 1822)

Southern Right Whale

Balaena australis Desmoulins, 1822a: 161.

TYPE LOCALITY: Algoa Bay, Cape of Good Hope, South Africa.
COMMENTS: Placed in *Eubalaena* at the species rank by J. Gray (1864c: 202) and Flower, 1865a: 394). Recognised as a subspecies of *Eubalaena glacialis* by Hershkovitz (1966: 186), or of *Balaena glacialis* by Rice (1998: 64), but recognised at species rank by Wakefield (1967b: 273), Mead and Brownell (1993: 349; 2005: 723), Rosenbaum *et al.* (2000: 1793) and Gaines *et al.* (2005: 537). Reviewed by Cummings (1985a: 275).

Balaenoptera antarctica Lesson, 1828d: 391.

TYPE LOCALITY: New Zealand.
COMMENTS: Synonymised within *australis* by J. Gray (1866b: 92), Hershkovitz (1966: 188) and Mead and Brownell (1993: 349; 2005: 723).

Balaena mysticetus antarctica Schlegel, 1841: 37.

TYPE LOCALITY: Unknown.
COMMENTS: Synonymised within *australis* by Hershkovitz (1966: 188).

Balaena antipodum J. Gray, 1843d: 183; Plate 1.

TYPE LOCALITY: New Zealand.
COMMENTS: Name spelt '*Antipodum*' on page 183 and as '*antipodarum*' in the text for Figure 1. Taxon recognised by Iredale and Troughton (1934: ix, 56), but synonymised within *australis* by Hershkovitz (1966: 186) and Mead and Brownell (1993: 349; 2005: 723).
HOMONYMS:
Balaena antipodarum J. Gray, 1864c, the Pygmy Right Whale of the Class Mammalia (Order Artiodactyla, Family Neobalaenidae). Name is a junior synonym of *Caperea marginata* (J. Gray, 1846c). See individual entry.
Balaena antipodarum Tomilin, 1957, the Fin Whale of the Class Mammalia (Order Artiodactyla, Family Neobalaenidae). Name is a junior synonym of *Balaenoptera physalus* (Linnaeus, 1758). See individual entry.

Hunterus Temminckii J. Gray, 1864d: 349.

TYPE LOCALITY: Cape of Good Hope, South Africa.
COMMENTS: Recognised by J. Gray (1866b: 98). Synonymised within *australis* by Hershkovitz (1966: 188) and Mead and Brownell (1993: 349; 2005: 723).

Macleayius australiensis J. Gray, 1865e: 588; Figs. 1–2.

TYPE LOCALITY: South Pacific, New Zealand.
COMMENTS: Synonymised within *antipodum* by Iredale and Troughton (1934: 56), and within *australis* by Hershkovitz (1966: 189) and Mead and Brownell (1993: 349; 2005: 723).

Family Balaenopteridae J. Gray, 1864

Family Balaenopteridae J. Gray, 1864c: 203.

TYPE GENUS: *Balaenoptera* Lacépède, 1804.
COMMENTS: When originally proposed, this rank was placed in the Section Mysticete (Flower, 1865a [=Mysticeti (J. Gray, 1864c)]) and included the tribes Megapterina (J. Gray, 1864c [=Balaenopteridae (J. Gray, 1864c)]) and Physalina (J. Gray, 1864d [=Balaenopteridae (J. Gray, 1864c)]). Recognised by most authors since its description. McKenna and Bell (1997: 377, 378) recognised two subfamilies, the Balaenopterinae and Eschrichtiinae, whereas Ellerman and Morrison-Scott (1951: 713), and Mead and Brownell (2005: 724, 725) recognised these as distinct families. Superfamily

rank (Balaenopteroidea) has been recognised by Mitchell (1989: 2231) and Steeman (2007: 880) and Rice (2009: 235). Molecular data by authors including McGowen *et al.* (2009: 896, 897) and Hassanin *et al.* (2012: 37, 39) indicate that *Eschrichtius* J. Gray, 1864d: 350 is nested within the Family Balaenopteridae, rather than separated in its own subfamily or family, and is most closely related to *Megaptera novaeangliae* and *Balaenoptera physalus*. The subfamilies Megapterinae and Balaenopterinae were recognised by Rice (1998: 67, 69) but this does not appear to have been widely adopted, and indeed molecular data indicate that *Megaptera* forms a clade within *Balaenoptera* (Nishida *et al.*, 2007: 727, 730; McGowen *et al.*, 2009: 896, 897; Hassanin *et al.*, 2012: 37, 39). Hassanin *et al.* (2012: 37, 39) resurrected the genera *Pterobalaena* Eschricht, 1849a and *Rorqualus* F. Cuvier, 1836b. Taxonomic and evolutionary history of fossil and modern balaenopteroid mysticetes reviewed by Deméré *et al.* (2005: 99).

FUTURE TAXONOMIC RESEARCH: The phylogeny of the Family Balaenopteridae needs to be thoroughly reviewed, using morphology and nuclear sequences to test whether the genera *Pterobalaena* (for the species *acutorostrata* and *bonaerensis*) and *Rorqualus* (for the species *borealis*, *brydei*, *edeni*, *musculus* and *omurai*), leaving only *B. physalus* in the genus *Balaenoptera*, should be applied as proposed by Hassanin *et al.* (2012: 37).

Tribe Megapterina J. Gray, 1864c: 205.

TYPE GENUS: *Megaptera* J. Gray, 1846b.

COMMENTS: When originally proposed, this rank was placed in the Family Balaenopteridae (J. Gray, 1864c) and included the genera *Megaptera* J. Gray, 1846b; *Benedenia* J. Gray, 1864c [=*Balaenoptera* (Lacépède, 1804)]; *Physalus* J. Gray, 1821 [=*Balaenoptera* Lacépède, 1804]; *Sibbaldus* J. Gray, 1864c [=*Balaenoptera* Lacépède, 1804]; and *Balaenoptera* Lacépède, 1804. Subfamily rank recognised by Flower (1865a: 391), Zeigler *et al.* (1997: 115), Rice (1998: 67) and Bisconti (2008: 174). Synonymised within the Subfamily Balaenopterinae by McKenna and Bell (1997: 377).

Tribe? Physalina J. Gray, 1864c: 211.

TYPE GENUS: *Physalus* J. Gray, 1821 [=*Balaenoptera* Lacépède, 1804].

COMMENTS: When originally proposed, this rank was placed in the Family Balaenopteridae (J. Gray, 1864c) and included the genera *Benedenia* J. Gray, 1864d [=*Balaenoptera* Lacépède, 1804]; *Physalus* J. Gray, 1821 [=*Balaenoptera* Lacépède, 1804]; *Sibbaldus* J. Gray, 1864c [=*Balaenoptera* Lacépède, 1804]; and *Balaenoptera* Lacépède, 1804. Synonymised within the Subfamily Balaenopterinae by McKenna and Bell (1997: 377).

Subfamily Megapterinae Flower, 1865a: 391.

TYPE GENUS: *Megaptera* J. Gray, 1846b.

COMMENTS: When originally proposed, this rank was placed in the Family Balaenopteridae (J. Gray, 1864c) and included the genus *Megaptera* J. Gray, 1846b. Subfamily rank recognised by Fordyce and Barnes (1994: 428) who have the author as J. Gray (1866b). Synonymised within the Subfamily Balaenopterinae by McKenna and Bell (1997: 377).

Suborder Balaenopteroidea J. Gray, 1868c: 2.

COMMENTS: When originally proposed, this rank was placed in the Section Mysticete (Flower, 1865a [=Mysticeti (J. Gray, 1864c)]) and families Megapteridae (J. Gray, 1868c [=Balaenopteridae (J. Gray, 1864c)]), Physalinidae (J. Gray, 1868c [=Balaenopteridae (J. Gray, 1864c)]) and Balaenopteridae (J. Gray, 1864c). Superfamily rank recognised by Mitchell (1989: 2231) and Steeman (2007: 880), who all give the author as J. Gray (1868c: 2), and Rice (2009: 235), but not recognised by McKenna and Bell (1997: 377), who placed it within the family Balaenopteridae.

Family Physalinidae J. Gray, 1868c: 2.

TYPE GENUS: *Physalus* J. Gray, 1821 [=*Balaenoptera* Lacépède, 1804].

COMMENTS: When originally proposed, this rank was placed in the Suborder Balaenopteroidea (J. Gray, 1868c) and included the genera *Benedenia* J. Gray, 1864d [=*Balaenoptera* Lacépède, 1804]; *Physalus* J. Gray, 1821 [=*Balaenoptera* Lacépède, 1804]; *Cuvierius* J. Gray, 1866b [=*Balaenoptera* Lacépède, 1804]; *Rudolphius* J. Gray, 1866b [=*Balaenoptera* Lacépède, 1804]; and *Sibbaldius* Flower, 1865a [=*Balaenoptera* Lacépède, 1804].

Family Megapteridae J. Gray, 1868c: 2.

TYPE GENUS: *Megaptera* J. Gray, 1846b.

COMMENTS: When originally proposed, this rank was placed in the Suborder Balaenopteroidea (J. Gray, 1868c) and included the genera *Megaptera* J. Gray, 1846b; *Poescopia* J. Gray, 1864c [=*Megaptera* J. Gray, 1846b]; and *Eschrichtius* J. Gray, 1864d: 350. Synonymised within Balaenopteridae by McKenna and Bell (1997: 377).

Family Agaphelidae J. Gray, 1870b: 391.

TYPE GENUS: *Agaphelus* Cope, 1868b: 159 [=*Balaenoptera* Lacépède, 1804].

COMMENTS: When originally proposed, this rank was placed in the Family Balaenidae (J. Gray, 1821) and included the genera *Agaphelus* Cope, 1868b: 159 [=*Balaenoptera* Lacépède, 1804]; and *Rhachianectes* Cope, 1869: 15 [=*Eschrichtius* J. Gray, 1864d: 350]. Synonymised within Balaenopteridae by McKenna and Bell (1997: 377).

Subfamily Balaenopterinae Brandt, 1872: 116.

TYPE GENUS: *Balaenoptera* Lacépède, 1804.

COMMENTS: When originally proposed, this rank was placed in the Family Balaenidae (J. Gray, 1821) and included the genera *Pterobalaena* Eschricht, 1849a [=*Balaenoptera* Lacépède, 1804] and *Kyphobalaena* Eschricht, 1849a [=*Megaptera* J. Gray, 1846b].

Family Ogmobalaenae (Balaenopterida) Haeckel, 1895: 566.

TYPE GENUS: *Balaenoptera* Lacépède, 1804.

COMMENTS: When originally proposed, this rank was placed in the Order Mysticeta (Haeckel, 1895 [=Mysticeti (J. Gray, 1864c)]) and included the genera † *Cetotherium* Brandt, 1843a: 241, 1843b: 270; 1843c: 148; and *Balaenoptera* Lacépède, 1804.

Family Rhachianectidae Weber, 1904: 575.

TYPE GENUS: *Rhachianectes* Cope, 1869: 15 [=*Eschrichtius* J. Gray, 1864d: 350].

COMMENTS: When originally proposed, this rank was placed in the Suborder Mystacoceti (Flower, 1867 [=Mysticeti (J. Gray, 1864c)]) and included the genus *Rhachianectes* Cope, 1869: 15 [=*Eschrichtius* J. Gray, 1864d: 350]. Synonymised within Balaenopteridae by McKenna and Bell (1997: 377).

Family Eschrichtiidae Ellerman & Morrison-Scott, 1951: 713.

TYPE GENUS: *Eschrichtius* J. Gray, 1864d: 350.

COMMENTS: When originally proposed this rank was placed in the Order Cetacea (Brisson, 1762) and included the genus *Eschrichtius* J. Gray, 1864d: 350. Some authors place *Eschrichtius* J. Gray, 1864d: 350 within the family Balaenopteridae but others recognise it in a separate family, Eschrichtiidae, in a clade that also contains the families Neobalaenidae and Balaenopteridae (e.g. Sasaki *et al.*, 2005: 82; Gatesy *et al.*, 2012: 488).

Balaenoptera Lacépède, 1804

Balaenoptera Lacépède, 1804: xxxvi, 114.

TYPE SPECIES: *Balaenoptera gibbar* Lacépède, 1804 [=*Balaenoptera physalus* (Linnaeus, 1758)] by monotypy.

COMMENTS: Recognised by J. Gray (1847d: 89; 1850: 31; 1864d: 226; 1866b: 186). Genus reviewed by Hershkovitz (1966: 151).

Balenoptera Duméril, 1806b: 28.

TYPE SPECIES: Emendation of *Balaenoptera* Lacépède, 1804.

COMMENTS: Synonymised within *Balaenoptera* by Iredale and Troughton (1934: 56) and Hershkovitz (1966: 151).

Catoptera Rafinesque, 1815: 61.

TYPE SPECIES: *Nomen novum* for *Balaenoptera* Lacépède, 1804.

COMMENTS: Synonymised within *Balaenoptera* by Iredale and Troughton (1934: 56), Hershkovitz (1966: 151), and Mead and Brownell (1993: 349; 2005: 724).

Cetoptera Rafinesque, 1815: 219.

TYPE SPECIES: Correction of spelling for *Catoptera* Rafinesque, 1815.

COMMENTS: Emendation of *Captoptera* Rafinesque, 1815. Synonymised within *Balaenoptera* by Iredale and Troughton (1934: 56), and Hershkovitz (1966: 151).

Physalus J. Gray, 1821: 310.

TYPE SPECIES: *Balaena physalus* Linnaeus, 1758 [=*Balaenoptera physalus* (Linnaeus, 1758)] by monotypy.

COMMENTS: Recognised by J. Gray (1866b: 114, 139). Synonymised within *Balaenoptera* by Hershkovitz (1966: 151), and Mead and Brownell (1993: 349; 2005: 724).

HOMONYMS:

Physalus Lacépède, 1804, the Sperm Whale of the Class Mammalia (Order Artiodactyla, Family Physeteridae). Genus is a synonym of *Physeter* Linnaeus, 1758. See individual entry.

Physalus de Blainville, 1830: 103, 'blue-bottle' jelly-fish of the Phylum Cnidaria (Class Hydrozoa, Order Siphonophorae, Family Physalidae). An unjustified emendation of *Physalia* Lamarck, 1801: 355. See Sherborn (1929: 4934) and ICZN (1941: 17).

Boops J. Gray, 1821: 310.

TYPE SPECIES: *Balaena Boops* Linnaeus, 1758 [=*Balaenoptera physalus* (Linnaeus, 1758)] by monotypy.

COMMENTS: Synonymised within *Balaenoptera* by Iredale and Troughton (1934: 56) and Hershkovitz (1966: 151).

HOMONYMS:

Boops Goüan, 1770: 27, fish. *Incertae sedis*.

Boops G. Cuvier, 1814: 91, sea bream and porgy fish of the Class Actinopterygii (Order Perciformes, Family Sparidae). Author sometimes cited as F. Cuvier (1817: 8) or G. Cuvier (1816b: 270). Genus currently recognised. See WoRMS (2012).

Boops Gronow, 1854: 58, bigeye fish of the Class Actinopterygii (Order Perciformes, Family Priacanthidae). Genus is a synonym of *Priacanthus* Oken, 1817: 1182. See Starnes (1988: 154).

Physalis Fleming, 1822b: 206.

TYPE SPECIES: *'Finner'* [=*Balaenoptera physalus* (Linnaeus, 1758)] by monotypy.

COMMENTS: Synonymised within *Balaenoptera* by Hershkovitz (1966: 151).

HOMONYMS:

Physalis Lamarck, 1816: 480, 'blue-bottle' jelly-fish of the Phylum Cnidaria (Class Hydrozoa, Family Physalidae). An unjustified emendation of *Physalia* Lamarck, 1801: 355. See ICZN (1957b: 237).

Rorqual G. Cuvier, 1829a: 298.

TYPE SPECIES: *Balaena boops* Linnaeus, 1758 [=*Balaenoptera physalus* (Linnaeus, 1758)] and *Balaena musculus* Linnaeus, 1758 [=*Balaenoptera musculus* (Linnaeus, 1758)]. See Palmer (1904: 612).

COMMENTS: Synonymised within *Balaenoptera* by McKennna and Bell (1997: 377).

Balenopterus F. Cuvier, 1829b: 518.

TYPE SPECIES: Emendation of *Balaenoptera* Lacépède, 1804.

COMMENTS: Synonymised within *Balaenoptera* by Palmer (1904: 132) and Hershkovitz (1966: 151).

Mysticetus Wagler, 1830: 33.

TYPE SPECIES: *Balaena boops* Linnaeus, 1758 [=*Balaenoptera physalus* (Linnaeus, 1758)] by subsequent designation. See Palmer (1904: 444).

COMMENTS: Synonymised within *Balaenoptera* by Iredale and Troughton (1934: 56) and Hershkovitz (1966: 151).

Rorqual Voight, 1831: 342.

TYPE SPECIES: *Balaena boops* Voigt, 1831 [=*Balaenoptera musculus* (Linnaeus, 1758), not *Balaena boops* Linnaeus, 1758] and *Balaenoptera musculus* (Linnaeus, 1758).

COMMENTS: Described as a subgenus of *Balaena* Linnaeus, 1758: 75. Synonymised within *Balaenoptera* by Iredale and Troughton (1934: 56) and Hershkovitz (1966: 152).

Rorqualus F. Cuvier, 1836b: 303.

TYPE SPECIES: *Balaena boops* Linnaeus, 1758 [=*Balaenoptera musculus* (Linnaeus, 1758), as indicated by Hershkovitz (1966: 152)].

COMMENTS: Synonymised within *Balaenoptera* by Iredale and Troughton (1934: 56), Hershkovitz (1966: 152), and Mead and Brownell (1993: 349; 2005: 724). The species *boops* is the first of the three listed by Cuvier for this genus, and Hershkovitz (1966: 152) indicated that, at least in his usage, the name is a synonym of *musculus* Linnaeus, 1758. Name resurrected by Hassanin *et al.* (2012: 37, 43) for the species *musculus*, *omurai*, *borealis*, *brydei* and *edeni*; the finding that these species form a monophyletic clade was pre-empted by the observations of McGowen *et al.* (2009: 896).

Rorqualis R. Hamilton, 1837: 125.

TYPE SPECIES: *Balaena musculus* Linnaeus, 1758 [=*Balaenoptera musculus* (Linnaeus, 1758)] by tautotypy.

COMMENTS: Synonymised within *Balaenoptera* by Iredale and Troughton (1934: 56).

Ptychocētus Gloger, 1841: xxxiv, 174.

TYPE SPECIES: *Balaenoptera musculus* (Linnaeus, 1758) by subsequent designation. See Hershkovitz (1966: 152).

COMMENTS: Synonymised within *Balaenoptera* by Iredale and Troughton (1934: 57) and Hershkovitz (1966: 152).

Ogmobalaena Eschricht, 1849a: 108.

TYPE SPECIES: 'Furehvaler eller Rorhvaler', a name apparently including all rorquals; Hershkovitz (1966: 152) equated this with *Balaenoptera physalus* (Linnaeus, 1758); although the concept was much broader than this, his designation of *physalus* may be accepted for convenience.

COMMENTS: Synonymised within *Balaenoptera* by Iredale and Troughton (1934: 57) and Hershkovitz (1966: 152).

Pterobalaena Eschricht, 1849a: 108.

TYPE SPECIES: *Pterobalaena minor groenlandica* and *bergensis* Eschricht, 1849a: 109 [=*Balaenoptera acutorostrata* Lacépède, 1804]. When the name is first introduced, on p. 108, *Pterobalaena* is characterised as 'Finhvaler', and Hershkovitz (1966: 152) interpreted this as the Fin Whale, *Balaenoptera physalus* (Linnaeus, 1758: 75), but *Pterobalaena* is part of a dichotomous classification of rorquals *(Ogmobalaena)*, in which it is contrasted with *Kyphobalaena*, 'Pukkelhvaler' (meaning 'humped whales', probably the Humpback Whale, *Megaptera*). That is to say, *Pterobalaena* was designed to include all those rorquals that have noticeable fins; and as the only species named under the heading of the genus is *Pterobalaena minor* with its two 'subspecies' [=*Balaenoptera acutorostrata* Lacépède, 1804], it is clear that the type species of the genus is *acutorostrata*.

COMMENTS: Name also used described by Eschricht (1849b: xi, 56). Synonymised within *Physalus* by J. Gray (1864c: 215) and within *Balaenoptera* by Iredale and Troughton (1934: 57), Hershkovitz (1966: 152), and Mead and Brownell (1993: 349; 2005: 724). Name resurrected by Hassanin *et al.* (2012: 37, 43), for the species *acutorostrata* and *bonaerensis*, who give the type species as *Balaenoptera acutorostrata* Lacépède, 1804 (see above). The recognition of this taxon was predicted by the observations of McGowen *et al.* (2009: 896).

Benedenia J. Gray, 1864c: 211.

TYPE SPECIES: *Benedenia knoxii* J. Gray, 1864c (as *Benedia knoxii*) [=*Balaenoptera physalus* (Linnaeus, 1758)] by monotypy.

COMMENTS: Recognised by J. Gray (1866b: 114, 135). Synonymised within *Balaenoptera* by Iredale and Troughton (1934: 57) and Hershkovitz (1966: 152).

HOMONYMS:

Benedenia Diesing, 1858: 363, parasitic worms of the Phylum Platyhelminthes (Class Monogenea, Order Monopisthocotylea, Family Capsalidae). Currently recognised genus. See Gibson and Bray (2010).

Benedenia Schneider, 1875: xliv, coccidian of the Phylum Myzozoa, Class Conoidasida (Order Eucoccidiorida, Family

Aggregatidae). Genus is a synonym of *Aggregata* Frenzel, 1885: 560. See Upton (2000: 322).

Benedenia Foettinger, 1881: 346, of the Kingdom Protista. *Incertae sedis*.

Benedenia Lehrer, 1976: 197, fresh flies of the Class Insecta (Order Diptera, Family Sarcophagidae). Genus is a synonym of *Leclercqiomyia* Lehrer, 1976: 195.

Sibbaldus J. Gray, 1864c: 222.

TYPE SPECIES: *Balaenoptera laticeps* J. Gray, 1846c (as *Sibbaldus laticeps*) [=*Balaenoptera borealis* Lesson, 1828] by subsequent designation. See Palmer (1904: 630) and Iredale and Troughton (1934: 57).

COMMENTS: Synonymised within *Balaenoptera* by Iredale and Troughton (1934: 57), Hershkovitz (1966: 152), and Mead and Brownell (1993: 349; 2005: 724).

Sibbaldius Flower, 1865a: 391.

TYPE SPECIES: Emendation of *Sibbaldus* J. Gray, 1864c.

COMMENTS: Synonymised within *Sibbaldus* by Palmer (1904: 630), but Barnes and McLeod (1984: 25) recognised the genus *Sibbaldius* for the blue whales, with substantially different cranial morphology from the other four living species of balaenopterines, which are all placed in the genus *Balaenoptera*. They used the spelling *Sibbaldius* (Flower, 1865a: 391) in preference to *Sibbaldus* (Gray, 1864c), an emendation with the proper patronymic suffix later accepted by Gray (1866b: 169).

Cuvierius J. Gray, 1866b: 114, 164.

TYPE SPECIES: *Physalus latirostris* Flower, 1865a [=*Balaenoptera musculus* (Linnaeus, 1758)] by monotypy.

COMMENTS: Synonymised within *Balaenoptera* by Iredale and Troughton (1934: 57), Hershkovitz (1966: 152), and Mead and Brownell (1993: 349; 2005: 724).

HOMONYMS:

Cuvieria Péron, 1807: Plate 30, jellyfish of the Phylum Cnidaria (Class Scyphozoa, Order Leptothecatae, Family Dipleurosomatidae). Currently accepted name. See Schuchert (2010).

Cuvieria Leach, 1814: 435, amphipods of the Subphylum Crustacea (Order Amphipoda, Family Leucothoidae). Genus is a synonym of *Leucothoe* Leach, 1814: 403. See Lowry and Stoddart (2003: 154).

Cuvieria G. Cuvier, 1816c: 22, worms of the Phylum Echinodermata (Class Holothuroidea, Order Dendrochirotida, Family Psolidae). Taxon is a synonym of *Psolus* Jäger, 1833: 9, 20. See Hansson (2010).

Cuvieria Rang, 1827: 322, sea snails of the Class Gastropoda (Order Thecosomata, Family Cavoliniidae). Genus is a synonym of *Cuvierina* Boas, 1886: 131.

Cuvieria Roberts, 1922: 210, falcons of the Class Aves (Order Falconiformes, Family Falconidae). Genus is a synonym of *Falco* Linnaeus, 1758: 83, 88.

Rudolphius J. Gray, 1866b: 170.

TYPE SPECIES: *Balaenoptera laticeps* J. Gray, 1846c [=*Balaenoptera borealis* Lesson, 1828d] by monotypy.

COMMENTS: Described as a subgenus of *Sibbaldius* Flower, 1865a. Synonymised within *Balaenoptera* by Iredale and Troughton (1934: 57) and Hershkovitz (1966: 153).

Swinhoia J. Gray, 1866b: 382.

TYPE SPECIES: *Balaenoptera swinhoii* J. Gray, 1866g [=*Balaenoptera physalus* (Linnaeus, 1758)] by tautonomy.

COMMENTS: Subgenus of *Balaenoptera* Lacépède, 1804. Synonymised within *Balaenoptera* by Iredale and Troughton (1934: 57) and Hershkovitz (1966: 153).

Fabricia J. Gray, 1866b: 382.

TYPE SPECIES: *Balaena rostrata* Fabricius, 1780 [=*Balaenoptera acutorostrata* Lacépède, 1804] by monotypy.

COMMENTS: Described as a subgenus of *Balaenoptera*. Synonymised within *Balaenoptera* by Iredale and Troughton (1934: 57) and Hershkovitz (1966: 153).

HOMONYMS:

Fabricia de Blainville, 1828: 439, worms of the Phylum Annelida (Class Polychaeta, Order Sabellida, Family Sabellidae). Currently accepted name. See Fauchald and Bellan (2012).

Fabricia Latreille, 1829: 510, tachinid flies of the Class Insecta (Order Diptera, Family Tachinidae). Genus is a synonym of *Tachina* Meigen, 1803: 280. See O'Hara and Wood (2004: 325).

Fabricia Meigen, 1838: 250, flies of the Class Insecta (Order Diptera, Family Tachinidae). Genus is a synonym of *Zaira* Robineau-Desvoidy, 1830: 150. See O'Hara and Wood (2004: 111).

Flowerius Lilljeborg, 1867: 11.

TYPE SPECIES: *Balaenoptera gigas* Reinhardt, 1857 [=*Balaenoptera musculus* (Linnaeus, 1758)] by monotypy.

COMMENTS: Synonymised within *Balaenoptera* by Iredale and Troughton (1934: 57) and Hershkovitz (1966: 153).

Agaphelus Cope, 1868b: 159.

TYPE SPECIES: *Balaena gibbosa* Erxleben, 1777 [=*Balaenoptera acutorostrata* Lacépède, 1804] by original designation. See Hershkovitz (1966: 153).

COMMENTS: Synonymised within *Balaenoptera* by Hershkovitz (1966: 153). There appears to be some confusion over the allocation of the type species of this genus. It was based on 'the Scrag Whale of Dudley, *Balaena gibbosa* of Erxleben'. Hershkovitz (1966: 153) give the type as '*Balaena gibbosa* Erxleben, 1777 (=*Balaenoptera acutorostrata* Lacépède)', but Cope also included '"the gray whale" of the coast of California', which he called *Agaphelus glaucus*. Later in the same year, Cope (1868c:

221–223) gave a much more detailed description, explaining that the Scrag Whale had been previously described by Dudley in his 1725 description of the whales found off New England, north-eastern United States of America. More recently Mead and Brownell (2005: 726) suggested that Dudley's 'scrag whale' was the extinct North Atlantic Gray Whale [=*Eschrichtius robustus* (Lilljeborg, 1861a: 602)], although Cope's description seems to suggest a minke whale.

Stenobalaena J. Gray, 1874b: 305.

TYPE SPECIES: *Stenobalaena xanthogaster* J. Gray, 1874b [=*Balaenoptera physalus* (Linnaeus, 1758)] by original designation.

COMMENTS: Synonymised within *Balaenoptera* by Iredale and Troughton (1934: 57) and Hershkovitz (1966: 153).

Dactylaena J. Gray, 1874c: 449.

TYPE SPECIES: *Balaenoptera huttoni* J. Gray, 1874c [=*Balaenoptera bonaerensis* Burmeister, 1867a] by monotypy.

COMMENTS: Described as a subgenus of *Balaenoptera* Lacépède, 1804. Synonymised within *Balaenoptera* by Iredale and Troughton (1934: 57) and Hershkovitz (1966: 153).

Eubalaenoptera Aclogue, 1899: 83.

TYPE SPECIES: *Physalus (Rorqualus) Sibbaldii* J. Gray, 1847d [=*Balaenoptera musculus* (Linnaeus, 1758)] by monotypy.

COMMENTS: Subgenus of *Balaenoptera* Lacépède, 1804. Synonymised within *Balaenoptera* by Hershkovitz (1966: 153).

Balaenoptera acutorostrata Lacépède, 1804

Common Minke Whale

Φ ***Balaenoptera acutorostrata acutorostrata*** Lacépède, 1804

North Atlantic Minke Whale

Φ *Balaenoptera acuto-rostrata* Lacépède, 1804: xxxvii, 134; Plate 4, Fig. 2; Plate 8.

TYPE LOCALITY: 'pris aux environs de la rade de Cherbourg, Mancha, France'.

COMMENTS: *Nomen novum* for *Balaena rostrata* Fabricius, 1780. Reviewed by Omura (1975: 1), Rice (1977: 6), and Stewart and Leatherwood (1985: 91). Nominate subspecies recognised from North Atlantic by Bannister *et al*. (1996: 178). Species name spelt *acutorostrata*, without the hyphen, by Oliver (1922: 563). Two forms have been described from the Antarctic by Williamson (1961: 133) and in SW Pacific waters by Arnold *et al*. (1987: 1), who recognised a dwarf form, that may represent two or three species (e.g. Wada & Numachi, 1991: 125). The occurrence of three unusual specimens from the Antarctic was also discussed by Williamson (1959: 135). Similarly Best (1985: 1) and Kato (1992: 61) described two different forms, including a dwarf form, of minke whales from the Southern Hemisphere (based on baleen colouration, external coloration and morphometric differences). The recognition of a dwarf form was accepted by authors including Arnold *et al*. (1987: 1), Zerbini *et al*. (1996: 333) and subsequent authors. Rice (1998: 70) recognised three subspecies: *acutorostrata* in the North Atlantic, *scammoni* from the North Pacific and an unnamed dwarf subspecies from the lower latitudes of the Southern Hemisphere referred to as the Dwarf Minke Whale. The recognition of a dwarf subspecies from the Southern Hemisphere was supported by Arnold *et al*. (2005: 277). Taxon placed in *Pterobalaena* by Hassanin *et al*. (2012: 37, 43), which is supported by the earlier observations of McGowen *et al*. (2009: 896).

FUTURE TAXONOMIC RESEARCH: Perrin *et al*. (2009: 3) concluded that there is good evidence for the existence of an undescribed dwarf Antarctic subspecies of this species; it is distinct morphologically and genetically, and there is some evidence for geographic overlap in the South Atlantic – if this is so, then presumably a new, distinct species of minke whale awaits description.

Φ *Balaena gibbosa* Erxleben, 1777: 610.

TYPE LOCALITY: Unknown.

COMMENTS: The description referred to it as the 'Scrag Whale of Dudley'. The placement of this taxon is uncertain as it was associated with *acutorostrata* by Hershkovitz (1966: 153), but Mead and Brownell (2005: 726) suggested that Dudley's 'scrag whale' was the extinct North Atlantic J. Gray Whale [=*Eschrichtius robustus* (Lilljeborg, 1861a: 602)], although Cope's description seems to suggest a minke whale.

HOMONYMS:

Balaena gibbosa J. Gray, 1843c, the Humpback Whale of the Class Mammalia (Order Artiodactyla, Family Balaenopteridae). Name is a synonym of *Megaptera novaeangliae* (Borowski, 1781). See individual entry.

Balaena gibbosa Cope, 1868b, the Dwarf Minke Whale of the Class Mammalia (Order Artiodactyla, Family Balaenopteridae). Name appears to be a synonym of *Balaenoptera acutorostrata* Lacépède, 1804. See individual entry.

Φ *Balaena rostrata* Fabricius, 1780: 40.

TYPE LOCALITY: North Atlantic, Greenland Seas.

COMMENTS: Synonymised within *acutorostrata* by Hershkovitz (1966: 153), and Mead and Brownell (2005: 724).

HOMONYMS:

Balaena rostrata O. Müller, 1776: 7, the Northern Bottle-nosed Whale of the Class Mammalia (Order Artiodactyla,

Family Ziphiidae). Taxon is a synonym of Φ *Hyperoodon ampullatus* (Forster, 1770: 18). See Mead and Brownell (2005: 740).

Balaena rostrata Rudolphi, 1822, the Sei Whale of the Class Mammalia (Order Artiodactyla, Family Balaenopteridae). Taxon is a synonym of *Balaenoptera borealis* Lesson, 1828d. See individual entry.

Φ *R.[orqualus] Boops* F. Cuvier, 1836b: 321; Plate 20, Fig. 1.

TYPE LOCALITY: Unknown.

COMMENTS: Included within '*balaena*' on page 308 and '*R.* [*orqualis*]' on page 321. Synonymised within *acutorostrata* by Hershkovitz (1966: 155).

Φ *Balaena minima* Rapp, 1837: 52.

TYPE LOCALITY: *Nomen novum* for *Balaena rostrata* Fabricius, 1780.

COMMENTS: Synonymised within *acutorostrata* by Hershkovitz (1966: 155), and Mead and Brownell (2005: 724).

Φ *Rorqualus Minor* R. Hamilton, 1837: 142; Plate 7.

TYPE LOCALITY: North Atlantic, Firth of Forth, Scotland.

COMMENTS: Synonymised within *acutorostrata* by Hershkovitz (1966: 152).

Φ *Balaena minimus borealis* F. Knox, 1838: 14.

TYPE LOCALITY: North Atlantic, Firth of Forth, Scotland.

COMMENTS: Synonymised within *acutorostrata* by Hershkovitz (1966: 155).

Φ *Balaenoptera Eschrichtii* Rasch, 1845: 123.

TYPE LOCALITY: North Atlantic, Sweden.

COMMENTS: Synonymised within *acutorostrata* by Hershkovitz (1966: 155).

Φ *Pterobalaena minor* Eschricht, 1849a: 109.

TYPE LOCALITY: Unknown.

COMMENTS: Also described by Eschricht (1849b: xv, 169). Synonymised within *acutorostrata* by Hershkovitz (1966: 155).

Φ *Pterobalaena minor groenlandica* Eschricht, 1849a: 109.

TYPE LOCALITY: North Atlantic, Greenland.

COMMENTS: Also described by Eschricht (1849b: Plates 5, 8). Synonymised within *acutorostrata* by Hershkovitz (1966: 155).

Φ *[Pterobalaena minor] bergensis* Eschricht, 1849a: 109.

TYPE LOCALITY: North Atlantic, Norway.

COMMENTS: Synonymised within *acutorostrata* by Hershkovitz (1966: 155).

Φ *Balaenoptera microcephala* J. Gray, 1850: 32.

TYPE LOCALITY: Unknown.

COMMENTS: Based on a manuscript of Brandt, but placed within *rostrata* by J. Gray (1850: 32). Synonymised within *acutorostrata* by Hershkovitz (1966: 156).

Φ *P. [terobalaena] N. [ana] pentadactyla* Barkow, 1862: 17.

TYPE LOCALITY: Unknown.

COMMENTS: Synonymised within *acutorostrata* by Hershkovitz (1966: 156).

Φ *P. [terobalaena] N. [ana] tetradactyla* Barkow, 1862: 17.

TYPE LOCALITY: *Nomen novum* for *Pterobalaena minor* Eschricht, 1849a.

COMMENTS: Synonymised within *acutorostrata* by Hershkovitz (1966: 156).

Φ *Pterobalaena pentadactyla* Flower, 1865a: 394, footnote.

TYPE LOCALITY: Unknown.

COMMENTS: Synonymised within *acutorostrata* by Hershkovitz (1966: 156).

Φ *Balaena gibbosa* Cope, 1868b: 159.

TYPE LOCALITY: North Atlantic.

COMMENTS: Synonymised within *acutorostrata* by Hershkovitz (1966: 157).

HOMONYMS:

Balaena gibbosa J. Gray, 1843c, the Humpback Whale of the Class Mammalia (Order Artiodactyla, Family Balaenopteridae). Name is a synonym of *Megaptera novaeangliae* (Borowski, 1781). See individual entry.

Balaena gibbosa Erxleben, 1777, the Dwarf Minke Whale of the Class Mammalia (Order Artiodactyla, Family Balaenopteridae). Name appears to be a synonym of *Balaenoptera acutorostrata* Lacépède, 1804. See individual entry.

Φ *Agaphelus gibbosus* Cope, 1868c: 224.

TYPE LOCALITY: North Atlantic, Long Beach, New Jersey, United States of America.

COMMENTS: Synonymised within *acutorostrata* by Hershkovitz (1966: 157).

Φ *Sibbaldius mondinii* Capellini, 1877: 423.

TYPE LOCALITY: Adriatic Sea, North Atlantic.

COMMENTS: Synonymised within *acutorostrata* by Hershkovitz (1966: 156).

Φ *Balaenoptera rostrata* Van Bénéden & Gervais, 1880: 146.

TYPE LOCALITY: Graham Land, Southern Shetlands, South Atlantic.

COMMENTS: Synonymised within *acutorostrata* by Hershkovitz (1966: 153).

Φ *Neobalaena marginata* E. Wilson, 1907: 4; Fig. 2.

TYPE LOCALITY: Unknown.
COMMENTS: Synonymised within *acutorostrata* by Hershkovitz (1966: 157).

Φ *Balaena microcephala* Tomilin, 1957: 239.

TYPE LOCALITY: Greenland.
COMMENTS: Synonymised within *acutorostrata* by Hershkovitz (1966: 157).

Φ *Balaenoptera acutorostrata bonaerensis* Deraniyagala, 1960: 84.

TYPE LOCALITY: Derived from *Balaenoptera bonaëerensis* Burmeister, 1868: 707, but apparently does not belong to that species.
COMMENTS: Synonymised within *acutorostrata* by Hershkovitz (1966: 156).

Φ *Balaenoptera acutorostrata scammoni* Deméré, 1986

North Pacific Minke Whale

Balaenoptera acutorostrata scammoni Deméré, 1986: 277, 288.

TYPE LOCALITY: Substitute name for *Balaenoptera davidsoni* Scammon, 1872, which was pre-occupied by *Balaenoptera davidsonii* Cope, 1872b: 30.
COMMENTS: Deméré (1986: 277) placed *Eschrichtius davidsoni* (Cope, 1872b: 30) in *Balaenoptera*, making *B. davidsoni* Scammon, 1872 a junior synonym and erected *Balaenoptera acutorostrata scammoni* as a substitute. Subspecies recognised by Rice (1998: 70) but only tentatively by Perrin *et al.* (2009: 2) as it has only been weakly defined.

Balaenoptera Davidsoni Scammon, 1872: 269.

TYPE LOCALITY: Admiralty Inlet, Puget Sound, Washington, United States of America.
COMMENTS: Changed to subspecies by Tomilin (1957: 274). Synonymised within *acutorostrata* by Hershkovitz (1966: 156). Recognised as a subspecies from North Pacific by Bannister *et al.* (1996: 178) but not Mead and Brownell (2005: 724).
HOMONYMS:
† *Eschrichitus [sic = Eschrichtius] davidsonii* Cope, 1872b: 30, extinct whale of the Class Mammalia (Order Artiodactyla, Family Balaenopteridae).

Physalus antarcticus Hutton, 1874: 316; Plate 16.

TYPE LOCALITY: Otago Head, New Zealand.
COMMENTS: Synonymised within *acutorostrata* by Hershkovitz (1966: 157).

B. [alaenoptera] Racovitzai Lahille, 1905: 74.

TYPE LOCALITY: Antarctica.
COMMENTS: Name based on two Antarctic specimens of *acutorostrata* described by Racovitza (1903: 57). Synonymised within *acutorostrata* by Hershkovitz (1966: 157).

Balaenoptera acutorostrata thalmaha Deraniyagala, 1963a: 82.

TYPE LOCALITY: Sri Lanka.
COMMENTS: Synonymised within *acutorostrata* by Hershkovitz (1966: 157).

Balaenoptera bonaerensis Burmeister, 1867

Antarctic Minke Whale

B. [alaenoptera] bonaërensis Burmeister, 1867a: xxiv.

TYPE LOCALITY: South Atlantic, near Belgrano, Río de La Plata at mouth of Ricachuelo Medrano, Buenos Aires, Argentina.
COMMENTS: Taxon also described by Burmeister (1867b: 310; 1868: 707). Synonymised within *acutorostrata* by Hershkovitz (1966: 156). Recognised as a subspecies from southern hemisphere by Bannister *et al.* (1996: 178) and at the species rank by Rice (1998: 71), Mead and Brownell (2005: 724), and Perrin and Brownell (2009: 733). Taxon placed within *Pterobalaena* by Hassanin *et al.* (2012: 37) and supported by the observations of McGowen *et al.* (2009: 896).

Balaenoptera Huttoni J. Gray, 1874c: 448, 449, 450.

TYPE LOCALITY: Otago Heads, New Zealand.
COMMENTS: On page 449, this species was made the type of a new 'section', *Dactylaena*. Synonymised within *acutorostrata* by Iredale and Troughton (1934: 58), Hershkovitz (1966: 157), and Mead and Brownell (1993: 349), and within *bonaerensis* by Mead and Brownell (2005: 724).

Balaenoptera borealis Lesson, 1828

Sei Whale

Φ *Balaenoptera borealis borealis* Lesson, 1828

Northern Sei Whale

Φ *Balaenoptera borealis* Lesson, 1828d: 342.

TYPE LOCALITY: Coast of Holstein near Gromitz, Schleswig-Holstein, Germany.
COMMENTS: History of description given by Gamble (1985b: 155). Species reviewed by Gambell (1985a). Taxon placed in *Rorqualis* by Hassanin *et al.* (2012: 37, 43), which is supported by the observations of McGowen *et al.* (2009: 896).

HOMONYMS:

Balaena borealis J. Fischer, 1829, Blue Whales of the Class Mammalia (Order Artiodactyla, Family Balaenopteridae). Taxon is a synonym of *Balaenoptera musculus* (Linnaeus, 1758). See individual entry.

Rorqualis borealis Hamilton, 1837, Blue Whales of the Class Mammalia (Order Artiodactyla, Family Balaenopteridae). Taxon is a synonym of *Balaenoptera musculus* (Linnaeus, 1758). See individual entry.

Φ *Balaena rostrata* Rudolphi, 1822: 27.

TYPE LOCALITY: Gromitz, Schleswig-Holstein, West Germany.

COMMENTS: Synonymised within *borealis* by Hershkovitz (1966: 160) and Mead and Brownell (2005: 724).

HOMONYMS:

Balaena rostrata O. Müller, 1776: 7, the Northern Bottle-nosed Whale of the Class Mammalia (Order Artiodactyla, Family Ziphiidae). Taxon is a synonym of *Hyperoodon ampullatus* (Forster, 1770: 18). See Mead and Brownell (2005: 740).

Balaena rostrata Fabricius, 1780, the Dwarf Minke Whale of the Class Mammalia (Order Artiodactyla, Family Balaenopteridae). Taxon is a synonym of *Balaenoptera acutorostrata* Lacépède, 1804. See individual entry.

Φ *Balaenoptera arctica* Temminck, 1844b: 26.

TYPE LOCALITY: North Pacific, coast of Kii, Japan.

COMMENTS: Synonymised within *borealis* by Hershkovitz (1966: 161).

Φ *Balaenoptera laticeps* J. Gray, 1846c: 20.

TYPE LOCALITY: *Nomen novum* for *Balaena rostrata* Rudolphi, 1822.

COMMENTS: Synonymised within *borealis* by Hershkovitz (1966: 161).

Φ *Balaenoptera Iwasi* J. Gray, 1846c: 20.

TYPE LOCALITY: Japan.

COMMENTS: Name based on *Balaenoptera arctica* or 'Iwasi Kuzira' of Temminck, 1844b. Recognised within *Physalus* with caution by J. Gray (1866b: 163). Synonymised within *borealis* by Hershkovitz (1966: 161).

Balaenoptera borealis schlegelii (Flower, 1865)

Southern Sei Whale

Sibbaldius schlegelii Flower, 1865a: 390.

TYPE LOCALITY: Java Sea, Indonesia.

COMMENTS: Recognised as a species within *Sibbaldius* by J. Gray (1866b: 178) and *Balaenoptera* by Van Bénéden and Gervais (1880: 220). Synonymised within *Balaenoptera borealis* by Hershkovitz (1966: 162). Subspecies within *Balaenoptera borealis* recognised by Rice (1998: 76) and followed by Mead and Brownell (2005: 724) and Burbidge *et al.* (2014: 26, 32), but not by Van Dyck and Strahan (2008: 797), and Perrin *et al.* (2009: 4) who suggested genetic support was weak.

Sibbaldus Schlegellii J. Gray, 1864d: 352.

TYPE LOCALITY: *Nomen nudum*.

COMMENTS: Synonymised within *borealis* by Hershkovitz (1966: 162).

Balaenoptera alba Giglioli, 1870: 74.

TYPE LOCALITY: Java Sea, Indonesia.

COMMENTS: Synonymised within *borealis* by Hershkovitz (1966: 162).

Balaenoptera brydei Olsen, 1913

Bryde's Whale

Balaenoptera brydei Olsen, 1913: 1073, 1074; Plates 109–111.

TYPE LOCALITY: Saldanha Bay, Western Cape Province, South Africa.

COMMENTS: Accepted with caution by R. Andrews (1918: 105) who said that further information is awaited with interest since it is not improbable that it was synonymous with *edeni* or that both species are synonymous with *borealis*. But it has subsequently been recognised as distinct, typically with hesitation, by Lönnberg (1931: 1), Soot-Ryan (1961: 323), Best (1977: 34; 2007: v, 58), Rice (1998: 71) and Sasaki *et al.* (2006: 40). Synonymised within *edeni* by Junge (1950: 24), Omura (1959: 31), Best (1960: 203), Hershkovitz (1966: 159), and Mead and Brownell (1993: 350; 2005: 725). Rice (1998: 71) thoroughly reviewed the evidence for the heterogeneity of the species, concluding that there are at least two different taxa which had, up to that time, been confused under the name *Balaenoptera edeni*. It seems evident that, as he concludes, the larger, 'true' Bryde's whale is different from the smaller, 'true' *edeni* (and Wada *et al.*, 2003: 278, showed that these smaller ex-Bryde's whales themselves constitute two species: see below under *B. omurai*). More recently this taxon has been recognised as a subspecies of *edeni* by authors including Kershaw *et al.* (2013: 755), who recognised it as a larger offshore form compared to the smaller coastal *edeni* form, Burbidge *et al.* (2014: 32) and the Society for Marine Mammalogy (Committee on Taxonomy, 2014). While there are no specific records of *B. brydei* from Australian waters, they do occur in warm oceanic waters between about 40°N and 40°S, and in the south-western Pacific they are known from New Zealand, between New Zealand and New Caledonia, south of Fiji, and south of Java, so it is probable that they will be found to exist marginally in Australian waters, as for example Christmas Island or the

Cocos Islands, or around Norfolk Island. Taxon placed in *Rorqualis* by Hassanin *et al.* (2012: 37, 43), along with *B. musculus* and others which is supported by the observations of McGowen *et al.* (2009: 896).

Balaenoptera edeni Anderson, 1879

Eden's Whale

Balaenoptera edeni Anderson, 1879: 551; Plate 44.

TYPE LOCALITY: Thaybyoo Creek Beach, Gulf of Marataban, Burma.

COMMENTS: Rice (1998: 72) noted that the date of publication is usually, but incorrectly, cited as 1878 (as this appears on the front page of the book) but that the date of publication is actually 1879 as this is recorded on the 'Corrigenda' located on the unnumbered page xi. *Balaenoptera edeni* was recognised as being distinct from *B. brydei* by Rice (1998: 74–75), although only *B. edeni* was recognised by Hershkovitz (1966: 158), and Mead and Brownell (2005: 725), the latter because they felt it was not clear if *edeni* and *brydei* differ in size, and they noted that only *edeni* has a type specimen. Species reviewed by Cummings (1985b: 137). Wada and Numachi (1991: 125) revealed that the smaller form occurring off the coast of the Solomon Islands and Java was genetically distinct from those of other *edeni* using allozymes. These results were supported by Yoshida and Kato (1999: 1269). More recently Kato and Perrin (2009: 159) indicated that they followed Rice's (1998: 74) lead of suggesting that there may indeed be two species of Bryde's whale (aside from *Balaenoptera omurai*) and suggested that genetic analysis should be carried out on the holotype of *B. edeni* to resolve the situation (genetic analysis has, however, been carried out on a specimen from P. Sugi, Sumatra, said to be near-identical to the holotype). Taxon placed in *Rorqualis* by Hassanin *et al.* (2012: 37, 43), which is supported by the observations of McGowen *et al.* (2009: 896).

FUTURE TAXONOMIC RESEARCH: The type specimen of *Balaenoptera edeni* located in the Indian Museum in Kolkata needs to be studied genetically to determine if the specimen from P. Sugi, from which DNA has been extracted, is indeed part of this species, or possibly *B. omurai* (see Kato and Perrin (2009: 159).

Balaenoptera musculus (Linnaeus, 1758)

Blue Whale

Φ ***Balaenoptera musculus musculus*** (Linnaeus, 1758)

Northern Blue Whale

Φ *[Balaena] Musculus* Linnaeus, 1758: 76.

TYPE LOCALITY: Firth of Forth, Scotland.

COMMENTS: Included within *Sibbaldus* by Kellogg (1929: 483). Included within *Balaenoptera* by Racovitza (1903: 33, 54). Reviewed by Yochem and Leatherwood (1985: 193). Taxon placed in *Rorqualis* by Hassanin *et al.* (2012: 37, 43), which is supported by the observations of McGowen *et al.* (2009: 896).

HOMONYMS:

Rorqualus musculus F. Cuvier, 1836b, Fin Whales of the Class Mammalia (Order Artiodactyla, Family Balaenopteridae). Taxon is a synonym of *Balaenoptera physalus* (Linnaeus, 1758). See individual entry.

Balaenoptera musculus Van Bénéden & Gervais, 1880, Fin Whales of the Class Mammalia (Order Artiodactyla, Family Balaenopteridae). Taxon is a synonym of *Balaenoptera physalus* (Linnaeus, 1758). See individual entry.

Φ *Balaenoptera jubartes* Lacépède, 1804: xxxvii, 120; Plate 4.

TYPE LOCALITY: Greenland Seas, between Greenland and Iceland.

COMMENTS: Synonymised within *musculus* by Hershkovitz (1966: 172).

Φ *B. [alaena] borealis* J. Fischer, 1829: 524.

TYPE LOCALITY: North Sea, Coast of Yorkshire, England.

COMMENTS: Synonymised within *musculus* by Hershkovitz (1966: 172).

HOMONYMS:

Balaenoptera borealis Lesson, 1828d, Sei Whales of the Class Mammalia (Order Artiodactyla, Family Balaenopteridae). Taxon is a synonym of *Balaenoptera borealis* Lesson, 1828d. See individual entry.

Rorqualis borealis Hamilton, 1837, Blue Whales of the Class Mammalia (Order Artiodactyla, Family Balaenopteridae). Taxon is a synonym of *Balaenoptera musculus* (Linnaeus, 1758). See individual entry.

Φ *Balaena boops* Voigt, 1831: 342.

TYPE LOCALITY: none given

COMMENTS: This name is based on the 'Jubarte of the Basques', which may refer to the blue whale or some other species, as earlier on the same page the author says that *B. physalus* is the 'Gibber of the Basques'.

HOMONYMS:

Balaena boops Fabricius, 1780, the Humpback Whale of the Class Mammalia (Order Artiodactyla, Family Balaenopteridae). Name is a junior synonym of *Megaptera novaeangliae* (Borowski, 1781). See Clapham and Mead (1999: 1).

Balaena boops Linnaeus, 1758, the Fin Whale of the Class Mammalia (Order Artiodactyla, Family Balaenopteridae).

Name is a junior synonym of *Balaenoptera physalus* (Linnaeus, 1758). See individual entry.

Φ *R. [orqualus] Boops* F. Cuvier, 1836b: 321; Plate 20, Fig. 1.

TYPE LOCALITY: North Sea, Coast of Yorkshire, England.

COMMENTS: Included within '*balaena*' on page 308 and '*R. [orqualis]*' on page 321. Not the same as *Balaena boops* Linnaeus, 1758 [= *Balaenoptera physalus*]. Synonymised within *musculus* by Hershkovitz (1966: 173).

Φ *Rorqualis borealis* Hamilton, 1837: 125; Plate 5.

TYPE LOCALITY: North Berwick, Firth of Forth, North Atlantic, Scotland.

COMMENTS: Synonymised within *musculus* by Hershkovitz (1966: 172).

HOMONYMS:

B[alaena] borealis J. Fischer, 1829, Blue Whales of the Class Mammalia (Order Artiodactyla, Family Balaenopteridae). Taxon is a synonym of *Balaenoptera musculus* (Linnaeus, 1758). See individual entry.

Balaenoptera borealis Lesson, 1828d, Sei Whale of the Class Mammalia (Order Artiodactyla, Family Balaenopteridae). Taxon is a currently recognised. See individual entry.

Φ *Physalus (Rorqualus) Sibbaldii* J. Gray, 1847d: 92.

TYPE LOCALITY: Coast of Yorkshire, England.

COMMENTS: Recognised by J. Gray (1866b: 160). Synonymised within *musculus* by True (1898: 633), Iredale and Troughton (1934: 57), Hershkovitz (1966: 173), and Mead and Brownell (2005: 725).

Φ *Balaenoptera gigas* Reinhardt, 1857: 10.

TYPE LOCALITY: Greenland Seas.

COMMENTS: Synonymised within *musculus* by Hershkovitz (1966: 173).

Φ *Pterobalaena gigas* Van Bénéden, 1861: 37.

TYPE LOCALITY: Belgium.

COMMENTS: Synonymised within *musculus* by Hershkovitz (1966: 173), and Mead and Brownell (2005: 725).

Φ *[Physalus] latirostris* Flower, 1865a: 414.

TYPE LOCALITY: Katwijk-aan-zee, 6 miles from Leiden.

COMMENTS: Recognised within *Cuvierius* by J. Gray (1866b: 165). Synonymised within *musculus* by Iredale and Troughton (1934: 57) and Hershkovitz (1966: 174).

Φ *Balaenoptera Carolinae* Malm, 1866: 10.

TYPE LOCALITY: West coast, Sweden.

COMMENTS: Synonymised within *musculus* by Hershkovitz (1966: 174).

Φ *Sibbaldius sulfureus* Cope, 1869: 20; Fig. 11.

TYPE LOCALITY: North Atlantic, north west coast, United States of America.

COMMENTS: Synonymised within *musculus* by Hershkovitz (1966: 174), and Mead and Brownell (2005: 725), who recorded the subspecies as 'unassigned'.

Φ *Rorqualus major* Knox, 1870: 21, 23; Plate 2, Fig. 1.

TYPE LOCALITY: Redescription and new name for the type of *Rorqualus borealis* R. Hamilton, 1837. See Hershkovitz (1966: 173).

COMMENTS: Synonymised within *musculus* by Hershkovitz (1966: 173), and Mead and Brownell (2005: 725).

Φ *Pterobalaena Gryphus* Münter, 1877: 1.

TYPE LOCALITY: Baltic Sea, Coast of Pomerania.

COMMENTS: Synonymised within *musculus* by Hershkovitz (1966: 175).

Φ *Balaenoptera musculus indica* Blyth, 1859

Indian Ocean Blue Whale

Φ *Balaenoptera indica* Blyth, 1859b: 488.

TYPE LOCALITY: Sordip, Bay of Bengal.

COMMENTS: Synonymised within *musculus* by Hershkovitz (1966: 173). Elevated to subspecies rank by Rice (1998: 78) and followed by Perrin *et al.* (2009: 5), Mead and Brownell (2005: 725) and the Committee on Taxonomy (2011) of the Society for Marine Mammalogy, but not Van Dyck and Strahan (2008: 800), Sears and Perrin (2009: 121) or Burbidge *et al.* (2014: 26, 32). Rice (1998: 78) noted that the type specimen was 25.6 m long, and that another individual 27.4 m long has been reported; these Indian Ocean whales cannot, therefore, be Pygmy Blue whales, but it is possible that they could be Antarctic Blue whales, which do grow to such lengths and are known at least as far north as Diego Garcia in the Indian Ocean (Stafford *et al.* 2004: 1342); the name *indicus* has priority over *intermedius* if they prove to be identical.

Balaenoptera musculus intermedia Burmeister, 1871

Antarctic Blue Whale

Balaenoptera intermedia Burmeister, 1871: xii.

TYPE LOCALITY: Near mouth of Rio Jujan, Buenos Aires, Argentina.

COMMENTS: Paper translated in Burmeister (1872: 413). Synonymised within *musculus* by Hershkovitz (1966: 174), but recognised as a subspecies by Rice (1998: 79) and followed by Mead and Brownell (2005: 725), Van Dyck and Strahan (2008: 800), Sears and Perrin (2009: 121) and

Burbidge *et al.* (2014: 26, 32). Perrin (2009: 6) suggested the taxon needs to be confirmed, but Gilpatrick and Perryman (2008: 9, 15) gave good evidence that adult Antarctic blue whales are consistently larger than those from the Northern Pacific, which presumably represent nominotypical *B. musculus musculus*. If this is so, then consideration should be given to whether these large blue whales actually represent two different species, respectively characteristic of the two hemispheres. The type specimen described by Burmeister was said to be young, and 58 feet (17.65 m) in length, which in fact makes it difficult to judge whether it truly was a specimen of the giant Antarctic Blue Whale or the Pygmy Blue Whale; for the moment, we continue to use the name for the giant species, but restudy of the specimen would be valuable, in case the name actually refers to the Pygmy species, antedating *brevicaudatus* by 93 years.

Baleinoptère de Miramar Lahille, 1899: 79; Plate 1.

TYPE LOCALITY: Miramar, Buenos Aires, Argentina.

COMMENTS: Synonymised within *musculus* by Hershkovitz (1966: 175).

Balaenoptera musculus brevicauda Zemsky and Boronin, 1964

Pygmy Blue Whale

Balaenoptera musculus brevicauda Zemsky and Boronin, 1964: 310.

TYPE LOCALITY: Subantarctic Ocean. (Approx. 49°59′S, 28°25′E)

COMMENTS: Zemsky and Boronin's (1964: 310) use of the name is not a *nomen nudum* (*pace* Rice, 1977: 6; Mead & Brownell 1993: 350; 2005: 725), and predates Ichihara's (1966: 79) first publication of the name by two years, which is unfortunate because Ichihara gave several detailed descriptions of Pygmy Blue Whales prior to naming it (e.g. Ichihara, 1961: 1, 19; 1963: 128), and his paper finally awarding it a name was given at a conference whose publication seems to have been inordinately delayed. Zemsky and Boronin (1964: 310) described the Pygmy Blue Whale under the name *brevicauda*, but without mentioning Ichihara's work. Taxon synonymised within *musculus* by Hershkovitz (1966: 175), but recognised by Yochem and Leatherwood (1985: 195), Rice (1998: 79), Mead and Brownell (1993: 350; 2005: 725), Bannister *et al.* (1996: 188), Van Dyck and Strahan (2008: 800) and Burbidge *et al.* (2014: 26, 32). Perrin *et al.* (2009: 7) suggested there is strong support for the taxon. There is good evidence for the existence of a very distinct, non-migratory 'pygmy blue whale' in the southern Oceans, including Western Australian waters, and there is even a suggestion that it may overlap with much larger, migratory blue whales (Stafford *et al.* 2004: 1342–1343; LeDuc *et al.* 2007: 76), in which case we have to do with two separate species even under the biological species concept. Branch and Mikhalev (2008: 697) indicate that length at sexual maturity for Pygmy and Antarctic Blue Whales does not overlap, and LeDuc *et al.* (2007) found complete separation between Pygmy and Antarctic Blue Whales using the mitochondrial Control Region and seven microsatellite loci, but, intriguingly, found considerable difference also between Pygmy populations from the Indian Ocean and the south-eastern Pacific. The evidence is that Pygmy and large Blue Whales, both northern and southern hemisphere, are 100% distinct, and specific distinction could be upheld on these grounds alone. Perrin *et al.* (2009: 5–6), however, warned that the status of the earlier names *indica* and *intermedia* is unclear, and that one of them may conceivably represent the same 'pygmy' species; this seems out of the question in the case of *indica*, but as noted above it is possible that the type specimen of *intermedia* may actually have been a Pygmy blue whale.

Balaenoptera omurai Wada *et al.*, 2003

Omura's Whale

Balaenoptera omurai Wada *et al.*, 2003: 278.

TYPE LOCALITY: Tsunoshima Island, The Sea of Japan, Japan. (34°21′03″N, 130°53′09″E)

COMMENTS: Separated from *Balaenoptera brydei*. Synonymised within *edeni* by Mead and Brownell (2005: 725) with the provision that it may be recognised as a full species when the genus *Balaenoptera* is revised. Species rank confirmed by Sasaki *et al.*, (2006: 40) who showed it has a separate and ancient lineage. Subsequently recognised as a distinct species by Van Dyck and Strahan (2008: 801) and Burbidge *et al.* (2014: 26, 32). Taxon placed in *Rorqualis* by Hassanin *et al.* (2012: 37, 43), along with B. musculus and the Sei/Bryde's group, which placement supported by the observations of McGowen *et al.* (2009: 896). See also comments above under *Balaenoptera edeni*.

Balaenoptera physalus (Linnaeus, 1758)

Fin Whale

Φ *Balaenoptera physalus physalus* (Linnaeus, 1758)

Northern Fin Whale

Φ *[Balaena] Physalus* Linnaeus, 1758: 75.

TYPE LOCALITY: 'Habitat in Oceano Europeao', but restricted to near Svalbard, Spitzbergen Sea, Norway by Thomas (1911b: 156).

COMMENTS: History of description given by Gamble (1985a: 171). Transferred to *Balaenoptera* by Racovitza

(1903: 55). Reviewed by Gambell (1985b: 171) and Rice (1998: 76), who recognised *B. b. quoyi* [sic], and Mead and Brownell (2005: 725). The phylogenetic research of F. Archer *et al.* (2013: 1) strongly suggests the need for revision of the global taxonomy of this species.

Φ *[Balaena] Boops* Linnaeus, 1758: 76.

TYPE LOCALITY: 'Oceano Septentrionali', but restricted to Firth of Forth, Scotland by Thomas (1911b: 156).

COMMENTS: Synonymised within *physalus* by True (1898: 633), Hershkovitz (1966: 164), and Mead and Brownell (2005: 725).

HOMONYMS:

Balaena boops Fabricius, 1780, the Humpback Whale of the Class Mammalia (Order Artiodactyla, Family Balaenopteridae). Name is a junior synonym of *Megaptera novaeangliae* (Borowski, 1781). See Clapham and Mead (1999: 1).

Balaena boops Voigt, 1831, the Blue Whale of the Class Mammalia (Order Artiodactyla, Family Balaenopteridae). Name is a junior synonym of *Balaenoptera musculus* (Linnaeus, 1758). See individual entry.

Φ *Balaenoptera gibbar* Lacépède, 1804: xxxvi, 114.

TYPE LOCALITY: Arctic Seas, near Greenland.

COMMENTS: Synonymised within *physalus* by Hershkovitz (1966: 165), and Mead and Brownell (2005: 725).

Φ *Balaenoptera rorqual* Lacépède, 1804: xxxvii, 126; Plate 1, Fig. 2.

TYPE LOCALITY: Scottish Seas.

COMMENTS: Synonymised within *physalus* by Hershkovitz (1966: 165).

Φ *Balaena sulcata* Neill, 1811: 212.

TYPE LOCALITY: North Atlantic, Firth of Forth, above Alloa, Scotland.

COMMENTS: Synonymised within *physalus* by Hershkovitz (1966: 165).

Φ *Balaena rostrata major* Rosenthal, 1827: Plate 1.

TYPE LOCALITY: West coast Rugen, Baltic Sea, Germany.

COMMENTS: Synonymised within *physalus* by Hershkovitz (1966: 165).

Φ *balaenoptera mediterraneensis* Lesson, 1828d: 442.

TYPE LOCALITY: Sainte Marguerite Island, Mediterranean Sea.

COMMENTS: Synonymised within *physalus* by Hershkovitz (1966: 165).

Φ *P.[hysalis (sic) vulgaris* Fleming, 1828: 32.

TYPE LOCALITY: *Nomen novum* for *Balaena physalus* Linnaeus, 1758.

COMMENTS: Synonymised within *physalus* by Hershkovitz (1966: 166).

Φ *B.[alaena] Antiquorum* J. Fischer, 1829: 525.

TYPE LOCALITY: Mediterranean Sea, North Atlantic.

COMMENTS: Species recognised by J. Gray (1866b: 144). Synonymised within *physalus* by Hershkovitz (1966: 166), and Mead and Brownell (2005: 725).

Φ *Balaenoptera Aragous* Farines & Carcassone, 1829: Plate.

TYPE LOCALITY: Coast of Saint Cyprien, Pyrénées Orientales, Mediterranean Sea, France.

COMMENTS: Synonymised within *physalus* by Hershkovitz (1966: 166).

Φ *Balaenoptera tenuirostris* Sweeting, 1840: 343.

TYPE LOCALITY: North Atlantic, Charmouth Beach, England.

COMMENTS: Synonymised within *physalus* by Hershkovitz (1966: 166).

Φ *Balaena sulcata arctica* Schlegel, 1841: 38; Plate 6.

TYPE LOCALITY: North Atlantic, Netherlands.

COMMENTS: Synonymised within *physalus* by Hershkovitz (1966: 166).

HOMONYMS:

Balaena mysticetus arctica Schlegel, 1841: 36, the Bowhead Whale of the Class Mammalia (Order Artiodactyla, Balaenidae). Name is a synonym of *Balaena mysticetus* Linnaeus, 1758: 75. See Hershkovitz (1966: 194).

Φ *Physalus Duguidii* Heddle, 1856: 187; Plate 44.

TYPE LOCALITY: North Atlantic, Island of Laman, or Lambholm Orkneys, Scotland.

COMMENTS: Synonymised within *physalus* by Hershkovitz (1966: 167).

Φ *Pterobalaena communis* Van Bénéden, 1857: 403.

TYPE LOCALITY: North Atlantic, near Vlieland Island, Netherlands.

COMMENTS: Synonymised within *physalus* by Hershkovitz (1966: 167).

Φ *Pterobalaena Gigantea michrochira* Barkow, 1862: 17.

TYPE LOCALITY: North Sea, Coast of Wales.

COMMENTS: Synonymised within *physalus* by Hershkovitz (1966: 167).

Φ *Benedenia Knoxii* J. Gray, 1864c: 212; Fig. 8.

TYPE LOCALITY: North Sea, Coast of Wales.

COMMENTS: Synonymised within *physalus* by Hershkovitz (1966: 167).

Φ *Balaenoptera swinhoii* J. Gray, 1866g: 725; Figs. 1–6.

TYPE LOCALITY: North Pacific Ocean, Formosa.

COMMENTS: Synonymised within *physalus* by Hershkovitz (1966: 168).

Φ *Swinhoia chinensis* J. Gray, 1868c: 3.

TYPE LOCALITY: Formosa, North Pacific Ocean.
COMMENTS: Synonymised within *physalus* by Hershkovitz (1966: 168).

Φ *B. [alaenoptera] velifera* Cope, 1869: 16.

TYPE LOCALITY: Oregon Coasts, North Pacific Ocean, United States of America.
COMMENTS: Synonymised within *physalus* by Hershkovitz (1966: 168), and Mead and Brownell (2005: 725).

Φ *B. [alaenoptera] swinhoei* Cope, 1869: 16.

TYPE LOCALITY: Formosa, North Pacific Ocean.
COMMENTS: Synonymised within *physalus* by Hershkovitz (1966: 168).

Φ *S. [ibbaldius] tuberosus* Cope, 1869: 16.

TYPE LOCALITY: North Atlantic, eastern North American coast.
COMMENTS: Synonymised within *physalus* by Hershkovitz (1966: 169).

Φ *S. [ibbaldius] tectirostris* Cope, 1869: 17.

TYPE LOCALITY: North Atlantic, Sinepuxent Bay, Maryland Peninsula, United States of America.
COMMENTS: Synonymised within *physalus* by Hershkovitz (1966: 169).

Φ *B. [alaenoptera] blythii* Anderson, 1879: 564.

TYPE LOCALITY: Calcutta, India.
COMMENTS: Synonymised within *physalus* by Hershkovitz (1966: 169).

Φ *Balaenoptera musculus* Van Bénéden & Gervais, 1880: 167; Plates 12–13, Figs. 11–24.

TYPE LOCALITY: Not given.
COMMENTS: Mis-citation for *Balaena rostrata major* Rosenthal, 1827 in synonymy of *B. musculus* (see Hershkovitz (1966: 169). Synonymised within *physalus* by Hershkovitz (1966: 169).
HOMONYMS:

Balaena musculus Linnaeus, 1758, the Blue Whale of the Class Mammalia (Order Artiodactyla, Family Balaenopteridae). Taxon is a synonym of *Balaenoptera musculus* (Linnaeus, 1758). See individual entry.

Rorqualus musculus F. Cuvier, 1836b, the Fin Whale of the Class Mammalia (Order Artiodactyla, Family Balaenopteridae). Taxon is a synonym of *Balaenoptera physalus* (Linnaeus, 1758). See individual entry.

Φ *[Balaenoptera] [velifera] copei* Elliot, 1901: 13.

TYPE LOCALITY: North Pacific, Shumagen Islands, Alaska, United States of America.
COMMENTS: Synonymised within *physalus* by Hershkovitz (1966: 169).

Φ *Balaenopteris [sic] guibusdam [sic]* Tomilin, 1957: 93.

TYPE LOCALITY: Not given.
COMMENTS: Mis-citation for *Balaena rostrata major* Rosenthal, 1827 in synonymy of *B. musculus* (see Hershkovitz 1966: 169). Synonymised within *physalus* by Hershkovitz (1966: 169).

Φ *Balaena antipodarum* Tomilin, 1957: 130.

TYPE LOCALITY: Not given.
COMMENTS: Name erroneously attributed to J. Fischer, 1829, evidently a lapsus for *antiquorum* J. Fischer, 1829. Synonymised within *physalus* by Hershkovitz (1966: 170).
HOMONYMS:

Balaena antipodarum J. Gray, 1843d, the Southern Right Whale of the Class Mammalia (Class Artiodactyla, Family Balaenidae). Name is a junior synonym of *Eubalaena australis* (Desmoulins, 1822). See individual entry.

Balaena antipodarum J. Gray, 1864c, the Pygmy Right Whale of the Class Mammalia (Class Artiodactyla, Family Neobalaenidae). Name is a junior synonym of *Caperea marginata* (J. Gray, 1846c). See individual entry.

Φ *Dubertus rhodinsulensis* Trumbull, 1884: 29.

TYPE LOCALITY: Unknown.
COMMENTS: *Nomen nudum*. Synonymised within *physalus* by Tomilin (1957: 131) and Hershkovitz (1966: 170).

Φ ***Balaenopterus physalus patachonica*** Burmeister, 1865

Pygmy Fin Whale

Φ *Balaenopterus patachonica* Burmeister, 1865a: 190.

TYPE LOCALITY: South Atlantic Ocean, Río de La Plata, near Mouth, Argentina.
COMMENTS: Synonymised within *physalus* by Hershkovitz (1966: 168), Mead and Brownell (2005: 725), and not recognised by Aguilar (2009: 434). Elevated to a subspecies of *physalus* by Clarke (2004: 329), to represent a pygmy form of *physalus* in the southern hemisphere based on its distribution, size and colouration. This subspecies rank was subsequently adopted by the Society for Marine Mammalogy (Committee on Taxonomy, 2014). If there really is a 'pygmy fin whale', with a range in the South Atlantic abutting or overlapping that of the giant Southern Fin Whale *Balaenoptera physalus quoyi*, then it can hardly be a subspecies of *B. physalus*.

Balaenoptera physalus quoyi (J. Fischer, 1829)

Southern Fin Whale

B. [alaena] Quoyi J. Fischer, 1829: 526.

TYPE LOCALITY: South Atlantic, Falkland Islands.
COMMENTS: Transferred to *Balaenoptera* by Lönnberg (1906b: 28). Synonymised within *physalus* by Hershkovitz

(1966: 166). Subspecies rank recognised, as *quoyii*, by Tomilin (1946: 468), and as *quoyi* by Tomilin (1957: 200), Mead and Brownell (2005: 725), Aguilar (2009: 434) and Burbidge *et al*. (2014: 26, 32), but not Van Dyck and Strahan (2008: 804). Perrin *et al*. (2009: 5) suggested there is no morphological justification for its recognition.

Balaena rostrata australis Desmoulins, 1822a: 164.

TYPE LOCALITY: Shores of the Falkland Islands.

COMMENTS: Based on an animal observed by Quoy. Synonymised within *physalus* by Hershkovitz (1966: 165).

HOMONYMS:

Balaena rostrata australis Desmoulins, 1822a: 161, the Southern Right Whale of the Class Mammalia (Order Artiodactyla, Family Balaenidae). Taxon currently known as *Eubalaena australis*. See Mead and Brownell (2005: 723).

R. [orqualus] musculus F. Cuvier, 1836b: 334.

TYPE LOCALITY: South Atlantic, Falkland Islands.

COMMENTS: Included within '*balaena*' on page 308 and '*R*[*orqualis*]' on page 334. Name based on 'rorqual de la Méditeranée' of Lacépède, 1804 stranded on Sainte Marguerite Island. Synonymised within *physalus* by Hershkovitz (1966: 166).

HOMONYMS:

Balaena musculus Linnaeus, 1758, the Blue Whale of the Class Mammalia (Order Cetacea, Family Balaenopteridae). Currently recognised species. See individual entry.

Balaenoptera musculus Van Bénéden & Gervais, 1880, the Fin Whale of the Class Mammalia (Order Artiodactyla, Family Balaenopteridae). Taxon is a synonym of *Balaenoptera physalus* (Linnaeus, 1758). See individual entry.

Balaenoptera australis J. Gray, 1846c: 51.

TYPE LOCALITY: Unknown.

COMMENTS: Synonymised within *physalus* by Hershkovitz (1966: 165).

Balaenoptera antarctica J. Gray, 1846c: 51.

TYPE LOCALITY: South Pacific, New Zealand.

COMMENTS: Synonymised within *physalus* by Iredale and Troughton (1934: 57) and Hershkovitz (1966: 167).

Balaenoptera Brasiliensis J. Gray, 1846c: 51.

TYPE LOCALITY: South Atlantic, Bahia, Brazil.

COMMENTS: Synonymised within *physalus* by Hershkovitz (1966: 167).

Physalus? fasciatus J. Gray, 1850: vii, 42.

TYPE LOCALITY: South Pacific, Peru.

COMMENTS: Synonymised within *physalus* by Hershkovitz (1966: 167).

Physalus? australis J. Gray, 1850: vii, 43.

TYPE LOCALITY: South Atlantic, Falklands.

COMMENTS: Recognised by J. Gray (1866b: 161). Synonymised within *physalus* by Hershkovitz (1966: 165).

Physalus antarcticus J. Gray, 1850: vii, 43.

TYPE LOCALITY: New Zealand.

COMMENTS: Synonymised within *physalus* by Hershkovitz (1966: 167).

Physalus Brasiliensis J. Gray, 1850: vii, 43.

TYPE LOCALITY: South Atlantic, Brazil.

COMMENTS: Synonymised within *physalus* by Hershkovitz (1966: 167).

Physalus Grayi McCoy, 1867b: 177.

TYPE LOCALITY: Victoria, Australia.

COMMENTS: Synonymised within *physalus* by Iredale and Troughton (1934: 57) and Hershkovitz (1966: 168).

Physalus australis J. Gray, 1870b: 393.

TYPE LOCALITY: South Atlantic, Patagonia.

COMMENTS: Synonymised within *physalus* by Hershkovitz (1966: 165).

Stenobalaena xanthogaster J. Gray, 1874b: 305.

TYPE LOCALITY: South Pacific, New Zealand.

COMMENTS: Synonymised within *physalus* by Iredale and Troughton (1934: 58) and Hershkovitz (1966: 169).

Megaptera J. Gray, 1846

Megaptera J. Gray, 1846b: 83.

TYPE SPECIES: *Megaptera longipinna* J. Gray, 1846b (derived from *Balaena longimana* Rudolphi, 1832) [=*Megaptera novaeangliae* (Borowski, 1781)] by monotypy.

COMMENTS: Described as a Genus listed within J. Gray, 1846c: 16, with the type species names as *Balaena nodosa* Bonnaterre, 1789 [=*Megaptera novaeangliae* (Borowski, 1781)]. Genus recognised by Iredale and Troughton (1934: ix, 58), Hershkovitz (1966: 176) and subsequent authors. The genus is sister to *B. musculus* according to mtDNA studies (McGowen *et al*., 2009: 896; Hassanin *et al*., 2011:37, 43), but separation time between the two is well above that suggested here for generic separation.

HOMONYMS:

Megaptera Meek and Worthen, 1868: 22, bivalves of the Phylum Mollusca (Class Bivalva, Order Pterioida, † Family Myalinidae). Genus is a synonym of † *Opisthoptera* Meek, 1872: 320, which was proposed as a replacement name. See Spamer *et al*. (1995: 271) and Vokes (1980: 30, 96).

Megaptera Modell, 1964: 95, bivalves of the Phylum Mollusca (Class Bivalva, Superfamily Unionacea). Incorrect subsequent spelling of *Metaptera* Rafinesque, 1820b: 299. Name has been placed as a junior synonym of *Potamilus*

Rafinesque, 1818b: 107 and *Proptera* Rafinesque, 1819a: 426. See Modell (1964: 95).

Megapteron J. Gray, 1846c: 51.

TYPE SPECIES: Apparently in *lapsus* for *Megaptera* J. Gray, 1846b. See Hershkovitz (1966: 176).

COMMENTS: Name also used by Wagner (1847: 38) as '*Megaptera* s. *Megapteron*'. Synonymised within *Megaptera* by Palmer (1904: 406), J. Gray (1864c: 205), Iredale and Troughton (1934: 58) and Hershkovitz (1966: 176).

Perqualus J. Gray, 1846c: Plate 32.

TYPE SPECIES: *Balaena boops* Fabricius, 1780 (as *Balaenoptera boops*) [=*Megaptera novaeangliae* (Borowski, 1781)] by monotypy.

COMMENTS: Subgenus of *Balaenoptera*. Taxon synonymised within *Balaenoptera* by Iredale and Troughton (1934: 57), and within *Megaptera* by Hershkovitz (1966: 176), Clapham and Mead (1999: 1), and Mead and Brownell (2005: 725).

Kyphobalaena Eschricht, 1849a: 108.

TYPE SPECIES: *Balaena boops* Fabricius, 1780 [=*Megaptera novaeangliae* (Borowski, 1781)] by monotypy.

COMMENTS: Taxon also described by Eschricht (1849b: xv, 56, 146). Synonymised within *Balaenoptera* by Iredale and Troughton (1934: 57) and within *Megaptera* by J. Gray (1864c: 205), Hershkovitz (1966: 176), Clapham and Mead (1999: 1), and Mead and Brownell (2005: 725).

Poescopia J. Gray, 1864c: 207.

TYPE SPECIES: *Balaena lalandii* J. Fischer, 1829; *Megaptera novae-zelandiae* J. Gray, 1864c [=*Megaptera novaeangliae* (Borowski, 1781)].

COMMENTS: Described as a subgenus of *Megaptera* J. Gray, 1846b and raised to full genus by J. Gray (1866b: 113, 125). Synonymised within *Megaptera* by Iredale and Troughton (1934: 58), Hershkovitz (1966: 176), Clapham and Mead (1999: 1), and Mead and Brownell (2005: 725).

Poeskopia Gervais, 1871: 88.

TYPE SPECIES: Emendation of *Poescopia* J. Gray, 1864c.

COMMENTS: Synonymised within *Poescopia* by Palmer (1904: 554) and Hershkovitz (1966: 176).

Cyphobalaena de Marschall, 1873: 5

TYPE SPECIES: Emendation of *Kyphobalaena* Eschricht, 1849a.

COMMENTS: Synonymised within *Kyphobalaena* by Palmer (1904: 359), *Balaenoptera* by Iredale and Troughton (1934: 57) and *Megaptera* by Hershkovitz (1966: 176), and Mead and Brownell (2005: 725).

Megapterina Tomilin, 1957: 274.

TYPE SPECIES: Not given.

COMMENTS: Erroneously listed as a generic name in synonymy of *Megaptera* J. Gray, 1846b: 16 by Tomilin (1957: 274; see Hershkovitz (1966: 176). Recognised at suprageneric rank by J. Gray (1864c: 205). Synonymised within *Megaptera* by Hershkovitz (1966: 176).

Megaptera novaeangliae (Borowski, 1781)

Humpback Whale

Φ ***Megaptera novaeangliae novaeangliae*** (Borowski, 1781)

North Atlantic Humpback Whale

Φ *Balaena Novae Angliae* Borowski, 1781: 21.

TYPE LOCALITY: Coast of New England, United States of America.

COMMENTS: The specific name is always nowadays treated as a single word, *novaeangliae* (see Hershkovitz 1966: 177, Clapham and Mead 1999: 1). Reviewed by Winn and Reichley (1985: 241). Taxon reviewed by Jackson *et al.* (2014: 1, 8) who recognised three subspecies that include *novaeangliae, australis and kuzira*, which is followed here.

Φ *Balaena boops* Fabricius, 1780: 36.

TYPE LOCALITY: North Atlantic, Greenland.

COMMENTS: Synonymised within *novaeangliae* by Hershkovitz (1966: 177) and Clapham and Mead (1999: 1).

HOMONYMS:

Balaena boops Linnaeus, 1758, the Fin Whale of the Class Mammalia (Order Artiodactyla, Family Balaenopteridae). Name is a junior synonym of *Balaenoptera physalus* (Linnaeus, 1758). See individual entry.

Balaena boops Voigt, 1831, the Blue Whale of of the Class Mammalia (Order Artiodactyla, Family Balaenopteridae). Name is a junior synonym of *Balaenoptera musculus* (Linnaeus, 1758). See individual entry.

Φ *B. [alaena] Nodosa* Bonnaterre, 1789: 5; Plates 1–12.

TYPE LOCALITY: North Atlantic, New England Coast, United States of America.

COMMENTS: Synonymised within *novaeangliae* by Hershkovitz (1966: 177), Mean and Brownell (1993: 350), and Clapham and Mead (1999: 1).

Φ *Balaena longimana* Rudolphi, 1832: 133.

TYPE LOCALITY: North Atlantic, Mouth of Elbe River, Germany.

COMMENTS: Recognised within *Megaptera* by J. Gray (1866b: 119). Synonymised within *novaeangliae* by Hershkovitz (1966: 179), Mead and Brownell (1993: 350; 2005: 725), and Clapham and Mead (1999: 1).

Φ *Balaenoptera leucopteron* Lesson, 1842: 202.

TYPE LOCALITY: Unknown.

COMMENTS: Synonymised within *novaeangliae* by Hershkovitz (1966: 181), and Clapham and Mead (1999: 1).

Φ *Balaena gibbosa* J. Gray, 1843c: 183; Plate 1.

TYPE LOCALITY: Unknown.

COMMENTS: Synonymised within *novaeangliae* by Hershkovitz (1966: 183), and Clapham and Mead (1999: 1).

HOMONYMS:

Balaena gibbosa Erxleben, 1777, the Dwarf Minke Whale of the Class Mammalia (Order Artiodactyla, Family Balaenopteridae). Name appears to be a synonym of *Balaenoptera acutorostrata* Lacépède, 1804. See individual entry.

Balaena gibbosa Cope, 1868b, the Dwarf Minke Whale of the Class Mammalia (Order Artiodactyla, Family Balaenopteridae). Name is a synonym of *Balaenoptera acutorostrata* Lacépède, 1804. See individual entry.

Φ *Megaptera longipinna* J. Gray, 1846b: 83.

TYPE LOCALITY: Unknown.

COMMENTS: Synonymised within *novaeangliae* by Hershkovitz (1966: 180), Mean and Brownell (1993: 350) and Clapham and Mead (1999: 1).

Φ *Balaena Allamack* J. Gray, 1846c: 17.

TYPE LOCALITY: Unknown.

COMMENTS: Synonymised within *novaeangliae* by Hershkovitz (1966: 181), and Clapham and Mead (1999: 1).

Φ *Megaptera Americana* J. Gray, 1846c: 17.

TYPE LOCALITY: Unknown.

COMMENTS: Synonymised within *novaeangliae* by Hershkovitz (1966: 181), and Clapham and Mead (1999: 1).

Φ *Megaptera longimana* J. Gray, 1846c: 17.

TYPE LOCALITY: Unknown.

COMMENTS: Recognised by J. Gray (1864c: 207) and synonymised within *novaeangliae* by Hershkovitz (1966: 179) and Hershkovitz (1966: 179).

Φ *Balaenoptera syncondylus* A. Müller, 1863: 38, 48; Plates 1–3.

TYPE LOCALITY: 'Ostsee und die Kurische Nehrung' [=Baltic Sea, near Courland Spit].

COMMENTS: Synonymised within *novaeangliae* by Hershkovitz (1966: 181) and Clapham and Mead (1999: 1).

Φ *M. [egaptera gigas]* Cope, 1865a: 179.

TYPE LOCALITY: 'North Atlantic Species'.

COMMENTS: Synonymised within *novaeangliae* by Hershkovitz (1966: 182) and Clapham and Mead (1999: 1).

Φ *Megaptera osphyia* Cope, 1865a: 180.

TYPE LOCALITY: North Atlantic, 40 miles off Pettit Menan Lighthouse, Maine, United States of America.

COMMENTS: Synonymised within *novaeangliae* by Hershkovitz (1966: 182), Clapham and Mead (1999: 1) and Mead and Brownell (2005: 725).

Φ *[Megaptera longimanna]* var. *moorei* J. Gray, 1866b: 122.

TYPE LOCALITY: North Atlantic, estuary of the Dee River, England.

COMMENTS: Synonymised within *novaeangliae* by Hershkovitz (1966: 182) and Clapham and Mead (1999: 1).

Φ *Kyphobalaena Keporkak* Van Bénéden, 1868: 109.

TYPE LOCALITY: Davis Strait.

COMMENTS: Synonymised within *novaeangliae* by Hershkovitz (1966: 182), and Clapham and Mead (1999: 1).

Φ *Megaptera bellicosa* Cope, 1871: 103.

TYPE LOCALITY: 'San Domingo', Haiti. Given as North Atlantic, French West Indies, St. Barthélemy Island by Hershkovitz (1966: 182).

COMMENTS: Synonymised within *novaeangliae* by Hershkovitz (1966: 182), and Clapham and Mead (1999: 1).

Φ *Megaptera boops* Van Bénéden & Gervais, 1880: 120; Plates 10–11, Figs. 1–8.

TYPE LOCALITY: North Atlantic, Grenadine Islands, Lesser Antilles.

COMMENTS: Synonymised within *novaeangliae* by Hershkovitz (1966: 177).

Φ *Balaena atlanticus* Hurdis, 1897: 330, 339.

TYPE LOCALITY: North Atlantic, Bermudas.

COMMENTS: Synonymised within *novaeangliae* by Hershkovitz (1966: 183), and Clapham and Mead (1999: 1).

Φ *Megaptera novaeangliae* Kellogg, 1932: 148.

TYPE LOCALITY: 'New England', North Atlantic, United States of America.

COMMENTS: Synonymised within *novaeangliae* by Hershkovitz (1966: 177).

Megaptera novaeangliae australis (Lesson, 1828)

Southern Humpback Whale

Balaenoptera australis Lesson, 1828d: 372.

TYPE LOCALITY: South Atlantic, Cape of Good Hope, South Africa.

COMMENTS: Recognised as a distinct species within *Megaptera* by Iredale and Troughton (1934: ix, 58), but synonymised within *novaeangliae* by Hershkovitz (1966: 179), and Clapham and Mead (1999: 1). Taxon recognised as a subspecies of *novaeangliae* by Jackson *et al.* (2014: 8).

B. [alaena] Lalandii J. Fischer, 1829: 525

TYPE LOCALITY: South Atlantic, Cape of Good Hope, South Africa.

COMMENTS: Recognised and placed in *Megaptera* by J. Gray (1864c: 207). Synonymised within *Megaptera australis* (Lesson, 1828d) by Iredale and Troughton (1934: 58) and within *novaeangliae* by Hershkovitz (1966: 179), Mean and Brownell (1993: 350; 2005: 725), and Clapham and Mead (1999: 1).

Balaenoptera Capensis A. Smith, 1834: 242

TYPE LOCALITY: South Atlantic, Cape of Good Hope, South Africa.

COMMENTS: Synonymised within *novaeangliae* by Hershkovitz (1966: 180), and Clapham and Mead (1999: 1).

R. [orqualus] antarcticus F. Cuvier, 1836b: 347; Plate 20, Figs. 2–4.

TYPE LOCALITY: Unknown.

COMMENTS: Synonymised within *novaeangliae* by Hershkovitz (1966: 180), and Clapham and Mead (1999: 1).

Balaena sulcata antarctica Schlegel, 1841: 43.

TYPE LOCALITY: South Atlantic, Cape of Good Hope, South Africa.

COMMENTS: Synonymised within *novaeangliae* by Hershkovitz (1966: 180).

Megaptera Poeskop J. Gray, 1846c: 17.

TYPE LOCALITY: South Atlantic, Cape of Good Hope, South Africa.

COMMENTS: Synonymised within *novaeangliae* by Hershkovitz (1966: 181), and Clapham and Mead (1999: 1).

Megaptera antarctica J. Gray, 1846c: 17.

TYPE LOCALITY: Unknown.

COMMENTS: Synonymised within *novaeangliae* by Hershkovitz (1966: 180).

Balaenoptera Astrolabe Pucheran, 1853: 42; Plate 24, Fig. 1.

TYPE LOCALITY: 'Southern oceans'.

COMMENTS: Synonymised within *novaeangliae* by Hershkovitz (1966: 181) and Clapham and Mead (1999: 1).

Megaptera novae-zelandiae J. Gray, 1864c: 207; Fig. 4.

TYPE LOCALITY: South Pacific, Otago, New Zealand.

COMMENTS: Recognised by J. Gray (1866b: 128). Synonymised within *Megaptera australis* (Lesson, 1828d) by Iredale and Troughton (1934: 58) and within *novaeangliae* by Hershkovitz (1966: 181) and Clapham and Mead (1999: 1).

Megaptera? Burmeisteri J. Gray, 1866b: 129.

TYPE LOCALITY: South Atlantic, Island between Paraná and Guazu and Panama de las Palmas, Mouth of the Panama River, Buenos Aires, Argentina.

COMMENTS: Synonymised within *novaeangliae* by Hershkovitz (1966: 182), Mean and Brownell (1993: 350; 2005: 725), and Clapham and Mead (1999: 1).

Megaptera braziliensis Cope, 1867: 32.

TYPE LOCALITY: Near Bahia, Brazil.

COMMENTS: Synonymised within *novaeangliae* by Hershkovitz (1966: 182), Mean and Brownell (1993: 350; 2005: 725), and Clapham and Mead (1999: 1).

Megaptera nodosa Lahille, 1905: 72.

TYPE LOCALITY: South Atlantic, Punta Indio, Río de La Plata, Argentina.

COMMENTS: Synonymised within *novaeangliae* by Hershkovitz (1966: 178).

Megaptera nodosa novae-zealandiae Ivashin, 1958: 77.

TYPE LOCALITY: Unknown.

COMMENTS: Synonymised within *novaeangliae* by Hershkovitz (1966: 182).

Φ ***Megaptera novaeangliae kuzira*** J. Gray, 1850

North Pacific Humpback Whale

Φ *Megaptera Kuzira* J. Gray, 1850: vii, 30.

TYPE LOCALITY: North Pacific, southern coast of Japan.

COMMENTS: Recognised by J. Gray (1866b: 130). Synonymised within *novaeangliae* by Hershkovitz (1966: 181) and Clapham and Mead (1999: 1).

Φ *Megaptera versabilis* Cope, 1869: 15.

TYPE LOCALITY: North Pacific.

COMMENTS: Synonymised within *novaeangliae* by Hershkovitz (1966: 182), Mean and Brownell (1993: 350; 2005: 725), and Clapham and Mead (1999: 1).

Φ *Megaptera indica* Gervais, 1883: 1567.

TYPE LOCALITY: Persian Gulf.

COMMENTS: Synonymised within *novaeangliae* by Hershkovitz (1966: 183), Clapham and Mead (1999: 1) and Mead and Brownell (2005: 725).

Parvorder Odontoceti Flower, 1867

Suborder Odontoceti Flower, 1867: 110, 115.

COMMENTS: When originally proposed, this rank was placed in the Order Cetacea (Brisson, 1762) and included the families Physeteridae (J. Gray, 1821), Platanistidae (J. Gray, 1846c: 25) and Delphinidae (J. Gray, 1821). McKenna and Bell (1997: 379) suggest the date of publication is 1869.

Order Pinnata Storr, 1780: Table C.

COMMENTS: When originally proposed, this rank was placed in the Mammalia (Linnaeus, 1758) and included the genera *Delphinus* Linnaeus, 1758; *Diodon* Storr, 1780: 42, Table C [=*Monodon* Linnaeus, 1758: 75]; *Physeter* Linnaeus, 1758; and *Balaena* Linnaeus, 1758: 75.

Tribe Carnivora Lesson, 1842:197.

TYPE GENUS: Not based on a genus group name.

COMMENTS: When originally proposed, this rank was placed in the Order Cétacés (Lesson, 1842 [=Cetacea (Brisson, 1762)]) and included the families Delphinusideae (Lesson, 1842 [=Delphinidae (J. Gray, 1821)]) and Physetereae (Lesson, 1842 [=Physeteridae (J. Gray, 1821)]).

HOMONYM:

Order Carnivora Bowdich, 1821, carnivores of the Class Mammalia. See individual entry.

Section Denticete J. Gray, 1864c: 231.

COMMENTS: When originally proposed, this rank was placed in the Suborder Cete (Linnaeus, 1758 [=Cetacea (Brisson, 1762)]) and included the families Catodontidae (F. Cuvier, 1836a [=Physeteridae (J. Gray, 1821)]) and Delphinidae (J. Gray, 1821). Does not appear to have been recognised previously.

Suborder Odontocete Flower, 1865a: 388.

COMMENTS: When originally proposed, this rank was placed in the Order Cetacea (Brisson, 1762) and included the families Physeteridae (J. Gray, 1821), Platanistidae (J. Gray, 1846c: 25) and Delphinidae (J. Gray, 1821). Synonymised within Odontoceti by Simpson (1945: 100) and McKenna and Bell (1997: 379).

Suborder Delphinoidea Flower, 1865a: 388, 389.

COMMENTS: When originally proposed, this rank was placed in the Order Cetacea (Brisson, 1762) and included the families Physeteridae (J. Gray, 1821), Platanistidae (J. Gray, 1846c: 25) and Delphinidae (J. Gray, 1821). Delphinoidea recognised at Superfamily rank by Rice (1984: 466). Proposed as an alternative name to the suborder. Synonymised within Odontoceti and Delphinoidea by McKenna and Bell (1997: 379, 383).

Suborder Denticeti Cope, 1869: 14, 20.

COMMENTS: When originally proposed placed in the Order Cetacea (Brisson, 1762) and included the families Delphinidae (J. Gray, 1821), Physeteridae (J. Gray, 1821) and Platanistidae (J. Gray, 1846c: 25). Both Denticeti and Denticete are mentioned.

Group Delphinoidea Huxley, 1872: 336.

COMMENTS: When originally proposed, this rank was placed in the Cetacea (Brisson, 1762) with lower ranks unknown. Synonymised within the Infraorder Autoceta (Haeckel, 1866) and Superfamily Delphinoidea (J. Gray, 1821) by McKenna and Bell (1997: 371, 383).

Family Hypognathodontidae Brandt, 1873b: 575.

TYPE GENUS: Not based on a genus group name.

COMMENTS: When originally proposed, this rank was placed in the Odontoceti (Flower, 1867) and included the subfamilies Physeterinae (J. Gray, 1821) and Ziphiinae (J. Gray, 1865d). Synonymised within the families Physeteridae and Ziphiidae by Rice (1998: 82, 85).

Megazoophaga A. Scott, 1873: 63.

COMMENTS: When originally proposed, this rank was placed in the Suborder Odontocete (Flower, 1865a [=Odontoceti (Flower, 1867)]) and included families Platanistidae (J. Gray, 1846c: 25), Pontoporiadae (J. Gray, 1870b: 393 [=Pontoporiidae (J. Gray, 1870b: 393)]), † Champsodelphidae (A. Scott, 1873: 63), Delphinidae (J. Gray, 1821) and Delphinapteridae (Gill, 1871b: 124).

Order Odontoceta Ameghino, 1889: xxvi, 883.

COMMENTS: When originally proposed, this rank was placed in the Grand Seccion Cetacea (Ameghino, 1889 [=Cetacea (Brisson, 1762)]) and included the families Platanistidae (J. Gray, 1846c: 25), Delphinidae (J. Gray, 1821), Monodontidae (J. Gray, 1821: 310), Physeteridae (J. Gray, 1821) and Ziphiidae (J. Gray, 1865d).

Order Denticeta Haeckel, 1895: 566, 569.

COMMENTS: When originally proposed, this rank was placed in the Legion Cetomorpha (Haeckel, 1895 [=Cetacea (Brisson, 1762)]) and included the families † Archidelphines (Haeckel, 1895: 566), † Zeuglododontida (Haeckel, 1895: 566 [=† Archaeoceti (Flower, 1883: 182)]),† Squalodontida (Mesoceta) (Haeckel, 1895: 566 [=† Squalodontidae (Brandt, 1873b: 576)]); Delphinida (Delphinoceta) (Haeckel, 1866 [=Delphinidae (J. Gray, 1821)]), Monodontida (Haeckel, 1895: 566 [=Monodontidae (J. Gray, 1821: 310)]), Ziphioida (Hyperoodontida) (J. Gray, 1868c [=Ziphiidae (J. Gray, 1865d)]) and Physeterida (Catodontida) (Haeckel, 1866 [=Odontoceti (Flower, 1867)]). Synonymised within Odontoceti by McKenna and Bell (1997: 379).

Infraorder Physeterida de Muizon, 1988a: 6, 65.

COMMENTS: When originally proposed as a new rank it was placed in the Suborder Odontoceti (Flower, 1867) and included the Superfamily Physeteroidea (J. Gray, 1821).

Order Odontocetiformes Kinman, 1994: 38.

COMMENTS: When originally proposed, this rank was placed in the Class Mammalea (Kinman, 1994 [=Mammalia

(Linnaeus, 1758)]). Synonymised within Odontoceti by McKenna and Bell (1997: 379).

Clade Synrhina Geisler *et al.*, 2011: 5, 29.

COMMENTS: When originally proposed, this clade was placed in the Odontoceti (Flower, 1867) and included the Delphinidae (J. Gray, 1821), Phocoenidae (J. Gray, 1825a), Monodontidae (J. Gray, 1821: 310), Inioidea (Gray, 1846c: 25), Lipotidae (Zhou *et al.* 1979: 72, 74 [=Iniidae (Gray, 1846c: 25)]), Ziphiidae (J. Gray, 1865d), Platanistidae (J. Gray, 1846c: 25), † Squalodelphinidae (Dal Piaz, 1917: 32), † Eurhinodelphinidae (Abel, 1901: 60), † *Kentriodon* Kellogg, 1927: 1; † *Atocetus* de Muizon, 1988a: 7, 129; and † *Albireo* Barnes, 1984: 1, 31.

Superfamily Physeteroidea J. Gray, 1821 *sensu* Bianucci & Landini, 2006

Family Physeteridae J. Gray, 1821: 310.

TYPE GENUS: *Physeter* Linnaeus, 1758.

COMMENTS: When originally proposed, this rank was placed in the Order Carnivorae (J. Gray, 1821 =[Cetacea (Brisson, 1762)]) and included the genera *Physeter* Linnaeus, 1758; and *Catodon* Linnaeus, 1761 [=*Physeter* Linnaeus, 1758]. Superfamily rank Physeteroidea recognised by Fordyce and Barnes (1994: 428), McKenna and Bell (1997: 379) and Rice (2009: 235). The scope of the superfamily was discussed by Bianucci and Landini (2006: 125) as a result of the discovery of a new basal physeteroid from late Miocene deposits of Italy. An updated review of phylogeny of Physeteroidea was provided by Lambert (2008: 277) including the crown and stem Physeteridae and Kogiidae.

Family Physeterida Haeckel, 1866: clix.

TYPE GENUS: *Physeter* Linnaeus, 1758.

COMMENTS: When originally proposed, this rank was placed in the Suborder Autoceta (Haeckel, 1866 [=Cetacea (Brisson, 1762)]) and included the genus *Physeter* Linnaeus, 1758. Name synonymised within Odontoceti by McKenna and Bell (1997: 379).

HOMONYMS:

Infraorder Physeterida de Muizon, 1988a, killer whales of the Class Mammalia. Name is a synonym of Physeteroidea J. Gray, 1821. See individual entry.

Suborder Physeteroidea J. Gray, 1871a: iv, 57.

COMMENTS: When originally proposed, this rank was placed in the Section Denticete (J. Gray, 1866b [=Odontoceti (Flower, 1867)]) and included the families Catodontidae (F. Cuvier, 1836a [=Physeteridae (J. Gray, 1821)]) and Physeteridae (J. Gray, 1821). Recognised as a superfamily by Fordyce and Barnes (1994: 428) who gave the author as J. Gray (1821: 428) and refer to the usage of Gill (1872: 15).

Superfamily? Physeteroidea Gill, 1872: 15.

TYPE GENUS: *Physeter* Linnaeus, 1758.

COMMENTS: When originally proposed, this rank was placed in the Suborder Denticete (J. Gray, 1866b [=Odontoceti (Flower, 1867)]) and included the Family Physeteridae (J. Gray, 1821). Synonymised within Physeteridae by McKenna and Bell (1997: 379).

Infraorder Physeterida de Muizon, 1988a: 65.

COMMENTS: When originally proposed as new, this rank included the Superfamily Physeteroidea (Gray, 1821).

HOMONYMS:

Family Physeterida Haeckel, 1866, killer whales of the Class Mammalia. Name is a synonym of Physeteroidea J. Gray, 1821. See individual entry.

Family Physeteridae J. Gray, 1821

Family Physeteridae J. Gray, 1821: 310.

TYPE GENUS: *Physeter* Linnaeus, 1758.

COMMENTS: When originally proposed, this rank was placed in the Order Carnivorae (J. Gray, 1821 =[Cetacea (Brisson, 1762)]) and included the genera *Physeter* Linnaeus, 1758; and *Catodon* Linnaeus, 1761 [=*Physeter* Linnaeus, 1758]. Superfamily rank Physeteroidea also recognised by McKenna and Bell (1997: 379). Extant subfamilies Physeterinae and Kogiinae were also recognised by McKenna and Bell (1997: 380).

Tribe Physeterina J. Gray, 1825a: 340

TYPE GENUS: *Physeter* Linnaeus, 1758.

COMMENTS: When originally proposed, this rank was placed in the Family Balaenidae (J. Gray, 1821) and included the genera *Physalus* Lacépède, 1804 [=*Physeter* Linnaeus, 1758]; *Physeter* Linnaeus, 1758; and *Catodon* Linnaeus, 1761 [=*Physeter* Linnaeus, 1758]. Synonymised within Physeteridae by McKenna and Bell (1997: 379).

Family Catodontidae F. Cuvier, 1836a: 564.

TYPE GENUS: *Catodon* Linnaeus, 1761 [=*Physeter* Linnaeus, 1758].

COMMENTS: When originally proposed, this rank was placed in the Tribe Zoophaga (F. Cuvier, 1836a [=Cetacea (Brisson, 1762)]) and included the genera *Catodon* Linnaeus, 1761 [=*Physeter* Linnaeus, 1758] and *Physeter* Linnaeus, 1758. Family name recognised by Wall (1851: 63). Synonymised within Physeteridae by McKenna and Bell (1997: 379).

Family Physetereae Lesson, 1842: 201.

TYPE GENUS: *Physeter* Linnaeus, 1758.

COMMENTS: When originally proposed, this rank was placed in the Order Cetaces (Lesson, 1842 [=Cetacea (Brisson, 1762)]) and included the genus *Physeter* Linnaeus,

1758. Does not appear to have been recognised by other authors.

Subfamily Physeterinae Flower, 1867: 114.

TYPE GENUS: *Physeter* Linnaeus, 1758.

COMMENTS: When originally proposed, this rank was placed in the Family Physeteridae (J. Gray, 1821) and included the genera *Physeter* Linnaeus, 1758; and *Kogia* J. Gray, 1846c. Subfamily rank recognised by Rice (1984: 490). Synonymised within the Subfamily Physeterinae (J. Gray, 1821: 310) by McKenna and Bell (1997: 379).

Family Physodontidae Lydekker, 1894b: 4.

TYPE GENUS: † *Physodon* Gervais, 1872: 101 [=*Scaldicetus* Du Bus, 1867: 568].

COMMENTS: When originally proposed, this rank included the genus † *Physodon* Gervais, 1872: 101. Synonymised within Physeteridae by McKenna and Bell (1997: 379).

Physeter Linnaeus, 1758

Physeter Linnaeus, 1758: 76.

TYPE SPECIES: *Physeter macrocephalus* Linnaeus, 1758 by subsequent designation. See discussion below under *P. macrocephalus*.

COMMENTS: See Husson and Holthuis (1974: 210) and Schevill (1986: 156). Taxonomic decision of Hershkovitz (1966: 116) to accept *Physeter* in preference to *Catodon*.

HOMONYMS:

Physeter de Blainville, 1838, Pygmy and Dwarf Sperm Whales of the Class Mammalia (Infraorder Cetacea, Family Kogiidae). Genus is a synonym of *Kogia* J. Gray, 1846c. See individual entry.

Catodon Linnaeus, 1761: 18.

TYPE SPECIES: *Physeter macrocephalus* Linnaeus, 1758 (as *Physeter catodon* Linnaeus, 1758) by virtual tautonomy.

COMMENTS: Recognised at genus rank by Lacépède (1804: xxxix), Tiedemann (1808: 575) and J. Gray (1864c: 231; 1866b: 195, 196). Synonymised within *Physeter* by Iredale and Troughton (1934: 58), Hershkovitz (1966: 116), and Mead and Brownell (1993: 359; 2005: 737).

HOMONYMS:

Catodon Duméril & Bibron, 1844: 319, blind snakes of the Class Reptilia (Order Squamata, Family Leptotyphlopidae). Genus is a synonym of *Leptotyphlops* Fitzinger, 1843: 24. See Giraudo and Scrocchi (2002: 4).

Physalus Lacépède, 1804: xl, 219.

TYPE SPECIES: *Phiseter cylindricus* (as *Physalus cylindricus*) Bonnaterre, 1789 [=*Physeter macrocephalus* Linnaeus, 1758] by monotypy.

COMMENTS: Genus recognised by J. Gray (1866b: 139). Synonymised within *Physeter* by Hershkovitz (1966: 116), and Mead and Brownell (1993: 359; 2005: 737).

HOMONYMS:

Physalus J. Gray, 1821, baleen whales of the Class Mammalia (Order Artiodactyla, Family Balaenopteridae). Genus is a synonym of *Balaenoptera* Lacépède, 1804. See individual entry.

Physalus de Blainville, 1830: 103, 'blue-bottle' jellyfish of the Phylum Cnidaria (Class Hydrozoa, Order Siphonophorae, Family Physalidae). An unjustified emendation of *Physalia* Lamarck, 1801: 355. See Sherborn (1929: 4934) and ICZN (1941: 17).

Physeterus Duméril, 1806b: 28.

TYPE SPECIES: 'les physeteres'.

COMMENTS: Also described by Duméril (1806a: 29). Synonymised within *Physeter* by Palmer (1904: 536), Iredale and Troughton (1934: 58), Hershkovitz (1966: 116) and McKenna and Bell (1997: 380).

Physelus Rafinesque, 1815: 60.

TYPE SPECIES: Incorrect subsequent spelling of *Physalus* Lacépède, 1804.

COMMENTS: Synonymised within *Physeter* by Palmer (1904: 536) and McKenna and Bell (1997: 380).

Cetus Oken, 1816: xiii, 674.

TYPE SPECIES: *Phiseter cylindricus* Bonnaterre, 1789 (as *Cetus cylindricus*) [=*Physeter macrocephalus* Linnaeus, 1758] by monotypy.

COMMENTS: Synonymised within *Physeter* by Hershkovitz (1966: 116), and Mead and Brownell (1993: 359; 2005: 737) who all give the author as Billberg (1828: 38) because the work of Oken (1816: 674) has been rejected for nomenclatural purposes by Opinion 417 of the ICZN (1956: 1).

HOMONYMS:

Cetus Brisson, 1762: 217, 225, whales of the Class Mammalia (Order Artiodactyla). Rejected for nomenclatural purposes by Opinion 1894 of the ICZN (1998: 64).

Cetus Billberg, 1828, the Sperm Whale of the Class Mammalia (Order Artiodactyla, Family Physeteridae). Genus is a synonym of *Physeter* Linnaeus, 1758. See individual entry.

Cetus Wagler, 1830, pilot whales of the Class Mammalia (Order Artiodactyla, Family Delphinidae). Genus is a synonym of *Globicephala* Lesson, 1828d. See individual entry.

Tursio Fleming, 1822b: 211.

TYPE SPECIES: *Delphinus vulgaris* Lacépède, 1804 (as *Tursio vulgaris*) [=*Delphinus delphis* Linnaeus, 1758] and *Physeter microps* Linnaeus, 1758 (as *T. microps*) [=*Physeter macrocephalus* Linnaeus, 1758].

COMMENTS: Synonymised within *Physeter* by Iredale and Troughton (1934: 59), Hershkovitz (1966: 116), and McKenna and Bell (1997: 380).

HOMONYMS:

Tursio Wagler, 1830, the Southern Rightwhale Dolphin of the Class Mammalia (Order Artiodactyla, Family Delphinidae). Genus is a synonym of *Lissodelphis* Gloger, 1841. See individual entry.

Tursio J. Gray, 1843a, bottle-nosed dolphins of the Class Mammalia (Order Artiodactyla, Family Delphinidae). Genus is a synonym of *Tursiops* Gervais, 1855a. See individual entry.

Megistosaurus Harlan, 1828: 186.

TYPE SPECIES: *Physeter macrocephalus* Linnaeus, 1758. See Hershkovitz (1966: 116).

COMMENTS: Synonymised within *Physeter* by Iredale and Troughton (1934: 59), Hershkovitz (1966: 116), and Mead and Brownell (1993: 359; 2005: 737).

Cetus Billberg, 1828: 38.

TYPE SPECIES: *Cetus cylindricus* Billberg, 1828 [=*Physeter macrocephalus* Linnaeus, 1758] by monotypy.

COMMENTS: Name synonymised within *Physeter* by Mead and Brownell (2005: 737).

HOMONYMS:

Cetus Brisson, 1762: 217, 225, whales of the Class Mammalia (Order Artiodactyla). Rejected for nomenclatural purposes by Opinion 1894 of the ICZN (1998: 64).

Cetus Oken, 1816, the Sperm Whale of the Class Mammalia (Order Artiodactyla, Family Physeteridae). Genus is a synonym of *Physeter* Linnaeus, 1758. See individual entry.

Cetus Wagler, 1830, pilot whales of the Class Mammalia (Order Artiodactyla, Family Delphinidae). Genus is a synonym of *Globicephala* Lesson, 1828d. See individual entry.

Physeteres F. Cuvier, 1829b: 518.

TYPE SPECIES: Appears to be an emendation of *Physeter* Linnaeus, 1758.

COMMENTS: Synonymised within *Physeter* by Palmer (1904: 536) and McKenna and Bell (1997: 380).

Cachelot Hamilton, 1937: 154.

TYPE SPECIES: *Physeter catodon* Linnaeus, 1758 [=*Physeter macrocephalus* Linnaeus, 1758] by tautotypy. See Iredale and Troughton (1934: 59).

COMMENTS: Synonymised within *Physeter* by Iredale and Troughton (1934: 59).

Meganeuron J. Gray, 1865f: 439, 440.

TYPE SPECIES: *Catodon (Meganeuron) krefftii* J. Gray, 1865f [=*Physeter macrocephalus* Linnaeus, 1758] by monotypy.

COMMENTS: Described as a subgenus of *Catodon*, and raised to generic rank by J. Gray (1866b: 387). Synonymised within *Physeter* by Iredale and Troughton (1934: 59), Hershkovitz (1966: 116), and Mead and Brownell (1993: 359; 2005: 737).

Physeter macrocephalus Linnaeus, 1758

Sperm Whale

Physeter macrocephalus Linnaeus, 1758: 76.

TYPE LOCALITY: 'Oceano Europeao'.

COMMENTS: Linnaeus (1758: 76) used both *catodon* and *macrocephalus*; of these, *P. catodon* has line priority. The use of *macrocephalus* or *catodon* has been under considerable debate since their description. The specific name *macrocephalus* was given preference by various authors including J. Gray (1866b: 202), A. Murray (1866: 210), Boschma (1938: 166), Holthius (1987: 87), Bannister *et al.* (1996: 39) and more recently by Van Dyck and Strahan (2007: 808) and Whitehead (2009: 1091). However other authors including Schevill (1987a: 89), who replied to Holthius (1987), strongly argued for *catodon* to be the senior synonym. Synonymised within *catodon* by Bannister (1988b: 199), and Mead and Brownell (1993: 359; 2005: 737). Rice (1998: 82) noted that *P. macrocephalus* takes precedence over P. *catodon* because of the Principle of the First Reviser, as decreed in Article 24 of the Code (ICZN, 1985a: 53) with Husson and Holthuis (1974: 214) acting as first reviser, and this was supported by Holthuis (1987: 87) and the others mentioned above. More recently this has been supported by the Committee on Taxonomy (2011) of the Society for Marine Mammalogy.

Physeter catodon Linnaeus, 1758: 76.

TYPE LOCALITY: 'Oceano septentrionali' but given as Kairston, Orkney, Scotland by Thomas (1911b: 157).

COMMENTS: Neotype designated by Husson and Holthuis (1974: 212). Linnaeus (1758: 76) used both *catodon* and *macrocephalus*; *P. catodon* has line priority. G. Cuvier (1823a: 334) was the first author to suggest that all sperm whales were conspecific. Taxonomic decision of Thomas (1911b: 157), Hershkovitz (1966: 116) and Scheville (1986: 156) to recognise *catodon* as the senior name. Since the description numerous authors have used both *catodon* and *macrocephalus* (e.g. see Boschma, 1938: 161). Type designation by Boschma (1938: 161) and Schevill (1986: 156; 1987a: 89) who both reviewed the taxonomic history of the species. However the review of Schevill (1986: 153) was critically assessed by Holthius (1987: 87) who proposed that it was not strictly correct and suggested that *Physeter macrocephalus* is the valid name. This was in turn disputed by Schevill (1987a: 89). Subsequently Rice (1989a: 178) reviewed the taxon and concurred that the correct name was

Physeter macrocephalus. Despite this the name *catodon* was given priority by Mead and Brownell (2005: 737).

Physeter microps Linnaeus, 1758: 76.

TYPE LOCALITY: 'Oceano septentrionali' but given as Kairston, Orkney, Scotland by Thomas (1911b: 157).

COMMENTS: Synonymised within *Physeter catodon* Linnaeus, 1758 by Hershkovitz (1966: 118).

Physeter Andersoni Borowski, 1780: 33.

TYPE LOCALITY: Iceland and Greenland.

COMMENTS: Name based on the 'cachalot a dents pointues' of Brisson (1756: 362). Synonymised within *catodon* Linnaeus, 1758 by Hershkovitz (1966: 119).

P. [hiseter (sic)] Trumpo Bonnaterre, 1789: 14; Plate 8, Fig. 1.

TYPE LOCALITY: Bayonne, N. Atlantic, France.

COMMENTS: Synonymised within *Physeter catodon* Linnaeus, 1758 by Hershkovitz (1966: 119).

P. [hiseter (sic)] Cylindricus Bonnaterre, 1789: 16; Plate 7, Fig. 1.

TYPE LOCALITY: North Atlantic, Greenland.

COMMENTS: Synonymised within *catodon* by Hershkovitz (1966: 119).

Physalus cylindricus Lacépède, 1804: xl, 219; Plate 9, Fig. 3.

TYPE LOCALITY: Moluccas and New Zealand.

COMMENTS: Synonymised within *catodon* by Hershkovitz (1966: 119).

Physeterus [sic] sulcatus Lacépède, 1818: 474.

TYPE LOCALITY: Name based on a Japanese drawing of a sperm whale.

COMMENTS: Synonymised within *catodon* by Hershkovitz (1966: 120).

physeter polycyphus Quoy & Gaimard, 1824: 77; Plate 12.

TYPE LOCALITY: Name based on an observed animal at sea.

COMMENTS: Species placed within *Catodon* by Lesson (1827a: 422). Synonymised within *australasianus* by Iredale and Troughton (1934: 59), and within *catodon* by Hershkovitz (1966: 120).

Physeter australasianus Desmoulins, 1822b: 618.

TYPE LOCALITY: Moluccas and New Zealand.

COMMENTS: Taxon recognised by Iredale and Troughton (1934: 59), but synonymised within *catodon* by Hershkovitz (1966: 120), Mead and Brownell (1993: 359; 2005: 737) and other authors.

D. [elphinus] Bayeri Risso, 1826b: 22.

TYPE LOCALITY: Nice, Mediterranean.

COMMENTS: Synonymised within *catodon* by Hershkovitz (1966: 120).

Cetus Cylindricus Billberg, 1828: 38.

TYPE LOCALITY: Unknown.

COMMENTS: Synonymised within *catodon* by Hershkovitz (1966: 119).

P. [hyseter] australis J. Gray, 1846c: 22.

TYPE LOCALITY: Port Jackson, Sydney, New South Wales, Australia.

COMMENTS: Lapsus for *Physeter australasianus*. Synonymised within *catodon* by Hershkovitz (1966: 120), and Mead and Brownell (1993: 359; 2005: 737).

Catodon Colneti J. Gray, 1850: vii, 52.

TYPE LOCALITY: Point Angeles, Mexico.

COMMENTS: Synonymised within *catodon* by Hershkovitz (1966: 120).

Catodon australis Wall, 1851: 1.

TYPE LOCALITY: Port Jackson, Sydney, New South Wales, Australia.

COMMENTS: Synonymised within *australasianus* by Iredale and Troughton (1934: 59), and within *catodon* by Hershkovitz (1966: 120) and Bannister (1988b: 199).

Catodon (Meganeuron) krefftii J. Gray, 1865f: 440.

TYPE LOCALITY: New South Wales, Australia.

COMMENTS: Synonymised within *australasianus* by Iredale and Troughton (1934: 59), and within *catodon* by Hershkovitz (1966: 120) and Bannister (1988b: 199).

Family Kogiidae Gill, 1871

Subfamily Kogiinae Gill, 1871c: 732.

TYPE GENUS: *Kogia* J. Gray, 1846c.

COMMENTS: When originally proposed, this rank was placed in the Family Physeteridae (J. Gray, 1821) and included the genus *Kogia* J. Gray, 1846c. de Muizon (1988a: 65; 1991: 297) recognised two subfamilies, the † Subfamily Scaphokogiinae (de Muizon, 1988a: 66) for a Miocene species from Peru and the Kogiinae for the living genus and a Miocene fossil genus from Mexico. The Subfamily Kogiinae was subsequently recognised by Rice (1984: 490). Synonymised within Physeteridae by Simpson (1945: 102) and Mead and Brownell (2005: 737).

Family Kogiidae Miller, 1923: 33, 45.

TYPE GENUS: *Kogia* J. Gray, 1846c.

COMMENTS: When originally proposed, this rank was placed in the Order Cetacea (Brisson, 1762) and included the genus *Kogia* J. Gray, 1846c. Synonymised within the Subfamily Kogiinae (Gill, 1871c: 732) by McKenna and Bell (1997: 380).

Kogia J. Gray, 1846

Kogia J. Gray, 1846c: 22.

TYPE SPECIES: *Physeter breviceps* de Blainville, 1838 [=*Kogia breviceps* (de Blainville, 1838)] by monotypy.

COMMENTS: Genus recognised by Iredale and Troughton (1934: ix, 59), Hershkovitz (1966: 113), and subsequent authors. Genus reviewed by Handley (1966: 62).

Physeter de Blainville, 1838: 337.

TYPE SPECIES: *Physeter breviceps* de Blainville, 1838 [=*Kogia breviceps* (de Blainville, 1838)] by monotypy.

COMMENTS: Synonymised within *Kogia* by Handley (1966: 67) and Nagorsen (1985: 1).

HOMONYMS:

Physeter Linnaeus, 1758, the Sperm Whale of the Class Mammalia (Infraorder Cetacea, Family Physeteridae). Currently accepted genus. See individual entry.

Euphysetes Wall, 1851: 37, 46.

TYPE SPECIES: *Euphysetes grayii* Wall, 1851 [= *Kogia breviceps* (de Blainville, 1838)] by monotypy.

COMMENTS: Synonymised within *Kogia* by Iredale and Troughton (1934: 59), Handley (1966: 67), Hershkovitz (1966: 113), Nagorsen (1985: 1), and Mead and Brownell (1993: 359; 2005: 737).

Euphycetes J. Gray, 1866b: 391.

TYPE SPECIES: *Euphysetes grayii* Wall, 1851 [=*Kogia breviceps* (de Blainville, 1838)] by monotypy.

COMMENTS: Proposed as alternative spelling to *Euphysetes* but not adopted. Synonymised within *Kogia* by McKenna and Bell (1997: 381).

Callignathus Gill, 1871c: 738.

TYPE SPECIES: *Physeter (Euphysetes) simus* Owen, 1866b (as *Callignathus simus*) [=*Kogia sima* (Owen, 1866b)] by original designation. See Hershkovitz (1966: 113).

COMMENTS: Synonymised within *Kogia* by Iredale and Troughton (1934: 59), Handley (1966: 67), Hershkovitz (1966: 113), Nagorsen (1985: 1), and Mead and Brownell (1993: 359; 2005: 737).

HOMONYMS:

Callignathus Agassiz, 1846: 58, darkling beetles of the Class Insecta (Order Coleoptera, Family Tenebrionidae). Incorrect subsequent spelling of *Calognathus* Guérin, 1837: Text to Plate 172.

Cogia Wallace, 1876: 208.

TYPE SPECIES: Invalid emendation of *Kogia* J. Gray, 1846c.

COMMENTS: Recognised by Flower and Lydekker (1891: 250). Synonymised within *Kogia* by Palmer (1904: 358), Handley (1966: 67), Hershkovitz (1966: 113), Nagorsen (1985: 1), and Mead and Brownell (1993: 359; 2005: 737).

Callignathula Strand, 1928: 61.

TYPE SPECIES: Substitute name for *Callignathus* Gill, 1871c.

COMMENTS: Often stated as published in 1926 but this appears to be in error (McKenna & Bell, 1997: 381). Synonymised within *Kogia* by Iredale and Troughton (1934: 59), Hershkovitz (1966: 113) and Handley (1966: 67).

Gallignathus Tomilin, 1957: 418.

TYPE SPECIES: Misprint for *Callignathus* Gill, 1871c.

COMMENTS: Synonymised within *Kogia* by Hershkovitz (1966: 113).

Kogia breviceps (de Blainville, 1838)

Pygmy Sperm Whale

physeter [sic] *breviceps* de Blainville, 1838: 337; Plate 10.

TYPE LOCALITY: Cape of Good Hope, South Africa.

COMMENTS: Placed within *Kogia* by J. Gray (1846c: 22) and followed by most subsequent authors including Hershkovitz (1966: 113). Reviewed by Ross (1984: 247), and Caldwell and Caldwell (1989: 235).

Euphysetes Grayii Wall, 1851: 37; Plate 2.

TYPE LOCALITY: Maroubra Beach, New South Wales, Australia.

COMMENTS: Recognised within *Kogia* by J. Gray (1866b: 218) and Gill (1871c: 738). Synonymised within *breviceps* by Iredale and Troughton (1934: 59), Handley (1966: 67), Hershkovitz (1966: 114), and Mead and Brownell (1993: 359; 2005: 737).

Euphysetes macleayi Krefft, 1866b: 713; Figs. 1–6.

TYPE LOCALITY: Manly Beach, Sydney, Australia.

COMMENTS: Species recognised within *Kogia* by Iredale and Troughton (1934: 59), J. Gray (1966c: 391). Synonymised within *breviceps* by Iredale and Troughton (1934: 59) and Hershkovitz (1966: 114).

Kogia Floweri Gill, 1871c: 738; Figs. 168–172.

TYPE LOCALITY: North Pacific, Mazatlán, Sinaloa, Mexico.

COMMENTS: Synonymised within *breviceps* by Handley (1966: 67), Hershkovitz (1966: 115), and Mead and Brownell (1993: 359; 2005: 737).

E. [uphysetes] pottsii Haast, 1874: 97, 100; Plate 15.

TYPE LOCALITY: Governor Bay, near Ohinitahi, New Zealand.

COMMENTS: Synonymised within *breviceps* by Iredale and Troughton (1934: 59), Handley (1966: 68) and Hershkovitz (1966: 115).

Kogia Goodei True, 1884: 630, 641.

TYPE LOCALITY: *Nomen nudum*.

COMMENTS: Synonymised within *breviceps* by Hershkovitz (1966: 115), and Mead and Brownell (1993: 359; 2005: 737).

Kogia sima (Owen, 1866)

Dwarf Sperm Whale

Physeter (Euphysetes) simus Owen, 1866b: 30; Plates 10–14.

TYPE LOCALITY: Waltair, Madras, India.

COMMENTS: Taxon placed within *Kogia* by Yamada (1954: 37). Synonymised within *Kogia breviceps* by Hershkovitz (1966: 114), but recognised by Handley (1966: 67, 68). The feminine form of the specific name, *sima*, fixed to replaced *simus* or *simum* by Rice (1998: 84). Reviewed by Nagorsen (1985: 1) and Caldwell and Caldwell (1989: 235). Genetic evidence of Chivers *et al*. (2006: 619) gives evidence for the existence of two species within *Kogia sima*, one from the Indo-Pacific Ocean and one in the Atlantic Ocean.

FUTURE TAXONOMIC RESEARCH: Research is required to confirm the existence of a second species within *Kogia sima*, and if confirmed, should be formally described as a new species.

Superfamily Ziphioidea J. Gray, 1865 *sensu* Rice, 2009

Family Ziphiidae J. Gray, 1865d: 528.

TYPE GENUS: *Ziphius* G. Cuvier, 1823a.

COMMENTS: When originally proposed, this rank included the tribes Hyperoodontina (J. Gray, 1846c [=Ziphiidae (J. Gray, 1865d)]), Epiodontina (J. Gray, 1865d [=Ziphiidae (J. Gray, 1865d)]) and Ziphiina (J. Gray, 1863d [=Ziphiidae (J. Gray, 1865d)]). Synonymised within the Superfamily Hyperoodontoidea by McKenna and Bell (1997: 381) but superfamily rank recognised by Fordyce and Barnes (1994: 428) and Rice (2009: 235). Not recognised by Mead (2009: 94) who only recognised the subfamilies Ziphiinae and Hyperoodontinae.

Suborder Ziphioidea J. Gray, 1868c: 9.

COMMENTS: When originally proposed, this rank was placed in the Section Denticete (J. Gray, 1866b [=Odontoceti (Flower, 1867)]) and included the families Hyperoodontidae (J. Gray, 1846c [=Ziphiidae (J. Gray, 1865d)]) and Epiodontidae (J. Gray, 1868 [=Ziphiidae (J. Gray, 1865d)]) and Ziphiidae (J. Gray, 1865d). Taxon recognised by F. Fraser and Purves (1960: 88) and as a superfamily by Fordyce and Barnes (1994: 428). Synonymised within Hyperoodontoidea J. Gray, 1846c: 24 by McKenna and Bell (1997: 381).

Family Ziphiidae J. Gray, 1865

Family Ziphiidae J. Gray, 1865d: 528.

TYPE GENUS: *Ziphius* G. Cuvier, 1823a.

COMMENTS: When originally proposed, this rank included the tribes Hyperoodontina (J. Gray, 1846c [=Ziphiidae (J. Gray, 1865d)]), Epiodontina (J. Gray, 1865d [=Ziphiidae (J. Gray, 1865d)]) and Ziphiina (J. Gray, 1863d [=Ziphiidae (J. Gray, 1865d)]). Mead and Brownell (1993: 361; 2005: 739) noted that although Hyperoodontidae (J. Gray, 1846c) has priority over Ziphiidae, they chose the latter name following Article 23(b) (1985) and Article 23.12 (1999) of the Code (ICZN, 1985a: 47; 1999: 29) because Ziphiidae has been the name of choice for more than 100 years. Family reviewed by Moore (1968: 209) and Dalebout *et al*. (2004: 459) who used mitochondrial DNA to confirm each species. Contra Mead (2009: 94), we recognise no subfamilies, as the DNA work of McGowen *et al*. (2009: 891) indicates a comparatively rapid successive separation of the genera of the family beginning about 22 million years ago, and their interrelationships do not conform to the postulated subfamilies.

Tribe Hyperoodontina J. Gray, 1846c: 24.

TYPE GENUS: *Hyperoodon* Lacépède, 1804.

COMMENTS: When originally proposed, this rank was placed in the Family Delphinidae (J. Gray, 1821) and included the genera *Hyperoodon* Lacépède, 1804; *Ziphius* G. Cuvier, 1823a; and *Delphinorhynchus* de Blainville, 1817: 151 *nomen dubium*. Recognised as a 'section' within the Family Delphinidae. Spelt Hyperodontidae by J. Gray (1866b: 62). Recognised as the Family Hyperoodontidae by Hershkovitz (1966: 122) and Superfamily Hyperoodontoidea by McKenna and Bell (1997: 381). Synonymised within Ziphiidae by Mead and Brownell (1993: 361; 2005: 739). The rank Hyperooodontina was synonymised within Hyperoodontoidea and Hyperoodontidae (J. Gray, 1846c) by McKenna and Bell (1997: 381). Tribe rank recognised by Moore (1968: 276) as a new rank, with family rank recognised by McKenna and Bell (1997: 381). The technical priority of the name Hyperoodontidae over the name Ziphiidae was discussed by Rice (1998: 85), who noted that apart from Iredale and Troughton (1934: ix, 60), Hershkovitz (1966: vii, 122) and Moore (1968: 276), the name Ziphiidae has been in universal use for over a century. As a result Rice (1998: 85) followed Mead and Brownell (1993: viii; 361) and other authors who retained it under Article 23b of the Code (ICZN, 1985a: 47).

Family Heterodontidae Girard, 1852: 324.

TYPE GENUS: *Heterodon* de Blainville, 1817.

COMMENTS: When originally proposed, this rank was placed in the Order Odontoceti (Flower, 1867) and included the genus *Heterodon* de Blainville, 1817 [=*Hyperoodon*

Lacépède, 1804]; and *Monodon* Linnaeus, 1758: 75. Synonymised within Ziphiidae by Rice (1998: 85).

Tribe? Ziphiina J. Gray, 1863d: 200.

TYPE GENUS: *Ziphius* G. Cuvier, 1823a.

COMMENTS: When originally proposed, this rank was placed in the Family Delphinidae (J. Gray, 1821) and included the genera *Bernardus* [sic =*Berardius*] Duvernoy, 1851; *Ziphius* G. Cuvier, 1823a; *Delphinorhynchus* de Blainville, 1817: 151 *nomen dubium*; and *Dioplodon* Gervais, 1850a [=*Mesoplodon* Gervais, 1850b]. Rank unknown but equivalent to subfamily or tribe. Tribe rank recognised by Moore (1968: 276) as a new rank. Reference given as J. Gray (1850: 59, 51) by McKenna and Bell (1997: 381).

Tribe Epiodontina J. Gray, 1865d: 528.

TYPE GENUS: *Epiodon* Rafinesque, 1814a [=*Mesoplodon* Gervais, 1850b].

COMMENTS: When originally proposed, this rank was placed in the Family Ziphiidae (J. Gray, 1865d) and included the genera *Aliama* J. Gray, 1864c [=*Ziphius* G. Cuvier, 1823a]; *Epiodon* Rafinesque, 1814a [=*Mesoplodon* Gervais, 1850b]; and *Petrorhynchus* J. Gray, 1865d [=*Ziphius* G. Cuvier, 1823a]. Synonymised within the Family Ziphiidae by Rice (1998: 85).

Family Hyperodontidae J. Gray, 1866b: 62.

TYPE GENUS: *Hyperoodon* Lacépède, 1804.

COMMENTS: When originally proposed, this rank was placed in the Section Denticete (J. Gray, 1866b [=Odontoceti (Flower, 1867)]). Recognised as a subfamily Hyperodontina and placed in the Family Ziphiidae on page 327. Synonymised within Hyperoodontidae by McKenna and Bell (1997: 381).

Family Hyperoodonta Haeckel, 1866: clix.

TYPE GENUS: *Hyperoodon* Lacépède, 1804.

COMMENTS: When originally proposed, this rank was placed in the Suborder Autoceta (Haeckel, 1866 [=Cetacea (Brisson, 1762)]) and included the genera *Hyperoodon* Lacépède, 1804; *Ziphius* G. Cuvier, 1823a; and *Dioplodon* Gervais, 1850a [=*Mesoplodon* Gervais, 1850b]. Synonymised within Hyperoodontidae by McKenna and Bell (1997: 381).

Family Epiodontidae J. Gray, 1868c: 9.

TYPE GENUS: *Epiodon* Rafinesque, 1814a [=*Mesoplodon* Gervais, 1850b].

COMMENTS: When originally proposed, this rank was placed in the Suborder Ziphioidea (J. Gray, 1868c [=Ziphiidae (J. Gray, 1865d)]) and included the genera *Epiodon* Rafinesque, 1814a [=*Mesoplodon* Gervais, 1850b]; and *Petrorhynchus* J. Gray, 1865d [=*Ziphius* G. Cuvier, 1823a]. Synonymised within the Family Ziphiidae by Rice (1998: 85).

Subfamily Ananarcinae Gill, 1871b: 124.

TYPE GENUS: *Anarnak* Lacépède, 1804 [*incertae sedis*].

COMMENTS: When originally proposed, this rank was placed in the Family Ziphiidae (J. Gray, 1865d) and included the genus *Anarnacus* [=*Anarnak* Lacépède, 1804; *incertae sedis*]. Subfamily name spelt Anarnacinae on page 126. Subfamily name is an incorrect subsequent spelling of the Subfamily Anarnacinae. Synonymised within Hyperoodontidae by McKenna and Bell (1997: 381) and within Ziphiidae by Rice (1998: 85).

Family Xiphidae Ameghino, 1889: xxvi, 895.

TYPE GENUS: *Xiphius* Agassiz, 1846 [=*Ziphius* G. Cuvier, 1823a].

COMMENTS: When originally proposed, this rank was placed in the Order Odontoceta (Ameghino, 1889 [=Odontoceti (Flower, 1867)]) and included the genus *Xiphius* Agassiz, 1846 [=*Ziphius* G. Cuvier, 1823a]. Synonymised within the Family Ziphiidae by Rice (1998: 85).

Xiphiini Winge, 1918 [p. 11 of 1921 English Edition].

TYPE GENUS: *Xiphius* Agassiz, 1846 [=*Ziphius* G. Cuvier, 1823a].

COMMENTS: When originally proposed, this rank was placed in the Family Physeteridae (J. Gray, 1821) and included the genera † *Argyrodelphis* Lydekker, 1894b: 12 [=† *Notocetus* Moreno, 1892: 397]; *Mesoplodon* Gervais, 1850b; † *Xiphirostrum* Du Bus, 1868: 622; † *Chonoxiphius* [=† *Choneziphius*] Duvernoy, 1851: 43; *Xiphius* Agassiz, 1846 [=*Ziphius* G. Cuvier, 1823a]; *Berardius* Duvernoy, 1851; and *Hyperoodon* Lacépède, 1804. Synonymised within the Family Ziphiidae by Rice (1998: 85).

Subfamily Ziphiinae F. Fraser & Purves, 1960: Table adjacent to page 108.

TYPE GENUS: *Ziphius* G. Cuvier, 1823a.

COMMENTS: When originally proposed, this rank was placed in the Family Ziphiidae (J. Gray, 1865d) and included the genera *Hyperoodon* Lacépède, 1804; *Ziphius* G. Cuvier, 1823a; *Berardius* Duvernoy, 1851; and *Mesoplodon* Gervais, 1850b. Synonymised within Hyperoodontidae by McKenna and Bell (1997: 381).

Family Hyperoodontidae Hershovitz, 1966: vii, 122.

TYPE GENUS: *Hyperoodon* Lacépède, 1804.

COMMENTS: When originally proposed, this rank was placed in the Suborder Odontoceti (Flower, 1867) and included the genera *Tasmacetus* Oliver, 1937; *Berardius* Duvernoy, 1851; *Mesoplodon* Gervais, 1850b; *Ziphius* G. Cuvier, 1823a; and *Hyperoodon* Lacépède, 1804. Name also recognised by Moore (1968: 276).

Superfamily Hyperoodontoidea Moore, 1968: 276.

TYPE GENUS: *Hyperoodon* Lacépède, 1804.

COMMENTS: When originally proposed as a new rank it was placed in the Suborder Odontoceti (Flower, 1867) and included the Family Hyperoodontidae (J. Gray, 1866b [=Ziphiidae (J. Gray, 1865d)]). Synonymised within Hyperoodontoidea (J. Gray, 1846c: 24), by McKenna and Bell (1997: 381).

Tribe Ziphiini Moore, 1968: 276.

TYPE GENUS: *Ziphius* G. Cuvier, 1823a.

COMMENTS: When originally proposed as a new rank it was placed in the Family Hyperoodontidae (J. Gray, 1866b [=Ziphiidae (J. Gray, 1865d)]) and included the subtribes Berardiina (Moore, 1968 [=Ziphiidae (J. Gray, 1865d)]) and Ziphiina (J. Gray, 1863d [=Ziphiidae (J. Gray, 1865d)]). Synonymised within Hyperoodontidae by McKenna and Bell (1997: 381).

Subtribe Berardiina Moore, 1968: 276.

TYPE GENUS: *Berardius* Duvernoy, 1851.

COMMENTS: When originally proposed as a new rank it was placed in the Tribe Ziphini (Moore, 1968 [=Ziphiidae (J. Gray, 1865d)]) and included the genus *Berardius* Duvernoy, 1851. Synonymised within Hyperoodontidae by McKenna and Bell (1997: 381) and within the Family Ziphiidae by Rice (1998: 85).

Tribe Hyperoodontini Moore, 1968: 276.

TYPE GENUS: *Hyperoodon* Lacépède, 1804.

COMMENTS: When originally proposed as a new rank it was placed in the Family Hyperoodontidae (J. Gray, 1866b [=Ziphiidae (J. Gray, 1865d)]) and included the subtribes Tasmacetina (Moore, 1968 [=Ziphiidae (J. Gray, 1865d)]), Indopacetina (Moore, 1968 [=Ziphiidae (J. Gray, 1865d)]) and Hyperoodontina (J. Gray, 1846c [=Ziphiidae (J. Gray, 1865d)]). Synonymised within Hyperoodontidae by McKenna and Bell (1997: 381).

Subtribe Tasmacetina Moore, 1968: 276.

TYPE GENUS: *Tasmacetus* Oliver, 1937.

COMMENTS: When originally proposed as a new rank it was placed in the Tribe Hyperoodontini (Moore, 1968 [=Ziphiidae (J. Gray, 1865d)]) and included the genus *Tasmacetus* Oliver, 1937. Synonymised within Hyperoodontidae by McKenna and Bell (1997: 381) and within the Family Ziphiidae by Rice (1998: 85).

Subtribe Indopacetina Moore, 1968: 277.

TYPE GENUS: *Indopacetus* Moore, 1968.

COMMENTS: When originally proposed as a new rank it was placed in the Tribe Hyperoodontini (Moore, 1968 [=Ziphiidae (J. Gray, 1865d)]) and included the genus *Indopacetus* Moore, 1968. Synonymised within Hyperoodontidae by McKenna and Bell (1997: 381) and within the Family Ziphiidae by Rice (1998: 85).

Subfamily Hyperoodontinae Mead, 2009: 94.

TYPE GENUS: *Hyperoodon* Lacépède, 1804.

COMMENTS: When originally proposed this rank was placed in the Family Ziphiidae (J. Gray, 1865d) and included the genera *Hyperoodon* Lacépède, 1804; *Indopacetus* Moore, 1968 and *Mesoplodon* Gervais, 1850b.

Berardius Duvernoy, 1851

Berardius Duvernoy, 1851: 41, 52.

TYPE SPECIES: *Berardius arnuxii* Duvernoy, 1851 by original designation.

COMMENTS: Genus recognised by J. Gray (1866b: 327, 348) and most subsequent authors.

Berardus J. Gray, 1863d: 200.

TYPE SPECIES: Invalid emendation of *Berardius* Duvernoy, 1851.

COMMENTS: Incorrect subsequent spelling of *Berardius* Duvernoy, 1851. Synonymised within *Berardius* by Palmer (1904: 136).

Rostrifer Zenkovicz, 1947: 15.

TYPE SPECIES: *Rostrifer nestoresmirnovi* Zenkovicz, 1947.

COMMENTS: *Nomen nudum*. Synonymised within *Berardius* by McKenna and Bell (1997: 383).

Berardius arnuxii Duvernoy, 1851

Arnoux's Beaked Whale

Berardius arnuxii Duvernoy, 1851: 52; Fig. 1.

TYPE LOCALITY: Akaroa, near Banks Peninsula, New Zealand.

COMMENTS: Species recognised by J. Gray (1866b: 348) and Flower (1872: 203). Species reviewed by Hershkovitz (1966: 122). Reviewed by Balcomb (1989: 261).

Hyperoodon Lacépède, 1804

Hyperoodon Lacépède, 1804: xliv, 319.

TYPE SPECIES: Φ *Hyperoodon butskopf* Lacépède, 1804: xliv, 319 [=Φ *Hyperoodon ampullatus* (Forster, 1770: 18)] by monotypy.

COMMENTS: Genus recognised by J. Gray (1866b: 327, 328). Genus reviewed by Hershkovitz (1966: 142).

HOMONYMS:

Hyperoodon Philippi, 1902: 1, frogs of the Class Amphibia (Order Anura, Family Cycloramphidae). Genus is a synonym of *Odontophrynus* Reinhardt and Lütken, 1862: 159, 162. See Frost (2011).

Uranodon Illiger, 1811: 143.

TYPE SPECIES: Φ *Delphinus Butzkopf* [sic =*Butskopf*] Bonnaterre, 1789: 25 [sic] [=Φ *Hyperoodon ampullatus* (Forster, 1770: 18)] by monotypy.

COMMENTS: *Nomen novum* for *Hyperoodon* Lacépède, 1804. Synonymised within *Hyperoodon* by Iredale and Troughton (1934: 60), Hershkovitz (1966: 142), McKenna and Bell (1997: 382), and Mead and Brownell (1993: 361; 2005: 740).

Bidens G. Fischer, 1814: 686.

TYPE SPECIES: Φ *Delphinus diodon* Lacépède, 1804: xliii, 309 (via bottle-nosed whale of Hunter, 1787: 373) [=Φ *Hyperoodon ampullatus* (Forster, 1770: 18)] by monotypy. See Hershkovitz (1966: 147).

COMMENTS: Synonymised within *Hyperoodon* by Iredale and Troughton (1934: 60), Hershkovitz (1966: 142) and McKenna and Bell (1997: 382).

Heterodon de Blainville, 1817: 151, 175.

TYPE SPECIES: Φ *Delphinus Butzkopf* Bonnaterre, 1789: 25 (as *Delphinus Butskode* and *Dauphin butskopf*) [=Φ *Hyperoodon ampullatus* (Forster, 1770: 18)] by subsequent designation. See Hershkovitz (1966: 142).

COMMENTS: Described as a subgenus of *Delphinus*. Synonymised within *Hyperoodon* by Hershkovitz (1966: 142), McKenna and Bell (1997: 382), and Mead and Brownell (1993: 361, 2005: 740).

HOMONYMS:

Heterodon Latreille, 1801: 32, hognose snakes of the Class Reptilia (Order Squamata, Family Colubridae). Currently recognised name.

Heterodon Lund, 1838: 11, extinct relative of modern armadillos of the Class Mammalia (Order Cingulata, † Family Glyptodontidae).

Heterodon J. Gray, 1840b: 150, of the Phylum Mollusca (Order Goniopoda?, Family Unionidae?). *Incertae sedis*.

Heterodon Bleeker, 1845: 523, threadfin bream fish of the Superclass Pisces (Order Perciformes, Family Nemipteridae). Genus is a synonym of *Pentapodus* Quoy and Gaimard, 1824: 294. See Russell (1990: 78).

Hyperdordon J. Gray, 1821: 310

TYPE SPECIES: Typographical error of *Heterodon* de Blainville, 1817.

COMMENTS: Synonymised within *Hyperoodon* by Hershkovitz (1966: 143).

Ceto-diodon Jacob, 1825: 72.

TYPE SPECIES: Φ *Delphinus hunteri* Desmarest, 1822b: 520 (as *Ceto-diodon Hunteri*) [=Φ *Hyperoodon ampullatus* (Forster, 1770: 18)] by monotypy.

COMMENTS: Synonymised within *Hyperoodon* by Iredale and Troughton (1934: 60), Hershkovitz (1966: 143) and McKenna and Bell (1997: 382), but not considered by Mead and Brownell (1993: 361, 2005: 740).

Anodon Wagler, 1830: 34 footnote 3.

TYPE SPECIES: Φ *Delphinus edentulus* Schreber, 1792: Plate 347 [=Φ *Hyperoodon ampullatus* (Forster, 1770: 18)] by subsequent designation. See Hershkovitz (1966: 147).

COMMENTS: Appears to be an incorrect subsequent spelling of *Aodon* Lesson, 1828d [=*Mesoplodon* Gervais, 1850b]. Synonymised within *Hyperoodon* by Hershkovitz (1966: 143), and Mead and Brownell (1993: 361; 2005: 740).

HOMONYMS:

Anodon Oken, 1815: 236, freshwater mussels of the Phylum Mollusca (Class Bivalvia, Order Unionoida, Family Unionidae). An unjustified emendation of *Anodonta* Lamarck, 1799: 87. See Melville and Smith (1987: 46).

Anodon A. Smith, 1829: 443, African egg-eating snakes of the Class Reptilia (Order Squamata, Family Colubridae). Genus is a synonym of *Dasypeltis* Wagler, 1830: 178.

Anodon Agassiz, 1846: 24, 27, fish of the Superclass Pisces (Order Rajiformes, Family Myliobatidae). An unjustified emendation of *Aodon* Lacépède, 1798: 297.

Anodon J. Gray, 1850, beaked whales of the Class Mammalia (Order Artiodactyla, Family Ziphiidae). Genus is a junior synonym of *Mesoplodon*, Gervais, 1850b. See individual entry.

Anodon Fairmaire, 1871: 36, scarab beetles of the class Insecta (Order Coleoptera, Family Scarabaeidae). Genus is a synonym of *Paranodon* Cockerell, 1905: 104.

Orca Wagler, 1830: 34.

TYPE SPECIES: Φ *Delphinus bidentatus* Bonnaterre, 1789: 25 (as *Orca bidentatus*) [=Φ *Hyperoodon ampullatus* (Forster, 1770: 18)] by subsequent designation. See Iredale and Troughton (1934: 60).

COMMENTS: Synonymised within *Ziphius* by Hershkovitz (1966: 137) and within *Hyperoodon* by Iredale and Troughton (1934: 60), and McKenna and Bell (1997: 382).

HOMONYMS:

Orca J. Gray, 1846b: 84, the Killer Whale of the Class Mammalia (Order Artiodactyla, Family Delphinidae). Genus is a synonym of *Orcinus* Fitzinger, 1860b. See individual entry.

Chaenodelphinus Eschricht, 1843: 655.

TYPE SPECIES: *Nomen novum* for *Hyperoodon* Lacépède, 1804.

COMMENTS: Synonymised within *Hyperoodon* by Iredale and Troughton (1934: 60), Hershkovitz (1966: 143), and Mead and Brownell (1993: 361; 2005: 740).

Uperoodon J. Gray, 1843a: xxiii.

TYPE SPECIES: Misprint of either *Hyperoodon* or *Uranodon*.

COMMENTS: Synonymised within *Hyperoodon* by Iredale and Troughton (1934: 60), Palmer (1904: 340), Hershkovitz (1966: 143), and McKenna and Bell (1997: 382).

Chaenocetus Eschricht, 1845: 17.

TYPE SPECIES: *Nomen novum* for *Hyperoodon* Lacépède, 1804.

COMMENTS: Synonymised within *Hyperoodon* by Iredale and Troughton (1934: 60), Hershkovitz (1966: 143), and McKenna and Bell (1997: 382).

Chenocetus J. Gray, 1846c: 52.

TYPE SPECIES: Emendation of *Chaenocetus* Eschricht, 1845.

COMMENTS: Placed within *Hyperoodon* by J. Gray (1866b: 328), but synonymised within *Chaenocetus* by Palmer (1904: 176) and within *Hyperoodon* by Hershkovitz (1966: 143) and McKenna and Bell (1997: 382).

Hyperhoodon Gervais, 1850b: 6.

TYPE SPECIES: Emendation of *Hyperoodon* Lacépède, 1804.

COMMENTS: Synonymised within *Hyperoodon* by Palmer (1904: 340), Hershkovitz (1966: 143) and McKenna and Bell (1997: 382).

Chenodelphinus Duvernoy, 1851: 45.

TYPE SPECIES: Incorrect subsequent spelling of *Chaenodelphinus* Eschricht, 1845.

COMMENTS: Synonymised within *Chaenodelphinus* by Palmer (1904: 176).

Lagenocetus J. Gray, 1863d: 200.

TYPE SPECIES: Φ *Hyperoodon latifrons* J. Gray, 1846c: 27 (as *Lagenocetus latifrons*) [=Φ *Hyperoodon ampullatus* (Forster, 1770: 18)] by monotypy.

COMMENTS: Genus recognised by J. Gray (1866b: 327, 336). Synonymised within *Hyperoodon* by Iredale and Troughton (1934: 60), Hershkovitz (1966: 143), McKenna and Bell (1997: 382), and Mead and Brownell (1993: 361; 2005: 740).

Hyperodon J. Gray, 1863d: 200.

TYPE SPECIES: Φ *Balaena rostata* O. Müller, 1776: 7 (as *Hyperodon rostratum*) [= Φ *Hyperoodon ampullatus* (Forster, 1770: 18)] by monotypy.

COMMENTS: Synonymised within *Hyperoodon* by Palmer (1904: 340), and McKenna and Bell (1997: 382).

Lagocetus J. Gray, 1866b: 82, 104.

TYPE SPECIES: Emendation of *Lagenocetus* J. Gray, 1863d.

COMMENTS: Synonymised within *Lagenocetus* by Palmer (1904: 360) and *Hyperoodon* by McKenna and Bell (1997: 382).

Hyperaodon Cope, 1869: 31.

TYPE SPECIES: Incorrect subsequent spelling of *Hyperoodon* Lacépède, 1804.

COMMENTS: Synonymised within *Hyperoodon* by Palmer (1904: 340), Hershkovitz (1966: 143), and McKenna and Bell (1997: 382).

Hyperoodus E. Schulze, 1897: 78.

TYPE SPECIES: Incorrect subsequent spelling of *Hyperoodon* Lacépède, 1804.

COMMENTS: Synonymised within *Hyperoodon* by Palmer (1904: 340), and McKenna and Bell (1997: 382).

Frasercetus Moore, 1968: 274.

TYPE SPECIES: *Hyperoodon planifrons* Flower, 1882 by original designation.

COMMENTS: Described as a subgenus of *Hyperoodon*. Synonymised within *Hyperoodon* by Mead and Brownell (1993: 361; 2005: 740).

Hyperoodon planifrons Flower, 1882

Southern Bottle-nosed Whale

Hyperoodon planifrons Flower, 1882: 393, 395.

TYPE LOCALITY: Lewis Island, Dampier Archipelago, north Western Australia, Australia.

COMMENTS: Placed in the subgenus *Frasercetus* by Moore (1968: 274) but typically placed within *Hyperoodon*. Reviewed by Mead (1989a: 321).

Hyperoodon Burmeisteri Moreno, 1895: 5.

TYPE SPECIES: *Nomen nudum* attributed to 'Moreno, 1888'.

COMMENTS: Synonymised within *planifrons* by Hershkovitz (1966: 145), and Mead and Brownell (1993: 361; 2005: 740).

Indopacetus Moore, 1968

Indopacetus Moore, 1968: 254.

TYPE SPECIES: *Mesoplodon pacificus* Longman, 1926 [=*Indopacetus pacificus* (Longman, 1926)] by original designation.

COMMENTS: Mead and Brownell (1993: 362) noted that this genus is considered by many authors to be more correctly included within *Mesoplodon*, to which it is the sister genus. Included within *Mesoplodon* by Reyes *et al.* (1995: 37) and Bannister *et al.* (1996: 39). Separated from *Mesoplodon* by Mead and Brownell (1993: 362; 2005: 740).

Indopacetus pacificus (Longman, 1926)

Longman's Beaked Whale

Mesoplodon pacificus Longman, 1926: 269; Plate 43.

TYPE LOCALITY: Mackay, eastern Queensland, Australia.

COMMENTS: Only known from ten skulls (Pitman, 2009: 601). Reviewed by Mead (1989b: 368) and Dalebout (2003: 421) who confirmed its status within *Indopacetus*. Recognised within *Mesoplodon* by Iredale and Troughton (1934: 62), Mead (1989b: 351), Reyes *et al.* (1995: 37) and

Bannister *et al.* (1996: 39). Synonymised within *Hyperoodon planifrons* by Hershkovitz (1966: 145) but transferred to *Indopacetus* and recognised as a distinct species by Moore (1968: 254), and followed by subsequent authors including Mead and Brownell (1993: 362; 2005: 740).

Mesoplodon Gervais, 1850

Mesoplodon Gervais, 1850b: 16.

TYPE SPECIES: Φ *Delphinus sowerbensis* de Blainville, 1817: 177 [= Φ *Mesoplodon bidens* (Sowerby, 1804: 310)] by monotypy.

COMMENTS: Genus reviewed by Hershkovitz (1966: 125). Iredale and Troughton (1934: 61) also considered other genera as synonyms of *Mesoplodon* including *Mesodiodon* Duvernoy, 1851 and *Callidon* J. Gray, 1871b. Rice (1980: 30) petitioned the ICZN to place *Nodus* on the list of rejected names and *Mesoplodon* on the list of valid names. The ICZN (1985b: 19) rejected *Nodus*, *Micropteron* and *Mikropteron*. Genus reviewed by Mead (1989b: 349) and Dalebout *et al.* (2004: 459). All species included within this genus appear to represent long-separated lines (apparently Late Miocene). Some of the generic synonyms listed here may, therefore, prove to merit resurrection as full genera, if further study supports these early separation dates. The taxon *Mesoplodon hotaula* Deraniyagala, 1963b: 13 from the tropical Indo-Pacific was resurrected by Dalebout *et al.* (2014), but it remains to be seen if this species occurs within Australian waters.

Aodon Lesson, 1828d: 155, 440; Plate 3, Fig. 1.

TYPE SPECIES: Φ *Aodon dalei* Lesson, 1828d: 440 [= Φ *Mesoplodon bidens* (Sowerby, 1804: 310)] by monotypy.

COMMENTS: Synonymised within *Mesoplodon* by Iredale and Troughton (1934: 61), Hershkovitz (1966: 125), and Mead and Brownell (1993: 362; 2005: 740).

HOMONYMS:

Aodon Lacépède, 1798: 297 or Anon, 1798b: 675, fish of the Superclass Pisces (Order Rajiformes, Family Myliobatidae). Name is recognised as described.

Nodus Wagler, 1830: 34.

TYPE SPECIES: Φ *Delphinus edentulus* Schreber, 1792: Plate 347 [= Φ *Hyperoodon ampullatus* (Forster, 1770: 18)]. Name for *Aodon* Lesson, 1828d.

COMMENTS: *Nomen oblitum* as not used in the primary literature in the last 50 years (e.g. Hershkovitz, 1966: 125; Mead, 1989b: 349), although *nomen oblitum* is redefined by the Glossary of the Fourth Edition of the Code as a name which has not been used since 1899. Suppressed by the ICZN (1985b: 19). Included as a synonym of *Hyperoodon* by Iredale and Troughton (1934: 60), and a synonym of *Mesoplodon* by Hershkovitz (1966: 125), and Mead and Brownell (1993: 362; 2005: 740).

Micropterus Wagner, 1846: 281, 352.

TYPE SPECIES: Φ *Delphinus micropterus* G. Cuvier, 1829a: 288. [= Φ *Mesoplodon bidens* (Sowerby, 1804: 310)] by tautonomy and monotypy.

COMMENTS: Described as a subgenus of *Delphinus*. Synonymised within *Mesoplodon* by Hershkovitz (1966: 125), and Mead and Brownell (1993: 362; 2005: 740).

HOMONYMS:

Micropterus Lacépède, 1802b: 324, sunfish of the Superclass Pisces (Order Perciformes, Family Centrarchidae). Genus currently recognised (Near *et al.*, 2003: 1610).

Micropterus Lesson, 1828e: 416, ducks of the class Aves (Order Anseriformes, Family Anatidae). Genus is a synonym of *Tachyeres* Owen, 1875b: 254.

Micropterus Chevrolat, 1842: 277, beetles of the Class Insecta (Order Coleoptera, Family Cleridae). Genus was replaced by *Micropteroclerus* Chapin, 1920: 51.

Micropteron Eschricht, 1849a: 97.

TYPE SPECIES: Φ *Delphinus micropterus* G. Cuvier, 1829b: 288 [= Φ *Mesoplodon bidens* (Sowerby, 1804: 310)] by tautonomy.

COMMENTS: Recognised as an incorrect subsequent spelling of *Micropterus* Wagner, 1846 by the ICZN (1985b: 19). A *nomen oblitum* as not used in the primary literature in the last 50 years (Mead, 1989b: 349), although *nomen oblitum* is redefined by the Glossary of the Fourth Edition of the Code as a name which has not been used since 1899. Synonymised within *Micropterus* by Palmer (1904: 423) and *Mesoplodon* by Hershkovitz (1966: 125) and Mead (1989b: 349).

Mikropteron Eschricht, 1849a: 98.

TYPE SPECIES: Incorrect subsequent spelling of *Micropteron* Eschricht, 1849a.

COMMENTS: Recognised as an incorrect subsequent spelling of *Micropterus* Wagner, 1846 by Opinion 1289 of the ICZN (1985b: 19).

Anodon J. Gray, 1850: 71.

TYPE SPECIES: Incorrect subsequent spelling of *Aodon* Lesson, 1828d.

COMMENTS: Genus synonymised within *Aodon* by Palmer (1904: 110) and *Mesoplodon* by McKenna and Bell (1997: 381).

HOMONYMS:

Anodon A. Smith, 1829: 443, African egg-eating snakes of the Class Reptilia (Order Squamata, Family Colubridae). Genus is a synonym of *Dasypeltis* Wagler, 1830: 178.

Anodon Oken, 1815: 236, freshwater mussels of the Phylum Mollusca (Class Bivalvia, Order Unionoida, Family Unionidae). An unjustified emendation of *Anodonta* Lamarck, 1799: 87. See Melville and Smith (1987: 46).

Anodon Wagler, 1830: 34, the Southern Bottle-nosed Whale of the class Mammalia (Order Artiodactyla, Family Ziphiidae). Genus is a synonym of *Hyperoodon* Lacépède, 1804. See individual entry.

Anodon Agassiz, 1846: 24, 27, fish of the Superclass Pisces (Order Rajiformes, Family Myliobatidae). An unjustified emendation of *Aodon* Lacépède, 1798: 297. See Whitley (1936: 165).

Anodon Fairmaire, 1871: 36, scarab beetles of the class Insecta (Order Coleoptera, Family Scarabaeidae). Genus is a synonym of *Paranodon* Cockerell, 1905: 104.

Dioplodon Gervais, 1850a: 512.

TYPE SPECIES: *Delphinus densirostris* de Blainville, 1817 [=*Mesoplodon densirostris* (de Blainville, 1817)] by original designation.

COMMENTS: Described as a subgenus. An extract of this paper, including the genus name, appeared in Gervais (1950c: 16). Ranked as a *nomen oblitum*, as not used in the primary literature in the last 50 years by Mead (1989b: 349), although *nomen oblitum* is redefined by the Glossary of the Fourth Edition of the Code as a name which has not been used since 1899, and it is specified that such a name still remains available. Genus recognised by J. Gray (1866b: 327, 355) and Iredale and Troughton (1934: 61) and Mead (1989b: 350). Synonymised within *Mesoplodon* by Hershkovitz (1966: 125), and Mead and Brownell (1993: 362; 2005: 740), and is available to be used for *M. densirostris*, and presumably the closely related *M. stejnegeri*, should the genus *Mesoplodon* be split up in future.

Mesodiodon Duvernoy, 1851: 41.

TYPE SPECIES: Φ *Mesodiodon Sowerbyi* Duvernoy, 1851: 41, 69 [= Φ *Mesoplodon bidens* (Sowerby, 1804: 310)] by subsequent designation.

COMMENTS: Synonymised within *Mesoplodon* by Iredale and Troughton (1934: 60), Hershkovitz (1966: 126) and McKenna and Bell (1997: 381).

Mesiodon J. Gray, 1866b: 349.

TYPE SPECIES: Incorrect subsequent spelling of *Mesodiodon* Duvernoy, 1851.

COMMENTS: Originally placed within *Ziphius* by J. Gray (1866b: 349), but subsequently synonymised within *Mesodiodon* by Palmer (1904: 414) and within *Mesoplodon* by Hershkovitz (1966: 126).

Diplodon J. Gray, 1866b: 349.

TYPE SPECIES: Incorrect subsequent spelling of *Dioplodon* Gervais, 1850a.

COMMENTS: Originally placed within *Ziphius* by J. Gray (1866b: 349), but subsequently synonymised within *Mesoplodon* by Hershkovitz (1966: 126).

Dolichodon J. Gray, 1866b: 353.

TYPE SPECIES: *Ziphius layardii* J. Gray, 1865g [=*Mesoplodon layardii* (J. Gray, 1865g)] by monotypy.

COMMENTS: Described as a subgenus of *Ziphius*. Synonymised within *Mesoplodon* by Iredale and Troughton (1934: 61), Hershkovitz (1966: 126), Mead (1989b: 350), and Mead and Brownell (1993: 362; 2005: 740). This name would be available for use for the *Mesoplodon layardi* group (presumably including *M. bowdoini* R. Andrews, 1908; *M. carlhubbsi* Moore, 1963: 396, 422; *M. traversii* (Gray, 1874d: 96)), should the genus *Mesoplodon* be split up in the future.

HOMONYMS:

Dolicodon Fanzago, 1874: 22, centipedes of the Subphylum Myriapoda (Order Lithobiomorpha, Family Lithobiidae). Name corrected to *Dolichodon* by the Cambridge (1875: 262). Genus is a junior synonym of *Lithobius* Leach, 1814: 387.

Dolichodon Parr, 1931: 45, deep-sea fish of the Class Actinopterygii (Order Perciformes, Family Chiasmodontidae). Genus is a synonym of *Kali* Lloyd, 1909: 154. See Melo (2008: 3).

Callidon J. Gray, 1871b: 368.

TYPE SPECIES: *Mesoplodon guntheri* Krefft, 1871c [=*Mesoplodon layardii* (J. Gray, 1865g)] by monotypy.

COMMENTS: Objective synonym of *Dolichodon*. Synonymised within *Mesoplodon* by Iredale and Troughton (1934: 60), Hershkovitz (1966: 126) and Mead (1989b: 350), but not discussed by Mead and Brownell (2005: 740).

Neoziphius J. Gray, 1871a: vi, 101.

TYPE SPECIES: Φ *Diplodon europaeus* Gervais, 1855a: 320 (as *Neoziphius europaeus*) [= Φ *Mesoplodon europaeus* (Gervais, 1855a: 320)] by monotypy.

COMMENTS: Synonymised within *Mesoplodon* by McKenna and Bell (1997: 381), but the name would be available for *M. europaeus*, a species of isolated affinities, should the genus *Mesoplodon* be split up in future.

Diplodon de Marschall, 1873: 5.

TYPE SPECIES: Incorrect subsequent spelling of *Dioplodon* Gervais, 1850a.

COMMENTS: Synonymised within *Dioplodon* by Palmer (1904: 235).

Oulodon Haast, 1876a: 457.

TYPE SPECIES: *Mesoplodon grayi* Haast, 1876b by monotypy.

COMMENTS: Synonymised within *Mesoplodon* by Iredale and Troughton (1934: 61), Hershkovitz (1966: 126), Mead (1989b: 350), and Mead and Brownell (1993: 362; 2005: 740); the name would be available for *M. grayi* should the genus *Mesoplodon* be split up in the future.

Calliodon Trouessart, 1898 [1898–1899]: 1067.

TYPE SPECIES: Incorrect subsequent spelling of *Callidon* J. Gray, 1871b.

COMMENTS: Synonymised within *Callidon* by Palmer (1904: 151).

Paikea Oliver, 1922: 574.

TYPE SPECIES: *Berardius hectori* J. Gray, 1871c [=*Mesoplodon hectori* J. Gray, 1871c] by original designation.

COMMENTS: Synonymised within *Berardius* by Hershkovitz (1966: 122). Synonymised within *Mesoplodon* by Harmer (1924: 555), Mead (1989b: 349), and Mead and Brownell (1993: 362; 2005: 740), although the name would be available for *M. hectori*, a species of isolated affinities, should the genus *Mesoplodon* be split up in future.

Mesoplodon bowdoini R. Andrews, 1908

Andrews' Beaked Whale

Mesoplodon bowdoini R. Andrews, 1908: 203; Plate 13, Figs. 1–5.

TYPE LOCALITY: New Brighton Beach, Canterbury Province, New Zealand.

COMMENTS: Historically often not recognised by authors including Orr (1953: 243) and McCann (1962: 18; 1964: 124; 1976: 107) who felt that *M. bowdoini* was synonymous with *Mesoplodon stejnegeri* (True, 1885: 584). *Mesoplodon bowdoini* was separated from *stejnegeri* by Moore (1963: 396) on the basis that *stejnegeri* represented a 'subarctic lot' and *bowdoini* represented a 'south temperate lot'. This recognition was followed by subsequent authors including Hershkovitz (1966: 136), Mead (1989b: 363), Baker (2001: 473), Dalebout *et al.* (2004: 466–468) and subsequent authors.

Mesoplodon densirostris (de Blainville, 1817)

Blainville's Beaked Whale

Delphinus densirostris de Blainville, 1817: 178.

TYPE LOCALITY: Unknown.

COMMENTS: Placed within *Mesoplodon* by Flower (1878a: 684) and most subsequent authors. Species reviewed by Hershkovitz (1966: 134), Ross (1984: 215) and Mead (1989b: 356).

Ziphius Sechellensis J. Gray, 1846c: 28; Plate 6, Fig. 1–2.

TYPE LOCALITY: Seychelles, India.

COMMENTS: Synonymised within *densirostris* by Iredale and Troughton (1934: 61), Hershkovitz (1966: 135), and Mead and Brownell (1993: 362; 2005: 741) who spelt the name '*seychellensis*'.

Mesoplodon ginkgodens Nishiwaki & Kamiya, 1958

Ginkgo-toothed Beaked Whale

Mesoplodon ginkgodens Nishiwaki & Kamiya, 1958: 77; Fig. 13.

TYPE LOCALITY: Oiso Beach, Sagami Bay, near Tokyo, Japan.

COMMENTS: Species recognised by Hershkovitz (1966: 132), and Mead and Brownell (1993: 363; 2005: 741).

Mesoplodon grayi Haast, 1876

Gray's Beaked Whale

Mesoplodon Grayi Haast, 1876b: 9, 13.

TYPE LOCALITY: Waitangi Beach, Chatham Island, New Zealand.

COMMENTS: Reviewed by Hershkovitz (1966: 130), Ross (1984: 198) and Mead (1989b: 358).

M. [esoplodon] australis Flower, 1878b: 417, footnote; Plates 71–73.

TYPE LOCALITY: South Pacific, Lyall Bay, New Zealand.

COMMENTS: Synonymised within *grayi* by Hershkovitz (1966: 131), and Mead and Brownell (1993: 363; 2005: 742).

Mesoplodon haasti Flower, 1878a: 684.

TYPE LOCALITY: South Pacific, North Island, New Zealand.

COMMENTS: Synonymised within *grayi* by Iredale and Troughton (1934: 62), Hershkovitz (1966: 131), and Mead and Brownell (1993: 363; 2005: 742).

Mesoplodon hectori (J. Gray, 1871)

Hector's Beaked Whale

Berardius Hectori J. Gray, 1871c: 117.

TYPE LOCALITY: Titai Bay, Cook Strait, New Zealand.

COMMENTS: Placed within *Mesoplodon* by Flower (1878b: 416) and Ross (1970: 195). Synonymised within *Berardius arnuxi* by Hershkovitz (1966: 123). Species recognised by Mead and Brownell (1993: 363; 2005: 742). Reviewed by Mead (1989b: 361).

Mesoplodon knoxi Hector, 1873a: 166, 167; Plate 6, Figs. 4a-b.

TYPE LOCALITY: South Pacific, Titai Bay, near Porirua, New Zealand.

COMMENTS: Synonymised within *Berardius arnuxi* by Hershkovitz (1966: 124) and within *Mesoplodon hectori* by Mead and Brownell (1993: 363; 2005: 742). Appears to be of uncertain status as Reeves *et al.* (2002: 274) suggests it is restricted to southern oceans.

Mesoplodon layardii (J. Gray, 1865)

Strap-toothed Beaked Whale

Ziphius layardii J. Gray, 1865g: 358.

TYPE LOCALITY: Probably off the Cape of Good Hope, South Africa.

COMMENTS: Placed in *Dolichodon* by J. Gray (1871a: 101) and *Mesoplodon* by Hector (1873b: 106), Flower (1878a: 684) and followed by most subsequent authors. Reviewed by Mead (1989b: 366).

Mesoplodon Güntheri Krefft, 1871c: 368.

TYPE LOCALITY: Little Bay, New South Wales, Australia.

COMMENTS: Synonymised within *Mesoplodon layardi* by Iredale and Troughton (1934: 61), Hershkovitz (1966: 133), and Mead and Brownell (1993: 363; 2005: 742).

Mesoplodon longirostris J. Gray, 1873b: 145.

TYPE LOCALITY: Unknown.

COMMENTS: Synonymised within *Mesoplodon layardi* by Iredale and Troughton (1934: 62), Hershkovitz (1966: 134), and Mead and Brownell (1993: 363; 2005: 742).

Mesoplodon floweri Haast, 1876c: 478; Plates 45–46.

TYPE LOCALITY: Saltwater Creek, Canterbury, New Zealand.

COMMENTS: Synonymised within *Mesoplodon layardi* by Iredale and Troughton (1934: 62), Hershkovitz (1966: 134), and Mead and Brownell (1993: 363; 2005: 742).

M. [esoplodon] thomsoni J. Ogilby, 1892: 71.

TYPE LOCALITY: Little Bay, near Sydney, New South Wales, Australia.

COMMENTS: Ogilby's paper refers to a manuscript of Krefft and states 'In Mr Krefft's MSS occurs a notice of a Ziphiid Whale, which he names *M. thomsoni*, but is probably a female of this species [*Mesoplodon layardii*]'. Not considered by Iredale and Troughton (1934: 61–62) or Mead and Brownell (1993: 363; 2005: 742). Synonymised within *layardii* by Hershkovitz (1966: 134).

Mesoplodon mirus True, 1913

True's Beaked Whale

Mesoplodon mirum True, 1913: 1.

TYPE LOCALITY: North Atlantic, Bird Island Shoal, Beaufort Harbour, North Carolina, United States of America.

COMMENTS: Spelling of the species name changed to *mirus* by Ulmer (1941: 107). Reviewed by Ross (1984: 202) and Mead (1989b: 366).

Tasmacetus Oliver, 1937

Tasmacetus Oliver, 1937: 371, 372.

TYPE SPECIES: *Tasmacetus shepherdi* Oliver, 1937, by original designation.

COMMENTS: Taxon reviewed by Hershkovitz (1966: 122).

Tasmacetus shepherdi Oliver, 1937

Tasman Beaked Whale

Tasmacetus shepherdi Oliver, 1937: 371, 373; Plates 1–5.

TYPE LOCALITY: Beach at Ohawe, Taranaki, west coast of North Island, New Zealand.

COMMENTS: Taxon reviewed by Hershkovitz (1966: 122) and Mead (1989c: 309).

Ziphius G. Cuvier, 1823a

ziphius G. Cuvier, 1823a: 350, 352.

TYPE SPECIES: *Ziphius cavirostris* G. Cuvier, 1823a by original designation.

COMMENTS: Genus recognised by J. Gray (1866b: 327, 348). Taxon reviewed by Hershkovitz (1966: 137).

Diodon Lesson, 1828d: 123, 440.

TYPE SPECIES: *Delphinus desmaresti* Risso, 1826b [=*Ziphius cavirostris* G. Cuvier, 1823a] by original designation.

COMMENTS: Synonymised within *Ziphius* by Iredale and Troughton (1934: 60), Hershkovitz (1966: 137), and Mead and Brownell (1993: 364; 2005: 743).

HOMONYMS:

Diodon Linnaeus, 1758: 334, porcupine fishes of the Class Actinopterygii (Order Tetraodontiformes, Family Diodontidae). See Bray *et al.* (2006: 1933).

Diodon Storr, 1780: 42, Table C, the Narwhal of the Class Mammalia (Order Artiodactyla, Family Monodontidae). Genus is a synonym of *Monodon* Linnaeus, 1758: 75. See Mead and Brownell (2005: 735).

Diodon Lesson, 1830: 95, falcons of the Class Aves (Order Falconiformes, Family Falconidae). Genus is a synonym of *Falco* Linnaeus, 1758: 83, 88.

Hypodon Haldeman, 1841: 127.

TYPE SPECIES: *Nomen novum* for *Diodon* Lesson, 1828d.

COMMENTS: Replacement name for *Diodon* Lesson, 1828d. Synonymised within *Ziphius* by Iredale and Troughton (1934: 60), Hershkovitz (1966: 137), and Mead and Brownell (1993: 364; 2005: 743).

Xiphias Murchison, 1843: 560.

TYPE SPECIES: Incorrect subsequent spelling of *Ziphius* G. Cuvier, 1823a.

COMMENTS: *Nomen novum* for *Diodon* Lesson, 1828d. Synonymised within *Ziphius* by Palmer (1904: 716) and McKenna and Bell (1997: 382).

HOMONYMS:

Xiphias Linnaeus, 1758: 248, swordfish of the Class Actinopterygii (Order Percifomes, Family Xiphiidae). Currently recognised name. See Bray and Hoese (2006: 1783).

Xiphius Agassiz, 1846: 389, 392.

TYPE SPECIES: Emendation of *Ziphius* G. Cuvier, 1823a.

COMMENTS: Synonymised within *Ziphius* by Palmer (1904: 716) and Hershkovitz (1966: 137), and McKenna and Bell (1997: 382).

Ziphices Duvernoy, 1851: 42.

TYPE SPECIES: Incorrect subsequent spelling of *Ziphius* G. Cuvier, 1823a.

COMMENTS: Synonymised within *Ziphius* by Iredale and Troughton (1934: 60).

Aliama J. Gray, 1864c: 242.

TYPE SPECIES: *Delphinus desmaresti* Risso, 1826b: (as *Aliama desmarestii*) [=*Ziphius cavirostris* G. Cuvier, 1823a] by monotypy.

COMMENTS: Synonymised within *Ziphius* by Iredale and Troughton (1934: 60), Hershkovitz (1966: 137), and McKenna and Bell (1997: 382).

Petrorhynchus J. Gray, 1865d: 524.

TYPE SPECIES: *Hyperoodon capensis* J. Gray, 1865g (as *Petrorhynchus capensis*) [=*Ziphiurus cavirostris* G. Cuvier, 1823a] by monotypy.

COMMENTS: Genus recognised by J. Gray (1866b: 327, 342). Synonymised within *Ziphius* by Iredale and Troughton (1934: 60), Hershkovitz (1966: 137), and Mead and Brownell (1993: 364; 2005: 743) give the year as 1875.

Ziphiorrhynchus Burmeister, 1866a: 94; Plate 3.

TYPE SPECIES: *Ziphiorhynchus cryptodon* Burmeister, 1866a [=*Ziphiurus cavirostris* G. Cuvier, 1823a] by monotypy.

COMMENTS: Synonymised within *Ziphius* by Iredale and Troughton (1934: 60), Hershkovitz (1966: 137), and Mead and Brownell (1993: 364; 2005: 743).

HOMONYMS:

Ziphorhynchus Swainson, 1837: 313, scythebills of the Class Aves (Order Passeriformes, Family Furnariidae). Genus is a synonym of *Campylorhamphus* Bertoni, 1901: 70.

Ziphiorhynchus Van Bénéden, 1868: 96.

TYPE SPECIES: Incorrect subsequent spelling of *Ziphiorrhynchus* Burmeister, 1866a.

COMMENTS: Synonymised within *Ziphius* by McKenna and Bell (1997: 382).

Mesoodon Brandt, 1873a: 220.

TYPE SPECIES: † *Ziphius longirostris* G. Cuvier, 1823b: 245.

COMMENTS: Described as a subgenus of *Ziphius*.

HOMONYMS:

Mesodon Rafinesque, 1819b: 66, snails of the Phylum Mollusca (Class Gastropoda). *nomen nudum*. Described further by Rafinesque (1831: 3).

Mesodon Férussac, 1821: 33, snails of the Phylum Mollusca (Class Gastropoda). Name is a junior synonym of *Helix* Say, 1817: 7 (in Conchology). See Rosenberg and Emberton (1990: 204).

Mesodon Wagner, 1851: 54, 56, ray-fined fish of the († Order Pycnodontiformes, Family Pycnodontidae). Genus is a synonym of *Typodus* Quenstedt, 1858: 781.

Mesodon Ameghino, 1882: 41, edentates of the Class Mammalia (Order Pilosa, † Family Scelidotheriidae). Genus is a synonym of † *Mylodon* Owen, 1839b: 63, 68, which is a currently accepted name (e.g. see Fernicola *et al*. 2009: 152).

Ziphius cavirostris G. Cuvier, 1823a

Cuvier's Beaked Whale

ziphius cavirostris G. Cuvier, 1823a: 352.

TYPE LOCALITY: Fossilised skull, from between Fos and the mouth of the Galegeon, Bouches-du-Rhone, France.

COMMENTS: Reviewed by Ross (1984: 227) and Heyning (1989a: 54; 1989b: 289).

D. [elphinus] Desmaresti Risso, 1826b: 24; Plate 2, Fig. 3.

TYPE LOCALITY: North Atlantic, Mediterranean Sea.

COMMENTS: Recognised as *Hyperoodon Desmarestii* by J. Gray (1850: 69), *Aliama desmarestii* by J. Gray (1864c: 242) and *Epiodon desmarestii* by Carus (1893: 716). Synonymised within *cavirostris* by Hershkovitz (1966: 139), and not considered by Mead and Brownell (1993: 364; 2005:743).

Delphinus Philippii Cocco, 1846: 104.

TYPE LOCALITY: Italy, North Atlantic.

COMMENTS: Synonymised within *cavirostris* by Hershkovitz (1966: 139).

Hyperoodon Doumetii J. Gray, 1850: vii, 68.

TYPE LOCALITY: Corsica, Mediterranean Sea.

COMMENTS: Synonymised within *cavirostris* by Hershkovitz (1966: 139).

Hyperoodon Gervaisii Duvernoy, 1851: 49, 67.

TYPE LOCALITY: Aresquiers, near Frontignan, Herault, North Atlantic, France.

COMMENTS: Synonymised within *cavirostris* by Hershkovitz (1966: 139).

Ziphius indicus Van Bénéden, 1864: 23; Plate 1.

TYPE LOCALITY: Cape of Good Hope, South Atlantic, South Africa.
COMMENTS: Recognised as *Aliama indica* by J. Gray (1865d: 528). Synonymised within *cavirostris* by Hershkovitz (1966: 139), and Mead and Brownell (1993: 364; 2005: 743).

Delphinorhynchus australis Burmeister, 1865b: 262.

TYPE LOCALITY: South Atlantic, Coast of Buenos Aires, Argentina.
COMMENTS: Synonymised within *cavirostris* by Hershkovitz (1966: 140), and Mead and Brownell (1993: 364; 2005: 743).

Hyperoodon Capensis J. Gray, 1865g: 359; Figure.

TYPE LOCALITY: South Atlantic, possibly off Cape of Good Hope, South Australia, Australia.
COMMENTS: Synonymised within *cavirostris* by Hershkovitz (1966: 140), and Mead and Brownell (1993: 364).

Ziphiorrhynchus cryptodon Burmeister, 1866a: 94; Plate 3.

TYPE LOCALITY: Buenos Aires, Argentina.
COMMENTS: Synonymised within *Ziphius* by Hershkovitz (1966: 140).

Epiodon chathamiensis Hector, 1873b: 105.

TYPE LOCALITY: South Pacific, Chatham Islands, New Zealand.
COMMENTS: Recognised as a species of *Ziphius* by Iredale and Troughton (1934: 61). Synonymised within *cavirostris* by Hershkovitz (1966: 141), and Mead and Brownell (1993: 364).

Ziphius novae-zealandiae Haast, 1876d: 466; Plates 45–46.

TYPE LOCALITY: Lyttelton Harbor, New Zealand.
COMMENTS: Synonymised within *Ziphius chathamensis* Hector, 1873b [=*Ziphius cavirostris* G. Cuvier, 1823a] by Iredale and Troughton (1934: 61). Synonymised within *cavirostris* by Hershkovitz (1966: 142).

Ziphius grebnitzkii Stejneger, 1883: 77.

TYPE LOCALITY: Komandorskye Island, Bering Sea, eastern Siberia, Russia.
COMMENTS: Synonymised within *cavirostris* by Hershkovitz (1966: 142).

Superfamily Delphinoidea J. Gray, 1821 *sensu* Rice, 2009

Family Delphinidae J. Gray, 1821: 310.

TYPE GENUS: *Delphinus* Linnaeus, 1758.

COMMENTS: When originally proposed, this rank was placed in the Order Carnivorae (J. Gray, 1821 =[Cetacea (Brisson, 1762)]) and included the genera *Delphinus* Linnaeus, 1758; *Phocena* J. Gray, 1821 [=*Phocoena* G. Cuvier, 1816a]; *Delphinaster* [sic =*Delphinapterus*] Lacépède, 1804: xli, 243; and *Hyperdordons* [sic =*Hyperoodon*] Lacépède, 1804. Recognised at superfamily/suborder rank by Flower (1865a: 388), Huxley (1872: 336), F. Fraser (1966: 7), Mead (1975: 64), Fordyce and Barnes (1994: 429) and McKenna and Bell (1997: 383). Composition of the superfamily follows Rice (2009: 235) and Le Duc (2009: 299).

Suborder Delphinoidea Flower, 1865a: 388.

COMMENTS: When originally proposed, this rank was placed in the Order Cetacea (Brisson, 1762) and included the families Physeteridae (J. Gray, 1821), Platanistidae (J. Gray, 1846c: 25) and Delphinidae (J. Gray, 1821). Delphinoidea recognised at Superfamily rank by Rice (1984: 466). Proposed as an alternative name to the suborder. Synonymised within Odontoceti and Delphinoidea by McKenna and Bell (1997: 379, 383).

Family Delphinidae J. Gray, 1821

Family Delphinidae J. Gray, 1821: 310.

TYPE GENUS: *Delphinus* Linnaeus, 1758.
COMMENTS: When originally proposed, this rank was placed in the Order Carnivorae (J. Gray, 1821 [=Cetacea (Brisson, 1762)]) and included the genera *Delphinus* Linnaeus, 1758; *Phocena* J. Gray, 1821 [=*Phocoena* G. Cuvier, 1816a]; *Delphinaster* [sic =*Delphinapterus*] Lacépède, 1804: xli, 243; and *Hyperdordons* [sic =*Hyperoodon*] Lacépède, 1804. The subfamilies Orcininae, Delphininae, Globicephalinae and Cephalorhynchinae were recognised by McKenna and Bell (1997: 384–386). Family reviewed by Kasuya (1973: 32), Barnes (1978: 1), Perrin (1989: iv) and Rice (1998: 97), with LeDuc *et al.* (1999: 619) and Caballero *et al.* (2008: 263) reviewing the recognised subfamilies and their constituents. Subfamilies were proposed by LeDuc (2009: 299), that were derived from LeDuc (1999: 639), and the results of Beasley *et al.* (2005: 365) and Caballero *et al.* (2007: 358); see also Nishida *et al.* (2007: 723), McGowen *et al.* (2009: 891), and Hassanin *et al.* (2012: 37). The evolutionary relationships among the Delphinidae are not well understood and relationships of the various genera are almost certainly not as traditionally envisaged (Cunha *et al.*, 2011: 3; McGowen, 2011: 349–354; Perrin *et al.*, 2013: 567), and therefore we recognise no subfamilies. Within the typically recognised Subfamily Delphininae, Le Duc *et al.* (1999: 619, 641) suggested that the genera *Stenella* and *Tursiops* are polyphyletic and that these genera, along with *Sousa* and *Lagenodelphis* should be synonymised within *Delphinus* as that is the oldest genus name. This potential taxonomic grouping was also noted by other studies

including Caballero *et al.* (2008: 260), McGowen *et al.* (2009), McGowen (2011), and Hassanin *et al.* (2012: 43).

The option to synonymise the species of the genera *Stenella*, *Tursiops*, *Sousa* and *Lagenodelphis* within *Delphinus* was considered by Perrin *et al.* (2013: 567, 581), who suggested the status-quo does not appeal to them but could understood why it may prevail, and concluded that should further work using next generation molecular technologies still prove unable to robustly identify relationships, it may then be time to revisit the taxonomy and consider a single-genus approach. Pending that work (which will, we hope, provide consensus to the shape of delphinid phylogeny), in the interests of stability of taxa that are currently well recognised and defined we have, with considerable reluctance, not followed these at this time, even if the phylogenetics currently implies that substantial taxonomic nomenclatorial revision is warranted.

FUTURE TAXONOMIC RESEARCH: All recent studies have indicated that the species customarily assigned to the five putative genera *Delphinus*, *Tursiops*, *Sousa*, *Lagenodelphis* and *Stenella* are intertwined, and that they are the result of a rapid radiation during the Pleistocene (see especially McGowen, 2011: 349–354; Cunha *et al.*, 2011: 3). Taxonomic studies are needed to resolve the interrelationships of the different taxa, to determine the validity of these and some recently recognised genera, and of several disputed species, and to identify and describe the undescribed species.

Tribe Delphinina J. Gray, 1825a: 340.

TYPE GENUS: *Delphinus* Linnaeus, 1758.

COMMENTS: When originally proposed, this rank was placed in the Family Delphinidae (J. Gray, 1821) and included the genera *Delphinus* Linnaeus, 1758; and *Delphinorhynchus* de Blainville, 1817: 151 *nomen dubium*. Recognised as an infraorder by de Muizon (1984: 68). Synonymised within Delphinidae by McKenna and Bell (1997: 386).

Family Delphinusideae Lesson, 1842: 197.

TYPE GENUS: *Delphinus* Linnaeus, 1758.

COMMENTS: When originally proposed, this rank was placed in the Order Cetaces (Lesson, 1842 [=Cetacea (Brisson, 1762)]) and included the genera *Delphinus* Linnaeus, 1758; *Inia* D'Orbigny, 1834: 31; *Heterodon* de Blainville, 1817 [=*Hyperoodon* Lacépède, 1804]; and *Monodon* Linnaeus, 1758: 75. Synonymised within Delphinidae by McKenna and Bell (1997: 384).

Tribe? Orcini Wagner, 1846: 292.

TYPE GENUS: *Orca* J. Gray, 1846b [=*Orcinus* Fitzinger, 1860b].

COMMENTS: Subfamily rank recognised by McKenna and Bell (1997: 384) and LeDuc (2009: 299). Rice (1998: 98) noted that the name is unavailable because the type genus is a junior homonym (Article 39 of the Code; ICZN, 1985a: 79): see below under generic synonymy.

Subfamily? Orcadina J. Gray, 1846c: 24.

TYPE GENUS: *Orca* J. Gray, 1846b [=*Orcinus* Fitzinger, 1860b].

COMMENTS: When originally proposed, this rank was placed in the Family Delphinidae (J. Gray, 1821) and included the genus *Orca* J. Gray, 1846b [=*Orcinus* Fitzinger, 1860b]. Synonymised within the Subfamily Orcininae (Wagner, 1846) by McKenna and Bell (1997: 384), and within the Family Delphinidae by Rice (1998: 97).

Tribe? Lagenorhynchi Wagner, 1846: 317.

TYPE GENUS: *Lagenorhynchus* J. Gray, 1846b.

COMMENTS: When originally proposed, this rank included the genus *Lagenorhynchus* J. Gray, 1846b. Synonymised within the Subfamily Orcininae by McKenna and Bell (1997: 384).

Family Delphinodea Giebel, 1855: ix, 86.

TYPE GENUS: *Delphinus* Linnaeus, 1758.

COMMENTS: When originally proposed, this rank was placed in the Order Cetacea (Brisson, 1762) and included the genera *Physeter* Linnaeus, 1758; *Phocaena* G. Cuvier, 1817; *Delphinus* Linnaeus, 1758; *Inia* D'Orbigny, 1834: 31; *Platanista* Wagler, 1830: 35; *Hyperoodon* Lacépède, 1804; *Berardius* Duvernoy, 1851; *Ziphius* G. Cuvier, 1823a; and *Delphinapterus* Lacépède, 1804: xli, 243. Synonymised within Delphinidae by McKenna and Bell (1997: 384).

Tribe? Globiocephalina J. Gray, 1863d: 201.

TYPE GENUS: *Globiocephalus* J. Gray, 1846c [=*Globicephala* Lesson, 1828d].

COMMENTS: When originally proposed, this rank was placed in the Family Delphinidae (J. Gray, 1821) and included the genus *Globiocephalus* J. Gray, 1846c [=*Globicephala* Lesson, 1828d]. The year of publication is often given as J. Gray, 1850 (e.g. McKenna and Bell, 1997: 385; Mead & Brownell, 2005: 726; Aguirre-Fernández *et al.*, 2009: 248), but this is incorrect as this rank does not occur within this publication. Subfamily rank recognised by Rice (1984: 481), McKenna and Bell (1997: 385), LeDuc *et al.* (1999: 639) and LeDuc (2009: 299).

Family Globiocephalidae J. Gray, 1866b: 62, 313.

TYPE GENUS: *Globiocephalus* J. Gray, 1846c [=*Globicephala* Lesson, 1828d].

COMMENTS: When originally proposed, this rank was placed in the Section Denticete (J. Gray, 1866b [=Odontoceti (Flower, 1867)]) and included the genera *Globiocephalus* J. Gray, 1846c [=*Globicephala* Lesson, 1828d]; and *Sphaerocephalus* J. Gray, 1864c [*Globicephala* Lesson, 1828d]. Citation for description sometimes given as J. Gray (1950), but this citation only mentions the genus *Globiocephalus* and not the family name. Family rank recognised by Nishiwaki (1963: 98), and at subfamily rank by McKenna and Bell (1997: 385) but synonymised within Delphinidae by Rice (1998: 98), and Mead and Brownell (2005: 726).

Family Delphinida Haeckel, 1866: clix.

TYPE GENUS: *Delphinus* Linnaeus, 1758.

COMMENTS: When originally proposed, this rank was placed in the Suborder Autoceta (Haeckel, 1866 [=Cetacea (Brisson, 1762)]) and included the genera *Delphinus* Linnaeus, 1758; *Phocaena* G. Cuvier, 1817; † *Arionius* H. Meyer, 1841: 315 [=† *Squalodon* Grateloup, 1840a: 208, 1840b: 346]; and † *Stereodelphis* Gervais, 1848: 152. Synonymised within Delphinidae by McKenna and Bell (1997: 384).

HOMONYMS:

Infraorder Delphinida de Muizon, 1984, dolphins of the Class Mammalia (Infraorder Cetacea, Parvorder Odontoceti). Name is a synonym of Delphinidae (J. Gray, 1821). See individual entry.

Subfamily Delphininae Flower, 1867: 115.

TYPE GENUS: *Delphinus* Linnaeus, 1758.

COMMENTS: When originally proposed, this rank was placed in the Family Delphinidae (J. Gray, 1821) and included the genera *Phocaena* G. Cuvier, 1817; *Neomeris* J. Gray, 1846c: 30 [=*Neophocaena* Palmer, 1899d: 23]; *Grampus* J. Gray, 1828; *Orca* J. Gray, 1846b [=*Orcinus* Fitzinger, 1860b]; *Pseudorca* Reinhardt, 1862; *Lagenorhynchus* J. Gray, 1846b; *Delphinus* Linnaeus, 1758; *Delphinapterus* Lacépède, 1804: xli, 243; and *Globiocephalus* J. Gray, 1846c [=*Globicephala* Lesson, 1828d]. Synonymised within Delphinidae by McKenna and Bell (1997: 384).

Tribe Stenonina J. Gray, 1868c: 5.

TYPE GENUS: *Steno* J. Gray, 1846c.

COMMENTS: When originally proposed, this rank was placed in the Family Delphinidae (J. Gray, 1821) and included the genera *Pontoporia* J. Gray, 1846c: 45; and *Steno* J. Gray, 1846c. Synonymised within Delphinidae by Rice (1998: 98), and McKenna and Bell (1997: 384), but it was recognised as a subfamily by Rice (1984: 480), LeDuc *et al.* (1999: 639) and LeDuc (2009: 299).

Tribe Lagenorhynchina J. Gray, 1868c: 7.

TYPE GENUS: *Lagenorhynchus* J. Gray, 1846b.

COMMENTS: When originally proposed, this rank was placed in the Family Delphinidae (J. Gray, 1821) and included the genera *Electra* J. Gray, 1866b [*Lagenorhynchus* J. Gray, 1846b]; *Leucopleurus* J. Gray, 1866h [=*Lagenorhynchus* J. Gray, 1846b]; and *Lagenorhynchus* J. Gray, 1846b. Synonymised within Delphinidae by Rice (1998: 98), and McKenna and Bell (1997: 384).

Tribe Pseudorcaina J. Gray, 1871a: v, 79.

TYPE GENUS: *Pseudorca* Reinhardt, 1862.

COMMENTS: When originally proposed, this rank was placed in the Family Delphinidae (J. Gray, 1821) and included the genera *Pseudorca* Reinhardt, 1862; and *Orcaella* J. Gray, 1866b. Synonymised within the Subfamily Orcininae by McKenna and Bell (1997: 384), and the Family Delphinidae by Rice (1998: 98).

Family Grampidae J. Gray, 1871a: v, 82.

TYPE GENUS: *Grampus* J. Gray, 1828.

COMMENTS: When originally proposed, this rank was placed in the Suborder Physeteroidea (J. Gray, 1871a [=Physeteridae (J. Gray, 1821)]) and included the genus *Grampus* J. Gray, 1828. Synonymised within the Family Delphinidae and Subfamily Globicephalinae by McKenna and Bell (1997: 384, 385), and within the Family Delphinidae by Rice (1998: 98).

Family Orcadae J. Gray, 1871a: vi, 85.

TYPE GENUS: *Orca* J. Gray, 1846b [=*Orcinus* Fitzinger, 1860b].

COMMENTS: When originally proposed, this rank was placed in the Suborder Physeteroidea (J. Gray, 1871a [=Physeteridae (J. Gray, 1821)]) and included the genera *Orca* J. Gray, 1846b [=*Orcinus* Fitzinger, 1860b] and *Ophysia* J. Gray, 1868c [=*Orcinus* Fitzinger, 1860b]. Rice (1998: 98) noted that the name is unavailable because the type genus is a junior homonym (Article 39 of the Code; ICZN, 1985a: 79). Synonymised within the Subfamily Orcininae by McKenna and Bell (1997: 384) and within the Family Delphinidae by Rice (1998: 98).

Subfamily Globiocephalinae Gill, 1872: 15, 95.

TYPE GENUS: *Globiocephalus* J. Gray, 1846c [=*Globicephala* Lesson, 1828d].

COMMENTS: When originally proposed, this rank was placed in the Family Delphinidae (J. Gray, 1821) and included the genera *Globiocephalus* J. Gray, 1846c [=*Globicephala* Lesson, 1828d]; and *Grampus* J. Gray, 1828. Synonymised within the Subfamily Globicephalinae (J. Gray, 1850: 313 [=J. Gray, 1863d: 201 or J. Gray, 1866b: 62, 313]) by McKenna and Bell (1997: 385). Subfamily recognised by van Bree and Cadenat (1968: 193), van Bree (1972: 212) and LeDuc *et al.* (1999: 639). Synonymised within Globicephalidae 'J. Gray, 1850' by McKenna and Bell (1997: 385).

Family Holoodontidae Brandt, 1873b: 575.

TYPE GENUS: Not based on a genus group name.

COMMENTS: When originally proposed, this rank included the subfamilies Orcinae (Brandt, 1873b [=Delphinidae

(J. Gray, 1821)]); Phocaeninae [=Phocoenidae] (J. Gray, 1825a); Delphininae (J. Gray, 1821); and Platanistinae (J. Gray, 1846c: 45). Synonymised within the Family Delphinidae by Rice (1998: 98).

Subfamily Orcinae Brandt, 1873b: 576.

TYPE GENUS: *Orcinus* Fitzinger, 1860b.

COMMENTS: When originally proposed this rank was placed in the Family Holoodontidae (Brandt, 1873b) and included the genus *Orcinus* Fitzinger, 1860b.

Family Delphinoidae Guérin, 1874: 62.

TYPE GENUS: Not based on a genus group name.

COMMENTS: When originally proposed, this rank was placed in the 'Cétodontes' and included the genera *Lagenorhynchus* J. Gray, 1846b; *Delphinorhynchus* de Blainville, 1817: 151 *nomen dubium*; *Tursio* J. Gray, 1843a [=*Tursiops* Gervais, 1855a] and 'Dauphins divers'.

Family Delphinorhynchidae W. Sclater, 1887: 60.

TYPE GENUS: *Delphinorhynchus* de Blainville, 1817: 151 *nomen dubium*.

COMMENTS: Family placed on the Cetaceae and included the genera † *Chamsodelphis* True, 1907: 104; † *Squalodon* Grateloup, 1840a: 208, 1840b: 346]; † *Schizodelphis* Gervais, 1861: 126; *Delphinorhynchus* de Blainville, 1817: 151 *nomen dubium*.

Tribe? Globicipites Winge, 1918: 36 [in 1921 English Edition].

TYPE GENUS: *Globiceps* Flower, 1884 [=*Globicephala* Lesson, 1828d].

COMMENTS: When originally proposed, this rank was placed in the Family Delphinidae (J. Gray, 1821) and included the genera *Orca* J. Gray, 1846b [=*Orcinus* Fitzinger, 1860b]; *Orcella* Anderson, 1871 [=*Orcaella* J. Gray, 1866b]; *Grampus* J. Gray, 1828; *Pseudorca* Reinhardt, 1862; and '*Globiceps* [*Globicephala*]' Flower, 1884 [=*Globicephala* Lesson, 1828d].

Subfamily Orcinae Slijper, 1936: 556.

TYPE GENUS: *Orcinus* Fitzinger, 1860b.

COMMENTS: When originally proposed, this rank was placed in the Family Delphinidae (J. Gray, 1821) and included the genera *Orca* J. Gray, 1846b [=*Orcinus* Fitzinger, 1860b], *Pseudorca* Reinhardt, 1862; *Orcella* Anderson, 1871 [=*Orcaella* J. Gray, 1866b]; and *Globicephalus* Lesson, 1828d. Synonymised within the Subfamily Orcininae by McKenna and Bell (1997: 384).

Family Stenidae F. Fraser & Purves, 1960: 59.

TYPE GENUS: *Steno* J. Gray, 1846c.

COMMENTS: When originally proposed, this rank was placed in the Delphinoidea (Flower, 1865a [=Odontoceti (Flower, 1867)]) and included the genera *Steno* J. Gray, 1846c; *Stenella* J. Gray, 1866h; and *Tursiops* Gervais, 1855a. Synonymised within Delphinidae by Mead and Brownell (2005: 726). Rank is an incorrect subsequent spelling of Stenonina (J. Gray, 1868c) (see Steyskal, 1980: 583; Rice (1998: 98). Steyskal (1980: 583) suggested that the correct spelling should be Stenonidae. Rice (1998: 98) noted that because of the incorrect spelling of its stem, the name becomes a homonym of the currently used subfamily name Stenidae [=Stenides Rey, 1883: 175], based on *Stenus* Latreille, 1796: 77, a genus of beetles [Coleoptera: Staphylinidae, See Shun-Ichiro (2006: 39)].

HOMONYMS:

Stenidae Rey, 1883: 175, rove beetles of the Class Insecta (Order Coleoptera, Family Staphylinidae). Name derived from the *Stenus* Latreille, 1796: 77, which is currently recognised as the Subfamily Steninae. See Shun-Ichiro (2006: 39).

Subfamily Orcinae F. Fraser and Purves, 1960: 94.

TYPE GENUS: *Orcinus* Fitzinger, 1860b.

COMMENTS: When originally proposed, this rank was placed in the Family Phocaenidae (Bravard, 1885 [=Phocoenidae (J. Gray, 1825a)]) and included the genera *Orcinus* Fitzinger, 1860b; and *Pseudorca* Reinhardt, 1862. Synonymised within Delphinidae by Rice (1998: 98), and Mead and Brownell (2005: 726).

Subfamily Lissodelphinae F. Fraser and Purves, 1960: 108.

TYPE GENUS: *Lissodelphis* Gloger, 1841.

COMMENTS: When originally proposed as a new rank it was placed in the Family Phocaenidae (Bravard, 1885 [=Phocoenidae (J. Gray, 1825a)]) and included the genus *Lissodelphis* Gloger, 1841. Synonymised within Delphinidae by Rice (1998: 98), McKenna and Bell (1997: 384), and Mead and Brownell (2005: 726). Subfamily recognised by Mead (1975: iii, 29), Rice (1984: 481), LeDuc *et al.* (1999: 639) who spelt it Lissodelphininae, and followed by LeDuc (2009: 299), but Rice (1998: 98) noted that the subfamily name is an incorrect original spelling under Article 32(c)(iii) of the Code (ICZN, 1985a: 69), because the grammatical stem of *delphis* is *delphin-*. Strong support for the subfamily was made by Harlin-Cognato and Honeycutt (2006: 1), who also suggested the genus *Lagenorhynchus* was polyphyletic.

Subfamily Cephalorhynchinae F. Fraser and Purves, 1960: 108.

TYPE GENUS: *Cephalorhynchus* J. Gray, 1846c: 36.

COMMENTS: When originally proposed as a new rank it was placed in the Family Phocaenidae (Bravard, 1885 [=Phocoenidae (J. Gray, 1825a)]) and included the genus *Cephalorhynchus* J. Gray, 1846c: 36. Subfamily recognised by Mead (1975: iii, 28), Rice (1984: 481) and McKenna and Bell (1997: 386), but synonymised within Delphinidae by

Rice (1998: 98), and Mead and Brownell (2005: 726), but not recognised by LeDuc *et al.* (1999: 639) or LeDuc (2009: 299).

Family Orcaellidae Nishiwaki, 1963: 98.

TYPE GENUS: *Orcaella* J. Gray, 1866b.

COMMENTS: When originally proposed as a new rank it included the genus *Orcaella* J. Gray, 1866b. Recognised as the Subfamily by McKenna and Bell (1997: 389), but synonymised within the Family Delphinidae by Rice (1998: 99).

Family Globicephalidae Nishiwaki, 1963: 98.

TYPE GENUS: *Globicephala* Lesson, 1828d.

COMMENTS: When originally proposed as a new rank it included the genera *Globicephala* Lesson, 1828d; *Feresa* J. Gray, 1870e; *Pseudorca* Reinhardt, 1862; *Grampus* Iredale & Troughton, 1933; and *Grampidelphis* Iredale and Troughton, 1933 [=*Grampus* J. Gray, 1828]. Name is a justified emendation of Globiocephalidae. Synonymised within the Family Delphinidae by Rice (1998: 99).

Family Globidelphinidae Nishiwaki, 1963: 98.

TYPE GENUS: Not based on a genus group name.

COMMENTS: When originally proposed as a new rank it included the genus *Grampidelphis* Iredale and Troughton, 1933 [=*Grampus* J. Gray, 1828]. It appears that the spelling of this family name was incorrectly given since it is based on *Grampidelphis* Iredale and Troughton, 1933. Synonymised within Delphinidae by Mead and Brownell (2005: 726).

Subfamily Orcininae Rice, 1967: 324.

TYPE GENUS: *Orcinus* Fitzinger, 1860b.

COMMENTS: When originally proposed, this rank was placed in the Family Delphinidae (J. Gray, 1821) and included the genera *Feresa* J. Gray, 1870e; *Globicephala* Lesson, 1828d; *Orcaella* J. Gray, 1866b; *Orcinus* Fitzinger, 1860b; *Peponocephala* Nishiwaki and Norris, 1966 and *Pseudorca* Reinhardt, 1862. Rice (1998: 99) argued that the name is a justified emendation of the Subfamily Orcinae (Fraser and Purves, 1960) in accordance with Article 29, 32(c)(iii), and 32(d) of the Code (ICZN, 1985a: 55, 69, 71).

Family Orcaelidae Nishiwaki, 1972: 111.

TYPE GENUS: *Orcaella* J. Gray, 1866b.

COMMENTS: When originally proposed, this rank was placed in the Suborder Odontoceti (Flower, 1867) and included the genus *Orcaella* J. Gray, 1866b. Incorrect subsequent spelling of Orcaellidae (Nishiwaki, 1963). Synonymised within the Family Delphinidae by Rice (1998: 99).

Subfamily Sotaliinae Kasuya, 1973: 32.

TYPE GENUS: *Sotalia* J. Gray, 1866b: 401.

COMMENTS: When originally proposed, this rank was placed in the Family Delphinidae (J. Gray, 1821) and included the genera *Sotalia* J. Gray, 1866b: 401; *Sousa* J. Gray, 1866h; and *Cephalorhynchus* J. Gray, 1846c: 36. Synonymised within Delphinidae by McKenna and Bell (1997: 384).

Subfamily Orcininae Kasuya, 1973: 40.

TYPE GENUS: *Orcinus* Fitzinger, 1860b.

COMMENTS: When originally proposed, this rank was placed in the Family Delphinidae (J. Gray, 1821) and included the genera *Pseudorca* Reinhardt, 1862; and *Orcinus* Fitzinger, 1860b. Subfamily was tentatively recognised by LeDuc *et al.* (1999: 639). Synonymised within the Subfamily Orcininae (Wagner, 1846) by McKenna and Bell (1997: 384).

Subfamily Orcaellinae Kasuya, 1973: 61.

TYPE GENUS: *Orcaella* J. Gray, 1866b.

COMMENTS: When originally proposed, this rank was placed in the Family Delphinapteridae (Gill, 1871b: 124) and included the genus *Orcaella* J. Gray, 1866b. Synonymised within the Subfamily Orcaellinae Nishiwaki (1972: 111) by McKenna and Bell (1997: 389).

Family Steninae Mead, 1975: iii, 26.

TYPE GENUS: *Steno* J. Gray, 1846c.

COMMENTS: When originally proposed, this rank was placed in the Family Delphinidae (J. Gray, 1821) and included the genera *Steno* J. Gray, 1846c; and *Sotalia* J. Gray, 1866b: 401. Synonymised within the Family Delphinidae by McKenna and Bell (1997: 384).

Family Stenonidae Steyskal, 1980: 583.

TYPE GENUS: *Steno* J. Gray, 1846c.

COMMENTS: Name is a correction of the spelling used in the Family Stenidae (F. Fraser & Purves, 1960). See also the discussion of Rice (1998: 98).

Infraorder Delphinida de Muizon, 1984: 68.

COMMENTS: When originally used, this rank was placed in the Odontoceti and included the superfamilies Delphinoidea (J. Gray, 1821) and Inioidea (Gray, 1846c: 25). Name subsequently recognised and further defined by de Muizon (1988b: 159, 160) and as an unranked clade within the Odontoceti by Geisler *et al.* (2011: 6).

HOMONYMS:

Family Delphinida Haeckel, 1866, dolphins of the Class Mammalia (Infraorder Cetacea, Parvorder Odontoceti). Name is a synonym of Delphinidae (J. Gray, 1821). See individual entry.

Subfamily Stenoninae Rice, 1984: 480.

TYPE GENUS: *Steno* J. Gray, 1846c.

COMMENTS: When originally proposed, this rank was placed in the Family Delphinidae (J. Gray, 1821) and included the genera *Sotalia* J. Gray, 1866b: 401; *Sousa* J.

Gray, 1866h; and *Steno* J. Gray, 1846. Subfamily tentatively recognised by LeDuc *et al.* (1999: 639) but not typically by subsequent authors.

Subfamily Lissodelphininae Rice, 1984: 481.

TYPE GENUS: *Lissodelphis* Gloger, 1841.

COMMENTS: When originally proposed, this rank was placed in the Family Delphinidae (J. Gray, 1821) and included the genus *Lissodelphis* Gloger, 1841. Rice (1998: 99) suggested this name is a justified emendation of the Subfamily Lissodelphinae in accordance with Article 29, 32(c)(iii), and 32(d) of the Code (ICZN, 1985a: 55, 69, 71).

Tribe Delphinini Pavlinov & Rossolimo, 1987: 92.

TYPE GENUS: *Delphinus* Linnaeus, 1758.

COMMENTS: When originally proposed, this rank was placed in the Subfamily Delphininae (J. Gray, 1821) and included the genera *Stenella* J. Gray, 1866h; *Delphinus* Linnaeus, 1758; *Tursiops* Gervais, 1855a; and *Lagenorhynchus* J. Gray, 1846b. Synonymised within Delphinidae by McKenna and Bell (1997: 384).

Tribe Grampini Pavlinov & Rossolimo, 1987: 94.

TYPE GENUS: *Grampus* J. Gray, 1828.

COMMENTS: When originally proposed, this rank was placed in the Subfamily Delphininae (J. Gray, 1821) and included the genera *Grampus* J. Gray, 1828 and *Lissodelphis* Gloger, 1841. Synonymised within Delphinidae by McKenna and Bell (1997: 385).

Family Grampidelphidae Mead and Brownell, 2005: 726.

TYPE GENUS: *Grampidelphis* Iredale and Troughton, 1933 [=*Grampus* J. Gray, 1828].

COMMENTS: When originally used this rank appears to be a correction of the Family Globidelphinidae (Nishiwaki, 1963), but potentially a correction of the type genus name *Grampidelphis* Iredale and Troughton, 1933. Synonymised within Delphinidae by Mead and Brownell (2005: 726).

Delphinus Linnaeus, 1758

Delphinus Linnaeus, 1758: 77.

TYPE SPECIES: *Delphinus delphis* Linnaeus, 1758 by Linnaean tautonomy.

COMMENTS: Only one species of *Delphinus*, *D. delphis*, was typically recognised until Heyning and Perrin (1994: 1, 29) also recognised *D. capensis* Gray, 1828 from the eastern North Pacific Ocean. Other studies by LeDuc (1999: 631–632) and Jefferson and Waerebeek (2002: 808) have suggested that there may be other valid taxa recognised in the future including *Delphinus bairdii* Dall, 1873 or possibly *Delphinus microps* Burmeister, 1866b. A review of 211 *Delphinus* skulls from throughout southern Australia revealed the presence of only one species (Bell *et al.*, 2002: 1), which is supported by the observations of Jefferson and Waerebeek (2002: 787, 794) who tentatively suggested that *D. delphis* occurs in southern Australian waters and *D. capensis* appears to occur in western and northern waters of Australia.

Epiodon Rafinesque, 1814a: 13.

TYPE SPECIES: *Epiodon urganantus* Rafinesque, 1814a: 13 [=*Incertae sedis*].

COMMENTS: Genus recognised by J. Gray (1866b: 327, 340), and allied to *Delphinus*, by Hershkovitz (1966: 199).

Rhinodelphis Wagner, 1846: 281, 316.

TYPE SPECIES: *Delphinus delphis* Linnaeus, 1758 by subsequent designation.

COMMENTS: Described as a subgenus of *Delphinus*. Synonymised within *Delphinus* by Hershkovitz (1966: 42), and Mead and Brownell (2005: 727).

Delphis J. Gray, 1864c: 236.

TYPE SPECIES: *Delphinus delphis* Linnaeus, 1758 by tautonomy.

COMMENTS: Described as a subgenus of *Delphinus*. Synonymised within *Delphinus* by Hershkovitz (1966: 42).

HOMONYMS:

Delphis Forskål, 1775: xviii, whales of the Class Mammalia (Order Artiodactyla). Name appears to be a *nomen nudum*.

Delphis Wagler, 1830: 34, the Beluga Whale of the Class Mammalia (Order Artiodactyla, Family Delphinidae). Taxon is a synonym of *Delphinapterus* Lacépède 1804: xli, 243.

Eudelphinus Van Bénéden & Gervais, 1880: 600.

TYPE SPECIES: *Delphinus delphis* Linnaeus, 1758 by original designation.

COMMENTS: Synonymised within *Delphinus* by Hershkovitz (1966: 42), and Mead and Brownell (2005: 727).

HOMONYMS:

Eudelphis Du Bus, 1872: 500, extinct dolphins of the Class Mammalia (Order Artiodactyla, Family Delphinidae). Taxon is a synonym of † *Scaldicetus* Du Bus, 1867: 568. See Hampe (2006: 63).

Mamdelphinus Herrera, 1899: 27.

TYPE SPECIES: *Mamdelphinus delphis* [=*Delphinus delphis* Linnaeus, 1758] by original designation.

COMMENTS: Article 1(b)(8) of the Code (ICZN, 1985a: 5) excludes modifications of available names by addition of a standard prefix or suffix in order to indicate that the taxa named are members of that group. Synonymised within *Delphinus* by McKenna and Bell (1997: 384).

Delphinus capensis Gray, 1828

Long-beaked Common Dolphin

Φ *Delphinus capensis capensis* Gray, 1828

Long-beaked Common Dolphin

Φ *Delphinus Capensis* Gray, 1828: 2.

TYPE LOCALITY: Cape of Good Hope, South Africa.

COMMENTS: Synonyms reviewed by Heyning and Perrin (1994: 21–24, 29). Species historically not recognised as occurring within Australian waters until it was added to the list of Australian mammals by Burbidge *et al.* (2014: 32). The addition of this species to the Australian mammal fauna was based on Brewer *et al.* (2009: 60) who suggested it occurs on the Scott Reef on the continental slope waters. It may also be the *Delphinus* that occurs around the Cocos (Keeling) Islands and Christmas Island in Australian territorial waters in north-eastern Indian Ocean (Woinarski *et al.*, 2014: 788). Jefferson and Van Waerebeek (2002: 787, 792–793) proposed that any Indian Ocean records should be attributed to *D. capensis*, or *tropicalis*, though the Australian forms in their study did not fit neatly with descriptions of either *capensis* or *delphis* and were not associated with the *tropicalis*-form. Jefferson and Van Waerebeek (2002: 807) proposed that this species generally does not occur around oceanic islands, so confirmation is needed for the occurrence of this species near Christmas Island and Cocos (Keeling) Islands. Woinarski *et al.* (2014: 789) indicated the distribution of this species is not well defined due to confusion with *D. delphis*.

FUTURE TAXONOMIC RESEARCH: A broadscale study of the distribution and taxonomic position of *Delphinus capensis*, and *tropicalis*, needs to be confirmed to determine if additional taxa should be recognised.

HOMONYMS:

Delphinus capensis G. Cuvier, 1829: 289, Heaviside's Dolphin of the Class Mammalia (Order Artiodactyla, Family Delphinidae). Name appears to be a synonym of *Cephalorhynchus heavisidii* (J. Gray, 1828: 2). See Hershkovitz (1966: 76).

Delphinus capensis Rapp, 1837, Pantropical Spotted Dolphin of the Class Mammalia (Order Artiodactyla, Family Delphinidae). Name is a synonym of *Stenella attenuata* (J. Gray, 1846). See individual entry.

Φ *Delphinus major* Gray, 1866b: 396.

TYPE LOCALITY: Unknown.

COMMENTS: Taxon synonymised within *capensis* by Heyning and Perrin (1994: 29), which was supported by the observations of Banks and Brownell (1969: 269).

Φ *Delphinus moorii* [sic] Gray, 1866i: 736, Fig. 1.

TYPE LOCALITY: Southwest of Cape of Good Hope, South Africa. (34°S, 7°3′W)

COMMENTS: Synonymised within *capensis* by Heyning and Perrin (1994: 29).

Φ *Delphinus microps* Burmeister, 1866b: 101.

TYPE LOCALITY: South Atlantic, Brazil.

COMMENTS: Synonymised within *delphis* by Hershkovitz (1966: 46) and synonymised within *capensis* by Heyning and Perrin (1994: 29).

HOMONYMS:

Delphinus microps J. Gray, 1846c, Spinner Dolphin of the Class Mammalia (Order Artiodactyla, Family Delphinidae). Name is a junior synonym of *Stenella longirostris* (J. Gray, 1828). See individual entry.

Φ *Delphinus Bairdii* Dall, 1873: 12.

TYPE LOCALITY: Coast of California, United States of America.

COMMENTS: Neotype allocated by Heyning and Perrin (1994: 1, 29), who synonymised it within *capensis*, which was followed by Mead and Brownell (2005: 728).

Delphinus capensis tropicalis van Bree, 1971

Indo-Pacific Common Dolphin

Delphinus tropicalis van Bree, 1971: 345.

TYPE LOCALITY: Malabar Coast, India.

COMMENTS: *Novum nomen* for *Delphinus longirostris* G. Cuvier, 1829a and *Delphinus dussumieri* Blanford, 1891 that are both preoccupied. Heyning and Perrin (1994: 1) suggested that further work was required to resolve whether *tropicalis* was a valid species. Taxon not recognised by Mead and Brownell (2005: 728), who placed it within *capensis*, but recognised as a subspecies of it by Jefferson and Van Waerebeek (2002: 787), Burbidge *et al.* (2014: 32), Wang *et al.* (2014: 502) and the Society for Marine Mammalogy (Committee on Taxonomy, 2014). It has also been recognised as a distinct species by van Bree and Gallagher (1978: 1) and Rice (1998: 112). Considered to be the subspecies most likely to occur within Australian waters by Jefferson and Van Waerebeek (2002: 793) and Woinarski *et al.* (2014: 788).

Delphinus longirostris G. Cuvier, 1829a: 288.

TYPE LOCALITY: Malabar Coast, India.

COMMENTS: Name pre-occupied by *Delphinus longirostris* Gray, 1828. Synonymised within *delphis* by Hershkovitz (1966: 43).

HOMONYMS:

Delphinus longirostris Gray, 1828, Spinner Dolphin of the Class Mammalia (Order Artiodactyla, Family Delphinidae). Name is currently recognised as *Stenella longirostris* (J. Gray, 1828).

Delphinus Sao J. Gray, 1850: viii, 125.

TYPE LOCALITY: Madagascar, Indian Ocean.

COMMENTS: Synonymised within *delphis* by Hershkovitz (1966: 44), but synonymised within *capensis* by Heyning and Perrin (1994: 29).

D. Frithii Blyth, 1859b: 492.

TYPE LOCALITY: 'procured during the voyage from England to India'.

COMMENTS: Synonymised within *capensis* by Heyning and Perrin (1994: 29).

Delphinus dussumieri Blanford, 1891: 588.

TYPE LOCALITY: *Novum nomen* for *Delphinus longirostris* G. Cuvier, 1829a.

COMMENTS: *Delphinus longirostris* G. Cuvier, 1829a is preoccupied by *D. longirostris* J. Gray, 1828 [= *Stenella longirostris* (J. Gray, 1828)] (see Hershkovitz, 1966: 46). Taxon synonymised within *delphis* by Hershkovitz (1966: 46) and within *tropicalis* by Pilleri and Gihr (1972: 406). Two additional skulls were discussed by van Bree (1971a: 169). Taxonomic history reviewed by Heyning and Perrin (1994: 24).

HOMONYMS:

Delphinus dussumieri J. Fischer, 1829: 456, Heaviside's Dolphin of the Class Mammalia (Order Artiodactyla, Family Delphinidae). Name proposed as a replacement for *Delphinus capensis* G. Cuvier, 1829: 289, which is preoccupied by *D. capensis* Gray, 1828. Name appears to be a synonym of *Cephalorhynchus heavisidii* (J. Gray, 1828: 2). See Hershkovitz (1966: 76).

Delphinus delphis Linnaeus, 1758

Short-beaked Common Dolphin

Delphinus delphis delphis Linnaeus, 1758

Short-beaked Common Dolphin

[Delphinus] Delphis Linnaeus, 1758: 77.

TYPE LOCALITY: 'Oceano Europaeo'.

COMMENTS: Based on description of an animal by Artedi (1738: 105). Taxonomic decision of Hershkovitz (1966: 42) to recognise only one species, which was followed by most subsequent authors including Mitchell (1975: 931). Evidence for two species of the traditional (restricted) *Delphinus* was proposed by Miller (1936: 146), Heyning and Perrin (1994: 1) when they proposed separating *D. bairdii* (Dall, 1873: 12); most authors recognise also *D. capensis* (e.g. Mead & Brownell, 2005: 727).

Delphinus vulgaris Lacépède, 1804: xlii, 250; Plates 13, Fig. 1; 14, Fig. 1.

TYPE LOCALITY: Unknown.

COMMENTS: Synonymised within *delphis* by Hershkovitz (1966: 43), Heyning and Perrin (1994: 28), and Mead and Brownell (2005: 728).

Delphinus Novae-Zelandiae Quoy & Gaimard, 1830: 149; Plate 28.

TYPE LOCALITY: South Pacific, Near Cape Gable, not far from Tolga Bay, New Zealand. Location given as Cook Straits by J. Gray (1843d: 183).

COMMENTS: Synonymised within *delphis* by Hershkovitz (1966: 44), and Heyning and Perrin (1994: 28).

Delphinus Zelandiae J. Gray, 1843d: 183.

TYPE LOCALITY: Incorrect subsequent spelling of *Delphinus Novae-Zelandiae* Quoy & Gaimard, 1830.

COMMENTS: Synonymised within *delphis* by Hershkovitz (1966: 44), and Heyning and Perrin (1994: 28).

Delphinus fulvifasciatus Wagner, 1846: 427; Plate 361, Fig. 1.

TYPE LOCALITY: Hobart, Tasmania, Australia.

COMMENTS: Synonymised within *delphis* by Hershkovitz (1966: 44), Bannister (1988c: 201), Heyning and Perrin (1994: 28), and Mead and Brownell (2005: 728).

Delphinus novae zeelandiae Wagner, 1846: 338, Plate 357.

TYPE LOCALITY: Incorrect subsequent spelling of *Delphinus Novae-Zelandiae* Quoy & Gaimard, 1830.

COMMENTS: Synonymised within *delphis* by Hershkovitz (1966: 44), and Heyning and Perrin (1994: 28).

Delphinus loriger Wiegmann, 1846: Plate 362.

TYPE LOCALITY: No locality.

COMMENTS: See comments under *Delphinus pseudodelphis* (below) for discussion on the date of publication. Taxon synonymised within *Stenella dubia* by Hershkovitz (1966: 32) and E. Hall (1981: 883). Perrin *et al.* (1987: 113) suggested the figure of external appearance that comprises the description is clearly of *Delphinus delphis* Linnaeus, 1758, which was confirmed by Heyning and Perrin (1994: 22, 28). Name not discussed by Mead and Brownell (2005).

Delphinus Janira J. Gray, 1846c: 41; Plate 23.

TYPE LOCALITY: North Atlantic, Newfoundland, Canada.

COMMENTS: Recognised within *Delphinus* by J. Gray (1866b: 245), but synonymised within *delphis* by Hershkovitz (1966: 44), Heyning and Perrin (1994: 28), and Mead and Brownell (2005: 728).

Delphinus Forsteri J. Gray, 1846c: 42; Plate 24.

TYPE LOCALITY: South Pacific, between New Caledonia and Norfolk Island.

COMMENTS: Synonymised within *delphis* by Hershkovitz (1966: 44), and Heyning and Perrin (1994: 28).

Delphinus albimanus Peale, 1848: 33; Plate 5, Fig. 1.

TYPE LOCALITY: South Pacific, off coast Chile. (27°16′S, 75°30′W)

COMMENTS: Recognised within *Delphinus* by J. Gray (1866b: 247). Synonymised within *delphis* by Hershkovitz (1966: 44), and Heyning and Perrin (1994: 28).

Delphinus novae zealandiae Gray, 1850: 123.

TYPE LOCALITY: Incorrect subsequent spelling of *Delphinus Novae-Zelandiae* Quoy & Gaimard, 1830.

COMMENTS: Synonymised within *delphis* by Hershkovitz (1966: 44), and Heyning and Perrin (1994: 28).

Delphinus algeriensis Loche, 1860: 474; Plate 22, Fig. 1.

TYPE LOCALITY: North Atlantic, Algeria, along coast.

COMMENTS: Synonymised within *delphis* by Hershkovitz (1966: 45), and Heyning and Perrin (1994: 28).

Delphinus fulvofasciatus True, 1889: 45.

TYPE LOCALITY: Incorrect subsequent spelling of *Delphinus fulvifasciatus* Wagner, 1846.

COMMENTS: Synonymised within *delphis* by Hershkovitz (1966: 44) and Bannister (1988c: 201). Not considered by Heyning and Perrin (1994) or Mead and Brownell (2005).

Φ ***Delphinus delphis ponticus*** Barabash-Nikiforov, 1935

Black Sea Common Dolphin

Φ *Delphinus delphis ponticus* Barabash-Nikiforov, 1935: 246, 249.

TYPE LOCALITY: Yalta region, Black Sea, Russia.

COMMENTS: Recognised as a subspecies of *delphis* by Hershkovitz (1966: 46), Rice (1998: 111), and Mead and Brownell (2005: 728), but not by Van Dyck and Strahan (2008: 846).

Delphinus Incertae sedis

Delphinus pomeegra Owen, 1866b: 23; Plate 6, Fig. 3.

TYPE LOCALITY: Off coast of Madras, India.

COMMENTS: Synonymised within *delphis* by Hershkovitz (1966: 45), and Mead and Brownell (2005: 728), but considered to be a *nomen dubium* by Heyning and Perrin (1994: 28) because the specimen is a juvenile and the rostrum is damaged.

Delphinus walkeri Gray, 1866i: 737, Fig. 2.

TYPE LOCALITY: Southwest of Cape of Good Hope, South Africa. 35°38′ S., 0°10′ E.

COMMENTS: Appears to be very similar to *moorei* [= *moorii*] according to Heyning and Perrin (1994: 23).

E. [udelphinus] Tasmaniensis Van Beneden & Gervais, 1880: 604, Plate 39, fig. 9.

TYPE LOCALITY: Hobart, Tasmania, Australia.

COMMENTS: Synonymised within delphis by Hershkovitz (1966: 45), but considered a *nomen dubium* by Heyning and Perrin (1994: 24).

Feresa J. Gray, 1870

Feresa J. Gray, 1870e: 77.

TYPE SPECIES: *Delphinus intermedia* J. Gray, 1827b (as *Orca intermedia*) [=*Feresa attenuata* J. Gray, 1874e] by monotypy.

COMMENTS: Described as a subgenus of *Orca* and raised to generic rank by J. Gray (1871a: v, 78). *Delphinus intermedius* J. Gray, 1827b was occupied by *Delphinus intermedius* Harlan, 1827 [=*Globicephala melas* (Traill, 1809)].

Feresia Flower, 1884: 510.

TYPE SPECIES: Emendation of *Feresa* J. Gray, 1870e.

COMMENTS: Synonymised within *Feresa* J. Gray, 1870e by Palmer (1904: 285) and Hershkovitz (1966: 98).

Feresa attenuata J. Gray, 1874

Pygmy Killer Whale

Feresa attenuata J. Gray, 1874e: 238.

TYPE LOCALITY: 'South Seas'.

COMMENTS: Described further by J. Gray (1875: 184). Molecular clock dates (McGowen *et al*. 2009: 898; see also Nishida *et al*. 2007: 727) for the separation of this genus and *Globicephala* seem to be well within the Pliocene; the generic status of *Feresa* seems thus to be very insecure.

Delphinus intermedius J. Gray, 1827b: 376.

TYPE LOCALITY: Unknown.

COMMENTS: Name attributed to J. Gray (1827b: 376) by Hershkovitz (1966: 98) and to J. Gray (1843a: 106) by Mead and Brownell (1993: 734; 2005: 728). Synonymised within *attenuata* by Hershkovitz (1966: 98).

HOMONYMS:

Delphinus intermedius Harlan, 1827, the Long-finned Pilot Whale of the Class Mammalia (Order Artiodactyla, Family Delphinidae). Taxon is a synonym of *Globicephala melas* (Traill, 1809). See individual entry.

Feresa occulta Jones & Packard, 1956: 167.

TYPE LOCALITY: North Pacific, Taiji, Honshu, Japan.

COMMENTS: Synonymised within *attenuatus* by Hershkovitz (1966: 99) and Mead and Brownell (1993: 352; 2005: 728).

Globicephala Lesson, 1828

globicephala [sic] Lesson, 1828d: 276, 441.

TYPE SPECIES: *Delphinus globiceps* G. Cuvier, 1812 [=*Globicephala melas* (Traill, 1809)] by original designation.

COMMENTS: See van Bree (1971c: 79) for a further review of *Globicephala* taxonomy.

Globicephalus Lesson, 1828f: 116.

TYPE SPECIES: Emendation of *Globicephala* Lesson, 1828d.

COMMENTS: Synonymised within *Globicephala* Lesson, 1828d by Palmer (1904: 296), Hershkovitz (1966: 90) and McKenna and Bell (1997: 385).

Cetus Wagler, 1830: 33.

TYPE SPECIES: *Nomen novum* for *Globicephala* Lesson, 1828d.

COMMENTS: Proposed as a *Nomen novum* for *Globicephala* Lesson, 1828d. Synonymised within *Globicephala* by Hershkovitz (1966: 90) and Mead and Brownell (1993: 352; 2005: 728).

HOMONYMS:

Cetus Brisson, 1762: 217, 225, whales of the Class Mammalia (Order Artiodactyla). Rejected for nomenclatural purposes by Opinion 1894 of the ICZN (1998: 64).

Cetus Oken, 1816, the Sperm Whale of the Class Mammalia (Order Artiodactyla, Family Physeteridae). Genus is a synonym of *Physeter* Linnaeus, 1758. See individual entry.

Cetus Billberg, 1828, the Sperm Whale of the Class Mammalia (Order Artiodactyla, Family Physeteridae). Genus is a synonym of *Physeter* Linnaeus, 1758. See individual entry.

Globiocephalus J. Gray, 1846c: 32.

TYPE SPECIES: Emendation of *Globicephalus* Lesson, 1828d.

COMMENTS: Spelling recognised by J. Gray (1866b: 313) and other authors until synonymised within *Globicephala* Lesson, 1828d by Hershkovitz (1966: 90).

Sphaerocephalus J. Gray, 1864c: 244.

TYPE SPECIES: *Delphinus globiceps* G. Cuvier, 1812 [=*Globicephala melas* (Traill, 1809)] by original designation.

COMMENTS: Described as subgenus of *Globiocephalus* J. Gray, 1846c, and raised to generic rank by J. Gray (1866b: 323). Synonymised within *Globicephala* by Iredale and Troughton (1936: 64), Hershkovitz (1966: 90), McKenna ad Bell (1997: 385), and Mead and Brownell (1993: 352; 2005: 728).

Globicephalus Van Bénéden & Gervais, 1880: 554.

TYPE SPECIES: *Delphinus globiceps* G. Cuvier, 1812 [=*Globicephala melas* (Traill, 1809)] by original designation.

COMMENTS: Synonymised within *Globicephalus* by Palmer (1904: 296) and McKenna and Bell (1997: 385).

Globiceps Flower, 1884: 508.

TYPE SPECIES: *Nomen novum* for *Globicephala* Lesson, 1828d.

COMMENTS: Flower was the first to select a type species (McKenna and Bell, 1997: 385), but his *Globiceps* was preoccupied by *Globiceps* Le Pelletier and Audinet-Serville, 1825, an insect. Synonymised within *Globicephala* by Palmer (1904: 296), Iredale and Troughton (1936: 64), Hershkovitz (1966: 90), McKenna and Bell (1997: 385), and Mead and Brownell (1993: 352; 2005: 728).

HOMONYMS:

Globiceps Le Peletier and Audinet-Serville, 1825: 326, bugs of the Class Insecta (Order Hemiptera, Family Miridae). Currently recognised name.

Globiceps Ayres, 1854: 193, polyps of the Class Hydrozoa (Order Anthoathecata, Family Pennariidae). Genus is a synonym of *Pennaria* Goldfuss, 1820b: xii, 89. See Schuchert (2006: 364).

Globiceps Van Bénéden, 1868: 93, whales of the Class Mammalia (Order Artiodactyla, Family Delphinidae). Genus is a *nomen nudum*.

Globiceps Felder, 1869: 31, butterflies of the Class Insecta (Order Lepidoptera, Family Pieridae). Genus is a synonym of *Pseudopontia* Plötz, 1870: 348.

Globicephala macrorhynchus (J. Gray, 1846)

Short-finned Pilot Whale

Globiocephalus macrorhynchus J. Gray, 1846c: 33.

TYPE LOCALITY: 'South Seas'.

COMMENTS: Transferred to *Globicephala* by F. Fraser (1950: 49). Synonymised within *Globicephala melaena melaena* (Traill, 1809) by Hershkovitz (1966: 94). Species recognised and synonyms reviewed by van Bree (1971c: 85), Ross (1984: 282), and Mead and Brownell (1993: 352; 2005: 728).

Globiocephalus Sieboldii J. Gray, 1846c: 32.

TYPE LOCALITY: Not known.

COMMENTS: Recognised as a subspecies within *melaena* by Hershkovitz (1966: 97) and synonymised within *macrorhynchus* by van Bree (1971c: 79), and Mead and Brownell (1993: 352; 2005: 728).

Globicephalus scammonii Cope, 1869: 21; Figs. 12–13.

TYPE LOCALITY: North Pacific, 10 miles off Baja California.

COMMENTS: Synonymised within *sieboldi* by Hershkovitz (1966: 97) and within *macrorhynchus* by van Bree (1971c: 85), and Mead and Brownell (1993: 352; 2005: 728).

Globicephalus ventricosus Iredale and Troughton, 1933: 35.

TYPE LOCALITY: See below.

COMMENTS: Erroneously applied to the pilot whale by Iredale & Troughton (1933: 35) who suggested that *Delphinus ventricosus* was based on the pilot whale or blackfish, which they believed must therefore be called *Globicephalus ventricosus*. Fraser (1951: 943) showed that *Delphinus ventricosus* was in fact *Phocoena phocoena* (Linnaeus, 1758: 77).

HOMONYMS:

Globiocephalus ventricosus Lacépède, 1804: xliii, 311, the Harbour Porpoise of the Class Mammalia (Order Artiodactyla, Family Phocoenidae). Name is a synonym of *Phocoena phocoena* (Linnaeus, 1758: 77). See Fraser (1951: 943).

Globicephala melas (Traill, 1809)

Long-finned Pilot Whale

Φ *Globicephala melas melas* (Traill, 1809)

North Atlantic Long-finned Pilot Whale

Φ *Delphinus melas* Traill, 1809: 81; Plate 3.

TYPE LOCALITY: Scapay Bay, Pomona, Orkney Islands, Scotland.

COMMENTS: Placed in *Globiocephalus* by Murie (1873: 235), *Globiceps* by Flower (1884: 509), *Globicephalus* by Flower (1885: v, 19) and *Globicephala* by Thomas (1898b: 99) and Sapin-Jaloustre (1953: 253). Type designation by Flower (1885: 19). Synonyms reviewed by van Bree (1971c: 83), and Mead and Brownell (1993: 352; 2005: 728). The spelling *melas* (the male form) has been used by various others including Scott and Lord (1920b: 13), Rice (1989b: 210; 1990: 360), and Mead and Brownell (1993: 352; 2005: 728). Article 31.2.3 of the fourth edition of the Code (ICZN, 1999: 38) gave *melas* as an example of a Greek adjective that does not change its ending when transferred to a genus of another gender. Kasuya (1975: 95) referred to historical records of an extinct North Pacific long-finned pilot whale, which was recognised as an unnamed subspecies by Rice (1998: 119).

Φ *Globicephala melaena* Thomas, 1898b: 99.

TYPE LOCALITY: Introduced as a feminine form of *melas*.

COMMENTS: There has previously been a lot of debate of the correct spelling of this taxon and whether to use *melas* or *melaena* (e.g. Schevill, 1990a: 169; 1990b: 360). The feminine form *melaena/melaina* has been used by various authors including Hershkovitz (1966: 91). See comments under *melas* above.

Φ *Globicephalus brachypterus* Cope, 1876: 129; Figs. 1–3.

TYPE LOCALITY: North Atlantic, east coast of Delaware Bay at Mouth of Maurice River, United States of America.

COMMENTS: Synonymised within *melaena* by Hershkovitz (1966: 95), van Bree (1971c: 85), and Mead and Brownell (1993: 352; 2005: 728).

Φ *delphinus [sic] globiceps* G. Cuvier, 1812: 14; Plate 1.

TYPE LOCALITY: 'le dauphin de St. Brieux', France.

COMMENTS: Synonymised within *melas* by Hershkovitz (1966: 92), van Bree (1971c: 83), and Mead and Brownell (1993: 352; 2005: 728).

Φ *Delphinus intermedius* Harlan, 1827: 51; Plate 1, Fig. 3.

TYPE LOCALITY: North Atlantic, Salem Harbour, Massachusetts, United States of America.

COMMENTS: Recognised within *Globiocephalus* by J. Gray (1866b: 318). Synonymised within *melaena* by Hershkovitz (1966: 93), and *melas* by Mead and Brownell (1993: 352; 2005: 728).

HOMONYMS:

Delphinus intermedius J. Gray, 1827b, the Pygmy Killer Whale of the Class Mammalia (Order Artiodactyla, Family Delphinidae). Taxon is a synonym of *Feresa attenuata* J. Gray, 1874e. See individual entry.

Φ *Globiocephalus Svineval* J. Gray, 1846c: 32.

TYPE LOCALITY: North Atlantic, Coast of North America.

COMMENTS: Synonymised within *melas* by Hershkovitz (1966: 94), van Bree (1971c: 85), and Mead and Brownell (1993: 352; 2005: 728).

Globicephala melas edwardii (Smith, 1834)

Southern Long-finned Pilot Whale

Phocaena Edwardii A. Smith, 1834: 239.

TYPE LOCALITY: Slang-kop, near Cape of Good Hope, South Africa.

COMMENTS: Placed within *Globiocephalus* at species rank by J. Gray (1866b: 320) and within *Globicephala* as a subspecies of *melaena* by Davies (1960: 34). Synonymised within *melas* by Hershkovitz (1966: 93), van Bree (1971c: 85), and Mead and Brownell (1993: 352). Recognised as a subspecies by Mead and Brownell (2005: 728), Clayton *et al.* (2006: 115), and Van Dyck and Strahan (2008: 864). Perrin *et al.* (2009: 16) considered that there is no strong morphological evidence supporting this taxon, although the separation of the northern and southern whales by a vast region of tropical and temperate waters does suggest that this is a case where reconsideration of distribution as a primary criterion might be justified.

Globicephala leucosagmaphora Rayner, 1939: 543.

TYPE LOCALITY: South Atlantic, South of Cape of Good Hope, South Africa.

COMMENTS: Synonymised within *melas* by Hershkovitz (1966: 96), van Bree (1971c: 85), and Mead and Brownell (1993: 352) and *edwardii* by Mead and Brownell (2005: 728).

Grampus J. Gray, 1828

Grampus J. Gray, 1828: 2.

TYPE SPECIES: *Delphinus griseus* G. Cuvier, 1812 [= *Grampus griseus* (G. Cuvier, 1812)] by virtual selection and formal subsequent selection. See J. Gray (1846c: 30) and Hershkovitz (1966: 87).

COMMENTS: Described as a subgenus of *Delphinus* Linnaeus, 1758 but elevated to generic rank by J. Gray (1843a: xxiii, 106) and followed by J. Gray (1866b: 295) and most subsequent authors. Iredale and Troughton (1933: 35) outlined the confusion in the genus name *Grampus*; they suggested that the name does not belong to Risso's dolphin but to the killer whale, and proposed that the common name and genus name should be *Grampus orca* rather than *Orcinus orca*. See review of Hershkovitz (1966: 81ff).

HOMONYMS:

Grampus Iredale & Troughton, 1933, the Killer Whale of the Class Mammalia (Order Artiodactyla, Family Delphinidae). Genus is synonym of *Orcinus* Fitzinger, 1860b. See individual entry. See individual entry.

Grayius A. Scott, 1873: 104

TYPE SPECIES: *Nomen novum* for *Grampus* J. Gray, 1828.

COMMENTS: Synonymised within *Grampidelphis* by Iredale and Troughton (1934: 64). Synonymised within *Grampus* by Hershkovitz (1966: 87), and Mead and Brownell (1993: 353; 2005: 729).

Grampidelphis Iredale and Troughton, 1933: 31.

TYPE SPECIES: *Nomen novum* for *Grampus* J. Gray, 1846b: 85.

COMMENTS: According to Hershkovitz (1966: 87) this name is an incorrect homonym of *Grampus* J. Gray, 1828. Synonymised within *Grampus* by Ellerman and Morrison-Scott (1951: 741), Schevill (1954: 124), Hershkovitz (1966: 87), and Mead and Brownell (1993: 353; 2005 729) and McKenna and Bell (1997: 386).

Grampus griseus (G. Cuvier, 1812)

Risso's Dolphin

delphinus [sic] griseus G. Cuvier, 1812: 13.

TYPE LOCALITY: Brest, France.

COMMENTS: Transferred to *Phocaena* by F. Cuvier (1836b: 182), *Grampus* by R. Hamilton (1837: 233) and *Grampidelphis* by Miller and Kellogg (1955: 662). Taxonomic decision of Hershkovitz (1966: 87) to return it to *Grampus*. Bannister *et al.* (1996: 68) noted morphological difference between regions, and suggested there are several races worldwide.

[Delphinus] aries G. Cuvier, 1812: 12; Plate 1.

TYPE LOCALITY: North Atlantic, Mediterranean at Nice, France.

COMMENTS: Synonymised within *griseus* by Hershkovitz (1966: 88).

delphinus Rissoanus Desmarest, 1822b: 519.

TYPE LOCALITY: North Atlantic, Mediterranean at Nice, France.

COMMENTS: *Nomen novum* for *Delphinus aries* Cuvier, 1812. Recognised within *Grampus* by J. Gray (1866b: 298). Synonymised within *griseus* by Hershkovitz (1966: 88), and Mead and Brownell (2005: 729).

Grampus Stearnsii Dall, 1873: 13.

TYPE LOCALITY: North Pacific, Monterey, California, United States of America.

COMMENTS: Synonymised within *griseus* by True (1889: 125), Hershkovitz (1966: 89), and Mead and Brownell (2005: 729).

Grampus Souverbianus P. Fischer, 1881: 210.

TYPE LOCALITY: Unknown. Skull is in the Bordeaux Museum (Hershkovitz, 1966: 89).

COMMENTS: Synonymised within *griseus* by Hershkovitz (1966: 89).

Grampidelphis exilis Iredale & Troughton, 1933: 32; Plate 10, Figs. 1–5.

TYPE LOCALITY: Ocean Beach, Manly, Sydney, New South Wales.

COMMENTS: Synonymised within *griseus* by Hershkovitz (1966: 89).

Lagenodelphis F. Fraser, 1956

Lagenodelphis F. Fraser, 1956: 496.

TYPE SPECIES: *Lagenodelphis hosei* F. Fraser, 1956 by original designation.

COMMENTS: Molecular data place the type species *hosei* well within the *Delphinus/Tursiops/Stenella* radiation (McGowen, 2011: 349; Hassanin *et al.*; 2012: 43).

Lagenodelphis hosei F. Fraser, 1956

Fraser's Dolphin

Lagenodelphis hosei F. Fraser, 1956: 496.

TYPE LOCALITY: Lutong River, Baram, Sarawak, Malaysia.

COMMENTS: No subspecies recognised by Bannister *et al.* (1996: 91). Species rediscovered in 1971 after not having been seen since its description (Perrin *et al.*, 1973: 345).

Lagenorhynchus J. Gray, 1846

Lagenorhynchus J. Gray, 1846b: 84.

TYPE SPECIES: Φ *Delphinus albirostris* J. Gray, 1846b: 84 [= Φ *Lagenorhynchus albirostris* (J. Gray, 1846b: 84)] by monotypy.

COMMENTS: Also described by J. Gray (1846c: 30, 34). Genus reviewed by Hershkovitz (1966: 60) and F. Fraser (1966: 16). This genus is considered polyphyletic and containing morphologically convergent species (Cipriano, 1997: 305; LeDuc *et al.*, 1999: 631; McGowan, 2011: 353; Committee on Taxonomy, 2011).

FUTURE TAXONOMIC RESEARCH: Further research required to resolve whether *Lagenorhynchus* is paraphyletic and, if required, make the necessary taxonomic changes.

Leucopleurus J. Gray, 1866h: 216.

TYPE SPECIES: Φ *Delphinus leucopleurus* Rasch, 1843: 112 [= Φ *Lagenorhynchus acutus* (J. Gray, 1828: 2)] by monotypy.

COMMENTS: Described as a subgenus of *Lagenorhynchus*. Synonymised within *Lagenorhynchus* by Hershkovitz (1966: 60), and Mead and Brownell (1993: 353; 2005: 729).

Sagmatias Cope, 1866: 294.

TYPE SPECIES: Φ *Sagmatias amblodon* Cope, 1866 [= Φ *Lagenorhynchus australis* (Peale, 1848: 33)] by monotypy.

COMMENTS: Synonymised within *Lagenorhynchus* by Hershkovitz (1966: 60), and Mead and Brownell (1993: 353; 2005: 729). Due to the apparent polyphyletic nature of *Lagenorhynchus*, the Society for Marine Mammalogy (Committee on Taxonomy, 2011) noted this genus has been resurrected by some authors (including Harlin-Cognato, 2010: 9) because of the disagreement about the composition of the genus.

Lagenorhynchus cruciger (Quoy & Gaimard, 1824)

Hourglass Dolphin

delphinus cruciger Quoy & Gaimard, 1824: 87; Plate 11, Figs. 3–4.

TYPE LOCALITY: Pacific Ocean, 49°S, between Cape Horn, South Africa and Australia.

COMMENTS: Transferred to *Lagenorhynchus* by Van Bénéden and Gervais (1880: 598, footnote), *Phocaena*, as *crucigera*, by Philippi (1893: 11). Formerly included *australis* and *obscurus*.

delphinus albigena Quoy & Gaimard, 1824: 87; Plate 11, Fig. 2.

TYPE LOCALITY: South Pacific, observed at sea between Cape Horn, South Africa and Australia.

COMMENTS: Synonymised within *cruciger* by Hershkovitz (1966: 63), and Mead and Brownell (1993: 353; 2005: 730).

delphinus bivittatus Lesson, 1827c: 178; Plate 9, Fig. 3.

TYPE LOCALITY: South Atlantic, observed 140 leagues west of the Falkland Islands, en route from Cape Horn, South Africa.

COMMENTS: Synonymised within *cruciger* by Hershkovitz (1966: 63), and Mead and Brownell (1993: 353; 2005: 730).

Lagenorhynchus clanculus J. Gray, 1846c: Plate 35.

TYPE LOCALITY: No type locality specified.

COMMENTS: Synonymised within *cruciger* by Hershkovitz (1966: 66), and Mead and Brownell (1993: 353; 2005: 730).

Sagmatias amblodon Cope, 1866: 294

TYPE LOCALITY: Unknown, but possibly Cape Horn to Lima, or in Australia or New Zealand.

COMMENTS: Synonymised within cruciger by Hershkovitz (1966: 68).

[Lagenorhynchus] wilsoni Lillie, 1915: 85, 123.

TYPE LOCALITY: Antarctic, between 54° and 65°S, including 55–60°S, 135°E.

COMMENTS: Synonymised within *cruciger* by Hershkovitz (1966: 68), and Mead and Brownell (1993: 353; 2005: 730).

Lagenorhynchus obscurus (J. Gray, 1828)

Dusky Dolphin

Lagenorhynchus obscurus obscurus (J. Gray, 1828)

African Dusky Dolphin

Φ *Delphinus (Grampus) obscurus* J. Gray, 1828: 2; Plate 2, Figs. 2–5.

TYPE LOCALITY: South Atlanta, Cape of Good Hope, South Africa.

COMMENTS: Transferred to *Tursio* by J. Gray (1866b: 264), *Clymenia* by J. Gray (1868c: 6), *Prodelphinus* by Flower (1885: vi, 28) and *Lagenorhynchus* by True (1889: 62). Synonymised within *cruciger* by Hershkovitz (1966: 64), but considered a distinct species by Rice (1977: 9), Brownell (1974: 14) and Mitchell (1975: 920). Some doubt exists over the placement of this taxon in *Lagenorhynchus* (Bannister *et al.*, 1996: 62). Confirmation of its occurrence along the coast of Australia was made by Gill *et al.* (2000: 452). Rice (1989: 114) refers to an unnamed subspecies from the east coast of New Zealand from Whitianga on the North Island south to Stewart Island, Campbell Island, Auckland Islands and the Chatham Islands.

Φ *Clymene similis* J. Gray, 1868d: 146, 147; Fig. 2.

TYPE LOCALITY: South Atlantic, Cape of Good Hope, South Africa.

COMMENTS: Synonymised within *Stenella coeruleoalba* by Hershkovitz (1966: 30), and F. Archer and Perrin (1999: 1) and *Lagenorhynchus obscurus* by Mead and Brownell (1993: 354; 2005: 730).

delphinus superciliosus Lesson, 1827c: 181; Plate 9, Fig. 2.

TYPE LOCALITY: South Cape, Tasmania, Australia.

COMMENTS: Synonymised within *cruciger* by Hershkovitz (1966: 64) and within *obscurus* by Mead and Brownell (1993: 354; 2005: 730). Taxon recognised as a subspecies of *obscurus* by Harlin-Cognato (2010: 9) and but this was not followed by the Society for Marine Mammalogy (Committee on Taxonomy, 2011) who raised doubt over the figure of Lesson (1827c) and referred to it as the 'unnamed New Zealand subspecies'.

Φ *Lagenorhynchus obscurus fitzroyi* (Waterhouse, 1838)

Fitzroy's Dusky Dolphin

Φ *Delphinus Fitzroyi* Waterhouse, 1838f: 23.

TYPE LOCALITY: South Atlantic, Golfo San Jose, Chebutt, coast of Patagoia, Argentina. (Approx. 42°30S)

COMMENTS: Transferred to *Lagenorhynchus* by Flower (1885: vi, 23) and *Phocaena* by Philippi (1893: 13). Synonymised within *Lagenorhynchus cruciger* by Hershkovitz (1966: 66) and within obscurus by Mead and Brownell (1993: 354). Subspecies within *Lagenorhynchus obscurus* recognised by Rice (1998: 114), Mead and Brownell (2005: 730), Harlin-Cognato (2010: 9), the Society for Marine Mammalogy (Committee on Taxonomy, 2011). Taxon not recognised by Van Dyck and Strahan (2008: 850). Subspecies discussed by Perrin *et al.* (2009a: 14).

Φ *Delphinus breviceps* Wagner, 1846: 427; Plate 368, Fig. 1.

TYPE LOCALITY: South Atlantic, Río de La Plata, Buenos Aires, Argentina.

COMMENTS: Synonymised within *cruciger* by Hershkovitz (1966: 67), and Mead and Brownell (1993: 354) and within *fitzroyi* by Mead and Brownell (2005: 730).

Φ *Tursio? panope* Philippi, 1896: 14; Plates 4–6.

TYPE LOCALITY: South Pacific, Chile.

COMMENTS: Synonymised within *Cephalorhynchus eutropica* J. Gray, 1846c: Plate 34 by Hershkovitz (1966: 75), but synonymised within *obscurus* by Mead and Brownell (1993: 354; 2005: 730) and within *fitzroyi* by Mead and Brownell (2005: 730).

Lagenorhynchus obscurus posidonia (Philippi, 1893)

Peruvian/Chilean Dusky Dolphin

Phocaena posidonia Philippi, 1893: 9.

TYPE LOCALITY: Southern tip of South America. Harpooned at '48° 10' latit. sur i 77° lonjit. oeste, i era hembra'.

COMMENTS: Name not typically considered by other authors until recognised as a subspecies of *obscurus* by Harlin-Cognato (2010: 9) and followed by the Society for Marine Mammalogy (Committee on Taxonomy, 2011).

Lissodelphis Gloger, 1841

Lissodelphis Gloger, 1841: xxxiv, 169.

TYPE SPECIES: *Delphinus peronii* Lacépède, 1804 [=*Lissodelphis peronii* (Lacépède, 1804)] by monotypy.

COMMENTS: Taxonomic decision of Hershkovitz (1966: 56) to recognise this taxon.

delphinapterus Lesson, 1827c: 179.

TYPE SPECIES: *Delphinus peronii* Lacépède, 1804 [=*Lissodelphis peronii* (Lacépède, 1804)] by monotypy.

COMMENTS: Taxon recognised by J. Gray (1846c: 36; 1866b: 276). Synonymised within *Lissodelphis* by Hershkovitz (1966: 56), McKenna and Bell (1997: 385), and Mead and Brownell (1993: 354; 2005: 730).

HOMONYMS:

Delphinapterus Lacépède, 1804: xli, 243, the Beluga Whale of the Class Mammalia (Order Artiodactyla, Family Monodontidae). Genus is currently recognised.

Delphinaptera Bowdich, 1821: 86, the Beluga Whale of the Class Mammalia (Order Artiodactyla, Family Monodontidae). Emendation of *Delphinapterus* Lacépède, 1804.

Tursio Wagler, 1830: 34.

TYPE SPECIES: *Delphinus peronii* Lacépède, 1804 [=*Lissodelphis peronii* (Lacépède, 1804)] by monotypy.

COMMENTS: Synonymised within *Lissodelphis* by Hershkovitz (1966: 56), and Mead and Brownell (1993: 354; 2005: 730).

HOMONYMS:

Tursio Fleming, 1822b, the Sperm Whale of the Class Mammalia (Order Artiodactyla, Family Physeteridae). Genus is a synonym of *Physeter* Linnaeus, 1758. See individual entry.

Tursio J. Gray, 1843a, bottle-nosed dolphins of the Class Mammalia (Order Artiodactyla, Family Delphinidae). Genus is a synonym of *Tursiops* Gervais, 1855a. See individual entry.

Leucorhamphus Lilljeborg, 1861b: 4, 5.

TYPE SPECIES: *Nomen novum* for *Delphinapterus* Lesson 1827c.

COMMENTS: Synonymised within *Lissodelphis* by Hershkovitz (1966: 56), and Mead and Brownell (1993: 354; 2005: 730).

Pachypleurus Brandt, 1873a: 234.

TYPE SPECIES: *Delphinus peronii* Lacépède, 1804 [=*Lissodelphis peronii* (Lacépède, 1804)] by monotypy.

COMMENTS: Described as a subgenus of *Delphinapterus*. Synonymised within *Lissodelphis* by McKenna and Bell (1997: 385).

HOMONYMS:

Pachypleura White, 1853: 27, longhorn beetles of the Class Insecta (Order Coleoptera, Family Cerambycidae). Genus is a synonym of *Megopis* Audinet-Serville, 1832: 161.

Pachypleura Cornalia, 1854: 45, reptiles of the Class Reptilia (Order Sauropterygia, Family Pachypleurosauridae). Taxon is a homonym of *Neusticosaurus* Seeley, 1882: 350. See Sander (1989: 576) for synonymy.

Archaeocetus Sinzow, 1898: 125.

TYPE SPECIES: *Delphinus peronii* Lacépède, 1804 [=*Lissodelphis peronii* (Lacépède, 1804)] by monotypy.

COMMENTS: Unnecessary replacement name for *Pachypleurus*. Synonymised within *Lissodelphis* by McKenna and Bell (1997: 385).

Pristinicetus Trouessart, 1898 [1898–1899]: 1071.

TYPE SPECIES: *Delphinus peronii* Lacépède, 1804 [=*Lissodelphis peronii* (Lacépède, 1804)] by monotypy.

COMMENTS: Unnecessary replacement name for *Pachypleurus* Brandt (1873a), which is preoccupied by *Pachypleurus* White (1953: 27) a genus of coleopteran beetles. Synonymised within *Lissodelphis* by McKenna and Bell (1997: 385).

Lissodelphis peronii (Lacépède, 1804)

Southern Rightwhale Dolphin

Delphinus Peronii Lacépède, 1804: xliii, 316.

TYPE LOCALITY: South of Tasmania, Australia. (Approx. 44°S, 144°E)

COMMENTS: Transferred to *Delphinapterus* by Lesson (1827c: 179) and J. Gray (1866b: 276), *Tursio* by True (1889: 78) and *Lissodelphis* by Palmer (1899d: 24).

Delphinus leucoramphus Lacépède, 1804: 316, footnote.

TYPE LOCALITY: Observed at sea north and east of 44°S, 141°W.

COMMENTS: *Nomen nudum*. Péron's manuscript name for *Delphinus peronii* Lacépède, 1804 (see Hershkovitz, 1966: 58). Synonymised within *peronii* by Hershkovitz (1966: 58), and Mead and Brownell (1993: 354; 2005: 731).

Orcaella J. Gray, 1866

Orcaella J. Gray, 1866b: 285.

TYPE SPECIES: *Orca brevirostris* J. Gray, 1866b: 285 [=*Orcaella brevirostris* (Owen in J. Gray, 1866b: 285)] by monotypy.

COMMENTS: Described as a subgenus of *Orca* and raised to generic rank by J. Gray (1868c: 7). Taxon included within Delphinidae by Hershkovitz (1966: 77), Heyning (1989a: 54), Lint *et al.* (1990: 17), and Mead and Brownell (2005: 731). *Orcaella* has been placed in the Subfamily Orcaellinae, within the Family Monodontidae, by Barnes *et al.* (1985: 26) and McKenna and Bell (1997: 389).

Orcella Anderson, 1871: 142, footnote.

TYPE SPECIES: *Nomen novum* for *Orcaella* J. Gray, 1866b.

COMMENTS: Synonymised within *Orcaella* by Palmer (1904: 478), Hershkovitz (1966: 77) and Stacey and Arnold (1999: 1).

Orcaella heinsohni Beasley *et al.*, 2005

Australian Snub-finned Dolphin

Orcaella heinsohni Beasley *et al.*, 2005: 365, 378.

TYPE LOCALITY: Horseshoe Bay, Magnetic Island, Queensland, Australia.

COMMENTS: Separated from *O. brevirostris* (J. Gray, 1866b: 285). The phylogenetic position of this species (as *O. brevirostris*) was explored by Arnold and Heinsohn (1996: 141) from a specimen collected in Queensland.

Orcinus Fitzinger, 1860

Orcinus Fitzinger, 1860b: 204.

TYPE SPECIES: *Delphinus orca* Linnaeus, 1758 [=*Orcinus orca* (Linnaeus, 1758)] by monotypy.

COMMENTS: Taxon reviewed by Hershkovitz (1966: 81).

Orca J. Gray, 1846b: 84.

TYPE SPECIES: *Delphinus gladiator* Bonnaterre, 1789 (as *Orca gladiator*) [=*Orcinus orca* (Linnaeus, 1758)] by virtual tautonomy.

COMMENTS: Also described by J. Gray (1846b: 33). Synonymised within *Orcinus* by Hershkovitz (1966: 81), Heyning and Dahlheim (1988: 1), and Mead and Brownell (1993: 354; 2005: 731).

HOMONYMS:

Orca Wagler, 1830, bottle-nosed whales of the Class Mammalia (Order Artiodactyla, Family Ziphiidae). Genus is a synonym of *Hyperoodon* Lacépède, 1804. See individual entry.

Ophysia J. Gray, 1868c: 8.

TYPE SPECIES: *Orca capensis* J. Gray, 1846c [=*Orcinus orca* (Linnaeus, 1758)] by monotypy.

COMMENTS: Described as a subgenus of *Orca*. Synonymised within *Orca* by Hershkovitz (1966: 81), Heyning and Dahlheim (1988: 1), and Mead and Brownell (1993: 354; 2005: 731).

Gladiator J. Gray, 1870e: 71.

TYPE SPECIES: *Orca stenorhyncha* J. Gray, 1870e [=*Orcinus orca* (Linnaeus, 1758)] by monotypy.

COMMENTS: Described as a subgenus of *Orca*. Synonymised within *Orca* by Hershkovitz (1966: 81), Heyning and Dahlheim (1988: 1), and Mead and Brownell (1993: 354; 2005: 731).

HOMONYMS:

Gladiator Gistl (in Gistel & Bromme), 1850 [1847–1850]: 381, beetles of the Class Insecta (Order Coleoptera).

Grampus Iredale & Troughton, 1933: 28.

TYPE SPECIES: '*Delphinus grampus* Linn. =Hunter' [=*Delphinus grampus* de Blainville, 1817] [=*Orcinus orca* (Linnaeus, 1758)] by monotypy.

COMMENTS: Synonymised within *Orcinus* by Hershkovitz (1961: 549; 1966: 81), Ellerman and Morrison-Scott (1951: 739), Bannister (1988c: 206), Heyning and Dahlheim (1988: 1), and Mead and Brownell (1993: 354; 2005: 731).

HOMONYMS:

Grampus J. Gray, 1828, Risso's Dolphin of the Class Mammalia (Order Artiodactyla, Family Delphinidae). Currently accepted name. See individual entry.

Orcinus orca (Linnaeus, 1758)

Killer Whale

[*Delphinus*] *orca* Linnaeus, 1758: 77.

TYPE LOCALITY: 'Oceano Europaeo'.

COMMENTS: Transferred to *Orcinus* by Palmer (1899d: 24) and *Grampus* by Iredale and Troughton (1933: 30). Reviewed by Hershkovitz (1966: 82), Ross (1984: 259), and Heyning and Dahlheim (1988: 1). No widely recognised subspecies according to Bannister *et al.* (1996: 107). It now seems evident that there are several undescribed species in this genus. Pitman and Ensor (2003: 131) distinguished three 'forms' in Antarctic waters, and Pitman *et al.* (2007: 43) described a fourth dwarf 'form' from Antarctica. Jefferson *et al.* (2008: 158–161) also described the different forms. LeDuc *et al.* (2008: 426) found that the existence of 'multiple' species seemed to be confirmed by mtDNA, and Morin *et al.* (2010: 908) produced convincing evidence of this, finding that three Antarctic species could be separated by complete mitochondrial genomes. Two unnamed 'subspecies' were recognised by Krahn *et al.* (2004: 16) and supported by Hoelzel *et al.* (2007: 1407), Perrin *et al.* (2009: 17) and Pitman *et al.* (2011: 303). Despite this accumulating evidence, Ford (2009: 650) suggested that it was unclear at present whether killer whale ecotypes represent a single species, multiple species, or subspecies. In fact, the evidence seems quite clear; Morin *et al.* (2010: 908) discussed in some detail the various 'ecotypes' that have been revealed over the last 30 years, supporting them using their complete mitochondrial genome sequences, and proposed that each 'ecotype' be elevated to species with some other types being recognised as subspecies pending additional data. Colour plates by Gorter (2011: 34) illustrated males and females of four 'ecotypes and forms' (plus a 'small type' of one of them) from the Southern Hemisphere, as well as five from the Northern Hemisphere. Each of these has its own preferred habitat, social organisation and diet, further corroborating the idea that they are distinct species. At least two names are available for Southern Hemisphere killer whales: certainly *glacialis* and *nanus*, and very likely also the names *capensis*, *victorini*, *magellanica*, *africana* and *tasmanica*. The name *antarctica* was based on a drawing, and may not be identifiable, although the name itself suggests it is another of these Southern Hemisphere killer whales. The name *nanus* was considered a *nomen nudum* by Rice (1998: 118), which was followed by Mead and Brownell (2005: 731), but the name was accompanied by a satisfactory description, so is available, and the stipulation that a type specimen be designated was not brought into the Code until the fourth edition (1999: 20, Art.16.4).

FUTURE TAXONOMIC RESEARCH: Further research is need to confirm the identity of the different 'forms' of *Orcinus* and which of the available names corresponds to which of them; new species should be described for those presently without a name, after due consideration of the names presently assigned to synonymy.

Delphinus serra Borowski, 1781: 38.

TYPE LOCALITY: North Atlantic, Davis Strait, Spitzbergen.

COMMENTS: Synonymised within *orca* by Hershkovitz (1966: 83), and Heyning and Dahlheim (1988: 1).

D. [elphinus] Gladiator Bonnaterre, 1789: 23.

TYPE LOCALITY: North Atlantic, Spitzbergen, Davis Strait and New England Coast.

COMMENTS: Recognised within *Orca* by J. Gray (1866b: 279) and others including Scott and Lord (1920b: 9). Synonymised within *orca* by Hershkovitz (1966: 83), Heyning and Dahlheim (1988: 1), and Mead and Brownell (1993: 355; 2005: 731).

Delphinus Duhamelii Lacépède, 1804: xliii, 314.

TYPE LOCALITY: France.

COMMENTS: Synonymised within *orca* by Hershkovitz (1966: 83), and Heyning and Dahlheim (1988: 1).

Delphinus grampus de Blainville, 1817: 168.

TYPE LOCALITY: North Atlantic.
COMMENTS: Synonymised within *orca* by Hershkovitz (1966: 83), Heyning and Dahlheim (1988: 1).

Orca Capensis J. Gray, 1846c: 34.

TYPE LOCALITY: South Atlantic, Cape of Good Hope, South Africa.
COMMENTS: Recognised by J. Gray (1866b: 283). Synonymised within *orca* by Hershkovitz (1966: 83), Heyning and Dahlheim (1988: 1), and Mead and Brownell (1993: 355; 2005: 731).

Delphinus victorini Grill, 1858: 21; Plate 1, Figs. 1–2.

TYPE LOCALITY: South Atlantic, Knysua River, west Cape Town, South Africa.
COMMENTS: Synonymised within *orca* by Hershkovitz (1966: 83), Heyning and Dahlheim (1988: 1).

O. [rca] Schlegelii Lilljeborg, 1866b: 235.

TYPE LOCALITY: North Atlantic, west coast of Norway.
COMMENTS: Synonymised within *orca* by Hershkovitz (1966: 84), Heyning and Dahlheim (1988: 1).

Orca magellanica Burmeister, 1866b: 99; Plate 9, Fig. 5.

TYPE LOCALITY: South Atlantic, Arroyo de Cristiano Muerto, south of Cabo Corrientes, Buenos Aires, Argentina.
COMMENTS: Synonymised within *orca* by Hershkovitz (1966: 84), Heyning and Dahlheim (1988: 1).

Orca Eschrichtii Reinhardt, 1866: 188; Fig. page 187.

TYPE LOCALITY: North Atlantic, Kollefjord of Strömö, Faeroe Islands, Great Britain
COMMENTS: Synonymised within *orca* by Hershkovitz (1966: 84), Heyning and Dahlheim (1988: 1).

Orca ater Cope, 1869: 22.

TYPE LOCALITY: North Pacific, north west coasts from Oregon to Aleuian Islands.
COMMENTS: Synonymised within *orca* by Hershkovitz (1966: 84), Heyning and Dahlheim (1988: 1), and Mead and Brownell (1993: 355; 2005: 731).

Orca rectipinna Cope, 1869: 22; Figs. 15–16.

TYPE LOCALITY: North Pacific, Coast of California, United States of America.
COMMENTS: Synonymised within *orca* by Hershkovitz (1966: 84), Heyning and Dahlheim (1988: 1), and Mead and Brownell (1993: 355; 2005: 731).

Orca stenorhyncha J. Gray, 1870e: 71; Figs. 1–3.

TYPE LOCALITY: North Atlantic, English coast.
COMMENTS: Synonymised within *orca* by Hershkovitz (1966: 85), Heyning and Dahlheim (1988: 1).

Orca pacifica J. Gray, 1870e: 71, 76.

TYPE LOCALITY: North Pacific.
COMMENTS: Synonymised within *orca* by Hershkovitz (1966: 85), Heyning and Dahlheim (1988: 1).

Orca latirostris J. Gray, 1870e: 76.

TYPE LOCALITY: North Atlantic, coast of Essex, England.
COMMENTS: Synonymised within *orca* by Hershkovitz (1966: 85), Heyning and Dahlheim (1988: 1).

Orca africana J. Gray, 1871a: vi, 91.

TYPE LOCALITY: Indian Ocean, Algoa Bay, Cape of Good Hope, South Africa.
COMMENTS: Synonymised within *orca* by Hershkovitz (1966: 85), Heyning and Dahlheim (1988: 1).

Orca tasmanica J. Gray, 1871a: vi, 92.

TYPE LOCALITY: Tasmania, Australia.
COMMENTS: Synonymised within *orca* by Hershkovitz (1966: 85), Bannister (1988c: 206), Heyning and Dahlheim (1988: 1).

Orca minor Malm, 1871: 80, 81.

TYPE LOCALITY: North Atlantic, Sweden.
COMMENTS: Synonymised within *orca* by Hershkovitz (1966: 85), Heyning and Dahlheim (1988: 1).

Orca antarctica P. Fischer, 1876: 146.

TYPE LOCALITY: Unknown.
COMMENTS: Based on a drawing. Synonymised within *orca* by Heyning and Dahlheim (1988: 1).

[Grampus] orca Iredale & Troughton, 1933: 30.

TYPE LOCALITY: Identification of *nomen nudum Delphinus grampus* 'Linn.' (J. Gray, 1828: 2).
COMMENTS: Name synonymised within *orca* by Bannister (1988c: 206) and Hershkovitz (1966: 82).

Orca nanus Mikhalev *et al.*, 1981: 564.

TYPE LOCALITY: Antarctic waters.
COMMENTS: Synonymised within *orca* by Heyning and Dahlheim (1988: 1) and Mead and Brownell (2005: 731). As no holotype was designated for this taxon it was recognised as a *nomen nudum* by the International Whaling Commission (1982: 625), which was followed by Rice (1998: 118), and Pitman and Ensor (2003: 131), but this is incorrect: until 2000, in the Fourth Edition of the Code (ICZN, 1999), a type specimen was not required in order for a name to be available. Careful comparison of the description, the localities and other data should enable this name to be identified with one of the four or five Southern Hemisphere 'forms'.

O. [rcinus] glacialis Berzin & Vladimirov, 1982: 31.

TYPE LOCALITY: Indian Ocean sector of Antarctic.

COMMENTS: Described again in greater detail by Berzin and Vladimirov (1983: 288). Doubt over the validity of this taxon as a separate species was made by Miyazaki (1992: 48) and it was synonymised within *orca* by Heyning and Dahlheim (1988: 1), and Mead and Brownell (2005: 731). Taxon discussed by Pitman and Ensor (2003: 131) who noted that much more descriptive detail was provided in the description of *glacialis* and that the authors designated a holotype and five paratypes at the Pacific Research Institute of Fisheries and Oceanography, TINRO, Vladivostok, although Pitman and Ensor (2003: 131) advised that apparently all of the specimens were subsequently discarded.

Peponocephala Nishiwaki & Norris, 1966

Peponocephala Nishiwaki & Norris, 1966: 95.

TYPE SPECIES: *Nomen novum* for *Electra* J. Gray, 1866b.

COMMENTS: Formerly included within *Lagenorhynchus* (e.g. Mead & Brownell, 2005: 731). Molecular clock dates (McGowen *et al.* 2009: 898; Nishida *et al.* 2007: 727; McGowen 2011: 349) for the separation of this genus and *Globicephala* seem to be well within the Pliocene and its separation from *G. melas* postdates that of *Feresa*; the generic status of *Peponocephala* seems thus very insecure.

Electra J. Gray, 1866b: 268.

TYPE SPECIES: *Lagenorhynchus electra* J. Gray, 1846c [=*Peponocephala electra* (J. Gray, 1846c)] by tautonomy.

COMMENTS: Also named by J. Gray (1866h: 216). Described as a subgenus of *Lagenorhynchus*, but raised to generic rank by J. Gray (1868c: 7). Formerly synonymised in *Lagenorhynchus* by Hershkovitz (1966: 60), and Mead and Brownell (1993: 355), but placed within *Peponocephala* by Mead and Brownell (2005: 731).

HOMONYMS:

Electra Lamouroux, 1816: 120, bryozoans of the Phylum Bryozoa (Class Gymnolaemata, Order Cheilostomata, Family Electridae). Currently accepted name. See Hayward (2011).

Electra Stephens, 1829a: 44, 1829b: 238, 1829c: 135, moths of the Class Insecta (Order Lepidoptera, Family Geometridae). Genus is a synonym of *Pelurga* Hübner, 1825 [1816–1826]: 335.

Electra Curtis, 1836: 603, geometer moths of the Class Insecta (Order Lepidoptera, Family Geometridae). Genus is a synonym of *Electrophaes* Prout, 1923: 197. See Prout (1923: 197).

Electra Loew, 1845: 57, flies of the Class Insecta (Order Diptera, Family Rachiceridae). Genus is a synonym of *Paleorachicerus* Nagatomi, 1970: 420.

Electra Albers, 1850: 194, land snails of the Phylum Mollusca (Class Gastropoda, Family Subulinidae). Genus is a synonym of *Glessula* Martens, 1860: 254.

Peponocephala electra (J. Gray, 1846)

Melon-headed Whale

Lagenorhynchus Electra J. Gray, 1846c: 35; Plate 13.

TYPE LOCALITY: Unknown.

COMMENTS: Included in *Lagenorynchus* by Hershkovitz (1966: 69). Transferred to the newly created genus *Peponocephala* by Nishiwaki and Norris (1966: 95), and followed by subsequent authors including van Bree and Cadenat (1968: 193) and Bryden *et al.* (1977: 180).

Lagenorhynchus Asia J. Gray, 1846c: 35; Plate 14.

TYPE LOCALITY: Unknown.

COMMENTS: Synonymised within *electra* by Hershkovitz (1966: 70), and Mead and Brownell (2005: 731).

Delphinus (Lagenorhynchus) fusiformis Owen, 1866b: 22; Plate 5, Fig. 1.

TYPE LOCALITY: Indian Ocean, Madras, India.

COMMENTS: Synonymised within *electra* by Hershkovitz (1966: 70), and Mead and Brownell (1993: 355; 2005: 731).

Delphinus pectoralis Peale, 1848: 32; Plate 5, Fig. 2.

TYPE LOCALITY: North Pacific, Hilo Bay, Hawaii, United States of America.

COMMENTS: Synonymised within *electra* by Hershkovitz (1966: 70), and Mead and Brownell (1993: 355; 2005: 731).

Pseudorca Reinhardt, 1862

Pseudorca Reinhardt, 1862: 151

TYPE SPECIES: *Phocaena crassidens* Owen, 1846 (as *Pseudorca crassidens*) [=*Pseudorca crassidens* (Owen, 1846)] by original designation.

COMMENTS: Recognised by J. Gray (1866b: 290), Hershkovitz (1966: 78) and subsequent authors, but molecular clock dates (McGowen *et al.* 2009: 898; Nishida *et al.* 2007: 727; McGowen 2011: 353) for the separation of this genus and *Globicephala* seem to be well within the Pliocene.

Neoorca J. Gray, 1871e: 80.

TYPE SPECIES: *Orca meridionalis* Flower, 1865b (as *Pseudorca meridionalis* J. Gray, 1866b: 291) [=*Pseudorca crassidens* (Owen, 1846)] by monotypy.

COMMENTS: Described as a subgenus of *Pseudorca*. Synonymised within *Pseudorca* by Iredale and Troughton (1934: 63), Hershkovitz (1966: 78), Bannister (1988c: 207), and Mead and Brownell (1993: 355; 2005: 731).

Pseudorca crassidens (Owen, 1846)

False Killer Whale

Phocaena crassidens Owen, 1846: 516; Fig. 213.

TYPE LOCALITY: Lincolnshire fens, near Stanford, England. (subfossil).

COMMENTS: Transferred to *Pseudorca* by Reinhardt (1862: 151), and followed by J. Gray (1866b: 290), Hershkovitz (1966: 79) and most subsequent authors.

Orca meridionalis Flower, 1865b: 420; Figs. 1–2.

TYPE LOCALITY: South Pacific, Tasmania, Australia.

COMMENTS: Recognised within *Pseudorca* by J. Gray (1866b: 291), but synonymised within *crassidens* by Iredale and Troughton (1934: 63), Hershkovitz (1966: 80), Bannister (1988c: 207), Mead and Brownell (1993: 355; 2005: 732) and Stacey *et al.* (1994: 1).

Orca destructor Cope, 1866: 293.

TYPE LOCALITY: South Pacific, Piura, off Paita, Peru.

COMMENTS: Synonymised within *crassidens* by Hershkovitz (1966: 80), Mead and Brownell (1993: 355; 2005: 732) and Stacey *et al.* (1994: 1).

Globicephalus Grayi Burmeister, 1869: 367.

TYPE LOCALITY: South Atlantic, Golfo de Somborombón, Buenos Aires, Argentina.

COMMENTS: Not considered by Mead and Brownell (1993: 355; 2005: 732). Synonymised within *crassidens* by Hershkovitz (1966: 80) and Stacey *et al.* (1994: 1).

Pseudorca? mediterranea Giglioli, 1882: 289.

TYPE LOCALITY: North Atlantic, Mediterranean Sea.

COMMENTS: Not considered by Mead and Brownell (1993: 355; 2005: 732). Synonymised within *crassidens* by Hershkovitz (1966: 80) and Stacey *et al.* (1994: 1).

Sousa J. Gray, 1866

Sousa J. Gray, 1866h: 213.

TYPE SPECIES: *Delphinus (Steno?) lentiginosus* Owen, 1866b: 20 (as *Steno lentiginosus*) [=*Sousa chinensis* (Osbeck, 1765: 337)] by subsequent designation. See Iredale and Troughton (1934: 67).

COMMENTS: Described as a subgenus of *Steno* J. Gray, 1846c. Genus synonymised within *Sotalia* J. Gray, 1866b: 401 by Hershkovitz (1966: 18) but recognised by various authors including Bannister (1988c: 207) and Bannister *et al.* (1996: 58) who suggested the genus is in need of review, and that there may be several species in addition to *Sousa chinensis* (Osbeck, 1765: 337). Rice (1998: 102–103) recognised three species in the genus including *Sousa chinensis* (Osbeck, 1765: 337), *Sousa plumbea* (G. Cuvier, 1829a: 288) and *Sousa teuszii* (Kükenthal, 1892: 442). Jefferson and Van Waerebeek (2004: 3) undertook morphological studies and found that *chinensis* and *teuszii* were clearly distinct and that *plumbea* may also be valid. Mead and Brownell (2005: 732), however, took a more conservative approach and only recognised *chinensis* and *teuszii* as distinct species. More recently there has been increasing clarity of the different taxa that should be recognised with support for the recognition of three species discussed above and an undescribed species from Australian waters. Studies in support of this arrangement include those of Frère *et al.* (2008: 259, 263), who used mitochondrial DNA, and Mendez *et al.* (2013: 5936) using both genetic sequence and morphologic data. Subsequently Jefferson and Rosenbaum (2014: 1494) formally revised the genus and recognised *chinensis*, *plumbea* and *teuszii* as distinct species, and named the species that occurs from northern Australia to southern New Guinea as *Sousa sahulensis*. With respect to the genus *Sousa* we are reluctant to recognise it because it is nested within the *Delphis/Tursiops/Stenella* radiation according to molecular data (Cunha *et al.*, 2011: 3; McGowen, 2011: 349; Hassanin *et al.*, 2012: 37), and we do so only because (as indicated above, under Delphinidae) the entire small-dolphin complex is in urgent need of revision.

Sousa sahulensis Jefferson & Rosenbaum, 2014

Australian Hump-backed Dolphin

Sousa sahulensis Jefferson & Rosenbaum, 2014: 1501, 1526.

TYPE LOCALITY: Saunders Beach, near Townsville, North Queensland, Australia. (19°09′20″S, 146°36′26″E)

COMMENTS: The Australian species has historically been recognised as *Sousa chinensis* (Osbeck, 1765: 337) but see the discussion above under *Sousa*.

Sousa queenslandensis Gaskin, 1972: 124.

TYPE LOCALITY: Not provided.

COMMENTS: Taxon named by Gaskin (1972) who suggested the animals found in Queensland may be a species or subspecies new to science, for which the name *Sousa queenslandensis* might be considered. However, as no biological description of the animal was provided, the name is a *nomen nudum* (Jefferson & Rosenbaum, 2014: 1498, 1526).

Stenella J. Gray, 1866

Stenella J. Gray, 1866h: 213.

TYPE SPECIES: *Steno attenuatus* J. Gray, 1846c [=*Stenella attenuatus* (J. Gray, 1846c)] by monotypy.

COMMENTS: Described as a subgenus of *Steno* J. Gray, 1846c. Reviewed by Perrin (1975: 1) and Perrin *et al.* (1981: 583; 1987: 99). Opinion 1660 of the ICZN (1991: 277) conserved *Stenella* J. Gray, 1866h. As all recent studies indicate that *Stenella* is polyphyletic (see LeDuc *et al.*, 1999:

619; Charlton *et al.*, 2011: 11, 15; McGowan, 2011: 353; Hassanin *et al.* 2012: 37), we are very reluctant to recognise the genus here, but do so, as explained earlier, pending an urgently needed total revision of the small-dolphin complex.

FUTURE TAXONOMIC RESEARCH: Further research required to resolve the interrelationships between the species formerly ascribed to *Stenella*.

Clymene J. Gray, 1864c: 237.

TYPE SPECIES: *Delphinus euphrosyne* J. Gray, 1846c [=*Stenella coeruleoalba* (Meyen, 1833)] by monotypy.

COMMENTS: Described as a subgenus of *Delphinus* Linnaeus, 1758. Raised to generic rank by J. Gray (1866h: 214). Synonymised within *Stenella* by Iredale and Troughton (1934: 65), Hershkovitz (1966: 25), McKenna and Bell (1997: 385), and Mead and Brownell (2005: 733).

HOMONYMS:

Clymene Oken, 1807: 1168, worms of the Class Polychaeta. *Nomen dubium* (arbitrary confused usages). See Read (2012).

Clymene Oken, 1815: 378, worms of the Class Polychaeta. *Nomen dubium* (arbitrary confused usages). See Read (2012).

Clymene Savigny, in Lamarck, 1818: 339, bamboo worms of the Class Polychaeta (Subclass Scolecida, Family Maldanidae). *Nomen dubium*. See Read and Fauchald (2012).

Clymene Savigny, 1822: 63, worms of the Phylum Annelida (Class Polychaeta, Family Maldanidae).

Clymene Chambers, 1873: 114, caddis flies of the Class Insecta (Order Trichoptera, Family Hydroptilidae). Genus is a synonym of *Orthotrichia* Eaton, 1873: 141.

Euphrosyne J. Gray, 1866h: 214.

TYPE SPECIES: *Delphinus euphrosyne* J. Gray, 1846c [=*Stenella coeruleoalba* (Meyen, 1833)] by tautonomy.

COMMENTS: Described as a subgenus of *Clymene* J. Gray, 1864c. Synonymised within *Stenella* by Iredale and Troughton (1934: 65), Hershkovitz (1966: 25), and Mead and Brownell (2005: 733).

HOMONYMS:

Euphrosyne Meigen, 1800: 16, flies of the Class Insecta (Order Diptera, Family Mycetophilidae). *Macrocera* Meigen, 1803: 261 was a change of name. See Coquillett (1910: 542, 564).

Euphrosyne Buchecker, 1876: Plate 19, moths of the Class Insecta (Order Lepidoptera, Family Castniidae). Name is a junior synonym of *Xanthocastnia* Houlbert, 1918: 63, 71, 257, 262.

Euphrosyne Savigny, 1822: 63, worms of the Phylum Annelida (Class Polychaeta, Order Aciculata, Family Euphrosinidae).

Clymenia J. Gray, 1868c: 6.

TYPE SPECIES: Emendation of *Clymene* J. Gray, 1864c.

COMMENTS: Synonymised within *Clymene* by Palmer (1904: 191) and within *Stenella* by Iredale and Troughton (1934: 66) and Hershkovitz (1966: 25).

HOMONYMS:

Clymenia Savi, 1817, a genus of 'vermes' that was an obsolete taxon used by Linnaeus.

Clymenia Münster, 1832: 489, ammonites of the Phylum Mollusca (Class Cephalopoda, Order Clymeniida, Family Clymeniidae). Incorrect subsequent spelling of *Clymenea* Münster, 1830: 78.

† *Clymenia* Bronn, 1835: 334, of the Class Cephalopoda. *Incertae sedis*.

Clymenia Örsted, 1844: 79, worms of the Class Polychaeta (Order Sabellida, Family Oweniidae). See Fauchald (2012).

Micropia J. Gray, 1868c: 6.

TYPE SPECIES: *Delphinus stenorhynchus* J. Gray, 1866b (as *Clymenia stenorhyncha*) [=*Stenella longirostris* (J. Gray, 1828)] by virtual tautonomy and monotypy.

COMMENTS: Described as a subgenus of *Clymenia* (J. Gray, 1868c). Synonymised within *Stenella* by Iredale and Troughton (1934: 66), Hershkovitz (1966: 25), and Mead and Brownell (2005: 733).

Fretidelphis Iredale & Troughton, 1934: ix, 65.

TYPE SPECIES: *Delphinus roseiventris* Wagner, 1946 [=*Stenella longirostris roseiventris* (Wagner, 1846)] by original designation.

COMMENTS: Proposed as subgenus of *Steno* J. Gray, 1846c. Synonymised within *Stenella* by Hershkovitz (1966: 26), and Mead and Brownell (2005: 733).

Prodelphinus Gervais, 1880: 604; Plate 38.

TYPE SPECIES: *Nomen novum* for *Clymenia* J. Gray, 1868c.

COMMENTS: Synonymised within *Stenella* by Hershkovitz (1966: 26), and Mead and Brownell (2005: 733).

Stenella attenuata (J. Gray, 1846)

Pantropical Spotted Dolphin

Stenella attenuata attenuata (J. Gray, 1846)

Offshore Pantropical Spotted Dolphin

Steno attenuatus J. Gray, 1846c: 44.

TYPE LOCALITY: Not given, unknown (possibly India, see J. Gray, 1843a: 105).

COMMENTS: Placed in the genus *Stenella* by Iredale and Troughton (1934: 66). Synonymised within *Delphinus dubia* G. Cuvier, 1812: 14, by Hershkovitz (1966: 32). Reviewed by Ross (1984: 352) and Perrin *et al.* (1987: 149). Opinion 1660 of the ICZN (1991: 277) conserved *attenuata* J. Gray, 1846c and suppressed *velox* G. Cuvier, 1829a, *pseudodelphis* Schlegel, 1841 and *brevimanus* Wagner, 1846. Specific name *attenuata* added to the Official List of Specific Names (J. Smith, 2001: 8). Recognised as the subspecies that occurs within Australian waters by Van Dyck and Strahan (2008: 841).

Delphinus attenuatus J. Gray, 1843: 105.

TYPE LOCALITY: ? India.

COMMENTS: *Nomen nudum*. Synonymised within *dubia* by Hershkovitz (1966: 32).

D.[elphinus] dubius G. Cuvier, 1812: 14.

TYPE LOCALITY: There is no holotype specimen.

COMMENTS: Recognised at the species rank by Hershkovitz (1966: 31). The name *Delphinus dubius* was considered a *nomen nudum* by Perrin *et al.* (1987: 111).

delphinus malayanus Lesson, 1827c: 184; Plate 9, Fig. 5.

TYPE LOCALITY: Karaimata Strait, between Java and Borneo. There is no holotype specimen.

COMMENTS: Species name recognised within *Steno* by J. Gray (1866b: 232) and within *Prodelphis* by True (1889: 67). Subsequently synonymised within *Stenella dubia* (G. Cuvier, 1812: 14) by Hershkovitz (1966: 32), reduced to a *nomen nudum* by Perrin *et al.* (1987: 111) and it synonymised within *Stenella attenuata* by Mead and Brownell (2005: 733).

D.[elphinus] velox G. Cuvier, 1829a: 288.

TYPE LOCALITY: Ceylon [=Sri Lanka].

COMMENTS: Suppressed by Opinion 1660 of the ICZN (1991: 277). Synonymised *dubia* by Hershkovitz (1966: 32) and *attenuata* by Perrin *et al.* (1987: 149), Perrin (2001: 1), and Mead and Brownell (2005: 733).

Delphinus capensis Rapp, 1837: 31.

TYPE LOCALITY: South Atlantic, Cape of Good Hope, South Africa.

COMMENTS: Synonymised within *malayanus* by J. Gray (1866b: 232). Synonymised within *Delphinus dubius* G. Cuvier, 1812 [=*Stenella attenuatus* (J. Gray, 1846c)] by Hershkovitz (1966: 34) and within *attenuata* by Perrin *et al.* (1987: 149), and Mead and Brownell (2005: 733).

HOMONYMS:

Delphinus capensis J. Gray, 1828, the Long-beaked Common Dolphin of the Class Mammalia (Order Artiodactyla, Family Delphinidae). Recognised as described. See Mead and Brownell (2005: 727). See individual entry.

Delphinus capensis G. Cuvier, 1829: 289, Heaviside's Dolphin of the Class Mammalia (Order Artiodactyla, Family Delphinidae). Name appears to be a synonym of *Cephalorhynchus heavisidii* (J. Gray, 1828: 2). See Hershkovitz (1966: 76).

Delphinus pseudodelphis Wiegmann in Wagner, 1846: Plate 358.

TYPE LOCALITY: Not given.

COMMENTS: The year of publication of the volume in which this name exists has been unstable with authors including Sherborn (1892: 592; 1927: 3674, see '*loriger*' entry), Heyning and Perrin (1994: 22 – see '*loriger*' entry), and Mead and Brownell (2005: 733) giving the publication date as 1846. In contrast Hershkovitz (1966: 32) and Perrin *et al.* (1987: 112) gave the year of publication as 1840. Also note that when Schlegel (1841: 22) referred to this taxon he gave the author as 'Wiegmann im Wagner' and referred to Plate 358 so the name must have already been published. Taxon synonymised within *dubia* by Hershkovitz (1966: 32) and within *attenuatus* by True (1889: 67), Perrin *et al.* (1987: 113, 149), Perrin (2001: 1), and Mead and Brownell (2005: 733).

Delphinus pseudodelphis Schlegel, 1841: 22.

TYPE LOCALITY: Not given.

COMMENTS: Suppressed by Opinion 1660 of the ICZN (1991: 277) and discussed by Perrin *et al.* (1987: 112–113).

Delphinus Rappii L. Reichenbach, 1845: 12; Plate 18, Fig. 57.

TYPE LOCALITY: South Atlantic, Cape of Good Hope.

COMMENTS: It was described earlier by Rapp (1837: 31) under '*Delphinus capensis* J. Gray [1828]' but it is not the same species as the holotype of *D. capensis* J. Gray, 1828, which was a specimen of *D. delphis* (Hershkovitz, 1966: 34). Synonymised within *malayanus* by J. Gray (1866b: 232). Perrin *et al.* (1987: 113) believed that Hershkovitz (1966: 34) was not justified in including the name in the synonymy of *Stenella dubia* and suggested that until the holotype skull can be critically examined the species should remain *incertae sedis*. Perrin *et al.* (1987: 113) also suggested that as the name has not been used since 1846 it should be considered a *nomen oblitum*, although the Fourth Edition of the Code recalls that such a name is still potentially available.

D. [elphinus] brevimanus Wagner, 1846: 427; Plate 361, Fig. 2.

TYPE LOCALITY: Indian Ocean, near Singapore.

COMMENTS: Recognised within *Steno* by J. Gray (1866b: 236). Considered a *nomen oblitum* by Perrin *et al.* (1987: 114). Suppressed by Opinion 1660 of the ICZN (1991: 277). Synonymised within *malayanus* by True (1889: 67), *dubia* by Hershkovitz (1966: 33) and *attenuatus* by Perrin *et al.* (1987: 149), Perrin (2001: 1), and Mead and Brownell (2005: 733).

Delphinus albirostratus Peale, 1848: 34.

TYPE LOCALITY: Pacific Ocean, near Phoenix Islands (2°47′5″S, 174°13′W)

COMMENTS: Synonymised within *Stenella coeruleoalba* by Hershkovitz (1966: 29) and within *attenuatus* by Perrin *et al*. (1987: 150), Perrin (2001: 1), and Mead and Brownell (2005: 733).

Delphinus? microbrachium J. Gray, 1850: viii, 119.

TYPE LOCALITY: South Sea?

COMMENTS: Synonymised within *Delphinus dubius* G. Cuvier, 1812 [=*Stenella attenuatus* (J. Gray, 1846c)] by Hershkovitz (1966: 34). Considered a *nomen oblitum* by Perrin *et al*. (1987: 114). Synonymised within *attenuatus* by Perrin *et al*. (1987: 149) and Perrin (2001: 1).

Steno capensis J. Gray, 1865d: 522.

TYPE LOCALITY: South Pacific, Cape of Good Hope, South Africa.

COMMENTS: Synonymised within *Delphinus dubius* G. Cuvier, 1812 [=*Stenella attenuatus* (J. Gray, 1846c)] by Hershkovitz (1966: 34) and *attenuatus* by True (1889: 67), Perrin *et al*. (1987: 150) and Perrin (2001: 1).

Clymene punctata J. Gray, 1866i: 738.

TYPE LOCALITY: Near Cape Verde, West Africa. (16°40′N., 21°W)

COMMENTS: Synonymised within *Delphinus dubius* G. Cuvier, 1812 [=*Stenella attenuatus* (J. Gray, 1846c)] by Hershkovitz (1966: 34), *attenuatus* by True (1889: 67), Perrin *et al*. (1987: 150), Perrin (2001: 1), and Mead and Brownell (2005: 733).

Steno consimilis Malm, 1871: 104; Plate 6, Fig. 53.

TYPE LOCALITY: Indian Ocean, Madagascar.

COMMENTS: Synonymised within *longirostris* by Hershkovitz (1966: 40) and *attenuatus* by Perrin *et al*. (1987: 150), Perrin (2001: 1), and Mead and Brownell (2005: 733).

Φ ***Stenella attenuata graffmani*** (Lönnberg, 1934)

Coastal Pantropical Spotted Dolphin

Φ *Prodelphinus graffmani* Lönnberg, 1934: 1; Plate 1.

TYPE LOCALITY: 20 miles north of the port of Acapulco, Mexico.

COMMENTS: Recognised at the species rank within *Stenella* by Hershkovitz (1966: 37) but placed within *attenuatus* by Perrin *et al*. (1987: 150). Recognised as a subspecies of *attenuatus* by Rice (1998: 108), Perrin (2001: 1), Mead and Brownell (2005: 733), and Perrin *et al*. (2009: 12). Further research is still needed to determine the status of this taxon.

Stenella coeruleoalba (Meyen, 1833)

Striped Dolphin

Delphinus coeruleo-albus Meyen, 1833: 609; Plate 43.

TYPE LOCALITY: Near Río de la Plata, east coast of South America.

COMMENTS: Included within *Lagenorhynchus* by J. Gray (1866b: 268) and within *Stenella* by Hershkovitz (1966: 27) and many recent authors. Reviewed by Hershkovitz (1966: 27), Perrin *et al*. (1981: 584, 592), Ross (1984: 337), and F. Archer and Perrin (1999: 1). No subspecies formally recognised (Bannister *et al*., 1996: 84).

Delphinus Styx J. Gray, 1846c: 39; Plate 21.

TYPE LOCALITY: South Atlantic, South Africa.

COMMENTS: Recognised within *Delphinus* by J. Gray (1866b: 250). Synonymised within *coeruleoalba* by Hershkovitz (1966: 28), Mead and Brownell (1993: 356; 2005: 733), and F. Archer and Perrin (1999: 1).

Delphinus Euphrosyne J. Gray, 1846c: 40; Plate 22.

TYPE LOCALITY: Unspecified. Hershkovitz (1966: 28) give locations as North Atlantic, England and South Atlantic, Río de La Plata, mouth Buenos Aires, Argentina.

COMMENTS: Recognised within *Delphinus* by J. Gray (1866b: 251). Synonymised within *coeruleoalba* by Hershkovitz (1966: 28), Mead and Brownell (1993: 356; 2005: 733), and F. Archer and Perrin (1999: 1).

Del. [phinus] Holböllii Nilsson, 1847: 595.

TYPE LOCALITY: Greenland.

COMMENTS: Synonymised within *coeruleoalba* by van Bree (1973: 129), and F. Archer and Perrin (1999: 1).

Delphinus lateralis Peale, 1848: 35; Plate 8, Fig. 1.

TYPE LOCALITY: North Pacific. (13°58′N, 161°22′W)

COMMENTS: Recognised within *Delphinus* by J. Gray (1866b: 254). Synonymised within *coeruleoalba* by Hershkovitz (1966: 29), and F. Archer and Perrin (1999: 1).

Delphinus Tethyos Gervais, 1853: 150; Plate 1, Fig. 1.

TYPE LOCALITY: 'Mediterranee'. Location given as Valreas, mouth of the Orb River, Herault, North Atlantic, France by Hershkovitz (1966: 30).

COMMENTS: Recognised within *Delphinus* by J. Gray (1866b: 251). Synonymised within *coeruleoalba* by Hershkovitz (1966: 30), Mead and Brownell (1993: 356; 2005: 733), and F. Archer and Perrin (1999: 1).

Delphinus marginatus Pucheran, 1856: 545; Plate 25.

TYPE LOCALITY: Atlantic Ocean, 'Dieppe', France.

COMMENTS: Recognised within *Delphinus* by J. Gray (1866b: 245). Synonymised within *coeruleoalba* by Hershkovitz (1966: 30), and F. Archer and Perrin (1999: 1).

Delphinus mediterraneus Loche, 1860: 475; Plate 22, Fig. 2.

TYPE LOCALITY: Algeria.

COMMENTS: Placed in synonymy with *Delphinus frontalis* (G. Cuvier, 1829: 288) by Hershkovitz (1966: 36) and by E. Hall (1981: 880), although Perrin *et al.* (1987: 114) considered that Loche's description of the colour pattern identifies the type specimen as a striped dolphin, *Delphinus coeruleoalba* Meyen, 1833. Synonymised within *coeruleoalba* by F. Archer and Perrin (1999: 1).

[Delphinus] asthenops Cope, 1865b: 200.

TYPE LOCALITY: Unknown.
COMMENTS: Synonymised within *coeruleoalba* by Mead and Brownell (1993: 356; 2005: 733), and F. Archer and Perrin (1999: 1).

Delphinus crotaphiscus Cope, 1865b: 200.

TYPE LOCALITY: Unknown.
COMMENTS: Recognised as a distinct species within *Stenella* by Hershkovitz (1966: 31). Synonymised within *coeruleoalba* by Mead and Brownell (1993: 356; 2005: 733), and F. Archer and Perrin (1999: 1).

Tursio Dorcides J. Gray, 1866b: 400.

TYPE LOCALITY: Unknown.
COMMENTS: Synonymised within *coeruleoalba* by Hershkovitz (1966: 30), and F. Archer and Perrin (1999: 1).

Clymene dorides J. Gray, 1866h: 214.

TYPE LOCALITY: Renaming of *Tursio dorcides* J. Gray, 1866b.
COMMENTS: Synonymised within *coeruleoalba* by Hershkovitz (1966: 30), and F. Archer and Perrin (1999: 1).

Clymenia euphrosynoides J. Gray, 1868c: 6; Plate 31.

TYPE LOCALITY: Renaming of *Delphinus euphrosyne* J. Gray, 1846c.
COMMENTS: Synonymised within *coeruleoalba* by Hershkovitz (1966: 30), and F. Archer and Perrin (1999: 1).

Clymenia Burmeisteri Malm, 1871: 63; Plate 6, Fig. 54.

TYPE LOCALITY: South Atlantic, Brazil.
COMMENTS: Synonymised within *coeruleoalba* by Hershkovitz (1966: 30), and F. Archer and Perrin (1999: 1).

Clymenia Novae Zealandiae Hector, 1873a: 159; Plate 2.

TYPE LOCALITY: Waikanae, New Zealand.
COMMENTS: Synonymised within *coeruleoalba* by F. Archer and Perrin (1999: 1).

Prodelphis Petersii Lütken, 1889: 40, 43; Figure.

TYPE LOCALITY: Indian Ocean.
COMMENTS: Synonymised within *coeruleoalba* by Hershkovitz (1966: 30), and F. Archer and Perrin (1999: 1).

Delphinus amphitriteus Philippi, 1893: 7; Plate 1, Fig. 3.

TYPE LOCALITY: Atlantic Ocean. (29°15′S)
COMMENTS: Synonymised within *coeruleoalba* by Hershkovitz (1966: 31), and F. Archer and Perrin (1999: 1).

Stenella longirostris (J. Gray, 1828)

Spinner Dolphin

Stenella longirostris longirostris (J. Gray, 1828)

Gray's Spinner Dolphin

Delphinus longirostris J. Gray, 1828: 1.

TYPE LOCALITY: Unknown.
COMMENTS: When originally proposed, this rank included *Fretidelphis roseiventris* (Wagner, 1846). Recognised within *Delphinus* by J. Gray (1866b: 241) and transferred to *Clymenia* by Flower (1884: 499) and *Prodelphinus* by Flower (1885: vi, 31). Species placed within *Stenella* by Iredale and Troughton (1934: 66), Hershkovitz (1966: 37) and Mead and Brownell (1993: 356; 2005: 734). Reviewed by Ross (1984: 364), Perrin (1990: 453; 1998: 1) and Perrin *et al.* (1999: 1029) who recognised a dwarf form from the Gulf of Thailand. The supposed subspecies of this species are badly in need of revision.
HOMONYMS:
Delphinus longirostris G. Cuvier, 1829a, the Indo-Pacific Common Dolphin of the Class Mammalia (Order Artiodactyla, Family Delphinidae). Name is a junior synonym of *Delphinus capensis tropicalis* van Bree, 1971. See individual entry.

Delphinus Alope J. Gray, 1846c: Plate 32.

TYPE LOCALITY: Unknown.
COMMENTS: Recognised within *Delphinus* by J. Gray (1850: 118; 1866b: 252). Synonymised within *longirostris* by Hershkovitz (1966: 38), Mead and Brownell (1993: 356; 2005: 734), and Perrin (1998: 1).

Delphinus microps J. Gray, 1846c: 42.

TYPE LOCALITY: Unknown.
COMMENTS: Recognised by J. Gray (1866b: 240). Synonymised within *longirostris* by Hershkovitz (1966: 39), Mead and Brownell (1993: 356; 2005: 734), and Perrin (1998: 1).
HOMONYMS:
Delphinus microps Burmeister, 1866b, Long-beaked Common of the Class Mammalia (Order Artiodactyla, Family Delphinidae). Name is a synonym of *Delphinus capensis* Gray, 1828. See individual entry.

Delphinus stenorhynchus J. Gray, 1866b: 396.

TYPE LOCALITY: Unknown.
COMMENTS: Synonymised within *longirostris* by Hershkovitz (1966: 40), Mead and Brownell (1993: 356) and Perrin (1998: 1). Not discussed by Mead and Brownell (1993; 2005).

Prodelphis longirostris kunitomoi Ogawa, 1950: 73, 85.

TYPE LOCALITY: Goto Islands, southwest Kyushu, Japan.

COMMENTS: Recognised as a *nomen nudum* by Nishiwaki (1957: 151). Synonymised within *longirostris* by Hershkovitz (1966: 40) and Perrin (1998: 1). Not discussed by Mead and Brownell (1993; 2005).

Φ *Stenella longirostris centroamericana* Perrin, 1990

Central American Spinner Dolphin

Φ *Stenella longirostris centroamericana* Perrin, 1990: 461.

TYPE LOCALITY: Gulf of Nicoya, Pacific coast of Costa Rica. (9°47′N, 85°42′W)

COMMENTS: Synonymised within *longirostris* by Mead and Brownell (1993: 356). Recognised as a subspecies by Perrin (1998: 1), Mead and Brownell (2005: 734), and Perrin *et al.* (2009: 13).

Φ *Stenella longirostris orientalis* Perrin, 1990

Eastern Spinner Dolphin

Φ *Stenella longirostris orientalis* Perrin, 1990: 457.

TYPE LOCALITY: Approx. 120 km off the coast of Guatemala in the eastern tropical Pacific.

COMMENTS: Synonymised within *longirostris* by Mead and Brownell (1993: 356). Recognised as a subspecies by Perrin, 1998: 1), Mead and Brownell (2005: 734) and Perrin *et al.* (2009a: 12). Not recognised by Van Dyck and Strahan (2008: 844).

Stenella longirostris roseiventris (Wagner, 1846)

Dwarf Spinner Dolphin

Delphinus roseiventris Wagner, 1846: Plate 360.

TYPE LOCALITY: Torres Strait.

COMMENTS: Recognised within *Steno* by J. Gray (1866b: 233). Synonymised within *longirostris* by Hershkovitz (1966: 39), Mead and Brownell (1993: 356; 2005: 734), and Perrin (1998: 1). Recognised as a subspecies of *longirostris* by Perrin (1999: 1029, 1048), who redescribed this taxon, and by Van Dyck and Strahan (2008: 844), Perrin *et al.* (2009: 13), and the Society for Marine Mammalogy (2011).

Steno J. Gray, 1846

Steno J. Gray, 1846c: 30, 43.

TYPE SPECIES: *Delphinus rostratus* G. Cuvier, 1812: 9 (as *Steno rostratus*) [=*Steno bredanensis* (G. Cuvier, 1828)]. See Hershkovitz (1966: 15).

COMMENTS: Taxon reviewed by Hershkovitz (1966: 15).

Glyphidelphis Gervais, 1859: 301.

TYPE SPECIES: *Delphinus rostratus* G. Cuvier, 1823a: 278 [=*Steno bredanensis* (G. Cuvier, 1828)] by monotypy.

COMMENTS: Synonymised within *Steno* by Hershkovitz (1966: 15) and McKenna and Bell (1997: 385).

Stenopontistes de Miranda-Ribeiro, 1936: 3, 19, 42.

TYPE SPECIES: *Stenopontistes zambezicus* de Miranda-Ribeiro, 1936 [=*Steno bredanensis* (G. Cuvier, 1828)] by original designation.

COMMENTS: Synonymised within *Steno* by McKenna and Bell (1997: 385) and within *Sousa* by Mead and Brownell (2005: 732).

Steno bredanensis (G. Cuvier, 1828)

Rough-toothed Dolphin

Delphinus bredanensis G. Cuvier in Lesson, 1828d: 206.

TYPE LOCALITY: Coast of France.

COMMENTS: *Nomen nova pro Delphinus rostratus* G. Cuvier, 1812. See Schevill (1987b: 78) for further taxonomic notes. No subspecies recognised by Bannister *et al.* (1996: 55) or Mead and Brownell (2005: 734).

[Delphinus] rostrata G. Cuvier, 1812: 9.

TYPE LOCALITY: Coast of France.

COMMENTS: Name recognised by Desmarest, 1817f: 160, who recognised the author as G. Cuvier and give the type locality as Paimpol, Brittany, France. Name, with Desmarest (1817f: 160) as the author, synonymised within *bredanensis* by Hershkovitz (1966: 15), and Mead and Brownell (1993: 357; 2005: 734).

HOMONYMS:

Delphinus Rostratus Shaw, 1801: 514, the Ganges River Dolphin of the Class Mammalia (Order Artiodactyla, Family Platanistidae). Currently recognised as *Platanista gangetica* (Roxburgh, 1803: 171). See Hershkovitz (1966: 14), and Mead and Brownell (2005: 738).

[Delphinus] frontatus G. Cuvier, 1823a: 278.

TYPE LOCALITY: Unknown.

COMMENTS: Recognised within *Steno* by J. Gray (1866b: 233). Synonymised within *bredanensis* by Hershkovitz (1966: 17), and Mead and Brownell (1993: 357; 2005: 734).

Delphinus compressus J. Gray, 1843a: 105.

TYPE LOCALITY: Unknown.

COMMENTS: *nomen nudum*. Recognised within *Steno* by J. Gray (1866b: 234). Synonymised within *bredanensis* by Hershkovitz (1966: 17), and Mead and Brownell (1993: 357; 2005: 734).

Steno compressus J. Gray, 1846: 43; Plate 27.

TYPE LOCALITY: Unknown.

COMMENTS: Synonymised within *bredanensis* by Hershkovitz (1966: 17).

Delphinus (Steno) perspicillatus Peters, 1877a: 360; Plates 2–3.

TYPE LOCALITY: South Atlantic. (32°29′S, 2°1′W)

COMMENTS: Synonymised within bredanensis by Hershkovitz (1966: 17), and Mead and Brownell (1993: 357; 2005: 734).

Tursiops Gervais, 1855

Tursiops Gervais, 1855a: 323.

TYPE SPECIES: *Nomen novum* for *Tursio* J. Gray, 1843a.

COMMENTS: Taxonomic history of Australian species assigned to *Tursiops* was discussed by Bannister (1988c: 209) and Ross and Cockcroft (1990: 101), who concluded that only one species, *T. truncatus*, occurs in Australian waters. Mead and Brownell (1993: 357) also proposed that only one species be recognised. More recent morphological studies by Hale *et al.* (2000: 101) concluded that two distinct forms were present in Australian waters, a large unspotted dolphin (*T. truncatus*) living in waters deeper than 30m and a small, spotted one (*T.* cf. *aduncus*) living in waters shallower than 30m and this was subsequent confirmed by genetic tests by Möller and Bererearay (2001: 249) and further osteological examination by Kemper (2004: 29), though Hale in personal communications to Kemper (2004: 30) referred them to cf. *aduncus* and cf. *truncatus*, pending further research. Two species were provisionally recognised by Mead and Brownell (2005: 734). It seems evident that there are undescribed species (Natoli *et al.*, 2004: 363; Charleton *et al.*, 2006: 173, 177; Möller *et al.*, 2008: 678). As all recent studies indicate that *Tursiops* is polyphyletic (see LeDuc *et al.*, 1999: 619; Charlton *et al.*, 2011: 11, 15; Hassanin *et al.* 2012: 37), we recognise the genus here with considerable reluctance, as explained earlier under *Delphinus*.

Tursio J. Gray, 1843a: xxiii, 105.

TYPE LOCALITY: *Delphinus truncatus* Montagu, 1821 [=*Tursiops truncatus* (Montagu, 1821)] by monotypy.

COMMENTS: Synonymised within *Tursiops* by Iredale and Troughton (1934: 68), Hershkovitz (1966: 47), and Mead and Brownell (2005: 734).

HOMONYMS:

Tursio Fleming, 1822b, the Sperm Whale of the Class Mammalia (Order Artiodactyla, Delphinidae). Genus is a synonym of *Physeter* Linnaeus, 1758: 76. See individual entry.

Tursio Wagler, 1830: 34, the Southern Rightwhale Dolphin of the Class Mammalia (Order Artiodactyla, Family Delphinidae). Genus is a synonym of *Lissodelphis* Gloger, 1841. See individual entry.

Gudamu J. Gray, 1868c: 6.

TYPE SPECIES: *Delphinus gadamu* Owen, 1866b: 17 (as *Clymenia gudamu* [sic]) [=*Tursiops aduncus* (Ehrenberg, 1833)] by monotypy.

COMMENTS: Taxon described as a subgenus of *Clymenia* by J. Gray (1868c: 6). Synonymised within *Tursiops* by Hershkovitz (1966: 48), and Mead and Brownell (2005: 734).

Tursiops aduncus (Ehrenberg, 1833)

Indo-Pacific Bottle-nosed Dolphin

Delphinus aduncus Ehrenberg, 1833: 6, footnote 1 of *Herpestes leucurus* entry.

TYPE LOCALITY: Belhosse Island, Dahlak Archipelago, Eritrea.

COMMENTS: Synonymised within *truncatus* by many authors including Ross and Cockcroft (1990: 124), and Mead and Brownell (1993: 357). Hershkovitz (1966: 48) and Bannister *et al.* (1996: 71) recognised this as a subspecies of *truncatus*. Species recognised by LeDuc and Curry (1997: 393), Rice (1998: 106), Mead and Brownell (2005: 734) and most subsequent authors. Also supported by morphological studies of Wang *et al.* (1999: 1603; 2000a: 1157; 2000b: 147) who clearly distinguished this species from *Tursiops truncatus*. These observations were also supported by Hale *et al.* (2000: 101). In contrast to these conclusions Natoli *et al.* (2004: 363) suggested *aduncus* was polytypic, which has not been widely accepted, and LeDuc *et al.* (1999: 619) who suggested that *aduncus* is most closely related to *Stenella frontalis* (G. Cuvier, 1829a: 288), which is inconsistent with the osteological characters. The presence of *aduncus* off eastern Australia was confirmed by Möller and Beheregaray (2001: 249).

Tursiops nuuanu R. Andrews, 1911a: 233; Plate 10.

TYPE LOCALITY: North Pacific, Santa Catalina Island, Gulf of California, United States of America. (12°N, 120°W)

COMMENTS: Synonymised within *aduncus* by Hershkovitz (1966: 50), and Mead and Brownell (1993: 357).

Tursiops tursio Ihering, 1893: 104.

TYPE LOCALITY: Brazil, Canal do Norte, near the Rio Grande; Rio Grande do Sul.

COMMENTS: Synonymised within *aduncus* by Hershkovitz (1966: 51), but not considered by Mead and Brownell (1993: 734).

HOMONYMS:

Delphinus tursio Fabricius, 1780: 49, dolphins of the Class Mammalia (Order Artiodactyla, Family Delphinidae?). *Incertae sedis*. See Hershkovitz (1966: 199).

Delphinus tursio Bonnaterre, 1789, the Common Bottle-nosed Dolphin of the Class Mammalia (Order Artiodactyla,

Family Delphinindae). Taxon is a synonym of *Delphinus truncatus* Montagu, 1821. See individual entry.

Tursiops tursio Scott & Lord, 1920a (part), the Burrunan Dolphin of the Class Mammalia (Order Artiodactyla, Family Delphinidae). Taxon is a synonym of *Tursiops australis* Charlton-Robb *et al.*, 2011. See individual entry.

Tursiops tursio Ihering, 1893, the Indo-Pacific Bottle-nosed Dolphin of the Class Mammalia (Order Artiodactyla, Family Delphinidae). Name is a synonym of *Delphinus aduncus* Ehrenberg, 1833. See individual entry.

Tursiops australis Charlton-Robb *et al.*, 2011

Burrunan Bottle-nosed Dolphin

Tursiops australis Charlton-Robb *et al.*, 2011: 1, 13; Figs. 4–5.

TYPE LOCALITY: Hobblers Bridge, North Esk River 5 km upstream from Tamar River, Tasmania, Australia.

COMMENTS: Previously published as the female of the southern form of *Tursiops tursio* by Scott and Lord (1920a) and later ascribed to *Tursiops maugeanus* Iredale and Troughton, 1934: 68. Further support for the recognition of this species was given by Moura *et al.* (2013: 874), who showed that it represented a sister group of all other *Tursiops* lineages. Despite this serious doubt has been cast over the validity of this species by authors including Burbidge *et al.* (2014, 26, 32) and the Society for Marine Mammalogy (Committee on Taxonomy, 2014) who suggested that a rigorous re-evaluation of the relevant data and arguments is needed. As a result of these concerns this taxon is recognised here with caution pending formal publication of its acceptance or rejection.

FUTURE TAXONOMIC RESEARCH: The validity of this taxon needs to be assessed to confirm whether or not it should be recognised.

Tursiops tursio Scott & Lord, 1920a: 96 (part).

TYPE LOCALITY: Hobblers Bridge, North Esk River 5 km upstream from Tamar River, Tasmania, Australia.

COMMENTS: Based on the female of the 'southern form'. Synonymised within *australis* by Charlton-Robb *et al.* (2011: 13).

HOMONYMS:

Delphinus tursio Fabricius, 1780: 49, dolphins of the Class Mammalia (Order Artiodactyla, Family Delphinidae?). *Incertae sedis*. See Hershkovitz (1966: 199).

Delphinus tursio Bonnaterre, 1789, the Common Bottle-nosed Dolphin of the Class Mammalia (Order Artiodactyla, Family Delphinindae). Taxon is a synonym of *Delphinus truncatus* Montagu, 1821. See individual entry.

Tursiops tursio Ihering, 1893, the Indo-Pacific Bottle-nosed Dolphin of the Class Mammalia (Order Artiodactyla, Family Delphinidae). Name is a synonym of *Delphinus aduncus* Ehrenberg, 1833. See individual entry.

Tursiops truncatus (Montagu, 1821)

Common Bottle-nosed Dolphin

Tursiops truncatus truncatus (Montagu, 1821)

Common Bottle-nosed Dolphin

Delphinus Truncatus Montagu, 1821: 75; Plate 3.

TYPE LOCALITY: Duncannon Pool, near Stoke Gabriel, River Dart, Devonshire, England.

COMMENTS: Recognised within *Tursio* by J. Gray (1843a: 105; 1866b: 258) and *Tursiops* by True (1903: 314) and most subsequent authors. Has often been placed with '*Tursiops*' *aduncus*. Synonymised within *nesarnack* by E. Hall (1981: 886). This species name was conserved and *nesarnack* suppressed by Opinion 1413 of the ICZN (1986: 256). Only this species was recognised in the genus by Bannister (1988c: 209) and Ross and Cockcroft (1990: 101). Bannister *et al.* (1996: 71) tentatively recognise two subspecies, *T. t. truncatus* and *T. t. aduncus*, to occur in Australian waters.

D. [elphinus] Tursio Bonnaterre, 1789: 21; Plate 11, Fig. 1.

TYPE LOCALITY: Description of specimen in the Veterinary School of Alford, France.

COMMENTS: Synonymised within *truncatus* by Hershkovitz (1966: 54).

HOMONYMS:

Delphinus tursio Fabricius, 1780: 49, dolphins of the Class Mammalia (Order Artiodactyla, Family Delphinidae?). *Incertae sedis*. See Hershkovitz (1966: 199).

Tursiops tursio Ihering, 1893, the Indo-Pacific Bottle-nosed Dolphin of the Class Mammalia (Order Artiodactyla, Family Delphinidae). Name is a synonym of *Delphinus aduncus* Ehrenberg, 1833. See individual entry.

Tursiops tursio Scott & Lord, 1920a (part), the Burrunan Dolphin of the Class Mammalia (Order Artiodactyla, Family Delphinidae). Taxon is a synonym of *Tursiops australis* Charlton-Robb *et al.*, 2011. See individual entry.

Delphinus nesarnack Lacépède, 1804: xliii, 307; Plate 15, Fig. 2.

TYPE LOCALITY: North Atlantic.

COMMENTS: Recognised at the species rank by E. Hall (1981: 885). Synonymised within *truncatus* by Hershkovitz (1966: 52), and Mead and Brownell (1993: 357; 2005: 734).

Delphinus catalania J. Gray, 1862b: 143, 144.

TYPE LOCALITY: Cape Melville, north east Queensland, Australia, Australia.

COMMENTS: Synonymised within *aduncus* by Hershkovitz (1966: 49) but not considered by Mead and Brownell (1993: 357; 2005: 734).

Tursiops Gillii Dall, 1873: 13.

TYPE LOCALITY: Monterey, California, United States of America.

COMMENTS: Recognised at the species rank by Hershkovitz (1966: 55) and E. Hall (1981: 885–887). Synonymised within *truncatus* by Mead and Brownell (1993: 357). Subspecies rank recognised by Mead and Brownell (2005: 724), but not by Van Dyck and Strahan (2008: 840). Rice (1998: 106) mentioned this taxon in the text but did not recognise it systematically as a subspecies (with a separate paragraph) as he did for subspecies of the other taxa.

Tursiops gephyreus Lahille, 1908: 347; Plates 3–4, Figs. 1–2.

TYPE LOCALITY: South Atlantic, Punta Lara, Río de La Plata; Quilmes, Río de La Plata, Argentina.

COMMENTS: Synonymised within *aduncus* by Hershkovitz (1966: 50) and *truncatus* by Mead and Brownell (1993: 357; 2005: 734).

Tursiops maugeanus Iredale & Troughton, 1934: ix, 68.

TYPE LOCALITY: Tamar River, Tasmania, Australia.

COMMENTS: *Nomen novum* for the unnamed southern form of *Tursiops tursio* described by Scott and Lord (1920a). Synonymised within *aduncus* by Hershkovitz (1966: 50), and within *truncatus* by Van Dyck and Strahan (2008: 840). Morphometric and genetic studies by Charleton-Robb *et al.* (2011: 11) revealed that the syntytpes, a male and female, allocated to this species comprise two different species, of which the female syntype was described as a new species and given the name *Tursiops australis*, while the male syntype was designated the lectotype of *maugeanus*.

FUTURE TAXONOMIC RESEARCH: Although this form is generally regarded as simply the southern form of *T. truncatus*, Charleton-Robb *et al.* (2011: 11) made the case that it is likely to be a valid species as long as it can be confirmed as different from *truncatus*. Therefore further research is needed to confirm if this taxon is distinct from *truncatus*.

Φ *Tursiops truncatus ponticus* (Barabash-Nikiforov, 1940)

Black Sea Bottle-nosed Dolphin

Φ *Delphinus delphis ponticus* Barabash-Nikiforov, 1940: 56; Figs. 14–15.

TYPE LOCALITY: Black Sea at Novorosisk.

COMMENTS: Synonymised within *truncatus* by Hershkovitz (1966: 54), but subsequently recognised as a subspecies of *truncatus* by Rice (1998: 106), Mead and Brownell (2005: 724), and Perrin *et al.* (2009: 11), but not by Van Dyck and Strahan (2008: 840).

Family Phocoenidae J. Gray, 1825

Tribe Phocaenina J. Gray, 1825a: 340.

TYPE GENUS: *Phocoena* G. Cuvier, 1816a.

COMMENTS: When originally proposed, this rank was placed in the Family Delphinidae (J. Gray, 1821) and included the genera *Phocoena* G. Cuvier, 1816a; *Delphinapterus* Lacépède, 1804: xli, 243; *Heterodon* de Blainville, 1817 [=*Hyperoodon* Lacépède, 1804]; and *Monodon* Linnaeus, 1758: 75. Family rank not recognised by Hershkovitz (1966). Formerly considered a subfamily of Delphinidae (see Mead & Brownell, 2005: 735). Family rank recognised by McKenna and Bell (1997: 390), Barnes (1971: 79), and Mead and Brownell (2005: 735). Comprised the subfamilies Phocoenoidinae and Phocoeninae by McKenna and Bell (1997: 390–391).

Tribe Phocaenina J. Gray, 1868c: 8.

TYPE GENUS: *Phocaena* G. Cuvier, 1817.

COMMENTS: When originally proposed, this rank was placed in the Family Delphinidae (J. Gray, 1821) and included the genera *Pseudorca* Reinhardt, 1862; *Phocaena* G. Cuvier, 1817 [=*Phocoena* G. Cuvier, 1816a]; *Acanthodelphis* J. Gray, 1866b [=*Phocoena* G. Cuvier, 1816a]; and *Neomeris* J. Gray, 1846c: 30 [=*Neophocaena* Palmer, 1899d: 23]. Does not appear to have been previously considered.

Family Phocaenidae Bravard, 1885: 144; Plates 2–3.

TYPE GENUS: *Phocaena* G. Cuvier, 1817.

COMMENTS: Synonymised within the Family Phocoenidae by McKenna and Bell (1997: 390).

Subfamily Phocoenoidinae Barnes, 1984: 1, 17.

TYPE GENUS: *Phocoenoides* R. Andrews, 1911b: 31.

COMMENTS: When originally proposed as a new rank it was placed in the Family Phocoenidae (J. Gray, 1825a) and included the genera *Phocoenoides* R. Andrews, 1911b: 31; † *Loxolithax* Kellogg, 1931: 361, 390; and † *Piscolithax* de Muizon, 1983: 1203. Recognised as a subfamily within by McKenna and Bell (1997: 390).

Phocoena G. Cuvier, 1816

Phocoena G. Cuvier, 1816a: 279.

TYPE SPECIES: Φ *Delphinus phocoena* Linnaeus, 1758: 77 [= Φ *Phocoena phocoena* (Linnaeus, 1758: 77)] by monotypy.

COMMENTS: Mead and Brownell (2005: 736) pointed out that *Phocaena* and *Phocena* are later spellings.

Phocaena G. Cuvier, 1817: 163.

TYPE SPECIES: Φ *Delphinus phocoena* Linnaeus, 1758: 77 [= Φ *Phocoena phocoena* (Linnaeus, 1758: 77)] by monotypy.

COMMENTS: Replacement name for *Phocoena* G. Cuvier, 1816a. Spelling used by G. Cuvier (1829a: 289) and used in preference to *Phocoena* by Palmer (1904: 532). Mead and Brownell (2005: 736) pointed out that the correct spelling is *Phocoena* and that *Phocaena* and *Phocena* are later spellings.

HOMONYMS:

Phocaena Brisson, 1762: 234, cetaceans of the Class Mammalia (Order Artiodactyla, Family Delphinidae). Rejected for nomenclatural purposes by Opinion 1894 of the ICZN (1998: 64).

Phocena J. Gray, 1821: 310.

TYPE SPECIES: Φ *Delphinus phocoena* Linnaeus, 1758: 77 [= Φ *Phocoena phocoena* (Linnaeus, 1758: 77)] by monotypy.

COMMENTS: Emendation or misprint of *Phocoena* Cuvier, 1816a. Synonymised within *Phocoena* by Hershkovitz (1966: 100), and Mead and Brownell (2005: 736).

Phocaena J. Gray, 1828: 2.

TYPE SPECIES: Φ *Delphinus phocoena* Linnaeus, 1758: 77 [= Φ *Phocoena phocoena* (Linnaeus, 1758: 77)] by monotypy.

COMMENTS: Described as a subgenus of *Delphinus*. Synonymised within *Phocoena* by Hershkovitz (1966: 100).

Acanthodelphis J. Gray, 1866b: 304.

TYPE SPECIES: Φ *Phocaena* [sic] *spinipinnis* Burmeister, 1865c: 228 [= Φ *Phocoena spinipinnis* Burmeister, 1865c: 228] by monotypy.

COMMENTS: Described as a subgenus of *Phocaena*, but raised to generic rank by J. Gray (1868c: 8). Synonymised within *Phocoena* by Hershkovitz (1966: 100), and Mead and Brownell (2005: 736).

Australophocaena Barnes, 1985: 149, 153.

TYPE SPECIES: *Phocoena dioptrica* Lahille, 1912 by original designation.

COMMENTS: Recognised by Mead and Brownell (1993: 358), but synonymised within *Phocoena* by Rosel *et al.* (1995: 463) and Mead and Brownell (2005: 736).

Phocoena dioptrica Lahille, 1912

Spectacled Porpoise

Phocoena dioptrica Lahille, 1912: 269.

TYPE LOCALITY: Punta Colares, near Quilmes, Río de la Plata, Argentina.

COMMENTS: Taxonomic decision of Hershkovitz (1966: 102) to recognise this species. Included in the genus *Phocoena* by Brownell (1975: 1) and Bannister (1988d: 211). Placed in the genus *Australophocaena* by Mead and Brownell (1993: 358) and Goodall and Schiavini (1995: 411), but transferred back to *Phocoena* by Mead and Brownell (2005: 736).

Phocaena Stornii Marelli, 1922: 229; Figs. 1–5.

TYPE LOCALITY: South Atlantic, Tierra del Fuego, Argentina.

COMMENTS: Synonymised within *dioptrica* by Hershkovitz (1966: 102), Brownell (1975: 1), and Mead and Brownell (2005: 736).

Incertae Sedis

Anarnak Lacépède, 1804: xxxviii, 164.

TYPE SPECIES: *Anarnak groenlandicus* Lacépède, 1804: xxxviii, 164 by monotypy.

COMMENTS: Considered *incertae sedis* by Hershkovitz (1966: 195). Synonymised with doubt within *Hyperoodon* by McKenna and Bell (1997: 382) and does not appear to have been considered by other authors.

anarcus Duméril, 1806b: 28.

TYPE SPECIES: Emendation of *Anarnak* Lacépède, 1804.

COMMENTS: Synonymised within *Anarnak* by Palmer (1904: 100) and Hershkovitz (1966: 195), and within *Hyperoodon* by McKenna and Bell (1997: 382) with a query.

anarcus Froriep, 1806: 29.

TYPE SPECIES: *Anarnak groenlandicus* Lacépède, 1804: xxxviii, 164.

COMMENTS: Synonymised within *Anarnak* by Palmer (1904: 100).

Anarnacus Tiedemann, 1808: 575.

TYPE SPECIES: *Anarnak groenlandicus* Lacépède, 1804: xxxviii, 164.

COMMENTS: Synonymised within *Anarnak* by Palmer (1904: 100).

Ancylodon Illiger, 1811: 142.

TYPE SPECIES: *Monodon spurius* Fabricius, 1780: 31.

COMMENTS: Synonymised within *Hyperoodon* by McKenna and Bell (1997: 382).

Ananareus J. Gray, 1843a: xxiii.

TYPE SPECIES: *Anarnak groenlandicus* Lacépède, 1804: xxxviii, 164.

COMMENTS: Incorrect subsequent spelling of *Anarnak*. Synonymised within *Anarnak* Lacépède, 1804 by Palmer (1904: 100).

Anarmacus Zittel, 1893: 178.

TYPE SPECIES: *Anarnak groenlandicus* Lacépède, 1804: xxxviii, 164.

COMMENTS: Incorrect subsequent spelling of *Anarnak* Lacépède, 1804. Synonymised within *Anarnak* by Palmer (1904: 100).

References

Abbie AA (1937) Some observations on the major subdivisions of the Marsupialia, with special reference to the position of the Peramelidae and Caenolestidae. *Journal of Anatomy* 71, 424 436.

Abel O (1901) Les dauphins longirostres du Boldérien (Miocène supérieur) des environs d'Anvers. *Extrait des Mémoires du Mémoirs Museum Royal Histoire Naturelle Belgique* 1, 1–95.

Abel O (1913) Die eocänen Sirenen der Mittelmeerregion. Erster Tiel: Der Schädel von Eotherium aegyptiacum. *Palaeontographica* 59, 289–360.

Abel O (1914) *Die Vorzeitlichen Säugetiere.* Gustav Fischer, Jena.

Acloque A (1899) *Faune de France, contenant la description des espèces indigénes disposées en tableaux analytiques et illustré de figures représentant les types caractéristiques des genres.* Volume 1. *Mammifères, Oiseaux, Poissons, Reptiles, Batraciens, Protochordes.* Librairie J.B. Ballière et fils, Paris.

Adam PJ (2005) *Lobodon carcinophaga. Mammalian Species* 772, 1–4. [15 Jul 2005]

Adams M, Baverstock PR, Watts CHS, Reardon T (1987a) Electrophoretic resolution of species boundaries in Australian Microchiroptera, II. The *Pipistrellus* group (Chiroptera: Vespertilionidae). *Australian Journal of Biological Sciences* 40, 163–170.

Adams M, Baverstock PR, Watts CHS, Reardon T (1987b) Electrophoretic resolution of species boundaries in Australian Microchiroptera. I. *Eptesicus* (Chiroptera: Vespertilionidae). *Australian Journal of Biological Sciences* 40, 143–162.

Adams M, Reardon PR, Baverstock PR, Watts CHS (1988) Electrophoretic resolution of species boundaries in Australian Microchiroptera IV. The Molossidae (Chiroptera). *Australian Journal of Biological Sciences* 41, 315–326.

Adams M, Cooper N, Armstrong J (2000) Revision of *Dasycercus* systematics. A Report to the South Australian Department of Environment and Heritage. 34pp. [Jun 2000]

Adkins RM, Honeycutt RL (1991) Molecular phylogeny of the superorder Archonta. *Proceedings of the National Academy of Sciences of the United States of America* 88, 10317–10321. [15 Nov 1991]

Adkins RM, Walton AH, Honeycutt RL (2003) Higher-level systematics of rodents and divergence time estimates based on two congruent nuclear genes. *Molecular Phylogenetics and Evolution* 26, 409–420. [Mar 2003]

Agassiz L (1842) *Nomina Systematica, Generum Mammalium, tam viventium quam fossilium, secundum ordinum alphabeticum disposita, adjectis auctoribus, libris in quibus reperiuntur anno editionis, etymologia et familis, ad quas pertinent.* pp. 1–38 + 1–8 (Addenda), in L Agassiz (1842–1846) *Nomenclator Zoologicus, continens nomina systematica generum animalium tam viventium quam fossilium, secundum ordinem alphabeticum disposita, adjectis auctoribus, libris, in quibus reperijuntur anno editionis, etymologia et familis, ad quas pertinent, in singulis classibus.* Jent et Gassmann, Soloduri.

Agassiz L (1846) *Nomenclatoris Zoologici, Index Universalis, continens nomina systematica classium, ordinum familiarum et generum animalium omnium, tam viventium quam fossilium, secundum ordinem alphabeticum unicum disposita, adjectis homonymiis plantarum, nec non variis adnotationibus et emendationibus.* pp. 1–393, in L Agassiz (1842–1846) *Nomenclator Zoologicus, continens nomina systematica generum animalium tam viventium quam fossilium, secundum ordinem alphabeticum disposita, adjectis auctoribus, libris, in quibus reperijuntur anno editionis, etymologia et familis, ad quas pertinent, in singulis classibus.* Jent et Gassmann, Soloduri.

Agassiz L (1859) *An Essay on Classification.* Longman, Brown, Green, Longmans & Roberts, London.

Agnarsson I, May-Collado LJ (2008) The phylogeny of Cetartiodactyla: the importance of dense taxon sampling, missing data, and the remarkable promise of cytochrome b to provide reliable species-level phylogenies. *Molecular Phylogenetics and Evolution* 48, 964–985. [Sep 2008]

Aguilar A (2009) Fin whale *Balaenoptera physalus.* pp. 433–437, in WF Perrin, B Würsig & JGM Thewissen (eds.) *Encyclopedia of Marine Mammals.* Academic Press, Amsterdam. Second Edition.

Aguirre-Fernández G, Barnes LG, Aranda-Manteca FJ, Fernández-Rivera JR (2009) *Protoglobicephala mexicana,* a new genus and species of Pliocene fossil dolphin (Cetacea; Odontoceti; Delphinidae) from the Gulf of California, Mexico. *Boletín de la Sociedad Geológica Mexicana* 61, 245–265.

Aitken PF (1968) Observations on *Notomys fuscus* (Wood Jones) (Muridae-Pseudomyinae) with notes on a new synonym. *South Australian Naturalist* 43, 37–45. [Dec 1968]

Aitken PF (1971) Rediscovery of the large desert sminthopsis (*Sminthopsis psammophila* Spencer) on Eyre Peninsula, South Australia. *Victorian Naturalist* 88, 103–111. [Apr 1971]

Aitken PF (1972a) *Planigale gilesi* (Marsupialia, Dasyuridae); a new species from the interior of south eastern Australia. *Records of the South Australian Museum* 16(10), 1–14. [12 Jun 1972]

Aitken PF (1972b) *Sminthopsis murina* (Waterhouse) 1838: A new record from Kangaroo Island, South Australia. *South Australian Naturalist* 46, 36–37. [Mar 1972]

Aitken PF (1976) Vertebrate type-specimens in the South Australian Museum. V. Mammals. *Records of the South Australian Museum* 17, 197–202. [Sep 1976]

Alberico MS, Orejuela JE (1982) Diversidad específica de dos communidades de murciélagos Nariño, Columbia. *Cespedesia* 3(suppl.), 31–40.

Albers JC (1850) *Die Heliceen, nach natürlicher verwandtschaft systematisch geordnet*. Th. Crr. Fr. Enslin, Berlin.

Albrecht CMP (1879) Bericht und Vorträge. *Königsberg Physikalisch-Oekonomische Gesellschaft Schriften* 20(1), 22.

Alexander JSA (1981) The status of the Squirrel Glider, *Petaurus norfolcensis* (Marsupialia: Petauridae) in Victoria. *BSc Honours Thesis*. Latrobe University, Melbourne.

Alexander WB (1919) A new species of marsupial of the Subfamily Phalanginae. *Journal of the Royal Society of Western Australia* 4, 31–36.

Alfken D (1891) Erster Beitrag zur Insekten-Fauna der Nordsee-Insel Juist. *Abhandlungen Herausgegeben vom Naturwissenschaftlichen Vereine zu Bremen* 12, 97–130. [Apr 1891]

Allen GM (1912) *Zaglossus*. *Memoirs of the Museum of Comparative Zoology* 40, 253–307.

Allen GM (1933) Two new bats from Australia. *Journal of Mammalogy* 14, 149–151. [May 1933]

Allen GM (1935) A fruit bat, *Dobsonia*, new to Australia. *Australian Zoologist* 8, 151.

Allen GM (1936) The status of *Vespertilio pilosus* Peters. *Journal of Mammalogy* 17, 168–169. [May 1936]

Allen GM (1938) The Mammals of China and Mongolia. Part 1. pp. 1–620, in W Granger (ed.) *Natural History of Central Asia*. Volume 11. American Museum of Natural History, New York.

Allen GM (1939) A checklist of African mammals. *Bulletin of the Museum of Comparative Zoology. Harvard University* 83, 1–763. [Feb 1939]

Allen GM, Barbour T (1909) A new marsupial from Netherlands New Guinea. *Proceedings of the New England Zoölogical Club* 4, 43–46. [12 Jul 1909]

Allen GR, McCosker JE, Cross NJ, Bray DJ, Hoese DF (2006) Muraenidae. pp. 243–259, in DF Hoese, DJ Bray, GR Allen, JR Paxton, A Wells & PL Beesley (eds.) *Zoological Catalogue of Australia. Volume 35. Part 1. Fishes*. CSIRO Publishing / Australian Biological Resources Study (ABRS), Melbourne and Canberra.

Allen H (1861) Description of new pteropine bats from Africa. *Proceedings. Academy of Natural Sciences of Philadelphia* 13, 156–160.

Allen H (1862) Descriptions of two new species of Vespertilionidae, and some remarks on the genus *Antrozous*. *Proceedings. Academy of Natural Sciences of Philadelphia* 14, 246–248.

Allen H (1864) Monograph of the bats of North America. *Smithsonian Miscellaneous Collections* 7(165), 1–85. [Jun 1864]

Allen H (1865) On a new genus of Vespertilionidae. *Proceedings. Academy of Natural Sciences of Philadelphia* 17, 173–175. [Sep-Oct 1965]

Allen H (1866) Notes on the Vespertilionidae of tropical America. *Proceedings. Academy of Natural Sciences of Philadelphia* 18, 279–288.

Allen H (1891) A new genus of Vespertilionidae. *Proceedings. Academy of Natural Sciences of Philadelphia* 43, 467–470.

Allen JA (1870) On the eared seals (Otariadae), with detailed descriptions of the North Pacific species by J.A. Allen. Together with an account of the habits of the northern fur seal (*Callorhinus ursinus*), by Charles Bryant. *Bulletin of the Museum of Comparative Zoology. Harvard University* 2, 1–108.

Allen JA (1880) History of North American pinnipeds: A monograph of the walruses, sea-lions, sea-bears and seals of North America. *Miscellaneous Publications, United States Geological and Geographical Survey of the Territories*. Government Printing Office, Washington. Volume 12, 1–785.

Allen JA (1892) A synopsis of the pinnipeds, or seals and walruses, in relation to their commercial history and products. pp. 367–391, in *Fur seal Arbitration, Appendix to the case of the United States before the tribunal of arbitration to convene at Paris*. United States Government Printing Office, Washington, D.C. Volume 1.

Allen JA (1905) Mammalia of Southern Patagonia. pp. 1–210, in WB Scott (ed.) Reports of the Princeton University Expeditions to Patagonia, 1896–1899. Princeton University, New Jersey & E. Schweizerbart'Sche Verlagshandlung. Volume 3. Zoölogy.

Allen JA (1908) The case of *Trix* vs. *Aluco*. *The Auk* 25, 288–291. [Jul 1908]

Allen JA (1917a) The skeletal characters of *Scutisorex* Thomas. *Bulletin of the American Museum of Natural History* 37, 769–784.

Allen JA (1917b) In Allen, J.A., Lang, H. & Chapin, J.P. The American Museum Congo Expedition collection of bats. *Bulletin of the American Museum of Natural History* 37, 405–563.

Allen JA, Coues E (1877) Appendix A. Synoptical List of the Fossil Rodentia of North America. pp. 943–949, in E Coues & JA Allen (eds.) *Monographs of North American Rodentia*. Report of the United States Geological Survey of the Territories. Volume XI. Government Printing Office, Washington.

Allison FR (1989) Molossidae. pp. 892–909, in DW Walton & BJ Richardson (eds.) *Fauna of Australia. Volume 1B. Mammalia*. Australian Government Publishing Service, Canberra.

Alston ER (1876) On the classification of the Order Glires. *Proceedings of the Zoological Society of London* 44, 61–98. [Jun 1876]

Alston ER (1877) On the rodents and marsupials collected by the Rev. G. Brown in Duke-of-York Island, New Britain, and New Ireland. *Proceedings of the Zoological Society of London* 45, 123–127. [Jun 1877]

Alston ER (1879a) On the *Acanthomys* of Gray. *Proceedings of the Zoological Society of London* 47, 645–647. [Oct 1879]

Alston ER (1879b) Untitled. *Nature* 20(504), 210–211. [26 Jun 1879]

Amadon D (1949) The seventy-five percent rule for subspecies. *The Condor* 51, 250–258.

Ameghino F (1882) Catálogo de las colecciones de Antropología prehistórica y de paleontología de Florentino Ameghino, Partido de Mercedes. In *Catálogo de la Sección de la Provincia de Buenos Aires (República Argentina), en la Exposición Continental Sudamericana. Anexo A*: 35–42. Buenos Aires. Reference not seen.

Ameghino F (1889) Contribución al Conocimiento de los Mamíferos Fósiles de la República Argentina. *Actas de la Academia Nacional de Ciencias de la República Argentina en Córdoba* 6, 1–1027.

Ameghino F (1891) Nuevos restos de mamíferos fósiles descubiertos por Cárlos Ameghino en el eoceno inferior de la Patagonia austral. — Especies nuevas, adiciones y correcciones. *Revista Argentina Historia Natural* 1, 289–328.

Ameghino F (1893) Enumération synontique des espèces de mammifères fossils des formations éocènes de Patagonie. *Boletin de la Academia Nacional de Ciencias en Córdoba. República Argentina* 13, 259–445.

Ameghino F (1895) Première contribution à connaissance de la fauna mammalogique de couches à *Pyrotherium*. *Boletín del Instituto Geográfico Argentino* 15, 603–660.

Ameghino F (1916) *Obras completas y correspondencia científica*. Volumen 6. *Los Mamíferos Fósiles de la República Argentina*. Taller de Impresiones Oficiales, LaPlata.

Amerling C (1861) Naturökonomie der von ihm beobachteten Milben, insbesondere der Trombidieen. *Sitzungsberichte der Königlichen Böhmischen Gesellschaft der Wissenschaften, Prague* 1861, 54–56.

Ammerman LK, Lee DN, Tips TM (2012) First molecular phylogenetic insights into the evolution of free-tailed bats in the subfamily Molossinae (Molossidae, Chiroptera). *Journal of Mammalogy* 93, 12–28. [Feb 2012]

Amrine-Madsen H, Scally M, Westerman M, Stanhope MJ, Krajewski C, Springer MS (2003a) Nuclear gene sequences provide evidence for the monophyly of australidelphian marsupials. *Molecular Phylogenetics and Evolution* 28, 186–196. [Aug 2003]

Amrine-Madsen H, Koepfli KP, Wayne RK, Springer MS (2003b) A new phylogenetic marker, apolipoprotein B, provides compelling evidence for eutherian relationships. *Molecular Phylogenetics and Evolution* 28, 225–240. [Aug 2003]

Amyot CJB, Audinet-Serville JG (1843) *Histoire Naturelle des Insectes: Hémiptères*. Libraire Encyclopédique de Roret, Paris.

Andersen K (1905a) On some bats of the genus *Rhinolophus*, with remarks on their mutual affinities, and descriptions of twenty-six new forms. *Proceedings of the Zoological Society of London* 75(105)(2), 75–145. [7 Oct 1905]

Andersen K (1905b) On *Hipposideros diadema* and its closest allies. *Annals & Magazine of Natural History* (7)16, 497–507. [1 Nov 1905]

Andersen K (1906) On some new or little known bats of the genus *Rhinolophus* in the collection of the Museo Civico, Genoa. *Annali del Museo Civico di Storia Naturale di Genova* (3)2(42), 173–195.

Andersen K (1907) Chiropteran notes. *Annali del Museo Civico di Storia Naturale di Genova* (3)3(43), 5–45.

Andersen K (1908) Twenty new forms of *Pteropus*. *Annals & Magazine of Natural History* (8)2, 361–370. [1 Oct 1908]

Andersen K (1909a) Notes on the genus *Acerodon*, with a synopsis of its species and subspecies, and descriptions of four new forms. *Annals & Magazine of Natural History* (8)3, 20–29. [1–2 Jan 1909]

Andersen K (1909b) Two new bats from the Solomon Islands. *Annals & Magazine of Natural History* (8)3, 266–270. [1–2 Mar 1909]

Andersen K (1911) Six new fruit-bats of the genera *Macroglossus* and *Syconycteris*. *Annals & Magazine of Natural History* (8)7, 641–643. [1 Jun 1911]

Andersen K (1912) *Catalogue of the Chiroptera in the Collection of the British Museum. Volume 1. Megachiroptera*. Second Edition. British Museum, London.

Andersen K (1918) Diagnoses of new bats of the families Rhinolophidae and Megadermatidae. *Annals & Magazine of Natural History* (9)2, 374–384. [1 Oct 1918]

Anderson J (1802) Notices of a singular animal, lately discovered in New South Wales, with a figure. *Recreations in Agriculture, Natural-history. Arts and Miscellaneous Literature* 2(2), 562–565. [= Volume 6 of combined series]

Anderson J (1871) Description of a new cetacean from the Irrawaddy River, Burma. *Orcella fluminalis* Anderson. *Proceedings of the Zoological Society of London* 39, 142–144. [Jun 1871]

Anderson J (1879) *Anatomical and Zoological Research*; comprising and account of zoological results of two expeditions to western Yunnan in 1868 and 1875; and a monograph of the two cetacean genera, *Platanista* and *Orcella*. Bernard Quaritch, London. Volume 1. [title page says 1878 but Corrigenda p. xii says 1879]

Andrew P (2008) Checklist of the mammals of Australia and its territories. pp. 629–638, in L Vogelnest & R Woods (eds.) *Medicine of Australian Mammals*. CSIRO Publishing, Melbourne.

Andrews CW (1900) *A Monograph of Christmas Island (Indian Ocean)*. British Museum (Natural History), London.

Andrews CW (1902) Preliminary note on some recently discovered extinct vertebrates from Egypt. (Part III). *Geological Magazine* 9, 291–295. [Decade IV]

Andrews CW (1906) A suggested change in nomenclature. *Nature* 73(1888), 224. [4 Jan 1906]

Andrews RC (1908) Description of a new species of *Mesoplodon* from Canterbury Province, New Zealand. *Bulletin of the American Museum of Natural History* 24, 203–215.

Andrews RC (1911a) Description of an apparently new porpoise of the genus *Tursiops*, with remarks upon a skull of *Tursiops gillii* Dall. *Bulletin of the American Museum of Natural History* 30, 233–237.

Andrews RC (1911b) A new porpoise from Japan. *Bulletin of the American Museum of Natural History* 30, 31–51.

Andrews RC (1918) A note on the skeleton of *Balaenoptera edeni* in the Indian Museum, Calcutta. *Records of the Indian Museum* 15, 105–107. [30 Aug 1918]

Angus GF (1861) Notes on the broad-fronted wombat of South Australia (*Phascolomys latifrons* Owen). *Proceedings of the Zoological Society of London* 29, 268–271. [Sep 1861]

Anon (1790) in J White. *Journal of a Voyage to New South Wales with sixty-five plates of non descript animals, birds, lizards, serpents, curious cones of tees and other natural productions*. J. Debrett, London.

Anon (1791) In A Phillip. *Voyage du Gouverneur Phillip à Botany-Bay*, avec une description de l'éstablissement des colonies du Port Jackson et de l'île Norfolk; faite sur des papiers authentiques, obtenus des divers départemens, auxquels on a ajouté les journaux des lieutenans Shortland,

Watts, Ball, et du Capitane Marshall, avec un récit de leurs nouvelles découvertes. Traduit de l'Anglais, Buisson, Paris. Reference not seen.

Anon (1798a) Compendiöse Bibliothek der gemeinnützigen Kenntnisse für alle Stände 21. Der Zoologe V-VIII. *Neue Allgemeine Deutsche Bibliothek* 39, 380–381.

Anon (1798b) Histoire naturelle des poissons par le Cit. Lacépède. *Allgemeine Literatur-Zeitung*. Expedition deifer Zeitung, Jena. Volume 3 (24 Sep 1798) Numéro 287 col., 673–678.

Anon (1820) Notices respecting new books: *The Cyclopædia; or, Universal Dictionary of Arts, Sciences, and Literature*, by Abraham Rees. *The Philosophical Magazine and Journal* (1) 56 (269), 218–224. [Jul 1820]

Anon (1825) *Monographies de Mammalogie*, ou description de quelques genres de Mammifères, dont les espèces ont eté observées dans les différens Musées de l'Europe par C.J. Temminck. *The Zoological Journal, London* 1, 574–576.

Anon (1838) *A Bibliographic Account and Collation of La Description de l'Egypt*, presented to the Library of the London Institution by Sir Thomas Baring, Baronet, President. London.

Anon (1839) Marsupia'lia or Marsupiata. pp. 450–469, in *The Penny Cyclopaedia of the Society for the Diffusion of Useful Knowledge*. Charles Knight, London. Volume 14. Limonia-Massachusetts.

Anon (1845) Zoology. pp. 728–784, in *London Encyclopaedia or Universal Dictionary of Science, Art, Literature, and Practical Mechanics, comprising a popular view of the present state of knowledge, illustrated by numerous engravings, a general atlas and appropriate diagrams*. Volume 22. THALES-ZYPAEUS. T. Gegg & Son, Sydney.

Anon (1854) Bandicoot. p. 382, in C Knight (ed.) *The English Cyclopaedia. A New dictionary of Universal Knowledge*. Natural History. Bradbury and Evans, London. Volume 1.

Anon (1859) *Descriptive Catalogue of the Specimens of Natural History in Spirit Contained in the Museum of the Royal College of Surgeons of England. Vertebrata: Pisces, Reptila, Aves, Mammalia*. Royal College of Surgeons of England, London.

Anon (1903) Catalogue of the Books, Manuscripts, and Drawings in the British Museum (Natural History). Volume 1. *A—D*. British Museum, London.

Anon (1913) *Catalogue of the Books, Manuscripts, Maps and Drawings in the British Museum (Natural History). British Museum, London. Volume 4. P—SN*. British Museum, London.

Anon (1915) *Procès-verbaux des séances de l'Académie tenues depuis la fondation de l'Institut jusqu à la fin du mois d'août 1835, 6, and 1816–1819*. Académie des Sciences, Institut de France. [p. 114]

Anon (1966) A relict marsupial. *Nature* 212(5059), 225. [15 Oct 1966]

Anthony HE (1923) Mammals from Mexico and South America. *American Museum Novitates* 43, 1–10. [17 Jan 1923]

Aplin KP (1987) In KP Aplin & M Archer. Recent advances in marsupial systematics with a new syncretic classification. Volume 1. pp. xv-lxxii, in M Archer (ed.) *Possums & Opossums*. Surrey Beatty, Sydney. [Nov 1987]

Aplin KP, Archer M (1987) Recent advances in marsupial systematics with a new syncretic classification. Volume 1. pp. xv-lxxii, in M Archer (ed.) *Possums & Opossums: Studies in Evolution*. Surrey Beatty and Sons, Sydney. [Nov 1987]

Aplin KP, Rhind SG, Chesser RT, ten Havem J (2015) Taxonomic revision of *Phascogale tapoatafa* (Meyer, 1793) (Dasyuridae; Marsupialia), including descriptions of two new subspecies and confirmation of *P. pirata* Thomas, 1904 as 'Top End' endemic. *Zootaxa*, in press.

Aplin KP, Suzuki H, Chinen AA, Chesser T, ten Have J, Donnellan SC, Austin J, Frost A, Gonzalez JP, Herbreteau V, Catzeflis F, Soubrier J, Fang Y-P, Robins J, Matisoo-Smith E, Bastos A, Maryanto I, Sinaga MH, Denys C, Van Den Bussche R, Conroy C, Rowe KC, Cooper A (2011) Multiple geographic origins of commensalism and complex dispersal history of Black Rats. *PLoS ONE* 6(11), 1–20. [2 Nov 2011]

Appleton BR, McKenzie JA, Christidis L (2004) Molecular systematics and biogeography of the bent-wing bat complex *Miniopterus schreibersii* (Kuhl, 1817) (Chiroptera: Vespertilionidae). *Molecular Phylogenetics and Evolution* 31, 431–439. [May 2004]

Arbogast BS, Aplin KP, Armstrong KN, Ogawa LM, Taylor AC (2011) Cryptic diversity in an iconic Australian marsupial: The Greater Glider, *Petauroides volans*. Abstract. Joint meeting of the 57th Scientific Meeting of the Australian Mammal Society with the 91st Annual Meeting of the American Society of Mammalogists. Portland State University, Oregon. USA. 24–28 June 2011.

Archer FI, Perrin WF (1999) *Stenella coeruleoalba*. *Mammalian Species* 603, 1–9. [5 May 1999]

Archer FI, Morin PA, Hancock-Hanser BL, Robertson KM, Leslie MS, Bérubé M, Panigada S, Taylor BL (2013) Mitogenomic phylogenetics of Fin Whales (*Balaenoptera physalus* spp.): Genetic evidence for revision of subspecies. *PLoS ONE* 8(5), e63396. [17 May 2013]

Archer M (1975) *Ningaui*, a new genus of tiny dasyurids (Marsupialia) and two new species. *N. timealeyi* and *N. ridei*, from arid Western Australia. *Memoirs of the Queensland Museum* 17, 237–249. [31 Jul 1975]

Archer M (1976a) The basicranial region of marsupial carnivores (Marsupialia), inter-relationships of carnivorous marsupials, and affinities of the insectivorous marsupial peramelids. *Zoological Journal of the Linnean Society* 59, 217–322. [Nov 1976]

Archer M (1976b) A revision of the marsupial genus *Planigale* Troughton (Dasyuridae). *Memoirs of the Queensland Museum* 17, 341–365. [30 Jun 1976]

Archer M (1976c) Application for the suppression of the name *Sminthopsis murina* var. *constricta* Spencer, 1896 (Marsupialia, Dasyuridae). Z.N. (S.) 2080. *Bulletin of Zoological Nomenclature* 33, 127–128. [30 Sep 1976]

Archer M (1977a) Revision of the dasyurid marsupial genus *Antechinomys* Krefft. *Memoirs of the Queensland Museum* 18, 17–29. [31 Mar 1977]

Archer M (1977b) In JAW Kirsch. The comparative serology of Marsupialia, and a classification of marsupials. *Australian Journal of Zoology. Supplement Series* 52, 1–152. [26 Sep 1977]

Archer M (1978) Quaternary vertebrate faunas from the Texas Caves of southeastern Queensland. *Memoirs of the Queensland Museum* 19, 61–109. [Jul 1978]

Archer M (1979a) The status of Australian dasyurids, thylacinids and myrmecobiids. pp. 29–43, in MJ Tyler (ed.) *The Status of Endangered Australasian Wildlife*. Royal Zoological Society of South Australia, Adelaide.

Archer M (1979b) Two new species of *Sminthopsis* Thomas (Dasyuridae: Marsupialia) from northern Australia, *S. butleri* and *S. douglasi*. *Australian Zoologist* 20, 327–345.

Archer M (1979c) *Wabularoo naughtoni* gen. et sp. nov., an enigmatic kangaroo (Marsupialia) from the middle Tertiary Carl Creek Limestone of northwestern Queensland. Results of the Ray. E. Lemley Expeditions, Part 4. *Memoirs of the Queensland Museum* 19, 299–307. [Jun 1979]

Archer M (1981) Results of the Archbold Expeditions. No. 104. Systematic revision of the marsupial dasyurid genus *Sminthopsis* Thomas. *Bulletin of the American Museum of Natural History* 168, 61–224. [10 Apr 1981]

Archer M (1982a) Review of the dasyurid (Marsupialia) fossil record, integration of data bearing on phylogenetic interpretation of suprageneric classification. pp. 397–443, in M Archer (ed.) *Carnivorous Marsupials*. Volume 2. Royal Zoological Society of New South Wales, Sydney. [Feb 1982]

Archer M (1982b) A review of the Miocene thylacinids (Thylacinidae, Marsupialia), the phylogenetic position of the Thylacinidae and the problem of apriorisms in character analysis. pp. 445–476, in M Archer (ed.) *Carnivorous Marsupials*. Volume 2. Royal Zoological Society of New South Wales, Sydney. [Feb 1982]

Archer M (1984) The Australian marsupial radiation. pp. 633–808, in M Archer & G Clayton (eds.) *Vertebrate Evolution and Zoogeography in Australasia - animals in space and time*. Hesperian Press, Perth.

Archer M (1989) Emendations to the nomenclature of recently established palaeontological taxa. *Australian Zoologist* 25, 67.

Archer M, Bartholomai A (1978) Tertiary mammals of Australia: a synoptic review. *Alcheringa* 2, 1–19.

Archer M, Flannery TF (1985) Revision of the extinct gigantic rat kangaroos (Potoroidae: Marsupialia), with description of a new Miocene genus and species and a new Pleistocene species of *Propleopus*. *Journal of Paleontology* 59, 1331–1349. [Nov 1985]

Archer M, Kirsch JAW (1977) The case for Thylacomyidae and Myrmecobiidae, Gill, 1872, or why are marsupial families so extended? *Proceedings of the Linnean Society of New South Wales* 102, 18–25. [28 Sep 1977]

Archer M, Plane MD, Pledge NS (1978) Additional evidence for interpreting the Miocene *Obdurodon insignis* Woodburne and Tedford, 1975, to be a fossil platypus (Ornithorhynchidae: Monotremata) and a reconsideration of the status of *Ornithorhynchus agilis* De Vis, 1885. *Australian Zoologist* 20, 9–27.

Archer M, Clayton G, Hand S (1984) A checklist of Australasian fossil mammals. pp. 1027–1087, in M Archer & G Clayton (eds.) *Vertebrate Evolution and Zoogeography in Australasia - animals in space and time*. Hesperian Press, Perth.

Archer M, Hand SJ, Godthelp HJ (1988) A new order of Tertiary zalambdodont marsupials. *Science* 239(4847), 1528–1531. [25 Mar 1988]

Archer M, Godthelp H, Hand SJ, Megirian D (1989) Fossil mammals of Riversleigh, northwestern Queensland: preliminary overview of biostratigraphy, correlation and environmental change. *Australian Zoologist* 25, 29–65.

Archibald JD (1998) Archaic ungulates (Condylartha). pp. 292–331, in CS Janis, KM Scott & LL Jacobs (eds.). *Evolution of Tertiary Mammals of North America. Vol. 1. Terrestrial Carnivores, Ungulates, and Ungulate like Mammals*. Cambridge University Press, Cambridge.

Archibald JD (2003) Timing and biogeography of the eutherian radiation: fossils and molecules compared. *Molecular Phylogenetics and Evolution* 28, 350–359. [Aug 2003]

Archibald JD, Averianov AO, Ekdale EG (2001) Late Cretaceous relatives of rabbits, rodents, and other extant eutherian mammals. *Nature* 414(6859), 62–65. [1 Nov 2001]

Ardalan A, Oskarsson M, Natanaelsson C, Wilton AN, Ahmadian A, Savolainen P (2012) Narrow genetic basis for the Australian dingo confirmed through analysis of paternal ancestry. *Genetica* 140, 65–73.

Armstrong KN (2006a) Resolving the correct nomenclature of the orange leaf-nosed bat *Rhinonicteris aurantia* (Gray, 1845) (Hipposideridae). *Australian Mammalogy* 28, 125–130. [10 May 2006]

Armstrong KN (2006b) Phylogeographic structure in *Rhinonicteris aurantia* (Chiroptera: Hipposideridae): implications for conservation. *Acta Chiropterologica* 8, 63–81.

Armstrong KN, Reardon T (2006) Standardising the common names of Australian bats – an update. *The Australasian Bat Society Newsletter* 26, 37–42. [Apr 2006]

Arnason U, Bodin K, Gullberg A, Ledje C, Mouchaty S (1995) A molecular view of pinniped relationships with particular emphasis on the true seals. *Journal of Molecular Evolution* 40, 78–85.

Arnason U, Gullberg A, Janke A (1999) The mitochondrial DNA molecule of the aardvark, *Oryctopus afer*, and the position of the Tubulidenta in the eutherian tree. *Proceedings of the Royal Society of London. Series B, Biological Sciences* 266, 339–345.

Arnason U, Adegoke JA, Bodin K, Born EW, Esa YB, Gullberg A, Nilsoon M, Short RV, Xu X, Janke A (2002) Mammalian mitogenomic relationships and the root of the eutherian tree. *Proceedings of the National Academy of Sciences of the United States of America* 99, 8151–8156. [11 June 2002]

Arnason U, Gullberg A, Janke A, Kullberg M, Lehman N, Petrov EA, Väinölä R (2006) Pinniped phylogeny and a new hypothesis for their origin and dispersal. *Molecular Phylogenetics and Evolution* 41, 345–354. [Nov 2006]

Arnason U, Adegoke JA, Gullberg A, Harley EH, Janke A, Kullberg M (2008) Mitogenomic relationships of placental mammals and molecular estimates of their divergences. *Gene* 421, 37–51.

Arnett RH Jr, Thomas MC, Skelley PE, Frank JH (2002) American Beetles – Polyphaga: Scarabaeoidea through Curculionoidea. CRC Press, Boca Raton. Volume 2.

Arnold P, Heinsohn G (1996) Phylogenetic status of the Irrawaddy dolphin *Orcaella brevirostris* (Owen in Gray): a cladistic analysis. *Memoirs of the Queensland Museum* 39, 141–204. [20 Jul 1996]

Arnold P, Marsh H, Heinsohn G (1987) The occurrence of two forms of minke whales in east Australian waters with a description of external characters and skeleton of the diminutive or dwarf form. *Scientific Reports of the Whales Research Institute, Tokyo* 38, 1–46.

Arnold PW, Birtles RA, Dunstand A, Lukoschek V, Matthews M (2005) Colour patterns of the dwarf minke whales *Balaenoptera acutorostrata* sensu lato: description, cladistic analysis and taxonomic implications. *Memoirs of the Queensland Museum* 51, 277–307. [31 Dec 2005]

Arredondo O (1981) Nuevos género y especie de mamífero (Carnivora: Canidae) de) Holoceno de Cuba. *Poeyana* 218, 1–28. [20 Oct 1981]

Artaud S (1803) *Manuel d'Histoire Naturelle*, traduit de allemand, de J. Fr. Blumenbach, Professeur a l'Université de Gottingue. Volume 1. Collignon, Metz.

Artedi P (1738) *Ichthyologia sive opera omnia de Piscibus scilicet: Bibliotheca ichthyologica. Philosophia ichthyologica. Genera piscium. Synonymia specierum. Descriptiones specierum. Omnia in hoc genere perfectiora, quam antea ulla.* Part 4. Conradum Wishoff, Lugduni Batavorum.

Asher RJ (2005) Insectivoran-grade placentals. pp. 50–70, in KD Rose & JD Archibald (eds.) *The Rise of Placental Mammals: Origin and Relationships of the Major Clades*. Johns Hopkins University Press, Baltimore.

Asher RJ, Helgen KM (2010) Nomenclature and placental mammal phylogeny. *BMC Evolutionary Biology* 10(102), 1–9. [20 Apr 2010]

Asher RJ, Horovitz I, Sánchez-Villagra MR (2004) First combined cladistic analysis of marsupial mammal interrelationships. *Molecular Phylogenetics and Evolution* 33, 240–250. [7 Jul 2004]

Asher RJ, Meng J, Wible JR, McKenna MC, Rougier GW, Dashzeveg D, Novacek MJ (2005) Stem Lagomorpha and the antiquity of Glires. *Science* 307(5712), 1091–1094. [18 Feb 2005]

Audinet-Serville A (1832) Nouvelle classification de la famille des longicornes. *Annales de la Société Entomologique de France, Paris* 1, 118–201.

Aymard A (1848) Essai monographique sur un nouveau genre de mammifère fossile trouvé dans la Haute-Loire, et nommé Entélodon. *Annales de la Société d'Agriculture, Sciences, Arts et Commerce du Puy* 12(1842–1846), 227–267. Date of publication given as 1846 but confirmed as 1848 by Palmer (1904, 211).

Ayoub N, McGowen M, Clark C, Springer M, Gatesy J (2009) Evolution and phylogenetic utility of the melanocortin-1 receptor gene (MC1R) in Cetartiodactyla. *Molecular Phylogenetics and Evolution* 52, 550–557. [Aug 2009]

Ayres WO (1854) A description of a new species of polyp from Long Island, allied to *Tubullaria* under the name *Globiceps tiarella* Ayres. *Proceedings of the Boston Society of Natural History* 4, 193–195.

Baird SF (1857) Mammals: General Report upon the Zoology of the Several Pacific Railroad Routes. *Reports, explorations and surveys for a railroad route from Mississippi River to the Pacific Ocean, Washington, D.C.* 8(1), 1–757.

Baker AM, Van Dyck S (2013a) Taxonomy and redescription of the Fawn Antechinus, *Antechinus bellus* (Thomas) (Marsupialia: Dasyuridae). *Zootaxa* 3613, 201–228. [11 Feb 2013]

Baker AM, Van Dyck S (2013b) Taxonomy and redescription of the Yellow-footed Antechinus, *Antechinus flavipes* (Waterhouse) (Marsupialia: Dasyuridae). *Zootaxa* 3649, 1–62. [14 May 2013]

Baker AM, Van Dyck S (2013c) Taxonomy and redescription of the Atherton Antechinus, *Antechinus godmani* (Thomas) (Marsupialia: Dasyuridae). *Zootaxa* 3670, 401–439. [14 Jun 2013]

Baker AM, Mutton TY, Van Dyck S (2012) A new dasyurid marsupial from eastern Queensland, Australia: the buff-footed antechinus, *Antechinus mysticus* sp. nov. (Marsupialia: Dasyuridae). *Zootaxa* 3515, 1–37. [12 Oct 2012]

Baker AM, Mutton TY, Hines HB (2013) A new dasyurid marsupial from Kroombit Tops, south-east Queensland, Australia: the Silver-headed Antechinus, *Antechinus argentus* sp. nov. (Marsupialia: Dasyuridae). *Zootaxa* 3746, 201–239. [11 Dec 2013]

Baker AM, Mutton TY, Hines HB, Van Dyck S (2014) The Black-tailed Antechinus, *Antechinus arktos* sp. nov.: a new species of carnivorous marsupial from montane regions of the Tweed Volcano caldera, eastern Australia. *Zootaxa* 3765, 101–133. [17 Feb 2014]

Baker AN (1985) Pygmy right whales *Caperea marginata* (Gray, 1846). pp. 345–354, in SH Ridgway & R Harrison (eds.) *Handbook of Marine Mammals. Volume 3. The Sirenians and Baleen Whales*. Academic Press, London.

Baker AN (2001) Status, relationships, and distribution of *Mesoplodon bowdoini* Andrews, 1908 (Cetacea: Ziphiidae). *Marine Mammal Science* 17, 473–493. [Jul 2001]

Baker RJ, Bradley RD (2006) Speciation in mammals and the genetic species concept. *Journal of Mammalogy* 87, 643–662. [Aug 2006]

Balcomb KC (1989) Baird's beaked whale, *Berardius bairdii* Stejneger, 1883: Arnoux's beaked whale *Berardius arnouxii* Duvernoy, 1851. pp. 261–288, in SH Ridgway & R Harrison (eds.) *Handbook of Marine Mammals. Volume 4. River Dolphins and Larger Toothed Whales*. Academic Press, London.

Banks RC, Brownell RL (1969) Taxonomy of the common dolphins of the eastern Pacific Ocean. *Journal of Mammalogy* 50, 262–271.

Bannister JL (1988a) Balaenidae. pp. 215–216, in DW Walton (ed.) *Zoological Catalogue of Australia. 5. Mammalia*. Australian Government Publishing Service, Canberra. [13 Apr 1988]

Bannister JL (1988b) Physeteridae. pp. 198–200, in DW Walton (ed.) *Zoological Catalogue of Australia. 5. Mammalia*. Australian Government Publishing Service, Canberra. [13 Apr 1988]

Bannister JL (1988c) Delphinidae. pp. 201–210, in DW Walton (ed.) *Zoological Catalogue of Australia. 5. Mammalia*. Australian Government Publishing Service, Canberra. [13 Apr 1988]

Bannister JL (1988d) Phocoenidae. pp. 211, in DW Walton (ed.) *Zoological Catalogue of Australia. 5. Mammalia*. Australian Government Publishing Service, Canberra. [13 Apr 1988]

Bannister JL, Kemper CM, Warneke RM (1996) *The Action Plan for Australian Cetaceans*. Endangered Species Program. Project Number 30. Wildlife Australia, Canberra. [Sep 1996]

Barabash-Nikiforov II (1935) *Delphinus delphis ponticus* subsp. n. *Bulletin de la Société Impériale des Naturalistes de Moscou* 44, 246–249.

Barabash-Nikiforov II (1940) *Fauna kitsobraznykh Chernogo Moria, ee sostav i proiskhozhdenie*. [Whales of the Black Sea] Izdatel'stvo Voronezhkogo Gosudarstvennogo Universitera, Voronezh. Reference not seen.

Barber AD, Minelli A (2012) *Henia* C. Koch, 1847. In AD Barber (2012) World database of littoral Myriapoda. Accessed through: J van der Land (ed). (2008) UNESCO-IOC Register of Marine Organisms (URMO) at http://www.marinespecies.org/urmo/aphia.php?p=taxdetails&id=105480 on 2012–06–01.

Barghoorn SF (1977) New material of *Vespertiliavus* Schlosser (Mammalia: Chiroptera) and suggested relationships of emballonurid bats based on cranial morphology. *American Museum Novitates* 2618, 1–29. [28 Apr 1977]

Barkow HCL (1862) *Das Leben der Walle in seiner Begishung zum Athmen und zum Blutlauf: Nebst Bemerkungen über die Benennung der Finnwalle*. Breslau.

Barnard KH (1950) The dates of issue of the '*Illustrations of the Zoology of South Africa*' and the '*Marine Investigations in South Africa*'. *Journal of the Society for the Bibliography of Natural History* 2, 187–189.

Barnes LG (1971) Comments on the Phocenidae. *Abstracts with Programs -. Proceedings of the Geological Society of America* 3, 79.

Barnes LG (1978) A review of *Lophocetus* and *Liolithax* and their relationships to the delphinoid family Kentriodontidae (Cetacea: Odontoceti). *Bulletin of the Los Angeles County Museum of Natural History* 28, 1–35.

Barnes LG (1979) Fossil enaliarctine pinnipeds (Mammalia: Otariidae) from Pyramid Hill, Kern County, California. *Contributions in Science* 318, 1–41.

Barnes LG (1984) Fossil odontocetes (Mammalia: Cetacea) from the Almejas Formation, Isla Cedros, Mexico. *PaleoBios* 42, 1–46. [23 Jul 1984]

Barnes LG (1985) Evolution, taxonomy and antitropical distributions of the porpoises (Phocoenidae, Mammalia). *Marine Mammal Science* 1, 149–165. [Apr 1985]

Barnes LG (1989) A new Enaliarctine pinniped from the Astoria formation, Oregon, and a classification of the Otariidae (Mammalia: Carnivora). *Contributions in Science* 403, 1–26.

Barnes LG, McLeod SA (1984) The fossil record and phyletic relationships of gray whales. pp. 3–32, in ML Jones, SL Swartz & S Leatherwood (eds.) *The Gray Whale*. Academic Press, New York.

Barnes LG, Domning DP, Ray CE (1985) Status of studies on fossil marine mammals. *Marine Mammal Science* 1, 15–53. [Jan 1985]

Barrett-Hamilton GEH (1902) The natural history collections of the 'Southern Cross'. 1. Mammalia. pp. 1–66, in *Report on the Collections of Natural History made in the Antarctic Regions during the Voyage of the 'Southern Cross'*. British Museum, London.

Bartholomai A (1967) *Troposodon*, a new genus of fossils Macropodinae (Marsupialia). *Memoirs of the Queensland Museum* 15, 21–33. [23 Jun 1967]

Bartholomai, A. (1973a) The genus *Protemnodon* Owen (Marsupialia: Macropodidae) in the Upper Cainozoic deposits of Queensland. *Memoirs of the Queensland Museum* 16, 309–363. [30 Jun 1973]

Bartholomai A (1973b) *Fissuridon pearsoni*, a new fossil macropodid (Marsupialia) from Queensland. *Memoirs of the Queensland Museum* 16, 365–368. [30 Jun 1973]

Bartholomai A (1977) The fossil vertebrate fauna from Pleistocene deposits at Cement Mills, Gore, southeastern Queensland. *Memoirs of the Queensland Museum* 18, 41–51. [31 Mar 1977]

Bartholomai A, Marshall LG (1973) The identity of the supposed dasyurids marsupial, *Sarcophilus prior* De Vis, 1883, with comments on the reported 'Pliocene' occurrences of *Sarcophilus*. *Memoirs of the Queensland Museum* 16, 369–374. [30 Jun 1973]

Bastos AD, Nair D, Taylor PJ, Brettschneider H, Kirsten F, Mostert E, von Maltitz E, Lamb JM, van Hooft P, Belmain SR, Contrafatto G, Downs S, Chimimba CT (2011) Genetic monitoring detects an overlooked cryptic species and reveals the diversity and distribution of three invasive *Rattus* congeners in south Africa. *BMC Genetics* 12(26), 1–18.

Bates PJJ, Harrison DL (1997) *Bats of the Indian Subcontinent*. Harrison Zoological Museum Publications, Kent.

Bauer AM, Gunther R, Klipfel M (1995) *The herpetological contributions of Wilhelm C.H. Peters (1815–1883)*. Society for the Study of Amphibians and Reptiles in association with the Deutsche fur Herpetologie und Terrarienkunde.

Baverstock PR (1984a) The molecular relationships of Australasian possums and gliders. pp. 1–8, in A Smith & I Hume (eds.) *Possums and Gliders*. Surrey Beatty, Sydney. [Dec 1984]

Baverstock P (1984b) Australia's living rodents: A restrained explosion. pp. 913–919, in M Archer & G Clayton (eds.) *Vertebrate Zoogeography and Evolution in Australasia*. Hesperian Press, Carlisle, Western Australia.

Baverstock PR, Hogarth JT, Cole S, Covacevich J (1976) Biochemical and karyotypic evidence for the specific status of the rodent *Leggadina lakedownensis* Watts. *Transactions of the Royal Society of South Australia* 100, 109–112. [31 May 1976]

Baverstock PR, Watts CHS, Cole SR (1977) Electrophoretic comparisons between allopatric populations of five Australian pseudomyine rodents (Muridae). *Australian Journal of Biological Sciences* 30, 471–485.

Baverstock PR, Watts CHS, Adams M, Gelder M (1980) Chromosomal and electrophoretic studies of Australian *Melomys* (Rodentia: Muridae). *Australian Journal of Zoology* 28, 553–574.

Baverstock PR, Watts CHS, Adams M, Cole SR (1981) Genetical relationships among Australian rodents (Muridae). *Australian Journal of Zoology* 29, 289–303.

Baverstock PR, Archer M, Adams M, Richardson BJ (1982) Genetic relationships among 32 species of Australian dasyurid marsupials. pp. 641–650, in M Archer (ed.) *Carnivorous Marsupials*. Volume 2. Royal Zoological Society of NSW, Sydney. [Feb 1982]

Baverstock PR, Adams M, Archer M (1984) Electrophoretic resolution of species boundaries in the *Sminthopsis murina* complex (Dasyuridae). *Australian Journal of Zoology* 32, 823–832.

Baverstock PR, Adams M, Reardon T, Watts CHS (1987) Electrophoretic resolution of species boundaries in Australian Microchiroptera. III. The Nycticeiini - *Scotorepens* and *Scoteanax* (Chiroptera: Vespertilionidae). *Australian Journal of Biological Sciences* 40, 417–433.

Baverstock PR, Richardson BJ, Birrell J, Krieg M (1989) Albumin immunologic relationships of the Macropodidae (Marsupialia). *Systematic Zoology* 38, 38–50. [Mar 1989]

Baverstock PR, Krieg M, Birrell J (1990a) Evolutionary relationships of Australian marsupials as assessed by albumin immunology. *Australian Journal of Zoology* 37, 131–145.

Baverstock KPR, Flannery T, Aplin K, Birrell J, Krieg M (1990b) Albumin immunologic relationships of the bandicoots (Perameloidea Marsupialia) a preliminary report. pp. 13–18, in JH Seebeck, PR Brown, RL Wallis & CM Kemper (eds.) *Bandicoots and Bilbies*. Surrey Beatty & Sons, Sydney. [Dec 1990]

Baverstock PR, Krieg M, Birrell J, McKay GM (1990c) Albumin immunologic relationships of Australian marsupials. II. The Pseudocheiridae. *Australian Journal of Zoology* 38, 519–526.

Baverstock PR, Krieg M, Birrell J (1990d) Evolutionary relationships of Australian marsupials as assessed by albumin immunology. pp. 131–145, in JAM Graves, RM Hope & D Cooper (eds.) *Mammals from Pouches and Eggs: Genetics, Breeding and Evolution of Marsupials and Monotremes*. CSIRO Publishing, Melbourne.

Beasley I, Robertson KM, Arnold P (2005) Description of a new dolphin, the Australian Snubfin Dolphin *Orcaella heinsohni* sp. n. (Cetacea, Delphinidae). *Marine Mammal Science* 21, 365–400. [Jul 2005]

Beaver RA (1998) New synonymy, new combinations and taxonomic notes on Scolytidae and Platypodidae (Insecta: Coleoptera). *Annalen des Naturhistorischen Museums in Wien* 100B, 179–192.

Bechstein JM (1799) *Thomas Pennant's allgemeine Übersicht der Vierfüssigen Thiere. Aus dem englischen übersetzt aud mit Anmerkungen und Zusätzen versehen von J.M. Bechstein*. Industrie-Comptoir's, Weimar. Volume 1.

Bechstein JM (1800) *Thomas Pennant's allgemeine Übersicht der Vierfüssigen Thiere. Aus dem englischen übersetzt aud mit Anmerkungen und Zusätzen versehen von J.M. Bechstein*. Industrie-Comptoir's, Weimar. Volume 2.

Beck RMD (2008) A dated phylogeny of marsupials using a molecular supermatrix and multiple fossil constraints. *Journal of Mammalogy* 89, 175–189. [Feb 2008]

Beck RMD (2012) An 'ameridelphian' marsupial from the early Eocene of Australia supports a complex model of Southern Hemisphere marsupial biogeography. *Naturwissenschaften* 99, 715–729. [5 Aug 212]

Beck RMD, Bininda-Emonds ORP, Cardillo M, Liu F-GR, Purvis A (2006) A higher-level MRP supertree of placental mammals. *BMC Evolutionary Biology* 6(93), 1–14. [13 Nov 2006]

Beck RMD, Godthelp H, Weisbecker V, Archer M, Hand SJ (2008) Australia's oldest marsupial fossils and their biogeographical implications. *PLoS ONE* 3(3), 1–8. e1858. [26 Mar 2008]

Beck RMD, Travouillon KJ, Aplin KP, Godthelp H, Archer M (2014) The osteology and systematics of the enigmatic Australian Oligo-Miocene metatherian Yalkaparidon (Yalkaparidontidae; Yalkaparidontia; ?Australidelphia; Marsupialia). *Journal of Mammalian Evolution* 21, 127–172. [Published online 24 Sep 2013]

Becker L (1858) A small Australian bat. *The Argus*. [Melbourne], No. 3785, p. 5, col. 5. [29 Jul 1858]

Becker L (1859) On an Australian bat. No. 1. *Transactions of the Philosophical Institute of Victoria* 3, 38–40.

Becker L (1860) Notes on Australian bats, No. II. *Transactions of the Philosophical Institute of Victoria* 4, 41–43.

Beddard FE (1890) On the structure of Hooker's Sea Lion (*Arctocephalus hookeri*). *Transactions of the Zoological Society of London* 12, 369–380. [Jan 1890]

Beeton B, Grigg G, Harrison P, How R, Humphreys B, McKenzie N, Woinarski J (2010) Final report of the Christmas Island Expert Working Group to the Minister of the Environment Protection, Heritage and the Arts. Canberra.

Bell CH, Kemper CM, Conran JG (2002) Common dolphins *Delphinus delphis* in southern Australia: A morphometric study. *Australian Mammalogy* 24, 1–10. [16 Sep 2002]

Bell T (1829) Description of a new species of *Phalangista*. *Transactions of the Linnean Society of London* (1)16, 121–128. [12–19 Feb 1829]

Bell T (1836) Cheiroptera. pp. 594–600, in RB Todd (ed.) *The Cyclopaedia of Anatomy and Physiology*. Sherwood, Gilbert and Piper, London. Volume 1. A-DEA.

Bellamy CL (1997) Clarification of authorship of certain generic names of Buprestidae (Coleoptera). *African Entomology* 5, 217–224.

Benda P, Vallo P (2009) Taxonomic revision of the genus *Triaenops* (Chiroptera: Hipposideridae) with description of a new species from southern Arabia and definitions of a new genus and tribe. *Folia Zoologica* 58 (Monograph), 1–45.

Bennett AF, Lumsden LF (1995) Little Pygmy-possum *Cercartetus lepidus*. pp. 99–100, in PW Menkhorst (ed.) *Mammals of Victoria: Distribution, Ecology and Conservation*. Oxford University Press, Melbourne.

Bennett DK (1980) Stripes do not make a zebra, Part 1: A cladistic analysis of *Equus*. *Systematic Zoology* 29, 272–287. [Sep 1980]

Bennett DK, Hoffmann RS (1999) *Equus caballus*. *Mammalian Species* 628, 1–14. [3 Dec 1999]

Bennett ET (1833) On the Family Chinchillidae, and on a new genus referrible. [sic] to it. *Proceedings of the Zoological Society of London* 1, 57–60. [5 Jul 1833]

Bennett ET (1835) On a new species of kangaroo. *Proceedings of the Zoological Society of London* 2(1834), 151–152. [3 Apr 1835]

Bennett ET (1836) On a remarkable pteropine bat from the. [sic] Gambia. *Proceedings of the Zoological Society of London* 3(1835), 149. [12 Feb 1836]

Bennett G (1837) *A Catalogue of the Specimens of Natural History and Miscellaneous Curiosities Deposited in the Australian Museum*. Sydney.

Benshemesh J (2004) Recovery Plan for Marsupial Moles *Notoryctes typhlops* and *N. caurinus*. 2005–2010. Northern

Territory Department of Infrastructure, Planning and Environment, Alice Springs.

Bensley BA (1903) On the evolution of the Australian Marsupialia; with remarks on the relationships of the marsupials in general. *Transactions of the Linnean Society of London* (2)9, 83–217.

Benton MJ (2005) *Vertebrate Palaeontology*. Blackwell Publishing, Oxford. Third Edition.

Berg C (1898) *Lobodon carcinophagus* (H.J.) Gr. en el Rio de la Plata. *Communicaciones del Museo Nacional de Buenus Aires* 1, 15.

Bergmans W (2001) Notes on distribution and taxonomy of Australasian bats. 1. Pteropodinae and Nyctimeninae (Mammalia, Megchiroptera, Pteropodidae). *Beaufortia* 51, 119–152. [10 Dec 2001]

Bergmans W, Sarbini S (1985) Fruit bats of the genus *Dobsonia* Palmer, 1898 from the islands of Biak, Owii, Numfoor and Yapen, Irian Jaya (Mammalia; Megachiroptera). *Beaufortia* 34, 181–189. [31 Jan 1985]

Berkenhout J (1769) *Outlines of the Natural History of Great Britain and Ireland*. Containing a systematic arrangement and concise description of all the animals, vegetables, and fossils which have hitherto been discovered in these kingdoms. Comprehending the Animal Kingdom. P. Elmsly, London. Volume 1.

Berta A (1994) New specimens of the pinnipediform *Pteronarctos* from the Miocene of Oregon. *Smithsonian Contributions to Palaeontology* 78, 1–30.

Berta A (2009a) Pinniped evolution. pp. 861–868, in WF Perrin, B Würsig & JGM Thewissen (eds.) *Encyclopedia of Marine Mammals*. Academic Press, Amsterdam. Second Edition.

Berta A (2009b) Pinnipedia, Overview. pp. 878–885, in WF Perrin, B Würsig & JGM Thewissen (eds.) *Encyclopedia of Marine Mammals*. Academic Press, Amsterdam. Second Edition.

Berta A, Churchill M (2012) Pinniped taxonomy: review of currently recognised species and subspecies, and evidence used for their description. *Mammal Review* 42, 207–234. [On line 13 Sep 2011]

Berta A, Wyss AR (1994) Pinniped phylogeny. *Proceedings of the San Diego Society of Natural History* 29, 33–56.

Berta A, Ray CE, Wyss AR (1989) Skeleton of the oldest known pinniped *Enaliarctos mealsi*. *Science* 244(4900), 60–62. [7 Apr 1989]

Berthold AA (1827) In PA Latreille (ed.) *Natürliche Familien des Thierreichs* / Aus dem Französischen mit Anmerkungen und Zusätzen von Dr Arnold Adolph Berthold. Landes-Industrie-Compoirs, Weimar.

Bertoni A de W (1901) Aves Nuevas del Paraguay: Continuación á Azara. Talleres Nacionales de H. Kraus, Asunción.

Bertuch FJ (1801) *Bilderuch für Kinder*. Verlage des Industrie-Comtoirs, Weimer. Volume 1.

Berzin AA, Vladimirov VL (1982) Novyi vid kosatok iz Antarktiki. *Prioroda* 6, 31. [= A new species of killer whale from the Antarctic]

Berzin AA, Vladimirov VL (1983) Novyi vid kosatki (Cetacea, Delphinidae) iz vod Antarktiki. *Зоологический журнал* 62, 287–295. [= A new species of killer whale (Cetacea, Delphinidae) from the Antarctic waters]

Best PB (1960) Further information on Bryde's whale (*Balaenoptera edeni* Anderson) from Saldanha Bay, South Africa. *Norsk Hvalfangst-Tidende* 49, 201–215.

Best PB (1977) Two allopatric forms of Bryde's whales off South Africa. *Scientific Reports of the Whales Research Institute* (Special Issue 1), 10–38.

Best PB (1985) External characters of the southern minke whales and the existence of a diminutive form. *Scientific Reports of the Whales Research Institute* 36, 1–33.

Best PB (2007) *Whales and Dolphins of the Southern African Subregion*. Cambridge University Press, Cambridge.

Bianchi V (1917) Notes préliminaires sur les chauve-souris ou chiroptères de la Russie. Annuaire du Musée Zoologique de L'Académie des Sciences. [*Ezhegodnik Zoologicheskogo Muzeia Imperatorskoi Akademii Nauk - Annuaire du Musée zoologique de l'Académie impériale des sciences de St. Petersbourg*] 21 (1916), lxxiii–lxxxii.

Bianucci G, Landini W (2006) Killer sperm whale: a new basal physeteroid (Mammalia, Cetacea) from the late Miocene of Italy. *Zoological Journal of the Linnean Society* 148, 103–131.

Bianucci G, Landini W (2007) Fossil History. pp. 35–93, in DL Miller (ed.) *Reproductive Biology and Phylogeny of Cetacea: Whales, Dolphins and Porpoises*. Science Publishers, Enfield, NH. [6 Jan 2007]

Billberg GJ (1828) *Synopsis Faunae Scandinaviae*. Tome 1. Part 1. Mammalia. C. Doleen, Holmiae.

Bininda-Emonds ORP, Russell AP (1996) A morphological perspective on the phylogenetic relationships of the extant phocid seals (Mammalia: Carnivora: Phocidae). *Bonner Zoologische Monographien* 41, 1–256.

Bino R (1996) Notes on behaviour of New Guinea singing dogs (*Canis lupus dingo*). *Science in New Guinea* 22, 43–47.

Birula AA (1893) Beitrage zur acariden-fauna russlands. I *Rhyncholophus* (*Macropus*) *plumifer* n. sp. *Horae Societatis Entomologicae Rossicae* 27, 387–390.

Bisconti M (2008) Morphology and phylogenetic relationships of a new eschrichtiid genus (Cetacea: Mysticeti) from the early Pliocene of northern Italy. *Zoological Journal of the Linnean Society* 153, 161–186. [25 Apr 2008]

Bisconti M (2012) Comparative osteology and phylogenetic relationships of *Miocaperea pulchra*, the first fossil pygmy right whale genus and species (Cetacea, Mysticeti, Neobalaenidae). *Zoological Journal of the Linnean Society* 166, 876–911. [26 Nov 2012]

Black K (1999) Diversity and relationships of living and extinct koalas (Phascolarctidae, Marsupialia). *Australian Mammalogy* 21, 16–17. [24 Sep 1999]

Black KH, Archer M, Hand SJ, Godthelp H (2012) The rise of Australian marsupials: A synopsis of biostratigraphic, phylogenetic, palaeoecologic and palaeobiogeographic understanding. pp. 983–1078, in JA Talent (ed.) *Earth and Life: International Year of Planet Earth*. Springer, City.

Blacket MJ, Krajewski C, Labrinidis A, Cambron B, Cooper S, Westerman M (1999) Systematic relationships within the dasyurid marsupial tribe Sminthopsini – a multigene approach. *Molecular Phylogenetics and Evolution* 12, 140–155. [Jul 1999]

Blacket MJ, Adams M, Krajewski C, Westerman M (2000) Genetic variation within the dasyurid marsupial genus *Planigale*. *Australian Journal of Zoology* 48, 443–459.

Blacket MJ, Adams M, Cooper S, Krajewski C, Westerman M (2001) Systematics and evolution of the dasyurid marsupial genus *Sminthopsis:* I. The *Macroura* group. *Journal of Mammalian Evolution* 8, 149–170. [Jun 2001]

Blacket MJ, Cooper S, Krajewski C, Westerman M (2006) Systematics and evolution of the dasyurid marsupial genus *Sminthopsis:* II. The Murina species group. *Journal of Mammalian Evolution* 13, 125–138. [online 25 May 2006]

Blanford WT (1888) Critical notes on the nomenclature of Indian mammals. *Proceedings of the Zoological Society of London* 55(1887), 620–638. [Apr 1888]

Blasius JH (1853) Beschreibung zweier neuer deutscher fledermausarten. *Archiv für Naturgeschichte* 19, 35–57.

Bleeker P (1845) Bijdragen tot de geneeskundige Topographie van. Batavia. Generisch Overzicht Fauna. *Natuur- en Geneeskundig Archief voor Neerland's Indië* 2, 497–534.

Bleeker P (1858) Scripta Ichthyologiae Varia. Volume III. Scripta Ex Actis Societatis Regiae Scientiarum Indo-Neerlandicae, Vol-1–6. Ichthyologie Archipelagi Indici Prodromus. Siluri. Bataviae. Volume 1.

Bloch JI, Silcox MT, Boyer DM, Sargis EJ (2007) New Paleocene skeletons and the relationship of plesiadapiformes to crown-clade primates. *Proceedings of the National Academy of Sciences, USA* 104, 1159–1164. [17 Jan 2007]

Bloch ME (1788) Ueber zwey merkwürdige Fischarten. *Abhandlungen der Böhmischen Gesellschaft der Wissenschaften* (1787)3, 278–282.

Bloch ME (1797) *Systema Ichthyologiae iconibus* cx illustratum. Post obitum auctoris opus inchoatum absoluit, correxit, interpolavit Jo. Gottlob Schneider, Saxo, Berolini. Volume 12. Reference not seen.

Blochii ME, Schneider JG (1801) *Systema Ichthyologiae iconibus cx illustratum*. Post obitum auctoris opus inchoatum absolvit, correxit, interpolavit Jo. Gottlob Schneider, Saxo. Sumtibus Auctoris Impressum et Bibliopolio Sanderiano Commissum, Berolini.

Blumenbach JF (1779) *Handbuch der Naturgeschichte*. Johann Christian Dieterich, Göttingen. Volume 1.

Blumenbach JF (1791) *Handbuch der Naturgeschichte*. Johann Christian Dieterich, Göttingen. Fourth Edition.

Blumenbach JF (1800a) Über das Schnabelthier (*Ornithorhynchus paradoxus*) ein neuentdecktes Geschlecht von Säugthieren des fünften Welttheils. *Magazin für den Neuesten Zustand Naturkunde mit Rücksicht auf die dazu Gehörigen Hülfswissenschaften* 2, 205–214.

Blumenbach, JF (1800b) 19 April 1800. *Göttingische Anzeigen von gelehrten Sachen* (= *Göttingische Gelehrte Anzeigen*) 1, 609–612.

Blumstein DT, Daniel JC (2003) Foraging behavior of three Tasmanian macropodid marsupials in response to present and historical predation threat. *Ecography* 26, 585–594.

Blyth E (1840) Mammalia. pp. 38–152 in E Blyth, R Mudie, G Johnston & JO Westwood (eds.) *Cuvier's Animal Kingdom*, arranged according to its organization; forming the basis for a natural history of animals, and an introduction to comparative anatomy. W.M.S. Orr & Co., London.

Blyth E (1843) in Hodgson, B.H. Description of a new genus of Falconidae. *Journal of the Asiatic Society of Bengal, Calcutta* 12, 127–128.

Blyth E (1844) Notices of various Mammalia, with descriptions of many new species. *Journal of the Asiatic Society of Bengal, Calcutta* 13, 463–494.

Blyth E (1848) Report of Curator Zoological Department. *Journal of the Asiatic Society of Bengal, Calcutta* 17, 247–255.

Blyth E (1859a) Report of Curator, Zoological Department, for May, 1858. *Journal of the Asiatic Society of Bengal, Calcutta* 27, 267–290.

Blyth E (1859b) On the great rorqual of the Indian Ocean, with notices of other cetals, and of the Syrenia or marine pachyderms. *Journal of the Asiatic Society of Bengal, Calcutta* 28, 481–498.

Blyth E (1863) *Catalogue of the Mammalia in the Museum of the Asiatic Society*. Asiatic Society of Bengal, Calcutta.

Boas JEV (1886) Spolia Atlantica: Bidrag til pteropodernes. Morfologi og Systematik samt til Kundskaben om deres geografiske Udbredelse. *Det Kongelige Danske Videnskabernes Selskabs Skrifter. Naturvidenskabelig og Mathematisk Afdeling* 4(1), 1–231.

Boddaert P (1783) *Table des Planches Enluminéez d'Histoire Naturelle, de M. d'Aubenton / Avec les denominations de M.M. de Buffon, Brisson, Edwards, Linnaeus et Latham. Precedé d'une notice des principeux ouvrages zoologiques enluminés, &c.* Utrecht.

Boddaert P (1785) *Elenchus Animalium*. Volume 1. Sistens Quadrupedia huc usque nota eorumque varietates. C.R. Hake, Rotterdam.

Boeck AJ (1876) *De Skandinaviske og Arktiske Amphipoder*. Christiana. Fasc 2., 229, 712 (as Acanthozoma; also as Acanthoscelides; also as Acanthonotozoma p. 237).

Boetticher H von (1934) Die geographische Verbreitung der Robben (Pinnipedia). *Zeitschrift fur Saugetierkunde* 9, 359–368.

Bogdanowicz W, Owen RD (1992) Phylogenetic analyses of the bat family Rhinolophidae. *Journal of Zoological Systematics and Evolutionary Research* 30, 142–160.

Boie F (1830) Naturgeschichtliche Beyträge II *Leuconoë* Boie - Eine neue Fledermausgattung. *Isis von Oken* 23, 256–257.

Boie F (1831) Bemerkungen über species und einige ornithologische familien. *Isis von Oken* 24, 538–547.

Boitard P (1841) *Le Jardin des Plantes* description et moeurs des mammifères de la Ménagerie et du Muséum d'Histoire Naturelle. J.J. Dubochet, Paris. [dated 1842 = Aug 1841]

Boitard P (1842) *Le Jardin des Plantes* description et moeurs des mammifères de la Ménagerie et du Muséum d'Histoire Naturelle. Gustav Barba, Paris.

Bonaccorso FJ (1998) *Bats of Papua New Guinea*. Conservation International Tropical Field Guide Series. Conservation International, Washington, D.C.

Bonaparte CL (1831) *Saggio di una Distribuzione Metodica degli Animali Vertebrati*. Presso Antonio Boulzaler, Roma.

Bonaparte CL (1832) *Saggio d'una Distribuzione Metodica degli Animali Vertebrati a Sangue Freddo*. Presso Antonio Boulzaler, Roma.

Bonaparte CL (1832–1841) *Iconografia della Fauna Italica* per le Quattro classi degli animali vertebrati. Tomo 1. Mammiferi e Uccelli. Dalla Tipografia Salviucci, Roma.

Bonaparte CL (1838) Synopsis vertebratorum systematis. *Nuovi Annali delle Scienze Naturali, Bologna* 2(1), 105–133.

Bonaparte CL (1840) A new systematic arrangement of vertebrated animals. *Transactions of the Linnean Society of London* (1)18, 247–304. [25 Jun 1840]

Bonaparte CL (1845) *Catalogo Metodico dei Mammiferi Europei*. Giacomo Pirola, Milan.

Bonaparte, CL (1850a) *Conspectus systematis: Mastozoologiae*. Editio Altera Reformata. E.J. Brill, Batavia. Unpaginated table.

Bonaparte, CL (1850b) *Conspectus generum avium*. E.J. Brill, Lugduni Batavorum. Volume 1.

Bonaparte JF (1986) Sobre *Mesungulatum houssayi* y nuevos mamíferos Cretácicos de Patagonia, Argentina. *Actas IV Congreso Argentino Paleontologia, y Bioestratigrafía* 2, 48–61.

Bonelli FA (1810) Observations Entomologique. Première partie (cicindélètes et portion des carabiques). Tabula Synoptica exhibens genera Carabicorum in Sectiones et Stirpes disposita. *Mémoires de l'Académie Impériale des Sciences. Littérature et Beaux-Arts de Turin* 18, 21–78.

Bonnaterre JP (1789) *Tableau Encyclopédique et Méthodique des trois règnes de la nature: Cetologie*. Panckoucke, Paris. [29 Aug 1789]

Bonner WN (1988) What shall we call the weddell and ross seals? *Marine Mammal Science* 4, 75–77. [Jan 1988]

Bonnet R (1892) Über hypotrichosis congenita universalis. *Anatomische Hefte: Referate und Beiträge zur Anatomie und Entwickelungsgeschichte. Erste Abteilung Arbeiten aus Anatomischen Instituten* 1, 233–273.

Borchard P, Wight IA (2010) Using camera-trap data to model habitat use by bare-nosed wombats (*Vombatus ursinus*) and cattle (*Bos taurus*) in a south-eastern Australian agricultural riparian ecosystem. *Australian Mammalogy* 32, 16–22.

Borkhausen MB (1797) Deutsche Fauna, order kurzgefasste Naturgeschichte der Thiere Deutschlands. Erster Theil. Saugthiere und Vogel. Varrentrapp & Wenner, Frankfurt am Mayn. Volume 1.

Borowski GH (1781) *Gemeinnüzzige Naturgeschichte des Thierreichs : darinn die merkwürdigsten und nüzlichsten Thiere in systematischer Ordnung beschrieben und alle Geschlechter in Abbildungen nach der Natur vorgestellet werden*. Gottlieb August Lange, Berlin. Volume 2.

Boschma H (1938) On the teeth and some other particulars of the sperm whale (*Physeter macrocephalus* L.). *Temminckia* 3, 151–278.

Bouetel V, de Muizon C (2006) The anatomy and relationships of *Piscobalaena nanna* (Cetacea, Mysticeti), a Cetotheriidae s.s. from the early Pliocene of Peru. *Geodiversitas* 28, 319–395.

Bourret R (1951) Une nouvelle chauve-souris du Tonkin, *Rhinomegalophus paradoxolophus*. *Bulletin du Muséum National d'Histoire Naturelle, Paris* (2)33, 607–609.

Bowdich TE (1821) *An Analysis of the Natural Classifications of Mammalia, for the Use of Students and Travellers*. J. Smith, Paris.

Bowdich TE (1825) *Excursions in Madeira and Porto Santo*, during the autumn of 1823, while on his third voyage to Africa. By the late T. Edward Bowdich, Esq. conductor of the mission to Ashantee; to which is added, by Mrs. (S.) Bowdich. I. A narrative of the continuance of the voyage to its completion, together with subsequent occurrence from Mr. Bowdich's arrival in Africa to the period of his death. II. A description of the English settlements on the River Gambia. III. Appendix: containing zoological and botanical descriptions, and translations from the Arabic. George B. Whittaker, London.

Braby MF, Eastwood R, Murray N (2012) The subspecies concept in butterflies: has its application in taxonomy and conservation biology outlived its usefulness? *Biological Journal of the Linnean Society. Linnean Society of London* 106, 699–716. [Aug 2012]

Bradley RD, Baker RJ (2001) A test of the genetic species concept: Cytochrome-b sequences and mammals. *Journal of Mammalogy* 82, 960–973. [Nov 2001]

Branch TA, Mikhalev YA (2008) Regional differences in length at sexual maturity for female blue whales based on recovered Soviet whaling data. *Marine Mammal Science* 24, 690–703. [Jul 2008]

Brandle R (2010) *A Biological survey of the Eyre Peninsula, South Australia*. Department for Environment and Heritage, South Australia.

Brandt JF (1833a) Über den Zahnbau der Stellerschen Seekuh (*Rhytina stelleri*) nebst Bemerkungen zur Characteristik der in zwei Unterfamilien zu zerfällanden Familie der Pflanzenfressenden Cetacean. *Mémoires de l'Académie Impériale des Sciences de St. Pétersbourg* (6)2, 103–118.

Brandt JF (1833b) De solenodonte novo mammalium insectivorum genere. *Mémoires de L'Académie Impérial des Sciences de St Pétersbourg* 2(6), 459–478.

Brandt JF (1843a) In Zoologie et Physiology. *L'Institute, Journal Universal des Sciences et des Sociétés savants en France et a l'étranger. l'ère Section. Sciences mathématiques, physiques et naturelles* 499, 241. [20 Jul 1843]

Brandt JF (1843b) In Palaéontologie. *L'Institute, Journal Universal des Sciences et des Sociétés savants en France et à l'étranger. L'ère Section. Sciences mathématiques, physiques et naturelles* 502, 270. [10 August 1843]

Brandt JF (1843c) De Cetotherio, novo balaenarum familiae genere, in Rossia meridionali anti aliquot annos effoso. *Bulletin de la Classe Physico-Mathématique de L'Académie Impérial des Sciences de Saint-Pétersberg* 1, 145–148.

Brandt JF (1855) Beiträge zur nähern Kenntniss der Säugethiere Russland's. *Mémoires de l'Académie Impériale des Sciences de Saint Pétersbourg* 7(6), 1–365.

Brandt JF (1872) Über eine neue Classification der Bartenwale (Balaenoidea) mit Berucksichtigung der untergegangenen Gattungen derselben. *Bulletin de l'Académie Impériale des Sciences de St. Pétersbourg* 17, 113–130.

Brandt JF (1873a) Untersuchungen über die fossilen und subfossilen cetaceen Europa's. *Mémoires de l'Académie Impériale des Sciences de Saint-Pétersbourg* 20(7), 1–372.

Brandt JF (1873b) Einige Worte über die Eintheilung der Zahnwale (Odontoceti). *Bulletin de l'Académie Impériale des Sciences de St. Pétersbourg* 18, 575–577.

Brass LJ (1953) Results of the Archbold Expeditions. No. 68. Summary of the 1848 Cape York (Australia) Expedition. *Bulletin of the American Museum of Natural History* 102, 135–206. [16 Jun 1953]

Bravard DA (1885) Exámen crítico de los mamíferos y reptiles fósiles. *Anales del Museo Nacional de Historia Natural de Buenos Aires* 3(4), 95–174.

Bray DJ, Hoese DF (2006) Xiphiidae. pp. 1783–1784, in DF Hoese, DJ Bray, GR Allen, JR Paxton, A Wells & PL Beesley (eds.) *Zoological Catalogue of Australia. Volume 35. Part 3. Fishes*. CSIRO Publishing / Australian Biological Resources Study (ABRS), Melbourne and Canberra.

Bray DJ, Hoese DF, Paxton JR (2006) Diodontidae. pp. 1930–1935, in DF Hoese, DJ Bray, GR Allen, JR Paxton, A Wells & PL Beesley (eds.) *Zoological Catalogue of Australia. Volume 35. Part 3. Fishes*. CSIRO Publishing / Australian Biological Resources Study (ABRS), Melbourne and Canberra.

Brazenor CW (1932) A re-examination of *Gymnobelideus leadbeateri* McCoy. *Australian Zoologist* 7, 106–109.

Brazenor CW (1934a) A revision of the Australian jerboa mice. *Memoirs of the National Museum of Victoria* 8, 74–89. [Sep 1934]

Brazenor CW (1934b) A new species of mouse, *Pseudomys* (*Gyomys*), and a record of broad-toothed rat, *Mastacomys*, from Victoria. *Memoirs of the National Museum of Victoria* 8, 158–161. [Sep 1934]

Brazenor CW (1936a) Muridae recorded from Victoria. *Memoirs of the National Museum of Victoria* 10, 62–85. [Nov 1936]

Brazenor CW (1936b) Two new rats from Central Australia. *Memoirs of the National Museum of Victoria* 9, 5–8. [Nov 1936]

Brazenor CW (1936c) An extended description of *Pseudomys novae-hollandiae* with remarks on its affinity to *P. hermannsburgensis*. *Memoirs of the National Museum of Victoria* 9, 9–13. [Nov 1936]

Breed B, Ford F (2007) *Native Mice and Rats*. CSIRO Publishing, Melbourne.

Breed WG (2000) Taxonomic implications of variation in sperm head morphology of the Australian delicate mouse, *Pseudomys delicatulus*. *Australian Mammalogy* 21, 193–199. [15 Jun 2000]

Brehm AE (1877) *Brehms thierleben. Allgemeine kune des Thierreich. Erste Abtheilung – Säugethiere*. Verlag des Bibliographischen Instituts, Leipzig. Volume 3.

Brehm CL (1824a) *Lehrbuch der Naturgeschichte aller Europäischen Vögel*. August Schmid, Jena. Volume 2.

Brehm CL (1824b) *Ornis, oder das Neueste und Wichtigste der Vögelkunde in Verbindung mit mehreren Naturforschern*. August Schmid, Jena. Volume 1.

Brewer DT, Potter A, Skewes TD, Lyne V, Andersen J, Davies C, Taranto T, Heap AD, Murphy NE, Rochester WA, Fuller M, Donovan A (2009) Conservation Values in Commonwealth Waters of the Christmas and Cocos (Keeling) Islands Remote Australian Territories. CSIRO, Cleveland.

Briggs KT, Morejohn VG (1976) Dentition, cranial morphology and evolution in the elephant seals. *Mammalia* 40, 199–222. [Jan 1976]

Bright DE, Skidmore RE (2002) A Catalogue of Scolytidae and Platypodidae (Coleoptera), Supplement 2 (1995–1999). National Research Council, Ottawa.

Brisbin IL Jr, Coppinger RP, Feinstein MH, Austad SN (1994) The New Guinea singing dog: taxonomy, captive studies and conservation priorities. *Science in New Guinea* 20, 27–38.

Briscoe DA, Fox BJ, Ingleby S (1981) Genetic differentiation between *Pseudomys pilligaensis* and related *Pseudomys* (Rodentia: Muridae). *Australian Mammalogy* 4, 89–92. [13 May 1981]

Briscoe DA, Calaby JH, Close RL, Maynes GM, Murtagh CE, Sharman GB (1982) Isolation, introgression and genetic variation in rock-wallabies. pp. 73–87, in RH Groves & WDL Ride (eds.) *Species at Risk: Research in Australia*. Australian Academy of Science, Canberra.

Brisson MJ (1756) *Regnum Animale in Classes* IX Distributum, sive Synopsis Methodica Sistens generalem Animalium distributionem in Classes IX, & duarum primarum Classium, Quadrupedum scilicet & Cetaceorum, particularem divisionem in Ordines, Sectiones, Genera & Species. Cum brevi cujusque Species. Paris.

Brisson MJ (1760) *Ornithologie ou méthode contenant la division des oiseaux en ordres, sections, genres, especes & leurs variétés*. A laquelle on a joint une description exacte de chaque espèce, avec les citations des auteurs qui en ont traité, les noms qu'ils leur ont donnés, ceux que leur ont donnés les différentes nations, & les noms vulgaires. Ouvrage enrichi de figures en taille-douce. Quay des Augustins, Paris. Volume 1.

Brisson MJ (1762) *Regnum Animale in Classes* IX. Distributum, sive Synopsis Methodica. Sistens generalem Animalium distributionem in Classes IX, & duarum primarum Classium, Quadrupedum scilicet & Cetaceorum, particularem divisionem in Ordines, Sectiones, Genera & Species. Cum brevi cujusque Speciei Descriptione, Cicationibus Auctorum de iis tractantium, Nominibus eis ab ipsis & Nationibus impositis, Nominibusque vulgaribus. Theodorum Haak, Lugduni Batavorum.

Bronn (1835) No title. *Neues Jahrbuch für Mineralogie, Geognosie, Geologie und Petrefaktenkunde, Stuttgart* 1835, 334.

Brookes J (1827) In JE Gray. Synopsis of the species of the class Mammalia. pp. 1–391, Volume 5, in E Griffith, CH Smith & E Pidgeon (eds.)(16 vols: 1824–1835) *The Animal Kingdom arranged in conformity with its organisation, by the Baron Cuvier, member of the Institute of France etc, with additional descriptions of all the species hitherto named, and of many not before noticed*. G.B. Whittaker & Co., London. [Sep 1827] Author of volume given as J.E. Gray by Anon (1903: 410).

Brookes J (1828) *A Catalogue of the Anatomical and Zoological Museum of Joshua Brookes, Esq., F.R.S. F.L.S. &c.* Richard Taylor, London.

Brookes J (1829) On a new genus of the Order Rodentia. *Transactions of the Linnean Society of London* (1)16, 95–104. [12–19 Feb 1829]

Broom R (1895a) On a new fossil marsupial allied to *Hypsiprymnus*, but resembling in some points the Plagiaulacidae. *Abstract of Proceedings of the Linnean Society of New South Wales* 26, ii. [for 26 Jun 1895]

Broom R (1895b) Linnean Society of New South Wales. *Zoologischer Anzeiger* 18, 371–372. [16 Sep 1895]

Broom R (1895c) On a new fossil mammal allied to *Hypsiprymnus*, but resembling in some points the Plagiaulacidae. *Proceedings of the Linnean Society of New South Wales* 10(2), 373. [18 Nov 1895]

Broom R (1896a) Report on a bone breccia deposit near Wombeyan Caves, NSW: with descriptions of some new species of marsupials. *Proceedings of the Linnean Society of New South Wales* 21, 48–61. [16 Jul 1896]

Broom R (1896b) Linnean Society of New South Wales. *Zoologischer Anzeiger* 19, 47–48. [30 Jan 1896]

Broom R (1896c) On a small fossil marsupial with large grooved premolars. *Proceedings of the Linnean Society of New South Wales* (2)10, 563–567. [29 Apr 1896]

Broom R (1896d) On a small marsupial allied to *Petaurus*. *Proceedings of the Linnean Society of New South Wales* (2)10, 568–570. [31 Jan 1896]

Broom R (1898) On the affinities and habits of *Thylacoleo*. *Proceedings of the Linnean Society of New South Wales* 23, 57–74. [23 Jun 1898]

Broom R (1915) On the organ of Jacobsen and its relations in the 'Insectivora'. Part II. *Talpa, Centetes* and *Chrysochloris*. *Proceedings of the Zoological Society of London* 85, 347–354. [17 Sep 1915]

Broom R (1935) A further contribution to our knowledge of the structure of the mammalian basicranial axis. *Annals of the Transvaal Museum, Pretoria* 18, 33–36.

Brothers DJ (1983) Nomenclature at the ordinal and higher ranks. *Systematic Zoology* 32, 34–42. [Mar 1983]

Broun T (1880) *Manual of the New Zealand Coleoptera*. James Hughes, Wellington. Volume 1.

Brown M, Cooksley H, Carthew SM, Coopers SJB (2006) Conservation units and phylogenetic structure of an arboreal marsupial, the yellow-bellied glider (*Petaurus australis*). *Australian Journal of Zoology* 54, 305–317.

Brown T (1832) *The Zoologist's Text-Book*. Embracing the Characters of the Classes, Orders, and Genera of Almost the Whole Animal Kingdom with an example of a species of nearly each genus, and a complete glossary of technical terms. Archbold Fullarton & Co., Glasgow. Volume 1.

Brown WL (1973) A comparison of the hylean and Congo-west African rain forest ant faunas. pp. 161–185, in BJ Meggers, ES Ayensu & WD Duckworth (ed.) *Tropical Forest Ecosystems in Africa and South America: A Comparative Review*. Smithsonian Institution Press, Washington, D.C.

Brownell RL Jr (1974) Small odontocetes of the Antarctic. pp. 13–19, in SG Brown, R Brownell Jn, AW Erickson, RJ Hofman, GA Llano & NA Mackintosh (eds.) *Antarctic Mammals*. American Geographic Society.

Brownell RL Jr (1975) *Phocoena dioptrica*. *Mammalian Species* 66, 1–3. [21 Nov 1975]

Brunner S (1998) A cranial morphometrics of the southern fur seals *Arctocephalus fosteri* and *A. pusillus* (Carnivora: Otariidae). *Australian Journal of Zoology* 46, 67–108.

Brunner S (2004) Fur seals and sea lions (Otariidae): identification of species and taxonomic review. *Systematics and Biodiversity* 1, 339–439.

Brünnich MT (1772) *Zoologiae fundamenta praelectionibus academicis accomodata. Grunde I dyrelaeren*. F.C. Pelt, Hafnae et Lipsiae.

Bryden MM (1968) Control of growth in two populations of elephant seal. *Nature* 217(5134), 1106–1108. [23 Mar 1968]

Bryden MM, Dawbin WH, Heinsohn GE, Bron DH (1977) Melon-headed whale, *Peponocephala electra*, on the east coast of Australia. *Journal of Mammalogy* 58, 180–187. [May 1977]

Buchannan F (1800) Description of the *Vespertilio plicatus*. *Transactions of the Linnean Society of London* (1)5, 261–263. [20–22 Feb 1800]

Buchecker H (1876–1880) *Systema Entomologiae sistens insectorum classes, genera, species*. Volume 1. *Lepidoptera*. München.

Buckton GB (1901) *A Monograph of the Membracidae*. Lovell Reeve & Co., London.

Burbidge AA, McKenzie NL, Brennan KEC, Woinarski JCZ, Dickman CR, Baynes A, Gordon G, Menkhorst PW, Robinson AC (2008) Conservation status and biogeography of Australia's terrestrial mammals. *Australian Journal of Zoology* 56, 411–422.

Burbidge AA, Eldridge MDB, Groves C, Harrison PL, Jackson SM, Reardon TB, Westerman M, Woinarski JCZ (2014) A list of native Australian mammal species and subspecies. pp. 15–32, in JCZ Woinarski, AA Burbidge & PL Harrison (eds.) *The Action Plan for Australian Mammals 2012*. CSIRO Publishing, Melbourne. [2 Jun 2014]

Burk A, Springer MS (2000) Intergeneric relationships among Macropodoidea (Metatheria: Diprotodontia) and the chronicle of kangaroo evolution. *Journal of Mammalian Evolution* 7, 213–237. [Dec 2000]

Burk A, Westerman M, Springer M (1998) The phylogenetic position of the musky rat-kangaroo and the evolution of bipedal hopping in kangaroos. (Macropodidae: Diprotodontidae). *Systematic Biology* 47, 457–474. [Sep 1998]

Burmeister CHC (1837) *Handbuch der Naturgeschichte zum Gebrauch bei Vorlesungen. II. Zoologie*. Berlin.

Burmeister H (1854) *Systematisch Uebersicht der Thiere Brasiliens*, welche während einer Reise durch die Provinzen von Rio de Janeiro und Minas Geraës gesammelt oder beobachtet wurden. Ester theil, Säugethiere (Mammalia). G. Reimer, Berlin.

Burmeister H (1856) *Erläuterungen zur Fauna Brasiliens, enthaltend Abbildungen und ausfuhrliche Beschreibungen neuer oder ungenügend bekannter Thier-Arten*. G. Reimer, Berlin.

Burmeister H (1865a) Description of a new species of whale. *Proceedings of the Zoological Society of London* 33, 190–191. [Jun 1865]

Burmeister H (1865b) *Delphinorhynchus australis* n. sp. *Zeitschrift fur die Gesammten Naturwissenschaften, Berlin* 26, 262–263.

Burmeister H (1865c) Description of a new species of porpoise in the museum of Buenos Ayres. *Proceedings of the Zoological Society of London* 33, 228–231. [Jun 1865]

Burmeister H (1866a) Preliminary account of a new cetacean captured on the shore of Buenos Ayres. *Annals & Magazine of Natural History* (3)17, 94–98. [1 Feb 1866]

Burmeister H (1866b) On some cetaceans. *Annals & Magazine of Natural History* (3)18, 99–103. [1 Aug 1866]

Burmeister H (1867a) No title. *Actas de la Sociedad Paleontologica de Buenos Aires* 1867, xxiv–xxv.

Burmeister H (1867b) Fauna Argentina: Segunda Part. Mammifera Pinnata Argentina. *Anales del Museo Público de Buenos Aires* 1, 301–311.

Burmeister H (1868) Preliminary description of a new species of finner whale (*Balaenoptera bonaërensis*). *Proceedings of the Zoological Society of London* 35(1867), 707–713. [Apr 1868]

Burmeister H (1869) Descripcion de cuatro espèces de Delphinides. *Anales del Museo Publico de Buenos Aires* 1, 367–445.

Burmeister H (1871) Sobre *Balaenoptera intermedia, B. patachonica* y *B. bonaerensis* y sus caracteres distintivos. *Boletín del Museo Publico de Buenos Aires* 2, xi–xv.

Burmeister H (1872) On *Balaenoptera patachonica* and *B. intermedia. Annals & Magazine of Natural History* (4)10, 413–418. [1 Dec 1872]

Burnett GT (1829) Illustrations of the Alipeda (alipeds), or bats and their allies; being the arrangement of the Cheiroptera, Volitantia, or wing-footed beasts. *Quarterly Journal of Science, Literature, and Art* Jan–Jun 1829, 262–269.

Burnett GT (1830a) Illustrations of the Herpornitherae; or the arrangement of the *Ornithorhynchus* and *Echidna*, indicated in outline. *Quarterly Journal of Science, Literature, and Art* Jan–Jun 1830, 362–367.

Burnett GT (1830b) Illustrations of the Quadrupeda, or quadrupeds; being the arrangement of the true four-footed beasts indicated in outline. *Quarterly Journal of Science, Literature, and Art* Jul–Dec 1830, 336–353.

Burnett GT (1830c) Illustrations of the Cetetherae, including the Loripeda, Semipeda and Pinnipeda or loripeds, semipeds and pinnipeds: being the arrangement of the seals, dugong, whales and their allies. *Quarterly Journal of Science, Literature, and Art* Jan–Jun 1830, 355–361.

Burns JJ, Fay FH (1970) Comparative morphology of the skull of the ribbon seal, *Histriophoca fasciata*, with remarks on systematics of Phocidae. *Journal of Zoology* 161, 363–394.

Butler JRA, du Toit JT, Bingham J (2004) Free-ranging domestic dogs (*Canis familiaris*) as predators and prey in rural Zimbabwe: threats of competition and disease to large wild carnivores. *Biological Conservation* 115, 369–378.

Butler PM (1939) The teeth of the Jurassic mammals. *Proceedings of the Zoological Society of London* 109B, 329–356.

Butler PM (1978) A new interpretation of the mammalian teeth of tribosphenic pattern from the Albion of Texas. *Museum of Comparative Zoology. Breviora* 446, 1–27.

Caballero S, Trujillo F, Vianna JA, Barrios-Garrido H, Montiel MG, Beltrán-Pedreros S, Marmontel M, Santos MC, Rossi-Santos M, Santos FR, Baker CS (2007) Taxonomic status of the genus *Sotalia*: Species-level ranking for 'tucuxi' (*Sotali fluviatilis*) and 'costero' (*Sotalia guianensis*) dolphins. *Marine Mammal Science* 23, 358–386. [Apr 2007]

Caballero S, Jackson J, Mignucci-Giannoni AA, Barrios-Garrido H, Beltrán-Pedreros S, Montiel-Villalobos MA, Robertson KM, Baker CS (2008) Molecular systematics of South American dolphins *Sotalia*: Sister taxa determination and phylogenetic relationships, with insights into a multi-locus phylogeny of the Delphinidae. *Molecular Phylogenetics and Evolution* 46, 252–268.

Cabrera A (1919) *Genera Mammalium. Monotremata, Marsupialia*. Museo Nacional de Ciencias Naturales, Madrid.

Cabrera A (1922) *Manual de Mastozoologia*. Calpe, Madrid.

Cabrera A (1931) On some South American canine genera. *Journal of Mammalogy* 12, 54–67. [Feb 1931]

Cabrera A (1958) Catálogo de los mamíferos de América del Sur. Catálogo de los mamíferos de América del Sur. *Revista del Museo Argentino de Ciencias Naturales 'Bernardino Rivadavia', e Instituto Nacional de Investigación de las Ciencias Naturales. Ciencias Zoológicas* 4(1), 1–307. [27 Mar 1958]

Cahn P (1906) Eine neue Form des Roten Riesenkänguruh (*Macropus occidentalis*). *Zoologischer Beobachter* 47, 381.

Cahn P (1907) Känguruhs im Frankfurter Zoologichen Garten. *Zoologische Beobachter der Zoologische Garten* 48, 1–6.

Calaby JH (1954) Comments on Gilbert's note-book on marsupials. *Western Australian Naturalist* 4, 147–148. [15 Sep 1954]

Calaby JH (1966) Mammals of the Upper Richmond and Clarence Rivers, New South Wales. *Technical Papers of the Division of Wildlife Survey, CSIRO* 10, 1–55.

Calaby J (1971) The current status of Australian Macropodidae. *Australian Zoologist* 16, 17–29.

Calaby JH, Richardson BJ (1988a) Potoroidae. pp. 53–59, in DW Walton (ed.) *Zoological Catalogue of Australia. 5. Mammalia*. Australian Government Publishing Service, Canberra. [13 Apr 1988]

Calaby JH, Richardson BJ (1988b) Macropodidae. pp. 60–80, in DW Walton (ed.) *Zoological Catalogue of Australia. 5. Mammalia*. Australian Government Publishing Service, Canberra. [13 Apr 1988]

Calaby JH, Taylor JM (1974) Type locality of *Rattus villosissimus. Australian Mammalogy* 1, 267–268. [Oct 1974]

Calaby JH, Taylor M (1981) Reproduction in two marsupial-mice *Antechinus bellus* and *Antechinus bilarni* (Dasyuridae), of tropical Australia. *Journal of Mammalogy* 62, 329–341. [May 1981]

Calaby JH, Mack G, Ride WDL (1962) The application of the generic name *Macropus* Shaw, 1790 and of other names commonly referred to the grey kangaroos. *Memoirs of the Queensland Museum* 14, 25–31. [4 Jul 1962]

Calaby JH, Mack G, Ride WDL (1963) The generic name *Macropus* Shaw, 1790 (Mammalia). Z.N. (S.) 1584. *Bulletin of Zoological Nomenclature* 20, 376–379. [21 Oct 1963]

Calaby JH, Horner BE, Taylor JM (1966) *Acanthomys leucopus* Gray 1867 (Mammalia): proposed validation under the plenary powers. Z.N.(S.) 1724. *Bulletin of Zoological Nomenclature* 22, 330–331. [31 Jan 1966]

Calaby JH, Corbett LK, Sharman GB, Johnston PG (1974) The chromosomes and systematic position of the marsupial mole, *Notoryctes typhlops. Australian Journal of Biological Sciences* 27, 529–532.

Caldwell DK, Caldwell MC (1989) Pygmy sperm whale *Kogia breviceps* (de Blainville, 1838): Dwarf sperm whale *Kogia simus* Owen, 1866. pp. 235–260, in SH Ridgway & R Harrison (eds.) *Handbook of Marine Mammals. Volume 4. River Dolphins and the Larger Toothed Whales*. Academic Press, London.

Cambridge OP (1875) Myriopoda. *Zoological Research* 12, 262–264.

Cambridge OP (1898) Note on the generic name *Eatonia. Proceedings of the Zoological Society of London* 66, 348. [Aug 1898]

Campbell AJ (1888) Field Naturalist's Club of Victoria expedition to King Island, November 1887. *Victorian Naturalist* 4, 129–164. [Jan 1888]

Canestrini G, Fanzago F (1878 [1877–1878]) Intorno agli acari Italiani. *Atti del Reale Istituto Veneto di Scienze, Lettere ed Arti* (5)4, 69–208. Reference not seen.

Cantor T (1846) Catalogue of Mammalia inhabiting the Malayan Peninsula and Islands. *Journal of the Asiatic Society of Bengal* 15, 171–203.

Capellini G (1877) Sulla balenottera di Mondini: rorqual de la mer Adriatique di G. Cuvier. *Memorie Accademia delle Scienze dell'Instituto di Bologna* 3(7), 413–448.

Cardillo M, Bininda-Edmonds ORP, Boakes E, Purvis A (2004) A species level phylogenetic supertree of marsupials. *Journal of Zoology* 264, 11–31.

Cardinal BR, Christidis L (2000) Mitochondrial DNA and morphology reveal three geographically distinct lineages of the large bentwing bat (*Miniopterus schreibersii*) in Australia. *Australian Journal of Zoology* 48, 1–19.

Carleton MD (1984) Introduction to Rodents. pp. 255–265, in S Anderson & JK Jones (eds.) *Orders and Families of Recent Mammals of the World*. John Wiley & Sons, New York.

Carleton MD, Musser GG (1984) Muroid rodents. pp. 289–379, in S Anderson & JK Jones (eds.) *Orders and Families of Recent Mammals of the World*. John Wiley & Sons, New York.

Carpenter PP (1864a) Supplementary report on the present state of our knowledge with regard to the Mollusca of the west coast of North America. pp. 517–686, in *Report of the Thirty-Third Meeting of the British Association for the Advancement of Science*. John Murray, London. Volume 33. 1863.

Carpenter PP (1864b) On new Californian marine shells. No. II. *Proceedings of the California Academy of Natural Science* 3, 175–177. [Dec 1874]

Carpenter SM, McKean JL, Richards GC (1978) Multivariate morphometric analysis of *Eptesicus* (Mammalia: Chiroptera) in Australia. *Australian Journal of Zoology* 26, 629–638.

Carr SG, Robinson AC (1997) The present status and distribution of the desert rat-kangaroo *Caloprymnus campestris* (Marsupialia: Potoroidae). *South Australian Naturalist* 72, 4–27. [Sep–Dec 1997]

Carrick R, Csordas SE, Ingham SE (1962) Studies on the southern elephant seal *Mirounga leonina* (L.). IV. Breeding and development. *Wildlife Research* 7, 161–197.

Carter TD, Hill JE, Tate GHH (1945) *Mammals of the Pacific World*. New York.

Carus JV (1868) *Handbuch der Zoologie. Volume 1. Wirbelthiere, Mollusken und Molluscoiden*. Leipzig, Wilhelm Engelmann.

Carus JV (1889–1893) *Prodromus Faunae Mediterraneae sive descriptio animalium maris Mediterranei incolarum. Volume 2: Brachiostomata. Mollusca. Tunicata. Vertebrata*. E. Schweitzerbart, Stuttgart.

Casari SA (2008) A phylogenetic study of the subtribe Deicrepidiina (Elateridae, Elaterinae, Ampedini). *Revista Brasileira de Entomologia* 52, 182–260.

Case JA (1984) A new genus of Potoroinae (Marsupialia: Macropodidae) from the Miocene Ngapakaldi Local Fauna, South Australia, and a definition of the Potoroinae. *Journal of Paleontology* 58, 1074–1086. [Jul 1984]

Cassin J (1858) *United States Exploring Expedition during the years 1838, 1839, 1840, 1841*, under the command of Charles Wilkes, United States Navy. Volume 8. Mammalogy and Ornithology. C. Sherman & Son, Philadelphia. Reference not seen.

Caughley G, Sheppard N, Short J (1987) *Kangaroos: Their Ecology and Management in the Sheep Rangelands of Australia*. Cambridge University Press, Cambridge.

Chambers VT (1873) Micro-Lepidoptera. *Canadian Entomologist* 5, 110–115.

Chapin EA (1920) New American Cleridae, with note on the synonymy of *Micropterus* Chevr (Coleoptera). *Proceedings of the Entomological Society of Washington* 22(3), 50–54. [Mar 1920]

Chapskii KK (1955) Opyt peresmotra sistemy I diagnostiki tyulenei podsemeistva Phocinae. [An attempt at revision of the systematics and diagnoses of seals of the Subfamily Phocinae] *Trudy Zoologicheskovo Instituta. Akademiya Nauk, Leningrad* 17, 160–199.

Charlton-Robb K, Taylor AC, McKechnie SW (2006) A note on divergent mtDNA lineages of bottlenose dolphins from coastal waters of southern Australia. *The Journal of Cetacean Research and Management* 8, 173–179.

Charlton-Robb K, Gershwin LA, Thompson R, Austin J, Owen K, McKechnie S (2011) A new dolphin species, the Burrunan Dolphin *Tursiops australis* sp. nov., endemic to Southern Australian coastal waters. *PLoS ONE* 6(9), e24047.

Chasen FN (1939) The mammals of the Netherlands Indian Mt. Leuder Expedition 1937 to north Sumatra. *Treubia* 17(3), 479–502.

Chasen FN (1940) A handlist of Malaysian mammals: A Systematic List of the Mammals of the Malay Peninsula, Sumatra, Borneo and Java, including the adjacent small islands. *Bulletin of the Raffles Museum* 15, 1–209.

Chevrolat A (1842) Descriptions de quelque térédiles de l'Afrique australe du voyage de M. Drege. *Revue Zoologique par La Société Cuvierienne* 1842, 276–278.

Chimimba CT, Kitchener DJ (1991) A systematic revision of Australian Emballonuridae (Mammalia: Chiroptera). *Records of the Western Australian Museum* 15, 203–265. [31 Jan 1991]

Chivers SJ, Leduc RG, Robertson KM, Barros NB, Dizon AE (2005) Genetic variation of *Kogia* spp. With preliminary evidence for two species of *Kogia sima*. *Marine Mammal Science* 21, 619–634.

Chow M, Rich TH (1982) *Shuotherium dongi*, n. gen. and sp., a therian with Pseudo-Tribosphenic molars from the Jurassic of Sichuan, China. *Australian Mammalogy* 5, 127–142. [10 May 1982]

Chruszcz B, Barclay RMR (2002) *Chalinolobus gouldii*. *Mammalian Species* 690, 1–4. [5 Jul 2002]

Churchill S (1998) *Australian Bats*. Reed New Holland, Sydney.

Churchill S (2008) *Australian Bats*. Reed New Holland, Sydney. Second Edition.

Clague CI, Coles RB, Whybird OJ, Spencer HJ, Flemons P (1999) The occurrence and distribution of the tube-nosed insectivorous bat (*Murina florium*) in Australia). *Proceedings of the Linnean Society of New South Wales* 121, 175–191. [28 Dec 1999]

Clapham PJ, Mead JG (1999) *Megaptera novaeangliae*. *Mammalian Species* 604, 1–9. [5 May 1999]

Clark JW (1874) On the eared seals of the Auckland Islands. *Proceedings of the Zoological Society of London* 41(1873), 750–760. [Apr 1874]

Clarke R (2004) Pygmy fin whales. *Marine Mammal Science* 20, 329–334. [Apr 2004]

Clayton M, Wombey JC, Mason IJ, Chesser RT, Wells A (2006) *CSIRO List of Australian Vertebrates: A Reference with Conservation Status*. SCIRO Publishing, Melbourne.

Close RL, Murray JD, Briscoe DA (1990) Electrophoretic and chromosome surveys of the taxa of short-nosed bandicoots within the genus *Isoodon*. pp. 19–27, in JH Seebeck, PR Brown, RL Wallis & CM Kemper (eds.) *Bandicoots and Bilbies*. Surrey Beatty, Sydney. [Dec 1990]

Clutton-Brock J, Corbet GB, Hills M (1976) A review of the family Canidae, with a classification by numerical methods. *Bulletin of the British Museum (Natural History). Zoology Series* 29, 117–199.

Cocco A (1846) Euber cinen in der meerenge von messina gefnndenen Delphin. *Archiv für Naturgeschichte* 1(12), 104–108.

Cockerell TDA (1905) Preoccupied names in Coleoptera. *The Entomologist* 38, 104.

Cohn L (1910) Die papuasischen *Perameles*-Arten. *Zoologischer Anzeiger* 35, 718–728.

Colgan DJ, Flannery TF (1992) Biochemical systematics studies in the genus *Petaurus* (Marsupialia: Petauridae). *Australian Journal of Zoology* 40, 245–256.

Colgan DJ, Flannery TF, Trimble J, Aplin K (1993) Electrophoretic and morphological analysis of the systematics of the *Phalanger orientalis* species complex in Papua New Guinea and the Solomon Islands. *Australian Journal of Zoology* 41, 355–378.

Collett R (1884a) *Echidna acanthion*, en sandsynligvis ubeskreven art myre-pindsvin fra Nord-Queensland. *Forhandlinger Videnskabs – Selskabot Kristiana* 1884(13), 1–12.

Collett R (1884b) On some apparently new marsupials from Queensland. *Proceedings of the Zoological Society of London* 52, 381–389. [Oct 1884]

Collett R (1887a) On *Phascogale virginiae*, a rare pouched mouse from Northern Queensland. *Proceedings of the Zoological Society of London* 54(1886), 548–549. [Apr 1887]

Collett R (1887b) On a collection of mammals from central and northern Queensland. *Zoologische Jahrbücher – Zeitschrift für Systematik. Geographie und Biologie der Thiere* 2, 829–940.

Collett R (1895) On a new *Pseudochirus* from N.W. Australia. *Zoologischer Anzeiger* 18, 464–468. [2 Dec 1895]

Collett R (1897) On a collection of mammals from north and north-west Australia. *Proceedings of the Zoological Society of London* 65, 317–336. [Aug 1897]

Collins BB (2003) Phylogeography of brushtail possums (*Trichosurus*). B.Sc. (Hons.) thesis. School of Tropical Biology, James Cook University, Townsville.

Collins D (1802) *An Account of the English Colony in New South Wales: from its first settlement in January 1788, to August 1801, with remarks on the disposition, customs, manners, &c. of the native inhabitants of that country*. Cadell Jun. & W. Davies, London. Volume 2.

Collins LD (1973) *Monotremes and Marsupials: A Reference for Zoological Institutions*. Smithsonian Institution Press, Washington D.C.

Committee on Taxonomy (2011) List of marine mammal species and subspecies. Society for Marine Mammalogy, www.marinemammalscience.org, consulted on 23 November 2011.

Committee on Taxonomy (2014) List of marine mammal species and subspecies. Society for Marine Mammalogy, www.marinemammalscience.org, consulted on 19 November 2014.

Common IFB, Moulds MS (1973) The date of publication of Captain Phillip King's narrative of a survey of the intertropical and western coasts of Australia. *Journal of the Society for the Bibliography of Natural History* 6, 257–259. [Apr 1973]

Condon T (1906) A new fossil pinniped (*Desmatophoca oregonensis*) from the Miocene of the Oregon coast. *University of Oregon Bulletin*, supplement 3(3), 1–14. [May 1906]

Conisbee LR (1953) *A List of the Names Proposed for Genera and Subgenera of Recent Mammals from the Publication of T.S. Palmer's 'Index Generum Mammalium' 1904 to the end of 1951*. British Museum Natural History, London.

Conroy CJ, Rowe KC, Rowe KMC, Kamath PL, Aplin KP, Hui L, James DK, Moritz C, Patton JL (2012) Cryptic genetic diversity in *Rattus* of the San Francisco Bay region, California. *Biological Invasions* 15, 741–758. [12 Sep 2012]

Cook J, King J (1784) *A Voyage to the Pacific Ocean, undertaken by Command of His Majesty, for making discoveries in the Northern Hemisphere*. Performed under the Direction of Captains Cook, Clerk, and Gore, in the years 1776, 1777, 1778, and 1780. Being a copious, comprehensive, and satisfactory abridgement of the Voyage. London. Third Voyage.

Cook OF (1914) Terms relating to generic types. *American Naturalist* 48, 308–314. [May 1914]

Cooper CE (2011) *Myrmecobius fasciatus* (Dasyuromorphia: Myrmecobiidae). *Mammalian Species* 43(881), 129–140. [28 Jul 2011]

Cooper DW, Woolley PA (1983) Confirmation of a new species of small dasyurid marsupials by electrophoretic analysis of enzymes and proteins. *Australian Journal of Zoology* 31, 743–751.

Cooper GA (1942) New genera of North American brachiopods. *Journal of the Washington Academy of Sciences* 32(8), 228–235.

Cooper NK, Aplin KP, Adams M (2000) A new species of false antechinus (Marsupialia: Dasyuridae) from the Pilbara region, Western Australia. *Records of the Western Australian Museum* 20, 115–136. [22 Aug 2000]

Cooper NK, Adams M, Anthony C, Schmitt LH (2003a) Morphological and genetic variation in *Leggadina* (Thomas, 1910) with special reference to Western Australian populations. *Records of the Western Australian Museum* 21, 333–351. [27 Jun 2003]

Cooper NK, Bertozzi T, Baynes A, Teale RJ (2003b) The relationship between eastern and western populations of the heath rat, *Pseudomys shortridgei* (Rodentia: Muridae). *Records of the Western Australian Museum* 21, 367–370. [27 Jun 2003]

Cooper SJB, Reardon T, Skilins J (1998) Molecular systematics of Australasian rhinolophid bats (Chiroptera: Rhinolophidae). *Australian Journal of Zoology* 46, 203–220.

Cooper SJB, Adams M, Labrinidis A (2000) Phylogeography of the Australian dunnart *Sminthopsis crassicaudata* (Marsupialia: Dasyuridae). *Australian Journal of Zoology* 48, 461–473.

Cooper SJB, Day PR, Reardon TB, Shulz M (2001) Assessment of species boundaries in Australian *Myotis* (Chiroptera: Vespertilionidae) using Mitochondrial DNA. *Journal of Mammalogy* 82, 328–338. [May 2001]

Cope ED (1865a) Note on a species of hunchback whale. *Proceedings. Academy of Natural Sciences of Philadelphia* 1865, 178–181. [Sep–Oct 1865]

Cope ED (1865b) A contribution to a knowledge of the Delphinidae. *Proceedings. Academy of Natural Sciences of Philadelphia* 1865, 198–204. [Sep–Oct 1865]

Cope ED (1866) Third contribution to the history of the Balaenidae and Delphinidae. *Proceedings. Academy of Natural Sciences of Philadelphia* 1866, 293–300.

Cope ED (1867) A young species of whale, known as the Bahia finner. *Proceedings. Academy of Natural Sciences of Philadelphia* 1867, 32. [19 Mar 1867]

Cope ED (1868a) An addition to the vertebrate fauna of the Miocene period, with a synopsis of the extinct Cetacea of the United States. *Proceedings. Academy of Natural Sciences of Philadelphia* 19, 138–156.

Cope ED (1868b) Some remains of extinct Cetacea from the Miocene bed in Maryland. *Proceedings. Academy of Natural Sciences of Philadelphia* 20, 159–160. [23 Jun 1868]

Cope ED (1868c) On *Agaphelus*, a genus of toothless Cetacea. *Proceedings. Academy of Natural Sciences of Philadelphia* 20, 221–227.

Cope ED (1869) In C.M. Scammon. On the cetaceans of the western coast of North America. *Proceedings. Academy of Natural Sciences of Philadelphia* 21, 13–63.

Cope ED (1871) On *Megaptera bellicosa*. *Proceedings of the American Philosophical Society* 12, 103–108.

Cope ED (1872a) Second account of the new vertebrate from the Bridger Eocene. *Proceedings of the American Philosophical Society* 12, 466–468.

Cope ED (1872b) On an extinct whale from California. *Proceedings. Academy of Natural Sciences of Philadelphia* 24, 29–30. [Jan–Apr 1872]

Cope ED (1873) On the short-footed Ungulata of the Eocene of Wyoming. *Proceedings of the American Philosophical Society* 13, 38–74.

Cope ED (1875) On the supposed Carnivora of the Eocene of the Rocky Mountains. *Proceedings. Academy of Natural Sciences of Philadelphia* 27, 444–448. [30 Nov 1875]

Cope ED (1876) Fourth contribution to the history of the existing Cetacea. *Proceedings. Academy of Natural Sciences of Philadelphia* 28, 129–139.

Cope ED (1880) On the extinct cats of America. *American Naturalist* 14, 833–858. [Dec 1880]

Cope ED (1882a) Contribution to the history of the vertebrata of the lower Eocene of Wyoming and New Mexico, made during 1881. *Proceedings of the American Philosophical Society* 20, 139–197.

Cope ED (1882b) Notes on Eocene Mammalia. *American Naturalist* 16, 522. [Jun 1882]

Cope ED (1882c) On the systematic relation of the Carnivora Fissipedia. *Proceedings of the American Philosophical Society* 20, 471–475.

Cope ED (1884) The Tertiary Marsupialia. *American Naturalist* 18, 686–697. [Jul 1884]

Cope ED (1889a) The Edentata of North America. *American Naturalist* 23, 657–664. [Aug 1889]

Cope ED (1889b) Synopsis of the families of Vertebrata. *American Naturalist* 23, 849–877. [Oct 1889]

Cope ED (1891) *Syllabus of Lectures on Geology and Paleontology. Part 3. Paleontology of the Vertebrata*. Ferris Brothers, Philadelphia.

Cope ED (1898) Syllabus of Lectures on the Vertebrata. *Publications of the University of Pennsylvania*. University of Pennsylvania, Philadelphia.

Coquillett DW (1910) The type-species of the North American genera of Diptera. *Proceedings of the United States National Museum* 37, 499–647. [4 Aug 1910]

Corbet GB (1978) *The Mammals of the Palaearctic Region: a Taxonomic Review*. British Museum, London.

Corbet GB, Clutton-Brock J (1984) Appendix: Taxonomy and nomenclature. pp. 434–438, in IL Mason (ed.) *Evolution of Domesticated Animals*. Longman, London.

Corbet GB, Hill JE (1980) *A World List of Mammalian Species*. British Museum, London. First Edition.

Corbet GB, Hill JE (1986) *A World List of Mammalian Species*. British Museum, London. Second Edition.

Corbet GB, Hill JE (1992) *The Mammals of the Indomalayan Region: A Systematic Review*. Oxford University Press, Oxford.

Corbet GB, Hill JE (1994) Book Review – *Mammal Species of the World: A Taxonomic and Geographic Reference* (2nd Edition) edited by DE Wilson and DM Reeder. *Journal of Mammalogy* 75, 239–243. [Feb 1994]

Corbett L (1995) *The Dingo in Australia and Asia*. J.B. Books, Marleston, South Australia. First Edition.

Corbett L (2004) Dingo. pp. 223–230, in C Sillero-Zubiri, M Hoffmann & DW Macdonald (eds.) *Canids: Foxes, Wolves, Jackals and Dogs*. International Union for Conservation of Nature and Natural Resources.

Corbett L (2006) The Australasian Dingo. pp. 745–751, in JR Merrick, M Archer, GM Hickey & MSY Lee (eds.) *Evolution and Biogeography in Australia*. Auscipub Pty Ltd, Sydney.

Cornalia E (1854) Notizie zoologiche sul *Pachypleura Edwardsii* Cor. Nuovo sauro acrodonte degli strati triasici di Lombardia. *Giornale dell'I. R. Istituto Lombardo di Scienze, Lettere ed Arti E Biblioteca Italiana* 6, 45–58.

Cory CB, Hellmayr CH (1924) Catalogue of the birds of the Americas: Part III. Pteroptochidae – Conophagidae – Formicariidae. *Field Museum of Natural History. Zoological Series* 13(223), 1–369.

Costello M, Bellan-Santini D (2011) *Acanthonotozoma* Boeck, 1876. In J Lowry. World Amphipoda database. Accessed through: World Register of Marine Species at http://www.marinespecies.org/aphia.php?p=taxdetails&id=101443 on 2011-10-10.

Coues E (1895) The genus *Zaglossus*. *Science* 1(22), 610. [31 May 1895]

Coues E, Yarrow HC (1875) Report upon the collections of mammals made in portions of Nevada, Utah, California,

Colorado, New Mexico and Arizona during the years 1871, 1872, 1873, and 1874. *Report of the United States Geological and Geographic Explorer Survey*. 5, 35–129.

Courtenay JD (1993) The systematics of the hare-wallabies *Lagorchestes* Gould, 1841 and *Lagostrophus* Thomas, 1887. Ph.D. Thesis. Australian National University, Canberra.

Courtney J (1963) King Island, Bass Strait – the remarkable faunal unit (with description of a new form of *Potorous* rat-kangaroo). *Australian Aviculture* 17, 18–20, 92.

Cowan CF (1969a) Cuvier's Règne Animal, first edition. *Journal of the Society of Bibliography and Natural History* 5(3), 219. [1 Oct 1969] Gives date as 7 December 1816.

Cowan CF (1969b) Notes on Griffith's Animal Kingdom of Cuvier (1824–1835). *Journal of the Society for the Bibliography of Natural History* 5, 137–140.

Cowan CF (1973) *Proceedings of the Zoological Society of London*, publication dates. *Journal of the Society of Bibliography and Natural History* 9(4), 293–294.

Cramb J, Hocknull S (2010a) Two new species of *Antechinus* Macleay (Dasyuridae: Marsupialia) from mid-Pleistocene cave deposits in eastern central Queensland. *Australian Mammalogy* 32, 127–144.

Cramb J, Hocknull S (2010b) New Quaternary records of *Conilurus* (Rodentia: Muridae) from eastern and northern Australia with the description of a new species. *Zootaxa* 2634, 41–56. [4 Oct 2010]

Cramb J, Hocknull S, Webb GE (2009) High diversity Pleistocene rainforest dasyurids assemblages with implications for the radiation of the Dasyuridae. *Austral Ecology* 34, 663–669.

Cretella M (2010) The complete collation and dating of the section Zoologie of the Coquille voyage. *Bollettino Malacologico* 46, 83–103.

Cretzschmar J (1826–1831) Säugethiere, pp. 1–78, in E Rüppell (ed.) *Atlas zu der Reise im nördlichen Afrika von Eduard Ruppell*. 1. Abth. Zoologie. Hrsg. von der Senkenbergischen Naturforschenden Gesellschaft. H. L. Brönner, Frankfurt am Main.

Crisp E (1862) On the situation, form, and capacity of the gall bladder in the Vertebrata; on its absence in certain animals; and on the colour of the bile. *Proceedings of the Zoological Society of London* 30, 132–139. [Sep 1862]

Crook N, Cairns SC, Vernes K (2013) Bare-nosed wombats (*Vombatus ursinus*) use drainage culverts to cross roads. *Australian Mammalogy* 35, 23–29.

Crosby K, Bassarova M, Archer M, Carbery K (2004) Fossil Possums in Australasia: Discovery, diversity and evolution. pp. 161–176. In RL Goldingay & SM Jackson (eds.) *The Biology of Australian Possums and Gliders*. Surrey Beatty & Sons, Sydney.

Crowcroft P (1967) Studies on the hairy-nosed wombat *Lasiorhinus latifrons* (Owen 1845). I. Measurements and taxonomy. *Records of the South Australian Museum* 15, 383–398. [21 Dec 1967]

Crowther MS, Dickman CR, Lynam AJ (1999) *Sminthopsis griseoventer boullangerensis* (Marsupialia: Dasyuridae), a new subspecies in the *S. murina* complex from Boullanger Island, Western Australia. *Australian Journal of Zoology* 47, 215–243.

Crowther MS, Spencer PBS, Alpers D, Dickman CR (2002) Taxonomic status of the Mardo, *Antechinus flavipes leucogaster* (Marsupialia: Dasyuridae): A morphological and molecular, reproductive and bioclimatic approach. *Australian Journal of Zoology* 50, 627–647.

Crowther MS, Sumner J, Dickman CR (2003) Speciation of *Antechinus stuartii* and *A. subtropicus* (Marsupialia: Dasyuridae) in eastern Australia: molecular and morphological evidence. *Australian Journal of Zoology* 51, 443–462.

Crowther MS, Fillios M, Colman N, Letnic M (2014) An updated description of the *Australian dingo* (*Canis dingo* Meyer, 1793). *Journal of Zoology* 293, 192–203. [online 28 Mar 2014]

Csorba G, Ujhelyi P, Thomas N (2003) *Horseshoe Bats of the World (Chiroptera: Rhinolophidae)*. Alana Books, Thropshire, United Kingdom.

Cummings WC (1985a) Right whales *Eubalaena glacialis* (Müller, 1776) and *Eubalaena australis* (Desmoulins, 1822). pp. 275–304, in SH Ridgway & R Harrison (eds.) *Handbook of Marine Mammals. Volume 3. The Sirenians and Baleen Whales*. Academic Press, London.

Cummings WC (1985b) Bryde's whale *Balaenoptera edeni* Anderson, 1878. pp. 137–154, in SH Ridgway & R Harrison (eds.) *Handbook of Marine Mammals. Volume 3. The Sirenians and Baleen Whales*. Academic Press, London.

Cunha HA, Moraes LC, Medeiros BV, Lailson-Brito J Jr, da Silva VMF, Solé-Cava AM, Schrago CG (2011) Phylogenetic status and timescale for the diversification of *Steno* and *Sotalia* dolphins. *PLoS ONE* 6(12), 1–7. [e 28297]

Curtis J (1836) British Entomology; being illustrations and descriptions of the genera of insects found in Great Britain and Ireland: Containing coloured figures from nature of the most rare and beautiful species, and in many instances of the plants upon which they are found. T. Hurst, London. Volume 13.

Cuvier F (1807) Du genre *Paca. Coelogenus* (*Cavia paca*, Lin.). *Annales du Muséum d'Histoire Naturelle, Paris* 10, 203–209. [Sep? 1807]

Cuvier F (1813) Essai sur de nouveaux caractères pour les genres des Mammifères. *Annales du Muséum d'Histoire Naturelle, Paris* 19, 268–295. [Ante Jul 1813]

Cuvier F (1817) *Dictionnaire des sciences naturelles, dans lequel on traite méthodiquement des différens êtres de la nature, considérés soit en eux-mêmes, d'après l'état actuel de nos connoisances, soit relativement à l'utilité qu'en peuvent retirer la médicine, l'agriulture, le commerce et les arts. Suivi d'une biographie des plus célèbres naturalistes.. / par plusieurs professeurs du Muséum National d'Histoire Naturelle et des autres principales écoles de Paris*. F.G. Levrault, Strasbourg, Paris. Edition 2. Volume 5 BOA-BYT. (Supplement)

Cuvier F (1821) Martre des Palmiers. Plate 186; 5 pp. unpaginated. In É. Geoffroy Saint-Hilaire and F. Cuvier (eds.). [1824–1842] *Histoire Naturelle des Mammifères, avec des figures originales, coloriées, dessinées d'après des animaux vivants; publiée sous l'autorité de l'Administration du Muséum d'Histoire Naturelle*. A Belin, Paris. Volume 3. Livr. 24.

Cuvier F (1821–1825) *Des Dents des Mammifères, considérées comme caractères zoologiques*. Levrault, Paris.

Cuvier F (1824) De quelques espèces de phoques et des groupes géneriques entre lesquels ils se partagent. *Mémoires du Muséum d'Histoire Naturelle, Paris* 11, 175–214. [Oct 1824]

Cuvier F (1825) Péramèle; *Perameles*, Geoff. (Mamm.). pp. 414–416, in FG Cuvier (ed.) *Dictionnaire des sciences naturelles*, dans lequel on traite méthodiquement des differens êtres de la nature.. suivi d'une biographie des plus célèbres naturalistes.. / par plusieurs professeurs du Muséum National d'Histoire Naturelle et des autres principales écoles de Paris. F.G. Levrault, Strasbourg, Paris. Volume 38. PARN-PERRON.

Cuvier F (1826a) Phoque. (Mamm.). pp. 540–559, in FG Cuvier (ed.) *Dictionnaire des Sciences Naturelles*, dans lequel on traite méthodiquement des différens êtres de la nature, considérés soit en eux – mêmes, d'après l'état actuel de nos connaissances, soit relativement à l'utilité qu'en peuvent retirer la médicine, l'agriculture, le commerce et les arts. Suive d'une Biographie des Plus Célèbres Naturalistes. Ouvrage destiné aux médecins, aux agriculteurs, aux commercans, aux artistes, aux manufacturiers, et à tous ceux qui ont intérêt à connaître les productions de la nature, leurs caractères génériques et spécifiques, leur lieu natal, leurs propriétés et leurs usages. Paris. Volume 39. PERROQ-PHOQ.

Cuvier F (1826b) Phacochoeres, *Phacochoerus*. (Mamm.). pp. 383–386, in FG Cuvier (ed.) *Dictionnaire des Sciences Naturelles*, dans lequel on traite méthodiquement des différens êtres de la nature, considérés soit en eux – mêmes, d'après l'état actuel de nos connoissances, soit relativement a l'utilité qu'en peuvent retirer la médicine, l'agriculture, le commerce et les arts. Suive D'Une Biographie des Plus Célèbres Naturalistes. Ouvrage destiné aux médecins, aux agriculteurs, aux commercans, aux artistes, aux manufacturiers, et à tous ceux qui ont intérêt à connoître les productions de la nature, leurs caractères génériques et spécifiques, leur lieu natal, leurs propriétés et leurs usages. Paris. Volume 39. PERROQ-PHOQ.

Cuvier F (1828) Description d'un nouveau genre de chauve-souris sous le nom de Furie. *Mémoires du Muséum National d'Histoire Naturelle* 16, 149–155. [Oct 1828]

Cuvier F (1829a) Kanguroo, ou Halmature Thétis. Plate 225; 2 pp. unpaginated. In É Geoffroy Saint-Hilaire & F Cuvier (eds.). [1824–1842] *Histoire Naturelle des Mammifères, avec des figures originales, coloriées, dessinées d'après des animaux vivants; publiée sous l'autorité de l'Administration du Muséum d'Histoire Naturelle*. A Belin, Paris. Volume 6. Livr. 56. [Oct 1829]

Cuvier F (1829b) Zoologie = Mammalogie. pp. 357–519, in FG Cuvier (ed.) *Dictionnaire des sciences naturelles, dans lequel on traite méthodiquement des différens êtres de la nature, considérés soit en eux-mêmes, d'après l'état atuel de nos connoisances, soit relativement a l'utilité qu'en peuvent retirer la médicine, l'agriulture, le commerce et les arts. Suivi d'une biographie des plus célèbres naturalistes.. / par plusieurs professeurs du Muséum National d'Histoire Naturelle et des autres principales écoles de Paris*. F.G. Levrault, Strasbourg, Paris. Edition 2. Volume 59. WAA-ZOOP.

Cuvier F (1832) Essai de classification naturelle des vespertiliones, et descriptions de plusieurs espèces de ce genre. *Nouvelles Annales du Muséum d'Histoire Naturelle, Paris* 1, 1–21. [May 1832]

Cuvier F (1836a) Cetacea. pp. 562–594, in RB Todd (ed.) *The Cyclopaedia of Anatomy and Physiology*. Sherwood, Gilbert & Piper, London. Volume 1. A-DEA.

Cuvier F (1836b) *De 'Histoire Naturelle des Cétacés*, ou recueil et examen des faits dont se compose l'histoire naturelle de ces animaux. Librairie Encyclopedique de Roret, Paris.

Cuvier F (1837a) Sarcophile Oursin. Plate 113; 6 pp. unpaginated. In É Geoffroy Saint-Hilaire & F Cuvier (eds.). [1824–1842] *Histoire Naturelle des Mammifères, avec des figures originales colorées, dessinées d'après des animaux vivants; publiée sous l'authorité de l'Administration du Muséum d'Histoire Naturelle*. Paris. Volume 7. Livr. 70.

Cuvier F (1837b) Memoir on the genera of *Dipus* and *Gerbillus*. *Proceedings of the Zoological Society of London* 4(1836), 141–142. [27 Jun 1837]

Cuvier F (1837c) Du genre Eligmodonte et de l'Eligmodonte de Bueonis-Ayres. *Eligmodontia typus*. *Annales des Sciences Naturelles. Zoologie* (2)7, 168–171.

Cuvier F (1837d) Rapport sur un mémoire de M. Jourdan, de Lyon, convernant quelques mammifères nouveaux. *Annales des Sciences Naturelles. Zoologie* (2)8, 367–374.

Cuvier F (1838) Rapport sur un mémoire de M. Jourdan, de Lyon, concernant quelques mammifères nouveaux. *Comptes Rendus Hebdomadaires des Séances de l'Académie des Sciences, Paris* 6, 2–6.

Cuvier G (1796) Notice sur le squelette d'une très-grande espèce de quadrupède inconnue jusqu'à présent, trouvé au Paraguay, et déposé au cabinet d'histoire naturelle de Madrid. *Magasin Encyclopédique, ou Journal des Sciences, des Lettres et des Arts* 7: 303–310.

Cuvier G (1797) *Tableau Élémentaire de l'Histoire Naturelle des Animaux*. Baudouin, Paris.

Cuvier G (1800) *Leçons d'Anatomie Comparée. Recueillies et publiées sous ses yeux par C. Duméril, chef des travaux anatomiques de l'École de Médecine de Paris*. National Des Sciences et des Arts, Paris. Volume 1.

Cuvier G (1809) Extrait des premiers Mémoires de M.F. Cuvier, sur les dents des mammifères considérés comme charactères génerics. *Nouveau Bulletin des Sciences, par la Société Philomathique* 1(24), 393–395.

Cuvier G (1812) Rapport fait à la classe des sciences mathématiques et physiques, sur divers cétacés pris sur les côtes de france, principalement sur ceux qui sont échoués près de Paimpol, le 7 janvier 1812. *Annales du Muséum d'Histoire Naturelle, Paris* 19, 1–16. [Aug 1812]

Cuvier G (1814) Observations et recherches critiques sur différens poissons de la Méditerranée et, à leur occasion, sur des Poissons des autres mers plus ou moins liés avec eux; par M. G. Cuvier. *Bulletin de la Société Philomatique, Paris* 1814, 80–92. [In a report by A. Desmarest.]

Cuvier G (1816a) *Le Règne Animal distribué d'après son organisation pour servir de base à l'histoire naturelle des*

Animaux et d'introduction à l'Anatomie Comparée. Chez Déterville, Paris. Volume 1. [Nov–17 Dec 1816]

Cuvier G (1816b) *Le Règne Animal distribué d'après son organisation pour servir de base à l'histoire naturelle des Animaux et d'introduction à l'Anatomie Comparée.* Chez Déterville, Paris. Volume 2. [Nov–17 Dec 1816]

Cuvier G (1816c) *Le Règne Animal distribué d'après son organisation pour servir de base à l'histoire naturelle des Animaux et d'introduction à l'Anatomie Comparée.* Chez Déterville, Paris. Volume 4. [Nov–17 Dec 1816]

Cuvier G (1817) In Desmarest, A.G. Dauphin. pp. 146–179, in C.S. Sonnini (ed.). *Nouveau Dictionnaire d'Histoire Naturelle*, appliquée aux arts, à l'agriculture, à l'economie rurale et domestique, à la medecine, etc. Par une société de natualistes et d'agriculteurs. Nouv. Édn. Presqu' entièrement refondue et considérablement augmentée. Déterville, Paris. Volume 9. CUN-DZW.

Cuvier G (1823a) *Recherches sur les Ossements Fossiles de Quadrupèdes, où l'on rétablit les caractères de plusieurs espèces d'Animaux que les révolutions du Globe paraissent avoir détruit les espèces.* Nouvelle Édition. Tome 5. Partie 1. Dufor et E. D'Ocagne, Paris.

Cuvier G (1823b) Recherches sur les Ossements Fossiles de Quadrupèdes, où l'on rétablit les caractères de plusieurs espèces d'Animaux que les révolutions du Globe paraissent avoir détruit les espèces. Nouvelle Édition. Tome 8. Partie 2. Dufor et E. D'Ocagne, Paris. Reference not seen.

Cuvier G (1824) Phalanger de Cook. Plate 224; 4 pp. unpaginated. In É Geoffroy Saint-Hilaire & F Cuvier (eds.). [1824–1842] *Histoire Naturelle des Mammifères, avec des figures originales colorées, dessinées d'après des animaux vivants; publiée sous l'authorité de l'Administration du Muséum d'Histoire Naturelle.* A Belin, Paris. Volume 6. Livr. 45.

Cuvier G (1828) In Lesson, R.P. *Histoire Naturelle*, Générale et Particulière des Mammifères et des Oiseaux Découverts depuis 1788 jusqu'à nos Jours: Complément des oeuvres de Buffon, ou histoire naturelle des animaux rares découverts par les naturalistes et les voyageurs depuis la mort de Buffon. Tom. 1. Cétacés. Baudoin Frères, Paris.

Cuvier G (1829a) *Le règne animal distribué d'après son organisation, pour servir de base à l'histoire naturelle des animaux et d'introduction à l'anatomie comparée.* Chez Déterville Libraire, Paris. Second Edition. Volume 1.

Cuvier G (1829b) *Le règne animal distribué d'après son organisation, pour servir de base à l'histoire naturelle des animaux et d'introduction à l'anatomie comparée.* Chez Déterville Libraire, Paris. Second Edition. Volume 2.

d'Alton JWE (1823) In CH Pander & E D'Alton. *Vergleichende Osteologie. Lief. 4. Die Skelete der Wiederkäuer, abgebildet und verglichen.* E. Weber, Bonn.

D'Orbigny A (1834) Notice sur un nouveau genre de cétacé, des rivières du centre de l'Amérique méridionale. *Nouvelle Annales du Muséum d'Histoire Naturelle, Paris* 3, 31–36.

Dahl G (1823) *Coleoptera und Lepidoptera : Ein Systematisches Verzeichniss, mit begesetzten Preisen der Vorräthe.* J.E. Akkermann, Wien.

Dahl K (1897) Biological notes on north-Australian Mammalia. *Zoologist* 1(4), 189–216.

Dal Piaz G (1917) Gli Odontoceti del Miocene Bellunese. *Parte Terza. Squalodelphis fabianii. Memorie dell' Istituto Geologico della R. Università di Padova* 5, 1–34.

Dalebout ML, Ross GJB, Baker CS, Anderson RC, Best PB, Cockcroft VG, Hinsz HL, Peddemors V, Pitman RL (2003) Appearance, distribution and genetic distinctiveness of Longman's beaked whale, *Indopacetus pacificus. Marine Mammal Science* 19, 421–461. [Jul 2003]

Dalebout M, Baker CS, Cockcroft VG, Mead JG, Yamada TK (2004) A comprehensive and validated molecular taxonomy of beaked whales, Family Ziphiidae. *The Journal of Heredity* 95, 459–473.

Dalebout ML, Baker CS, Steel D, Thompson K, Robertson KM, Chivers SJ, Perrin WF, Goonatilake M, Anderson RC, Mead JG, Potter CW, Thomson L, Jupiter D, Yamada TK (2014) Resurrection of *Mesoplodon hotaula* Deraniyagala 1963: A new species of beaked whale in the tropical Indo Pacific. *Marine Mammal Science* 30, 1081–1108. [online 5 Feb 2014]

Dall WH (1873) Description of three species of Cetacea, from the coast of California. *Proceedings of the California Academy of Sciences* (1)5, 12–14. [29 Jan 1873]

Daniels MJ, Corbett L (2003) Redefining introgressed protected mammals: when is a wildcat a wildcat and a dingo a wild dog? *Wildlife Research* 30, 213–218.

Darragh TA (1988) A revision of the Tertiary Volutidae (Mollusca: Gastropoda) of South-Eastern Australia. *Memoirs of the Museum of Victoria* 49, 195–307. [Nov 1988]

Dasmahapatra KK, Hoffman JI, Amos W (2009) Pinniped phylogenetic relationships inferred using AFLP markers. *Heredity* 103, 168–177.

Davidson MEM (1929) Notes on the northern elephant seal. *Proceedings of the California Academy of Sciences* 18(4), 229–242. [5 Apr 1929]

Davie P, Türkay M (2010) *Stenorhynchus* Lamarck, 1818. Accessed through: World Register of Marine Species at http://www.marinespecies.org/aphia.php?p= taxdetails&id=106914 on 2011–10–10.

Davie P, Türkay M (2011) *Uca* Leach, 1814. Accessed through: World Register of Marine Species at: http://www.marinespecies.org/aphia.php?p=taxdetails&id= 106971 on 2011–10–10.

Davies JL (1960) The southern form of the pilot whale. *Journal of Mammalogy* 41, 29–34. [Feb 1960]

Davis CS, Delisle I, Stirling I, Siniff DB, Strobeck C (2004) A phylogeny of the extant Phocidae inferred from complete mitochondrial DNA coding regions. *Molecular Phylogenetics and Evolution* 33, 363–377. [Nov 2004]

Davison A (1991) A new subspecies of dusky antechinus, *Antechinus swainsonii* (Marsupialia: Dasyuridae) from western Victoria. *Australian Mammalogy* 14, 103–113. [28 Jun 1991]

Dawson L (1982a) Taxonomic status of fossil devils (*Sarcophilus*, Dasyuridae, Marsupialia) from late Quaternary eastern Australian localities. pp. 517–525, in M Archer (ed.) *Carnivorous Marsupials.* Royal Zoological Society of New South Wales, Sydney. Volume 2. [Feb 1982]

Dawson L (1982b) Taxonomic status of fossil thylacines (*Thylacinus*, Thylacinidae, Marsupialia) from late Quaternary deposits in eastern Australia. pp. 527–536, in M

Archer (ed.) *Carnivorous Marsupials*. Royal Zoological Society of New South Wales, Sydney. Volume 2. [Feb 1982]

Dawson L (1983) The taxonomic status of small fossil wombats (Vombatidae: Marsupialia) from quaternary deposits, and of related modern wombats. *Proceedings of the Linnean Society of New South Wales* 107, 99–121. [Dec 1983]

Dawson LJ (1988) Vombatidae. pp. 48–50, in DW Walton (ed.) *Zoological Catalogue of Australia. 5. Mammalia*. Australian Government Publishing Service, Canberra. [13 Apr 1988]

Dawson L, Flannery T (1985) Taxonomic and phylogenetic status of living and fossil kangaroos and wallabies of the genus *Macropus* Shaw (Macropodidae: Marsupialia), with a new subgeneric name for the larger wallabies. *Australian Journal of Zoology* 33, 473–498.

Day F (1889) *Fauna of British India. Fishes*. Volume 1. Taylor and Francis, London.

de Beaufort F (1966) Catalogue des types de mammifères du Muséum National d'Histoire Naturelle. Paris. VI. Monotremata. - VII Marsupialia. *Bulletin du Muséum National d'Histoire Naturelle. Section B. Botanique. Biologie et Écologie Végétables Phytochimie, Paris* (2)38, 509–553.

de Beaufort LF (1911) Die Säugetiere der Aru- und Kei-Inseln. *Abhandlungen der Senckenbergischen Naturforschenden Gesellschaft* 34, 99–115.

de Blainville HMD (1816a) Prodrome d'une nouvelle distribution systématique du règne animal. *Bulletin des Sciences par la Société Philomathique de Paris* 1816, 113–124. [Jul 1816]

de Blainville HMD (1816b) Prodrome d'une nouvelle distribution systématique du règne animal. *Journal de Physique, de Chimie, d'Histoire Naturelle et des Arts* 83, 244–267. [July 1816]

de Blainville HMD (1816c) Sur plusieurs espèces d'animaux mammifères, de l'ordre des ruminants. *Bulletin des Sciences par la Société Philomathique de Paris* 1816, 73–82.

de Blainville HMD (1817) In Desmarest, AG Dauphin. pp. 146–179, in CS Sonnini (ed.). *Nouveau Dictionnaire d'Histoire Naturelle*, appliquée aux arts, à l'agriculture, à l'economie rurale et domestique, à la médecine, etc. Par une société de natualistes et d'agriculteurs. Nouv. Édn. Presqu'entièrement refondue et considérablement augmentée. Déterville, Paris. Volume 9. CUN-DZW.

de Blainville HMD (1820) Sur quelques cranes de phoques. *Journal de Physique, de Chimie, d'Histoire Naturelle et des Arts* 91, 286–300.

de Blainville HMD (1822a) *De l'Organisation des animaux, ou Principes d'anatomie comparée*. Volume 1. F.G. Levrault, Paris.

de Blainville HMD (1822b) Sur les caractères distinctifs des espèces de cerfs. *Journal de Physique, de Chimie, d'Histoire Naturelle et des Arts* 94, 254–283.

de Blainville HMD (1828) Vers. (Entomoz.), pp. 372–625, in F Cuvier (ed.). *Dictionnaire des sciences naturelles, dans lequel on traite méthodiquement des differents êtres de la nature.. suivi d'une biographie des plus célèbres naturalistes.. / par plusieurs professeurs du Muséum National d'Histoire Naturelle et des autres principales écoles de Paris*. F.G. Levrault, Strasbourg, Paris. Volume 57. VEA-VERS.

de Blainville HMD (1830) Zoophytes, Zoophyta. pp. 1–546, in FG Cuvier (ed.) *Dictionnaire des sciences naturelles, dans lequel on traite méthodiquement des differents êtres de la nature.. suivi d'une biographie des plus célèbres naturalistes.. / par plusieurs professeurs du Muséum National d'Histoire Naturelle et des autres principales écoles de Paris*. F.G. Levrault, Strasbourg, Paris. Volume 60. Zoo-Zyt.

de Blainville HMD (1834) *Cours de la faculté des sciences*. According to Gregory (1910: 82) and Simpson (1954: 356) the classification used by de Blainville was copied in Gervais (1836: 619). Reference not seen.

de Blainville HMD (1838) Sur les Cachalots. *Annales Françaises Étrangères Anatomie et de Physiologie* 2, 335–337.

de Boisduval JBAD (1833) Mémoire sur les Lépidoptères de Madagascar. Bourbon et Maurice. *Nouvelles Annales du Muséum d'Histoire Naturelle, Paris* 2, 149–270. [Mid 1833]

de Bonvouloir H (1870) Monographie de la famille des Eucnémides. *Annales de la Société Entomologique de France* 10(4, Supplement), 5–907.

de Jong N, Bergmans W (1981) A revision of the fruit bats of the genus *Dobsonia* Palmer, 1898 from Sulawesi and some nearby islands (Mammalia, Megachiroptera, Pteropodinae). *Zoologische Abhandlungen / Staatliches Museum fur Tierkunde in Dresden* 37, 209–224.

de Laporte FL (1840) In E Blanchard, Compte de Castelnau & Brulle. *Histoire Naturelle des Insectes Coléoptères*. P. Duménil, Paris. Volume 2(Part 1).

de Marschall AF (1873) *Nomenclator zoologicus continens nomina systematica generum animalium tam viventium quam fossilium, secundum ordinem alphabeticum disposita, sub auspiciis et sumptibus C. R. Societatis zoologico-botanicae*. Vindobonae. Mammalia.

de Mello-Leitão CF (1923) Sobre uma Aranha parasite de Saúva. *Revista do Museu Paulista, São Paulo* 13, 523–525.

de Mello-Leitão CF (1925) Pequenas notas arachnologicas. I. nomes emendados. *Boletim do Museu Nacional, Rio de Janeiro* 1(6), 455–463.

de Miklouho-Maclay N (1884) Notes on zoology of the Maclay-Coast in New Guinea. *Proceedings of the Linnean Society of New South Wales* (1)9, 713–720. [29 Nov 1884]

de Miranda-Ribiero A (1936) Notas cetologicas (os generos *Steno, Sotalia* e *Stenopontistes*). *Boletim do Museu Nacional, Rio de Janeiro* 12(1), 3–23.

de Muizon C (1978) *Arctocephalus* (*Hydrarctos*) *lomasiensis*, subgen, nov. et nov. sp., un nouvel Otariidae du Mio-Pliocene de Sacaco (Perou). *Bulletin de l'Institute Francais d'Etudes Andines* 7(3–4), 169–189.

de Muizon C (1982) Phocid phylogeny and dispersal. *Annals of the South African Museum* 89, 175–213.

de Muizon C (1983) Un nouveau Phocoenidae (Cetacea) du Pliocène inférieur du Pérou. *Comptes-Rendus Hebdomadaires des Séances de l'Académie des Sciences Ser. II.* 296, 1203–1206.

de Muizon C (1984) Les vertébrés fossiles de la Formation Pisco (Pérou). Deuxieme partie: Les Odontocètes (Cetacea, Mammalia) du Pliocène inférieur de Sud-Sacaco. *Institut Français d'Étude Andines. Mémoire* 50, 1–188.

de Muizon C (1988a) Les vertébrés fossiles de la Formation Pisco (Pérou). Troisième partie. Les odontocètes (Cetaceae,

Mammalia) du Miocène. *Institut Français d'Etudes Andines, Editions Recherche sur les Civilisations. Mémoire* 78, 1–246.

de Muizon C (1988b) Les relations phylogenetiques des Delphinida (Cetacea, Mammalia). *Annales de Paléontologie* 74, 159–227.

de Muizon C (1991) A new Ziphiidae (Cetacea) from the Early Miocene of Washington State (USA) and phylogenetic analysis of the major groups of odontocetes. *Bulletin du Muséum National d'Histoire Naturelle, Paris* 12(4) (sect C (3–4)), 279–326.

De Queiroz K (2007) Species concepts and species delimitation. *Systematic Biology* 56, 879–886.

de Rochebrune AT (1883) Faune de la Sénégambie. Mammifères. *Actes de la Société Linnéenne de Bordeaux* 37(4)7, 49–203.

de Saussure H (1860) Note sur quelques Mammifères du Mexique. *Revue et Magasin de Zoologie Pure et Appliquée, Paris* 12(2), 281–293.

de Serres M (1838) *Essai sur les Cavernes à Ossements, et sur les causes qui les y ont accumulés*. Montpellier, Paris. Third Edition.

de Spix JB (1823) *Simiarum et Vespertilionum Brasiliensium species novae ou Histoire Naturelle des Espèces Nouvelles de Singes et de Chauves-Souris Observees et Recueillies Pendant le Voyage dans l'interieur du Brésil*. Francisci Seraphici Hübschmanni, Munich.

de Spix JB (1824) *Avium Species Novae, quas in itinere per Brasiliam annis MDCCCXVII - MDCCCXX jussu et auspiciis Maximiliani Josephi I Bavariae Regis*. Francisci Seraphi Hybschmanni, Monachii. Volume 1.

de Tarragon L (1847) Description du mâle adulte du *Colobe guereza* et du *Phascogale Virginiae. Revue Zoologique par La Société Cuvierienne* 1847, 177–178.

De Vis C (1883a) On tooth-marked bones of extinct marsupials. *Proceedings of the Linnean Society of New South Wales* (1)8, 187–190. [17 Jul 1883]

De Vis CW (1883b) Description of a new *Belideus* from northern Queensland. *Abstract of Proceedings of the Linnean Society of New South Wales* 7(1), 619–620. [Apr. 1883]

De Vis CW (1883c) Description of a new *Belideus* from northern Queensland, in The Linnean Society of New South Wales. *The Southern Science Record and Magazine of Natural History* 2, 26–29. [Jan 1883]

De Vis CW (1883d) On *Brachalletes palmeri* an extinct marsupial. *Proceedings of the Linnean Society of New South Wales* (1)8, 190–193. [17 Jul 1883]

De Vis CW (1883e) Notes on a lower jaw of *Palorchestes azael. Proceedings of the Linnean Society of New South Wales* (1)8, 221–224. [17 Jul 1883]

De Vis CW (1884a) On apparently new species of *Halmaturus. The Proceedings of the Royal Society of Queensland* 1(1883), 107–112.

De Vis CW (1884b) Notes on the fauna of the Gulf of Carpentaria. *The Proceedings of the Royal Society of Queensland* 1(1883), 154–160.

De Vis CW (1885a) In Anon, *The Daily Observer*. [Brisbane], no. 549, vol. 2. 11 April 1885. p. 2, column 6. Reference not seen.

De Vis CW (1885b) On an extinct monotreme, *Ornithorhynchus agilis. The Proceedings of the Royal Society of Queensland* 2(1884), 35–38.

De Vis CW (1886a) On new or rare vertebrates from the Herbert River, North Queensland. *Abstract of Proceedings of the Linnean Society of New South Wales* (1886), vi. [for 29 Dec 1886]

De Vis CW (1886b) On a probably new species of tree-kangaroo from North Queensland. *Abstract of Proceedings of the Linnean Society of New South Wales* (1886), v. [for 27 Oct 1886]

De Vis CW (1887a) On a new rare vertebrate from the Herbert River, North Queensland. *Proceedings of the Linnean Society of New South Wales* (2)1, 1129–1137. [22 Feb 1887]

De Vis CW (1887b) Notice of a probable new species of *Dendrolagus. The Proceedings of the Royal Society of Queensland* 3(1886), 11–14. [Aug 1887]

De Vis CW (1888) On a third species of Australian tree kangaroo. *The Proceedings of the Royal Society of Queensland* 4(1887), 132–134.

De Vis CW (1889) On a new genus of extinct mammals. *The Proceedings of the Royal Society of Queensland* 5(1888), 158–160.

De Vis CW (1893) A thylacine of the earlier nototherian period in Queensland. In Anon. *Abstracts of the Proceedings of the Linnean Society of New South Wales*, 29 November 1893, v.

De Vis CW (1894) A thylacine for the earlier nototherian period in Queensland. *Proceedings of the Linnean Society of New South Wales* (2)8, 443–447. [5 Jun 1894]

De Vis CW (1895) A review of the fossil jaws of the Macropodidae in the Queensland Museum. *Proceedings of the Linnean Society of New South Wales* (2)10, 75–133. [9 Sep 1895]

De Vis CW (1900) A new species of hairy-nosed wombat. *Annals of the Queensland Museum* 5, 14–16.

De Vis CW (1905) Bats. *Annals of the Queensland Museum* 6, 36–40.

De Vis CW (1907) An eccentric rat. *Annals of the Queensland Museum* 7, 8–9.

de Winton WE (1898) On the hares of Western Europe and North Africa. *Annals & Magazine of Natural History* (7)1, 149–158. [1 Feb 1898]

de Zigno A (1887) Quelques observations sur les Siréniens fossiles. *Bulletin de la Société Géologique de France* 3(15), 728–732. [Dec 1887]

Deberer PH (1909) Comparison of *Caenolestes* with Polyprotodonta and Diprotodonta. *American Naturalist* 43, 614–618. [Oct 1909]

Decher J, Fahr J (2005) *Hipposideros cyclops. Mammalian Species* 763, 1–7. [15 Jul 2005]

Dejean PFMA (1821) *Catalogue de la Collection de Coléoptères*. Chez Crevot, Paris. [May 1821]

Dejean PFMA (1833) *Catalogue des Coléoptères de la collection de M. le Comte Dejean*. Méquignon-Marvis, Paris. Second Edition. Part 1: 1–96.

Dejean PFMA (1835) *Catalogue des Coléoptères de la collection de M. le Comte Dejean*. Livraison 4. Méquignon-Marvis, Paris, 257–360. Third Edition. Reference not seen.

Delson E, Napier PH (1976) Request for the determination of the generic names of the baboon and the mandrill (Mammalia: Primates, Cercopithecidae). Z.S. (S.) 2093. *Bulletin of Zoological Nomenclature* 33, 46–60. [Jun 1976]

Deméré TA (1986) The fossil whale, *Balaenoptera davidsoni* (Cope, 1872), with a review of the other Neogene species of *Balaenoptera* (Cetacea: Mysticeti). *Marine Mammal Science* 2, 277–298. [Oct 1986]

Deméré TA, Berta A, Adam PJ (2003) Pinnipedimorph evolutionary biogeography. *Bulletin of the American Museum of Natural History* 279, 32–76.

Deméré TA, Berta A, McGowan MR (2005) The taxonomic and evolutionary history of fossil and modern balaenopteroid mysticetes. *Journal of Mammalian Evolution* 12, 99–143. [Jun 2005]

Dennis AJ, Storch D (1998) Conservation and Taxonomic Status of the Bramble Cay Melomys, *Melomys rubicola*. Progress Report for Endangered Species Program. Canberra. Environment Australia. [Dec 1998]

Deraniyagala P (1960) Some southern temperate zone snakes, birds and whales that enter the Ceylon area. *Spolia Zeylanica: Bulletin of the National Museums of Sri Lanka: geology, zoology, anthropology* 29, 79–85.

Deraniyagala PEP (1963a) Mass mortality of a new subspecies of little piked whale *Balaenoptera acutorostrata thalmaha* and a new beaked whale *Mesoplodon hotaula* from Ceylon. *Spolia Zeylanica: Bulletin of the National Museums of Sri Lanka: geology, zoology, anthropology* 30, 80–84.

Deraniyagala PEP (1963b) A new beaked whale from Ceylon. *Ceylon Today* 12, 13–14.

Desmarest AG (1803a) Chauve-Souris. pp. 107–111, in CS Sonnini (ed.) Société de naturalistes et d'agriculteurs (eds.) *Nouveau Dictionnaire d'Histoire Naturelle, appliquée aux arts, principalement à l'agriculture et à l'economie rurale et domestique*. Deterville, Paris. Volume 5. CHA-COC.

Desmarest AG (1803b) Rhinolophe. pp. 383–384, in CS Sonnini (ed.) Société de naturalistes et d'agriculteurs (eds) *Nouveau Dictionnaire d'Histoire Naturelle, appliquée aux arts, principalement à l'agriculture et à l'economiie rurale et domestique*. Deterville, Paris. Volume 19. QOT-RYZ.

Desmarest AG (1804a) Tableau Méthodique des Mammifères. pp. 5–38, in CS Sonnini (ed.) in *Nouveau Dictionnaire d'Histoire Naturelle*, appliquée aux arts, principalement à l'agriculture et à l'economie rurale et domestique: par une société de naturalistes et d'agricltureurs: avec des figures tirées des trois règnes de la nature. Déterville, Paris. First Edition. Volume 24.

Desmarest AG (1804b) Dasyure Gutté. pp. 10, in CS Sonnini (ed.) *Nouveau Dictionnaire d'Histoire Naturelle* appliquée aux arts, principalement à l'agriculture et à l'économie rurale et domestique: par une société de naturalistes et d'agricltureurs: avec des figures tirées des trois règnes de la nature. Déterville, Paris. First Edition. Volume 24.

Desmarest AG (1804c) Kanguroo ou Kangurou. pp. 354–358, in CS Sonnini (ed.) *Nouveau Dictionnaire d'Histoire Naturelle*, appliquée aux arts, principalement à l'agriculture, à l'économie rurale et domestique: par une société de naturalistes et d'agricltureurs: avec des figures tirées des trois règnes de la nature. Déterville, Paris. First Edition. Volume 12. INA-LAT.

Desmarest AG (1817a) *Isoodon*. pp. 409–410, in CS Sonnini (ed.) *Nouveau Dictionnaire d'Histoire Naturelle*, appliquée aux arts, l'agriculture, à l'economie rurale et domestique, à la medecine, etc. Par une société de naturalistes et d'agriculteurs. Nouv. Édn. Presqu'entièrement refondue et considérablement augmentée. Deterville, Paris. Tome 16. IAS-JYN.

Desmarest AG (1817b) Petauriste. pp. 398–406, in CS Sonnini (ed.) *Nouveau Dictionnaire d'Histoire Naturelle*, appliquée aux arts, à l'agriculture, à l'économie rurale et domestique, à la medecine, etc. Par une société de natualistes et d'agriculteurs. Nouv. Édn. Presqu'entièrement refondue et considérablement augmentée. Déterville, Paris. Volume 25. PAS-PHO.

Desmarest AG (1817c) Phalanger. pp. 470–477, in CS Sonnini (ed.) *Nouveau Dictionnaire d'Histoire Naturelle*, appliquée aux arts, àl'agriculture, à l'économie rurale et domestique, à la medecine, etc. Par une société de natualistes et d'agriculteurs. Nouv. Édn. Presqu'entièrement refondue et considérablement augmentée. Déterville, Paris. Volume 25. PAS-PHO.

Desmarest AG (1817d) Kanguroo. pp. 28–44, in CS Sonnini (ed.) *Nouveau Dictionnaire d'Histoire Naturelle*, appliquée aux arts, à l'agriculture, à l'économie rurale et domestique, à la medecine, etc. Par une société de natualistes et d'agriculteurs. Nouv. Édn. Presqu'entièrement refondue et considérablement augmentée. Déterville, Paris. Volume 17. KAA-LIG.

Desmarest AG (1817e) Phoque. pp. 544–604, in CS Sonnini (ed.) *Nouveau Dictionnaire d'Histoire Naturelle*, appliquée aux arts, àl'agriculture, à l'économie rurale et domestique, à la medecine, etc. Par une société de natualistes et d'agriculteurs. Nouv. Édn. Presqu'entièrement refondue et considérablement augmentée. Déterville, Paris. Volume 25. PAS-PHO.

Desmarest AG (1817f) Dauphin. pp. 146–179, in CS Sonnini (ed.). *Nouveau Dictionnaire d'Histoire Naturelle*, appliquée aux arts, àl'agriculture, à l'économie rurale et domestique, à la medecine, etc. Par une société de natualistes et d'agriculteurs. Nouv. Édn. Presqu'entièrement refondue et considérablement augmentée. Déterville, Paris. Volume 9. CUN-DZW.

Desmarest AG (1819) Vespertilion. pp. 461–481, in CS Sonnini (ed.) *Nouveau Dictionnaire d'Histoire Naturelle*, appliquée aux arts, à l'agriculture, à l'économie rurale et domestique, à la medecine, etc. Par une société de natualistes et d'agriculteurs. Nouv. Édn. Presqu'entièrement refondue et considérablement augmentée. Deterville, Paris. Volume 35.

Desmarest AG (1821) *Mammalogie, ou description des espèces de mammifères. Premier Partie, contenant les orders des bimanes, des quadrumanes et des carnassiers*. V. Agasse, Paris. [6 Jan 1821]

Desmarest, A.G. (1822a) Mémoire sur un nouveau genre de mammifères de l'ordre des rongeurs, nommé *Capromys*. *Bulletin des Sciences par la Societe Philomathique de Paris* 1822, 185–188.

Desmarest AG (1822b) *Mammalogie, ou description des espèces de mammifères. Seconde Partie, contenant les orders des bimanes, des quadrumanes et des carnassiers*. V. Agasse, Paris. [6 Jul 1822]

Desmarest E (1847) Péramèle. *Pelameles*. pp. 577–579, in C. D'Orbigny (ed.) *Dictionnaire Universal d'Histoire Naturelle*. Paris. Second Edition. Volume 9. OIE-PHO.

Desmoulins A (1822a) Baleinas ou Balènas. Mam. pp. 155–165, in JBGM Bory de Saint-Vincent (ed.) *Dictionnaire Classique d'Histoire Naturelle*. J. Tastu, Paris. Tome 2. ASA-CAC.

Desmoulins A (1822b) Cachalot. Physeter. pp. 613–619, in JBGM Bory de Saint-Vincent (ed.) *Dictionnaire Classique d'Histoire Naturelle*. J. Tastu, Paris. Volume 2. ASA-CAC.

Dickinson EC (2005) The *Proceedings of the Zoological Society of London*, 1859–1900: an exploration of break between calendar years of publication. *Journal of Zoology* 266, 427–430.

Dickinson EC, Bruce MD, Dowsett RJ (2006) Vivarium naturae or the naturalist's miscellany (1789–1813) by George Shaw: an assessment of the dating of the parts and volumes. *Archives of Natural History* 33, 322–343.

Dickman CR, King DH, Adams M, Baverstock PM (1988) Electrophoretic identification of a new species of *Antechinus* (Marsupialia: Dasyuridae) in south-eastern Australia. *Australian Journal of Zoology* 36, 455–463.

Dickman CR, Parnaby HE, Crowther MS, King DH (1998) *Antechinus agilis* (Marsupialia: Dasyuridae), a new species from the *A. stuartii* complex in south-eastern Australia. *Australian Journal of Zoology* 46, 1–26.

Diesing KM (1858) Revision der Myzhelminthen. Abtheilung: Trematoden. *Mathematisch-Naturwissenschaftliche Classe der Kaiserlichen Akademie der Wissenschaften in Wien* 32, 307–390.

Dieterlen F (1986) *Kurzfassungen der Vorträge und Posterdemonstrationen, Deutsche Gesellschaft für Säugetierkunde Hauptversammlung, Stuttgart*. Parey, Hamburg and Berlin. Reference not seen.

Dieterlen F (1989) Revocation of name. p. 65. WFH Ansell (ed.) *African Mammals 1938–1988*. Trendrine, Zennor.

Dieterlen F (2005) Family Petetidae. pp. 1535, in DE Wilson & DA Reeder (eds.) *Mammal Species of the World. A Taxonomic and Geographic Reference*. Johns Hopkins University Press, Baltimore. Third Edition.

Dixon JM (1970) Catalogue of mammal types (Class Mammalia) in the National Museum of Victoria. *Memoirs of the National Museum of Victoria* 31, 105–114. [18 May 1970]

Dixon JM (1981) Selection of a neotype for the southern short-nosed (brown) bandicoot, *Isoodon obesulus* (Shaw & Nodder, 1797). *Victorian Naturalist* 98, 130–135. [May–Jun 1981]

Dobson GE (1871a) On a new genus and species of Rhinolophidae, with description of a new species of *Vesperus*, and notes on some other species of insectivorous bats from Persia. *Journal of the Asiatic Society of Bengal, Calcutta* 40, 455–461.

Dobson GE (1871b) Description of four new species of Malayan bats, from the collection of Dr. Stoliczka. *Journal of the Asiatic Society of Bengal, Calcutta* 40, 260–267.

Dobson GE (1871c) Notes on nine new species of Indian and Indo-Chinese Vespertilionidae, with remarks on the synonymy and classification of some other species of the same family. *Proceedings of the Asiatic Society of Bengal, Calcutta* 1871, 210–215.

Dobson GE (1872) Brief descriptions of five new species of rhinolophine bats. *Journal of the Asiatic Society of Bengal, Calcutta* 41, 336–338.

Dobson GE (1873a) Notes on the Pteropi of India and its islands, with descriptions of some new or little known species. *Proceedings of the Asiatic Society of Bengal, Calcutta* 1873, 147–148.

Dobson GE (1873b) Description of a remarkable new species of *Molossus* (*Nyctinomus*) from Johore in the Malay Peninsula. *Proceedings of the Asiatic Society of Bengal, Calcutta* 1873, 22–23.

Dobson GE (1874) On the Asiatic species of *Molossi*. *Journal of the Asiatic Society of Bengal, Calcutta* 43, 142–144.

Dobson GE (1875a) Conspectus of the suborders, families, and genera of Chiroptera arranged according to their natural affinities. *Annals & Magazine of Natural History* (4)16, 345–357. [1 Nov 1875]

Dobson GE (1875b) Description of a new species of *Taphozous* from Labuan. *Annals & Magazine of Natural History* (4)16, 232. [1 Sep 1875]

Dobson GE (1875c) Description of new or little-known species of bats of the genus *Vesperugo*. *Proceedings of the Zoological Society of London* 43, 470–474. [Oct 1875]

Dobson GE (1875d) On the genus *Chalinolobus*, with descriptions of new or little-known species. *Proceedings of the Zoological Society of London* 43, 381–388. [Oct 1875]

Dobson GE (1875e) On the genus *Scotophilus*, with description of a new genus and species allied thereto. *Proceedings of the Zoological Society of London* 43, 368–373. [Oct 1875]

Dobson GE (1876a) A monograph of the genus *Taphozous*, Geoffr. *Proceedings of the Zoological Society of London* 43(1875), 546–556. [Apr 1876]

Dobson GE (1876b) Description of a new species of *Chalinolobus* from Australia. *Annals & Magazine of Natural History* (4)17, 289–290. [1 Apr 1876]

Dobson GE (1877a) On a collection of Chiroptera from Duke-of-York Island and the adjacent parts of New Ireland and New Britain. *Proceedings of the Zoological Society of London* 45, 114–123. [Jun 1877]

Dobson GE (1877b) A monograph of the Group Molossi. *Proceedings of the Zoological Society of London* 44(1876), 701–735. [Apr 1877]

Dobson GE (1878) *Catalogue of the Chiroptera in the Collections of the British Museum*. British Museum Trustees, London.

Dobson GE (1880) On some new or rare species of Chiroptera in the collection of the Gottingham Museum. *Proceedings of the Zoological Society of London* 48, 461–465. [Oct 1880]

Dobson GE (1889) In O Thomas. On the mammals of Christmas Island. *Proceedings of the Zoological Society of London* 56(1888), 532–534. [April 1889]

Dobzhansky T (1937) *Genetics and the Origin of Species*. Columbia University Press, New York.

Doescher RA (1981) Living and fossil brachiopod genera 1775–1979 lists and bibliography. *Smithsonian Contributions to Paleobiology* 42, 1–238.

Domning DP (1994) A phylogenetic analysis of the Sirenia. *Proceedings of the San Diego Society of Natural History* 29, 177–189.

Domning DP (1996) Bibliography and index of the Sirenia and Desmostylia. *Smithsonian Contributions to Paleobiology* 80, 1–611.

Donndorff JA (1798) *Amphibiologisch und Ichthyologische Beyträge* zur XIII. Ausgabe des Linnéischen Natursystems. Weidmannschen Buchhandlung, Leipzig.

Donovan E (1825) *Naturalist's Repository or Monthly Miscellany*. London. Volume 3. Plate 73. Text unnumbered.

Douady CJ, Chatelier PI, Madsen O, de Jong WW, Catzeflis F, Springer MS, Stanhope MJ (2002) Molecular phylogenetic evidence confirming the Eulipotyphla concept and in support of hedgehogs as the sister group to shrews. *Molecular Phylogenetics and Evolution* 25, 200–209. [Oct 2002]

Douady CJ, Catzeflis F, Raman J, Springer MS, Stanhope MJ (2003) The Sahara as a vicariant agent, and the role of Miocene climatic events, in the diversification of the mammalian order Macroscelidea (elephant shrews). *Proceedings of the National Academy of Sciences of the United States of America* 100, 8325–8330. [8 Jul 2003]

Douglas AM (1962) *Macroderma gigas saturata* (Chiroptera, Megadermatidae) a new subspecies from the Kimberley division of Western Australia. *Western Australian Naturalist* 8(3), 59–61. [31 Jan 1962]

Douzery EJP, Huchon D (2004) Rabbits, if anything, are likely Glires. *Molecular Phylogenetics and Evolution* 33, 922–935. [Dec 2004]

Du Bus B (1867) Sur quelques mammifères du crag d'Anvers. *Bulletins de L'Academie Royale des Sciences, des Lettres et des Beaux-Arts* (2)24, 562–577.

Du Bus B (1868) Sur différents ziphiides nouveaux du crag d'Anvers. *Bulletins de L'Academie Royale des Sciences, des Lettres et des Beaux-Arts* (2)25, 621–630.

Du Bus B (1872) Mammifères nouveaux du crag d'Anvers. *Bulletins de L'Académie Royale des Sciences, des Lettres et des Beaux-arts de Belgique* 34(2), 491–509.

Dubois A (1882) Remarques sur l'*Acathoglossus bruynii*. *Bulletin de la Société Zoologique de France* 6, 266–270.

Duckworth JW, Pine RH (2003) English names for a world list of mammals, exemplified by species of Indochina. *Mammal Review* 33, 151–173.

Ducroz JF, Volobouev V, Granjon L (2001) An assessment of the systematics of Arvicanthine rodents using mitochondrial DNA sequences: Evolutionary and biogeographical implications. *Journal of Mammalian Evolution* 8, 173–206. [Sep 2001]

Duméril AMC (1806a) *Analytische Zoologie*, ou Méthode Naturelle de Classification des Animaux, rendue plus facile à l'aide de Tableaux Synoptiques. Allais, Weimar.

Duméril AMC (1806b) *Zoologie Analytique*, ou Méthode Naturelle de Classification des Animaux, rendue plus facile à l'aide de Tableaux Synoptiques. Allais, Weimar.

Duméril AMC, Bibron G (1839) *Erpétologie Générale ou Histoire Naturelle Complète des Reptiles*. Volume 5. Libr. Encyclopédique Roret, Paris.

Duméril AMC, Bibron G (1844) *Erpétologie Générale ou Histoire Naturelle Complète des Reptiles*. Volume 6. Libr. Encyclopédique Roret, Paris.

Duncan FM (1937) On the dates of publication of the Society's 'Proceedings', 1859–1926. With an appendix containing the dates of 'Proceedings', 1830–1858, compiled by the late F.H. Waterhouse, and of the 'Transactions', 1833–1869, compiled by the late Henry Peavot, originally published in P.Z.S. 1893, 1913. *Proceedings of the Zoological Society of London* 107A, 71–84.

Duvernoy GL (1851) Mémoire sur les caractères ostéologiques des genres nouveaux ou des espèces nouvelles de cétacés vivants ou fossiles, dont les squelettes entiers, ou les têtes seulement, sont conservés dans les galéries d'anatomie comparée du Muséum d'Histoire Naturelle. *Annales des Sciences Naturelles. Zoologie* 15(3), 5–71.

Dybowski B (1929) Przyczynek do znajomości fok syberyjskich – Zur Kenntnis der Sibirischen Seehunde. *Bulletin International de l'Académie des Sciences de Cracovie (B)* 1929, 405–415.

Eaton AE (1873) On the Hydroptilidae, a family of the Trichoptera. *Transactions of the Entomological Society of London* 1873, 125–151.

Edwards D, Westerman M (1992) DNA-DNA hybridization and position of Leadbeater's possum (*Gymnobelideus leadbeateri* McCoy) in the family Petauridae (Marsupialia: Diprotodontia). *Australian Journal of Zoology* 40, 563–571.

Edwards D, Westerman M (1995) The molecular relationships of possum and glider families as revealed by DNA-DNA hybridisations. *Australian Journal of Zoology* 43, 231–240.

Ehrenberg CG (1830) In FG Hemprich & CG Ehrenberg, 1828–1845. *Symboláe Physicae seu Icones et descriptiones Mammalium* quae ex itinere Africam Borealem et Asiam Occidentalem. Decas Secunda. Berolini ex Officina Academica.

Ehrenberg CG (1833) In FG Hemprich & CG Ehrenberg, 1828–1845. *Symboláe Physicae seu Icones et descriptiones Mammalium* quae ex itinere Africam Borealem et Asiam Occidentalem. Decas Secunda. Berolini ex Officina Academica. Decas II, Folio K.

Eichwald E von (1831) *Zoologia specialis quam expositus animalibus tum vivis, tum fossilibus potissimum Rossiae in universum et Poloniae in specie*. 3 Parts in 2 volumes. Vilnae. 1829–1831.

Eick GN, Jacobs DS, Matthee CA (2005) A nuclear DNA phylogenetic perspective on the evolution of echolocation and historical biogeography of extant bats (Chiroptera). *Molecular Biology and Evolution* 22, 1869–1886. [Sep 2005]

Eldridge MDB (1997) Taxonomy of rock-wallabies, *Petrogale* (Marsupialia: Macropodidae). II. An historical overview. *Australian Mammalogy* 19, 113–122. [30 Sep 1997]

Eldridge MDB, Close RL (1992) Taxonomy of rock wallabies, *Petrogale* (Marsupialia: Macropodidae. I. A revision of the eastern *Petrogale* with the description of three new species. *Australian Journal of Zoology* 40, 605–625.

Eldridge MDB, Dollin A, Johnston PG, Close RL, Murray JD (1988) Chromosomal rearrangements in rock wallabies, *Petrogale* (Marsupialia: Macropodidae. I. The *Petrogale assimilis* species complex. G-banding and synaptonemal complex analysis. *Cytogenetics and Cell Genetics* 48, 228–232.

Eldridge MDB, Johnston PG, Close RL, Lowry PS (1989) Chromosomal rearrangements in rock wallabies, *Petrogale* (Marsupialia: Macropodidae. II. G-banding analysis of *Petrogale godmani*. *Genome* 32, 935–940.

Eldridge MDB, Close RL, Johnston PG (1990) Chromosomal rearrangements in rock wallabies, *Petrogale* (Marsupialia: Macropodidae. III. G-banding analysis of *P. inornata and P. penicillata. Genome* 33, 798–802.

Eldridge MDB, Close RL, Johnston PG (1991a) Chromosomal rearrangements in rock wallabies, *Petrogale* (Marsupialia: Macropodidae. IV. G-banding analysis of *Petrogale lateralis group. Australian Journal of Zoology* 39, 621–627.

Eldridge MDB, Johnston PG, Close RL (1991b) Chromosomal rearrangements in rock wallabies, *Petrogale* (Marsupialia: Macropodidae. V. Chromosomal phylogeny of the lateralis/ penicillata group. *Australian Journal of Zoology* 39, 629–641.

Eldridge MDB, Wilson ACC, Metcalfe CJ, Dollin AE, Bell JN, Johnson PM, Johnston PG, Close RL (2001) Taxonomy of rock-wallabies, *Petrogale* (Marsupialia: Macropodidae). III. Molecular data confirms the species status of the purple-necked rock-wallaby (*Petrogale purpureicollis* le Souef). *Australian Journal of Zoology* 49, 323–343.

Eldridge MDB, Kinnear JE, Zengler KR, McKenzie LM, Spencer PBS (2004) Genetic diversity in remnant mainland and 'pristine' island populations of three endemic Australian macropodids (Marsupialia): *Macropus eugenii, Lagorchestes hirsutus* and *Petrogale lateralis. Conservation Genetics* 5, 325–338. [Jun 2004]

Eldridge MDB, Johnson RN, Meek PD (2009) The taxonomy of the Christmas Island shrew *Crocidura attenuata trichura*. Report to the Department of the Environment, Heritage, Water and the Arts. Australian Museum, Sydney. [Apr 2009]

Eldridge MDB, Heckenberg K, Neaves LE, Metcalfe CJ, Hamilton S, Johnson PM, Close RL (2011) Genetic differentiation and introgression amongst *Thylogale* (pademelons) taxa in eastern Australia. *Australian Journal of Zoology* 59, 103–117.

Eldridge MDB, Meek PD, Johnson RN (2013) Taxonomic uncertainty and the loss of biodiversity on Christmas Island, Indian Ocean. *Conservation Biology* **28**, 572–579. [27 Nov 2013]

Elledge AE, Leung LK-P, Allen LE, Firestone K, Wilton AN (2006) Assessing the taxonomic status of dingoes *Canis familiaris dingo* for conservation. *Mammal Review* 36, 142–156.

Elledge AE, Allen LE, Carlsson BL, Wilton AN, Leung LK-P (2008) An evaluation of genetic analyses, skull morphology and visual appearance for assessing dingo purity: implications for dingo conservation. *Wildlife Research* 35, 812–820.

Ellerman JR (1940) *The Families and Genera of Living Rodents*. With a list of named forms (1758–1936) by RW Hayman and GWC Holt. Volume 1. Rodents other than the Muridae. British Museum, London.

Ellerman JR (1941) *The Families and Genera of Living Rodents*. With a list of named forms (1758–1936) by RW Hayman and GWC Holt. Volume 2. Family Muridae. British Museum, London.

Ellerman JR (1947) Notes on some Asiatic rodents in the British Museum. *Proceedings of the Zoological Society of London* 117, 259–271.

Ellerman JR (1949) *The Families and Genera of Living Rodents*. With a list of named forms (1758–1936) by RW Hayman and GWC Holt. Volume 3. Part 1. British Museum, London.

Ellerman JR (1954) In EMO Laurie & JE Hill. *List of Land Mammals of New Guinea, Celebes and Adjacent Islands 1758–1952*. British Museum, London.

Ellerman JR, Morrison-Scott TCS (1951) *Checklist of Palearctic and Indian Mammals*. British Museum, London.

Elliot DG (1901) A synopsis of the mammals of North America and the adjacent seas. *Publications of the Field Columbian Museum. Zoological Series* 2, 1–519.

Elliot DG (1907) A catalogue of the collection of mammals in the Field Columbian Museum. *Publications of the Field Columbian Museum. Zoological Series* 8, 1–694.

Elliot W (1839) A catalogue of the species of Mammalia found in the Southern Mahratta Country; with their synonymies in the native languages in use there. *The Madras Journal of Literature and Science* 10, 207–234.

Erridge N (1988) Case 275. *Camelus* Linnaeus, 1758 (Mammalia, Artiodactyla): Proposed designation of *Camelus bactrianus* Linnaeus, 1758 as type species. *Bulletin of Zoological Nomenclature* 45, 141–142. [24 Jun 1988]

Erxleben JCP (1777) *Systema Regni Animalis* per classes, ordines, genera, species, varietates cum synonymia et historia animalium. Classis 1 Mammalia. Impensis Weygandianis, Lipsiae.

Eschricht DF (1843) Iagttagelser paa nebbehvalen eller Islaendernes Andarnefia, Faeröernes Dögling. *Förhandlinger vid de Skandinaviske Naturforskeres tredje Möte, Stockholm* 3 (July 1842), 651–658.

Eschricht DF (1845) No title. *Oversigt over det Kongelige Danske Videnskabernes Selskabs Forhandlinger. Kjøbenhavn* 1845, 15–18.

Eschricht DF (1849a) Undersögelser over hvaldyrene. *Det Kongelige Danske Videnskabernes Selskabs Skrifter. Naturvidenskabelig og Mathematisk* 1(5), 87–138.

Eschricht DF (1849b) *Zoologisch-Anatomisch-Physiologisch Untersuchungen über die Nordischen Wallthiere*. Verlag Von Leopold Voss, Leipzig. Volume 1.

Eschscholtz JF von (1829) *Zoologischer Atlas, enthaltend Abbildungen und Beschreibungen neuer Thierarten, während des Flottcaauf der Russisch –Kaiserlichen Kriegsshlupp Predpriaetie in en Jahren 1823–1826*. Gebruckt und Verlegt, Berlin. Volume 1.

Eschscholtz JF von (1836) In Silbermann, G. *Revue Entomologique*. Bureau de la Revue Entomologique, Strasbourg and Lequien Fils et Roret, Paris. Volume 4.

Etheridge R (1889) General Zoology of Lord Howe Island. *Australian Museum Memoirs* 2, 1–42. [31 Dec 1889]

Evenhuis NL (1990) Dating of the livraisons and volumes of d'Orbigny's *Dictionnaire Universel d'Histoire Naturelle. Bishop Museum Occasional Papers* 30: 219–225.

Evenhuis NL (2003a) Publication and dating of the journals forming the *Annals and Magazine of Natural History* and the *Journal of Natural History. Zootaxa* 385, 1–68. [16 Dec 2003]

Evenhuis NL (2003b) Dating and publication of the *Encyclopédie Méthodique* (1782–1832), with special reference to the parts of the *Histoire Naturelle* and details on the *Histoire Naturelle des Insectes. Zootaxa* 166, 1–48. [17 Mar 2003]

Eyton TC (1838) *A Monograph on the Anatidae or Duck Tribe*. London: Longman, Orme, Brown, Green & Longman. Two unpaginated appendices.

Fabricius JC (1794) *Genera et Ordines Insectorum Linnaei, quae, in Entomologia Systematica Dni. Fabricii, tum quoad totum, tum quoad partem, mutata, reformata, atque nova & adaucta exponuntur*. Hedemora. Volume 4.

Fabricius, JC (1798) *Supplementum entomologiae systematicae*. Hafniae.

Fabricius O (1780) *Fauna Groenlandica*, systematice sistens animalia Groenlandiae occidentalis hactenus indagata, quoad nomen specificum, triviale, vernaculumque; synonyma auctorumplurium, descriptionem, locum, victum, generationem, mores, usum, capturamque singuli, proust detegendi occasio fuit, maximaque parte secundum proprias observations Othonis Fabricii, Ministri Evangeli, quondam Groenlandis ad Coloniam Friderichshab, posthac Norvagus Drangedeliae, nunc vero Danis hopunti iutiae, Memri Societatis Scientarum quae est Hafniac. Hafniae & Lipsiae, Copenhagen and Leipzig.

Fagge CH, Pye-Smith PH (1902) *The Principles and Practice of Medicine*. London. Fourth Edition. Reference not seen.

Fairmaire L (1871) Notes sur les coléopteres recueillis par Charles Coquerel à Madagascar et sur les côtes d'Afrique. *Annales de la Société Entomologique de France* 1(5), 36.

Falcimagne A (1854) *Mémoires Historiques sur l'Australie Par Mgr Rudesindo Salvado*. Alphonse Pringuet, Paris.

Falconer H (1857) Description of two species of fossil mammalian genus *Plagiaulax*. *Quarterly Journal of the Geological Society of London* 13, 261–282.

Fanzago F (1874) I Chilopoda Italiani Monografia. *Atti della Accademia Scientifica Veneto-Trentino-Istriana* 3, 17–26.

Farines JN, Carcassone C (1829) *Mémoire sur un Cétacé échoué, le 27 Novembre 1828, sur la côte dépendante de la commune de Saint-Cyprien, Pyrénées-Orientales, &c*. Perpignan. Reference not seen.

Fatio V (1869) *Faune des Vertébrés de la Suisse. Volume 1. Histoire Naturelle des Mammifères*. H. Georg, Genève et Bâle.

Fauchald K (2012) *Clymenia*. In G Read & K Fauchald (eds.) World Polychaeta database. Accessed through: World Register of Marine Species at: http://www.marinespecies.org/aphia.php?p=taxdetails&id=324788. [10 Nov 2012]

Fauchald K, Bellan G (2012) *Fabricia* Blainville, 1828. In G Read & K Fauchald (eds.) World Polychaeta database. Accessed through: World Register of Marine Species at http://www.marinespecies.org/aphia.php?p=taxdetails&id=129529. [5 Jun 2012]

Faxon W (1878) On the presence of *Demodex folliculum* in the skin of the ox. *Bulletin of the Museum of Comparative Zoology. Harvard University* 5, 11–16.

Fedorowski J, Bamber EW, Stevens CH (2007) *Lower Permian Collonial Rugose Corals, Western and Northwestern Pangea: Taxonomy and Distribution*. National Research Council Research Press, Ottawa.

Feiler A (1978a) Üeber artliche Abgrenzung und innerartliche Ausformung bei *Phalanger maculatus*. *Zoologische Abhandlungen / Staatliches Museum fur Tierkunde in Dresden* 35, 1–30.

Feiler A (1978b) Bemerkungen über Phalanger der '*orientalis*'-Gruppe' nach Tate (1945). *Zoologische Abhandlungen / Staatliches Museum fur Tierkunde in Dresden* 34, 385–395.

Felder G-R (1869) Lépidoptères, *Globiceps paradoxa*. *Petites Nouvelles Entomologiques* 1(8), 4, 30–31.

Feldhamer GA, Farris-Renner KC, Barker CM (1988) *Dama dama*. *Mammalian Species* 317, 1–8. [27 Dec 1988]

Felten H (1964) Zur taxonomie indo-australischer Fledermäuse der Gattung *Tadarida* (Mammalia, Chiroptera). *Senckenbergiana Biologica* 45, 1–13.

Fernicola JC, Vizcaino SF, De Iuliis G (2009) The fossil mammals collected by Charles Darwin in South America during his travels on board the HMS Beagle. *Revista de la Asociación Geológica Argentina* 64, 147–159.

Ferraris CJ Jr (2007) Checklist of catfishes, recent and fossil (Osteichthyes: Siluriformes), and catalogue of siluriform primary types. *Zootaxa* 1418, 1–628. [8 Mar 2007]

Férussac JBL de A (1821) *Tableau systématique des animaux mollusques classés en familles naturelles, dans lesquelles on a établi la concordance de tous les systèmes; suivis d'un prodrome général pour tous les mollusques terrestres oufluviatiles vivants ou fossiles*. Arthus Bertrand, Paris.

Filewood LW (1983) The possible presence in New Guinea of the ghost bat (*Macroderma gigas*: Chiroptera, Megadermatidae). *Australian Mammalogy* 6, 35–36. [1 May 1983]

Filhol H (1882) Étude des mammifères fossiles de Ronzon. *Annales des Sciences Géologiques de Paris* 12(Art 3), 1–270.

Finarelli JA (2008) A total evidence phylogeny of the Arctoidea (Carnivora: Mammalia): Relationships among basal taxa. *Journal of Mammalian Evolution* 15, 231–259. [29 Jan 2008]

Findley JS (1972) Phenetic relationships among bats of the genus *Myotis*. *Systematic Zoology* 21, 31–52. [Mar 1972]

Finlayson HH (1927) Observations on the South Australian members of the subgenus, '*Wallabia*. *Transactions of the Royal Society of South Australia* 51, 363–377. [23 Dec 1927]

Finlayson HH (1930) Observations on the South Australian members of the subgenus, 'Wallabia'. Part II. *Transactions of the Royal Society of South Australia* 54, 47–56. [20 Dec 1930]

Finlayson HH (1931) On mammals from the Dawson Valley, Queensland. Part 1. *Transactions of the Royal Society of South Australia* 55, 67–89. [24 Dec 1931]

Finlayson HH (1932) Preliminary descriptions of two new mammals from South Australia. *Transactions of the Royal Society of South Australia* 56, 168–171. [24 Dec 1932]

Finlayson HH (1933a) On mammals from the Lake Eyre Basin. Part 1. Dasyuridae. *Transactions of the Royal Society of South Australia* 57, 195–202. [23 Dec 1933]

Finlayson HH (1933b) On the Eremian representative of *Myrmecobius fasciatus* (Waterhouse). *Transactions of the Royal Society of South Australia* 57, 203–205. [23 Dec 1933]

Finlayson HH (1933c) On *Mastacomys fuscus* (Thomas). *Transactions of the Royal Society of South Australia* 57, 125–129. [23 Dec 1933]

Finlayson HH (1934) On mammals from the Dawson and Fitzroy Valleys; central coastal Queensland. Part 2. *Transactions of the Royal Society of South Australia* 58, 218–231. [22 Dec 1934]

Finlayson HH (1935a) On the mammals of the Lake Eyre Basin. Part II. The Peramelidae. *Transactions of the Royal Society of South Australia* 59, 227–236. [23 Dec 1935]

Finlayson HH (1935b) Notes on some Victorian mammals. *Transactions of the Royal Society of South Australia* 59, 221–226. [23 Dec 1935]

Finlayson HH (1938) On a new species of *Potorous* (Marsupialia) from a cave deposit on Kangaroos Island, South Australia. *Transactions of the Royal Society of South Australia* 62, 132–140. [22 Jul 1938]

Finlayson HH (1939a) On mammals from the Lake Eyre Basin. Part IV. The Monodelphia. *Transactions of the Royal Society of South Australia* 63, 88–118. [28 Jul 1939]

Finlayson HH (1939b) Records and descriptions of Muridae from Ooldea, South Australia. *Transactions of the Royal Society of South Australia* 63, 354–364. [22 Dec 1939]

Finlayson HH (1940) On central Australian mammals. Part I. The Muridae. *Transactions of the Royal Society of South Australia* 64, 125–136. [26 Jul 1940]

Finlayson HH (1941) On central Australian mammals. Part II. The Muridae. *Transactions of the Royal Society of South Australia* 65, 215–232. [19 Dec 1941]

Finlayson HH (1942) A new *Melomys* from Queensland with notes of two other Queensland rats. *Transactions of the Royal Society of South Australia* 66, 243–247. [18 Dec 1942]

Finlayson HH (1943) A new species of *Lagorchestes* (Marsupialia). *Transactions of the Royal Society of South Australia* 67, 319–321. [30 Jul 1943]

Finlayson HH (1957) Preliminary description of two new forms of *Bettongia* (Marsupialia). *Annals & Magazine of Natural History* (12)10, 552–554. [29 Oct 1957]

Finlayson HH (1958a) A case of duplex convergent resemblance in Australian mammals, with a review of some aspects of the morphology *Phascogale (Antechinus) swainsonii* Waterhouse and *Phascogale (Antechinus) flavipes* Waterhouse. *Transactions of the Royal Society of South Australia* 81, 141–151. [Mar 1958]

Finlayson HH (1958b) On central Australian mammals (with notice of related species from adjacent tracts). Part III. Potoroinae. *Records of the South Australian Museum* 13, 235–302. [15 Aug 1958]

Finlayson HH (1960a) Nomenclature of *Notomys* (Muridae) in the Lake Eyre Basin. *Transactions of the Royal Society of South Australia* 83, 79–82. [Mar 1960]

Finlayson HH (1960b) *Rattus greyi* Gray and its derivatives. *Transactions of the Royal Society of South Australia* 83, 123–147. [Mar 1960]

Finlayson HH (1961a) On central Australian mammals. Part IV. The distribution and status of central Australian species. *Records of the South Australian Museum* 14, 141–191. [8 Aug 1961]

Finlayson HH (1961b) Mitchell's wombat in South Australia. *Transactions of the Royal Society of South Australia* 85, 207–215. [Nov 1961]

Finlayson HH (1963) The brush-tailed opossum of Kangaroo Island, South Australia. *Transactions of the Royal Society of South Australia* 87, 17–22. [Dec 1963]

Firestone KB (2000) Phylogenetic relationships among quolls revisited: the mtDNA control region as a useful tool. *Journal of Mammalian Evolution* 7, 1–22. [Mar 2000]

Firestone KB, Elphinstone MS, Sherwin WB, Houlden BA (1999) Phylogeographical population structure of tiger quoll *Dasyurus maculatus* (Dasyuridae: Marsupialia), an endangered carnivorous marsupial. *Molecular Ecology* 8, 1613–1625. [Oct 1999]

Fischer G (1803) *Das National Museum der Naturgeschichte zu Paris*. Von seinem ersten Ursprunge bis zu seinem jetzigen Glanze geschildert. Frankfurt-am-Main, Paris. Volume 2.

Fischer G (1809) Sur l'Elasmotherium et le Trogonthérium deux animaux fossiles et inconnus de la Russie. *Mémoires de la Société Imperiale des Naturalistes de Moscou* 2, 250–268.

Fischer G (1811) Note de quelques animaux rares qui se trouvent au Muséum Impérial d'Histoire de Moscou décrits. *Mémoires de la Société Impériale des Naturalistes de Moscou* (second edition) 1, 9–13.

Fischer G (1813a) *Zoognosia tabulis synopticis illustrata* : in usum praelectionum Academiae imperialis medico-chirugicae mosquensis edita. Typis Nicolai S. Vsevolozsky, Mosquae. Third Edition. Volume 1.

Fischer G (1813b) *Zoognosia tabulis synopticis illustrata* : in usum praelectionum Academiae imperialis medico-chirugicae mosquensis edita. Typis Nicolai S. Vsevolozsky, Mosquae. Third Edition. Volume 2.

Fischer G (1814) *Zoognosia Tabulis Synopticis Illustrata, in usum praelectionum Academiae Imperialis Medico-Chirurgicae Mosquensis*. Nicolai Sergeidis Vsevolozsky, Mosquae. Third Edition. Volume 3.

Fischer G (1817) Adversaria Zoologica. *Mémoires de la Société Impériale des Naturalistes de Moscou* 5, 357–428.

Fischer G (1821) *Entomographia Imperii Russici. Auctoritate Societatis Caesareae Mosquesis Naturae Scrutatorum Collecta et in Lucum Edita. Genéra Insectorum Systematice Exposiita et Analysi Iconographica Instructa*. Volume Primum Genera Coleopterorum. Augusti Semen Excusa, Mosquae.

Fischer JB (1829) *Synopsis Mammalium*. (Addenda, emendanda et index.). J.G. Cottae, Stuttgardtiae.

Fischer P (1876) Sur une espèce de Cétacé (*Orca Antarctica*). *Journal de Zoologica* 5, 146–151.

Fischer, P (1881) Cétacés du sud-ouest de la France. *Actes de la Société Linnéenne de Bordeaux* 35(4)5, 5–219.

Fitzgerald EMG (2010) The morphology and systematics of *Mammalodon colliveri* (Cetacea: Mysticeti), a toothed mysticete from the Oligocene of Australia. *Zoological Journal of the Linnean Society* 158, 367–476. [published online 21 Dec 2009]

Fitzgerald EMG (2012) Possible neobalaenid from the Miocene of Australia implies a long evolutionary history for the pygmy right whale *Caperea marginata* (Cetacea, Mysticeti). *Journal of Vertebrate Paleontology* 32, 976–980. [published online 26 Jun 2012]

Fitzinger LI (1826) *Neue Classification der Reptilien nach ihren natürlichen Verwandtschaften : Nebst einer Verwandtschafts-Tafel und einem Verzeichnisse der Reptilien-Sammlung des K. K. Zoologischen Museum's zu Wien*. Von J.G. Heubner, Wien.

Fitzinger LJ (1843) *Systema Reptilium*. Fasciculus primus. Amblyglossae. Braumüller et Seidel, Wien.

Fitzinger LJ (1860a) *Wissenschaftliche-populäre. Naturgeschichte der Säugethiere* in ihren sammtlichen Hauptformen. Nebst einer Einleitung in die Naturgeschichte uberhaupt und in die Lehre von den Thieren insbesondere. Wien. Volume 5.

Fitzinger LJ (1860b) *Wissenschaftlich-populäre Naturgeschichte der Säugethiere* in ihren sammtlichen hauptformen. Nebst einer ein leitung in die naturgeschichte uberhaupt und in die lehre von den thieren insbesondere. Aus der Kaiserlich-koniglichen hafund staatsdruckerei, Wien. Volume 6.

Fitzinger LJ (1861) Die ausbeute der österreichischen näturforscher an säugethieren und reptilien während der weltumsegelung Sr. Majestät Fregatte Novara. *Sitzungsberichte der Kaiserlichen Akademie der Wissenschaften Mathematisch-Naturwissenschaftliche Classe, Wien* 42, 383–416.

Fitzinger LJ (1870a) Kritische durchsicht der ordung der flatterthiere oder handflüger (Chiroptera). Familie der fledermäuse (Vespertiliones). *Sitzungsberichte Mathematisch-Naturwissenschaftliche Classe der Kaiserlichen Akademie der Wissenschaften, Wien* 62, 13–144.

Fitzinger LJ (1870b) Kritische Durchsicht der Ordung der Flatterthiere oder handflüger (Chiroptera). Familie der fledermäuse (Vespertiliones). *Sitzungsberichte Mathematisch-Naturwissenschaftliche Classe der Kaiserlichen Akademie der Wissenschaften, Wien* 62, 353–438.

Fitzinger LJ (1870c) Kritische Durchsicht der Ordung der Flatterthiere oder handflüger (Chiroptera). Familie der fledermäuse (Vespertiliones). *Sitzungsberichte Mathematisch-Naturwissenschaftliche Classe der Kaiserlichen Akademie der Wissenschaften, Wien* 62, 527–579.

Flannery TF (1983) Revision of the macropod subfamily Sthenurinae (Marsupialia: Macropodidae) and the relationships of the *Troposodon* and *Lagostrophus*. *Australian Mammalogy* 6, 15–28. [Mar 1983]

Flannery TF (1984) Kangaroos: 15 million years of Australian bounders. pp. 817–835, in M Archer & G Clayton (eds.) *Vertebrate Evolution and Zoogeography in Australasia*. Hesperian Press, Perth.

Flannery TF (1989) Phylogeny of the Macropodoidea: A study in convergence. pp. 1–46, in G Grigg, P Jarman & I Hume (eds.) *Kangaroos, Wallabies and Rat-Kangaroos*. Surrey Beatty, Sydney. Volume 1.

Flannery T (1990) *Mammals of New Guinea*. Australian Museum, Robert Brown & Associates, Carina, Queensland.

Flannery TF (1994) *Possums of the World: A Monograph of the Phalangeroidea*. GEO, Sydney.

Flannery T (1995a) *Mammals of New Guinea*. Australian Museum/Reed Books, Sydney. Second Edition.

Flannery T (1995b) *Mammals of the South-West Pacific & Moluccan Islands*. Reed Books, Sydney.

Flannery TF, Archer M (1984) The macropodoids (Marsupialia) of the early Pliocene Bow Local Fauna, central eastern New South Wales. *Australian Zoologist* 21, 357–383.

Flannery TF, Archer M (1987) *Bettongia moyesi*, A new and plesiomorphic kangaroo (Marsupialia: Potoroidae) from Miocene sediments of northwestern Queensland. Volume 2. pp. 759–767, in M Archer (ed.) *Possums & Opossums: Studies in Evolution*. Surrey Beatty and Sons, Sydney. [Nov 1987]

Flannery TF, Groves CP (1998) A revision of the genus *Zaglossus* (Monotremata, Tachyglossidae), with description of new species and subspecies. *Mammalia* 62, 367–396. [29 Dec 1998]

Flannery TF, Hann LM (1984) A new macropodine genus and species (Marsupialia: Macropodidae) from the early Pleistocene of southwestern Victoria. *Australian Mammalogy* 7, 193–204. [Dec 1984]

Flannery TF, Rich THR (1986) Macropodoids of the middle Miocene Namba Formation, South Australia, and the homology of some dental structures in kangaroos. *Journal of Paleontology* 60, 418–447. [Mar 1986]

Flannery T, Schouten P (2001) *A Gap in Nature: Discovering the World's Extinct Animals*. Text Publishing, Melbourne.

Flannery TF, Szalay FS (1982) *Bohra paulae*, a new giant fossil tree kangaroo (Marsupialia: Macropodidae) from New South Wales, Australia. *Australian Mammalogy* 5, 83–94. [18 May 1982]

Flannery TF, Archer M, Plane M (1983) Middle Miocene kangaroos (Macropodoidea: Marsupialia) from three localities in northern Australia, with a description of two new subfamilies. *BMR Journal of Australian Geology and Geophysics* 7, 287–302.

Flannery T, Archer M, Plane M (1984) Phylogenetic relationships and a reconsideration of higher level systematics within the Potoroidae (Marsupialia). *Journal of Paleontology* 58, 1087–1097. [Jul 1984]

Flannery TF, Archer M, Maynes G (1987) The phylogenetic relationships of living phalangerids with a suggested new taxonomy. Volume 2. pp. 477–506, in M Archer (ed.) *Possums & Opossums: Studies in Evolution*. Surrey Beatty and Sons, Sydney. [Nov 1987]

Flannery TF, Martin R, Szalay A (1996) *Tree Kangaroos: A Curious Natural History*. Reed Books, Melbourne.

Fleay D (1947) *Gliders of the Gum Trees*. Bread and Cheese Club, Melbourne.

Fleay D (1954) The squirrel glider. *Victorian Naturalist* 70, 208–210. [4 Mar 1954]

Fleming J (1822a) *The Philosophy of Zoology; or a general view of the structure, function, and classification of animals*. Archbold Constable and Co., Edinburgh. Volume 1.

Fleming J (1822b) *The Philosophy of Zoology; or a general view of the structure, function, and classification of animals*. Archbold Constable and Co., Edinburgh. Volume 2.

Fleming J (1828) *History of British Animals*. Bell & Bradfute, Edinburgh.

Fletcher JJ (1882) On the existence after parturition of a direct communication between the median vaginal cul-de-sac so called, and the urogenital canal, in certain species of kangaroo. *Proceedings of the Linnean Society of New South Wales* (1)6, 796–811. [20 Mar 1882]

Fletcher JJ (1896) On the dates of publication of the early volumes of the Society's Proceedings. *Proceedings of the*

Linnean Society of New South Wales (2)10, 533–536. [29 Apr 1896]

Flood J (1989) *Archaeology of the Dreamtime*. Angus & Robertson, Sydney. Revised Edition.

Flourens MJP (1853) Eloge historique d'Ètienne Geoffroy Saint-Hilaire. *Mémoires de l'Académie des Sciences de l'Institut de France* 23, i–lxxi.

Flower WH (1865a) Notes on the skeletons of whales in the principal museums of Holland and Belgium, with descriptions of two species apparently new to science. *Proceedings of the Zoological Society of London* 32(1864), 384–420. [May 1865]

Flower WH (1865b) A new species of Grampus (*Orca meridionalis*) from Tasmania. *Proceedings of the Zoological Society of London* 32(1864), 420–426. [May 1865]

Flower WH (1867) Description of the skeleton of *Inia geoffrensis* and the skull of *Pontoporia blainvillii*, with remarks on the systematic position of these animals in the Order Cetacea. *Transactions of the Zoological Society of London* 6, 87–116. [28 Mar 1867]

Flower WH (1869) On the value of the characters of the base of the cranium in the classification of the Order Carnivora, and on the systematic position of *Bassaris* and other disputed forms. *Proceedings of the Zoological Society of London* 37, 4–37. [Jun 1869]

Flower WH (1872) On the recent ziphoid whales, with a description of the skeleton of *Berardius arnouxi*. *Transactions of the Zoological Society of London* 8, 203–234. [Sep 1872]

Flower WH (1878a) A further contribution to our knowledge of the existing ziphoid whales, of the genus *Mesoplodon*. *Proceedings of the Zoological Society of London* 45(1877), 684. [Mar 1878]

Flower WH (1878b) A further contribution to the knowledge of the existing ziphoid whales, genus *Mesoplodon*. *Transactions of the Zoological Society of London* 10, 415–437.

Flower WH (1881) On the elephant seal, *Macrorhinus leoninus* Linn. *Proceedings of the Zoological Society of London* 49, 145–162. [Jun 1881]

Flower WH (1882) On the cranium of a new species of *Hyperoodon* from the Australian seas. *Proceedings of the Zoological Society of London* 50, 392–396. [Oct 1882]

Flower WH (1883) On the arrangement of the orders and families of existing Mammalia. *Proceedings of the Zoological Society of London* 51, 178–186. [Aug 1883]

Flower WH (1884) On the characters and divisions of the family Delphinidae. *Proceedings of the Zoological Society of London* 51(1883), 466–513. [Apr 1884]

Flower WH (1885) *List of Specimens of Cetacea in the Zoological Department of the British Museum*. British Museum, London.

Flower WH, Garson JG (1884) *Catalogue of the Specimens Illustrating the Osteology and Dentition of Vertebrated Animals, Recent and Extinct, Contained in the Museum of the Royal College of Surgeons of London. Part II. Class Mammalia, other than Man*. Royal College of Surgeons, London.

Flower WH, Lydekker R (1891) *An Introduction to the Study of Mammals Living and Extinct*. Adam & Charles Black, London.

Flynn JJ, Wesley-Hunt GD (2005) Carnivora. pp. 175–198, in KD Rose & JD Archibald (eds.) *The Rise of Placental Mammals*. The John Hopkins University Press, Baltimore, Maryland.

Flynn JJ, Parrish JM, Rakotosamimanana B, Simpson WF, Wyss AR (1999) A Middle Jurassic mammal from Madagascar. *Nature* 401(6748), 57–60. [2 Sep 1999]

Flynn JJ, Finarelli JA, Zehr S, Hsu J, Nedbal MA (2005) Molecular phylogeny of the Carnivora (Mammalia): Assessing the impact of increased sampling on resolving enigmatic relationships. *Systematic Biology* 54, 317–337.

Flynn JJ, Finarelli JA, Zehr S, Hsu J, Nedba MA (2005) Molecular phylogeny of the Carnivora (Mammalia): Assessing the impact of increased sampling on resolving enigmatic relationships. *Systematic Biology* 54, 317–337. [Apr 2005]

Flynn TT (1911) Notes on marsupialian anatomy. 1. On the condition of the median vaginal septum in the Trichosuridae. *Papers and Proceedings of the Royal Society of Tasmania* 1911, 120–123.

Foettinger A (1881) Recherches sur quelques infusoires nouveaux, parasites des céphalopodes. *Archives de Biologie* 2, 345–378.

Foley NM, Thong VD, Soisook P, Goodman SM, Armstrong KN, Jacobs D, Peuchmaille SJ, Teeling EC (2015) How and why overcome the impediments to resolution: Lessons from rhinolophid and hipposiderid bats. *Molecular Biology and Evolution* 32, 313–333.

Ford F (2003) *Conilurine Rodent Evolution: The role of ecology in modifying evolutionary consequences of environmental change*. PhD Thesis. James Cook University, Townsville.

Ford F (2006) A splitting headache: relationships and generic boundaries among Australian murids. *Biological Journal of the Linnean Society. Linnean Society of London* 89, 117–138.

Ford JKB (2009) Killer whale *Orcinus orca*. pp. 650–657, in WF Perrin, B Würsig & JGM Thewissen (eds.) *Encyclopedia of Marine Mammals*. Academic Press, Amsterdam. Second Edition.

Fordyce RE (2009) Neoceti. pp. 758–763, in WF Perrin, B Würsig & JGM Thewissen (eds.) *Encyclopedia of Marine Mammals*. Academic Press, Amsterdam. Second Edition.

Fordyce RE, Barnes LG (1994) The evolutionary history of whales and dolphins. *Annual Review of Earth and Planetary Sciences* 22, 419–455.

Fordyce RE, de Muizon C (2001) Evolutionary history of the cetaceans: a review. pp. 169–233, in J-M Mazinand & V Buffrénil (eds.) *Secondary Adaptations of Tetrapods to Life in Water*. Verlag Dr. Friedrich Pfeil, München, Germany.

Fordyce RE, Marx FG (2012) The pygmy right whale *Caperea marginata*: the last of the cetotheres. *Proceedings of the Royal Society B Biological Sciences* 280, 1–6.

Forskål P (1775) *Descriptiones animalium avium, amphibiorum, piscium, insectorum, vermium*. Heineck & Faber, Hauniae.

Forster JR (1770) In P Kalm (ed.) *Travels into North America*; containing its natural history, and a circumstantial account of its plantations and agriculture in general, with the civil, ecclesiastical and commercial state of the country, the manners of the inhabitants, and several curious and important remarks on various subjects. Translated into English by John

Reinhold Forster. Enriched with a map, several cuts for the illustration of natural history, and some additional notes. William Eyres, Warrington. Volume 1.

Forster JR (1778) Beskrifning på djuret *Yerbua capensis*, med anmarkningar om genus *Yerbuae. Kongliga Svenska Vetenskaps Academiens Handlingar* 39, 108–119.

Forster JR (1788) *Enchiridion Historiae Naturali Inserviens, quo, termini et delineationes ad avium, piscium, insectorum et plantarum adumbrationes intelligendas et concinnandas, secundum methodum systematis Linnaeani continentur.* Apud Hemmerde et Schwetschke, Halae.

Forster JR (1844) *Descriptiones Animalium* quae in Itinere ad Maris Austalis Terras per Annos 1772, 1773 et 1774 suscepto collegit observavit et delineavit. Berolini Ex Officina Academica. Written in about 1775 but not published until 1844.

Fox BJ, Briscoe DA (1980) *Pseudomys pilligaensis*, a new species of murid rodent from the Pilliga scrub, northern New South Wales. *Australian Mammalogy* 3, 109–126. [May 1980]

Fox S (2011) The spectacled flying fox – a review of past and present knowledge. pp. 136–145, in B Law, P Eby, D Lunney & L Lumsden (eds.) *The Biology and Conservation of Australasian Bats*. Royal Zoological Society of NSW, Sydney. [Sep 2011]

Frankham GJ, Handasyde KA, Eldridge MDB (2012) Novel insights into the phylogenetic relationships of the endangered marsupial genus *Potorous. Molecular Phylogenetics and Evolution* 64, 592–602.

Fraser FC (1950) Two skulls of *Globicephala macrorhyncha* (Gray) from Dakar. pp. 49–60, in AF Brunn (ed.) *Atlantide Report No. 1: Scientific Results of the Danish Expedition to the Coasts of Tropical West Africa, 1945–1946*. Danish Science Press, Copenhagen.

Fraser FC (1951) The specific name of the northern pilot whale or blackfish. *Annals & Magazine of Natural History* (12)4, 942–944. [1–5 Sep 1951]

Fraser FC (1956) A new Sarawak dolphin. *Sarawak Museum Journal* 7(8), 478–503.

Fraser FC (1966) Comments on the Delphinoidea. pp. 7–31, in KS Norris (ed.) *Whales, Dolphins and Porpoises*. University of California Press, Berkeley.

Fraser FC, Purves PE (1960) Hearing in cetaceans: Evolution of the accessory air sacs and the structure and function of the outer and middle ear in recent cetaceans. *Bulletin of the British Museum (Natural History). Zoology Series* 7(1), 1–140. [4 Nov 1960]

Fraser L (1843) Exhibition of a specimen of the *Galago senegalensis* and description of a new species of shrew from Fernando Po. *Proceedings of the Zoological Society of London* 10(1842), 200–201. [Feb 1843]

Frechkop S (1932) Notes sur les mammifères XI. Un grand rat de l'île Waïgeu (Nouvelle Guinée). *Bulletin du Musée Royal d'Histoire Naturelle de Belgique* 28(8), 1–11.

Freedman AH, Gronau I, Schweizer RM, Ortega-Del Vecchyo D, Han E, Silva PM, Galaverni M, Fan Z, Marx P, Lorente-Galdos B, Beale H, Ramirez O, Hormozdiari F, Alkan C, Vilà C, Squire K, Geffen E, Kusak J, Boyko AR, Parker HG, Lee C, Tadigotla V, Siepel A, Bustamante CD, Harkins TT, Nelson SF, Ostrander EA, Marques-Bonet T, Wayne RK & Novembre J (2014) Genome sequencing highlights the dynamic early history of dogs. *PLoS Genetics* 10(1), e1004016. [16 Jan 2014]

Freedman L, Rightmire GP (1971) Skull and tooth variation in Australian bandicoots (Peramelidae, Marsupialia): the genus *Isoodon* and multivariate comparisons with *Perameles. Journal of the Royal Society of Western Australia* 54, 21–31.

Freeman PW (1981) A multivariate study of the Family Molossidae (Mammalia, Chiroptera): morphology, ecology, evolution. *Fieldiana Zoology*, new series 7, 1–173. [31 Mar 1981]

Frenzel J (1885) Ueber einige in seethiren lebende gregarinen. *Archiv für Mikroskopische Anatomie* 24, 545–588.

Frère CH, Hale PT, Porter L, Cockcroft VG, Dalebout ML (2008) Phylogenetic analysis of mtDNA sequences suggests revision of humpback dolphin (*Sousa* spp.) taxonomy is needed. *Marine and Freshwater Research* 59, 259–268.

Frère CH, Seddon J, Palmer C, Porter L, Parra GJ (2011) Multiples lines of evidence for an Australasian geographic boundary in the Indi-Pacific humpback dolphin (*Sousa chinensis*): population or species divergence? *Conservation Genetics* 12, 1633–1638. [Dec 2011]

Freyman B, Schuchmann KL (2005) Collecting history of the hummingbird genera *Chaetocercus* Gray, 1855 and *Lophornis* Lesson, 1829. *Journal fur Ornithologie* 146, 61–64.

Frick C (1926) The Hemicyoninae and an American Tertiary bear. *Bulletin of the American Museum of Natural History* 56, 1–119.

Friend JA (1989) Myrmecobiidae. pp. 583–590, in DW Walton & BJ Richardson (eds.) *Fauna of Australia. Volume 1B. Mammalia*. Australian Government Publishing Service, Canberra.

Frisch JL (1775) *Das Natur-System der Vierfüssigen Thiere*, in Tabellen, darinnen alle Ordnungen, Geschlechte und Arten, nicht nur mit bestimmenden Benennungen, sondern beygesetzten underscheidenden Kennzeichen angezeigt werden, zum Nutzen der erwachsenen Schuljugend. C.F. Gunter, Glogau.

Frith HJ, Calaby JH (1969) *Kangaroos*. Cheshire, Melbourne.

Froriep LF (1806) In C Duméril's *Analytische Zoologie*, ou Méthode Naturelle de Classification des Animaux, rendue plus facile à l'aide de Tableaux Synoptiques. Allais, Weimar.

Frost DR (1985) *Amphibian Species of the World: A Taxonomic and Geographic Reference*. Allen Press Inc. & The Association of Systematics collections, Kansas.

Frost DR (2011) Amphibian Species of the World: an Online Reference. Version 5.5 (31 January, 2011). Electronic Database accessible at http://research.amnh.org/vz/herpetology/amphibia/. American Museum of Natural History, New York, USA.

Fulton TL, Strobeck C (2006) Molecular phylogeny of the Arctoidea (Carnivora): Effect of missing data on supertree and matrix analyses of multiple gene data sets. *Molecular Phylogenetics and Evolution* 41, 165–181.

Fulton TL, Strobeck C (2010a) Multiple fossil calibrations, nuclear loci and mitochondrial genomes provide new insight into biogeography and divergence timing for true seals (Phocidae, Pinnipedia). *Journal of Biogeography* 37, 814–829.

Fulton TL, Strobeck C (2010b) Multiple markers and phylogenetic individuals refine true seal phylogeny and bring molecules and morphology back in line. *Proceedings of the Royal Society of London. Series B, Biological Sciences* 277, 1065–1070.

Furlan E, Umina PA, Mitrovski PJ, Gust N, Griffiths J, Weeks AR (2010) High levels of genetic divergence between Tasmanian and Victorian platypuses, *Ornithorhynchus anatinus*, as revealed by microsatellite loci. *Conservation Genetics* 11, 319–323.

Gabriel, SI, Stevens, MI, Mathias, MdL., Searle, JB (2011) Of mice and 'convicts': Origin of the Australian House Mouse, Mus musculus. *PLoS ONE* 6(12), e28622.

Gadow H (1892) On the systematic position of *Notoryctes typhlops*. *Proceedings of the Zoological Society of London* 60, 361–370. [Oct 1892]

Gaimard P (1823) Sur une nouvelle espèce de kanguroo lanieux (*Kangurus lanigeri*). *Bulletin des Sciences par la Societe Philomatique de Paris* 1823, 138–139.

Gaines CA, Hare MP, Beck SE, Rosenbaum HC (2005) Nuclear markers confirm taxonomic status and relationships among highly endangered and closely related right whale species. *Proceedings of the Royal Society of London. Series B, Biological Sciences* 272, 533–542. [7 Mar 2005]

Gales N, Adams M, Burton HR (1989) Genetic relatedness of two populations of the southern elephant seal. *Marine Mammal Science* 5, 57–67. [Jan 1989]

Gamble R (1985a) Fin whale *Balaenoptera physalus* (Linnaeus, 1758). pp. 171–192, in SH Ridgway & R Harrison (eds.) *Handbook of Marine Mammals. Volume 3. The Sirenians and Baleen Whales*. Academic Press, London.

Gamble R (1985b) Sei whale *Balaenoptera borealis* Lesson, 1828. pp. 155–170, in SH Ridgway & R Harrison (eds.) *Handbook of Marine Mammals. Volume 3. The Sirenians and Baleen Whales*. Academic Press, London.

Gardiner BG (1982) Tetrapod classification. *Zoological Journal of the Linnean Society* 74, 207–232. [Mar 1982]

Gardner AL (2005a) Order Pilosa. pp. 100–103, in DE Wilson & DA Reeder (eds.) *Mammal Species of the World. A Taxonomic and Geographic Reference*. Johns Hopkins University Press, Baltimore. Third Edition.

Gardner AL (2005b) Order Didelphimorphia. pp. 3–18, in DE Wilson & DA Reeder (eds.) *Mammal Species of the World. A Taxonomic and Geographic Reference*. Johns Hopkins University Press, Baltimore. Third Edition.

Gardner AL (2007) *Mammals of South America. Volume 1. Marsupials, Xenarthrans, Shrews, and Bats*. Chicago University Press, Chicago.

Gardner AL, Hayssen V (2004) A guide to constructing and understanding synonymies for mammalian species. *Mammalian Species* 739, 1–17. [13 Jul 2004]

Gardner AL, Robbins CB (1998) Generic names of northern and southern fur seals (Mammalia: Otariidae). *Marine Mammal Science* 14, 544–551. [Jul 1998]

Gardner AL, Robbins CB (1999) Case 3058. *Arctocephalus* F. Cuvier, 1826 and *Callorhinus* Gray, 1859 (Mammalia, Pinnipedia): proposed conservation by the designation of *Phoca pusilla* Schreber,. [1775] as the type species of *Arctocephalus*; and *Otaria* Péron, 1816 and *Eumetopias* Gill, 1866: proposed conservation by the designation of *Phoca leonina* Molina, 1782 as the type species of *Otaria*. *Bulletin of Zoological Nomenclature* 56, 136–141. [30 Jun 1999]

Garrison EG, McFall G, Cherkinsky A, Scott EN (2012) Discovery of a Pleistocene mysticete whale, Georgia Bight (USA). *Palaeontologia Electronica* 15(3), 1–10. [Nov 2012]

Garrod AH (1875) On the kangaroo called *Halmaturus luctuosus* by D'Albertis and its affinities. *Proceedings of the Zoological Society of London* 43, 48–59. [Jun 1875]

Gaskin DE (1972) *Whales, Dolphins, and Seals, with special reference to the New Zealand region*. St. Martin's Press, New York.

Gaskin LJP, Lewis E (1956) On the '*Tabula Synoptica*' and the '*Observations Entomologiques*' of F.A. Bonelli. *Journal of the Society for the Bibliography of Natural History* 3, 158–164. [Jan 1956]

Gatesy J, Hayashi C, Cronin M, Arcander P (1996) Evidence from milk casein genes that cetaceans are close relatives of hippopotamid artiodactyls. *Molecular Biology and Evolution* 13, 954–963. [Sep 1996]

Gatesy J, Milinkovitch M, Waddell V, Stanhope M (1999) Stability of cladistic relationships between Cetacea and higher-level artiodactyl taxa. *Systematic Biology* 48, 6–20. [Mar 1999]

Gatesy J, Matthee C, DeSalle R, Hayashi C (2002) Resolution of a supertree/supermatix paradox. *Systematic Biology* 51, 652–664. [Aug 2002]

Gatesy J, Geisler JH, Chang J, Buell C, Berta C, Meredith RW, Springer MS, McGowna MR (2012) A phylogenetic blueprint for a modern whale. *Molecular Phylogenetics and Evolution* 66, 479–506. [Feb 2013]

Gattolliat J-L (2002) Two new genera of Baetidae (Ephemeroptera; Insecta) from Madagascar. *Aquatic Insects* 24, 143–159.

Gattolliat J-L, Jacobus LM (2010) *Madaechinopus*, a replacement name for *Echinopus* Gattolliat, 2002 (Ephemeroptera: Baetidae), nec Fisher, 1814 (Monotremata: Tachyglossidae) and nec Schenherr, 1836 (Coleoptera: Curculionidae). *Aquatic Insects* 32, 159–160. [online 8 Jun 2010]

Geisler JH, Sanders AE (2003) Morphological evidence for the phylogeny of Cetacea. *Journal of Mammalian Evolution* 10, 23–129.

Geisler JH, Theodor JM (2009) Hippopotamus and whale phylogeny. *Nature* 458(7236), E1–E4. [19 Mar 2009]

Geisler JH, Uhen MD (2005) Phylogenetic relationships of extinct cetartiodactyls: results of simultaneous analyses of molecular, morphological, and stratigraphic data. *Journal of Mammalian Evolution* 12, 145–160. [Jun 2005]

Geisler JH, Theodor JM, Uhen MD, Foss SE (2007) Phylogenetic relationships of cetaceans to terrestrial artiodactyls. pp. 19–31, in DR Prothero & SE Foss (eds.) *The Evolution of Artiodactyls*. Johns Hopkins University Press, Baltimore.

Geisler JH, McGowen MR, Yang G, Gatesy J (2011) A supermatrix analysis of genomic, morphological, and paleontological data from crown Cetacea. *BMC Evolutionary Biology* 11(112), 1–33.

Genov PV (1999) A review of the cranial characteristics of the wild boar (*Sus scrofa* Linnaeus 1758), with systematic conclusions. *Mammal Review* 29, 205–238.

Gentry AW (1992) The subfamilies and tribes of the Bovidae. *Mammal Review* 22, 1–32.

Gentry A, Clutton-Brock J, Groves CP (1996) Case 3010. Proposed conservation of usage of mammal specific names based on wild species which are antedated by or contemporary with those based on domestic animals. *Bulletin of Zoological Nomenclature* 53, 28–37. [29 Mar 1996]

Gentry A, Clutton-Brock J, Groves CP (2004) The naming of wild animal species and their domestic derivatives. *Journal of Archaeological Science* 31, 645–651.

Geoffroy ÉL (1762) *Histoire Abregée des Insectes* qui se trouvent aux environs de Paris; Dans laquelle ces animaux sont rangés suivant un ordre méthodique. Durand, Paris. Volume 2.

Geoffroy Saint-Hilaire É (1795) Extrait d'un mémoire sur une nouveau genre de quadrupèdes, de l'ordre des rongeurs (Glires L.), lu à la Société d'Histoire-Naturelle. *La Décade Philosophique. Littéraire et Politique* 4(28), 193–206.

Geoffroy Saint-Hilaire É (1796a) Extrait d'un mémoire sur le *Myrmecophaga capensis*. Gme. *Bulletin des Sciences par la Société Philomatique de Paris* 1, 102–103.

Geoffroy Saint-Hilaire É (1796b) Dissertation sur les animaux à bourse (*Didelphis*, L.). *Magazin Encyclopédique: ou journal des sciences, des letters et des arts* 2(3), 445–472.

Geoffroy Saint-Hilaire É (1796c) Extrait d'une dissertation sur les animaux à bourse. *Bulletin des Sciences par la Société Philomatique de Paris* 1, 106.

Geoffroy Saint-Hilaire É (1796d) Mammifères. Mémoire sur les rapports naturels des Makia lemur, L. et description d'une espèce nouvelle de mammifère. *Magazin Encyclopédique: ou Journal des Sciences, des Letters et des Arts* 2(1), 20–50.

Geoffroy Saint-Hilaire É (1803a) Extrait des observations anatomiques de M. Home, sur l'echidné. *Bulletin des sciences par la Société Philomatique de Paris* 3(77), 125–127. Misprinted as should be pages 225–227.

Geoffroy Saint-Hilaire É (1803b) Note sur une nouveau mammifère découvert à la Nouvelle Hollande, par M. Bass, voyageur anglais. *Bulletin des sciences par la Société Philomatique de Paris* 3(72), 185–186. [Mar 1803]

Geoffroy Saint-Hilaire É (1803c) *Catalogue des mammifères du Muséum National d'Histoire Naturelle*. Muséum National d'Histoire Naturelle, Paris. (printed draft but not published)

Geoffroy Saint-Hilaire É (1803d) Note sur les genres *Phascolomis* et *Perameles*, nouveaux genres d'animaux à bourse. *Bulletin des Sciences par la Société Philomatique de Paris* (3)80, 149–150. Misprinted as should be pages 249–250. [23 Oct-20 Nov 1803]

Geoffroy Saint-Hilaire É (1803e) Note sur les espèces du genre dasyure. *Bulletin des sciences par la Société Philomathique de Paris* 3(81), 158–159. Misprinted as should be pages 258–259.

Geoffroy Saint-Hilaire É (1803f) Notice sur une nouvelle espèce de mammifère apportée vivante par le vaisseau. *Annales du Muséum d'Histoire Naturelle, Paris* 2, 364–367. [Aug 1803]

Geoffroy Saint-Hilaire É (1804a) Note sur un nouveau genre de mammifères, de l'ordre des rongeurs, sous le nom d'*Hydromys*. *Bulletin des sciences par la Société Philomathique de Paris* 3(93), 253–254. (Misprinted as should be page 353–354)

Geoffroy Saint-Hilaire É (1804b) Mémoire sur les espèces du genre Dasyure. *Annales du Muséum d'Histoire Naturelle, Paris* 3, 353–363. [Feb 1804]

Geoffroy Saint-Hilaire É (1804c) Mémoire sur un nouveau genre de mammifères à bourse, nommé *Péramèles*. *Annales du Muséum d'Histoire Naturelle, Paris* 4, 56–64. [Mar 1804]

Geoffroy Saint-Hilaire É (1805a) Sur un nouveau genre de mammifères nommé *Hydromis*. *Annales du Muséum d'Histoire Naturelle, Paris* 6, 81–90. [Mar 1805]

Geoffroy Saint-Hilaire É (1805b) Sur quelques chauve-souris Amérique formant une petite famille sous le nom de molossus. *Annales du Muséum d'Histoire Naturelle, Paris* 6, 150–156. [May 1805]

Geoffroy Saint-Hilaire É (1806) Mémoire sur le genre et les espèces de vespertilion, l'un des genres de la famille des chauve-souris. *Annales du Muséum d'Histoire Naturelle, Paris* 8, 187–205. [By Sep 1806]

Geoffroy Saint-Hilaire É (1810a) Sur les phyllostomes et les mégadermes, deux genres de la famille des Chauve-souris. *Annales du Muséum d'Histoire Naturelle, Paris* 15, 187–198. [Apr–Sep 1810]

Geoffroy Saint-Hilaire É (1810b) Description des rousettes et des céphalotes, deux nouxeaux genres de la famille des chauve-souris. *Annales du Muséum d'Histoire Naturelle, Paris* 15, 86–108. [Apr–Sep 1810]

Geoffroy Saint-Hilaire É (1812) Suite au tableau des quadrumanes. Seconde famille. Lemuriens. Strepsirrhini. *Annales du Muséum d'Histoire Naturelle, Paris* 19, 156–170. [Oct 1812]

Geoffroy Saint-Hilaire É (1813) Sur un genre de chauve-souris, sous le nom de rhinolophiles. *Annales du Muséum d'Histoire Naturelle, Paris* 20, 254–266. [Aug 1813]

Geoffroy Saint-Hilaire É (1818) Description des mammifères qui se trouvent en Égypte. pp. 99–114. in EF Jomard (ed.) *Description de l'Égypte, ou, recueil des observations et des recherches qui ont été faites en Égypte pendant l'expédition de l'armée Française. [1798–1801], &c. Histoire Naturelle*. Tome. 2. Volume 9. Histoire Naturelle. L'Imprimerie Imperiale, Paris.

Geoffroy Saint-Hilaire É (1828) *Cours de l'Histoire Naturelle des Mammifères*. Partie comprenant quelques vues préliminaires de philosophie naturelle, et l'histoire des singes, des makis, des chauve-souris et de la taupe; pouvant servir de complément à l'histoire naturelle des quadrupèdes de Buffon. 13e leçon-27 Juin 1828. Pichon et Didier, Paris. Dated from Andersen 1912, 448. [27 Jun 1828] [footnote)]

Geoffroy Saint-Hilaire É, Cuvier G (1795) Mémoire sur une nouvelle division des mammifères, et sur les principes qui doivent servir de base dans cette sorte de travail, lu à la société d'histoire naturelle, le premier floréal de l'an troisième. *Magazin Encyclopédique, ou journal des sciences, des lettres et des arts* 1(2), 164–190.

Geoffroy Saint-Hilaire I (1824) Sur les vespertilions du Brésil. *Annales des Sciences Naturelles, Paris* 3(1), 440–447.

Geoffroy Saint-Hilaire I (1827) Ornithorhynque. pp. 393–411, in JV Audouin & JBGM Bory de Saint-Vincent (ed.) *Dictionnaire Classique d'Histoire Naturelle*. Rey et Gravier, Paris. Volume 12. NUA-PAM.

Geoffroy Saint-Hilaire I (1828) Roussette. Mam. pp. 695–708, in JV Audouin & JBGM Bory de Saint-Vincent (eds.)

Dictionnaire Classique d'Histoire Naturelle. Rey et Gravier, Paris. Volume 14. PLA-ROY.

Geoffroy Saint-Hilaire I (1838) Notice sur les rongeurs épineaux désignés par les auteurs sous les noms *d'Échimys, Loncheres, Heteromys* et *Netomys. Annales des Sciences Naturelles. Zoologie* 10(2), 122–127.

Geoffroy Saint-Hilaire I (1847) Vie, Travaux et Doctrine Scientifique d'Étienne Geoffroy Saint-Hilaire par son fils M. Isidore Geoffroy Saint-Hilaire. P. Bertrand, Paris.

George GG (1979) The status of endangered Papua New Guinea mammals. pp. 93–100, in MJ Tyler (ed.) *The Status of Endangered Australasian Wildlife.* Royal Zoological Society of South Australia, Adelaide.

Gerdtz WD, Archbold NW (2003) An early occurrence of *Sarcophilus laniarius harrisii* (Marsupialia, Dasyuridae) from the Early Pleistocene of Nelson Bay, Victoria. *Proceedings of the Royal Society of Victoria* 115, 45–54.

Germar EF (1839) *Zeitschrift für die Entomologie.* Friedrich Fleischer, Leipzig. Volume 1. Part 2.

Gerrard E (1862) *Catalogue of the Bones of Mammalia in the Collection of the British Museum.* British Museum, London.

Gervais P (1835a) Echidné, Echina. pp. 622–623, in *Dictionnaire Pittoresque d'Histoire Naturelle et des Phénomènes de la Nature.* Legier, Paris. Volume 2.

Gervais P (1835b) Didelphes ou marsupiaux. pp. 527–534, in F-E Guérin-Méneville (ed.) *Dictionnaire Pittoresque d'Histoire Naturelle et des Phénomènes de la Nature.* Bureau de Sousription, Paris. Volume 2.

Gervais P (1836) Mammalogie ou Mastologie. pp. 614–640, in F-E Guérin-Méneville (ed.) *Dictionnaire Pittoresque d'Histoire Naturelle et des Phénomènes de la Nature.* Bureau de Sousription, Paris. Volume 4. HOLO-MAMM.

Gervais P (1842a) Mammalogie: Nouveau genre de Didelphe. *Extraits des Procès-Verbaux des Séances. Societé Philomatique de Paris* 7, 19. [after 3 Mar 1842]

Gervais P (1842b) Société Philomatique de Paris. *Revue Zoologique* 1842, 94. [before 9 Oct 1842]

Gervais P (1842c) Zoologie. Nouveau genre de mammifères Didelphes. *L'Echo du Monde Savant*, no 27. Columns 631–632. [9 Oct 1842] Reference not seen.

Gervais P (1847) Phalanger. *Phalangista.* pp. 700–706, in C d'Orbigny (ed.) *Dictionnaire Universel d'Histoire Naturelle.* Paris. Second Edition. Volume 9. OIE-PHO.

Gervais P (1848) [1848–1852] *Zoologie et Paléontologie Françaises (Animaux Vertébrés) ou Nouvelles Recherches sur les Animaux Vivants et Fossiles de la France.* Arthus Bertrand, Paris. First Edition. Volume 1.

Gervais P (1849a) Rongeurs. pp. 198–204, in C D'Orbigny (ed.) *Dictionnaire Universel d'Histoire Naturelle.* Volume 11. Renard, Martinet et Cie, Paris. [28 Feb 1849]

Gervais P (1849b) Vespertilion. pp. 211–215, in C d'Orbigny (ed.) *Dictionnaire Universel d'Histoire Naturelle.* Second Edition. Volume 13. VAN-ZY.

Gervais P (1850a) Recherches sur les Cétacés du genre *Ziphius*, de Cuvier et plus particulièrement sur le *Ziphius cavirostris. Comptes Rendus Hebdomadaires des Séances de l'Académie des Sciences, Paris* 31, 510–512.

Gervais P (1850b) Mémoire sur la famille des cétacés Ziphioides, et plus particulièrement sur le *Ziphius cavirostris* de la Méditerranée. *Annales des Sciences Naturelles. Zoologie* 14(3), 5–17.

Gervais P (1853) Remarques sur les mammifères marins qui fréqentent les côtes de la France et plus particulièrement sur une nouvelle espèce de Dauphins propre à la Méditerranée. *Bulletin de la Société Centrale d'Agriculture du Departement de l'Hérault, Montpellier* 40, 140–156.

Gervais P (1854) *Histoire Naturelle des Mammifères : avec l'indication de leurs mœurs et de leur rapports avec les arts, le commerce et l'agriculture.* L. Curmer, Paris.

Gervais P (1855a) *Histoire Naturelle des Mammifères*, avec l'indication de leurs moeurs, et de leur rapports avec les arts, le commerce et l'agriculture. Tom. 2. Carnivores, Proboscidiens, Jumentés, Bisulques, Ėdentes, Marsupiaux, Monotrèmes, Phoques, Sirénides, et Cétacés. L. Curmer, Paris.

Gervais P (1855b) Mammifères. pp. 1–116, in F de Castelnau (1855–1859) *Animaux nouveaux ou rares. Recueillis Pendant L'Expédition dans les Parties Centrales de l'Amérique du Sud, de Rio de Janeiro à Lima, et de Lima au Para; exécutée par ordre du gouvernment Français pendant les années 1843 à 1847.* Sous la Direction du Comte Francis de Castelnau. Partie 7. Volume 1. Part 2. Zoologie, anatomie, mammifères. Paris. [23 Jul 1855]

Gervais P (1859) *Zoologie et Paléontologie Françaises.* Nouvelles Recherches sur les Animaux Vertébrés dont on trouve les ossements enfouis dans le sol de la France et sur leur comparaison avec les espèces propres aux autres régions du globe. Arthus Bertrand, Paris. Second Edition.

Gervais P (1861) Sur différentes espèces de vertébrés fossils observées pour la plupart dans le midi de la France. *Mémoires de la Section des Sciences Académie des Sciences et Lettres de Montpellier* 5, 117–132.

Gervais P (1869) Phalanger. pp. 568–574, in C D'Orbigny (ed.) *Dictionaire Universal d'Histoire Naturelle.* Second Edition. Volume 10. Abel Pilon, Paris.

Gervais P (1871) Remarques sur l'anatomie des Cétacés de la division des Balénidés. *Nouvelles Archives du Muséum d'Histoire Naturelle, Paris* 7, 65–146.

Gervais P (1872) Coup d'oeil sur les mammifères fossiles de l'Italie. *Bulletin de la Société Géologique de France* 29(2), 92–103.

Gervais P (1877a) L'Échidné de la Nouvelle-Guinée. *Comptes Rendus Hebdomadaires des Séances de l'Académie des Sciences, Paris* 85, 837–838. [5 Nov 1877]

Gervais P (1877b) L'Échidné de la Nouvelle-Guinée. Deuxième Note. *Comptes Rendus Hebdomadaires des Séances de l'Académie des Sciences, Paris* 85, 990–991. [30 Nov 1877]

Gervais P (1877c) L'Échidné de la Nouvelle-Guinée. *Journal de Zoologie Comprenant les Différentes Branches de Cette Science* 6, 375–379.

Gervais P (1877d [1877–1878]) *Ostéographie des monotrèmes vivants et fossiles comprenant la description et l'iconographie du squelette et du système dentaire de ces animaux ainsi que des documents relatifs à leur histoire naturelle.* Arthus Bertrand, Paris. [30 Nov 1877] Reference not seen.

Gervais P (1880) Genre *Prodelphinus.* pp. 604–605, in P-J Van Bénéden & P Gervais (eds.) *Ostéographie des Cétacés Vivants et Fossiles*, comprenant la desciption et l'iconographie de squelette et du système dentaire de ces animaux. Arthus Bertrand, Paris.

Gervais P (1883) Sur une nouvelle espèce du genre mégaptère provenant de la baie de Bassora (Golfe Persique). *Comptes Rendus Hebdomadaires des Séances l'Academie des Sciences, Paris* 97, 1566–1569.

Gervais P, Verreaux J (1842a) On a new genus of marsupial animals, *Tarsipes rostratus. Proceedings of the Zoological Society of London* 10, 1–5. [Jun 1842]

Gervais P, Verreaux J (1842b) Mammalogie: Nouveau genre de didelphe. *L'Institute, Journal Universal des Sciences et des Sociétés savants en France et à l'étranger. 1'ère Section. Sciences mathématiques, physiques et naturelles* 427, 75–76. [3 Mar 1842]

Gervais P, Verreaux J (1842c) Description du *Tarsipes rostratus*, nouveau mammifère didelphe de la Nouvelle-Holland. *Magasin de Zoologie, d'Anatomie Comparée et de Palaeontologie. Paris (Classe 1, Mammifères)*, 1–12. [before Oct 1842]

Gheerbrant E, Domning DP, Tassy P (2005) Paenungulata (Sirenia, Proboscidea, Hyracoidea, and relatives). pp. 84–105, in KD Rose & JD Archibald (eds.) *The Rise of Placental Mammals: Origins and Relationships of the Major Extant Clades*. Johns Hopkins University Press, Baltimore.

Giannini NP, Simmons AB (2007) The chiropteran premaxilla: A reanalysis of the morphological variation and its phylogenetic interpretation. *American Museum Novitates* 3585, 1–44. [6 Sep 2007]

Gibson D, Bray R (2010) *Benedenia* Diesing, 1858. Accessed through: World Register of Marine Species at: http://www.marinespecies.org/aphia.php?p=taxdetails&id=119261 on 2012–06–05.

Giebel CGA (1855) *Die Säugethiere in Zoologischer, anatomischer und palaeontologischer Beziehung umfassend dargestellt*. Verlag Von Ambrosius Abel, Leipzig.

Giebel CG (1874) Säugethiere (= Mammalia). Plates to Volume 6; Abtheilung 5; Lieferung 1–60 (Part 2), in HG Bronn. (ed.) *Klassen und Ordnungen des Thier-Reichs Wissenschaften Dargestellt in Wort und Bild*. C.F. Winter'sche Verlagshandlung, Leipzig

Giglioli EH (1870) *Note Intorno alla Distribuzione della Fauna Vertebrata nell' occeano prese durante un viaggio intorno al Globo 1865–68 : [in the 'Magenta']* Florence.

Giglioli H (1882) Note intorno un nuovo Cetaceo nel Mediterraneo da riferirsi probabilmente al genere *Pseudorca. Zoologischer Anzeiger* 5, 288–290. [5 Jun 1882]

Gilbert CH (1890) Scientific results of explorations by the U.S. Fish Commission Steamer Albatross. No. XII. A preliminary report on the fishes collected by the steamer albatross on the Pacific coast of North America during the year 1889, with descriptions of twelve new genera and ninety-two new species. *Proceedings of the United States National Museum* 13, 49–126.

Gilbert CH (1903) Correspondence, in D.S. Jordon. *American Naturalist* 37, 360. [May 1903]

Gill PC, Ross GJB, Dawbin WH, Wapstra H (2000) Confirmed sightings of dusky dolphins (*Lagenorhynchus obscurus*) in southern Australian waters. *Marine Mammal Science* 16, 452–459. [Apr 2000]

Gill T (1866) Prodrome of a monograph of the pinnipedes. *Communications of the Essex Institute, Salem* 5, 3–13.

Gill T (1871a) On the characteristics of the primary groups of the class of mammals. *American Naturalist* 5, 526–533. [Sep 1871]

Gill T (1871b) Synopsis of the primary subdivisions of the cetaceans. *Proceedings of the Essex Institute* 6, 121–126.

Gill T (1871c) The sperm whales, giant whales and pygmy. *American Naturalist* 4, 725–743. [Feb 1871]

Gill T (1872) Arrangement of the families of mammals with analytical tables. *Smithsonian Miscellaneous Collections* 11(1), 1–98. [Nov 1872]

Gill T (1873a) On the status of Aristotle in systematic zoology. *American Naturalist* 7, 458–463. [Aug 1873]

Gill T (1873b) The ribbon seal of Alaska. *American Naturalist* 7, 178–179. [Mar 1873]

Gill T (1877) Vertebrate zoology. *Annual Record of Science and Industry* May: clxvi–clxxv. [5 May 1877]

Gill T (1883) On the classification of the insectivorous mammals. *Bulletin of the Philosophical Society of Washington* 25, 118–120.

Gill T (1885) The species of tachyglossids. *Annual Report of the Board of Regents of the Smithsonian Institution* 1884, 642–643.

Gill T (1901) The bat genus *Pteronotus* renamed *Dermonotus. Proceedings of the Biological Society of Washington* 14, 177. [25 Sep 1901]

Gillham NW (1956) Geographic variation and the subspecies concept in butterflies. *Systematic Zoology* 5, 110–120. [Sep 1956]

Gilpatrick JW, Perryman WL (2008) Geographic variation in external morphology of North Pacific and Southern Hemisphere blue whales (*Balaenoptera musculus*). *The Journal of Cetacean Research and Management* 10, 9–21.

Ginsburg L (1982) Sur la position systématique du petit panda, *Ailurus fulgens* (Carnivora, Mammalia). *Geobios* 15(Supplement), 247–258.

Ginsburg L, Janvier P (1971) Les mammifères marins des faluns miocènes de la Touraine et de l'Anjou. *Bulletin du Museum National d'Histoire Naturelle. Sciences de la Terre* (22)6, 161–195.

Girard C (1852) On the classification of Mammalia. *Proceedings of the American Association for the Advancement of Science* 6, 319–335.

Giraudo AR, Scrocchi GJ (2002) Argentinian Snakes: An Annotated Checklist. *Smithsonian Herpetological Information Service* 132, 1–53.

Gistel J (1848) *Naturgeschichte des Thierreichs für höhere Schulen*. Hoffmannsche, Stuttgart.

Gistel J (1857) *Vacuna oder die Geheimnisse aus der organischen und leblosen Welt*. Straubing. Volume 2.

Gistel J, Bromme T (1847–1850) *Handbuch der Naturgeschichte aller drei Reiche für Lehrer und Lernende, für Schule und Haus*. Hoffmann, Stuttgart.

Glauert L (1926) A list of Western Australian fossils. Supplement. No. 1. *Geological Survey of Western Australia. Bulletin* 88, 36–71.

Glauert L (1950) The development of our knowledge of the marsupials of Western Australia. *Journal of the Royal Society of Western Australia* 34, 115–134.

Gloger C (1827) Etwas über einige ornithologische Gattungs-benennungen. *Notizen aus dem Gebiete der Natur- und Heilkunde. Froriep* 16, 275–280.

Gloger C (1841) *Gemeinnütziges hand-und Hilfsbuch der Naturgeschichte.* Für gebildete Leser aller Stände, besonders für die reifere Jugend und ihre Lehrer. Erster Band, enthaltend die erste Halfte der Thiere, nebst erfahrungsmassigen Andeutungen über den gegewartigen Zustand und Erfolfg des Unterrichts in dieser Wissenschaft, namentlich auf Gymnasien, und Vorschlagen über fernere Einrichtung desselben im Verhaltnisse zu seinen wirklichen Zwicke. Hefte 1–2. A. Schultz & Co. Breslau. Volume 1. See also Thomas 1895a, 189. [Published before May 1841]

Gmelin JF (1788) *Caroli a Linné, Systema naturae per regna tria naturae secundum classes, ordines, genera, species, cum characteribus, differentiis, synonymis, locis.* Editio decima tertia, aucta, reformata. Georg Emanuel Beer, Lipsiae. Volume 1.

Göbbel L (2002) Ontogenetic and phylogenetic transformations of the lacrimal-conducting apparatus among Microchiroptera. *Mammalian Biology –. Zeitschrift fur Saugetierkunde* 67, 338–357.

Godman FD (1900) In FD Godman & O Salvin (1887–1901) *Biologia Centrali-Americana. Insecta. Lepidoptera-Rhopalocera.* Volume 2. R.H. Porter, London.

Goeldi EA (1894) Critical gleanings on the Didelphyidae of the Serra dos Orgãos, Brazil. *Proceedings of the Zoological Society of London* 62, 457–467. [Oct 1894]

Goldfuss GA (1809) *Vergleichende Naturbeschreibung der Säugethiere.* Waltherschen Kunst- und Buchhandlung, Erlangen. Reference not seen.

Goldfuss GA (1817) *Die Säugethiere*, in Abbildungen nach Natur mit Beschreibungen von Johan Christian Daniel von Schreiber. Part 65. No text. Plate 155, Aa, Ab.

Goldfuss GA (1818) Schrebers Säugthiere. *Isis Von Oken* 2–3, 1081–1083.

Goldfuss GA (1819a) Über *Lipurus cinereus* und die Ordnung der Beutelthiere uberhaupt. *Isis von Oken* 4, 271–274.

Goldfuss GA (1819b) Über die verschiedenen Arten der Känguruh. *Isis von Oken* 4, 266–270.

Goldfuss GA (1820a) *Handbuch der Zoologie.* J.L. Schrag, Nurnberg. Abteilung II. Nürnberg, Johan Leonhard Schrag.

Goldfuss GA (1820b) *Handbuch der Zoologie.* J.L. Schrag, Nurnberg. Abteilung I. Nürnberg, Johan Leonhard Schrag.

Gollan K (1984) The Australian dingo: in the shadow of man. pp. 921–927, M Archer & G Clayton, (eds.) *Vertebrate Zoogeography and Evolution in Australasia.* Hesperian Press, Carlisle.

Gongora J, Swan AB, Chong AY, Ho SYW, Damayanti CS, Kolomyjec S, Grant T, Miller E, Blair D, Furlan E, Gust N (2012) Genetic structure and phylogeography of platypuses revealed by mitochondrial DNA. *Journal of Zoology* 286, 110–119. [online 7 Oct 2011]

González JC, Fabián ME (1995) Una nueva especie de murciélago para estado de Rio Grande do Sul, Brasil: *Myotis riparius* Handley, 1960 (Chiroptera, Vespertilionidae). *Comunicações do Museu de Ciência e Tecnologia, PUCRS. Série Zoologia* 8, 55–59.

Goodall RNP, Schiavini ACM (1995) On the biology of the spectacled porpoise, *Australophocoena dioptrica. Report of the International Whaling Commission* 16(Special Issue), 411–543.

Goode GB (1881) *Notacanthus phasganorus*, a new species of Notacanthidae from the Grand Banks of Newfoundland. *Proceedings of the United States National Museum* 3, 535–537.

Goodman M, Porter CA, Czelusniak J, Page SL, Schneider H, Shoshani J, Gunnell G, Groves CP (1998) Toward a phylogenetic classification of Primates based on DNA evidence complemented by fossil evidence. *Molecular Physiology and Evolution* 9, 585–598.

Goodrich ES (1930) *Studies on the Structure and Development of the Vertebrates.* Macmillan, London. Reference not seen.

Goodwin RE (1979) The bats of Timor: systematics and ecology. *Bulletin of the American Museum of Natural History* 163, 73–122. [30 Apr 1979]

Gordon G, Lawrie BC (1980) The rediscovery of the bridled nail-tailed wallaby, *Onychogalea fraenata* (Gould) (Marsupialia; Macropodidae) in Queensland. *Australian Wildlife Research* 7, 339–345.

Gordon M (1984) *Notes on a Search for Bats on Norfolk Island.* Australian National Parks and Wildlife Service, Canberra. Unpublished Report.

Gorter U (2011) Ecotypes and forms. [of Killer Whales] drawn to scale. *Journal of the American Cetacean Society* 40, 34–35.

Goüan A (1770) *Historia Piscium: sistens ipsorum anatomen externam, internam, atque genera in classes & ordines redacta : accedunt vocabularium locupletissimum, indices latini ac gallici, experimenta circa motum natatorium & muscularem, respirationis mechanismum, auditas & generationis organa.* Amandi König, Argentorati.

Gould J (1836) Characters of several new species of insessorial birds, including a new genus (*Stenorhynchus*). *Proceedings of the Zoological Society of London* 3(1835), 185–187. [8 Apr 1836]

Gould J (1837) Observations on the raptorial birds in Mr. Darwins collection, with characters of the new species. *Proceedings of the Zoological Society of London* 5, 9–11. [Oct 1837]

Gould J (1840a) No title. *The Athenaeum* 670, 685. [29 Aug 1840]

Gould J (1840b) In Zoological Society. *The Literary Gazette* 1232, 561. [29 Aug 1840]

Gould J (1840c) Zoological Society. *The Athenaeum* 679, 876. [31 Oct 1840]

Gould J (1841a) Observations on *Dasyurus maugei* and *D. viverrinus* of Geoffroy, and description of a new species. *Proceedings of the Zoological Society of London* 8(1840), 151. [Jul 1841]

Gould J (1841b) Description of a new species of *Hypsiprymnus. Proceedings of the Zoological Society of London* 8(1840), 178–179. [Jul 1841]

Gould J (1841c) On a new species of *Hypsiprymnus* from King George's Sound. *Proceedings of the Zoological Society of London* 9, 14. [Sep 1841]

Gould J (1841d) On two new species of kangaroos (*Macropus manicatus* and *M. brachyotis*) from Swan River. *Proceedings of the Zoological Society of London* 8(1840), 127–128. [Jul 1841]

Gould J (1841e) On five new species of kangaroos. *Proceedings of the Zoological Society of London* 8(1840), 92–94. [Apr 1841]

Gould J (1841–1842) *A Monograph of the Macropodidae, or Family of Kangaroos*. John Gould, London. [1 Aug 1841 & 1 May 1842]

Gould J (1842a) Characters of a new species of *Perameles*, and a new species of *Dasyurus*. *Proceedings of the Zoological Society of London* 10, 41–42. [Nov 1842]

Gould J (1842b) On some new species of Australian mammals. *Proceedings of the Zoological Society of London* 10, 10–14. [Jun 1842]

Gould J (1842c) On two new species of kangaroo (*Petrogale concinna* and *Halmaturus binoë*). *Proceedings of the Zoological Society of London* 10, 57–58. [Nov 1842]

Gould J (1842d) On a New Species of *Petrogale,—P. inornata*. *Proceedings of the Zoological Society of London* 10, 5–6. [Jun 1842]

Gould J (1842e) Descriptions of four new species of kangaroos. *Proceedings of the Zoological Society of London* 9(1841), 80–83. [Mar 1842]

Gould J (1842f) Description of two new species of kangaroos from Western Australia. *Annals & Magazine of Natural History* (1)10, 1–3. [1 Sep 1842]

Gould J (1843a) New Australian mammals. *Annals & Magazine of Natural History* (1)10, 404–406. [1 Jan 1843]

Gould J (1843b) On a new species of kangaroo rat. *Proceedings of the Zoological Society of London* 11, 81. [Dec 1843]

Gould J (1844a) Exhibition and character of a number of animals, &c. transmitted from Australia by Mr. Gilbert. *Proceedings of the Zoological Society of London* 12, 103–107. [Oct 1844]

Gould J (1844b) Description of three species of *Halmaturus* and *Lagorchestes*. *Proceedings of the Zoological Society of London* 12, 31–33. [Sep 1844]

Gould J (1845a) Descriptions of five new species of mammals. *Proceedings of the Zoological Society of London* 13, 77–79. [Oct 1845]

Gould J (1845b) On three new species of birds from China; and on a small mammal and new grallatorial bird from western Australia. *Proceedings of the Zoological Society of London* 13, 1–3. [Apr 1845]

Gould J (1848) Appendix. In C Sturt. *Narrative of an Expedition into Central Australia Performed Under the Authority of Her Majesty's Government, During the Years 1844, 5, 6, Together with a Notice of the Province of South Australia in 1847*. T & W. Boone, London.

Gould J (1850) On new species of Mammalia and birds from Australia. *Proceedings of the Zoological Society of London* 17(1849), 109–112. [Jan–Jun 1850]

Gould J (1851) *The Literary Gazette* 1823, 923. [27 Dec 1851] Reference not seen.

Gould J (1853) Remarks on the genus *Hapalotis*. *Proceedings of the Zoological Society of London* 19(1851), 126–127. [29 Apr 1853]

Gould J (1854) Description of two new species of Mammalia of the genus *Antechinus*. *Proceedings of the Zoological Society of London* 19(1851), 284–285. [14 June 1854]

Gould J (1858) On four new species of *Mus* and one of *Hapalotis* from Australia. *Proceedings of the Zoological Society of London* 25(1857), 241–243. [12 Jan 1858]

Gould J (1861a) On a new species of kangaroo, of the genus *Halmaturus*. *Proceedings of the Zoological Society of London* 28(1860), 375. [Mar 1861]

Gould J (1861b) Remarks on a kangaroo living in the Society's gardens. *Proceedings of the Zoological Society of London* 28(1860), 373. [Mar 1861]

Gould J (1845–1863) *The Mammals of Australia*. John Gould, London.

Grant J (1831) Notice of the Van Diemen's land tiger. *Gleanings in Science* 3, 175–177.

Grateloup JPS (1840a) Description d'un fragment de mâchoire fossile, d'un genre nouveau de reptile (saurien), de taille gigantesque, voisin de *l'Iguanodon*, trouvé dans le Grès Marin, à Léognan, près Bordeaux (Gironde). *Actes de l'Academie national des sciences, belles-lettres et artes de Bordeaux* 2, 201–210.

Grateloup Le D (1840b) Considérations générale sur la géologie et la zoologie fossile de la commune de Léognan, près Bordeaux. *Actes de la Société Linnéenne de Bordeaux* 11, 335–346.

Graur D, Duret L, Gouy M (1996) Phylogenetic position of the Order Lagomorpha (rabbits, hares and allies). *Nature* 379(6563), 333–335. [25 Jan 1996]

Graur D, Gouy M, Duret L (1997) Evolutionary affinities of the order Perissodactyla and the phylogenetic status of the superordinal Ungulata and Altungulata. *Molecular Phylogenetics and Evolution* 7, 195–200. [Apr 1997]

Gray GR (1840) *A List of the Genera of Birds, with an indication of the typical species of each genus*. Richard and John E. Taylor, London.

Gray GR (1855) *Catalogue of the Genera and Subgenera of Birds in the British Museum*. British Museum, London.

Gray JE (1821) On the natural arrangement of vertebrose animals. *London Medical Repository* 15, 296–310. [1 Apr 1821]

Gray JE (1825a) An outline of an attempt at the disposition of the Mammalia into tribes and families with a list of the genera apparently appertaining to each tribe. *Annals of Philosophy* 10(2), 337–344.

Gray JE (1825b) An attempt at a division of the family Vespertilionidae into groups. *The Zoological Journal, London* 2, 242–243.

Gray JE (1826a) The third order of the Mammalia, continued. Supplement to the Marsupiata. pp. 1–60, Volume 3, in E Griffith, CH Smith & E Pidgeon (eds.) (16 vols: 1824–1835) *The Animal Kingdom arranged in conformity with its organisation, by the Baron Cuvier, member of the Institute of France etc, with additional descriptions of all the species hitherto named, and of many not before noticed*. G.B. Whittaker & Co., London. Author of volume given as E Griffith, CH Smith and E Pidgeon by Anon (1903: 410). [Mar–Sep 1826]

Gray JE (1826b) Vertebrata. Mammalia. pp. 412–415. Appendix B in part, in PP King (ed.) *Narrative of a Survey of the Intertropical and Western Coasts of Australia*. Performed between the years 1818 and 1822. With an Appendix, containing various subjects relating to hydrography and natural history. J. Murray, London. Volume 2. [15 Apr 1826]

Gray JE (1827a) Synopsis of the species of the class Mammalia. pp. 1–391, Volume 5, in E Griffith, CH Smith & E Pidgeon (eds.) (16 vols: 1824–1835) *The Animal Kingdom arranged in conformity with its organisation, by the Baron Cuvier, member of the Institute of France etc, with additional descriptions of all the species hitherto named, and of many not before noticed.* G.B. Whittaker & Co., London. Author of volume given as J.E. Gray by Anon (1903: 410). [Sep 1827]

Gray JE (1827b) Description of the skulls of two apparently undescribed species of dolphins, which are in the British Museum. *The Philosophical Magazine and Annals of Philosophy* (2) 2 (2), 375–376.

Gray JE (1828) *Spicilegia Zoologica*; or original figures and short systematic descriptions of new and unfigured animals. Treuttel, Wurtz & Co., London. Part 1.

Gray JE (1830a [1830–1835]) *Illustrations of Indian Zoology; chiefly selected from the collection of Major-General Hardwicke, F.R.S.* Treuttel, Wurtz, Jun. & Richter, London. Volume 1.

Gray JE (1830b) *Spicilegia Zoologica*; or original figures and short systematic descriptions of new and unfigured animals. Treuttel, Wurtz & Co., London. Part 2.

Gray JE (1831a) A synopsis of the species of the class Reptilia. pp. 1–110, Volume 9, in E Griffith & E Pidgeon (eds.) (16 vols: 1824–1835) *The Animal Kingdom arranged in conformity with its organisation, by the Baron Cuvier, member of the Institute of France etc, with additional descriptions of all the species hitherto named, and of many not before noticed.* G.B. Whittaker & Co., London. Author of volume given as E. Griffith and E. Pidgeon by Anon (1903: 410). [Sep 1830–Mar 1831]

Gray JE (1831b) Descriptions of some new genera and species of bats. pp. 37–38, in *The Zoological Miscellany*. Treuttel, Wurtz & Co., London.

Gray JE (1832) Characters of a new genus of Mammalia, and of a new genus and two new species of lizards from New Holland. *Proceedings of the Committee of Science and Correspondence of the Zoological Society of London* 2, 39–40.

Gray JE (1834) A specimen of bat captured in New Holland by George Bennett, Esq. *Proceedings of the Zoological Society of London* 2, 52–53. [26 Sep 1834]

Gray JE (1837) Description of some new or little known Mammalia, principally in the British Museum collection. *Magazine of Natural History and Journal of Zoology, Botany, Mineralogy, Geology, and Meteorology* 1(2), 577–587. [1 Nov 1837]

Gray JE (1838a) In TL Mitchell. *Three Expeditions into the Interior of Eastern Australia; with descriptions of the recently explored region of Australia Felix, and of the present colony of New South Wales*. T. & W. Boone, London. Volume 1.

Gray JE (1838b) Notes on the above, with descriptions of two new species. *Annals of Natural History; or. Magazine of Zoology, Botany, and Geology* 1(1), 106–109. [1 Apr 1838]

Gray JE (1838c) A revision of the genera of bats (Vespertilionidae), and the descriptions of some new genera and species. *Magazine of Zoology and Botany* 2, 483–505. [1 Feb 1838]

Gray JE (1839a) Catalogue of the slender-tongued saurians, with descriptions of many new genera and species. *Annals of Natural History; or. Magazine of Zoology, Botany, and Geology* 2(1), 331–337. [1 Jan 1839]

Gray JE (1839b) Descriptions of some Mammalia discovered in Cuba by W.S. Macleay, Esq. With some account of their habits, extracted from Mr. Macleay's notes. *Annals of Natural History; or. Magazine of Zoology, Botany, and Geology* 4(1), 1–7. [1 Sep 1839]

Gray JE (1840a) A new marsupial animal. *Annals of Natural History; or. Magazine of Zoology, Botany, and Geology* 5(1), 150. [1 April 1840]

Gray JE (1840b) *Synopsis of the Contents of the British Museum*. British Museum, London. 42nd Edition.

Gray JE (1841) Contributions towards the geographical distribution of the Mammalia of Australia, with notes on some recently discovered species, a letter addressed to the Author. Appendix C. pp. 397–414, in G Grey (ed.) *Journals of Two Expeditions of Discovery in north-west and Western Australia, During the Years 1837, 38, and 39, Under the Authority of Her Majesty's Government. Describing many newly discovered, important, and fertile districts, with observations on the moral and physical condition of the aboriginal inhabitants, &c.* T. & W. Voone, London. Volume 2. [Nov 1841]

Gray JE (1842a) Monographic synopsis of the vipers of the family Viperidae. *Zoological Miscellany* 2, 68–71.

Gray JE (1842b) Description of some new genera and fifty unrecorded species of Mammalia. *Annals & Magazine of Natural History* (1)10, 255–267. [1 Dec 1842]

Gray JE (1842c) A new species of tapering-tailed Phascogale in the collection of the British Museum. *Annals & Magazine of Natural History* (1)9, 518. [1 Aug 1842]

Gray JE (1842d) Description of two new species of Mammalia discovered in Australia by Captain George Grey, Governor of South Australia. *Annals & Magazine of Natural History* (1)9, 39–42. [8 Mar. 1842]

Gray JE (1842e) *Synopsis of the Contents of the British Museum*. British Museum, London. 44nd Edition.

Gray JE (1843a) *List of the Specimens of Mammalia in the Collection of the British Museum*. British Museum, London.

Gray JE (1843b) On two new species of bats and a porcupine (*Hystrix subspinosus*, Licht.) in the British Museum. *Proceedings of the Zoological Society of London* 11, 20–22. [Jul 1843]

Gray JE (1843c) Description of some new genera and species of Mammalia in the British Museum collection. *Annals & Magazine of Natural History* (1)11, 117–119. [1 Feb 1843]

Gray JE (1843d) Fauna of New Zealand. pp. 177–295, in E Dieffenback (ed.) *Travels in New Zealand: with contributions to the geography, geology, botany, and natural history of that country*. John Murray, London. Volume 2.

Gray JE (1844a) Miscellanea. pp. 12a–12d, in J Richardson & JE Gray (eds.) (1844–1875) *The Zoology of the Voyage of HMS Erebus and Terror*, under the Command of Captain Sir James Clark Ross, N.N., F.R.S., during the years 1839 to 1843. Mammalia, birds. E.W. Janson, London. Volume 1. Beasts.

Gray JE (1844b) Seals of the southern hemisphere. pp. 1–12, in J Richardson & JE Gray (eds.) (1844–1875) *The Zoology of the Voyage of HMS Erebus and Terror*, under the Command of Captain Sir James Clark Ross, N.N., F.R.S., during the years 1839 to 1843. Volume 1. Mammalia, Birds. E.W. Janson, London.

Gray JE (1845a) *Catalogue of the Specimens of Lizards in the Collection of the British Museum*. British Museum, London.

Gray JE (1845b) Description of some new Australian animals. pp. 405–411, in EJ Eyre (ed.) *Journal of Expeditions of Discovery into Central Australia, and Overland from Adelaide to King George Sound, in the years 1840–1841*, sent by the colonists of South Australia, with the sanction and support of the Government: including an account of the manners and customs of the aborigines and the state of their relations with Europeans. T. & W. Boone, London. Volume 1.

Gray JE (1846a) *Catalogue of the Specimens and Drawings of Mammalia and Birds of Nepal and Thibet presented by B.H. Hodgson, Esq. to the British Museum*. British Museum, London.

Gray JE (1846b) On the British Cetacea. *Annals & Magazine of Natural History* (1)17, 82–85. [1 Feb 1846]

Gray JE (1846c) On the cetaceous animals. pp. 13–53, in J Richardson & JE Gray (eds.) *The Zoology of the Voyage of H.M.S. Erebus and Terror under the Command of Captain Sir James Clark Ross*, R.N., F.R.S., during the years 1839 to 1843. Volume 1. Mammalia, Birds. E.W. Janson, London.

Gray JE (1847a) Description of a new rat from South Australia. *Proceedings of the Zoological Society of London* 15, 5–6. [29 Mar 1847]

Gray JE (1847b) Characters of six new genera of bats not hitherto distinguished. *Proceedings of the Zoological Society of London* 15, 14–16. [13 Apr 1847]

Gray JE (1847c) Characters of six new genera of bats not hitherto distinguished. *Annals & Magazine of Natural History* (1)19, 406–408. [1 Jun 1847]

Gray JE (1847d) On the finner whales, with the description of a new species. *Proceedings of the Zoological Society of London* 15, 88–93. [20 Jul 1857]

Gray JE (1849) Vertebrata. pp. 1–43, in E Belcher (ed.) *The Zoology of the Voyage of H.M.S. 'Samarang' under the Command of Captain Sir Edward Belcher during the years 1843–46*. Paris.

Gray JE (1850) *Catalogue of the Specimens of Mammalia in the Collection of the British Museum. Part 1. Cetacea*. British Museum, London.

Gray JE (1854) On the painted pig of the Camaroons (*Potamochoerus penicillatus*). *Proceedings of the Zoological Society of London* 20(1852), 129–132. [27 Jun 1854]

Gray JE (1855) Description of a new species of *Petrogale*. *Proceedings of the Zoological Society of London* 22(1854), 249. Plate 39. [11 Apr 1855]

Gray JE (1858a) List of species of Mammalia sent from the Aru Islands by Mr. A.R. Wallace to the British Museum. *Proceedings of the Zoological Society of London* 26, 106–113. [27 Apr 1858]

Gray JE (1858b) Observations on the genus *Cuscus*, with description of a new species. *Proceedings of the Zoological Society of London* 26, 100–105. [27 Apr 1858]

Gray JE (1859a) Notice of *Notopteris*, a New Guinea of Pteropine bat from the Feejee Islands. *Proceedings of the Zoological Society of London* 27(1859), 36–38. Plate 77. [Feb–Jun 1859]

Gray JE (1859b) On the sea-lions, or lobos marino of the Spaniards, on the coast of California. *Proceedings of the Zoological Society of London* 27, 357–361. [Oct 1859]

Gray JE (1859c) On the eared seals of the Cape of Good Hope (*Otaria delalandii*). *Proceedings of the Zoological Society of London* 27, 107–110. [Feb–Jun 1859]

Gray JE (1862a) Additional observations on the genus *Cuscus*. *Proceedings of the Zoological Society of London* 29(1861), 314–321. [Apr 1862]

Gray JE (1862b) Notice of a new species of dolphin (*Delphinus catalania*). *Proceedings of the Zoological Society of London* 30, 143–145. [Sep 1862]

Gray JE (1863a) Revision of the species of lemuroid animals, with the description of some new species. *Proceedings of the Zoological Society of London* 31, 129–152. [Oct 1863]

Gray JE (1863b) Notice of three wombats in the zoological gardens. *Annals & Magazine of Natural History* (3)11, 457–459. [1 June 1863]

Gray JE (1863c) Description of some new species of Mammalia. *Proceedings of the Zoological Society of London* 30(1862), 261–263. [Apr 1863]

Gray JE (1863d) On the arrangement of the cetaceans. *Proceedings of the Zoological Society of London* 31, 197–202. [Oct 1863]

Gray JE (1864a) Notes on some Mammalia, with the description of a new *Golunda*, from western Africa. *Proceedings of the Zoological Society of London* 32, 55–58. [Jul 1864]

Gray JE (1864b) Notes on seals (Phocidae), including the description of a new seal (*Halcyon richardii*) from the West Coast of North America. *Proceedings of the Zoological Society of London* 32, 27–34. [Jul 1864]

Gray JE (1864c) On the Cetacea which have been observed in the seas surrounding the British Islands. *Proceedings of the Zoological Society of London* 32, 195–248. [Nov 1864]

Gray JE (1864d) Notes on the whalebone-whales; with a synopsis of the species. *Annals & Magazine of Natural History* (3)14, 345–353. [1 Nov 1864]

Gray JE (1865a) Revision of the genera and species of Mustelidae contained in the British Museum. *Proceedings of the Zoological Society of London* 33, 100–154. [Jun 1865]

Gray JE (1865b) Revision of the genera and species of entomophagous edentata, founded on the examination of the specimens in the British Museum. *Proceedings of the Zoological Society of London* 33, 359–386. [Oct 1865]

Gray JE (1865c) A revision of the genera and species of viverrine animals (Viverridae) founded on the collection in the British Museum. *Proceedings of the Zoological Society of London* 32(1864), 502–579. [May 1865]

Gray JE (1865d) Notices of a new genus of delphinoid whales from the Cape of Good Hope, and of other cetaceans from the same seas. *Proceedings of the Zoological Society of London* 33, 522–529. [Oct 1865]

Gray JE (1865e) Notice of the atlas and other cervical vertebrae of a right whale in the museum of Sydney, New South Wales. *Proceedings of the Zoological Society of London* 32(1864), 587–594. [May 1865]

Gray JE (1865f) Notice of a new species of Australian sperm whale (*Catodon krefftii*) in the Sydney Museum. *Proceedings of the Zoological Society of London* 33, 439–442. [Oct 1865]

Gray JE (1865g) In EL Layard. Notes on the whales of the Cape. With Descriptions of two new species. *Proceedings of the Zoological Society of London* 33, 357–359. [Oct 1865]

Gray JE (1866a) Note on some Mammalia from Port Albany (Cape York Peninsula), north Australia, with the descriptions of some new species. *Proceedings of the Zoological Society of London* 34, 219–221. [Sep 1866]

Gray JE (1866b) *Catalogue of Seals and Whales in the British Museum*. British Museum, London. Second Edition.

Gray JE (1866c) Synopsis of the genera of Vespertilionidae and Noctilionidae. *Annals & Magazine of Natural History* (3)17, 89–93. [1 Feb 1866]

Gray JE (1866d) A revision of the genera of pteropine bats (Pteropidae), and the description of some apparently undescribed species. *Proceedings of the Zoological Society of London* 34, 62–67. [May 1866]

Gray JE (1866e) A revision of the genera of Rhinolophidae, or horseshoe bats. *Proceedings of the Zoological Society of London* 34, 81–83. [May 1866]

Gray JE (1866f) Notes on the skulls of sea-bears and sea-lions (Otariadae) in the British Museum. *Annals & Magazine of Natural History* (3)18, 228–237. [1 Sep 1866]

Gray JE (1866g) Short account of part of a skeleton of a finner whale, sent by Mr. Swinhoe from the coast of Formosa. *Proceedings of the Zoological Society of London* 33(1865), 725–728. [Apr 1866]

Gray JE (1866h) Notes on the skulls of dolphins, or bottlenose whales, in the British Museum. *Proceedings of the Zoological Society of London* 34, 211–216. [Sep 1866]

Gray JE (1866i) Description of three species of dolphins in the free museum of Liverpool. *Proceedings of the Zoological Society of London* 33(1865), 735–739. [Apr 1866]

Gray JE (1867) Notes on the variegated or yellow-tailed rats of Australasia. *Proceedings of the Zoological Society of London* 35, 597–600. [Oct 1867]

Gray JE (1868a) Observations on the fur-seals of the Antarctic seas and the Cape of Good Hope, with the description of a new species. *Annals & Magazine of Natural History* (4)1, 215–219. [1 Mar 1868]

Gray JE (1868b) Synopsis of the species of pigs (Suidae) in the British Museum. *Proceedings of the Zoological Society of London* 36, 17–49. [May 1868]

Gray JE (1868c) *Synopsis of the Species of Whales and Dolphins in the Collection of the British Museum*. Bernard Quartitch, London.

Gray JE (1868d) Notice of *Clymene similes*, a new dolphin sent from the Cape by Mr Layard. *Proceedings of the Zoological Society of London* 36, 146–149. [May 1868]

Gray JE (1869a) *Catalogue of Carnivorous, Pachydermatous, and Edentate Mammalia in the British Museum*. British Museum (Natural History), London.

Gray JE (1869b) Additional notes on sea-bears (Otariadae). *Annals & Magazine of Natural History* (4)4, 264–270. [1 Oct 1869]

Gray JE (1869c) Notes on seals (Phocidae) and the changes in the form of their lower jaw during growth. *Annals & Magazine of Natural History* (4)4, 342–346. [1 Nov 1869]

Gray JE (1870a) *Catalogue of Monkeys, Lemurs and Fruit-eating Bats in the Collection of the British Museum*. British Museum, London.

Gray JE (1870b) The geographical distribution of the Cetacea. *Annals & Magazine of Natural History* (4)6, 387–394. [1 Nov 1870]

Gray JE (1870c) Observations on the whales described in the '*Ostéographie des Cétacés*' of MM. Van Beneden and Gervais. *Annals & Magazine of Natural History* (4)6, 193–204. [1 Sep 1870]

Gray JE (1870d) Notes on the skull of *Balaena marginata*, the type of the genus, *Neobalaena*. *Annals & Magazine of Natural History* (4)6, 154–157. [1 Aug 1870]

Gray JE (1870e) Notes on the skulls of the genus *Orca* in the British Museum, and a notice of a specimen of the genus from the Seychelles. *Proceedings of the Zoological Society of London* 38, 70–77. [Jun 1870]

Gray JE (1871a) *Supplement to the Catalogue of Seals and Whales in the British Museum*. British Museum (Natural History), London.

Gray JE (1871b) In G. Krefft. Notice of a new Australian ziphioid whale. *Annals & Magazine of Natural History* (4)7, 368. [1 May 1871]

Gray JE (1871c) Notes on the *Berardius* of New Zealand. *Annals & Magazine of Natural History* (4)8, 115–117. [1 Aug 1871]

Gray JE (1872) On the sea-bears of New Zealand (*Arctocephalus cinereus*) and the north-Australian sea bear (*Gypsophoca tropicalis*). *Proceedings of the Zoological Society of London* 40, 653–662. [Nov 1872]

Gray JE (1873a) Notes on the rats; with the description of some new species from Panama and the Aru Islands. *Annals & Magazine of Natural History* (4)12, 416–419. [1 Nov 1873]

Gray JE (1873b) Notice of the skeleton of the New Zealand right whale (*Macleayius australiensis*) and other whales, and other New Zealand marine Mammalia. *Proceedings of the Zoological Society of London* 41, 129–145. [Jun 1873]

Gray JE (1873c) Remarks on some of the species in the foregoing paper. *Annals & Magazine of Natural History* (4)11, 107–112. [1 Feb 1873]

Gray JE (1874a) *Hand-list of Seals, Morses, Sea-Lions, and Sea-bears in the British Museum*. British Museum, London.

Gray JE (1874b) Note in J. Hector. Notes on the sulphur-bottom whale of the New-Zealand Whalers Bay. *Annals & Magazine of Natural History* (4)14, 304–305. [1 Oct 1874]

Gray JE (1874c) On the skeleton of the New-Zealand pike whale, *Balaenoptera Huttoni* (*Physalus antarticus*, Hutton). *Annals & Magazine of Natural History* (4)13, 448–452. [1 Jun 1874]

Gray JE (1874d) Notes on Dr. Hector's paper on the whales and dolphins of the New Zealand seas. *Transactions and Proceedings of the New Zealand Institute* 6, 93–97.

Gray JE (1874e) Description of the skull of a new species of Dolphin (*Feresa attenuata*). *Annals & Magazine of Natural History* (4)14, 238–239. [1 Sep 1874]

Gray JE (1875) Feresa attenuata. *Journal des Museum Godeffroy* 8, 184.

Green B (1978) Estimation of food consumption in the dingo, *Canis familiaris dingo*, by means of 22Na turnover. *Ecology* 59, 207–210. [Mar 1978]

Gregory WK (1910) The orders of mammals. *Bulletin of the American Museum of Natural History* 27, 1–524. [Feb 1910]

Gregory WK (1947) The monotremes and the Palimpsest Theory. *Bulletin of the American Museum of Natural History* 88, 1–52. [27 Jan 1947]

Gregory WK, Hellman M (1939) On the evolution and major classification of the civets (Viverridae) and allied fossil and recent Carnivora: a phylogenetic study of the skull and dentition. *Proceedings of the American Philosophical Society* 81, 309–392. [31 Aug 1939]

Gressitt JL (1951) *Longicorn beetles of China*. Longicornia, Paris. Volume 2.

Griffin FJ (1932) On the '*Catalogues*' of the collection of coleoptera of Dejean. *The Annals and Magazine of Natural History* (10)9: 177–178. [Feb 1932]

Griffiths M (1978) *The Biology of Monotremes*. Academic Press, New York.

Griffiths TA, Smith AL (1991) Systematics of emballonuroid bats (Chiroptera: Emballonuridae and Rhinopomatidae) based on the hyoid morphology. *Bulletin of the American Museum of Natural History* 206, 62–83. [12 Sep 1991]

Grill JW (1858) Zoologiska Anteckingar under en resa I sodra delarne af Caplandete aren 1853–1855. Af J.F. Victorin. Ur den a flidnes papper samlade och ordnade at J.W. Grill. Kong. Svenska Vetenska. *Kongliga Svenska Vetenskaps-Akademiens Handlingar, Stockholm* 2(2), 15–62.

Gromov IM, Baranova GI (eds.) (1981) Katalog mlekopitayushchikh SSSR. [Catalogue of Mammals of the USSR] Nauka, Leningrad.

Gronow LT (1854) In JE Gray. *Catalogue of fish collected and described by Laurence Theodor Gronow: now in the British Museum*. British Museum, London.

Groves CP (1974) *Horses, Asses and Zebras in the Wild*. David and Charles, Newton Abbot, London.

Groves CP (1981) Ancestors for the pigs: taxonomy and phylogeny of the genus *Sus*. *Department of Prehistory, Research School and Pacific Studies, Australian National University. Technical Bulletin* 3, 1–96.

Groves CP (1982) The systematics of tree kangaroos (*Dendrolagus*; Marsupialia, Macropodidae). *Australian Mammalogy* 5, 157–186. [20 Dec 1982]

Groves C (1993a) Order Monotremata. pp. 13, in DE Wilson & DA Reeder (eds.) *Mammal Species of the World. A Taxonomic and Geographic Reference*. Smithsonian Institution Press, Washington. Second Edition.

Groves C (1993b) Order Dasyuromophia. pp. 29–37, in DE Wilson & DA Reeder (eds.) *Mammal Species of the World. A Taxonomic and Geographic Reference*. Smithsonian Institution Press, Washington. Second Edition.

Groves C (1993c) Order Notoryctemorphia. pp. 43, in DE Wilson & DA Reeder (eds.) *Mammal Species of the World. A Taxonomic and Geographic Reference*. Smithsonian Institution Press, Washington. Second Edition.

Groves C (1993d) Order Peramelemorphia. pp. 39–42, in DE Wilson & DA Reeder (eds.) *Mammal Species of the World. A Taxonomic and Geographic Reference*. Smithsonian Institution Press, Washington. Second Edition.

Groves C (1993e) Order Diprotodontia. pp. 45–62, in DE Wilson & DA Reeder (eds.) *Mammal Species of the World. A Taxonomic and Geographic Reference*. Smithsonian Institution Press, Washington. Second Edition.

Groves CP (1995) On the nomenclature of domestic animals. *Bulletin of Zoological Nomenclature* 52, 137–141. [30 Jun 1995]

Groves CP (2001) *Primate Taxonomy*. Smithsonian Institution Press, Washington DC.

Groves C (2005a) Order Notoryctemorphia. pp. 22, in DE Wilson & DA Reeder (eds.) *Mammal Species of the World. A Taxonomic and Geographic Reference*. Johns Hopkins University Press, Baltimore. Third Edition.

Groves C (2005b) Order Diprotodontia. pp. 43–70, in DE Wilson & DA Reeder (eds.) *Mammal Species of the World. A Taxonomic and Geographic Reference*. Johns Hopkins University Press, Baltimore. Third Edition.

Groves C (2005c) Order Monotremata. pp. 1–2, in DE Wilson & DA Reeder (eds.) *Mammal Species of the World. A Taxonomic and Geographic Reference*. Johns Hopkins University Press, Baltimore. Third Edition.

Groves C (2005d) Order Dasyuromophia. pp. 23–37, in DE Wilson & DA Reeder (eds.) *Mammal Species of the World. A Taxonomic and Geographic Reference*. Johns Hopkins University Press, Baltimore. Third Edition.

Groves C (2005e) Order Peramelemorphia. pp. 38–42, in D.E. Wilson & D.A. Reeder (eds.) *Mammal Species of the World. A Taxonomic and Geographic Reference*. Johns Hopkins University Press, Baltimore. Third Edition.

Groves C (2005f) Order Primates. pp. 111–184, in DE Wilson & DA Reeder (eds.) *Mammal Species of the World. A Taxonomic and Geographic Reference*. Johns Hopkins University Press, Baltimore. Third Edition.

Groves CP, Flannery TF (1990) Revision of the families and genera of bandicoots. pp. 1–11, in JH Seebeck, RL Wallis, PR Brown & CM Kemper (eds.) *Bandicoots and Bilbies*. Surrey Beatty and Sons, Sydney. [Dec 1990]

Groves CP, Flannery TF (1994) A revision of the genus *Uromys* Peters, 1867 (Muridae: Mammalia) with descriptions of two new species. *Records of the Australian Museum* 46, 145–169. [28 Jul 1994]

Groves CP, Grubb P (1987) Relationships of the living deer. pp. 21–59, in CM Wemmer (ed.) *Biology and Management of the Cervidae*. Smithsonian Institution Press, Washington.

Groves C, Grubb P (2011) *Ungulate Taxonomy*. Johns Hopkins University Press, Baltimore.

Groves CP, Willoughby DP (1981) Studies on the taxonomy and phylogeny of the genus *Equus*. 1. Subgeneric classification of the recent species. *Mammalia* 45, 321–354. [15 Nov 1981]

Grubb P (2001) Review of family-group names of living bovids. *Journal of Mammalogy* 82, 374–388. [May 2001]

Grubb P (2005a) Order Artiodactyla. pp. 637–722, in DE Wilson & DA Reeder (eds.) *Mammal Species of the World. A Taxonomic and Geographic Reference*. Johns Hopkins University Press, Baltimore. Third Edition.

Grubb P (2005b) Order Perissodactyla. pp. 629–636, in DE Wilson & DA Reeder (eds.) *Mammal Species of the World.*

A Taxonomic and Geographic Reference. Johns Hopkins University Press, Baltimore. Third Edition.

Guenée A (1852) *Histoire Naturelle des Insectes. Species Général des Lépidoptéres*. Volume 7. Librairie Encyclopédique de Roret, Paris.

Guérin R (1837) *Magasin de Zoologie* 7, Plate 172.

Guérin R (1874) Études zoologiques et paléontologiques sur la famille des cétacés. Thèse présentée et soutenue devant l'Ecole Supérieure de Pharmacie de Paris. Paris.

Güldenstädt AJ (1770) *Spalax*, novum glirium genus. *Novi Commentarii Academiae Scientiarum Imperialis Petropolitanae* 14(2), 409–440.

Güldenstädt AJ (1777) Beschreibung des *Desmans*. pp. 107–137, in *Beschäftigungen Berlinischen Gesellschaft Naturforschender Freunde*. Berlin. Volume 3.

Gunn RC (1846) In Minutes of the Tasmanian Society, July 2, 1845. *Tasmania Journal* 2, 458–464.

Gunn RC (1851) On the introduction and naturalisation of *Petaurus sciureus* in Tasmania. *Papers and Proceedings of the Royal Society of Van Diemen's Land* 1, 253–255.

Gunnell GF, Simmons NB (2005) Fossil evidence and the origin of bats. *Journal of Mammalian Evolution* 12, 209–246. [Jun 205]

Gunson MM, Sanson GB, Thomson JA (1968) The affinities of *Burramys* (Marsupialia: Phalangeroidea) as revealed by its chromosomes. *Australian Journal of Science* 31, 40–41.

Günther A (1861) *Catalogue of the Fishes in the British Museum*. British Museum, London. Volume 3.

Günther A (1862) Description of new species of reptiles and fishes in the collection of the British Museum. *Proceedings of the Zoological Society of London* 30, 188–194. [Sep 1862]

Günther A (1864) *The Reptiles of British India*. Robert Hardwicke, London.

Günther A (1868) *Catalogue of the Fishes in the British Museum*. British Museum, London. Volume 7.

Günther A (1875) Description of a new species of kangaroo. *Proceedings of the Zoological Society of London* 42(1874), 653. [Apr 1875]

Gureev AA (1971) *Zemleroikii (Soricidae) fauny mira*. [*Shrew Fauna (Soricidae) World Wide*] Nauka, Leningrad. (in Russian)

György Z (2006) Checklist of Hungarian Anthribidae and Urodontidae (Coleoptera). *Folia Entomologica Hungarica Rovartani Közlemények* 67, 63–67.

Haast J von (1874) On the occurrence of a new species of *Euphytes* (*E. pottsii*), a remarkably small catodont whale, on the coast of New Zealand. *Transactions and Proceedings of the New Zealand Institute* 6(1873), 97–102.

Haast J von (1876a) Further notes on *Oulodon*, a new genus of ziphioid whale from the New Zealand seas. *Proceedings of the Zoological Society of London* 44, 457–458. [Oct 1876]

Haast J von (1876b) On a new ziphoid whale. *Proceedings of the Zoological Society of London* 44, 7–13. [Jun 1876]

Haast J von (1876c) Notes on *Mesoplodon floweri*. *Proceedings of the Zoological Society of London* 44, 478–485. [Oct 1876]

Haast J von (1876d) Notes on the skeleton of *Ziphius novaezealandiae*. *Proceedings of the Zoological Society of London* 44, 466–477. [Oct 1876]

Habersetzer J, Storch G (1987) Klassification und funktionelle Flügelmorphologie paläogener Fledermäuse (Mammalia, Chiroptera). *Courier Forschunginstitut Senckenberg* 91, 117–150. Reference not seen.

Haeckel E (1866) *Generelle Morphologie der Organismen allgemeine Grundzüge der organischen Formen-Wissenchaft: mechanisch begründet durch die von Charles Darwin reformirte Descendenz-Theorie*. George Reimer, Berlin. Volume 2.

Haeckel E (1895) *Systematische Phylogenie : Entwurf eines natürlichen Systems der Organismen auf Grund ihrer Stammesgeschichte. Tome 3. Systematische Phylogenie der Wirbelthiere (Vertebrata)*. Verlag von Georg Reimer, Berlin.

Haeckel E, Lankester E (1876) The history of creation, or, The development of the earth and its inhabitants by the action of natural causes : doctrine of evolution in general, and of that of Darwin, Goethe, and Lamarck in particular. Translation revised by ER Lankester. Henry S. King, London. Volume 2.

Hagen HA (1866) On some aberrant genera of Psocina. *Entomologist's Monthly Magazine* 2, 170–172. [Jan 1866]

Haldeman SS (1841) Written communication. *Proceedings. Academy of Natural Sciences of Philadelphia* 1, 127. [23 Nov 1841]

Hale PT, Barreto AS, Ross GJB (2000) Comparative morphology and distribution of the *aduncus* and *truncatus* forms of the bottlenose dolphin *Tursiops* in the Indian and western Pacific Oceans. *Aquatic Mammals* 26, 101–110.

Hall ER (1981) *The Mammals of North America*. Volume II. John Wiley & Sons, New York.

Hall J, Clarke JM (1894) *Palaeontology of New York. Volume 8. Palaeontology. Part 2. An Introduction to the Study of the Genera of Palaeozoic Brachiopoda*. Charles Van Benthuysen & Sons, Albany, New York.

Hall LS (1987) Identification, distribution and taxonomy of Australian flying-foxes (Chiroptera: Pteropodidae). *Australian Mammalogy* 10, 75–79. [29 Jun 1987]

Hall LS, Richards GC (1979) Bats of Eastern Australia. *Queensland Museum Booklet* 12, 1–66.

Hall LS, Richards GC (1991) Flying fox camps. *Wildlife Australia* 28(1), 19–22.

Hall LS, Richards G (2003) Flying around underground: cave bats. pp. 111–126, in B Finlayson & E Hamilton (eds.) *Beneath the Surface: A Natural History of Australian Caves*. University of New South Wales Press, Sydney.

Haltenorth T (1958) Klassifikation der Säugetiere. 1. (1, Ordnung Kloakentiere, Monotremata Bonaparte, 1838, und 2, Ordnung Beuteltiere, Marsupialia Illiger, 1811. [= Didelphia de Blainville, 1816]). *Handbuch der Zoologie, Berlin* 8(16), 1–40. [May 1858]

Haltenorth T (1963) Klassifikation der Säugetiere Artiodactyla. *Handbuch der Zoologie, Berlin* 8(32), 1–167.

Hamilton A (1972) Aboriginal man's best friend? *Mankind* 8, 287–295.

Hamilton JE (1940) On the history of the elephant seal *Mirounga leonina* (Linn.). *Proceedings of the Linnean Society of London* 1939–1940, 33–37. [9 Feb 1940]

Hamilton R (1837) *The Naturalist's Library* conducted by Sir William Jardine. Mammalia 6. [On the Ordinary Cetacea or Whales. W.H. Lizars, Edinburgh.]

Hamilton R (1839) Amphibious carnivora, including the walrus and seals, also the herbivorous cetacean, in W Jardine (ed.) *The Naturalist's Library*. W.H. Lizars, Edinburgh. Volume 8. Mammalia.

Hampe O (2006) Middle/late Miocene hoplocetine sperm whale remains (Odontoceti: Physeteridae) of North Germany with an emended classification of Hoplocetinae. *Fossil Record* 9, 61–86.

Hand SJ, Archer M (2005) A new hipposiderid genus (Microchiroptera) from an early Miocene bat community in Australia. *Palaeontology* 48, 371–383.

Hand SJ, Kirsch JAW (1998) A southern origin for the Hipposideridae (Microchiroptera). Evidence from the Australian fossil record. pp. 72–90, in TH Kunz & PA Racey (ed.) *Bat Biology and Conservation*. Smithsonian Institution Press, Washington, D.C.

Hand SJ, Kirsch JAW (2003) *Archerops*, a new annectant hipposiderid genus (Mammalia: Microchiroptera) from the Australian Miocene. *Journal of Paleontology* 77, 1139–1151.

Hand SJ, Novacek MJ, Godthelp HJ, Archer M (1994) First Eocene bat from Australia. *Journal of Vertebrate Paleontology* 14, 375–381.

Hand SJ, Mackness B, Wilkinson C, Wilkinson D (1999) First Australian Pliocene molossid bat: *Micronomus* sp. from the Chinchilla Local Fauna, southeastern Queensland. *Records of the Western Australian Museum* 57, 291–298.

Handley CO (1966) A synopsis of the genus *Kogia* (Pygmy Sperm whales). pp. 62–69 in KS Norris (ed.) *Whales, Dolphins and Porpoises*. University of California Press, Berkley.

Handley CO (1980) Inconsistencies in formation of family group names in Chiroptera. pp. 9–13, in DE Wilson & AL Gardner (1980) *Proceedings of the Fifth International Bat Research Conference*. Texas Tech Press, Lubbock, Texas, USA.

Hansen BD, Taylor AC (2008) Isolated remnant or recent introduction? Estimating the provenance of Yellingbo Leadbeater's possums by genetic analysis and bottleneck simulation. *Molecular Ecology* 17, 4039–4052.

Hansson H (2010) *Cuvieria*. Accessed through: World Register of Marine Species at http://www.marinespecies.org/aphia.php?p=taxdetails&id=123504 on 2012–06–05.

Harlan R (1827) Description of a new species of *Grampus*, (*Delphinus*, Cuv.) inhabiting the coast of New England. *Journal of the Academy of Natural Sciences of Philadelphia* 6, 51–53.

Harlan R (1828) Note on the examination of the large bones disinterred at the mouth of the Mississippi River, and exhibited in the city of Baltimore. *The American Journal of Science and Arts* 14, 186–187.

Harlin-Cognato AD (2010) The dusky dolphin's place in the delphinid family tree. pp. 1–20, in B Würsig & M Würsig (eds). *The Dusky Dolphin: Master Acrobat Off Different Shores*. Academic Press, Amsterdam.

Harlin-Cognato AD, Honeycutt RL (2006) Multi-locus phylogeny of dolphins in the subfamily Lissodelphininae: character synergy improves phylogenetic resolution. *BMC Evolutionary Biology* 6(87), 1–16. [1 Nov 2006]

Harmer SF (1924) On *Mesoplodon* and other beaked whales. *Proceedings of the Zoological Society of London* 94, 541–587. [9 Jul 1924]

Harris GP (1808) Description of two new species of *Didelphis* from Van Diemen's Land. *Transactions of the Linnean Society of London* (1)9, 174–178. [23 Nov 1808]

Harris JM (2006) The discovery and early history of the eastern pygmy possum, *Cercartetus nanus* (Geoffroy and Desmarest, 1817). *Proceedings of the Linnean Society of New South Wales* 127, 107–124. [23 Feb 2006]

Harris JM (2008) *Cercartetus nanus* (Diprotodontia: Burramyidae). *Mammalian Species* 815, 1–10. [6 Jun 2008]

Harris JM (2009a) *Cercartetus concinnus* (Diprotodontia: Burramyidae). *Mammalian Species* 831, 1–11. [27 May 2009]

Harris JM (2009b) *Cercartetus lepidus* (Diprotodontia: Burramyidae). *Mammalian Species* 842, 1–8. [22 Sep 2009]

Harris JM, Maloney KS (2010) *Petauroides volans* (Diprotodontia: Pseudocheiridae). *Mammalian SpeciesJanuary* 42(866), 207–219. [24 Sep 2010]

Hartlaub C (1886) Ueber *Manatherium delheidi*, eine Sirene aus dem Oligocän Belgiens. *Zoologische Jahrbucher. Abteilung fur Anatomie und Ontogenie der Tiere* 1, 369–378.

Harvey JB (1841) A sketch of the natural history of Port Lincoln. *South Australian Magazine* 1, 210–212.

Harvey JB (1904) A sketch of the natural history of Port Lincoln. *Proceedings of the Royal Geographical Society of Australasia, South Australian Branch* 7, 176–180.

Hassanin A, Delsuc F, Ropiquet A, Hammer C, van Vuuren BJ, Matthee C, Ruiz-Garcia M, Catzeflis F, Areskoug V, Nguyen TT, Couloux A (2012) Pattern and timing of diversification of Cetartiodactyla (Mammalia, Laurasiatheria), as revealed by a comprehensive analysis of mitochondrial genomes. *Comptes Rendus Biologies* 335, 32–50.

Hawkesworth J (1773) *An account of the voyages undertaken by the order of His Majesty for making discoveries in the Southern Hemisphere, and successively performed by Commander Byron, Captain Wallis, Captain Carteret, and Captain Cook, in the Dolphin, the Swallow and the Endeavour*. W. Strahan & T. Cadell, London. Volume 3.

Hay OP (1923) Characteristics of sundry fossil vertebrates. *The Pan-American Geologist* 39, 101–120.

Hay OP (1930) *Second Bibliography and Catalogue of the Fossil Vertebrata of North America*. Carnegie Institute of Washington. Publication No. 390. Volume 2.

Hayman RW (1936) On a new subspecies of *Mesembriomys hirsutus* from Melville Island. *Annals & Magazine of Natural History* (10)17, 366–368. [1 Mar 1936]

Hayman RW (1946) A new genus of fruit-bat and a new squirrel, from Celebes. *Annals & Magazine of Natural History* (11)12, 569–578. [20 Aug 1946]

Hayman RW, Hill JE (1971) Order Chiroptera. Part 2. pp. 1–73, in J Meester & HW Setzer (eds.) *The Mammals of Africa: An Identification Manual*. Smithsonian Institution Press, Washington.

Hayward P (2011) *Electra* Lamouroux, 1816. In P Bock (ed.) *World list of Bryozoa*. Accessed through *World Register of*

Marine Species at http://www.marinespecies.org/aphia.php?p=taxdetails&id=110904 on 2012–06–05.

Heaney LR, Peterson RL (1984) A new species of tube-nosed fruit bat (*Nycimene*) from Negros island, Philippines (Mammalia: Pteropodidae). *Occasional Papers of the Museum of Zoology, University of Michigan* 708, 1–16.

Heaney LR, Rabor DS (1982) Mammals of Dinagat and Siargao Islands, Philippines. *Occasional Papers of the Museum of Zoology, University of Michigan* 699, 1–30.

Hector J (1873a) On the whales and dolphins of the New Zealand seas. *Transactions and Proceedings of the New Zealand Institute* (1872) 5, 154–170.

Hector J (1873b) Notes on the whales and dolphins of New Zealand seas with remarks by J.E. Gray. *Annals & Magazine of Natural History* (4)11, 104–107. [1 Feb 1873]

Heddle R (1856) On a whale of the genus *Physalus*, Gray, captured in Orkney. *Proceedings of the Zoological Society of London* 24, 187–198. [11 Dec 1856]

Heffern DJ (2005) *Catalog and Bibliography of Longhorned Beetles from Borneo (Coleoptera: Cerambycidae).* Electronic Version, 2005.1.

Heim de Balsac H, Bourlière F (1955) Ordre des insectivores. Systématique. pp. 1653–1697, in Grassé (ed.) *Traité de Zoologie : Anatomie, Systématique, Biologie.* Volume 17. Part 2. Mammifères. Les Ordres: Anatomie, Éthologie, Systématique. Masson & Cie, Paris.

Heim de Balsac H, Meester J (1977) Order Insectivora. Part 1. pp. 1–29, in J Meester & HW Setzer (eds.) *The Mammals of Africa: An Identification Manual.* Smithsonian Institution Press, Washington D.C.

Heine F, Reichenow A (1890) *Nomenclator Musei Heineani Ornithologici: Verzeichniss der Vogel-Sammlung des.* Berlin.

Helgen KM (2003a) The feather-tailed glider (*Acrobates pygmeus.* [sic]) in New Guinea. *Treubia* 33, 107–111.

Helgen KM (2003b) Major mammalian clades: a review under consideration of molecular & palaeontological evidence. *Mammalian Biology* 68, 1–15.

Helgen KM (2004) On the identity of flying-foxes, genus *Pteropus* (Mammalia: Chiroptera), from islands in the Torres Strait, Australia. *Zootaxa* 780, 1–14. [20 Dec 2004]

Helgen KM (2007a) Taxonomic and geographic overview of the mammals of Papua. pp. 689–749, in AJ Marshall & BM Beehler (eds.) *The Ecology of Papua. Ecology of Indonesia Series.* Volume VI. Periplus Editions, Singapore.

Helgen KM (2007b) *A reassessment of taxonomic diversity and geographic patterning in the Melanesian mammal fauna.* PhD Thesis. University of Adelaide, Adelaide.

Helgen KM, Flannery TF (2003) Taxonomy and historical distribution of the wallaby genus *Lagostrophus. Australian Journal of Zoology* 51, 199–212.

Helgen KM, Flannery TF (2004) Notes on the phalangerid marsupial genus *Spilocuscus*, with description of a new species from Papua. *Journal of Mammalogy* 85, 825–833. [Oct 2004]

Helgen KM, Armstrong KN, Guzinski J, How RA, Donnellan SC (2009) Taxonomic status of the Christmas Island Pipistrelle, *Pipistrellus murrayi* Andrews, 1900, as assessed by morphometric and molecular phylogenetic investigations of Indo-Australian *Pipistrellus*'. Report to the Commonwealth Department of Environment, Water, Heritage and the Arts. [Oct 2009]

Helgen KM, Miguez RP, Kohen JL, Helgen LE (2012) Twentieth century occurrence of the Long-Beaked Echidna *Zaglossus bruijnii* in the Kimberley region of Australia. *Zookeys* 255, 103–132. [28 Dec 2012]

Heller E (1910) New species of insectivores from British East Africa, Uganda, and the Sudan. *Smithsonian Miscellaneous Collections* 56(15), 1–8. [23 Dec 1910]

Heller KM (1897) Zwei neue Beutelthiere aus Deutsch Neu Guinea nebst einer Aufzählung der bekannten papuanischen Säugethiere. *Abhandlungen und Berichte des Königlichen Zoologischen und Anthropologisch-Ethnologischen Museums zu Dresden* 6(8), 1–7.

Hellmayr CE, Conover B (1948) Catalogue of birds of the Americas and the adjacent islands. *Field Museum of Natural History, Zoological Series* (13)1(2), 1–434.

Hellwig JCL (1795) *Fauna Etrusca, sistens Insecta quae in provinciis Florentina et Pisana.* Second Edition. Volume 1.

Hemmer H (1983) *Domestikation: Verarmung der Merkwelt. Braunschweig, Wiesbaden: Vieweg* (translated (1990) as *Domestication: the Decline of Environmental Appreciation*). Cambridge University Press, Cambridge.

Hemmer H (1990) *Domestication: The Decline of Environmental Appreciation.* Cambridge University, Cambridge.

Hemming F (1939) On five genera in the Lepidoptera Rhopalocera at present with valid names. *Proceedings of the Royal Entomological Society of London* 8(3), 39. [B]

Hemming F (1955) Proposed acceptance of the specific name 'australis' Kerr, 1792, as published in the combination '*Canis australis*' as the name for the Falkland Islands wolf. *Bulletin of Zoological Nomenclature* 11, 122–123. [28 Feb 1955]

Hemprich W (1820) *Grundriss der Naturgeschichte für hohere Lehranstalten.* Entworfen von Dr. W. Hemprich. Berlin, August Rucker, and Vienna, Friedrich Volke. Reference not seen.

Herbst JFW (1792) In CG Jablonsky. Natursystem aller bekannten in- und ausländischen Insekten, als eine Fortsezzung der von Büffonschen Naturgeschichte. Paulischen Buchhandlung, Berlin. Volume 3.

Herbst JFW (1793) In CG Jablonsky. Natursystem aller bekannten in- und ausländischen Insekten, als eine Fortsezzung der von Büffonschen Naturgeschichte. Paulischen Buchhandlung, Berlin. Volume 4.

Hermann J (1780) In EAW von Zimmermann. *Geographische Geschichte des Menschen,* und der vierfüssigen Thiere. Enthalt ein vollstandiges Verzeichniss aller bekannten Quadrupen. Weygandschen Buchhandlung, Leipzig. Volume 2.

Hermann J (1783) *Tabula Affinitatum Animalium olim academico specimine edita nunc uberiore commentario illustrate cum annotationibus ad historiam naturalem animalium augendam facientibus.* Joh. Georgii Treuttel, Agentorati.

Herrera AL (1899) *Sinonimia Vulgar y Cientifica de los Principales Vertebrados Mexicanos.* Oficina Tipografica de la Secretaria de Fomento, Mexico.

Hershkovitz P (1957) The type locality of *Bison bison* Linnaeus. *Proceedings of the Biological Society of Washington* 70, 31–32.

Hershkovitz P (1959) Nomenclature and taxonomy of the neotropical mammals described by Olfers, 1818. *Journal of Mammalogy* 40, 337–353. [Aug 1959]

Hershkovitz P (1961) On the nomenclature of certain whales. *Fieldiana Zoology* 39, 547–565. [25 Jul 1961]

Hershkovitz P (1966) Catalogue of living whales. *Bulletin of the United States National Museum* 246, 1–259. [28 Feb 1966]

Hesse M (1881) Crustacés rares ou nouveaux des côtes de France. *Annales des Sciences Naturelles. Zoologie* 11(6), 1–19.

Heude PM (1896) *Mémoires Concernant l'Histoire Naturelle de l'Empire Chinois par des pères de la Compagnie de Jésus* 3(1), 1–52.

Heude PM (1897) *Mémoires Concernant l'Histoire Naturelle de l'Empire Chinois par des pères de la Compagnie de Jésus* 3(2–4), 53–198.

Heuglin MT von (1861) Beiträge zur fauna der säugethiere N.O.-Afrika's. *Novorum Actorum Academiae Caesarea Leopoldino–Carolinae Germanicae Naturae Curiosorum : Verhandlungen der Kaiserlichen Leopoldinisch-Carolinischen Deutschen Akademie der Naturforscher* 29(8), 1–24.

Heuglin MT von (1877) *Reise in Nordost-Afrika*. Schilderungen aus dem Gebiete der Beni Amer und Habab, nebst zoologischen Skizzen und einem Fürher für Jagdreisende. Westermann, Braunschweig. Volume 2.

Heyden CHG von (1826) Versuch empr system atischen Emtheilung der Acanden. *Isis Von Oken* 19, 608–613.

Heyning JE (1989a) Comparative facial anatomy of beaked whales (Ziphiidae) and a systematic revision among the families of extant Odontoceti. *Contributions in Science* 405, 1–64.

Heyning JE (1989b) Cuvier's beaked whale – *Ziphius cavirostris* G. Cuvier, 1823. pp. 289–308, in SH Ridgway & R Harrison (eds.) *Handbook of Marine Mammals: Volume 4 River Dolphins and the Larger Toothed Whales*. Academic Press, London.

Heyning JE, Dahlheim ME (1988) *Orcinus orca*. *Mammalian Species* 304, 1–9. [15 Jan 1988]

Heyning JE, Perrin WF (1994) Evidence for two species of common dolphins (Genus *Delphinus*) from the eastern Pacific Ocean. *Contributions in Science* 442, 1–35.

Higdon JF, Bininda-Emonds ORP, Beck RMD, Ferguson SH (2007) Phylogeny and divergence of the pinnipeds (Carnivora: Mammalia) assessed using a multigene dataset. *BMC Evolutionary Biology* 7(216), 1–19. [9 Nov 2007]

Higgins ET, Petterd WF (1883) Descriptions of hitherto undescribed Antechini and Muridae inhabiting Tasmania. *Papers and Proceedings of the Royal Society of Tasmania* 1882, 171–176.

Higgins ET, Petterd WF (1884a) New species of Tasmanian antechini and *Mus*. *Papers and Proceedings of the Royal Society of Tasmania* 1883, 184–186. [before 25 Jun 1884]

Higgins ET, Petterd WF (1884b) Descriptions of new Tasmanian animals. *Papers and Proceedings of the Royal Society of Tasmania* 1883, 181–184. [before 25 Jun 1884]

Higgins ET, Petterd WF (1884c) Description of a new species of *Mus*, with a list of terrestrial animals of Tasmania. *Papers and Proceedings of the Royal Society of Tasmania* 1883, 195–198.

Hill JE (1958) The mammals of Rennell Island. pp. 73–84, in T Wolff (ed.) *The Natural History of Rennell Island, British Solomon Islands*. Volume 1 (Vertebrates). Danish Sciences Press, Copenhagen.

Hill JE (1961) Indo-Australian bats of the genus *Tadarida*. *Mammalia* 25, 29–56. [Mar 1961]

Hill JE (1963) A revision of the genus *Hipposideros*. *Bulletin of the British Museum (Natural History). Zoology Series* 11(1), 1–129. [25 Oct 1963]

Hill JE (1965) Asiatic bats of the genera *Kerivoula* and *Phoniscus* (Vespertilionidae), with a note on *Kerivoula aerosa* Tomes. *Mammalia* 29, 524–556. [Dec 1965]

Hill JE (1966) The status of *Pipistrellus regulus* Thomas (Chiroptera, Vespertilionidae). *Mammalia* 30, 302–307. [Jun 1966]

Hill JE (1967) The bats of Andaman and Nicobar Islands. *Journal of the Bombay Natural History Society* 64, 1–9. [Apr 1967]

Hill JE (1971a) A note on *Pteropus* (Chiroptera: Pteropodidae) from the Andaman Islands. *Journal of the Bombay Natural History Society* 68, 1–8.

Hill JE (1971b) Bats from the Solomon Islands. *Journal of Natural History* 5, 573–581.

Hill JE (1974) A new family, genus, and species of bat (Mammalia: Chiroptera) from Thailand. *Bulletin of the British Museum (Natural History). Zoology Series* 27(7), 301–336. [24 Dec 1974]

Hill JE (1982) A review of the leaf-nosed bats *Rhinonycteris, Cloeotis* and *Triaenops* (Chiroptera: Hipposideridae). *Bonner Zoologische Beitrage* 33, 165–186.

Hill JE (1983) Bats (Mammalia: Chiroptera) from Indo-Australia. *Bulletin of the British Museum (Natural History). Zoology Series* 45(3), 103–208. [25 Aug 1983]

Hill JE (1990) A memoir and bibliography of Michael Rogers Oldfield Thomas, F.R.S. *Bulletin of the British Museum (Natural History). History Series* 18(1), 25–113. [31 May 1990]

Hill JE, Harrison DL (1987) The baculum in the Vespertilioninae (Chiroptera: Vespertilionidae) with a systematic review, a synopsis of *Pipistrellus* and *Eptesicus*, and the description of a new genus and subgenus. *Bulletin of the British Museum (Natural History). Zoology Series* 52(7), 225–305. [30 July 1987]

Hill JE, Koopman KF (1981) The status of *Lamingtona lophorhina* McKewan and Calaby, 1968 (Chiroptera: Vespertilionidae). *Bulletin of the British Museum (Natural History). Zoology Series* 41(5), 275–278. [17 Dec 1981]

Hinton MAC (1943) Preliminary diagnosis of five new murine rodents from New Guinea. *Annals & Magazine of Natural History* (10)11, 552–557. [1 Sep 1943]

Hitchcock E (1865) *Supplement to The Ichnology of New England: A Report to the Government of Massachusetts in 1863*. Wright & Potter, Boston.

Hitchcock G (1997) First record of the spectacled hare-wallaby, *Lagorchestes conspicillatus* (Marsupialia: Macropodidae), in New Guinea. *Science in New Guinea* 23, 47–51.

Hitchcock G (1998) First record of the false water-rat, *Xeromys myoides* (Rodentia: Muridae), in New Guinea. *Science in New Guinea* 23, 141–144.

Hodgson BH (1835) Synopsis of the Vespertilionidae of Nipal. *Journal of the Asiatic Society of Bengal, Calcutta* 4, 699–701. [sic]

Hodgson BH (1838) Classified catalogue of Nepalese Mammalia. *Annals of Natural History; or. Magazine of Zoology, Botany, and Geology* 1(1), 152–154. [1 Apr 1838]

Hodgson BH (1843) Description of a new genus of Falconidae. *Journal of the Asiatic Society of Bengal, Calcutta* 12, 127–128.

Hodgson BH (1845) On the rats, mice, and shrews of the central region of Nepal. *Annals & Magazine of Natural History* (1)15, 266–270. [1 Apr 1845]

Hodgson BH (1847) On a new form of the hog kind or Suidae. *Journal of the Asiatic Society of Bengal, Calcutta* 16, 423–428.

Hoelzel AR, Campagna C, Arnbom T (2001) Genetic and morphometric differentiation between island and mainland southern elephant seal populations. *Proceedings. Biological Sciences* 268, 325–332.

Hoelzel AR, Hey J, Dahlheim ME, Nicholson C, Burkanov V, Black N (2007) Evolution of population structure in a highly social top predator, the killer whale. *Molecular Biology and Evolution* 24, 1407–1415. [Jun 2007]

Hoffmann RS, Smith AT (2005) Order Lagomorpha. pp. 185–211, in DE Wilson & DA Reeder (eds.) *Mammal Species of the World. A Taxonomic and Geographic Reference.* Johns Hopkins University Press, Baltimore. Third Edition.

Hoffmann RS, Anderson CG, Thorington RW, Heaney LR (1993) Family Sciuridae. pp. 419–465, in DE Wilson & DM Reeder (eds.) *Mammal Species of the World: A Taxonomic and Geographic Reference.* Smithsonian Institution Press, Washington, D.C.

Holden ME (2005) Family Gliridae. pp. 819–841, in DE Wilson & DA Reeder (eds.) *Mammal Species of the World. A Taxonomic and Geographic Reference.* Johns Hopkins University Press, Baltimore. Third Edition.

Holden ME, Musser GG (2005) Family Dipodidae. pp. 871–893, in DE Wilson & DA Reeder (eds.) *Mammal Species of the World. A Taxonomic and Geographic Reference.* Johns Hopkins University Press, Baltimore. Third Edition.

Hollister N (1916) The type species of *Rattus. Proceedings of the Biological Society of Washington* 29, 206–207. [22 Sep 1916]

Hollister N (1918) East African mammals in the United States National Museum. Part I. Insectivora, Chiroptera, and Carnivora. *Bulletin of the United States National Museum* 99(1), 1–194. [after 12 Jun 1918]

Holthuis LB (1954) C. S. Rafinesque as a carcinologist, an annotated compilation of the information on Crustacea contained in the works of that author. *Zoologische Verhandelingen* 25, 1–43.

Holthuis LB (1987) The scientific name of the sperm whale. *Marine Mammal Science* 3, 87–89. [Jan 1987]

Hombron JB, Jacquinot H (1842) In H Jacquinot & J Pucheran (eds.) Mammifères et oiseaux. Volume 3. In JSC Dumont d'Urville (ed.) (1842–1854). *Voyage au Pôle Sud et dans l'Océanie sur les corvettes l'Astrolobe et la Zélée*: exécuté par ordre roi pendant les années 1837–1838–1839–1840 – sous le commandement de M.J. Dumont d'Urville, Capitaine de Vaisseau, publié par ordonnance de Sa majesté; Histoire du Voyage par M. Dumont D'Urville, Paris. Atlas Mammifères. [1842, Atlas Mammifères Plate 10; Tome 3. Mammifères et Oiseaux, 1853]

Home E (1802) Description of the anatomy of the *Ornithorhynchus hystrix. Philosophical Transactions of the Royal Society* 92, 348–364.

Honacki JH, Kinman KE, Koeppl JW (1982) *Mammal Species of the World – a Taxonomic and Geographic Reference.* Allen Press Inc. & The Association of Systematics Collections, Lawrence.

Honeycutt RL, Adkins RM (1993) Higher level systematics of eutherian mammals: An assessment of molecular characters and phylogenetic hypotheses. *Annual Review of Ecology and Systematics* 24, 279–305.

Hoofer SR, Van Den Bussch RA (2003) Molecular phylogenetics of the chiropteran family Vespertilionidae. *Acta Chiropterologica* 5(suppl.), 1–63.

Hoofer SR, Reeder SA, Hansem EW, Van Den Busshe RA (2003) Molecular phylogenetics and taxonomic revisions of the infraorder Yangochiroptera (Chiroptera: Mammalia). *Journal of Mammalogy* 84, 809–821. [Aug 2003]

Hooijer DA (1956) The valid name of the banteng: *Bibos javanicus* d'Alton. *Zoologische Mededeelingen* 34, 223–226. [28 Feb 1956]

Hooijer DA (1957) Three new giant prehistoric rats from Flores, Lesser Sunda Islands. *Zoologische Mededeelingen* 35, 299–314. [23 Dec 1957]

Hooker JJ (1996) A primitive emballonurid bat (Chiroptera, Mammalia) from the earliest Eocene of England. *Palaeovertebrata* 25, 284–300. [Reference not seen.]

Hope JH (1972) Mammals of the Bass Strait Islands. *Proceedings of the Royal Society of Victoria* 85, 163–195.

Hope RM, Godfrey GK (1988) Genetically determined variation of pelage colour and reflectance in natural and laboratory populations of the marsupial *Sminthopsis crassicaudata* (Gould). *Australian Journal of Zoology* 36, 441–454.

Hopson JA (1970) The classification of non-therian mammals. *Journal of Mammalogy* 51, 1–9. [Feb 1970]

Hopson JA (1994) Synapsid evolution and the radiation of non-eutherian mammals. pp. 190–219, in RS Spencer (ed.) *Major Features of Vertebrate Evolution.* Paleontological Society Short Courses in Paleontology. No. 7.

Hopwood AT (1947) The generic names of the mandrill and baboons, with notes on some of the genera of Brisson, 1762. *Proceedings of the Zoological Society of London* 117, 533–536.

Horáček I, Hanák V (1985) Generic status of *Pipistrellus savii* and comments on the genus *Pipistrellus* (Chiroptera, Vespertilionidae). *Myotis, Bonn* 23–24, 9–16.

Horner BE, Taylor JM (1959) Results of the Archbold Expeditions. No. 80. Observations on the biology of the yellow-footed marsupial mouse *Antechinus flavipes flavipes. American Museum Novitates* 1972, 1–24. [29 Oct 1959]

Horsfield T (1821–1824) *Zoological Researches in Java, and the Neighboring Islands.* Kingbury, Parbury & Allen, London. [No 1, Jul 1821; No. Nov 1821; No. 3 Feb 1822; No. 4. Jun 1822; No. 5. Oct 1822; No. 6. Apr 1823; No. 7. Oct 1923; No. 8. Apr 1824]

Höslin P (1781) *Des Ritters Carl von Linné Lehr-Buch über das Natur-System so weit es das Thierreich angehet : in einem vollständigen Auszuge der Müllerischen Ausgabe*. Gabriel Nicolaus Raspe, Nürnberg. Volume 1.

Houart R (2010) *Pteronotus* Swainson, 1833. Accessed through: World Register of Marine Species at: http://www.marinespecies.org/aphia.php?p=taxdetails&id =404597. Accessed on 2012–05–31.

Houlbert CV (1918) Révision monographique de la sous-famille des Castniinae. pp. 5–730 in C Oberthür (ed) *Études de Lépidoptérologie Comparée*. Volume 15.

Houlden BA, Costello BH, Sharkey D, Fowler EV, Melzer A, Ellis W, Carrick F, Baverstock PR, Elphinstone MS (1999) Phylogeographic differentiation in the mitochondrial control region in the koala, *Phascolarctos cinereus* (Goldfuss 1817). *Molecular Ecology* 8, 999–1011. [Jun 1999]

How RA, Cooper NK, Bannister JL (2001) Checklist of the mammals of Western Australia. *Records of the Western Australian Museum* 63(Supplement), 91–98.

Hoye G (2011) The status of microbats on Norfolk Island, southwest Pacific. pp. 297–307, in B Law, P Eby, D Lunney & L Lumsden (eds.) *The Biology and Conservation of Australasian Bats*. Royal Zoological Society of New South Wales, Sydney. [Sep 2011]

Hrabe S (1964) On *Peloscolex svirenkoi* (Jarošenko) and some other species of the genus *Peloscolex*. *Spisy Prirodovedecke Fakulty University J.E. Purkyne v Brne* **450**, 101–112.

Hübner J (1806) *Tentamen determinationis digestionis atque denominationis singularum stirpium Lepidopterorum, peritis ad inspiciendum et dijudicandum communicatum*. Jacobo Hübner.

Hübner J (1807) *Sammlung Exotischer Schmetterlinge*. Verlag der Hübnerschen Werke bey C. Geyer. Augsburg. Volume 1.

Hübner J (1818) *Zuträge zur Sammlung exotischer Schmetterlinge, bestehend in Bekundigung einzelner Fliegmuster neuer oder rarer nichteuropäischer Gattungen*. Augsburg. Volume 1.

Hübner J (1816–1826) *Verzeichniss Bekannter Schmetterlinge*. Jacob Hübner, Augsburg.

Huchon D, Madsen O, Sibbald MJJB, Ament K, Stanhope MJ, Catzeflis F, de Long WW, Douzery EJP (2002) Rodent phylogeny and a timescale for the evolution of Glires: evidence from an extensive taxon sampling using three nuclear genes. *Molecular Biology and Evolution* 19, 1053–1065. [Jul 2002]

Hudson WS, Wilson DE (1986) *Macroderma gigas*. *Mammalian Species* 260, 1–4. [16 Jun 1986]

Hulbert AJ, Gordon G, Dawson TJ (1971) Rediscovery of the marsupial *Echymipera rufescens* in Australia. *Nature* 231(5301), 330–331. [4 Jun 1971]

Hunt RM (2002) New amphicyonid carnivorans (Mammalia, Daphoeninae) from the early Miocene of southeastern Wyoming. *American Museum Novitates* 3385, 1–41. [27 Dec 2002]

Hunter J (1787) Observations of the structure and oeconomy. [sic] of whales. *Philosophical Transactions of the Royal Society of London* 77(2), 371–450.

Hurdis JL (1897) *Rough Notes and Memoranda Relating to the Natural History of the Bermudas*. R.H. Porter, London.

Husar SL (1978) *Dugong dugon*. *Mammalian Species* 88, 1–7. [6 Jan 1978]

Husson AM (1962) The bats of Suriname. *Zoologische Verhandelingen* 58, 1–282.

Husson AM, Holthius LB (1953) On the early editions of Lacépède's Tableaux des Mammifères et des Oiseaux', with remarks on two hitherto overlooked species: *Lori bengalensis* Lacépède, 1800, and *Ornithorhynchus novaehollandiae* Lacépède, 1800. *Zoologische Mededeelingen* 32, 211–219. [30 Dec 1953]

Husson AM, Holthius LB (1955) The dates of publication of '*Verhandelingen over de natuurlijke Geschiedenis der Nederlandsche overzeesche Bezittingen*' edited by CJ Temminck. *Zoologische Mededelingen* 34, 17–24. [28 Sep 1955]

Husson AM, Holthius LB (1974) *Physeter macrocephalus* Linnaeus, 1758, the valid name for the sperm whale. *Zoologische Mededeelingen* 48, 205–217. [24 Oct 1974]

Hutcheon JM, Kirsch JAW (2004) Camping in a different tree: Results of molecular systematic studies of bats using DNA-DNA hybridisation. *Journal of Mammalian Evolution* 11, 17–47. [Mar 2004]

Hutcheon JM, Kirsch JAW (2006) A moveable face: deconstructing the Microchiroptera and a new classification of extant bats. *Acta Chiropterologica* 8, 1–10.

Hutterer R (1986) African shrews allied to *Crocidura fischeri*: taxonomy, distribution and relationships. *Cimbebasia. Series A. Natuurwetenskappe* 8(4), 23–33. [14 Feb 1986]

Hutterer R (1993) Order Insectivora. pp. 69–130, in DE Wilson & DA Reeder (eds.) *Mammal Species of the World. A Taxonomic and Geographic Reference*. Smithsonian Institution Press, Washington. Second Edition.

Hutterer R (2005) Order Soricomorpha. pp. 220–311, in DE Wilson & DA Reeder (eds.) *Mammal Species of the World. A Taxonomic and Geographic Reference*. Johns Hopkins University Press, Baltimore. Third Edition.

Hutton FW (1874) In Gray, J.E. On a New-Zealand whale (*Physalus antarcticus*, Hutton), with notes. *Annals & Magazine of Natural History* (4)13, 316–318. [1 Apr 1874]

Hutton FW (1904) *Index Faunae Novae Zealandiae*. Dulau & Co., London.

Huxley TH (1869) *An Introduction to the Classification of Animals*. John Churchill & Sons, London.

Huxley TH (1872) *A Manual of the Anatomy of Vertebrated Animals*. D. Appleton, New York.

Huxley TH (1881) On the application of the laws of evolution to the arrangement of the vertebrate and more particularly of the Mammalia. *Proceedings of the Zoological Society of London* 48(1880), 649–662. [Apr 1881]

Hyett J, Shaw N (1980) *Australian Mammals. A Field Guide for New South Wales, Victoria, South Australia & Tasmania*. Nelson, Melbourne.

Ichihara T (1961) Blue whales in the waters around Kerguelen Island. *Norsk Hvalfangst-Tidende* 1, 1–20.

Ichihara T (1963) Identification of the pygmy blue whale in Antarctica. *Norsk Hvalfangst-Tidende* 5, 128–130.

Ichihara T (1966) The pygmy blue whale. *Balaenoptera musculus brevicauda*, a new subspecies from the Antarctic. pp. 79–113, in KS Norris (ed.) *Whales, Dolphins and Porpoises*. University of California Press, Berkeley.

Ihering, H von (1893) Os Mamíferos do Rio Grande do Sul. *Annuario do Estado do Rio Grande do Sul para o Anno 1893 de Graciano A de Azambuja* 1893, 96–123. Reference not seen.

Illiger JKW (1798) In JG Kugelann (ed.) Verzeichniss der Käfer Preussens. Johann Jacob Gebauer, Halle.

Illiger JKW (1807) Vorschlag zur aufnahme im Fabricischen System fehlender Käfergattungen. *Magazin für Insektenkunde* 6, 318–349. [p. 326 *Eurhin*]

Illiger JKW (1811) *Prodromus Systematis Mammalium et Avium additis terminis zoographicis utriusque classis, eorumque versione Germanica*. Sumptibus C. Salfeld, Berlin.

Illiger JKW (1815) Ueberblick der Säugthiere nach ihrer Verteilung über die Welttheile. *Abhandlungen der Physikalischen Klasse der Königlich-Preussischen Akademie der Wissenschaften zu Berlin* 1804–1811, 39–159.

Imamura T (1961) *Shin d butsu bunruihy . [New Classification Table of Animals]* Hokuryu-Kan Co. Ltd., Tokyo. Reference not seen.

Ingleby S, Colgan D (2003) Electrophoretic studies of the systematic and biogeographic relationships of the Fijian bat genera *Pteropus, Pteralopex, Chaerephon* and *Notopteris. Australian Mammalogy* 25, 13–29.

International Commission on Zoological Nomenclature (1910) Opinion 16. The status of prebinomial names (published prior to 1758) under Art. 30d. pp. 31–39, in Opinions Rendered by the International Commission on Nomenclature. Opinions 1–25. Smithsonian Publications 1838. Smithsonian Institution, Washington, D.C.

International Commission on Zoological Nomenclature (1926a) Rules and recommendations. *Proceedings of the Biological Society of Washington* 39, 75–103. [30 Jul 1926]

International Commission on Zoological Nomenclature (1926b) Opinion 97. Did Hübners Tentamen, 1806, create monotypic genera. pp. 19–30 in Opinions rendered by the International Commission on Zoological Nomenclature. Opinions 91–97. *Smithsonian Miscellaneous Collections* 73(4), 1–30. [8 Oct 1926]

International Commission on Zoological Nomenclature (1926c) Opinion 91. Thirty-five generic names of mammals placed in the Official List of Generic Names. pp. 1–2 in Opinions rendered by the International Commission on Zoological Nomenclature. Opinions 91–97. *Smithsonian Miscellaneous Collections* 73(4), 1–30. [8 Oct 1926]

International Commission on Zoological Nomenclature (1941) Opinion 80. Suspension of rules in the case of *Holothuria* and *Physalia. Smithsonian Miscellaneous Collections* 73, 17–18. [9 Feb 1924]

International Commission on Zoological Nomenclature (1954a) Opinion 278. Addition to the Official List of Generic Names in Zoology of the names of ten genera of the Sub-Order Rhopalocera of the Order Lepidoptera (Class Insecta). *Opinions and Declarations Rendered by the International Commission on Zoological Nomenclature* 6(10), 135–175. [1 Oct 1954]

International Commission on Zoological Nomenclature (1954b) Opinion 257. Rejection for nomenclatorial purposes of the work by Zimmermann (E.A.W. von) published in 1777 under the title *Specimen Zoologiae Geographicae Quadrupedum, Domicilia et Migrationes sistens*, and acceptance for the same purposes of the work by the same author published in the period 1778–1783 under the title Geographische Geschichte de Menschen, und der allgemein verbreiteten vierfussigen Thiere. *Opinions and Declarations Rendered by the International Commission on Zoological Nomenclature* 5(18), 231–243. [10 Aug 1954]

International Commission on Zoological Nomenclature (1954c) Opinion 258. Rejection for nomenclatorial purposes of the work by Frisch (J.L.) published in 1775 under the title '*Das Natur-System der vierfussigen Thiere*'. *Opinions and Declarations Rendered by the International Commission on Zoological Nomenclature* 5(19), 245–252. [10 Aug 1954]

International Commission on Zoological Nomenclature (1955) Direction 24. Completion of the entries relative to the names of certain genera in the Class Mammalia made in the 'Official List of Generic Names in Zoology' in the period up to the end of 1936. *Opinions and Declarations Rendered by the International Commission on Zoological Nomenclature* 1(c), 219–246. [Nov 1955]

International Commission on Zoological Nomenclature (1956) Opinion 417. Rejection for nomenclatorial purposes of Volume 3. (Zoologie) of the work by Lorenz Oken entitled Lehrbuch der Naturgeschichte published in 1815–1816. *Opinions and Declarations Rendered by the International Commission on Zoological Nomenclature* 14(1), 1–42. [1 Sep 1956]

International Commission on Zoological Nomenclature (1957a) Opinion 451. Use of plenary powers to secure that the specific name *dingo* Meyer, 1793, as published in the combination *Canis dingo* shall be the oldest available name for the dingo in Australia (Class Mammalia). *Opinions and Declarations Rendered by the International Commission on Zoological Nomenclature* 15(17), 329–338. [8 Mar 1957]

International Commission on Zoological Nomenclature (1957b) Direction 76. Completion and in certain cases correction of entries relating to the names of genera belonging to the Phyla Protozoa, Coelenterata, Platyhelminthes, Nemathelminthes and Annelida, made on the *Official List of Generic Names in Zoology* by Rulings given in *Opinions* rendered in the period up to the end of 1936. *Opinions and Declarations Rendered by the International Commission on Zoological Nomenclature* 1E(14), 227–264. [1 Nov 1957]

International Commission on Zoological Nomenclature (1958a) *Official List of Generic Names in Zoology*. International Commission on Zoological Nomenclature, London. [10 Jun 1958]

International Commission on Zoological Nomenclature (1958b) Direction 98. Interpretation under the Plenary Powers of the nominal species *Vespertilio murinus* Linnaeus, 1758, and insertion in the Official List of Generic Names in Zoology of a revised entry relating to the generic name *Vespertilio* Linnaeus, 1758 (Class Mammalia) (Direction supplementary to Opinion 91). *Opinions and Declarations Rendered by the International Commission on Zoological Nomenclature*. Volume 1. Section F, Part F9, 127–160. [16 May 1958]

International Commission on Zoological Nomenclature (1960) Opinion 581. Determination of the generic name of the fallow deer of Europe and the Virginia deer of America (Class Mammalia). *Bulletin of Zoological Nomenclature* 17, 267–275. [16 Sep 1960]

International Commission on Zoological Nomenclature (1961) *International Code of Zoological Nomenclature* adopted by the XV International Congress of Zoology. International Trust for Zoological Nomenclature, London. First Edition.

International Commission on Zoological Nomenclature (1963) Opinion 678. The suppression under the plenary powers of the pamphlet published by Meigen, 1800. *Bulletin of Zoological Nomenclature* 20, 339–342. [21 Oct 1963]

International Commission on Zoological Nomenclature (1964) *International Code of Zoological Nomenclature* adopted by the XV International Congress of Zoology. International Trust for Zoological Nomenclature, London. Second Edition.

International Commission on Zoological Nomenclature (1965) Opinion 730. *Yerbua* Forster, 1778 (Mammalia): suppressed under plenary powers. *Bulletin of Zoological Nomenclature* 22, 84–85. [18 May 1965]

International Commission on Zoological Nomenclature (1966) Opinion 760. *Macropus* Shaw, 1790 (Mammalia): Addition to the official list together with the validation under the plenary powers of *Macropus giganteus* Shaw, 1790. *Bulletin of Zoological Nomenclature* 22, 292–295. [31 Jan 1966]

International Commission on Zoological Nomenclature (1979a) Opinion 1116. *Acanthomys leucopus* Gray, 1867 (Mammalia): validated under the plenary powers. *Bulletin of Zoological Nomenclature* 35, 180–183. [1 Feb 1979]

International Commission on Zoological Nomenclature (1979b) Opinion 1129. *Vulpes* Frisch, 1775 (Mammalia) conserved under the plenary powers. *Bulletin of Zoological Nomenclature* 36, 76–78. [1 Aug 1979]

International Commission on Zoological Nomenclature (1981) Opinion 1198. *Sminthopsis murina* var. *constricta* Spencer, 1896 (Mammalia, Marsupialia) suppressed. *Bulletin of Zoological Nomenclature* 38, 274–275. [30 Nov 1981]

International Commission on Zoological Nomenclature (1985a) *International Code of Zoological Nomenclature*. International Trust for Zoological Nomenclature and the British Museum of Natural History London and University of California Press, Berkley. Third Edition. [Feb 1985]

International Commission on Zoological Nomenclature (1985b) Opinion 1289. *Mesoplodon* Gervais, 1850 (Mammalia, Cetacea); Conserved. *Bulletin of Zoological Nomenclature* 42, 19–20. [2 Apr 1985]

International Commission on Zoological Nomenclature (1986) Opinion 1413. *Delphinus truncatus* Montagu, 1821 (Mammalia, Cetacea): Conserved. *Bulletin of Zoological Nomenclature* 43, 256–257. [6 Oct 1986]

International Commission on Zoological Nomenclature (1990) Opinion 1607. *Mus musculus domesticus* Schwarz & Schwarz, 1943 (Mammalia, Rodentia): Specific name conserved. *Bulletin of Zoological Nomenclature* 47, 171–172. [29 Jun 1990]

International Commission on Zoological Nomenclature (1991) Opinion 1660. *Stenella attenuatus* Gray, 1846 (currently *Stenella attenuata*; Mammalia, Cetacea): Specific name conserved. *Bulletin of Zoological Nomenclature* 48, 277–278. [30 Sep 1991]

International Commission on Zoological Nomenclature (1998) Opinion 1894. *Regnum Animals*, Ed. 2 (M.J. Brisson, 1762): Rejected for nomenclatural purposes with the conservation of the mammalian generic names *Philander* (Marsupialia), *Pteropus* (Chiroptera), *Glis, Cuniculus* and *Hydrochoerus* (Rodentia), *Meles, Lutra* and *Hyaena* (Carnivora), *Tapirus* (Perrissodactyla), *Tragulus* and *Giraffa* (Artiodactyla). *The Bulletin of Zoological Nomenclature* 55, 64–71. [31 Mar 1998]

International Commission on Zoological Nomenclature (1999) *International Code of Zoological Nomenclature*. International Trust for Zoological Nomenclature and the British Museum of Natural History, London. Fourth Edition. [Effective 1 Jan 2000]

International Commission on Zoological Nomenclature (2000) Opinion 1962. *Arctocephalus* F. Cuvier, 1826 and *Callorhinus* Gray, 1859 (Mammalia, Pinnipedia): conserved by the designation of *Phoca pusilla* Schreber, 1775 as the type species of *Arctocephalus*; and *Otaria* Péron, 1816 and *Eumetopias* Gill, 1866: conserved by the designation of *Phoca leonina* Molina, 1782 as the type species of Otaria. *Bulletin of Zoological Nomenclature* 57, 193–195. [29 Sep 2000]

International Commission on Zoological Nomenclature (2001) Case 2661 Macropodidae Liem, 1963 (Osteichthyes, Perciformes): proposed emendation of spelling to Macropodusine, so removing the homonymy with Macropodidae Gray, 1821 (Mammalia, Marsupialia). *Bulletin of Zoological Nomenclature* 58, 297–299. [19 Dec 2001]

International Commission on Zoological Nomenclature (2002) Opinion 1994 (Case 3095). *Mystacina* Gray, 1843, *Chalinolobus* Peters, 1866, *M. tuberculata* Gray, 1843 and *Vespertilio tuberculatus* J.R. Forster, 1844 (currently *C. tuberculatus*) (Mammalia, Chiroptera): usage of the generic and specific names conserved. *Bulletin of Zoological Nomenclature* 59, 63–64. [27 Mar 2002]

International Commission on Zoological Nomenclature (2003a) Opinion 2058 (Case 2661). Macropodinae Hoedeman, 1948 (Osteichthyes, Perciformes): Spelling emended to Macropodusinae so removing the homonymy with Macropodinae Gray, 1821 (Mammalia, Marsupialia). *Bulletin of Zoological Nomenclature* 60, 253–254. [30 Sep 2003]

International Commission on Zoological Nomenclature (2003b) Opinion 2027. Usage of 17 specific names based on wild species which are pre-dated by or contemporary with those based on domestic animals (Lepidoptera, Osteichthyes, Mammalia): Conserved. *Bulletin of Zoological Nomenclature* 60, 81–84. [31 Mar 2003]

International Whaling Commission (1982) Report on the Workshop on the Identity, Structure and Vital Rates of Killer Whale Populations, Cambridge, England, June 23–25 1981. *Reports of the International Whaling Commission* 32, 617–631.

Iredale T (1937) The last letters of John Macgillivray. *Australian Zoologist* 9, 40–63.

Iredale T (1947) The scientific name of the dingo. *Proceedings of the Royal Zoological Society of New South Wales* 67(1946–1947), 35–36.

Iredale T, Troughton EleG (1925) Captain Cook's kangaroo. *Australian Zoologist* 3, 311–316.

Iredale T, Troughton EleG (1933) The correct generic name for the Grampus or Killer whale, and the so-called Grampus or Risso's dolphin. *Records of the Australian Museum* 19, 28–36. [2 Aug 1933]

Iredale T, Troughton EleG (1934) A check-list of the mammals recorded from Australia. *Memoirs of the Australian Museum* 6, 1–122. [4 May 1934]

Iredale T, Troughton EleG (1937) The identity of Captain Cook's kangaroo. *Records of the Australian Museum* 20, 67–71. [15 May 1937]

Iredale T, Troughton E (1963) The actual identity of Captain Cook's kangaroo. *Proceedings of the Linnean Society of New South Wales* 87, 177–184. [10 Jan 1963]

Irwin DM, Wilson AC (1993) Limitations of molecular methods for establishing the phylogeny of mammals, with special reference to the position of elephants. pp. 257–267, in FS Szalay, MJ Novacek & MC McKenna (eds.) Mammal Phylogeny: Placentals. Springer-Verlag, New York.

Ivashin MV (1958) O sistematicheskom polozhenii gorbatogo kita (*Megaptera nodosa lalandii* Fischer) yuzhnogo polushariya. [On the systematic position of the humpbacked whale is the southern hemisphere] *Informatsionnyi Byulleten'. Sovetskoi Antarkticheskoi Ekspeditsii* 3, 77–78.

Iwata A (1983) A revision of the cottid fish genus *Vellitor*. *Japanese Journal of Ichthyology* 30, 1–9.

Jackson BD (1896) The Dates of Rees's Cyclopædia. *Journal of Botany. British and Foreign* 34, 307–311. [Jul 1896]

Jackson JA, Steel DJ, Beerli P, Congdon BC, Olavarria C, Leslie MS, Pomilla C, Rosenbaum H, Baker CS (2014) Global diversity and oceanic divergence of humpback whales (*Megaptera novaeangliae*). *Proceedings of the Royal Society B* 241, 20133222, 1–10.

Jackson SM (2011) *Petaurus gracilis*. *Mammalian Species* 43(882), 141–148. [28 Jul 2011]

Jackson SM, Thorington R (2012) Gliding mammals: Taxonomy of living and extinct species. *Smithsonian Contributions to Zoology* 638, 1–107. [28 Mar 2012]

Jacob A (1825) On the generic characters and anatomical structure of the whale, entitled *Delphinus Diodon* by Hunter, and *Hyperoodon* by La Cepede. *Dublin Philosophical Journal and Science Review* 1, 58–73.

Jacobson G (1895) Chrysomelidae palaearcticae novae vel parum cognitae. *Horae Societatis Entomologicae Rossicae* 29, 529–558.

Jaekel OMJ (1911) *Die Wirbeltiere. Eine Überzichte über die Fossilen und Lebenden Formen*. Gebrüder Borntraeger, Berlin.

Jäger GF (1833) *De Holothuriis. Dissertatio Inauguralis quam consensu et auctoritate gratiosi medicorum ordinis in alma universiate litteraria Turicensi ut summi in medicina et chirurgia honores*. Gessnerianis. Turici.

Jäger G (1850) Übersicht der fossilen säugethiere, weche in würtemberg in verschiedenen formationen aufgefunden worden sind, und nähere beschreibung und abbildung einzelner derselben. *Novorum Actorum Academiae Caesareae Leopoldino-Carolinae Naturae Curiosorum : Verhandlungen der Kaiserlichen Leopoldinisch-Carolinischen Akademie der Naturforscher* 22(2), 665–934.

James DJ, Dale GJ, Retallick K, Orchard K (2007) Christmas Island Flying-fox *Pteropus natalis* Thomas 1887: An Assessment of Conservation Status and Threats. Parks Australia North Christmas Island Biodiversity Monitoring Programme. Report to the Department of Finance and Administration and the Department of Environment and Water Resources, Canberra. [Jan 2007]

Jan G (1863) *Elenco sistematico degli ofidi descritti e disegnati per l'Iconografia generale*. A. Lombardi, Milano.

Janke A, Gemmell NJ, Feldmaier-Fuchs G, von Haesele A, Pääbo S (1996) The mitochondrial genome of a monotreme – the platypus (*Ornithorhynchus anatinus*). *Journal of Molecular Evolution* 42, 153–159.

Janke A, Xu X, Arnason U (1997) The complete mitochondrial genome of the wallaroo (*Macropus robustus*) and the phylogenetic relationship among Monotremata, Marsupialia and Eutheria. *Proceedings of the National Academy of Sciences of the United States of America* 94, 1276–1281. [18 Feb 1997]

Janke A, Magnell O, Wieczorek G, Westerman M, Arnason U (2002) Phylogenetic analysis of 18S rRNA and the mitochondrial genomes of the wombat, *Vombatus ursinus*, and the spiny anteater, *Tachyglossus aculeatus*: Increased support for the Marsupionta hypothesis. *Journal of Molecular Evolution* 54, 71–80.

Jefferson TA, Rosenbaum HC (2014) Taxonomic revision of the humpback dolphins (*Sousa* spp.), and description of a new species from Australia. *Marine Mammal Science* 30, 1494–1541. [Oct 2014]

Jefferson TA, Van Waerebeek K (2002) The taxonomic status of the nominal dolphin species *Delphinus tropicalis* van Bree, 1971. *Marine Mammal Science* 18, 787–818. [Oct 2002]

Jefferson TA, Van Waerebeek K (2004) Geographic variation in skull morphology of humpback dolphins (*Sousa* spp.). *Aquatic Mammals* 30, 3–17.

Jefferson TA, Webber MA, Pitman RL (2008) *Marine Mammals of the World: A Comprehensive Guide to their Identification*. Academic Press, Amsterdam.

Jenkins PD (1976) Variation in Eurasian shrews of the genus *Crocidura* (Insectivora: Soricidae). *Bulletin of the British Museum (Natural History). Zoology Series* 30(7), 269–309. [25 Nov 1976]

Jenkins PD (1982) A discussion of the Malayan and Indonesian shrews of the genus *Crocidura* (Insectivora: Soricidae). *Zoologische Mededeelingen* 56, 267–279. [3 Dec 1982]

Jenkins PD, Hill JE (1981) The status of *Hipposideros galeritus* Cantor, 1846 and *Hipposideros cervinus* (Gould, 1854) (Chiroptera: Hipposideridae). *Bulletin of the British Museum (Natural History). Zoology Series* 41(5), 279–294. [17 Dec 1981]

Jenkins PD, Knutson L (1983) *A Catalogue of the Type Specimens of the Monotremata and Marsupialia in the British Museum (Natural History)*. British Museum (Natural History), London.

Jennings J (1828) *Ornithologia, or the Birds : A poem, in two parts; with an introduction to their natural history; with copious notes*. Poole and Edwards Stationers Court, London.

Jentink FA (1879) On a new genus and species of rat from Celebes. *Notes from the Leyden Museum* 1, 117–119.

Jentink FA (1881) Description of a new African bat, *Leiponyx büttikoferi*. *Notes from the Leyden Museum* 3, 59–62.

Jentink FA (1885) On *Didelphis caudivolvula* Kerr and *Didelphis vulpecular* Kerr. *Notes from the Leyden Museum* 7, 21–28.

Jentink FA (1887) *Catalogue ostéologique des mammifères. Muséum d'Histoire Naturelle des Pays-Bas*. Volume 9. E.J. Brill, Leiden.

Jentink FA (1889) On a new genus and a new species in the macroglossine-group of bats. *Notes from the Leyden Museum* 11, 209–212.

Jentink FA (1890) Mammalia from the Malay Archipelago. pp. 93–130, in MCW Weber (ed.) *Zoologische Ergebnisse einer Reise in Niederländische Ost-Indien*. Leiden.

Jentink FA (1902) Revision of the genera *Macroglossus* and *Syconycteris* and description of a new genus and species, *Odontonycteris meijeri*. *Notes from the Leyden Museum* 23, 131–142.

Jentink FA (1907) Mammals collected by the members of the Humboldt Bay- and the Merauke River- Expeditions. *Nova Guinea* 5, 364–374.

Jentink FA (1909) Mammals collected by the Dutch New Guinea expedition 1907. *Nova Guinea* 9, 1–14.

Jentink FA (1910) *Chryopteron bartelsii*, novum genus et species, from Java. *Notes from the Leyden Museum* 32, 73–77.

Jentink FA (1911) Mammals, collected by the Dutch New Guinea Expedition 1909/10. *Nova Guinea* 9, 165–183.

Jerdon TC (1867) *The Mammals of India: Natural History of all the Animals Known to Inhabit Continental India*. Thomason College Press, Roorkee, India.

Johnson DH (1954) A new marsupial of the genus *Antechinus* from northern Australia. *Proceedings of the Biological Society of Washington* 67, 77–79. [22 Mar 1954]

Johnson DH (1959) Four new mammals from the Northern Territory of Australia. *Proceedings of the Biological Society of Washington* 72, 183–187. [4 Nov 1959]

Johnson DJ (1964) Mammals of the Arnhem Land expedition. pp. 427–515, in RL Specht (ed.) *Records of the American-Australian Scientific Expedition to Arnhem Land*. Volume 4. Melbourne University Press, Melbourne.

Johnson KA, Roff AD (1980) Discovery of ningauis (*Ningaui* sp.: Dasyuridae: Marsupialia) in the Northern Territory, Australia. *Australian Mammalogy* 3, 127–129. [May 1980]

Johnson KA, Walton DW (1989) Notoryctidae. pp. 591–602, in DW Walton & BJ Richardson (eds.) *Fauna of Australia. Volume 1B. Mammalia*. Australian Government Publishing Service, Canberra.

Johnston PG, Sharman GB (1976) Studies on populations of *Potorous* Desmarest (Marsupialia). I. Morphological variation. *Australian Journal of Zoology* 24, 573–588.

Johnston PG, Sharman GB (1977) Studies on populations of *Potorous* Desmarest (Marsupialia). II. Electrophoretic, chromosomal and breeding studies. *Australian Journal of Zoology* 25, 733–747.

Jones E (2009) Hybridisation between the dingo, *Canis lupus dingo*, and the domestic dog, *Canis lupus familiaris*, in Victoria: a critical review. *Australian Mammalogy* 31, 1–7.

Jones JK, Packard RL (1956) *Feresa intermedia* (Gray) preoccupied. *Proceedings of the Biological Society of Washington* 69, 167. [31 Dec 1956]

Jones ME, Paetkau D, Geffen E, Moritz C (2003) Microsatellites for the Tasmanian devil (*Sarcophilus laniarius*). *Molecular Ecology Notes* 3, 277–279. [26 Mar 2003]

Jordan K (1938) In memory of Lord Rothschild, Ph.D. F.R.S., J.P. born the 8th February, 1868, died the 27th August, 1937. *Novitates Zoologicae* 41, 1–41.

Jordan DS, Starks EC (1904) A review of the Cottidae or sculpins found in the waters of Japan. *Proceedings of the United States National Museum* 27, 231–335. [28 Jan 1904]

Jourdan C (1837a) Mémoire sur quelques mammifères nouveaux. *Comptes Rendus Hebdomadaires des Séances de l'Académie des Sciences, Paris* 5, 521–524. [Oct 1837]

Jourdan C (1837b) *L'Echo du Monde Savant* 4(275), 156. [14 Oct 1837] Reference not seen.

Julien-Laferrière D (1994) Catalogue des types de marsupiaux (Marsupialia) du Muséum National d'Histoire Naturelle, Paris. *Mammalia* 58, 1–39.

Junge GCA (1950) On a specimen of the rare fin whale, *Balaenoptera edeni* Anderson, stranded on Pulu Sugi near Singapore. *Zoologische Verhandelingen* 9, 1–26.

Kalandadze NN, Rautian SA (1992) The system of mammals and historical zoogeography. [Systema mlekopitayushchikh i istorygeskaya zoogeographei] In OL Rossolimo (ed.) Filogenetika mlekopitaûŝih: issledovaniâ po faune. *Sbornik Trudov Zoologičeskogo Muzeâ Moskovskogo Gosudarstvennogo Universiteta* 29, 44–152.

Kammerer CF (2006) Notes on some preoccupied names in Arthropoda. *Acta Zootaxonomica Sinica* 31, 269–271. [Apr 2006]

Karawajew W (1926) Ameisen aus dem Indo-Australischen Gebiet. *Treubia* 8, 413–445.

Kasuya T (1973) Systematic consideration of recent toothed whales based on the morphology of the tympano-periotic bone. *Scientific Reports of the Whales Research Institute, Tokyo* 25, 1–103.

Kasuya T (1975) Past occurrence of *Globicephala melaena* in the western North Pacific. *Scientific Reports of the Whales Research Institute, Tokyo* 27, 95–110.

Kato H (1992) Diminutive minke whales in southern hemisphere. *IBI Reports* 3, 61–62.

Kato H, Perrin WF (2009) Bryde's whales *Balaenoptera edeni/brydei*. pp. 158–163, in WF Perrin, B Würsig & JGM Thewissen (eds.) *Encyclopedia of Marine Mammals*. Academic Press, Amsterdam. Second Edition.

Kaup JJ (1829) *Skizzirte Entwickelungs-Geschichte und natürliches System der Europäischen Thierwelt : Erster Theil welcher die Vogelsäugethiere und Vögel, nebst Andeutung der Entstehung der letzteren aus Amphibien enthält*. Darmstadt und Leipzig.

Kaup JJ (1838a) Ueber Zaihnen von *Halytherium* und *Pugmeodon* aus Flonheim. *Neues Jahrbuch ftir Mineralogie, Geognosie. Geologie und Petrefaktenkunde* 1838, 318–320.

Kaup JJ (1838b) Bei aufstellung meines Geschlechts *Halitherium* (Jahbr. 1838). *Neues Jahrbuch ftir Mineralogie, Geognosie. Geologie und Petrefaktenkunde* 1838, 536.

Kavanagh JR, Burk-Herrick A, Westerman M, Springer MS (2004) Relationships among families of Diprotodontia (Marsupialia) and the phylogenetic position of the autapomorphic honey possum (*Tarsipes rostratus*). *Journal of Mammalian Evolution* 11, 207–222. [Dec 2004]

Kear BP, Cooke BN (2001) A review of macropodoid (Marsupialia) systematics with the inclusion of a new family.

Memoirs of the Association of Australasian Palaeontologists 25, 83–101.

Kelaart EF (1850a) Description of additional mammals. *Journal of the Ceylon Branch of the Royal Asiatic Society* 2, 216–217. (= pp. 329–330 in reprint of 1890). Reference not seen.

Kelaart EF (1850b) Description of new species and varieties of mammals found in Ceylon. *Journal of the Ceylon Branch of the Royal Asiatic Society* 2, 208–215. (= pp. 321–328 in reprint of 1890).

Kellogg R (1922) Pinnipeds from Miocene and Pleistocene deposits of California. *University of California Publications. Bulletin of the Department of Geological Sciences* 13(4), 23–132. [14 Apr 1922]

Kellogg R (1925) New pinnipeds from the Miocene diatomaceous earth near Lompoc, California. *Carnegie Institution of Washington Publication* 348, 71–96. [Apr 1925]

Kellogg R (1927) *Kentriodon pernix*, a Miocene porpoise from Maryland. *Proceedings of the United States National Museum* 69(19), 1–55.

Kellogg R (1929) What is known of the migrations of some of the whalebone whales. *Smithsonian Institution Annual Report* 1928, 467–496.

Kellogg R (1931) Pelagic mammals from the Temblor Formation of the Kern River region, California. *Proceedings of the California Academy of Sciences* 19(4), 217–397. [30 Jan 1931]

Kellogg R (1932) New names for mammals proposed by Borowski in 1780 and 1781. *Proceedings of the Biological Society of Washington* 45, 147–148. [9 Sep 1932]

Kellogg R (1945) A new naked-tailed rat (*Melomys*). *Proceedings of the Biological Society of Washington* 58, 69–71. [7 May 1945]

Kemp TS (1983) The relationships of mammals. *Zoological Journal of the Linnean Society* 77, 353–384.

Kemper CM (2004) Osteological variation and taxonomic affinities of bottlenose dolphins, *Tursiops* spp., from South Australia. *Australian Journal of Zoology* 52, 29–48.

Kemper CM, Schmitt LH (1992) Morphological variation between populations of the brush-tailed tree rat (*Conilurus penicillatus*) in northern Australia and New Guinea. *Australian Journal of Zoology* 40, 437–452.

Kemper CM, Cooper SJB, Medlin GC, Adams M, Stemmer D, Saint KM, McDowell MC, Austin JJ (2011) Cryptic grey-bellied dunnart (*Sminthopsis griseoventer*) discovered in South Australia: genetic, morphological and subfossil analyses show the value of collecting voucher material. *Australian Journal of Zoology* 59, 127–144. [online 23 Nov 2011]

Kerbert C (1913) Über *Zaglossus*. *Zoologischer Anzeiger* 42, 162–167. [20 Jun 2013]

Kerle JA, McKay GM, Sharman GB (1991) A systematic analysis of the brushtail possum, *Trichosurus vulpecula* (Kerr, 1792) (Marsupialia: Phalangeridae). *Australian Journal of Zoology* 39, 313–331.

Kermack DM, Kermack KA, Mussett F (1968) The Welsh Pantothere *Kuehneotherium praecursoris*. *Journal of the Linnean Society of London. Zoology (Jena, Germany)* 47, 407–423.

Kermack KA, Mussett F, Rigney HW (1973) The lower jaw of Morganucodon. *Zoological Journal of the Linnean Society* 53, 87–175.

Kerr R (1792) *The Animal Kingdom*, or zoological system, of the celebrated Sir Charles Linnaeus; Class1. Mammalia: containing a complete systematic description, arrangement, and nomenclature, of all the known species and varieties of the Mammalia, or animals which give suck to their young; being a translation of that part of the Systema Naturae, as lately published, with great improvements, by Professor Gmelin of Goettingen. Together with numerous additions from more recent zoological writers, and illustrated with copper plates. J. Murray & R. Faulder, London. Volume 1.

Kerr SE (2011) Divergence of a mammal along a habitat gradient: a study of the coppery brushtail possum, *Trichosurus vulpecula johnsonii*. [sic] PhD Thesis. James Cook University Townsville.

Kershaw F, Leslie, MS, Collins T, Mansur RM, Smith BD, Minton G, Baldwin R, LeDuc RG, Anderson RC, Brownell Jr. RL, Rosenbaum HC (2013) Population differentiation of 2 forms of Bryde's whales in the Indian and Pacific Oceans. *Journal of Heredity* 104, 755–764. [30 Sep 2013]

Keyserling A, Blasius IH (1839) Uebersicht der Gattungs-und Artcharaktere der europäischer Flermäuse. *Archiv für Naturgeschichte* 5(1), 293–331.

Killian KJ, Buckley TR, Stewart N, Munday BL, Jirtle RL (2001) Marsupials and eutherians reunited: genetic evidence from the Theria hypothesis of mammalian evolution. *Mammalian Genome* 12, 513–517.

King JE (1959a) Notes on the specific name of the Kerguelen fur seal. *Mammalia* 23, 381. [Sep 1959]

King JE (1959b) The northern and southern populations of *Arctocephalus gazella*. *Mammalia* 23, 19–40. [Mar 1959]

King JE (1960) Sea-lions of the genera *Neophoca* and *Phocarctos*. *Mammalia* 24, 445–456. [Sep 1960]

King JE (1961) Notes on the pinnipèdes from Japan described by Temminck in 1844. *Zoologische Mededeelingen* 13, 211–224. [28 Sep 1961]

King JE (1966) Relationships of the hooded and elephant seals (genera *Cystopora* and *Mirounga*). *Journal of Zoology* 148, 385–398.

King JE (1968) The identity of fur seals of Australia. *Nature* 219(5154), 632–633. [10 Aug 1968]

King JE (1969) The identity of the fur seals of Australia. *Australian Journal of Zoology* 17, 841–853.

King JE (1983) *Seals of the World*. British Museum (Natural History), London. Second Edition.

Kinman KE (1994) *The Kinman System: Toward A Stable Cladisto-Eclectic Classification of Organisms (Living and Extinct: 48 Phyla, 269 Classes, 1,719 Orders*. K. Kinman, Hays, Kansas.

Kirby W (1837) The insects. In J Richardson (ed.). *Fauna Boreali-Americana. Insects. Coleoptera*. Part 4. Norwich, London.

Kirby WF (1882) List of Hymenoptera, with descriptions and figures of the typical specimens in the British Museum. British Museum, London. Volume 1. Tenthredinidae and Siricidae.

Kirkaldy GW (1908) Bibliographical and nomenclatorial notes on the Hemiptera.—No. 8. *The Entomologist* 41, 123–125.

Kirkpatrick TH (1963) A note on the dental eruption of some Macropodinae, with particular reference to Captain Cook's kangaroo. *Queensland Journal of Agricultural Science* 20, 539–541.

Kirkpatrick TH, Woods JT (1964) Comments on the proposed stabilization of *Macropus* Shaw, 1790. *Bulletin of Zoological Nomenclature* 21, 249–250. [16 Oct 1964]

Kirsch JAW (1968a) Prodromus of the comparative serology of Marsupialia. *Nature* 217(5127), 418–420. [3 Feb 1968]

Kirsch JAW (1968b) The serological affinities of *Burramys* and related possums (Marsupialia: Phalangeroidea). *Australian Journal of Science* 31, 43–45.

Kirsch JAW (1977a) The comparative serology of Marsupialia, and a classification of marsupials. *Australian Journal of Zoology, Supplement Series* 52, 1–152. [26 Sep 1977]

Kirsch JAW (1977b) The classification of the marsupials with special reference to karyotypes and serum proteins. pp. 1–50, in D Hunsaker II. (ed.) *The Biology of the Marsupials*. Academic Press, New York.

Kirsch JAW, Archer M (1982) Polythetic cladistics, or, when parsimony's not enough: The relationships of carnivorous marsupials. pp. 595–619, in M Archer (ed.) *Carnivorous Marsupials*. Volume 2. Royal Zoological Society of New South Wales, Sydney. [Feb 1982]

Kirsch JW, Calaby JH (1977) The species of living marsupials: an annotated list. pp. 9–26, in B Stonehouse & D Gilmore (eds.) *The Biology of Marsupials*. Macmillan Press, London.

Kirsch JAW, Mayer GC (1998) The platypus is not a rodent: DNA hybridization, amniote phylogeny and the palimpsest theory. *Philosophical Transactions of the Royal Society of London. Series B, Biological Sciences* 353, 1221–1237.

Kirsch JAW, Poole WE (1967) Serological evidence for speciation in the grey kangaroo, *Macropus giganteus* Shaw 1790 (Marsupialia: Macropodidae). *Nature* 215(5105), 1097–1098. [2 Sep 1967]

Kirsch JAW, Poole WE (1972) Taxonomy and distribution of the grey kangaroos, *Macropus giganteus* and *Macropus fuliginosus* (Desmarest), and their subspecies (Marsupialia: Macropodidae). *Australian Journal of Zoology* 20, 315–339.

Kirsch JAW, Wolman MA (2001) Molecular relationships of the bear cuscus, *Ailurops ursinus* (Marsupialia: Phalangeridae). *Australian Mammalogy* 23, 23–30.

Kirsch JAW, Krajewski C, Springer MS, Archer M (1990a) DNA-DNA hybridisation studies of the carnivorous marsupials. II. Relationships among dasyurids (Marsupialia: Dasyuridae). *Australian Journal of Zoology* 38, 673–696.

Kirsch JAW, Springer MS, Krajewski C, Archer M, Aplin K, Dickerman AW (1990b) DNA/DNA hybridisation studies of the carnivorous marsupials. I. The intergeneric relationships of bandicoots (Marsupialia: Perameloidea). *Journal of Molecular Evolution* 30, 434–448.

Kirsch JAW, Lapointe F-J, Foeste A (1995) Resolution of portions of the kangaroo phylogeny (Marsupialia: Macropodidae) using DNA hybridization. *Biological Journal of the Linnean Society. Linnean Society of London* 55, 309–328.

Kirsch JAW, Lapointe J-F, Springer MS (1997) DNA-hybridisation studies of marsupials and their implications for metatherian classification. *Australian Journal of Zoology* 45, 211–280.

Kitchener DJ (1976) *Eptesicus douglasi*, a new vespertilionid bat from the Kimberley, Western Australia. *Records of the Western Australian Museum* 4, 295–301. [15 Oct 1976]

Kitchener DJ (1980a) A new species of *Pseudomys* (Rodentia: Muridae) from Western Australia. *Records of the Western Australian Museum* 8, 405–414. [30 Jun 1980]

Kitchener DJ (1980b) *Taphozous hilli* sp. nov. (Chiroptera: Emballonuridae), a new sheath-tailed bat from Western Australia and Northern Territory. *Records of the Western Australian Museum* 8, 161–169. [30 Jun 1980]

Kitchener DJ (1985) Description of a new species of *Pseudomys* (Rodentia: Muridae) from Northern Territory. *Records of the Western Australian Museum* 12, 207–221. [30 Jul 1985]

Kitchener DJ (1988) A new species of false antechinus (Marsupialia: Dasyuridae) from the Kimberley, Western Australia. *Records of the Western Australian Museum* 14, 61–71. [10 Jun 1988]

Kitchener D (1989) Taxonomic appraisal of *Zyzomys* (Rodentia, Muridae) with descriptions of two new species from Northern Territory, Australia. *Records of the Western Australian Museum* 14, 331–373. [29 Sep 1989]

Kitchener DJ (1991) *Pseudantechinus mimulus* (Thomas, 1906) (Marsupialia, Dasyuridae): rediscovery and redescription. *Records of the Western Australian Museum* 15, 191–202. [31 Jan 1991]

Kitchener DJ, Caputi N (1985) Systematic revision of Australian *Scoteanax* (Chiroptera: Vespertilionidae), with remarks on relationships to other Nycticeiini. *Records of the Western Australian Museum* 12, 85–146. [2 Apr 1985]

Kitchener DJ, Caputi N (1988) A new species of false antechinus (Marsupialia: Dasyuridae) from Western Australia, with remarks on the generic classification within the Parantechini. *Records of the Western Australian Museum* 14, 35–59. [10 Jun 1988]

Kitchener DJ, Halse SA (1978) Reproduction in female *Eptesicus regulus* (Thomas) (Vespertilionidae) in South-western Australia. *Australian Journal of Zoology* 26, 257–267.

Kitchener DK, Humphreys WF (1986) Description of a new species of *Pseudomys* (Rodentia: Muridae) from the Kimberley Region, Western Australia. *Records of the Western Australian Museum* 12, 419–434. [21 May 1986]

Kitchener DK, Humphreys WF (1987) Description of a new subspecies of *Pseudomys* (Rodentia: Muridae) from the Northern Territory. *Records of the Western Australian Museum* 13, 285–295. [13 Apr 1987]

Kitchener DJ, Sanson G (1978) *Petrogale burbidgei* (Marsupialia, Macropodidae), a new rock wallaby from Kimberley, Western Australia. *Records of the Western Australian Museum* 6, 269–285. [30 Jun 1978]

Kitchener DJ, Suyanto A (2002) Morphological variation in *Miniopterus pusillus* and *M. australis* (*sensu* Hill, 1992) in southeastern Asia, New Guinea, and Australia. *Records of the Western Australian Museum* 21, 9–33. [1 May 2002]

Kitchener DJ, Vicker E (1981) *Catalogue of Modern Mammals in the Western Australian Museum 1895–1981*. Western Australian Museum, Perth.

Kitchener DJ, Stoddart J, Henry J (1983) A taxonomic appraisal of the genus *Ningaui* Archer (Marsupialia: Dasyuridae), including description of a new species. *Australian Journal of Zoology* 31, 361–379.

Kitchener DJ, Stoddart J, Henry J (1984a) A taxonomic revision of the *Sminthopsis murina* complex (Marsupialia, Dasyuridae) in Australia, including description of four new species. *Records of the Western Australian Museum* 11, 201–247. [2 Oct 1984]

Kitchener DJ, Adams M, Baverstock P (1984b) Redescription of *Pseudomys bolami* Troughton, 1932 (Rodentia: Muridae). *Australian Mammalogy* 7, 149–159. [Dec 1984]

Kitchener DJ, Caputi N, Jones B (1986) Revision of Australo-Papuan *Pipistrellus* and *Falsistrellus* (Microchiroptera: Vespertilionidae). *Records of the Western Australian Museum* 12, 435–495. [21 May 1986]

Kitchener DJ, Jones B, Caputi N (1987) Revision of Australian *Eptesicus* (Microchiroptera: Vespertilionidae). *Records of the Western Australian Museum* 13, 427–500. [30 Nov 1987]

Kitchener DJ, How RA, Maharadatunkamsi (1991) A new species of *Nyctophilus* (Chiroptera: Vespertilionidae) from Lembata Island, Nusa Tenggara, Indonesia. *Records of the Western Australian Museum* 15, 97–107. [31 Jan 1991]

Kitchener DJ, How RA, Cooper NK, Suyanto A (1992) *Hipposideros diadema* (Chiroptera: Hipposideridae) in the Lesser Sunda Islands, Indonesia: Taxonomy and geographic morphological variation. *Records of the Western Australian Museum* 16, 1–60. [18 Sep 1992]

Kitchener DJ, Packer WC, Maryanto I (1994a) Morphological variation in Maluku populations of *Syconycteris australis* (Peters, 1867) (Chiroptera: Pteropodidae). *Records of the Western Australian Museum* 16, 485–498. [27 Jun 1994]

Kitchener DJ, Adams M, Boeadi (1994b) Morphological and genetic relationships among populations of *Scotorepens sanborni* (Chiroptera: Vespertilionidae) from Papua New Guinea, Australia and Indonesia. *Australian Mammalogy* 17, 31–42. [30 Jun 1994]

Kitchener DJ, Cooper N, Maryanto I (1995) The *Myotis adversus* (Chiroptera: Vespertilionidae) species complex in Eastern Indonesia, Australia, Papua New Guinea and the Solomon Islands. *Records of the Western Australian Museum* 17, 191–212. [27 Jun 1995]

Kittlitz FH von (1830) Über einie vögel von Chili. *Mémoires Présentés à l'Académie Impériale des Sciences de St. Pétersbourg* 1(2), 173–194.

Knox E (1978) A note on the identification of *Melomys* species (Rodentia: Muridae) in Australia. *Journal of Zoology* 185, 276–277.

Knox FJ (1838) *Catalogue of Anatomical Preparations Illustrative of the Whale*, particularly the great rorqual (*Balaena maximus borealis*), now exhibiting in the pavilion, North College Street. Neill and Co., Edinburgh.

Knox FJ (1870) On the Balaenidae or whales with Baleen. With notes on the Cetacea, in the Colonial Museum, Wellington. *Transactions and Proceedings of the New Zealand Institute* 2 (1869): 2, 21–28. [April 1870]

Koch CL von (1839) *Übersicht des Arachnidensystems.*: C.H. Zeh'schen Buchhandlung. Nürnberg. Volume 2.

Koch CL von (1847) *System der Myriapoden, mit den verzeichnissen und berichtigenungen zu Deutschlands crustacean, myriapoden und arachniden.* Volume 3. pp. 1–270, in GWF Panzer & GAW Herrich-Schäffer (eds.) *Kritische Revision der Insectenfaune Deutschlans*. Friedrich Pustet, Regensburg.

Koch CL von (1860) Die fledermäuse (Chiropteren) oberhessens und der angrenzenden Ländertheile. pp. 25–62, in *Achter Bericht der Oberhessischen Gesellschaft für Natur- und Heilkunde*. Giefsen im Mai 1860.

Kohlbrugge JHF (1896) Bijdragen tot de natuurlijke geschiedenis van menschen en dieren IV. Zoogdieren van den Tengger. *Natuurkundig Tijdschrift voor Nederlandsch-Indië* 55, 261–298.

Kohler-Rollefson IU (1991) *Camelus dromedarius*. *Mammalian Species* 375, 1–8. [12 Apr 1991]

Kolenati FA (1856) Europa's Chiroptern. 1. Synonopsis der Europäischen Chiroptera. *Allgemeine Deutsche Naturhistorische Zeitung, Dresden* 2, 121–133, 161–192.

Kolenati FA (1858) Eine neue österreichische fledermaus. *Sitzungsberichte der Mathematisch-Naturwissenschaftliche Classe der Kaiserlichen Akademie der Wissenschaften, Wien* 29, 250–256.

Koler-Matznick J, Brisbin IL, Feinstein M, Bulmer S (2003) An updated description of the New Guinea singing dog (*Canis hallstromi*, Troughtoni 1957). *Journal of Zoology* 261, 109–118.

Koler-Matznick J, Yates BC, Bulmer S, Brisbin IL (2007) The New Guinea singing dog: its status and scientific importance. *Australian Mammalogy* 29, 47–56.

Kompanje EJO, Moeliker CW (2001) Some fruit bats from the remote Moluccan and West-Papuan islands, with the description of a new subspecies of *Macroglossus minimus* (Megachiroptera: Pteropodidae). *Deinsea* 8, 143–167.

Koopman KF (1971) Taxonomic notes on *Chalinolobus* and *Glauconycteris* (Chiroptera, Vespertilionidae). *American Museum Novitates* 2451, 1–10. [11 Feb 1971]

Koopman KF (1973) Systematics of the Indo-Australian *Pipistrellus*. *Periodicum Biologorum* 75, 113–116.

Koopman KF (1978) The genus *Nycticeius* (Vespertilionidae), with special reference to tropical Australia. pp. 165–171, in RJ Olembo, JB Castelino & FA Mutere (eds.) *Proceedings of the Fourth International Bat Research Conference*. Kenya Literature Bureau, Nairobi.

Koopman KF (1979) Zoogeography of mammals from islands of the northeastern coast of New Guinea. *American Museum Novitates* 2690, 1–17. [2 Nov 1979]

Koopman KF (1982) Results of the Archbold Expeditions. No. 109. Bats from Eastern Papua and the East Papuan Islands. *American Museum Novitates* 2747, 1–34. [23 Sep 1982]

Koopman KF (1984a) Taxonomic and distributional notes on tropical Australian bats. *American Museum Novitates* 2778, 1–48. [31 Jan 1984]

Koopman KF (1984b) Bats. pp. 145–186, in S Anderson & JK Jones (eds.) *Orders and Families of Recent Mammals of the World*. John Wiley & Sons, New York.

Koopman K (1985) A synopsis of the families of bats. Part VII. *Bat Research News* 25, 25–27. [Dated 1984 but issued 1985]

Koopman KF (1989) Distributional patterns of Indo-Malayan bats (Mammalia: Chiroptera). *American Museum Novitates* 2942, 1–19. [26 Jun 1989]

Koopman KF (1993) Order Chiroptera. pp. 137–241, in DE Wilson & DA Reeder (eds.) *Mammal Species of the World. A Taxonomic and Geographic Reference*. Smithsonian Institution Press, Washington. Second Edition.

Koopman KF (1994) Chiroptera: Systematics. *Handbook of Zoology*. Volume VIII. Part 60. Mammalia. 1–217.

Koopman KF, Cockrum EL (1967) Bats. pp. 109–150, in S Anderson & JK Jones Jr (eds.) *Recent Mammals of the World: A Synonopsis of Families*. The Ronald Press Company, New York.

Koopman KF, Jones JK Jr (1970) Classification of bats. pp. 22–28, in BH Slaughter & DW Walton (eds.) *About Bats : a chiropteran biology symposium (held at the 1969 meeting of the Texas Academy of Sciences*. Southern Methodist University Press, Dallas.

Kooyman GL (1981) *Weddell Seal: Consummate Diver*. Cambridge University Press, Cambridge.

Koretsky IA, Grigorescu D (2002) The Fossil Monk Seal *Pontophoca sarmatica* (Alekseev) (Mammalia: Phocidae: Monachinae) from the Miocene of Eastern Europe. *Smithsonian Contributions to Paleobiology* 93, 149–162.

Koretsky IA, Holec P (2002) A primitive seal (Mammalia: Phocidae) from the early middle Miocene of central Paratethys. *Smithsonian Contributions to Paleobiology* 93, 163–178.

Kossovaya OL (1997) Rugose corals from standard sections of the Gzhelian-Artinskian stages of northern Timan and the western slope of the Urals. In Stukalina, G.A. (ed.) *Atlas of reference complexes of Palaeozoic benthic fauna of northeastern European Russia. Ostracods, brachiopods, rugosans. Izdatelstvo Vserossiiskii Nauchno-Issledovatelskii Geologicheskii Institut im. A.P. Karpinskogo (VSEGEI)*, St. Petersburg. 3: 53–96, 106–115, 128–154. Reference not seen.

Kowarzik R (1909) Resultatae einer zusammenfassenden Bearbeitung der Monotremen. *Zoologischer Anzeiger* 35, 213–215. [7 Dec 1909]

Kraatz G (1859) Die Staphylinen-Fauna von Ostindien, insbesondere der Insel Ceylan. *Archiv für Naturgeschichte* 25(1), 1–194.

Krahn M, Ford MJ, Perrin WF, Wade PR, Angliss RP, Hanson MB, Taylor BL, Ylitalo GM, Dahlheim ME, Stein JE & Waples RS (2004) Status Review of Southern Resident Killer Whales (*Orcinus orca*) under the Endangered Species Act. NOAA Technical Memorandum NMFS-NWFSC-62. 73 pp. [Dec 2004]

Krajewski C, Westerman M (2003) Molecular systematics of Dasyuromorphia. pp. 3–20, in M Jones, C Dickman & M Archer (eds.) *Predators with Pouches: The Biology of Carnivorous Marsupials*. CSIRO Publishing, Melbourne.

Krajewski C, Painter J, Driskell AC, Buckley L, Westerman M (1993) Molecular systematics of New Guinean dasyurids (Marsupialia: Dasyuridae). *Science in New Guinea* 19, 157–166.

Krajewski C, Painter J, Buckley L, Westerman M (1994) Phylogenetic structure of the marsupial family Dasyuridae based on cytochrome b DNA sequences. *Journal of Mammalian Evolution* 2, 25–35. [Mar 1994]

Krajewski C, Buckley L, Woolley PA, Westerman M (1996) Phylogenetic analysis of cytochrome b sequences in the dasyurid marsupial subfamily Phascogalinae: Systematics and the evolution of reproductive strategies. *Journal of Mammalian Evolution* 3, 81–91. [Mar 1996]

Krajewski C, Young J, Buckley L, Woolley PA, Westerman M (1997a) Reconstructing the evolutionary radiation of dasyurine marsupials with cytochrome *b*, 12S rRNA, and protamine P1 gene trees. *Journal of Mammalian Evolution* 4, 217–236. [Sep 2007]

Krajewski C, Blacket MJ, Buckley L, Westerman M (1997b) A multigene assessment of phylogenetic relationships within the dasyurid marsupial Subfamily Sminthopsinae. *Molecular Phylogenetics and Evolution* 8, 236–248. [Oct 1997]

Krajewski C, Buckley L, Westerman M (1997c) DNA phylogeny of the marsupial wolf resolved. *Proceedings of the Royal Society of London. Series B, Biological Sciences* 264, 911–917.

Krajewski C, Blacket MJ, Westerman M (2000a) DNA sequence analysis of familial relationships among Dasyuromorphian marsupials. *Journal of Mammalian Evolution* 7, 95–108. [Jun 2000]

Krajewski C, Wroe S, Westerman M (2000b) Molecular evidence for the timing of cladogenesis in dasyurid marsupials. *Zoological Journal of the Linnean Society* 130, 375–404. [Nov 2000]

Krajewski C, Woolley PA, Westerman M (2000c) The evolution of reproductive strategies in dasyurid marsupials: implications of molecular phylogeny. *Biological Journal of the Linnean Society. Linnean Society of London* 71, 417–435.

Krajewski C, Anderson FE, Woolley PA, Westerman M (2012) Molecular evidence for a deep clade of dunnarts (Marsupialia: Dasyuridae: *Sminthopsis*). *Journal of Mammalian Evolution* 19, 265–276. [Online 13 Jun 2012]

Krefft G (1862) The vertebrated animals of the lower Murray and Darling, their habits, economy, and geographical distribution. *The Sydney Morning Herald*. Volume 46. No. 7608. Page 2. Columns 1–5. [24 Oct 1862]

Krefft G (1863) Description of a new species of the genus *Dromicia* discovered in the neighborhood of Sydney. *Proceedings of the Zoological Society of London* 31, 49–50. [May 1863]

Krefft G (1864) *Catalogue of the Mammals in the Collection of the Australian Museum*. Australian Museum, Sydney.

Krefft G (1865) Description of a new species of rock kangaroo from New South Wales. *Proceedings of the Zoological Society of London* 33, 324–325. [Jun 1865]

Krefft G (1866a) On the vertebrated animals of the lower Murray and Darling: their habits, economy, and geographical distribution. *Transactions of the Philosophical Society of New South Wales* 1862–1865, 1–33.

Krefft G (1866b) Notice of a new species of sperm-whale belonging to the genus *Euphysetes* of Macleay. *Proceedings of the Zoological Society of London* 33(1865), 708–713. [Apr 1866]

Krefft G (1867a) On the classification of the small Dasyuridae of Australia, with descriptions of two new genera and one new species. *Proceedings of the Zoological Society of London* 34(1866), 431–435. [Apr 1867]

Krefft G (1867b) *Halmaturus mastersii* in Appendix. Australian vertebrata, (Recent and Fossils) representing all the genera known up to the present time. pp. 91–110, in *Catalogue of the Natural and Industrial Products of New South Wales Forwarded to the Paris Universal Exhibition of 1867*, by the New South Wales Exhibition Commissioners. Government Printer, Sydney.

Krefft G (1867c) Notes on the mammals and birds of Cape York, with description of two new rodents of the genus *Hapalotis*. *Proceedings of the Zoological Society of London* 35, 316–319. [Oct 1867]

Krefft G (1868a) Notes on the fauna of Tasmania. *Papers and Proceedings of the Royal Society of Tasmania* 1868, 93–105.

Krefft G (1868b) Notes on Australian zoology. *Proceedings of the Zoological Society of London* 36, 2–4. [May 1868]

Krefft G (1868c) Description of a new species of thylacine (*Thylacinus breviceps*). *Annals & Magazine of Natural History* (4)2, 296–297. [1 Oct 1868]

Krefft G (1871a) *The Mammals of Australia*. Thomas Richards, Government Printer, Sydney.

Krefft G (1871b) *Australian Vertebrata: Fossil and Recent*. From the Industrial Progress of N.S.W. being a Report of the Intercolonial. Exhibition. Sydney.

Krefft G (1871c) Notice of a new Australian ziphoid whale: *Mesoplodon guntheri*. *Annals & Magazine of Natural History* (4)7, 368. [1 May 1871]

Krefft G (1872a) Natural History. Section Monotremata - continued. *The Sydney Mail and New South Wales Advertiser* 14(652), 808. [28 Dec 1872]

Krefft G (1872b) Natural History. The native cat family, or Dasyuridae (continued). *The Sydney Mail and New South Wales Advertiser* (new series), XIV (645), 598. [9 Nov 1872]

Krefft G (1873a) On the species of wombat (*Phascolomys*). *Proceedings of the Zoological Society of London* 40(1872), 795–796. [Apr 1873]

Krefft G (1873b) The bat tribe (continued). *The Sydney Mail and New South Wales Advertiser* Volume 16. No. 700. [29 Nov 1873]

Kretzoi M (1929) Materialen zur phylogenetischen klassifikation der Aeruroideen. *The Tenth International Congress of Zoology, Budapest* (for the year 1927). Part 2. pp. 1293–1355. [8 Sep 1927] Reference not seen.

Kretzoi M (1941) *Sirenavus hungaricus* n. g., n. sp., ein neuer Prorastomide aus dem Mitteleozän (Lutetium) von Felsőgalla in Ungarn. *Annales Historico-Naturales Musei Nationalis Hungarici, Pars Mineralogica. Geologica et Palaeontologica* 34, 146–156.

Kretzoi M (1943) *Kochictis cenntenii* n. g. n. sp., ein alterümlicher Creodonte aus dem Oberoligozän Sienbenbürgens. *Földtany Közlöny* 52, 10–17, 190–195.

Kretzoi M (1945) Bemerkungen über das Raubtiersystem. *Annales Historico-Naturales Musei Nationalis Hungarici* 38, 59–83. [Budapest]

Kretzoi M, Kretzoi M (2000) *Index generum et subgenerum mammalium*. Fossilium Catalogus Animalia, Pars 137, Section 1: xvi + 1–433; Section 2: 434–726. Backhuys Publishers, Leiden.

Kriegs JO, Churakov G, Jurka J, Brosius J, Schmitz J (2007) Evolutionary history of 7SL RNA-derived SINEs in Supraprimates. *Trends in Genetics* 23, 158–161. [online 20 Feb 2007]

Krikken J (1984) A new key to the suprageneric taxa in the beetle Family Cetoniidae, with annoted lists of the known genera. *Zoologische Verhandelingen* 210, 1–75.

Kroh A (2010) *Clypeaster* Lamarck, 1801. In A Kroh & R Mooi (2010) World Echinoidea Database. Accessed through: World Register of Marine Species at http://www.marinespecies.org/aphia.php?p=taxdetails&id=205242 on 2011–10–10.

Kuhl H (1817) *Die Deutschen Fledermäuse*. Hanau.

Kuhl H (1819) Die deutschen fledermäuse. *Annalen der Wetterauischen Gesellschaft für die Gesammte Naturkunde* 4, 11–49, 185–215.

Kuhl H (1820) *Beiträge zur Zoologie und vergleichenden Anatomie*. Hermannsche Buchhandlung, Frankfurt.

Kuhn H-J, Zeller U (1987) The cavum epiptericum in monotremes and therian mammals. *Mammalia Depicta* 13, 51–70.

Kühne WG (1958) Rhaetische Triconodonten aus Glamorgan, ihre Stellug zwischen den Klassen Reptilia und Mammalia und ihre Bedeutung für die Reichart,sche Theorie. *Paläontologische Zeitschrift* 32, 197–235.

Kühne WG (1973) The systematic position of monotremes reconsidered (Mammalia). *Zeitschrift für Morphologie der Tiere* 75, 59–64.

Kühne WG (1975) Marsupium and marsupial bone in Mesozoic mammals and in the marsupionta. *Colloques Internationaux du Centre National de la Recherche Scientifique* 218, 585–590.

Kühne WG (1977) On the Marsupionta, a reply to Dr. Parrington. *Journal of Natural History* 11, 225–228.

Kükenthal W (1892) *Sotalia teuszii* n. sp. ein pflanzenfressender (?) delphin aus Kamerun. *Zoologische Jahrbücher, Abteilung für Systematik* 6, 442–446.

Kullberg M, Hallstrom BM, Arnason U, Jamke A (2008) Phylogenetic analysis of 1.5 Mbp and platypus EST data refute the Marsupionta hypothesis and unequivocally support Monotremata as sister group to Marsupialia/Placentalia. *Zoologica Scripta* 37, 115–127.

Kuntner M, May-Collado LJ, Agnarsson I (2010) Phylogeny and conservation priorities of afrotherian mammals (Afrotheria, Mammalia). *Zoologica Scripta* 40, 1–15.

Kutt AS (2004) Clarification of the distribution of the long-eared horseshoe bat *Rhinolophus philippinensis* complex in Australia. *Australian Zoologist* 32, 629–631.

Labrinidis A, Cooper SJB, Adams M, Baczocha N (1998) Systematic affinities of island and mainland populations of the dunnart *Sminthopsis griseoventer* in Western Australia: data from allozymes and mitochondrial DNA. *Pacific Conservation Biology* 4, 289–295.

Lacépède BGÉ (1798) *Histoire Naturelle des Poissons, Par le citoyen Lacépède, Membre de l'Institut National, et Professeur au Muséum d'Histore Naturelle*. Plassan, Imprimeur-Libraire, Paris. Volume 1.

Lacépède BGÉ (1799a) *Tableau des divisions, sous-divisions, ordres et genres des mammifères*. pp. 1–20, in BGE Lacépède (ed.) *Discours d'Ouverture et de Clôture du cours d'Histoire Naturelle donné dans le Muséum National d'Histoire Naturelle : l'an VII de la République, et Tableaux Méthodiques des Mammifères et des Oiseaux*. Plassan, Paris.

Lacépède BGÉ (1799b) *Tableau des divisions, sous-divisions, ordres et genres des oiseaux*. pp. 1–20, in BGE Lacépède (ed.) *Discours d'Ouverture et de Clôture du cours d'Histoire Naturelle donné dans le Muséum National d'Histoire Naturelle : l'an VII de la République, et Tableaux Méthodiques des Mammifères et des Oiseaux*. Plassan, Paris.

Lacépède BGÉ (1800) Nouvelle table méthodique de la class des mammifères. pp. 66–86, in Anon. *Séances des Écoles*

Normales, recueilles par des sténographes, et revues par les professeurs. l'Imprimerie du Cercle-socia, Nouvelle Édition. L'Imprimerie du Cercle-Social, Paris. Volume 8.

Lacépède BGÉ (1801) Mémoire sur une nouvelle table méthodique des animaux a mammelles. *Mémoires de l'Institut des Sciences et Arts, Paris* 3, 469–502.

Lacépède BGÉ (1802a) *Histoire Naturelle des Poissons, Par le citoyen* Lacépède, *Membre de l'Institut National, et Professeur au Muséum d'Histore Naturelle*. Plassan, Imprimeur-Libraire, Paris. Volume 3.

Lacépède BGÉ (1802b) *Histoire Naturelle des Poissons, Par le citoyen* Lacépède, *Membre de l'Institut National, et Professeur au Muséum d'Histore Naturelle*. Plassan, Imprimeur-Libraire, Paris. Volume 4.

Lacépède BGÉ (1804) *Histoire Naturelle des Cétacées*. pp. i–xliv, 1–329, in GLL de Buffon (ed.) *Histoire Naturelle, Générale et Particulière*. Nouv. Édn. C.S. Sonnini, Paris. Volume 37.

Lacépède BGÉ (1818) Note sur des cétacées des mers voisins du Japon. *Mémoires du Muséum d'Histoire Naturelle, Paris* 4, 467–475.

Lack JB, Greene DU, Conroy CJ, Hamilton MJ, Braun JK, Mares MA, Van Den Bussche RA (2012) Invasion facilitates hybridization with introgression in the *Rattus rattus* species complex. *Molecular Ecology* 21, 3545–3561. [Jul 2012]

Lacordaire JT (1866) *Histoire Naturelle des Insectes. Genera des Coléoptères ou exposé méthodique et critique de tous les genres proposés jusqu'ici dans cet ordre d'insectes*. La Librarie Encyclopédique de Roret, Paris. Volume 7.

Lahille F (1899) Ostéologie du baleinoptère de Miramar. *Revista del Museo de La Plata. Sección Zoologia* 9, 79–120.

Lahille F (1905) Las ballenas de nuestro mares. *Revista del Jardín Zoológico de Buenos Aires* 2(1), 28–82.

Lahille F (1908) Nota sobre un delfin. (*Tursiops gephyreus* Lah). *Anales del Museo Nacional de Buenos Aires* 9(3), 347–365.

Lahille F (1912) Nota preliminar sobre una nueva especie de marsopa del rio dela Plata. (*Phocaena dioptrica*). *Anales del Museo Nacional de Historia Natural de Buenos Aires* 23, 269–278.

Lamarck J-B (1799) Prodrome d'une nouvelle classification des coquilles, comprenant une rédaction appropriée des caractères génériques, et l'éstablissement d'un grand nombre de genres nouveaux. *Mémoires de la Société d'Histoire Naturelle de Paris* 1, 63–91.

Lamarck J-B (1801) *Systême des Animaux sans Vertèbres*, ou, tableau général des classes, des ordres, et des genres de ces animaux : présentant leurs caractères essentiels et leur distribution, d'après la considération de leurs rapports naturels et de leur organisation, et suivant l'arrangement établi dans les galeries du Muséum d'Hist. Naturelle, parmi leurs dépouilles conservées. Précédé du discours d'ouverture du cours de zoologie, donné dans le Muséum National d'Histoire Naturelle l'an 8 de la République. Auteur et Déterville, Paris.

Lamarck J-B (1816) *Histoire naturelle des animaux sans vertèbres : présentant les caractères généraux et particuliers de ces animaux, leur distribution, leurs classes, leurs familles, leurs genres, et la citation des principales espèces qui s'y rapportent : précédée d'une introduction offrant la détermination des caractères essentiels de l'animal, sa distinction du végétal et des autres corps naturels, enfin, l'exposition des principes fondamentaux de la zoologie*. Verdière, Paris. Volume 2.

Lamarck J-B (1818) *Histoire naturelle des animaux sans vertèbres : présentant les caractères généraux et particuliers de ces animaux, leur distribution, leurs classes, leurs familles, leurs genres, et la citation des principales espèces qui s'y rapportent : précédée d'une introduction offrant la détermination des caractères essentiels de l'animal, sa distinction du végétal et des autres corps naturels, enfin, l'exposition des principes fondamentaux de la zoologie*. Verdière, Paris. Volume 5.

Lambert AB (1807) Description of a new species of *Macropus* from New Holland. *Transactions of the Linnean Society of London* (1)8, 318–319. [9 Mar 1807]

Lambert O (2008) Sperm whales from the Miocene of the North Sea: a re-appraisal. *Bulletin De L Institut Royal Des Sciences Naturelles de Belgique-Sciences de la Terre* 78, 277–316.

Lambert O, Bianucci G, Post K, Muizon C, Salas-Gismondi R, Urbina M, Reumer J (2010) The giant bite of a new raptorial sperm whale from the Miocene epoch of Peru. *Nature* 466(7302), 105–108. [1 Jul 2010]

Lamouroux JVF (1816) *Histoire des polypiers coralligènes flexibles, vulgairement nommés Zoophytes*. F Poiisson, A Caen.

Lampe E (1900) Catalog der Säugetier-Sammlung des Naturhistorischen Museums zu Wiesbaden. *Jahrbücher des Nassauischer Vereins Für Naturkunde* 53(Appendix), 1–39.

Lancaster ML, Bradshaw JA, Goldsworthy S, Sunnock P (2006) Ménage à trois on Macquarie Island: hybridisation among three species of fur seal (*Arctocephalus* spp.) following historical population extinction. *Molecular Ecology* 15, 3681–3692.

Landry SO Jr (1999) A proposal for a new classification and nomenclature for the Glires (Lagomorpha and Rodentia). *Mitteilungen aus dem Museum für Naturkunde in Berlin. Zoologische Reihe* 75, 283–316.

Lariviere S, Pasitschniak-Arts M (1996) *Vulpes vulpes*. *Mammalian Species* 537, 1–11. [27 Dec 1996]

Lartet E (1866) Notes sure deux nouveaux Siréniens fossiles des terrains tertiaires du basin de la Garonne. *Bulletin de la Société Géologique de France* 2(23), 673–686.

Latreille PA (1796) Précis des caractères génériques des Insectes disposés dans un ordre naturel, &c. de l'Imprimerie de F. Bourdeaux, Brive.

Latreille PA (1801) XII Genre Hétérodon, *Heterodon*. pp. 32–37, in CS Sonnini and PA Latreille (eds.) *Histoire Naturelle des Reptiles, avec figures dessinées d'après nature*. Chez Deterville, Paris. Volume 4.

Latreille PA (1802) *Histoire naturelle, générale et particulière des crustacés et des insectes* : ouvrage faisant suite à 'l'Histoire Naturelle générale et particulière, composée par Leclerc de Buffon, et régidée par CS Sonnini. L'Imprimerie de F Dufart, Paris. Volume 3.

Latreille PA (1817) Gelasime. pp. 517–520, in *Nouveau Dictionnaire d'Histoire Naturelle appliquée aux arts.. par une Société de naturalistes et d'agriculteurs*. Edition 2. Volume 12.

Latreille PA (1818) *Nouveau Dictionnaire d'Histoire Naturelle*, Appliquée Aux Arts, l'Agriculture, a l'Economie Rurale et Domestique, à la Medecine, etc. Par une Société de Naturalites et d'Agriculteurs. Nouv. Edn. Presqu' entierement refondue et considerablement augmentee. Déterville, Paris. Volume 23. NIL-ORC.

Latreille PA (1825) *Familles naturelles du règne animal, exposées succinctement et dans un ordre analytique, avec l'indiction de leurs genres*. J. Bailliere, Paris.

Latreille PA (1827) *Natürliche Familien des Thierreichs* / Aus dem Französischen mit Anmerkungen und Zusätzen von A.A. Berthold, &c. Weimar.

Latreille PA (1829) Suite et fin des insects, in Cuvier, G. (ed.) *Le règne animal distribué d'après son organisation, pour servir de base à l'histoire naturelle des animaux et d'introduction à l'anatomie comparée*. Chez Déterville Libraire, Paris. Second Edition. Volume 5.

Laurenti JN (1768) *Specimen medicum, exhibens synopsin reptilium emendatam cum experimentis circa venena et antidota reptilium austracorum, quod authoritate et consensu*. Joan Thomae, Vienna.

Laurie EMO, Hill JE (1954) *List of Land Mammals of New Guinea, Celebes and Adjacent Islands 1758–1952*. British Museum, London. [30 Jun 1954]

Laurie EMO, Hill JE (1957) Renaming of *Miniopterus australis minor* Laurie and Hill. *Journal of Mammalogy* 38, 128. [Feb 1957]

Lawrence B (1939) Collections from the Philippine Islands. Mammals. *Bulletin of the Museum of Comparative Zoology. Harvard University* 86, 28–73. [Nov 1939]

Laws RM (1994) History and present status of southern elephant seal populations. pp. 49–65, in BJ Le Boeuf & RM Laws (eds.) *Elephant Seals: Population Biology, Behavior and Physiology*. University of California Press, Berkeley, California.

Lay GT (1825) in J Wilkes (ed.) Zoology. pp. 728–784. *Encyclopaedia Londiniensis*, or universal dictionary of arts, sciences and literature. London. Volume 22. THALES-ZYPAEUS.

Lay GT (1829) Observations on a species of *Pteropus* from Bonin. *The Zoological Journal, London* 4, 457–459.

Le Conte JL (1856) Synopsis of the Melolonthidae of the United States. *Journal of the Academy of Natural Sciences of Philadelphia* 3(2), 225–288. [Nov 1856]

Le Peletier de Saint Fargeau A, Audinet-Serville JG (1825) Salde. pp. 321–326, in P-A Latreille *et al.* (eds.) *Encyclopédie Méthodique. Histoire Naturelle. Entomologie, ou Histoire Naturelle des Crustacés, des Arachnides et des Insects*. Volume 10. PAR-Z. Agasse, Paris. [1 Oct 1825]

Le Souef AS (1916) Notes on colour variation of opossums of the genus *Trichosurus*. *Australian Zoologist* 1, 62–64. [13 Mar 1916]

Le Souef AS (1923) The grey kangaroos (*Macropus giganteus*) and its allies. *Australian Zoologist* 3, 145–147.

Le Souef AS (1924) notes on some rock wallabies, genus *Petrogale*, with descriptions of two new species. *Australian Zoologist* 3, 272–276.

Le Souef AS (1929) Notes on some mammals from Bass Strait Islands, including a new subspecies of *Pseudochirus*. *Australian Zoologist* 5, 329–332. [24 Mar 1929]

Le Souef AS, Burrell H (1926) *The Wild Animals of Australasia*. Harrap and Co., London.

Leach WE (1814) Crustacealogy. pp. 383–437, in D Brewster (ed.) *The Edinburgh Encyclopaedia*. William Blackwood, Edinburgh. Volume 7(2). COL-DIV.

Leach WE (1815) *The Zoological Miscellany*; being descriptions of new, or interesting animals, illustrated with coloured figures, drawn from nature, by RP Nodder. B. McMillan, London. Volume 2.

Leach WE (1816) *Systematic Catalogue of the Specimens of the Indigenous Mammalia and Birds that are Preserved in the British Museum with their localities and authorities*. British Museum, London.

Leach WE (1819) Notice of some animals from the Arctic regions. *Annals of Philosophy; or Magazine of Chemistry, Mineralogy, Mechanics, Natural History, Agriculture, and the Arts* (1)13, 60–61.

Leach WE (1821a) The characters of seven genera of bats with foliaceous appendages to the nose. *Transactions of the Linnean Society of London* (1)13, 73–82. [23 May–21 Jun 1821]

Leach WE (1821b) The characters of three new genera of bats without foliaceous appendages to the nose. *Transactions of the Linnean Society of London* (1)13, 69–72. [23 May–21 Jun 1821]

Leach WE (1825) Description of the *Vespertilio pygmaeus*, a new species, recently discovered in Devonshire by Dr. Leach. *Zoological Journal* 1(4), 559–561.

Leche W (1884) On some species of Chiroptera from Australia. *Proceedings of the Zoological Society of London* 52, 49–54. [Jun 1884]

Leche W (1886) Über die Säugethiergattung *Galeopithecus*: Eine morphologische Untersuchung. *Kongliga Svenska Vetenskaps-Akademiens Handlingar, Stockholm* 21(11), 1–92.

Lecompte E, Aplin K, Denys C, Catzeflis F, Chades M, Chevret P (2008) Phylogeny and biogeography of African Murinae based on mitochondrial and nuclear gene sequences, with a new tribal classification of the subfamily. *BMC Evolutionary Biology* 8(199), 1–21. [10 Jul 2008]

LeDuc R (2009) Delphinids, overview. pp. 298–302, in WF Perrin, B Würsig & JGM Thewissen (eds.) *Encyclopedia of Marine Mammals*. Academic Press, Amsterdam. Second Edition.

LeDuc RG, Curry BE (1997) Mitochondrial DNA sequence analysis indicates need for revision of the genus *Tursiops*. *International Whaling Commission. Reports* 47, 393.

LeDuc RG, Perrin WF, Dizon AE (1999) Phylogenetic relationships among the delphinid cetaceans based on full cytochrome B sequences. *Marine Mammal Science* 15, 619–648. [Jul 1999]

LeDuc RG, Dizon AE, Goto M, Pastene LA, Kato H, Nishiwaki H, LeDuc CEA, Brownell PL (2007) Patterns of genetic variation in Southern Hemisphere blue whales and the use of assignment test to detect mixing in the feeding grounds. *The Journal of Cetacean Research and Management* 9, 73–80.

LeDuc RG, Robertson KM, Pitman RL (2008) Mitochondrial sequence divergence among Antarctic killer whale ecotypes is consistent with multiple species. *Biology Letters* 4, 426–429. [4 Jun 2008]

Lee AK (1995) *The Action Plan for Australian Rodents*. Australian Nature Conservation Agency, Endangered Species Program. Project 130. Canberra. [Apr 1995]

Lee AK, Martin R (1988) *The Koala - A Natural History*. New South Wales University Press, Sydney.

Lee AK, Baverstock PR, Watts CHS (1981) Rodents – The late invaders. pp. 1523–1553, in A Keast (ed.) *Ecological Biogeography of Australia*. Junk, The Hague.

Legendre S (1984) Étude odontologique des représentants actuels du groupe *Tadarida* (Chiroptera, Molossidae). Implications phylogéniques, systématiques et zoogéographiques. *Revue Suisse de Zoologie* 91, 399–442.

Lehrer AZ (1976) Leclercqiomyia: Genre nouveau de sarcophagines palearcttiques (Diptera: Sarcophagidae). *Bulletin et Annales de la Société Royale Belge d'Entomologie* 112, 195–203.

Leidy J (1869) Extinct Mammalia of Dakota and Nebraska, including an account of some allied forms from other localities. *Journal of the Academy of Natural Sciences of Philadelphia* 7(2), 23–472.

Lekagul B, McNeely JA (1988) *Mammals of Thailand*. Association for the Conservation of Wildlife, Bangkok. Second Edition.

Lenz HO (1831) *Naturgeschichte der Säugethiere, nach Cuvier's Systeme bearbeitet, &c*. Bedersche Buchhamdlung, Gotha.

Leonard JA, Wayne RK, Wheeler J, Valadez R, Guillén S, Vilà C (2002) Ancient DNA evidence for old world origin of new world dogs. *Science* 298(5598), 1613–1616. [22 Nov 2002]

Lesèble L (1890) Les dingos au chenil du Jardin Zoologique d'Acclimatisation. *Revue des Sciences Naturelles Appliquées: Bulletin Bimensuel de la Société National d'Acclimatisation de France* 37, 681–684.

Leske NG (1779) *Anfangsgründe der Naturgeschichte*. Erster Teil. Siegfried Lebrecht Crusius, Leipzig.

Lesson RP (1826) *Sur le Phoque Léopard de Mer (Sea Leopard) des Orcades Australes*; par James Weddell. (Voy. towards the South Pole, etc., in-8°. 1825. *Bulletin des Sciences Naturelles et de Géologie, Paris* 7, 437–438.

Lesson RP (1827a) *Manuel de Mammalogie, ou histoire naturelle des Mammifères*. Roret, Paris.

Lesson RP (1827b) Description des mammifères. pp. 137–176, in RP Lesson & P Garnot (eds.) (1826–1829) Zoologie. Volume 1. Part 1, in LI Duperry (ed.) (1826–1834) *Voyage autour du monde, Exécuté par Ordre du Roi, sur la Corvette de La Majesté, La Coquille, pendant les années 1822, 1823, 1824 et 1825, Sous le Ministère et conformément aux instructions de S. E. M. le Marquis Clermont-Tonnerre, Minister de la Marine; et publié sous les auspices de son Excellence Mgr le Cte de Chabrol, Ministre de la Marine et des Colonies*. Arthus Bertrand, Libraire-Editeur, Imprimerie de Firmin Didot, Paris. Volume 1. [p. 137–168, 25 Jul 1827, 169–216, 17 Oct 1827]

Lesson RP (1827c) Observations générales sur quelques cétacées. pp. 177–186, in RP Lesson & P Garnot (eds.) Zoologie. Volume 1. Part 1, in LI Duperry (ed.)(1826–1834) *Voyage autour du monde, Exécuté par Ordre du Roi, sur la Corvette de La Majesté, La Coquille, pendant les années 1822, 1823, 1824 et 1825, Sous le Ministère et conformément aux instructions de S. E. M. le Marquis Clermont-Tonnerre, Minister de la Marine; et publié sous les auspices de son Excellence Mgr le Cte de Chabrol, Ministre de la Marine et des Colonies*. Arthus Bertrand, Libraire-Editeur, Imprimerie de Firmin Didot, Paris. [17 Oct 1827]

Lesson RP (1828a) Péramèle. pp. 197–200, in JV Audouin & JBGM Bory de Saint-Vincent (eds.) *Dictionnaire Classique d'Histoire Naturelle*. Rey et Gravier, Paris. Volume 13. PAN-PIV.

Lesson RP (1828b) Phalanger. pp. 326–335, in JV Audouin & JBGM Bory de Saint-Vincent (eds.) *Dictionnaire Classique d'Histoire Naturelle*. Rey et Gravier, Paris. Volume 13. PAN-PIV.

Lesson RP (1828c) Phoque. pp. 400–426, in JV Audouin & JBGM Bory de Saint-Vincent (eds.) *Dictionnaire Classique d'Histoire Naturelle*. Rey et Gravier, Paris. Volume 13. PAN-PIV.

Lesson RP (1828d) *Histoire Naturelle*, générale et particulière des mammifères et des oiseaux découverts depuis 1788 jusqu'à nos jours. Complément des oeuvres de Buffon, ou histoire naturelle des animaux rares découverts par les naturalistes et les voyageurs depuis la mort de Buffon. Tom. 1. Cétacés. Baudoin Frères, Paris.

Lesson RP (1828e) *Manuel d'ornithologie, ou description des genres et des principales espèces d'oiseaux, &c*. Roret, Paris. Volume 2.

Lesson RP (1828f) Spilecia zoological, etc. Figures originales et descriptions systematiques d'animaux nouveaux et non décrits. *Bulletin des Sciences Naturelles et de Géologie, Paris* 16, 115–117.

Lesson RP (1830 [1830–1831]) *Traité d'Ornithologie, ou tableau méthodique des ordres, sous-ordres, familles, tribus, genres, sous-genres et races d'oiseaux*. F.G. Levrault, Paris. Volume 1.

Lesson RP (1831 [1830–1831]) *Traité d'Ornithologie, ou tableau méthodique des ordres, sous-ordres, familles, tribus, genres, sous-genres et races d'oiseaux*. F.G. Levrault, Paris. Volume 2.

Lesson RP (1835) *Illustrations de Zoologie ou Recueil de Figures D'Animaux*. Arthus Bertrand, Paris.

Lesson RP (1836 [1834–1836]) *Histoire Naturelle*, générale et particulière des mammifères et des oiseaux découverts depuis la mort de Buffon. Volume 5. Mammifères. Baudoin Frères, Paris.

Lesson RP (1842) *Nouveau Tableau du Règne Animal. Mammifères*. Arthus Bertrand, Paris.

Lesueur CA, Petit SL (1807) Atlas in F Péron & LCD de Freycinet (1807–1816) *Voyage de Découvertes aux Terres Australes, Exécuté par Ordre de Sa Majesté l'Empereur et roi, sur les Corvettes le Géographe, le Naturaliste, et la Goelette le Casuarina, Pendant les Années 1800, 1801, 1802, 1803 et 1804*. L'Imprimerie Impériale, Paris.

Lewin J (1808) *Birds of New Holland with their Natural History*. White & Bagster, London.

Lichtenstein AAH (1796) *Catalogus Musei zoologici ditissimi Hamburgi*, dIII Februar 1796. Auctionis lege distrahendi. Sectio Teria. Continens Insecta. Hamburg.

Lichtenstein MHC (1814) Die Gattung Antilope. *Magazin der Gesellschaft Naturforschender Freunde zu Berlin* 6, 147–160, 163–182.

Lichtenstein MHC (1818) *Das Zoologische Museum der Universitat*. Berlin. Second Edition.

Lichtenstein MHC (1829 [1827–1834]) *Darstellung neuer oder wenig bekannter Säugethiere in Abbildungen und Beschreibungen von fünf und sechzig Arten auf* funfzig colorirten Steindrucktafeln nach den Originalen des Zoologischen Museums der Universität zu Berlin. Luderitz, Berlin. Hft 6.

Lichtenstein MHC (1831 [1827–1834]) *Darstellung neuer oder wenig bekannter Säugethiere in Abbildungen und Beschreibungen von fünf und sechzig Arten auf* funfzig colorirten Steindrucktafeln nach den Originalen des Zoologischen Museums der Universität zu Berlin. Luderitz, Berlin. Hft 7.

Lidicker WZ Jr (1983) Dasyurids and their kin. *Science* 219(4590), 1316–1317. [18 Mar 1983]

Lidicker WZ Jr, Brylski PV (1987) The conilurine rodent radiation of Australia, analyzed on the basis of phallic morphology. *Journal of Mammalogy* 8, 617–641. [Aug 1987]

Lidicker WZ Jr, Follett WI (1968) *Isoodon* Desmarest, 1817, rather than *Thylacis* Illiger, 1811, as the valid generic name of the short nosed bandicoots (Marsupialia: Peramelidae). *Proceedings of the Biological Society of Washington* 81, 251–256. [30 Aug 1968]

Lidicker WZ Jr, Marlow BJ (1970) A review of the dasyurids marsupial genus *Antechinomys* Krefft. *Mammalia* 34, 212–227. [Jun 1970]

Lidicker WZ Jr, Ziegler AC (1968) Report on a collection of mammals from eastern New Guinea, including species keys for fourteen genera. *University of California Publications in Zoology* 87, 1–60. [18 Dec 1968]

Liem KF (1963) The comparative osteology and phylogeny of the Anabantoidei (Teleostei, Pisces). *Illinois Biological Monographs* 30, 1–149.

Lillie DG (1915) Cetacea. pp. 85–124, in *British Antarctic (Terra Nova) Expedition, 1910*. [under Capt. RF Scott] Natural History Report. London. Volume 1.

Lilljeborg W (1861a) Hvalben funna i jorden på Gräsön I Roslagen i Sverige. *Forhandlinger ved de Skandinaviske Naturforskeres. Ottende Møde, Kopenhagen* 8(1660), 599–616.

Lilljeborg W (1861b) Ofversigt af de inom Skandinavien (Sverige und Norrige) anträffade Hvalartade Däggdjur (Cetacea). *Upsala Universitets, Årsskrift* 1861, 1–38.

Lilljeborg W (1866a) *Systematisk Öfversigt af de Gnagande Däggdjuren, Glires*. Kongl. Akad Boktryckeriet, Upsala.

Lilljeborg W (1866b) Synopsis of the cetaceans Mammalia of Scandinavia (Sweden and Norway). pp. 219–309, in WH Flower (ed.) *Recent Memoirs on the Cetacea*. Royal Society of London, London.

Lilljeborg W (1867) On two subfossil whales discovered in Sweden. *Nova Acta Regiae Societatis Scientiarum Upsaliensis* 6(3), 1–48.

Lilljeborg W (1874) Sveriges och Norges Ryggradsdjur. 1. Däggdjuren, jemte inledning till ryggradsdjuren. Förra afdlelringen. W. Schultz, Upsala.

Lin YH, Waddell, PJ, Penny D (2002) Pika and vole mitochondrial genomes increase support for both rodent monophyly and glires. *Gene* 294, 119–129.

Lindenmayer DB, Dubach J, Viggers KL (2002) Geographical dimorphism in the mountain brushtail possum (*Trichosurus caninus*): the case for a new species. *Australian Journal of Zoology* 50, 369–393.

Ling JK (1988a) Otariidae. pp. 223–225, in DW Walton (ed.) *Zoological Catalogue of Australia. 5. Mammalia*. Australian Government Publishing Service, Canberra. [13 Apr 1988]

Ling JK (1988b) Phocidae. pp. 226–229, in DW Walton (ed.) *Zoological Catalogue of Australia. 5. Mammalia*. Australian Government Publishing Service, Canberra. [13 Apr 1988]

Ling JK (1992) *Neophoca cinereus*. *Mammalian Species* 392, 1–7. [5 Jun 1992]

Ling JK, Bryden MM (1992) *Mirounga leonina*. *Mammalian Species* 391, 1–8. [5 Jun 1992]

Link HF (1794) *Beyträge zur Naturgeschichte*. Karl Stiller, Rostock and Leipzig. Vol. 1.

Link HF (1795) *Beyträge zur Naturgeschichte*. Karl Stiller, Rostock and Leipzig. Volume 2.

Link HF (1806) Ueber die in Meklenburg einheimischen gistigen Schlangen. *Magazin für den neuesten Zustand der Naturkunde* (ed. J. H. Voight) 12, 289–300.

Linnaeus C (1758) *Systema Naturae per regna tria naturae*, secundum classes, ordines, genera, species, cum characteribus, differentiis, synonymis, locis. Editio decima, reformata. Laurentii Salvii, Holmiae. Volume 1. [1 Jan 1758]

Linnaeus C (1761) *Fauna Svecica sistens Animalia Sveciae regni*: mammalia, aves, amphibia, pisces, insecta, vermes. Distributa per classes & ordines, genera & species, cum differentiis specierum, synonymis auctorum, nominibus incolarum, locis natalium, descriptionibus insectorum. Editio Altera, Auctior. Laurentii Salvii, Stockholm. Second Edition.

Linnaeus C (1766) *Systema naturae per regna tria naturae. Editio duodecima, reformata*. Volume 1. Regnum Animale. Stockholm: Laurentii Salvii.

Lint DW, Clayton JW, Lillie WR, Postma L (1990) Evolution and systematics of the Beluga whale, *Delphinapterus leucas*, and other odontocetes: A molecular approach. *Canadian Bulletin of Fisheries and Aquatic Sciences* 224, 7–22.

Liu F-GR, Miyamoto MM, Freire NP, Ong PQ, Tennant MR, Young TS, Gugel KF (2001) Molecular and morphological supertrees for eutherian (placental) mammals. *Science* 291(5509), 1786–1789. [2 Mar 2001]

Lloyd RE (1909) A description of the deep-seafish caught by the R.I.M.S. Ship 'Investigator' since the year 1900, with supposed evidence of mutation in *Malthopsis*. *Memoirs of the Indian Museum* 2, 139–180.

Loche GG (1860) Description de deux nouvelles espèces du genre Dauphin. *Revue et Magasin de Zoologie Pure et Appliquée, Paris* 12(2), 473–479.

Loew H (1845) In GC Berendt (ed.) Die im Bernstein befindlichen Organischen Reste der Vorwelt gesammelt in Verbindung mit Mehreren bearbeitit und herausgegeben. Berlin. Volume 1. Part 1.

Long JL (2003) *Introduced Mammals of the World: Their History, Distribution and Influence*. CSIRO Publishing, Melbourne.

Long J, Archer M, Flannery T, Hand S (2002) *Prehistoric Mammals of Australia and New Guinea: One Hundred Million Years of Evolution*. University of New South Wales Press, Sydney.

Longman HA (1915) A new phalanger from North Queensland. *Memoirs of the Queensland Museum* 3, 22–23. [28 Jan 1915]

Longman HA (1916) *Notes on Classification of Rodents with a List of Australian Species*. Albert J. Mullet, Government Printer, Melbourne.

Longman HA (1921) A new *Nyctimene* from south Queensland. *Memoirs of the Queensland Museum* 7, 179–181. [4 Nov 1921]

Longman HA (1922) South Queensland marsupials. *Memoirs of the Queensland Museum* 7, 297–300. [19 Dec 1922]

Longman HA (1924a) Untitled abstract. *Proceedings of the Royal Society of Queensland* 36, ix.

Longman HA (1924b) Some Queensland fossil vertebrates. *Memoirs of the Queensland Museum* 8, 16–28. [30 Jan 1924]

Longman HA (1926) New records of Cetacea, with a list of Queensland species. *Memoirs of the Queensland Museum* 8, 266–278. [31 Mar 1926]

Longman HA (1939) A central Queensland wombat. *Memoirs of the Queensland Museum* 11, 283–287. [17 Mar 1939]

Lönnberg E (1906a) On the geographical races of red deer in Scandenavia. *Arkiv för Zoologi* 3(9), 1–19.

Lönnberg E (1906b) Contributions to the fauna of South Georgia. I. Taxonomic and biological notes on vertebrates. *Kungliga Svenska Vetenskaps-Akademiens Handlingar* 40(5), 1–104.

Lönnberg E (1910) On the variation of the sea-elephants. *Proceedings of the Zoological Society of London* 80, 580–588. [23 Aug 1910]

Lönnberg E (1913) Results of Dr. E. Mjöberg's Swedish Scientific Expeditions to Australia 1910–1913. 1. Mammals. *Kungliga Svenska Vetenskaps-Akademiens Handlingar* 52(1), 1–10.

Lönnberg E (1916) Results of Dr. E. Mjöberg's Swedish Scientific Expeditions to Australia 1910–1913. II. Mammals from Queensland. 1. List of mammals. *Kungliga Svenska Vetenskaps-Akademiens Handlingar* 52(2), 3–11. [18 Jan 1916]

Lönnberg E (1931) The skeleton of *Balaenoptera brydei* O. Olsen. *Arkiv för Zoologi* 23(1), 1–23.

Lönnberg E (1934) *Prodelphis graffmani* n. sp. a new dolphin from the Pacific coast of Mexico. *Archiv för Zoologi* 26A, 1–11.

Lord CE (1919) Notes on the mammals of Tasmania. *Papers and Proceedings of the Royal Society of Tasmania* 1918, 16–52. [1 Apr 1919]

Lord CE (1923) On the Tasmanian Muridae. *Papers and Proceedings of the Royal Society of Tasmania* 1922, 74–76. [23 Feb 1923]

Lord CE, Scott HH (1924) *A Synopsis of Vertebrate Animals of Tasmania*. Oldham, Beddome & Meredith, Hobart.

Lowry JK, Stoddart HE (2003) *Zoological Catalogue of Australia. Volume 19, Part 2. Crustacea: Malacostraca: Peracarida: Amphipoda, Cumacea, Mysidacea*. Commonwealth of Australia, Canberra.

Lucas AHS, Le Souëf WH (1909) *The Animals of Australia: Mammals Reptiles and Amphibians*. Whitcombe and Tombs, Melbourne.

Lucas FA (1899) The main divisions of the Pinnipedia. pp. 1–2, in DS Jordan (ed.). *The Fur Seals and Fur-seal Islands of the North Pacific Ocean*. Government Printing Office, Washington, D.C. Part 3.

Lucas HP (1846 [1846–1849]) Histoire naturelle des animaux articulés. Deuxième Partie. Insectes. Coleoptères. pp. i–xxxv, 1–590, in *Exploration scientifique de l'Algérie pendant les années 1840, 1841, 1842 publiée par ordre du Gouvernement et avec le concours d'une Commission Académique*. Imprimerie Nationale, Paris. Volume 2. Zoologie. Sciences Physique, Zoologie. Imprimerie Nationale, Paris. Reference not seen.

Lucas SG (1992) Extinction and the definition of the Class Mammalia. *Systematic Biology* 41, 370–371. [Sep 1992]

Luckett WP, Zeller U (1989) Development evidence for dental homologies in the monotreme *Ornithorhynchus anatinus* and its systematic implications. *Zeitschrift fur Saugetierkunde* 54, 193–204.

Lull RS (1904) Fossil footprints of the Jura-Trias of North America. *Memoirs of the Boston Society of Natural History* 5(11), 461–557.

Lumholtz C (1884) Notes upon some mammals recently discovered in Queensland. *Proceedings of the Zoological Society of London* 52, 406–409. [Oct 1884]

Lund PW (1831) Lettre sur les habitudes de quelques fourmis de Brésil. *Annales des Sciences Naturelles, Paris* 23(1), 113–138.

Lund PW (1838) Blik paa Brasiliens Dyreverden för den sidste Jordomvaeltning. *Oversigt over det Kongelige Danske Videnskabernes Selskabs Forhandlinger. Kjøbenhavn* 1838, 7–15.

Lund PW (1839) Coup-d'oeil sur les espèces éteintes de mammifères du Brésil; extrait de quelques mémoires présentés à l'Académie Royale des Sciences de Copenhague. *Annales des Sciences Naturelles. Zoologie* 11(2), 214–234.

Lund PW (1843) Utog af en Afhandling isaer over Brasiliens Rovdyr I nuvaerende og forrige Jordperiode. *Oversigt over det Kongelige Danske Videnskabernes Selskabs Forhandlinger. Kjøbenhavn* 1843, 77–82.

Lundelius EL, Turnbull WD (1984) The mammalian fauna of Madura Cave, Western Australia. Part IV. Macropodidae: Potoroidae. *Fieldiana Geology*, new series 14, 1–63. [30 Apr 1984]

Luo Z (2000) In search of the whales' sister. *Nature* 404(6775), 235–239. [16 Mar 2000]

Luo Z-X, Cifelli RL, Kielan-Jaworoska Z (2001a) Dual origin of tribosphenic mammals. *Nature* 409(6816), 53–57. [4 Jan 2001]

Luo Z-X, Yuan CX, Meng Q-J, Jo Q (2001b) A Jurassic eutherian mammal and divergence of marsupials and placentals. *Nature* 476(7361), 442–445. [25 Aug 2011]

Luo Z-X, Kielan-Jaworoska Z, Cifelli RL (2002) In quest for a phylogeny of Mezozoic mammals. *Acta Palaeontologica Polonica* 47, 1–78.

Luo Z-X, Ji Q, Wible JR, Yuan C-X (2003) An Early Cretaceous tribosphenic mammal and metatherian evolution. *Science* 302(5652), 1934–1940. [12 Dec 2003]

Lütken CF (1889) Spolia Atlantica. Bidrag til Kundskab om de tre pelagiske Tandhval-Slaegter *Steno, Delphinus* og *Prodelphinus*. *Kongelige Danske Videnskabs-selskabet skrifter 1 Christiania, Mathematisk-naturvidenskabelig* 6(5)(1), 1–61.

Lydekker R (1883) Indian Tertiary and Post-Tertiary Vertebrata. Part 5: Siwalik selenodont suina. *Memoirs of the Geological Survey of India, Palaeontologica Indica* (10)2(5), 143–177. [Feb 1883]

Lydekker R (1887) *Catalogue of the Fossil Mammalia in the British Museum (Natural History), Cromwell Road, S. W. Part V. Containing The Group Tillodontia, the Orders Sirenia, Cetacea, Edentata, Marsupialia, Monotremata, and Supplement*. British Museum (Natural History), London.

Lydekker R (1891) In WH Flower & R Lydekker. *An Introduction to the Study of Mammals Living and Extinct*. Adam & Charles Black, London.

Lydekker R (1894a) *A Handbook to the Marsupialia and Monotremata*. Allan & Co., London.

Lydekker R (1894b) Contributions to a knowledge of the fossil vertebrates of Argentina. Part 2. Cetacean skulls from Patagonia. *Anales del Museo de La Plata. Paleontologia* 2, 1–13. [uncertainty over publication date – ranging from 1891–1894]

Lydekker R (1895) *Royal Natural History*. Frederick Warne & Co., London. Volume 3.

Lydekker R (1906) Mammalia. pp. 1–47, in D Sharp (ed.) *The Zoological Record* being records of zoological literature relating chiefly to the year 1905. Zoological Society of London, London. Volume 42.

Lydekker R (1909) On the skull-characters of the southern sea-elephant. *Proceedings of the Zoological Society of London* 79, 600–606. [9 Oct 1909]

Lydekker R (1915) *Catalogue of the Ungulate Mammals in the British Museum* (Natural History). Volume 4. Artiodactyla, Families Cervidae (deer), Tragulidae (chevtotains), Camelidae (camels and llamas), Suidae (pigs and peccaries), and Hippopotamidae (hippopotamuses). British Museum, London.

Lydekker R (1916) *Catalogue of the Ungulate Mammals in the British Museum* (Natural History). Volume 5. Perissodactyla (Horses, Tapirs, Rhinoceroses), Hyracoidea (Hyraxes) Proboscidea (Elephants). With addenda to the earlier volumes. British Museum, London.

Lynam AJ (1987) Inbreeding and juvenile dispersal in insular populations of two dasyurid marsupials; the dibbler, *Parantechinus apicalis*, and the grey-bellied dunnart, *Sminthopsis griseoventer*. BSc Hons Thesis. University of Western Australia, Perth. Reference not seen.

Lyne AG (1967) *Marsupials and Monotremes*. Angus and Robertson, Sydney.

Lyne AG, Mort PA (1981) A comparison of skull morphology in the marsupial bandicoot genus *Isoodon*: its taxonomic implications and notes on a new species, *Isoodon arnhemensis*. *Australian Mammalogy* 4, 107–133. [13 May 1981]

Lyon MW (1914) *Tadarida* Rafinesque versus *Nyctinomus* Geoffroy. *Proceedings of the Biological Society of Washington* 27, 217–218. [31 Oct 1914]

Lyon NW, Osgood WH (1909) Catalogue of the type-specimens of mammals in the United States National Museum, including the Biological Survey collection. *Bulletin of the United States National Museum* 62, 1–325. [after 15 Jan 1909]

Lysaght A (1957) Captain Cook's Kangaroo. *New Scientist* 3(14), 17–19. [14 Mar 1957]

Macgillivray W (1827) Description of a new species of *Ornithorhynchus*. *Memoirs of the Wernerian Natural History Society* 6, 127–132.

Macintosh NWG (1975) The origin of the dingo: an enigma. pp. 87–106, in MW Fox (ed.) *The Wild Canids: Their Systematics, Behavioural Ecology and Evolution*. Van Nostrand Reinhold, New York.

Mack G (1961) Mammals from south-western Queensland. *Memoirs of the Queensland Museum* 13, 213–229. [28 Mar 1961]

Mackerras IM, Mackerras MJ (1960) Taxonomy of the common short-nosed marsupial bandicoot of eastern Queensland. *Australian Journal of Science* 23, 51–53.

Mackness BS, Wroe S, Muirhead J, Wilkinson C, Wilkinson D (2000) First fossil bandicoot from the Pliocene Chinchilla local fauna. *Australian Mammalogy* 22, 133–136. [Dec 2000]

Macleay WJ (1872) Notes on a collection of insects from Gayndah. Second paper. *Transactions of the Entomological Society of New South Wales* 2(4), 239–318.

Macleay WS (1841) Notice of a new genus of Mammalia discovered by J. Stuart, Esq., in New South Wales. *Annals & Magazine of Natural History* (1)8, 241–243. [1 Dec 1841]

Macqueen P, Seddon JM, Austin JJ, Hamilton S, Goldizen AW (2010) Phylogenetics of the pademelons (Macropodidae: *Thylogale*) and historical biogeography of the Australo-Papuan region. *Molecular Phylogenetics and Evolution* 57, 1134–1148.

Madge RB (1988) The publication dates of Dejean's catalogues. *Archives of Natural History* 15, 317–321.

Madsen O, Scally M, Douady CJ, Kao DJ, DeBry RW, Adkins R, Amrine HM, Stanhope MJ, de Jong WW, Springer MS (2001) Parallel adaptive radiations in two major clades of placental mammals. *Nature* 409(6820), 610–614. [1 Feb 2001]

Maeda K (1982) Studies on the classification of the *Miniopterus* in Eurasia, Australia, and Melanesia. *Honyurui Kagaku* 1(Supplement), 1–176. [Mammalian Science]

Mahoney JA (1964) The taxonomic status of *Dasyurus affinis* McCoy. [1865] (Dasyuridae) and *Hypsiprymnus trisulcatus* McCoy. [1865] (Macropodidae), two marsupials from a Holocene cave deposit near Gisborne, Victoria. *Proceedings of the Royal Society of Victoria* 77, 525–533.

Mahoney JA (1968) *Baiyankamys* Hinton, 1943 (Muridae, Hydromyinae) a New Guinea rodent genus named for an incorrectly associated skin and skull (Hydromyinae, *Hydromys*) and mandible (Murinae, *Rattus*). *Mammalia* 32, 64–71. [Mar 1968]

Mahoney JA (1972) The identity of *Hapalotis personata* Krefft, 1867 (Muridae, *Rattus*) from cape York, Queensland. *Australian Mammalogy* 1, 14–19. [Dec 1972]

Mahoney JA (1974) The Australian rodent specimens (Muridae) of J.E. Grays List of the specimens of Mammalia in the collection of the British Museum (1843). *Australian Mammalogy* 1, 231–242. [Oct 1974]

Mahoney JA (1975) The identity and status of Thomas' 'lectotype' of *Leporillus apicalis* (Gould, 1853).

Transactions of the Royal Society of South Australia 99, 101–104. [Rodentia: Muridae] [30 Aug 1975]

Mahoney JA (1977) Skull characters and relationships of *Notomys mordax* Thomas (Rodentia: Muridae), a poorly known Queensland hopping-mouse. *Australian Journal of Zoology* 25, 749–754.

Mahoney JA (1981) The specific name of the honey possum (Marsupialia; Tarsipedidae: *Tarsipes rostratus* Gervais & Verreaux, 1842). *Australian Mammalogy* 4, 135–138. [13 May 1981]

Mahoney JA (1982) Identities of the rodents (Muridae) listed in TL Mitchell's 'Three expeditions into the interior of eastern Australia, with descriptions of the recently explored region of Australia Felix, and of the present colon of New South Wales' (1st edn., 1838; 2nd edn., 1839). *Australian Mammalogy* 5, 15–36. [10 May 1982]

Mahoney JA (1988a) Ornithorhynchidae. pp. 7–10, in DW Walton (ed.) *Zoological Catalogue of Australia. 5. Mammalia*. Australian Government Publishing Service, Canberra. [13 Apr 1988]

Mahoney JA (1988b) Tachyglossidae. pp. 4–6, in DW Walton (ed.) *Zoological Catalogue of Australia. 5. Mammalia*. Australian Government Publishing Service, Canberra. [13 Apr 1988]

Mahoney JA, Richardson BJ (1988a) Muridae. pp. 154–192, in DW Walton (ed.) *Zoological Catalogue of Australia. 5. Mammalia*. Australian Government Publishing Service, Canberra. [13 Apr 1988]

Mahoney JA, Richardson BJ (1988b) Canidae. pp. 217–220, in DW Walton (ed.) *Zoological Catalogue of Australia. 5. Mammalia*. Australian Government Publishing Service, Canberra. [13 Apr 1988]

Mahoney JA, Richardson BJ (1988c) Cervidae. pp. 238–243, in DW Walton (ed.) *Zoological Catalogue of Australia. 5. Mammalia*. Australian Government Publishing Service, Canberra. [13 Apr 1988]

Mahoney JA, Ride WDL (1975) Index of the genera and species of fossil Mammalia described from Australia and New Guinea between 1838 and 1968 (including citations of type species and primary type species). *Special Publications of the Western Australian Museum* 6, 1–250.

Mahoney JA, Ride WDL (1984) The identity of Captain Cook's quoll *Mustella quoll* Zimmermann, 1783 (Marsupialia: Dasyuridae). *Australian Mammalogy* 7, 57–62. [May 1984]

Mahoney JA, Ride WDL (1986) *Dasyurus hallucatus* Gould, 1842 (Mammalia, Marsupialia): proposed conservation by the suppression of *Mustella quoll* Zimmermann, 1783. Z.N.(S.)2472. *Bulletin of Zoological Nomenclature* 43, 50–54. [9 Apr 1986]

Mahoney JA, Ride WDL (1988a) Dasyuridae. pp. 14–33, in DW Walton (ed.) *Zoological Catalogue of Australia. 5. Mammalia*. Australian Government Publishing Service, Canberra. [13 Apr 1988]

Mahoney JA, Ride WDL (1988b) Myrmecobiidae. pp. 34–35, in DW Walton (ed.) *Zoological Catalogue of Australia. 5. Mammalia*. Australian Government Publishing Service, Canberra. [13 Apr 1988]

Mahoney JA, Ride WDL (1988c) Thylacinidae. pp. 11–13, in DW Walton (ed.) *Zoological Catalogue of Australia. 5. Mammalia*. Australian Government Publishing Service, Canberra. [13 Apr 1988]

Mahoney JA, Ride WDL (1988d) Peramelidae. pp. 36–42, in DW Walton (ed.) *Zoological Catalogue of Australia. 5. Mammalia*. Australian Government Publishing Service, Canberra. [13 Apr 1988]

Mahoney JA, Ride WDL (1988e) Thylacomyidae. pp. 43–45, in DW Walton (ed.) *Zoological Catalogue of Australia. 5. Mammalia*. Australian Government Publishing Service, Canberra. [13 Apr 1988]

Mahoney JA, Walton DW (1988a) Pteropodidae. pp. 105–113, in DW Walton (ed.) *Zoological Catalogue of Australia. 5. Mammalia*. Australian Government Publishing Service, Canberra. [13 Apr 1988]

Mahoney JA, Walton DW (1988b) Megadermatidae. pp. 119–120, in DW Walton (ed.) *Zoological Catalogue of Australia. 5. Mammalia*. Australian Government Publishing Service, Canberra. [13 Apr 1988]

Mahoney JA, Walton DW (1988c) Rhinolophidae. pp. 121–123, in DW Walton (ed.) *Zoological Catalogue of Australia. 5. Mammalia*. Australian Government Publishing Service, Canberra. [13 Apr 1988]

Mahoney JA, Walton DW (1988d) Hipposideridae. pp. 124–127, in DW Walton (ed.) *Zoological Catalogue of Australia. 5. Mammalia*. Australian Government Publishing Service, Canberra. [13 Apr 1988]

Mahoney JA, Walton DW (1988e) Emballonuridae. pp. 114–118, in DW Walton (ed.) *Zoological Catalogue of Australia. 5. Mammalia*. Australian Government Publishing Service, Canberra. [13 Apr 1988]

Mahoney JA, Walton DW (1988f) Molossidae. pp. 146–150, in DW Walton (ed.) *Zoological Catalogue of Australia. 5. Mammalia*. Australian Government Publishing Service, Canberra. [13 Apr 1988]

Mahoney JA, Walton DW (1988g) Vespertilionidae. pp. 128–145, in DW Walton (ed.) *Zoological Catalogue of Australia. 5. Mammalia*. Australian Government Publishing Service, Canberra. [13 Apr 1988]

Mahoney JA, Smith MJ, Medlin GC (2008) A new species of hopping-mouse, *Notomys robustus* (Rodentia: Muridae) from cave deposits in the Flinders and Davenport Ranges, South Australia. *Australian Mammalogy* 29, 117–135. [1 Apr 2008]

Malekian M (2007) Molecular systematics and conservation genetics of gliding petaurids (Marsupialia: Petauridae). Ph.D. Thesis, The University of Adelaide.

Malekian M, Cooper SJB, Norman JA, Christidis L, Carthew SM (2010a) Molecular systematics and evolutionary origins of the genus *Petaurus* (Marsupialia: Petauridae) in Australia and New Guinea. *Molecular Phylogenetics and Evolution* 54, 122–135. [Jan 2010]

Malekian M, Cooper SJB, Carthew SM (2010b) Phylogeography of the Australian sugar glider (*Petaurus breviceps*): evidence for a new divergent lineage in eastern Australia. *Australian Journal of Zoology* 58, 165–181.

Malia MJ, Adkins RM, Allard MW (2002) Molecular support for Afrotheria and the polyphyly of Lipotyphla based on analyses of the growth hormone receptor gene. *Molecular Phylogenetics and Evolution* 24, 91–101. [Jul 2002]

Malm AW (1866) *Några blad on Hvaldjur I Allmänhet och Balaenoptera carolinae* I Synnerhet. Handdelstidningens Bolags Tryckeri, Göteborg.

Malm AW (1871) Hvaldjur I Sveriges år 1869. *Konglinga Svenska Vetenskaps Akademiens Handlingar Stockholm* 9(2), 1–104.

Maloney KS, Harris JM (2008) Early natural history of the greater glider *Petauroides volans* (Kerr, 1792). *Proceedings of the Linnean Society of New South Wales* 129, 39–55. [19 Mar 2008]

Manwell C, Baker CMA (1983) Origin of the dog: from the wolf or wild *Canis familiaris? Speculations in Science and Technology* 6, 213–224. [Reference not seen.]

Manwell C, Baker CM (1984) Domestication of the dog: hunter, food, bed-warmer, or emotional object? *Zeitschrift für Tierzüchtung und Züchtunsbiologie –. Journal of Animal Breeding and Genetics* 101, 241–256.

Marelli CA (1922) *Phocaena stornii* sp. n., una nueva espécie de marsopa del mar austral Argentino. *Anales de la Sociedad Científica Argentina, Buenos Aires* 94, 229–240.

Marlow BJ (1958) A survey of the marsupials of New South Wales. *CSIRO Wildlife Research* 3(2), 71–114.

Marlow BJ (1965) *Marsupials of Australia*. Jacaranda Press, Brisbane. Second Ed.

Marlow BJ (1981) *Marsupials of Australia*. Jacaranda Press, Brisbane. Third Ed.

Marlow BJ, King J (1974) Sea lions and fur seals of Australia and New Zealand – the growth of knowledge. *Australian Mammalogy* 1, 117–135. [Oct 1974]

Marsh OC (1873) On the gigantic fossil mammals of the Order Dinocerata. *American Journal of Science* 5(3), 117–122.

Marsh OC (1875) New order of Eocene mammals. *The American Journal of Science and Arts* (3)9(51), 221.

Marsh OC (1886) Dinocerata: A monograph of an extinct order of gigantic mammals. *Monographs of the United States Geological Survey* 10, 1–237.

Marsh OC (1887) American Jurassic mammals. *The American Journal of Science and Arts* (3)33(196), 327–348.

Marsh OC (1889) Discovery of Cretaceous Mammalia. *American Journal of Science* 38(3), 81–92.

Marshall JT Jr (1998) Identification and scientific names of Eurasian house mice and their European allies, subgenus *Mus* (Rodentia: Muridae). Privately Printed. Springfield, Virginia.

Marshall LG (1977) Cladistic analysis of borhyaenoid, dasyuroid, the thylacinid (Marsupialia: Mammalia) affinity. *Systematic Zoology* 26, 410–425. [Dec 1977]

Marshall LG (1979) Evolution of metatherian and eutherian (mammalian) characters: a review based on cladistic methodology. *Zoological Journal of the Linnean Society* 66, 369–410. [Aug 1979]

Marshall LG (1981) The families and genera of Marsupialia. *Fieldiana Geology*, new series 8, 1–65. [20 Jul 1981]

Marshall LG (1984) Monotremes and Marsupials. pp. 59–115, in S Anderson & JK Jones (1984) *Orders and Families of Recent Mammals of the World*. John Wiley & Sons, New York.

Marshall LG, Hope JH (1973) A revaluation of *Dasyurus bowlingi* Spencer and Kershaw 1910 (Marsupialia, Dasyuridae) from King Island, Bass Straight. *Proceedings of the Royal Society of Victoria* 85, 225–234.

Marshall LG, Case JA, Woodburne MO (1990) Phylogenetic relationships of the families of marsupials. pp. 433–505, in HH Genoways (ed.) *Current Mammalogy*. Plenum Press, New York.

Marshall TA (1868) A few more words on bad spelling. *Entomologist's Monthly Magazine* 4, 280–282.

Martens E von (1860) *Die Heliceen, nach natürlicher Verwandtschaft systematisch geordnet von J.C, Albers*. Wilhelm Engelmann, Leipzig. Second Ed.

Martin JHD (1976) *Karyotypes of native Muridae in Queensland*. MSc Thesis University of Queensland.

Martin R, Handasyde K (1999) *The Koala: Natural History, Conservation and Management*. Australian Natural History Series, NSWU Press, Sydney.

Martin T, Rauhut OWM (2005) Mandible and dentition of *Asfalotomys patagonicus* (Australosphenida, Mammalia) and the evolution of tribosphenic teeth. *Journal of Vertebrate Paleontology* 25, 414–425. [27 Jun 2005]

Martinez RC, Lidicker WZ Jr (1971) Description of a new genus and species of fossil rodent from Australia. *Journal of Mammalogy* 52, 775–781. [Nov 1971]

Mason GE (1908) On the fruit bats of the genus *Pteropus* inhabiting the Andaman and Nicobar Archipelago. *Records of the Indian Museum* 2, 159–166. [Jul 1908]

Mathews GM, Iredale T (1912) Perry's Arcana' – an overlooked work. *Victorian Naturalist* 24, 7–16.

Mathews GM, Iredale T (1922) An extraordinary bird book. *The Austral Avian Record* 4, 172–175. [7 Mar 1922]

Matschie P (1894) Neue säugethiere aus den sammlungen der Herren Zenker, Neumann, Stuhlmann und Emin. *Sitzungs-berichte der Gesellschaft Naturforschender Freunde zu Berlin* 8, 194–206.

Matschie P (1899a) *Die Fledermäuse des Berliner Museums für Naturkunde*. 1. Lieferung. *Die Megachiroptera des Berliner Museums für Naturkunde*. George Reimer, Berlin.

Matschie P (1899b) Die tierwelt New-Guineas. pp. 73–112, in M Krieger (ed.) *Neu-Guinea*. Alfred Schall, Berlin.

Matschie P (1901a) Die Säugetiere von W. Kükenthal auf Halmahera, Batjan und Nord-Celebes gemachten Ausbeute. *Abhandlungen der Senckenbergischen Naturforschenden Gesellschaft* 25, 247–296.

Matschie P (1901b) Rumänische Säugethiere. *Sitzungs-berichte der Gesellschraft Naturforschender Freunde zu Berlin* 1901, 220–238.

Matschie P (1903) Die Chiropteren, Insectivoren und Muriden der Semon'schen Forschungsreise. pp. 129–136. [= 771–778 in book], in RW Semon (ed.) *Zoologische Forschungsreisen in Australien und dem Malayischen Archipel*. Mit unterstützung des Herrn Dr Paul von Ritter ausgefuhrt in dem Jahren 1891–1893 von Richard Semon. Volume 5. Systematik, Tiergeographie, Anatomie Wirbelloser Thiere. Lg. VI. (Des ganzen Werkes lfg. 21). G. Fischer, Jena.

Matschie P (1915a) Einige Beiträge zur Kenntnis der Gattung *Pseudochirus* Ogilby. *Sitzungs-berichte der Gesellschaft Naturforschender Freunde zu Berlin* 4, 83–95. [15 Apr 1915]

Matschie P (1915b) Der dingo-hund des Macdonnell-Gebirges. *Sitzungs-berichte Der Gesellschraft Naturforschender Freunde zu Berlin* 1915, 101–107. [15 Apr 1915]

Matschie P (1916) Die verbreitung der Beuteltiere auf Neuguinea mit einigen Bemerkungen über ihre Einteilung in Untergattungen. *Mitteilungen aus dem Zoologischen Museums, Berlin* 8, 257–308. [June 1916]

Matschie P (1917) Die von O. Finsch bei Port Moresby in Sudostneuguinea gesammelten kängurus. *Sitzungs-berichte der Gesellschraft Naturforschender Freunde zu Berlin* 1916, 43–61.

Matthew WD (1918) A revision of the lower Eocene Wasatch and Wind River faunas. Part 5. Insectivora (continued) Glires, Edentata. *Bulletin of the American Museum of Natural History* 38, 565–657.

Matthew WD (1929) Reclassification of the artiodactyl families. *Bulletin of the Geological Society of America* 40, 403–408.

Matthews GM, Iredale T (1912) Perry's Arcana – an overlooked work. *Victorian Naturalist* 19, 7–16. [May 1912]

Maxwell S, Burbidge AA, Morris K (eds.) (1996) *The 1996 Action Plan for Australian Marsupials and Monotremes.* Wildlife Australia. Endangered Species Program. Project Number 500. Canberra.

Mayden RL (1997) A hierarchy of species concepts: The denouement in the saga of the species problem. pp. 381–424, in MF Claridge, HA Dawah & MR Wilson (eds.) *Species: The Units of Biodiversity.* Chapman and Hall, New York.

Maynes GM (1982) A new species of rock-wallaby, *Petrogale persephone* (Marsupialia, Macropodidae), from Proserpine, ventral Queensland. *Australian Mammalogy* 5, 47–58. [10 May 1982]

Mayr E (1942) *Systematics and the Origin of Species.* Columbia University Press, New York.

Mayr E (1982) Of what use are subspecies? *The Auk* 99, 593–595.

Mazak V (1967) Notes on Siberian long-haired tiger, *Panthera tigris altaica* (Temminck, 1844), with remark on Temminck's mammal volume of the *Fauna Japonica. Mammalia* 31, 537–573.

McAllan IAW, Bruce MD (1989) Some problems in vertebrate nomenclature. I. Mammals. *Dal Bollettino del Museo Regionale di Scienze Naturali, Torino* 7, 443–460.

McCann C (1962) Key to the family Ziphiidae beaked whales. *Tuatara* 10, 13–18.

McCann C (1964) A coincidental distribution pattern of some of the larger marine animals. *Tuatara* 12, 119–124.

McCann C (1976) Note on the foetal skull of *Mesoplodon stejnegeri. Scientific Reports of the Whales Research Institute* 28, 107–117.

McCoy F (1865) Note. Quarter Sheet 7 NW. Mt. Aitken. Geological Survey of Victoria.

McCoy F (1866a) On a *Halmaturus. The Australasian.* p. 583. [11 Aug 1866]

McCoy F (1866b) On a new species of *Halmaturus* from east Australia. *Annals & Magazine of Natural History* (3)18, 322–323. [1 Nov 1866]

McCoy F (1867a) On a new genus of *Phalanger. Annals & Magazine of Natural History* (3)20, 287–288. [1 Oct 1867]

McCoy F (1867b) On the recent zoology and palaeontology of Victoria. *Annals & Magazine of Natural History* (3)20, 175–202. [1 Sep 1867]

McCoy F (1877) *Prodromus of the Palaentology of Victoria; or figures and descriptions of the Victorian organic remains.* George Robertson, Melbourne. Decade 5.

McCoy F (1879 [1878–1885]) *Prodromus of the Zoology of Victoria.* George Robertson, Melbourne. Decade IV.

McDonald JN (1981) *North American Bison, Their Classification and Evolution.* University of California Press, Berkeley.

McDowell MC (2013) *Late Quaternary faunal responses to environmental change and isolation on a large Australian land-bridge island.* Unpublished PhD Thesis, Flinders University, Adelaide. Reference not seen.

McGlashan ND, Obendorf DL, Harrington JS (2006) Aspects of the fatal malignant disease among the Tasmanian devil population (*Sarcophilus laniarius*). *European Journal of Oncology* 11, 95–102.

McGowen MR (2011) Toward the resolution of an explosive radiation - A multilocus phylogeny of oceanic dolphins (Delphinidae). *Molecular Phylogenetics and Evolution* 60, 345–357. [Sep 2011]

McGowen MR, Spaulding M, Gatesy J (2009) Divergence date estimation and a comprehensive molecular tree of extant cetaceans. *Molecular Phylogenetics and Evolution* 53, 891–906.

McKay GM (1982) Nomenclature of the gliding possum genera *Petaurus* and *Petauroides* (Marsupialia: Petauridae). *Australian Mammalogy* 5, 37–39. [10 May 1982]

McKay GM (1984) Cytogenic relationships of possums and gliders. pp. 9–16, in A Smith & ID Hume (eds.) *Possums and Gliders.* Surrey Beatty and Sons, Sydney.

McKay GM (1988a) Phascolarctidae. pp. 51–52, in DW Walton (ed.) *Zoological Catalogue of Australia. 5. Mammalia.* Australian Government Publishing Service, Canberra. [13 Apr 1988]

McKay GM (1988b) Burramyidae. pp. 98–102, in DW Walton (ed.) *Zoological Catalogue of Australia. 5. Mammalia.* Australian Government Publishing Service, Canberra. [13 Apr 1988]

McKay GM (1988c) Petauridae. pp. 87–97, in DW Walton (ed.) *Zoological Catalogue of Australia. 5. Mammalia.* Australian Government Publishing Service, Canberra. [13 Apr 1988]

McKay GM (1988d) Tarsipedidae. pp. 103–104, in DW Walton (ed.) *Zoological Catalogue of Australia. 5. Mammalia.* Australian Government Publishing Service, Canberra. [13 Apr 1988]

McKay GM (1988e) Phalangeridae. pp. 81–86, in DW Walton (ed.) *Zoological Catalogue of Australia. 5. Mammalia.* Australian Government Publishing Service, Canberra. [13 Apr 1988]

McKean JL (1966) Some new distributional records of the broad-nosed bats (*Nycticeius* spp.). *Victorian Naturalist* 83, 25–30. [10 Feb 1966]

McKean JL (1970) A new subspecies of the horseshoe bat, *Hipposideros diadema* from the Northern Territory, Australia. *Western Australian Naturalist* 11, 138–140. [22 Apr 1970]

McKean JL (1972) Notes on some collections of bats (Order Chiroptera) from Papua-New Guinea and Bougainville Island. *Technical Paper, Division of Wildlife Research. CSIRO Australia* 26, 1–35.
McKean JL (1975) The bats of Lord Howe Island with the description of a new nyctophiline bat. *Australian Mammalogy* 1, 329–332. [Dec 1978]
McKean JL, Calaby JH (1968) A new genus and two new species of bats from New Guinea. *Mammalia* 32, 372–378. [Sep 1968]
McKean JL, Friend GR (1979) *Taphozous kapalgensis*, a new species of sheath-tailed bat from the Northern Territory, Australia. *Victorian Naturalist* 96, 239–241. [Nov–Dec 1979]
McKean JL, Price WJ (1967) Notes on some Chiroptera from Queensland, Australia. *Mammalia* 31, 101–119. [Mar 1967]
McKean JL, Richards GC, Price WJ (1978) A taxonomic appraisal of *Eptesicus* (Chiroptera, Mammalia) in Australia. *Australian Journal of Zoology* 26, 529–537.
McKenna MC (1975) Toward a phylogenetic classification of the Mammalia. pp. 21–46, in WP Luckett & FS Szalay (eds.) *Phylogeny of the Primates – A Multidisciplinary Approach*. Plenum Press, New York.
McKenna MC (1993) In RK Stucky & MC McKenna. Mammalia. pp. 739–771, in MJ Benton (ed.) *The Fossil Record 2*. Chapman & Hall, London.
McKenna MC, Bell SK (1997) *Classification of Mammals: Above the Species Level*. Columbia University Press, New York.
McKenzie NL, Archer A (1982) *Sminthopsis youngsoni* (Marsupialia: Dasyuridae) the lesser hairy-footed dunnarts, a new species from Arid Australia. *Australian Mammalogy* 5, 267–279. [20 Dec 1982]
McLaren IA (1960) Are the Pinnipedia biphyletic. *Systematic Zoology* 9, 18–28. [Mar 1960]
McMurtrie H (1834) *Cuvier's Animal Kingdom arranged according to its organisation: translated from the French and abridged for the use of students*. Orr & Smith, London.
McNamara JA (1997) Some smaller macropod fossils of South Australia. *Proceedings of the Linnean Society of New South Wales* 117, 97–105. [1 Mar 1997]
Mead JG (1975) Anatomy of the external nasal passages and facial complex in the Delphinidae (Mammalia: Cetacea). *Smithsonian Contributions to Zoology* 207, 1–72.
Mead JG (1989a) Bottlenose whales *Hyperoodon ampullatus* (Forster, 1770) and *Hyperoodon planifrons* Flower, 1882. pp. 321–348, in SH Ridgway & R Harrison (eds.) *Handbook of Marine Mammals. Volume 4. River Dolphins and Larger Toothed Whales*. Academic Press, London.
Mead JG (1989b) Beaked whales of the genus *Mesoplodon*. pp. 349–430, in SH Ridgway & R Harrison (eds.) *Handbook of Marine Mammals. Volume 4. River Dolphins and Larger Toothed Whales*. Academic Press, London.
Mead JG (1989c) Shepherd's Beaked Whale *Tasmacetus shepherdi* Oliver, 1937. pp. 309–320, in SH Ridgway & R Harrison (eds.) *Handbook of Marine Mammals. Volume 4. River Dolphins and Larger Toothed Whales*. Academic Press, London.
Mead JG (2009) Beaked whales, overview Ziphiidae. pp. 94–97, in WF Perrin, B Würsig & JGM Thewissen (eds.) *Encyclopedia of Marine Mammals*. Academic Press, Amsterdam. Second Edition.
Mead JG, Brownell RL (1993) Order Cetacea. pp. 349–364, in DE Wilson & DA Reeder (eds.) *Mammal Species of the World. A Taxonomic and Geographic Reference*. Smithsonian Institution Press, Washington. Second Edition.
Mead JG, Brownell RL (2005) Order Cetacea. pp. 723–743, in DE Wilson & DA Reeder (eds.) *Mammal Species of the World. A Taxonomic and Geographic Reference*. Johns Hopkins University Press, Baltimore. Third Edition.
Meagher M (1986) *Bison bison*. *Mammalian Species* 266, 1–8. [16 Jun 1986]
Mearns EA (1905) Descriptions of new genera and species of mammals from the Philippine Islands. *Proceedings of the United States National Museum* 28, 425–460. [13 May 1905]
Meckel JF (1826) *Ornithorhynchi paradoxi descriptio anatomica*. Gerhardum Fleischerum, Lipsiae.
Medlin GC (1993) *Field Guide to Chambers Gorge Flingers Ranges*. South Australian Museum, Adelaide.
Medlin GC (2008) Broad-cheeked hopping mouse *Notomys robustus*. pp. 609–610, in S Van Dyck and R Strahan (eds.) *The Mammals of Australia*. Reed New Holland, Sydney. Third Edition. March 2008.
Medway L (1977) Mammals of Borneo: Field keys and an annoted checklist. Second Edition. *Monograph, Malay Branch of the Royal Asiatic Society* 7, 1–172.
Meek FB (1872) Descriptions of new western Palaeozoic fossils, mainly from the Cincinnati Group of the lower Silurian Series of Ohio. *Proceedings of the Academy of Natural Sciences of Philadelphia* 23(1871), 308–336. [13 Feb 1872]
Meek P (2000) The decline and current status of the Christmas Island shrew *Crocidura attenuata trichura* on Christmas Island, Indian Ocean. *Australian Mammalogy* 22, 43–49. [Aug 2000]
Meek FB, Worthen AH (1868) Descriptions of Palaeozoic fossils from the Silurian, Devonian and Carboniferous rocks of Illinois, and other western States. *Proceedings of the Chicago Academy of Sciences* 1, 11–23.
Meigen JW (1800) *Nouvelle Classification des Mouches à deux ailes, (Diptera L.), d'après un plan tout nouveau, &c*. Paris.
Meigen JW (1803) Versuch einer neuen GattungsEintheilung der europäischen zweiflügligen Insekten. *Magazin für Insektenkunde* 2, 259–281. [Illiger]
Meigen JW (1830) Systematische Beschreibung der bekannten Europäischen zweiflügeligen Insekten. Schulzische Buchhandlung, Hamm. Volume 6.
Meigen JW (1838) *Systematische Beschreibung der bekannten Europäischen zweiflügeligen Insekten*. Schulzische Buchhandlung, Hamm. Volume 7.
Mein P, Tupinier Y (1977) Formule dentaire et position systématique de Miniooptère (Mammalia, Chiroptera). *Mammalia* 41, 207–211. [30 Sep 1977]
Meinert F (1870) Myriapoda Musaei Hauniensis: bidrag til myriapodernes morphologi og systematik. *Naturhistorisk Tidsskrift* 7(3), 1–128.
Melo MRS (2008) The genus *Kali* Lloyd (Chiasmodontidae: Teleostei) with description of new two species, and the

revalidation of *K. kerberti* Weber. *Zootaxa* 1747, 1–33. [11 Apr 2008]

Melville RV, Smith JDD (eds.) (1987) *Official Lists and Indexes of Names and Works in Zoology*. International Commission on Zoological Nomenclature, London.

Mendez M, Jefferson TA, Kolokotronis SO, Krützen M, Parra GJ, Collins T, Minton G, Baldwin R, Berggren P, Särnblad A, Amir OA, Peddemors VM, Karczmarski L, Guissamulo A, Smith B, Sutaria D, Amato G, Rosenbaum HC (2013) Integrating multiple lines of evidence to better understand the evolutionary divergence of humpback dolphins along their entire distribution range: a new dolphin species in Australian waters. *Molecular Ecology* 22, 5936–5948.

Menegaux MA (1901) Description d'une variété et d'une espèce nouvelles de chiroptères rapportées du Mexique par M. Diguet. *Bulletin du Muséum d'Histoire Naturelle, Paris* 7, 321–327.

Meng J (2004) Phylogeny and divergence of basal Glires. *Bulletin of the American Museum of Natural History* 285, 93–109.

Meng J, Wyss AR (2001) The morphology of *Tribosphenomys* (Rodentiaformes, Mammalia): phylogenetic implications for basal Glires. *Journal of Mammalian Evolution* 8, 1–71. [Mar 2001]

Meng J, Wyss AR (2005) Glires (Lagomorpha, Rodentia). pp. 145–158, in KD Rose and JD Archibald (eds.) *The Rise of Placental Mammals: Origins and Relationships of the Major Extant Clades*. The Johns Hopkins University Press, Baltimore.

Meng JA, Wyss AR, Dawson MR, Zhal RJ (1994) Primitive fossil rodent from Inner Mongolia and its implications for mammalian phylogeny. *Nature* 370(6485), 134–136. [14 Jul 1994]

Meng J, Hu Y, Li C (2003) The osteology of *Rhombomylus* (Mammalia, Glires): Implications for phylogeny and evolution of Glires. *Bulletin of the American Museum of Natural History* 275, 1–247. [28 Feb 2003]

Menkhorst P, Knight F (2011) *A Field Guide to the Mammals of Australia*. Oxford University Press, Melbourne.

Menu H (1984) Révision du statut de *Pipistrellus subflavus* (F. Cuvier, 1832). Proposition d'un taxon générique nouveau: *Perimyotis* nov. gen. *Mammalia* 48, 409–416. [18 Oct 1984]

Menu H (1987) Morphotypes dentaires actuels et fossils des chiroptères vespertilioninés. 2ème partie: implications systématiques et phylogéniques. *Palaeovertebrata* 17, 77–150.

Menzies JI (1990) A systematic revision of *Pogonomelomys* (Rodentia:Muridae) of New Guinea. *Science in New Guinea* 16, 118–137.

Menzies JI (1996) A systematic revision of *Melomys* (Rodentia: Muridae) of New Guinea. *Australian Journal of Zoology* 44, 367–426.

Menzies JI, Pernetta JC (1986) A taxonomic revision of cuscuses allied to *Phalanger orientalis* (Marsupialia: Phalangeridae). *Journal of Zoology* 1, 551–618.

Meredith RW, Westerman M, Case JA, Springer MS (2008a) A phylogeny and timescale for marsupial evolution based on sequences for five nuclear genes. *Journal of Mammalian Evolution* 15, 1–36. [online 7 Nov 2007]

Meredith RW, Westerman M, Springer MS (2008b) A timescale and phylogeny for 'bandicoots' (Peramelemorphia: Marsupialia) based on sequences for five nuclear genes. *Molecular Phylogenetics and Evolution* 47, 1–20. [Apr 2008]

Meredith RW, Westerman M, Springer MS (2008c) A phylogeny and timescale for the living genera of kangaroos and kin (Macropodiformes: Marsupialia) based on nuclear DNA sequences. *Australian Journal of Zoology* 56, 395–410.

Meredith RW, Westerman M, Springer MS (2009a) A phylogeny of Diprotodontia (Marsupialia) based on sequences for five nuclear genes. *Molecular Phylogenetics and Evolution* 51, 554–571. [Jun 2009]

Meredith RW, Krajewski C, Westerman M, Springer MS (2009b) Relationships and divergence times among the orders and families of Marsupialia. *Museum of Northern Arizona Bulletin* 65, 383–406.

Meredith RW, Menzoza MA, Roberts KK, Westerman M, Springer MS (2010) A phylogeny and timescale for the evolution of Pseudocheiridae (Marsupialia: Diprotodontia) in Australia and New Guinea. *Journal of Mammalian Evolution* 17, 75–99. [online 2 Mar 2010]

Merriam CH (1895) Brisson's genera of mammals, 1762. *Science* new series 1(14), 375–376. [5 Apr 1895]

Merrilees D (1973) Fossiliferous deposits at Lake Tandou, New South Wales, Australia. *Memoirs of the National Museum of Victoria* 34, 177–182. [9 May 1973]

Mertens R (1925) Verzeichnis der Säugetier-Typen des Senckenbergischen Museums. *Senckenbergiana* 7, 18–37.

Meschinelli L (1903) Un nuovo chiropterro fossile *(Archaeopterus transiens Mesch.)*, Delle Ligniti Di Monteviale, Contribuzione alla Paleontologia Vicentia. *Atti Del Reale Instituto Veneto di Scienze ed Arti* 62(2), 1329–1344.

Meyen FJF (1833) Beiträge zur Zoologie, gesammelt auf einer Reise um die Erde: Säugethiere. *Nova Acta Physico-Medica Academiae Caesareae Leopoldino-Carolinae Naturae Curiosorum* 16(2), 549–610.

Meyer AB (1876) On the habitat of *Uromys aruensis* (Gray) and its allies. *Annals & Magazine of Natural History* (4)17, 145–146. [1 Feb 1876]

Meyer AB (1899) Säugethiere vom Celébes- und Philippinen-Archipel, II. *Abhandlungen und Berichte des Königlichen Zoologischen und Anthropologisch-Ethnologischen Museums zu Dresden* 7(7), viii, 1–55.

Meyer FAA (1793) *Systematische-summarische Uebersicht der neuesten zoologischen Entdeckungen in Neuholland und Afrika. Nebst zwey andern zoologischen Abhandlungen*. Dykischen Buchhandlung, Leipzip.

Meyer H von (1841) *Arionus servatus* ein den Delphinen verwandtes Meeres-Säugethier aus der Molasse von Baltringen in Württemberg. *Neues Jahrbuch für Mineralogie, Geognosie. Geologie und Petrefakten-Kunde* 1841, 315–331.

Miescher F (1843) Über einen neuen parasiten der menschlichen haut. *Bericht über die Verhandlungen der Naturforschenden Gesellschaft in Basel* 5(1840–1842), 191–198.

Mikhalev YA, Ivashin MV, Savusin VP, Zelenaya FE (1981) The distribution and biology of killer whales in the southern hemisphere. *Report of the International Whaling Commission* 31, 551–566.

Miller GS Jr (1897a) Revision of North American bats of the family Vespertilionidae. *North American Fauna* 13, 1–135. [16 Oct 1897]

Miller GS Jr (1897b) The nomenclature of some European bats. *Annals & Magazine of Natural History* (6)20, 379–385. [1 Oct 1897]

Miller GS Jr (1900) Seven new rats collected by W.L. Abbott in Siam. *Proceedings of the Biological Society of Washington* 13, 137–150. [21 Apr 1900]

Miller GS Jr (1902a) The mammals of the Andaman and Nicobar Islands. *Proceedings of the United States National Museum* 24, 751–795. [29 May 1902]

Miller GS Jr (1902b) Twenty new American bats. *Proceedings. Academy of Natural Sciences of Philadelphia* 54, 389–412. [May-Sep 1902]

Miller GS (1902c) Two new tropical old world bats. *Proceedings of the Biological Society of Washington* 15, 245–246. [20 Nov 1902]

Miller GS Jr (1905) A new genus of bats from Sumatra. *Proceedings of the Biological Society of Washington* 18, 229–230. [9 Dec 1905]

Miller GS Jr (1906a) Seven new Malayan bats. *Proceedings of the Biological Society of Washington* 19, 61–65. [1 May 1906]

Miller GS Jr (1906b) Twelve new genera of bats. *Proceedings of the Biological Society of Washington* 19, 83–85. [4 Jun 1906]

Miller GS Jr (1907) The families and genera of bats. *Bulletin of the United States National Museum* 57, 1–282. [29 Jun 1907]

Miller GS Jr (1910) Description of two new genera and sixteen new species of mammals from the Philippine Islands. *Proceedings of the United States National Museum* 38, 391–404. [19 Aug 1910]

Miller GS Jr (1912) *Catalogue of the Mammals of Western Europe (Europe exclusive of Russia) in the Collection of the British Museum*. British Museum, London. [after Oct 1912]

Miller GS Jr (1923) The telescoping of the cetacean skull. *Smithsonian Miscellaneous Collections* 76(5), 1–70. [31 Aug 1923]

Miller GS Jr (1936) The status of *Delphinus bairdii* Dall. *Proceedings of the Biological Society of Washington* 49, 145–146. [22 Aug 1936]

Miller GS Jr, Gidley JW (1918) Synopsis of supergeneric groups of rodents. *Journal of the Washington Academy of Sciences* 8(13), 431–448. [19 Jul 1918]

Miller GS Jr, Kellogg R (1955) List of North American recent mammals. *Bulletin of the United States National Museum* 205, 1–954.

Miller GS Jr, Rehn JAG (1901) Systematic results of the study of North American land mammals to the close of the year 1900. *Proceedings of the Boston Society of Natural History* 30, 1–352.

Miller-Butterworth CM, Murphy JM, O'Brien SJ, Jacobs DS, Springer MS, Teeling EC (2007) A family matter: Conclusive resolution of the taxonomic position of the long-fingered bats, *Miniopterus*. *Molecular Biology and Evolution* 24, 1553–1561. [Jul 2007]

Milne DJ, Jackling FC, Sidhu M, Appleton BJ (2009) Shedding new light on old species identifications: morphological and genetic evidence suggest a need for conservation status review of the critically endangered bat *Saccolaimus saccolaimus*. *Wildlife Research* 36, 496–508.

Milne-Edwards A (1867) Mémoire sur le type d'une nouvelle famille de l'ordre des Rongeurs. *Nouvelles Archives du Muséum d'Histoire Naturelle, Paris* 3, 81–116.

Milne-Edwards A (1872 [1868–1874]) *Recherches pour servir à l'histoire des mammifères*. Masson, Paris. Volumes 1. Text.

Milne-Edwards A (1877) Note sur quelques mammifères nouveaux provenant de la Nouvelle-Guinée. *Comptes Rendus Hebdomadaires des Séances de l'Académie des Sciences, Paris* 85, 1079–1081.

Milne Edwards A (1888) Note sur une espèce nouvelle du genre *Dactylopsila*. *Mémoires publiés par la Société Philomathique à l'occasion du Centenaire de sa Fondation 1788–1888*, 1888, 173–177.

Minkoff EC (1976) Mammalian superorders. *Zoological Journal of the Linnean Society* 58, 147–158. [Mar 1976]

Misawa K, Janke A (2003) Revisiting the Glires concept--phylogenetic analysis of nuclear sequences. *Molecular Phylogenetics and Evolution* 28, 320–327. [Aug 2003]

Misonne X (1969) African and Indo-Australian Muridae: Evolutionary trends. *Annales de Musée Royal Afrique Centrale. Tervuren* 172, 1–219.

Mitchell ED (1968) The Mio-Pliocene pinniped *Imagotaria*. *Journal of the Fisheries Research Board of Canada* 25, 1843–1900.

Mitchell ED (1975) Report on the meeting on smaller cetaceans. Montreal, April 1–11, 1974. Review of Biology and Fisheries for Smaller Cetaceans. *Journal of the Fisheries Research Board of Canada* 32, 889–983.

Mitchell ED (1989) A new cetacean from the late Eocene La Meseta Formation, Seymour Island, Antarctic Peninsula. *Canadian Journal of Fisheries and Aquatic Sciences* 46(12), 2219–2235.

Mitchell ED, Tedford RH (1973) The Enaliarctinae, a new group of extinct aquatic Carnivora and a consideration of the origin of the Otariidae. *Bulletin of the American Museum of Natural History* 151, 201–284. [27 Jun 1973]

Mitchell TL (1838a) *Three Expeditions into the Interior of Eastern Australia; with descriptions of the recently explored region of Australia Felix, and of the present colony of New South Wales*. T. & W. Boone, London. Volume 1.

Mitchell TL (1838b) *Three Expeditions into the Interior of Eastern Australia; with descriptions of the recently explored region of Australia Felix, and of the present colony of New South Wales*. T. & W. Boone, London. Volume 2. Note: 2nd Edition is 1839.

Mivart SG (1885) Notes on the Pinnipedia. *Proceedings of the Zoological Society of London* 53, 484–501. [Oct 1 885]

Miyazaki N (1992) Sightings of killer whales in the Southern Hemisphere. *IBI Reports. International Marine Biological Research Institute* 3, 47–59.

Mjöberg E (1916) Results of Dr. E. Mjöberg's Swedish Scientific Expeditions to Australia 1910–1913. II. Mammals from Queensland. 2. On a new genus and species of marsupial. *Kungliga Svenska Vetenskaps-Akademiens Handlingar* 52(2), 13–20. [16 Jan. 1916]

Modell H (1964) Das natürliche System der Najaden. 3. *Archiv fuer Molluskenkunde* 93(3/4), 71–126. [20 Jul 1964]

Moeller H (1973) Zur Kenntnis der Schnädelgestalt großer Raubbeutler (Dasyuridae Waterhouse, 1838) - Eine allometrische Formanalyse. *Zoologische Jahrbucher. Abteilung fur Anatomie und Ontogenie der Tiere* 91, 257–303.

Molina GI (1782) *Saggio sulla storia naturale del Chili.* Stamperia di S. Tommaso d'aquino, Bologna.

Möller LM, Beheregaray LB (2001) Coastal bottlenose dolphins from southeastern Australia are *Tursiops aduncus* according to sequences of the mitochondrial DNA control region. *Marine Mammal Science* 17, 249–263. [Apr 2001]

Möller LM, Bilgmann K, Charlton-Robb K, Beheregaray L (2008) Multi-gene evidence for a new bottlenose dolphin species in southern Australia. *Molecular Phylogenetics and Evolution* 49, 674–681.

Montagu G (1808) An account of the larger and lesser species of horse-shoe bats, proving them to be distinct; together with a description of *Vespertilio barbastellus*, taken in the south of Devonshire. *Transactions of the Linnean Society of London* (1)9, 162–173. [23 Nov 1808]

Montagu G (1821) Description of a *Delphinus* (*D. truncatus*) which appears to be new. *Memoirs of the Wernerian Natural History Society* 3, 75–82.

Montgelard C, Catzeflis FM, Douzery E (1997) Phylogenetic relationships of artiodactyls and cetaceans as deduced from the comparison of cytochrome b and 125 rRNA mitochondrial sequences. *Molecular Biology and Evolution* 14, 550–559. [May 1997]

Moore JC (1963) Recognising certain species of beaked whales of the Pacific Ocean. *American Midland Naturalist* 70, 396–428. [Oct 1963]

Moore JC (1968) Relationships among the living genera of beaked whales with classification, diagnoses and keys. *Fieldiana Zoology* 53, 209–298. [13 Dec 1968]

Mörch OAL (1852) *Catalogus Conchyliorum quae reliquit D. Alphonso D'Aguirra & Gadea Comes de Yoldi, regis Daniae cubiculorum princeps, ordinis Dannebrogici in prima classe & ordinis Caroli Tertii Eques. Volume 1. Cephalophora.* Hafniae (Copenhagen). pp. 1–170.

Moreno FP (1892) Lijeros apuntes sobre dos generos de cetaceos fosiles de la Republica Argentina. *Revista del Museo de La Plata* 3, 393–400.

Moreno FP (1895) Nota sobre los restos de Hyperoodontes conservados en el Museo de la Plata. *Anales del Museo de la Plata. Zoologica* 3, 1–8.

Morin PA, Archer FI, Foote AD, Vilstrup J, Allen EE, Wade P, Durban J, Parsons K, Pitman R, Li L, Bouffard P, Abel Nielsen SC, Rasmussen M, Willerslev E, Gilbert MTP, Harkins T (2010) Complete mitochondrial genome phylogeographic analysis of killer whales indicates multiple species. *Genome Research* 20, 908–916. [22 Apr 2010 online]

Moritz C (1994) Defining 'evolutionarily significant units' for conservation. *Trends in Ecology & Evolution* 9, 373–375. [Oct 1994]

Morris FO (1837) A new system of nomenclature, illustrated by a list of British Birds. pp. 122–127, in N Wood (ed.) *The Naturalist; Illustrative of the Animal, Vegetable, and Mineral Kingdoms*. Volume 2. Part 9.

Morrison-Scott TCS (1955) Support for Tate's proposed use of the plenary powers to validate the specific name '*dingo*' Meyer, 1793, as published in the combination '*Canis dingo*' as the name for the dingo (Class Mammalia). *Bulletin of Zoological Nomenclature* 11, 168. [31 May 1955]

Morrison-Scott TCS, Sawyer FC (1950) The identity of Captain Cook's kangaroo. *Bulletin of the British Museum (Natural History). Zoology (Jena, Germany)* 1(3), 45–50. [31 Mar 1950]

Morton SR, Alexander F (1982) Geographic variation in the external morphology of *Sminthopsis crassicaudata* (Dasyuridae, Marsupialia). pp. 695–698, in M Archer (ed.) *Carnivorous Marsupials*. Volume 2. Royal Zoological Society of New South Wales, Sydney. [Feb 1982]

Moseley HN (1879) Notes by a naturalist on the 'Challenger': being an account of various observations made during the voyage of H.M.S. Challenger round the world, in the years 1872–1876, under the commands of Capt. G. S. Nares and Capt. F. T. Thomson. Macmillan, London.

Mostert ME (2009) Molecular and morphological assessment of invasive, inland *Rattus* (Rodentia: Muridae) congenerics in South Africa and their reservoir host potential with respect to *Helicobacter* and *Bartonella*. MSc Thesis. University of Pretoria, South Africa.

Moura AE, Nielsen SCA, Vilstrup JT, Moreno-Mayer JV, Gilbert MTP, Gray HWI, Natoli A, Möller L, Hoelzel AR (2013) Recent diversification of a marine genus (*Tursiops* spp.) tracks habitat preference and environmental change. *Systematic Biology* 62, 865–877. [8 Aug 2013]

Mudie R (1829) *The Picture of Australia*. Whittaker, Treacher, & Co., London.

Muirhead J (2000) Yaraloidea (Marsupialia, Peramelemorphia), a new superfamily of marsupial and a description and analysis of the cranium of the Miocene *Yarala burchfieldi*. *Journal of Paleontology* 74, 512–523.

Müller A (1863) Über das Bruchstuck vom Schädel eines Finnwales, *Balaenoptera syncondylus*, welches im Jahre 1860 von der Ostsee an die kurische Nehrung geworfen wurde. *Schriften der Königlichen Physikalisch-Öekonomische Gesellschaft zu Königsberg. Schriften* 4, 38–78.

Müller J (1836) Jahresbericht über die Fortschritte der anatomisch-physiologisehen Wissenschaften im Jahre 1835, pp. I–CCXXXV in J. Müller (ed.) *Archiv für Anatomie, Physiologie und Wissenschaftliche Medecin, in Verbindung mit Mehreren Gelehrten Herausgegeben*. Verlsag von G. Eichler, Berlin. (for 1835)

Müller OF (1776) *Zoologiae Danicae. Prodromus seu animalium Daniae et Norvegiae ingenarum characteres, nomina et synonyma impris popularium*. Hallageriis, Havniæ.

Müller PLS (1776) *Des Ritters Carl von Linné* Vollständiges Natursystem: nach der zwölften lateinischen Ausgabe, und nach Anleitung des Holländischen Houttuynischen Werks. Gabriel Nicolaus Raspe, Nürnberg. Supplements und Register-Band.

Müller S (1839–1841) Bijdragen tot de Kennis Van Nieuw-Guinea. pp. 1–80, in CJ Temminck (ed.). [1839–1847] *Verhandelingen over de Natuurlijke Geschiedenis der Nederlandsche overzeesche bezittingen*, door de Leden der Natuurkundige Commissie in Indië en andere Schrijvers.

Land-en Volkenkunde. S. and J. Luchtmans and C.C. van der Hoek, Leiden. [p. 1–8, 18 Dec 1839; p. 9–56, 13 Mar 1841; p. 57–84, 11 Oct 1841]

Müller S, Schlegel H (1845a) Beschrijving eener nieuwe sort van vleeschetende buideldieren, *Phascogale melas*. Plate XXV; pp. 149–152, in CJ Temminck (ed.). [1839–1847] *Verhandelingen over de Natuurlijke Geschiedenis der Nederlandsche overzeesche bezittingen*, door de Leden der Natuurkundige Commissie in Indië en andere Schrijvers. Zoologie. S. and J. Luchtmans and C.C. van der Hoek, Leiden. [Plate 10 May 1843; Text 26 Jun 1845]

Müller S, Schlegel H (1845b) Over de herten van den Indischen Archipel. Plate XLII-XLV. pp. 209–228, in CJ Temminck (ed.). [1839–1847] *Verhandelingen over de Natuurlijke Geschiedenis der Nederlandsche overzeesche bezittingen*, door de Leden der Natuurkundige Commissie in Indië en andere Schrijvers. Zoologie. S. and J. Luchtmans and C.C. van der Hoek, Leiden. [26 Jun 1845]

Munemasa M, Nikaido M, Donnellan S, Austin CC, Okada N, Hasegawa M (2006) Phylogenetic analysis of diprotodontian marsupials based on complete mitochondrial genomes. *Genes & Genetic Systems* 81, 181–191.

Münster G (1830) *Bemerkungen zur nähern Kenntniss der Belemniten*, &c. Reference not seen.

Münster G (1832) In G. Goldfuss, *Naturhistorischer* Atlas. Volume 4. Reference not seen.

Münter J (1877) Über zwei im Jahrhunderte bei Greifswald zur section gelangte männliche individen von *Balaenoptera Sibbaldii* v. Bened. s. *Cuvierius Sibbaldii* Gray s. *Pterobalaena Gryphus* Mtr. *Mitteilungen aus dem Naturwissenschaftlichen Verein von Neu-Vorpommern und Rügen* (Greifswald), Berlin 9, 1–174.

Murchison RI (1843) Proceedings of the Geological Society: Mr Murchison's Anniversary Address, February 1843. *The London, Edinburgh and Dublin Philosophical Magazine and Journal of Science* (3)22, 511–567.

Murie J (1873) On the organisation of the caaing whale, *Globiocephalus melas*. *Transactions of the Zoological Society of London* 8, 235–301. [Feb 1873]

Murphy WJ, Eizirik E, O'Brien SJ, Madsen O, Scally M, Douady CJ, Teeling E, Ryder OA, Stanhope MJ, de Jong WW, Springer MS (2001a) Resolution of the early placental mammal radiation using Bayesian phylogenetics. *Science* 294(5550), 2348–2351. [14 Dec 2001]

Murphy WJ, Eizirik E, Johnson WE, Zhang YP, Ryder OA, O'Brien SJ (2001b) Molecular phylogenetics and the origins of placental mammals. *Nature* 409(6820), 614–618. [1 Feb 2001]

Murray A (1860) Contributions to the fauna of Old Calabar – Mammals. *Proceedings of the Royal Physical Society of Edinburgh* 2, 156–159.

Murray A (1866) *The Geographical Distribution of Mammals*. Day & Son, London.

Murray JD, Sharman G, McKay GM, Calaby JH (1980) Karyotypes, constitutive heterochromatin and taxonomy of ringtail possums of the genus *Pseudocheirus* (Marsupialia: Petauridae). *Cytogenetics and Cell Genetics* 27, 73–81.

Murray JD, McKay GM, Winter JW, Ingleby S (1989) Cytogenetics of the Herbert River ringtail possum, *Pseudocheirus herbertensis* (Diprotodontia: Pseudocheiridae): evidence for two species. *Genome* 32, 1119–1123.

Murray P (1991) The Pleistocene Megafauna of Australia. pp. 1071–1164, in P Vickers-Rich, JM Monaghan, RF Baird & TH Rich (eds.) *Vertebrate Palaeontology of Australasia*. Monash University, Melbourne.

Musser AM (2003) Review of the monotreme fossil record and comparison of palaeontological and molecular data. *Comparative Biochemistry and Physiology. A. Comparative Physiology* 136, 927–942.

Musser GG (1981) A new genus of arboreal rat from West Java, Indonesia. *Zoologische Verhandelingen* 189, 1–35.

Musser GG (1982) Results of the Archbold Expeditions. No. 107. A new genus of arboreal rat from Luzon Island in the Philippines. *American Museum Novitates* 2730, 1–23. [7 Jun 1982]

Musser GG (1986) In GG Musser, A van de Weerd & E Strasser. *Paulamys*, a replacement name for *Floresomys* Musser, 1981 (Muridae), and new material of that taxon from Flores, Indonesia. *American Museum Novitates* 2850, 1–10. [30 Jun 1986]

Musser GG, Boeadi (1980) A new genus of murid rodent from the Komodo islands in Nusatenggara, Indonesia. *Journal of Mammalogy* 61, 395–413. [Aug 1980]

Musser GG, Carleton MD (1993) Family Muridae. pp. 501–755, in DE Wilson & DA Reeder (eds.) *Mammal Species of the World. A Taxonomic and Geographic Reference*. Smithsonian Institution Press, Washington. Second Edition.

Musser GG, Carleton MD (2005) Superfamily Muroidea. pp. 894–1531, in DE Wilson & DA Reeder (eds.) *Mammal Species of the World. A Taxonomic and Geographic Reference*. Johns Hopkins University Press, Baltimore. Third Edition.

Musser GG, Newcomb C (1983) Malaysian murids and the giant rat of Sumatra. *Bulletin of the American Museum of Natural History* 174, 327–598. [29 Jun 1983]

Nagatomi A (1970) Rachiceridae (Diptera) from the Oriental and Palaearctic regions. *Pacific Insects* 12(2), 417–466.

Nagorsen D (1985) *Kogia simus*. *Mammalian Species* 239, 1–6. [24 May 1985]

Nalepa A (1889) Beitrage zur systematik der phytopten. *Sitzungsberichte der Kaiserlichen Akademie der Wissenschaften in Wien. Mathematisch-Naturwissenschaftliche Classe* 98, 112–156.

Nalepa A (1890) Neue phytoptiden. *Anzeiger der Kaiserlichen Akademie der Wissenschaften in Wien* 28, 212–213.

Nathusius H (1839) In A de Demidoff (ed.) *Voyage dans la Russie méridionale et la Crimée par la Hongrie, la Valachie et la Moldavie*. Ernest Bourdin, Paris. Volume 3.

Natoli A, Peddermors VM, Hoelzel AR (2004) Population structure and speciation in the genus *Tursiops* based on microsatellite and mitochondrial DNA analyses. *Journal of Evolutionary Biology* 17, 363–375.

Near TJ, Kassler TW, Koppelman JB, Dillman CB, Philipp DP (2003) Speciation in North American black basses, *Micropterus* (Actinopterygii: Centrarchidae). *Evolution* 57, 1610–1621. [Jul 2003]

Neave SA (1939–1940) *Nomenclator Zoologicus: A list of the names of genera and subgenera in zoology from the tenth edition of Linnaeus 1758 to the end of 1935*. Zoological Society, London. Volumes 2 and 3.

Neaves LE, Zenger KR, Prince RIT, Eldridge MDB (2012) Impact of Pleistocene aridity oscillations on the population history of a widespread, vagile Australian mammal, *Macropus fuliginosus*. *Journal of Biogeography* 39, 1545–1563. [Aug 2012]

Nehring A (1887a) Ueber eine Pelzrobben-Art von der Küste 1887 Süd-Brasiliens. *Archiv für Naturgeschichte* 1(53), 75–94.

Nehring A (1887b) Über die südbrasilienische Pelzrobbe. *Sitzungs-berichte der Gesellschraft Naturforschender Freunde zu Berlin* 1887, 142–143.

Neill P (1811) Some account of a fin-whale stranded near Alloa. *Memoirs of the Wernerian Natural History Society* 1, 201–214.

Nelson J (1988) Where is the type locality of *Macroderma gigas? Macroderma* 4, 70–72. [Sep 1988]

Nelson JE, Stephan H (1982) Encephalisation in Australian Marsupials. pp. 699–706, in M Archer (ed.) *Carnivorous Marsupials*. Volume 2. Royal Zoological Society of New South Wales, Sydney. [Feb 1982]

Newbound CN, Hisheh S, Suyanto A, How RA, Schmitt LH (2008) Markedly discordant mitochondrial DNA and allozyme phylogenies of tube-nosed fruit bats, *Nyctimene*, at the Australian Oriental biogeographical interface. *Biological Journal of the Linnean Society. Linnean Society of London* 93, 589–602. [Mar 2008]

Newman E (1843) *The System of Nature: an essay*. John Van Voorst, London. Second Edition.

Newport G (1850a) Further observations on the habits of Monodontomerus, with some accounts of a new Acarus, *Heteropus ventricosus*, a parasite in the nests of *Anthophora retusa. Proceedings of the Linnean Society of London* 2(42), 70–71.

Newport G (1850b) Further observations on the habits of *Monodontomerus*, with some account of a new *Acarus, Heteropus ventricosus*, a parasite in the nests of *Anthophora retusa. Annals & Magazine of Natural History* (2)6, 395–396. [1 Nov 1850]

Newsome TM, Stephens D, Ballard G-A, Dickman CR, Fleming PJS (2013) Genetic profile of dingoes (*Canis lupus dingo*) and free-roaming domestic dogs (*C. l. familiaris*) in the Tanami Desert. *Australian Wildlife Research* 40, 196–206.

Newton AF Jr, Chandler DS (1989) World catalogue of the genera of Psephalidae (Coleoptera). *Fieldiana Zoology*, new series 53, 1–93. [28 Apr 1989]

Ng PLL, Guinot D, Davie PJF (2008) Systema brachyurorum: Part 1. An annoted checklist of extant brachyuran crabs of the world. *The Raffles Bulletin of Zoology* 17, 1–286.

Nicholson HA (1880) *Manual of Zoology for the Use of Students: With a general introduction on the principles of Zoology*. William Blackwood and Sons, Edinburgh & London.

Nikaido M, Rooney AP, Okada N (1999) Phylogenetic relationships among cetartiodactyls based on insertions of short and long interspersed elements: Hippopotamuses are the closes extant relatives of whales. *Proceedings of the National Academy of Sciences of the United States of America* 96(18), 10261–10266. [31 Aug 1999]

Nikolaev S, Montoya-Burgos JI, Margulies EH, NISC Comparative Sequencing Program, Rougemont J, Nyffeler B, Antonarakis SE (2007) Early history of mammals is elucidated with the ENCODE Multiple Species Sequencing Data. *PLOS Genetics* 3(1), 3–8. [Jan 2007]

Nilsson MA, Arnason U, Spencer PBS, Janke A (2004) Marsupial relationships and a timeline for marsupial relation in South Gondwana. *Gene* 340, 189–196.

Nilsson MA, Churakov G, Sommer M, Tran NV, Zemann A, Brosius J, Schmitz J (2010) Tracking marsupial evolution using archaic genomic retroposon insertions. PLoS Biology 8(7), 1–9. e1000436. [27 Jul 2010]

Nilsson S (1820) *Skandinavisk Fauna: en hanbok för jagare och zoologer*. Berlingska boktryckeriest, Lund. 1, 1–419.

Nilsson S (1838) Utkast till en systematisk indelning af phocacéerna. *Kongelige Vetenskaps-Academiens Handlingar, Stockholm* 1837, 235–240.

Nilsson S (1847) *Skandinavisk Fauna I – Däggdjuren*. C.W.K. Glerups Forlag, Lund.

Nishida S, Goto M, Pastene LA, Kanda N, Koike H (2007) Phylogenetic relationships among cetaceans revealed by Y-chromosome sequences. *Zoological Science* 24, 723–732.

Nishihara H, Hasegawa M, Okada N (2006) Pegasoferae, an unexpected mammalian clade revealed by tracking ancient retroposon insertions. *Proceedings of the National Academy of Sciences of the United States of America* 103(26), 9929–9934. [27 Jun 2006]

Nishiwaki M (1957) A list of marine mammals found in the seas adjacent to Japan. pp. 149–156, in Y Suehiro, Y Ohshima & Y Hiyama (eds.) *Collected Papers on Fisheries*. University of Tokyo Press, Tokyo.

Nishiwaki M (1963) Taxonomical consideration on genera of Delphinidae. *Scientific Reports of the Whales Research Institute, Tokyo* 17, 93–103.

Nishiwaki M (1972) General biology. pp. 3–204, in SH Ridgway (ed.) *Mammals of the Sea: Biology and Medicine*. Charles C Thomas, Springfield, IL.

Nishiwaki M, Kamiya T (1958) A beaked whale *Mesoplodon* stranded at iso Beach, Japan. *Scientific Reports of the Whales Research Institute, Tokyo* 13, 53–83.

Nishiwaki M, Norris KS (1966) A new genus, *Peponocephala*, for the odontocete cetacean species *Electra electra. Scientific Reports of the Whales Research Institute, Tokyo* 20, 95–99.

Nitzsch CL (1818) Die familien und gattungen der thierinsekten (Insecta Epizoica); als prodromus einer naturgeschichte der selben. *Magazin der Entomologie* 3, 261–316.

Norris CA (1994) The periotic bones of possums and cuscuses: Cuscus polyphyly and the division of the marsupial family Phalangeridae. *Zoological Journal of the Linnean Society* 111, 73–98. [May 1994]

Norris CA (1999) *Phalanger lullulae. Mammalian Species* 620, 1–4. [3 Dec 1999]

Norris CA, Musser GG (2001) Systematic revision of the *Phalanger orientalis* complex (Diprotodontia, Phalangeridae): A third species of lowland gray cuscus from New Guinea and Australia. *American Museum Novitates* 3356, 1–20. [31 Dec 2001]

Norton E (1872) Notes on North American Tenthredinidae, with descriptions of new species. *Transactions of the American Entomological Society* 4, 77–86.

Novacek MJ (1986) The skull of leptictid insectivorans and the higher-level classification of eutherian mammals. *Bulletin of the American Museum of Natural History* 183, 1–111. [29 Apr 1986]

Novacek MJ, Wyss AR (1986) Higher-level relationships of Recent eutherian orders: morphological evidence. *Cladistics* 2, 257–287.

Nuttal T (1834) *A Manual of the Ornithology of the United States and of Canada. The water birds*. Hilliard, Gray, & Co., Boston.

Nyakatura K, Birinda-Emonds ORP (2013) Updating the evolutionary history of Carnivora (Mammalia): a new species-level supertree complete with divergence time estimates. *BMC Biology* 10(12), 1–31.

O'Brien CW, Wibmer GJ (1982) Annotated checklist of the weevils (Curculionidae *sensu lato*) of North America, Central America, and the West Indies (Coleoptera: Curculionoidea). *Memoirs of the American Entomological Institute (Gainesville)* 34, 1–382.

O'Hara JE, Wood DM (2004) Catalogue of the Tachinidae (Diptera) of America North of Mexico. *Memoirs on Entomology. International* 18, 1–410.

O'Leary MA, Gatesy J (2008) Impact of increased character sampling on the phylogeny of Cetartiodactyla (Mammalia): Combined analysis including fossils. *Cladistics* 24, 397–442.

Ochsenheimer F (1810) *Die Schmetterlinge von Europa*. Gerhard. Fleischer dem Jüngern, Leipzig. Volume 3.

Ogawa T (1950) *Kujira no hanashi*. [*A narrative on cetaceans*] Chukoron Sha, Yokyo.

Ogilby JD (1892) *Catalogue of Australian Mammalia, with introductory notes on general mammalogy*. Australian Museum, Sydney. Catalogue No. 16.

Ogilby W (1831) On two new species of *Phalangista*, Cuv. *Proceedings of the Committee of Science and Correspondence of the Zoological Society of London* 1, 135–136. [Oct 1831]

Ogilby W (1832) On two new species of Mammalia from New Holland. *Proceedings of the Committee of Science and Correspondence of the Zoological Society of London* (1831) 1, 149–151. [22 Nov 1832]

Ogilby W (1836a) Observations upon the opposable power of the thumb in certain Mammals, considered as a zoological character, and on the natural affinities which subsist between the Bimana, Quadrumana, and Pedimana. *Proceedings of the Zoological Society of London* 4, 25–28. [9 Jun 1836]

Ogilby W (1836b) Remarks on some marsupials from the interior of New South Wales. *Proceedings of the Zoological Society of London* 3(1835), 191–192. [8 Apr 1836]

Ogilby W (1837a) Observations on the opposable power of the thumb in certain mammals, considered as a zoological a character; and on the natural affinities which subsist between the Bímana, Quadrùmana, and Pedímana. *The Magazine of Natural History, and Journal of Zoology, Botany, Mineralogy, Geology, and Meteorology* 1(2), 449–459. [1 Sep 1837]

Ogilby W (1837b) Zoological Society. *The Athenaeum* 527, 883. [2 Dec 1837]

Ogilby W (1837c) in Anon. Abstract untitled. P. 208. cols. 1–2 in *Zoologie. L'Echo du Monde Savant et L'Hermès*. No. 294. Volume 3. (2e division – Sciences naturelles et géographiques – no. 102). [23 Dec. 1837] Reference not seen.

Ogilby W (1837d) On the generic characters of ruminants. *Proceedings of the Zoological Society of London* 4(1836), 131–139. [27 Jun 1837]

Ogilby W (1838a) On a new species of marsupial animal by Major Mitchell on the banks of the Murray River in New South Wales. *Proceedings of the Zoological Society of London* 6, 25–27. [July 1838]

Ogilby W (1838b) On a New Phalanger (*Phalangista viverrina*), from Van Diemen's Land. *Proceedings of the Zoological Society of London* 5(1837), 131. [14 June 1838]

Ogilby W (1838c) Observations upon some recent communications of Mr. J. E. Gray, of the British Museum, to the *Annals of Natural History*; with description of two new kangaroos from Van Dieman's Land. *The Annals of Natural History; or. Magazine of Zoology, Botany, and Geology* 1(1), 216–221. [1 May 1838]

Ogilby W (1838d) Description of various species of the genus *Hypsiprymnus*. *Proceedings of the Zoological Society of London* 6, 62–63. [Jul 1838]

Ogilby W (1838e) On a new species of kangaroo. *Proceedings of the Zoological Society of London* 6, 23. [Jul 1838]

Ogilby W (1838f) Notice of certain Australian quadrupeds belonging to the Order Rodentia. *The London and Edinburgh Philosophical Magazine and Journal of Science* 12(3), 95–96.

Ogilby W (1838g) Notice on certain Australian quadrupeds, belonging to the Order Rodentia. *Transactions of the Linnean Society of London* (1)18, 121–132. [21 Jun 1838]

Ognev SI (1940) Zveri SSSR i prilezhashchikh stran: Gryzuny (Zveri vostochnoi Evropy i severnoi Azii). *Akademiya Nauk SSSR* 4, 1–615. [Mammals of the USSR and Adjacent Countries: Rodents (Mammals of Eastern Europe and Northern Asia).] [Reference not seen.]

Oken L (1807) No. Title. *Göttingische Gelehrte Anzeigen* 2, 1161–1168.

Oken L (1815) *Lehrbuch der Naturgeschichte*. Jena. Volume 3. Zoologie. 1.

Oken L (1816) *Lehrbuch der Naturgeschichte*. Jena. Volume 3. Zoologie. 2.

Oken L (1817) Cuviers und Okens zoologien neben einander gestellt. *Isis von Oken* 1, 1145–1185.

Oken L (1845) Abhandlunden der böhmischen Gesellschaft der Wissenschaften. *Isis Von Oken* 38, 866–871.

Oken L (1847) *Elements of Physiophilosophy*. From the Germ by Alfred Tulk. Ray Society, London.

Olfers I von (1818) Bemerkungen zu Illiger's Ueberblick der Säugthiere nach ihrer Verteilung uber die Welttheile, rücksichtlich der Südamericanischenn Arten (species). pp. 192–237, in WL von Eschwege (ed.) *Journal von Brasilien, Oder Vermischte Nachrichten aus Brasilien, auf Wissenschaftlichen Reisen Gesammelt*. Verlage des Gr. H. S. Landes-Industries-Comptoirs, Weimar, Germany. Volume 15. Part 2. Reference not seen.

Oliver WBR (1922) A review of the Cetacea of the New Zealand seas. *Proceedings of the Zoological Society of London* 92, 557–585. [28 Sep 1922]

Oliver WRB (1937) *Tasmacetus shepherdi*: a new genus and species of beaked whale from New Zealand. *Proceedings of the Zoological Society of London* 107B, 371–381. [Sep 1937]

Olsen O (1913) On the external characteristics and biology of Bryde's whale (*Balaenoptera brydei*) a new rorqual from the coast of South Africa. *Proceedings of the Zoological Society of London* 83, 1073–1090. [17 Dec 1913]

Olsen SJ (1985) *Origins of the Domestic Dog: The Fossil Record*. University of Arizona Press. Reference not seen.

Omura H (1959) Bryde's whale from the coast of Japan. *Scientific Reports of the Whales Research Institute, Tokyo* 14, 1–33.

Omura H (1975) Osteological study of the minke whale from the Antarctic. *Scientific Reports of the Whales Research Institute, Tokyo* 27, 1–36.

Orr RT (1953) Beaked whale (*Mesoplodon*) from California, with comments on taxonomy. *Journal of Mammalogy* 34, 239–249. [May 1953]

Örsted AS (1844) De regionibus marinis: Elementa Topographiae Historiconaturalis Freti Oeresund. Respondente ornatissimo E. Petit; publice defendere studebit A.S. Ørsted. J.C. Scharling, Copenhagen.

Osbeck P (1765) *Reise nach Ostindien und China* (Anleitung, &c.): nebst O. Toreens Reise nach Suratte und C. G. Ekebergs Nachricht von der Landwirthschaft der Chineser. Verlegts Johann Christian Koppe, Rostock.

Osborn HF (1888) On the structure and classification of the Mesozoic Mammalia. *Journal of the Academy of Natural Sciences* (2)9, 186–265.

Osborn HF (1910) *The Age of Mammals in Europe, Asia and North America*. Macmillan Co., New York.

Osborne MJ, Christidis L (2001) Molecular phylogenetics and Australo-Papuan possums and glider (Family Petauridae). *Molecular Phylogenetics and Evolution* 20, 211–224. [Aug 2001]

Osborne MJ, Christidis L (2002) Systematics and biogeography of pygmy possums (Burramyidae: *Cercartetus*). *Australian Journal of Zoology* 50, 25–37.

Osgood WH (1921) A monographic study of the American marsupial, *Caenolestes*. *Field Museum of Natural History. Zoological Series* 14, 1–162.

Owen R (1834) On the generation of the marsupial animals, with a description of the impregnated uterus of the kangaroo. *Philosophical Transactions of the Royal Society of London* 124, 333–364.

Owen R (1838a) On the jaws of *Thylacotherium Prevostii* (Valenciennes) from Stonesfield. *Proceedings of the Geological Society of London* 3, 5–9.

Owen R (1838b) In TL Mitchell. *Three Expeditions into the Interior of Eastern Australia*; with descriptions of the recently explored region of Australia Felix, and of the present colony of New South Wales. T. & W. Boone, London. Volume 2.

Owen R (1838c) Description of the osteology of the Marsupialia. *The Athenaeum* 1838, 747. [13 Oct 1838]

Owen R (1839a) Outlines of a classification of the Marsupialia. *Proceedings of the Zoological Society of London* 7, 5–19. [Jul 1839]

Owen R (1839b [1838–1840]) Part 1. Fossil Mammalia. In C Darwin (ed.) (1838–1843) *Zoology of the Voyage of H.M.S Beagle, under the command of Captain Fitzroy, during the years 1832 to 1836*. Smith, Elder & Co., London. [May 1839]

Owen R (1840) Outlines of a classification of the Marsupialia. *Transactions of the Zoological Society of London* 2, 315–333. [6 Apr 1840]

Owen R (1840–1845) *Odontography : or, a treatise on the Comparative Anatomy of the Teeth, their physical relations, mode of development and microscopic structure in the Vertebrate Animals*. Hippolyte, London. Volume 2.

Owen R (1841) On the osteology of the Marsupialia. *Transactions of the Zoological Society of London* 2, 379–408. [15 May 1841]

Owen R (1842) *Description of the Skeleton of an Extinct Gigantic Sloth, Mylodon robustus, Owen, with observations on the osteology, natural affinities, and probable habits of the megatheroid quadrupeds in general*. John Van Voost, London.

Owen R (1843) Notice of a new species of seal (*Stenorhynchus serridens*). *Annals & Magazine of Natural History* (1)12, 331–332. [1 Nov 1843]

Owen R (1845a) *Descriptive and illustrated catalogue of the fossil organic remains of Mammalia and Aves contained in the Museum of the Royal College of Surgeons of England*. Richard and John E. Taylor, England.

Owen R (1845b) Exhibited wombats. *Proceedings of the Zoological Society of London* 13, 82–83. [Oct 1845]

Owen R (1846) *A History of British Fossil Mammals and Birds*. John Van Voorst, London.

Owen R (1847a) Marsupialia. pp. 257–330, in RB Todd (ed.) *The Cyclopaedia of Anatomy and Physiology*. Sherwood, Gilbert & Piper, London. Volume 3. INS-PLA.

Owen R (1847b) Appendix No. VI. Notes on the characters of the skeleton of a Dugong. (*Halicore Australis*), from the north coast of Australia, indicative of its specific distinctness from the *Halicore indicus* and *Halicore Tabernaculi*. pp. 323–331, in JB Jukes (ed.) *Narrative of the Surveying Voyage of HMS Fly*, commanded by Captain FP Blackwood, R.N. in Torres Strait, New Guinea, and other islands of the Eastern Archipelago, during the years 1842–1846. T. & W. Boone, London. Volume 2.

Owen R (1848) Description of teeth and portions of jaws of two extinct anthracotheroid quadrupeds (*Hyopotamus vectianus* and *Hyop. bovinus*) discovered by the Marchioness of Hastings in the Eocene deposits on the N.W. coast of the Isle of Wight: with an attempt to develop Cuvier's idea of the classification of the pachyderms by the number of toes. *Quarterly Journal of the Geological Society of London* 4, 103–141.

Owen R (1849 [1849–1852]) Teeth. Comparative anatomy. pp. 864–935, in RB Todd (ed.) *The Cyclopaedia of Anatomy and Physiology*. Longman, Brown, Green, Longmans and Roberts, London. STA-WRI. Volume 4. Part 2.

Owen R (1853a) *Descriptive Catalogue of the Osteological Series Contained in the Museum of the Royal College of Surgeons of England*. Volume 1. Pisces, Reptilia, Aves, Marsupialia. Taylor and Francis, London.

Owen R (1853b) Description of some species of the extinct genus *Nesodon*, with remarks on the primary group (Toxodontia) of the hoofed quadrupeds to which that genus is referable. *Philosophical Transactions of the Royal Society of London* 143, 291–309.

Owen R (1858a) On the characters, principles of division, and primary groups of the Class Mammalia. *Journal of the*

Proceedings of the Linnean Society of London. Zoology (Jena, Germany) 2, 1–37.

Owen R (1858b) Odontology. pp. 407–484, in *Encyclopedia Britannica, or Dictionary of Arts, Sciences and General Literature*. 8th Edition. Volume 16. Reference not seen.

Owen R (1859) *On the Classification and Geographical Distribution of the Mammalia* being the lecture on Sir Robert Reads Foundation, delivered before the University of Cambridge, in the Senate-House, May 10, 1859. To which is added an appendix 'On the Gorilla,' and 'On the extinction and transmutation of species.' John W. Parker, London.

Owen R (1866a) *On the Anatomy of Vertebrates. Volume 2. Birds and Mammals*. Longmans, Green & Co., London.

Owen R (1866b) On some Indian Cetacea collected by Walter Eliot Esq. *Transactions of the Zoological Society of London* 6, 17–47. [15 Aug 1866]

Owen R (1868) *On the Anatomy of Vertebrates. Volume 3. Mammals*. Longmans, Green & Co., London.

Owen R (1872) On the fossil mammals of Australia. Part IV. Genus *Phascolomys* Geoffroy. *Philosophical Transactions of the Royal Society of London* 162, 173–196.

Owen R (1873a) On the fossil mammals of Australia. Part VIII. Family Macropodidae; genera *Macropus, Osphranter, Phascolagus, Sthenurus*, and *Protemnodon. Proceedings of the Royal Society of London* 21, 128. [23 Jan 1873]

Owen R (1873b) On the fossil mammals of Australia. Part VIII. Family Macropodidae: Genera *Macropus, Osphranter, Phascolagus, Sthenurus*, and *Protemnodon. Nature* 7(170), 255. [30 Jan 1873]

Owen R (1873c) On the fossil mammals of Australia. Part IX, Macropodidae: Genera: *Macropus, Pachysiagon, Leptosiagon, Procoptodon* and *Palorchestes. Proceedings of the Royal Society of London* 21, 386–387.

Owen R (1874a) On the fossil mammals of Australia. Part VIII. Family Macropodidae: Genera *Macropus, Osphanter, Phascolagus, Sthenurus* and *Protemnodon. Philosophical Transactions of the Royal Society of London* 164, 245–287.

Owen R (1874b) On the fossil mammals of Australia. Part IX. Family Macropodidae. Genera: *Macropus, Pachysaigon, Leptosaigon, Procoptodon* and *Palorchestes. Philosophical Transactions of the Royal Society of London* 164, 783–803.

Owen R (1875a) On *Prorastomus sirenoides* (Ow.). Part II. *Quarterly Journal of the Geological Society of London* 31, 559–567.

Owen R (1875b) On *Dinornis* (Part XX): containing a restoration of the skeleton of *Cnemiornis calcitrans*, Ow., with remarks on its affinities in the Lamellirostral group. *Transactions of the Zoological Society of London* 9, 253–272. [May 1875]

Owen R (1877a) *Researches on the fossil remains of the extinct mammals of Australia; with a notice of the extinct marsupials of England*. J. Erxleben, London. Volume 1.

Owen R (1877b) Bibliographical Notice - Researches on the fossil remains of the extinct mammals of Australia; with a notice of the extinct marsupials of England. *Annals & Magazine of Natural History* (4)20, 359–361. [1 Oct 1877]

Owen R (1877c) On a new marsupial from Australia. *Annals & Magazine of Natural History* (4)20, 542. [1 Dec 1877]

Owen R (1878a) On *Hypsiprymnodon*, Ramsay, a genus indicative of a distinct family (Pleopodidae) in the diprotodont section of the Marsupialia. *Transactions of the Linnean Society of London. Zoology (Jena, Germany)* (2)1, 573–582.

Owen R (1878b) On a new marsupial from Australia. *Annals & Magazine of Natural History* (5)1, 103. [1 Jan 1878]

Özdikmen H (2008) Nomenclatural changes for nine crustacean genera (Crustacea: Copepoda). *Munis Entomology & Zoology* 3(1), 265–274.

Paclt J (1960) The 'Classes' of Rafinesque and the modern biosystematics. *Taxon* 9(2), 47–49. [Feb 1960)]

Pagenstecher HA (1885a) In GA Fischer, Auftrage der geographischen Gesellschaft in Hamburg unternommenen reise in das Massai-Land gesammelten Säugethiere. *Jahrbuch der Hamburgischen Wissenschaftlichen Anstalten* 2, 29–46.

Pagenstecher HA (1885b) *Megaloglossus woermanni* nov. gen. et. spec. *Zoologischer Anzeiger* 8, 245. [28 Apr 1885]

Pages M, Chaval Y, Herbreteau V, Waengsothorn S, Cosson J-F, Hugot J-P, Morand S, Michaux J (2010) Revisiting the taxonomy of the Rattini tribe: a phylogeny-based delimitation of species boundaries. *BMC Evolutionary Biology* 10(184), 1–27.

Painter J, Krajewski C, Westerman M (1995) Molecular phylogeny of the marsupial genus *Planigale* (Dasyuridae). *Journal of Mammalogy* 76, 406–413. [May 1995]

Palisot de Beauvois A-M-F-J (1820 [-1805–1821]) *Insectes Recueillis en Afrique et en Amérique, dans les royaumes d'Oware et de Benin, à Saint-Dominique et dans les États-Unis, pendant les années 1786–1797*. De Fain et Compagnie, Paris.

Pallas PS (1766) *Miscellanea Zoologica, quibus novæ imprimis atque obscuræ animalium species describuntur et observationibus iconibusque illustrantur*. Petrun van Cleef, Hagae Comitum, Hagae Comitum.

Pallas PS (1767) Vespertiliones in genre. pp. 1–35, in *Spicilegia Zoologica quibus novae imprimis et obscurae animalium species iconibus, descriptionibus atque commentariis illustrantur*. Tomus 1. Fascicle Tertius. G. A. Lange, Berolini, Germany.

Pallas PS (1778–1779) *Novae Species Quadrupedum e Glirium Ordine* cum illustrationibus variis complurium ex hoc ordine animalium. Fasciculus 1. Wolfgang Waltheri, Erlangae. pp. 1–69 (1778), 70–388 (1779).

Pallas PS (1780a) *Galeopithecus volans*, camellii Descriptus. *Acta Academiae Scientiarum Imperialis Petropolitanae* 1, 208–222.

Pallas PS (1780b) *Spicilegia Zoologica quibus novae imprimis et obscurae animalium species iconibus, descriptionibus atque commentariis illustrantur*. Fasciculus XIV. Joachimum Pauli, Berolini, Germany.

Pallas PS (1811) In Tilesius, W. G. von. Piscium Camtschaticorum descriptiones et icones. *Mémoires de l'Académie Impériale des Sciences de St. Pétersbourg* 3, 225–285.

Pallas PS (1831) *Zoographia rosso-asiatica, sistens omnium animalium in extenso Imperio rossico, et adjacentibus maribus observatorum recensionem, domicilia, mores et descriptiones, anatomen atque icones plurimorum*. Academiae scientiarum impress, Petropoli. Volume 3.

Palmer TS (1892) A new generic name for the Bering Sea fur-seal. *Proceedings of the Biological Society of Washington* 7, 156. [Jul 1892]

Palmer TS (1895a) The generic names of the three-toed echidna. *Science* 1(19), 518–519. [10 May 1895]

Palmer TS (1895b) The earliest name of Steller's sea cow and dugong. *Science* (2)2(40), 449–450. [4 Oct 1895]

Palmer TS (1897a) A list of the generic and family names of rodents. *Proceedings of the Biological Society of Washington* 11, 241–270. [17 Dec 1897]

Palmer TS (1897b) Notes on the nomenclature of four genera of tropical American mammals. *Proceedings of the Biological Society of Washington* 11, 173–174. [9 Jun 1897]

Palmer TS (1898) Random notes on the nomenclature of the Chiroptera. *Proceedings of the Biological Society of Washington* 12, 109–114. [30 Apr 1898]

Palmer TS (1899a) On *Thylacomys*, Owen. *Annals & Magazine of Natural History* (7)4, 300–302. [1–2 Oct 1899]

Palmer TS (1899b) Review of Dr. E. L. Trouessart's *Catalogus Mammalium. Science* (2)10(249), 491–495. [6 Oct 1899]

Palmer TS (1899c) The family name of the dormice. *Science* 10(247), 412–413. [22 Sep 1899]

Palmer TS (1899d) Notes on three genera of dolphins. *Proceedings of the Biological Society of Washington* 13, 23–24. [31 Jan 1899]

Palmer TS (1904) *Index Generum Mammalium: A list of the genera and families of mammals (Index Genera Mammalium). North American Fauna*. No. 23. U.S. Dept. of Agriculture, Washington. [23 Jan 1904]

Palmer TS (1906) *Ammomys* and other compounds of *Mys*. *Proceedings of the Biological Society of Washington* 19, 97. [4 Jun 1906]

Panzer GW (1813) *Index entomologicus : sistens omnes insectorum species. Fauna insectorum Germanica descriptas atque delineatas secundum methodum Fabricianam: adiectis emendationibus, observationibus. Pars I. Eleutherata*. Norimbergae.

Pardini AT, O'Brien PCM, Fu B, Bonde RK, Elder FFB, Ferguson-Smith MA, Yang F, Robinson TJ (2007) Chromosome painting among Proboscidea, Hyracoidea and Sirenia: support for Paenungulata (Afrotheria, Mammalia) but not Tethytheria. *Proceedings of the Royal Society of London. Series B, Biological Sciences* 274, 1333–1340. [Online 20 Mar 2007]

Parker TJ, Haswell WA (1897) *A Text-Book of Zoology*. Macmillan & Co., London. Volume 2.

Parkes KC (1982) Subspecific taxonomy: unfashionable does not mean irrelevant. *The Auk* 99, 596–598.

Parnaby HE (1987) Distribution and taxonomy of the long-eared bats, *Nyctophilus gouldi* Tomes, 1858 and *Nyctophilus bifax* Thomas, 1915 (Chiroptera: Vespertilionidae) in eastern Australia. *Proceedings of the Linnean Society of New South Wales* 109, 153–174. [22 Dec 1987]

Parnaby HE (1992) An interim guide to the identification of bats in south-eastern Australia. *Technical Reports of the Australian Museum* 8, 1–33. [12 Oct 1992]

Parnaby H (1995) Identification criteria and taxonomic clarification of some problematic bat species in north-eastern New South Wales. Appendix A. *In* NSW National Parks and Wildlife Service *Vertebrates of Upper North East New South Wales*. NSW National Parks and Wildlife Service, Hurstville. [Aug 1995]

Parnaby HE (1996) The common name and common misconceptions about the common bent-wing bat. *Australian Bat Society Newsletter* 7, 35–38. [Oct 1996]

Parnaby HE (2002) A new species of long-eared bat (*Nyctophilus*: Vespertilionidae) from New Caledonia. *Australian Mammalogy* 23, 115–124. [30 Apr 2002]

Parnaby HE (2009) A taxonomic review of the Australian greater long-eared bat previously known as *Nyctophilus timoriensis* (Chiroptera: Vespertilionidae) and some associated taxa. *Australian Zoologist* 35, 39–81.

Parnaby H, Ingleby S, Divljan A (2015) Taxonomic status of *Podabrus albocaudatus* Krefft, 1872 and declaration of *Sminthopsis granulipes* Troughton, 1932 (Marsupialia: Dasyuridae) as a protected name for the White-tailed Dunnart from Western Australia. *Zootaxa* 3904, 283–292. [6 Jan 2015].

Parr AE (1931) Deepsea fishes from off the western coast of North and Central America. With keys to the genera *Stomias, Diplophos, Melamphaes* and *Bregmaceros*, and a revision of the *Macropterus* group of the genus *Lampanyctus*. *Bulletin of the Bingham Oceanographic Collection, Yale University* 2(4), 1–53.

Parrington FR (1974) The problem of the origin of the monotremes. *Journal of Natural History* 8, 421–426.

Partington CF (1837) *The British Cyclopaedia of Natural History: Combining a scientific classification of animals, plants, and minerals; with a popular view of their habits, economy and structure*. Volume 3. W.M.S. Orr & Co., London.

Partington CF (1838) *The British Cyclopaedia of Arts, Sciences, History, Geography, Literature, Natural History, and Biogeography. Volume 3. Literature, Geography, and History*. AA to FEA.W.M.S. Orr & Co., London.

Parton GF (1952) Encounter with native cat. *Western Australian Naturalist* 3, 93. [15 Mar 1952]

Pasitschniak-Arts M, Marineli L (1998) *Ornithorhynchus anatinus*. *Mammalian Species* 585, 1–9. [1 Jun 1998]

Pavlinov IYa, Rossolimo OL (1987) *Sistematika mlekopitayushchikh SSSR*. [*Taxonomy of the Mammals of the USSR*] Moscow University Press, Moscow.

Paxton JR, Gates JE, Bray DJ, Hoese DF (2006) Notacanthidae. pp. 232–233, in DF Hoese, DJ Bray, GR Allen, JR Paxton, A Wells & PL Beesley (eds.) *Zoological Catalogue of Australia. Volume 35. Part 1. Fishes*. CSIRO Publishing / Australian Biological Resources Study (ABRS), Melbourne and Canberra.

Payne RB (2005) *The Cuckoos*. Oxford University Press, Oxford.

Peale TR (1848) The *United States Exploring Expedition during the years 1838, 1839, 1840, 1841, 1842 under the command of Charles Wilkes, U.S.N. Volume 8. Mammals and Ornithology*. C. Sherman, Philadelphia, Pennsylvania. + Atlas Mammalogy.

Pearson J (1950) The relationships of the Potoroidae to the Macropodidae (Marsupialia). *Papers and Proceedings of the Royal Society of Tasmania* 1949, 211–229. [15 Sep 1950]

Peavot H (1937) List of dates of publication of the early parts of the Society's 'Transactions'. *Proceedings of the Zoological Society of London* 107(Series A), 83–84.

Pelham JP (2008) A catalogue of the butterflies of the United States and Canada, with a complete bibliography of the descriptive and systematic literature. *Journal of Research on the Lepidoptera* 40, 1–658.

Pennant T (1781) *History of the Quadrupeds*. B. White, London. 2 Volumes. Second Edition.

Pennant T (1783) *History of the Quadrupeds*. B. White, London. 2 Volumes. Third Edition.

Penny D, Hasegawa M (1997) The platypus in its place. *Nature* 387(6633), 549–550. [5 Jun 1997]

Pernety AJ (1769) Journal historique d'un voyage fait aux Îles Malouïnes en 1763 & 1764, pour les reconnoître, & y former un établissement; et de deux voyages au Détroit de Magellan, avec une rélation sur les Patagons. Etienne de Bourdeaux, Berlin. Volume 2.

Péron F (1807) *Voyage de Découvertes aux Terres Australes*, Exécuté par Ordre de Sa Majesté l'Empereur et roi, sur les Corvettes le Géographe, le Naturaliste, et la Goelette le Casuarina, Pendant les Années 1800, 1801, 1802 et 1804. l'Imprimerie Royale, Paris. Plates.

Péron F (1816) *Voyage de Découvertes aux Terres Australes*, Exécuté par Ordre de Sa Majesté l'Empereur et roi, sur les Corvettes le Géographe, le Naturaliste, et la Goelette le Casuarina, Pendant les Années 1800, 1801, 1802 et 1804. l'Imprimerie Royale, Paris. Volume 2. Historique.

Péron F, Lesueur CA (1807) in F Péron & LCD de Freycinet (1807–1816) *Voyage de Découvertes aux Terres Australes*, Exécuté sur les corvettes le Géographe, le Naturaliste, et la goelette le Casuarina, pendant les années 1800, 1801, 1802 et 1804. l'Imprimerie Royale, Paris. Volume 1. Historque.

Perrin WF (1975) Variation of spotted and spinner porpoise (genus *Stenella*) in the eastern tropical Pacific and Hawaii. *Bulletin of the Scripps Institution of Oceanography* 21, 1–206.

Perrin WF (1989) *Dolphins, Porpoises, and Whales - An Action Plan for the Conservation of Biological Diversity: 1988–1992*. IUCN, Gland, Switzerland. Second Edition.

Perrin WF (1990) Subspecies of *Stenella longirostris* (Mammalia: Cetacea: Delphinidae). *Proceedings of the Biological Society of Washington* 103, 453–463. [21 Mar 1990]

Perrin WF (1998) *Stenella longirostris*. *Mammalian Species* 599, 1–7. [4 Dec 1998]

Perrin WF (2001) *Stenella attenuata*. *Mammalian Species* 683, 1–8. [26 Dec 2001]

Perrin WF, Brownell RL (2009) Minke whales *Balaenoptera acutorostrata* and *B. bonaerensis*. pp. 733–735, in WF Perrin, B Würsig & JGM Thewissen (eds.) *Encyclopedia of Marine Mammals*. Academic Press, Amsterdam. Second Edition.

Perrin WF, Best PB, Dawbin WH, Balcomb KC, Gambell R, Ross GJB (1973) Rediscovery of Fraser's dolphin *Lagenodelphis hosei*. *Nature* 241(5388), 345–350. [2 Feb 1973]

Perrin WF, Mitchell ED, Mead JG, Caldwell DK, van Bree PJH (1981) *Stenella clymene* a rediscovered tropical dolphin of the Atlantic. *Journal of Mammalogy* 62, 583–598. [Aug 1981]

Perrin WF, Mitchell ED, Mead JG, Caldwell DK, Caldwell MC, van Bree PJH, Dawbin WH (1987) Revision of the spotted dolphins, *Stenella* spp. *Marine Mammal Science* 3, 99–170. [Apr 1987]

Perrin WF, Dolar MLL, Robineau D (1999) Spinner dolphins (*Stenella longirostris*) of the western Pacific and Southeast Asia: pelagic and shallow-water forms. *Marine Mammal Science* 15, 1029–1053. [Oct 1999]

Perrin WF, Mead JG, Brownell RL Jr (2009) Review of the evidence used in the description of currently recognized cetacean subspecies. NOAA-TM-NMFS-SWFSC-450. [Dec 2009]

Perrin WF, Rosel PE, Cipriano F (2013) How to contend with paraphyly in the taxonomy of the delphinine cetaceans? *Marine Mammal Science* 29, 567–588.

Perry G (1810–1811) *Arcana;* or the Museum of Natural History, British Museum, London. Volume 1. [See Petit (2009: 12–16) for specific dates of individual species]

Peters JL (1964) *Check-List of Birds of the World*. Harvard University Press, Cambridge. Volume 4(1).

Peters W (1846) Über neue säugethiergattungen aus den Ordnungen der insectenfresser und nagetheriere. *Bericht über die zur Bekanntmachung geeigneten Verhandelingen der Königlich-Preussischen Akademie der Wissenschaften zu Berlin* 11, 257–259.

Peters W (1847) Neue säugethieregattung aus der ordung der insektenfresser. *Bericht über die zur Bekanntmachung geeigneten Verhandelingen der Königlich-Preussischen Akademie der Wissenschaften zu Berlin* 12, 36–38.

Peters W (1852a) Einige neue säugethiere und flussfische aus Mossambique vor. *Bericht über die zur Bekanntmachung geeigneten Verhandlungen der Königlich Preussischen Akademie der Wissenschaften zu Berlin* for 1852 (1853), 273–276.

Peters W (1852b) *Naturwissenschaftliche Reise nach Mossambique. Zoologie. I. Säugethiere*. Druck und Verlag von Georg Reimer, Berlin.

Peters W (1859a) Über ein neues Flugbeutelthier, *Petaurus* (*Belideus*), aus dem südlichen Theile von Neuholland. *Monatsberichte der Königlichen Preussischen Akademie der Wissenschaften zu Berlin* for 1859 (1860), 14–15.

Peters W (1859b) Neue Beiträge zur Kenntniss der Chiropteren. *Monatsberichte der Königlichen Preussischen Akademie der Wissenschaften zu Berlin* for 1859 (1860), 222–225.

Peters W (1860a) Über einige merkwürdige Nagethiere (*Spalacomys indicus, Mus tomentosus* und *Mus squamipes*) des Königl. zoologischen Museums. *Abhandlungen der Königlichen Akademie der Wissenschaften zu Berlin* for 1860 (1861), 139–156.

Peters W (1860b) Über die chiropterengattung *Nyctophilus*. *Physikalische Abhandlungen der Königlichen Akademie der Wissenschaften zu Berlin* for 1860 (1861), 123–137.

Peters W (1861a) Über den *Pteropus stramineus* Geoffroy. *Monatsberichte der Königlichen Preussischen Akademie der Wissenschaften zu Berlin* for 1861 (1862), 423–424.

Peters W (1861b) Über die von Hrn. F. Jagor bisher auf Malacca, Borneo, Java und den Philippinen gesammelten Säugethiere. *Monatsberichte der Königlichen Preussischen Akademie der Wissenschaften zu Berlin* for 1861 (1862), 706–712.

Peters W (1862) Über einen neuen Flederhund, *Pteropus scapulatus*, aus Neuholland. *Monatsberichte der Königlichen Preussischen Akademie der Wissenschaften zu Berlin* for 1862 (1863), 574–576.

Peters W (1863) Über die Säugethiergattung *Solenodon*. *Abhandlungen der Königlichen Akademie Wissenschaften zu Berlin* for 1863 (1864), 1–22.

Peters W (1865a) Abbildungen zu einer monographie der chiropteren vor und gab eine Übersicht der von ihm befolgten systematischen Ordnung der hieher gehörigen Gattungen. *Monatsberichte der Königlichen Preussischen Akademie der Wissenschaften zu Berlin* for 1865 (1866), 256–258.

Peters W (1865b) Über die zu den *Vampyri* gehörigen Flederthiere und über die natürliche Stellung der Gattung *Antrozous*. *Monatsberichte der Königlichen Preussischen Akademie der Wissenschaften zu Berlin* for 1865 (1866), 503–524.

Peters W (1865c) Über Flederthiere (*Vespertilio soricinus* Pallas, *Choeronycteris* Lichtenst., *Rhinophylla pumilio* nov. gen., *Artibeus fallax* nov. sp., *A. concolor* nov. sp., *Dermanura quadrivittatum* nov. sp., *Nycteris grandis* n. sp.). *Monatsberichte der Königlich Preussischen Akademie der Wissenschaften zu Berlin* for 1865 (1866), 351–359.

Peters W (1865d) Über flederthiere brasilianischen, von Spix beschriebenen Flederthiere. *Monatsberichte der Königlich Preussischen Akademie der Wissenschaften zu Berlin* for 1865 (1866), 568–588.

Peters W (1865e) Über einige weniger bekannte Flederthiere (*Phyllostoma brachystoma, Coelops, Furia, Lasionycteris*). *Monatsberichte der Königlichen Preussischen Akademie der Wissenschaften zu Berlin* for 1865 (1866), 641–648.

Peters W (1866a) Über einige neue oder weniger bekannte Flederthiere. *Monatsberichte der Königlich Preussischen Akademie der Wissenschaften zu Berlin* for 1866 (1867), 16–25.

Peters W (1866b) Über die Ohrenrobben (Seelöwen und Seebären). Otariae, insbesondere über in den sammlungen zu Berlin benfindlichen Arten. *Monatsberichte der Königlich Preussischen Akademie der Wissenschaften zu Berlin* for 1866 (1867), 261–281.

Peters W (1867a) Über eine neue Gattung von Nagern, *Uromys*, aus Nordaustralien. *Monatsberichte der Königlichen Preussischen Akademie der Wissenschaften zu Berlin* for 1867 (1868), 343–345.

Peters W (1867b) Herpetologische Notizen. *Monatsberichte der Königlich Preussischen Akademie der Wissenschaften zu Berlin* for 1867 (1868), 13–37.

Peters W (1867c) Über Flederthiere (*Pteropus gouldii, Rhinolophus deckenii, Vespertilo lobipes, Vesperugo temminckii*) und Amphibien (*Hypsilurus godeffroyi, Lygosoma scutatum, Stenostoma narirostre, Onychocephalus unguirostris, Ahaetulla polylepis, Pseudechis scutellatus, Hoplobatrachus reinhardtii, Hyla coriacea*). *Monatsberichte der Königlich Preussischen Akademie der Wissenschaften zu Berlin* for 1867 (1868), 703–712.

Peters W (1867d) Über die Flederhunde, *Pteropi*, und insbesondere über die Arten der Gattung *Pteropus*. *Monatsberichte der Königlichen Preussischen Akademie der Wissenschaften zu Berlin* for 1867 (1868), 319–333.

Peters W (1867e) On some Mammalia collected by Capt. A.C. Beaven, C.M.Z.S., at Moulmein, Burmah. *Proceedings of the Zoological Society of London* 34(1866), 426–430. [Apr 1867]

Peters W (1867f) Über die zu den gattungen *Mimon* und *Saccopteryx* gehörigen flederthiere. *Monatsberichte der Königlich Preussischen Akademie der Wissenschaften zu Berlin for* 1867, 469–481.

Peters W (1867g) On *Taphozous flaviventris*, Gould, a new species of bat from Australia. *Proceedings of the Zoological Society of London* 34(1866), 430. [Apr 1867]

Peters W (1867h) Fernere Mittheilungen zur Kenntniss der Flederthiere, namentlich über Arten des Leidener und Britischen museums. *Monatsberichte der Königlich Preussischen Akademie der Wissenschaften zu Berlin for* 1866, 672–681.

Peters W (1867i) Nachtrag zu seiner Abhandlung uber die Ohrenrobben, Otariae. *Monatsberichte der Königlich Preussischen Akademie der Wissenschaften zu Berlin for* 1866, 665–672.

Peters W (1868a) Mittheilung über eine neue Nagergattung, *Chiropodomys penicillatus*, so wie über einige neue oder weniger bekannte Amphibien und Fische. *Monatsberichte der Königlich Preussischen Akademie der Wissenschaften zu Berlin* for 1868 (1869), 448–460.

Peters W (1868b) Mittheilung über die von dem Hrn. Marsquis Giacomo Doria in Sarawak auf Borneo gesammelten Flederthiere. *Monatsberichte der Königlich Preussischen Akademie der Wissenschaften zu Berlin* for 1868 (1869), 626–627.

Peters W (1869) Über neue oder weniger bekannte Flederthiere, besonders des Pariser Museums. *Monatsberichte der Königlich Preussischen Akademie der Wissenschaften zu Berlin* for 1869 (1870), 391–406.

Peters W (1870) Eine monographische übersicht der chiropterengattungen *Nycteris* und *Atalapha* vor. *Monatsberichte der Koniglich Preussischen Akademie der Wissenschaften zu Berlin* for 1870 (1871), 900–914.

Peters W (1871) Über die Gattungen und Arten der Hufeisennasen, Rhinolophi. *Monatsberichte der Königlich Preussischen Akademie der Wissenschaften zu Berlin for* 1871, 301–332.

Peters W (1872a) Mittheilung über neue Flederthiere (*Phyllorhina micropus, Harpyiocephalus Huttoni, Murina grisea, Vesperugo micropus, Vesperus* (*Marsipoloemus*) *albigularis, Vesperus propinquus, tenuipinnis*. *Monatsberichte der Königlich Preussischen Akademie der Wissenschaften zu Berlin* for 1872 (1873), 256–264.

Peters W (1872b) Über den *Vespertilio calcaratus* Prinz zu Wied und cine neue gattung der flederthiere, *Tylonycteris*. *Monatsberichte der Koniglich Preussischen Akademie der Wissenschaften zu Berlin* for 1872 (1873), 699–706.

Peters W (1874a) Diagnosi di tre nuovi mammiferi della Nuova Guinea ed Isole Key. *Annali del Museo Civico di Storia Naturale di Genova* 6(1), 303.

Peters W (1874b) Über eine neue art von flederthieren, *Promops bonariensis* und uber *Lophuromys*, eine Nagergattung von Westafrika. *Monatsberichte der Koniglich Preussischen Akademie der Wissenschaften zu Berlin* for 1874 (1875), 232–234.

Peters W (1875a) Abbildungen von zwei neuen Beuteltieren aus Neu-Guinea vor, welche dem Museo civico von Genoa. *Sitzungs-Berichte der Gesellschraft Naturforschender Freunde zu Berlin* 1875, 73.

Peters W (1875b) Über Dasymys, eine neue gattung von murinen nagethieren aus Sudafrika. *Monatsberichte der Koniglich*

Preussischen Akademie der Wissenschaften zu Berlin for 1875 (1876), 12–14.

Peters W (1875c) Über eine neue art von seebären, *Arctophoca gazella*, von den Kerguelen-Inseln. *Monatsberichte der Königlichen Preussischen Akademie den Wissenschaften zu Berlin* for 1875 (1876), 393–399.

Peters W (1876a) Über die von S.M.S. Gazelle gesammelten Säugethiere aus den Abtheilungen der Nager, Hufthiere, Sirenen, Cetacean und Beutelthiere. *Monatsberichte der Königlichen Preussischen Akademie der Wissenschaften zu Berlin* for 1876 (1877), 355–366.

Peters W (1876b) Mittheilung über die Pelzrobbe von den Inseln St. Paul und Amsterdam und über die von S.M.S. Gazelle mitgebrachten flederthiere. *Monatsberichte der Koniglich Preussischen Akademie der Wissenschaften zu Berlin* for 1876 (1877), 315–319.

Peters W (1877) Über die Ohrenrobben, Otariae, ais Nachtra zu seiner im vorigen Jahre über diese Thiere gelesenen Abhandlung. *Monatsberichte der Koniglich Preussischen Akademie der Wissenschaften zu Berlin* for 1877 (1878), 505–507.

Peters W (1881) Über die Chiropterengattung *Mormopterus* und die dahin gehörigen Arten. *Monatsberichte der Königlich Preussischen Akademie der Wissenschaften zu Berlin* for 1881 (1882), 482–485.

Peters W (1906) *Museum Zoologicum Berolinzuse. Chiroptera.* Plates only. Berlin. No text was produced for this title as the author died.

Peters W, Doria G (1875) Diagnosi di alcune nuove specie di Marsupiali appartenenti all fauna Papuana. *Annali del Museo Civico di Storia Naturale di Genova* 7(1), 541–544.

Peters W, Doria G (1876) *Tachyglossus bruijnii*, n. sp. *Annali del Museo Civico di Storia Naturale di Genova* 9(1), 183–187.

Peters W, Doria G (1880) Enumerazione dei mammiferi raccolti da O. Beccari, L.M. D'Albertis ed A.A. Bruijn nella Nuova Guinea propriamente detta. *Annali del Museo Civico di Storia Naturale di Genova* 16(1), 665–707.

Peterson RL (1965) A review of the flat-headed bats of the family Molossidae from South America and Africa. *Life Sciences Contributions. Royal Ontario Museum* 64, 1-32.

Peterson RL (1981) Systematic variation in the tristis group of the bent-winged bats of the genus *Miniopterus* (Chiroptera: Vespertilionidae). *Canadian Journal of Zoology* 59, 828–843.

Peterson RL (1985) A systematic review of the molossid bats allied with the genus *Mormopterus* (Chiroptera: Molossidae). *Acta Zoologica Fennica* 170, 205–208.

Peterson RL, Eger JL, Mitchell L (1995) Chiroptères. *Faune de Madagascar* 84, 1–204.

Petit RE (2009) *Perry's Arcana: A facsimile edition with collation and systematic review.* Temple University Press, Philadelphia.

Philippi RA (1892) Las focas chilenas del Museo Nacional. *Anales del Museo Nacional de Chile. Primera Seccion Zoolojía. Entr* 1892, 1–52.

Philippi RA (1893) Los delfines de la punta austral de la América del sur. *Anales del Museo Nacional de Chile. Zoolojía* 6, 1–17.

Philippi RA (1896) Los cráneos de los delphines Chilenos. *Anales del Museo Nacional de Chile. Zoolojía* 12, 1–18.

Philippi RA (1902) *Suplemento a los Batraquios Chilenos descritos en la Historia Física i Política de Chile de don Clauido Gay.* Santiago de Chile.

Phillip A (1789) *The Voyage of Governor Phillip to Botany Bay*, with an account of the establishment of the colonies of Port Jackson & Norfolk Island compiled from authentic papers, which have been obtained from the several departments, to which are added the journals of Lieuts. Shortland, Watts, Ball & Capt. Marshall; with an account of their discoveries, embellished with LV copper plates, the maps and charts taken from actual surveys, & the plans and views drawn on the spot, by Capt. Hunter, Lieuts. Shortland, Watts, Dawes, Bradley, Capt. Marshall. John Stockdale, London.

Phillips CJ (1967) A new subspecies of horseshoe bat (*Hipposideros diadema*) from the Solomon Islands. *Proceedings of the Biological Society of Washington* 80, 35–39. [24 Mar 1967]

Phillips CJ, Birney EC (1968) Taxonomic status of the vespertilionid genus *Anamygdon* (Mammalia: Chiroptera). *Proceedings of the Biological Society of Washington* 81, 491–498. [30 Dec 1968]

Phillips MJ, Penny D (2003) The root of the mammalian tree inferred from whole mitochondrial genomes. *Molecular Phylogenetics and Evolution* 28, 171–185. [Aug 2003]

Phillips MJ, Pratt RC (2008) Family-level relationships among the Australasian marsupial 'herbivorous' (Diprotodontia: koala, wombats, kangaroos and possums). *Molecular Phylogenetics and Evolution* 46, 594–605. [Feb 2008]

Phillips MJ, McLenachan PA, Down C, Gibb GC, Penny D (2006) Combined mitochondrial and nuclear DNA sequences resolve the interrelations of the major Australasian marsupial radiations. *Systematic Biology* 55, 122–137.

Phillips MJ, Haouchar D, Pratt RC, Gibb GC, Bunce M (2013) Inferring kangaroo phylogeny from incongruent nuclear and mitochondrial genes. *PLoS ONE* 2(2), 1–12.

Pilleri G, Gihr M (1972) A rare species of dolphin *Delphinus tropicalis* van Bree 1971 (= *dussumieri* Blanford, 1891) from the coast of Pakistan. *Mammalia* 36, 406–413. [Sep 1972]

Pine RH (1980) Taxonomic notes on '*Monodelphis dimidiata itatiayae* (Miranda-Ribeiro),' *Monodelphis domestica* (Wagner) and *Monodelphis maraxina* Thomas (Mammalia: Marsupialia: Didelphidae). *Mammalia* 43, 495–499. [15 Jan 1980]

Piper KJ (2007) Early Pleistocene mammals from the Nelson Bay Local Fauna, Portland, Victoria, Australia. *Journal of Vertebrate Paleontology* 27, 492–503. [Jun 2007]

Pitman R (2009) Indo-Pacific Beaked Whale: *Indopacetus pacificus*. pp. 600–602, in WF Perrin, B Würsig & JGM Thewissen (eds.) *Encyclopedia of Marine Mammals.* Academic Press, London. Second Edition.

Pitman RL, Ensor P (2003) Three forms of killer whales (*Orcinus orca*) in Antarctic waters. *The Journal of Cetacean Research and Management* 5, 131–139.

Pitman RL, Perryman WL, LeRoi D, Eilers E (2007) A dwarf form of killer whale in Antarctica. *Journal of Mammalogy* 88, 43–48. [Feb 2007]

Pitman RL, Durban JW, Greenfelder M, Guinet C, Jorgensen M, Olson PA, Plana J, Tixier P, Towers JR (2011) Observations of a distinctive morphotypes of killer whale (*Orcinus orca*), type D, from subantarctic waters. *Polar Biology* 34, 303–306. [online 7 Aug 2010]

Pizzey G, Knight F (1998) *Field Guide to the Birds of Australia*. Angus and Robertson, Sydney.

Platnick NI (2013) The World Spider Calalogue. Version 13.5. American Museum of Natural History, New York. Obtained at: http://research.amnh.org/iz/spiders/catalog/THOMISIDAE.html.

Platnick NI, Baptista RLC (1995) On the spider genus *Attacobius* (Araneae, Dionycha). *American Museum Novitates* 3120, 1–9. [27 Jan 1995]

Plötz C (1870) *Pseudopontia calabarica* n. gen. et. n. sp. *Stettiner Entomologische Zeitung* 31, 348–349.

Poche F von (1906) Zur nomenklatur der muriden. *Zoologischer Anzeiger* 30, 326. [19 Jun 1906]

Pocock RI (1921) The external characters and classification of the Procyonidae. *Proceedings of the Zoological Society of London* 1921, 389–422. [8 July 1921]

Pocock RI (1926a) The external characters of *Thylacinus, Sarcophilus*, and some related marsupials. *Proceedings of the Zoological Society of London* 96, 1037–1084.

Pocock RI (1926b) The external characters of the flying lemur (*Galeopterus temminckii*). *Proceedings of the Zoological Society of London* 96, 429–444.

Pocock RI (1951) Catalogue of the Genus *Felis*. British Museum, London.

Poeppig E (1835) Ueber den Cucurrito Chile's (*Psammoryctes noctivagus* Poepp.). *Archiv für Naturgeschichte* 1(1), 252–255.

Poole RW (1989) Noctuidae. *Lepidopterorum Catalogus*. Taylor & Francis, London. Fascicle 118. Reference not seen.

Poole RW, Gentili P (1996) Nomina Insecta Nearctica. A Check List of the Insects of North America. Volume 3. Diptera, Lepidoptera, Siphonaptera. Entomological Information Services. Rockville.

Poole WE (1979) The status of the Australian Macropodidae. pp. 13–27, in MJ Tyler (ed.) *The Status of Endangered Australian Wildlife*. Royal Zoological Society of South Australia, Adelaide.

Poole WE (1982) *Macropus giganteus*. *Mammalian Species* 187, 1–8. [23 Nov 1982]

Poole WE, Carpenter SM, Simms NG (1990) Subspecific separation in the western grey kangaroo, *Macropus fuliginosus*: a morphometric study. *Australian Wildlife Research* 17, 159–168.

Poole WE, Wood JT, Simms NG (1991) Distribution of the tammar, *Macropus eugenii*, and the relationships of populations as determined by cranial morphometrics. *Wildlife Research* 18, 625–639.

Pope L, Storch D, Adams M, Moritz C, Gordon G (2001) A phylogeny for the genus *Isoodon* and a range extension for *I. obesulus peninsulae* based on mtDNA control region and morphology. *Australian Journal of Zoology* 49, 411–434.

Popper K (1963) *Conjectures and Refutations: The Growth of Scientific Knowledge*. Routledge and Keagan Paul, London.

Potter S, Eldridge MD, Cooper S, Taggart DA (2010) Phylogeography of the *brachyotis* group with recognition of two species within *Petrogale brachyotis*. pp. 56, in Abstracts for the Rock-wallaby Symposium of the Australian Mammal Society. pp. 52–71.

Potter S, Eldridge MDB, Taggart DA, Cooper SJB (2012a) Multiple biogeographic barriers identified across the monsoon tropics of northern Australia: phylogeographic analysis of the *brachyotis* group of rock-wallabies. *Molecular Ecology* 21, 2254–2269. [May 2012]

Potter S, Cooper SJB, Metcalfe CJ, Taggart DA, Eldridge MDB (2012b) Phylogenetic relationships of rock-wallabies, *Petrogale* (Marsupialia: Macropodidae) and their biogeographic history within Australia. *Molecular Phylogenetics and Evolution* 62, 640–652. [online 20 Nov 2011]

Potter S, Close R, Taggart D, Cooper S, Eldridge M (2014) Taxonomy of rock-wallabies, Petrogale (Marsupialia: Macropodidae). IV. Multifaceted study of the *brachyotis* group identifies additional taxa. *Australian Journal of Zoology* 62, 401–414.

Poux C, Madsen O, Glos J, de Jong WW, Vences M (2008) Molecular phylogeny and divergence times of Malagasy tenrecs: influence of data partitioning and taxon sampling on dating analyses. *BMC Evolutionary Biology* 8(102), 1–16. [31 Mar 2008]

Prashad B (1936) Animal Remains from Harappa. *Memoirs of the Archaeological Survey of India* 51, 1–64.

Price SA, Bininda-Edmonds ORP, Gittleman JL (2005) A complete phylogeny of the whales, dolphins and even-toed hoofed mammals (Cetartiodactyla). *Biological Reviews of the Cambridge Philosophical Society* 80, 445–473. [Aug 2005]

Prideaux GJ (2004) Systematics and Evolution of the Sthenurine Kangaroos. *University of California Publications in Geological Sciences* 146, 1–622. [May 2004]

Prideaux GJ, Warburton NM (2010) An osteology-based appraisal of the phylogeny and evolution of the kangaroos and wallabies (Macropodidae: Marsupialia). *Zoological Journal of the Linnean Society* 159, 954–987. [26 Jul 2010]

Prothero DR (1981) New Jurassic mammals from Como Bluff, Wyoming, and the interrelationships of the non-tribosphenic Theria. *Bulletin of the American Museum of Natural History* 167, 277–362. [26 Feb 1981]

Prothero DR, Schoch RM (1989) *The Evolution of Perissodactyls*. pp. 504–529, in DR Prothero & RM Schoch. Oxford University Press, New York.

Prout LB (1923) New Geometridae in the Tring Museum. *Novitates Zoologicae* 30, 191–215.

Pucheran J (1853) In H Jacquinot & J Pucheran (ed.) Mammifères et oiseaux. Volume 3. In JSC Dumont d'Urville (ed.) (1841–1854) *Voyage au Pôle Sud et dans l'Océanie sur les corvettes l'Astrolobe et la Zélée*: Exécuté par ordre roi pendant les années 1837–1838–1839–1840 – sous le commandement de M.J. Dumont d'Urville, Capitaine de Vaisseau, publié par ordonnance de Sa majesté; Histoire du Voyage par M. Dumot D'Urville, Paris. [1842, Atlas Mammifères Plate 10; Tome 3. Mammifères et Oiseaux, 1853]

Pucheran D (1856) Travaux inedits. Notices mammalogiques. *Revue et Magasin de Zoologie pure et Appliquée* (2)8, 545–553.

Pumo DE, Finamore PS, Franek WR, Phillips CJ, Tarsami S, Balzaroano D (1998) Complete mitochondrial genome of a neotropical fruit bat, *Artibeus jamaicensis*, and a new hypothesis of the relationships of bats to other Eutherian mammals. *Journal of Molecular Evolution* 47, 709–717.

Purcell B (2010) *Dingo*. CSIRO Publishing, Canberra.

Queale LF (1997) Field identification of female little brown bats *Vespadelus* spp. (Chiroptera: Vespertilionidae) in South Australia. *Records of the South Australian Museum* 30, 29–33. [8–17 Apr 1997]

Qumsiyeh MB, Owen RD, Chesser RK (1988) Differential rates and chromosomal evolution in bats of the family Rhinolophidae. *Genome* 30, 326–335. [Jun 1988]

Quoy JRC, Gaimard JP (1824) Zoologie. pp. 1–712, in LCD de Freycinet (ed.) (1824–1826) *Voyage Autour du Monde*, entrepris par ordre du Roi, sous le Ministère et conformément aux instructions de S. Exc. M. le Vicomte du Bouchage, Secrétaire d'Etat au Département de la marine, exécuté sur les corvettes de S.M. l'Uranie et la Physicienne, pendant les anées 1817, 1818, 1819, et 1820; publié sous les auspices de S.E.M. le Comte Corbière. Secrétaire d'État de l'Intérieur, pour la partie historique et les sciences naturelles, et de S.E.M. le Marquis de Clermont-Tonnerre, Secrétaire d'État de la Marine et des Colonies, pour la partie nautique, par M. Louis Freycinet, etc. Pillet Aîné, Paris. Volume 1. [See Sherborn & Woodward, 1901a: 392 for dates]

Quoy JRC, Gaimard JP (1825) Notice sur les mammifères et les oiseaux de la Baie de Chiens-Marins et de la Nouvelle-Galles du Sud: sur leurs moeurs et leur distribution géographiqie. *Annales des Sciences Naturelles, Paris* 5(1), 476–491.

Quoy JRC, Gaimard JP (1830) Zoologie. Volume 1. pp. 1–268, in JSC Dumont d'Urville (1830–1835) *Voyage de Découvertes de l'Astrolabe. Exécuté par ordre du Roi pendant les années* 1826–1827–1828–1829 sous le commandement de M.J. Dumont d'Urville. J. Tastu, Paris.

Racovitza ÉG (1903) *Expédition Antarctique Belge : Résultats du Voyage du S. Y. Belgica en 1897–1898 - 1899 sous le commandement de A. de Gerlache de Gomery. Rapports Scientifiques. Zoologie. Cétacéa*. J.-E. Buschmann, Anvers.

Radford CD (1943) Genera and species of parasitic mites (Acarina). *Parasitology* 35, 58–81. [Feb 1943]

Radford CG, Letnic M, Fillios M, Crowther MS (2012) An assessment of the taxonomic status of wild canids in south-eastern New South Wales: phenotypic variation in dingoes. *Australian Journal of Zoology* 60, 73–80.

Raffles TS (1821) Descriptive Catalogue of a Zoological Collection, made on account of the Honourable East India Company, in the Island of Sumatra and its Vicinity, under the Direction of Sir Thomas Stamford Raffles, Lieutenant-Governor of Tort Marlborough; with additional Notices illustrative of the Natural History of those Countries. *Transactions of the Linnean Society of London* (1)13, 239–274. [23 May–21 Jun 1821]

Rafinesque CS (1814a) *Précis des découvertes et travaux somiologiques. Entre 1800 et 1814. Ou choix raisonné de ses principales découvertes en Zoologie et en Botanique, pour servir d'introduction à ses ouvrages futurs*. Royale Typographie Militaire, Palerme.

Rafinesque CS (1814b) *Principes fondamentaux de Somiologie ou les loix de la nomenclature et de la classification des corps organisés*. Palerme.

Rafinesque CS (1815) *Analyse de la Nature; ou Tableau de l'Univers et de Corps Organisés*, Palerme.

Rafinesque CS (1817) New species of mammifers, noticed in the notes to the (Tableau Methodique des Mammifères) Methodical Picture of the Mammifers, by D. Desmarest, in the 24th and last volume of the French New Dictionary of Natural History. Paris, 1804. *American Monthly Magazine and Critical Review* 1, 361–363.

Rafinesque CS (1818a) Further discoveries in natural history, made during a journey through the western region of the United States. *American Monthly Magazine and Critical Review* 3, 445–447.

Rafinesque CS (1818b) General account of the discoveries made in the zoology of the Western States. *American Monthly Magazine and Critical Review* 4, 106–107.

Rafinesque CS (1819a) De nouveaux genres d'animaux découverts dans l'intérieur des États – unis d'Amérique, durant l'année 1818. *Journal de Physique, de Chimie et d'Histoire Naturelle* 88, 417–429.

Rafinesque CS (1819b) Conchology. pp. 65–66, in H McMurtrie (ed.) *Sketches of Louisville and its Environs*. Louisville, Kentucky.

Rafinesque CS (1820a) *Annals of Nature or Annual Synopsis of New Genera and Species of Animals, Plants, and Discoveries in North America*. First Annual Number, for 1820 Thomas Smith. Lexington, Kentucky.

Rafinesque CS (1820b) Monographie des coquilles bivalves fluviatiles de la rivière Ohio, contenant douze genres et soixant-huit espèces. *Annales Générales des Sciences Physiques* 5, 287–322.

Rafinesque CS (1831) *Enumeration and account of some remarkable natural objects in the cabinet of Prof. Rafinesque in Philadelphia: being animals, shells, plants, and fossils, collected by him in North America, between 1816 and 1831*. Philadelphia. [Nov 1831]

Rafinesque CS (1832) *Atlantic Journal and Friend of Knowledge. Philadelphia.* 3(3), 91–122.

Ramme W von (1914) *Dactylopsila hindenburgi*, ein neuer Streifenbeutler aus Kaiser-Wilhelms-Land (Mamm. Marsup.). *Sitzungs-berichte der Gesellschraft Naturforschender Freunde zu Berlin* 1914, 412–419.

Ramsay EP (1876a) Description of a new genus and species of rat-kangaroo, allied to the genus *Hypsiprymnus*, proposed to be called *Hypsiprymnodon moschatus*. *Proceedings of the Linnean Society of New South Wales* (1)1, 33–35. [Feb 1876]

Ramsay EP (1876b) Description of a new species of kangaroo, from New Guinea. *Proceedings of the Linnean Society of New South Wales* (1)1, 162–164. [Jul 1876]

Ramsay EP (1876c) Description of a supposed new species of bat, from Stanwell, near Bulli, N.S.W. *Proceedings of the Linnean Society of New South Wales* (1)1, 81–82. [Feb 1876]

Ramsay EP (1877a) Note of a species of echidna (*Tachyglossus*), from Port Moresby, New Guinea. *Proceedings of the Linnean Society of New South Wales* (1)2, 31–33. [Jul 1877]

Ramsay EP (1877b) Zoology of the 'Chevert'. Mammals. Part 1. *Proceedings of the Linnean Society of New South Wales* (1)2, 7–15. [Jul 1877]

Ramsay EP (1877c) Notes on a collection of birds from the Norman River, Gulf of Carpentaria, with descriptions of some new species. *Proceedings of the Linnean Society of New South Wales* (1)1, 379–395. [Mar 1877]

Ramsay EP (1877d) Description of a supposed new species of rock-wallaby from the Palm Island; on the north-east coast of Australia, proposed to be called *Petrogale assimilis*. *Proceedings of the Linnean Society of New South Wales* (1)1, 359–361. [Mar 1877]

Ramsay EP (1879) Contributions to the Zoology of New Guinea. Parts I and II. *Proceedings of the Linnean Society of New South Wales* (1)3, 241–306.

Ramsay EP (1882) Description of a supposed new species of rat from the interior of New South Wales. *Proceedings of the Linnean Society of New South Wales* (1)6, 763–765. [20 Mar 1882]

Ramsay EP (1883) Contributions to the Zoology of New Guinea, Part VII. *Proceedings of the Linnean Society of New South Wales* (1)8, 15–29. [19 Jun 1883]

Ramsay EP (1886) Description of a new species of *Hapalotis*, (*H. Boweri*) from north west Australia. *Abstract of Proceedings of the Linnean Society of New South Wales* (1886), vi. [for 29 Dec 1886]

Ramsay EP (1887a) Description of two new species of marsupials (*Perameles* and *Antechinus*), and of a new species of *Mus* (*M. burtoni*), from the neighborhood of Derby, N.W.A. *Proceedings of the Linnean Society of New South Wales* (2)2, 551–553. [30 Nov 1887]

Ramsay EP (1887b) Description of three new species of mammals from north west Australia. *Abstract of Proceedings of the Linnean Society of New South Wales* 1887, vi. [for 31 Aug 1887]

Ramsay EP (1887c) Description of a new species of *Hapalotis*, (*H. boweri*) from north west Australia. *Proceedings of the Linnean Society of New South Wales* (2)1, 1153–1154. [22 Feb 1887]

Ramsay EP (1888) Notes on the fauna of the Bellenden-Ker Ranges. *Proceedings of the Linnean Society of New South Wales* (2)3, 1295–1299. [7 Dec 1888]

Ramsay EP (1890a) On a new species of *Petaurides* from the Bellenden-Kerr Range, Queensland. *Records of the Australian Museum* 1, 77–78. [30 Sep 1890]

Ramsay EP (1890b) Notes and exhibits. *Proceedings of the Linnean Society of New South Wales* (2)4, 1030. [3 Feb 1890]

Randi E, Mucci N, Claro-Hergueta F, Bonnet A, Douzery EJP (2001) A mitochondrial DNA control region phylogeny of the Cervinae: Speciation in *Cervus* and implications for the conservation. *Animal Conservation* 4, 1–11.

Rang PCAL (1827) Description de deux genres nouveau (*Cuvieria* et *Euribia*) appartenant à la class des Ptérorodes. *Annales des Sciences Naturelles, Paris* 12(1), 320–329.

Ranzani C (1839) De Chamaeleontibus. *Novi Commentarii Academiae Scientiarum Instituti Bononiensis* 3, 213–235.

Raphael S (1970) The publication dates of the *Transactions of the Linnean Society of London*, Series 1, 1791–1875. *Biological Journal of the Linnean Society. Linnean Society of London* 2, 61–76.

Rapp WL von (1837) *Die Cetaceen : zoologisch-anatomisch dargestellt*. J.G. Cotta'schen Buchhandlung, Stuttgart & Tübingen.

Rasch H (1845) Beskrivelse over en i Christinaiafjorden fanget nye Delphinart. *Nyt Magazin for Naturvidenskaberne* 4, 97–125.

Rathbun GB (1984) Sirenians. pp. 537–547, in S Anderson & JK Jones (eds.) *Orders and Families of Recent Mammals of the World*. John Wiley & Sons, New York.

Raven HC (1929 [1929–1933]) Kangaroo. pp. 254–255, Volume 13, in JL Garvin & FH Hooper in *Encyclopaedia Britannica*. London. 14th Edition.

Raven HC (1935) Wallace's line and the distribution of Indo-Australian mammals. *Bulletin of the American Museum of Natural History* 68, 179–293. [5 Apr 1935]

Raven HC (1939) The identity of Captain Cook's kangaroo. *Journal of Mammalogy* 20, 50–57. [Feb 1939]

Raven HC, Gregory WK (1946) Adaptive branching of the kangaroo family in relation to habitat. *American Museum Novitates* 1309, 1–15. [8 Mar 1946]

Ray J (1693) *Synopsis methodica animalium quadrupedum et serpentini generis*. Vulgarium natas characteristicas, rariorum descriptiones integras exhibens: cum historiis & observationibus anatomicis perquam curiosis. Præmittuntur nonnulla de animalium in genere. S. Smith & B. Walford, Londini.

Rayner GW (1939) *Globicephala leucosagmaphora*, a new species of the genus *Globicephala*. *Annals & Magazine of Natural History* (11)4, 543–544. [1 Nov 1939]

Read G (2012) *Clymene* Oken, 1807, 1815, in G Read & K Fauchald (eds.) *World Polychaeta* database. Accessed through: *World Register of Marine Species* at http://www.marinespecies.org/aphia.php?p=taxdetails&id=596529 on 2012–06–03.

Read G, Fauchald K (2012) *Clymene* Savigny, in Lamarck, 1818. In G Read & K Fauchald (eds) *World Polychaeta database*. Accessed through: *World Register of Marine Species* at http://www.marinespecies.org/aphia.php?p=taxdetails&id=146846 on 2012–06–03.

Reardon TB (1999a) Taxonomy and selection of taxa for this action plan. pp. 6–14, in A Duncan, GB Baker & N Montgomery (eds.) *Action Plan for Australian Bats*. Environment Australia, Canberra. [Aug 1999]

Reardon TB (1999b) Nomenclature of *Tadarida australis* (Gray, 1838). *The Australasian Bat Society Newsletter* 12, 22–24. [Mar 1999]

Reardon TB (2009a) Mega Micro, Yino Yango; whats going on? And who do bats relate to now? *The Australasian Bat Society Newsletter* 32, 39–40. [Mar 2009]

Reardon TB (2009b) A note on the taxonomic treatment used in the second edition of Australian bats by Sue Churchill. *The Australasian Bat Society Newsletter* 32, 52–53. [Mar 2009]

Reardon TB, Flavel SC (1987) *A Guide to the Bats of South Australia.* South Australian Museum and Field Naturalists Society of South Australia, Adelaide.

Reardon TB, McKenzie N (2008) *Mormopterus eleryi* in Reardon, T., Adams, M., McKenzie, N. & Jenkins, P. A new species of Australian freetail bat *Mormopterus eleryi* sp. nov. (Chiroptera: Molossidae) and a taxonomic reappraisal of *M. norfolkensis* (Gray). *Zootaxa* 1875, 1–31. [12 Sep 2008]

Reardon TB, Thomson B (2002) Taxonomy and conservation status of Troughton's sheathtail bat (*Taphozous troughtoni*). Report to Natural Heritage Trust, Canberra.

Reardon TB, Adams M, McKenzie N, Jenkins P (2008) A new species of Australian freetail bat *Mormopterus eleryi* sp. nov. (Chiroptera: Molossidae) and a taxonomic reappraisal of *M. norfolkensis* (Gray). *Zootaxa* 1875, 1–31. [12 Sep 2008]

Reardon TB, Robson SKA, Parsons JG, Inkster T (2010) *Review of the threatened status of the microchiropteran bat species on Cape York Peninsula.* South Australian Museum, Adelaide.

Reardon TB, McKenzie NL, Cooper SJB, Appleton B, Carthew S, Adams M (2014) A molecular and morphological investigation of species boundaries and phylogenetic relationships in Australian free-tailed bats *Mormopterus* (Chiroptera: Molossidae). *Australian Journal of Zoology* 62, 109–136. [online 19 Jun 2014]

Rees A (1807 [1802–1820]) *The Cyclopædia; or, Universal Dictionary of Arts, Sciences, and Literature.* Longman, Hurst, Rees, Orme & Brown, London. Volume 8. Part 1. [18 May 1807 or Jan 1807]

Reeves RR, Stewart BS, Clapham PJ, Powell JA (2002) *Sea Mammals of the World.* A & C Black, London.

Reeves RR, Perrin WF, Taylor BL, Baker CS, Mesnick SL (eds.) (2004) Report of the Workshop on Shortcomings of Cetacean Taxonomy in Relation to Needs of Conservation and Management. NOAA Technical Memorandum NOAA-TM-NMFS-SWFSC-363.

Reichenbach AB (1835) *Bildergalerie der Thierwelt, oder Abbildungen des interessantesten aus dem Thierreiche, mit ausführlicher Beschreibung.* Leipzig. Part 6. Reference not seen.

Reichenbach HGL (1837) *Deutschlands fauna; oder, Praktisch-gemeinnützige naturgeschichte der thiere des inlandes mit naturgetreuen abbildungen aller arten. Für gebildete leser aller stände, so wie für schulen.* I. Die Säugthiere. Wagner, Leipzig. Reference not seen.

Reichenbach HGL (1845) *Synopsis Mammalium Iconibus Illustrata.* A.F. Hofmeister, Leipzig. Volume 1.

Reichenbach HGL (1862) *Die vollständigste Naturgeschichte der Affen.* Dresden, Leipzig.

Reid J (1837) Description of a new species of the genus *Perameles* (*P. lagotis*). *Proceedings of the Zoological Society of London* 4(1836), 129–131. [7 Jun 1837]

Reinhardt J (1857) Fortegnelse over Grønlands Pattedyr, Fugle og Fiske. pp. 3–27, in J Reinhardt, JMC Schiødte, OAL Mørch, CF Lütfen, J Lange & H Rink (eds.) *Naturhistoriske Bidrag til en Beskrivelse af Grønland.* Saerskilt aftryk af tillaeggene til 'Grønland, geographisk og statistisk beskrevet,' af H. Rink. Kjøbenhavn.

Reinhardt J (1862) Om en for den Danske fauna ny Delphinart (*Pseudorca crassidens* Owen). *Oversigt over det Kongelige Danske Videnskabernes Selskabs Forhandlinger. Kjøbenhavn* 1862, 103–152.

Reinhardt J (1866) In DF Eschricht. On the species of the genus *Orca* inhabiting the northern seas. pp. 151–188, in WH Flower (ed.) *Recent Memoirs of the Cetacea.* Royal Society of London, London.

Reinhardt J, Lütken C (1862) Bidrag til Kundskab om Brasiliens padder og Krybdyr. *Videnskabelige Meddelelser fra den Naturhistoriske Forening i Kjøbenhavn* (Nr. 10–15) 1861, 143–242.

Reinhold L (1997) Taxonomy of bent-winged bats (genus *Miniopterus*) in northern Australia and New Guinea. BSc (Hons) Thesis. University of Queensland, Brisbane. Reference not seen.

Renfree MB (1980) Embryonic diapause in the honey possum *Tarsipes spencerae. Search* 11, 81.

Repenning CA (1967) Subfamilies and genera of the Soricidae. *Geological Survey Professional Paper* 565, 1–69.

Repenning CA, Tedford RH (1977) Otarioid seals of the Neogene. *U.S. Geological Survey Professional Paper* 992, 1–93.

Repenning CA, Peterson RS, Hubbs CL (1971) Contributions to the systematics of the southern fur seals, with particular reference to the Juan Fernandez and Guadalupe species. *Antarctic Research* 18, 1–34.

Retzius AJ (1794) Anmärkningar vid Genus *Trichechi. Kongliga Svenska Vetenskaps Akademiens Nya Handlingar, Stockholm* 15, 286–300.

Reumer JWF (1989) Speciation and evolution in the Soricidae (Mammalia: Insectivora) in relation with the paleoclimate. *Revue Suisse de Zoologie* 96, 81–90. [Mar 1989]

Reuvens CL (1890) *Die Myoxidae Oder Schlaffer ein Beitrag zur Osteologie und Systematik der Nagethiere.* P.W.M. Trap, Leiden.

Revilliod P (1914) Les mammifères de la Nouvalle-Calédonie et des îles Loyalty. pp. 339–365, in F Sarasin & J Roux (eds.) *Nova Caledonia, Recherches Scientifiques en Nouvelle-Calédonie et aux Iles Loyalty. A. Zoologie.* C.W. Kreidels Verlag, Wiesbaden, Germany. Volume 1.

Revilliod P (1917) Fledermäuse aus der Braunkohle von Messel bei Darmstadt. *Abhandlungen der Grossherzoglich Hessischen Geologischen Landesanstalt zu Darmstadt* 7, 157–201.

Revilliod P (1922) Contribution à l'étude des Chiroptères des terrains Tertiaires. *Mémoires de la Société Paléontologique Suisse* 45, 133–195. [Reference not seen.]

Rey C (1883) Tribu des Brévipennes. *Annales de la Société Linnéenne de Lyon* 30, 153–415.

Reyes JC, Van Waerebeek K, Cardenas JC, Yanez JL (1995) *Mesoplodon bahamondi* sp. n. (Cetacea, Ziphiidae), a new living beaked whale from the Juan Fernandez Archipelago, Chile. *Boletin del Museo Nacional de Historia Natural, Chile* 45, 31–44.

Reynolds JE, Odell DK, Rommell SA (1999) Marine mammals of the world. pp. 1–14, in JE Reynolds & SA Rommel (eds.) *Biology of Marine Mammals.* Melbourne University Press, Melbourne.

Rhind SG, Bradley JS, Cooper NK (2001) Morphometric variation and taxonomic status of brush-tailed phascogales, *Phascogale tapoatafa* (Meyer, 1793)(Marsupialia: Dasyuridae). *Australian Journal of Zoology* 49, 345–368.

Rice DW (1967) Cetaceans. pp. 291–324, in S Anderson & JK Jones Jr (eds.) *Recent Mammals of the World: A Synopsis of Families*. The Ronald Press Company, New York.

Rice DW (1977) A List of Marine Mammals of the World. *NOAA Technical Report, NMFS SSRF* 711, 1–75.

Rice DW (1980) *Mesoplodon* Gervais, 1850 (Mammalia: Cetacea): proposed conservation. Z.N. 2081. *Bulletin of Zoological Nomenclature* 37, 30–33. [8 May 1980]

Rice DW (1984) Cetaceans. pp. 447–490, in S Anderson & JK Jones (eds.) *Orders and Families of Recent Mammals of the World*. John Wiley & Sons, New York.

Rice DW (1989a) Sperm whale *Physeter macrocephalus* Linnaeus, 1758. pp. 177–233, in SH Ridgway & R Harrison (eds.) *Handbook of Marine Mammals. Volume 4. River Dolphins and Larger Toothed Whales*. Academic Press, London.

Rice DW (1989b) Scientific correspondence. *Marine Mammal Science* 5, 210.

Rice DW (1990) The scientific name of the pilot whale – a rejoiner to Schevill. *Marine Mammal Science* 6, 359–360. [Oct 1990]

Rice DW (1998) *Marine Mammals of the World: Systematics and Distribution*. Society for Marine Mammalogy, Special Publication 4.

Rice DW (2009) Classification (Overall). pp. 234–238, in WF Perrin, B Würsig & JGM Thewissen (eds.) *Encyclopedia of Marine Mammals*. Academic Press, Amsterdam. Second Edition.

Rich TH, Vickers-Rich P, Constantine A, Flannery TF, Kool L, Klaveren N (1997) A tribosphenic mammal from the Mesozoic of Australia. *Science* 278(5342), 1438–1442. [21 Nov 1997]

Rich TH, Flannery TF, Trusler P, Kool L, van Klaveren NA, Vickers-Rich P (2002) Evidence that monotremes and ausktribosphenids are not sister groups. *Journal of Vertebrate Paleontology* 22, 466–469. [8 Jul 2002]

Richards GC, Hall LS (2002) A new flying-fox of the genus *Pteropus* (Chiroptera: Pteropodidae) from Torres Strait, Australia. *Australian Zoologist* 32, 69–75.

Richards GC, Hall LS, Helman P, Churchill SK (1982) First discovery of a species of the rare tube-nosed insectivorous bat (*Murina*) in Australia. *Australian Mammalogy* 5, 149–151. [10 May 1982]

Richards JD, Short J, Prince RIT, Friend JA, Courtenay JM (2001) The biology of banded (*Lagostrophus fasciatus*) and rufous (*Lagorchestes hirsutus*) hare-wallabies (Diprotodontia: Macropodidae) on Dorre and Bernier Islands, Western Australia. *Wildlife Research* 28, 311–322.

Richardson BJ, Sharman GB (1976) Biochemical and morphological observations on the wallaroos (Macropodidae: Marsupialia) with a suggested new taxonomy. *Journal of Zoology* 179, 499–513.

Richardson J (1837) Report on North American zoology. *Report of the British Association for the Advancement of Science for* 1836(5), 121–224.

Richardson J (1848) Fishes. pp. 1–28, in A Adams (ed.) *The Zoology of the Voyage of H.M.S. Samarang*, under the command of Captain Sir E. Belcher, C.B., F.R.A.S., F.G.S. during the years 1843–1846. Reeve, Benham, and Reeve, London.

Richardson JD (2007) Western barred bandicoot *Perameles bougainville*, burrowing bettong *Bettongia lesueur* and banded hare-wallaby *Lagostrophus fasciatus* recovery plan 2007–2011. Wildlife Management Program No. 49. Department of Environment and Conservation, Western Australia. [Dec 2007]

Richardson J, Dallas WS, Cobbold TS, Baird W, White A (1862) *The Museum of Natural History*; being a popular account of the structure, habits, and classification of the various departments of the animal kingdom: quadrupeds, birds, reptiles, fishes, shells, and insects, including the insects destructive to agriculture. W. Mackenzie, Glasgow. Volume 1.

Richmond CW (1899) On the date of Lacépède's 'Tableaux'. *The Auk* 16, 325–329.

Ride WDL (1956a) The affinities of *Burramys parvus* Broom a fossil phalangeroid marsupial. *Proceedings of the Zoological Society of London* 127, 413–429.

Ride WDL (1956b) A new fossil *Mastacomys* (Muridae) and a revision of the genus. *Proceedings of the Zoological Society of London* 127, 431–439.

Ride WDL (1957) *Protemnodon parma* (Waterhouse) and the classification of related wallabies (*Protemnodon, Thylogale* and *Setonix*). *Proceedings of the Zoological Society of London* 128, 327–346. [Jun 1957]

Ride WDL (1962) On the use of generic names for kangaroo and wallabies. *Australian Journal of Science* 24, 367–372.

Ride WDL (1963) The effect of the suppression by the International Commission on Zoological Nomenclature of Zimmermann 1777 upon the stability of the generic name *Macropus* Shaw 1790. *Journal of the Royal Society of Western Australia* 46, 126–128.

Ride WDL (1964a) A review of Australian fossil marsupials. *Journal and Proceedings of the Royal Society of Western Australia* 47, 97–131.

Ride WDL (1964b) *Antechinus rosamondae*, a new species of dasyurid marsupial from the Pilbara District of Western Australia; with remarks on the classification of Antechinus. *Western Australian Naturalist* 9, 58–65. [13 Mar 1964]

Ride WDL (1970) *A Guide to the Native Mammals of Australia*. Oxford University Press, Melbourne.

Ride WDL (1993) *Jackmahoneya* gen. nov. and the genesis of the macropodiform molar. *Memoir of the Association of Australasian Palaeontologists* 15, 441–459.

Ride WDL, Calaby JH (1964) Comments on the proposed stabilization of *Macropus* Shaw, 1790. *Bulletin of Zoological Nomenclature* 21, 250–255. [16 Oct 1964]

Riedman M (1990) *The Pinnipeds: Seals, Sea Lions and Walruses*. University of California Press, Berkeley.

Risso A (1826a) *Histoire Naturelle des Principales Productions de l'Europe Méridionale et particulièrement de celles des environs de Nice et des Alpes Maritimes*. F.G. Levrault, Paris. Volume 5.

Risso A (1826b) *Histoire Naturelle des Principales Productions de l'Europe Méridionale et particulièrement de celles des environs de Nice et des Alpes Maritimes*. F.G. Levrault, Paris. Volume 3.

Robbins LW, Sarich VM (1988) Evolutionary relationships in the family Emballonuridae (Chiroptera). *Journal of Mammalogy* 69, 1–13. [25 Feb 1988]

Roberts A (1922) A review of the nomenclature of South African birds. *Annals of the Transvaal Museum, Pretoria* 8, 187–272.

Roberts A (1926) Some new S. African mammals and some changes in nomenclature. *Annals of the Transvaal Museum, Pretoria* 11, 245–263.

Roberts A (1932) Preliminary description of fifty-seven new forms of South African Mammals. *Annals of the Transvaal Museum, Pretoria* 15, 13–19.

Roberts A (1946) Descriptions of numerous new subspecies of mammals. *Annals of the Transvaal Museum, Pretoria* 20, 303–328.

Roberts KK, Archer M, Hand SJ, Godthelp H (2007) New genus and species of extinct Miocene ringtail possums (Marsupialia: Pseudocheiridae). *American Museum Novitates* 3560, 1–15. [8 Mar 2007]

Roberts RG, Flannery TF, Ayliffe LK, Yoshida H, Olley JM, Prideaux GJ, Laslett GM, Baynes A, Smith MA, Jones R, Smith BL (2001) New ages for the last Australian megafauna: Continent-wide extinction about 46,000 years ago. *Science* 292(5523), 1888–1892. [8 Jun 2001]

Roberts TR (1989) The freshwater fishes of western Borneo (Kalimantan Barat, Indonesia). *Memoirs of the California Academy of Sciences* 14, 1–210.

Robineau-Desvoidy JB (1830) Essai sur les myodaires. *Mémoires Présentés par Divers Savants à l'Académie des Sciences de l' Institut de France* 2, 1–813.

Robins JH, Hingston M, Matisoo-Smith E, Ross HA (2007) Identifying *Rattus* species using mitochondrial DNA. *Molecular Ecology Notes* 7, 717–729.

Robinson AC, Kemper CM, Medlin GC, Watts CHS (2000) The rodents of South Australia. *Wildlife Research* 27, 379–404.

Robinson HC, Kloss CB (1916) Preliminary diagnoses of some new species and subspecies of mammals and birds obtained in Korinchi, West Sumatra. *Journal of the Straits Branch of the Royal Asiatic Society* 73, 269–278. [Jul 1916]

Robinson NA, Murray ND, Sherwin WB (1993) VNTR loci reveal differentiation between and structure within populations of eastern barred bandicoot *Perameles gunnii. Molecular Ecology* 2, 195–207. [Aug 1993]

Rode P (1945) Catalogue des types de mammifères du Muséum National d'Histoire Naturelle. Ordre des rongeurs (suite). *Bulletin du Muséum National d'Histoire Naturelle, Paris* (2)17, 201–208.

Röderer JG (1761) Göttingische Anzeigen von gelehrten Sachen, unter der Aufsicht der Königl. *Gesellschaft der Wissenschaften* 10(October), 241–243.

Roewer CF (1929) On a collection of Indian palpatores (Phalangiidae) with a revision of the continental genera and species of the Sub-Family Gagrellinae Thorell. *Records of the Indian Museum* 31, 107–159. [Jul 1929]

Roger O (1887) Verzeichness der bisher bekannten fossilen Säugethiere. *Neunundzwanzigster Bericht des Naturwissenschaftlichen Vereins für Schwaben und Neuburg* 1887, 1–162.

Roig-Juñent S (2000) The subtribes and genera of the tribe Broscini (Coleoptera: Carabidae): Cladistic analysis, taxonomic treatment, and biogeographical considerations. *Bulletin of the American Museum of Natural History* 255, 1–90. [25 Sep 2000]

Rose RW, Rose RK (1998) *Bettongia gaimardi. Mammalian Species* 584, 1–6. [1 Jun 1998]

Rosel PE, Haygood MG, Perrin WF (1995) Phylogenetic relationships among the true porpoises (Cetacea: Phocoenidae). *Molecular Phylogenetics and Evolution* 4, 463–474. [Dec 1995]

Rosenbaum H, Brownell RL Jr, Brown MW, Schaeff C, Portway Y, White BN, Malik S, Pastene LA, Patenaude NJ, Baker CS, Goto M, Best PB, Clapham PJ, Hamilton P, Payne R, Rowntree V, Tynan CT, Bannister JL, DeSalle R (2000) World-wide genetic differentiation of *Eubalaena* questioning the number of right whale species. *Molecular Ecology* 9, 1793–1802. [Nov 2000]

Rosenberg G, Emberton KC (1990) Case 2642. Comment on the proposed precedence of Polygyridae Pilsbry, 1894 over Mesodontidae Tryon, 1866 (Mollusca, Gastropoda). *Bulletin of Zoological Nomenclature* 47, 204–205. [28 Sep 1990]

Rosenberg CBH von (1867) Reis Naar de Zuidoostereilanden Gedaan in 1865 op last der Regering van Nederlandsch-Indie. Martinus Nijhoff, S Gravenhage.

Rosenthal FC (1827) *Einige Naturhistorische* Bemerkungen *über die Walle nebst einer Abbildung. Greifswald*. Reference not seen.

Ross GJB (1970) The occurrence of Hector's beaked whale *Mesoplodon hectori* (Gray) in South African waters. *Annals of the Cape Provincial Museums. Natural History* 8, 195–204.

Ross GJB (1984) The smaller cetaceans of the south east coast of southern Africa. *Annals of the Cape Provincial Museums. Natural History* 15, 173–410.

Ross GJB, Cockcroft VG (1990) Comments on Australian bottlenose dolphins and the taxonomic status of *Tursiops aduncus* (Ehrenberg, 1832). pp. 101–128, in S Leatherwood & RR Reeves (ed.). *The Bottlenose Dolphin*. Academic Press, San Diego.

Ross J (1835) *Narrative of a Second Voyage in search of a North-West Passage*, and of a residence in the Arctic Regions during the years 1829, 1830, 1831, 1832, 1833. Includes the Reports of Commander, now Captain James Clark Ross, R.N., F.R.S., F.L.S., &c. and the Discovery of the Northern Magnetic Pole. A.W. Webster, London. Volume 2.

Rothschild W (1893) Descriptions of two new mammals from New Guinea. *Proceedings of the Zoological Society of London* 60(1892), 545–546. [Apr 1893]

Rothschild W (1903) Preliminary diagnosis of a new genus and species of kangaroo. *Novitates Zoologicae* 10, 414.

Rothschild W (1904) Note on *Dendrodorcopsis woodwardi. Novitates Zoologicae* 10, 543.

Rothschild W (1905a) Notes on *Zaglossus* and description of a new species of *Echidna hystrix. Novitates Zoologicae* 12, 305–306.

Rothschild W (1905b) Notes on two kangaroos from 'Northern Territory of South Australia' with descriptions of a new species. *Novitates Zoologicae* 12, 509–510.

Rothschild W (1905c) Note on *Macropus rufus* Desm., with description of a new subspecies. *Novitates Zoologicae* 12, 508.

Rothschild W (1907) Further notes on *Macropus magnus. Novitates Zoologicae* 14, 333.

Rothschild W (1910) Notes on sea-elephants (*Mirounga*). *Novitates Zoologicae* 17, 445–446.

Rothschild W (1913) Some notes on the genera *Zaglossus* and *Tachyglossus*. *Novitates Zoologicae* 20, 188–191.

Rothschild W, Dollman G (1936) The genus *Dendrolagus*. *Transactions of the Zoological Society of London* 21, 477–548. [Mar 1936]

Rothschild W, Rothschild NC (1898) Descriptions of three new kangaroos, and notes on the skulls of *Dendrolagus bennettianus*. *Novitates Zoologicae* 5, 511–513.

Rougier GW (1993) *Vincelestes neuquenianus* Bonaparte (Mammalia, Theria) un primitive mamífero del Cretácico Inferior de la Cuenca Neuquina. Unpublished PhD Thesis. University of Buenos Aires, Buenos Aires.

Rougier GW, Wible JR, Hopson JA (1996) Basicranial anatomy of *Priacodon fruitaensis* (Triconodontidae, Mammalia) from the Late Jurassic of Colorado, and a reappraisal of mammaliaform interrelationships. *American Museum Novitates* 3183, 1–40. [22 Nov 1996]

Rougier GW, Wible JR, Novacek MJ (1998) Implications of *Deltatheridium* specimens for early marsupial history. *Nature* 396(6710), 459–463. [3 Dec 1998]

Rougier GW, Martinelli AG, Forasiepi AM, Novacek MJ (2007) New Jurassic mammals from Patagonia, Argentina: A reappraisal of Australosphenidan morphology and interrelationships. *American Museum Novitates* 3566, 1–10. [16 May 2007]

Roux C (1976) On the dating of the first edition of Cuvier's Règne Animal. *Journal of the Society of Bibliography and Natural History* 8(1): 31. [Nov 1976]

Roux P (1797) Livres nouveax. *Journal Typographique et Bibliographique* 1, 81–86.

Rowe T (1988) Definition, diagnosis and origin of Mammalia. *Journal of Vertebrate Paleontology* 8, 241–264. [23 Sep 1988]

Rowe T, Gauthier J (1992) Ancestry, paleontology, and the definition of the name Mammalia. *Systematic Biology* 41, 372–378. [Sep 1992]

Roxburgh W (1803) An account of a new species of *Delphinus*, an inhabitant of the Ganges. *Asiatic Researches; or, Transactions of the Society Instituted in Bengal, for Inquiring into the History and Antiquities, the Arts. Sciences, and Literature of Asia* 7, 170–174. [Calcutta Edition]

Rudolphi CA (1822) Einige Anatomische Bemerkungen über *Balaena rostrata. Abhandlungen der Physikalischen Klasse der Königlich-Preussischen Akademie der Wissenschaften zu Berlin* 1820–1821, 27–40.

Rudolphi DKA (1832) Über *Balaena longimana. Abhandlungen der Königlich-Preussischen Akademie der Wissenschaften zu Berlin* 1829–1832, 133–144.

Ruedas LA, Morales JC (2005) Evolutionary relationships among genera of Phalangeridae (Metatheria: Diprotodontia) inferred from mitochondrial DNA. *Journal of Mammalogy* 86: 353–365. [Apr 2005]

Ruedi M (1995) Taxonomic revision of shrews of the genus *Crocidura* from the Sunda Shelf and Sulawesi with description of two new species (Mammalia: Soricidae). *Zoological Journal of the Linnean Society* 115, 211–265. [Nov 1995]

Rümmler H (1934) Eine neue Muridengattung aus dem Hochgebirge Neuguineas. *Zeitschrift fur Saugetierkunde* 9, 47–48.

Rümmler H (1936) Die formen der papuanischen Muridengattung *Melomys. Zeitschrift fur Saugetierkunde* 11, 247–253.

Rümmler H (1938) Die systematik und verbreitung der muriden Neuguineas. *Mitteilungen aus der Zoologischen Sammlung des Museums für Berlin* 23, 1–297.

Rüppell E (1834) Beschreibung des im rothen mere vorkommenden Dugong (*Halicore*). *Museum Senkenbergianum: Abhandlungen aus dem Gebiete der Beschreibenden Naturgeschichte, Frankfurt* 1, 95–114.

Russell BC (1990) FAO species catalogue. Volume 12. Nemipterid fishes of the world. (Threadfin breams, whiptail breams, monocle breams, dwarf monocle breams, and coral breams). Family Nemipteridae. An annotated and illustrated catalog of Nemipterid species known to date. FAO (Food and Agriculture Organization of the United Nations). *Fisheries Synopsis* 12(125), 1–149.

Rutty J (1772) *An Essay Towards a Natural History of the County of Dublin*. W. Sleater, Dublin. Volume 1.

Ryan RM (1965a) The type locality of the Australian horseshoe bat, *Rhinolophus megaphyllus. Australian Journal of Science* 27, 259–260.

Ryan RM (1965b) Taxonomic status of the vespertilionid genera *Kerivoula* and *Phoniscus. Journal of Mammalogy* 46, 517–518. [Aug 1965]

Ryan RM (1966) A new and some imperfectly known Australian *Chalinolobus* and the taxonomic status of African *Glauconycteris. Journal of Mammalogy* 47, 86–91. [Feb 1966]

Rydell J, Baagøe HJ (1994) *Vespertilio murinus. Mammalian Species* 467, 1–6. [2 Jun 1994]

Saban R (1954) Phylogénie des insectivores. *Bulletin de le Muséum Nationale d'Histoire Naturelle* 26(2), 419–432.

Salvadori T (1888) Le date della publicazione della '*Iconografia Della Fauna Italica*' del Bonaparte ed Indice delle specie illustrate in detta opera. *Bollettino dei Musei di Zoologia ed Anatomia Comparata della Regia Universita di Torino* 3(48), 1–25.

Sander PM (1989) The Pachypleurosaurids (Reptilia: Nothosauria) from the Middle Triassic of Monte San Giorgio (Switzerland) with the description of a new species. *Philosophical Transactions of the Royal Society of London. Series B, Biological Sciences* 325(1230), 561–666.

Sapin-Jaloustre J (1953) L'identification des cétacés Antarctiques a la mer. *Mammalia* 17, 221–259.

Sargis EJ (2002) Primate origins nailed. *Science* 298(5598), 1564–1565. [22 Nov 2002]

Sargis EJ (2004) New views on tree shrews: The role of tupaiids in Primate supraordinal relationships. *Evolutionary Anthropology* 13, 56–66.

Sarich VM (1969) Pinniped phylogeny. *Systematic Zoology* 18, 416–422. [Dec 1969]

Sasaki T, Nikaido M, Hamilton H, Goto M, Kato H, Kanda N, Pastene LA, Cao Y, Fordyce RE, Hasegawa M, Okada N (2005) Mitochondrial phylogenetics and evolution of Mysticete whales. *Systematic Biology* 54, 77–90.

Sasaki T, Nikaido M, Wada S, Yamada TK, Cao Y, Hasegawa M, Okada N (2006) *Balaenoptera omurai* is a newly discovered baleen whale that represents an ancient evolutionary lineage. *Molecular Phylogenetics and Evolution* 41, 40–52. [Oct 2006]

Sato JJ, Wolsan M, Suzuki H, Hosoda T, Yamaguchi Y, Hiyama K, Kobayashi M, Minami S (2006) Evidence from nuclear DNA sequences sheds light on the phylogenetic relationships of Pinnipedia: single origin with affinity to Musteloidea. *Zoological Science* 23, 125–146.

Saunders E (1871) *Catalogus Buprestidarum Synonymicus et Systematicus*. J. Janson, London. [Sep 1871]

Savi P (1817) Syst. Annel. Reference not seen.

Savigny P (1818) In J-B Lamarck. *Histoire naturelle des animaux sans vertèbres : présentant les caractères généraux et particuliers de ces animaux, leur distribution, leurs classes, leurs familles, leurs genres, et la citation des principales espèces qui s'y rapportent : précédée d'une introduction offrant la détermination des caractères essentiels de l'animal, sa distinction du végétal et des autres corps naturels, enfin, l'exposition des principes fondamentaux de la zoologie*. Verdière, Paris. Volume 5.

Savigny J-C (1822) Système des annélids, principalement de celles des côtes de l'Égypte et de la Syrie, offrant les caractères tant distinctifs que natuels des ordres, familles et genres, avec la description des espèces. pp. 1–128, in J-C Savigny (1809–1829) *Description de L'Égypte, ou Recueil des observations et des Recherches qui ont été faites en Égypte pendant l'expédition de l'armée Francaise, publié par les ordres de la Majesté l'Empereur Napoléon le Grand. Histoire Naturelle*. Volume 1. Part 3.

Savolainen P, Leitner T, Wilton AN, Matisoo-Smith E, Lundeberg J (2004) A detailed picture of the origin of the Australian dingo, obtained from the study of mitochondrial DNA. *Proceedings of the National Academy of Sciences of the United States of America* 101, 12387–12390. [17 Aug 2004]

Sawyer FC (1953) The dates of issue of J. E. Gray's '*Illustrations of Indian Zoology*' (London, 1830–1835). *Journal of the Society for the Bibliography of Natural History* 3, 48–55.

Say T (1817) Conchology. pp. A-3 - C-6. [= 1–20], in W. Nicholson (ed.) *American Edition of the British Encyclopedia, or, dictionary of arts and sciences comprising an accurate and popular view of the present improved state of human knowledge*. First Edition. Volume 4.

Say T (1822) On a quadruped, belonging to the order Rodentia. *Journal of the Academy of Natural Sciences of Philadelphia* 2(2), 330–343.

Say T, Ord G (1825a) A new genus of Mammalia proposed, and a description of the species upon which it is founded. *Journal of the Academy of Natural Sciences of Philadelphia* 4(2), 345–349.

Say T, Ord G (1825b) Description of a new species of Mammalia, where-on a genus is proposed to be founded. *Journal of the Academy of Natural Sciences of Philadelphia* 4(2), 352–354.

Scaliger GC (1557) *Exotericarum Exercitationum liber quintus decimus, de subtilitate, ad Hieronymum Cardanum*. Michaelis Vascosani, Lutetiae. *Exercitatio* CCVI, 10. [Animalia crumenata.]

Scally M, Madsen O, Douady CJ, de Jong WW, Stanhope MJ, Springer MS (2002) Molecular evidence for the major clades of placental mammals. *Journal of Mammalian Evolution* 8, 239–277. [Dec 2001]

Scammon CM (1872) On a new species of *Balaenoptera*. *Proceedings of the California Academy of Sciences* 4, 269–270. [4 Oct 1872]

Schaub S (1937) Ein neuer Muridae von Timor. *Verhandlungen der Naturforschenden Gesellschaft in Basel* 48, 1–6.

Scheffer VB (1958) *Seals, Sea Lions and Walruses. A Review of the Pinnipedia*. Standford University Press, Standford.

Schein H (1956) Neue afrikanische hoplien (Coleoptera, Scarabaeidae, Hopliinae). *Mitteilungen der Münchner Entomologischen Gesellschaft* 46, 27–41.

Schevill WE (1954) Sight records of the gray grampus, *Grampus griseus* (Cuvier). *Journal of Mammalogy* 35, 123–124. [Feb 1954]

Schevill WE (1986) The International Code of Zoological Nomenclature and a paradigm: the name *Physeter catodon* Linnaeus, 1758. *Marine Mammal Science* 2, 153–157. [Apr 1986]

Schevill WE (1987a) Reply to Holthius, 1987. *Marine Mammal Science* 3, 89–90. [Jan 1987]

Schevill WE (1987b) Note by William E. Schevill. In W.A Watkins, P. Tyak & K.E. Moore. *Steno bredanensis* in the Mediterranean Sea. *Marine Mammal Science* 3, 78–82. [Jan 1987]

Schevill WE (1990a) On stability in zoological nomenclature. *Marine Mammal Science* 6, 168–169. [Apr 1990]

Schevill WE (1990b) Reply to D.W. Rice's rejoiner. *Marine Mammal Science* 6, 360. [Oct 1990]

Schinz HR (1821) *Das Tierreich eingeheilt nach dem Bau der Thiere als Grundlage ihrer Naturgeschichte und der Vergleichenden Anatomie von dem Herrn Ritter von Cuvier staatsgrah von Fransreich und beständiger Secretar der Academie der Wissenschaften U.S.W. Aus dem Französischen frey übersetzt und mit vielen Zusäten versehen. Säugethiere und Vögel*. J.G. Cotta'schen, Stuttgart. Volume 1.

Schinz HR (1824) *Naturgeschichte und Abbildungen der Saügethiere*. Brodtmanus Lithographischer, Zürich.

Schinz HR (1825) *Das Tierreich eingeheilt nach dem Bau der Thiere als Grundlage ihrer Naturgeschichte und der Vergleichenden Anatomie von dem Herrn Ritter von Cuvier staatsgrah von Fransreich und beständiger Secretär der Academie der Wissenschaften U.S.W. Aus dem Französischen frey ubersetzt und mit vielen Zusätzen versehen. Säugethiere und Vogel*. J.G. Cotta'schen, Stuttgart. Volume 4.

Schinz HR (1844) *Systematisches vertzeichniss aller bis jetzt bekannten Säugethiere oder Synopsis mammalium nach dem Cuvier'schen System*. Volume 1.

Schlegel H (1841) *Abhandlungen aus dem Gebiete der Zoologie und Vergleichende Anatomie*. A. Arz & Comp., Leiden. Volume 1.

Schlegel H, Müller S (1845) Over drie buideldieren uit de familie der kengoeroe's. Plates XIX–XXIV. pp. 129–148, in CJ Temminck (ed.) (1839–1847) *Verhandelingen over de Natuurlijke Geschiedenis der Nederlandsche overzeesche bezittingen*, door de Leden der Natuurkundige Commissie in Indië en andere Schrijvers. Zoologie. S. and J. Luchtmans and C.C. van der Hoek, Leiden. [26 Jun 1845]

Schlitter DA (2005) Order Macroscelida. pp. 82–85, in DE Wilson & DA Reeder (eds.) *Mammal Species of the World. A Taxonomic and Geographic Reference*. Johns Hopkins University Press, Baltimore. Third Edition.

Schlosser M (1887) Die affen, lemuren, chiropteren, insectivoren, marsupialier, creodonten und carnivoren des Europaischen Tertiars. *Beitrage zur Paleontologie Osterreich-Ungarns un des Orients* 6, 1–224.

Schlosser M (1899) F. Ameghino, 1896. Sur l'Evolution des Dents des Mammifères. *Neues Jahrbuch für Mineralogie. Geologie und Paläontologie* 1, 345–353.

Schmeltz JDE (1879) *Museum Godeffroy Catalog VII. Wirbelthiere (Animalia Vertebrata) und Nachträge zu Catalog V & VI aus den übrigen Thierklassen.* L. Friederichsen & Co, Hamburg.

Schneider A (1875) Note sur la psorospermie oviforme du poulpe. *Archives de Zoologie Expérimentale et Générale* 4, xl–xlv.

Schneider DH (1791) *Neuestes Magazin für die Liebhaber der Entomologie.* Volume 1.

Schneider JG (1800) *Vespertilio speoris* Plate 59b, in JCD von Schreber (ed.) *Die Säugthiere in Abbildungen nach der Natur.* Erlangen.

Schönherr CF (1825) Continuation. Tabulae synopticae Familiae Curculionidum. *Isis von Oken* 16–17, 581–588.

Schönherr CJ (1826) *Curculionidum Dispositio Methodica cum generum characteribus, descriptionibus atque observationibus variis, seu prodromus ad synonymiae insectorum.* Fridericum Fleischer, Lipsiae. Part 4.

Schönherr CJ (1833) *Genera et Species Curculionidum, cum synonymia hujus familae.* Species novae aut hactenus minus cognitae, descriptionibus a Dom. Leonardo Gyllenhal, C. H. Boheman, et entomologis aliis illustratæ. Roret, Paris. Volume 1. Part 1.

Schönherr CJ (1836) *Genera et Species Curculionidum, cum synonymia huius familae. Species novae aut hactenus minus cognitae, descriptionibus a Dom. Leonardo Gyllenhal, C. H. Boheman, O.J. Fahraeus et entomologis aliis illustratæ.* Roret, Paris. Volume 3. Part 1.

Schönherr CJ (1845) *Genera et Species Curculionidum, cum synonymia hujus familae.* Species novae aut hactenus minus cognitae, descriptionibus a Dom. Leonardo Gyllenhal, C. H. Boheman, et entomologis aliis illustratæ. Roret, Paris. Volume 8. Part 2.

Schönherr CJ (1848) Mantissa secunda familiae curculionidum seu descriptions novorum quorundum generum curculionidum. *Kongliga Vetenskaps-Academiens Handlingar* 1846–1847, 51–136.

Schrank F von Paula (1798) *Fauna Boica : durchgedachte Geschichte der in Baiern einheimischen und zahmen Thiere.* Stein'schen Buchhandlung, Nürnberg. Volume 1. Part 1.

Schreber JCD von (1774) *Die Säugethiere in Abbildungen nach der Natur*, mit Beschreibungen. Walfgang Walther, Erlangen. Volume 1, Parts 7–9. [p. 166, 189; Plate 52; p. 167, 189, Plate 54; p. 174, 190, Plate 62]

Schreber JCD von (1775–1776) *Die Säugethiere in Abbildungen nach der Natur*, mit Beschreibungen. Walfgang Walther, Erlangen. Plate 85. Volume 3. Part 13 - 1775; Text p. 314, Volume 3, Part 18 - 1776.

Schreber JCD von (1777) *Die Säugethiere in Abbildungen nach der Natur*, mit Beschreibungen. Walfgang Walther, Erlangen. Volume 3, Part 23. [Plate 154 - 1777; p. 552 - 1878]

Schreber JCD von (1778) Die Säugethiere in Abbildungen nach der Natur, mit Beschreibungen. *Wolfgang Walther, Erlangen.* 4(Part 27), 551.

Schreber JCD von (1780) *Die Säugethiere in Abbildungen nach der Natur mit Beschreibungen.* Wolfgang Walther, Erlangen. Volume 4. Parts 31–32. Plate 207.

Schreber JCD von (1784) *Die Säugthiere in Abbildungen nach der Natur mit Beschreibungen.* Wolfgang Walther, Erlangen. Volume 4. Parts 40–41. Plate 255.

Schreber JCD von (1792) *Die Säugthiere in Abbildungen nach der Natur mit Beschreibungen.* Wolfgang Walther, Erlangen. Volume 4. Parts 53–54.

Schuchert P (2006) The European athecate hydroids and their medusa (Hydrozoa, Cnidaria): Capitata Part 1. *Revue Suisse de Zoologie* 113, 325–410.

Schuchert P (2010) *Cuvieria* Péron, 1807. In P Schuchert, World Hydrozoa database. Accessed through: World Register of Marine Species at: http://www.marinespecies.org/aphia.php?p=taxdetails&id=117077 on 2012–06–05.

Schulze E (1897) Mammalia Europaea, in *Helios, Abhandlungen und Vortrage aus dem Gesammtgebiete der Naturwissenschaften* 14, 75–100.

Schultz W (1969) Zur kenntis des Hallstromhunden (*Canis hallstromi*, Troughton 1957). *Zoologischer Anzeiger* 183, 47–72. [Jul-Aug 1969]

Schürhoff PN (1933) Beiträge zur Kenntnis der Cetoniden II. *Entomologisches Nachrichtenblatt* 7, 89–96.

Schwarz E (1909) Über zwei mit *Trichosurus vulpecula* verwanddte Kuscus. *Zoologischer Anzeiger* 34, 625–626. [10 Aug 1909]

Schwarz E (1910a) On the wallaby usually referred to as *Macropus agilis*, Gould. *Annals & Magazine of Natural History* (8)5, 164–166. [1 Feb 1910]

Schwarz E (1910b) Die Grossen Känguruhs und ihre geographischen Formen. *Novitates Zoologicae* 17, 86–108.

Schwarz E, Schwarz HK (1943) The wild and commensal stocks of the house mouse, *Mus musculus*. *Journal of Mammalogy* 24, 59–72. [Feb 1943]

Sclater PL (1870) On additions to the menagerie in February 1870. *Proceedings of the Zoological Society of London* 38, 125–127. [Jun 1870]

Sclater PL (1876) Footnote to a letter addressed to the Secretary by M.L.M.D. D'Albertis. *Proceedings of the Zoological Society of London* 43(1875), 530–532. [Apr 1876]

Sclater PL (1893) List of dates of receipt from the printers of the sheets of the Society's Proceedings from 1831 to 1859 inclusive. *Proceedings of the Zoological Society of London* 61, 435–440. [Aug 1893]

Sclater WL (1887) Mammalia. pp. 37–62, in FE Beddard (ed.) *Zoological Record for 1886; Being Volume the Twenty-Third of the Record of Zoological Nomenclature.* Gurney & Jackson, London. Volume 23.

Sclater WL (1901) *The Mammals of South Africa. Volume 2. Rodentia, Chiroptera, Insectivora, Cetacea and Edentata.* R.H. Porter, London.

Scopoli GA (1777) *Introductio ad Historiam naturalem, sistens genera Lapidum, Plantarum et Animalium hactenus detecta, caracteribus essentialibus donata, in tribus divisa, subinde ad leges Naturæ.* Prague.

Scott AW (1873) *Mammalia, Recent and Extinct: An Elementary Treatise for the Use of the Public Schools of New South Wales*. Section B - Pinnata Seals, dugongs, whales, &c. Thomas Richards, Sydney.

Scott GG (1988) *Taxonomy and diagnostic osteology of extant hairy nosed and common wombats*. MSc thesis, Australian National University, Canberra. [18 Nov 1988]

Scott HH, Lord CE (1920a) Studies on Tasmanian Cetacea. Part 3. *Tursiops tursio. Papers and Proceedings of the Royal Society of Tasmania* 1919, 96–109. [6 Mar 1920]

Scott HH, Lord CE (1920b) Studies on Tasmanian Cetacea. Part 1. *Papers and Proceedings of the Royal Society of Tasmania* 1919, 1–17. [6 Mar 1920]

Scott HH, Lord CE (1926) Studies on Tasmanian mammals, living and extinct. 14. The eared seals of Tasmania. *Papers and Proceedings of the Royal Society of Tasmania* 1925, 187–194. [5 Mar 1926]

Scudder SH (1872) A systematic revision of some of the American butterflies; with brief notes on those that occur in Essex County, Mass. Annual Report of the Trustees of the Peabody Academy of Science 4, 24–83.

Scudder SH (1890) The Tertiary insects of North America. *Report of the United States Geological Survey of the Territories* 13, 1–734.

Sears R, Perrin WF (2009) Blue whale *Balaenoptera musculus*. pp. 120–124, in WF Perrin, B Würsig & JGM Thewissen (eds.) *Encyclopedia of Marine Mammals*. Academic Press, Amsterdam. Second Edition.

Seebeck J (1989) Sciuridae. pp. 932–938, in DW Walton & BJ Richardson (eds.) *Fauna of Australia. Volume 1B. Mammalia*. Australian Government Publishing Service, Canberra.

Seebeck JH (1992) Sub-fossil potoroos in south-eastern Australia; with a record of *Potorous longipes* from New South Wales. *Victorian Naturalist* 109, 173–176. [Oct 1992]

Seebeck J, Johnson PG (1980) *Potorous longipes* (Marsupialia: Macropodidae); a new species from eastern Victoria. *Australian Journal of Zoology* 28, 119–134.

Seebeck JH, Rose RW (1989) Potoroidae. pp. 716–739, in DW Walton & BJ Richardson (eds.) *Fauna of Australia. Volume 1B. Mammalia*. Australian Government Publishing Service, Canberra.

Seebeck JH, Bennett AF, Scott DJ (1989) Ecology of the Potoroidae – a review. pp. 67–88, in G Grigg, P Jarman & I Hume (eds.) *Kangaroos, Wallabies and Rat Kangaroos*. Surrey Beatty, Sydney. Volume 1.

Seeley HG (1882) On *Neusticosaurus pusillus* (Fraas), an amphibious reptile having affinities with terrestrial Nothosauria and with marine Plesiosauria. *Quarterly Journal of the Geological Society of London* 38, 350–366.

Serena M, Soderquist TR, Morris KD (1991) The Chuditch (*Dasyurus geoffroii*). Wildlife Management Program No. 7. Department of Conservation and Land Management, Perth.

Serventy DL (1954) The recent increase of the rarer native mammals. *Western Australian Naturalist* 4, 128–141. [15 Sep 1954]

Sevastianof A (1809) Déscription de quelques nouvelles espèces d'animaux, du Musée Académique. *Mémoires de l'Académie Impériale des Sciences de St. Petersbourg* 1, 443–449.

Shamel HH (1940) The insectivorous bats collected by H.C. Raven in Celebes. *Journal of Mammalogy* 21, 352–354. [Aug 1940]

Sharman GB (1961) The mitotic chromosomes of the marsupials and their bearing on taxonomy and phylogeny. *Australian Journal of Zoology* 9, 38–60.

Sharman GB, Robinson ES, Walton SM, Weaver CM (1980) The chromosomes of rat-kangaroo attributable to Bettongia tropica (Marsupialia: Macropodidae). *Australian Journal of Zoology* 28, 59–63.

Sharman GB, Close RL, Maynes GM (1990) Chromosome evolution, phylogeny and speciation of rock-wallabies (*Petrogale*: Macropodidae). *Australian Journal of Zoology* 37, 351–363.

Shaughnessy PD (1992) New mammals recorded for Australia – Antarctic and subantarctic fur seals *Arctocephalus* species. *Australian Mammalogy* 15, 77–80. [28 Jun 1992]

Shaw G (1790) *The Naturalist's Miscellany: or coloured figures of natural objects; drawn and described immediately from nature*. F.P. Nodder & Co., London. Volume 1. Text to Plate 33. [Jun 1790]

Shaw G (1791) *The Naturalist's Miscellany: or coloured figures of natural objects, drawn and described immediately from nature*. F.P. Nodder & Co., London. Volume 2. Text to Plate 60. [March 1791]

Shaw G (1792) *The Naturalist's Miscellany: or coloured figures of natural objects, drawn and described immediately from nature*. F.P. Nodder & Co., London. Volume 3. Text to Plate 109. [Jun 1792]

Shaw G (1794) *Zoology of New Holland*. J. Sowerby, London.

Shaw G (1797) *The Naturalist's Miscellany: or coloured figures of natural objects, drawn and described immediately from nature*. F.P. Nodder & Co., London. Volume 8. Text to Plate 298. [Jun 1797]

Shaw G (1799) *The Naturalist's Miscellany: or coloured figures of natural objects, drawn and described immediately from nature*. F.P. Nodder & Co., London. Volume 10. Text to Plates 385–386. [Jun 1799]

Shaw G (1800) *General Zoology*, or systematic natural history. Quadrupeds. G. Kearsley, London. Volume 1. Part 2.

Shaw G (1801) *General Zoology*, or systematic natural history. Quadrupeds. G. Kearsley, London. Volume 2. Part 2.

Sheiko BA, Mecklenburg CW (2004) Family Agonidae Swainson 1839 - poachers. *California Academy of Sciences Annotated Checklists of Fishes* 30, 1–27. [Feb 2004]

Sheldon J (1992) *Wild Dogs: The Natural History of the Nondomestic Canidae*. Academic Press, Inc, San Diego.

Shepherd NC (1981) Predation of red kangaroos, *Macropus rufus*, by dingoes, *Canis familiaris dingo* (Blumenbach) in north-western New South Wales. *Wildlife Research* 8, 255–262.

Sherborn CD (1891) Dates of the Parts of P.S. Pallas's 'Icones Insect. p. Ross. Sibir.' and 'Nov. spec. Quadr, Glirium.'. *Annals & Magazine of Natural History* (6)7, 236. [1 Feb 1891]

Sherborn CD (1892) On the dates of the parts, plates, and text of Schreber's 'Saugthiere'. *Proceedings of the Zoological Society of London* (1891) 59, 587–592. [Apr 1892]

Sherborn CD (1895) On the dates of Shaw and Nodder's 'Naturalist's Miscellany'. *Annals & Magazine of Natural History* (6)15, 375–376. [1 Apr 1895]

Sherborn CD (1897a) On the dates of the natural history portion of Savigny's 'Description de l'Egypt'. *Proceedings of the Zoological Society of London* 65, 285–288. [Jun 1897]

Sherborn CD (1897b) Note on the dates of 'The Zoology of the Beagle'. *Annals & Magazine of Natural History* (6)20, 483. [1 Nov 1897]

Sherborn CD (1898) Dates of Blainville's '*Ostéographie*.'. *Annals & Magazine of Natural History* (7)2, 76. [1 Jul 1898]

Sherborn CD (1899) Lacépède's 'Tableau..des mammifères et des oiseaux,' 1799. *Natural Science* 15, 406–409. [Dec 1899]

Sherborn CD (1902) *Index animalium sive index nominum quae ab A.D. MDCCLVIII generibus et speciebus animalium imposta sunt*. J. and C. F. Clay, Cambridge University Press, London.

Sherborn CD (1914) An attempt at fixation of the dates of issue of the parts of the publications of the Musee d'Histoire Naturelle of Paris, 1802–1850. *Annals & Magazine of Natural History* (8)13, 365–368. [1 Apr 1914]

Sherborn CD (1922) *Index Animalium Sive Index Nominum Quae AB A.D. MDCCLVIII generibus et speciebus animalium imposita sunt*. Societatibus Eruditorum Adiuvantibus. Part 1. Introduction, Bibliography and Index A-Aff. pp. 1–128. 1801–1850. British Museum, London.

Sherborn CD (1927) *Index Animalium Sive Index Nominum Quae AB A.D. MDCCLVIII generibus et speciebus animalium imposita sunt*. Societatibus Eruditorum Adiuvantibus. Part 14. Index *laminella-Liyzzia*. pp. 3393–3746. 1801–1850. British Museum, London. December 1927.

Sherborn CD (1928) *Index Animalium Sive Index Nominum Quae AB A.D. MDCCLVIII generibus et speciebus animalium imposita sunt*. Societatibus Eruditorum Adiuvantibus. Part 15. Index *M-A-melanophysa*. pp. 3747–3970. 1801–1850. British Museum, London. [May 1828]

Sherborn CD (1929) *Index Animalium Sive Index Nominum Quae AB A.D. MDCCLVIII generibus et speciebus animalium imposita sunt*. Societatibus Eruditorum Adiuvantibus. Part 20. Index *phyllochroma-Pratincola*. pp. 4931–5138. 1801–1850. British Museum, London. [Oct 1929]

Sherborn CD, Jentink FA (1895) On the dates of the parts of Siebold's '*Fauna Japonica*' and Giebel's '*Allgemeine Zoologie*' (first edition). *Proceedings of the Zoological Society of London* 63(1895), 149–150.

Sherborn CD, Woodward BB (1893) On the dates of the '*Encyclopédie Méthodique*' (Zoology). *Proceedings of the Zoological Society of London* 61, 582–584. [Oct 1893]

Sherborn CD, Woodward BB (1901a) Dates of publication of the zoological and botanical portions of some French voyages. Part II. Ferret and Galinier's '*Voyage en Abyssinie*'; Lefebvre's '*Voyage en Abyssinie*'; '*Exploration scientifique de l'Algérie*'; Castelnau's '*Amérique du Sud*'; Dumont d'Urville's '*Voyage de l'Astrolabe*'; Laplace's '*Voyage sur la Favorite*'; Jacquemont's '*Voyage dans l'Inde*'; Tréhouart's '*Commission scientifique d'Islande*'; Cailliaud, '*Voyage a` Méroe*''; '*Expédition scientifique de Morée*'; Fabre, '*Commission scientifique du Nord*'; Du Petit-Thouars, '*Voyage de la Vénus*'; and on the dates of the '*Faune Française*.' *The Annals and Magazine of Natural History* (7)8,161–164, 333–336, and 491–494. [1 Aug 1901, 1 Oct 1901, 1 Nov 1901]

Sherborn CD, Woodward BB (1901b) Notes on the dates of publication of the natural history portions of some French voyages. Part I. '*Amérique méridionale*'; '*Indes orientales*': '*Pôle Sud*' ('*Astrolabe*' and '*Zélée*'); '*La Bonite*'; '*La Coquille*'; and '*L'Uranie et Physicienne*'. *Annals & Magazine of Natural History* (7)7, 388–392. [1 Apr 1901]

Sherborn CD, Woodward BB (1906) On the dates of publication of the natural history portion of the '*Encyclopedie Methodique*'. *Annals & Magazine of Natural History* (7)17, 577–582. [1 Jun 1906]

Shortridge GC (1910) An account of the geographical distribution of the marsupials and monotremes of south-west Australian, having special reference to the specimens collected during the Balston Expedition of 1904–1907. *Proceedings of the Zoological Society of London* 79(1909), 803–848. [8 Apr 1910]

Shoshani J (1992) Cuvier vis-à-vis Huxley on the relationship of Hyracoidea and an update on an old controversy. pp. 103–112, in F Spitz (ed.) *Ongules / Ungulates 91*: Proceedings of the International Symposium 'Ongulés / Ungulates 91', Toulouse, France, September 2–6, 1991. Société Française pour l'Etude et la Protection des Mammifères, Paris.

Shoshani J (2005) Order Sirenia. pp. 92–93, in DE Wilson & D.A. Reeder (eds.) *Mammal Species of the World. A Taxonomic and Geographic Reference*. Johns Hopkins University Press, Baltimore. Third Edition.

Shoshani J, McKenna MC (1998) Higher taxonomic relationships among extant mammals based on morphology, with selected comparisons of results from molecular data. *Molecular Phylogenetics and Evolution* 9, 572–584.

Shun-Ichiro N (2006) Taxonomic revision of the genus *Stenus* Latreille, 1797 (Coleoptera, Staphylinidae, Steninae) of Japan: Species Group of *S. indubius* Sharp. *Japanese Journal of Systematic Entomology* 12, 39–120.

Sickenberg O (1928) Eine sirene aus dem Leithakalk des Burgenlandes. *Denkschriften, Akademie der Wissenshcaften in Wien. Mathematisch-Naturwissenschafliche* 101, 293–323.

Sigé B (1968) Les chiroptères du Miocene inférieur de Bouzigues. I. Étude systématique. *Palaeovertebrata* 1, 65–133. [Apr 1968]

Silcox MT, Bloch JL, Sargis EJ, Boyer DM (2005) Euarchonta (Dermoptera, Scandentia, Primates. pp. 127–144, in KD Rose & JD Archibald (eds.) *The Rise of Placental Mammals: Origins and Relationships of the Major Extant Clades*. Johns Hopkins University Press, Baltimore.

Sillero-Zubiri C (2009) Family Canidae (Dogs). pp. 352–466, in DE Wilson & RA Mittermeier (eds.) *Handbook of the Mammals of the World. Volume 1. Carnivores*. Lynx Editions, Barcelona.

Simmons NB (1993) The importance of methods, archontan phylogeny and cladistic analysis of morphological data. pp. 1–61, in RDE MacPhee (ed.) *Primates and their Relatives in Phylogenetic Perspective*. Plenum Publishing, New York.

Simmons NB (1995) Bat relationships and the origin of flight. *Symposia of the Zoological Society of London* 67, 27–43.

Simmons NB (1998) A reappraisal of interfamilial relationships of bats. pp. 3–26, in TH Kunz & PA Racey (eds.) *Bat Biology and Conservation*. Smithsonian Institution Press, Washington, D.C.

Simmons NB (2005a) Order Chiroptera. pp. 312–529, in DE Wilson & DM Reeder (eds.) *Mammal Species of the World: A Taxonomic and Geographic Reference*. Johns Hopkins University Press, Baltimore. Third Edition.

Simmons NB (2005b) Chiroptera. pp. 159–174, in KD Rose & JD Archbold (eds.) *The Rise of Placental Mammals: Origins and Relationships of the Major Extant Clades*. Johns Hopkins University Press, Baltimore.

Simmons NB, Geisler JH (1998) Phylogenetic relationships of *Icaronycteris, Archaeonycteris, Hassianycteris*, and *Palaeochiropteryx* to extant bat lineages, with comments on the evolution of echolocation and foraging strategies in Microchiroptera. *Bulletin of the American Museum of Natural History* 235, 1–182.

Simon ME (1886) Espèces et genres nouveaux de la famille des Thomisidae. *Actes de la Société Linnéenne de Bordeaux* 40(4)10, 167–187.

Simpson GC (1928) *A Catalogue of the Mesozoic Mammalia in the Geological Department of the British Museum*. London.

Simpson GG (1925) Mesozoic mammals. 3. Preliminary comparisons of Jurassic mammals except multituberculates. *American Journal of Science* 10(5), 559–569.

Simpson GG (1930) Post-Mesozoic Marsupialia. Part 47. pp. 5–69, in JF Pompeckj (ed.) *Fossilium Catalogus. 1. Animalia*. W. Junk, Berlin.

Simpson GG (1931) A new classification of Mammals. *Bulletin of the American Museum of Natural History* 59, 259–293.

Simpson GG (1932) Fossil Sirenia of Florida and the evolution of the Sirenia. *Bulletin of the American Museum of Natural History* 59, 419–503.

Simpson GG (1945) The principles of classification and a classification of mammals. *Bulletin of the American Museum of Natural History* 85, 1–350. [5 Oct 1945]

Simpson GG (1954) Tendencies actuelles de la systematique des mammifères. *Mammalia* 18, 337–357.

Simpson GG (1961) Historical zoogeography of Australian mammals. *Evolution* 15, 431–446.

Simpson GG (1970) The Argyrolagidae, extinct South American marsupials. *Bulletin of the Museum of Comparative Zoology. Harvard University* 139, 1–86.

Sinclair EA, Westerman M (1997) Phylogenetic relationships within the genus *Potorous* (Marsupialia: Potoroidae) based on allozyme electrophoresis and sequence analysis of cytochrome-b gene. *Journal of Mammalian Evolution* 4, 147–161. [Sep 1997]

Sinclair EA, Danks A, Wayne AF (1996) Rediscovery of Gilbert's potoroo, *Potorous tridactylus*, in Western Australia. *Australian Mammalogy* 19, 69–72. [23 Nov 1996]

Sinclair EA, Murch AR, Renzo MD, Palermo M (2000) Chromosome morphology in Gilbert's Potoroo, *Potorous gilberti* (Marsupialia: Potoroidae). *Australian Journal of Zoology* 48, 281–287.

Sinzow J Von (1898) Über ein neues genus der neogenen Cetaceen. *Verhandlungen der Russisch-Kaiserlichen Mineralogischen Gesellschaft zu St. Petersburg* 35, 117–134.

Sivertsen E (1954) A survey of the eared seals (Family Otariidae) with remarks on the Antarctic seals collected by M/K 'Norvegica' in 1928–1929. *Det Norske Videnkap-Akademi I oslo. Scientific Results of the Norwegian Antarctic Expedition* 36, 1–76.

Skinner JD, Chimimba CT (2005) *Mammals of the Southern African Subregion*. Cambridge University Press, Cambridge. Second Edition.

Slade RW, Moritz C, Hoelzel AR, Burton HR (1998) Molecular population genetics of the southern elephant seal *Mirounga leonina*. *Genetics* 149, 1945–1957. [1 Aug 1998]

Slaughter BH (1965) A therian from the Lower Cretaceous (Albian) of Texas. *Postilla* 93, 1–18.

Slijper EJ (1936) Die Cetaceen. Vergleichend-anatomisch und systematisch. *Capita Zoologica* 7, i–xv, 1–590.

Smeenk C (2009) Has one of Captain Cook's possums landed in Leiden? The possible holotype of *Pseudocheirus peregrinus* (Boddaert, 1785). *Zoologische Mededelingen, Leiden* 83(19), 723–740.

Smirnov N (1908) Revue des pinnipèdes Russes. *Mémoires de L'Académie Imperiale des Sciences de St Petersbourg* (8)23(4), 1–75. In Russian.

Smith A (1829) Contributions to the natural history of South Africa. *The Zoological Journal, London* 4, 433–444.

Smith A (1834) African zoology. *South African Quarterly Journal* (2)3(3), 233–256.

Smith A (1849 [1838–1849]) *Illustrations of the Zoology of South Africa*: consisting chiefly of figures and descriptions of the objects of natural history collected during an expedition into the interior of South Africa, in the years 1834, 1835, and 1836; fitted out by 'The Cape of Good Hope Association for Exploring Central Africa.' Reptilia. Smith, Elder and Co., London. Appendix.

Smith ABT (2002) *Checklist of the Scarabaeoidea of the Nearctic Realm*. Version 2. From http://www.bio-nica.info/biblioteca/Smith2002ScarabaeoideaNearctica.pdf. Accessed 1 June 2012. [30 Aug 2002]

Smith AMA (1983) The subspecific biochemical taxonomy of *Antechinus minimus, A. swainsonii* and *Sminthopsis leucopus* (Marsupialia: Dasyuridae). *Australian Journal of Zoology* 32, 753–762.

Smith AP (1984) The species of living possums and gliders. pp. xiii–xv, in AP Smith & ID Hume (eds.) *Possums and Gliders*. Surrey Beatty, Sydney.

Smith CH (1827) Order VII. Ruminatia. Pecora, Lin. pp. 296–376, in JE Gray (ed.) Synopsis of the species of the class Mammalia. pp. 1–391, Volume 5. In E Griffith, CH Smith & E Pidgeon (eds.) (16 vols: 1824–1835) *The Animal Kingdom arranged in conformity with its organisation, by the Baron Cuvier, member of the Institute of France etc, with additional descriptions of all the species hitherto named, and of many not before noticed*. G.B. Whittaker & Co., London. Author of volume given as JE Gray by Anon (1903: 410). [Sep 1827]

Smith CH (1842) *An Introduction to the Mammalia*. In W. Jardine (ed.) *Naturalist's Library*. Volume 15. Lizars, Edinburgh.

Smith JC (1993) *Georges Cuvier: an annotated bibliography of his published works*. Smithsonian Institution Press, Washington, D.C.

Smith JDD (2001) *Official Lists and Indexes of Names and Works in Zoology. Supplement 1986–2000*. International Commision of Zoological Nomenclature, London.

Smith JD, Hood CS (1983) A new species of tube-nosed fruit bat (*Nyctimene*) from the Bismarck Archipelago, Papua New Guinea. *Occasional Papers of the Museum, Texas Tech University* 81, 1–14.

Smith MJ (1973) *Petaurus breviceps*. *Mammalian Species* 30, 1–5. [13 Jun 1973]

Smith WP (1991) *Odocoilus virginianus*. *Mammalian Species* 388, 1–13. [6 Nov 1991]

Smuts J (1832) *Dissertatio zoologica, enumerationem mammalium capensium continens*. Apud J.C. Cyfveer, Leidae.

Sody HJV (1930) On some new or insufficiently known mammals from Java, Borneo and Celebes. *Natuurkundig Tijdschrift voor Nederlandsch-Indië* 90, 258–273.

Sody HJV (1931) Six new mammals from Sumatra, Java, Bali and Borneo. *Natuurkundig Tijdschrift voor Nederlandsch-Indie* 91, 349–360.

Sody HJV (1936) Seventeen new generic, specific and subspecific names for Dutch East Indian mammals. *Natuurkundig Tijdschrift voor Nederlandsch-Indie* 96, 42–55.

Sody HJV (1941) On a collection of rats from the Indo-Malayan and Indo Australian regions with descriptions of 43 new genera, species and subspecies. *Treubia* 18, 255–325.

Sonnini CS (1804) Wombat. pp. 480–482, in *Nouveau Dictionnaire d'Histoire Naturelle*, appliquée aux arts, principalement à l'agriculture et à l'économie rurale et domestique: par une société de naturalistes et d'agricltureurs: avec des figures tirées des trios règnes de la nature. Déterville, Paris. First Edition. Volume 23. VAL-ZYZ.

Sonntag CF (1922) On the myology and classification of the wombat, koala and phalangers. *Proceedings of the Zoological Society of London* 92(2), 863–896.

Soot-Ryan T (1961) On a Bryde's whale stranded on Curacao. *Norsk Hvalfangst Tid* 50(8), 323–332. [Reference not seen.]

Sowerby J (1804) Extracts from the Minute-Book of the Linnean Society of London. *Transactions of the Linnean Society of London* (1)7, 309–310. [8–21 Nov 1804]

Spamer EE, Daescher E, Vostreys-Shapiro LG (1995) *A study of fossil vertebrate types in the Academy of Natural Sciences Philadelphia*. Academy of Natural Sciences Philadelphia. Special Publication 16. [5 Nov 1995]

Spaulding M, O'Leary MA, Gatesy J (2009) Relationships of Cetacea (Artiodactyla) among mammals: increased taxon sampling alters interpretations of key fossils and character evolution. *PLoS ONE* 4(9), 1–14. [23 Sep 2009]

Spencer B (1895) Preliminary notice of two new species of marsupials from central Australia. *Proceedings of the Royal Society of Victoria* 7, 222–224.

Spencer B (1896a) Mammalia. pp. 1–52, in B Spencer (ed.) *Report on the Work of the Horn Scientific Expedition to Central Australia. Part 2. Zoology*. Melville, Mullen & Slade, Melbourne. [Feb 1896]

Spencer B (1896b) Preliminary description of certain new marsupials from central Australia, together with remarks upon the occurrence and identity of *Phascogale cristicauda*. *Proceedings of the Royal Society of Victoria* new series 8, 5–13.

Spencer B (1896c) Through Larapinta Land: a narrative of the Horn Expedition to Central Australia. pp. 1–136, in B Spencer (ed.) *Report on the Work of the Horn Scientific Expedition to Central Australia. Part 1. Introduction, Narrative, Summary of Results, Supplementto Zoological Report, Map*. Melville, Mullin and Slade, Melbourne.

Spencer B (1897) Description of two new species of marsupials from Central Australia. *Proceedings of the Royal Society of Victoria* 9, 5–11.

Spencer B (1909) Description of a new species of *Sminthopsis*. *Proceedings of the Royal Society of Victoria* 21, 449–451.

Spencer B, Kershaw, JA (1910a) A collection of sub-fossil bird and marsupial remains from King Island, Bass Strait. *Memoirs of the National Museum of Melbourne* 3, 5–35. [Feb 1910]

Spencer B, Kershaw JA (1910b) On the existing species of the genus *Phascolomys*. *Memoirs of the National Museum of Melbourne* 3, 37–65. [Feb 1910]

Spencer PBS, Rhind SG, Eldridge MDB (2001) Phylogenetic structure within *Phascogale* (Marsupialia: Dasyuridae) based on partial cytochrome *b* sequence. *Australian Journal of Zoology* 49, 369–377.

Spinola M (1837) *Essai sur les Genres d'Insectes appartenants à l'ordre des Hémiptères, Lin., ou Rhyngotes, Fab., et la section des Hétéroptères, Dufour*. Yves Gravier, Gènes.

Springer MS (1997) Molecular clocks and the timing of the placental and marsupial radiations in relation to the Cretaceous-Tertiary boundary. *Journal of Mammalian Evolution* 4, 285–302. [Dec 1997]

Springer MS, de Jong WW (2001) Phylogenetics. Which mammalian supertree to bark up? *Science* 291(5509), 1709, 1711. [2 Mar 2001]

Springer MS, Westerman M, Kirsch JAW (1994) Relations among orders and families of marsupials based on 12S ribosomal DNA sequences and the timing of the marsupial radiation. *Journal of Mammalian Evolution* 2, 85–115. [Jun 1994]

Springer MS, Teeling EC, Madsen O, Stanhope MJ, de Jong WW (2001) Integrated fossil and molecular data reconstruct bat echolocation. *Proceedings of the National Academy of Sciences of the United States of America* 98, 6241–6246. [22 May 2001]

Springer MS, Murphy WJ, Eizirik E, O'Brien SJ (2005) Molecular evidence for major placental clades. pp. 37–49, in KD Rose & JD Archibald (eds.) *The Rise of Placental Mammals: Origins and Relationships of the Major Extant Clades*. Johns Hopkins University Press, Baltimore.

Springer MS, Murphy WJ, Eizirik E, Madsen O, Scally M, Douady CJ, Teeling EC, Stanhope MJ, de Jong WW, O'Brien SJ (2007) A molecular classification for the living orders of placental mammals and the phylogenetic placement of primates. pp. 1–28, in M Ravosa & M Dagosto (eds.) *Primate Origins: Adaptations and Evolution*. Plenum, New York.

St Mivart G (1868) Notes on the osteology of the Insectivora. *Journal of Anatomy and Physiology* 2, 117–154.

Stacey PJ, Arnold PW (1999) *Orcaella brevirostris*. *Mammalian Species* 616, 1–8. [5 May 1999]

Stacey PJ, Leatherwood S, Baird RW (1994) *Pseudorca crassidens*. *Mammalian Species* 456, 1–6. [2 Jun 1994]

Stafford BJ, Szalay FS (2000) Craniodental functional morphology and taxonomy of dermopterans. *Journal of Mammalogy* 81, 360–385. [May 2000]

Stafford BJ, Thorington RW (1998) Carpal development and morphology in archontan mammals. *Journal of Morphology* 235, 135–155.

Stafford KM, Bohnenstiehl D, Tolstoy M, Chapp E, Mellinger DK, Moore SE (2004) Antarctic type blue whale calls recorded at low latitudes in the Indian and eastern Pacific oceans. *Deep-sea Research. Part I, Oceanographic Research Papers* 51, 1337–1346.

Stains HJ (1984) Carnivores. pp. 491–521, in S Anderson & JK Jones (eds.) *Orders and Familes of Recent Mammals of the World*. John Wiley & Sons, New York.

Stål C (1869) Bidrag till membracidernas kännedom. *Öfversigt af Kongliga Vetenskaps-Akademiens Förhandlingar, Stockholm* 26, 231–300.

Stanhope MJ, Waddell VG, Madsen O, de Jong W, Hedges SB, Cleven GC, Kao D, Springer MS (1998) Molecular evidence for multiple origins of Insectivora and for a new order of endemic African insectivore mammals. *Proceedings of the National Academy of Sciences of the United States of America* 95(17), 9967–9972. [18 Aug 1998]

Starnes WC (1988) Revision, phylogeny and biogeographic comments on the circumtropical marine percoid fish family Priacanthidae. *Bulletin of Marine Science* 43, 117–203.

Start AN, Moro D, Adams M, Bencini R (2006) Dunnarts from Boullanger Island: new evidence and reassessment of a taxonomic issue with resource implications. *Australian Mammalogy* 28, 51–58. [10 May 2006]

Start AN, Burbidge AA, McDowell MC, McKenzie NL (2012) The status of non-volant mammals along a rainfall gradient in the south-west Kimberley, Western Australia. *Australian Mammalogy* 34, 36–48.

Start T, Burbidge A, Sinclair E, Wayne A (1995) Lost & Found: Gilbert's Potoroo. *Landscope* 10, 29–33.

Steeman ME (2007) Cladistic analysis and revised classification fossil and recent mysticetes. *Zoological Journal of the Linnean Society* 150, 875–894. [2 Aug 2007]

Steeman ME (2010) The extinct baleen whale fauna from the Miocene-Pliocene of Belgium and the diagnostic cetacean ear bones. *Journal of Systematic Palaeontology* 8, 63–80.

Stein G (1932) Eine neue Beuteliere aus Neuguinea. *Zeitschrift fur Saugetierkunde* 7, 254–257.

Stein G (1933) Weitere mitteilungen zur systematik papuanischer Säuger. *Zeitschrift fur Saugetierkunde* 8, 87–95.

Stejneger L (1883) Contributions to the history of the Commander Islands. No. 1. Notes on the natural history, including descriptions of new cetaceans. *Proceedings of the United States National Museum* 6, 58–89. [20–22 Jun 1883]

Stephens D (2011) The molecular ecology of Australian wild dogs: hybridisation, gene flow and genetic structure at multiple geographic scales. PhD Thesis. University of Western Australia, Perth.

Stephens JF (1829a) *The Nomenclature of British Insects; being a compendious list of such species as are contained in the Systematic Catalogue of British Insects, and forming a guide to their classification, &c.* Baldwin & Cradock, London.

Stephens JF (1829b) *Illustrations of British Entomology : or, A synopsis of Indigenous Insects : containing their generic and specific distinctions; with an account of their metamorphoses, times of appearance, localities, food, and economy, as far as practicable.* Baldwin and Cradock, London. Volume 3.

Stephens JF (1829c) *Systematic Catalogue of British Insects*. Baldwin & Cradock, London.

Stephenson NG (1963) Growth gradients among fossil monotremes and marsupials. *Palaeontology* 6, 615–624.

Stewart BS, Leatherwood S (1985) Minke whale *Balaenoptera acutorostrata* (Lacépède, 1804). pp. 91–136, in SH Ridgway & R Harrison (eds.) *Handbook of Marine Mammals. Volume 3. The Sirenians and Baleen Whales*. Academic Press, London.

Steyskal GC (1980) [Review] Gotch. A.F. Mammals – their Latin names explained. *Journal of Mammalogy* 61, 581–583. [Aug 1980]

Stirling EC (1888) A new Australian mammal. *Nature* 38(990), 588–589. [18 Oct 1888]

Stirling EC (1889a) Minutes. *Transactions of the Royal Society of South Australia* 12, 157–158.

Stirling EC (1889b) Preliminary notes on a new Australian mammal. *Transactions of the Royal Society of South Australia* 11, 21–24. [Apr 1889]

Stirling EC (1891a) Description of a new genus and species of Marsupialia, *Notoryctes typhlops*. *Transactions of the Royal Society of South Australia* 14, 154–187. [Jul 1891]

Stirling EC (1891b) Further notes on the habits and anatomy of *Notoryctes typhlops*. *Transactions of the Royal Society of South Australia* 14, 283–291. [Dec 1891]

Stirling EC (1894) Supplemental note on the osteology of *Notoryctes typhlops*. *Transactions of the Royal Society of South Australia* 18, 1–2. [Nov 1894]

Stirling I (1971) *Leptonychotes weddelli*. *Mammalian Species* 6, 1–5. [19 Jan 1971]

Stirton RA (1955) Late Tertiary marsupials from South Australia. *Records of the South Australian Museum* 11, 247–268. [28 Feb 1955]

Stirton RA (1963) A review of the macropod genus *Protemnodon*. *University of California Publications in Geological Sciences* 44, 97–161. [28 Feb 1963]

Stirton RA, Tedford RH, Woodburne MO (1967) A new Tertiary formation and fauna from the Tirari Desert, South Australia. *Records of the South Australian Museum* 15, 427–462. [21 Dec 1967]

Stokes JL (1846) *Discoveries in Australia*; with an account of the coasts and rivers explored and surveyed during the voyage of H.M.S. Beagle, in the years 1837–38–39–40–41–42–43, by command of the Lords Commissioners of the Admiralty, also a narration of Captain Owen Stanley's visits to the islands in the Arafura Sea. Volume 1.

Stoll C (1787) *Natuurlijke en naar het leven nauwkeurig gekleurde afbeeldingen en beschrijvingen der Spoken, Wandelende Bladen, Zabel springhanen, Krekels, Trekspringhanen en Kakkerlakken. In alle vier deelen der wereld Europia, Asia, Afrika en Amerika, Huishoudende. Représentation exactement colorée d'après nature des Spectres ou Phasmes, des Mantes, des Sauterelles, des Grillions, des Criquets et des Blattes. Qui se trouvent dans les quatre parties du monde, &c.* Amsterdam. Introduction. J.C. Sepp et Fils, Amsterdam.

Storr GCC (1780) *Prodromus Methodi Mammalium.. inauguralem disputationem propositus*. Tubingœ.

Strahan R (1980a) Recommended common names of Australian mammals. *Australian Mammal Society Bulletin* 6(2), 13–23.

Strahan R (1980b) Lack of provenance of *Tadarida norfolkensis*. *Australian Bat Research News* 16, 7–10.

Strahan R (ed.) (1983) *The Complete Book of Australian Mammals*. Angus and Robertson, Sydney.

Strahan R (1987) *What Mammal is That*. Angus and Robertson, Sydney.

Strahan R (1992) *Encyclopedia of Australian Animals: Mammals*. Angus & Robertson, Sydney.

Strahan R (ed.) (1995) *Mammals of Australia*. Reed Books, Sydney.

Strand E (1828) Miscellanea nomenclatorica zoologica et palaeontologica. *Archiv für Naturgeschichte* 92, 30–75.

Strand E (1934) Miscellanea nomenclatorica zoologica et palaeontologica. *Folia Zoologica et Hydrobiologica* 6, 271–277.

Stucky RK, McKenna MC (1993) Mammalia. pp. 739–771, in MJ Benton (ed.) *The Fossil Record 2*. Chapman & Hall, London.

Sturm J (1826) *Catalog meiner Insecten-Sammlung*. Erster Theil. Käfer. Nürnberg.

Sturt C (1848) *Narrative of an Expedition into Central Australia Performed Under the Authority of Her Majesty's Government, During the Years 1844, 5, 6, Together with a Notice of the Province of South Australia in 1847*. T & W. Boone, London.

Sundevall CJ (1843) Om Professor J. Hedenborgs insamlingar af Däggdjur i Nordöstra Africa och Arabien. *Kongliga Vetenskaps-Akademiens Handlingar, Stockholm* 1842, 189–282.

Sundevall CJ (1845) Öfversigt af däggdjursordningen pecora. *Öfversigt af Kongliga Vetenskaps-Akademiens Förhandlingar, Stockholm* 2(3), 26–32.

Sundevall CJ (1846a) Nya Mammalia från Sydafrika. *Öfversigt af Kongliga Vetenskaps-Akademiens Förhandlingar, Stockholm* 3(5), 118–121.

Sundevall CJ (1846b) Methodisk öfversigt af Idislade djurten, Linnés Pecora. *Kongliga Vetenskaps-Akademiens Handlingar, Stockholm* 1844, 121–210.

Suzuki H, Nunome M, Kinoshita G, Aplin KP, Vogel P, Kryukov AP, Jin M-L, Han S-H, Maryanto I, Tsuchiya K, Ikeda H, Shiroishi T, Yonekawa H, Moriwaki K (2013) Evolutionary and dispersal history of Eurasian house mice *Mus musculus* clarified by more extensive geographic sampling of mitochondrial DNA. *Heredity* 111, 375–390. [3 Jul 2013]

Svihla A (1936) The Hawaiian rat. *The Murrelet – A Journal of the Northwestern Ornithology and Mammalogy* 17, 3–14.

Swainson W (1831 [1831–1832]) *Zoological Illustrations, or original figures and descriptions of new, rare, or interesting Animals, selected chiefly from the classes of Ornithology, Entomology, and Conchology, &c*. Second Series. Volume 2.

Swainson W (1832 [1832–1833]) *Zoological Illustrations, or original figures and descriptions of new, rare, or interesting Animals, selected chiefly from the classes of Ornithology, Entomology, and Conchology, &c*. Second Series. Volume 3.

Swainson W (1833 [1832–1833]) *Zoological Illustrations, or Original Figures and Descriptions of New, Rare, or Interesting Animals, Selected Chiefly from the Classes of Ornithology, Entomology and Conchology, and Arranged According to their Natural Affinities*. Baldwin & Cradock, London. Volume 3. Second Series.

Swainson W (1835) *On the Natural History and Classification of Quadrupeds. The Cabinet Cyclopaedia, conducted by the Rev. Dionysius Lardner*. Longman, London.

Swainson W (1836) *On the Natural History and Classification of Birds*. Volume 1. Lardner's Cabinet Cyclopedia. Longman, Rees, Orme, Brown, Green, and Longman, London.

Swainson W (1837) *On the Natural History and Classification of Birds*. Lardner's Cabinet Cyclopedia. Longman, Rees, Orme, Brown, Green, and Longman, London. Volume 2.

Swainson W (1839) *The Natural History and Classification of Fishes, Amphibians, & Reptiles, or Monocardian Animals*. Longman, Orme, Brown, Green & Longmans, and John Taylor. London. Volume 2.

Swainson W, Richardson J (1832) *Fauna Boreali-Americana; or the Zoology of the northern parts of British America: Containing descriptions of the objects of natural history collected on the late northern land expeditions under the command of Captain Sir John Franklin, R.N.* Volume 2. The Birds. John Murray, London.

Sweeting RH (1840) Dimensions and description of a supposed new species of *Balaenoptera*, stranded on Charmouth Beach, February 5, 1840. *The Magazine of Natural History* 4(2), 341–343. [1 Jul 1840]

Szalay FS (1982) A new appraisal of marsupial phylogeny and classification. pp. 621–640, in M Archer (ed.) *Carnivorous Marsupials*. Royal Zoological Society of New South Wales, Sydney. Volume 2. [Feb 1982]

Szalay FS (1993) Metatherian taxon Phylogeny: evidence and interpretation from cranioskeletal system. pp. 216–242, in FS Szalay, MJ Novacek & MC McKenna (eds.) *Mammal Phylogeny: Mesozoic Differentiation, Multituberculates, Monotremes, Early Therians and Marsupials*. Springer-Verlag, New York.

Szalay FS (1994) *Evolutionary History of the Marsupials and an Analysis of Osteological Characters*. Cambridge University Press, Cambridge.

Szalay FS, Lucas SG (1993) Cranioskeletal morphology of archontans and diagnoses of Chiroptera, Volitantia and Archonta. pp. 187–226, in RDE MacPhee (ed.) *Primates and Their Relatives in Phylogenetic Perspective*. Plenum Publishing, New York.

Szalay FS, Lucas SG (1996) The postcranial morphology of Paleocene *Chriacus* and *Mixodectes* and the phylogenetic relationships of archontan mammals. *New Mexico Museum of Natural History and Science Bulletins* 7, 1–47.

Taczanowski L (1872) Les aranéides de la Guyane Française. *Horae Societatis Entomologicae Rossicae* 9, 64–150.

Takami K, Yoshida M, Yamamoto Y, Harada M, Furuyama J (1998) Genetic variation of mitochondrial cytochrome b genes among the subspecies of koala, *Phascolarctos cinereus*. *The Journal of Veterinary Medical Science* 60, 1161–1163.

Tate GHH (1935) Rodents of the genera *Rattus* and *Mus* from the Pacific Islands, collected by the Whitney South Sea Expedition, with a discussion of the origin and races of the Pacific island rat. *Bulletin of the American Museum of Natural History* 68, 145–178. [11 Feb 1935]

Tate GHH (1938) New or little known marsupials: A new species of Phascogalinae, with notes upon *Acrobates pulchellus* Rothschild. *Novitates Zoologicae* 41, 58–60.

Tate GHH (1940) Notes on the types of certain early described species of monotremes, marsupials, Muridae and bats from the Indo-Australian region. *American Museum Novitates* 1061, 1–10. [1 May 1940]

Tate GHH (1941a) Results of the Archbold Expeditions. No. 40. Notes on vespertilionid bats of the subfamilies Miniopterinae, Murininae, Kerivoulinae, and Nyctophilinae. *Bulletin of the American Museum of Natural History* 78, 567–597. [31 Dec 1941]

Tate GHH (1941b) Results of the Archbold Expeditions. No. 36. Remarks on some old world leaf-nosed bats. *American Museum Novitates* 1140, 1–11. [20 Aug 1941]

Tate GHH (1941c) Results of the Archbold Expeditions. No. 35. A review of the genus *Hipposideros* with special reference to Indo-Australian species. *Bulletin of the American Museum of Natural History* 78, 353–393. [19 Aug 1941]

Tate GHH (1941d) Results of the Archbold Expeditions. No. 37. Notes on oriental *Taphozous* and allies. *American Museum Novitates* 1141, 1–5. [20 Aug 1941]

Tate GHH (1941e) Results of the Archbold Expeditions. No. 38. Molossid bats of the Archbold Collections. *American Museum Novitates* 1142, 1–4. [20 Aug 1941]

Tate GHH (1941f) Results of the Archbold Expeditions. No. 39. A review of the genus *Myotis* (Chiroptera) of Eurasia, with special reference to species occurring in the East Indies. *Bulletin of the American Museum of Natural History* 78, 537–565. [29 Dec 1941]

Tate GHH (1942a) Results of the Archbold Expeditions. No. 48. Pteropodidae (Chiroptera) of the Archbold collections. *Bulletin of the American Museum of Natural History* 80, 331–347. [31 Dec 1942]

Tate GHH (1942b) Results of the Archbold Expedition. No. 46. A new genus and species of fruit bats, allied to *Nyctimene*. *American Museum Novitates* 1204, 1–2. [23 Oct 1942]

Tate GHH (1942c) Results of the Archbold Expeditions. No. 47. Review of the Vespertilioninae bats, with special attention to genera and species of the Archbold collections. *Bulletin of the American Museum of Natural History* 80, 221–297. [27 Nov 1942]

Tate GHH (1945a) Results of the Archbold Expeditions. No. 55. Notes on the squirrel-like and mouse-like possums (Marsupialia). *American Museum Novitates* 1305, 1–12. [29 Dec 1945]

Tate GHH (1945b) Results of the Archbold Expeditions. No. 54. The marsupial genus *Pseudocheirus* and its subgenera. *American Museum Novitates* 1287, 1–30. [11 June 1945]

Tate GHH (1945c) Results of the Archbold Expeditions. No. 52. The marsupial genus *Phalanger*. *American Museum Novitates* 1283, 1–31. [27 Apr 1945]

Tate GHH (1947) Results of the Archbold Expeditions. No. 56. On the anatomy and classification of the Dasyuridae (Marsupialia). *Bulletin of the American Museum of Natural History* 88, 97–155. [20 Feb 1947]

Tate GHH (1948a) Results of the Archbold Expeditions. No. 60. Studies in the Peramelidae (Marsupialia). *Bulletin of the American Museum of Natural History* 92, 313–346. [25 Nov 1948]

Tate GHH (1948b) Results of the Archbold Expeditions. No. 59. Studies on the anatomy and phylogeny of the Macropodidae (Marsupialia). *Bulletin of the American Museum of Natural History* 91, 233–351. [1 Nov 1948]

Tate GHH (1950) The Muridae of the Cocos-Keeling Islands. *Bulletin of the Raffles Museum* 22, 271–277.

Tate GHH (1951a) The Banded Anteater, *Myrmecobius* Waterhouse (Marsupialia). *American Museum Novitates* 1521, 1–8. [18 Jun 1951]

Tate GHH (1951b) The wombats (Marsupialia, Phascolomyidae). *American Museum Novitates* 1525, 1–18. [13 Jul 1951]

Tate GHH (1951c) Results of the Archbold Expeditions. No. 65. The rodents of Australia and New Guinea. *Bulletin of the American Museum of Natural History* 97, 183–430. [30 Oct 1951]

Tate GHH (1952a) Results of the Archbold Expeditions. No. 66. Mammals of Cape York Peninsula, with notes on the occurrence of rainforest in Queensland. *Bulletin of the American Museum of Natural History* 98, 563–616. [25 Mar 1952]

Tate GHH (1952b) Results of the Archbold Expeditions. No. 67. A new *Rhinolophus* from Queensland (Mammalia, Chiroptera). *American Museum Novitates* 1578, 1–3. [30 Jul 1952]

Tate GHH (1955) Proposed use of the plenary powers to validate the specific name '*dingo*' Meyer, 1793, as published in the combination '*Canis dingo*' as the name for the dingo (Class Mammalia). *Bulletin of Zoological Nomenclature* 11, 121. [28 Feb 1955]

Tate GHH, Archbold R (1935a) Results of the Archbold Expeditions. No. 5. Seven apparently new forms of Phalangeridae from the New Guinea region. *American Museum Novitates* 810, 1–8. [19 Jul 1935]

Tate GHH, Archbold R (1935b) Results of the Archbold Expeditions. No. 4. An apparently new race of wallabies from southern New Guinea. *American Museum Novitates* 804, 1–2. [18 May 1935]

Tate GHH, Archbold R (1935c) Results of the Archbold Expeditions. No. 2. Twelve apparently new forms of *Rattus* from the Indo-Australian region. *American Museum Novitates* 802, 1–10. [18 May 1935]

Tate GHH, Archbold R (1936) Results of the Archbold Expeditions. No. 8. Four apparently new polyprotodont marsupials from New Guinea. *American Museum Novitates* 823, 1–4. [9 Mar 1936]

Tate GHH, Archbold R (1937) Results of the Archbold Expeditions. No. 16. Some marsupials of New Guinea and Celebes. *Bulletin of the American Museum of Natural History* 73, 331–476. [11 June 1937]

Tate GHH, Archbold R (1938) Results of the Archbold Expeditions. No. 18. Two new Muridae from the Western Division of Papua. *American Museum Novitates* 982, 1–2. [12 May 1938]

Tate GHH, Archbold R (1939) Results of the Archbold Expeditions. No. 24. Oriental *Rhinolophus*, with special reference to material from the Archbold collections. *American Museum Novitates* 1036, 1–12. [11 Aug 1939]

Tate GHH, Archbold R (1941) Results of the Archbold Expeditions. No. 31. New rodents and marsupials from New

Guinea. *American Museum Novitates* 1101, 1–9. [30 Jan 1941]

Taylor AC, Cooper DW (1999) Microsatellites identify introduced New Zealand tammar wallabies (*Macropus eugenii*) as an 'extinct' taxon. *Animal Conservation* 2, 41–49.

Taylor AC, Foulkes J (2004) Molecules and morphology: a taxonomic analysis of the common brushtail possum *Trichosurus vulpecula* with an emphasis on the central Australian form. pp. 455–470. In RL Goldingay & SM Jackson (eds.) *The Biology of Australian Possums and Gliders*. Surrey Beatty & Sons, Sydney.

Taylor EH (1934) Philippine land mammals. *Monographs of the Philippine Bureau of Science* 30, 1–548.

Taylor JM, Calaby JH (1988a) *Rattus fuscipes*. *Mammalian Species* 298, 1–8. [15 Jan 1988]

Taylor JM, Calaby JH (1988b) *Rattus lutreolus*. *Mammalian Species* 299, 1–7. [15 Jan 1988]

Taylor JM, Horner BE (1967) Results of the Archbold Expeditions. No. 88. The historical misapplication of the name *Mus fuscipes* and a systematic re-evaluation of *Rattus lacus* (Rodentia, Muridae). *American Museum Novitates* 2281, 1–14. [30 Jan 1967]

Taylor JM, Horner BE (1973) Results of the Archbold Expeditions. No. 98. Systematics of native Australian *Rattus* (Rodentia, Muridae). *Bulletin of the American Museum of Natural History* 150, 1–130. [8 Jan 1973]

Taylor JM, Calaby JH, Van Deusen HM (1982) A revision of the genus *Rattus* (Rodentia, Muridae) in the New Guinean region. *Bulletin of the American Museum of Natural History* 173, 177–336. [30 Jul 1982]

Taylor RJ, O'Neill MG, Reardon T (1987) Tasmanian bats: Identification, distribution and natural history. *Papers and Proceedings of the Royal Society of Tasmania* 121, 109–119. [Jun 1987]

Tedford RH (1966) A review of the macropodid genus *Sthenurus*. *University of California Publications in Geological Sciences* 57, 1–156. [18 Feb 1966]

Tedford RH (1976) Relationship of pinnipeds to other carnivores (Mammalia). *Systematic Zoology* 25, 363–374. [Dec 1976]

Tedford RH, Woodburne MO (1987) The Ilariidae, a new family of vombatiform marsupials from Miocene strata of South Australia and an evaluation of the homology of molar cusps in the Diprotodontia. Volume 2. pp. 401–418, in M Archer (ed.) *Possums & Opossums: Studies in Evolution*. Surrey Beatty and Sons, Sydney. [Nov 1987]

Teeling EC, Scally M, Kao DJ, Romagnoll ML, Springer MS, Stanhope MJ (2000) Molecular evidence regarding the origins of evolution and flight in bats. *Nature* 403(6766), 188–192. [13 Jan 2000]

Teeling EC, Madsen O, Van Den Bussche RA, de Jong WW, Stanhope MJ, Springer MS (2002) Microbat paraphyly and the convergent evolution of a key innovation in old world rhinolophid microbats. *Proceedings of the National Academy of Sciences of the United States of America* 99, 1431–1436. [5 Feb 2002]

Teeling EC, Madsen O, Murphy WJ, Springer MS, O'Brien SJ (2003) Nuclear gene sequences confirm an ancient link between New Zealand short-tailed bats and South American noctilionoid bats. *Molecular Phylogenetics and Evolution* 28, 308–319. [Aug 2003]

Teeling EC, Springer MS, Madsen O, Bates P, O'Brien SJ, Murphy WJ (2005) A molecular phylogeny for bats illuminates biogeography and the fossil record. *Science* 307(5709), 580–584. [28 Jan 2005]

Teeling EC, Dool S, Springer MS (2012) Phylogenies, fossils and functional genes: the evolution of echolocation in bats. pp. 1–22, in GF Gunnell & NB Simmons (eds.) *Evolutionary History of Bats: Fossils, Molecules and Morphology*. Cambridge University Press, Cambridge.

Temminck CJ (1824 [1824–1827]) Couscous oursin *Phalangista ursina*. pp. 10–20. [and] Sur le genre Sarigue - *Didelphis* (Linn.). pp. 21–54, pls 5–6. [and] Sur les mammifères du genre Dasyure, et sur deux genres voisins, les Thylacynes et les Phascogales. pp. 55–72, pls 7–8. *Monographies de Mammalogie, ou description de quelques genres de mammifères, dont les espèces ont été observées dans les différents musées de l'Europe. Ouvrage accompagné de planches d'Ostéologie, pouvant servir de suit et de complément aux notices sur les animaux vivants, publiées par M. le Baron G. Cuvier, dans ses recherches sur les ossements fossils*. G. Dufour et E. D'Ocagne, Paris. Tome 1. [pp. i-xxxii, 1–72= 1824; pp. 73–204 = 1825; pp. 205–244 = 1826; pp. 245–268 = 1827]

Temminck CJ (1825 [1824–1827]) Sur l'ordre des cheiropteres. pp. 157–203, in CJ Temminck (1824–1827) *Monographies de Mammalogie, ou description de quelques genres de Mammifères, dont les espèces ont eté observées dans les différens Musées de l'Europe, &c. Ouvrage accompagné de planches d'Osteologie, pouvant servir de suite et de complement aux notices sur les animaux vivants, publiées par M. le Baron G. Cuvier, dans ses recherches sur les ossements fossils*. G. Dufour et E. D'Ocagne, Paris. Tome 1.

Temminck CJ (1827 [1824–1827]) Tableau méthodique des mammifères, répartis en orders, genres et sections, avec une énumération approximative des espèces comprises dans les groupes, suivant le relevé le plus récent dans cette classe du règne animal. pp. xiii-xxxii and Table des chapitres du premier volume. pp. 266–268, in CJ Temminck (1824–1827) *Monographies de Mammalogie, ou description de quelques genres de Mammifères, dont les espèces ont eté observées dans les différens Musées de l'Europe, &c. Ouvrage accompagné de planches d'Osteologie, pouvant servir de suite et de complement aux notices sur les animaux vivants, publiées par M. le Baron G. Cuvier, dans ses recherches sur les ossements fossils*. G. Dufour et E. D'Ocagne, Paris. Tome 1.

Temminck CJ (1834) Over een geslachk der vleugelhandige zoogdieren, bladneusgenaamd (*Rhinolophus* Geoff., Cuv., Illig., Desm.; *Vespertilio* Linn., Erxleb., *Noctilio* Kuhl). *Tijdschrift voor Natuurlijke Geschiedenis en Physiologie* 1, 1–30.

Temminck CJ (1836) Coup-d'oeil sur la faune des Iles de la Sonde et de l'Empire du Japon. Discours préliminaire destiné a servir d'introduction a la Faune du Japon. pp. I–XXX, in PT von Siebold (ed.) (1833–1850) *Fauna Japonica*, sive descriptio animalium, quae in itinere per Japoniam, jussu et auspiciis superiorum, qui summum in India Batava imperium tenent, suscepto, annis 1823–1830 collegit, notis, observationibus et adumbrationibus illustravit Ph. Fr. de Siebold. Conjunctis studiis C. J. Temminck et H. Schlegel pro vertebratis atque W. de Haan pro invertebratis elaborata. Regis Auspicii Edita. Apud Arnz et Socios, Lugduni Batavorum.

Temminck CJ (1837) Les cheiroptères frugivores. Additions aux vues générales sur l'ordre des cheiroptères, révision de la monographie du genre Rousette et monographies des genres Pachysome, Macroglosse, Harpie et Céphalote. pp. 49–112, in CJ Temminck (1835–1841) *Monographies de Mammalogie, ou description de quelques genres de Mammifères, dont les espèces ont été observées dans les différents Musées de l'Europe. Ouvrage accompagné de planches 'Ostéologie purvant server de suite et de complément aux Notices sur les animaux vivants, publiées par le Baron G. Cuvier, dans ses Recherches sur les ossements fossils.* C.C. van der Hoek, Leiden. Volume 2.

Temminck CJ (1838a) Over de geslachten *Taphozous, Emballonura, Urocryptus* en *Diclidurus. Tijdschrift voor Natuurlijke Geschiedenis en Physiologie* 5, 1–34.

Temminck CJ (1838b) Over de kennis en de verbreiing der zoogdiern van Japan. *Tijdschrift voor Natuurlijke Geschiedenis en Physiologie* 5, 273–293.

Temminck CJ (1839–1840) Sur les cheiroptères vespertilionides formant les genres Nyctice, Vespertilion et Furie. pp. 141–272, in CJ Temminck (1835–1841) *Monographies de Mammalogie, ou description de quelques genres de Mammifères, dont les espèces ont été observées dans les différens Musées de l'Europe. Ouvrage accompagné de planches 'Ostéologie purvant server de suite et de complément aux Notices sur les animaux vivants, publiées par le Baron G. Cuvier, dans ses Recherches sur les ossements fossils.* C.C. van der Hoek, Leiden. Tome 2.

Temminck CJ (1844a) Apercu général et spécifique sur les mammifères qui habitant le Japon et les îles qui en dèpendent. pp. 1–59, in PT von Siebold (ed.) (1833–1850) *Fauna Japonica*, sive descriptio animalium, quae in itinere per Japoniam, jussu et auspiciis superiorum, qui summum in India Batava imperium tenent, suscepto, annis 1823 - 1830 collegit, notis, observationibus et adumbrationibus illustravit Ph. Fr. de Siebold. Conjunctis studiis C. J. Temminck et H. Schlegel pro vertebratis atque W. de Haan pro invertebratis elaborata. Regis Auspicii Edita. Apud Arnz et Socios, Lugduni Batavorum. [pp. 1–24, 11 Feb 1843; pp. 25–59, 18 Dec 1844]

Temminck CJ (1844b) Les mammifères marins. pp. 1–26, in PT von Siebold (ed.) (1833–1850) *Fauna Japonica*, sive descriptio animalium, quae in itinere per Japoniam, jussu et auspiciis superiorum, qui summum in India Batava imperium tenent, suscepto, annis 1823–1830 collegit, notis, observationibus et adumbrationibus illustravit Ph. Fr. de Siebold. Conjunctis studiis C. J. Temminck et H. Schlegel pro vertebratis atque W. de Haan pro invertebratis elaborata. Regis Auspicii Edita. Apud Arnz et Socios, Lugduni Batavorum. [18 Dec 1844]

Temminck CJ (1846) *Coup-d'Oeil Général sur les Possessions Néerlandaises dans l'Indie Archipélagique.* A. Arnz & Comp., Liden. Volume 1.

Temminck CJ (1853) *Esquisses Zoologiques sur la Côte de Guinée: 1ère Partie, les Mammifères.* E.J. Brill, Leiden.

Templeton R (1848) In E Blyth. Report of Curator Zoological Department. *Journal of the Asiatic Society of Bengal, Calcutta* 17, 247–255.

Thaler L (1966) Les rongeurs fossils du Bas-Languedoc dans leur rapports avec l'histoire des faunes et la stratigraphic du Tertiaire d'Europe. *Mémoires du Muséum National d'Histoire Naturelle, Série C* 7, 1–295.

Thewissen JGM, Babcock SK (1991) Distinctive cranial and cervical innervation of wing muscles: new evidence for bat monophyly. *Science* 251(4996), 934–936. [22 Feb 1991]

Thewissen JGM, Babcock SK (1993) The implications of the propatagial muscles of flying and gliding mammals for archontan systematics. pp. 91–109, in RDE MacPhee (ed.) *Primates and Their Relatives in Phylogenetic Perspective.* Plenum Press, New York.

Thewissen JGM, Cooper LN, Clementz MT, Bajpai S, Tiwari BN (2007) Whales originated from aquatic artiodactyls in the Eocene epoch of India. *Nature* 450(7173), 1190–1195. [20 Dec 2007]

Thomas O (1879) On Robert Kerr's translation of the '*Systema Naturae*' of Linnaeus. *Annals & Magazine of Natural History* (5)4, 396–397. [1 Nov 1879]

Thomas O (1882) On two new Muridae from Tasmania. *Annals & Magazine of Natural History* (5)9, 413–416. [1 Jun 1882]

Thomas O (1883) Mammalia. pp. 1–40, in E Caldwell (ed.) *The Zoological Record for 1882.* Zoological Society of London, London. Volume 19.

Thomas O (1885) Notes on the characters of the different races of *Echidna. Proceedings of the Zoological Society of London* 53, 329–339. [Aug 1885]

Thomas O (1887a) On the specimens of *Phascogale* in the Museo Civico, Genoa, with notes on the allied species of the genus. *Annali del Museo Civico di Storia Naturale di Genova* (2)4(24), 502–511.

Thomas O (1887b) Description of a second species of rabbit-bandicoot (*Peragale*). *Annals & Magazine of Natural History* (5)19, 397–399. [1 Jun 1887]

Thomas O (1887c) On the wallaby commonly known as *Lagorchestes fasciatus. Proceedings of the Zoological Society of London* 54(1886), 544–547. [Apr 1887]

Thomas O (1887d) Report on a zoological collection made by the officers of H.M.S. 'Flying Fish' at Christmas Island. I. Mammalia. *Proceedings of the Zoological Society of London* 55, 511–514. [Oct 1887]

Thomas O (1887e) Diagnoses of two new fruit-eating bats from the Solomon Islands. *Annals & Magazine of Natural History* (5)19, 147. [1 Feb 1887]

Thomas O (1888a) *Catalogue of the Marsupialia and Monotremata in the Collection of the British Museum (Natural History).* British Museum, London. [3 Nov. 1888]

Thomas O (1888b) Description of a new genus and species of rat from New Guinea. *Proceedings of the Zoological Society of London* 56, 237–240. [Aug 1888]

Thomas O (1888c) Diagnoses of six new mammals from the Solomon Islands. *Annals & Magazine of Natural History* (6)1, 155–158. [1 Feb 1888]

Thomas O (1888d) Description of a new bat of the genus *Nyctophilus. The Annals and Magazine of Natural History* (6)2, 226. [1 Sept 1888]

Thomas O (1889a) Description of a new genus of Muridae allied to *Hydromys*. *Proceedings of the Zoological Society of London* 57, 247–250. [Oct 1889]

Thomas O (1889b) Description of a new species of *Mus* from South Australia. *Annals & Magazine of Natural History* (6)3, 433–435. [1 May 1889]

Thomas O (1889c) On the mammals of Christmas Island. *Proceedings of the Zoological Society of London* 56(1888), 532–534. [Apr 1889]

Thomas O (1892) Description of a third species of the genus *Nyctophilus*. *Annals & Magazine of Natural History* (6)9, 405–406. [1 May 1892]

Thomas O (1894) Description of a new species of *Vespertilio* from China. *Annals & Magazine of Natural History* (6)14, 300–301. [1 Oct 1894]

Thomas O (1895a) An analysis of the mammalian generic names given in Dr. C.W.L. Gloger's '*Naturgeschichte*' (1841). *Annals & Magazine of Natural History* (6)15, 189–193. [1 Feb 1895]

Thomas O (1895b) On some mammals collected by Mr. Albert Meek on Ferguson Island, d'Entrecasteaux Group. *Novitates Zoologicae* 2, 163–166.

Thomas O (1895c) Preliminary diagnoses of new mammals from northern Luzon, collected by Mr. John Whitehead. *Annals & Magazine of Natural History* (6)16, 160–164. [1 Aug 1895]

Thomas O (1896a) On *Caenolestes*, a still existing survivor of the Epanorthidae of Ameghino, and the representative of a new family of recent marsupials. *Proceedings of the Zoological Society of London* 63(1895), 870–878. [Apr 1896]

Thomas O (1896b) On mammals from Celebes, Borneo, and the Philippines recently received at the British Museum. *Annals & Magazine of Natural History* (6)18, 241–250. [1 Sep 1896]

Thomas O (1897a) Viaggio Di Lamberto Loria Nella Papuasia Oriental. XIX. On the mammals collected in British New Guinea by Dr. Lamberto Loria. *Annali del Museo Civico di Storia Naturale di Genova* (2)18(38), 606–622. [14 Dec 1897]

Thomas O (1897b) On the genera of rodents: An attempt to bring up to date the current arrangement of the order. *Proceedings of the Zoological Society of London* 64(1896), 1012–1028. [Apr 1897]

Thomas O (1898a) Description of three new mammals from the East Indian Archipelago and Australia. *Novitates Zoologicae* 5, 1–4.

Thomas O (1898b) The technical names of British mammals. *Zoologist* 2(4), 97–103.

Thomas O (1900a) Exhibition of, and remarks upon, a skin of a new species of kangaroo, proposed to be called *Macropus bedfordi*. *Proceedings of the Zoological Society of London* 69, 112. [Jun 1900]

Thomas O (1900b) Exhibition of, and remarks upon, a skin of a new sub-species of kangaroo from Western Australia, proposed to be named *Macropus robustus cervinus*. *Proceedings of the Zoological Society of London* 69, 113. [Jun 1900]

Thomas O (1900c) A new bat from the Key Islands. *Annals & Magazine of Natural History* (7)5, 145. [1 Jan 1900]

Thomas O (1901a) On some kangaroos and bandicoots from Barrow Island, N.W. Australia, and the adjoining mainland. *Novitates Zoologicae* 8, 394–396.

Thomas O (1901b) Some new African bats (including one from the Azores) and a new galago. *Annals & Magazine of Natural History* (7)8, 27–34. [1 Jul 1901]

Thomas O (1901c) On a new genus and species of vespertilionine bat from East Africa. *Annals & Magazine of Natural History* (7)7, 460–462. [1 May 1901]

Thomas O (1901d) A new scotophiline bat from British East Africa, with the description of a new genus of the group. *Annals & Magazine of Natural History* (7)7, 263–265. [1 Mar 1901]

Thomas O (1901e) New *Myotis, Artibeus, Sylvilagus*, and *Metachirus* from central and South America. *Annals & Magazine of Natural History* (7)7, 541–545. [1 Jun 1901]

Thomas O (1902a) Two new Australian small mammals. *Annals & Magazine of Natural History* (7)10, 491–492. [1 Dec 1902]

Thomas O (1902b) New forms of *Saimiri, Oryzomys, Phyllotis, Coendou*, and *Cyclopes*. *Annals & Magazine of Natural History* (7)10, 246–250. [1 Sep 1902]

Thomas O (1902c) On the generic names *Notophorus, Alces, Dama*, and *Cephalotes*, with remarks on the 'one-letter rule' in nomenclature. *Proceedings of the Biological Society of Washington* 15, 197–198. [10 Oct 1902]

Thomas O (1902d) On a new genus of vespertilionine bats from New Guinea. *Annals & Magazine of Natural History* (7)9, 220–222. [1 Mar 1902]

Thomas O (1903a) Note on the technical name of the Tasmanian devil. *Annals & Magazine of Natural History* (7)11, 289. [1–2 Mar 1903]

Thomas O (1903b) On a collection of mammals from Abyssinia, including some from Lake Tsana, collected by Mr. Edward Degen. *Proceedings of the Zoological Society of London* 71(1902)(2), 308–316. [Aug 1903]

Thomas O (1904a) On a collection of mammals made by Mr. J.T. Tunney in Arnhem Land, Northern Territory of South Australia. *Novitates Zoologicae* 11, 222–229.

Thomas O (1904b) On a new rock-wallaby from north-west Australia. *Novitates Zoologicae* 11, 365–366.

Thomas O (1904c) Exhibition of specimens and descriptions of new species of mammals from New Guinea. *Proceedings of the Zoological Society of London* 73(1903)(2), 196–202. [1 Apr 1904]

Thomas O (1904d) New species of *Pteropus, Mus*, and *Pogonomys* from the Australian region. *Novitates Zoologicae* 11, 597–600.

Thomas O (1904e) New bats and rodents from West Africa, the Malay Peninsula, and Papuasia. *Annals & Magazine of Natural History* (7)14, 196–202. [31 Aug 1904]

Thomas O (1904f) In G Schneider. Ergebnisse zoologischer Forschungsreisen in Sumatra. *Zoologischer Anzeiger* 27, 722–724. [12 Jul 1904]

Thomas O (1904g) On the osteology and systematic position of the rare Malagasy bat *Myzopoda aurita*. *Proceedings of the Zoological Society of London* 74(1904)(2), 2–6. [1 Oct 1904]

Thomas O (1904h) Mammals obtained in the island of Fernando Po by Mr E. Seimund. *Abstract of the Proceedings of the*

Zoological Society of London 74(10) (1904)(2), 12. [1 Oct 1904]

Thomas O (1904i) A new bat from the United States, representing the European *Myotis (Leuconoe) daubentoni*. *Annals & Magazine of Natural History* (7)13, 382–384. [1–7 May 1904]

Thomas O (1905a) On some Australasian mammals. *Annals & Magazine of Natural History* (7)16, 422–428. [1–2 Oct 1905]

Thomas O (1905b) The generic names given by Frisch in 1775. *Annals & Magazine of Natural History* (7)16, 461–464. [1–2 Oct 1905]

Thomas O (1906a) On mammals collected in south-west Australia for Mr. W.E. Balston. *Abstract of the Proceedings of the Zoological Society of London* 76(31)(1906), 1–2. [1 May 1906]

Thomas O (1906b) New mammals from the Australian region. *Annals & Magazine of Natural History* (7)17, 324–332. [1–3 Mar 1906]

Thomas O (1906c) On mammals collected in south-west Australia for Mr. W.E. Balston. *Proceedings of the Zoological Society of Lond*on 76(1906)(2), 468–478. [10 Oct 1906]

Thomas O (1906d) On mammals from northern Australia presented to the National Museum by Sir Wm. Ingram, Bt., and the Hon. John Forrest. *Proceedings of the Zoological Society of Lond*on 76(1906)(2), 536–543. [10 Oct 1906]

Thomas O (1906e) On a collection of mammals made by Mr. W. Stalker in the Northern Territory of South Australia, and presented to the National Museum by Sir William Ingram, Bart., and the Hon. John Forrest. *Abstract of the Proceedings of the Zoological Society of Lond*on 76(32)(1906)(2), 6. [15 May 1906]

Thomas O (1906f) On the generic arrangement of the Australian rats hitherto referred to *Conilurus*, with remarks on the structure and evolution of their molar cusps. *Annals & Magazine of Natural History* (7)17, 81–85. [1 Jan 1906]

Thomas O (1906g) New African mammals of the genera *Cercopithecus, Scotophilus, Miniopterus, Crocidura, Georychus*, and *Heliophobius*. *Annals & Magazine of Natural History* (7)17, 173–179. [1–3 Feb 1906]

Thomas O (1907a) On the occurrence of *Acanthoglossus* in British New Guinea. *Annals & Magazine of Natural History* (7)20, 293–294. [1–2 Sep 1907]

Thomas O (1907b) List of further collections of mammals from Western Australia, including a series from Bernier Island, obtained for Mr. W.E. Balston; with field notes by the collector, Mr. G.C. Shortridge. *Proceedings of the Zoological Society of London* 76(1906), 763–777. [11 Apr 1907]

Thomas O (1907c) On three new mammals from British New Guinea. *Annals & Magazine of Natural History* (7)20, 70–74. [1–3 Jul 1907]

Thomas O (1908a) The species of the genus *Dactylopsila*. *Annals & Magazine of Natural History* (8)1, 122–124. [1 Jan 1908]

Thomas O (1908b) New bats and rodents in the British Museum collection. *Annals & Magazine of Natural History* (8)2, 370–375. [1–5 Oct 1908]

Thomas O (1909a) Some mammals from N.E. Kimberley, northern Australia. *Annals & Magazine of Natural History* (8)3, 149–152. [1–4 Feb 1909]

Thomas O (1909b) Two new mammals from northern Australia. *Annals & Magazine of Natural History* (8)4, 197–198. [1–2 Sep 1909]

Thomas O (1909c) On the N. Australian rats referred to the genus *Mesembriomys*. *Annals & Magazine of Natural History* (8)3, 372–374. [1–2 Apr 1909]

Thomas O (1910a) A new genus for *Dactylopsila palpator*. *Annals & Magazine of Natural History* (8)6, 610. [1–3 Dec 1910]

Thomas O (1910b) New African mammals. *Annals & Magazine of Natural History* (8)5, 83–92. [1–4 Jan 1910]

Thomas O (1910c) The generic arrangement of the Australian murines hitherto referred to as '*Mus*'. *Annals & Magazine of Natural History* (8)6, 603–607. [1–3 Dec 1910]

Thomas O (1910d) New Australian Muridae of the genus *Pseudomys*. *Annals & Magazine of Natural History* (8)6, 607–610. [1–3 Dec 1910]

Thomas O (1910e) New genera of Australasian Muridae. *Annals & Magazine of Natural History* (8)6, 506–508. [1–3 Nov 1910]

Thomas O (1910f) Three new Asiatic mammals. *Annals & Magazine of Natural History* (8)5, 534–536. [1 Jun 1910]

Thomas O (1911a) A new kangaroo from the Northern Territory of Western Australia. *Annals & Magazine of Natural History* (8)7, 609–610. [1 Jun 1911]

Thomas O (1911b) The mammals of the tenth edition of Linnaeus; an attempt to fix the types of the genera and the exact bases and localities of the species. *Proceedings of the Zoological Society of London* 81, 120–158. [22 Mar 1911]

Thomas O (1911c) Two new eastern bats. *Annals & Magazine of Natural History* (8)8, 378–380. [29 Aug 1911]

Thomas O (1912a) The technical name of the Tasmanian Devil. *Proceedings of the Biological Society of Washington* 25, 116. [29 Jun 1912]

Thomas O (1912b) On a collection of small mammals from the Tsin-ling Mountains, Central China, presented by Mr. G. Fenwick Owen to the National Museum. *Annals & Magazine of Natural History* (8)10, 395–403. [1–2 Oct 1912]

Thomas O (1913a) Some new species of *Uromys*. *Annals & Magazine of Natural History* (8)12, 212–217. [1–2 Aug 1913]

Thomas O (1913b) On new mammals obtained by the Utakawa Expedition to Dutch New Guinea. *Annals & Magazine of Natural History* (8)12, 205–212. [1–2 Aug 1913]

Thomas O (1913c) On African bats and shrews. *Annals & Magazine of Natural History* (8)11, 314–321. [1 Mar 1913]

Thomas O (1913d) On a remarkable new free-tailed bat from southern Bombay. *Journal of the Bombay Natural History Society* 22, 87–91.

Thomas O (1913e) Two new Australian mammals. *Annals & Magazine of Natural History* (8)11, 79–80. [1–2 Jan 1913]

Thomas O (1914a) New Asiatic and Australasian bats and a new bandicoot. *Annals & Magazine of Natural History* (8)13, 439–444. [1 Apr 1914]

Thomas O (1914b) Report on the mammals collected by the British Ornithologists Union Expedition and the Wollaston

Expedition in Dutch New Guinea. *Transactions of the Zoological Society of London* 20, 315–324. [May 1914]
Thomas O (1914c) On mammals from Manus Island Admiralty Group, and Ruk Island, Bismarck Archipelago. *Annals & Magazine of Natural History* (8)13, 434–439. [1 Apr 1914]
Thomas O (1914d) A new genus of bats allied to *Nyctophilus*. *Annals & Magazine of Natural History* (8)14, 381–383. [1–3 Nov 1914]
Thomas O (1915a) List of mammals (exclusive of Ungulata) collected on the Upper Congo by Dr. Christy for the Congo Museum, Tervueren. *Annals & Magazine of Natural History* (8)16, 465–481. [1 Dec 1915]
Thomas O (1915b) New African rodents and insectivores, mostly collected by Dr. C. Christy of the Congo Museum. *Annals & Magazine of Natural History* (8)16, 146–152. [1–3 Aug 1915]
Thomas O (1915c) Notes on *Taphozous* and *Saccolaimus*. *Journal of the Bombay Natural History Society* 24, 57–63.
Thomas O (1915d) A special genus for the Himalayan bat known as *Murina grisea*. *Annals & Magazine of Natural History* (8)16, 309–310. [1–2 Oct 1915]
Thomas O (1915e) Notes on the genus *Nyctophilus*. *Annals & Magazine of Natural History* (8)15, 493–499. [1 May 1915]
Thomas O (1915f) Two new species of *Leuconoe*. *Annals & Magazine of Natural History* (8)15, 170–172. [1–4 Jan 1915]
Thomas O (1916a) On Muridae from Darjiling and the Chin Hills. *Journal of the Bombay Natural History Society* 24, 404–415.
Thomas O (1916b) On the generic names *Rattus* and *Phyllomys*. *Annals & Magazine of Natural History* (8)18, 240. [1–2 Oct 1916]
Thomas O (1916c) On *Rattus* as a generic name, with a note on the nomenclature of *Echimys* and *Loncheres*. *Annals & Magazine of Natural History* (8)18, 70–72. [1–3 Jul 1916]
Thomas O (1920a) *Notoryctes* in north-west Australia. *Annals & Magazine of Natural History* (9)6, 111–113. [1 Jul 1920]
Thomas O (1920b) On mammals from Ceram. *Annals & Magazine of Natural History* (9)6, 422–431. [1 Oct 1920]
Thomas O (1920c) On mammals from near Tinogata, Catamarca, collected by Sr. Budin. *Annals & Magazine of Natural History* (9)6, 116–120. [1 Jul 1920]
Thomas O (1920d) A further collection of mammals from July. *Annals & Magazine of Natural History* (9)5, 188–198. [3 Feb 1920]
Thomas O (1921a) Notes on Australasian rats, with a selection of lectotypes of Australasian Muridae. *Annals & Magazine of Natural History* (9)8, 425–433. [1 Oct 1921]
Thomas O (1921b) On three new Australian rats. *Annals & Magazine of Natural History* (9)8, 618–622. [1–2 Dec 1921]
Thomas O (1921c) Notes on the species of *Notomys*, the Australian jerboa-rats. *Annals & Magazine of Natural History* (9)8, 536–541. [31 Oct 1921]
Thomas O (1922a) A selection of lectotypes of the typical Australian marsupials in the British Museum collection. *Annals & Magazine of Natural History* (9)10, 127–128. [1–3 Jul 1922]
Thomas O (1922b) New mammals from New Guinea and neighbouring islands. *Annals & Magazine of Natural History* (9)9, 261–265. [1–4 Mar 1922]
Thomas O (1922c) The bandicoots of Nuyts Archipelago, S. Australia, and of Cape York, N. Queensland. *Annals & Magazine of Natural History* (9)9, 677–679. [30 May 1922]
Thomas O (1922d) On bandicoots allied to *Perameles bougainvillei*. *Annals & Magazine of Natural History* (10)9, 143–145. [1–3 Jul 1922]
Thomas O (1922e) Two new subspecies of *Phalanger orientalis*. *Annals & Magazine of Natural History* (9)9, 680–681. [30 May 1922]
Thomas O (1922f) A new rock wallaby (*Petrogale*) from the islands off South Australia. *Annals & Magazine of Natural History* (9)9, 681–683. [30 May 1922]
Thomas O (1922g) A subdivision of the genus *Uromys*. *Annals & Magazine of Natural History* (9)9, 260–261. [1–4 Mar 1922]
Thomas O (1922h) A new species of *Mastacomys* from a cave in South Australia. *Annals & Magazine of Natural History* (9)10, 550–551. [31 Oct 1922]
Thomas O (1922i) Two new jerboa-rats (*Notomys*). *Annals & Magazine of Natural History* (9)9, 315–317. [1–8 Apr 1922]
Thomas O (1922j) The generic classification of the *Taphozous* group. *Annals & Magazine of Natural History* (9)9, 266–267. [1–4 Mar 1922]
Thomas O (1922k) A new bat of the genus *Miniopterus* from N. Australia. *Annals & Magazine of Natural History* (9)10, 616–617. [31 Nov 1922]
Thomas O (1923a) The Godman Exploration Fund: List of mammals from north Queensland collected by Mr T.V. Sherrin. *Annals & Magazine of Natural History* (9)11, 170–178. [1–3 Jan 1923]
Thomas O (1923b) On some Queensland Phalangeridae. *Annals & Magazine of Natural History* (9)11, 246–250. [1 Feb 1923]
Thomas O (1923c) On the ring-tailed phalanger of South Australia, and a new rat from north Queensland. *Annals & Magazine of Natural History* (9)12, 158–160. [1 Jul 1923]
Thomas O (1923d) Exhibition of a new rock kangaroo from North Queensland. *Abstract of the Proceedings of the Zoological Society of London* 93 (235), 13. [6 Feb 1923]
Thomas O (1923e) Exhibition of a new rock kangaroo from northern Australia. *Proceedings of the Zoological Society of London* 93, 177–178. [12 Apr 1923]
Thomas O (1923f) The native rat of Pearson's Islands, S. Australia. *Annals & Magazine of Natural History* (9)11, 601–602. [1–2 May 1923]
Thomas O (1923g) On some small mammals, chiefly bats, from the East Indian Archipelago. *Annals & Magazine of Natural History* (9)11, 250–255. [31 Jan 1923]
Thomas O (1924a) A new pouched mouse (*Phascogale*) from northern New South Wales. *Annals & Magazine of Natural History* (9)14, 528–529. [1–3 Nov 1924]
Thomas O (1924b) Some new Australasian Muridae. *Annals & Magazine of Natural History* (9)13, 296–299. [1 Mar 1924]
Thomas O (1924c) A new subspecies of *Nyctinomus australis*. *Annals & Magazine of Natural History* (9)14, 455–456. [1–2 Oct 1924]
Thomas O (1924d) A new *Scoteinus* from Queensland. *Annals & Magazine of Natural History* (9)13, 540. [1–2 May 1924]

Thomas O (1925) A new Australian mouse of the genus *Pseudomys*. *Annals & Magazine of Natural History* (9)15, 669–671. [1–5 Jun 1925]

Thomas O (1926a) The local races of *Dasyurus hallucatus*. *Annals & Magazine of Natural History* (9)18, 543–544. [1–2 Nov 1926]

Thomas O (1926b) On various mammals obtained during Capt. Wilkins's expedition in Australia. *Annals & Magazine of Natural History* (9)17, 625–635. [1 Jun 1926]

Thomas O (1926c) Two new rock-wallabies (*Petrogale*) discovered by Capt. G.W. [sic G.H.] Wilkins in northern Australia. *Annals & Magazine of Natural History* (9)17, 184–187. [1–2 Jan 1926]

Thomas O (1926d) Two new Australian Muridae. *Annals & Magazine of Natural History* (9)18, 308–310. [1–4 Sep 1926]

Thomas O, Dollman G (1909) On mammals from Inkerman, North Queensland, presented to the National Museum by Sir William Ingram, Bt., and the Hon. John Forrest. *Proceedings of the Zoological Society of London* 78(1908), 788–794. [8 Apr 1909]

Thomson J (1857) *Archives Entomologiques ou Recueil Contenant des Illustrations d'Insectes Nouveau ou Rares*. Bureau du Trésorier de la Société Entomologique de France, Paris. Volume 1.

Thomson J (1864) *Systema Cerambycidarum : ou exposé de tous les genres compris dans la famille des Cérambycides et familles limitrophes*. H. Dessain, Liège.

Thonglongya K (1973) First record of *Rhinolophus paradoxolophus* (Bouret, 1951) from Thailand, with the description of a new species of the *Rhinolophus philippinensis* group (Chiroptera: Rhinolophidae). *Mammalia* 37, 587–597. [Dec 1973]

Thorell TTT (1876) Descrizione di alcune specie di Opilioni dell' Arcipelago Malese appartenenti al Museo Civico di Genova. *Annali del Museo Civico di Storia Naturale di Genova* 9(1), 111–138.

Thorington RW, Hoffman RS (2005) Family Sciuridae. pp. 754–818, in DE Wilson & DA Reeder (eds.) *Mammal Species of the World. A Taxonomic and Geographic Reference*. Johns Hopkins University Press, Baltimore. Third Edition.

Thunberg CP (1805) [Four new genera] No title. *Göttingischen Gelehrte Anzeigen* 1, 281–282.

Thunberg CP (1811) Mammalia capensia recensita et illustrata. *Mémoires de l'Académie Impériale des Sciences de St. Pétersbourg* 3, 299–323.

Thunberg CP (1815) *Hemipterorum maxillosorum* genera illustrata plurimisque novis speciebus ditata ac descripta. *Mémoires de l'Académie Impériale des Sciences de St. Pétersbourg* 5, 211–301.

Tian L, Liang B, Maeda K, Metzner W, Zhang S (2004) Molecular studies on the classification of *Miniopterus schreibersii* (Chiroptera: Vespertilionidae) inferred from mitochondrial cytochrome b sequences. *Folia Zoologica* 53, 303–311.

Tickell SR (1888) In F Day. *Supplement to the Fishes of India: being a natural history of the fishes known to inhabit the seas and fresh waters of India, Burma and Ceylon*. pp. 779–816.

Tidemann CR (1986) Morphological variation in Australian and island populations of Gould's wattled bat, *Chalinolobus gouldi* (Gray) (Chiroptera: Vespertilionidae). *Australian Journal of Zoology* 34, 503–514.

Tidemann CR (1987a) Notes on the flying-fox, *Pteropus melanotus* (Chiroptera: Pteropodidae), on Christmas Island, Indian Ocean. *Australian Mammalogy* 10, 89–91. [29 Jun 1987]

Tidemann CR (1987b) Recent sightings of bats on Lord Howe Island and Norfolk Island: recolonisers or survivors? *Macroderma* 3, 33–35. [Sep 1987]

Tiedemann F (1808) *Zoologie. Zu seinen Vorlesungen entworfen. Allgemeine Zoologie, Mensch und Säugthiere*. Weberschen Buchhandlung, Landshut. Volume 1.

Tokuda M (1941) A revised monograph of the Japanese and Manchou-Korean Muridae. *Transactions of the Biogeographical Society of Japan* 4(1), 1–155. Plates 1–10.

Toldt K (1905) Über das genus *Proechidna*. *Verhandlungen der Kaiserlich-Königlichen Zoologisch-Botanischen in Wien* 55, 5–11. [28 Jan 1905]

Toldt K (1906) Über des Haar- und Stachelkleid von *Zaglossus* Gill. *Annalen des K.K. Naturhistorischen Hofmuseums in Wien* 21, 1–21.

Tomes RF (1857) On two species of bats inhabiting New Zealand. *Proceedings of the Zoological Society of London* 25, 134–142. [21 Oct 1857]

Tomes RF (1858a) A monograph of the genus *Miniopteris*. *Proceedings of the Zoological Society of London* 26, 115–128. [13 May 1858]

Tomes RF (1858b) A monograph of the genus *Nyctophilus*. *Proceedings of the Zoological Society of London* 26, 25–37. [9 Mar 1858]

Tomes RF (1859a) Notice of five species of bats in the collection of L.L. Dillwyn, Esq., M.P.; collected in the Labuan by Mr. James Motley. *Proceedings of the Zoological Society of London* 26(1858): 536–540. [Jan–May 1859]

Tomes RF (1859b) Description of six hitherto undescribed species of bats. *Proceedings of the Zoological Society of London* 27, 68–79. [Feb–Jun 1859]

Tomes RF (1861) Notes on a collection of bats made by Mr. Anderson in the Damara Country, south-western Africa, with notices of some other African species. *Proceedings of the Zoological Society of London* 29, 31–40. [May 1861]

Tomilin AG (1946) Thermoregulation and the geographical races of cetaceans. *Comptes rendus (Doklady) de l'Académie des Sciences de l'URSS. Moscou* 54, 465–468.

Tomilin AG (1957) *Mammals of Eastern Europe and Northern Asia*. Volume 9 Cetacea. Israel Program for Scientific Translations, Jerusalem.

Tomiya S (2011) A new basal caniform (Mammalia: Carnivora) from the middle Eocene of North America and remarks on the phylogeny of early carnivorans. *PLoS ONE* 6(9), 1–14. e24146.

Toyosawa S, O'hUigin C, Figueroa F, Klein J (1998) Identification and characterization of amelogenin genes in monotremes, reptiles and amphibians. *Proceedings of the National Academy of Sciences, USA* 95, 13056–13061. [27 Oct 1998]

Traill TS (1809) Description of a new species of whale *Delphinus melas*. pp. 81–83, in WA Nicholson (ed.) *A Journal of Natural Philosophy, Chemistry and the Arts*. Volume 22.

Trask JB (1856) On some new microscopic organisms. *Proceedings of the California Academy of Natural Sciences* 1, 99–103. [7 Jan 1856]

Triggs B (2009) *Wombats*. CSIRO Publishing, Melbourne. Third Edition.

Triggs SJ (1990) Population genetics of the brushtail possum *Trichosurus vulpecula* (Marsupialia: Phalangeridae) in southeastern Australia. *Australian Journal of Zoology* 37, 545–551.

Trouessart E-L (1878) Catalogue des mammifères vivants et fossils. *Revue et Magasin de Zoologie Pure et Appliquée* 6(3), 201–254.

Trouessart E-L (1879) Catalogue des mammifères vivants et fossils. *Revue et Magasin de Zoologie Pure et Appliquée* 1878 (3)6, 108–140.

Trouessart E-L (1881) Catalogue des mammifères vivants et fossiles. *Bulletin de la Société d'Études Scientifiques d'Angers* 10(2), 58–212.

Trouessart E-L (1885) Catalogue des mammifères vivants et fossils (Carnivores). *Bulletin de la Société d'Études Scientifques d'Angers* (supplément to the year 1884) 14, 1–108.

Trouessart E-L (1891) Le *Notoryctes typhlops*: Nouveau type de marsupiaux fouisseurs originaire du désert Australien. *La Nature – Revue des Sciences et de Leurs Applications aux arts et à 'Industrie Journal Hebdomadaire Illustré. Dix-Neuvième Année* 8, 290–294.

Trouessart E-L (1897) *Catalogus Mammalium tam Viventium quam Fossilium*. Nova Editio (Prima Completa). Tomus I. Fasciculus 1. pp. 1–218; Fasciculus 2. pp. 219–452. *Fasciculus* 3, 45–664. [Primates Prosimiae, Chiroptera, Insectivora, Riodentia, Pinnipedia. R. Friedländer & Sohn, Berolini.]

Trouessart E-L (1898–1899) *Catalogus Mammalium tam Viventium quam Fossilium*. Nova editio (Prima completa). Tomus II. 1898 - Fasciculus 4, pp. 665–998; Fasciculus 5, 999–1264; 1899 - Fasciculus 6, pp. 1265–1469. (Paginae 665 ad finem) Tillodontia, Ungulata, Sirenia, Cetacea, Edentata, Marsupialia, Allotheria, Monotremata, Appendix (Addenda et Corrigenda). Index Alphabeticus. R. Friedländer & Sohn, Berolini.

Trouessart E-L (1904–1905) *Catalogus Mammalium tam Viventium quam Fossilium*. Quinquennale supplementum, anno 1904. 1904 - Fasciculus 1–2, pp. 1–546; 1905 - Fasciculus 3–4, pp. 547–929. R. Friedländer & Sohn, Berolini.

Trouessart E-L (1907) *Mammifères Pinnipèdes*. pp. 1–27, in EL Trouessart & A Menegaux (eds.) *Expédition Antarctique Française (1903–1905) Commandée par le Dr Jean Charcot*. Masson, Paris.

Trouessart E-L (1910) *Conspectus Mammalium Europae. Faune des mammifères d'Europe*. R. Friedlander & Sohn, Berlin.

Trouessart E-L (1915) Nouvelle espèce du genre *Nyctophilus* (*N. geayi*). *Bulletin du Muséum National d'Histoire Naturelle. Section B. Botanique. Biologie et Écologie Végétables Phytochimie, Paris* 21, 146–147.

Troughton EG (1920) Notes on Australian Mammals. *Records of the Australian Museum* 13, 118–122. [4 Dec 1920]

Troughton EG (1922) Notes on Australian bats, and the occurrence of *Chalinolobus gouldi*, Gray at Norfolk Island. *Australian Zoologist* 3, 39–41.

Troughton EG (1923a) The 'honey mouse', *Tarsipes spenserae* Gray. *Australian Zoologist* 3, 148–156.

Troughton EG (1923b) A revision of the rats of the genus *Leporillus* and the status of *Hapalotis personata* Krefft. *Records of the Australian Museum* 14, 23–41. [28 Feb 1923]

Troughton EG (1925) A revision of the genera *Taphozous* and *Saccolaimus* (Chiroptera) in Australia and New Guinea, including a new species, and a note on two Malayan forms. *Records of the Australian Museum* 14, 313–341. [9 Apr 1925]

Troughton EG (1926) The bats of Australia and New Guinea. pp. 21–88, in AS Le Souef & H Burrell (eds.) *The Wild Animals of Australasia*. Harrap and Co., London.

Troughton EG (1928) A new genus, species, and subspecies of marsupial mice (Family Dasyuridae). *Records of the Australian Museum* 16, 281–288. [11 June 1928]

Troughton EG (1929) A new genus and species of bat (Kerivoulinae) from the Solomons, with a review of the genera of the sub-family. *Records of the Australian Museum* 17, 85–99. [26 Jun 1929]

Troughton EG (1931) Three new bats of the genera *Pteropus, Nyctimene* and *Chaerephon* from Melanesia. *Proceedings of the Linnean Society of New South Wales* 56, 204–209. [15 Jul 1931]

Troughton EG (1932a) A new species of fat-tailed marsupial mouse, and the status of *Antechinus froggatti* Ramsay. *Records of the Australian Museum* 18, 349–353. [20 Apr 1932]

Troughton EG (1932b) A revision of the rabbit-bandicoots. Family Peramelidae, genus *Macrotis*. *Australian Zoologist* 7, 219–236.

Troughton EG (1932c) On five new rats of the genus *Pseudomys*. *Records of the Australian Museum* 18, 287–294. [20 Apr 1932]

Troughton EG (1935a) The southern race of the koala. *Australian Naturalist* 9, 137–140. [Sep 1935]

Troughton EG (1935b) Five new rats of the genera *Hydromys* and *Melomys* from northern Australia. *Records of the Australian Museum* 19, 251–258. [19 Sep 1935]

Troughton EG (1936) Description of new rats and mice from Queensland. *Memoirs of the Queensland Museum* 11, 14–22. [17 Apr 1936]

Troughton EG (1937a) Description of some New Guinea mammals. *Records of the Australian Museum* 20, 117–127. [27 Aug 1937]

Troughton EG (1937b) The status of '*Mus*' *novaehollandiae* Waterhouse, and allied forms. *Records of the Australian Museum* 20, 185–190. [27 Aug 1937]

Troughton EG (1937c) On new forms of the eastern swamp rat, and the relationship of *Mastacomys*. *Australian Zoologist* 8, 281–286.

Troughton EG (1937d) Six new bats (Microchiroptera) from the Australasian region. *Australian Zoologist* 8, 274–281.

Troughton EG (1939) Queensland rats of economic importance, and new forms of *Rattus* and *Thetomys*. *Records of the Australian Museum* 20, 278–281. [31 Mar 1939]

Troughton EG (1941) *Furred Animals of Australia*. Angus & Robertson, Sydney. First Edition. [Nov 1941]

Troughton EG (1944) *Furred Animals of Australia*. Angus and Robertson, Sydney. Second Edition. [Nov 1944] Publication date via Mahoney and Walton (1988f: 147).

Troughton EG (1945a) The kangaroo family. The nail-tail wallabies. *Australian Museum Magazine* 8, 346–349.

Troughton EG (1945b) Diagnoses of new mammals from the south-west Pacific. *Records of the Australian Museum* 21, 373–375. [25 Jun 1945]

Troughton EG (1946) Diagnoses of new rats from the New Guinea area. *Records of the Australian Museum* 21, 406–410. [24 Jun 1946]

Troughton EG (1957) A new native dog from the Papuan highlands. *Proceedings of the Royal Zoological Society of New South Wales* 76 (1955–1956), 93–94.

Troughton EG (1965) A review of the marsupial genus *Sminthopsis* (Phascogalinae) and diagnoses of new forms. *Proceedings of the Linnean Society of New South Wales* 89, 307–321. [7 May 1965]

Troughton EG (1967) *Furred Animals of Australia*. Angus and Robertson, Sydney. Ninth Edition.

Troughton EG (1971) The early history and relationships of the New Guinea Highland dog (*Canis hallstromi*). *Proceedings of the Linnean Society of New South Wales* 96, 93–98.

Troughton EG, Le Souef AS (1929a) A new species of ring-tailed phalanger (*P. laniginosus* group) from the Bunya Mountains, S.E. Queensland. *Records of the Australian Museum* 17, 291–296. [28 Nov 1929]

Troughton EG, Le Souef AS (1929b) New forms of mosaic-tailed rats (*Melomys* and *Uromys*) from Hinchinbrook Island, Queensland. *Australian Zoologist* 6, 96–99.

True FW (1884) Catalogue of the aquatic mammals exhibited by the United States National Museum. *Bulletin of the United States National Museum* 27, 623–644.

True FW (1885) Contributions to the history of the Commander Islands No 5. Description of a new species of *Mesoplodon, M. stejnegeri, obtained* by Dr. Leonard Stejneger, in Bering Island. *Proceedings of the United States National Museum* 8, 584–585.

True FW (1889) Contributions to the natural history of the cetaceans, a review of the Family Delphinidae. *Bulletin of the United States National Museum* 36, 1–191.

True FW (1898) The nomenclature of the whalebone whales of the tenth edition of Linnaeus's Systema Naturae. *Proceedings of the United States National Museum* 21, 617–635. [4 Nov 1898]

True FW (1903) A note on the common bottlenosed porpoise of the North Atlantic, *Tursiops truncatus* (Montagu). *Proceedings. Academy of Natural Sciences of Philadelphia* 55, 313–314.

True FW (1907) Observations on the type specimen of the fossil cetacean *Anoplonassa forcipata* Cope. *Bulletin of the Museum of Comparative Zoology. Harvard University* 51, 97–106.

True FW (1913) Diagnosis of a new beaked whale of the genus *Mesoplodon* from the coast of North Carolina. *Smithsonian Miscellaneous Collections* 60(25), 1–2. [14 Mar 1913]

Trumbull JH (1884) In GB Goode. pp. 7–32, Part 1, A. The whales and porpoises. In Section 1, Natural history of useful aquatic animals with an atlas of two hundred and seventy-seven plate. *The Fisheries and Fishery Industries of the United States: prepared through the co-operation of the Commissioner of Fisheries and the Superintendent of the Tenth Census*. Government Printing Office, Washington.

Tschudi JJ von (1844) *Fauna Peruana. Part 1. Mammalium conspectus quae in Republica Peruana reperiuntur et pleraque observata vel collecta sunt in itinere*. Berolini.

Tullberg T (1899) Ueber das system der Nagethiere. Eine phylogenetische studie. *Nova Acta Regiae Societatis Scientiarium Upsaliensis* 18(3), 1–514.

Tunbridge D (1991) *The Story of the Flinders Ranges Mammals*. Kangaroo Press, Sydney.

Turnbull WD (1971) The trinity therians: their bearing on evolution in marsupials and other therians. pp. 151–179, in AA Dahlberg (ed.) *Dental Morphology and Evolution*. University of Chicago Press, Chicago.

Turnbull WD, Lundelius EL Jr (1970) The Hamilton Fauna, a late Pliocene mammalian Fauna from the Grange Burn, Victoria, Australia. *Fieldiana Geology* 19, 1–123. [28 Aug 1970]

Turner HN Jr (1849) Observations relating to some of the foramina at the base of the skull in Mammalia, and on the classification of the Order Carnivora. *Proceedings of the Zoological Society of London* 16(1848), 63–88. [30 Jan 1849]

Turner HN Jr (1850) On the evidence of affinity afforded by the skull in the ungulate Mammalia. *Proceedings of the Zoological Society of London* 17(1849), 147–158. [Jan–Jun 1850]

Turner WM (1868) On the cranium of an apparently new species of *Arcto-cephalus. Journal of Anatomy* 3, 113–117. [Nov 1868]

Turner W (1888) Report on the seals collected during the voyage of H.M.S. Challenger in the years 1873–76. pp. 1–240, in CW Thomson & J Murray (eds.) *Report on the Scientific Results of the Voyage of the H.M.S. Challenger during the years 1873–76 under the command of Captain George S. Nares, R.N., F.R.S. and the late Captain Frank Tourle Thomson, R.N.* Volume 26. Her Majesty's Government. Edinburgh, Scotland.

Uhen MD (1998) Middle to late Eocene basilosaurines and dorudontines. pp. 29–61, in JGM Thewissen (ed.) *The Emergence of Whales*. Plenum Press, New York.

Uhen MD (2004) Form, function, and anatomy of *Dorudon atrox* (Mammalia, Cetacea): an archaeocete from the middle to late Eocene of Egypt. *University of Michigan Museum of Paleontology Papers on Paleontology* 34, 1–222.

Uhen MD (2008a) New protocetid whales from Alabama and Mississippi, and a new cetacean clade, Pelagiceti. *Journal of Vertebrate Paleontology* 28, 589–593.

Uhen MD (2008b) A new Xenorophus-like odontocete cetacean form the Oligocene of North Carolina and a discussion of the basal odontocete radiation. *Journal of Systematic Palaeontology* 6, 433–452.

Uhen MD (2010) The origins of whales. *Annual Review of Earth and Planetary Sciences* 38, 189–219. [28 Jan 2010]

Ulber TM (1999) Sir Andrew Smith's 'Illustrations of the Zoology of South Africa' An Annotated Table of Contents for the 'Reptilia' Volume. *Smithsonian Herpetological Information Service* 120, 1–17.

Ulmer FA (1941) *Mesoplodon mirus* in New Jersey, with additional notes on the New Jersey *M. densirostris*, and a list and key to the ziphoid whales of the Atlantic coast of North America. *Proceedings. Academy of Natural Sciences of Philadelphia* 93, 107–122.

Upton SJ (2000) Suborder Eimeriorina Léger, 1911. pp. 318–339, in JJ Lee, GF Leedale & P Bradbury (eds.) *An Illustrated Guide to the Protozoa*. Second Edition. Society of Protozoologists, Lawrence, Kansas, U.S.A. Volume 1.

Uvarov BP (1940) Eleven new generic names in Orthoptera. *Journal of Natural History* 6(11), 377–380.

Van Bénéden P-J (1852 [1852–1855]) *Anatomie Comparée*. A. Jamar, Bruxelles.

Van Bénéden P-J (1857) Sur une baleine prise pres de l'île Vlieland et dont le squelette est monté au Jardin royal de zoologie d'Anvers. *Bulletins de L'Académie Royale des Sciences, des Lettres et des Beaux-arts de Belgique* 1(2), 390–403.

Van Bénéden P-J (1859) Rapport. *Bulletins de L'Académie Royale des Sciences, des Lettres et des Beaux-arts de Belgique* 8(2), 123–146.

Van Bénéden P-J (1861) Recherches sur la faune littorale de Belgique. *Mémoires de l'Académie Royale des Sciences, des Lettres et des Beaux-Arts de Belgique* 32, 1–39.

Van Bénéden P-J (1864) Mémoire sur une nouvelle espèce de *Ziphius* de la mer des Indes. *Mémoires Couronnés et Autres Mémoires publiés par l'Académie Royale des Sciences, des Lettres et des Beaux-Arts de Belgique* 16(1), 1–23. [Jun 1864]

Van Bénéden P-J (1868) Les squelettes de cétacés et les musées qui les renferment. *Bulletins de L'Académie Royale des Sciences, des Lettres et des Beaux-arts de Belgique* 25(2), 88–125.

Van Bénéden P-J (1872) Les baleines fossils d'Anvers. *Bulletins de L'Académie Royale des Sciences, des Lettres et des Beaux-arts de Belgique* 34(2), 6–20.

Van Bénéden P-J (1876) Les phoques fossiles du bassin d'Anvers. *Bulletins de L'Académie Royale des Sciences, des Lettres et des Beaux-arts de Belgique* 41(2), 783–802.

Van Bénéden P-J, Gervais P (1880) *Ostéographie des Cétacés Vivants et Fossiles*, comprenant la desciption et l'iconographie de squelette et du système dentaire de ces animaux. Arthus Bertrand, Paris.

van Bree PJH (1971a) On two skulls of *Delphinus dussumieri* Blanford, 1891. *Beaufortia* 18, 169–172. [10 Feb 1971]

van Bree PJH (1971b) *Delphinus tropicalis*, a new name for *Delphinus longirostris* G. Cuvier, 1829. *Mammalia* 35, 345–346. [Jun 1971]

van Bree PJH (1971c) On *Globicephala sieboldii* Gray, 1846, and other species of pilot whales. (Notes on Cetacea, Delphinoidea III). *Beaufortia* 19, 79–87. [17 Sep 1971]

van Bree PJH (1972) Sur la présence de *Pseudorca crassidens* (Owen, 1846) (Cetacea, Globicephalinae) au large des côtes d'Afrique occidentale. *Bulletin de l'Institut Français d'Afrique Noire* 34A, 212–218.

van Bree PJH (1973) On the description and taxonomic status of *Delphinus holboellii* Nilsson, 1847 (Notes on Cetacea, Delphinoidea VI). *Beaufortia* 20, 129–134. [6 Feb 1973]

van Bree PJH, Cadenat J (1968) On a skull of *Peponocephala electra* (Gray, 1846) (Cetacea, Globicephalinae) from Sénégal. *Beaufortia* 14, 193–202. [10 Jul 1968]

van Bree PJH, Gallagher M (1978) On the taxonomic status of *Delphinus tropicalis* van Bree 1971 (Notes on Cetacea, Delphinoidea IX). *Beaufortia* 28, 1–8. [28 Aug 1978]

Van Cleave HJ (1943) An index to the Opinions Rendered by the International Commission on Zoological Nomenclature. *American Midland Naturalist* 30, 223–240. [Jul 1943]

Van Den Bussche RA, Hoofer SR (2004) Phylogenetic relationships among recent chiropteran families and the importance of choosing appropriate out-group taxa. *Journal of Mammalogy* 85, 321–330. [Apr 2004]

Van Den Bussche RA, Hoofer SR, Hansen EW (2002) Characterization and phylogenetic utility of the mammalian protamine P1 gene. *Molecular Phylogenetics and Evolution* 22, 333–341. [Mar 2002]

Van der Hoeven J (1827) *Handboek der Dierkunde, of Grondbeginsels der natuurlijke Geschiedenis van het Dierenrijk*. Rotterdam. Volume 1.

Van der Hoeven J (1855) *Handbuch der Zoologie*. Volume 2. J.C.A Sulpke, Amsterdam.

Van der Hoeven J (1858) *Handbook of Zoology*. Volume 2. Longman, Brown, Green, Longmans, and Roberts, London. Translated from the second Dutch edition.

Van Deusen HM (1960) Notes on the marsupial feather-tail glider from Australia. *Journal of Mammalogy* 41, 263–264. [May 1960]

Van Deusen HM, George GG (1969) Results of the Archbold Expeditions. No. 90. Notes on the Echidnas (Mammalia, Tachyglossidae) of New Guinea. *American Museum Novitates* 2383, 1–23. [28 Jul 1969]

Van Deusen HM, Jones KJ Jr (1967) Marsupials. pp. 61–86, in S Anderson & JK Jones Jr (eds.) *Recent Mammals of the World*. Ronald Press, New York.

Van Deusen HM, Koopman KF (1971) Results from the Archbold Expeditions. No. 95. The genus *Chalinolobus* (Chiroptera, Vespertilionidae). Taxonomic review of *Chalinolobus picatus, C. nigrogriseus* and *C. rogersi*. *American Museum Novitates* 2468, 1–30. [13 Oct 1971]

Van Dyck S (1980) The cinnamon antechinus, *Antechinus leo* (Marsupialia: Dasyuridae) a new species from the vine-forests of cape York Peninsula. *Australian Mammalogy* 3, 5–17. [May 1980]

Van Dyck S (1982) The relationships of *Antechinus stuartii* and *A. flavipes* (Dasyuridae, Marsupialia) with special reference to Queensland. pp. 723–766, in M Archer (ed.) *Carnivorous Marsupials*. Volume 2. Royal Zoological Society of New South Wales, Sydney. [Feb 1982]

Van Dyck S (1985) *Sminthopsis leucopus* (Marsupialia: Dasyuridae) in north Queensland rainforest. *Australian Mammalogy* 8, 53–60. [Feb 1985]

Van Dyck S (1986) *Sminthopsis archeri* (Marsupialia: Dasyuridae), a new species from the savannah of Papua New Guinea and Cape York Peninsula, Australia. *Australian Mammalogy* 9, 111–124. [30 Jun 1986]

Van Dyck S (1988) The bronze quoll, *Dasyurus spartacus* (Marsupialia: Dasyuridae), a new species from the savannahs of Papua New Guinea. *Australian Mammalogy* 11, 145–156. [29 Aug 1988]

Van Dyck S (1990) *Belideus gracilis* – soaring problems for an old De Vis glider. *Memoirs of the Queensland Museum* 28, 329–336. [31 Mar 1990]

Van Dyck S (1991) The status of mammals. pp. 349–353, in GJ Ingram & RJ Raven (eds.) *An Atlas of Queensland's Frogs, Reptiles Birds and Mammals*. Queensland Museum, Brisbane.

Van Dyck S (1993) The taxonomy and distribution of *Petaurus gracilis* (Marsupialia: Petauridae), with notes on its ecology and conservation status. *Memoirs of the Queensland Museum* 33, 77–122. [30 Jun 1993]

Van Dyck S (1997) Queensland pebble-mound mice - up from the tailings. *Nature Australia* 25(10), 40–47.

Van Dyck S (2002) Morphology-based revision of *Murexia* and *Antechinus* (Marsupialia: Dasyuridae). *Memoirs of the Queensland Museum* 48, 239–330. [31 May 2002]

Van Dyck S, Crowther MS (2000) Reassessment of northern representatives of the *Antechinus stuartii* complex (Marsupialia: Dasyuridae): *A. subtropicus* sp. nov. and *A. adustus* new status. *Memoirs of the Queensland Museum* 45, 611–635. [30 Jun 2000]

Van Dyck S, Strahan R (2008) *The Mammals of Australia*. Reed New Holland, Sydney. Third Edition. [1 Mar 2008]

Van Dyck S, Woinarski JCZ, Press AJ (1994) The Kakadu dunnart, *Sminthopsis bindi* (Marsupialia: Dasyuridae), a new species from the stony woodlands of the Northern Territory. *Memoirs of the Queensland Museum* 37, 311–323. [1 Dec 1994]

Van Dyck S, Gynther I, Baker A (2013) *Field Companion to the Mammals of Australia*. New Holland, Sydney.

Van Gelder RG (1977) Mammalian hybrids and generic limits. *American Museum Novitates* 2635, 1–25. [12 Oct 1977]

Van Valen L (1967) New Paleocene insectivores and insectivore classification. *Bulletin of the American Museum of Natural History* 135, 217–284. [29 May 1967]

Van Valen L (1969) The multiple origins of the placental carnivores. *Evolution* 23, 118–130. [Mar 1969]

Van Valen L (1979) The evolution of bats. *Evolutionary Theory* 4, 103–121.

Vandebroek G (1961) The comparative anatomy of the teeth of the lower and non-specialised mammals. Part 1. pp. 215–320; Part 2. 1–181, in G Vandebroek (ed.) International Colloquium on the Evolution of Lower and Non-specialised Mammals. *Koninklijke Vlaamse Academie voor Wetenschappen, Letteren en Schone Kunsten van België, Brussels*.

Vaurie P (1958) A revision of the genus *Diplotaxis* (Coleoptera, Scarabaeidae, Melolonthinae). Part 1. *Bulletin of the American Museum of Natural History* 115, 269–396.

Vaz-Ferreira R (1982) *Arctocephalus australis* Zimmermann, South American Fur Seal. pp. 497–508, in *Mammals in the Seas, Volume IV: Small Cetaceans, Seals, Sirenians and Otters*. FAO Fisheries Series, No. 5.

Vejdovsky F (1876a) Beiträge zur Oligochaetenfauna Böhmens. *Sitzungsberichte der Königlichen Böhmischen Gesellschaft der Wissenschaften in Prag* 1875, 191–201.

Vejdovsky F (1876b) Üeber *Psammoryctes umbellifer* (*Tubifex umbellifer* E.R. Lank.) und ihm verwandte gattungen. *Zeitschrift fur Wissenschartliche Zoologie* 27, 137–154.

Vernes K, Dennis A, Winter J (2001) Mammalian diet and broad hunting strategy of the dingo (*Canis familiaris dingo*) in the wet tropical rainforests of northeastern Australia. *Biotropica* 33, 339–345.

Vicq-d'Azyr F (1792) *Encyclopédie Méthodique. Systême Anatomique. Quadrupèdes*. Panckoucke, Paris. Volume 2. [Vol 2(1), 3 Jun 1791; Vol 2(2), May 1792]

Vieillot LP (1816) *Analyse d'une Nouvelle Ornithologie Élémentaire*. Déterville, Paris.

Vilà CP, Savolainen P, Maldonado JE, Amorim IR, Rice JE, Honeycutt RL, Crandall KA, Lundenberg J, Wayne RK (1997) Multiple and ancient origins of the domestic dog. *Science* 276(5319), 1687–1689. [13 June 1997]

Villa A, Villa GB (1833) *Coleoptera Europae dupleta in collectione Villa quae pro mutua commutatione offerri Jossunt*. Mediolani. Reference not seen.

Voigt FS (1831) In G Cuvier, *Das Thierreich, geordnet nach seiner organisation : Als grundlage der naturgeschichte der thiere und einleitung in die vergleichende anatomie*. Volume 1. Brockhaus, Leipzig.

Voigt FS (1839) In G Cuvier, *Das Thierreich, geordnet nach seiner organisation : Als grundlage der naturgeschichte der thiere und einleitung in die vergleichende anatomie*. Volume 5. Brockhaus, Leipzig.

Voigt JH (1802) *Magazin für den neuesten Zustand der Naturkunde*, mit Rücksicht auf die dazu gehörigen Hilfswissenschaften, hrsg. Von Johann Heinrich Voigt. Industrie-Comptoir, Weimr. Volume 4.

Vokes HE (1980) *Genera of the Bivalva: A systematic and bibliographical catalogue* (revised and updated). Paleontological Research Institution, Ithaca, New York.

Volleth M, Heller K-G (1994) Phylogenetic relationships of vespertilionid genera (Mammalia: Chiroptera) as revealed by karyological analysis. *Journal of Zoological Systematics and Evolutionary Research* 32, 11–34. [Mar 1994]

Volleth M, Tidemann CR (1989) Chromosome studies in three genera of Australian vespertilionid bats and their systematic implications. *Zeitschrift fur Saugetierkunde* 54, 215–222.

Volleth M, Tidemann CR (1991) The origin of the Australian Vespertilioninae bats. As indicated by chromosomal studies. *Zeitschrift fur Saugetierkunde* 56, 321–330.

vonHoldt BM, Polliger JP, Lohmueller KE, Han E, Parker HG, Quignon P, Degenhardt JD, Boyko AR, Earl DA, Auton A, Reynolds A, Bryc K, Brisbin A, Knowles JC, Mosher DS, Spady TC, Elkahloun A, Geffen E, Pilot M, Jedrzejewski W, Greco C, Randi E, Bannasch D, Wilton A, Shearman J, Musiani M, Cargill M, Jones PG, Qian Z, Huang W, Ding Z-L, Zhang Y-P, Bustamante CD, Ostrander EA, Novembre J, Wayne RK (2010) Genome-wide SNP and haplotype analyses reveal a rich history underlying dog domestication. *Nature* 464(7290), 898–903. [8 Apr 2010]

Vose HM (1973) Feeding habits of the Western Australian honey possum, *Tarsipes spenserae*. *Journal of Mammalogy* 54, 245–247. [Feb 1973]

Voth I (1988) Social behaviour of New Guinea dingoes (*Canis lupus familiaris*): expressive behaviour, social organisation and rank relationship. PhD (Vet.) Thesis. Ludwig Maximilian University, Munich. Reference not seen.

Vrana PB, Milinkovitch MC, Powell JR, Wheeler WC (1994) Higher level relationships of the arctoid Carnivora based on sequence data and 'total evidence'. *Molecular Phylogenetics and Evolution* 3, 47–58.

Wada S, Numachi K (1991) Allozyme analyses of genetic differentiation among the populations and species of *Balaenoptera*. pp. 125–154, in AR Hoelzl (ed.) *Genetic Ecology of Whales and Dolphins. Report of the International Whaling Commission, Special Issue* 13, 1–311.

Wada S, Oishi M, Yamada TK (2003) A newly discovered species of living baleen whale. *Nature* 426(6964), 278–281. [20 Nov 2003]

Waddell PJ, Okada N, Hasegawa M (1999a) Towards resolving the interordinal relationships of placental mammals. *Systematic Biology* 48, 1–5. [Mar 1999]

Waddell PJ, Cao Y, Hasegawa M (1999b) Using novel methods to evaluate mammalian mtDNA and detect internal conflicts: Including AA invariant sites-LogDet and site stripping, with special reference to the position of hedgehog, armadillo, and elephant. *Systematic Biology* 48, 31–53. [Mar 1999]

Waddell PJ, Cao Y, Hasegawa M, Mindell DP (1999c) Assessing the Cretaceous superordinal divergence times within birds and placental mammals by using whole mitochondrial protein sequences and an extended statistical framework. *Systematic Biology* 48, 119–137. [Mar., 1999]

Waddell PJ, Kishino H, Ota R (2001) A phylogenetic foundation for comparative mammalian genomics. *Genome Informatics* 12, 141–154.

Wagler JG (1829a) Beyträge und Bemerkungen zu dem ersten Band seines Systema Avium (Fortsetzung I). *Isis von Oken* 22, 645–664.

Wagler JG (1829b) Beyträge und Bemerkungen zu dem ersten Band seines Systema Avium (Fortsetzung III). *Isis von Oken* 22, 736–762.

Wagler JG (1830) *Natürliches System der Amphibien, mit vorangehender Classification der Säugthiere und Vögel. Ein Beitrag zur vergleichenden Zoologie*. J.G. Cotta'schen Buchhandlung, München, Stuttgart und Tübingen. [Aug 1830]

Wagler JG (1832a) Neue Sippen und Gattungen der Saugthiere und vogel. *Isis von Oken* 25, 1218–1235.

Wagler J (1832b) Mittheilungen über einige merkwürdige Thiere. 1. Saugthiere. *Isis von Oken, Jena* 25, 275–281.

Wagner JA (1841a) Beschreibubng einer neuen Art von Bandikuts, *Perameles mysurus*. [sic], nebt Bemerkungen über *Perameles obesula. Archiv für Naturgeschichte* 7(1), 289–297.

Wagner JA (1841b) Berichte über die Leistungen in der Naturgeschichte der Säugethiere während der beiden Jahre 1839 und 1840. *Archiv für Naturgeschichte* 7(2), 1–58.

Wagner JA (1843) *Die Säugthiere in Abbildungen nach der natur mit Beschreibungen* Von Dr Johann Christian Daniel von Schreber. Erlangen. Supplementband. Volume 3.

Wagner JA (1844) *Die Säugthiere in Abbildungen nach der natur mit Beschreibungen* Von Dr Johann Christian Daniel von Schreber. Erlangen. Supplementband. Volume 4.

Wagner JA (1845) Diagnosen einiger neuen Arten von Nagern und Handflüglern. *Archiv für Naturgeschichte* 11(1), 145–149.

Wagner JA (1846) *Die Säugthiere in Abbildungen nach der natur mit Beschreibungen* Von Dr Johann Christian Daniel von Schreber. Weigel, Lepzig. Volume 7.

Wagner JA (1847) Bericht über die Leistungen in der Naturgeschichte der Säugthiere während des Jahres 1846. *Archiv für Naturgeschichte* 13(2), 1–42.

Wagner JA (1851) Beiträge zur Kenntniss der in den lithographischen Schiefern abgelagerten urweltichen Fische. *Abhandlungen der Mathemat.-. Physikalischen Classe der Königlich Bayerischen Akademie der Naturwissenschaftlichen* 6(1), 1–80.

Wagner JA (1855) *Die Säugthiere in Abbildungen nach der natur mit Beschreibungen* Von Dr Johann Christian Daniel von Schreber. Weigel, Lepzig. Supplementband. Volume 5.

Waite ER (1896) Muridae. pp. 93–409, in B Spencer (ed.) *Report of the Horn Scientific Expedition to Central Australia*. Part 2. Zoology. Melville, Mullen & Slade, Melbourne.

Waite ER (1898) Observations on Muridae from central Australia. *Proceedings of the Royal Society of Victoria* 10, 114–128.

Waite ER (1899) Description of a ring-tailed possum regarded as a variety of *Pseudochirus herbertensis*, Collett. *Records of the Australian Museum* 3, 91–93. [17 Apr. 1899]

Waite ER (1900) The generic name *Thylacomys. Annals & Magazine of Natural History* (5)7, 222–223. [1 Feb 1900]

Waite ER (1901) A description of *Macropus isabellinus. Records of the Australian Museum* 4, 131–134. [29 Jul 1901]

Waite ER (1904) Description of a new *Phascogale* from north Western Australia. *Records of the Australian Museum* 5, 122–124. [28 Jan 1904]

Waithman J (1979) A report on a collection of mammals from southwest Papua, 1972–1973. *Australian Zoologist* 20, 313–326.

Wakefield NA (1963a) The Australian pigmy-possums. *Victorian Naturalist* 80, 99–116. [8 Aug 1963]

Wakefield NA (1963b) Mammal remains from the Grampians, Victoria. *Victorian Naturalist* 80, 130–133. [5 Sep 1963]

Wakefield NA (1966) Mammals of the Blandowski Expedition to north western Victoria, 1856–57. *Proceedings of the Royal Society of Victoria* 79, 371–379.

Wakefield NA (1967a) Some taxonomic revision in the Australia marsupial genus *Bettongia* (Macropodidae), with description of a new species. *Victorian Naturalist* 84, 8–22. [5 Jan 1967]

Wakefield NA (1967b) Whales and dolphins recorded for Victoria. *Victorian Naturalist* 84, 273–281. [8 Sep 1967]

Wakefield NA (1972a) Paleoecology of fossil mammal assemblages from some Australian caves. *Proceedings of the Royal Society of Victoria* 85(1), 1–26.

Wakefield NA (1972b) Studies in Australian Muridae: Review of *Mastacomys fuscus*, and description of a new subspecies of *Pseudomys higginsi. Memoirs of the National Museum of Victoria* 33, 15–31. [7 Feb 1972]

Wakefield NA, Warneke RM (1963) Some revision in *Antechinus* (Marsupialia) – 1. *Victorian Naturalist* 80, 194–219. [Nov 1963]

Wakefield NA, Warneke RM (1967) Some revision in *Antechinus* (Marsupialia) – 2. *Victorian Naturalist* 84, 69–99. [9 Mar 1967]

Wall WB (1851) History and description of the skeleton of a new sperm whale, lately set up in the Australian Museum. Together with some account of a new genus of sperm whales called *Euphysetes. Australian Museum Memoir* 1, 1–66. [31 Dec 1851]

Wallace AR (1876) *The Geographical Distribution of Animals*. Harper, New York. Volume 2.

Walters B (1995) *The Company of Dingoes: Two Decades with our Native Dog*. Australian Native Dog Conservation Society, Bargo, NSW.

Walton DW (ed.) (1988a) *Zoological Catalogue of Australia. 5. Mammalia*. Australian Government Publishing Service, Canberra. [13 Apr 1988]

Walton DW (1988b) Notoryctidae. pp. 46–47, in DW Walton (ed.) *Zoological Catalogue of Australia. 5. Mammalia*. Australian Government Publishing Service, Canberra. [13 Apr 1988]

Wang JY, Chou LS, White BN (1999) Mitochondrial DNA analysis of sympatric morphotypes of bottlenose dolphins (genus: *Tursiops*) in Chinese waters. *Molecular Ecology* 8, 1603–1612. [Oct 1999]

Wang JY, Chou LS, White BN (2000a) Differences in external morphology of two sympatric species of bottlenose dolphins (genus *Tursiops*) in the waters of China. *Journal of Mammalogy* 81, 1157–1165. [Nov 2000]

Wang JY, Chou LS, White BN (2000b) Osteological differences between two sympatric forms of bottlenose dolphins (genus *Tursiops*) in Chinese waters. *Journal of Zoology* 252, 147–162.

Wang JY, Riehl KN, Dungan SZ (2014) Family Delphinidae (Ocean Dolphins. pp. 410–526, in DE Wilson & RA Mittermeier (eds.) *Handbook of the Mammals of the Word. Volume 4. Sea Mammals*. Lynx Edicions, Barcelona.

Warlow W (1833) Systematically arranged catalogue of the Mammalia and birds belonging to the Museum of the Asiatic Society, Calcutta. *Journal of the Asiatic Society of Bengal, Calcutta* 2, 97–100.

Warneke RM, Shaughnessy PD (1985) *Arctocephalus pusillus*, the South African and Australian fur seal: taxonomy, evolution, biogeography and life history. pp. 53–77, in JK Ling & MM Bryden (eds.) *Studies of Sea Mammals in South Latitudes*. South Australian Museum, Adelaide.

Warren WC, Hillier LW, Graves JAM, Birney E, Ponting CP, Grützner F, Belov K, Miller W, Clarke L, Chinwalla AT, Yang S-P, Heger A, Locke DP, Miethke P, Waters PD, Veyrunes F, Fulton L, Fulton B, Graves T, Wallis J, Puente XS, López-Otín C, Ordóñez GR *et al.* (2008) Genome analysis of the platypus reveals unique signatures of evolution. *Nature* 453(7192), 175–184. [8 May 2008]

Waterhouse CO (1902) *Index Zoologicus*: An alphabetical list of names of genera and sub- genera proposed for use in zoology as recorded in the 'Zoological Record' 1880–1900 together with other names not included within the Nomenclator Zoologicus of S.H. Scudder. Gurney & Jackson, London.

Waterhouse FG (1876) The fauna of South Australia. pp. 281–296, in W Harcus (ed.) *South Australia: Its History, Resources and Productions*. Sampson, Low, Marston, Searle & Ringington, London.

Waterhouse FH (1885) *The Dates of Publication of some of the Zoological Works of the late John Gould, F.R.S.* R.H. Porter, London.

Waterhouse FH (1937) List of the dates of delivery of the sheets of the 'Proceedings' of the Zoological Society of London, from the commencement to 1859 inclusive. *Proceedings of the Zoological Society of London* 107(Series A), 78–83.

Waterhouse GR (1836a) Description of a new genus (*Myrmecobius*) of mammiferous animals from New Holland, probably belonging to the marsupial type. *Proceedings of the Zoological Society of London* 4, 69–70. [18 Oct 1836]

Waterhouse GR (1836b) Description of a new genus of mammiferous animals from New Holland, which will probably be found to belong to the marsupial type. *Philosophical Magazine and Journal of Science* 9(3), 520–521.

Waterhouse GR (1837a) Exhibition of the small rodents, belong to the collection of Mr. Darwin to the Society. *Proceedings of the Zoological Society of London* 5, 27–32. [21 Nov 1837]

Waterhouse GR (1837b) Characters of new species of the genus *Mus*, from the collection of Mr Darwin. *Proceedings of the Zoological Society of London* 5, 15–22. [21 Nov 1837]

Waterhouse GR (1838a) *Catalogue of the Mammalia Preserved in the Museum of the Zoological Society of London*. Second Edition. Richard and John E. Taylor, London. Higher ranks *fide* Waterhouse (1841a: 60).

Waterhouse GR (1838b) Characters of some new species of the genera *Mus* and *Phascogale*. *Proceedings of the Zoological Society of London* 5(1837), 75–77. [22 Jan 1838]

Waterhouse GR (1838c) Description of a new genus of mammiferous animals from Australia, belonging probably to the Order Marsupialia. *Transactions of the Zoological Society of London* 2, 149–154. [4 May 1838]

Waterhouse GR (1838d) Some skulls of the flying opossums (*Petaurista*). *The Athenaeum* 580, 880. [8 Dec 1838]

Waterhouse GR (1838e) Description of a new species of kangaroo (*Macropus bennetti*) and a mouse (*Mus subspinosus*) from the Cape of Good Hope. *Proceedings of the Zoological Society of London* 5(1837), 103–105. [25 May 1838]

Waterhouse GR (1838f) On a new species of the genus *Delphinus*. *Proceedings of the Zoological Society of London* 6, 23–24. [Jul 1838]

Waterhouse GR (1839a) On the dentition of the flying opossums. *Proceedings of the Zoological Society of London* 6(1838), 149–156. [May 1839]

Waterhouse GR (1839b) On the geographical distribution of the Rodentia. *Proceedings of the Zoological Society of London* 7, 172–174. [Mar 1840]

Waterhouse GR (1839c [1838–1839]) Part II. Mammalia. In C Darwin (ed.) (1838–1843) *The Zoology of the Voyage of H.M.S. Beagle, Under the Command of Captain Fitzroy, R.N., During the Year 1832 to 1836*. Smith, Elder & Co., London. [Sep 1939]

Waterhouse GR (1839d) A new species of rodent which had been sent from the island of Luzon, one of the Philippines, by Hugh Cumming, Esq. *Proceedings of the Zoological Society of London* 7, 107–109. [Nov 1839]

Waterhouse GR (1840a) Description of a new marsupial mammal, belonging to the genus *Phascogale*. *The Magazine of Natural History* 4(2), 299–301. [1 May 1840]

Waterhouse GR (1840b) On a new genus of the Family Muridae and Order Rodentia. *Proceedings of the Zoological Society of London* 8, 1–3. [Jul 1840]

Waterhouse GR (1840c) On the geographical distribution of the Rodentia. *Proceedings of the Zoological Society of London* 7(1839), 172–174. [Mar 1840]

Waterhouse GR (1841a) *The Natural History of Marsupialia or Pouched Animals*. W.H. Lizars, Edinburgh. [Aug 1841]

Waterhouse GR (1841b) Observations of the Rodentia. *Annals & Magazine of Natural History* (1)8, 81–84. [1 Oct 1841]

Waterhouse GR (1842) On two new species of marsupial animals from South Australia. *Proceedings of the Zoological Society of London* 10, 47–48. [Nov 1842]

Waterhouse GR (1843a) Description of two new species of the genus *Mus*. *Proceedings of the Zoological Society of London* 10(1842), 145–146. [Feb 1843]

Waterhouse GR (1843b) Two new species of mouse. *Annals & Magazine of Natural History* (1)12, 134–135. [1 Aug 1843]

Waterhouse GR (1843c) On various species of bats collected by H. Cuming, Esq. in the Philippine Islands. *Proceedings of the Zoological Society of London* 11, 66–69. [Nov 1843]

Waterhouse GR (1845) Descriptions of new species of bats collected in the Philippine Islands, and presented to the Society by H. Cuming, Esq. *Proceedings of the Zoological Society of London* 13, 3–10. [Apr 1845]

Waterhouse GR (1846) *A Natural History of the Mammalia. Volume 1. Containing the Order Marsupiata, or Pouched Animals*. Bailliere, London.

Watts CHS (1976) *Leggadina lakedownensis*, a new species of murid rodent from North Queensland. *Transactions of the Royal Society of South Australia* 100, 105–108. [31 May 1976]

Watts CHS, Aslin HJ (1974) Notes on the small mammals of north-eastern South Australia and south-western Queensland. *Transactions of the Royal Society of South Australia* 98, 61–69. [31 May 1974]

Watts CHS, Aslin HJ (1981) *The Rodents of Australia*. Angus and Robertson, Sydney.

Watts CHS, Baverstock PR (1994) Evolution of New Guinean Muridae (Rodentia) assessed by microcomplement fixation of albumin. *Australian Journal of Zoology* 42, 295–306.

Watts CHS, Kemper CM (1989) Muridae. pp. 939–950, in DW Walton & BJ Richardson (eds.) *Fauna of Australia. Volume 1B. Mammalia*. Australian Government Publishing Service, Canberra.

Watts CHS, Baverstock PR, Birrell J, Krieg M (1992) Phylogeny of the Australian rodents (Muridae): a molecular approach using microcomplement fixation of albumin. *Australian Journal of Zoology* 40, 81–90.

Wayne RK (1993) Molecular evolution of the dog family. *Trends in Genetics* 9, 218–224. [Jun 1993]

Webb NJ, Tidemann CR (1995) Hybridisation between black (*Pteropus alecto*) and grey-headed (*P. poliocephalus*) flying foxes (Megachiroptera: Pteropodidae). *Australian Mammalogy* 18, 19–26. [21 Dec 1995]

Weber F (1795) *Nomenclator entomologicus secundum entomologiam systematicam ill. Fabricii : adjectis speciebus recens detectis et varietatibus*. Chilonii & Hamburgi.

Weber MCW (1904) *Die Säugetiere: Einführung in die Anatomie und Systematik der recenten und fossilen Mammalia*. Systematischer Teil. Gustav Fischer, Jena.

Weber MCW (1928) *Die Säugetiere: Einführung in die Anatomie und Systematik der recenten und fossilen Mammalia*. Volume 2. Systematischer Teil. Gustav Fischer, Jena.

Weddell J (1825) *A Voyage Toward the South Pole. Performed in the years 1822–24. Containing an examination of the Antarctic Sea, to the seventy-fourth degree of latitude: and a visit to Tierra del Fuego, with particular account of the inhabitants*. Longman, Hurst, Rees, Orme, Brown and Green, London.

Weise J (1886) In WF Erichson. *Naturgeschichte der Insecten Deutschlands*. Nicolaische Verlags-Buchhandlung, Berlin. Volume 6.

Werdelin L (1987) Some observations of *Sarcophilus laniarius* and the evolution of *Sarcophilus*. *Records of the Queen Victoria Museum, Launceston* 90, 1–27. [Feb 1987]

Westerman M (1991) Phylogenetic relationships of the marsupial mole, *Notoryctes typhlops* (Marsupialia: Notoryctidae). *Australian Journal of Zoology* 39, 529–537.

Westerman M, Krajewski C (2000) Molecular relationships of the Australian bandicoot genera *Isoodon* and *Perameles* (Marsupialia: Peramelina). *Australian Mammalogy* 22, 1–8. [Aug 2000]

Westerman M, Springer MS, Dixon J, Krajewski C (1999) Molecular relationships of the extinct pig-footed bandicoot *Chaeropus ecaudatus* (Marsupialia: Perameloidea) using 12S rRNA sequences. *Journal of Mammalian Evolution* 6, 271–288. [Sep 1999]

Westerman M, Springer MS, Krajewski C (2001) Molecular relationships of the New Guinean bandicoot genera *Microperoryctes* and *Echymipera* (Marsupialia: Peramelina). *Journal of Mammalian Evolution* 8, 93–105. [Jun 2001]

Westerman M, Burk A, Amrine-Madsen HM, Prideaux GJ, Case JA, Springer MS (2002) Molecular evidence for the last survivor of an ancient kangaroo lineage. *Journal of Mammalian Evolution* 9, 209–223. [Sep 2002]

Westerman M, Young J, Krajewski C (2008) Molecular relationships of species of *Pseudantechinus, Parantechinus* and *Dasykaluta* (Marsupialia; Dasyuridae). *Australian Mammalogy* 29, 201–212. [1 Apr 2008]

Westerman M, Kear BP, Aplin K, Meredith RW, Emerling C, Springer MS (2012) Phylogenetic relationships of living and recently extinct bandicoots based on nuclear and mitochondrial DNA sequences. *Molecular Phylogenetics and Evolution* 62, 97–108.

Westwood JO (1840) Synopsis of the genera of British insects. pp. 1–154, in W Westwood (ed.) *An Introduction to the Modern Classification of Insects*. Volume 2. Synopsis. Longman, Orme, Brown, Green and Longmans, London.

White A (1853) *Catalogue of the coleopterous insects in the collection of the British Museum*. British Museum (Natural History), London.

White J (1790) *Journal of a Voyage to New South Wales*; with sixty-five plates of non descript animals, birds, lizards, serpents, curious cones of trees and other natural productions. J. Debrett, London.

White J (1803) in Geoffroy Saint-Hilaire, É. Note sur les espèces du genre dasyure. *Bulletin des sciences par la Société Philomatique de Paris* 3(81), 158–159. Misprinted as should be pages 258–259.

White N (2006) The conservation significance of island versus mainland populations of the Western Chestnut Mouse (*Pseudomys nanus*). Honours Thesis. Murdoch University, Perth. Reference not seen.

Whitehead GK (1972) *Deer of the World*. Constable, London. Reference not seen.

Whitehead H (2009) Sperm whale *Physeter macrocephalus*. pp. 1091–1097, in WF Perrin, B Würsig & JGM Thewissen (eds.) Encyclopedia of Marine Mammals. Academic Press, Amsterdam. Second Edition.

Whitehead PJP (1967) The dating of the 1st Edition of Cuvier's Le Règne Animal Distribué d'Après son Organisation. *Archives of Natural History* 4, 300–301. [1 Jul 1967]

Whitley GP (1936) The Australian devil ray, *Daemomanta alfredi* (Krefft), with remarks on the Superfamily Mobuloidea (Order Batoidei). *Australian Zoologist* 8, 164–188. [29 Jun 1936]

Whittell HM (1954) John Gilbert's notebook on marsupials. *Western Australian Naturalist* 4, 104–114. [30 Jun 1954]

Wible JR, Martin JR (1993) Ontogeny of the tympanic floor and roof in archontans. pp. 111–148, in RDE MacPhee (ed.) *Primates and Their Relatives in Phylogenetic Perspective*. Plenum Publishing, New York.

Wible JR, Rougier GW, Novacek MJ, McKenna MC, Dashzeveg DD (1995) A mammalian petrosal from the Early Cretaceous of Mongolia: implications for the evolution of the ear region and mammaliamorph relationships. *American Museum Novitates* 3149, 1–19. [16 Nov 1995]

Wible JR, Rougier GW, Novacek MJ, Asher RJ (2007) Cretaceous eutherians and Laurasian origin for placental mammals near the K/T boundary. *Nature* 447(7147), 1003–1006. [21 Jun 2007]

Wied-Neuwied M Prince von (1820) *Diclidurus* Klappenschwanz. Bin neues genus der Chiropteren aus Brasilian. *Isis von Oken* 4–5, 1629–1630.

Wied-Neuwied M Prince von (1832) *Beiträge zur Naturgeschichte von Brasilien*. Verlage des Landes-Industrie-Comptoirs, Weimar. Volume 4.

Wied-Neuwied M Prince von (1839) *Reise in das Innere Nord-America*. Volume 1. Coblenz.

Wiedemann CRW (1800) Nachricht von einem äusserst fonderbaren, neu entdeckten säugethiere: *Platypus anatinus*. *Archiv für Zoologie und Zootomie* 1, 175–180.

Wiegmann AFA von (1832) In AFA von Wiegmann & JF Ruthe (eds.) *Handbuch der Zoologie*. Berlin.

Wiegmann AFA von (1834) *Herpetologia Mexicana*, seu Descriptio amphibiorum Novae Hispaniae, quae itineribus Comitis de Sack, Ferdinandi Deppe et Chr. Guil. Schiede in Museum Zoologicum Berolinense pervenerunt. Pars Prima, Sauroum Species. G. C. Lüderitz, Berlin.

Wiegmann AFA von (1839) Amphibia, aves, mammalia bearbeitet com Herausgeber. *Archiv für Naturgeschichte* 5(2), 386–428.

Wiegmann AFA von (1846) In JA Wagner. *Die Säugethiere*, in Abbildungen nach Natur mit Beschreibungen von Johan Christian Daniel von Schreiber. Weigel, Leipzig. Supplementband. Volume 7.

Williamson GR (1959) Three unusual rorqual whales from the Antarctic. *Proceedings of the Zoological Society of London* 133, 135–144. [Jul 1859]

Williamson GR (1961) Two kinds of minke whales in the Antarctic. *Norsk Hvalfangst-Tidende* 50, 133–141.

Wilson DE (1993) Order Sirenia. pp. 365–366, in DE Wilson & DA Reeder (eds.) *Mammal Species of the World. A Taxonomic and Geographic Reference*. Smithsonian Institution Press, Washington. Second Edition.

Wilson DE (2007) Genus *Myotis* Kaup, 1829. pp. 468–481, in AL Gardner (ed.) *Mammals of South America. Volume 1. Marsupials, Xenarthrans, Shrews, and Bats*. Chicago University Press, Chicago.

Wilson DE, Reeder DM (2005) *Mammal Species of the World: A Taxonomic and Geographic Reference*. Johns Hopkins University Press, Baltimore. Third Edition.

Wilson EA (1907) *National Antarctic Expedition. [of the S.S. 'Discovery', under Captain R.F. Scott], 1901–1904: Natural history. [including some specimens obtained by G.A. Davidson of the relief ship, the 'Morning']* Volume 2 - Zoology. Part 4 – Mammalia (Whales and Seals). British Museum (Natural History), London.

Wilson EO, Brown WL (1953) The subspecies concept and its taxonomic application. *Systematic Zoology* 2, 97–111. [Sep 1953]

Wilson P (1982) Metrical variation within and between populations of *Miniopterus australis* and *M. oceanensis* (Chiroptera: Vespertilionidae) from southeastern Australia. *Honours Thesis*, School of Zoology. University of New South Wales, Sydney. Reference not seen.

Wilson P (1985) Maeda's *Miniopterus* taxonomy. *Macroderma* 1, 29–36. [Sep 1985]

Wilton AN, Steward DJ, Zafiris K (1999) Microsatellite variation in the Australian dingo. *The Journal of Heredity* 90, 108–111.

Winge H (1887) Jordfunde og nulevende Gnavere (Rodentia) fra Lagoa Santa, Minas Geraes, Brasilien: Med udsigt over gnavernes indbyrdes slagtskab. *En Samling af Afhandlinger E Museo Lundii* 1(3), 1–178.

Winge H (1893a) Jordfunde og nulevende Pungdyr (Marsupialia) fra Lagoa Santa, Monas Geraes, Brasilien. *E Museo Lundii - En Samling af Afhandlinger* 2(2), 1–133.

Winge H (1893b) Jordfunde og nulevende Flagermus (Chiroptera) fra Lagoa Santa, Minas Geraes, Brasilien. *En Samling af Afhandlinger. E Museo Lundii* 2(1), 1–54.

Winge H (1917) Udsigt over Insektaedernes indbyrdes Slaegtskab. *Videnskabelige Meddelelser fra Dansk Naturhistorisk Forening, Copenhagen* 68, 83–203.

Winge H (1918) A review of the interrelationships of the Cetacea. *Smithsonian Miscellaneous Collections* 72(8), 1–97. [1921 Translated from the Danish by G.S. Miller, Jr.]

Winge H (1941) *The Inter-relationships of the Mammalian Genera. Volume 1. Monotremata, Marsupialia, Insectivora, Chiroptera, Edentata*. Translated from Danish by E. Deichmann and G.M. Allen. C.A. Reitzels Forlag, Copenhagen.

Winn HE, Reichley NE (1985) Humpback whale *Megaptera novaeangliae*. pp. 241–274, in SH Ridgway & R Harrison (eds.) *Handbook of Marine Mammals. The Sirenians and Baleen Whales*. Academic Press, London.

Winter JW (1983) Thornton Peak Melomys. pp. 379, in R Strahan (ed.) *The Complete Book of Australian Mammals*. Angus and Robertson, Sydney.

Winter J (1984) The Thornton Peak Melomys, *Melomys hadrourus* (Rodentia: Muridae): a new rainforest species from northeastern Queensland, Australia. *Memoirs of the Queensland Museum* 21, 519–539. [Mar 1984]

Winter JW, Allison FR (1980) The native mammals of Cape York Peninsula – changes in status since the Archbold Expeditions. pp. 31–44, in NC Stevens & A Baily (eds.) *Contemporary Cape York Peninsula*. Royal Society of Queensland, Brisbane.

Woinarski JCZ, Burbidge AA, Harrison P (eds.) (2014) *The Action Plan for Australian Mammals 2012*. CSIRO Publishing, Melbourne. [2 Jun 2014]

Wolsan M (1993) Phylogeny and classification of early European *Mustelida* (*Mammalia*: *Carnivora*). *Acta Theriologica* 38, 345–384.

Wood AE (1957) What, if anything, is a rabbit? *Evolution* 11, 417–425. [Dec 1957]

Wood HE (1937) Perissodactyl suborders. *Journal of Mammalogy* 18, 106. [Feb 1937]

Wood Jones F (1921) The status of the dingo. *Transactions of the Royal Society of South Australia* 45, 254–263. [21 Dec 1921]

Wood Jones F (1922a) External characters of pouch embryos of marsupials. No. 3. *Isoodon barrowensis*. *Transactions of the Royal Society of South Australia* 46, 39–45. [22 Dec 1922]

Wood Jones F (1922b) On the dental characters of certain Australian rats. *Proceedings of the Zoological Society of London* 92, 587–598. [28 Sep 1922]

Wood Jones F (1922c) The flora and fauna of Nuyts Archipelago and the Investigator group. No. 2. The monodelphian mammals. *Transactions of the Royal Society of South Australia* 46, 181–193. [22 Dec 1922]

Wood Jones F (1923) The marsupial genus *Thalacomys*. A review of the rabbit-bandicoots; with the description of a new species. *Records of the South Australian Museum* 2, 333–352. [30 Jun 1923]

Wood Jones F (1924) The flora and fauna of Nuyt's Archipelago and the Investigator Group. No. 15. The Pearson Island rat and the Flinder's Island wallaby. *Transactions of the Royal Society of South Australia* 48, 10–14. [24 Dec 1924]

Wood Jones F (1925a) A new South Australian dormouse opossum. *Transactions of the Royal Society of South Australia* 49, 96–98. [24 Dec. 1925]

Wood Jones F (1925b) A revision of South Australian jerboa mice, with the description of a new species. *Records of the South Australian Museum* 3, 1–7. [30 Jun 1925]

Wood Jones F (1925c) The eared seals of South Australia. *Records of the South Australian Museum* 3, 9–16. [30 Jun 1925]

Wood Jones F (1949) The study of a generalised marsupial (*Dasycercus cristicauda*) Krefft. *Transactions of the Zoological Society of London* 26, 409–501.

Wood Jones F (1923–1925) *The Mammals of South Australia*. Parts 1–3. Government Printer, Adelaide. [Part 1, 5 Apr 1923; Part 2, 8 Aug 1924; Part 3, 21 Dec 1925]

Woodburne MO (1967) The Alcoota Fauna, central Australia. An integrated palaeontological and geological study. *Bureau of Mineral Resources. Geology and Geophysics Bulletin* 87, 1–187.

Woodburne MO (1984a) Families of marsupials: Relationships, evolution and biogeography. pp. 48–71, in TW Broadhead (ed.) *Mammals: Notes for a Short Course. University of Tennessee Department of Geological Sciences, Studies in Geology*. Volume 8. [4 Nov 1984]

Woodburne MO (1984b) *Wakiewakie lawsoni*, a new genus and species of Potoroinae (Marsupialia: Macropodidae) of medial Miocene age, South Australia. *Journal of Paleontology* 58, 1062–1073. [Jul 1984]

Woodburne MO (2003) Monotremes as pretribosphenic mammals. *Journal of Mammalian Evolution* 10, 195–248. [Sep 2003]

Woodburne MO, Rich TH, Springer MS (2003) The evolution of tribospheny and the antiquity of mammalian clades. *Molecular Phylogenetics and Evolution* 28, 360–385. [Aug 2003]

Woods CA, Kilpatrick CW (2005) Infraorder Hystricognathi. pp. 1538–1600, in DE Wilson & DA Reeder (eds.) *Mammal Species of the World. A Taxonomic and Geographic Reference*. Johns Hopkins University Press, Baltimore. Third Edition.

Woods JT (1960) The genera *Proleopus* and *Hypsiprymnodon* and their position in the Macropodidae. *Memoirs of the Queensland Museum* 13, 199–212. [5 Sep 1960]

Woodward AS (1898) *Outlines of Vertebrate Palaeontology for Students of Zoology*. Cambridge University Press, Cambridge.

Woodward BH (1902) Fauna. pp. 211–85, in MAC Fraser (ed.) *Western Australian Yearbook for the year 1900–1901*. 12th Edition. Volume 1. Government Printer, Perth. Reference not seen.

Woolley PA (1971) Observations on the reproductive biology of the dibbler, *Antechinus apicalis* (Marsupialia: Dasyuridae). *Journal of the Royal Society of Western Australia* 54, 99–102.

Woolley PA (1974) The pouch of *Planigale subtillissima* and other dasyurid marsupials. *Journal of the Royal Society of Western Australia* 57, 11–15.

Woolley PA (1982) Phallic morphology of the Australian species of *Antechinus* (Marsupialia: Dasyuridae): A new taxonomic tool? pp. 767–781, in M Archer (ed.) *Carnivorous Marsupials*. Volume 2. Royal Zoological Society of New South Wales, Sydney. [Feb 1982]

Woolley PA (1984) Reproduction in *Antechinomys laniger* ('spenceri' form) (Marsupialia: Dasyuridae): Field and laboratory investigations. *Australian Wildlife Research* 11, 481–489.

Woolley PA (1995) Observations on reproduction in captive *Parantechinus bilarni* (Marsupialia: Dasyuridae). *Australian Mammalogy* 18, 83–85. [21 Dec 1995]

Woolley PA (2005) The species of *Dasycercus* Peters, 1875 (Marsupialia: Dasyuridae). *Memoirs of Museum Victoria* 62, 213–221.

Woolley PA (2006) Studies on the crest-tailed mulgara *Dasycercus cristicauda* and the brush-tailed mulgara *Dasycercus blythi* (Marsupialia: Dasyuridae). *Australian Mammalogy* 28, 117–120. [10 May 2006]

Woolley PA, Webb SJ (1977) The penis of dasyurid marsupials. pp. 307–323, in B Stonehouse & D Gilmore (eds.) *The Biology of Marsupials*. Macmillan Press, London.

Woolley PA, Westerman M, Krajewski C (2007) Interspecific affinities within the genus *Sminthopsis* (Dasyuromorphia: Dasyuridae) based on morphology of the penis: Congruence

with other anatomical and molecular data. *Journal of Mammalogy* 88, 1381–1392. [Dec 2007]
WoRMS (2012) *Boops* Cuvier, 1814. In R Froese & D Pauly (eds.) FishBase. Accessed through: World Register of Marine Species at: http://www.marinespecies.org/aphia.php?p=taxdetails&id=126073. [10 Nov 2012]
Worthington Wilmer J, Hall L, Barratt E, Moritz C (1999) Genetic structure and male-mediated gene flow in the ghost bat (*Macroderma gigas*). *Evolution* 53, 1582–1591. [Oct 1999]
Wozencraft WC (1993) Order Carnivora. pp. 279–348, in DE Wilson & DA Reeder (eds.) *Mammal Species of the World. A Taxonomic and Geographic Reference*. Smithsonian Institution Press, Washington. Second Edition.
Wozencraft WC (2005) Order Carnivora. pp. 532–628, in DE Wilson & DA Reeder (eds.) *Mammal Species of the World. A Taxonomic and Geographic Reference*. Johns Hopkins University Press, Baltimore. Third Edition.
Wroe S (1996) *Muribacinus gadiyuli*, (Thylacinidae: Marsupialia), a very plesiomorphic thylacinid from the Miocene of Riversleigh, northwestern Queensland, and the problem of paraphyly for the Dasyuridae (Marsupialia). *Journal of Paleontology* 70, 1032–1044. [Nov 1996]
Wroe S (1997) A re-examination of morphology-based synapomorphies for the families of Dasyuromorphia (Marsupialia). I. Dasyuridae. *Journal of Mammalian Evolution* 4, 19–52. [Mar 1997]
Wroe S (1999) The geologically oldest dasyurid, from the Miocene of Riversleigh, north-west Queensland. *Palaeontology* 42, 501–527.
Wroe S, Mackness BS (1998) Revision of the Pliocene dasyurid, *Dasyurus dunmalli* (Dasyuridae; Marsupialia). *Memoirs of the Queensland Museum* 42, 605–612. [29 Jun 1998]
Wroughton RC (1905) The common striped palm squirrel. *Journal of the Bombay Natural History Society* 16, 406–413.
Wynen LP, Goldsworthy SD, Insley SJ, Adams M, Bickman JW, Francis J, Pablo Gallo J, Hoelzel AR, Majluf P, White RWG, Slade R (2001) Phylogenetic relationships within the eared seals (Otariidae: Carnivora): Implications for the historical biogeography of the family. *Molecular Phylogenetics and Evolution* 21, 270–284. [Nov 2001]
Wyss AR (1988) Evidence from flipper structure for a single origin of pinnipeds. *Nature* 334(6181), 427–428. [4 Aug 1988]
Wyss AR, Flynn JJ (1993) A phylogenetic analysis and definition of the Carnivora. pp. 32–52, in FS Szalay, MJ Novacek & MC McKenna (eds.) *Mammal Phylogeny: Placentals*. Springer-Verlag, New York. Volume 2.
Wyss AR, Meng J (1996) Application of phylogenetic taxonomy to poorly resolved crown clades: a stem-modified node-based definition of Rodentia. *Systematic Biology* 45, 559–568. [Dec 1996]
Xiao-Feng X, Zhi-Qiang Z (2009) Eriophyoid mites (Acari: Prostigmata) in Southeast Asia: A Synopsis of 104 genera, with an illustrated key to genera and checklist of species. *Zootaxa* 2257, 1–128. [8 Oct 2009]
Xu X, Janke A, Arnason U (1996) The Complete mitochondrial DNA sequence of the greater Indian rhinoceros, *Rhinoceros unicornis*, and the phylogenetic relationship among Carnivora, Perissodactyla, and Artiodactyla (+ Cetacea). *Molecular Biology and Evolution* 13, 1167–1173. [Nov 1996]
Yabe H, Hayasaka I (1916) Palaeozoic corals from Japan, Korea and China. *Journal of Geological Society of Tokyo* 23, 57–75.
Yamada M (1954) Some remarks on the pygmy sperm whale *Kogia*. *Scientific Reports of the Whales Research Institute, Tokyo* 9, 37–58.
Yapa A, Ratnavira G (2013) *The Mammals of Sri Lanka*. Field Ornithology Group of Sri Lanka, Columbo.
Yates TL (1984) Insectivores, Elephant Shrews, Tree Shrews, and Dermopterans. pp. 117–144, in S Anderson & JK Jones (eds.) *Orders and Families of Recent Mammals of the World*. John Wiley & Sons, New York.
Yochem PK, Leatherwood S (1985) Blue whale *Balaenoptera musculus* (Linnaeus, 1758). pp. 193–240, in SH Ridgway & R Harrison (eds.) *Handbook of Marine Mammals. Volume 3. The Sirenians and Baleen Whales*. Academic Press, London.
Yonezawa T, Kohno N, Hasegawa M (2009) The monophyletic origin of sea lions and fur seals (Carnivora; Otariidae) in the Southern Hemisphere. *Gene* 441, 89–99.
Yoshida H, Kato H (1999) Phylogenetic relationships of Bryde's whale in the western North Pacific and adjacent waters inferred from mitochondrial DNA sequences. *Marine Mammal Science* 15, 1269–1286. [Oct 1999]
Yosida TH (1980) *Cytogenetics of the Black Rat. Karyotype Evolution and Species Differentiation*. Tokyo University, Tokyo. Reference not seen.
Young GE (1980) Geographic variation in the common wombat, *Vombatus ursinus* (Shaw, 1800). *Victorian Naturalist* 97, 200–204. [Sep–Oct 1980]
Zardoya R, Meyer A (1998) Complete mitochondrial genome suggests diapsid affinities of turtles. *Proceedings of the National Academy of Sciences of the United States of America* 95, 14226–14231. [24 Nov 1998]
Zeigler CV, Chan GL, Barnes LG (1997) A new Late Miocene balaenopterid whale (Cetacea: Mysticeti), *Parabalaenoptera baulinensis*, (new genus and species) from the Santa Cruz Mudstone, Point Reyes Peninsula, California. *Proceedings of the California Academy of Sciences* 50, 115–138. [9 Dec 1997]
Zelebor J (1869) Säugethiere. pp. 1–42, in B von Wüllerstorf-Ubair (ed.) *Reise der Österreichischen Fregatte Novara um die Erde, in den Jahren 1857, 1858, 1859: Zoologische Theil*. Kaiserlich-Königlichen hof-und Staatsdruckerel, Wien. Volume 1.
Zemsky VA, Boronin VA (1964) On the question of the pygmy blue whale taxonomic position. *Norsk Hvalfangst-Tidende* 53, 306–311.
Zenger KR, Eldridge MDB, Cooper DW (2003) Intraspecific variation, sex-biased dispersal and phylogeography of the eastern grey kangaroo (*Macropus giganteus*). *Heredity* 91, 153–162. [Aug 2003]
Zenger KR, Eldridge MDB, Johnston PG (2005) Phylogenetics, population structure and genetic diversity of the endangered southern brown bandicoot (*Isoodon obesulus*) in south-eastern Australia. *Conservation Genetics* 6, 193–204.
Zenkovicz (1947) Рыбное хоз-во СССР (= *Fish Households of the USSR*) 10, 15. Reference not seen.

Zerbini AN, Secchi ER, Siciliano S, Simöes-Lopes PC (1996) The dwarf form of the minke whale, *Balaenoptera acutorostrata* Lacépède, 1804, in Brazil. *Report of the International Whaling Commission* 46, 333–340.

Zhou K, Qian W, Li Y (1979) The osteology and the systematic position of the Baiji, *Liptes vexillifer. Acta Zoologica Sinica* 25, 58–74.

Zhou X, Xu S, Yang Y, Zhou K, Yang G (2011) Phylogenetic analyses and improved resolution of Cetartiodactyla. *Molecular Phylogenetics and Evolution* 61, 255–264. [Nov 2011]

Zhou X, Xu S, Xu J, Chen B, Zhou K, Yang G (2012) Phylogenomic analysis resolves the interordinal relationships and rapid diversification of the Laurasiatherian mammals. *Systematic Biology* 61, 150–164.

Ziegler AC (1982) The Australo-Papuan genus *Syconycteris* (Chiroptera: Pteropodidae) with the description of a new Papua New Guinea species. *Occasional Papers of the Bernice P. Bishop Museum* 25(5), 1–22. [30 Apr 1982]

Ziegler AC, Lidicker WZ Jr (1968) Keys to the genera of New Guinea recent land mammals. *Proceedings of the California Academy of Sciences* (4)36(2), 33–71. [25 Sep 1968]

Zietz ACH (1904) Notes by Mr. A. Zietz on the foregoing paper of Dr. Harvey. *Proceedings of the Royal Geographical Society of Australasia, South Australian Branch* 7, 180–184.

Zimmermann EAW von (1777) *Specimen Zoologiae Geographicae, Quadrupedum, Domicilia et Migrationes sistens. Dedit, tabulamque mundi zoographicam adjunxit. Theodorum Haak, et Socios*. Lugduni Batavorum.

Zimmermann EAW von (1780) *Geographische Geschichte des Menschen*, und der vierfüssigen Thiere. Enthalt ein vollstandiges Verzeichniss aller bekannten Quadrupen. Weygandschen Buchhandlung, Leipzig. Volume 2.

Zimmermann EAW von (1783) *Geographische Geschichte des Menschen, und der allgemein verbreiteten vierfussigen Thiere. Mit einer hiezu gehorigen Zoologischen Weltcharte*. Weygandschen Buchhlandlung, Leipzig. Volume 3.

Zittel KA von (1893) *Handbuch der Palaeontologie*. Abteilung I. Palaeozoic. Volume 4. Vertebrata (Mammalia). R. Oldenbourg, Munich.

Appendix 1. Secondary sources used to determine publication dates of particular journals and books

Publication	Secondary source for dates
Journals	
Annales du Muséum d'Histoire Naturelle, Paris.	Sherborn, 1914: 366.
Annals and Magazine of Natural History.	Evenhuis, 2003a: 14.
Annals of Natural History.	Evenhuis, 2003a: 13.
Encyclopédie Méthodique.	Evenhuis, 2003b: 38; Sherborn & Woodward, 1893: 584; Sherborn & Woodward, 1906: 580.
Histoire Naturelle, Paris.	Evenhuis, 2003b: 39.
Journal of Natural History.	Evenhuis, 2003a: 57.
Magazine of Natural History.	Evenhuis, 2003a: 10.
Magazine of Zoology and Botany.	Evenhuis, 2003a: 13.
Mémoires du Muséum d'Histoire Naturelle, Paris.	Sherborn, 1914: 367.
Monatsberichte der [Königlich] Preussischen Akademie der Wissenschaften zu Berlin.	Bauer *et al.* 1995: 22.
Nouvelles Annales du Muséum d'Histoire Naturelle, Paris.	Sherborn, 1914: 368.
Proceedings of the Zoological Society of London.	Sclater, 1893: 435; Duncan, 1937: 71; Waterhouse, 1937: 78; Cowan, 1973: 293; Dickinson, 2005: 427.
Proceedings of the Linnean Society of New South Wales.	Fletcher, 1896: 533.
Transactions of the Linnean Society of London.	Raphael, 1970: 61.
Transactions of the Zoological Society of London.	Peavot, 1937: 83.
Books	
Blainville, H.M.D. de (1839–1864) *Ostéographie ou description iconographique comparée du squelette et du système dentaire des mammifères.*	Sherborn, 1898: 76.
Boitard, P. (1841) *Le Jardin des Plantes description et moeurs des mammifères de la Ménagerie et du Muséum d'Histoire Naturelle.*	Sherborn, 1922: xxvi.
Bonaparte, C.L. (1832–1841) *Iconografia della Fauna Italica.*	Salvadori, 1888: 3.
Bonelli 1810 *Tabula Synoptica.*	Gaskin & Lewis 1956: 158.
Castelnau F. de. (1855–1859) *Animaux nouveaux ou rares. Recueillis Pendant L'Expédition dans les parties centrales de l'Amérique du Sud, de Rio de Janeiro à Lima, et de Lima au Para; exécutée par ordre du gouvernment Français pendant les années 1843 à 1847.*	Sherborn & Woodward, 1901a: 164.
Cuvier, G. (1797) *Tableau Élémentaire de l'Histoire Naturelle des Animaux.*	Roux, 1797: 81.
Cuvier, G. (1816) *Le Règne Animal.*	Anon, 1915: 114; Whitehead, 1967: 300; Cowan, 1969a: 219; Roux, 1976: 31.
Cuvier. G. Various.	Smith, J. 1993.
Darwin, C. (1838–1843) *The Zoology of the Voyage of H.M.S. Beagle.*	Sherborn, 1897b: 483.
Dejean, P.F.M.A. (1821) *Catalogue de la Collection de Coléoptères.*	Griffin, 1932: 177; Madge, 1988: 83.
Desmarest, A.G. (1804) Tableau Méthodique des Mammifères. Pp. 5–38, in C.S. Sonnini (ed.) in Nouveau Dictionnaire d'Histoire Naturelle. Vol. 24.	Rafinesque, 1817: 361.

Appendix 1. (*Continued*)

Publication	Secondary source for dates
Desmarest, A.G. (1821–1822) *Mammalogie, ou description des espèces de mammifères.*	Sherborn & Woodward, 1906: 580; Evenhuis, 2003b: 38.
D'Orbigny, C. (ed.)(1839–1849) *Dictionaire Universal d'Histoire Naturelle.* Paris.	Evenhuis, 1990: 222.
Dumont d'Urville, J.S.C. (1830–1835) *Voyage de Découvertes de l'Astrolabe. Exécuté par ordre du Roi pendant les années* 1826–1827–1828–1829.	Sherborn & Woodward, 1901a: 333.
Dumont d'Urville, J.S.C. (ed.)(1842–1854) *Voyage au Pole Sud et dans l'Océanie sur les corvettes l'Astrolobe et la Zélée.*	Sherborn & Woodward, 1901b: 390.
Duperry, L.I. (1826–1834) *Voyage autour du monde, Exécuté par Ordre du Roi, sur la Corvette de La Majesté, La Coquille, pendant les années 1822, 1823, 1824 et 1825.*	Sherborn & Woodward, 1901b: 391; Cretella, 2010: 83.
Gloger, C. (1841) *Gemeinnützige hand-und Hilfsbuch der Naturgeschichte.*	Thomas, 1895a: 189.
Gould, J. (1841–1842) *A Monograph of the Macropodidae, or Family of Kangaroos.* John Gould, London.	Waterhouse, F. 1885: 5.
Gould, J. (1845–1863) *The Mammals of Australia. John Gould, London.*	Waterhouse, F. 1885: 1.
Gray, J.E. (1830–1835) *Illustrations of Indian Zoology.*	Sawyer, 1953: 48.
Griffith (1824–1835) *The Animal Kingdom arranged in conformity with its organisation, by the Baron Cuvier.* Vol. 16.	Anon, 1903: 410; Cowan, 1969b: 137.
King, P.P. (ed.) (1826) *Narrative of a Survey of the Intertropical and Western Coasts of Australia.*	Common & Moulds, 1973: 257.
Lacépède, B.G.É. (1799–1802) *Tableau des divisions, sous-divisions, ordres et genres des mammifères.*	Sherborn, 1899: 406; Richmond, 1899: 325.
Lucas, H.P. (1846–1849) Histoire naturelle des animaux articulés. Deuxième Partie. Insectes. Coleoptères. Pp. i–xxxv, 1–590, in *Exploration scientifique de l'Algérie pendant les années 1840, 1841, 1842.*	Sherborn & Woodward, 1901a: 163.
Pallas, P.S. (1778–1779) *Novae Species Quadrupedum e Glirium Ordine.*	Sherborn, 1891: 236.
Perry, G. (1810–1811) Arcana; or the Museum of Natural History, British Museum, London.	Matthews & Iredale, 1912: 7; Petit, 2009: 12.
Quoy, J.R.C. & Gaimard, J.P. (1824–1826) Zoologie. Pp. 1–712, in L.C.D. de Freycinet (ed.) (1824–1830) *Voyage Autour du Monde.*	Sherborn & Woodward, 1901b: 392.
Rees, A. (1802–1820) *The Cyclopædia : or, universal dictionary of Arts, Sciences, and Literature.* London.	Anon, 1820: 222; Jackson, 1896: 310.
Richardson *et al.* (1862) 1862) *The Museum of Natural History.*	Anon, 1913: 1699.
Rothschild, W. Various.	Jordan, 1938: 17.
Savigny, J.-C. (1809–1829) *Description de L'Égypte, ou Recueil des observations et des Recherches qui ont été faites en Égypte pendant l'expédition de l'armée Francaise.*	Sherborn, 1897a: 285.
Schreber (1774–1855) *Die Säugthiere in Abbildungen nach der natur mit Beschreibungen.* Including volumes edited by G.A. Goldfuss and J.A. Wagner.	Sherborn, 1892: 587.
Shaw, G. (1789–1813) *The Naturalist's Miscellany.*	Sherborn, 1895: 375; Dickinson *et al.*, 2006: 322.
Smith, A. (1838–1849) *Illustrations of the Zoology of South Africa.*	Barnard, 1950: 187.
Siebold P.T. von (1833–1850) *Fauna Japonica.*	Sherborn & Jentink, 1895: 149; Mazak, 1967: 537.
Temminck, C.J. (1839–1847) *Verhandelingen over de natuurlijke Geschiedenis der Nederlandsche overzeesche Bezittingen.*	Husson & Holthuis, 1955: 17.
Temminck, C.J. (1824–1827) *Monographies de Mammalogie, ou description de quelques genres de mammifères.*	Anon, 1825: 574; Husson, 1962: 276.
Thomas, O. Various.	Hill, 1990: 36–82.
Various.	Palmer, 1904.
Various.	Sherborn, 1902; 1922; 1927; 1928; 1929.
Various.	Walton, 1988a.

Index of common names

Index of scientific names

www.ingramcontent.com/pod-product-compliance
Lightning Source LLC
Chambersburg PA
CBHW042013200926
51615CB00029B/7
9781486300129